Advances in Modal Logic

Volume 12

Advances in Modal Logic

Volume 12

Edited by

Guram Bezhanishvili

Giovanna D'Agostino

George Metcalfe

and

Thomas Studer

ISBN 978-1-84890-255-8

College Publications
Scientific Director: Dov Gabbay
Managing Director: Jane Spurr

http://www.collegepublications.co.uk

Printed by Lightning Source, Milton Keynes, UK

Contents

Preface

Advances in Modal Logic (AiML) is an initiative founded in 1995 and aimed at presenting an up-to-date picture of the state of the art in modal logic and its many applications. It consists of a conference series together with volumes based on the conferences. The conference series is the main international forum at which research on all aspects of modal logic is presented. The first installment was held in 1996 in Berlin, Germany, and since then it has been organized biennially, with meetings in 1998 in Uppsala, Sweden; in 2000 in Leipzig, Germany (jointly with ICTL-2000); in 2002 in Toulouse, France; in 2004 in Manchester, UK; in 2006 in Noosa, Australia; in 2008 in Nancy, France; in 2010 in Moscow, Russia; in 2012 in Copenhagen, Denmark; in 2014 in Groningen, The Netherlands; and in 2016 in Budapest, Hungary. Information about AiML and related events, including conference proceedings, is available at the website `www.aiml.net`.

The twelfth conference in the AiML series was held on August 27–31, 2018 at the University of Bern, Switzerland, co-located with the sixth edition of the conference "Logic, Algebra and Truth Degrees" (LATD 2018). AiML 2018 was organized by George Metcalfe and Thomas Studer (University of Bern, Switzerland) with the assistance of Bettina Choffat, Almudena Colacito, José Gil-Férez, Eveline Lehmann, Nenad Savić, Olim Tuyt, and Silvia Steila. The conference website can be found at `http://www.aiml2018.unibe.ch/`.

This volume contains abstracts of invited talks and contributed papers from the conference. The invited talks were given by

- Agata Ciabattoni (TU Wien, Austria),
- Rob Goldblatt (Victoria University of Wellington, New Zealand),
- Rosalie Iemhoff (Utrecht University, The Netherlands), joint AiML-LATD invited speaker, and
- Stanislav Kikot (Birkbeck College London and Institute for Information Transmission Problems, Moscow).

The Programme Committee received 64 regular paper submissions. Of these, 34 were selected for this volume by a reviewing process where every paper received three independent expert reviews. The volume includes papers on propositional modal logics, their products, predicate modal logics, temporal and epistemic reasoning, provability, intuitionistic, substructural, and paraconsistent logics, and other related logics. The topics include decidability and complexity results, proof theory, model theory, interpolation, as well as other related problems in algebraic logic.

In addition, there were 34 submissions for short presentations at the conference, and 24 were accepted for presentation.

The members of the Programme Committee for the conference were

- Natasha Alechina (University of Nottingham)
- Lev Beklemishev (Steklov Mathematical Institute of Russian Academy of Sciences in Moscow)
- Guram Bezhanishvili (New Mexico State University)
- Marta Bílková (Charles University)
- Patrick Blackburn (University of Roskilde)
- Giovanna D'Agostino (University of Udine)
- Stéphane Demri (CNRS, LSV, ENS Paris-Saclay)
- David Fernández-Duque (Ghent University)
- David Gabelaia (TSU Razmadze Mathematical Institute)
- Mai Gehrke (CNRS Paris Diderot)
- Silvio Ghilardi (Universitá degli Studi di Milano)
- Nina Gierasimczuk (Danish Technical University)
- Valentin Goranko (Stockholm University)
- Rajeev Goré (The Australian National University)
- Helle Hvid Hansen (Delft University of Technology)
- Ian Hodkinson (Imperial College)
- Wesley Holliday (UC Berkeley)
- Emil Jeřábek (Czech Academy of Sciences)
- Marcus Kracht (Bielefeld University)
- Clemens Kupke (University of Strathclyde)
- Agi Kurucz (King's College London)
- Roman Kuznets (TU Wien)
- Tamar Lando (Columbia University)
- Carsten Lutz (Universität Bremen)
- Larry Moss (Indiana University)
- Sara Negri (University of Helsinki)
- Eric Pacuit (University of Maryland)
- Valeria de Paiva (Nuance Communications, USA)
- Mark Reynolds (The University of Western Australia)
- Ilya Shapirovsky (Institute for the Information Transmission Problems)
- Valentin Shehtman (Institute for the Information Transmission Problems)
- Sonja Smets (Institute of Logic, Language and Computation)
- Viorica Sofronie-Stokkermans (Universität Koblenz-Landau)
- Thomas Studer (Universität Bern)
- Heinrich Wansing (Ruhr University Bochum)

- Frank Wolter (University of Liverpool)
- Michael Zakharyaschev (Birkbeck College London)

The Programme Committee was chaired by

- Guram Bezhanishvili (New Mexico State University)
- Giovanna D'Agostino (University of Udine)

The Steering Committee of AiML for 2016–2018 consisted of

- Lev Beklemishev (Steklov Mathematical Institute of Russian Academy of Sciences in Moscow)
- Guram Bezhanishvili (New Mexico State University)
- Giovanna D'Agostino (University of Udine)
- Stephane Demri (LSV, CNRS, ENS Cachan)
- Silvio Ghilardi (Universitá degli Studi di Milano)
- Rajeev Goré (Australian National University)
- Agi Kurucz (King's College London)
- George Metcalfe (Universität Bern) (local organizer AiML 2018)
- Larry Moss (Indiana University)
- Thomas Studer (Universität Bern) (local organizer AiML 2018)

Many other people assisted with the reviewing process, including: Juan Pablo Aguilera, Philippe Balbiani, Francesco Belardinelli, Gianluigi Bellin, Nick Bezhanishvili, Adam Bjorndahl, Annemarie Borg, Joseph Boudou, Wojciech Buszkowski, Inma P. Cabrera, Marcelo Coniglio, Willem Conradie, Giovanna Corsi, Fredrik Dahlqvist, Evgenij Dashkov, Jeremy Dawson, Martín Diéguez, Yifeng Ding, Wojciech Dzik, Harley Eades, Cristina Feier, Camillo Fiorentini, Melvin Fitting, Fredrik Nordvall Forsberg, Tim French, Peter Fritz, Andreas Fjellstad, Nick Galatos, Malvin Gattinger, Brunella Gerla, Meghdad Ghari, Marianna Girlando, Samuel J. van Gool, Giuseppe Greco, Gianluca Grilletti, Raul Hakli, Christopher Hampson, Eduardo Hermo Reyes, Nebojsa Ikodinovic, Andrzej Indrzejczak, Jean Christoph Jung, Stanislav Kikot, Kohei Kishida, Roman Kontchakov, Zofia Kostrzycka, Vladimir Krupski, Andrey Kudinov, Louwe Kuijer, Antti Kuusisto, Evgeny Kuznetsov, Frederik M. Lauridsen, Bjoern Lellmann, Tadeusz Litak, Sonia Marin, Sérgio Marcelino, Johannes Marti, Michel Marti, Manuel A. Martins, Renato Neves, Damian Niwinski, Andrés Occhipinti Liberman, Eugenio Orlandelli, Fedor Pakhomov, Dirk Pattinson, Edi Pavlovic, Elaine Pimentel, Sophie Pinchinat, Vaughan Pratt, Adam Prenosil, Vit Puncochar, James Raftery, Revantha Ramanayake, Rasmus K. Rendsvig, Grigore Rosu, Nenad Savic, Thomas Schneider, Lutz Schröder, Guido Sciavicco, Igor Sedlar, Dmitry Skvortsov, Luca Spada, Shawn Standefer, Lutz Straßburger, Levan Uridia, Moshe Vardi, Fan Yang, Junhua

Yu, Evgeny Zolin. We apologize to anyone whose name was inadvertently left off this list.

We thank the organizers of the conference for their dedicated efforts in bringing this meeting to fruition. We thank the members of the Programme Committee and all other reviewers for the time, professional effort and the expertise that they invested in ensuring the high scientific standards of the conference and its proceedings. We also thank the authors for their excellent contributions and Jane Spurr for bringing this volume to publication. Special thanks go to Lev Beklemishev and Stéphane Demri (Chairs of AiML 2016) who generously shared their knowledge and experience with us, which made our work so much easier.

We would like to thank University of Bern for hosting the conference. This is very much appreciated! Finally, we would like to thank the sponsors of the conference: Burgergemeinde Bern, Swiss National Science Foundation, Stadt Bern, University of Bern, the Association for Symbolic Logic, and the European Union Horizon 2020 research and innovation program.

July 15th, 2018

Guram Bezhanishvili
Giovanna D'Agostino
George Metcalfe
Thomas Studer

Abstracts of Invited Talks

Intermediate Logics: From Hypersequents to Concurrent Computation

Agata Ciabattoni

Vienna University of Technology

We provide a general proof-theoretic framework connecting logic and concurrent computation. We describe an algorithm for introducing analytic calculi in a uniform and systematic way for a large class of logics [4]. Our calculi use hypersequents, which are sequents working in parallell [3]. Reformulated first as natural deduction systems [7,5,6], the introduced hypersequent calculi are employed to provide a concurrent computational interpretation for many intermediate logics, classical logic included [1,2].

We use the CurryHoward correspondence to obtain new typed concurrent λ-calculi, each of which features a specific communication mechanism and implements techniques for handling and transmitting process closures.

References

[1] Aschieri, F., A. Ciabattoni and F. A. Genco, *Gödel logic: From natural deduction to parallel computation*, in: *Proceedings of Logic in Computer Science (LICS 2017)*, 2017, pp. 1–12.

[2] Aschieri, F., A. Ciabattoni and F. A. Genco, *Classical proofs as parallel programs* (2018), submitted.

[3] Avron, A., *Hypersequents, logical consequence and intermediate logics for concurrency*, Annals of Mathematics and Artificial Intelligence **4** (1991), pp. 225–248.

[4] Ciabattoni, A., N. Galatos and K. Terui, *From axioms to analytic rules in nonclassical logics*, in: *Proceedings of Logic in Computer Science (LICS 2008)*, 2008, pp. 229–240.

[5] Ciabattoni, A. and F. A. Genco, *Embedding formalisms: Hypersequents and two-level systems of rules*, in: *Proceedings of Advances in Modal Logic 2016 (AIML 2016)*, College Publications, **11** (2016), pp. 197–216.

[6] Ciabattoni, A. and F. A. Genco, *Hypersequents and systems of rules: Embeddings and applications*, ACM TOCL **19** (2018), pp. 1–27.

[7] Negri, S., *Proof analysis beyond geometric theories: From rule systems to systems of rules*, J. Logic Comput **26** (2016), pp. 513–537.

Canonicity Frameworks and Ultraproducts of Polarities

Robert Goldblatt

Victoria University of Wellington

The duality between modal algebras and Kripke frames depends on the formation of *canonical* extensions of algebras, a construction introduced in the 1940s by Jónsson and Tarski for Boolean algebras with additional operations preserving finite joins. After a hiatus of several decades, Gehrke and Harding [2] provided a theory of canonical extensions of lattice-ordered algebras in general. It builds a canonical extension as the lattice of all stable subsets of a polarity structure in the sense of Birkhoff.

This theory will be used to study a generalisation to lattice-ordered algebras of an influential result of Fine [1] from modal model theory. Fine's theorem states that if a modal logic is determined by a first-order definable class of frames, then it is valid in all its canonical frames. That was generalised in [3] to the fact that the complex algebras of an ultraproducts-closed class of relational structures generates a variety of Boolean algebras with operators that is closed under canonical extensions.

The talk will describe an axiomatisation of the structural relationships underlying this result by providing a list of conditions on an algebra/structure duality that suffice to ensure that if a class of polarity-based structures is closed under ultraproducts, then the stable set lattices of these structures generate a variety of algebras that is closed under canonical extensions.

In particular, situations to which this applies arise when the polarity-based structures have stable set lattices whose additional operations are first-order definable over the structures and preserve all joins, or all meets. That includes the residuated operations modelling the fusion and implication connectives of a number of substructural logics. A partial account of this work is given in [4].

References

[1] Fine, K., *Some connections between elementary and modal logic*, in: S. Kanger, editor, *Proceedings of the Third Scandinavian Logic Symposium*, North-Holland, 1975 pp. 15–31.

[2] Gehrke, M. and J. Harding, *Bounded lattice expansions*, Journal of Algebra **239** (2001), pp. 345–371.

[3] Goldblatt, R., *Varieties of complex algebras*, Annals of Pure and Applied Logic **44** (1989), pp. 173–242.

[4] Goldblatt, R., *Canonical extensions and ultraproducts of polarities*, arXiv:1709.09798 (2017).

The Existence of Proof Systems

Rosalie Iemhoff

Utrecht University, The Netherlands

During the last hundred years proof systems of all kinds have been developed for a great variety of logics. These proof systems can often be used to establish that the corresponding logics have nice properties, such as decidability, interpolation or Skolemization. Results stating that a logic does not have certain proof systems are less common. In this talk a method is introduced to prove such negative results. The method establishes a connection between the existence of certain proof systems for a logic and certain regular properties that the logic satisfies. The talk focusses on (intuitionistic) modal logics, although the method is applicable to other logics as well. The regular properties considered in this talk are variants of interpolation, and the developed method can be used not only to obtain the negative results, but also to prove uniform interpolation for several classical and intuitionistic modal logics. The method is in fact inspired by the syntactic proof that intuitionistic logic has uniform interpolation by Pitts.

The method makes use of sequent calculi, but in a very abstract form. The key property of rules that this method uses is that of being *focussed*, a property that expresses the structurality of a rule. Many of the standard sequent rules for connectives have this property and thus are focussed. In [2] it is shown that if a modal logic has a proof system that consists of focussed rules, then it has uniform interpolation, which implies that the many modal logics without uniform interpolation [1,3] cannot have focussed proof systems. The generality of the notions involved makes the method applicable to many other logics, for example to intermediate logics.

In how far other proof systems lend themselves to this approach is still not clear. Besides the technical results above, such unresolved issues as well as related conjectures will be addressed during the talk.

References

[1] Ghilardi, S. and M. Zawadowski, "Sheaves, Games, and Model Completions: A Categorical Approach to Nonclassical Propositional Logics," Trends in Logic (Book 14), Springer, 2002.

[2] Iemhoff, R., *Uniform interpolation and sequent calculi in modal logic*, Archive for Mathematical Logic Https://link.springer.com/article/10.1007/s00153-018-0629-0, to appear in print in 2018.

[3] Maksimova, L., *Craig's theorem in superintuitionistic logics and amalgamable varieties of pseudo-Boolean algebras*, Algebra Logika **16** (1977), pp. 643–681.

Kripke Completeness of Strictly Positive Modal Logics Over Meet Semilattices with Operators

Stanislav Kikot

Birkbeck College, UK Institute for Information Transmission Problems, Russia

This talk is about a connection between various consequence relations for the fragment of propositional multi-modal logic that comprises implications $\sigma \to \tau$, where σ and τ are *strictly positive modal formulas* constructed from propositional variables using conjunction $\wedge$, unary diamond operators $\Diamond_i$, and the constant 'truth' $\top$. We call such formulas SP-*implications* and we call sets of SP-implications SP-*theories*. These formulas can be interpreted on Kripke frames as well as on meet semilattices with unary monotone operators (SLOs). This gives rise to the following problem:

(***completeness***) identify SP-theories P that are *complete* in the sense that the two consequence relations $P \models_{\mathsf{Kr}}$ and $P \models_{\mathsf{SLO}}$ coincide, where for any SP-implication ι,

$$P \models_{\mathsf{Kr}} \iota \quad \text{iff} \quad \iota \text{ is valid in every Kripke frame validating } P;$$
$$P \models_{\mathsf{SLO}} \iota \quad \text{iff} \quad \iota \text{ is valid in every SLO validating } P.$$

SP-implications are Sahlqvist, so for every modal formula φ and SP-theory P

$$P \models_{\mathsf{Kr}} \varphi \quad \text{iff} \quad \varphi \in \mathbf{K} + P \quad \text{iff} \quad \varphi \approx \top \text{ is valid in every BAO validating } P, \quad (1)$$

where BAO stands for *Boolean algebras with normal and $\vee$-additive unary operators*. Note that, by (1), the completeness problem is equivalent to

(***conservativity***) the purely algebraic problem of whether the consequence relation $P \models_{\mathsf{BAO}}$ is *conservative* over $P \models_{\mathsf{SLO}}$ with respect to SP-implications, that is, $P \models_{\mathsf{SLO}} \iota$ iff $P \models_{\mathsf{BAO}} \iota$, for any ι; and also to

(***axiomatisability***) the problem whether P *axiomatises* the SP-implicational fragment of the normal modal logic $\mathbf{K} + P$ using the syntactic Birkhoff-type calculus corresponding to the algebraic consequence relation $P \models_{\mathsf{SLO}}$ (in other words, the problem whether P has a *modal companion*).

I am going to present two methods for proving completeness for SP-theories together with numerous sufficient conditions for their applicability. Note that incomplete SP-theories are easy to find, with two simplest ones being $\{\Diamond p \to p\}$ and $\{\Diamond p \to \Diamond q\}$.

This talk is based on a recent joint work with Agi Kurucz, Yoshihito Tanaka, Frank Wolter and Michael Zakharyaschev accessible at `https://arxiv.org/pdf/1708.03403.pdf`

Contributed Papers

The Worm Calculus

Ana de Almeida Borges [1]

Universitat de Barcelona
C. Montalegre 6
08001 Barcelona, Catalonia, Spain

Joost J. Joosten [2]

Universitat de Barcelona
C. Montalegre 6
08001 Barcelona, Catalonia, Spain

Abstract

We present a propositional modal logic WC, which includes a logical *verum* constant $\top$ but does not have any propositional variables. Furthermore, the only connectives in the language of WC are consistency-operators $\langle\alpha\rangle$ for each ordinal α. As such, we end up with a class-size logic. However, for all practical purposes, we can consider restrictions of WC up to a given ordinal. Given the restrictive signature of the language, the only formulas are iterated consistency statements, which are called worms. The theorems of WC are all of the form $A \vdash B$ for worms A and B. The main result of the paper says that the well-known strictly positive logic RC, called Reflection Calculus, is a conservative extension of WC. As such, our result is important since it is the ultimate step in stripping spurious complexity off the polymodal provability logic GLP, as far as applications to ordinal analyses are concerned. Indeed, it may come as a surprise that a logic as weak as WC serves the purpose of computing something as technically involved as the proof theoretical ordinals of formal mathematical theories.

Keywords: Provability logic, strictly positive logics, closed fragment, feasible fragments, Reflection Calculus, ordinal notations.

1 Introduction

Quite some interest has arisen in feasible fragments of modal logics recently. One of the common goals is to find fragments with good computational properties that still maintain a decent amount of expressibility. Description logics and their applications to database theory [1] are a good example of this.

The current paper also studies fragments of modal logic, but coming from a different tradition. Our starting point is GLP: a polymodal version of Gödel-

[1] ana.agvb@gmail.com

[2] jjoosten@ub.edu

Löb's provability logic as introduced by Japaridze [17]. The logic GLP is a propositional modal logic which in its simplest version has a modality for each natural number. Although this logic is known to be PSPACE-complete [20], it behaves rather ghastly. While complete with respect to topological semantics [8], GLP is easily seen to be frame-incomplete.

The logic GLP has received a substantial amount of interest due to its applications to ordinal analysis [2]. The variable-free fragment GLP^0 of GLP actually suffices for various purposes. Going from GLP to GLP^0 is then a first weakening leading up to our final system WC to be introduced below.

The reason why GLP^0 is still suitably expressible lies in the fact that terms in it can be read in various ways. One can conceive of these terms as consistency statements or reflection principles. Furthermore, natural fragments of arithmetic are denoted by terms. The simplest terms of GLP^0 are iterated consistency statements, and they are called *worms* due to their relation to the heroic worm battle [4]. The worms modulo provable equivalence can be ordered, so that they can also be conceived of as ordinals [14]. Apart from their interpretation as consistency statements, reflection principles, fragments of arithmetic, or ordinals, worms also stand in an intimate relation with Turing progressions [18]. All of these mathematical entities can be manipulated and reasoned about within the rather simple modal logic GLP^0.

Even though the logic GLP^0 is already a substantial simplification with respect to GLP, its decidability problem is still PSPACE-complete [19]. Furthermore, the problem of frame incompleteness is still there, but the logic GLP^0 does have a rather well behaved universal model [9].

A next step in simplifying GLP^0 arose by studying strictly positive fragments of GLP and GLP^0 by means of the so called *reflection calculi* RC and RC^0 [10], [5], [6]. The theorems of RC and RC^0 are of the form $\varphi \vdash \psi$, where the only connectives in φ and ψ are conjunctions and consistency modalities. GLP is conservative over RC, in the sense that for φ and ψ only using conjunctions and consistency operators, we have that $\varphi \vdash \psi$ is provable in RC if and only if $\varphi \to \psi$ is a theorem of GLP [10].

The reflection calculi are known to be very well-behaved. In particular, the problem of frame-incompleteness is no longer there, and the decision problem is decidable in polynomial time [10]. Yet, as far as applications to ordinal analysis are concerned, no essential expressive power has been lost. Thus, the second step in our simplification brings us from GLP^0 to RC^0.

Given the limited signature of RC^0, its formulas are just built from diamonds, conjunctions, and top. However, it is provable in RC^0 that each of its formulas is equivalent to a single worm [16]. As such, one may wonder if some decent axiomatization of the worm fragment of RC^0 exists that only uses worms and only proves statements of the form $A \vdash B$ with A and B being worms. The current paper settles this question in the positive, presenting a calculus WC that only manipulates worms, so that RC^0, and thus also GLP^0, are conservative extensions of WC.

In the last two sections of the paper we dwell on semantics for WC. In particular we see that although WC has the finite model property, any (moderately nice) universal model for WC inherits much of the intrinsic complexity of Ignatiev's universal model for GLP^0.

2 The Reflection Calculus

Given an ordinal Λ, the *Reflection Calculus* for Λ — we write RC_Λ — is a propositional sequent logic in a modal language that is strictly positive. The language is hence composed of $\top$, variables, and closed both under the binary connective $\wedge$, and the unary modal operators $\langle\alpha\rangle$ for each ordinal $\alpha < \Lambda$.

Definition 2.1 [Reflection Calculus, RC_Λ, [5]] Let φ, ψ and χ be formulas in the language of RC_Λ, and $\alpha, \beta < \Lambda$ be ordinals.

The axioms of RC_Λ are:

1. $\varphi \vdash_{\mathsf{RC}} \varphi$ and $\varphi \vdash_{\mathsf{RC}} \top$;
2. $\varphi \wedge \psi \vdash_{\mathsf{RC}} \varphi$ and $\varphi \wedge \psi \vdash_{\mathsf{RC}} \psi$;
3. $\langle\alpha\rangle\langle\alpha\rangle\varphi \vdash_{\mathsf{RC}} \langle\alpha\rangle\varphi$;
4. $\langle\alpha\rangle\varphi \vdash_{\mathsf{RC}} \langle\beta\rangle\varphi$ for $\alpha > \beta$;
5. $\langle\alpha\rangle\varphi \wedge \langle\beta\rangle\psi \vdash_{\mathsf{RC}} \langle\alpha\rangle(\varphi \wedge \langle\beta\rangle\psi)$ for $\alpha > \beta$.

The rules are:

1. If $\varphi \vdash_{\mathsf{RC}} \psi$ and $\psi \vdash_{\mathsf{RC}} \chi$, then $\varphi \vdash_{\mathsf{RC}} \chi$;
2. If $\varphi \vdash_{\mathsf{RC}} \psi$ and $\varphi \vdash_{\mathsf{RC}} \chi$, then $\varphi \vdash_{\mathsf{RC}} \psi \wedge \chi$;
3. If $\varphi \vdash_{\mathsf{RC}} \psi$, then $\langle\alpha\rangle\varphi \vdash_{\mathsf{RC}} \langle\alpha\rangle\psi$.

If $\varphi \vdash_{\mathsf{RC}} \psi$, we say that ψ follows from φ in RC. If both $\varphi \vdash_{\mathsf{RC}} \psi$ and $\psi \vdash_{\mathsf{RC}} \varphi$, we say that φ and ψ are equivalent in RC, and write $\varphi \equiv_{\mathsf{RC}} \psi$.

In this paper we are mainly interested in the closed fragment of RC_Λ, denoted by RC^0_Λ, which is the same as RC_Λ without variables in the language. Since the following results hold for any chosen Λ, we will omit it.

There are some inhabitants of RC^0 on which we take special interest: the worms. These are just the formulas of RC^0 that have no $\wedge$.

Definition 2.2 [Worms, $\mathbb{W}$ and $\mathbb{W}_\alpha$] Worms are inductively defined as follows: $\top$ is in $\mathbb{W}$; if A is in $\mathbb{W}$ and α is an ordinal, then $\langle\alpha\rangle A$ is in $\mathbb{W}$.

Worms whose modalities are all at least α — we write $\mathbb{W}_\alpha$ — are defined inductively in a similar manner: $\top$ is in $\mathbb{W}_\alpha$; if A is in $\mathbb{W}_\alpha$ and $\gamma \geq \alpha$ is an ordinal, then $\langle\gamma\rangle A$ is in $\mathbb{W}_\alpha$.

It is a known result [16] that any formula in the language of RC^0 is equivalent to a worm.

Lemma 2.3 *For each formula φ of RC^0 there is a worm A such that $\varphi \equiv_{\mathsf{RC}} A$.*

This makes one wonder whether it would be possible to work with a calculus that only involves worms as far as RC^0 is concerned. This paper settles the question in the positive.

3 The Worm Calculus

We propose a *Worm Calculus* — we write WC — which derives sequents of worms. Since the language of WC only includes $\top$ and diamonds $\langle\alpha\rangle$ for an ordinal α, we omit the $\langle\cdot\rangle$, obtaining formulas which are simply strings of ordinals ending in $\top$. To further simplify, for the worms $A\top$ and $B\top$, we will write A and B. When we write AB this is understood as $AB\top$.

Definition 3.1 [Worm Calculus, WC] Let A, B and C be worms, and α, β be ordinals.

The axioms of WC are:

A1. $A \vdash_{\mathsf{WC}} \top$;

A2. $\alpha\alpha A \vdash_{\mathsf{WC}} \alpha A$ (Transitivity);

A3. $\alpha A \vdash_{\mathsf{WC}} \beta A$ for $\alpha > \beta$ (Monotonicity).

The rules of WC are:

R1. If $A \vdash_{\mathsf{WC}} B$ and $B \vdash_{\mathsf{WC}} C$, then $A \vdash_{\mathsf{WC}} C$ (Cut);

R2. If $A \vdash_{\mathsf{WC}} B$, then $\alpha A \vdash_{\mathsf{WC}} \alpha B$ (Necessitation);

R3. If $A \vdash_{\mathsf{WC}} B$ and $A \vdash_{\mathsf{WC}} \alpha C$, then $A \vdash_{\mathsf{WC}} B\alpha C$, for $B \in \mathbb{W}_{\alpha+1}$.

If $A \vdash_{\mathsf{WC}} B$, we say that B follows from A in WC. If both $A \vdash_{\mathsf{WC}} B$ and $B \vdash_{\mathsf{WC}} A$, we say that A and B are equivalent in WC, and write $A \equiv_{\mathsf{WC}} B$.

To express recursion and the notion of simplicity, we use a simple measure on worms, their length. The length of a worm is the total number of symbols other than $\top$.

Definition 3.2 [Length] The *length* of a worm A — we write $|A|$ — is defined recursively as such: $|\top| := 0$, and $|\alpha A| := |A| + 1$.

We can immediately prove some facts about worms using the worm calculus.

Lemma 3.3 *For any worms A and B, we have that $AB \vdash_{\mathsf{WC}} A$.*

Proof. The proof goes by induction on the length of A. Starting from $B \vdash \top$ (base case), repeatedly apply Necessitation to build A up front. □

From this lemma we obtain a simple but useful corollary.

Corollary 3.4 *For any worm A, we have that $A \vdash_{\mathsf{WC}} A$.*

It is in general not true that $AB \vdash_{\mathsf{WC}} B$, but there is a special case.

Lemma 3.5 *For any ordinal α and worms A and B such that $A \in \mathbb{W}_{\alpha+1}$, we have that $A\alpha B \vdash_{\mathsf{WC}} \alpha B$.*

Proof. By induction on the length of A, with the help of Transitivity and Monotonicity. □

Lemma 3.6 *For any non-trivial worm $A \in \mathbb{W}_\alpha$, we have that $A \vdash_{\mathsf{WC}} \alpha$.*

Proof. By induction on the length of A. If $|A| = 1$, then $A = \beta$ for some $\beta \geq \alpha$. The result follows by Monotonicity and Corollary 3.4. For the induction step, consider $A = \beta A'$, where $\beta \geq \alpha$ and we already know $A' \vdash_{\mathsf{WC}} \alpha$. Then by Necessitation and Transitivity, $\alpha A' \vdash_{\mathsf{WC}} \alpha\alpha \vdash_{\mathsf{WC}} \alpha$. Since $\beta A' \vdash_{\mathsf{WC}} \alpha A'$, we are done. □

It is easy to see that RC extends WC. As we shall later see, RC is conservative over WC, which means that this extension is, in a sense, not proper. The first of these two claims is articulated in the following theorem.

Theorem 3.7 *For any two worms A and B we have that $A \vdash_{\mathsf{WC}} B$ implies $A \vdash_{\mathsf{RC}} B$.*

Proof. By an easy induction on the length of a WC proof. To see that Rule R3 is admissible in RC, we use induction on the length of B and Axiom 5. □

The proof of the converse is a bit more involved. We shall use the fact that an implication between worms can be recursively broken down into implications between simpler worms.

4 Decomposing worms

The notions of α-head and α-remainder are useful to break down worms into smaller ones.

Definition 4.1 [α-head, α-remainder] Let A be a worm and α be an ordinal.

The α-*head* of A — we write $h_\alpha(A)$ — is defined recursively as: $h_\alpha(\top) := \top$, $h_\alpha(\beta A) := \beta h_\alpha(A)$ if $\beta \geq \alpha$, and $h_\alpha(\beta A) := \top$ if $\beta < \alpha$.

Likewise, the α-*remainder* of A — we write $r_\alpha(A)$ — is defined recursively as: $r_\alpha(\top) := \top$, $r_\alpha(\beta A) := r_\alpha(A)$ if $\beta \geq \alpha$, and $r_\alpha(\beta A) := \beta A$ if $\beta < \alpha$.

Intuitively, the α-head of A is the greatest initial segment of A which is in $\mathbb{W}_\alpha$, and the α-remainder is what remains after cutting off the α-head. It then follows that $A = h_\alpha(A) r_\alpha(A)$, for every worm A and ordinal α. An immediate consequence is that the lengths of the α-head and of the α-remainder of a worm are always at most the length of the worm itself.

It is possible to prove that $A \equiv_{\mathsf{RC}} h_\alpha(A) \wedge r_\alpha(A)$ for every worm A and ordinal α. In WC we cannot state such a result due to the lack of the conjunction connective in the language. We can, however, obtain the same consequences.

Lemma 4.2 *Let A be a worm and α be an ordinal. Then:*

(i) *$A \vdash_{\mathsf{WC}} h_\alpha(A)$;*

(ii) *$A \vdash_{\mathsf{WC}} r_\alpha(A)$;*

(iii) *If $B \vdash_{\mathsf{WC}} h_\alpha(A)$ and $B \vdash_{\mathsf{WC}} r_\alpha(A)$, then $B \vdash_{\mathsf{WC}} A$.*

Proof. Note that $A = h_\alpha(A) r_\alpha(A)$, this is to say, they are syntactically the same. Thus, Part (i) follows from Lemma 3.3. Part (ii) is a consequence of Lemma 3.5, taking into consideration that $h_\alpha(A) \in \mathbb{W}_\alpha$ and that $r_\alpha(A)$ always starts with either $\top$ — making the result trivial — or with an ordinal less than α. Part (iii) follows from rule R3 unless $r_\alpha(A) = \top$, in which case it is trivial. □

There is another relevant part of a worm, the α-body. It is obtained from the $(\alpha + 1)$-remainder by dropping its leftmost modality (as long as said remainder is not trivial).

Definition 4.3 [α-body] Let A be a worm and α an ordinal. The *α-body* of A — we write $b_\alpha(A)$ — is defined from $r_{\alpha+1}(A)$ as follows: if $r_{\alpha+1}(A) = \top$ then $b_\alpha(A) := \top$, and if $r_{\alpha+1}(A) = \beta B$ then $b_\alpha(A) := B$.

The α-body of a non-trivial worm A is particularly useful because its length is always strictly smaller than the length of A. We can also prove a counterpart of Lemma 4.2 about the α-body.

Lemma 4.4 *Let α be an ordinal and A be a non-trivial worm in $\mathbb{W}_\alpha$. Then:*

(i) $A \vdash_{\mathsf{WC}} \alpha b_\alpha(A)$;

(ii) *If* $B \vdash_{\mathsf{WC}} h_{\alpha+1}(A)$ *and* $B \vdash_{\mathsf{WC}} \alpha b_\alpha(A)$, *then* $B \vdash_{\mathsf{WC}} h_{\alpha+1}(A)\alpha b_\alpha(A)$;

(iii) $h_{\alpha+1}(A)\alpha b_\alpha(A) \vdash_{\mathsf{WC}} A$;

(iv) $A \equiv_{\mathsf{WC}} h_{\alpha+1}(A)\alpha b_\alpha(A)$.

Proof. We make a case distinction on $r_{\alpha+1}(A)$ in order to prove Parts (i) to (iii) separately in each case.

Suppose that $r_{\alpha+1}(A) = \beta b_\alpha(A)$ for some ordinal β. Since $A \in \mathbb{W}_\alpha$, then $\beta \geq \alpha$. But since it is in the $(\alpha+1)$-remainder, $\beta < \alpha + 1$. We conclude that $\beta = \alpha$, and hence that $r_{\alpha+1}(A) = \alpha b_\alpha(A)$. Then Parts (i) to (iii) are just a corollary of Lemma 4.2.

However it can be the case that $r_{\alpha+1}(A) = \top$ and hence $b_\alpha(A) = \top$ as well. Then Part (i) becomes an instance of Lemma 3.6, Part (ii) follows from Rule R3 and Part (iii) is a consequence of Lemma 3.3.

Finally, Part (iv) is a corollary of all of the other parts put together. □

The following result is Lemma 3.15 of [14]. It describes part of a recursive decision procedure for provability in RC between worms.

Lemma 4.5 *For any two worms A and B and for any ordinal α we have that $A \vdash_{\mathsf{RC}} \alpha B$ if and only if both $h_\alpha(A) \vdash_{\mathsf{RC}} \alpha h_\alpha(B)$ and $A \vdash_{\mathsf{RC}} r_\alpha(B)$.*

Let us see that we can prove one of the implications in WC, which we will later use in the proof of our main theorem (Theorem 6.1). There is no *a priori* reason why the other implication can't also hold; in fact, we will see that it does, since the calculi are equivalent for worms. It just so happens that we have no use for it.

Lemma 4.6 *For any two worms A and B, and for any ordinal α, if we have that $h_\alpha(A) \vdash_{\mathsf{WC}} \alpha h_\alpha(B)$ and $A \vdash_{\mathsf{WC}} r_\alpha(B)$, then we have $A \vdash_{\mathsf{WC}} \alpha B$.*

Proof. Taking into consideration that $A = h_\alpha(A) r_\alpha(A)$ and similarly for B, consider two cases. In the first case, $r_\alpha(B) = \top$, and this is a consequence of Lemma 4.2. In the second case, $r_\alpha(B) = \beta C$ for some $\beta < \alpha$ and worm C. Then the result follows from Rule R3. □

We now want to prove that RC is conservative over WC using the following inductive strategy. If $A \vdash_{\mathsf{RC}} B$, we use Lemma 4.5 to recast this into a collection of provability statements in RC between worms with smaller lengths. We then translate them to WC using the induction hypothesis, and finally go back with the help of Lemma 4.6. However, depending on the worms A and B, it could be the case that these two theorems are not enough, since they don't always reduce the length of the provability statements. In what follows, we introduce some more useful notions and results, which will help us deal with that problem.

5 Well founded orders on worms

It is possible to define an order relation between worms as is standard in the literature.

Definition 5.1 [Ordering worms] We say that $A <_\alpha B$ if $B \vdash_{\mathsf{WC}} \alpha A$. Furthermore, we say that $A \leq_\alpha B$ if either $A <_\alpha$ or $A \equiv_{\mathsf{WC}} B$. The provability can be taken in RC to obtain $<_\alpha^{\mathsf{RC}}$ and $\leq_\alpha^{\mathsf{RC}}$, respectively.

It is well-known that $<_\alpha^{\mathsf{RC}}$ is irreflexive [7]. Since WC is embedded in RC, we also know that $<_\alpha$ is irreflexive. It is easy to see that both relations are transitive.

Our goal now is to show that $<_\alpha$ is a total relation over worms in $\mathbb{W}_\alpha$. This has been shown for $<_\alpha^{\mathsf{RC}}$ using worm normal forms [7], but here we follow a different strategy, proposed in [11]. We start by presenting a number of useful sufficient conditions to deduce $A <_\alpha B$, and one to deduce $A \equiv B$.

Lemma 5.2 *Let $A, B \in \mathbb{W}_\alpha$ such that $A, B \neq \top$. Then in* WC *(and hence in* RC*) we have the following:*

(i) *If $b_\alpha(B) \vdash A$, then $A <_\alpha B$;*

(ii) *If $A <_\alpha b_\alpha(B)$, then $A <_\alpha B$;*

(iii) *If $b_\alpha(A) <_\alpha B$ and $h_{\alpha+1}(A) <_{\alpha+1} h_{\alpha+1}(B)$, then $A <_{\alpha+1} B$ (and consequently $A <_\alpha B$);*

(iv) *If $b_\alpha(A) <_\alpha B$ and $b_\alpha(B) <_\alpha A$ and $h_{\alpha+1}(A) \equiv h_{\alpha+1}(B)$, then $A \equiv B$.*

Proof. For the first item, from $b_\alpha(B) \vdash A$ we get by Necessitation that $\alpha b_\alpha(B) \vdash \alpha A$. Since by Lemma 4.4 we know that $B \vdash \alpha b_\alpha(B)$, we can conclude that $B \vdash \alpha A$. The second item follows by the transitivity of $<_\alpha$, taking into account that $b_\alpha(B) <_\alpha B$ (Lemma 4.4.(i)). The third item follows from Lemma 4.6. Finally, for the fourth item we use $B \vdash \alpha b_\alpha(A)$ and $h_{\alpha+1}(B) \vdash h_{\alpha+1}(A)$ to get $B \vdash h_{\alpha+1}(A)\alpha b_\alpha(A)$, and hence $B \vdash A$. Then we obtain $A \vdash B$ in the same way. □

Now we are ready to prove the totality of $<_\alpha$ for worms in $\mathbb{W}_\alpha$.

Lemma 5.3 (Trichotomy, [11]) *Given worms $A, B \in \mathbb{W}_\alpha$, we have that either $A <_\alpha B$, or $A \equiv B$, or $B <_\alpha A$.*

Proof. We proceed by induction on the length of AB. If the length is zero, *i.e.*, if $A = B = \top$, then clearly $A \equiv B$.

Note that by Lemma 3.6, $\top <_\alpha C$ regardless of the worm $C \in \mathbb{W}_\alpha$, as long as $C \neq \top$. Then if exactly one of A, B is $\top$ we have also solved our problem.

Now for the induction step, take both A and B with positive length. Our induction hypothesis is:

> For any ordinal β and worms $C, D \in \mathbb{W}_\beta$ such that $|CD| < |AB|$, we have $C <_\beta D$, or $C \equiv D$, or $D <_\beta C$.

Let ξ be the minimum ordinal in AB, which means that $\alpha \leq \xi$. According to Lemma 5.2, if $A \leq_\xi b_\xi(B)$ or $B \leq_\xi b_\xi(A)$, we can conclude $A <_\xi B$ or $B <_\xi A$, respectively. Assume then that $A \not\leq_\xi b_\xi(B)$ and $B \not\leq_\xi b_\xi(A)$. Since we took $A \neq \top$, it is clear that $|b_\xi(A)| < |A|$, and analogously for B. Then by the induction hypothesis, we have $b_\xi(B) <_\xi A$ and $b_\xi(A) <_\xi B$.

Since we are assuming ξ is in AB, we also know that

$$|h_{\xi+1}(A)h_{\xi+1}(B)| < |AB|,$$

and thus by the induction hypothesis we have

$$\begin{aligned} h_{\xi+1}(A) <_{\xi+1} h_{\xi+1}(B), &\text{ or} \\ h_{\xi+1}(B) <_{\xi+1} h_{\xi+1}(A), &\text{ or} \\ h_{\xi+1}(A) \equiv h_{\xi+1}(B). & \end{aligned}$$

In the first two cases we use Lemma 5.2 again to conclude $A <_\xi B$ or $B <_\xi A$ respectively. In the last case we can use the same lemma to get $A \equiv B$.

As a final remark, we observe that since $\alpha \leq \xi$, we have that $C <_\xi D$ implies $C <_\alpha D$ for any worms $C, D \in \mathbb{W}_\xi$. □

With the totality of $<_\alpha$ at hand, we can already show that proofs in RC and WC are equivalent for certain worms.

Theorem 5.4 *If $A, B \in \mathbb{W}_\alpha$, then:*

$$\begin{aligned} A <_\alpha B &\iff A <_\alpha^{\mathsf{RC}} B; \\ A \equiv_{\mathsf{WC}} B &\iff A \equiv_{\mathsf{RC}} B. \end{aligned}$$

Proof. Both left-to-right implications are a consequence of RC extending WC (Theorem 3.7).

For the right-to-left implication of the first statement, we reason as follows: suppose that $A <_\alpha^{\mathsf{RC}} B$ but it is not the case that $A <_\alpha B$. Then by the totality of $<_\alpha$ for worms in $\mathbb{W}_\alpha$ (Lemma 5.3), either $B <_\alpha A$, or $A \equiv_{\mathsf{WC}} B$. By the inclusion of WC in RC, we then conclude that either $B <_\alpha^{\mathsf{RC}} A$, or $A \equiv_{\mathsf{RC}} B$. But neither of these two cases is possible, as they contradict the irreflexivity of $<_\alpha^{\mathsf{RC}}$. Then it must be the case that $A <_\alpha B$. The proof of the second statement is analogous. □

6 Conservativity of RC over WC

We are now ready to prove the main result of this paper.

Theorem 6.1 *For any worms A and B', if $A \vdash_{\mathsf{RC}} B'$, then also $A \vdash_{\mathsf{WC}} B'$.*

Proof. If $B' = \top$, the result is immediate. Assume then that $B' = \alpha B$ for some ordinal α and worm B.

The proof proceeds by complete induction on the length of $A\alpha B$. The minimum length of $A\alpha B$ is 1, and it occurs only when $A = B = \top$. But the premise $\top \vdash_{\mathsf{RC}} \alpha$ is absurd since it would contradict the irreflexivity of $<_\alpha^{\mathsf{RC}}$, and hence there is nothing left to prove.

For the induction step, note that our induction hypothesis is the following:

For any worms C, D such that $|CD| < |A\alpha B|$ and $C \vdash_{\mathsf{RC}} D$, we have that $C \vdash_{\mathsf{WC}} D$.

Assume that $A \vdash_{\mathsf{RC}} \alpha B$. From Lemma 4.5 we get $h_\alpha(A) \vdash_{\mathsf{RC}} \alpha h_\alpha(B)$ and $A \vdash_{\mathsf{RC}} r_\alpha(B)$. Consider the following cases:

(i) $r_\alpha(A) \neq \top$ or $r_\alpha(B) \neq \top$.
Then, since $|A| = |h_\alpha(A)| + |r_\alpha(A)|$ (and equivalently for B), we know that $|h_\alpha(A)| < |A|$ or $|h_\alpha(B)| < |B|$. Then $|h_\alpha(A)\alpha h_\alpha(B)| < |A\alpha B|$. Furthermore, $|Ar_\alpha(B)| < |A\alpha B|$. By using the induction hypothesis twice we get $h_\alpha(A) \vdash_{\mathsf{WC}} \alpha h_\alpha(B)$ and $A \vdash_{\mathsf{WC}} r_\alpha(B)$, which is enough to show $A \vdash_{\mathsf{WC}} \alpha B$ by Lemma 4.6.

(ii) $r_\alpha(A) = \top$ and $r_\alpha(B) = \top$.
In this case, we know that $A, B \in \mathbb{W}_\alpha$, and hence that $A \vdash_{\mathsf{WC}} \alpha B$ by Theorem 5.4.

□

The proof of the preceding result (together with the proofs of the results it uses) gives us a constructive algorithm to decide whether $A \vdash_{\mathsf{WC}} B$. Furthermore, if indeed $A \vdash_{\mathsf{WC}} B$, this algorithm provides a list of syntactical steps which form a formal proof. Since at each iteration of the recursion we may need to decide several different statements with only slightly smaller lengths, the algorithm is exponential. It is known that there is a polynomial procedure to decide RC [10] (and hence, as we've seen, WC), but it uses semantics. Finding a polynomial syntactical algorithm remains an open problem.

Combining Theorems 3.7 and 6.1, we obtain the promised result: RC is a conservative extension of WC.

Theorem 6.2 *For any worms A and B we have that $A \vdash_{\mathsf{RC}} B$ if and only if $A \vdash_{\mathsf{WC}} B$.*

Combining this theorem with Lemma 2.3 we obtain the following corollary.

Corollary 6.3 *Let φ and ψ be closed RC-formulas with corresponding worms A and B such that $\varphi \equiv_{\mathsf{RC}} A$ and $\psi \equiv_{\mathsf{RC}} B$. Then we have $\varphi \vdash_{\mathsf{RC}} \psi$ if and only if $A \vdash_{\mathsf{WC}} B$.*

7 Relational semantics and Ignatiev's model

Let us briefly recall how we arrived at the calculus WC while we comment on the relational semantics for the intermediate steps. Japaridze went from the regular provability logic GL to its polymodal version GLP. Whereas GL is frame-complete, it turned out that GLP is frame incomplete. Ignatiev intensively studied the closed fragment GLP^0 and — although the frame incompleteness is still salient — introduced a universal model $\mathcal{I}$ for it. Ignatiev's model $\mathcal{I}$ is essentially infinite, having fractal features.

Dashkov and Beklemishev studied reflection calculi and in particular the strictly positive fragments RC and RC^0 of GLP and GLP^0, respectively. Here the only connectives are the diamond modalities together with conjunctions. The reflection calculi are known to be frame complete and have the finite model property. Furthermore, linear frames (xRy and xRz imply $y = z$, yRz or zRy) suffice for the closed fragment [11].

In this paper we perform a final simplification on RC^0, getting rid of the conjunctions to end up with WC. Inspired by the finite model property of RC^0 and whence of WC, in the last section of this paper we question whether WC may have a universal model $\mathcal{U}$ that is significantly simpler than Ignatiev's model $\mathcal{I}$. We settle the answer to this question with a yes and a no.

Yes, $\mathcal{U}$ can be simpler in that we can bound the length of the strict chains of successors in $\mathcal{U}$ by ω. For this, it suffices to take the disjoint union of all finite RC^0 counter models for all statements for which $A \nvdash B$. This clearly defines a universal model for RC^0 with only finite strict R_α-chains whereas, as we shall see below, the model $\mathcal{I}$ has arbitrarily long strict R_α-chains. On the other hand, we shall see that for a large class of universal models $\mathcal{U}$, they inherit much of the intrinsic complexity of $\mathcal{I}$ in that for infinitely many essentially different points $x \in \mathcal{I}$ we can find corresponding points $y \in \mathcal{U}$ such that x and y have the same modal theory.

Before we can make this statement precise, we need a couple of technical definitions that allow us to describe Ignatiev's model $\mathcal{I}$. As a first step, we need to define the *end-logarithm* ℓ as a function from the ordinals to the ordinals by stipulating $\ell(0) := 0$ and $\ell(\alpha + \omega^\beta) := \beta$. Next, we need to define iterates of ℓ — the *hyper-logarithms* — and write ℓ^ξ to denote the *ξ-th iterate of ℓ*. We define:

(i) $\ell^0 := \mathsf{id}$,
(ii) $\ell^1 := \ell$ and,
(iii) $\ell^{\alpha+\beta} := \ell^\beta \circ \ell^\alpha$.

Clearly, these three properties do not tell us anything about ℓ^ξ for an additively indecomposable ξ. To fix that, we will use the notion of *initial function*. An initial function on the ordinals is a function that maps initial segments $[0, \ldots, \alpha]$ onto initial segments $[0, \ldots, \beta]$.

We now further require that each ℓ^ξ is point-wise maximal among all families of initial ordinal functions $\{f^\xi\}_{\xi \in \mathsf{On}}$ that satisfy the three properties. In this way, clearly each f^ξ defines an initial function. For the purposes of this paper, many of the exact details of the ℓ^ξ functions are irrelevant and we refer the interested reader to [13] or [12] for further details.

Let us, by way of example, compute the first couple of values of ℓ^ω. Recall that as always, ε_ζ denotes the ζ-th fixpoint of $x \mapsto \omega^x$. Since the initial segment

$[0]$ should be mapped onto an initial segment, it must be that $\ell^\omega(0) = 0$. If $\alpha < \varepsilon_0$ then it is easy to see that for some $n < \omega$ we have that $\ell^n(\alpha) = 0$. Consequently, $\ell^\omega(\alpha) = \ell^{n+\omega}(\alpha) = \ell^\omega \circ \ell^n(\alpha) = \ell^\omega(0) = 0$. Consequently, each initial segment $[0, \ldots, \alpha]$ for $\alpha < \varepsilon_0$ will be mapped by ℓ^ω to the initial segment $[0]$.

What about $\ell^\omega(\varepsilon_0)$? If we disregard the requirement on initiality, it is not hard to see that $\ell^\omega(\varepsilon_0)$ could be *any* value. However, since ℓ^ω should map the initial segment $[0, \ldots, \varepsilon_0]$ to an initial segment, the maximal possible value doing so requires that $\ell^\omega(\varepsilon_0) = 1$.

We observe that $\ell^\omega(\xi + \zeta) = \ell^{1+\omega}(\xi + \zeta) = \ell^\omega \circ \ell(\xi + \zeta) = \ell^\omega \circ \ell(\zeta) = \ell^{1+\omega}(\zeta) = \ell^\omega(\zeta)$ so that $\ell^\omega(\varepsilon_0 + \alpha) = 0$ for any $\alpha < \varepsilon_0$.

Following this kind of arguments, we can now see that the next value where ℓ^ω increases will be at ε_1 and $\ell^\omega(\varepsilon_1) = 2$. Fortunately, we do not need to prove tons of theorems any time we are required to know some value of $\ell^\alpha(\beta)$ and in [12,13] a recursive algorithm is presented to compute these values:

Proposition 7.1 *For ordinals ξ, ζ, the following recursion is well-defined and determines all the $\ell^\xi(\zeta)$ values:*

(i) $\ell^0(\alpha) = \alpha$,

(ii) $\ell^\xi(0) = 0$,

(iii) $\ell^1(\alpha + \omega^\beta) = \beta$,

(iv) $\ell^{\omega^\rho + \xi} = \ell^\xi \circ \ell^{\omega^\rho}$ *provided that* $\xi < \omega^\rho + \xi$,

(v) $\ell^{\omega^\rho}(\zeta) = \ell^{\omega^\rho}(\ell^\eta(\zeta))$ *if* $\rho > 0$ *and* $\eta < \omega^\rho$ *is such that* $\ell^\eta(\zeta) < \zeta$,

(vi) $\ell^{\omega^\rho}(\zeta) = \sup_{\delta \in [0,\zeta)} (\ell^{\omega^\rho}(\delta) + 1)$ *if* $\rho > 0$ *and* $\zeta = \ell^\eta(\zeta)$ *for all* $\eta < \omega^\rho$.

Now that the hyper-logarithms have been defined, we can specify the points of Ignatiev's model $\mathcal{I}$ which are the so-called *ℓ-sequences.*

Definition 7.2 [ℓ-sequence] An *ℓ-sequence* is a function $f : \mathsf{On} \to \mathsf{On}$ such that for each ordinal ζ we have $f(\zeta) \leq \ell^{-\xi+\zeta}(f(\xi))$ for $\xi < \zeta$ large enough.

At times we shall write f_ξ instead of $f(\xi)$. We note that for each ℓ-sequence f the inequality $f(\alpha + 1) \leq \ell(f(\alpha))$ holds. Furthermore, the requirement of $\xi < \zeta$ being large enough is important, as it means that $f = \langle \omega^{\varepsilon_0+1}, \varepsilon_0, \varepsilon_0, \ldots 1, 0 \ldots \rangle$ is an ℓ-sequence, where $f(0) = \omega^{\varepsilon_0+1}$, $f(i) = \varepsilon_0$ for $0 < i < \omega$, $f(\omega) = 1$ and $f(i) = 0$ for $i > \omega$. It is easy to see that $\ell^\omega(\omega^{\varepsilon_0+1}) = 0$ and $\ell^\omega(\varepsilon_0) = 1$. Then $f(\omega) \leq \ell^\omega(f(1))$ but it is not the case that $f(\omega) \leq \ell^\omega(f(0))$.

We can now define the class-size version of Ignatiev's model as the collection of all ℓ-sequences with suitable relations R_ξ to model each of the $\langle \xi \rangle$ modalities. For all practical purposes we can take sufficiently large set-size truncations of the class-size model.

Definition 7.3 [Ignatiev's model] Ignatiev's model is $\mathcal{I} := \langle I, \{R_\xi\}_{\xi \in \mathsf{On}} \rangle$, where I is the collection of all ℓ-sequences and $f R_\xi g$ if both $f(\alpha) = g(\alpha)$ for $\alpha < \xi$ and $f(\xi) > g(\xi)$.

For example, we can see that

$$\begin{array}{rl} & \langle \omega^{\varepsilon_0+1}, \varepsilon_0, \varepsilon_0, \ldots 0, 0 \ldots \rangle \ R_0 \ \langle \varepsilon_0, \varepsilon_0, \varepsilon_0, \ldots 0, 0 \ldots \rangle, \\ & \langle \omega^{\varepsilon_0+1}, \varepsilon_0, \varepsilon_0, \ldots 1, 0 \ldots \rangle \ R_0 \ \langle \varepsilon_0, \varepsilon_0, \varepsilon_0, \ldots 1, 0 \ldots \rangle, \\ & \langle \omega^{\varepsilon_0+1}, \varepsilon_0, \varepsilon_0, \ldots 1, 0 \ldots \rangle \ R_\omega \ \langle \omega^{\varepsilon_0+1}, \varepsilon_0, \varepsilon_0, \ldots 0, 0 \ldots \rangle, \\ \text{but} & \neg\Big(\langle \omega^{\varepsilon_0+1}, \varepsilon_0, \varepsilon_0, \ldots 1, 0 \ldots \rangle \ R_\omega \ \langle \varepsilon_0, \varepsilon_0, \varepsilon_0, \ldots 0, 0 \ldots \rangle \Big). \end{array}$$

The relation $\mathcal{I}, x \Vdash \varphi$ is defined as usual, where we omit the mention of the model $\mathcal{I}$: $x \Vdash \top$; $x \Vdash \varphi \wedge \psi :\Leftrightarrow x \Vdash \varphi$ and $x \Vdash \psi$; $x \Vdash \neg\varphi :\Leftrightarrow x \nVdash \varphi$; and finally, $x \Vdash \langle \xi \rangle \varphi :\Leftrightarrow$ there is a y s.t. $xR_\xi y$ and $y \Vdash \varphi$.

Theorem 7.4 ([13]) GLP^0 *is sound and complete with respect to Ignatiev's model, that is* $\mathsf{GLP}^0 \vdash \varphi$ *if and only if* $\forall x \in \mathcal{I}\ x \Vdash \varphi$.

We now define an important subset of $\mathcal{I}$ that has rather nice properties.

Definition 7.5 [Main axis, MA] By MA we denote the *main axis* of Ignatiev's model $\mathcal{I}$ and define it as such: $f \in \mathsf{MA} :\Leftrightarrow \forall \zeta\ \forall \xi{<}\zeta\ f(\zeta) = \ell^{-\xi+\zeta}(f(\xi))$.

For example, $\langle \omega^{\varepsilon_0+1}, \varepsilon_0, \varepsilon_0, \ldots 1, 0 \ldots \rangle$ is not on the main axis, whereas $\langle \omega^{\varepsilon_0+1}, \varepsilon_0 + 1, 0 \ldots 0, 0 \ldots \rangle$ is. One of the nice properties of the main axis is that each point on it is modally definable. We refer the reader to [13] for a proof of the following.

Lemma 7.6 *For each* $x \in \mathsf{MA}$ *there is a worm* A *such that* $\mathcal{I}, y \Vdash A \wedge [0]\neg A$ *if and only if* $x = y$. *Moreover, each worm* A *defines a point at the main axis via* $A \wedge [0]\neg A$.

8 On universal models for WC

The current proof of the main result of this section does not hold for any universal model but only for models that satisfy an additional natural condition. Let us recall that in the context of provability logics, a model is called *Euclidean* whenever $xR_\alpha y$ and $xR_\beta z$ imply $yR_\beta z$ for $\beta < \alpha$. Furthermore, by $\mathsf{Th}(x)$ we denote the collection of worms $\{A \mid x \Vdash A\}$. Now we are able to state the main result.

Theorem 8.1 *Let* $\mathcal{U}$ *be a Euclidean universal model for* WC. *We have that for each point* $x \in \mathcal{I}$ *with* $x \in \mathsf{MA}$, *there is some* $y \in \mathcal{U}$ *such that* $\mathsf{Th}(x) = \mathsf{Th}(y)$.

Proof. Let $x \in \mathsf{MA}$ be arbitrary and let A be the worm given by Lemma 7.6 such that $A \wedge [0]\neg A$ is true at x and nowhere else. Since $A \nvdash_{\mathsf{WC}} 0A$, we can find $y \in \mathcal{U}$ with $\mathcal{U}, y \Vdash A$ and $\mathcal{U}, y \nVdash 0A$. We shall show that for this particular choice of y we have $\mathsf{Th}(x) = \mathsf{Th}(y)$.

First, we assume that $\mathcal{I}, x \Vdash B$ for some worm B. By the definability of x and the completeness of $\mathcal{I}$, we know that $\mathsf{GLP} \vdash A \wedge [0]\neg A \to B$. By Lemma 8.3, which we prove below, and the conservativity of GLP over WC, we may conclude that actually $A \vdash_{\mathsf{WC}} B$, and hence $\mathcal{U}, y \Vdash B$.

Now assume that $\mathcal{I}, x \nVdash B$ for some worm B. Then $\mathsf{GLP} \nvdash A \to B$, which means that $A \nvdash_{\mathsf{WC}} B$. Let C be a worm equivalent to $A \wedge B$. Clearly, by the trichotomy of $<_0$ and since $A \nvdash_{\mathsf{WC}} B$ and $A \nvdash_{\mathsf{RC}} \langle 0\rangle(A \wedge B)$, we have that $A <_0 C$, whence $C \vdash_{\mathsf{WC}} 0A$. We assume for a contradiction that $\mathcal{U}, y \Vdash B$. In that case, since also $\mathcal{U}, y \Vdash A$, we may conclude by Lemma 8.5 below that $\mathcal{U}, y \Vdash C$. But since $C \vdash_{\mathsf{WC}} 0A$, this would mean that $\mathcal{U}, y \Vdash 0A$ which is a contradiction by our choice of y. □

We finish the paper by proving the two critical lemmas that were needed in the above proof, together with some auxiliary observations. First we define a relation $y \succeq x$ on $\mathcal{I}$ as y being point-wise at least x, that is, $y \succeq x$ if and only if for all ξ we have $y_\xi \geq x_\xi$. The following lemma tells us that this relation, together with the point x where a worm A is true for the first time, characterizes all the points where A holds.

Lemma 8.2 *Let A be a worm and $x, y \in \mathcal{I}$. We have that*

$$x \Vdash A \implies \Big(y \succeq x \implies y \Vdash A\Big);$$

$$x \Vdash A \wedge [0]\neg A \implies \Big(y \succeq x \iff y \Vdash A\Big).$$

Proof. During this proof we make use of a specific operation on ℓ-sequences: for an ordinal α and ℓ-sequences f and g, we define $\alpha(f,g)$ to be the ordinal sequence such that $\alpha(f,g)_\zeta = f_\zeta$ for $\zeta < \alpha$ and $\alpha(f,g)_\zeta = g_\zeta$ for $\zeta \geq \alpha$. Clearly, whenever $g_\alpha \leq f_\alpha$, we have that $\alpha(f,g)$ is again an ℓ-sequence.

The first item is proven by an easy induction on A. It was already observed as Lemma 2.4.3. of [15]. Note also that the $\implies$ direction of the second item follows from the first one.

For the $\Longleftarrow$ direction of the second item, we fix x such that $x \Vdash A \wedge [0]\neg A$, consider y such that $y \Vdash A$, and assume for a contradiction that $y \not\succeq x$. Let ξ be the smallest ordinal such that $y_\xi < x_\xi$. Then it is easy to see that $xR_\xi\xi(x,y)$.

We will set out to prove the following **claim**: for any worm B and any ℓ-sequences f and g with $f_\alpha > g_\alpha$ we have that if $f \Vdash B$ and $g \Vdash B$, then $\alpha(f,g) \Vdash B$.

Clearly the result follows from the claim, as it would imply that $\xi(x,y) \Vdash A$, and hence that $x \Vdash \langle\xi\rangle A$. Consequently, also $x \Vdash \langle 0\rangle A$, which contradicts the assumption that $x \Vdash [0]\neg A$.

We prove the claim by induction on B with the base case being trivial. Thus we consider the inductive case where $B = \langle\zeta\rangle C$ assuming $f \Vdash \langle\zeta\rangle C$ and $g \Vdash \langle\zeta\rangle C$.

In the case where $\zeta < \alpha$, we see that for any w such that $fR_\zeta w$, we also have $\alpha(f,g)R_\zeta w$, which tells us that $\alpha(f,g) \Vdash \langle\zeta\rangle C$.

In the case where $\zeta \geq \alpha$, we find w and w' such that $fR_\zeta w \Vdash C$ and $gR_\zeta w' \Vdash C$. By the induction hypothesis, we know that $\alpha(w,w') \Vdash C$. But since $\alpha(f,g)R_\zeta\alpha(w,w')$, we see that $\alpha(f,g) \Vdash \langle\zeta\rangle C$ as was to be shown. □

With this characterization lemma, we can easily prove the following admissible rule.

Lemma 8.3 *Let A and B be worms such that* $\mathsf{GLP} \vdash A \wedge [0]\neg A \to B$. *Then we have that* $\mathsf{GLP} \vdash A \to B$.

Proof. Given A, let x be the unique ℓ-sequence where $A \wedge [0]\neg A$ holds. Clearly, since $\mathsf{GLP} \vdash A \wedge [0]\neg A \to B$, we also have that $x \Vdash B$. We prove that for any y, if $y \Vdash A$, then $y \Vdash B$, from which the result follows by completeness. By the previous lemma, if $y \Vdash A$, then $y \succeq x$. But then, using the previous lemma again, we may conclude that $y \Vdash B$. □

The next two lemmas relate to conjunctions and models of WC.

Lemma 8.4 *Let $\mathcal{U}$ be a Euclidean universal model for* WC *and $x \in \mathcal{U}$. Then, if $x \Vdash A$ with $A \in \mathbb{W}_{\alpha+1}$ and $x \Vdash \alpha B$, it also holds that $x \Vdash A\alpha B$.*

Proof. By induction on the length of A with the base case being trivial. For the inductive case, suppose that $x \Vdash \gamma A$ for some $\gamma > \alpha$, and that $x \Vdash \alpha B$. Then there is $y \in \mathcal{U}$ such that $xR_\gamma y$ and $y \Vdash A$. Likewise, there is $z \in \mathcal{U}$ such that $xR_\alpha z$ and $z \Vdash B$. Since $\mathcal{U}$ is Euclidean, we know that $yR_\alpha z$, and hence by induction hypothesis that $y \Vdash A\alpha B$. Then clearly $x \Vdash \gamma A\alpha B$. □

With this lemma at hand we can show that although a Euclidean WC-model $\mathcal{U}$ cannot speak directly about conjunctions, it can indirectly do so.

Lemma 8.5 *Let $\mathcal{U}$ be a Euclidean universal model for* WC *and $x \in \mathcal{U}$. Let A, B and C be worms such that $A \wedge B \equiv_{\mathsf{RC}} C$. Then we have $x \Vdash A$ and $x \Vdash B$ if and only if $x \Vdash C$.*

Proof. The right-to-left direction is trivial. The left-to-right direction follows by induction on the number of different symbols of AB (the *width* of AB) following the standard proof that worms are closed under conjunctions (Lemma 9 of [3] and Corollary 4.13 of [7]). The base case is trivial. For the induction step, let α be the minimal modality of AB. We know by Lemmas 4.2 and 4.4 that $x \Vdash h_{\alpha+1}(A)$, $x \Vdash \alpha b_\alpha(A)$, $x \Vdash h_{\alpha+1}(B)$, and $x \Vdash \alpha b_\alpha(B)$. Let D be provably equivalent to $h_{\alpha+1}(A) \wedge h_{\alpha+1}(B)$. Then, since there is no α in the $(\alpha+1)$-heads of A and B, we know that $x \Vdash D$ by the induction hypothesis. By Corollary 4.12 of [7], we know that either $\alpha b_\alpha(A) \vdash \alpha b_\alpha(B)$ or $\alpha b_\alpha(B) \vdash \alpha b_\alpha(A)$. Let αE be the maximum. We obtain $x \Vdash D\alpha E$ by Lemma 8.4, and clearly $D\alpha E \equiv_{\mathsf{RC}} A \wedge B$. □

We conclude by observing that all the results proven about universal (Euclidean) models of WC also hold for universal (Euclidean) models of RC^0. It remains to see whether we can prove the same results for non-Euclidean models.

References

[1] Artale, A., D. Calvanese, R. Kontchakov and M. Zakharyaschev, *The DL-Lite family and relations*, Journal of Artificial Intelligence Research **36** (2009), pp. 1–69.

[2] Beklemishev, L. D., *Provability algebras and proof-theoretic ordinals, I*, Annals of Pure and Applied Logic **128** (2004), pp. 103–124.

[3] Beklemishev, L. D., *Veblen hierarchy in the context of provability algebras*, in: *Logic, Methodology and Philosophy of Science, Proceedings of the Twelfth International Congress* (2005), pp. 65–78.

[4] Beklemishev, L. D., *The worm principle*, in: Z. Chatzidakis, P. Koepke and W. Pohlers, editors, *Logic Colloquium 2002, Lecture Notes in Logic 27*, ASL Publications, 2006 pp. 75–95.

[5] Beklemishev, L. D., *Calibrating provability logic: From modal logic to Reflection Calculus*, in: T. Bolander, T. Braner, T. S. Ghilardi and L. Moss, editors, *Advances in Modal Logic 9* (2012), pp. 89–94.

[6] Beklemishev, L. D., *Positive provability logic for uniform reflection principles*, Annals of Pure and Applied Logic **165** (2014), pp. 82–105.

[7] Beklemishev, L. D., D. Fernández-Duque and J. J. Joosten, *On provability logics with linearly ordered modalities*, Studia Logica **102** (2014).

[8] Beklemishev, L. D. and D. Gabelaia, *Topological completeness of the provability logic GLP*, Annals of Pure and Applied Logic **164** (2013), pp. 1201–1223.

[9] Beklemishev, L. D., J. J. Joosten and M. Vervoort, *A finitary treatment of the closed fragment of Japaridze's provability logic*, Journal of Logic and Computation **15** (2005), pp. 447–463.

[10] Dashkov, E. V., *On the positive fragment of the polymodal provability logic GLP*, Mathematical Notes **91** (2012), pp. 318–333.

[11] Fernández-Duque, D., *Worms and spiders: Reflection calculi and ordinal notation systems*, IfCoLoG Journal of Logics and their Applications **4** (2017), pp. 3277–3356.

[12] Fernández-Duque, D. and J. J. Joosten, *Hyperations, Veblen progressions and transfinite iteration of ordinal functions*, Annals of Pure and Applied Logic **164** (2013), pp. 785–801.

[13] Fernández-Duque, D. and J. J. Joosten, *Models of transfinite provability logic*, The Journal of Symbolic Logic **78** (2013), pp. 543–561.

[14] Fernández-Duque, D. and J. J. Joosten, *Well-orders in the transfinite Japaridze algebra*, Logic Journal of the IGPL **22** (2014), pp. 933–963.

[15] Icard, T., "Models of the Polymodal Provability Logic," Master's thesis, Universiteit van Amsterdam (2008).

[16] Ignatiev, K. N., *On strong provability predicates and the associated modal logics*, The Journal of Symbolic Logic **58** (1993), pp. 249–290.

[17] Japaridze, G. K., *The polymodal provability logic*, in: *Intensional logics and logical structure of theories: material from the fourth Soviet-Finnish symposium on logic, Telavi, May 20–24, 1985*, Metsniereba, (1988), pp. 16–48, (In Russian).

[18] Joosten, J. J., *Turing-Taylor expansions for arithmetic theories*, Studia Logica **104** (2016), pp. 1225–1243.

[19] Pakhomov, F., *On the complexity of the closed fragment of Japaridze's provability logic*, Archive for Mathematical Logic **53** (2014), pp. 949–967.

[20] Shapirovsky, I., *PSPACE-decidability of Japaridze's polymodal logic*, in: C. Areces and R. Goldblatt, editors, *Advances in Modal Logic 7* (2008), pp. 289–304.

A Simple Cut-Free System for a Paraconsistent Logic Equivalent to S5

Arnon Avron [1]

Tel Aviv University, Israel

Ori Lahav [2]

Tel Aviv University, Israel

Abstract

NS5 is a paraconsistent logic in the classical language, which is equivalent to the well-known modal logic **S5**. We provide a particularly simple hypersequential system for the propositional **NS5**, and prove a strong cut-admissibility theorem for it. Our system is obtained from the standard hypersequential system for classical logic by just weakening its two rules for negation, and without introducing any new structural rule. We also explain how to extend the results to the natural first-order extension of **NS5**. The latter is equivalent to the Constant Domain first-order **S5**.

Keywords: **S5**, modal logic, paraconsistent logic, hypersequents, cut-elimination

It is well-known that the cut-admissibility theorem fails for the standard Gentzen-type system for the propositional modal logic **S5** . Therefore various alternative Gentzen-type systems for this logic that do enjoy cut-admissibility have been proposed in the literature. Among those systems, the simplest are those that employ *hypersequents*. Several such systems for propositional **S5** have been presented over the years, e.g. in [12,18,2,19,16,6]. (See also [20], [17], and [10] for further examples and references.) However, we have not been able to find in the literature any reasonable sound and complete hypersequential calculus for *first-order* **S5** which is known to be cut-free.

The main goal of this paper is to present a particularly simple hypersequential system for first-order **S5**, for which even a *strong* cut-admissibility theorem holds. What is more, our system is obtained from the standard hypersequential system for *classical* logic using only slight changes in its two rules for *negation*, and without introducing any new structural rules. For this we extend to the first-order level the presentation which was investigated in [4] of the propositional **S5** as a paraconsistent logic **NS5** in the standard classical propositional

[1] Supported by the Israel Science Foundation under grant agreement 817/15.

[2] Supported by Len Blavatnik and the Blavatnik Family foundation.

language. This method also demonstrates the usefulness of viewing **S5** as a paraconsistent logic, and shows at the same time how close it is as such to classical logic.

The structure of the paper is as follows. In its first half (Sections 1–3) we provide our hypersequential version of the propositional **NS5**, and prove cut-admissibility for it using a *semantic* method.[3] The results and method are then extended to the first-order level in its second half (Section 4).

1 Preliminaries

We assume that all propositional languages share the same set $\{P_1, P_2, ...\}$ of propositional variables, and use p, q, r to vary over this set. The set of formulas of a propositional language $\mathcal{L}$ is denoted by $\mathcal{WFF}(\mathcal{L})$, and φ, ψ, σ will vary over it.

1.1 Calculi of Hypersequents

Definition 1.1 A *hypersequent* is a finite set of ordinary sequents. The elements of this set are called its *components*. We denote by $s_1 \mid \cdots \mid s_n$ the hypersequent whose components are $s_1, \dots, s_n$, and use G, H as metavariables for (possibly empty) hypersequents.[4]

Note 1 Hypersequents are often taken to be *multisets* of ordinary sequents. In such a case it is necessary to add to the list of structural rules the rule [EC] of *external contraction*, which allows one to infer $H \mid s$ from $H \mid s \mid s$. This is a reasonable choice, since external contraction is the main source of problems for hypersequential calculi (both for proving cut-elimination and for producing efficient proof-search procedures), and this is hidden if it is built into the definition of a hypersequent (as we do here in order to get simpler and more economic systems).

Usually, most of the rules in hypersequential calculi are obtained from standard rules of ordinary sequential calculi by allowing also side components in applications of the rules (in addition to the presence of side formulas). A simple example is provided in Figure 1, which presents a hypersequential version LK^h of Gentzen's system LK for classical logic. The rules of LK^h are just the obvious hypersequential versions of the rules of LK. Therefore it is easy to see that a hypersequent is derivable in LK^h iff one of its components is classically valid.

1.2 The Propositional System NS5

The language of the propositional modal logic **S5** is usually taken to be the one induced by $\{\wedge, \vee, \supset, \mathsf{F}, \Box\}$. *Classical* negation $\sim$ can be defined by $\sim\varphi = \varphi \supset \mathsf{F}$. Following Béziau ([7]) and Batens ([5]), one may also define a *paraconsistent*

[3] Gentzen-style syntactic proofs of cut-elimination are notoriously error-prone and difficult to check. Accordingly, we continue in this paper our program of providing checkable semantic proofs for hypersequential calculi. (See e.g. [11,3])

[4] Semantically, the interpretation of '|' is usually taken to be disjunctive.

Structural rules:

$$[\text{id}]\ \frac{}{\varphi \Rightarrow \varphi} \qquad [\text{cut}]\ \frac{H \mid \Gamma \Rightarrow \Delta, \varphi \quad H \mid \varphi, \Gamma \Rightarrow \Delta}{H \mid \Gamma \Rightarrow \Delta}$$

$$[\text{IW}]\ \frac{H \mid \Gamma \Rightarrow \Delta}{H \mid \Gamma', \Gamma \Rightarrow \Delta, \Delta'} \qquad [\text{EW}]\ \frac{H}{H \mid \Gamma \Rightarrow \Delta}$$

Logical rules:

$$[\wedge\Rightarrow]\ \frac{H \mid \Gamma, \varphi, \psi \Rightarrow \Delta}{H \mid \Gamma, \varphi \wedge \psi \Rightarrow \Delta} \qquad [\Rightarrow\wedge]\ \frac{H \mid \Gamma \Rightarrow \Delta, \varphi \quad H \mid \Gamma \Rightarrow \Delta, \psi}{H \mid \Gamma \Rightarrow \Delta, \varphi \wedge \psi}$$

$$[\vee\Rightarrow]\ \frac{H \mid \Gamma, \varphi \Rightarrow \Delta \quad H \mid \Gamma, \psi \Rightarrow \Delta}{H \mid \Gamma, \varphi \vee \psi \Rightarrow \Delta} \qquad [\Rightarrow\vee]\ \frac{H \mid \Gamma \Rightarrow \Delta, \varphi, \psi}{H \mid \Gamma \Rightarrow \Delta, \varphi \vee \psi}$$

$$[\supset\Rightarrow]\ \frac{H \mid \Gamma \Rightarrow \Delta, \varphi \quad H \mid \Gamma, \psi \Rightarrow \Delta}{H \mid \Gamma, \varphi \supset \psi \Rightarrow \Delta} \qquad [\Rightarrow\supset]\ \frac{H \mid \Gamma, \varphi \Rightarrow \Delta, \psi}{H \mid \Gamma \Rightarrow \Delta, \varphi \supset \psi}$$

$$[\neg\Rightarrow]\ \frac{H \mid \Gamma \Rightarrow \Delta, \varphi}{H \mid \Gamma, \neg\varphi \Rightarrow \Delta} \qquad [\Rightarrow\neg]\ \frac{H \mid \Gamma, \varphi \Rightarrow \Delta}{H \mid \Gamma \Rightarrow \Delta, \neg\varphi}$$

$$[\forall\Rightarrow]\ \frac{H \mid \Gamma, \varphi\{t/a\} \Rightarrow \Delta}{H \mid \Gamma, \forall x(\varphi\{x/a\}) \Rightarrow \Delta} \qquad [\Rightarrow\forall]\ \frac{H \mid \Gamma \Rightarrow \Delta, \varphi}{H \mid \Gamma \Rightarrow \Delta, \forall x(\varphi\{x/a\})}$$

$$[\exists\Rightarrow]\ \frac{H \mid \Gamma, \varphi \Rightarrow \Delta}{H \mid \Gamma, \exists x(\varphi\{x/a\}) \Rightarrow \Delta} \qquad [\Rightarrow\exists]\ \frac{H \mid \Gamma \Rightarrow \Delta, \varphi\{t/a\}}{H \mid \Gamma \Rightarrow \Delta, \exists x(\varphi\{x/a\})}$$

The rules $[\Rightarrow\forall]$ and $[\exists\Rightarrow]$ must obey the eigenvariable condition: a must not occur in the lower hypersequent.

Fig. 1. The proof system LK^h

negation $\neg$ by $\neg\varphi = \Box\varphi \supset \mathsf{F}$, and in fact, we can take the language of **S5** to be the language of classical logic $\mathcal{L}_{CL} = \{\supset, \wedge, \vee, \neg\}$. The connectives F and $\Box$ of **S5** are definable in $\mathcal{L}_{CL}$: F is equivalent to $\neg(p \supset p)$, where p is some atomic formula; while $\Box\varphi$ is equivalent in **S5** to $\neg\neg\varphi$.[5] When formulated in this language, **S5** becomes a *paraconsistent* logic, which was called **NS5** in [4].[6] Its semantics is given in the next definitions.

Definition 1.2 A pair $\langle W, \nu \rangle$ is called an **NS5**-*frame* for $\mathcal{L}_{CL}$ if W is a nonempty (finite) set (of "worlds"), and $\nu : W \times \mathcal{WFF}(\mathcal{L}_{CL}) \to \{t, f\}$ satisfies the following conditions:[7]

[5] This implies that we could have restricted ourselves to the language of $\{\supset, \neg\}$, since the classical $\vee$ and $\wedge$ can of course be defined in terms of $\supset$ and F.

[6] What is called here **NS5** was independently introduced by Béziau ([7,8,9] and Batens ([5]). (**NS5** was called $\mathbb{Z}$ by Béziau, and **A** by Batens.) Further study of this system was done by Osorio, Carballido, Zepeda, and Castellanos in [14] and [15].

[7] For **S5/NS5** this suffices. For normal modal logics in general one should use triples

- $\nu(w, \psi \wedge \varphi) = t$ iff $\nu(w, \psi) = t$ and $\nu(w, \varphi) = t$.
- $\nu(w, \psi \vee \varphi) = t$ iff $\nu(w, \psi) = t$ or $\nu(w, \varphi) = t$.
- $\nu(w, \psi \supset \varphi) = t$ iff $\nu(w, \psi) = f$ or $\nu(w, \varphi) = t$.
- $\nu(w, \neg\psi) = t$ iff there exists $w' \in W$ such that $\nu(w', \psi) = f$.

Definition 1.3 Let $\langle W, \nu \rangle$ be an **NS5**-frame.

- A formula φ is *true* in a world $w \in W$ ($w \Vdash \varphi$) if $\nu(w, \varphi) = t$.
- Let $\mathcal{T} \cup \{\varphi\}$ be a set of formulas in $\mathcal{L}_{CL}$. φ *follows in* **NS5** from $\mathcal{T}$ ($\mathcal{T} \vdash_{\mathbf{NS5}} \varphi$) if for every **NS5**-frame $\langle W, \nu \rangle$ and every $w \in W$: if $w \Vdash \psi$ for every $\psi \in \mathcal{T}$ then $w \Vdash \varphi$.
- A sequent $s = \Gamma \Rightarrow \Delta$ is *true* in a world $w \in W$ ($w \Vdash s$) if $\nu(w, \varphi) = f$ for some $\varphi \in \Gamma$, or $\nu(w, \varphi) = t$ for some $\varphi \in \Delta$.
- A sequent s is *valid* in $\langle W, \nu \rangle$ ($\langle W, \nu \rangle \models s$) if it is true in every world $w \in W$.
- Let $S \cup \{s\}$ be a set of sequents in $\mathcal{L}_{CL}$. s *follows in* **NS5** from S ($S \vdash_{\mathbf{NS5}} s$) if for every **NS5**-frame $\mathcal{W}$, if $\mathcal{W} \models s'$ for every $s' \in S$, then $\mathcal{W} \models s$. s is **NS5**-*valid* if s follows in **NS5** from $\emptyset$ (that is, s is valid in every **NS5**-frame).
- A hypersequent H is *valid* in $\langle W, \nu \rangle$, or $\langle W, \nu \rangle$ is a *model* of H ($\langle W, \nu \rangle \models H$), if one of the components of H is valid in $\langle W, \nu \rangle$.
- Let $\mathcal{H} \cup \{H\}$ be a set of hypersequents in $\mathcal{L}_{CL}$. H follows from $\mathcal{H}$ in **NS5** ($\mathcal{H} \vdash_{\mathbf{NS5}} H$) if every model of $\mathcal{H}$ is also a model of H.

Note 2 Note that the consequence relation we use between formulas is the *local* one (or the "truth" consequence relation in the terminology of [1]), while those we use between sequents or hypersequents are the *global* ones (or the "validity" consequence relation in the terminology of [1]). This implies that if Γ is a finite set of formulas then $\Gamma \vdash_{\mathbf{NS5}} \varphi$ iff the sequent (which is also a hypersequent) $\Gamma \Rightarrow \varphi$ is **NS5**-valid.

2 The Hypersequential System $GNS5^h$

Our hypersequential system $GNS5^h$ for **NS5** differs from the propositional fragment of LK^h (given in Figure 1) only with respect to its two rules for $\neg$. Instead of the rules of LK^h, the system $GNS5^h$ employs the following two rules:

$$[\neg \Rightarrow]\ \frac{H \mid\ \Rightarrow \varphi}{H \mid \neg\varphi \Rightarrow} \qquad [\Rightarrow \neg]\ \frac{H \mid \Gamma, \varphi \Rightarrow \Delta}{H \mid \Gamma \Rightarrow \Delta \mid\ \Rightarrow \neg\varphi}$$

The rule $[\neg \Rightarrow\ ]$ of $GNS5^h$ is just a special case of the corresponding rule of LK^h. In contrast, the rule $[\ \Rightarrow \neg]$ of $GNS5^h$ is *stronger* than the corresponding rule of LK^h. Thus the latter is derivable from the former with the help of

$\langle W, R, \nu \rangle$, where R is a relation on W, which in the case of **NS5** should be an equivalence relation. Note also that in the literature on modal logics one usually means by a "frame" just the pair $\langle W, R \rangle$, while we find it convenient to follow [13], and use this technical term a little bit differently, so that the valuation ν is a part of it.

[IW]. On the other hand, the rule $[\Rightarrow\neg]$ of $GNS5^h$ allows the inference of $\Rightarrow\varphi \mid \Rightarrow\neg\varphi$ for every φ, while if p is atomic then $\Rightarrow p \mid \Rightarrow\neg p$ is not provable in LK^h (since neither $\Rightarrow p$ nor $\Rightarrow\neg p$ is classically valid).

Note 3 The rule $[\Rightarrow\neg]$ of $GNS5^h$ becomes derivable if we add to LK^h the following splitting rule: from $H \mid \Gamma_1, \Gamma_2 \Rightarrow \Delta_1, \Delta_2$ infer $H \mid \Gamma_1 \Rightarrow \Delta_1 \mid \Gamma_2 \Rightarrow \Delta_2$. It is easy to see that the extended system is again sound and complete for classical logic, but with a different semantic interpretation of hypersequents: A hypersequent H is provable in that system iff every classical valuation is a model of one of the components of H.

Example 2.1 Let *GNS*5 be the system which is obtained from the propositional fragment of *LK* (the ordinary, sequential Gentzen-type system for classical logic) by replacing its rule $[\neg\Rightarrow]$ by the rule:

$$[\neg\Rightarrow]_5 \quad \frac{\neg\Gamma \Rightarrow \psi, \neg\Delta}{\neg\Gamma, \neg\psi \Rightarrow \neg\Delta}$$

This system (which is the variant of the usual Gentzen-type system for $S5$ in which the present $\neg$ is used instead of $\Box$) is known to be sound and complete for $NS5$ ([4]). However, the cut-admissibility theorem fails for it. For example, the sequent $\neg\neg P_1 \Rightarrow P_1$ is derivable in *GNS*5 by applying the cut rule to the sequents $\Rightarrow P_1, \neg P_1$ and $\neg\neg P_1, \neg P_1 \Rightarrow$ (both of which have very short cut-free proofs in *GNS*5). It is easy to see that this cut cannot be eliminated. Here is a cut-free proof of this sequent in $GNS5^h$:

$$\cfrac{\cfrac{\cfrac{\cfrac{P_1 \Rightarrow P_1}{\Rightarrow P_1 \mid \Rightarrow \neg P_1}\,[\Rightarrow\neg]}{\Rightarrow P_1 \mid \neg\neg P_1 \Rightarrow}\,[\neg\Rightarrow]}{\neg\neg P_1 \Rightarrow P_1 \mid \neg\neg P_1 \Rightarrow}\,[IW]}{\neg\neg P_1 \Rightarrow P_1}\,[IW]$$

(Note that the last step includes a hidden application of [EC] — See Note 1.)

Note 4 Obviously, an equivalent hypersequential system $GS5^h$ for propositional $S5$ in its standard language ($\{\wedge, \vee, \supset, \mathsf{F}, \Box\}$) is obtained from $GNS5^h$ by adding all sequents of the form $H \mid \mathsf{F} \Rightarrow$ as axioms, and replacing the two rules for $\neg$ by their following counterparts for $\Box$:

$$[\Box\Rightarrow]\ \frac{H \mid \Gamma, \varphi \Rightarrow \Delta}{H \mid \Gamma \Rightarrow \Delta \mid \Box\varphi \Rightarrow} \qquad [\Rightarrow\Box]\ \frac{H \mid \Rightarrow \varphi}{H \mid \Rightarrow \Box\varphi}$$

The strong soundness and completeness theorem and the strong cut-admissibility theorem which are proved for $GNS5^h$ in the next section can be proved for $GS5^h$ by practically the same proofs. [8]

[8] The rules of the propositional $GS5^h$ are very close to those given by Restall in [19], which are themselves close to Poggiolesi's system in [16]. (We are indebted to an anonymous reviewer for bringing these papers, as well as [6], to our attention.)

3 Soundness, Completeness, Cut-Admissibility

In this section we prove the main properties of $GNS5^h$.

Proposition 3.1 (strong soundness of $GNS5^h$**)** *Let* $\mathcal{H} \cup \{H\}$ *be a set of hypersequents. If* $\mathcal{H} \vdash_{GNS5^h} G$ *then* $\mathcal{H} \vdash_{\mathbf{NS5}} H$.

Proof. It is easy to see that the axioms of $GNS5^h$, its structural rules, and its logical rules for the positive connectives, all preserve validity of hypersequents in frames. Now we show that the same applies to the two negation rules.

- Suppose that $H \mid \Rightarrow \varphi$ is valid in $\langle W, \nu \rangle$. Then either one of the components of H is valid in $\langle W, \nu \rangle$, or $\Rightarrow \varphi$ is. The second case holds iff φ is valid in $\langle W, \nu \rangle$, implying that $\neg\varphi \Rightarrow$ is valid in $\langle W, \nu \rangle$. Hence in both cases one of the components of $H \mid \neg\varphi \Rightarrow$ is valid in $\langle W, \nu \rangle$.
- Suppose that $H \mid \varphi, \Gamma \Rightarrow \Delta$ is valid in $\langle W, \nu \rangle$. Then either one of the components of H is valid in $\langle W, \nu \rangle$, or $\varphi, \Gamma \Rightarrow \Delta$ is. If the first case holds then obviously $H \mid \Gamma \Rightarrow \Delta \mid \Rightarrow \neg\varphi$ is valid in $\langle W, \nu \rangle$. So assume that $\varphi, \Gamma \Rightarrow \Delta$ is valid there. Then either $\nu(w, \varphi) = f$ for some $w \in W$, or $\Gamma \Rightarrow \Delta$ is valid in $\langle W, \nu \rangle$. In the first case $\Rightarrow \neg\varphi$ is valid in $\langle W, \nu \rangle$. Hence in either case $H \mid \Gamma \Rightarrow \Delta \mid \Rightarrow \neg\varphi$ is valid in $\langle W, \nu \rangle$.

□

We turn to the strong completeness of $GNS5^h$ and to the cut-admissibility theorem for it.

Notation. $\mathcal{H} \vdash^{cf}_{GNS5^h} H$ means that there is a proof in $GNS5^h$ of the hypersequent H from the set of hypersequents $\mathcal{H}$ in which each cut is on a formula φ such that $\varphi \in \Gamma \cup \Delta$ for some component $\Gamma \Rightarrow \Delta$ of some hypersequent in $\mathcal{H}$.

Proposition 3.2 (strong completeness of $GNS5^h$**)** *Let* $\mathcal{H} \cup \{H\}$ *be a finite set of hypersequents. If* $\mathcal{H} \vdash_{\mathbf{NS5}} H$ *then* $\mathcal{H} \vdash^{cf}_{GNS5^h} H$.

Proof. Suppose that $\mathcal{H} \nvdash^{cf}_{GNS5^h} H$. We construct a model of $\mathcal{H}$ which is not a model of H.

Let $\mathcal{F}$ be the set of subformulas of formulas in $\mathcal{H} \cup \{H\}$. We call a hypersequent H^* an *$\mathcal{F}$-hypersequent* if it has the following properties:

- Every formula which occurs in H^* belongs to $\mathcal{F}$.
- $\mathcal{H} \nvdash^{cf}_{GNS5^h} H^*$.
- If $\Gamma \cup \Delta \subseteq \mathcal{F}$ then either $\Gamma \Rightarrow \Delta \in H^*$, or $\mathcal{H} \vdash^{cf}_{GNS5^h} H^* \mid \Gamma \Rightarrow \Delta$.

Let $s_1, \dots, s_n$ be an enumeration of all the sequents $\Gamma \Rightarrow \Delta$ such that $\Gamma \cup \Delta \subseteq \mathcal{F}$. ($n$ is finite because $\mathcal{H}$ is finite, and so $\mathcal{F}$ is finite.) Let $H_0 = H$. Define a sequence $H_1, \dots, H_n$ of hypersequents by letting $H_i = H_{i-1} \mid s_i$ in case $\mathcal{H} \nvdash^{cf}_{GNS5^h} H_{i-1} \mid s_i$, and $H_i = H_{i-1}$ otherwise. Let $H^* = H_n$. Then H^* is an $\mathcal{F}$-hypersequent such that $H \subseteq H^*$. Call a component $\Gamma^* \Rightarrow \Delta^*$ of H^* *maximal* if it has no proper extension in H^* (i.e. if $\Gamma' \Rightarrow \Delta' \in H^*$, $\Gamma^* \subseteq \Gamma'$ and $\Delta^* \subseteq \Delta'$, then $\Gamma^* = \Gamma'$ and $\Delta^* = \Delta'$). Let W be the set of all maximal components of H^*. For a world $w \in W$, we denote by Γ_w and Δ_w the sets Γ^*

and Δ^* (respectively) such that $w = \Gamma^* \Rightarrow \Delta^*$. Let ν be the valuation defined by $\nu(w, p) = t$ iff $p \in \Gamma_w$ for every atomic variable p (its values for compound formulas are then uniquely determined following Definition 1.2).

We show by induction on the structure of formulas that the following hold for every $\varphi \in \mathcal{F}$ and every maximal component w of H^*:

(i) If $\varphi \in \Gamma_w$ then $\nu(w, \varphi) = t$.

(ii) If $\varphi \in \Delta_w$ then $\nu(w, \varphi) = f$.

- The case where φ is a propositional variable is immediate from the definition of ν, and the fact that if $\varphi \in \Delta_w$ then $\varphi \notin \Gamma_w$ (because $\mathcal{H} \not\vdash^{cf}_{GNS5^h} H^*$).
- Suppose that $\varphi = \varphi_1 \wedge \varphi_2$.
  - Suppose $\varphi \in \Gamma_w$. Assume for contradiction that $\{\varphi_1, \varphi_2\} \not\subseteq \Gamma_w$. Then the maximality of w implies that $\varphi_1, \varphi_2, \Gamma_w \Rightarrow \Delta_w$ is not a component of H^*. Since H^* is an $\mathcal{F}$-hypersequent, it follows that $\mathcal{H} \vdash^{cf}_{GNS5^h} H^* \mid \varphi_1, \varphi_2, \Gamma_w \Rightarrow \Delta_w$. By applying $[\wedge \Rightarrow]$ to this hypersequent we get that $\mathcal{H} \vdash^{cf}_{GNS5^h} H^* \mid w$, and so that $\mathcal{H} \vdash^{cf}_{GNS5^h} H^*$ (implicitly using [EC]). A contradiction. Hence $\{\varphi_1, \varphi_2\} \subseteq \Gamma_w$, and so $\nu(w, \varphi_1) = \nu(w, \varphi_2) = t$ by the induction hypothesis for φ_1, φ_2. Hence $\nu(w, \varphi) = t$.
  - Suppose $\varphi \in \Delta_w$. Assume for contradiction that $\{\varphi_1, \varphi_2\} \cap \Delta_w = \emptyset$. Then the maximality of w and the fact that H^* is an $\mathcal{F}$-hypersequent imply that both $\mathcal{H} \vdash^{cf}_{GNS5^h} H^* \mid \Gamma_w \Rightarrow \Delta_w, \varphi_1$ and $\mathcal{H} \vdash^{cf}_{GNS5^h} H^* \mid \Gamma_w \Rightarrow \Delta_w, \varphi_2$. By applying $[\Rightarrow \wedge]$ to these two hypersequents (together with an implicit use of [EC]) we get that $\mathcal{H} \vdash^{cf}_{GNS5^h} H^*$. A contradiction. Hence either $\varphi_1 \in \Delta_w$ or $\varphi_2 \in \Delta_w$. It follows by the induction hypothesis for φ_1 and φ_2 that either $\nu(w, \varphi_1) = f$ or $\nu(w, \varphi_2) = f$. In both cases $\nu(w, \varphi) = f$.
- The cases $\varphi = \varphi_1 \vee \varphi_2$ and $\varphi = \varphi_1 \supset \varphi_2$ are similar to the case $\varphi = \varphi_1 \wedge \varphi_2$, and are left for the reader.
- Suppose $\varphi = \neg\psi$.
  - Suppose $\varphi \in \Gamma_w$. Assume for contradiction that $\Rightarrow \psi \notin H^*$. Then $\mathcal{H} \vdash^{cf}_{GNS5^h} H^* \mid \;\Rightarrow \psi$, because H^* is an $\mathcal{F}$-hypersequent. By applying $[\neg \Rightarrow]$ to this hypersequent followed by internal weakenings, we get that $\mathcal{H} \vdash^{cf}_{GNS5^h} H^* \mid w$, and so that $\mathcal{H} \vdash^{cf}_{GNS5^h} H^*$. A contradiction. Hence $\Rightarrow \psi \in H^*$. It follows that there is a maximal component w' of H^* that extends it, i.e. $\psi \in \Delta_{w'}$. Therefore the induction hypothesis for ψ implies that $\nu(w', \psi) = f$. Hence $\nu(w, \varphi) = t$.
  - Suppose $\varphi \in \Delta_w$. Let $w' \in W$. We show that $\psi \in \Gamma_{w'}$. Assume otherwise. Then $\mathcal{H} \vdash^{cf}_{GNS5^h} H^* \mid \psi, \Gamma_{w'} \Rightarrow \Delta_{w'}$ (by the maximality of w' and the fact that H^* is an $\mathcal{F}$-hypersequent). By applying $[\Rightarrow \neg]$ to $H^* \mid \psi, \Gamma_{w'} \Rightarrow \Delta_{w'}$, we get that $\mathcal{H} \vdash^{cf}_{GNS5^h} H^* \mid w' \mid \;\Rightarrow \varphi$, and so (implicitly using [EC]) $\mathcal{H} \vdash^{cf}_{GNS5^h} H^* \mid \;\Rightarrow \varphi$. Since $\varphi \in \Delta_w$, by applying [IW] to $H^* \mid \;\Rightarrow \varphi$ we get $\mathcal{H} \vdash^{cf}_{GNS5^h} H^* \mid w$, and so $\mathcal{H} \vdash^{cf}_{GNS5^h} H^*$. A contradiction. It follows by the induction hypothesis for ψ that $\nu(w, \psi) = t$ for every $w \in W$. Hence $\nu(w, \varphi) = f$.

Suppose now that $\Gamma \Rightarrow \Delta$ is some component of H. Since $H \subseteq H^*$, there is a maximal component w of H^* such that $\Gamma \subseteq \Gamma_w$ and $\Delta \subseteq \Delta_w$. Therefore properties 1 and 2 above imply that if $\varphi \in \Gamma$ then $\nu(w,\varphi) = t$, while if $\varphi \in \Delta$ then $\nu(w,\varphi) = f$. It follows that $\Gamma \Rightarrow \Delta$ is not true in the world w, and so $\Gamma \Rightarrow \Delta$ is not valid in $\langle W, \nu\rangle$. Hence $\langle W, \nu\rangle$ is not a model of H.

Finally, we prove that $\langle W, \nu\rangle$ is a model of $\mathcal{H}$. So let $H' \in \mathcal{H}$. It is impossible that every component of H' is a subsequent of some component of H^*, because otherwise H^* can be derived from H' (and so from $\mathcal{H}$) using just internal and external weakenings ([IW] and [EW]), and this contradicts the fact that $\mathcal{H} \nvdash^{cf}_{GNS5^h} H^*$. Therefore there is a component $\Gamma \Rightarrow \Delta$ of H' which is not a subsequent of any component of H^*. We show that $\Gamma \Rightarrow \Delta$ is valid in $\langle W, \nu\rangle$. So let $w \in W$. Then either $\Gamma \nsubseteq \Gamma_w$, or $\Delta \nsubseteq \Delta_w$. Assume e.g. the former. (The proof in the second case is similar). Then $\varphi \notin \Gamma_w$ for some $\varphi \in \Gamma$. Hence $\mathcal{H} \vdash^{cf}_{GNS5^h} H^* \mid \varphi, \Gamma_w \Rightarrow \Delta_w$ (because $\varphi \in \mathcal{F}$, w is maximal, and H^* is an $\mathcal{F}$-hypersequent). Assume for contradiction that $\varphi \notin \Delta_w$. Then $\mathcal{H} \vdash^{cf}_{GNS5^h} H^* \mid \Gamma_w \Rightarrow \Delta_w, \varphi$ as well. By applying a cut on φ to these two hypersequents, we get that $\mathcal{H} \vdash^{cf}_{GNS5^h} H^* \mid w$, and so that $\mathcal{H} \vdash^{cf}_{GNS5^h} H^*$. A contradiction. It follows that $\varphi \in \Delta_w$, and so $\nu(w,\varphi) = f$ by property 2 above of maximal components of H^*. Since $\varphi \in \Gamma$, this means that $\Gamma \Rightarrow \Delta$ is true in the world w. It follows that $\Gamma \Rightarrow \Delta$ is valid in $\langle W, \nu\rangle$, and so H' is valid $\langle W, \nu\rangle$. □

Theorem 3.3 (strong soundness and completeness) *Let $\mathcal{H} \cup \{H\}$ be a finite set of hypersequents. Then $\mathcal{H} \vdash_{\mathbf{NS5}} H$ iff $\mathcal{H} \vdash_{GNS5^h} H$.*

Proof. Immediate from Propositions 3.1 and 3.2. □

Note 5 Theorem 3.3 can be extended to the case in which $\mathcal{H}$ is an arbitrary set of hypersequents (not necessarily finite). Like in the first-order case which is discussed below, this is done by using infinite hypersequents, letting such an infinite hypersequent H^* follow from a set $\mathcal{H}$ of finite hypersequents iff there is finite subset H of H^* such that $\mathcal{H} \vdash^{cf}_{GNS5^h} H$. We omit the details.

Theorem 3.4 (cut-admissibility) $GNS5^h$ *admits strong cut-admissibility: If $\mathcal{H} \vdash_{GNS5^h} H$ then $\mathcal{H} \vdash^{cf}_{GNS5^h} H$. In particular: If $\vdash_{GNS5^h} H$ then H has a cut-free proof in $GNS5^h$.*

Proof. Immediate from Propositions 3.1 and 3.2. □

4 The First-order Case

In this section we explain how the results of the previous sections can be extended to the first-order level.

4.1 Preliminaries

Let $\mathcal{L}$ be the version of the classical first-order language in which the set of free variables and the set of bounded variables are disjoint (thus in a well-formed formula, the use of the bound variables is always in the scope of a quantification of the same variables). We use the metavariables a, b to range over the free variables, x to range over the bounded variables, p to range over the predicate symbols of $\mathcal{L}$, c to range over its constant symbols, and f to range over its function symbols. The sets of $\mathcal{L}$-terms and $\mathcal{L}$-formulas are defined as usual, and are denoted by $trm_{\mathcal{L}}$ and $frm_{\mathcal{L}}$, respectively. We mainly use t as a metavariable standing for $\mathcal{L}$-terms, φ, ψ for $\mathcal{L}$-formulas, Γ, Δ for finite sets of $\mathcal{L}$-formulas, and $\mathcal{T}, \mathcal{U}$ for (possibly infinite) sets of $\mathcal{L}$-formulas.

Given an $\mathcal{L}$-term t, a free variable a, and another $\mathcal{L}$-term t', we denote by $t\{t'/a\}$ the $\mathcal{L}$-term obtained from t by replacing all occurrences of a by t'. This notation is extended to formulas, sets of formulas, etc. in the obvious way.

To improve readability we use square parentheses in the meta-language, and reserve round parentheses to the first-order language.

4.2 The Hypersequential System $GQNS5^h$

Let $GQNS5^h$ denote the extension of $GNS5^h$ with the rules for the quantifiers of LK^h. (See Figure 1. Note that again, $GQNS5^h$ differs from the classical system LK^h only with respect to its two rules for $\neg$.) For a set $\mathcal{H}$ of hypersequents and a hypersequent H, we write $\mathcal{H} \vdash_{GQNS5^h} H$ if there is a proof in $GQNS5^h$ of H from the set $\mathcal{H}$, and $\mathcal{H} \vdash^{cf}_{GQNS5^h} H$ if there is such a proof in which each cut is on a formula that belongs to some component $\Gamma \Rightarrow \Delta$ of some hypersequent in $\mathcal{H}$.

4.3 The Constant Domain Semantics of NS5

Definition 4.1 An $\mathcal{L}$-*algebra* is a pair $\langle D, I\rangle$ where D is a non-empty domain and I is an interpretation of constant and function symbols of $\mathcal{L}$ such that $I[c] \in D$ for every constant symbol c of $\mathcal{L}$, and $I[f] \in D^n \to D$ for every n-ary function symbol f of $\mathcal{L}$.

Definition 4.2 Let $M = \langle D, I\rangle$ be an $\mathcal{L}$-algebra. An $\langle \mathcal{L}, M\rangle$-evaluation is a function assigning an element in D to every free variable of $\mathcal{L}$. An $\langle \mathcal{L}, M\rangle$-evaluation e is naturally extended to $trm_{\mathcal{L}}$ as follows: $e[c] = I[c]$ for every constant symbol c; and $e[f(t_1, \dots, t_n)] = I[f][e[t_1], \dots, e[t_n]]$ for every $f(t_1, \dots, t_n) \in trm_{\mathcal{L}}$.

Notation. Given an $\langle \mathcal{L}, M\rangle$-evaluation e, a free variable a, and $d \in D$, we denote by $e_{[a:=d]}$ the $\langle \mathcal{L}, M\rangle$-evaluation which is identical to e except that $e_{[a:=d]}[a] = d$.

Definition 4.3 An $\mathcal{L}$-*frame* is a tuple $\mathcal{W} = \langle W, M, \mathcal{I}\rangle$, where W is a set (of "worlds"), $M = \langle D, I\rangle$ is an $\mathcal{L}$-algebra, and $\mathcal{I} = \{I_w\}_{w\in W}$, where for every $w \in W$, I_w is an interpretation of predicate symbols, i.e., a function assigning a subset of D^n to every n-ary predicate symbol of $\mathcal{L}$.

Definition 4.4 Let $\mathcal{W} = \langle W, M, \mathcal{I}\rangle$ be an $\mathcal{L}$-frame, where $M = \langle D, I\rangle$ and $\mathcal{I} = \{I_w\}_{w\in W}$. Let e be an $\langle \mathcal{L}, M\rangle$-evaluation. The satisfaction relation $\models$ is recursively defined as follows:

(i) $\mathcal{W}, w, e \models p(t_1, \dots, t_n)$ iff $\langle e[t_1], \dots, e[t_n]\rangle \in I_w[p]$.

(ii) $\mathcal{W}, w, e \models \varphi_1 \wedge \varphi_2$ iff $\mathcal{W}, w, e \models \varphi_1$ and $\mathcal{W}, w, e \models \varphi_2$.

(iii) $\mathcal{W}, w, e \models \varphi_1 \vee \varphi_2$ iff $\mathcal{W}, w, e \models \varphi_1$ or $\mathcal{W}, w, e \models \varphi_2$.

(iv) $\mathcal{W}, w, e \models \varphi_1 \supset \varphi_2$ iff $\mathcal{W}, w, e \not\models \varphi_1$ or $\mathcal{W}, w, e \models \varphi_2$.

(v) $\mathcal{W}, w, e \models \neg\varphi$ iff $\mathcal{W}, w', e \not\models \varphi$ for some $w' \in W$.

(vi) $\mathcal{W}, w, e \models \forall x(\varphi\{x/a\})$ iff $\mathcal{W}, w, e_{[a:=d]} \models \varphi$ for every $d \in D$.

(vii) $\mathcal{W}, w, e \models \exists x(\varphi\{x/a\})$ iff $\mathcal{W}, w, e_{[a:=d]} \models \varphi$ for some $d \in D$.

It is easy to see that $\models$ is well-defined, and in particular in vi and vii, the exact choice of the free variable a is immaterial.

We now define the consequence relation of **NS5** in semantic terms.

Definition 4.5 Let $\mathcal{T} \cup \{\varphi\}$ be a set of $\mathcal{L}$-formulas. $T \vdash_{\mathbf{NS5}} \varphi$ if $\mathcal{W}, w, e \models \mathcal{T}$ implies $\mathcal{W}, w, e \models \varphi$ for every $\mathcal{L}$-frame $\mathcal{W} = \langle W, M, \mathcal{I}\rangle$, world $w \in W$, and $\langle \mathcal{L}, M\rangle$-evaluation e.

4.4 Soundness, Completeness and Cut-Admissibility

Notation. Given an $\mathcal{L}$-frame $\mathcal{W} = \langle W, M, \mathcal{I}\rangle$, an $\langle \mathcal{L}, M\rangle$-evaluation e, $w \in W$, and a sequent $\Gamma \Rightarrow \Delta$, we write $\mathcal{W}, w, e \models \Gamma \Rightarrow \Delta$ if either $\mathcal{W}, w, e \not\models \varphi$ for some $\varphi \in \Gamma$, or $\mathcal{W}, w, e \models \varphi$ for some $\varphi \in \Delta$.

Definition 4.6 Let $\mathcal{W} = \langle W, M, \mathcal{I}\rangle$ be an $\mathcal{L}$-frame. $\mathcal{W}$ is a *model* of a hypersequent H if for every $\langle \mathcal{L}, M\rangle$-evaluation e, there exists a component $s \in H$ such that $\mathcal{W}, w, e \models s$ for every $w \in W$. $\mathcal{W}$ is a model of a set $\mathcal{H}$ of hypersequents if it is a model of every $H \in \mathcal{H}$.

Definition 4.7 Let $\mathcal{H} \cup \{H\}$ be a set of hypersequents. $\mathcal{H} \vdash^{hs}_{\mathbf{NS5}} H$ iff every $\mathcal{L}$-frame which is a model of $\mathcal{H}$ is also a model of H.

The proof of the next theorem is not difficult:

Theorem 4.8 (strong soundness of $GQNS5^h$) *$GQNS5^h$ is strongly sound with respect to $\vdash^{hs}_{\mathbf{NS5}}$: If $\mathcal{H} \vdash_{GQNS5^h} H$ then $\mathcal{H} \vdash^{hs}_{\mathbf{NS5}} H$.*

The completeness proof is of course more complicated. Essentially, it combines the ideas of the proof of Theorem 3.2 with the method of [11]. In particular: we have to use in it the notions of *extended sequents* and *extended hypersequents.*

Definition 4.9 An *extended sequent* is an ordered pair of (possibly infinite) sets of $\mathcal{L}$-formulas. Given two extended sequents $\mu_1 = \langle \mathcal{T}_1, \mathcal{U}_1\rangle$ and $\mu_2 = \langle \mathcal{T}_2, \mathcal{U}_2\rangle$, we write $\mu_1 \sqsubseteq \mu_2$ if $\mathcal{T}_1 \subseteq \mathcal{T}_2$ and $\mathcal{U}_1 \subseteq \mathcal{U}_2$. An extended sequent is called *finite* if it consists of finite sets of formulas.

Definition 4.10 An *extended hypersequent* is a (possibly infinite) set of extended sequents. Given two extended hypersequents Ω_1, Ω_2, we write $\Omega_1 \sqsubseteq \Omega_2$ (and say that Ω_2 *extends* Ω_1) if for every extended sequent $\mu_1 \in \Omega_1$, there exists $\mu_2 \in \Omega_2$ such that $\mu_1 \sqsubseteq \mu_2$. An extended hypersequent is called *finite* if it consists of finitely many finite extended sequents.

We use the same notations as above for extended sequents and extended hypersequents. For example, we write $\mathcal{T} \Rightarrow \mathcal{U}$ instead of $\langle \mathcal{T}, \mathcal{U}\rangle$, and $\Omega \mid \mathcal{T} \Rightarrow \mathcal{U}$ instead of $\Omega \cup \{\langle \mathcal{T}, \mathcal{U}\rangle\}$.

Definition 4.11 An extended sequent $\mathcal{T} \Rightarrow \mathcal{U}$ admits *the witness property* if the following hold:

(i) If $\forall x(\varphi\{x/a\}) \in \mathcal{U}$ then $\varphi\{b/a\} \in \mathcal{U}$ for some free variable b.

(ii) If $\exists x(\varphi\{x/a\}) \in \mathcal{T}$ then $\varphi\{b/a\} \in \mathcal{T}$ for some free variable b.

Definition 4.12 Let Ω be an extended hypersequent, and $\mathcal{H}$ be a set of (ordinary) hypersequents.

(i) Ω is called *$\mathcal{H}$-consistent* if $\mathcal{H} \not\vdash^{cf}_{GQNS5^h} H$ for every (ordinary) hypersequent $H \sqsubseteq \Omega$.

(ii) Ω is called *internally $\mathcal{H}$-maximal with respect to an $\mathcal{L}$-formula φ* if for every $\mathcal{T} \Rightarrow \mathcal{U} \in \Omega$:
   (a) If $\varphi \notin \mathcal{T}$ then $\Omega \mid \mathcal{T}, \varphi \Rightarrow \mathcal{U}$ is not $\mathcal{H}$-consistent.
   (b) If $\varphi \notin \mathcal{U}$ then $\Omega \mid \mathcal{T} \Rightarrow \mathcal{U}, \varphi$ is not $\mathcal{H}$-consistent.

(iii) Ω is called *internally $\mathcal{H}$-maximal* if it is internally $\mathcal{H}$-maximal with respect to any $\mathcal{L}$-formula.

(iv) Let s be a sequent. Ω is called *externally $\mathcal{H}$-maximal with respect to s* if either $\{s\} \sqsubseteq \Omega$, or $\Omega \mid s$ is not $\mathcal{H}$-consistent.

(v) Ω is called *externally $\mathcal{H}$-maximal* if it is externally $\mathcal{H}$-maximal with respect to any sequent of the form $\Rightarrow \varphi$.

(vi) Ω admits *the witness property* if every $\mu \in \Omega$ admits the witness property.

(vii) Ω is called *$\mathcal{H}$-maximal* if it is $\mathcal{H}$-consistent, internally $\mathcal{H}$-maximal, externally $\mathcal{H}$-maximal, and it admits the witness property.

Less formally, an extended hypersequent Ω is internally $\mathcal{H}$-maximal if every new formula added on some side of some component of Ω would make it $\mathcal{H}$-inconsistent. Similarly, Ω is externally $\mathcal{H}$-maximal if every new sequent of the form $\Rightarrow \varphi$ added to Ω would make it $\mathcal{H}$-inconsistent.

Obviously, every hypersequent is an extended hypersequent, and so all of these properties apply to (ordinary) hypersequents as well.

Next we list, without proofs, a sequence of lemmas which are needed for the proof of the completeness of $GQNS5^h$.

Lemma 4.13 *Let Ω be an extended hypersequent that is internally $\mathcal{H}$-maximal with respect to an $\mathcal{L}$-formula φ. For every $\mathcal{T} \Rightarrow \mathcal{U} \in \Omega$:*

(i) *If $\varphi \notin \mathcal{T}$, then $\mathcal{H} \vdash^{cf}_{GQNS5^h} H \mid \Gamma, \varphi \Rightarrow \Delta$ for some hypersequent $H \sqsubseteq \Omega$ and sequent $\Gamma \Rightarrow \Delta \sqsubseteq \mathcal{T} \Rightarrow \mathcal{U}$.*

(ii) *If $\varphi \notin \mathcal{U}$, then $\mathcal{H} \vdash^{cf}_{GQNS5^h} H \mid \Gamma \Rightarrow \Delta, \varphi$ for some hypersequent $H \sqsubseteq \Omega$ and sequent $\Gamma \Rightarrow \Delta \sqsubseteq \mathcal{T} \Rightarrow \mathcal{U}$.*

Lemma 4.14 *Let Ω be an extended hypersequent that is externally $\mathcal{H}$-maximal with respect to a sequent s. If $\{s\} \not\sqsubseteq \Omega$, then there exists a hypersequent $H \sqsubseteq \Omega$ such that $\mathcal{H} \vdash^{cf}_{GQNS5^h} H \mid s$.*

Lemma 4.15 *Let $\mathcal{H}$ be a set of hypersequents, and let $H = \Gamma_1 \Rightarrow \Delta_1 \mid ... \mid \Gamma_n \Rightarrow \Delta_n$ be a $\mathcal{H}$-consistent finite extended hypersequent. Then, there exists a $\mathcal{H}$-consistent finite extended hypersequent H' of the form $\Gamma'_1 \Rightarrow \Delta'_1 \mid ... \mid \Gamma'_n \Rightarrow \Delta'_n$, such that $\Gamma_i \subseteq \Gamma'_i$ and $\Delta_i \subseteq \Delta'_i$ for every $1 \leq i \leq n$, and H' admits the witness property.*

Lemma 4.16 *Let $\mathcal{H}$ be a set of hypersequents, and $H = \Gamma_1 \Rightarrow \Delta_1 \mid ... \mid \Gamma_n \Rightarrow \Delta_n$ be a $\mathcal{H}$-consistent finite extended hypersequent. Let φ be an $\mathcal{L}$-formula, and $\Gamma^* \Rightarrow \Delta^*$ be a sequent. Then, there exists a $\mathcal{H}$-consistent finite extended hypersequent H', such that:*

- *$H' = \Gamma'_1 \Rightarrow \Delta'_1 \mid ... \mid \Gamma'_{n'} \Rightarrow \Delta'_{n'}$, where $n' \in \{n, n+1\}$, $\Gamma_i \subseteq \Gamma'_i$ and $\Delta_i \subseteq \Delta'_i$ for every $1 \leq i \leq n$.*
- *H' is internally $\mathcal{H}$-maximal with respect to φ.*
- *H' is externally $\mathcal{H}$-maximal with respect to $\Gamma^* \Rightarrow \Delta^*$.*
- *H' admits the witness property.*

Lemma 4.17 *Let $\mathcal{H}$ be a set of hypersequents. Every $\mathcal{H}$-consistent hypersequent can be extended to a $\mathcal{H}$-maximal extended hypersequent Ω.*

Next we define the $\mathcal{L}$-algebra used in the completeness proof.

Definition 4.18 The *Herbrand $\mathcal{L}$-algebra* is an $\mathcal{L}$-algebra, $\langle D, I \rangle$, such that $D = trm_{\mathcal{L}}$ (the set of *all* $\mathcal{L}$-terms), $I[c] = c$ for every constant c, and $I[f][t_1, ..., t_n] = f(t_1, ..., t_n)$ for every n-ary function symbol f and $t_1, ..., t_n \in D$.

Note that the domain of the Herbrand $\mathcal{L}$-algebra contains also non-closed terms. However, recall that we assume that the set of free variables and the set of bounded variables are disjoint, so an $\mathcal{L}$-term cannot contain a bounded variable.

We are now ready to establish the main completeness theorem.

Theorem 4.19 (strong completeness of $GNS5^h$) *Let $\mathcal{H}_0$ be a set of hypersequents closed under substitutions, and H_0 be a hypersequent. If $\mathcal{H}_0 \vdash^{hs}_{\mathbf{NS5}} H_0$ then $\mathcal{H}_0 \vdash^{cf}_{GQNS5^h} H_0$.*

Outline of Proof: Assume that $\mathcal{H}_0 \nvdash^{cf}_{GQNS5^h} H_0$. We construct an $\mathcal{L}$-frame $\mathcal{W}$ that is a model of $\mathcal{H}_0$ but not of H_0. The availability of external and internal weakenings ensures that H_0 is $\mathcal{H}_0$-consistent. Thus by Lemma 4.17, there exists a $\mathcal{H}_0$-maximal extended hypersequent Ω such that $H_0 \sqsubseteq \Omega$. Using Ω, $\mathcal{W} = \langle W, M, \mathcal{I} \rangle$ is defined as follows:

- $W = \Omega$.
- $M = \langle D, I \rangle$ is the Herbrand $\mathcal{L}$-algebra.
- $\mathcal{I} = \{I_w\}_{w \in W}$ where $\langle t_1, ..., t_n \rangle \in I_{\mathcal{T} \Rightarrow \mathcal{U}}[p]$ iff $p(t_1, ..., t_n) \in \mathcal{T}$.

Now, let e be the identity $\langle \mathcal{L}, M \rangle$-evaluation (defined by $e[a] = a$ for every free variable a). We prove that the following hold for every $w = \mathcal{T} \Rightarrow \mathcal{U} \in W$:

(a) If $\psi \in \mathcal{T}$ then $\mathcal{W}, w, e \models \psi$.

(b) If $\psi \in \mathcal{U}$ then $\mathcal{W}, w, e \not\models \psi$.

(a) and (b) are proved together by induction on the complexity of ψ.

Once (a) and (b) are established, one shows that $\mathcal{W}$ is a model of $\mathcal{H}_0$ but not of H_0. □

The following corollary establishes the link to our logic (cf. Definition 4.5).

Corollary 4.20 *Let $\Gamma \cup \{\varphi\}$ be a finite set of $\mathcal{L}$-formulas. Then, $\Gamma \vdash_{\mathbf{NS5}} \varphi$ iff $\vdash_{GQNS5^h} \Gamma \Rightarrow \varphi$.*

Taken together, Theorems 4.8 and 4.19 naturally entail the following strong cut-admissibility result.

Corollary 4.21 *$\mathcal{H} \vdash_{GQNS5^h} H$ implies $\mathcal{H} \vdash^{cf}_{GQNS5^h} H$, for every set $\mathcal{H}$ of hypersequents closed under substitutions, and a hypersequent H. In particular, for every hypersequent H, $\vdash_{GQNS5^h} H$ implies that there exists a cut-free derivation of H in $GQNS5^h$.*

Note that it is necessary to require that the set of assumptions $\mathcal{H}$ is closed under substitutions. Indeed, $\Rightarrow p(a) \vdash_{GQNS5^h} \Rightarrow p(b)$, but if $a \neq b$ there is no derivation of $\Rightarrow p(b)$ from $\Rightarrow p(a)$ in $GQNS5^h$ with cuts only on $p(a)$.

References

[1] Avron, A., *Natural 3-valued logics: Characterization and proof theory*, Journal of Symbolic Logic **56** (1991), pp. 276–294.

[2] Avron, A., *The method of hypersequents in proof theory of propositional non-classical logics*, in: W. Hodges, M. Hyland, C. Steinhorn and J. Truss, editors, *Logic: Foundations to Applications*, Oxford Science Publications, 1996 pp. 1–32.

[3] Avron, A., *Cut-elimination in* **RM** *proved semantically*, IFCoLog Journal of Logics and their Applications **4** (2017), pp. 605–621.

[4] Avron, A. and A. Zamansky, *Paraconsistency, self-extensionality, modality*, to appear in a special issue of The Logic Journal of the IGPL.

[5] Batens, D., *On some remarkable relations between paraconsistent logics, modal logics, and ambiguity logics*, in: W. A. Carnielli, M. E. Coniglio and I. D'Ottaviano, editors, *Paraconsistency: The Logical Way to the Inconsistent*, number 228 in Lecture Notes in Pure and Applied Mathematics, Marcel Dekker, 2002 pp. 275–293.

[6] Bednarska, K. and A. Indrzejczak, *Hypersequent calculi for S5: The methods of cut elimination*, Logic and Logical Philosophy **24** (2015), pp. 277–311.

[7] Béziau, J. Y., *S5 is a paraconsistent logic and so is first-order classical logic*, Logical Investigations **8** (2002), pp. 301–309.

[8] Béziau, J. Y., *Paraconsistent logic from a modal viewpoint*, Journal of Applied Logic **3** (2005), pp. 6–14.

[9] Béziau, J. Y., *The paraconsistent logic Z. A possible solution to Jaśkowski's problem*, Logic and Logical Philosophy **15** (2006), pp. 99–111.

[10] Bimbó, K., "Proof Theory: Sequent Calculi and Related Formalisms," Discrete Mathematics and Its Applications, CRC Press, 2015.

[11] Lahav, O. and A. Avron, *A semantic proof of strong cut-admissibility for first-order Gödel logic*, Journal of Logic and Computation **23** (2013), pp. 59–86.

[12] Mints, G. E., *Some calculi of modal logic*, Proc. Steklov Inst. of Mathematics **98** (1968), pp. 97–122.

[13] Nerode, A. and R. A. Shore, "Logic for Applications," Springer, 1997.
[14] Osorio, M., J. L. Carballido and C. Zepeda, *Revisiting* $\mathbb{Z}$, Notre Dame Journal of Formal Logic **55** (2014), pp. 129–155.
[15] Osorio, M., J. L. Carballido, C. Zepeda and J. A. Castellanos, *Weakening and extending* $\mathbb{Z}$, Logica Universalis **9** (2015), pp. 383–409.
[16] Poggiolesi, F., *A cut-free simple sequent calculus for modal logic S5*, The Review of Symbolic Logic **1** (2008), p. 315.
[17] Poggiolesi, P., "Gentzen Calculi for Modal Propositional Logic," Trends in Logic **32**, Springer, 2010.
[18] Pottinger, G., *Uniform, cut-free formulations of T, S4 and S5 (abstract)*, Journal of Symbolic Logic **48** (1983), p. 900.
[19] Restall, G., *Proofnets for S5: Sequents and circuits for modal logic*, in: C. Dimitracopoulos, L. Newelski and D. Normann, editors, *Logic Colloquium 2005*, Cambridge: Cambridge University Press, 2007 pp. 151–172.
[20] Wansing, H., *Sequent systems for modal logics*, in: D. Gabbay and F. Guenther, editors, *Handbook of Philosophical Logic, 2nd edition, Vol. 8*, Kluwer, 2002 pp. 61–145.

A Hypersequent Calculus with Clusters for Linear Frames

David Baelde Anthony Lick Sylvain Schmitz

LSV, ENS Paris-Saclay & CNRS & Inria, Université Paris-Saclay

Abstract

The logic $\mathbf{K_t4.3}$ is the basic modal logic of linear frames. Along with its extensions, it is found at the core of linear-time temporal logics and logics on words. In this paper, we consider the problem of designing proof systems for these logics, in such a way that proof search yields decision procedures for validity with an optimal complexity—coNP in this case. In earlier work, Indrzejczak has proposed an ordered hypersequent calculus that is sound and complete for $\mathbf{K_t4.3}$ but does not yield any decision procedure. We refine his approach, using a hypersequent structure that corresponds to weak rather than strict total orders, and using annotations that reflect the model-theoretic insights given by small models for $\mathbf{K_t4.3}$. We obtain a sound and complete calculus with an associated coNP proof search algorithm. These results extend naturally to the cases of unbounded and dense frames, and to the complexity of the two-variable fragment of first-order logic over total orders.

Keywords: modal logics, proof systems, hypersequents, clusters.

1 Introduction

Modal logics are expressive and intuitive languages for describing properties of relational structures. Accordingly, when investigating properties of linear frames, it is often quite useful to express them using a tense logic [19] able to reason on temporal flows. For instance, LTL [17,20] and CTL [4] are widely used for verifying computer programs.

When studying a logic, a common approach is to design a proof system, such as a sequent calculus. Our own interest in (enriched) sequent calculi, compared to e.g. axiomatisations, is that their associated proof-search procedures often yield decidability and even complexity results for the satisfiability and validity problems. They are also modular, allowing them to be easily adapted to handle extensions or fragments of the logic at hand. However, basic sequent calculi are often ill-suited for modal logics, as the class of frames underlying the logic is typically difficult to capture. Therefore, more expressive variants of the

* Work funded by ANR grant ANR-14-CE28-0005 PRODAQ.

sequent calculus have been developed, such as labelled sequents [15], nested sequents [3,18], linear nested sequents [12] or hypersequents [1,7,8,9,10].

In this paper, we focus on $\mathbf{K_t4.3}$ [5,2], the tense logic of linear frames. Somewhat surprisingly for the logic lying at the heart of LTL with past modalities—which is largely studied in verification [13,11]—, to the best of our knowledge, a sound and complete sequent-style calculus for $\mathbf{K_t4.3}$ was only recently proposed by Indrzejczak [9]. This is an *ordered hypersequent* calculus, where the structure of the hypersequents reflects the linear structure of $\mathbf{K_t4.3}$ frames. However, this calculus does not yield a proof-search algorithm, even though $\mathbf{K_t4.3}$ satisfiability is known to be decidable and even NP-complete [16]. The issue here is that ordered hypersequents correspond to *strictly ordered* linear frames, which are arguably not the most adequate structures for the logic. Although every satisfiable $\mathbf{K_t4.3}$ formula has a model whose underling frame is a strict total order, there are examples of invalid formulæ (like $\mathsf{G}\perp \vee \mathsf{F}\,\mathsf{G}\perp$), whose strictly ordered counter-models are all infinite. On such invalid instances, the hypersequent calculus of Indrzejczak [9] yields a proof tree with some infinite failure branches, thus proof-search does not terminate.

The decidability of the satisfiability problem of $\mathbf{K_t4.3}$ comes from its finite model property, shown by Ono and Nakamura [16, Thm. 3]. But this property can only be obtained when working with *weak* total orders, i.e. allowing some worlds of the models to be equivalent for the order relation. Such groups of nodes are commonly called ‘clusters.’ Note that the logic itself is not able to distinguish between a weakly ordered frame and any of its ‘bulldozed’ strict orders [2, Thm. 4.56].

In the remainder of this paper, we capture the syntactic aspects of these model-theoretic results. In Section 3, we show how to enhance the hypersequent calculus of Indrzejczak [9] by capturing the model-theoretic ideas in hypersequents with *clusters* and *annotations*. This leads to a sound and complete proof system where proof search always terminates, furthermore with a coNP complexity—which is optimal for the validity problem. Moreover, this proof system is also modular: we consider some classical extensions of $\mathbf{K_t4.3}$ in Section 4, and provide new rules for our hypersequent calculus to handle these extensions; these new rules still yield an optimal coNP proof search. Finally, Manuel and Sreejith [14] have recently shown that validity in first-order logic with two variables over strict total orders is in coNEXP. The same statement can be derived from our results and further extended to *dense* linear orders, by first converting the first-order formulæ into equivalent exponential-sized $\mathbf{K_t4.3}$ formulæ [6]; see Section 5.

We start by recalling the definition of $\mathbf{K_t4.3}$ in Section 2.

2 Modal Logic on Weak Total Orders

We consider tense logics with two unary temporal operators, over a set Φ of propositional variables, with the following syntax:

$$\varphi ::= \perp \mid p \mid \varphi \supset \varphi \mid \mathsf{G}\,\varphi \mid \mathsf{H}\,\varphi \qquad \text{(where } p \in \Phi\text{)}$$

Formulæ $\mathsf{G}\,\varphi$ and $\mathsf{H}\,\varphi$ are called *modal formulæ*. Intuitively, $\mathsf{G}\,\varphi$ expresses that φ holds 'globally' in all future worlds reachable from the current one, while $\mathsf{H}\,\varphi$ expresses that φ holds 'historically' in all past worlds from which the current world is accessible. Other Boolean connectives may be encoded from $\bot$ and $\supset$, and we define, as is common, $\mathsf{F}\,\varphi = \neg\mathsf{G}\,\neg\varphi$ expressing that φ will hold 'in the future' and $\mathsf{P}\,\varphi = \neg\mathsf{H}\,\neg\varphi$ expressing that φ was true 'in the past.'

2.1 Semantics

As is standard, our formulæ shall be evaluated on *Kripke structures.* A *frame* is a pair $\mathfrak{F} = (W, \precsim)$, where W is a set of worlds, and $\precsim \subseteq W \times W$ is a binary relation over W. A *structure* is a pair $\mathfrak{M} = (\mathfrak{F}, V)$, where $\mathfrak{F} = (W, \precsim)$ is a frame, and $V : \Phi \to 2^W$ is a valuation function. Given such a structure, we define the *satisfaction* relation $\mathfrak{M}, w \models \varphi$, where $w \in W$ and φ is a formula, by structural induction on φ:

$$
\begin{array}{ll}
\mathfrak{M}, w \not\models \bot & \\
\mathfrak{M}, w \models p & \text{iff } w \in V(p) \\
\mathfrak{M}, w \models \varphi \supset \psi & \text{iff if } \mathfrak{M}, w \models \varphi \text{ then } \mathfrak{M}, w \models \psi \\
\mathfrak{M}, w \models \mathsf{G}\,\varphi & \text{iff } \forall w' \in W \text{ such that } w \precsim w',\ \mathfrak{M}, w' \models \varphi \\
\mathfrak{M}, w \models \mathsf{H}\,\varphi & \text{iff } \forall w' \in W \text{ such that } w' \precsim w,\ \mathfrak{M}, w' \models \varphi
\end{array}
$$

When $\mathfrak{M}, w \models \varphi$, we say that $(\mathfrak{M}, w)$ is a *model* of φ.

A formula that is satisfied in all worlds of all structures is said to be *valid.* In this paper, we shall not consider the validity problem in general, but only in restricted classes of structures. Namely, we will consider the logic of *weak total orders*, i.e., the formulæ that hold in all structures whose accessibility relation is transitive and total. This logic can be defined axiomatically, as shown next. Later in Section 4, we will study further restrictions of the logic.

The choice of the symbol $\precsim$ for our frames' accessibility relations is in line with our focus on weak total orders. When working on such orders, it is useful to define $x \prec y$ when $x \precsim y$ but not $y \precsim x$. Note that $\prec$ may not be a strict total order: it is transitive but not necessarily total.

2.2 Weak Total Orders

The logic $\mathbf{K_t 4.3}$ is defined as the set of theorems generated by necessitation, modus ponens and substitution from classical tautologies and the axioms:

$$\mathsf{G}\,(p \supset q) \supset (\mathsf{G}\,p \supset \mathsf{G}\,q) \tag{$\mathbf{K_r}$}$$

$$\mathsf{H}\,(p \supset q) \supset (\mathsf{H}\,p \supset \mathsf{H}\,q) \tag{$\mathbf{K_\ell}$}$$

$$p \supset \mathsf{G}\,\mathsf{P}\,p \tag{$\mathbf{t_r}$}$$

$$p \supset \mathsf{H}\,\mathsf{F}\,p \tag{$\mathbf{t_\ell}$}$$

$$\mathsf{F}\,\mathsf{F}\,p \supset \mathsf{F}\,p \tag{4}$$

$$\mathsf{F}\,p \wedge \mathsf{F}\,q \supset \mathsf{F}\,(p \wedge \mathsf{F}\,q) \vee \mathsf{F}\,(p \wedge q) \vee \mathsf{F}\,(q \wedge \mathsf{F}\,p) \tag{$.\mathbf{3_r}$}$$

$$\mathsf{P}\,p \wedge \mathsf{P}\,q \supset \mathsf{P}\,(p \wedge \mathsf{P}\,q) \vee \mathsf{P}\,(p \wedge q) \vee \mathsf{P}\,(q \wedge \mathsf{P}\,p) \tag{$.\mathbf{3_\ell}$}$$

The first two axioms are simply the Kripke schema, given for each modality. Next we find the **t** axioms, which are obviously satisfied in our setting since the two modalities are converses of each other.[1] The next axiom, dubbed **4**, corresponds to the transitivity of $\precsim$. More precisely, canonical models of **4** are transitive [2]. Similarly, canonical models of the *trichotomy* axioms **.3** have accessibility relationships that are non-branching to the left and to the right. All together, this implies the following completeness result:

Fact 2.1 ([2, p. 220]) *A formula is a theorem of* $\mathbf{K_t4.3}$ *iff it is valid in all structures whose relation is transitive and total, i.e., in* weak total orders.

The logic $\mathbf{K_t4.3}$ is perhaps better known for being complete wrt. the class of *strict* total orders [2, Thm. 4.56]. As we shall see, focusing on this characterisation would however be counterproductive for our purposes. As a simple illustration of when weak total orders could be beneficial, note that some formulæ admit finite weak total orders as models but only infinite strict total orders. It is the case, for example, of $(\mathsf{G}\,\mathsf{F}\,\top) \wedge (\mathsf{F}\,\top)$, which admits a single-world model that is a weak total order. The use of weak total orders is instrumental in order to derive decidability and complexity results.

3 Hypersequents with Clusters

Indrzejczak [9] proposed a complete calculus for $\mathbf{K_t4.3}$ using the framework of *ordered* hypersequents (aka. linear nested sequents [12]): his calculus works with *lists* of sequents rather than the usual *multisets* of sequents of hypersequent calculi. The semantics of ordered hypersequents relies on a mapping from ordered sequents to worlds that are ordered accordingly. This extension allows for a natural calculus, enjoying the subformula property and extending nicely to accommodate semantic restrictions such as unboundedness and density.

For example, the calculus of [9] allows the following inference:

$$\frac{\Gamma \vdash \Delta;\ \vdash \varphi}{\Gamma \vdash \Delta, \mathsf{G}\,\varphi}$$

It expresses that, if $w \not\models \mathsf{G}\,\varphi$ for an arbitrary world w, there must be a $w \precsim w'$ such that $w' \not\models \varphi$.

Unfortunately, Indrzejczak's completeness argument is quite complex, and does not yield a decision procedure. The argument is Hintikka-style: if a careful exhaustive proof search fails in his calculus, then some failed proof-search branch yields a counter-model of the conclusion hypersequent. In Indrzejczak's calculus, that failure branch may be infinite, in which case the extracted counter-model is obtained as a limit, and is itself infinite.

Finite Models and Hypersequents with Clusters. In fact, the counter-models extracted from failure branches of Indrzejczak's calculus are always *strictly* (and totally) ordered, hence they must be infinite in some cases. The

[1] In a standard bi-modal setting, we would have two a priori unrelated relations. The **t** axioms would then force the two relations to be converses of each other in canonical models.

model theory of $\mathbf{K_t4.3}$ provides us with a way to circumvent this problem. Indeed, $\mathbf{K_t4.3}$ enjoys a finite model property for weak total orders [16, Thm. 3]; the desired finite model is obtained from the original model by applying Lemmon's filtration (see Appendix A for details).

This insight leads us to consider ordered hypersequents with *clusters*, corresponding semantically to sets of worlds which are all equivalent with respect to the weak total order, i.e. where $w \precsim w'$ and $w' \precsim w$ for all w, w' in the cluster. In itself, this only complicates the calculus as it only creates more premises (and indeed some rules in our calculus have a large number of premises), and does not allow us to bound failure branches. For example, the inference shown above would be modified as follows:

$$\dfrac{\Gamma \vdash \Delta;\ \vdash \varphi \qquad \dfrac{\{\Gamma \vdash \Delta \parallel \vdash \varphi\} \quad \{\Gamma \vdash \Delta\};\ \vdash \varphi}{\{\Gamma \vdash \Delta, \mathsf{G}\,\varphi\}}}{\Gamma \vdash \Delta, \mathsf{G}\,\varphi}$$

The bottom inference expresses that, if $w \not\models \mathsf{G}\,\varphi$, then either there is $w \prec w'$ such that $w' \not\models \varphi$ (first premise) or w is reflexive (second premise). In the latter case, the next inference expresses that there must be a w' such that $w' \not\models \varphi$ satisfying either $w \precsim w' \precsim w$ (first premise) or $w \prec w'$ (second one).

Extremal Models and Annotations. Crucially, this new formalism allows us to benefit from another model-theoretic insight. It is known that satisfiability in $\mathbf{K_t4.3}$ is NP-complete because any satisfiable formula φ admits a model of size linear in the size of the formula [16, Thm. 5] (see also [2, Thm. 6.38]). We shall not exploit this result as such, but the construction behind it: the linear-sized model is obtained by keeping, for each subformula $\mathsf{F}\,\psi$ (resp. $\mathsf{P}\,\psi$) of φ, only the *rightmost* (resp. *leftmost*) world of the original model that satisfies ψ.

Viewing our hypersequent calculus as a search for counter-models, we constrain it to search for 'extremal' counter-models as above. Concretely, we annotate some sequents with modal formulæ, requiring that a modal formula occurs at most once as an annotation. For example, the previous inferences are enriched as follows (with the annotations between parentheses):

$$\dfrac{\Gamma \vdash \Delta;\ \vdash \varphi\,(\mathsf{G}\,\varphi) \qquad \dfrac{\{\Gamma \vdash \Delta \parallel \vdash \varphi\,(\mathsf{G}\,\varphi)\} \quad \{\Gamma \vdash \Delta\};\ \vdash \varphi\,(\mathsf{G}\,\varphi)}{\{\Gamma \vdash \Delta, \mathsf{G}\,\varphi\}}}{\Gamma \vdash \Delta, \mathsf{G}\,\varphi}$$

The annotation indicates a maximal sequent for the contradiction of the considered modal formula. This is then reflected by special inferences, for example

$$\dfrac{}{\ldots;\Gamma \vdash \Delta\,(\mathsf{G}\,\varphi)\,;\ldots;\Pi \vdash \Sigma, \mathsf{G}\,\varphi\,;\ldots}$$

which expresses that, for the (complete) class of counter-models that we are considering, there is no counter-model, since $\mathsf{G}\,\phi$ would have to be contradicted strictly after a rightmost contradicting world.

With this in place, we finally obtain a calculus where failure branches are finite. This allows for an elementary completeness argument, extracting finite weakly ordered counter-models from failure branches. For the soundness argument, we indirectly make use of the extremal counter-model construction of [16, Thm. 4]. Another consequence is that proof search in our calculus directly yields an optimal coNP procedure for validity.

3.1 Definitions and Basic Meta-Theory

We shall now formally describe our calculus. We first define hypersequents with clusters and their semantics in terms of embeddings into weak total orders. We then extend them with annotations, and present our system of deduction rules.

Hypersequents with Clusters. A *sequent* (denoted S) is a pair of two finite sets of formulæ, written $\Gamma \vdash \Delta$. It is satisfied in a world w of a model $\mathfrak{M}$ if, in that world, the conjunction of the formulæ of Γ implies the disjunction of the formulæ of Δ. In that case, we write $\mathfrak{M}, w \models \Gamma \vdash \Delta$.

In this paper, a *hypersequent* is a list of *cells*, each cell being either a sequent or a list of sequents called a (syntactic) *cluster*. We shall use the following abstract syntax, where both operators ';' and '$\|$' are associative with unit '$\bullet$':

$$H ::= \bullet \mid C\,;H \qquad \text{(hypersequents)}$$
$$C ::= S \mid \{S \parallel Cl\} \qquad \text{(cells)}$$
$$Cl ::= \bullet \mid S \parallel Cl \qquad \text{(clusters)}$$

The main feature of hypersequents with clusters is that their structures are weak total orders. The order of cells in a hypersequent *is* relevant, as it yields a strict ordering in the semantics. The order of sequents inside a cluster is semantically irrelevant; nevertheless, assuming an ordering as part of the syntactic structure of clusters is sometimes useful, as in the upcoming definition.

Underlying Frames and Embeddings. Let H be a hypersequent containing n sequents, counting both the sequents found directly in its cells and those in its clusters. We call a natural number $i \in [1;n]$ a *position* of H, and we write $H(i)$ for the i-th sequent of H. We define the *underlying frame* of H as $\mathfrak{F}(H) = ([1;n], \precsim)$ where $i \precsim j$ iff either the i-th and j-th sequents are in the same cluster, or the i-th sequent is in a cell that lies strictly to the left of the cell of the j-th sequent. In particular, a position can only be reflexive in the underlying frame of a hypersequent if it is in a cluster.

Let $\mathfrak{F} = (W, \precsim)$ and $\mathfrak{F}' = (W', \precsim')$ be two frames. We say that $\mu : W \to W'$ is an *embedding* of $\mathfrak{F}$ into $\mathfrak{F}'$ if, for all $(w_1, w_2) \in W^2$,

- $w_1 \precsim w_2$ implies $\mu(w_1) \precsim' \mu(w_2)$ and
- $w_1 \prec w_2$ implies $\mu(w_1) \prec' \mu(w_2)$.

In that case, we write $\mathfrak{F} \hookrightarrow_\mu \mathfrak{F}'$. We simply write $H \hookrightarrow_\mu \mathfrak{F}'$ when $\mathfrak{F}(H) \hookrightarrow_\mu \mathfrak{F}'$.

An example embedding is shown in Figure 1. Note that it is possible that $\mu(i)$ is reflexive when i is not. However, positions from distinct cells cannot be embedded into worlds of a same cluster. By contrast, distinct positions belonging to the same cluster may be mapped to the same (reflexive) world.

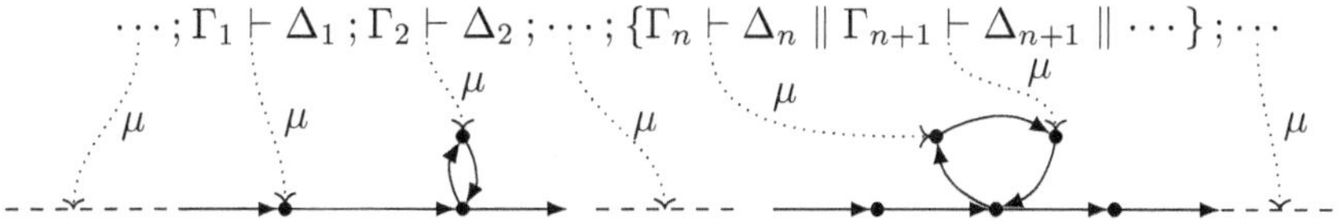

Fig. 1. Embedding of a hypersequent in a weak total order.

Definition 3.1 (semantics) Let $\mathfrak{M} = (\mathfrak{F}, V)$ be a structure. Given an embedding $H \hookrightarrow_\mu \mathfrak{F}$, we say that $(\mathfrak{M}, \mu)$ is a *model* of a hypersequent H, written $\mathfrak{M}, \mu \models H$, when there exists a position i of H such that $\mathfrak{M}, \mu(i) \models H(i)$. We say that a hypersequent is *valid* if for any weak total order $\mathfrak{M} = (\mathfrak{F}, V)$ and any embedding $H \hookrightarrow_\mu \mathfrak{F}$, we have $\mathfrak{M}, \mu \models H$. □

Annotations. We finally introduce annotations, and their semantics. An *annotated sequent* is a sequent that may be annotated with modal formulæ. We simply write $\Gamma \vdash \Delta$ for a sequent carrying no annotation, otherwise we write, e.g., $\Gamma \vdash \Delta\ (\mathsf{H}\,\varphi, \mathsf{G}\,\psi, \ldots)$. Then, *annotated hypersequents* are hypersequents whose sequents are annotated, with the constraint that an annotation may only occur once in an annotated hypersequent. Formally, we can see annotations as partial functions from the set of modal formulæ to the set of positions of the hypersequent. For instance, $\Gamma \vdash \Delta\ (\mathsf{G}\,\varphi) ; \{\Pi \vdash \Sigma\ (\mathsf{H}\,\varphi)\}$ is an annotated hypersequent but $\Gamma \vdash \Delta\ (\mathsf{H}\,\varphi, \mathsf{G}\,\varphi) ; \{\Pi \vdash \Sigma\ (\mathsf{H}\,\varphi)\}$ is not allowed because of the two occurrences of $\mathsf{H}\,\varphi$ as an annotation.

Since we use these annotations to guide the search for a *finite* counter-model, we only define a semantics for annotated hypersequents over finite structures.

Definition 3.2 (annotation semantics) Given an annotated hypersequent H and a *finite* structure $\mathfrak{M} = (\mathfrak{F}, V)$, an embedding $H \hookrightarrow_\mu \mathfrak{F}$ is *annotation-respecting* if, for all i such that $H(i)$ carries the annotation $(\mathsf{G}\,\varphi)$ (resp. $(\mathsf{H}\,\varphi)$), there is no $w \in W$ such that $\mathfrak{M}, w \models \neg\varphi$ and $\mu(i) \prec w$ (resp. $w \prec \mu(i)$).

An *annotation-respecting model* of H is a model $((\mathfrak{F}, V), \mu)$ of H where $H \hookrightarrow_\mu \mathfrak{F}$ is annotation-respecting. The sequent H is *annotation-respecting valid* if, for any finite weak total order $\mathfrak{M} = (\mathfrak{F}, V)$ and any annotation-respecting embedding $H \hookrightarrow_\mu \mathfrak{F}$, we have $\mathfrak{M}, \mu \models H$. □

A peculiarity of our system is that it is sound with respect to the annotation-respecting validity of Definition 3.2, but only complete with respect to the general validity of Definition 3.1: if a hypersequent is annotation-respecting valid but not valid in general, it might not have a derivation. Thus our annotation system may be seen as a proof search strategy over a more standard, annotation-free system. [2] In any case, our proof system is sound and complete for hypersequents without annotations, since the two notions of validity coincide in that case.

[2] This more standard system simply consists of the rules of figures 2 and 3 without annotations, ignoring the rules of Figure 4. It is obviously sound. For completeness, we conjecture that Indrzejczak's proof could be adapted to weak total orders. However that system is not interesting as it is subsumed by Indrzejczak's original calculus, only adding more branches.

$$\text{(ax)}\ \frac{}{H\,[\varphi,\Gamma\vdash\Delta,\varphi]} \qquad \frac{H\,[\varphi\supset\psi,\Gamma\vdash\Delta,\varphi] \qquad H\,[\varphi\supset\psi,\psi,\Gamma\vdash\Delta]}{H\,[\varphi\supset\psi,\Gamma\vdash\Delta]}\ (\supset\vdash)$$

$$(\bot)\ \frac{}{H\,[\Gamma,\bot\vdash\Delta]} \qquad \frac{H\,[\varphi,\Gamma\vdash\Delta,\psi,\varphi\supset\psi]}{H\,[\Gamma\vdash\Delta,\varphi\supset\psi]}\ (\vdash\supset)$$

Fig. 2. Propositional rules of the hypersequent calculus with clusters.

Rules of the Hypersequent Calculus. The rules are given in figures 2 to 4, making use of a few notations.

First, we use hypersequents with *holes.* One-placeholder hypersequents, cells, and clusters are defined by the syntax

$$H\,[] ::= H\,;C\,[]\,;H \qquad C\,[] ::= \star \mid \{\ Cl[]\ \} \qquad Cl[] ::= Cl \parallel \star \parallel Cl$$

Two-placeholder cells and hypersequents have two holes identified by $\star_1$ and $\star_2$:

$$H\,[]\,[] ::= H\,;C\,[]\,[]\,;H \mid H[\star_1]\,;H[\star_2]$$
$$C\,[]\,[] ::= \{\ Cl[\star_1] \parallel Cl[\star_2]\ \} \mid \{\ Cl[\star_2] \parallel Cl[\star_1]\ \}$$

As usual, $C\,[S]$ (resp. $C\,[Cl]$) denotes the same cell with S (resp. Cl) substituted for $\star$; two-placeholder cells and hypersequents with holes behave similarly. In terms of the frames underlying hypersequents with two holes, observe that the positions i and j associated resp. to $\star_1$ and $\star_2$ are such that $i \precsim j$.

Second, we do not write explicitly the annotations that sequents may carry in rule applications. These annotations are implicitly the same in a conclusion sequent and the corresponding sequents in premises, or updated by adding the explicit annotation; freshly created sequents always have an explicit annotation. Annotations can prevent a rule application if the addition of an annotation would break the single-annotation constraint.

Third, we use a convenient notation for *enriching* a sequent: if S is a sequent $\Gamma\vdash\Delta\ (A)$, then $S \ltimes (\Gamma'\vdash\Delta'\ (A'))$ is the sequent $\Gamma,\Gamma'\vdash\Delta,\Delta'\ (A,A')$. Moreover, we sometimes need to enrich an arbitrary sequent of a cluster C with a sequent S; then $C \ltimes S$ denotes the cluster with its leftmost sequent enriched.

Modal Rules. After the usual propositional rules of Figure 2, we give in Figure 3 the introduction rules for modalities. The left introduction rules are symmetric for our two modalities. The first two, $(\mathsf{G}\vdash)$ and $(\mathsf{G}\vdash')$, express that if $\mathsf{G}\,\varphi$ holds at some position, then φ must also hold at a position to its right in the underlying frame.

Regarding the right introduction rules for modalities, let us start with the particular case where these modalities occur in extremal cells. In rule $(\vdash\mathsf{G})$, we introduce a formula $\mathsf{G}\,\varphi$ to the right of a principal sequent that is in the rightmost cell of the hypersequent. The premises cover all the ways in which a world could occur to the right of (the embedding of) the principal sequent:

- We always have to consider a possible new cell strictly further to the right; in that case, the cell carries the (single) annotation $(\mathsf{G}\,\varphi)$.

$$(\mathsf{G}\vdash)\ \frac{H\,[\mathsf{G}\,\varphi,\Gamma\vdash\Delta]\,[\varphi,\Pi\vdash\Sigma]}{H\,[\mathsf{G}\,\varphi,\Gamma\vdash\Delta]\,[\Pi\vdash\Sigma]} \qquad \frac{H_1;\{\mathit{Cl}_1 \parallel \varphi,\mathsf{G}\,\varphi,\Gamma\vdash\Delta \parallel \mathit{Cl}_2\}\,;H_2}{H_1;\{\mathit{Cl}_1 \parallel \mathsf{G}\,\varphi,\Gamma\vdash\Delta \parallel \mathit{Cl}_2\}\,;H_2}\ (\mathsf{G}\vdash')$$

$$(\mathsf{H}\vdash)\ \frac{H\,[\varphi,\Pi\vdash\Sigma]\,[\mathsf{H}\,\varphi,\Gamma\vdash\Delta]}{H\,[\Pi\vdash\Sigma]\,[\mathsf{H}\,\varphi,\Gamma\vdash\Delta]} \qquad \frac{H_1;\{\mathit{Cl}_1 \parallel \varphi,\mathsf{H}\,\varphi,\Gamma\vdash\Delta \parallel \mathit{Cl}_2\}\,;H_2}{H_1;\{\mathit{Cl}_1 \parallel \mathsf{H}\,\varphi,\Gamma\vdash\Delta \parallel \mathit{Cl}_2\}\,;H_2}\ (\mathsf{H}\vdash')$$

$$\frac{\begin{array}{ll} H\,;C\,[\Gamma\vdash\Delta,\mathsf{G}\,\varphi]\,;\ \vdash\varphi\,(\mathsf{G}\,\varphi) & \\ H\,;\{\Gamma\vdash\Delta,\mathsf{G}\,\varphi\} & \text{if } C=\star \\ H\,;C\,[\Gamma\vdash\Delta,\mathsf{G}\,\varphi \parallel \vdash\varphi\,(\mathsf{G}\,\varphi)] & \text{if } C\neq\star \end{array}}{H\,;C\,[\Gamma\vdash\Delta,\mathsf{G}\,\varphi]}\ (\vdash\mathsf{G})$$

$$\frac{\begin{array}{ll} \vdash\varphi\,(\mathsf{H}\,\varphi)\,;C\,[\Gamma\vdash\Delta,\mathsf{H}\,\varphi]\,;H & \\ \{\Gamma\vdash\Delta,\mathsf{H}\,\varphi\}\,;H & \text{if } C=\star \\ C\,[\Gamma\vdash\Delta,\mathsf{H}\,\varphi \parallel \vdash\varphi\,(\mathsf{H}\,\varphi)]\,;H & \text{if } C\neq\star \end{array}}{C\,[\Gamma\vdash\Delta,\mathsf{H}\,\varphi]\,;H}\ (\vdash\mathsf{H})$$

$$\frac{\begin{array}{ll} H\,[C\,[\Gamma\vdash\Delta,\mathsf{G}\,\varphi]\,;\ \vdash\varphi\,(\mathsf{G}\,\varphi)\,;C'] & \\ H\,[C\,[\Gamma\vdash\Delta,\mathsf{G}\,\varphi]\,;C'\ltimes(\vdash\mathsf{G}\,\varphi)] & \\ H\,[C\,[\Gamma\vdash\Delta,\mathsf{G}\,\varphi]\,;C'\ltimes(\vdash\varphi\,(\mathsf{G}\,\varphi))] & \text{if } C' \text{ is not a cluster} \\ H\,[\{\Gamma\vdash\Delta,\mathsf{G}\,\varphi\}\,;C'] & \text{if } C=\star \\ H\,[C\,[\Gamma\vdash\Delta,\mathsf{G}\,\varphi \parallel \vdash\varphi\,(\mathsf{G}\,\varphi)]\,;C'] & \text{if } C\neq\star \end{array}}{H\,[C\,[\Gamma\vdash\Delta,\mathsf{G}\,\varphi]\,;C']}\ (\vdash\mathsf{G}')$$

$$\frac{\begin{array}{ll} H\,[C';\ \vdash\varphi\,(\mathsf{H}\,\varphi)\,;C\,[\Gamma\vdash\Delta,\mathsf{H}\,\varphi]] & \\ H\,[C'\rtimes(\vdash\mathsf{H}\,\varphi)\,;C\,[\Gamma\vdash\Delta,\mathsf{H}\,\varphi]] & \\ H\,[C'\rtimes(\vdash\varphi\,(\mathsf{H}\,\varphi))\,;C\,[\Gamma\vdash\Delta,\mathsf{H}\,\varphi]] & \text{if } C' \text{ is not a cluster} \\ H\,[C'\,;\{\Gamma\vdash\Delta,\mathsf{H}\,\varphi\}] & \text{if } C=\star \\ H\,[C'\,;C\,[\Gamma\vdash\Delta,\mathsf{H}\,\varphi \parallel \vdash\varphi\,(\mathsf{H}\,\varphi)]] & \text{if } C\neq\star \end{array}}{H\,[C'\,;C\,[\Gamma\vdash\Delta,\mathsf{H}\,\varphi]]}\ (\vdash\mathsf{H}')$$

Fig. 3. Modal rules of the hypersequent calculus with clusters.

- If the active sequent does not belong to a cluster, i.e., if $C=\star$, it may still be embedded in a cluster in a frame, so we have to consider a premise where the last cell is changed into a single-sequent cluster.
- Alternatively, if $C\neq\star$, the active sequent belongs to a cluster and we need the last premise when φ is falsified in an arbitrary world of that cluster.

Rule $(\vdash\mathsf{H})$ is, as expected, symmetric. Note that the $(\vdash\mathsf{G})$ and $(\vdash\mathsf{H})$ rules *cannot* apply when the principal formula already belongs to the annotations of some sequent of the hypersequent, since it would then create a new cell with that annotation. The rules $(\vdash\mathsf{G}')$ and $(\vdash\mathsf{H}')$, where the active sequent is not extremal, follow the same idea but have extra premises corresponding to the case where φ is falsified in the next cell C' or beyond.

Annotation Rules. Finally, the rules of Figure 4 allow special deduction steps based on the annotations, leveraging the annotation-respecting semantics.

$$((\mathsf{G}))\ \frac{}{H_1\,[\Gamma \vdash \Delta\ (\mathsf{G}\,\varphi)]\ ;\ H_2\,[\Pi \vdash \Sigma, \mathsf{G}\,\varphi]} \qquad \frac{H_1\ ;\ \{\Gamma \vdash \Delta, \mathsf{H}\,\varphi\ (\mathsf{H}\,\varphi)\}\ ;\ H_2}{H_1\ ;\ \Gamma \vdash \Delta, \mathsf{H}\,\varphi\ (\mathsf{H}\,\varphi)\ ;\ H_2}\ (\{(\mathsf{H})\})$$

$$((\mathsf{H}))\ \frac{}{H_1\,[\Pi \vdash \Sigma, \mathsf{H}\,\varphi]\ ;\ H_2\,[\Gamma \vdash \Delta\ (\mathsf{H}\,\varphi)]} \qquad \frac{H_1\ ;\ \{\Gamma \vdash \Delta, \mathsf{G}\,\varphi\ (\mathsf{G}\,\varphi)\}\ ;\ H_2}{H_1\ ;\ \Gamma \vdash \Delta, \mathsf{G}\,\varphi\ (\mathsf{G}\,\varphi)\ ;\ H_2}\ (\{(\mathsf{G})\})$$

Fig. 4. Annotation rules of the hypersequent calculus with clusters.

The $((\mathsf{G}))$ rule allows to derive any hypersequent where $\mathsf{G}\,\varphi$ occurs strictly to the right of a sequent carrying the annotation $(\mathsf{G}\,\varphi)$, and symmetrically for $((\mathsf{H}))$: such hypersequents cannot have annotation-respecting counter-models. The $(\{(\mathsf{G})\})$ and $(\{(\mathsf{H})\})$ rules express that, if a hypersequent features a sequent containing a modal formula both in its right hand side and in its set of annotations, then that sequent must occur in a cluster for the hypersequent to have an annotation-respecting counter-model.

Invertibility. Note that our rules are formulated in an invertible style, keeping the principal formula in the premises. This eases the proof of completeness, where proof search induces a form of saturation. The following weakening rules are admissible in our system, and we shall use them implicitly in examples to avoid carrying around useless formulæ:

$$\frac{H[\Gamma \vdash \Delta]}{H[\Gamma, \varphi \vdash \Delta]} \qquad \frac{H[\Gamma \vdash \Delta]}{H[\Gamma \vdash \varphi, \Delta]}$$

We prove invertibility with respect to Definition 3.1.

Lemma 3.3 (invertibility) *For any instance of a deduction rule where the conclusion hypersequent is valid, all premisses are also valid.*

Proof. Considering a rule instance with a counter-model $(\mathfrak{M}, \mu)$ of a premise H, we build a counter-model $(\mathfrak{M}, \mu')$ of the conclusion H'. Depending on the rule that is applied, H and H' will either have exactly the same structure, or H will have a new cell, or H will have a cluster cell where H' contains a simple sequent cell. Accordingly, we take μ' to be the restriction of μ to the positions of H' (and adapt it accordingly for the positions that have been shifted). It is indeed a proper embedding of H' into $\mathfrak{M}$. It is then easy to see that $(\mathfrak{M}, \mu')$ is a counter-model of H', since any sequent $H'(i)$ is contained in the corresponding sequent $H(j)$: $\mathfrak{M}, \mu(j) \not\models H(j)$ implies $\mathfrak{M}, \mu'(i) \not\models H'(i)$. □

Example 3.4 We provide on the next page a proof of the hypersequent $\{\mathsf{H}\,p, \mathsf{G}\,p, p \vdash \mathsf{G}\,\mathsf{H}\,p\}$ in our system. At each inference, the principal formula is underlined and weakenings are implicit.

Example 3.5 Consider the hypersequent $\mathsf{G}\,\neg\mathsf{G}\,\bot\ \vdash\ \mathsf{G}\,\bot$, which has finite counter-models with a weak total order, but no finite counter-models with a strict total order (a counter-model of this sequent must be unbounded to the right). When trying to prove this sequent with the calculus of Indrzejczak [9], the proof search strategy underlying its completeness argument unfolds the

$$
\dfrac{\mathcal{P} \qquad \dfrac{\dfrac{\overline{\{\mathsf{H}p, \mathsf{G}p, p \vdash \; \| \vdash \; (\mathsf{G}\,\mathsf{H}p) \; \| \; \underline{p} \vdash \underline{p}\; (\mathsf{H}p)\}}\;(\mathsf{ax})}{\{\mathsf{H}p, \underline{\mathsf{G}p}, p \vdash \; \| \vdash \; (\mathsf{G}\,\mathsf{H}p) \; \| \; \vdash p\; (\mathsf{H}p)\}}\;(\mathsf{G}\vdash) \qquad \dfrac{\overline{\underline{p} \vdash \underline{p}\; (\mathsf{H}p) \,;\, \{\mathsf{H}p, \mathsf{G}p, p \vdash \; \| \vdash \; (\mathsf{G}\,\mathsf{H}p)\}}\;(\mathsf{ax})}{\vdash p\; (\mathsf{H}p) \,;\, \{\underline{\mathsf{H}p}, \mathsf{G}p, p \vdash \; \| \vdash \; (\mathsf{G}\,\mathsf{H}p)\}}\;(\mathsf{H}\vdash)}{\{\mathsf{H}p, \mathsf{G}p, p \vdash \; \| \vdash \underline{\mathsf{H}p}\; (\mathsf{G}\,\mathsf{H}p)\}}\;(\vdash\mathsf{H})}{\{\mathsf{H}p, \mathsf{G}p, p \vdash \underline{\mathsf{G}\,\mathsf{H}p}\}}\;(\vdash\mathsf{G})
$$

where $\mathcal{P}$ is:

$$
\dfrac{\overline{\{\underline{\mathsf{H}p}, \mathsf{G}p, p \vdash \underline{\mathsf{H}p}\} \,;\; \vdash \; (\mathsf{G}\,\mathsf{H}p)}\;(\mathsf{ax}) \qquad \dfrac{\overline{\{\mathsf{H}p, \mathsf{G}p, p \vdash\} \,;\, \underline{p} \vdash \underline{p}\; (\mathsf{H}p) \,;\; \vdash \; (\mathsf{G}\,\mathsf{H}p)}\;(\mathsf{ax})}{\{\mathsf{H}p, \underline{\mathsf{G}p}, p \vdash\} \,;\; \vdash p\; (\mathsf{H}p) \,;\; \vdash \; (\mathsf{G}\,\mathsf{H}p)}\;(\mathsf{G}\vdash) \qquad \mathcal{P}'}{\{\mathsf{H}p, \mathsf{G}p, p \vdash\} \,;\; \vdash \underline{\mathsf{H}p}\; (\mathsf{G}\,\mathsf{H}p)}\;(\vdash\mathsf{H}')
$$

where $\mathcal{P}'$ is:

$$
\dfrac{\overline{\{\underline{\mathsf{H}p}, \mathsf{G}p, p \vdash \underline{\mathsf{H}p}\} \,;\, \{\vdash \; (\mathsf{G}\,\mathsf{H}p)\}}\;(\mathsf{ax}) \qquad \dfrac{\overline{\{\mathsf{H}p, \mathsf{G}p, p \vdash\} \,;\, \underline{p} \vdash \underline{p}\; (\mathsf{H}p) \,;\, \{\vdash \; (\mathsf{G}\,\mathsf{H}p)\}}\;(\mathsf{ax})}{\{\mathsf{H}p, \underline{\mathsf{G}p}, p \vdash\} \,;\; \vdash p\; (\mathsf{H}p) \,;\, \{\vdash \; (\mathsf{G}\,\mathsf{H}p)\}}\;(\mathsf{G}\vdash) \qquad \dfrac{\overline{\{\mathsf{H}p, \mathsf{G}p, p \vdash\} \,;\, \{\vdash \; (\mathsf{G}\,\mathsf{H}p) \; \| \; \underline{p} \vdash \underline{p}\; (\mathsf{H}p)\}}\;(\mathsf{ax})}{\{\mathsf{H}p, \underline{\mathsf{G}p}, p \vdash\} \,;\, \{\vdash \; (\mathsf{G}\,\mathsf{H}p) \; \| \; \vdash p\; (\mathsf{H}p)\}}\;(\mathsf{G}\vdash)}{\{\mathsf{H}p, \mathsf{G}p, p \vdash\} \,;\, \{\vdash \underline{\mathsf{H}p}\; (\mathsf{G}\,\mathsf{H}p)\}}\;(\vdash\mathsf{H}')
$$

following infinite derivation, by alternating the right and left introduction rules for G (with implicit uses of the left rules for $\supset$ and $\bot$):

$$\dfrac{\dfrac{\dfrac{\dfrac{\vdots}{\mathsf{G}\neg\mathsf{G}\bot \vdash \mathsf{G}\bot \;;\; \vdash \mathsf{G}\bot, \bot \;;\; \vdash \bot}}{\mathsf{G}\neg\mathsf{G}\bot \vdash \mathsf{G}\bot \;;\; \vdash \underline{\mathsf{G}\bot}, \bot}}{\underline{\mathsf{G}\neg\mathsf{G}\bot} \vdash \mathsf{G}\bot \;;\; \vdash \bot}}{\mathsf{G}\neg\mathsf{G}\bot \vdash \underline{\mathsf{G}\bot}}$$

Principal formulas underlined, useless formulas in gray.

In our calculus, a derivation of that same hypersequent would necessarily contain several branches. The analogue of the one shown above will quickly lead to a point where only $(\{(\mathsf{G})\})$ applies, after which no rule applies:

$$\dfrac{\cdots \quad \dfrac{\dfrac{\mathsf{G}\neg\mathsf{G}\bot \vdash \mathsf{G}\bot \,;\{\vdash \mathsf{G}\bot, \bot\ (\mathsf{G}\bot)\}}{\mathsf{G}\neg\mathsf{G}\bot \vdash \mathsf{G}\bot \;;\; \vdash \underline{\mathsf{G}\bot}, \bot\ (\mathsf{G}\bot)}(\{(\mathsf{G})\})}{\underline{\mathsf{G}\neg\mathsf{G}\bot} \vdash \mathsf{G}\bot \;;\; \vdash \bot\ (\mathsf{G}\bot)}(\mathsf{G}\vdash) \quad \cdots}{\mathsf{G}\neg\mathsf{G}\bot \vdash \underline{\mathsf{G}\bot}}(\vdash\mathsf{G})$$

In other words, it is a finite failure branch. As we shall see, we can extract from it a finite counter-model featuring a reflexive world. □

3.2 Soundness

We show two soundness statements, relative to definitions 3.2 and 3.1.

Lemma 3.6 (annotation-respecting soundness) *All the rules of our hypersequent calculus with clusters are sound with respect to the annotation-respecting semantics.*

Proof. We prove the contrapositive. Considering a rule instance whose conclusion H admits an annotation-respecting counter-model $(\mathfrak{M}, \mu)$, we show that one of its premises also admits an annotation-respecting counter-model $(\mathfrak{M}, \mu')$. Below, embeddings and counter-models are implicitly annotation-respecting.

We first consider the case of rule $(\vdash\mathsf{G}')$, applied on a principal sequent $\Gamma \vdash \Delta, \mathsf{G}\,\varphi$ at position i in H. Since $\mathfrak{M}, \mu(i) \not\models \mathsf{G}\,\varphi$, there exists w' such that $\mu(i) \precsim w'$ and $\mathfrak{M}, w' \models \neg\varphi$. Since $\mathfrak{M}$ is finite we can take w' to be a rightmost world invalidating φ, i.e., such that there is no $w' \prec w''$ such that $w'' \models \neg\varphi$.

- We first consider the case where $\mu(i)$ and w' are two worlds (distinct or not) of the same cluster. If i is not in a cluster in the underlying frame of H, i.e., if $C = \star$, then the rule has a premise $H\left[\{\Gamma \vdash \Delta, \mathsf{G}\,\varphi\}\,;C'\right]$ of which $(\mathfrak{M}, \mu)$ is a counter-model. Otherwise, the premise $H\left[C\left[\Gamma \vdash \Delta, \mathsf{G}\,\varphi \parallel \vdash \varphi\ (\mathsf{G}\,\varphi)\right];C'\right]$ is available. We extend μ into μ', mapping the new sequent, at position $i+1$, to the world w': $\mu'(k) = \mu(k)$ for all $k \leq i$, $\mu'(i+1) = w'$, and $\mu'(k+1) = \mu(k)$ for all $k > i$. Then $(\mathfrak{M}, \mu')$ is a counter-model of the premise. In particular, the annotation $(\mathsf{G}\,\varphi)$ at position $i+1$ is respected, as we have chosen $\mu'(i+1) = w'$ such that for any $\mu'(i+1) \prec w''$, $w'' \models \varphi$.
- Otherwise, $\mu(i) \prec w'$. Let j be the first position in the cell C'. If $w' \prec \mu(j)$, we obtain a counter-model of premise $H\left[C\left[\Gamma \vdash \Delta, \mathsf{G}\,\varphi\right];\ \vdash \varphi\ (\mathsf{G}\,\varphi)\,;C'\right]$

by adapting μ into an embedding μ' that assigns w' to the new position. If $\mu(j) \precsim w'$ then we have a counter-model of the second premise $H\,[C\,[\Gamma \vdash \Delta, \mathsf{G}\,\varphi]\,;C' \ltimes (\vdash \mathsf{G}\,\varphi)]$, with the same embedding μ. Otherwise, $\mu(j) = w'$ and $\mu(j)$ is not reflexive, hence the next premise is available, namely $H\,[C\,[\Gamma \vdash \Delta, \mathsf{G}\,\varphi]\,;C' \ltimes (\vdash \varphi\;(\mathsf{G}\,\varphi))]$. Our counter-model $(\mathfrak{M}, \mu)$ is a counter-model of that premise.

In the case of rule $(\{(\mathsf{G})\})$ applied on a principal sequent $\Gamma \vdash \Delta, \mathsf{G}\,\varphi\;(\mathsf{G}\,\varphi)$ at position i in H, i cannot be in a cluster. Since $(\mathfrak{M}, \mu)$ is a counter-model of H, we have $\mathfrak{M}, \mu(i) \models \neg\mathsf{G}\,\varphi$, i.e., there is a world w of $\mathfrak{M}$ such that $\mu(i) \precsim w$ and $\mathfrak{M}, w \models \neg\varphi$. Since $H(i)$ carries the annotation $(\mathsf{G}\,\varphi)$, we cannot have $\mu(i) \prec w$, hence $\mu(i) = w$ or $w \precsim \mu(i)$. Either way, $\mu(i)$ is reflexive, hence $(\mathfrak{M}, \mu)$ is still a counter-model of the premise of the rule, which creates a cluster at position i.

We finally consider the case of rule $((\mathsf{G}))$. We show that there cannot be a counter-model $(\mathfrak{M}, \mu)$ of the conclusion $H_1\,[\Gamma \vdash \Delta\;(\mathsf{G}\,\varphi)]\,;H_2\,[\Pi \vdash \Sigma, \mathsf{G}\,\varphi]$. Let $i \prec j$ be the respective positions of the two active sequents in the rule application. Since $\mathfrak{M}, \mu(j) \models \neg\mathsf{G}\,\varphi$, there exists w such that $\mu(j) \precsim w$ and $\mathfrak{M}, w \models \neg\varphi$. Since $\mu(i) \prec w$, this contradicts the fact that μ was assumed annotation-respecting.

The other rules are analogous, or easy to handle. □

Theorem 3.7 (soundness) *Our hypersequent calculus with clusters is sound: if an annotation-free hypersequent is provable, then it is valid.*

Proof. We prove the contrapositive. If an annotation-free hypersequent has a counter-model, then it has a finite counter-model $\mathfrak{M}$ as a consequence of the finite model property of $\mathbf{K_t4.3}$ [16] recalled in Lemma A.2, Appendix A. Since $\mathfrak{M}$ is finite and H does not carry any annotations, $\mathfrak{M}$ is also an annotation-respecting counter-model of H. So, by Lemma 3.6, H is not provable. □

3.3 Completeness and Complexity

We now turn to establishing completeness for our calculus, and to showing that proof search yields an optimal coNP procedure for deciding $\mathbf{K_t4.3}$ validity. These results follow from two properties of our calculus: deduction rules are invertible wrt. the (annotation-blind) semantics (recall Lemma 3.3), and proof search branches are polynomially bounded (as shown next in Lemma 3.8).

In this section, we call *partial proof* a finite open derivation tree: each node corresponds to a rule application, but some leaves may be left open. Partial proofs arise from (backward) proof search. We require that the conclusion hypersequent of any rule application differs from all of the premisses of that rule—this amounts to forbidding useless proof search steps.

In general, proof search may diverge by expanding partial proofs infinitely, or require backtracking due to (finite) choices in rule applications. Lemma 3.8 shows that divergence cannot happen with our calculus, regardless of the way rules are applied. Lemma 3.3 shows that backtracking is not necessary either. Hence, proof search in our calculus simply consists in expanding one proof attempt, either reaching a complete proof or obtaining a partial proof with at least one open leaf that cannot be derived by any rule application.

We define $|H|$ to be the maximum of the number of positions in H and the number of distinct subformulæ occurring in H.

Lemma 3.8 (small branch property) *For any partial proof of a hypersequent H, any branch of the proof is of length at most $4|H|^2 + 2|H|$.*

Proof. Let H be a hypersequent of size $|H|$, $\mathcal{P}$ a partial proof of it, and β a branch of $\mathcal{P}$. Remark that the number of positions in hypersequents of β is bounded by $2|H|$: we have at most $|H|$ positions initially, and a new position may only be created together with a new annotation among at most $|H|$ formulæ. Any rule application adds some subformula among $|H|$ to the left or to the right of the turnstile at a position among $2|H|$, hence with $4|H|^2$ choices, or changes a simple cell among $2|H|$ into a cluster. Thus β is of length at most $4|H|^2 + 2|H|$. □

Theorem 3.9 (completeness) *Our hypersequent calculus with clusters is complete: every annotation-free valid hypersequent H has a proof.*

Proof. Assume that a hypersequent H is not provable. Consider a partial proof $\mathcal{P}$ of H that cannot be expanded any more: its leaves cannot be obtained as the conclusion of a rule instance. Such a partial proof exists by Lemma 3.8. By invertibility, it suffices to exhibit a counter-model for an open leaf of $\mathcal{P}$ to obtain a counter-model of H as required.

We thus consider a leaf hypersequent H', which cannot be derived by any rule (excluding rule applications which would have H' itself as a premise). Let $\mathfrak{F} = (W, \precsim)$ be the underlying frame of H'. Let V be the valuation defined for all $i \in W$ by $V(p) = \{i \in W \mid p \text{ appears on the left-hand side of } H'(i)\}$. Finally, let $\mathfrak{M} = (\mathfrak{F}, V)$. We shall establish that $(\mathfrak{M}, \mu)$ is a counter-model of H', where μ is the identity embedding $H' \hookrightarrow_\mu \mathfrak{F}$. More precisely, we prove by structural induction on φ that, for every position i of H':

- If φ appears on the left of the turnstile in $H'(i)$, then $\mathfrak{M}, i \models \varphi$.
- If φ appears on the right of the turnstile in $H'(i)$, then $\mathfrak{M}, i \not\models \varphi$.

We reason by case analysis on φ. We only detail below the case where $\varphi = \mathsf{G}\,\varphi'$, since the other cases are either standard or analogous.

- If $\mathsf{G}\,\varphi'$ appears on the left-hand side of a sequent $H'(i)$, then, since rules $(\mathsf{G}\vdash)$ and $(\mathsf{G}\vdash')$ cannot be applied on H', φ' appears on the left-hand side of every sequent $H'(j)$ such that $i \precsim j$. By induction hypothesis, $\mathfrak{M}, j \models \varphi'$ for all $i \precsim j$. Hence $\mathfrak{M}, i \models \mathsf{G}\,\varphi'$.
- If $\mathsf{G}\,\varphi'$ appears on the right-hand side of $H'(i)$, there must be some position j such that $H'(j)$ carries the annotation $(\mathsf{G}\,\varphi')$, as otherwise, either $(\vdash\mathsf{G})$ or $(\vdash\mathsf{G}')$ would apply. Moreover, by inspection of our rules and since H was initially annotation-free, necessarily j must contain φ' on its right-hand side. By totality, we have $i \precsim j$, $j \precsim i$ or $i = j$. If $i = j$, since rule $(\{(\mathsf{G})\})$ does not apply, i is in a cluster. If $j \precsim i$, since rule $((\mathsf{G}))$ does not apply, i and j must be in the same cluster. So we have $i \precsim j$ in any case. By induction hypothesis on φ', we have $\mathfrak{M}, j \not\models \varphi'$, hence $\mathfrak{M}, i \not\models \mathsf{G}\,\varphi'$. □

Proposition 3.10 *Proof search in our hypersequent calculus is in* coNP.

Proof. Proof search can be implemented in an alternating Turing machine maintaining the current hypersequent on its tape, where existential states choose which rule to apply to which principal sequent(s) and formula, and universal states choose a premise of the rule. By Lemma 3.8, the computation branches are of length bounded by a polynomial. By Lemma 3.3, the non-deterministic choices in existential states can be replaced by arbitrary deterministic choices, thus this Turing machine has only universal states, hence is in coNP. □

4 Extensions

The logic $\mathbf{K_t4.3}$ can be extended by additional axioms to further restrict the class of frames. We consider here two examples of such extensions also considered by Indrzejczak [9]: density and unboundedness. For each extension, we show that our calculus can be adapted by adding new rules corresponding to the new axioms, and yields the same coNP upper bound. These new rules are rather different from Indrzejczak's, and exploit our use of hypersequents with clusters. Together, these rules extend our calculus into a sound and complete proof system with a coNP proof search algorithm for $\mathbf{K_tQ}$, the logic of *dense unbounded* linear frames, consisting of $\mathbf{K_t4.3}$ with both extensions.

Density. A frame $\mathfrak{F} = (W, \precsim)$ is *dense* if $\forall(x,y) \in W^2$, if $x \precsim y$ then $\exists z \in W$ such that $x \precsim z \precsim y$. Density is axiomatised by adding the following axiom:

$$\mathsf{F}\,p \supset \mathsf{F}\,\mathsf{F}\,p \qquad \textbf{(Den)}$$

This new logic also has a finite model property as well as a small model property [16]. Moreover, a finite weak total order is dense if and only if it never has two consecutive worlds that are not in clusters. This last property leads to the following new rule for our calculus to handle density:

$$\frac{H\,[\{S_1\}\,;S_2] \qquad H\,[S_1\,;\{\vdash\}\,;S_2] \qquad H\,[S_1\,;\{S_2\}]}{H\,[S_1\,;S_2]}\ (\mathsf{den})$$

Proposition 4.1 *Adding* (den) *to our calculus yields a sound and complete proof system for* $\mathbf{K_t4.3} \cup \textbf{(Den)}$, *where proof search is in* coNP.

Proof. Our rule is obviously sound, as it closely reflects the shape of dense finite weak total orders. It is also invertible, since the underlying frame of the conclusion of the rule is always a subframe of its premises. Hence Theorem 3.7 and Lemma 3.3 still hold.

To obtain that proof search is in coNP, it suffices to check that Lemma 3.8 carries over to our extension. This is true because the rule (den) can only be applied on two consecutive non-cluster cells, and whenever the rule (den) is applied on such a bad occurrence, this occurrence is no longer present in the premises. Hence, every time the rule (den) is applied, we reduce at least by one the number of bad occurrences, so we can only apply the rule (den) a finite number of times between applications of other rules creating new cells such as $(\vdash \mathsf{G})$ and $(\vdash \mathsf{H})$. Finally, since new cells can only be created polynomialy many times by those other rules thanks to our initial strategy, the new rule (den) can, in the end, only be applied polynomialy many times along a branch. So the branches of our proof tree are still polynomial.

Finally, completeness is obtained as in Theorem 3.9. It only remains to show that the underlying frame of a hypersequent found at the end of a failing branch is dense. Indeed, if its underlying frame was not dense, we could apply the rule (den) which would contradict the fact that no rules can be applied any more on this hypersequent. □

Unboundedness. A frame $\mathfrak{F} = (W, \precsim)$ is *unbounded to the right* if $\forall x \in W, \exists y \in W$ such that $x \precsim y$. Symmetrically, a frame $\mathfrak{F} = (W, \precsim)$ is *unbounded to the left* if $\forall x \in W$, $\exists y \in W$ such that $y \precsim x$. These frame properties can be axiomatised by adding the following axiom(s):

$$\mathsf{G}p \supset \mathsf{F}p \qquad (\mathbf{D_r})$$

$$\mathsf{H}p \supset \mathsf{P}p \qquad (\mathbf{D}_\ell)$$

The logics we obtain when adding these axioms still have a finite model property and a small model property [16]. Moreover, a finite weak total order is unbounded to the right (resp. left) if and only if its rightmost (resp. leftmost) world is in a cluster. This leads to the following new rules for our calculus to handle unboundedness:

$$(\mathsf{D_r})\ \frac{H\,;\{S\} \qquad H\,;S\,;\{\vdash\}}{H\,;S} \qquad\qquad \frac{\{S\}\,;H \qquad \{\vdash\}\,;S\,;H}{S\,;H}\ (\mathsf{D}_\ell)$$

Proposition 4.2 *Adding* $(\mathsf{D_r})$ *(resp.* (D_ℓ)*) yields a sound and complete proof system for* $\mathbf{K_t4.3} \cup (\mathbf{D_r})$ *(resp.* $\mathbf{K_t4.3} \cup (\mathbf{D}_\ell)$*), where proof search is in* coNP.

Proof. It is easy to check that rule $(\mathsf{D_r})$ is sound, as it reflects the shape of right-unbounded finite weak total orders. It is also invertible, since the underlying frame of the conclusion of the rule is always a subframe of its premises. Hence Theorem 3.7 and Lemma 3.3 still hold.

To obtain that proof search is in coNP, it suffices to check that Lemma 3.8 carries over to our extension. This is true because the rule $(\mathsf{D_r})$ can only be applied when the last cell of the hypersequent is not a cluster, and whenever the rule $(\mathsf{D_r})$ is applied, the last cell of its premises is always a cluster. Hence, the rule $(\mathsf{D_r})$ can only be applied once between applications of other rules creating new cells such as $(\vdash \mathsf{G})$ and $(\vdash \mathsf{H})$. Finally, since new cells can only be created polynomialy many times by those other rules thanks to our initial strategy, the new rule $(\mathsf{D_r})$ can, in the end, only be applied polynomialy many times along a branch. So the branches of our proof tree are still polynomial.

Finally, completeness is obtained as in Theorem 3.9. It only remains to show that the underlying frame of a hypersequent found at the end of a failing branch is unbounded to the right. Indeed, if its underlying frame was not unbounded to the right, we could apply the rule $(\mathsf{D_r})$ which would contradict the fact that no rules can be applied any more on this hypersequent. □

One can see that all rules can be taken together to form a sound and complete calculus for $\mathbf{K_tQ}$, with coNP proof search. Note that the rules proposed in this section differ from the ones proposed by Indrzejczak for density and unboundedness [9]. These rules would be sound but would break our polynomial bound on the length of proof branches.

5 First-Order Logic with Two Variables

We show here a coNEXP upper bound on the complexity of validity in the two-variable fragment of first-order logic over linear orders, re-proving and extending recent results by Manuel and Sreejith [14].

Syntax and Semantics. We consider first-order formulæ with two variables x and y over the signature $(=, <, (p)_{p\in\Phi})$ where $=$ and $<$ are binary relational symbols and each p is a unary relational symbol:

$$\psi ::= z = z' \mid z < z' \mid p(z) \mid \bot \mid \psi \supset \psi \mid \forall z.\psi \qquad \text{(first-order formulæ)}$$

where z, z' range over $\{x, y\}$ and p over Φ. We call this logic $\mathrm{FO}^2(<)$.

We interpret our formulæ over structures $\mathfrak{M} = (W, <, V)$ where $=$ is interpreted as the equality over W, $<$ as the strict total ordering of W, and each p as $V(p)$ for the valuation $V : \Phi \to 2^W$.

Equivalence with $\mathbf{K_t4.3}$. Given an $\mathrm{FO}^2(<)$ formula $\psi(z)$ with one free variable z, Etessami et al. [6] show how to construct a $\mathbf{K_t4.3}$ formula φ such that, for all strict totally ordered structures $\mathfrak{M} = (W, <, V)$, $\mathfrak{M}, [w/z] \models \psi$ if and only if $\mathfrak{M}, w \models \varphi$, where $[w/z]$ is the variable assignment mapping z to w.

Fact 5.1 ([6, Thm. 2]) *Every $\mathrm{FO}^2(<)$ formula $\psi(z)$ can be converted to an equivalent $\mathbf{K_t4.3}$ formula φ with $|\varphi| \in 2^{poly(|\psi|)}$.*

Although the proof of [6, Thm. 2] is given for the case of the strict total order ω—i.e., for ω-words over the alphabet 2^Φ—, it actually does not rely on this specific frame and applies similarly to arbitrary strict total orders.

We have therefore the following, where the NEXP upper bounds in items (i–iii) were already shown by Manuel and Sreejith [14, Thm. 15] using automata-based techniques. Let us reiterate that the complexity bounds on the satisfiability problem for the modal logics in question were already known [16], so the interest here lies in the use of proof search in our hypersequent proof system rather than a brutal enumeration of all potential models up to some bound.

Theorem 5.2 *The following problems are in* NEXP*: satisfiability of $\mathrm{FO}^2(<)$ over (i) arbitrary strict total orders, (ii) countable strict total orders, (iii) scattered strict total orders, and (iv) dense strict total orders.*

Proof. Regarding (i), given an $\mathrm{FO}^2(<)$ formula ψ, we first turn it into the equisatisfiable formula $\exists y.\psi$ with one free variable x. Fact 5.1 then allows to construct a $\mathbf{K_t4.3}$ formula φ of exponential size, which is equisatisfiable over strict total orders. By Fact 2.1, it is also equisatisfiable over weak total orders, and Theorem 3.10 shows that satisfiability can be checked in non-deterministic polynomial time in $|\varphi|$, hence in NEXP overall.

Regarding (ii) and (iii), by [16, Thm. 3], the above-constructed φ is satisfiable over weak total orders if and only if it is satisfiable over finite weak total orders. The bulldozing construction used to prove Fact 2.1 (see [2, Thm. 4.56]) consists essentially in turning each cluster into a direct product $\omega^* \cdot \omega$ (i.e., a copy of $\mathbb{Z}$), which shows that φ is satisfiable over finite weak total orders if and only if it is satisfiable over countable scattered strict total orders.

Finally, regarding (iv), by adapting [2, theorems 4.41 and 4.56] to bulldoze clusters over $\mathbb{Q}$ rather than $\mathbb{Z}$, ψ is satisfiable over dense strict total orders if and only if the above-constructed φ is satisfiable over dense weak total orders as a $\mathbf{K_t4.3} \cup (\mathbf{Den})$ formula. By Proposition 4.1, the latter can be checked in non-deterministic polynomial time in $|\varphi|$, hence in NEXP overall. □

6 Discussion

We have designed a sound and complete hypersequent calculus with clusters for the modal logic $\mathbf{K_t4.3}$ of linear temporal frames. The proof system relies on the finite model property of our logic in the presence of clusters to bound the length of branches during a proof search, which yields a proof search with optimal coNP complexity for the validity problem. Moreover, the approach is modular, as these results remain true when extending the proof system to handle density and unboundedness, yielding a sound and complete system for $\mathbf{K_tQ}$ with the same complexity, and a sound and complete system for $\mathrm{FO}^2(<)$ with coNEXP upper bounds. This coNEXP upper bound itself is hardly surprising, but from a proof-theoretic perspective, the two-variable fragment of first-order logic is an unusual beast—eigenvariables must be avoided—, hence our solution through a proof system for a modal logic is arguably a natural one.

An extension we would like to consider in future work is *well-foundedness*, by adding the Gödel-Löb axiom to our logic. Here, the logic of weak total orders well-founded to the left and unbounded to the right does not enjoy a finite model property.

Appendix

A Finite Model Property

We recall the result from Ono and Nakamura [16] which yields the finite model property for all logics considered in this paper. Finite models are obtained by using a filtration [2, Def. 2.36] on a structure to obtain a finite structure of the same 'shape.' The relevant filtration in this case is called the *Lemmon filtration.*

Lemmon Filtration. Let $\mathfrak{M} = (W, \precsim, V)$ be a Kripke structure. Let Ψ be a set of $\mathbf{K_t4.3}$ formulæ closed under taking subformulæ. We define a binary relation $\equiv$ on W by:

$$w \equiv w' \text{ iff } \forall \psi \in \Psi,\ \mathfrak{M}, w \models \psi \iff \mathfrak{M}, w' \models \psi$$

The relation $\equiv$ is an equivalence relation, and we note $[w]$ the equivalence class of a world $w \in W$. Note that, if Ψ is finite, then $\equiv$ has finite index. Moreover, if $w \equiv w'$, then $\forall p \in \Phi \cap \Psi$, $w \in V(p) \iff w' \in V(p)$. Hence, we can define the *Lemmon filtration* of $\mathfrak{M}$ by Ψ as $\mathfrak{M}^f = (W^f, \precsim^f, V^f)$ such that:

$$W^f = W/{\equiv} \qquad V^f(p) = V(p)/{\equiv}$$

$$[w] \precsim^f [w'] \text{ iff } \begin{cases} \forall \mathsf{G}\,\psi \in \Psi, \text{ if } \mathfrak{M}, w \models \mathsf{G}\,\psi \text{ then } \mathfrak{M}, w' \models \mathsf{G}\,\psi \text{ and } \mathfrak{M}, w' \models \psi \\ \forall \mathsf{H}\,\psi \in \Psi, \text{ if } \mathfrak{M}, w' \models \mathsf{H}\,\psi \text{ then } \mathfrak{M}, w \models \mathsf{H}\,\psi \text{ and } \mathfrak{M}, w \models \psi \end{cases}$$

Fact A.1 ([16, Thm. 3]) *Let $\mathfrak{M} = (W, \precsim, V)$ be a weak total order and Ψ a set of $\mathbf{K_t4.3}$ formulæ closed under taking subformulæ, and let $\mathfrak{M}^f = (W^f, \precsim^f, V^f)$ be the Lemmon filtration of $\mathfrak{M}$ by Ψ. Then (i) $[w] \precsim^f [w']$ if $w \precsim w'$, (ii) $\precsim^f$ is transitive and linear, (iii) $\precsim^f$ is unbounded to the right (resp. left) if $\precsim$ is unbounded to the right (resp. left), and (iv) $\precsim^f$ is dense if $\precsim$ is dense.*

Now, if $\mathfrak{M}$ is a model of a $\mathbf{K_t4.3}$ formula φ and Ψ is the set of subformulæ of φ, then $\mathfrak{M}^f$ is finite since Ψ is finite. Moreover, if we had $\mathfrak{M}, w \models \varphi$, then we also have $\mathfrak{M}^f, [w] \models \varphi$ since it is a filtration [2, Thm. 2.39]. Hence, all the logics presented in this paper have the finite model property.

Finally, we show that our hypersequents with clusters also enjoy the finite counter-model property; the following proof also captures our extensions to dense and unbounded frames.

Lemma A.2 *If H is an invalid annotation-free hypersequent with clusters, then H has a finite counter-model.*

Proof. Let $(\mathfrak{M}, \mu)$ be a counter-model of H and $\mathfrak{M}^f$ its Lemmon filtration for Ψ the set of subformulæ of H. For every position i of H, we have $\mathfrak{M}^f, [\mu(i)] \not\models H(i)$. But $\mu^f : i \mapsto [\mu(i)]$ might not be an embedding of H in $\mathfrak{M}^f$, as we could have two positions $i \prec j$ such that $[\mu(i)] \precsim^f [\mu(j)]$ and $[\mu(j)] \precsim^f [\mu(i)]$, i.e., $[\mu(i)] \sim^f [\mu(j)]$. We can avoid this problem by duplicating such clusters.

Formally, let $i \prec j$ and $[\mu(i)] \sim^f [\mu(j)]$. Let $C = \{w \in \mathfrak{M}^f \mid w \sim^f [\mu(i)]\}$ be the cluster containing $[\mu(i)]$ and $[\mu(j)]$. We define a modified model $\mathfrak{M}^f_1 = (W^f_1, \precsim^f_1, V^f_1)$ featuring two copies of C as follows:

$$W^f_1 = (W^f \setminus C) \cup \{(w, b) \mid w \in C, b \in \{0,1\}\}$$

$$V^f_1(w) = V^f(w)\ \forall w \in W^f \setminus C \qquad V^f_1((w,b)) = V^f(w)\ \forall (w,b) \in C \times \{0,1\}$$

$$(w,b) \precsim^f_1 (w',b)\ \forall (w,w',b) \in C^2 \times \{0,1\} \quad (w,0) \precsim^f_1 (w',1)\ \forall (w,w') \in C$$

$$w \precsim^f_1 (w',b) \text{ whenever } w \precsim^f w' \qquad (w,b) \precsim^f_1 w' \text{ whenever } w \precsim^f w'$$

$$w \precsim^f_1 w' \text{ whenever } w \precsim^f w'$$

We now have $([\mu(i)], 0) \prec^f_1 ([\mu(j)], 1)$ and we still have $\mathfrak{M}^f_1, ([\mu(i)], 0) \not\models H(i)$ and $\mathfrak{M}^f_1, ([\mu(j)], 1) \not\models H(j)$, because $[\mu(i)]$ and $([\mu(i)], 0)$ (resp. $[\mu(j)]$ and $([\mu(j)], 1)$)

are bisimilar [2, Thm. 2.20]. The mapping μ^f can be modified into μ^f_1 as follows:

$$\mu^f_1(k) = \begin{cases} ([\mu(k)], 0) & \text{if } k \in C \text{ and } k \precsim i \\ ([\mu(k)], 1) & \text{if } k \in C \text{ and } i \prec k \\ [\mu(k)] & \text{if } k \notin C \end{cases}$$

This fixes the failure of the second condition of embeddings on i and j, though not necessarily on other positions.

Finally, let $\mathfrak{M}'$ be the model obtained from $\mathfrak{M}^f$ after performing this duplication for all such $i \prec j$; $\mathfrak{M}'$ is finite, since $\mathfrak{M}^f$ is finite and we only did finitely many copies, and the resulting μ' is such that $(\mathfrak{M}', \mu')$ is a counter-model of H.□

References

[1] Avron, A., *Hypersequents, logical consequence and intermediate logics for concurrency*, Annals of Mathematics and Artificial Intelligence **4** (1991), pp. 225–248.

[2] Blackburn, P., M. de Rijke and Y. Venema, "Modal Logic," Cambridge Tracts in Theoretical Computer Science **53**, Cambridge University Press, 2001.

[3] Brünnler, K., *Deep sequent systems for modal logic*, Archiv für Mathematische Logik und Grundlagenforschung **48** (2009), pp. 551–577.

[4] Clarke, E. M. and E. A. Emerson, *Design and synthesis of synchronization skeletons using branching time temporal logic*, in: *Proc. Workshop on Logic of Programs*, Lecture Notes in Computer Science **131** (1981), pp. 52–71.

[5] Cocchiarella, N. B., "Tense and Modal Logic: a Study in the Topology of Temporal Reference," Ph.D. thesis, University of California, Los Angeles (1965).

[6] Etessami, K., M. Y. Vardi and T. Wilke, *First-order logic with two variables and unary temporal logic*, Information and Computation **179** (2002), pp. 279–295.

[7] Indrzejczak, A., *Cut-free hypersequent calculus for S4.3*, Bulletin of the Section of Logic **41** (2012), pp. 89–104.

[8] Indrzejczak, A., *Eliminability of cut in hypersequent calculi for some modal logics of linear frames*, Information Processing Letters **115** (2015), pp. 75–81.

[9] Indrzejczak, A., *Linear time in hypersequent framework*, Bulletin of Symbolic Logic **22** (2016), pp. 121–144.

[10] Indrzejczak, A., *Cut elimination theorem for non-commutative hypersequent calculus*, Bulletin of the Section of Logic **46** (2017), pp. 135–149.

[11] Laroussinie, F., N. Markey and Ph. Schnoebelen, *Temporal logic with forgettable past*, in: *Proc. LICS 2002* (2002), pp. 383–392.

[12] Lellmann, B., *Linear nested sequents, 2-sequents and hypersequents*, in: *Proc. Tableaux 2015*, Lecture Notes in Computer Science **9323** (2015), pp. 135–150.

[13] Lichtenstein, O., A. Pnueli and L. Zuck, *The glory of the past*, in: *Proc. Workshop on Logics of Programs*, Lecture Notes in Computer Science **193** (1985), pp. 196–218.

[14] Manuel, A. and A. V. Sreejith, *Two-variable logic over countable linear orderings*, in: *Proc. MFCS 2016*, Leibniz International Proceedings in Informatics **58** (2016), pp. 66:1–66:13.

[15] Negri, S., *Proof analysis in modal logic*, Journal of Philosophical Logic **34** (2005), pp. 507–544.

[16] Ono, H. and A. Nakamura, *On the size of refutation Kripke models for some linear modal and tense logics*, Studia Logica **39** (1980), pp. 325–333.

[17] Pnueli, A., *The temporal logic of programs*, in: *FOCS 1977* (1977), pp. 46–57.

[18] Poggiolesi, F., *The method of tree-hypersequents for modal propositional logic*, in: *Proc. Trends in Logic IV*, Trends in Logic **28** (2009), pp. 31–51.

[19] Prior, A. N., "Time and Modality," Oxford University Press, 1957.

[20] Sistla, A. P. and E. M. Clarke, *The complexity of propositional linear temporal logics*, Journal of the ACM **32** (1985), pp. 733–749.

Here and There Modal Logic with Dual Implication

Philippe Balbiani

Institut de recherche en informatique de Toulouse,
CNRS — Toulouse University

Martin Diéguez [1]

Lab-STICC, CERV, Ecole Nationale d'Ingénieurs de Brest

Abstract

We define and study an extension of the logic of Here and There with dual implication and modal operators of necessity and possibility. We provide a complete axiomatisation. We prove as well other results such as the interdefinability of modal operators and the Hennessy-Milner property. We give an upper bound to the complexity of the satisfiability problem.

Keywords: modal logic, Here and There logic, dual implication, axiomatization and completeness, Hennessy-Milner property.

1 Introduction

In the last twenty years, research on extensions of the logic of *Here and There* [16,21,38] (HT) have been very active due to the advent of *Equilibrium Logic* [29,30], which is considered the best-known logical characterisation of the *Stable Models Semantics* [15] and *Answer Sets Semantics* [6] in *Logic Programming* (LP). Recently, combinations of intermediate and modal logics [3,9,13,11,39] have caught the attention of the LP community since they can support the definition of non-monotonic modal logics [9,3]. Extending intermediate logics [25] (IL) with modalities is not new, since several semantics and properties have been studied about this topic in both philosophy and formal logic [4,7,28,35,37] and computer science [2,5,12,24,32].

Also related to IL, several types of negation were considered: for instance, Nelson's Constructive Logic [27] was used by [30] in order to characterise the strong negation in LP. However, other dual operators of IL [2] have not been considered in the HT setting. More precisely, we focused on the dual implication proposed by C. Rauszer [33].

---

[1] Martin Diéguez is funded by the ANR-12-ASTR-0020 Project STRATEGIC.

[2] See the discussion presented in [40].

Rauszer proposed an extension of intuitionistic logic equipped with a new implication (denoted by $\leftarrow$) in order to provide "a more elegant algebraic and model-theoretic theory than in ordinary intuitionistic logic" [33]. Later on, this new implication was further studied: in [41] this new operator is added to the intuitionistic modal language providing several results such as matrix and Kripke semantics or embeddings into (extended) tense logics. A display calculus unifying intuitionistic and dual-intuitionistic logic was presented in [18] and refined in [19]. Recently, in [20], a cut-free sequent calculi in terms of *derivations* and *refutations* have been introduced[3].

In this paper, we have considered the combination of propositional HT with dual implication and modal logic K. On it, we have defined the concept of *modal equilibrium model* and we study several interesting properties, which can serve as a starting point for future modal extensions. These properties are presented along this paper in the following way. In Section 2, we present syntax and two equivalent alternative semantics based on Kripke models. The former semantics (the "Here and There" semantics) is simulated by two valuation functions while the latter semantics possesses two accessibility relations to interpret implication, dual implication and modal operators. In Section 3 and Section 4, we present an axiomatisation of this logic and we prove its completeness with respect to the birelational semantics. In Section 5, we establish the complexity, in PSPACE, of the satisfiability problem in this logic. In Section 6 we define bisimulations for our BHT-modal extensions and we use them to prove the Hennessy-Milner property. In Section 7 we define the concept of modal equilibrium logic and shows that such definition is suitable for proving the theorem of *strong equivalence.*

2 Syntax and semantics

In this section, we present the syntax and the semantics of BHT.

2.1 Syntax

Let VAR be a countable set of propositional variables (denoted p, q, etc). The set FOR of all formulas (denoted φ, ψ, etc) is defined as follows:

$$\varphi, \psi ::= p \mid \bot \mid \top \mid (\varphi \vee \psi) \mid (\varphi \wedge \psi) \mid (\varphi \to \psi) \mid (\varphi \leftarrow \psi) \mid \Box\varphi \mid \Diamond\varphi \qquad (1)$$

We follow the standard rules for omission of the parentheses. As in [33], two negations can be defined: $\neg\varphi \overset{def}{=} \varphi \to \bot$ and $\llcorner\varphi \overset{def}{=} \top \leftarrow \varphi$. Let $|\varphi|$ denote the number of symbol occurrences in φ. A set Σ of formulas is *closed* iff it is closed under subformulas and for all formulas φ, if $\varphi \in \Sigma$ then $\neg\varphi \in \Sigma$ and $\llcorner\varphi \in \Sigma$.

[3] Derivation Calculi are used to reason about a syntactic derivability relation ($\vdash$). usually associated with the ordinary implication ($\to$). Conversely, Refutation Calculi [17] are thought for reasoning about a syntactic refutability relation ($\dashv$) and it comes from the use of the dual implication ($\leftarrow$).

The *modal degree* of a formula φ (in symbols $deg(\varphi)$) is defined as follows:

$$deg(\varphi) \overset{def}{=} \begin{cases} 0 & if\ \varphi = p\ (p \in VAR) \text{ or } \varphi = \bot \text{ or } \varphi = \top \\ max(deg(\psi), deg(\chi)) & if\ \varphi = \psi \odot \chi, \text{ with } \odot \in \{\vee, \wedge, \rightarrow, \leftarrow\} \\ 1 + deg(\psi) & if\ \varphi = \odot\psi, \text{ with } \odot \in \{\Box, \Diamond\} \end{cases}$$

A *theory* is a set of formulas. For all theories x, y, we define the theories $\Box x \overset{def}{=} \{\varphi \mid \Box\varphi \in x\}$ and $\Diamond y \overset{def}{=} \{\Diamond\varphi \mid \varphi \in y\}$.

2.2 *BHT* semantics

Given a nonempty set W and $H, T : VAR \rightarrow 2^W$, we say that H is *included* in T (in symbols $H \leq T$) iff for all $p \in VAR$, $H(p) \subseteq T(p)$. A *BHT-frame* is a structure (W, R) where W is a nonempty set and R is a binary relation on W. A *BHT-model* is a structure $\mathbf{M} = \langle W, R, H, T \rangle$ where (W, R) is a *BHT*-frame and $H, T : VAR \rightarrow 2^W$ are such that $H \leq T$. Given a *BHT*-model $\mathbf{M} = \langle W, R, H, T \rangle$, $x \in W$, and $\alpha \in \{h, t\}$, interpreting $\bot$, $\top$, $\vee$ and $\wedge$ as usual, the satisfaction of a formula φ at (x, α) in $\mathbf{M}$ (in symbols $\mathbf{M}, (x, \alpha) \vDash \varphi$) is defined as follows:

- $\mathbf{M}, (x, h) \vDash p$ iff $x \in H(p)$ and $\mathbf{M}, (x, t) \vDash p$ iff $x \in T(p)$,
- $\mathbf{M}, (x, \alpha) \vDash \varphi \rightarrow \psi$ iff for all $\alpha' \in \{\alpha, t\}$, $\mathbf{M}, (x, \alpha') \nvDash \varphi$, or $\mathbf{M}, (x, \alpha') \vDash \psi$,
- $\mathbf{M}, (x, \alpha) \vDash \varphi \leftarrow \psi$ iff there exists $\alpha' \in \{h, \alpha\}$ such that $\mathbf{M}, (x, \alpha') \vDash \varphi$ and $\mathbf{M}, (x, \alpha') \nvDash \psi$,
- $\mathbf{M}, (x, \alpha) \vDash \Box\varphi$ iff for all $y \in W$, if $xR\,y$ then $\mathbf{M}, (y, \alpha) \vDash \varphi$,
- $\mathbf{M}, (x, \alpha) \vDash \Diamond\varphi$ iff there exists $y \in W$ such that $xR\,y$ and $\mathbf{M}, (y, \alpha) \vDash \varphi$.

As a result, $\mathbf{M}, (x, \alpha) \vDash \neg\varphi$ iff for all $\alpha' \in \{\alpha, t\}$, $\mathbf{M}, (x, \alpha') \nvDash \varphi$ and $\mathbf{M}, (x, \alpha) \vDash \llcorner\varphi$ iff there exists $\alpha' \in \{h, \alpha\}$ such that $\mathbf{M}, (x, \alpha') \nvDash \varphi$. Notice that if $H = T$ then the satisfaction relation is essentially the same as the satisfaction relation used in classical modal logic [10]. We say that the formulas φ and ψ are *BHT-equivalent* (in symbols $\varphi \simeq \psi$) iff for all *BHT* models $\mathbf{M} = \langle W, R, H, T \rangle$, for all $x \in W$ and for all $\alpha \in \{h, t\}$, $\mathbf{M}, (x, \alpha) \vDash \varphi$ iff $\mathbf{M}, (x, \alpha) \vDash \psi$. The satisfaction of a theory Γ at (x, α) in $\mathbf{M}$ (in symbols $\mathbf{M}, (x, \alpha) \vDash \Gamma$) is defined as usual. Two theories Γ_1 and Γ_2 are *BHT-equivalent* (in symbols $\Gamma_1 \simeq \Gamma_2$) iff for all *BHT* models $\mathbf{M} = \langle W, R, H, T \rangle$, for all $x \in W$ and for all $\alpha \in \{h, t\}$, $\mathbf{M}, (x, \alpha) \vDash \Gamma_1$ iff $\mathbf{M}, (x, \alpha) \vDash \Gamma_2$.

Lemma 2.1 *Let φ be a formula. For all BHT-models $\mathbf{M} = \langle W, R, H, T \rangle$ and for all $x \in W$, if $\mathbf{M}, (x, h) \vDash \varphi$ then $\mathbf{M}, (x, t) \vDash \varphi$.*

As a result, for arbitrary $x \in W$ and $\alpha \in \{h, t\}$, $\mathbf{M}, (x, \alpha) \vDash \neg\varphi$ iff $\mathbf{M}, (x, t) \nvDash \varphi$ and $\mathbf{M}, (x, \alpha) \vDash \llcorner\varphi$ iff $\mathbf{M}, (x, h) \nvDash \varphi$. Hence, $\mathbf{M}, (x, t) \vDash \varphi \vee \neg\varphi$ and $\mathbf{M}, (x, h) \nvDash \varphi \wedge \llcorner\varphi$. Remark also that $\mathbf{M}, (x, \alpha) \vDash \neg\neg\varphi$ iff $\mathbf{M}, (x, t) \vDash \varphi$ and $\mathbf{M}, (x, \alpha) \vDash \llcorner\llcorner\varphi$ iff $\mathbf{M}, (x, h) \vDash \varphi$. A formula φ is said to be *satisfiable* iff there exists a *BHT* model $\mathbf{M} = \langle W, R, H, T \rangle$, there exists $x \in W$ and there exists $\alpha \in \{h, t\}$ such that $\mathbf{M}, (x, \alpha) \vDash \varphi$. A formula φ is said to be *valid* iff for all *BHT* models $\mathbf{M} = \langle W, R, H, T \rangle$, for all $x \in W$ and for all $\alpha \in \{h, t\}$, $\mathbf{M}, (x, \alpha) \vDash \varphi$. In order to grasp the differences between $\neg$ and $\llcorner$, let us notice that, although $p \vee \neg p$ is not valid and $p \wedge \llcorner p$ is satisfiable, we have $\varphi \vee \llcorner\varphi$ is valid and $\varphi \wedge \neg\varphi$ is not satisfiable for arbitrary formula φ. In other respect, by Lemma 2.1, one can readily conclude that φ is valid iff for all BHT models $\mathbf{M} = \langle W, R, H, T\rangle$ and for all $x \in W$, $\mathbf{M}, (x, h) \models \varphi$ and φ is not satisfiable iff for all BHT models $\mathbf{M} = \langle W, R, H, T\rangle$ and for all $x \in W$, $\mathbf{M}, (x, t) \not\models \varphi$. It can be easily checked that if a formula φ is not satisfiable then $\neg\varphi$ is valid and if φ is valid then $\llcorner\varphi$ is not satisfiable. Finally, remark that for all formulas φ, ψ, $\varphi \to \psi$ is valid iff $\varphi \leftarrow \psi$ is not satisfiable.

Lemma 2.2 *The following formulas are valid:*

1) *Standard axioms of Intuitionistic Propositional Calculus (IPC):*

- $\varphi \to (\psi \to \varphi)$,
- $(\varphi \to (\psi \to \chi)) \to ((\varphi \to \psi) \to (\varphi \to \chi))$,
- $(\varphi \to \chi) \to ((\psi \to \chi) \to (\varphi \vee \psi \to \chi))$,
- $\varphi \to \varphi \vee \psi$,
- $\psi \to \varphi \vee \psi$,
- $\varphi \wedge \psi \to \varphi$,
- $\varphi \wedge \psi \to \psi$,
- $\varphi \to (\psi \to \varphi \wedge \psi)$,
- $\bot \to \varphi$,

2) *Hosoi formula [22]:* $\varphi \vee (\varphi \to \psi) \vee \neg\psi$,

3) *Fisher Servi axioms:*

- $\Box(\varphi \to \psi) \to (\Box\varphi \to \Box\psi)$,
- $\Box(\varphi \to \psi) \to (\Diamond\varphi \to \Diamond\psi)$,
- $\Diamond(\varphi \vee \psi) \to \Diamond\varphi \vee \Diamond\psi$,
- $(\Diamond\varphi \to \Box\psi) \to \Box(\varphi \to \psi)$,
- $\neg \Diamond \bot$,

4) *Additional formulas:*

- $\varphi \to (\varphi \leftarrow \psi) \vee \psi$,
- $\varphi \vee \llcorner\varphi$,
- $\neg\varphi \vee \neg\neg\varphi$,
- $\llcorner \Diamond (\varphi \wedge \llcorner\varphi)$,
- $\neg\neg \Box \varphi \to \Box\neg\neg\varphi$ *and* $\Diamond\neg\neg\varphi \to \neg\neg \Diamond \varphi$,
- $\llcorner\llcorner \Box \varphi \to \Box\llcorner\llcorner\varphi$ *and* $\Diamond\llcorner\llcorner\varphi \to \llcorner\llcorner \Diamond \varphi$.

Lemma 2.3 *The following formulas are valid:*

1) *Negation-dual of standard axioms of Intuitionistic Logic:*

- $\neg((\varphi \leftarrow \psi) \leftarrow \varphi)$,
- $\neg(((\chi \leftarrow \varphi) \leftarrow (\psi \leftarrow \varphi)) \leftarrow ((\chi \leftarrow \psi) \leftarrow \varphi))$,
- $\neg(((\chi \leftarrow \psi \wedge \varphi) \leftarrow (\chi \leftarrow \psi)) \leftarrow (\chi \leftarrow \varphi))$,
- $\neg(\psi \wedge \varphi \leftarrow \varphi)$,
- $\neg(\psi \wedge \varphi \leftarrow \psi)$,
- $\neg(\varphi \leftarrow \psi \vee \varphi)$,
- $\neg(\psi \leftarrow \psi \vee \varphi)$,
- $\neg((\psi \vee \varphi \leftarrow \psi) \leftarrow \varphi)$,
- $\neg(\varphi \leftarrow \top)$.

2) *Negation-dual of Hosoi formula:* $\neg(\llcorner\psi \wedge (\psi \leftarrow \varphi) \wedge \varphi)$,

3) *Negation-dual of Fisher Servi axioms:*

- $\neg((\Diamond\psi \leftarrow \Diamond\varphi) \leftarrow \Diamond(\psi \leftarrow \varphi))$,
- $\neg((\Box\psi \leftarrow \Box\varphi) \leftarrow \Diamond(\psi \leftarrow \varphi))$,
- $\neg(\Box\psi \wedge \Box\varphi \leftarrow \Box(\psi \wedge \varphi))$,
- $\neg(\Diamond(\psi \leftarrow \varphi) \leftarrow (\Diamond\psi \leftarrow \Box\varphi))$,
- $\neg\llcorner\Box\top$,

4) Negation-dual of additional formulas:

- $\neg(\psi \wedge (\psi \rightarrow \varphi) \leftarrow \varphi)$,
- $\neg(\neg\varphi \wedge \varphi)$,
- $\neg(\llcorner\llcorner\varphi \wedge \llcorner\varphi)$,
- $\neg\neg\Box(\neg\varphi \vee \varphi)$,
- $\neg(\Diamond\llcorner\llcorner\varphi \leftarrow \llcorner\llcorner\Diamond\varphi)$ *and* $\neg(\llcorner\llcorner\Box\varphi \leftarrow \Box\llcorner\llcorner\varphi)$,
- $\neg(\Diamond\neg\neg\varphi \leftarrow \neg\neg\Diamond\varphi)$ *and* $\neg(\neg\neg\Box\varphi \leftarrow \Box\neg\neg\varphi)$.

From now on, the set of all valid formulas is denoted BHT.

Lemma 2.4 *In the following tables, the formulas on the left are BHT-equivalent to the corresponding formulas on the right.*

$\neg(\varphi \vee \psi)$	$\neg\varphi \wedge \neg\psi$
$\neg(\varphi \wedge \psi)$	$\neg\varphi \vee \neg\psi$
$\llcorner(\varphi \vee \psi)$	$\llcorner\varphi \wedge \llcorner\psi$
$\llcorner(\varphi \wedge \psi)$	$\llcorner\varphi \vee \llcorner\psi$
$\neg(\varphi \rightarrow \psi)$	$\neg\neg\varphi \wedge \neg\psi$
$\neg(\varphi \leftarrow \psi)$	$\neg\varphi \vee \llcorner\llcorner\psi \vee (\llcorner\varphi \wedge \neg\neg\psi)$
$\llcorner(\varphi \rightarrow \psi)$	$\neg\neg\varphi \wedge \llcorner\psi \wedge (\llcorner\llcorner\varphi \vee \neg\psi)$
$\llcorner(\varphi \leftarrow \psi)$	$\llcorner\varphi \vee \llcorner\llcorner\psi$

$\neg\llcorner\varphi$	$\llcorner\llcorner\varphi$
$\llcorner\neg\varphi$	$\neg\neg\varphi$

$\neg\neg\neg\varphi$	$\neg\varphi$
$\neg\neg\llcorner\varphi$	$\llcorner\varphi$
$\neg\llcorner\neg\varphi$	$\neg\varphi$
$\llcorner\neg\neg\varphi$	$\neg\varphi$
$\neg\llcorner\llcorner\varphi$	$\llcorner\varphi$
$\llcorner\neg\llcorner\varphi$	$\llcorner\varphi$
$\llcorner\llcorner\neg\varphi$	$\neg\varphi$
$\llcorner\llcorner\llcorner\varphi$	$\llcorner\varphi$

Lemma 2.5 *Let φ be a formula. The least closed set of formulas containing φ contains at most $5|\varphi|$ equivalence classes of formulas modulo $\simeq$.*

In most modal extensions of intuitionistic logic, $\Box$ and $\Diamond$ are non-interdefinable. Within the context of BHT, this is no longer the case.

Lemma 2.6 *In the following table, the formulas on the left are BHT-equivalent to the corresponding formulas on the right.*

$\Box\varphi$	$\Diamond(\varphi \wedge \llcorner\varphi) \vee \llcorner\Diamond\llcorner\varphi \leftarrow \Diamond\neg\varphi$
$\Box\varphi$	$(\neg\Diamond(\varphi \wedge \llcorner\varphi) \vee \Diamond(\varphi \wedge \llcorner\varphi) \vee \llcorner\Diamond\llcorner\varphi) \wedge \neg\Diamond\neg\varphi$
$\Diamond\varphi$	$\Box\llcorner\varphi \rightarrow \neg\Box\neg\varphi \wedge \Box(\varphi \vee \neg\varphi)$
$\Diamond\varphi$	$\llcorner\Box\llcorner\varphi \vee (\neg\Box\neg\varphi \wedge \Box(\varphi \vee \neg\varphi) \wedge \llcorner\Box(\varphi \vee \neg\varphi))$

2.3 Birelational semantics

A standard approach in the semantics of a modal intuitionistic logic is to consider structures based on a partial order and a binary relation [37]. A *birelational frame* is a structure $(W, \leq, R)$ where W is a nonempty set, $\leq$ is a partial order on W and R is a binary relation on W. A birelational frame $(W, \leq, R)$ is *normal* iff it satisfies the following conditions for all $x, y, z \in W$:

1) if $x \leq y$ and $x \leq z$ then $x = y$, or $x = z$, or $y = z$,
2) if $x \leq z$ and $y \leq z$ then $x = y$, or $x = z$, or $y = z$.

As a result, if $(W, \leq, R)$ is normal then for all $x \in W$, x is a maximal element with respect to $\leq$, or there exists exactly one $y \in W$ such that $x \leq y$ and $x \neq y$.

In the former case let $\widehat{x}$ denote x. In the latter case, let $\widehat{x}$ denote this y. From this definition, it follows that for all $x, y \in W$, $x \leq y$ iff $y = x$, or $y = \widehat{x}$. Similarly, if $(W, \leq, R)$ is normal then for all $x \in W$, x is a minimal element with respect to $\leq$, or there exists exactly one $y \in W$ such that $y \leq x$ and $x \neq y$. In the former case, let $\check{x}$ denote x. In the latter case, let $\check{x}$ denote this y. From this definition it follows that for all $x, y \in W$, $y \leq x$ iff $x = y$, or $\check{x} = y$. Obviously, for all $x \in W$, $\check{x} \leq x \leq \widehat{x}$. Moreover, $\widehat{\check{x}} = \widehat{x}$ and $\check{\widehat{x}} = \check{x}$. A normal birelational frame $(W, \leq, R)$ is *Cartesian* iff it satisfies the following conditions for all $x, y \in W$:

1) if $\check{x} R y$ then $\check{y} = y$ and $\widehat{x} R \widehat{y}$, 2) if $\widehat{x} R y$ then $\widehat{y} = y$ and $\check{x} R \check{y}$.

Lemma 2.7 *Let $(W, \leq, R)$ be a Cartesian birelational frame. For all $x, y \in W$, if xRy then $\widehat{x} R \widehat{y}$ and $\check{x} R \check{y}$.*

A *birelational model* is a structure $\langle W, \leq, R, V \rangle$ where $(W, \leq, R)$ is a birelational frame and $V : VAR \to 2^W$ is such that for all $x, y \in W$, if $x \leq y$ then for all $p \in VAR$, if $x \in V(p)$ then $y \in V(p)$. Given a birelational model $\mathbf{M} = \langle W, \leq, R, V \rangle$ and $x \in W$, interpreting $\bot$, $\top$, $\vee$ and $\wedge$ as usual, the satisfaction of a formula φ at x in $\mathbf{M}$ (in symbols $\mathbf{M}, x \models \varphi$) is defined as follows:

(i) $\mathbf{M}, x \models p$ iff $x \in V(p)$,

(ii) $\mathbf{M}, x \models \varphi \to \psi$ iff for all $y \in W$ if $x \leq y$ then $\mathbf{M}, y \not\models \varphi$, or $\mathbf{M}, y \models \psi$,

(iii) $\mathbf{M}, x \models \varphi \leftarrow \psi$ iff there exists $y \in W$ such that $y \leq x$, $\mathbf{M}, y \models \varphi$ and $\mathbf{M}, y \not\models \psi$,

(iv) $\mathbf{M}, x \models \Box\varphi$ iff for all $y, z \in W$, if $x \leq y$ and $yR\ z$ then $\mathbf{M}, z \models \varphi$,

(v) $\mathbf{M}, y \models \Diamond\varphi$ iff there exists $y \in W$ such that $xR\ y$ and $\mathbf{M}, y \models \varphi$.

Remark that the clause concerning $\Box$ imitates the clause for the quantifier $\forall$ in first-order intuitionistic logic. Nevertheless, it can be proved that in a Cartesian model $\mathbf{M} = \langle W, \leq, R, V \rangle$, replacing the clause concerning $\Box$ by the clause

$$\mathbf{M}, x \models' \Box\varphi \text{ iff for all } y \in W, \text{ if } xR\ y \text{ then } \mathbf{M}, y \models' \varphi$$

would define a satisfaction relation equivalent to the relation $\models$ defined above. Regarding the birelational semantics, a formula φ is said to be *satisfiable* iff there exists a birelational model $\mathbf{M} = \langle W, \leq, R, V \rangle$ and there exists $x \in W$ such that $\mathbf{M}, x \models \varphi$. Moreover, a formula φ is said to be *valid* iff for all birelational models $\mathbf{M} = \langle W, \leq, R, V \rangle$ and for all $x \in W$, $\mathbf{M}, x \models \varphi$.

2.4 Equivalence between the two semantics

In this section, we prove that a formula is satisfiable (respectively, valid) in the BHT semantics iff it is satisfiable (respectively, valid) in the birelational semantics. Let $\mathbf{M} = \langle W, R, H, T \rangle$ be a BHT model. We define the birelational model $\mathbf{M}' = \langle W', \leq', R', V' \rangle$ as follows:

1) $W' = W \times \{h, t\}$,

2) $(x, \alpha) \leq' (y, \beta)$ iff $x = y$ and $\alpha = h$, or $\beta = t$,

3) $(x, \alpha) R' (y, \beta)$ iff $xR\ y$ and $\alpha = \beta$,

4) $V'(p) = \{(x, h) :\ x \in H(p)\} \cup \{(x, t) :\ x \in T(p)\}$.

The reader can easily check that $\mathbf{M}'$ satisfies the conditions to be normal. Moreover, the reader can check that for all $(x,\alpha) \in W'$, $\overset{\smile}{(x,\alpha)} = (x,h)$, and $\overset{\frown}{(x,\alpha)} = (x,t)$. Let us prove that $\mathbf{M}'$ is Cartesian. Let us consider (x,α) and (y,β) in W' satisfying $\overset{\smile}{(x,\alpha)}R'(y,\beta)$. By definition, $(x,h)R'(y,\beta)$, so xRy and $\beta = h$. Again, by definition $\overset{\smile}{(y,\beta)} = (y,\beta)$. Assume that not $\overset{\frown}{(x,\alpha)}R'\overset{\frown}{(y,\beta)}$, By definition not $(x,t)R'(y,t)$. By definition we conclude not xRy: a contradiction. Therefore $\overset{\frown}{(x,\alpha)}R'\overset{\frown}{(y,\beta)}$. Let us consider now (x,α) and (y,β) in W' satisfying $\overset{\frown}{(x,\alpha)}R'(y,\beta)$. By definition, $(x,t)R'(y,\beta)$, so xRy and $\beta = t$. Again, by definition $\overset{\frown}{(y,\beta)} = (y,\beta)$. Assume that not $\overset{\smile}{(x,\alpha)}R'\overset{\smile}{(y,\beta)}$, By definition not $(x,h)R'(y,h)$. By definition we conclude not xRy: a contradiction. Therefore $\overset{\smile}{(x,\alpha)}R'\overset{\smile}{(y,\beta)}$. Finally, we can prove the following correspondence between $\mathbf{M}$ and $\mathbf{M}'$.

Lemma 2.8 *Let φ be a formula. For all $x \in W$ and for all $\alpha \in \{h,t\}$, $\mathbf{M},(x,\alpha) \models \varphi$ iff $\mathbf{M}',(x,\alpha) \models \varphi$.*

Proof. By induction on φ. In the case of a propositional variable p, if $\mathbf{M},(x,h) \models p$ then $x \in H(p)$ so, by definition, $(x,h) \in V(p)$, so $\mathbf{M}',(x,h) \models p$. If $\mathbf{M},(x,t) \models p$ then $x \in T(p)$ so, by definition, $(x,t) \in V(p)$, so $\mathbf{M}',(x,t) \models p$. The converse direction is proved in a similar way. Also, the cases of conjunction and disjunction are proved by using the induction hypothesis. We consider the operators $\rightarrow$ and $\Box$ below:

- Case $\varphi \rightarrow \psi$: from left to right, assume by contradiction that $\mathbf{M}',(x,\alpha) \not\models \varphi \rightarrow \psi$. Therefore, there exists $(y,\beta) \in W'$ such that $(x,\alpha) \leq' (y,\beta)$ and $\mathbf{M}',(y,\beta) \models \varphi$ and $\mathbf{M}',(y,\beta) \models \psi$. By induction hypothesis we get that $\mathbf{M},(y,\beta) \not\models \varphi \rightarrow \psi$. By definition $x = y$ and either $\alpha = h$ or $\beta = t$. If $\alpha = h$ then there exists $\beta \in \{h,t\}$, $\mathbf{M},(x,\beta) \not\models \varphi \rightarrow \psi$, so $\mathbf{M},(x,\alpha) \not\models \varphi \rightarrow \psi$: a contradiction. If $\beta = t$ then for all $\alpha \in \{h,t\}$, $\mathbf{M},(x,\alpha) \not\models \varphi \rightarrow \psi$: a contradiction. For the converse direction, let us consider $\mathbf{M},(x,\alpha) \not\models \varphi \rightarrow \psi$. Therefore there exists some $\beta \in \{\alpha,t\}$, $\mathbf{M},(x,\beta) \models \varphi$ and $\mathbf{M},(x,\beta) \not\models \psi$. By induction $\mathbf{M}',(x,\beta) \models \varphi$ and $\mathbf{M}',(x,\beta) \not\models \psi$. Therefore $\mathbf{M}',(x,\beta) \not\models \varphi \rightarrow \psi$. If $\beta = \alpha$ we get $\mathbf{M}',(x,\alpha) \not\models \varphi \rightarrow \psi$: a contradiction. If $\beta = t$ then $(x,\alpha) \leq' (x,\beta)$ and $\mathbf{M}',(x,\beta) \not\models \varphi \rightarrow \psi$: a contradiction.
- Case $\Box\psi$: from left to right, assume by contradiction that $\mathbf{M}',(x,\alpha) \not\models \Box\psi$. This means that there exists (x',β) and (y,γ) in W' such that $(x,\alpha) \leq' (x',\beta)R'(y,\gamma)$ and $\mathbf{M}',(y,\gamma) \not\models \varphi$. By induction $\mathbf{M},(x,\gamma) \not\models \psi$. By definition $x'Ry$ and $\gamma = \beta$ (so $\mathbf{M},(x',\beta) \not\models \Box\psi$). Again, by definition $x = x'$ and either $\beta = t$ or $\alpha = h$. Any of the cases leads to $\mathbf{M},(x,\alpha) \not\models \Box\psi$. Conversely, assume by contradiction that $\mathbf{M},(x,\alpha) \not\models \Box\psi$. Therefore $\mathbf{M},(y,\alpha) \not\models \psi$ for some xRy. By induction $\mathbf{M}',(y,\alpha) \not\models \psi$. By definition $(x,\alpha)R'(y,\alpha)$ and $(x,\alpha) \leq' (x,\alpha)$, so $\mathbf{M}',(x,\alpha) \not\models \Box\psi$.

□

Let $\mathbf{M} = \langle W, \leq, R, V\rangle$ be a Cartesian birelational model. We define the *BHT* model $\mathbf{M}' = \langle W', R', H', T'\rangle$ as follows:

1) $W' = \{(\check{x},\hat{x}) \mid x \in W\}$,
2) $(\check{x},\hat{x})R'(\check{y},\hat{y})$ iff $\check{x}R\check{y}$ and $\hat{x}R\hat{y}$,
3) $H'(p) = \{(\check{x},\hat{x}) : \check{x} \in V(p)\}$,
4) $T'(p) = \{(\check{x},\hat{x}) : \hat{x} \in V(p)\}$.

Take $(\check{x},\hat{x}) \in H'(p)$. By definition $\check{x} \in V(p)$. Since $\check{x} \leq \hat{x}$ then, by definition, $\hat{x} \in V(p)$. Finally, by definition, $(\check{x},\hat{x}) \in T'(p)$. Thus, $H' \leq T'$. Moreover, the following result relates birelational and BHT semantics.

Lemma 2.9 *Let φ be a formula. For all $x \in W$,*

1) $\mathbf{M},\check{x} \models \varphi$ *iff* $\mathbf{M}',((\check{x},\hat{x}),h) \models \varphi$,

2) $\mathbf{M},\hat{x} \models \varphi$ *iff* $\mathbf{M}',((\check{x},\hat{x}),t) \models \varphi$.

Proof. By induction on φ. For the case of a propositional variable p we get that if $\mathbf{M},\check{x} \models p$ then $\check{x} \in V(p)$ and, by definition, $(\check{x},\hat{x}) \in H'(p)$. Therefore, $\mathbf{M}',((\check{x},\hat{x}),h) \models p$. The converse direction follows a similar reasoning. If $\mathbf{M},\hat{x} \models p$ then $\hat{x} \in V(p)$ and, by definition, $(\check{x},\hat{x}) \in T'(p)$. Therefore, $\mathbf{M}',((\check{x},\hat{x}),t) \models p$. The converse direction follows a similar reasoning. The cases of conjunction and disjunction are proved by induction. We present the proof for the $\rightarrow$ and $\Box$ connectives below:

- Case $\varphi \rightarrow \psi$: from $\mathbf{M},\check{x} \models \varphi \rightarrow \psi$ then for all $x' \in \{\check{x},\hat{x}\}$, either $\mathbf{M},x' \not\models \varphi$ or $\mathbf{M},x' \models \psi$. From the induction hypothesis we get that for all $\alpha \in \{h,t\}$, $\mathbf{M}',((\check{x},\hat{x}),\alpha) \not\models \varphi$ or $\mathbf{M}',((\check{x},\hat{x}),\alpha) \models \psi$, so $\mathbf{M}',((\check{x},\hat{x}),\alpha) \models \varphi \rightarrow \psi$. The converse direction and the second part of the theorem are proved in a similar way.
- Case $\Box\psi$: in the first case, assume by contradiction that $\mathbf{M}',((\check{x},\hat{x}),h) \not\models \Box\psi$. Therefore, $\mathbf{M}',((\check{y},\hat{y}),h) \not\models \psi$ for some $(\check{y},\hat{y}) \in W'$ satisfying $(\check{x},\hat{x})R'(\check{y},\hat{y})$. By induction hypothesis $\mathbf{M},\check{y} \not\models \psi$. By definition, $\check{x}R\check{y}$. Therefore, $\mathbf{M},\check{x} \not\models \Box\psi$: a contradiction. Conversely, assume by contradiction that $\mathbf{M},\check{x} \not\models \Box\psi$. Therefore there exists $y \in W$ such that $\check{x} \leq x'Ry$ and $\mathbf{M},y \not\models \psi$. If $x' = \check{x}$, we use the first condition of being Cartesian to conclude that $\check{y} = y$ (so $\check{x}R\check{y}$) and $\hat{x}R\hat{y}$. By definition $(\check{x},\hat{x})R'(\check{y},\hat{y})$. By induction $\mathbf{M}',((\check{y},\hat{y}),h) \not\models \psi$. Therefore $\mathbf{M}',((\check{x},\hat{x}),h) \not\models \Box\psi$. If $x' = \hat{x}$, we use the second condition of being Cartesian to conclude that $\hat{y} = y$ (so $\hat{x}R\hat{y}$) and $\check{x}R\check{y}$. By definition $(\check{x},\hat{x})R'(\check{y},\hat{y})$. By induction $\mathbf{M}',((\check{y},\hat{y}),t) \not\models \psi$. Therefore $\mathbf{M}',((\check{x},\hat{x}),t) \not\models \Box\psi$. By Lemma 2.1, $\mathbf{M}',((\check{x},\hat{x}),h) \not\models \Box\psi$. The proof of the second item is similar.

□

Proposition 2.10 *For any modal formula φ, φ is satisfiable (respectively, valid) in the class of all BHT-frames iff φ is satisfiable (respectively, valid) in the class of all Cartesian birelational frames.*

3 Axiomatisation

The axiomatic system of BHT consists of the formulas considered in Lemmas 2.2 and 2.3 plus the following inference rules:

$MP\ \frac{\varphi\ \varphi\to\psi}{\psi}$, $MR_{\to}\ \frac{\chi\wedge\psi\to\varphi}{\chi\to(\psi\to\varphi)}$, $MR_{\Box}\ \frac{\varphi\to\psi\vee\chi}{\Box\varphi\to\Diamond\psi\vee\Box\chi}$,

$Nec\ \frac{\varphi}{\Box\varphi}$, $MR_{\leftarrow}\ \frac{\varphi\to\psi\vee\chi}{(\varphi\leftarrow\psi)\to\chi}$, $MR_{\Diamond}\ \frac{\chi\wedge\psi\to\varphi}{\Box\chi\wedge\Diamond\psi\to\Diamond\varphi}$.

The notion of *BHT-derivability* is defined as usual.

Lemma 3.1 *$\top$ is derivable in BHT.*

Proof. Notice that $\top$ is valid in *IPC* (in fact it is equivalent to $\neg\bot$). Since $IPC \subseteq BHT$ we conclude that $\top$ is derivable in *BHT*. □

Proposition 3.2 (Soundness) *Let φ be a formula. If φ is BHT-derivable then φ is valid in the class of all BHT-model.*

Proof. It suffices to check that all axioms are valid and the inference rules preserve validity. □

Let x, y be theories. We say that x *derives* y (in symbols $x \vdash y$) iff there exists $m, n \geq 0$, there exists formulas $\varphi_1, \ldots, \varphi_m \in x$ and there exists formulas $\psi_1, \ldots, \psi_n \in y$ such that $\varphi_1 \wedge \ldots \wedge \varphi_m \to \psi_1 \vee \ldots \vee \psi_n$ is *BHT*-derivable.

4 Completeness

We base our proof of completeness on the canonical model construction.

4.1 Tableaux

A *tableau* is a couple of theories. We say that a tableau (x, y) is *consistent* iff $x \nvdash y$. The tableau (x, y) is said to be *maximal* iff for all formulas φ, $\varphi \in x$, or $\varphi \in y$. We say that a tableau (x, y) is *disjoint* iff $x \cap y = \emptyset$. The tableau (x, y) is said to be *saturated* iff $\bot \in y$, $\top \in x$ and for all formulas φ, ψ,

(i) if $\varphi \vee \psi \in x$ then $\varphi \in x$, or $\psi \in x$,

(ii) if $\varphi \vee \psi \in y$ then $\varphi \in y$ and $\psi \in y$,

(iii) if $\varphi \wedge \psi \in x$ then $\varphi \in x$ and $\psi \in x$,

(iv) if $\varphi \wedge \psi \in y$ then $\varphi \in y$, or $\psi \in y$,

(v) if $\varphi \to \psi \in x$ then $\varphi \in y$, or $\psi \in x$,

(vi) if $\varphi \leftarrow \psi \in y$ then $\varphi \in y$, or $\psi \in x$.

Lemma 4.1 *Every consistent tableau is disjoint.*

Thus, if $(x, y), (x', y')$ are maximal consistent tableaux then $x \subseteq x'$ iff $y \supseteq y'$.

Lemma 4.2 (Lindenbaum Lemma) *Let (x, y) be a tableau. If (x, y) is consistent then there exists a maximal consistent tableau (x', y') such that $x \subseteq x'$ and $y \subseteq y'$.*

Lemma 4.3 *If (x, y) is a maximal consistent tableau then x contains the set of all BHT-derivable formulas and x is closed under the rule MP. Moreover, x and y constitute a partition of the set of all formulas.*

Lemma 4.4 *Every maximal consistent tableau is saturated.*

Proof. Let (x, y) be a maximal consistent tableau. We demonstrate (x, y) is saturated, which amounts to prove that conditions (i)-(vi) of the definition of saturated tableaux are satisfied. We only present the proof for conditions (v)

and (vi).
Suppose $\varphi \to \psi \in x$, $\varphi \notin y$ and $\psi \notin x$. Since (x,y) is maximal consistent, therefore $\varphi \to \psi \notin y$ and $\psi \in y$. Moreover, $\varphi \wedge (\varphi \to \psi) \leftarrow \psi$ is in y. Since $\psi \in y$, therefore $\varphi \wedge (\varphi \to \psi) \in y$ (otherwise, we would obtain $x \vdash y$ which contradicts the maximal consistency of (x,y)). Hence, $\varphi \in y$, or $\varphi \to \psi \in y$. Since $\varphi \notin y$, therefore $\varphi \to \psi \in y$: a contradiction.
Suppose $\varphi \leftarrow \psi \in y$, $\varphi \notin y$ and $\psi \notin x$. Since (x,y) is maximal consistent, therefore $\varphi \leftarrow \psi \notin x$ and $\varphi \in x$. Moreover, $\varphi \to (\varphi \leftarrow \psi) \vee \psi$ is in x. Since $\varphi \in x$, therefore $(\varphi \leftarrow \psi) \vee \psi \in x$ (otherwise, we would obtain $x \vdash y$ which contradicts the maximal consistency of (x,y)). Hence, $\varphi \leftarrow \psi \in x$, or $\psi \in x$. Since $\psi \notin x$, therefore $\varphi \leftarrow \psi \in x$: a contradiction. □

Lemma 4.5 (Hosoi Lemma) *Let (x,y), (x',y') and (x'',y'') be maximal consistent tableaux. if $x \subseteq x'$ and $x \subseteq x''$ then $x = x'$, or $x = x''$, or $x' = x''$.*

Proof. Suppose $x \subseteq x'$, $x \subseteq x''$, $x \neq x'$, $x \neq x''$ and $x' \neq x''$. Without loss of generality, suppose $x' \nsubseteq x''$. Let φ be a formula such that $\varphi \in x'$ and $\varphi \notin x''$. Since $x \subseteq x'$, therefore $\neg\varphi \notin x$ (otherwise, we would obtain $\varphi \wedge \neg\varphi \in x'$ which contradicts the maximal consistency of (x',y')). Since $x \subseteq x''$ and $x \neq x''$, therefore let ψ be a formula such that $\psi \notin x$ and $\psi \in x''$. Since (x,y) is maximal consistent, therefore $\psi \vee (\psi \to \varphi) \vee \neg\varphi$ (Hosoi axiom) is in x. Hence, $\psi \in x$, or $\psi \to \varphi \in x$, or $\neg\varphi \in x$. Since $\neg\varphi \notin x$ and $\psi \notin x$, therefore $\psi \to \varphi \in x$. Since $x \subseteq x''$, therefore $\psi \to \varphi \in x''$. Since $\psi \in x''$, therefore $\varphi \in x''$: a contradiction. □

Lemma 4.6 (Negation-dual of Hosoi Lemma) *Let (x,y), (x',y') and (x'',y'') be maximal consistent tableaux. if $x \supseteq x'$ and $x \supseteq x''$ then $x = x'$, or $x = x''$, or $x' = x''$.*

Proof. Suppose $x \supseteq x'$, $x \supseteq x''$, $x \neq x'$, $x \neq x''$ and $x' \neq x''$. Without loss of generality, suppose $x' \nsupseteq x''$. Let φ be a formula such that $\varphi \notin x'$ and $\varphi \in x''$. Since $x \supseteq x'$, therefore $\llcorner\varphi \in x$ (otherwise, we would obtain $\varphi \vee \llcorner\varphi \notin x'$ which contradicts the maximal consistency of (x',y')). Since $x \supseteq x''$ and $x \neq x''$, therefore let ψ be a formula such that $\psi \in x$ and $\psi \notin x''$. Since (x,y) is maximal consistent, therefore $\neg(\llcorner\varphi \wedge (\varphi \leftarrow \psi) \wedge \psi)$ (dual of Hosoi axiom) is in x and $\llcorner\varphi \wedge (\varphi \leftarrow \psi) \wedge \psi$ is not in x. Hence, $\llcorner\varphi \notin x$, or $\varphi \leftarrow \psi \notin x$, or $\psi \notin x$. Since $\llcorner\varphi \in x$ and $\psi \in x$, therefore $\varphi \leftarrow \psi \notin x$. Since $x \supseteq x''$, therefore $\varphi \leftarrow \psi \notin x''$. Since $\psi \notin x''$, therefore $\varphi \notin x''$: a contradiction. □

4.2 Canonical model

The canonical model $\mathbf{M}_c$ is defined as the structure $\mathbf{M}_c = \langle W_c, \leq_c, R_c, V_c \rangle$ where:

- W_c is the set of all maximal consistent tableaux,
- $\leq_c$ is defined by $(x,y) \leq_c (x',y')$ iff $x \subseteq x'$ and $y \supseteq y'$,
- R_c is defined by $(x,y)R_c(x',y')$ iff $\Box x \subseteq x'$ and $x \supseteq \Diamond x'$,
- $V_c : VAR \to 2^{W_c}$ is defined by $(x,y) \in V_c(p)$ iff $p \in x$,

Lemma 4.7 $\mathbf{M}_c$ *is normal.*

Lemma 4.8 $\mathbf{M}_c$ *is Cartesian.*

Proof. Suppose $\mathbf{M}_c$ is not Cartesian. Let $(x,y),(x',y') \in W_c$ be such that $\widecheck{(x,y)}R_c(x',y')$ and $\widecheck{(x',y')} \neq (x',y')$, or $\widecheck{(x,y)}R_c(x',y')$ and not $\widehat{(x,y)}R_c\widehat{(x',y')}$, or $\widehat{(x,y)}R_c(x',y')$ and $\widehat{(x',y')} \neq (x',y')$, or $\widehat{(x,y)}R_c(x',y')$ and not $\widecheck{(x,y)}R_c\widecheck{(x',y')}$. Let $x^{\downarrow}$, $x^{\uparrow}$, $y^{\downarrow}$, $y^{\uparrow}$, $x'^{\downarrow}$, $x'^{\uparrow}$, $y'^{\downarrow}$ and $y'^{\uparrow}$ be theories such that $\widecheck{(x,y)} = (x^{\downarrow},y^{\downarrow})$, $\widehat{(x,y)} = (x^{\uparrow},y^{\uparrow})$, $\widecheck{(x',y')} = (x'^{\downarrow},y'^{\downarrow})$ and $\widehat{(x',y')} = (x'^{\uparrow},y'^{\uparrow})$.

Suppose $\widecheck{(x,y)}R_c(x',y')$ and $\widecheck{(x',y')} \neq (x',y')$. Let φ be a formula such that $\varphi \in x'$ and $\varphi \notin x'^{\downarrow}$. Since $\varphi \vee \llcorner\varphi$ is derivable, therefore $\llcorner\varphi \in x'^{\downarrow}$. Hence, $\llcorner\varphi \in x'$. Since $\varphi \in x'$, therefore $\varphi \wedge \llcorner\varphi \in x'$. Since $\widecheck{(x,y)}R_c(x',y')$, therefore $\Diamond(\varphi \wedge \llcorner\varphi) \in x^{\downarrow}$. Thus, $\llcorner\llcorner\Diamond(\varphi \wedge \llcorner\varphi) \in x^{\downarrow}$. Consequently, $\llcorner\Diamond(\varphi \wedge \llcorner\varphi) \notin x^{\downarrow}$ (otherwise we would obtain $\llcorner\Diamond(\varphi\wedge\llcorner\varphi) \wedge \llcorner\llcorner\Diamond(\varphi\wedge\llcorner\varphi) \in x^{\downarrow}$ which contradicts the maximal consistency of $(x^{\downarrow},y^{\downarrow})$). Hence, $\llcorner\Diamond(\varphi \wedge \llcorner\varphi)$ is not derivable: a contradiction.

Suppose $\widecheck{(x,y)}R_c(x',y')$ and not $\widehat{(x,y)}R_c\widehat{(x',y')}$. Let φ be a formula such that $\Box\varphi \in x^{\uparrow}$ and $\varphi \notin x'^{\uparrow}$, or $\Diamond\varphi \notin x^{\uparrow}$ and $\varphi \in x'^{\uparrow}$. In the former case, $\neg\neg\Box\varphi \in x^{\downarrow}$. Since $\neg\neg\Box\varphi \to \Box\neg\neg\varphi$ is derivable, therefore $\Box\neg\neg\varphi \in x^{\downarrow}$. Since $\widecheck{(x,y)}R_c(x',y')$, therefore $\neg\neg\varphi \in x'$. Hence, $\varphi \in x'^{\uparrow}$: a contradiction. In the latter case, $\neg\neg\varphi \in x'$. Since $\widecheck{(x,y)}R_c(x',y')$, therefore $\Diamond\neg\neg\varphi \in x^{\downarrow}$. Since $\Diamond\neg\neg\varphi \to \neg\neg\Diamond\varphi$ is derivable, therefore $\neg\neg\Diamond\varphi \in x^{\downarrow}$. Thus, $\Diamond\varphi \in x^{\uparrow}$: a contradiction.

The cases when $\widehat{(x,y)}R_c(x',y')$ and $\widehat{(x',y')} \neq (x',y')$, or $\widehat{(x,y)}R_c(x',y')$ and not $\widecheck{(x,y)}R_c\widecheck{(x',y')}$ are addressed in a similar way. □

4.3 Truth Lemma

We now prepare ourselves for the prof of the Truth Lemma.

Lemma 4.9 *Let φ,ψ be formulas. Let (x,y) be a maximal consistent tableau. If $\varphi \to \psi \in y$ then there exists a maximal consistent tableau (x',y') such that $x \subseteq x'$, $\varphi \in x'$ and $\psi \in y'$.*

Proof. Suppose $\varphi \to \psi \in y$. Let $x' = x \cup \{\varphi\}$ and $y' = \{\psi\}$. Suppose the tableau (x',y') is not consistent. Let $n \geq 0$ and $\chi_1,\ldots,\chi_n \in x'$ be such that $\chi_1 \wedge \ldots \wedge \chi_n \to \psi$ is BHT-derivable. There are 2 cases: there exists a positive integer $i \leq n$ such that $\chi_i = \varphi$, or such integer does not exist. In the former case, since $\chi_1 \wedge \ldots \wedge \chi_n \to \psi$ is BHT-derivable, therefore $\chi \wedge \varphi \to \psi$ is BHT-derivable where χ is the conjunction of the formulas in $\chi_1,\ldots,\chi_n$ which are not equal to φ. Hence, by $MR_{\to}$, $\chi \to (\varphi \to \psi)$ is BHT-derivable: a contradiction with the consistency of (x,y). In the latter case, since $\chi_1 \wedge \ldots \wedge \chi_n \to \psi$ is BHT-derivable, therefore $\chi_1 \wedge \ldots \wedge \chi_n \to (\varphi \to \psi)$ is BHT-derivable: a contradiction with the consistency of (x,y). Consequently, the tableau (x',y') is consistent. By Lindenbaum Lemma, let (x'',y'') be a maximal consistent tableau such that $x' \subseteq x''$ and $y' \subseteq y''$. Obviously, $\varphi \in x''$ and $\psi \in y''$. Moreover, $x \subseteq x''$. □

Lemma 4.10 *Let φ, ψ be formulas. Let (x, y) be a maximal consistent tableau. If $\varphi \leftarrow \psi \in x$ then there exists a maximal consistent tableau (x', y') such that $x \supseteq x'$, $\varphi \in x'$ and $\psi \in y'$.*

Proof. Suppose $\varphi \leftarrow \psi \in x$. Let $x' = \{\varphi\}$ and $y' = y \cup \{\psi\}$. Suppose the tableau (x', y') is not consistent. Let $n \geq 0$ and $\chi_1, \ldots, \chi_n \in y'$ be such that $\varphi \rightarrow \chi_1 \vee \ldots \vee \chi_n$ is BHT-derivable. There are 2 cases: there exists a positive integer $i \leq n$ such that $\chi_i = \psi$, or such integer does not exist. In the former case, since $\varphi \rightarrow \chi_1 \vee \ldots \vee \chi_n$ is BHT-derivable, therefore $\varphi \rightarrow \chi \vee \psi$ is BHT-derivable where χ is the disjunction of the formulas in $\chi_1, \ldots, \chi_n$ which are not equal to ψ. Hence, by $MR_{\leftarrow}$, $(\varphi \leftarrow \psi) \rightarrow \chi$ is BHT-derivable: a contradiction with the consistency of (x, y). In the latter case, since $\varphi \rightarrow \chi_1 \vee \ldots \vee \chi_n$ is BHT-derivable, therefore $(\varphi \leftarrow \psi) \rightarrow \chi$ is BHT-derivable: a contradiction with the consistency of (x, y). Consequently, the tableau (x', y') is consistent. By Lindenbaum Lemma, let (x'', y'') be a maximal consistent tableau such that $x' \subseteq x''$ and $y' \subseteq y''$. Obviously, $\varphi \in x''$ and $\psi \in y''$. Moreover, $x \supseteq x''$. □

Lemma 4.11 *Let φ be a formula. Let (x, y) be a maximal consistent tableau. If $\Box\varphi \in y$ then there exists a maximal consistent tableau (x', y') such that $\Box x \subseteq x'$, $x \supseteq \Diamond x'$ and $\varphi \in y'$.*

Proof. Suppose $\Box\varphi \in y$. Let $x' = \Box x$ and $y' = \{\chi : \Diamond\chi \in y\} \cup \{\varphi\}$. Suppose the tableau (x', y') is not consistent. Let $m, n \geq 0$, $\psi_1, \ldots, \psi_m \in x'$ and $\chi_1, \ldots, \chi_n \in y'$ be such that $\psi_1 \wedge \ldots \wedge \psi_m \rightarrow \chi_1 \vee \ldots \vee \chi_n$ is BHT-derivable. There are 2 cases: there exists a positive integer $i \leq n$ such that $\chi_i = \varphi$, or such integer does not exist. In the former case, since $\psi_1 \wedge \ldots \wedge \psi_m \rightarrow \chi_1 \vee \ldots \vee \chi_n$ is BHT-derivable, therefore $\psi \rightarrow \chi \vee \varphi$ is BHT-derivable where ψ is the conjunction of the formulas in $\psi_1, \ldots, \psi_m$ and χ is the disjunction of the formulas in $\chi_1, \ldots, \chi_n$ which are not equal to φ. Hence, by $MR_{\Box}$, $\Box\psi \rightarrow \Diamond\chi \vee \Box\varphi$ is BHT-derivable: a contradiction with the consistency of (x, y). In the latter case, since $\psi_1 \wedge \ldots \wedge \psi_m \rightarrow \chi_1 \vee \ldots \vee \chi_n$ is BHT-derivable, therefore $\psi \rightarrow \chi \vee \varphi$ is BHT-derivable. Hence, $\Box\psi \rightarrow \Diamond\chi \vee \Box\varphi$ is BHT-derivable: a contradiction with the consistency of (x, y). Consequently, the tableau (x', y') is consistent. By Lindenbaum Lemma, let (x'', y'') be a maximal consistent tableau such that $x' \subseteq x''$ and $y' \subseteq y''$. Obviously, $\Box x \subseteq x''$ and $x \supseteq \Diamond x''$. Moreover, $\varphi \in y''$. □

Lemma 4.12 *Let φ be a formula. Let (x, y) be a maximal consistent tableau. If $\Diamond\varphi \in x$ then there exists a maximal consistent tableau (x', y') such that $\Box x \subseteq x'$, $x \supseteq \Diamond x'$ and $\varphi \in x'$.*

Proof. Suppose $\Diamond\varphi \in x$. Let $x' = \Box x \cup \{\varphi\}$ and $y' = \{\chi : \Diamond\chi \in y\}$. Suppose the tableau (x', y') is not consistent. Let $m, n \geq 0$, $\psi_1, \ldots, \psi_m \in x'$ and $\chi_1, \ldots, \chi_n \in y'$ be such that $\psi_1 \wedge \ldots \wedge \psi_m \rightarrow \chi_1 \vee \ldots \vee \chi_n$ is BHT-derivable. There are 2 cases: there exists a positive integer $i \leq m$ such that $\psi_i = \varphi$, or such integer does not exist. In the former case, since $\psi_1 \wedge \ldots \wedge \psi_m \rightarrow \chi_1 \vee \ldots \vee \chi_n$ is BHT-derivable, therefore $\psi \wedge \varphi \rightarrow \chi$ is BHT-derivable where ψ is the conjunction of the formulas in $\psi_1, \ldots, \psi_m$ which are not equal to φ and χ is the disjunction of the formulas in $\chi_1, \ldots, \chi_n$. Hence, by $MR_{\Diamond}$, $\Box\psi \wedge \Diamond\varphi \rightarrow \Diamond\chi$ is BHT-

derivable: a contradiction with the consistency of (x, y). In the latter case, since $\psi_1 \wedge \ldots \wedge \psi_m \rightarrow \chi_1 \vee \ldots \vee \chi_n$ is BHT-derivable, therefore $\psi \wedge \varphi \rightarrow \chi$ is BHT-derivable. Hence, $\Box\psi \wedge \Diamond\varphi \rightarrow \Diamond\chi$ is BHT-derivable: a contradiction with the consistency of (x, y). Consequently, the tableau (x', y') is consistent. By Lindenbaum Lemma, let (x'', y'') be a maximal consistent tableau such that $x' \subseteq x''$ and $y' \subseteq y''$. Obviously, $\Box x \subseteq x''$ and $x \supseteq \Diamond x''$. Moreover, $\varphi \in x''$. □

Lemma 4.13 (Truth Lemma) *For all formulas φ and for all $(x, y) \in W_c$, $\varphi \in x$ iff $\mathbf{M}_c, (x, y) \models \varphi$ and $\varphi \in y$ iff $\mathbf{M}_c, (x, y) \not\models \varphi$.*

Proposition 4.14 (Completeness) *Let φ be a formula. If φ is valid in the class of all BHT-frames then φ is BHT-derivable.*

Proof. Suppose φ is not BHT-derivable. Hence, the tableau $(\emptyset, \{\varphi\})$ is consistent. By Lindenbaum Lemma, let (x, y) be a maximal consistent tableau such that $\varphi \in y$. By the Truth Lemma, $\mathbf{M}_c, (x, y) \not\models \varphi$. By Proposition 2.10, φ is not valid in the class of all BHT-frames. □

5 Decidability/complexity

There is a classical method for building finite models, namely filtration, and this method could easily be adapted to the BHT setting. Nevertheless, it will fail to give us a tight upper bound for the complexity of the satisfiability problem in BHT. The truth is that, as shown in this section, see below Proposition 5.7, the satisfiability problem is in PSPACE. Our argument is based on a translation into modal logic. In order to define a translation from the BHT to modal logic K, we need for all formulas φ, two new variables: a_φ and b_φ. Let $\mathtt{h}()$ and $\mathtt{t}()$ be translations that are structure-preserving for $\bot$, $\top$, $\vee$, $\wedge$, $\Box$ and $\Diamond$ such that

- $\mathtt{h}(p) = a_p$,
- $\mathtt{h}(\varphi \rightarrow \psi) = (\mathtt{h}(\varphi) \rightarrow \mathtt{h}(\psi)) \wedge (b_\varphi \rightarrow b_\psi)$,
- $\mathtt{h}(\varphi \leftarrow \psi) = \mathtt{h}(\varphi) \wedge \neg\mathtt{h}(\psi)$,
- $\mathtt{t}(p) = b_p$,
- $\mathtt{t}(\varphi \rightarrow \psi) = \mathtt{t}(\varphi) \rightarrow \mathtt{t}(\psi)$,
- $\mathtt{t}(\varphi \leftarrow \psi) = (\mathtt{t}(\varphi) \wedge \neg\mathtt{t}(\psi)) \vee (a_\varphi \wedge \neg a_\psi)$.

Lemma 5.1 *For all formulas φ, $|\mathtt{h}(\varphi)| \leq 11|\varphi|$ and $|\mathtt{t}(\varphi)| \leq 11|\varphi|$.*

For all formulas φ, let $\mu(\varphi)$ be the conjunction of the following formulas:

- $a_p \rightarrow b_p$ for each φ's variable p,
- $a_\psi \leftrightarrow \mathtt{h}(\psi)$ for each φ's subformula ψ,
- $b_\psi \leftrightarrow \mathtt{t}(\psi)$ for each φ's subformula ψ.

Lemma 5.2 *For all formulas φ, $|\mu(\varphi)| = \mathcal{O}(|\varphi|^2)$.*

For all formulas φ, let $\nu(\varphi) = \mu(\varphi) \wedge \Box\mu(\varphi) \wedge \ldots \wedge \Box^{deg(\varphi)}\mu(\varphi)$.

Lemma 5.3 *For all formulas φ, $|\nu(\varphi)| = \mathcal{O}(|\varphi|^3)$.*

Given a BHT-model $\mathbf{M} = \langle W, R, H, T\rangle$, we define its associated model $\mathcal{M} = \langle W, R, V\rangle$ as follows:

1) $V(a_\psi) = \{x \in W : \mathbf{M}, (x,h) \models \psi\}$ for each φ's subformula ψ,

2) $V(b_\psi) = \{x \in W : \mathbf{M}, (x,t) \models \psi\}$ for each φ's subformula ψ.

This associated model $\mathcal{M}$ is considered as a model of modal logic K. Obviously, for all φ's variables p, $V(a_p) \subseteq V(b_p)$. Moreover,

Lemma 5.4 *For all φ's subformulas ψ and for all $x \in W$,*

1) $\mathbf{M}, (x,h) \models \psi$ iff $\mathcal{M}, x \models \mathtt{h}(\psi)$ iff $x \in V(a_\psi)$,
2) $\mathbf{M}, (x,t) \models \psi$ iff $\mathcal{M}, x \models \mathtt{t}(\psi)$ iff $x \in V(b_\psi)$.

Thus, for all $x \in W$, $\mathcal{M}, x \models \mu(\varphi)$. Consider now a generated model $\mathcal{M} = \langle W, R, V\rangle$ of modal logic K of depth at most $deg(\varphi)$ such that for all $x \in W$, $\mathcal{M}, x \models \mu(\varphi)$. We define the corresponding BHT-model $\mathbf{M} = \langle W, R, H, T\rangle$ as follows:

1) $H(p) = V(a_p)$, 2) $T(p) = V(b_p)$.

Obviously, $\mathbf{M}$ is a BHT-model. Moreover,

Lemma 5.5 *For all φ's subformulas ψ and for all $x \in W$,*

1) $\mathcal{M}, x \models \mathtt{h}(\psi)$ iff $\mathbf{M}, (x,h) \models \psi$, *2) $\mathcal{M}, x \models \mathtt{t}(\psi)$ iff $\mathbf{M}, (x,t) \models \psi$.*

Proposition 5.6 *For all formulas φ,*

- *$\llcorner\llcorner\varphi$ is satisfiable in a BHT-model iff $\mathtt{h}(\varphi) \wedge \nu(\varphi)$ is satisfiable in a model of modal logic K,*
- *$\neg\neg\varphi$ is satisfiable in a BHT-model iff $\mathtt{t}(\varphi) \wedge \nu(\varphi)$ is satisfiable in a model of modal logic K.*

Proposition 5.7 *The satisfiability problem in BHT is in* PSPACE.

6 Bisimulations

Bisimulations are binary relations that relate elements of models carrying the same modal information. We now adapt the definition of bisimulations to the BHT setting.

6.1 Bisimulations for BHT

Let $\mathbf{M}_1 = \langle W_1, R_1, H_1, T_1\rangle$ and $\mathbf{M}_2 = \langle W_2, R_2, H_2, T_2\rangle$ be BHT-models. Let $D_1 = W_1 \times \{h,t\}$ and $D_2 = W_2 \times \{h,t\}$. A binary relation $\mathcal{Z}$ between D_1 and D_2 is a *bisimulation* iff the following conditions are satisfied:

1) if $(x_1,\alpha_1)\mathcal{Z}(x_2,\alpha_2)$ then $\mathbf{M}_1, (x_1,\alpha_1) \models p$ iff $\mathbf{M}_2, (x_2,\alpha_2) \models p$ for all $p \in VAR$,

2) if $(x_1,\alpha_1)\mathcal{Z}(x_2,\alpha_2)$ then $(x_1,t)\mathcal{Z}(x_2,t)$,

3) if $(x_1,\alpha_1)\mathcal{Z}(x_2,\alpha_2)$ then $(x_1,h)\mathcal{Z}(x_2,h)$,

4) if $(x_1,\alpha_1)\mathcal{Z}(x_2,\alpha_2)$ and $x_1R_1y_1$ then there exists $y_2 \in W_2$ such that $x_2R_2y_2$ and $(y_1,\alpha_1)\mathcal{Z}(y_2,\alpha_2)$, or $(y_1,t)\mathcal{Z}(y_2,\alpha_2)$,

5) if $(x_1,\alpha_1)\mathcal{Z}(x_2,\alpha_2)$ and $x_2R_2y_2$ then there exists $y_1 \in W_1$ such that $x_1R_1y_1$ and $(y_1,\alpha_1)\mathcal{Z}(y_2,\alpha_2)$, or $(y_1,\alpha_1)\mathcal{Z}(y_2,t)$,

6) if $(x_1,\alpha_1)\mathcal{Z}(x_2,\alpha_2)$ and $x_2R_2y_2$ then there exists $y_1 \in W_1$ such that $x_1R_1y_1$ and $(y_1,\alpha_1)\mathcal{Z}(y_2,\alpha_2)$, or $(y_1,t)\mathcal{Z}(y_2,\alpha_2)$,

7) if $(x_1,\alpha_1)\mathcal{Z}(x_2,\alpha_2)$ and $x_1R_1y_1$ then there exists $y_2 \in W_2$ such that $x_2R_2y_2$ and $(y_1,\alpha_1)\mathcal{Z}(y_2,\alpha_2)$, or $(y_1,\alpha_1)\mathcal{Z}(y_2,t)$.

Lemma 6.1 (Bisimulation Lemma) *Let φ be a formula. For all $(x_1,\alpha_1) \in D_1$ and for all $(x_2,\alpha_2) \in D_2$, if $(x_1,\alpha_1)\mathcal{Z}(x_2,\alpha_2)$ then $\mathbf{M}_1,(x_1,\alpha_1) \vDash \varphi$ iff $\mathbf{M}_2,(x_2,\alpha_2) \vDash \varphi$.*

Obviously, the union of two bisimulations is also a bisimulation.

6.2 Hennessy-Milner property

In this section we show that BHT possesses the Hennessy-Milner property. Our proof follows the line of reasoning suggested in [24]. Let $\mathbf{M}_1 = \langle W_1,R_1,H_1,T_1\rangle$ and $\mathbf{M}_2 = \langle W_2,R_2,H_2,T_2\rangle$ be finite BHT models. Let $D_1 = W_1 \times \{h,t\}$ and $D_2 = W_2\times\{h,t\}$. We define the binary relation $\leftrightsquigarrow$ between D_1 and D_2 as follows: $(x_1,\alpha_1) \leftrightsquigarrow (x_2,\alpha_2)$ iff for all formulas φ, $\mathbf{M}_1,(x_1,\alpha_1) \vDash \varphi$ iff $\mathbf{M}_2,(x_2,\alpha_2) \vDash \varphi$.

Lemma 6.2 (Hennesy-Milner property) *The binary relation $\leftrightsquigarrow$ is a bisimulation between $\mathbf{M}_1$ and $\mathbf{M}_2$.*

Proof. Suppose $\leftrightsquigarrow$ is not a bisimulation. Hence, one of the conditions 1)-7) does not hold for some $(x_1,\alpha_1) \in D_1$ and some $(x_2,\alpha_2) \in D_2$.
Suppose Condition 1) is not satisfied. Hence, there exists a variable p such that, without loss of generality, $\mathbf{M}_1,(x_1,\alpha_1) \vDash p$ and $\mathbf{M}_2,(x_2,\alpha_2) \nvDash p$. Thus, $(x_1,\alpha_1)\not\leftrightsquigarrow(x_2,\alpha_2)$: a contradiction.
Suppose Condition 2) is not satisfied. Hence, $(x_1,\alpha_1) \leftrightsquigarrow (x_2,\alpha_2)$ but $(x_1,t)\not\leftrightsquigarrow (x_2,t)$. Let φ be a formula such that $\mathbf{M}_1,(x_1,t) \vDash \varphi$ and $\mathbf{M}_2,(x_2,t) \nvDash \varphi$, or $\mathbf{M}_1,(x_1,t) \nvDash \varphi$ and $\mathbf{M}_2,(x_2,t) \vDash \varphi$. Thus, $\mathbf{M}_1,(x_1,\alpha_1) \vDash \neg\neg\varphi$ and $\mathbf{M}_2,(x_2,\alpha_2) \nvDash \neg\neg\varphi$, or $\mathbf{M}_1,(x_1,\alpha_1) \nvDash \neg\neg\varphi$ and $\mathbf{M}_2,(x_2,\alpha_2) \vDash \neg\neg\varphi$. Consequently, $(x_1,\alpha_1) \not\leftrightsquigarrow (x_2,\alpha_2)$: a contradiction.
Suppose Condition 3) is not satisfied. Hence, $(x_1,\alpha_1) \leftrightsquigarrow (x_2,\alpha_2)$ but $(x_1,h)\not\leftrightsquigarrow (x_2,h)$. Let φ be a formula such that $\mathbf{M}_1,(x_1,h) \vDash \varphi$ and $\mathbf{M}_2,(x_2,h) \nvDash \varphi$, or $\mathbf{M}_1,(x_1,h) \nvDash \varphi$ and $\mathbf{M}_2,(x_2,h) \vDash \varphi$. Thus, $\mathbf{M}_1,(x_1,\alpha_1) \vDash \llcorner\llcorner\varphi$ and $\mathbf{M}_2,(x_2,\alpha_2) \nvDash \llcorner\llcorner\varphi$, or $\mathbf{M}_1,(x_1,\alpha_1) \nvDash \llcorner\llcorner\varphi$ and $\mathbf{M}_2,(x_2,\alpha_2) \vDash \llcorner\llcorner\varphi$. Consequently, $(x_1,\alpha_1) \not\leftrightsquigarrow (x_2,\alpha_2)$: a contradiction.
Suppose Condition 4) is not satisfied: Then $(x_1,\alpha_1) \leftrightsquigarrow (x_2,\alpha_2)$ and there exists $x_1 \in W_1$ such that $x_1R_1y_1$ and for all $y_2 \in W_2$, if $x_2R_2y_2$ then $(y_1,\alpha_1)\not\leftrightsquigarrow(y_2,\alpha_2)$ and $(y_1,t)\not\leftrightsquigarrow(y_2,\alpha_2)$. Let $R_2(x_2) \stackrel{def}{=} \{(y_2,\alpha_2) \in D_2 \mid x_2R_2y_2\}$ and $R_1(x_1) \stackrel{def}{=} \{(y_1,\alpha_1) \mid x_1R_1y_1\}$. Let $I \subseteq R_2(x_2)$, $J \subseteq R_2(x_2)$, $(y_2,\alpha_2) \in R_2(x_2)$ and for all $(y_2,\alpha_2) \in R_2(x_2)$, let $\varphi(y_2,\alpha_2)$ and $\psi(y_2,\alpha_2)$ be formulas such that

1) $\mathbf{M}_1,(y_1,\alpha_1) \vDash \varphi(y_2,\alpha_2)$ and $\mathbf{M}_2,(y_2,\alpha_2) \nvDash \varphi(y_2,\alpha_2)$ if $(y_2,\alpha_2) \in I$;

2) $\mathbf{M}_1,(y_1,\alpha_1) \nvDash \varphi(y_2,\alpha_2)$ and $\mathbf{M}_2,(y_2,\alpha_2) \vDash \varphi(y_2,\alpha_2)$ if $(y_2,\alpha_2) \in \overline{I}$;

3) $\mathbf{M}_1, (y_1, t) \models \psi(y_2, \alpha_2)$ and $\mathbf{M}_2, (y_2, \alpha_2) \not\models \psi(y_2, \alpha_2)$ if $(y_2, \alpha_2) \in J$;

4) $\mathbf{M}_1, (y_1, t) \not\models \psi(y_2, \alpha_2)$ and $\mathbf{M}_2, (y_2, \alpha_2) \models \psi(y_2, \alpha_2)$ if $(y_2, \alpha_2) \in \overline{J}$.

Let us define $\chi(y_2, \alpha_2)$ as the following formula:

$$\chi(y_2,\alpha_2) = \begin{cases} \varphi(y_2,\alpha_2) & \text{if } y_2 \in I; \\ \varphi(y_2,\alpha_2) \to \psi(y_2,\alpha_2) & \text{if } y_2 \in \overline{I} \cap J; \\ \neg\psi(y_2,\alpha_2) & \text{if } y_2 \in \overline{I} \cap \overline{J}. \end{cases}$$

It follows that $\mathbf{M}_1, (y_1, \alpha_1) \models \chi(y_2, \alpha_2)$ and $\mathbf{M}_2, (y_2, \alpha_2) \not\models \chi(y_2, \alpha_2)$, for all $(y_2, \alpha_2) \in R_2(x_2)$. Therefore $\mathbf{M}_1, (x_1, \alpha_1) \models \Diamond \bigwedge_{(y_2,\alpha_2) \in R_2(x_2)} \chi(y_2, \alpha_2)$ while $\mathbf{M}_2, (x_2, \alpha_2) \not\models \Diamond \bigwedge_{(y_2,\alpha_2) \in R_2(x_2)} \chi(y_2, \alpha_2)$: a contradiction.

Suppose Condition 5) is not satisfied. Then $(x_1, \alpha_1) \leftrightsquigarrow (x_2, \alpha_2)$, $x_2 R_2 y_2$ and for all $y_1 \in W_1$, if $x_1 R_1 y_1$ then $(y_1, \alpha_1) \not\leftrightsquigarrow (y_2, \alpha_2)$ and $(y_1, \alpha_1) \not\leftrightsquigarrow (y_2, t)$. Let $R_2(x_2) \stackrel{def}{=} \{(y_2, \alpha_2) \in D_2 \mid x_2 R_2 y_2\}$ and $R_1(x_1) \stackrel{def}{=} \{(y_1, \alpha_1) \mid x_1 R_1 y_1\}$. Let $I \subseteq R_1(x_1)$, $J \subseteq R_1(x_1)$ and for all $(y_1, \alpha_1) \in R_1(x_1)$, let $\psi(y_1, \alpha_1)$ and $\varphi(y_1, \alpha_1)$ be formulas such that:

1) $\mathbf{M}_1, (y_1, \alpha_1) \models \varphi(y_1, \alpha_1)$ and $\mathbf{M}_2, (y_2, \alpha_2) \not\models \varphi(y_1, \alpha_1)$ if $(y_1, \alpha_1) \in I$;

2) $\mathbf{M}_1, (y_1, \alpha_1) \not\models \varphi(y_1, \alpha_1)$ and $\mathbf{M}_2, (y_2, \alpha_2) \models \varphi(y_1, \alpha_1)$ if $(y_1, \alpha_1) \in \overline{I}$;

3) $\mathbf{M}_1, (y_1, \alpha_1) \models \psi(y_1, \alpha_1)$ and $\mathbf{M}_2, (y_2, t) \not\models \psi(y_1, \alpha_1)$ if $(y_1, \alpha_1) \in J$;

4) $\mathbf{M}_1, (y_1, \alpha_1) \not\models \psi(y_1, \alpha_1)$ and $\mathbf{M}_2, (y_2, t) \models \psi(y_1, \alpha_1)$ if $(y_1, \alpha_1) \in \overline{J}$.

Let us consider the formula $\chi(y_1, \alpha_1)$ defined as

$$\chi(y_1,\alpha_1) = \begin{cases} \varphi(y_1,\alpha_1) & \text{if } (y_1,\alpha_1) \in \overline{I}; \\ \varphi(y_1,\alpha_1) \to \psi(y_1,\alpha_1) & \text{if } (y_1,\alpha_1) \in I \cap \overline{J}; \\ \neg\psi(y_1,\alpha_1) & \text{if } (y_1,\alpha_1) \in I \cap J. \end{cases}$$

It follows that $\mathbf{M}_1, (y_1, \alpha_1) \not\models \chi(y_1, \alpha_1)$ and $\mathbf{M}_2, (y_2, \alpha_2) \models \chi(y_1, \alpha_1)$, for all $(y_1, \alpha_1) \in R_1(x_1)$. Therefore $\mathbf{M}_2, (x_2, \alpha_2) \models \Diamond \bigwedge_{(y_1,\alpha_1) \in R_1(x_1)} \chi(y_1, \alpha_1)$ while $\mathbf{M}_1, (x_1, \alpha_1) \not\models \Diamond \bigwedge_{(y_1,\alpha_1) \in R_1(x_1)} \chi(y_1, \alpha_1)$: a contradiction. The proof for Condition 6) is similar to the proof for Condition 5) but using

$$\chi(y_1,\alpha_1) \stackrel{def}{=} \begin{cases} \varphi(y_1,\alpha_1) & \text{if } (y_1,\alpha_1) \in I; \\ \varphi(y_1,\alpha_1) \to \psi(y_1,\alpha_1) & \text{if } (y_1,\alpha_1) \in \overline{I} \cap J; \\ \neg\psi(y_1,\alpha_1) & \text{if } (y_1,\alpha_1) \in \overline{I} \cap \overline{J}. \end{cases}$$

The proof for Condition 7) is similar to the proof for Condition 4) but using

$$\chi(y_2,\alpha_2) = \begin{cases} \varphi(y_2,\alpha_2) & \text{if } (y_2,\alpha_2) \in \overline{I}; \\ \varphi(y_2,\alpha_2) \to \psi(y_2,\alpha_2) & \text{if } (y_2,\alpha_2) \in I \cap \overline{J}; \\ \neg\psi(y_2,\alpha_2) & \text{if } (y_2,\alpha_2) \in I \cap J. \end{cases} \tag{2}$$

□

Remark how the formulas defining $\chi(y_1, \alpha_1)$ and $\chi(y_2, \alpha_2)$ above are related to the Hosoi Axiom.

7 Strong equivalence property

Pearce's Equilibrium logic [29] is the best-known logical characterization of the stable models semantics [15] and of Answer Sets [6]. It is defined in terms of the monotonic logic of Here and There [30] (HT) plus a minimisation criterion among the given models. This simple definition led to several modal extensions of *Answer Set Programming* [9,13]. All these extensions have their roots in the corresponding modal extensions of HT-logic defined as the combination of propositional HT and any modal logic [14]) that play an important role in the proof of several interesting properties of the resulting formalisms such as *strong equivalence* [8,13,23]. In this section, we define the concept of modal equilibrium model and prove the associated theorem of *strong equivalence.* A BHT-model $\mathbf{M} = \langle W, R, H, T\rangle$, is said to be *total* iff $H = T$. Given a BHT-model $\mathbf{M} = \langle W, R, H, T\rangle$, $x \in W$ and $k \in \mathbb{N}$, we say that H is strictly included in T with respect to x and k (in symbols $H <_x^k T$) iff there exists $y \in W$ such that $xR^{\leq k}y$ and $H(y) \neq T(y)$. A total BHT-model $M = \langle W, R, T, T\rangle$ is a *Modal Equilibrium Model* of a formula φ iff there exists $x \in W$ such that

1) $\mathbf{M}, (x, h) \models \varphi$;
2) For all $\mathbf{M}' = \langle W, R, H, T\rangle$, if $H <_x^{deg(\varphi)} T$ then $\mathbf{M}', (x, h) \not\models \varphi$.

The notion of modal equilibrium model of a theory is defined in a similar way. When dealing with non-monotonicity the relation of equivalence between theories depends on the context where they are considered. We say that two theories Γ_1 and Γ_2 are *strongly equivalent* (in symbols $\Gamma_1 \equiv_s \Gamma_2$) iff for all theories Γ, the equilibrium models of $\Gamma_1 \cup \Gamma$ and $\Gamma_2 \cup \Gamma$ coincide [23].

Proposition 7.1 *For all theories Γ_1 and Γ_2, $\Gamma_1 \equiv_s \Gamma_2$ iff Γ_1 and Γ_2 are BHT-equivalent.*

Proof. Suppose Γ_1 and Γ_2 are BHT-equivalent. Let Γ be an arbitrary theory. Thus $\Gamma_1 \cup \Gamma$ and $\Gamma_2 \cup \Gamma$ are BHT-equivalent. Therefore $\Gamma_1 \cup \Gamma$ and $\Gamma_2 \cup \Gamma$ have the same equilibrium models. Reciprocally, suppose that Γ_1 and Γ_2 are not BHT-equivalent.

- **First case:** Γ_1 and Γ_2 are not K-equivalent. Without loss of generality, there exists a total BHT-model $\mathbf{M} = \langle W, R, T, T\rangle$ and $x \in W$ such that $\mathbf{M}, (x, h) \models \Gamma_1$ but $\mathbf{M}, (x, h) \not\models \Gamma_2$. Let $\Gamma_0 \stackrel{def}{=} \{\Box^k (p \vee \neg p) \mid k \geq 0 \text{ and } p \in VAR\}$. It can be checked that $\mathbf{M}$ is an equilibrium model of $\Gamma_1 \cup \Gamma_0$ but not of $\Gamma_2 \cup \Gamma_0$.
- **Second case:** Γ_1 and Γ_2 are K-equivalent. Without loss of generality, there exists a BHT-model $\mathbf{M} = \langle W, R, H, T\rangle$ ($\widehat{\mathbf{M}} = \langle W, R, T, T\rangle$ denote its corresponding total model) such that
  (1) for all $y \in W$, $\mathbf{M}, (y, t) \models \Gamma_1$ iff $\mathbf{M}, (y, t) \models \Gamma_2$;
  (2) there exists $x \in W$ such that $\mathbf{M}, (x, h) \models \Gamma_1$ and $\mathbf{M}, (x, h) \not\models \Gamma_2$.

  Therefore there exists $\varphi \in \Gamma_2$ such that $\mathbf{M}, (x, h) \not\models \varphi$. Let $\Gamma \stackrel{def}{=} \{\varphi \to \Box^k (p \vee \neg p) \mid k \geq 0 \text{ and } p \in VAR\}$. It follows that $\mathbf{M}, (x, h) \models \Gamma_1 \cup \Gamma$, since $\mathbf{M}, (x, h) \not\models \varphi$ and $\mathbf{M}, (x, t) \models \Box^k (p \vee \neg p)$, for all $k \geq 0$ and for all $p \in VAR$.

Therefore $\widehat{M}$ is not an equilibrium model of $\Gamma_1 \cup \Gamma$. Since $\Gamma_1 \equiv_s \Gamma_2$, $\widehat{M}$ is not an equilibrium model of $\Gamma_2 \cup \Gamma$. Since $\widehat{\mathbf{M}}, (x,h) \vDash \Gamma_2 \cup \Gamma$, therefore there exists a BHT-model $\mathbf{M}' = \langle W, R, H', T \rangle$ such that $H' <_x^{deg(\Gamma_2 \cup \Gamma)} T$ and $\mathbf{M}', (x,h) \vDash \Gamma_2 \cup \Gamma$. However, from $\mathbf{M}', (x,h) \vDash \Gamma_2$ and $\mathbf{M}', (x,h) \vDash \Gamma$ we conclude that $\mathbf{M}', (x,h) \vDash \Gamma_0$, thus $H' = T$ and this is a contradiction. □

The theorem played a important role in the area of Answer Set Programming [6] since it allows, under ASP semantics, to exchange two logic programs (or theories) regardless the context in which they are considered. This theorem also justifies the use of BHT as a monotonic basis supporting non-monotonicity.

8 Conclusions

In this paper we have studied a combination of the modal logic of Here and There equipped with the dual implication [33]. For this new logic we have presented two alternative (and equivalent) semantics as well as several results concerning axiomatisation, bisimulation, Hennessy-Milner property, decidability and complexity. Finally we have considered the property of strong equivalence from Answer Set Programming [6,30] in our setting.

The reader might have noticed that the dual implication is not used in the proof of the strong equivalence theorem. This fact gives us the idea that this new operator would allow us to characterise, in terms of strong equivalence, new kinds of minimal models like the ones introduced in [1].

Another area of potential application of this logic could be *Inductive Logic Programming* [36,31] (ILP). Among other techniques used to infer rules from facts, called *Inverse Entailment* [26] (IE) reverse the ordinary semantical consequence ($\vDash$). This technique was revisited under the perspective of ASP in [34]. Thanks to the dual implication we can define an inverse entailment relation ($\dashv$) in a very natural way allowing us to investigate the application of ILP in modal contexts.

Finally, we would like to extend the results presented in this paper to general combinations of modal and Gödel Logics [16] as done, for the Hennessy-Milner property, in [24].

Appendix

Proof of Lemma 2.1. By induction on φ.
Proof of Lemma 2.5. By induction on φ.
Proof of Proposition 2.10. By Lemmas 2.8 and 2.9.
Proof of Lemma 4.2. This is a standard result [10].
Proof of Lemma 4.3. This is a standard result [10].
Proof of Lemma 4.7. By Lemmas 4.5 and 4.6.
Proof of Lemma 4.13. By induction on φ. While considering the cases for formulas $\psi \to \chi$, $\psi \leftarrow \chi$, $\Box\psi$ and $\Diamond\chi$, one has to respectively use Lemmas 4.9, 4.10, 4.11 and 4.12.
Proof of Lemma 5.1. By induction on φ.

Proof of Lemma 5.2. By Lemma 5.1.
Proof of Lemma 5.3. By Lemma 5.2.
Proof of Lemma 5.4. By induction on ψ.
Proof of Lemma 5.5. By induction on ψ.
Proof of Proposition 5.6. By Lemmas 5.4 and 5.5.
Proof of Proposition 5.7. By Lemmas 5.1 and 5.3, Proposition 5.6 and the fact that the satisfiability problem in modal logic K is in PSPACE.
Proof of Lemma 6.1. By induction on φ.

References

[1] Amendola, G., T. Eiter, M. Fink, N. Leone and J. Moura, *Semi-equilibrium models for paracoherent answer set programs*, Artificial Intelligence **234** (2016), pp. 219–271.

[2] Balbiani, P., J. Boudou, M. Diéguez and D. Fernández-Duque, *Bisimulations for intuitionistic temporal logics*, in: *Intuitionistic Modal Logic and Applications (IMLA)*, 2017, forthcoming.

[3] Balbiani, P. and M. Diéguez, *Temporal Here and There*, in: *JELIA'16*, 2016, pp. 81–96.

[4] Božić, M. and K. Došen, *Models for normal intuitionistic modal logics*, Studia Logica **43** (1984), p. 217–245.

[5] Boudou, J., M. Diéguez and D. Fernández-Duque, *A decidable intuitionistic temporal logic*, in: *26th EACSL Annual Conference on Computer Science Logic (CSL)*, 2017, pp. 14:1–14:17.

[6] Brewka, G., T. Eiter and M. Truszczyński, *Answer set programming at a glance*, Commun. ACM **54** (2011), pp. 92–103.

[7] Bull, R. A., *A modal extension of intuitionist logic*, Notre Dame Journal of Formal Logic **6** (1965), pp. 142–146.

[8] Cabalar, P. and M. Diéguez, *Strong equivalence of non-monotonic temporal theories*, in: *KR'14*, 2014.

[9] Cabalar, P. and G. Pérez, *Temporal Equilibrium Logic: A first approach*, in: *EUROCAST'07*, 2007, pp. 241–248.

[10] Chagrov, A. and M. Zakharyaschev, "Modal Logic," Oxford University Press, 1997.

[11] Diéguez, M., *Temporal answer set programming*, available at `http://www.dc.fi.udc.es/~cabalar/DieguezPhD-final.pdf`.

[12] Fairtlough, M. and M. Mendler, *An intuitionistic modal logic with applications to the formal verification of hardware*, in: *Computer science logic (Kazimierz, 1994)*, Lecture Notes in Comput. Sci. **933**, Springer, Berlin, 1995, pp. 354–368.

[13] Fariñas del Cerro, L., A. Herzig and E. Su, *Epistemic Equilibrium Logic*, in: *IJCAI'15*, 2015, pp. 2964–2970.

[14] Gabbay, D., A. Kurucz, F. Wolter and M. Zakharyaschev, "Many-Dimensional Modal Logics: Theory and Applications," North Holland, 2003.

[15] Gelfond, M. and V. Lifschitz, *The stable model semantics for logic programming*, in: *ICLP'88*, 1988, pp. 1070–1080.

[16] Gödel, K., *Zum intuitionistischen Aussagenkalkül*, Anzeiger der Akademie der Wissenschaften Wien, mathematisch, naturwissenschaftliche Klasse **69** (1932), pp. 65–66.

[17] Goranko, V., *Refutation systems in modal logic*, Studia Logica **53** (1994), pp. 299–324.

[18] Goré, R., "A Uniform Display System for Intuitionistic and Dual Intuitionistic Logic", TR-ARP-6-95, Automated Reasoning Project, Australian Nat. Uni., (1995).

[19] Goré, R., *Dual intuitionistic logic revisited*, in: R. Dyckhoff, editor, *Automated Reasoning with Analytic Tableaux and Related Methods* (2000), pp. 252–267.

[20] Goré, R. and L. Postniece, *Combining derivations and refutations for cut-free completeness in bi-intuitionistic logic*, J. Log. Comput. **20** (2010), pp. 233–260.

[21] Heyting, A., "Die Formalen Regeln der Intuitionistischen Logik," Sitzungsberichte der Preussischen Akademie der Wissenschaften. Physikalisch-mathematische Klasse, Deütsche Akademie der Wissenschaften zu Berlin, Mathematisch-Naturwissenschaftliche Klasse, 1930.
[22] Hosoi, T., *The axiomatization of the intermediate propositional systems S_2 of Gödel*, Journal of the Faculty of Science of the University of Tokyo **13** (1966), pp. 183–187.
[23] Lifschitz, V., D. Pearce and A. Valverde, *Strongly equivalent logic programs*, ACM Transactions on Computational Logic **2** (2001), pp. 526–541.
[24] Marti, M. and G. Metcalfe, *A Hennessy-Milner property for many-valued modal logics*, in: *Advances in Modal Logic* (2014), pp. 407–420.
[25] Mints, G., "A Short Introduction to Intuitionistic Logic," University Series in Mathematics, Springer, 2000.
[26] Muggleton, S., *Inverse entailment and progol*, New Generation Computing **13** (1995), pp. 245–286.
[27] Nelson, D., *Constructible falsity*, The Journal of Symbolic Logic **14** (1949), pp. 16–26.
[28] Ono, H., *On some intuitionistic modal logics*, Publications of the Research Institute for Mathematical Sciences **13** (1977), p. 687–722.
[29] Pearce, D., *A new logical characterisation of stable models and answer sets*, in: *NMELP'96*, 1996, pp. 57–70.
[30] Pearce, D., *Equilibrium logic*, Annals of Mathematics and Artificial Intelligence **47** (2006), pp. 3–41.
[31] Plotkin, G., "Automatic Methods of Inductive Inference," Ph.D. thesis, Edinburgh University (1971).
[32] Plotkin, G. D. and C. Stirling, *A framework for intuitionistic modal logics*, in: *TARK'86*, 1986, pp. 399–406.
[33] Rauszer, C., *An algebraic and Kripke-style approach to a certain extension of intuitionistic logic*, Dissertationes Math. (Rozprawy Mat.) **167** (1980).
[34] Sakama, C., *Inverse entailment in nonmonotonic logic programs*, in: *Proceedings of the 10th International Conference on Inductive Logic Programming*, ILP '00 (2000), pp. 209–224.
[35] Servi, G. F., *On modal logic with an intuitionistic base*, Studia Logica **36** (1977), pp. 141–149.
[36] Shapiro, E. Y., *Inductive inference of theories from facts*, Technical Report Research Report 192, Yale University, Department of Computer Science (1981).
[37] Simpson, A., "The Proof Theory and Semantics of Intuitionistic Modal Logic," Ph.D. thesis, University of Edinburgh (1994).
[38] Smetanich, Y. S., *On the completeness of the propositional calculus with additional operations in one argument*, Trudy Moskovskogo Matematicheskogo Obshchestva **9** (1960), pp. 357–371, (in Russian).
[39] Su, E. I., "Extensions of Equilibrium Logic by Modal Concepts," Ph.D. thesis, Université de Toulouse, Toulouse, France (2015), available at `https://tel.archives-ouvertes.fr/tel-01636791/document`.
[40] Wansing, H., *Constructive negation, implication, and co-implication*, Journal of Applied Non-Classical Logics **18** (2008), pp. 341–364.
[41] Wolter, F., *On logics with coimplication*, Journal of Philosophical Logic **27** (1998), pp. 353–387.

Frame-Validity Games and Absolute Minimality of Modal Axioms

Philippe Balbiani [1]

Institut de Recherche en Informatique de Toulouse
CNRS — Toulouse University

David Fernández-Duque [2]

Department of Mathematics
Ghent University

Andreas Herzig [3]

Institut de Recherche en Informatique de Toulouse
CNRS — Toulouse University

Petar Iliev [4]

Centre International de Mathématiques et d'Informatique de Toulouse
Toulouse University

Abstract

We introduce frame-equivalence games tailored for reasoning about the size, modal depth, number of occurrences of symbols and number of different propositional variables of modal formulae defining a given property. Using these games we prove that the Löb axiom $\Box(\Box p \to p) \to \Box p$ and the (m, n)-transfer axioms $\Diamond^m p \to \Diamond^n p$ are optimal among those defining their respective class of frames.

Keywords: modal logic, correspondence theory, formula-size games, lower bounds on formula-size.

1 Introduction

One of the key advantages of modal logics over first-order logic is that the former are often decidable. However, decidability is not sufficient for applications: efficiency plays a huge role in determining the usefulness of a formal system.

[1] Philippe.Balbiani@irit.fr
[2] David.FernandezDuque@UGent.be
[3] Andreas.Herzig@irit.fr
[4] Petar.Iliev@irit.fr

Typical measures of complexity revolve around problems such as satisfiability and model-checking, but the sometimes-overlooked *succinctness* plays a crucial role as well: there is little use in a PTIME logic if properties of interest can only be defined by exponentially large formulas.

The power of first-order logic and some of its extensions to succinctly define graph properties has been investigated extensively [7], as that of the modal language and natural extensions to define properties of relational *models* [3,9]. In contrast, it seems that the only study of how succinctly *frame* properties can be expressed in modal logic is [8], where the question of how many different propositional variables are needed to modally define certain classes of Kripke frames is being considered. To increase our understanding of the succinctness of modal languages, we develop in the present paper techniques for proving lower bounds on the complexity of modal formulas defining frame properties and apply them to some well-known classes of frames.

As usual, we say that a modal formula φ defines a class $\mathbf{F}$ of frames if $\mathbf{F}$ exactly consists of the frames on which φ is valid. If a class of frames is definable by a modal formula, it is natural to ask how *complex* any such formula must be, where the complexity of a formula may be measured according to the total number of symbols, the modal depth, the number of occurrences of symbols of a certain type, or the number of different variables needed.

The techniques we will employ are based on *frame equivalence games,* closely related to model-equivalence games as appeared in [4,5,6]. To demonstrate the applicability of the former to both first- and second-order semantic conditions, we prove that

(i) For each $m, n \geq 0$, the (m, n)-transfer axiom $\Diamond^m p \to \Diamond^n p$ is essentially the shortest modal formula defining the first-order condition

$$\forall x \forall y (x R^m y \to x R^n y), \tag{1}$$

where R^j denotes the j-fold composition of R.

(ii) The Löb axiom $\Box(\Box p \to p) \to \Box p$ is essentially the shortest modal formula defining transitivity plus the second-order property of converse well-foundedness.

Note that the former result applies to the well-studied axioms defining transitivity, reflexivity, and density.

2 Technical preliminaries

Our formula size games are based on formulas in negation normal form, i.e., negations appear only in front of propositional symbols. Fix a countably infinite set of *propositional variables* $P = \{p_1, p_2, \ldots\}$, and let $\mathcal{L}_\Diamond$ denote the uni-modal language that has as atomic formulas the *literals* p, $\overline{p}$ for each $p \in P$ as well as $\bot$, $\top$ and as primitive connectives $\vee$, $\wedge$, $\Diamond$, and $\Box$. The expressions $\neg\varphi$ and $\varphi \to \psi$ will be regarded as abbreviations defined using De Morgan's rules.

As usual a frame is a pair $\mathcal{A} = (W_\mathcal{A}, R_\mathcal{A})$ where $W_\mathcal{A}$ is a nonempty set and $R_\mathcal{A} \subseteq W_\mathcal{A} \times W_\mathcal{A}$, a *model based on* $(W_\mathcal{B}, R_\mathcal{B})$ is a tuple $\mathcal{B} = (W_\mathcal{B}, R_\mathcal{B}, V_\mathcal{B})$

consisting of a frame equipped with a *valuation* $V_{\mathcal{B}}\colon W_{\mathcal{B}} \to 2^P$, and a *pointed model* is a tuple $\boldsymbol{c} = (\mathcal{C}, c)$ consisting of a model $\mathcal{C}$ equipped with a designated point $c \in W_{\mathcal{C}}$; pointed models will always be denoted by $\boldsymbol{a}, \boldsymbol{b}, \ldots$ and frames or models by $\mathcal{A}, \mathcal{B}, \ldots$. For a pointed model $\boldsymbol{a} = (\mathcal{A}, a)$, we denote by $\Box\boldsymbol{a}$ the set $\{(\mathcal{A}, b) : a\ R_{\mathcal{A}}\ b\}$, i.e., the set of all pointed models that are *successors* of the pointed model $\boldsymbol{a}$ along the relation $R_{\mathcal{A}}$.

Given $\varphi \in \mathcal{L}_{\Diamond}$ and a pointed model $\boldsymbol{a}$, we define $\boldsymbol{a} \models \varphi$ according to standard Kripke semantics, and as usual if $\mathcal{A}$ is a model we write $\mathcal{A} \models \varphi$ if $(\mathcal{A}, a) \models \varphi$ for all $a \in W_{\mathcal{A}}$, and if $\mathcal{A}$ is a frame, $\mathcal{A} \models \varphi$ if $(\mathcal{A}, V) \models \varphi$ for every valuation V. We use *structure* as an umbrella term to denote either a model, a frame, or a pointed model. For a class of structures $\mathbf{A}$ and a formula φ, we write $\mathbf{A} \models \varphi$ when $\mathcal{X} \models \varphi$ for all $\mathcal{X} \in \mathbf{A}$, and say that the formulae φ and ψ are *equivalent* on $\mathbf{A}$ when for all $\mathcal{X} \in \mathbf{A}$, $\mathcal{X} \models \varphi$ if and only if $\mathcal{X} \models \psi$.

Our goal in this paper is to develop techniques to establish when a formula φ is of minimal complexity among those defining some class of frames. Here *complexity* could mean many things: by a *complexity measure* (or just *measure*) we simply mean a function $\mu\colon \mathcal{L}_{\Diamond} \to \mathbb{N}$. We are interested in the following measures: (i) the *length* of a formula φ, denoted $|\varphi|$ and defined as the number of nodes in its syntax tree (including leaves); (ii) the *number of ocurrences* of any connective, (iii) the *modal depth,* and (iv) the *number of variables.*

Note that these are a total of nine measures, as each connective gives rise to its own measure in (ii). We will show that several modal axioms of interest are minimal with respect to all of these measures simultaneously. To this end, given a set $\Gamma \subseteq \mathcal{L}_{\Diamond}$ and $\varphi \in \Gamma$, say that φ *is absolutely minimal among* Γ if for all $\psi \in \Gamma$ and any of the nine measures μ described above, $\mu(\varphi) \leq \mu(\psi)$.

3 A formula-bound game on models

The game described below is the modal analogue of the formula-size game developed in the setting of first-order logic in [1]. The general idea is that we have two competing players, *Hercules* and the *Hydra.* Given two classes of pointed models $\mathbf{A}$ and $\mathbf{B}$, Hercules is trying to show that there is a "small" $\mathcal{L}_{\Diamond}$-formula φ such that $\mathbf{A} \models \varphi$ but $\mathbf{B} \models \neg\varphi$ whereas the Hydra is trying to show that any such φ is "big". The players move by adding and labelling nodes on a game-tree T. For our purposes a *tree* is a finite set partially ordered by some order $\preccurlyeq$ such that if $\eta \in T$ then $\downarrow\eta = \{\nu : \nu \preccurlyeq \eta\}$ is linearly ordered; any set of the form $\downarrow\eta$ is a *branch* of T.

Definition 3.1 The $(\mathcal{L}_{\Diamond}, \langle\mathbf{A}, \mathbf{B}\rangle)$ *formula-complexity game on models* (denoted $(\mathcal{L}_{\Diamond}, \langle\mathbf{A}, \mathbf{B}\rangle)$-FGM) is played by two players, Hercules and the Hydra, who construct a game-tree T in such a way that each node $\eta \in T$ is labelled with a pair $\langle \mathsf{L}(\eta), \mathsf{R}(\eta)\rangle$ of classes of pointed models and either a literal or a symbol from the set $\{\bot, \top, \vee, \wedge, \Diamond, \Box\}$ according to the rules below.

Any leaf η can be declared either a *head* or a *stub.* Once η has been declared a stub, no further moves can be played on it. The construction of T begins with a root labelled by $\langle\mathbf{A}, \mathbf{B}\rangle$ that is declared a head. Afterwards, the game continues as long as there is at least one head. In each turn, Hercules goes first by choosing a head η labelled by $\langle \mathsf{L}(\eta), \mathsf{R}(\eta)\rangle$. Then, he plays one of the following moves.

LITERAL-MOVE: Hercules chooses a literal ι such that $\mathsf{L}(\eta) \models \iota$ and $\mathsf{R}(\eta) \models \neg\iota$. The node η is declared a stub and labelled with the symbol ι.

$\bot$-MOVE: Hercules can play this move only if $\mathsf{L}(\eta) = \varnothing$. The node η is declared a stub and labelled with the symbol $\bot$.

$\top$-MOVE: Hercules can play this move only if $\mathsf{R}(\eta) = \varnothing$. The node η is declared a stub and labelled with the symbol $\top$.

$\vee$-MOVE: Hercules labels η with the symbol $\vee$ and chooses two sets $\mathbf{L}_1, \mathbf{L}_2 \subseteq \mathbf{L}$ such that $\mathsf{L}(\eta) = \mathbf{L}_1 \cup \mathbf{L}_2$. Two new heads, labelled by $\langle \mathbf{L}_1, \mathsf{R}(\eta)\rangle$ and $\langle \mathbf{L}_2, \mathsf{R}(\eta)\rangle$, are added to T as daughters of η.

$\wedge$-MOVE: Dual to the $\vee$-move, except that in this case Hercules chooses $\mathbf{R}_1$, $\mathbf{R}_2$ such that $\mathbf{R}_1 \cup \mathbf{R}_2 = \mathsf{R}(\eta)$.

$\Diamond$-MOVE: Hercules labels η with the symbol $\Diamond$ and, for each pointed model $\boldsymbol{l} \in \mathsf{L}(\eta)$, he chooses a pointed model from $\Box\boldsymbol{l}$; if for some $\boldsymbol{l} \in \mathsf{L}(\eta)$ we have $\Box\boldsymbol{l} = \varnothing$, Hercules cannot play this move. All these new pointed models are collected in the set $\mathbf{L}_1$. For each pointed model $\boldsymbol{r} \in \mathsf{R}(\eta)$, the Hydra replies by picking a subset of $\Box\boldsymbol{r}$ [5]. All the pointed models chosen by the Hydra are collected in the class $\mathbf{R}_1$. A new head labelled by $\langle \mathbf{L}_1, \mathbf{R}_1\rangle$ is added as a daughter to η.

$\Box$-MOVE: Dual to the $\Diamond$-move, except that Hercules first chooses a successor for each $\boldsymbol{r} \in \mathbf{R}$ and Hydra chooses her successors for frames in $\mathbf{L}$.

The $(\mathcal{L}_\Diamond, \langle \mathbf{A}, \mathbf{B}\rangle)$-FGM concludes when there are no heads and we say in this case that T is a *closed game tree.*

Note that the Hydra has no restrictions on the number of pointed models she chooses on modal moves; in fact, she can choose all of them, and it is often convenient to assume that she always does so. To be precise, say that the Hydra *plays greedily* if (i) whenever Hercules makes a $\Diamond$-move on a node η and adds a new node η', then Hydra sets $\mathsf{R}(\eta') = \bigcup_{\boldsymbol{r}\in\mathsf{R}(\eta)} \Box\boldsymbol{r}$, and similarly (ii) whenever Hercules makes a $\Box$-move on a node η and adds a new node η', then Hydra sets $\mathsf{L}(\eta') = \bigcup_{\boldsymbol{l}\in\mathsf{L}(\eta)} \Box\boldsymbol{l}$.

The $(\mathcal{L}_\Diamond, \langle \mathbf{A}, \mathbf{B}\rangle)$-FGM can be used to give lower bounds on the length of formulae defining a given property [4,5,6]. Here we will generalize this to show that it can be used to give lower bounds on any complexity measure. For this, we need to view game-trees as formulae.

Definition 3.2 Given a closed $(\mathcal{L}_\Diamond, \langle \mathbf{A}, \mathbf{B}\rangle)$-FGM tree T, we define $\psi_T \in \mathcal{L}_\Diamond$ to be the unique formula whose syntax tree is given by T.

Formally speaking, ψ_T is defined by recursion on T starting from leaves: if T is a single leaf then it must be labelled by a literal ι, or by $\bot$, or by $\top$, so we

[5] In particular, if $\Box\boldsymbol{r} = \varnothing$ for some $\boldsymbol{r} \in \mathsf{R}(\eta)$, the Hydra does not add anything to $\mathbf{R}_1$ for the pointed model $\boldsymbol{r}$.

respectively set $\psi_T = \iota$, or $\psi_T = \bot$, or $\psi_T = \top$; if T has a root η labelled by $\vee$, then η has two daughters η_1, η_2. Letting T_1, T_2 be the respective generated subtrees, we define $\psi_T = \psi_{T_1} \vee \psi_{T_2}$. The cases for $\wedge$, $\Diamond$ and $\Box$ are all analogous. Then, given a complexity measure μ, we extend the domain of μ to include the set of closed game trees by defining $\mu(T) = \mu(\psi_T)$.

If $m \in \mathbb{N}$, $\mathbf{A}, \mathbf{B}$ are classes of models, and $\mu \colon \mathcal{L}_\Diamond \to \mathbb{N}$ a complexity measure, we say that Hercules has a *winning strategy for the* $(\mathcal{L}_\Diamond, \langle \mathbf{A}, \mathbf{B} \rangle)$-FGM *with* μ *below* m if Hercules has a strategy so that no matter how Hydra plays, the game terminates in finite time with a closed tree T so that $\mu(T) < m$.

Theorem 3.3 *Let* $\mathbf{A}$, $\mathbf{B}$ *be classes of models,* μ *any complexity measure, and* $m \in \mathbb{N}$. *Then, the following are equivalent:*

(i) *Hercules has a winning strategy for the* $(\mathcal{L}_\Diamond, \langle \mathbf{A}, \mathbf{B} \rangle)$-FGM *with* μ *below* m*;*

(ii) *there is an* $\mathcal{L}_\Diamond$*-formula* φ *with* $\mu(\varphi) < m$ *and* $\mathbf{A} \models \varphi$ *whereas* $\mathbf{B} \models \neg\varphi$.

We defer the proof of Theorem 3.3 to Appendix A, where we also establish some useful properties of the formula-complexity game. However, we remark that the proof is essentially the same as that of the special case where $\mu(\varphi) = |\varphi|$, which can be found in any of [4,5,6]. We will also use the following easy consequence of Theorem 3.3. We assume familiarity with bisimulations [2].

Corollary 3.4 *Let* $\mathbf{A}$ *and* $\mathbf{B}$ *be classes of pointed models such that there are* $\boldsymbol{a} \in \mathbf{A}$ *and* $\boldsymbol{b} \in \mathbf{B}$ *with* $\boldsymbol{a}$ *bisimilar to* $\boldsymbol{b}$. *For all complexity measures* μ *and for all nonnegative integers* m*, Hercules has no winning strategy for the* $(\mathbf{A}, \mathbf{B})$-FGM *with* μ *below* m.

4 A formula-complexity game on frames

We develop an analogous game to the one above that is played on frames instead of models in order to reason about the "resources" needed to modally define properties of frames.

Definition 4.1 Let $\mathbf{A}$, $\mathbf{B}$ be classes of frames. The $(\mathcal{L}_\Diamond, \langle \mathbf{A}, \mathbf{B} \rangle)$ formula-complexity game on frames (denoted $(\mathcal{L}_\Diamond, \langle \mathbf{A}, \mathbf{B} \rangle)$-FGF) is played by Hercules and the Hydra as follows.

HERCULES SELECTS MODELS: For each $\mathcal{B} \in \mathbf{B}$ Hercules chooses a model $\mathcal{B}^{\mathrm{M}}$ based on $\mathcal{B}$ and a point $\triangleright_{\mathcal{B}} \in W_{\mathcal{B}}$ and then sets $\mathbf{B}^{\mathrm{m}} = \{(\mathcal{B}^{\mathrm{M}}, \triangleright_{\mathcal{B}}) : \mathcal{B} \in \mathbf{B}\}$.

THE HYDRA SELECTS MODELS: The Hydra replies by choosing a class of pointed models $\mathbf{A}^{\mathrm{m}}$ of the form $(\mathcal{A}, V, a)$ with $\mathcal{A} \in \mathbf{A}$.

FORMULA GAME ON MODELS: Hercules and the Hydra play the $(\mathcal{L}_\Diamond, \langle \mathbf{A}^{\mathrm{m}}, \mathbf{B}^{\mathrm{m}} \rangle)$-FGM.

The game tree assigned to a match of the $(\mathcal{L}_\Diamond, \langle \mathbf{A}, \mathbf{B} \rangle)$-FGF is the game tree of the subsequent $(\mathcal{L}_\Diamond, \mathbf{A}^{\mathrm{m}}, \mathbf{B}^{\mathrm{m}} \rangle)$-FGM.

Remark 4.2 The Hydra is free to assign as many models as she wants to each $\mathcal{A} \in \mathbf{A}$, even no model at all. We say that the Hydra plays *functionally* if she chooses $\mathbf{A}^{\mathrm{m}}$ so that for each $\mathcal{A} \in \mathbf{A}$ there is exactly one pointed model $(\mathcal{A}^{\mathrm{M}}, \triangleright_{\mathcal{A}}) \in \mathbf{A}^{\mathrm{m}}$ with $\mathcal{A}^{\mathrm{M}}$ based on $\mathcal{A}$.

As was the case for the FGM, for $m \in \mathbb{N}$, classes of frames $\mathbf{A}$, $\mathbf{B}$, and $\mu\colon \mathcal{L}_\diamond \to \mathbb{N}$ a complexity measure, Hercules has a *winning strategy for the* $(\mathcal{L}_\diamond, \langle \mathbf{A}, \mathbf{B} \rangle)$-FGF *with* μ *below* m if no matter how Hydra plays, the game terminates in finite time with a closed tree T so that $\mu(T) < m$.

Theorem 4.3 *Let* $\mathbf{A}$, $\mathbf{B}$ *be classes of frames,* μ *any complexity measure, and* $m \in \mathbb{N}$. *Then, the following are equivalent:*

(i) *Hercules has a winning strategy for the* $(\mathcal{L}_\diamond, \langle \mathbf{A}, \mathbf{B} \rangle)$-FGF *with* μ *below* m;

(ii) *there is an* $\mathcal{L}_\diamond$*-formula* φ *with* $\mu(\varphi) < m$ *such that* $\mathbf{A} \models \varphi$ *but* φ *is not valid in any frame of* $\mathbf{B}$.

Proof. (ii) IMPLIES (i). Let φ be an $\mathcal{L}_\diamond$-formula with $\mu(\varphi) < m$ that is valid on all frames in $\mathbf{A}$ and not valid in any frame in $\mathbf{B}$. For each $\mathcal{B} \in \mathbf{B}$, Hercules can choose a pointed model $\mathcal{B}^{\mathrm{M}} = (\mathcal{B}, V, b)$ based on $\mathcal{B}$ so that $\mathcal{B}^{\mathrm{M}} \not\models \varphi$. The Hydra then responds with some set of pointed models $\mathbf{A}^{\mathrm{m}}$; since φ is valid on $\mathbf{A}$, for all $\mathcal{A} \in \mathbf{A}^{\mathrm{m}}$ we have $\mathcal{A} \models \varphi$. By Theorem 3.3, it follows that Hercules has a winning strategy with μ below m for the $(\mathcal{L}_\diamond, \mathbf{A}^{\mathrm{m}}, \mathbf{B}^{\mathrm{m}} \rangle)$-FGM and thus for $(\mathcal{L}_\diamond, \langle \mathbf{A}, \mathbf{B} \rangle)$-FGF.

(i) IMPLIES (ii). Assume that Hercules has such a strategy, and that he chooses $\mathbf{B}^{\mathrm{m}}$ according to this strategy. Then Hydra *opens greedily* by choosing every pointed model based on a frame in $\mathbf{A}$; in other words, she sets $\mathbf{A}^{\mathrm{m}}$ to be the set of all $(\mathcal{A}, V, a)$ with $\mathcal{A} \in \mathbf{A}$, V a valuation on $\mathcal{A}$ and $a \in W_{\mathcal{A}}$.

By playing according to his strategy, Hercules can win the $(\mathbf{A}^{\mathrm{m}}, \mathbf{B}^{\mathrm{m}})$-FGM with a closed game tree T such that $\mu(T) < m$; but this is only possible if his sub-strategy for the $(\mathbf{A}^{\mathrm{m}}, \mathbf{B}^{\mathrm{m}})$-FGM is a winning strategy with μ below m. Thus by Theorem 3.3, there is an $\mathcal{L}_\diamond$-formula φ with $\mu(\varphi) < m$ such that $\mathbf{A}^{\mathrm{m}} \models \varphi$ and $\mathbf{B}^{\mathrm{m}} \models \neg\varphi$. Since Hercules chose one pointed model for each $\mathcal{B} \in \mathbf{B}$ it follows that φ is not valid in any frame in $\mathbf{B}$, while since Hydra chose all possible pointed models, it follows that $\mathcal{A} \models \varphi$. □

5 The transfer axioms

We apply our formula-complexity games to prove the minimality of some modal axioms. We begin with what we call the *transfer axioms,* defined as $\mathrm{TA}(m, n) = \diamond^m p \to \diamond^n p$, where $m \neq n \in \mathbb{N}$; since we treat $\varphi \to \psi$ as an abbreviation, we can rewrite these axioms as $\Box^m \overline{p} \vee \diamond^n p$. It is well-known that $\mathrm{TA}(m, n)$ defines the first-order property of (m, n)*-transfer* (1) from the introduction. As special cases we have that $(2, 1)$-transfer is just transitivity and $(0, 1)$-transfer is reflexivity. Instead of (m, n)-transfer we write n*-reflexivity* when $m = 0$, m*-recurrence* when $n = 0$, (m, n)*-transitivity* when $m > n > 0$ and (m, n)*-density* when $0 < m < n$.

In this and the following sections, we define a number of pointed models using figures for ease of understanding. We follow the convention that such pointed models consist of the relevant Kripke model and a point that is denoted

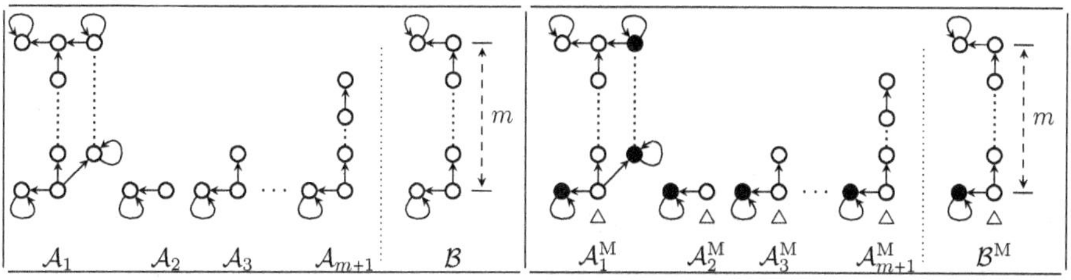

Fig. 1. The frames $\mathcal{A}_1, \ldots, \mathcal{A}_{m+1}$ and $\mathcal{B}$ and the pointed models based on them.

by the $\triangleright$ sign next to it. Our goal is to prove the following.

Theorem 5.1 *For any $n \neq m \in \mathbb{N}$, $\Box^m \overline{p} \vee \Diamond^n p$ is absolutely minimal among all formulas defining (m,n)-transfer.*

The proof that for each $m, n \geq 0$, $\Diamond^m p \to \Diamond^n p$ is essentially the shortest formula defining (m,n)-transfer is split in four parts according to the ordering between m and n. Cases where one of the two is zero are treated in Appendix B.

5.1 Generalized density axioms

First we consider the generalized density axioms, i.e. (m,n)-transfer when $0 < m < n$. We prove that Theorem 5.1 holds in this case by considering a suitable formula-complexity game. Specifically, Hercules and the Hydra play a $(\mathcal{L}_\Diamond, \langle \mathbf{A}, \mathbf{B} \rangle)$-FGF where $\mathbf{A} = \{\mathcal{A}_1, \ldots \mathcal{A}_{m+1}\}$ and $\mathbf{B}$ contains a single element $\mathcal{B}$. These frames are shown in the left rectangle in Figure 1 and separated by the dotted line. $\mathcal{A}_1$ is constructed so that the vertical path leading from the lowest non-reflexive point to the uppermost non-reflexive one consists of m steps whereas the rightmost path that starts and ends respectively with these two points consists of n steps (not counting the reflexive steps) and every point on this rightmost path is reflexive. The frame $\mathcal{B}$ is obtained from $\mathcal{A}_1$ by simply erasing the latter path. Each $\mathcal{A}_i$, for $2 \leq i \leq m+1$, contains a vertical path of $i-2$ steps. Obviously, $\Diamond^m p \to \Diamond^n p$ is valid in all frames in $\mathbf{A}$ and not valid on $\mathcal{B}$.

SELECTION OF THE MODELS ON THE RIGHT: If Hercules wishes to win the game, he must choose his pointed models with some care.

Lemma 5.2 *In any winning strategy for Hercules for an $(\mathcal{L}_\Diamond, \langle \mathbf{L}, \mathbf{R} \rangle)$-FGF in which $\mathcal{A}_1 \in \mathbf{L}$ and $\mathcal{B} \in \mathbf{R}$, Hercules must pick a pointed model $(\mathcal{B}^{\mathrm{M}}, \triangleright)$ based on the lowest irreflexive point in $\mathcal{B}$.*

Proof. It is easy to see that Hercules is not going to select a pointed model that is not based on the lowest non-reflexive point in $\mathcal{B}$ because the Hydra can always reply with a bisimilar pointed model based on $\mathcal{A}_1$. $\Box$

SELECTION OF MODELS ON THE LEFT: The Hydra replies with the pointed models shown on the left of the dotted line in the right rectangle in Figure 1. She has constructed them as follows. Using the fact that $\mathcal{B}$ is a sub-structure of $\mathcal{A}_1$, the Hydra makes sure that the same points in $\mathcal{A}_1^{\mathrm{M}}$ and $\mathcal{B}^{\mathrm{M}}$ satisfy the same literals; moreover, the black points in both models satisfy the same literals, too. The models $\mathcal{A}_i^{\mathrm{M}}$ for $2 \leq i \leq m+1$ receive valuations that make them initial segments of the vertical path in $\mathcal{B}^{\mathrm{M}}$, i.e., the lowest non-reflexive point in any $\mathcal{A}_i^{\mathrm{M}}$ and the lowest non-reflexive point in $\mathcal{B}^{\mathrm{M}}$ satisfy the same literals and similarly for their vertical successors. When the Hydra chooses her pointed models in this way, we say she *mimics* Hercules' choice.

FORMULA SIZE GAME ON MODELS: We consider the FGM starting with $(\mathcal{A}_1^{\mathrm{M}}, \triangleright), \ldots, (\mathcal{A}_{m+1}^{\mathrm{M}}, \triangleright)$ on the left and $(\mathcal{B}^{\mathrm{M}}, \triangleright)$ on the right. First we show that there are some constraints on the moves that Hercules may make.

Lemma 5.3 *Let* $\mathbf{L}$, $\mathbf{R}$ *be classes of models such that Hercules has a winning strategy for the* $(\mathcal{L}_\Diamond, \langle \mathbf{L}, \mathbf{R} \rangle)$*-FGM. Let* T *be any closed game tree on which the Hydra played greedily and* η *be any position of* T *such that* $(\mathcal{B}^{\mathrm{M}}, \triangleright) \in \mathsf{R}(\eta)$ *while* $(\mathcal{A}_i, \triangleright) \in \mathsf{L}(\eta)$ *for some* i *with* $1 \leq i \leq m+1$.

(i) *If Hercules played a* $\Diamond$*-move at* η *then he did not pick the left lowest reflexive point in* $\mathcal{A}_i^{\mathrm{M}}$, *and if* $i = 1$ *then he picked the bottom-right reflexive point on* $\mathcal{A}_1^{\mathrm{M}}$.

(ii) *If Hercules played a* $\Box$*-move at* η *then he did not pick the left lowest reflexive point in* $\mathcal{B}^{\mathrm{M}}$.

Proof. If Hercules picks the left lowest reflexive point when playing such a move, the Hydra is going to reply with the same point in $\mathcal{B}_1^{\mathrm{M}}$ and obtain bisimilar pointed models on each side. If $i = 1$ and Hercules picks the unique irreflexive successor on $\mathcal{A}_1^{\mathrm{M}}$, then Hydra can reply with the irreflexive successor on $\mathcal{B}^{\mathrm{M}}$, which means by Corollary 3.4 that Hercules cannot win. The second claim is symmetric. □

Lemma 5.4 *Suppose that* $\mathbf{L}$, $\mathbf{R}$ *are classes of models and Hercules has a winning strategy for the* $(\mathcal{L}_\Diamond, \langle \mathbf{L}, \mathbf{R} \rangle)$*-FGM. If* T *is any closed game tree in which the Hydra played greedily and* η *is any position of* T *such that* $(\mathcal{B}^{\mathrm{M}}, \triangleright) \in \mathsf{R}(\eta)$, *then*

(i) *if* $(\mathcal{A}_1^{\mathrm{M}}, \triangleright) \in \mathsf{L}(\eta)$, *then Hercules did not play a* $\Box$*-move on* η;

(ii) *if* $(\mathcal{A}_2^{\mathrm{M}}, \triangleright) \in \mathsf{L}(\eta)$, *then Hercules did not play a* $\Diamond$*-move on* η.

Proof. The first claim is immediate from the fact that if Hercules played a $\Box$-move, the Hydra can reply with the same point in $\mathcal{A}_1^{\mathrm{M}}$ and obtain bisimilar pointed models on each side. For the second, Hercules is forced to pick the reflexive point in $\mathcal{A}_2^{\mathrm{M}}$ when playing a $\Diamond$-move which contradicts Lemma 5.3.□

With this we can establish lower bounds on the number of moves of each type that Hercules must make, as established by the proposition below.

Proposition 5.5 *Let* $\mathbf{L}$, $\mathbf{R}$ *be classes of models such that Hercules has a winning strategy for the* $(\mathcal{L}_\Diamond, \langle \mathbf{L}, \mathbf{R} \rangle)$*-FGM and let* T *be a closed game tree in which the Hydra played greedily.*

(i) *If* $\{(\mathcal{A}_1, \triangleright), (\mathcal{A}_2, \triangleright)\} \subseteq \mathbf{L}$ *and* $(\mathcal{B}, \triangleright) \in \mathbf{R}$, *then Hercules made at least one* $\vee$*-move during the game.*

(ii) *If* $(\mathcal{A}_1^{\mathrm{M}}, \triangleright) \in \mathbf{L}$, *and* $(\mathcal{B}^{\mathrm{M}}, \triangleright) \in \mathbf{R}$, *then* T *has modal depth at least* n, *at least* n $\Diamond$*-moves and one literal.*

(iii) *If* $\{(\mathcal{A}_2^{\mathrm{M}}, \triangleright), \ldots, (\mathcal{A}_{m+1}^{\mathrm{M}}, \triangleright)\} \subseteq \mathbf{L}$ *and* $(\mathcal{B}^{\mathrm{M}}, \triangleright) \in \mathbf{R}$, *then Hercules made at least* m $\Box$*-moves during the game.*

Proof.
(i) By Lemma 5.4, Hercules cannot play a modality as long as $(\mathcal{A}_1, \triangleright), (\mathcal{A}_2, \triangleright)$ are on the left and $(\mathcal{B}, \triangleright)$ on the right, and the three satisfy the same literals, so that he cannot play a literal either. Playing a $\wedge$-move would lead to at least one new game position that is the same as the previous one. Hence, every winning strategy for Hercules must 'separate' $(\mathcal{A}_1, \triangleright)$, from $(\mathcal{A}_2, \triangleright)$ with an $\vee$-move.

(ii) Note that $(\mathcal{A}_1^{\mathrm{M}}, \triangleright)$ and $(\mathcal{B}^{\mathrm{M}}, \triangleright)$ satisfy the same literals and $\vee$ and $\wedge$-moves lead to at least one new game-position in which $(\mathcal{A}_1^{\mathrm{M}}, \triangleright)$ is on the left and $(\mathcal{B}^{\mathrm{M}}, \triangleright)$ is on the right. By Lemma 5.4.i, Hercules cannot play a $\Box$-move in any of these positions. Thus Hercules must perform a $\Diamond$-move in a position in which $(\mathcal{A}_1^{\mathrm{M}}, \triangleright)$ is on the left and $(\mathcal{B}^{\mathrm{M}}, \triangleright)$ is on the right. By Lemma 5.3 he is going to pick the first reflexive point on the rightmost path in $\mathcal{A}_1^{\mathrm{M}}$.

The Hydra replies with, among others, the left lowest reflexive point in $\mathcal{B}^{\mathrm{M}}$. Since this point satisfies the same literals as the reflexive points lying on the rightmost path in $\mathcal{A}_1^{\mathrm{M}}$, Hercules cannot play a literal-move; moreover, $\vee$, $\wedge$ and $\Box$-moves lead to at least one new game position that is essentially the same as the previous one. In the case of $\Box$-moves this is true because, when playing such a move, Hercules must stay in the lowest reflexive point in $\mathcal{B}^{\mathrm{M}}$ while the Hydra can stay in the current reflexive point on the rightmost path in $\mathcal{A}_1^{\mathrm{M}}$. Hence, he must make at least $n-1$ subsequent $\Diamond$-moves to reach a point in $\mathcal{A}_1^{\mathrm{M}}$ that differs on a literal from the lowest reflexive point in $\mathcal{B}^{\mathrm{M}}$. Finally he must play a literal, as no other move can close the tree.

(iii) Fix $i \in [2, m+1]$. Let $w_1, \ldots, w_{i-1}$ enumerate the vertical path of $\mathcal{A}_i$ starting at the root, and similarly let $v_1, \ldots, v_m$ enumerate the vertical path of $\mathcal{B}$. Let $\boldsymbol{w}_j = (\mathcal{A}_i^{\mathrm{M}}, w_j)$ and $\boldsymbol{v}_j = (\mathcal{B}^{\mathrm{M}}, v_j)$.

Say that a branch $\overrightarrow{\nu} = (\nu_0, \ldots, \nu_k)$ on T is *i-critical* if there exists $j \in [1, i)$ with $\boldsymbol{w}_j \in \mathsf{L}(\nu_k)$, $\boldsymbol{v}_j \in \mathsf{R}(\nu_k)$ and Hercules has played exactly $j-1$ modal moves on $\nu_1, \ldots, \nu_{k-1}$. Since T is finite and the singleton branch consisting of the root is i-critical, we can pick a maximal i-critical branch $\overrightarrow{\eta} = (\eta_0, \ldots, \eta_\ell)$ for some value of j.

We claim that $j = i-1$ and Hercules plays a $\Box$-move on η_ℓ. Since T is closed η_ℓ cannot be a head, but $\boldsymbol{w}_j$ and $\boldsymbol{v}_j$ share the same valuation so it cannot be a stub either, thus η_ℓ is not a leaf. If Hercules played a $\wedge$- or a $\vee$-move then η_ℓ would have a daughter giving us a longer i-critical branch. Thus Hercules played a modality on η_ℓ. If $j < i-1$ then for the unique daughter η' of η_ℓ we have that $\boldsymbol{w}_{j+1} \in \mathsf{L}(\eta')$ and $\boldsymbol{v}_{j+1} \in \mathsf{R}(\eta')$, where in the case of $j = 0$ we use Lemma 5.3 and otherwise there simply are no other options for Hercules; but this once again gives us a longer i-critical branch. Thus $j = i-1$; but then Hercules is not allowed to play $\Diamond$, as there is a pointed model on the left without successors, so he played a $\Box$-move on η_ℓ.

We conclude that for each $i \in [2, m+1]$ there is an instance of $\Box$ of modal depth exactly $i-1$, which implies that each instance is distinct. $\Box$

With this we prove Theorem 5.1 in the case $0 < m < n$.

Proof. If $0 < m < n$ we consider the $(\mathcal{L}_\Diamond, \langle \mathbf{A}, \mathbf{B} \rangle)$-FGF with $\mathbf{A}$, $\mathbf{B}$ as depicted in Figure 1. By Lemma 5.2 Hercules chooses some pointed model $\mathcal{B}^{\mathrm{M}}$ based on the irreflexive point at the bottom of $\mathcal{B}$, and Hydra replies by mimicking Hercules' pointed models. Then by Proposition 5.5 Hercules must play at least one disjunction, one literal, n $\Diamond$-moves, modal depth n, and at least m $\Box$-moves. By Theorem 4.3, any formula valid on every frame of $\mathbf{A}$ and no frame of $\mathbf{B}$ must satisfy these bounds; but the frames in $\mathbf{A}$ satisfy the (m, n)-transfer property while those in $\mathbf{B}$ do not. $\Box$

5.2 Generalized transitivity axioms

Next we treat Theorem 5.1 in the case where $0 < n < m$. As before, we do so by considering a suitable $(\mathcal{L}_\Diamond, \langle \mathbf{A}, \mathbf{B} \rangle)$-FGF where $\mathbf{A} = \{\mathcal{A}_1, \ldots \mathcal{A}_{m+1}\}$ and $\mathbf{B}$ contains a single element $\mathcal{B}$, but now using the frames shown in Figure 2. The frame $\mathcal{A}_1$ is based on a right-angled triangle in which the sum of the relation steps in the legs is m whereas the number of relation steps in the hypotenuse is n; moreover, each path on the left of the hypotenuse that shares nodes with it consist of n relation steps too. The frame $\mathcal{B}$ is obtained from $\mathcal{A}_1$ by "separating" the hypotenuse from the horizontal leg and erasing the points that do not lie either on the hypotenuse or on the legs of $\mathcal{A}_1$. Each $\mathcal{A}_i$, for $2 \le i \le m+1$, contains a vertical path of $i-2$ relation steps and a diagonal one of n relation steps. Obviously, $\Diamond^m p \to \Diamond^n p$ is valid in all frames in $\mathbf{A}$ and not valid on $\mathcal{B}$.

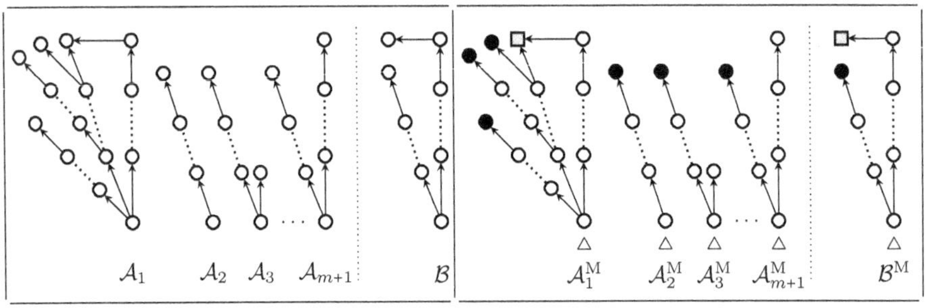

Fig. 2. The frames $\mathcal{A}_1, \ldots, \mathcal{A}_{m+1}$ and $\mathcal{B}$ and the pointed models based on them.

SELECTION OF THE MODELS ON THE RIGHT: In this case, Hercules must choose his models according to the following.

Lemma 5.6 *In any winning strategy for Hercules for an* $(\mathcal{L}_\Diamond, \langle \mathbf{L}, \mathbf{R} \rangle)$-FGF *in which* $\mathcal{A}_1 \in \mathbf{L}$ *and* $\mathcal{B} \in \mathbf{R}$*, Hercules picks a pointed model* $(\mathcal{B}^{\mathrm{M}}, \triangleright)$ *based on the lowest point in* $\mathcal{B}$*, and assigns different valuations to the two dead-end points of* $\mathcal{B}$*.*

Proof. Hercules is not going to select a pointed model that is not based on the lowest point in $\mathcal{B}$ because the Hydra can always reply with a bisimilar pointed model based on $\mathcal{A}_1$. Similarly, if Hercules assigns the same valuation to the two dead-ends the Hydra can choose a bisimilar model based on $\mathcal{A}_1$ by copying the valuations from the hypothenuse onto all paths of length n, and copying the valuations from the legs onto the path of length m; since the valuations coincide on the end-points, there is no clash at the top left of the triangle. □

To indicate that the two end-points of $\mathcal{B}$ receive different valuations, we have drawn one of them black while the other is shaped as a rectangle. The literals true in the rest of the points are immaterial. Thus, Hercules constructs the pointed model $(\mathcal{B}^{\mathrm{M}}, \triangleright)$ shown in the right rectangle in Figure 2.

SELECTION OF MODELS ON THE LEFT: The Hydra replies with the pointed models shown on the left of the dotted line in the right rectangle in Figure 2. The pointed model based on $\mathcal{A}_1$ is defined so that the set of literals true in the points on a diagonal path that shares points with the hypotenuse but do not coincide with it copy the respective sets of literals true in the points of the diagonal path in $\mathcal{B}$.

The models $\mathcal{A}_i$ for $2 \leq i \leq m+1$ receive valuations so that their diagonal paths coincide with the diagonal path in the model $\mathcal{B}$ whereas their vertical paths are 'initial segments' of the vertical path in $\mathcal{B}$, i.e., the lowest point in any $\mathcal{A}_i$ for $2 \leq i \leq m+1$ and the lowest point in $\mathcal{B}$ satisfy the same literals and similarly for their vertical successors. As before, if the Hydra chooses her models in this way, we say that she *mimics* Hercules' choice.

FORMULA SIZE GAME ON MODELS: We consider the FGM starting with $(\mathcal{A}_1^{\mathrm{M}}, \triangleright), \ldots, (\mathcal{A}_{m+1}^{\mathrm{M}}, \triangleright)$ on the left and $(\mathcal{B}^{\mathrm{M}}, \triangleright)$ on the right. These lemmas are analogous to those in Section 5.1.

Lemma 5.7 *Let* $\mathbf{L}$, $\mathbf{R}$ *be classes of models so that Hercules has a winning strategy for the* $(\mathcal{L}_\Diamond, \langle \mathbf{L}, \mathbf{R} \rangle)$*-FGM. Let* T *be any closed game in which the Hydra played greedily and* η *be a node on which Hercules played a* $\Diamond$*-move.*

(i) *If* $(\mathcal{A}_1^{\mathrm{M}}, \triangleright) \in \mathsf{L}(\eta)$ *and* $(\mathcal{B}^{\mathrm{M}}, \triangleright) \in \mathsf{R}(\eta)$*, then he picked a pointed model based on a point that lies on the hypotenuse of* $\mathcal{A}_1^{\mathrm{M}}$.

(ii) *If for some* $i \in [3, m+1]$ *we have that* $(\mathcal{A}_i^{\mathrm{M}}, \triangleright) \in \mathsf{L}(\eta)$ *and* $(\mathcal{B}^{\mathrm{M}}, \triangleright) \in \mathsf{R}(\eta)$*, then he picked the rightmost daughter as a successor of* $(\mathcal{A}_i^{\mathrm{M}}, \triangleright)$.

Proof. Both items hold because if Hercules picked a different point, the Hydra replied with the same point in $\mathcal{B}^{\mathrm{M}}$. In either case we obtain bisimilar models on each side, which by Corollary 3.4 means that Hercules cannot win. □

Lemma 5.8 *Suppose that* $\mathbf{L}$ *and* $\mathbf{R}$ *are classes of models and Hercules has a winning strategy for the* $(\mathcal{L}_\Diamond, \langle \mathbf{L}, \mathbf{R} \rangle)$*-FGM. Suppose that* T *is a closed game tree, the Hydra played greedily, and* η *is a node of* T.

(i) *If* $(\mathcal{A}_1^{\mathrm{M}}, \triangleright) \in \mathsf{L}(\eta)$ *and* $(\mathcal{B}^{\mathrm{M}}, \triangleright) \in \mathsf{R}(\eta)$*, then Hercules did not play a* $\Box$*-move at* η.

(ii) *If* $(\mathcal{A}_2^{\mathrm{M}}, \triangleright) \in \mathsf{L}(\eta)$ *and* $(\mathcal{B}^{\mathrm{M}}, \triangleright) \in \mathsf{R}(\eta)$*, then Hercules did not play a* $\Diamond$*-move at* η.

Proof. The first item is immedate from the fact that if Hercules played a $\Box$-move, the Hydra can reply with the same point in $\mathcal{A}_1^{\mathrm{M}}$, and similarly in the second case the Hydra would reply with the same pointed model based on $\mathcal{B}^{\mathrm{M}}$. □

As was the case for the generalized density axioms, Hercules must play at least one $\vee$-move to separate $\mathcal{A}_1^{\mathrm{M}}$ from $\mathcal{A}_2^{\mathrm{M}}$.

Proposition 5.9 *Let* $\mathbf{L}$ *and* $\mathbf{R}$ *be classes of models such that Hercules has a winning strategy for the* $(\mathcal{L}_\Diamond, \langle \mathbf{L}, \mathbf{R} \rangle)$*-*FGM*. Let* T *be a closed game tree in which the Hydra played greedily.*

(i) *If* $(\mathcal{A}_1, \triangleright), (\mathcal{A}_2, \triangleright) \in \mathbf{L}$ *and* $(\mathcal{B}, \triangleright) \in \mathbf{R}$*, then Hercules made at least one* $\vee$*-move during the game.*

(ii) *If* $(\mathcal{A}_1^{\mathrm{M}}, \triangleright) \in \mathbf{L}$*, then* T *has at least* n *nested* $\Diamond$*-moves and at least one literal move.*

(iii) *If* $\{(\mathcal{A}_2^{\mathrm{M}}, \triangleright), \ldots, (\mathcal{A}_{m+1}^{\mathrm{M}}, \triangleright)\} \subseteq \mathbf{L}$*, then* T *has at least* m $\Box$*-moves.*

Proof. The proof of the first item is analogous to that of Proposition 5.5.i, except that it uses Lemma 5.8, and the proof of the third item is essentially the same as the proof of Proposition 5.5.iii. Thus we focus on the second item.

Since $(\mathcal{A}_1^{\mathrm{M}}, \triangleright)$ and $(\mathcal{B}^{\mathrm{M}}, \triangleright)$ satisfy the same literals, $\vee$, and $\wedge$-moves lead to at least one new game-position in which $(\mathcal{A}_1^{\mathrm{M}}, \triangleright)$ is on the left and $(\mathcal{B}^{\mathrm{M}}, \triangleright)$ is on the right, Hercules must perform a $\Diamond$-move in a position in which $(\mathcal{A}_1^{\mathrm{M}}, \triangleright)$ is on the left and $(\mathcal{B}^{\mathrm{M}}, \triangleright)$ is on the right. It follows from Lemma 5.7, that he is going to pick the immediate successor along the hypotenuse of $\mathcal{A}_1^{\mathrm{M}}$. The Hydra replies, with among others, the immediate successor along the diagonal path in $\mathcal{B}^{\mathrm{M}}$. Since the new pointed models satisfy the same literals, Hercules cannot play a literal-move; moreover, $\vee$- and $\wedge$-moves lead to at least one new game position that is essentially the same as the previous one. If he decided to play a $\Box$-move and picked a pointed model based on a point along the diagonal path in $\mathcal{B}^{\mathrm{M}}$, the Hydra will reply with the same point along a path that is different from the hypotenuse because such paths are always available. Hence, he must make at least $n-1$ subsequent $\Diamond$-moves to reach the point in which the hypotenuse of $\mathcal{A}_1^{\mathrm{M}}$ and its horizontal leg meet. Finally, at this point Hercules must play a literal, as this is the only move that will lead to a closed game-tree. □

With this we conclude the proof of Theorem 5.1 in the case $0 < n < m$.

Proof. Similar to the proof for the case $0 < m < n$, except that we use the classes $\mathbf{A}$, $\mathbf{B}$ of Figure 2 and Proposition 5.9. □

6 The Löb axiom

Finally, we consider the Löb axiom, which defines the property of transitivity and converse-well-foundedness, i.e. that there are no infinite chains $w_0 \; R \; w_1 \; R \; \ldots$. Note that this is a second-order property, and cannot be defined in first-order logic.

Theorem 6.1 *The formula* $\Box\overline{p} \vee \Diamond(p \wedge \Box\overline{p})$ *is absolutely minimal among all formulas defining the class of transitive and converse well-founded frames.*

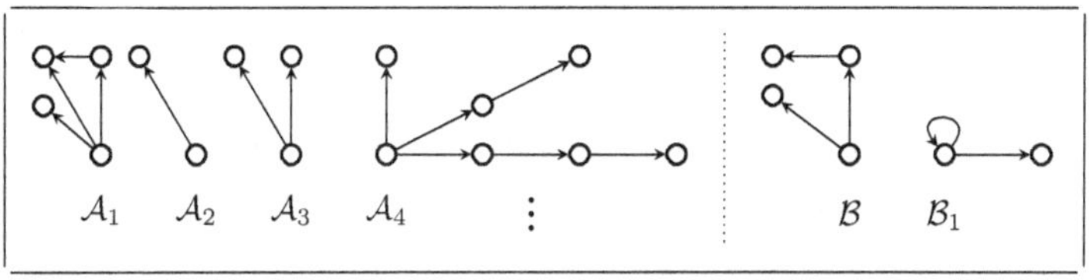

Fig. 3. The sets of frames $\mathbf{A} = \{\mathcal{A}_1, \mathcal{A}_2, \mathcal{A}_3, \mathcal{A}_4\}$ and $\mathbf{B} = \{\mathcal{B}, \mathcal{B}_1\}$.

We have already shown that $\Box\Box\overline{p} \vee \Diamond p$ is absolutely minimal among those formulas defining transitivity, so our strategy will be to expand on the frames and pointed models in Figure 2 to additionally force Hercules to play a conjunction. Since these models were already well-founded we can use previous results.

Let us consider an $(\mathcal{L}_\Diamond, \langle \mathbf{A}, \mathbf{B} \rangle)$-FGF played by Hercules and the Hydra with the frames shown in Figure 3. Obviously, $\mathcal{A}_1, \mathcal{A}_2, \mathcal{A}_3$, and $\mathcal{B}$ are obtained from the frames in Figure 2 for $m = 2$ and $n = 1$. Additionally, $\mathbf{A}$ contains the frame $\mathcal{A}_4$ that is a transitive tree with infinitely many branches such that, for every natural number $n > 0$, there is a branch for which the maximum number or relation steps from the root to its leaf is n. Similarly, $\mathbf{B}$ contains the frame $\mathcal{B}_1$ shown on the right of the dotted line in the same figure. Intuitively, we are going to use $\mathcal{A}_4$ and $\mathcal{B}_1$ in order to force Hercules to play an $\wedge$-move.

SELECTION OF THE MODELS ON THE RIGHT: We only consider the choice of pointed model for the frame $\mathcal{B}_1$. It is obvious that Hercules is not going to base a pointed model on the dead-end point in $\mathcal{B}_1$ because the Hydra would reply with a bisimilar pointed model based on one of the leaves of $\mathcal{A}_4$.

Lemma 6.2 *In any winning strategy for Hercules in the* $(\mathcal{L}_\Diamond, \langle \mathbf{A}, \mathbf{B} \rangle)$-FGF, *Hercules will choose a pointed model based on the reflexive point on* $\mathcal{B}_1$.

SELECTION OF MODELS ON THE LEFT: Hydra will choose her pointed models based on $\mathcal{A}_1$, $\mathcal{A}_2$ and $\mathcal{A}_3$ as before. For her pointed model based on $\mathcal{A}_4$, she picks a pointed model based on the root of the tree in which all leaves of $\mathcal{A}_4$ satisfy the same literals as the ones satisfied by the dead-end point in $\mathcal{B}_1$ whereas the rest of the points satisfy the same literals as the ones satisfied by the reflexive point in $\mathcal{B}_1$. Once again if Hydra plays in this way we say that she *mimics* Hercules' selection.

FORMULA SIZE GAME ON MODELS: The next lemmas will be used to prove that Hercules must play an $\wedge$-move.

Lemma 6.3 *Let* $\mathbf{L}$, $\mathbf{R}$ *be classes of models such that Hercules has a winning strategy for the* $(\mathcal{L}_\Diamond, \langle \mathbf{L}, \mathbf{R} \rangle)$-FGM. *If* T *is a closed game on which the Hydra played greedily, then for any game position* η *and any non-leaf point* w *of* $\mathcal{A}_4$, *if* $(\mathcal{A}_4^{\mathrm{M}}, w) \in \mathsf{L}(\eta)$, $(\mathcal{B}_1^{\mathrm{M}}, \triangleright) \in \mathsf{R}(\eta)$, *and Hercules played a* $\Box$-*move at* η, *then he selected* $(\mathcal{B}_1^{\mathrm{M}}, \triangleright)$ *again.*

Proof. If Hercules picked the dead-end point in $\mathcal{B}_1^{\mathrm{M}}$, the Hydra, using the transitivity of the relation, would reply with a bisimilar pointed model based on a leaf in $\mathcal{A}_4^{\mathrm{M}}$. □

Proposition 6.4 *Suppose that* $\mathbf{L}$, $\mathbf{R}$ *are classes of models for which Hercules has a winning strategy for the* $(\mathcal{L}_\Diamond, \langle \mathbf{L}, \mathbf{R}\rangle)$*-FGM and let* T *be a closed game tree on which the Hydra played greedily.*

(i) *If* $(\mathcal{A}_4^{\mathrm{M}}, \rhd) \in \mathbf{L}$ *and* $(\mathcal{B}_1^{\mathrm{M}}, \rhd) \in \mathbf{R}$, *Hercules played at least one* $\Diamond$*-move on a node* η *such that* $\mathsf{L}(\eta)$ *contains a pointed model based on* $\mathcal{A}_4^{\mathrm{M}}$ *whereas* $(\mathcal{B}_1^{\mathrm{M}}, \rhd) \in \mathsf{R}(\eta)$.

(ii) *If Hercules plays a* $\Diamond$*-move in a position* η *in which* $\mathsf{L}(\eta)$ *contains a pointed model based on* $\mathcal{A}_4^{\mathrm{M}}$ *while* $(\mathcal{B}_1^{\mathrm{M}}, \rhd)$ *is on the right, he must play at least one subsequent* $\wedge$*-move.*

Proof.
(i) Let us suppose that Hercules plays without $\Diamond$-moves. Since $(\mathcal{A}_4^{\mathrm{M}}, \rhd)$ and $(\mathcal{B}_1^{\mathrm{M}}, \rhd)$ satisfy the same literals, no literal move is possible in a game position η in which $(\mathcal{A}_4^{\mathrm{M}}, \rhd)$ is on the left and $(\mathcal{B}_1^{\mathrm{M}}, \rhd)$ on the right. Playing a $\wedge$- or a $\vee$-move results in at least one new position in which $(\mathcal{A}_4^{\mathrm{M}}, \rhd)$ is on the left and $(\mathcal{B}_1^{\mathrm{M}}, \rhd)$ is on the right. Hence a $\Box$-move is inevitable and by Lemma 6.3, he selected $(\mathcal{B}_1^{\mathrm{M}}, \rhd)$ again.

When Hercules plays such a move, the Hydra would reply with all infinitely many pointed models based on $\mathcal{A}_4^{\mathrm{M}}$ and an immediate successor of the root of the tree. From this new position on any finite number of $\vee$, $\wedge$ and $\Box$-moves are going to result in at least one new position that contains $(\mathcal{B}_1^{\mathrm{M}}, \rhd)$ on the right whereas on the left we have infinitely many pointed models based on $\mathcal{A}_4^{\mathrm{M}}$ and a non-leaf point. Obviously, none of the $\top$-, $\bot$-, and literal-moves are possible in such a position. Hence, Hercules has no winning strategy without $\Diamond$-moves.

(ii) Let us suppose that Hercules plays a $\Diamond$-move in such a position. The Hydra is going to respond with both $(\mathcal{B}_1^{\mathrm{M}}, \rhd)$ and a pointed model based on the dead-end point in $\mathcal{B}_1^{\mathrm{M}}$. Let us suppose now that Hercules is not going to play any subsequent $\wedge$-move. Obviously, $\bot$, $\top$, and literal moves are impossible; moreover, the presence of a dead-end pointed model on the right prevents $\Box$-moves. Clearly, playing an $\vee$-move would result in at least one new game position which is the same as the previous one. Therefore, Hercules can only play $\Diamond$-moves until he reaches a pointed model $(\mathcal{A}_4, v)$ such that the only successor of v is a leaf. Playing a $\Diamond$-move in such a position would lead to a loss in the next step because of the presence of bisimilar pointed models on the left and right. Since $(\mathcal{A}_4^{\mathrm{M}}, v)$ and $(\mathcal{B}_1^{\mathrm{M}}, \rhd)$ satisfy the same literals no literal moves are possible either. Therefore, Hercules has no winning strategy without playing at least one $\wedge$-move. □

With this we can prove Theorem 6.1.

Proof. Consider a $(\mathcal{L}_\Diamond, \langle \mathbf{A}, \mathbf{B}\rangle)$-FGF where $\mathbf{A} = \{\mathcal{A}_1, \mathcal{A}_2, \mathcal{A}_3, \mathcal{A}_4\}$ and $\mathbf{B} = \{\mathcal{B}, \mathcal{B}_1\}$ as given in Figures 2 and 3. Hercules must choose his pointed models according to Lemmas 5.6 and 6.2, and Hydra replies by mimicking Hercules.

Using Proposition 5.9 we see that if the Hydra plays greedily then any closed game tree must have modal depth at least two, contain two instances of $\Box$, one instance of each $\Diamond$ and $\vee$, and one variable. By Proposition 6.4, it also contains one conjunction, as required. $\Box$

7 Conclusion

The present work was motivated to a large degree by ideas and results from [8], where the notion of minimal modal equivalent of a first-order condition was introduced. Note however that the term minimal is used in [8] only with respect to the number of different variables needed to modally define a first-order condition which does not tell us much about the length, modal depth, or the number of Boolean connectives required and that is why we have extended the notion of minimality to cover these as well. With this we have shown that several familiar modal axioms are minimal with respect to all measures considered, including the Löb axiom, which is not first-order definable. It is obvious that once we have shown that a given frame property is modally definable, we can study its minimal modal complexity with respect to different complexity measures and therefore there are many natural open problems related to the present work. We would like to mention one in particular.

Question 1 *Is there a complexity measure μ and an infinite sequence of formulae $\varphi_1, \varphi_2, \ldots$ such that if $\psi_1, \psi_2, \ldots$ is a sequence of equivalent Sahlqvist formulae then $\mu(\psi_n)$ grows exponentially in $\mu(\varphi_n)$?*

Appendix

A Properties of the formula-complexity game on models

We have seen that a closed game tree T induces a formula ψ_T. Under certain conditions, we can also turn formulae into game trees.

Lemma A.1 *Let $\mathbf{A}, \mathbf{B}$ be classes of models and $\varphi \in \mathcal{L}_\Diamond$ be so that $\mathbf{A} \models \varphi$ and $\mathbf{B} \models \neg\varphi$. Then, Hercules has a strategy for the $(\mathcal{L}_\Diamond, \langle \mathbf{A}, \mathbf{B} \rangle)$-FGM so that any match terminates on a closed game tree T with $\psi_T = \varphi$.*

Proof. We proceed by induction on the structure of φ.

φ IS A LITERAL. If φ is a literal ι, then Hercules plays the literal-move by choosing ι and the game tree T is closed with $\psi_T = \iota$, as required.

φ IS $\bot$. If φ is $\bot$, then Hercules plays the $\bot$-move and the game tree T is closed with $\psi_T = \bot$, as required.

$\varphi = \varphi_1 \vee \varphi_2$. Hercules can play the $\vee$-move and add two nodes η_1, η_2 labelled by $\langle \mathbf{A}_1, \mathbf{B} \rangle$ and $\langle \mathbf{A}_2, \mathbf{B} \rangle$, respectively, where $\mathbf{A} = \mathbf{A}_1 \cup \mathbf{A}_2$, $\mathbf{A}_1 \models \varphi_1$ and $\mathbf{A}_2 \models \varphi_2$. Applying the induction hypothesis to each sub-game, for $i \in \{1, 2\}$ Hercules has a strategy for the $(\mathcal{L}_\Diamond, \langle \mathbf{A}_i, \mathbf{B}_i \rangle)$-FGM with resulting closed game trees T_i so that $\psi_{T_i} = \varphi_i$. This yields a closed game tree T for the original game with $\psi_T = \varphi$, as desired.

$\varphi = \Diamond\theta$. For each $\boldsymbol{a} \in \mathbf{A}$, Hercules chooses a pointed model from $\Box\boldsymbol{a}$ that satisfies θ and collects all these pointed models in the class $\mathbf{A}_1$. Hydra replies by choosing a subset of $\Box \boldsymbol{b}$ for each $\boldsymbol{b} \in \mathbf{B}$ and collects these pointed models in $\mathbf{B}_1$. A new node η labelled with $\langle \mathbf{A}_1, \mathbf{B}_1 \rangle$ is added to the game tree as a successors to the one labelled with $\langle \mathbf{A}, \mathbf{B} \rangle$. Obviously, $\mathbf{A}_1 \models \theta$ and $\mathbf{B}_1 \models \neg\theta$. Applying the induction hypothesis, we conclude that Hercules has a strategy for the sub-game starting at η so that the resulting game tree S is closed with $\psi_S = \theta$. This yields a closed tree T for the original game with $\psi_T = \Diamond\theta$.

OTHER CASES. Each of the remaining cases is dual to one discussed above. □

Next we show that if the Hydra plays greedily, then any closed game tree T for the $(\mathcal{L}_\Diamond, \langle \mathbf{A}, \mathbf{B} \rangle)$-FGM is such that $\mathbf{A} \models \psi_T$ and $\mathbf{B} \models \neg\psi_T$.

Lemma A.2 *Let* $\mathbf{A}$, $\mathbf{B}$ *be classes of models and let* T *be a closed game tree for the* $(\mathcal{L}_\Diamond, \langle \mathbf{A}_i, \mathbf{B}_i \rangle)$-FGM *on which the Hydra played greedily. Then,* $\mathbf{A} \models \psi_T$ *and* $\mathbf{B} \models \neg\psi_T$.

Proof. For a node η of T let T_η be the subtree with root η, and let $\psi_\eta = \psi_{T_\eta}$. By induction on η starting from the leaves we show that $\mathsf{L}(\eta) \models \psi_\eta$ and $\mathsf{R}(\eta) \models \neg\psi_\eta$. The base case is immediate since Hercules can only play a literal when it is true on the left but false on the right, and the inductive steps for $\bot$, $\top$, $\vee$ and $\wedge$ are straightforward. The critical case is when Hercules plays a modality on η, which is when we use that the Hydra plays greedily. If Hercules played a $\Diamond$-move on η with daughter η' then for each $\boldsymbol{l} \in \mathsf{L}(\eta)$ he chose $\boldsymbol{l}' \in \Box\mathsf{L}(\eta)$ and placed $\boldsymbol{l}' \in \mathsf{L}(\eta')$; by the induction hypothesis $\boldsymbol{l}' \models \psi_{\eta'}$, so that $\boldsymbol{l} \models \Diamond\psi_{\eta'} = \psi_\eta$ by the semantics of $\Diamond$. Meanwhile for $\boldsymbol{r} \in \mathsf{R}(\eta)$, if $\boldsymbol{r}' \in \Box\boldsymbol{r}$ then since the Hydra played greedily $\boldsymbol{r}' \in \mathsf{R}(\eta')$, and since $\boldsymbol{r}'$ was arbitrary we see that $\boldsymbol{r} \models \neg\Diamond\psi_{\eta'}$. The case for a $\Box$-move is symmetric. □

With this we prove Theorem 3.3

Proof. Let $\mathbf{A}$, $\mathbf{B}$ be classes of models, μ any complexity measure, and $m \in \mathbb{N}$. Recall that Theorem 3.3 states that the following are equivalent:

(i) Hercules has a winning strategy for the $(\mathcal{L}_\Diamond, \langle \mathbf{A}, \mathbf{B} \rangle)$-FGM with μ below m, and

(ii) there is an $\mathcal{L}_\Diamond$-formula φ with $\mu(\varphi) < m$ and $\mathbf{A} \models \varphi$ whereas $\mathbf{B} \models \neg\varphi$.

First assume that (i) holds, and let Hydra play the $(\mathcal{L}_\Diamond, \langle \mathbf{A}, \mathbf{B} \rangle)$-FGM greedily. By using his winning strategy, Hercules can ensure that the game terminates on some closed tree T with $\mu(T) < m$. But by definition this means that $\mu(\psi_T) < m$, and by Lemma A.2, $\mathbf{A} \models \psi_T$ while $\mathbf{B} \models \neg\psi_T$.

Conversely, if (ii) holds, by Lemma A.1 Hercules has a strategy so that no matter how the Hydra plays, any match ends with a closed tree T with $\psi_T = \varphi$, so that in particular $\mu(T) < m$. □

B Generalized reflexivity and recurrence

In this appendix we prove Theorem 5.1 in cases where one of the parameters is zero.

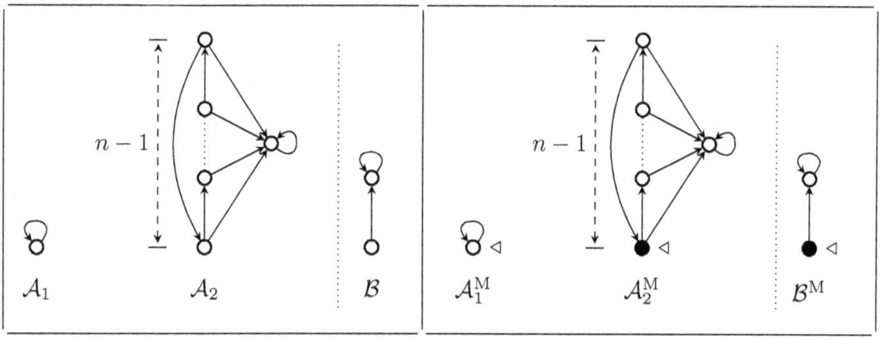

Fig. B.1. The frames $\mathcal{A}_1$, $\mathcal{A}_2$ and $\mathcal{B}$ and the pointed models based on them.

B.1 The generalized reflexivity axioms

Recall that we write *n-reflexivity* instead of $(0, n)$-transfer. In order to prove that Theorem 5.1 holds in this case, we consider a $(\mathcal{L}_\Diamond, \langle \mathbf{A}, \mathbf{B} \rangle)$-FGF where $\mathbf{A} = \{\mathcal{A}_1, \mathcal{A}_2\}$ and $\mathbf{B}$ contains a single element $\mathcal{B}$. These frames are shown in the left rectangle in Figure B.1 and separated by the dotted line. The "highest" point in $\mathcal{A}_2$ can be reached in $n-1$ relation steps from the lowest one and then we can return back to the latter in one additional relation step, i.e, the points in $\mathcal{A}_2$ that are different from the reflexive one form a cycle of length n. It is immediate that $p \to \Diamond^n p$ is valid on both $\mathcal{A}_1$ and $\mathcal{A}_2$ and not valid on $\mathcal{B}$.

Next we study Hercules' possible strategies. We begin with his choice of models on the right.

SELECTION OF THE POINTED MODELS ON THE RIGHT. If Hercules is to win the formula-complexity game, he must choose his models in a specific way.

Lemma B.1 *In any winning strategy for Hercules for an* $(\mathcal{L}_\Diamond, \langle \mathbf{L}, \mathbf{R} \rangle)$-FGF *in which* $\mathcal{A}_1 \in \mathbf{L}$ *and* $\mathcal{B} \in \mathbf{R}$,

(i) *Hercules chooses the valuation on* $\mathcal{B}$ *so that at least one literal is true in one point but not on the other, and*

(ii) *he picks the pointed model based on the irreflexive point in* $\mathcal{B}$.

The pointed model based on $\mathcal{B}$ and its irreflexive point chosen by Hercules is shown in the right half of Figure B.1. We indicate that the two points in $\mathcal{B}$ satisfy different sets of literals by making one of them black and the other white.

SELECTION OF THE POINTED MODELS ON THE LEFT. The Hydra can reply with the pointed models shown on the left of the dotted line in the right half in Figure B.1. She selects these pointed models so that two points in any two models satisfy the same set of literals iff they have the same colour. As usual, we say that she *mimics* Hercules if she chooses her pointed models in this way.

FORMULA SIZE GAME ON MODELS: Let us consider now the FGM starting with $(\mathcal{A}_1^{\mathrm{M}}, \triangleright), (\mathcal{A}_2^{\mathrm{M}}, \triangleright)$ on the left and $(\mathcal{B}^{\mathrm{M}}, \triangleright)$ on the right. We first note that the modal moves that Hercules may make have some restrictions. The following can be seen by observing that playing otherwise would produce bisimilar pointed models on each side.

Lemma B.2 *Let* $\mathbf{L}$, $\mathbf{R}$ *be classes of models so that Hercules has a winning strategy for the* $(\mathcal{L}_\Diamond, \langle \mathbf{L}, \mathbf{R} \rangle)$*-FGM and* T *a closed game tree in which the Hydra played greedily.*

(i) *If there is a game position* η *in which any pointed model based on either* $\mathcal{A}_1^{\mathrm{M}}$ *or* $\mathcal{A}_2^{\mathrm{M}}$ *is on the left and any pointed model based on* $\mathcal{B}^{\mathrm{M}}$ *is on the right, then Hercules did not play a* $\Box$*-move at* η.

(ii) *If there is a game position* η *in which* $(\mathcal{A}_1^{\mathrm{M}}, \triangleright)$ *is on the left and a pointed model based on* $\mathcal{B}^{\mathrm{M}}$ *is on the right, then Hercules did not play a* $\Diamond$*-move at* η.

From this it is easy to see that Hercules must play at least one variable.

Lemma B.3 *Suppose that* $\mathbf{L}$, $\mathbf{R}$ *are classes of models and that Hercules has a winning strategy for the* $(\mathcal{L}_\Diamond, \langle \mathbf{L}, \mathbf{R} \rangle)$*-FGM. Let* T *be a closed game tree in which the Hydra played greedily and such that there is a position* η *in which* $(\mathcal{A}_1^{\mathrm{M}}, \triangleright)$ *is on the left and* $(\mathcal{B}^{\mathrm{M}}, \triangleright)$ *is on the right. Then, the number of literal moves in* T *is at least one.*

Proof. By Lemma B.2 Hercules cannot play any $\Diamond$- or $\Box$-moves, and $\wedge$- or $\vee$-moves result in at least one new position with both of these pointed models. Since Hercules cannot play $\bot$ or $\top$, he must use at least one variable. $\Box$

With this we are ready to prove Theorem 5.1 in the case where $m = 0$.

Proof. Let $\mathbf{A}$ and $\mathbf{B}$ be as depicted in the left rectangle in Figure B.1; since the frames of $\mathbf{A}$ are n-reflexive but the ones in $\mathbf{B}$ are not, by Theorem 4.3 it suffices to show that the Hydra can play so that any closed game tree has at least one $\vee$-move, one literal move, and modal depth at least n.

Let $\mathbf{B}^{\mathrm{m}} = \{(\mathcal{B}^{\mathrm{M}}, \triangleright_{\mathcal{B}})\}$ be the singleton set of pointed models chosen by Hercules, which by Lemma B.2 must be so that the top and bottom points have different valuations, and let Hydra choose $\mathbf{A}^{\mathrm{m}}$ as depicted in the right-hand side of Figure B.1. Lemma B.2 implies that Hercules cannot begin the FGM starting with $(\mathcal{A}_1^{\mathrm{M}}, \triangleright), (\mathcal{A}_2^{\mathrm{M}}, \triangleright)$ on the left and $(\mathcal{B}^{\mathrm{M}}, \triangleright)$ on the right by playing either a $\Diamond$- or a $\Box$-move. Playing an $\wedge$-move will result in at least one new position that is the same as the previous one. Therefore, Hercules must play an $\vee$-move and he and the Hydra will have to compete in two new sub-games: the first one starting with $(\mathcal{A}_1^{\mathrm{M}}, \triangleright)$ on the left and $(\mathcal{B}^{\mathrm{M}}, \triangleright)$ on the right while the second starts with $(\mathcal{A}_2^{\mathrm{M}}, \triangleright)$ on the left and $(\mathcal{B}^{\mathrm{M}}, \triangleright)$ on the right.

By Lemma B.3 he can win the former only by playing a literal-move whereas the latter can be won only by playing a sequence of n $\Diamond$-moves that must be made in order to perform a cycle leading back to the black point in $\mathcal{A}_2$, giving us at least n ocurrences of $\Diamond$ and modal depth at least n. We can then use

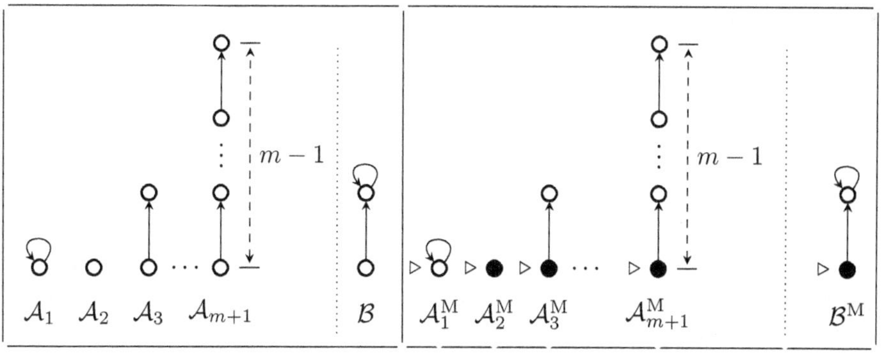

Fig. B.2. The frames $\mathcal{A}_1$, ..., $\mathcal{A}_{m+1}$ and $\mathcal{B}$ and the pointed models based on them.

Theorem 4.3 to conclude that $\overline{p} \vee \Diamond^n p$ is absolutely minimal. □

B.2 The generalized recurrence axioms

Now we treat the m-recurrence axioms, where $n = 0$. This time Hercules and the Hydra play a $(\mathcal{L}_\Diamond, \langle \mathbf{A}, \mathbf{B} \rangle)$-FGF where $\mathbf{A} = \{\mathcal{A}_1, \ldots \mathcal{A}_{m+1}\}$ while $\mathbf{B}$ contains a single element $\mathcal{B}$, as depicted in the left rectangle in Figure B.2. For $2 \leq i \leq m+1$, each $\mathcal{A}_i$ is a path of $i-2$ relation steps. Clearly, $\Diamond^m p \to p$ is valid in all the frames in $\mathbf{A}$ and it is not valid in the frame $\mathcal{B}$.

SELECTION OF THE MODELS ON THE RIGHT: It follows from Lemma B.1 that Hercules must pick the pointed model $(\mathcal{B}^{\mathrm{M}}, \triangleright)$ shown in the right half of Figure B.2. Again, to indicate that the two points of $\mathcal{B}^{\mathrm{M}}$ satisfy different sets of literals, we colour one of them black and the other white.

SELECTION OF THE POINTED MODELS ON THE LEFT: The Hydra replies with the pointed models shown on the left of the dotted line in the right half in Figure B.2. Again, she picks these pointed models so that points that satisfy the same set of literals have the same colour.

FORMULA SIZE GAME ON MODELS: Let us consider the FGM starting with $\mathbf{A}^{\mathrm{m}} = \{(\mathcal{A}_1^{\mathrm{M}}, \triangleright), \ldots, (\mathcal{A}_{m+1}^{\mathrm{M}}, \triangleright)\}$ on the left and $\mathbf{B}^{\mathrm{m}} = \{(\mathcal{B}^{\mathrm{M}}, \triangleright)\}$ on the right.

Lemma B.4 *In any closed game tree T for the $(\mathbf{A}^{\mathrm{m}}, \mathbf{B}^{\mathrm{m}})$-FGM in which the Hydra played greedily, Hercules played at least one $\vee$-move.*

Proof. Using Lemma B.2, we see that in order to win a FGM with a starting position η in which $(\mathcal{A}_1^{\mathrm{M}}, \triangleright)$ is on the left and $(\mathcal{B}^{\mathrm{M}}, \triangleright)$ is on the right, Hercules must not play either a $\Diamond$- or a $\Box$-move at η. On the other hand, for every game position ν in which there is some $(\mathcal{A}_i^{\mathrm{M}}, \triangleright)$ on the left for $2 \leq i \leq m+1$ and $(\mathcal{B}^{\mathrm{M}}, \triangleright)$ on the right, Hercules must play at least one $\Diamond$- or $\Box$-move at ν. This implies that in any FGM with a starting position in which the pointed models selected by the Hydra are on the left and $(\mathcal{B}^{\mathrm{M}}, \triangleright)$ is on the right, Hercules must play at least one $\vee$ to separate every $(\mathcal{A}_i^{\mathrm{M}}, \triangleright)$ for $2 \leq i \leq m+1$ from $(\mathcal{A}_1^{\mathrm{M}}, \triangleright)$. □

Lemma B.5 *Let* **L**, **R** *be classes of models so that Hercules has a winning strategy for the* $(\mathcal{L}_\diamond, \langle \mathbf{L}, \mathbf{R} \rangle)$*-*FGM*. Let T be a closed game tree in which the Hydra played greedily. If all $(\mathcal{A}_i^{\mathrm{M}}, \triangleright)$ for $2 \leq i \leq m{+}1$ are in* **L** *and* $(\mathcal{B}^{\mathrm{M}}, \triangleright) \in \mathbf{R}$*, Hercules must have played at least m $\Box$-moves and the modal depth of T must be at least m.*

We omit the proof, which is similar to that of Proposition 5.5.iii. With this we are ready to prove Theorem 5.1 for the case where $n = 0$.

Proof. Consider the $(\mathbf{A}, \mathbf{B})$-FGF where **A**, **B** are as depicted in Figure B.2 on the left: by Lemma B.1 Hercules must choose different valuations for the points of **B** and choose the bottom point. Let Hydra reply as depicted on the right-hand side of the figure.

By Lemma B.3 Hercules must play at least one variable, by Lemma B.4 he must play at least one $\vee$-move, by Lemma B.5 he must play at least m $\Box$-moves and modal depth at least m on the resulting FGM, and we can apply Theorem 4.3. $\Box$

References

[1] Adler, M. and N. Immerman, *An n! lower bound on formula size*, ACM Transactions on Computational Logic **4** (2003), pp. 296–314.

[2] Chagrov, A. and M. Zakharyaschev, "Modal Logic," Oxford logic guides **35**, Oxford University Press, 1997.

[3] Fernåndez-Duque, D. and P. Iliev, *Succinctness in subsystems of the spatial mu-calculus*, Journal of Applied Logics (2018).

[4] French, T., W. van der Hoek, P. Iliev and B. Kooi, *Succinctness of epistemic languages*, in: T. Walsh, editor, *Proceedings of IJCAI*, 2011, pp. 881–886.

[5] French, T., W. van der Hoek, P. Iliev and B. Kooi, *On the succinctness of some modal logics*, Artificial Intelligence **197** (2013), pp. 56–85.

[6] Hella, L. and M. Vilander, *The succinctness of first-order logic over modal logic via a formula size game*, in: *Advances in modal logic. Vol. 11*, Coll. Publ., [London], 2016 pp. 401–419.

[7] Immerman, N., "Descriptive Complexity," Graduate Texts in Computer Science, Springer-Verlag, New York, 1999.

[8] Vakarelov, D., *Modal definability in languages with a finite number of propositional variables and a new extension of the Sahlqvist's class*, Advances in Modal Logic **4** (2003), pp. 499–518.

[9] van Ditmarsch, H. and P. Iliev, *The succinctness of the cover modality*, Journal of Applied Non-Classical Logics **25** (2015), pp. 373–405.

Complexity of Dynamic Epistemic Logic with Common Knowledge

Tristan Charrier

Univ Rennes
IRISA
France

François Schwarzentruber

Univ Rennes
IRISA
France

Abstract

We consider the language of Dynamic epistemic logic with knowledge operators, common knowledge operators and dynamic operators based on event models. First, we prove that the model checking remains PSPACE-complete when common knowledge is added. Second, we prove that the satisfiability problem is 2EXPTIME-complete. We further address the model checking and the satisfiability problem for succinct inputs: we prove that complexities remain unchanged.

Keywords: Dynamic epistemic logic, common knowledge, complexity theory.

1 Introduction

Dynamic epistemic logic (DEL) [29] is a framework for reasoning about knowledge and complex actions (public announcement, public actions, private announcements, etc.). On top of that, *common knowledge* (everybody knows that everybody knows that...) is a condition for agents to act simultaneously in distributed systems [14] , in games [2] and more generally in artificial intelligence and computer science ([20], p. 45).

In this paper, we tackle both the model checking problem and the satisfiability problem of DEL with common knowledge. Both notions are relevant and have their advantages and drawbacks (see [16]).

- *Model checking.* It consists in checking whether a specification is true in a specific multi-agent system, described here by means of a Kripke model. Model checking allows for solving epistemic planning [5] but in its bounded version. For instance, one may check that formula $\varphi_{plan?}$ defined by $\langle\{e_1, e_2\}\rangle^n C_G p$ (there is a plan of length n made up of actions e_1 or

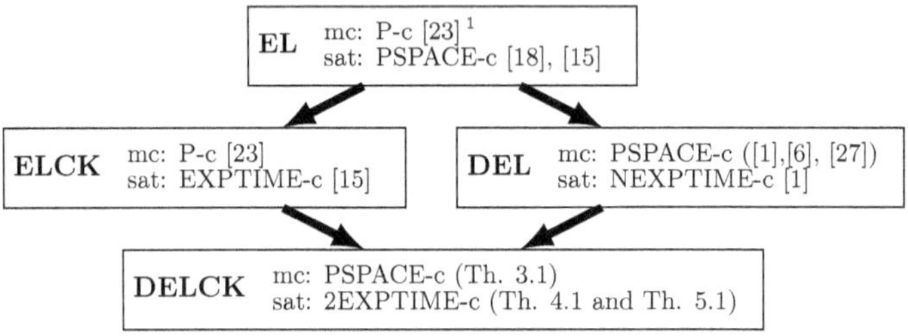

Figure 1. Complexities for the model checking (mc) and the satisfiability (sat) problem.

e_2 that leads to the common knowledge of p). Model checking has low complexity in general but we only reason about one fully-described state.

- *Satisfiability problem.* It consists in checking consistency of a specification. It allows for solving bounded planning/games not in only one initial state but in a *class* of initial states, described by a formula φ_{init}. Typically, we check that $\varphi_{init} \rightarrow \varphi_{plan?}$ is valid (by proving that the negation of it is unsatisfiable). Unfortunately, compared to model checking, the satisfiability problem is more computationally expensive in general.

The use of complex formulas is relevant for checking bounded games instead of bounded plans, by alternating diamond and box dynamic operators ($\langle\{e_1, e_2\}\rangle[\{e_3, e_4\}]C_p$). Furthermore, we can handle bounded implicit coordinated plans [12], where each agent a_i executing an event e_i knows that the rest of the plan will be correct, in the following sense: a formula of the form $K_{a_1}\langle e_1\rangle K_{a_2}\langle e_2\rangle \dots \varphi_G$ where φ_G is a goal formula, is true.

The exact complexities were first investigated in [1] but the problem was left open for common knowledge. Other papers address the model checking problem but always without the common knowledge operator ([6], [27]). Figure 1 shows the complexities of Dynamic epistemic logics (**EL**: epistemic logic without common knowledge, **ELCK**: epistemic logic with common knowledge, **DEL**: dynamic epistemic logic without common knowledge, **DELCK**: dynamic epistemic logic with common knowledge). Our paper solves open problems for the complexity of **DELCK**:

- We show that model checking remains in PSPACE when formulas contain common knowledge operators. The algorithm follows the same principle than the one given in [1]. Interestingly, common knowledge is treated via the *divide-and-conquer* paradigm [13]. Compared to [1], we also treat postconditions in event models. The contribution is a neat analysis of the complexity when event model update and common knowledge are mixed.
- We provide a double-exponential algorithm for the satisfiability problem of **DELCK** by using the method Pratt developed for Propositional Dynamic

[1] Hardness was proven for the EX, AX-fragment of temporal logic CTL, which is technically the same logic as epistemic logic K.

Logic [22].

- We show that the satisfiability problem with common-knowledge is actually 2EXPTIME-hard, already for public actions, and no constraint on the frames, by providing a reduction from the halting problem of an alternating Turing machine running in exponential space (see [7]).
- Succinctness is a key concept in complexity theory (see [21]) and is central in symbolic model checking. As shown in [10], a succinct representation for event models is also relevant for modeling complex actions such as attention-based announcements. Therefore, succinctness is even relevant for the satisfiability problem where formulas contain event models presented in a succinct way. We show that our complexity results for model checking and for the satisfiability problem still hold for succinct representations of Kripke models and event models. Incidentally, we simplify the formalism introduced in [10].

The results are proven for arbitrary models but they also hold for **S5** models, in which epistemic relations are equivalence relations. In Section 2, we define the logic **DELCK** and alternating Turing machines, which are relevant for the proofs on the other sections. In Section 3 we prove that the model checking problem against **DELCK** is PSPACE-complete. In Sections 4 and 5 we prove that the SAT problem of **DELCK** is in 2-EXPTIME and is 2-EXPTIME-hard respectively. In Section 6 we prove that the same complexity results hold when the input models are represented succinctly. Finally, in Section 7 we discuss related work and conclude.

2 Background

2.1 Background on dynamic epistemic logic

We consider a countable set of atomic propositions AP and a finite set of agents Ag. We recall that the language of epistemic logic with common knowledge (**ELCK**), denoted by $\mathcal{L}_{\mathbf{ELCK}}$, is defined by the following BNF:

$$\varphi ::= \top \mid p \mid \neg\varphi \mid (\varphi \vee \varphi) \mid K_a\varphi \mid C_G\varphi$$

with $p \in AP$, $a \in Ag$, $G \subseteq Ag$. Formula $K_a\varphi$ reads as "agent a knows that φ holds" and $C_G\varphi$ reads as "φ is common knowledge among the agents of G". As usual, we define the abbreviations $(\varphi_1 \wedge \varphi_2)$ for $\neg(\neg\varphi_1 \vee \neg\varphi_2)$, $\hat{K}_a\varphi$ for $\neg K_a\neg\varphi$ and $\hat{C}_G\varphi$ for $\neg C_G\neg\varphi$.

Definition 2.1 A *Kripke model* $\mathcal{M} = (W, (R_a)_{a\in Ag}, V)$ is defined by a non-empty set W of epistemic worlds, epistemic relations $(R_a)_{a\in Ag} \subseteq W \times W$ and a valuation function $V : W \to 2^{AP}$.

We write $R_a^{\mathcal{M}}$ for the epistemic relation for agent a in model $\mathcal{M}$. A pair $(\mathcal{M}, w)$ is called a *pointed epistemic model.* The left part of Figure 2 shows a pointed epistemic model $\mathcal{M}, w$ with two words w and u. As usual, formulas are interpreted in pointed epistemic models and we define $\mathcal{M}, w \models \varphi$ (φ is true in $\mathcal{M}, w$) by induction on φ (Boolean cases are omitted):

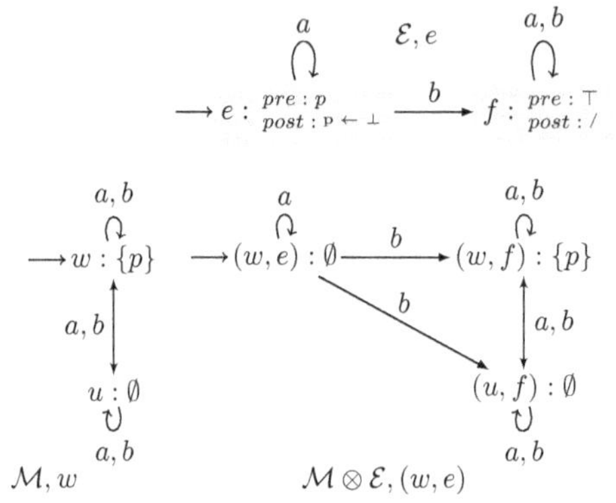

Figure 2. Example of product.

- $\mathcal{M}, w \models p$ if $p \in V(w)$;
- $\mathcal{M}, w \models K_a\varphi$ if for all $u \in W, wR_a u$ implies $\mathcal{M}, u \models \varphi$;
- $\mathcal{M}, w \models C_G\varphi$ if for all $u \in W, wR_G u$ implies $\mathcal{M}, u \models \varphi$ where R_G is the transitive and reflexive closure of $\bigcup_{a \in G} R_a$.

As usual, we define $\varphi_1 \wedge \varphi_2 = \neg(\neg\varphi_1 \vee \neg\varphi_2)$, $\varphi_1 \to \varphi_2 = \neg\varphi_1 \vee \varphi_2$ and $\varphi_1 \leftrightarrow \varphi_2 = (\varphi_1 \to \varphi_2) \wedge (\varphi_2 \to \varphi_1)$. The dynamic of the system is modeled by event models. An event model is like a Kripke model but epistemic worlds are replaced by events labeled by a precondition and a postcondition.

Definition 2.2 An *event model* $\mathcal{E} = (\mathsf{E}, (R_a^{\mathcal{E}})_{a \in Ag}, pre, post)$ is defined by a non-empty set of *events* E, epistemic relations $(R_a^{\mathcal{E}})_{a \in Ag} \subseteq \mathsf{E} \times \mathsf{E}$, a precondition function $pre : \mathsf{E} \to \mathcal{L}_{\mathbf{ELCK}}$ and a postcondition function $post : \mathsf{E} \times AP \to \mathcal{L}_{\mathbf{ELCK}}$.

A pair $(\mathcal{E}, e)$ with $e \in \mathcal{E}$ is called a *pointed event model*, where e represents the actual event. A pair $(\mathcal{E}, \mathsf{E}_0)$ with $\mathsf{E}_0 \subseteq \mathsf{E}$ is called a *multi-pointed event model*, where E_0 represents the set of possible actual events. Pointed event models correspond to deterministic actions and multi-pointed event models correspond non-deterministic actions. We may confuse $(\mathcal{E}, e)$ and $(\mathcal{E}, \{e\})$.

Example 2.3 The top of Figure 2 shows an example of a pointed event model with two events e and f. The actual event is e but agent b imagines event f as the sole possible event.

Definition 2.4 Let $\mathcal{M} = (W, (R_a)_{a \in Ag}, V)$ be a Kripke model. Let $\mathcal{E} = (\mathsf{E}, (R_a^{\mathcal{E}})_{a \in Ag}, pre, post)$ be an event model. The *product* of $\mathcal{M}$ and $\mathcal{E}$ is $\mathcal{M} \otimes \mathcal{E} = (W', (R_a)', V')$ where:

- $W' = \{(w, e) \in W \times \mathsf{E} \mid \mathcal{M}, w \models pre(e)\}$;
- $(w, e)R'_a(w', e')$ iff $wR_a w'$ and $eR_a^{\mathcal{E}} e'$;
- $V'((w, e)) = \{p \in AP \mid \mathcal{M}, w \models post(e, p)\}$.

Example 2.5 Figure 2 shows the product operation.

An event is e is said *executable* in a world w if its precondition $pre(e)$ holds in w. The language $\mathcal{L}_{\mathbf{DELCK}}$ extends $\mathcal{L}_{\mathbf{ELCK}}$ with dynamic modalities and is defined by the following BNF:

$$\varphi ::= \top \mid p \mid \neg\varphi \mid (\varphi \vee \varphi) \mid K_a\varphi \mid C_G\varphi \mid \langle \mathcal{E}, \mathsf{E}_0\rangle\varphi$$

with $p \in AP$, $a \in Ag$. Formula $\langle \mathcal{E}, \mathsf{E}_0\rangle\varphi$ reads as "There is an executable event in E_0 and φ holds after having executed it". In [29], the event models can feature any formula of $\mathcal{L}_{\mathbf{DELCK}}$, not just $\mathcal{L}_{\mathbf{ELCK}}$. The results of the paper still hold for this definition, but for the sake of simplicity, we consider event models to not feature dynamic constructions in preconditions and postconditions. We define the dual construction $[\mathcal{E}, \mathsf{E}_0]\varphi$ for $\neg\langle \mathcal{E}, \mathsf{E}_0\rangle\neg\varphi$, that is read as "For all executable events in E_0, φ holds after having executed it".

Definition 2.6 We extend the definition $\mathcal{M}, w \models \varphi$ to $\mathcal{L}_{\mathbf{DELCK}}$ with:

- $\mathcal{M}, w \models \langle \mathcal{E}, \mathsf{E}_0\rangle\varphi$ if there exists $e \in \mathsf{E}_0$ s.t. $\mathcal{M}, w \models pre(e)$ and $\mathcal{M} \otimes \mathcal{E}, (w, e) \models \varphi$.

A formula φ is satisfiable iff there exists a pointed epistemic model $\mathcal{M}, w$ such that $\mathcal{M}, w \models \varphi$. In the sequel, we suppose w.l.o.g. that, given a formula φ, there is a common $\mathcal{E}$ and dynamic operators $\langle \mathcal{E}, \mathsf{E}_0\rangle$ and $[\mathcal{E}, \mathsf{E}_0]$ are written $\langle \mathsf{E}_0\rangle$ and $[\mathsf{E}_0]$ respectively. The idea is to set $\mathcal{E}$ to be the disjoint union of all occurrences of event models in the formula. E.g. formula $\langle \mathcal{E}_1, e_1\rangle K_a p \wedge \langle \mathcal{E}_2, e_2\rangle K_a q$ is rewritten as $\langle e_1\rangle K_a p \wedge \langle e_2\rangle K_a q$ with a common event model $\mathcal{E}$ defined as the disjoint union of $\mathcal{E}_1$ and $\mathcal{E}_2$.

In the sequence, we take the abbreviation $w\overrightarrow{e}$ for $(w, e_1, \dots, e_n)$ and $\mathcal{M}\overrightarrow{\mathcal{E}}$ for $\mathcal{M}, \mathcal{E}_1, \dots, \mathcal{E}_n$. We write $\mathcal{M}\overrightarrow{\mathcal{E}}, w\overrightarrow{e} \models \varphi$ instead of $\mathcal{M} \otimes \mathcal{E}_1 \otimes \dots \otimes \mathcal{E}_n, (\dots(w, e_1), \dots, e_n) \models \varphi$. A sequence of events $\overrightarrow{e}$ is executable in w if $w\overrightarrow{e}$ is in $\mathcal{M}\overrightarrow{\mathcal{E}}$. The empty sequence of events is denoted by ϵ. The empty sequence is of course executable in all worlds.

2.2 Background on alternation

In Sections 3 and 5, we will make use of *alternation* [8]. Formally, an *alternating Turing machine* is a tuple $M = (Q, \Sigma, \Gamma, \delta, q_0, q_{acc}, q_{rej}, g)$ where:

Q is the finite set of states of M; Σ is the finite input alphabet; Γ is the finite tape alphabet with $\Sigma \subseteq \Gamma$; $\delta \subseteq Q \times \Gamma \times Q \times \Gamma \times \{-1, +1\}$ is the transition function; $q_0 \in Q$ is the initial state, $q_{acc} \in Q$ the accepting state and $q_{rej} \in Q$ is the rejecting state; $g : Q \to \{\exists, \forall\}$ is the quantification function for the states.

We suppose that only q_{acc} and q_{rej} are end states (i.e. have no outgoing transitions to other states). The execution of the machine is controlled by two players $\exists$ and $\forall$. When the current state q is existential ($g(q) = \exists$) (resp. universal ($g(q) = \forall$)), player $\exists$ (resp. $\forall$) chooses the transition to apply. A configuration is accepting if player $\exists$ has a winning strategy for reaching the accepting state. An input word is accepted by the machine if the corresponding initial configuration is *accepting*. Chandra and Stockmeyer [8] defined complexity classes with respect to alternating Turing machines. E.g. APTIME is the class of decision problems decided by an alternating Turing machine in polynomial time (the height of the computation tree is polynomial in the size of the input). They proved that APTIME = PSPACE and AEXPSPACE = 2-EXPTIME [8].

The original definition of alternating Turing machine [8] also allows for *negative states*. When the current state q is negative, the acceptance condition is negated. We can get rid off negative states without changing the definition of complexity classes (see [17], Lecture 7, Lemma 7.3, p. 47).

3 Model checking

The model checking for **DELCK**, given a pointed Kripke model $\mathcal{M}, w$, a formula Φ of $\mathcal{L}_{\mathbf{DELCK}}$, asks to decide whether $\mathcal{M}, w \models \Phi$. In this section, we prove the following theorem.

Theorem 3.1 *The model checking problem for* **DELCK** *is* PSPACE*-complete.*

Hardness comes directly from the PSPACE-hardness of the model checking of **DEL** without common knowledge [1] (actually, it is already PSPACE-hard for single-pointed event models [6], but actually even when the Kripke model is **S5** and event models are **S5** and single-pointed [27]). For the PSPACE-membership, Figure 3 provides the pseudo-code of an alternating Turing machine that decides the model checking problem for **DELCK** in polynomial time. In the pseudo-code in Figure 3, existential **(∃)** and keyword **or** (resp. universal **(∀)** choices and keyword **and**) corresponds to existential (resp. universal) states. Keyword **not** corresponds to a negated state. The upper bound is proven since PSPACE = APTIME. The machine starts by calling $mc(\mathcal{M}, w, \Phi)$. The specifications of the procedures mc, $inval$ in, rel and rel^* (see Figure 3) are given in the following proposition:

Proposition 3.2 *For all formulas φ, for all Kripke models $\mathcal{M}$, for all sequences of event models $\vec{\mathcal{E}}$, for all worlds $w\vec{e}, u\vec{f}$ of $\mathcal{M}\vec{\mathcal{E}}$, for all agents a, for all groups of agents G, for integers i that are powers of two,*

$mc(\mathcal{M}\vec{\mathcal{E}}, w\vec{e}, \varphi)$ *is accepting*	*iff* $\mathcal{M}\vec{\mathcal{E}}, w\vec{e} \models \varphi$,
$inval(p, w\vec{e}, \mathcal{M}\vec{\mathcal{E}})$ *is accepting*	*iff* $p \in V(w\vec{e})$,
$in(w\vec{e}, \mathcal{M}\vec{\mathcal{E}})$ *is accepting*	*iff* $w\vec{e} \in \mathcal{M}\vec{\mathcal{E}}$,
$rel(w\vec{e}, u\vec{f}, a, \mathcal{M}\vec{\mathcal{E}})$ *is accepting*	*iff* $(w\vec{e}, u\vec{f}) \in R_a$,
and $rel^*(w\vec{e}, u\vec{f}, G, i, \mathcal{M}\vec{\mathcal{E}})$ *is accepting*	*iff* $(w\vec{e}, u\vec{f}) \in \bigcup_{j \leq i} \left(\bigcup_{a \in G} R_a\right)^j$.

Proof The proposition is straightforwardly proven by induction since the pseudo-code directly reflects the semantics of **DELCK**. The only difficulties are:

- The induction works on the quantities given in Figure 3 and thanks to the following Lemma 3.4.
- $\bigcup_{j \leq B_{\mathcal{M},\Phi}} \left(\bigcup_{a \in G} R_a\right)^j = \left(\bigcup_{a \in G} R_a\right)^*$ since the number of worlds in $\mathcal{M}\vec{\mathcal{E}}$ in bounded by $B_{\mathcal{M},\Phi}$;
- The design of Procedure rel^* relies on the *divide and conquer* paradigm. For checking that $u\vec{f}$ is reachable by at most i $\bigcup_{a \in G} R_a$-steps from $w\vec{e}$,

procedure $mc(\mathcal{M}\overrightarrow{\mathcal{E}}, w\overrightarrow{e}, \varphi)$ $|\mathcal{M}\overrightarrow{\mathcal{E}}| + |\varphi|$

- **case** $\varphi = p$: $inval(p, \mathcal{M}\overrightarrow{\mathcal{E}}, w\overrightarrow{e})$
- **case** $\varphi = (\varphi_1 \vee \varphi_2)$: (∃) choose $i \in \{1,2\}$; $mc(\mathcal{M}\overrightarrow{\mathcal{E}}, w\overrightarrow{e}, \varphi_i)$
- **case** $\varphi = \neg\psi$: **not** $mc(\mathcal{M}\overrightarrow{\mathcal{E}}, w\overrightarrow{e}, \psi)$.
- **case** $\varphi = K_a\psi$:
  - (∀) choose $u\overrightarrow{f} \in \mathcal{M}\overrightarrow{\mathcal{E}}$
  - (∃) **not** $in(u\overrightarrow{f}, \mathcal{M}\overrightarrow{\mathcal{E}})$ **or not** $rel(w\overrightarrow{e}, u\overrightarrow{f}, a, \mathcal{M}\overrightarrow{\mathcal{E}})$
    or $mc(\mathcal{M}\overrightarrow{\mathcal{E}}, u\overrightarrow{f}, \psi)$
- **case** $\varphi = \langle \mathsf{E}_0 \rangle \psi$:
  - (∃) choose $e \in \mathsf{E}_0$; (∀) $mc(\mathcal{M}\overrightarrow{\mathcal{E}}, w\overrightarrow{e}, pre(e))$
    and $mc(\mathcal{M}\overrightarrow{\mathcal{E}} :: \mathcal{E}, w\overrightarrow{e} :: e, \psi)$.
- Case $\varphi = C_G\psi$:
  - (∀) choose $u\overrightarrow{f} \in \mathcal{M}\overrightarrow{\mathcal{E}}$
  - (∃) **not** $in(u\overrightarrow{f}, \mathcal{M}\overrightarrow{\mathcal{E}})$ **or not** $rel^*(w\overrightarrow{e}, u\overrightarrow{f}, G, B_{\mathcal{M},\Phi}, \mathcal{M}\overrightarrow{\mathcal{E}})$
    or $mc(\mathcal{M}\overrightarrow{\mathcal{E}}, u\overrightarrow{f}, \psi)$

procedure $inval(p, w\overrightarrow{e}, \mathcal{M}\overrightarrow{\mathcal{E}})$ $|\mathcal{M}\overrightarrow{\mathcal{E}}|$

- **case** $\overrightarrow{\mathcal{E}} = \epsilon$: **if** $p \in V(w)$ **then accept else reject**
- **case** $\overrightarrow{\mathcal{E}} = \overrightarrow{\mathcal{E}}'{::}\mathcal{E}$ and $w\overrightarrow{e} = w\overrightarrow{e}'{::}e$: $mc(\mathcal{M}\overrightarrow{\mathcal{E}}', w\overrightarrow{e}', post(e,p))$

Procedure $in(w\overrightarrow{e}, \mathcal{M}\overrightarrow{\mathcal{E}})$ $|\mathcal{M}\overrightarrow{\mathcal{E}}|$

- **case** $\overrightarrow{\mathcal{E}} = \epsilon$: **accept**
- **case** $\overrightarrow{\mathcal{E}} = \overrightarrow{\mathcal{E}}'{::}\mathcal{E}$ and $w\overrightarrow{e} = w\overrightarrow{e}'{::}e$: (∀) $mc(\mathcal{M}\overrightarrow{\mathcal{E}}', w\overrightarrow{e}', pre(e))$
  and $in(w\overrightarrow{e}', \mathcal{M}\overrightarrow{\mathcal{E}}')$

procedure $rel(w\overrightarrow{e}, u\overrightarrow{f}, a, \mathcal{M}\overrightarrow{\mathcal{E}})$ $|\mathcal{M}\overrightarrow{\mathcal{E}}|$

- **case** $\overrightarrow{\mathcal{E}} = \epsilon$: **if** $(w,u) \in R_a^{\mathcal{M}}$ **then accept else reject**
- **case** $\overrightarrow{\mathcal{E}} = \overrightarrow{\mathcal{E}}' :: \mathcal{E}$, $\overrightarrow{e} = \overrightarrow{e}' :: e$ and $\overrightarrow{f} = \overrightarrow{f}' :: f$:
  - (∀) $rel(w\overrightarrow{e}', u\overrightarrow{f}', a, \mathcal{M}\overrightarrow{\mathcal{E}}')$ **and if** $(e,f) \in R_a^{\mathcal{E}}$ **then accept else reject**

procedure $rel^*(w\overrightarrow{e}, u\overrightarrow{f}, G, i, \mathcal{M}\overrightarrow{\mathcal{E}})$ $|\mathcal{M}\overrightarrow{\mathcal{E}}| + \log i$

- **case** $i = 1$: **if** $u\overrightarrow{f} = w\overrightarrow{e}$ **then accept else** (∃) choose $a \in G$;
  $rel(w\overrightarrow{e}, u\overrightarrow{f}, a, \mathcal{M}\overrightarrow{\mathcal{E}})$
- **case** $i \geq 2$:
  - (∃) choose $v\overrightarrow{g} \in \mathcal{M}\overrightarrow{\mathcal{E}}$
  - (∀) $in(v\overrightarrow{g}, \mathcal{M}\overrightarrow{\mathcal{E}})$ **and** $rel^*(w\overrightarrow{e}, v\overrightarrow{g}, G, i/2, \mathcal{M}\overrightarrow{\mathcal{E}})$ **and** $rel^*(v\overrightarrow{g}, u\overrightarrow{f}, G, i/2, \mathcal{M}\overrightarrow{\mathcal{E}})$

Figure 3. Model checking procedures for **DELCK** (in gray: quantities associated to each procedure call).

we guess an intermediate world $v\vec{g} \in \mathcal{M}\vec{\mathcal{E}}$ and check that $v\vec{g}$ is reachable by at most $\frac{i}{2} \bigcup_{a\in G} R_a$-steps from $w\vec{e}$ and that $u\vec{f}$ is reachable by at most $i/2 \bigcup_{a\in G} R_a$-steps from $v\vec{g}$.

□

Better than giving a tedious and straightforward proof of Proposition 3.2, let us explain the algorithm on the example of $mc(\mathcal{M}, w, \langle\mathcal{E}, \mathsf{E}_0\rangle\neg C_G p)$.The procedure starts by choosing e in E_0. Then, we check that both $pre(e)$ holds in w and that $\neg C_G p$ holds in we. Checking that $\neg C_G p$ holds in we leads to a negated configuration: we negate the fact that $C_G p$ holds in we. It is followed by a universal choice of $u\vec{f} \in \mathcal{M}\vec{\mathcal{E}}$. For each choice of $uf \in \mathcal{M}\mathcal{E}$, we progress in an existential configuration that checks that either uf is not a world of $\mathcal{M}\mathcal{E}$, either wf is not reachable from we by at most $B_{\mathcal{M},\Phi} \bigcup_{a\in G} R_a^{\mathcal{M}\mathcal{E}}$-steps or that p in uf. Checking that p holds in uf is performed by the call of $inval(p, uf, \mathcal{M}\mathcal{E})$, which itself check that the postcondition $post(e, p)$ holds in u.

The quantities associated to each procedure call used for the proofs by induction require a careful definition. They depend on the input $(\mathcal{M}, w, \Phi)$ of the model checking problem. Let $B_{\mathcal{M},\Phi}$ be the smallest power of two that is greater than the number of worlds in Kripke model $\mathcal{M}\vec{\mathcal{E}}$ where $\vec{\mathcal{E}}$ is the list of all event models appearing in the formula Φ.

Definition 3.3 $|\mathcal{M}|$, $|\mathcal{E}|$ and $|\varphi|$ are defined by mutual induction. First, $|\mathcal{M}|$ (resp. $|\mathcal{E}|$) are the number of bits to encode Kripke model $\mathcal{M}$ (resp. event model $\mathcal{E}$). In particular, $|\mathcal{E}|$ takes into account the memory needed to store the precondition and the postcondition functions. Then, $|\mathcal{M}\vec{\mathcal{E}}|$ denotes $|\mathcal{M}| + \sum_{i=1}^{n} |\mathcal{E}_i|$. Second $|\varphi|$ denotes the length of φ, defined by induction as usual except for the two following cases:

- $|\langle\mathcal{E}, \mathsf{E}_0\rangle\varphi| := |\mathcal{E}| + 1 + |\varphi|$;
- $|C_G\varphi| := \log_2 B_{\mathcal{M},\Phi} + 1 + |\varphi|$.

Lemma 3.4 *The quantities given in gray in Figure 3 are strictly decreasing along a branch of the computation tree.*

Proof Let us discuss the following cases (the other ones are left to the reader):

- The quantity for $mc(\mathcal{M}\vec{\mathcal{E}}, w\vec{e}, C_G\varphi)$ is $|\mathcal{M}\vec{\mathcal{E}}| + |C_G\varphi| + 1 = |\mathcal{M}\vec{\mathcal{E}}| + \log_2 B_{\mathcal{M},\Phi} + |\varphi| + 1$ and is strictly greater than the quantity for $rel^*(w\vec{e}, u\vec{f}, G, B_{\mathcal{M},\Phi}, \mathcal{M}\vec{\mathcal{E}})$, which is $|\mathcal{M}\vec{\mathcal{E}}| + \log_2 B_{\mathcal{M},\Phi}$.
- The quantity for $mc(\mathcal{M}\vec{\mathcal{E}}, w\vec{e}, \langle\mathcal{E}, \mathsf{E}_0\rangle\varphi)$ is $|\mathcal{M}\vec{\mathcal{E}}| + |\mathcal{E}| + \varphi| + 1$ and is strictly greater than the quantity for $mc(\mathcal{M}\vec{\mathcal{E}}, w\vec{e}, pre(e))$, which is $|\mathcal{M}\vec{\mathcal{E}}| + |pre(e)| < |\mathcal{M}\vec{\mathcal{E}}| + |\mathcal{E}|$.
- The quantity for $inval(p, w\vec{e}, \mathcal{M}\vec{\mathcal{E}})$ is $|\mathcal{M}\vec{\mathcal{E}}'\mathcal{E}| = |\mathcal{M}\vec{\mathcal{E}}'| + |\mathcal{E}|$ and is strictly greater than the quantity for $mc(\mathcal{M}\vec{\mathcal{E}}', w\vec{e}', post(e,p))$ which $|\mathcal{M}\vec{\mathcal{E}}'| + post(e,p)$.

□

Proposition 3.5 *$mc(\mathcal{M}, w, \Phi)$ is executed in polynomial time in the size of the input $(\mathcal{M}, w, \Phi)$.*

Proof The time is bounded the height of the computation tree rooted in $mc(\mathcal{M}, w, \Phi)$. Thanks to Lemma 3.4, the height of the computation treeis bounded by the quantity associated to $mc(\mathcal{M}\overrightarrow{\mathcal{E}}, w\overrightarrow{e}, \varphi)$, that is $|\mathcal{M}| + |\varphi|$. This quantity is *not* the size of the input $(\mathcal{M}, w, \Phi)$: for instance the weight of C_G-modalities is $\log_2 B_{\mathcal{M},\Phi}$. However this quantity is polynomial in the the size of the input $(\mathcal{M}, w, \Phi)$.

At each node of the computation tree, the computation performed in a single node is polynomial. For instance, the instruction '**(∀)** choose $u\overrightarrow{f} \in \mathcal{M}\overrightarrow{\mathcal{E}}$' consists in choosing each bit of $u\overrightarrow{f}$, thus is polynomial in the size of the input.

To conclude, the execution time on each branch in the computation tree is polynomial. □

4 Upper bound of SAT

The satisfiability problem for **DELCK**, given a **DELCK**-formula Φ, asks to decide Φ is satisfiable. In this section, we prove the following upper bound result:

Theorem 4.1 *The satisfiability problem of* **DELCK** *is in* 2-EXPTIME.

In order to prove Theorem 4.1, we will proceed as for proving that Propositional Dynamic Logic is in EXPTIME and use the method of Pratt [22], but we will simulate tableau method rules of the same kind that in [1]. To ease the reading, we will w.l.o.g consider that formulas are in negative normal form, that is, negations are in front of atomic propositions, and we will use all connectives $\vee$, $\wedge$, K_a, $\hat{K}_a$, C_a, $\hat{C}_a$, $\langle e \rangle$, $[e]$. The *negation* of a formula φ is the formula in negative normal form obtained by negating all connectives, e.g. the negation of $C_G((\hat{K}_a \neg q) \wedge \langle e \rangle p)$ is formula $\hat{C}_G(K_a q \vee [e]\neg p)$. The dynamic modal depth of a formula Φ, noted $dmd(\Phi)$, is the modal depth by only counting the dynamic operators. E.g. the dynamic modal depth of $K_a[e][e']p \wedge [e']C_G q$ is 2.

Definition 4.2 The *closure*[2] of formula Φ is the set $Cl(\Phi)$ that contains elements $in_{\overrightarrow{e}}$ and $(\overrightarrow{e}, \psi)$ where $\overrightarrow{e}$ is a sequence of events in E of length at most $dmd(\Phi)$, and ψ is a subformula (or negation) of Φ or a subformula (or negation) of a precondition or postcondition in E, under the condition that $dmd(\varphi) + |\overrightarrow{e}| \leq dmd(\Phi)$.

The intended meaning of $in_{\overrightarrow{e}}$ is that the current world survives the sequence of events $\overrightarrow{e}$. The intended meaning of $(\overrightarrow{e}, \varphi)$ is that formula φ is true after having executed the sequence of events $\overrightarrow{e}$.

[2] The definition given here contains 'too many' formulas. We could have given a much more thorough definition, but the definition would have been more complicated to understand and the closure would have had the same asymptotic size.

Example 4.3 Let us take the event model $\mathcal{E}$ of Figure 2 and formula $\Phi := [e]K_a[f]q$. The closure $Cl(\Phi)$ is the set $\{in_\epsilon, in_e, in_{ef}, in_f, in_{fe}, (\epsilon, \Phi), (\epsilon, K_a[f]q), (e, K_a[f]q), \ldots\}$.

Proposition 4.4 *The size of the closure of Φ is exponential in $|\Phi|$.*

Proof There is a direct correspondence between a subformula of Φ and a node in the syntactic tree of Φ. Therefore, the number of subformulas of Φ is in $O(|\Phi|)$. The number of possible ψ is then bounded by $O(|\Phi|)$ (the size of Φ is the number of memory cells needed to write down Φ, all the information of the event model $\mathcal{E}$ included). The number of possible sequences $\overrightarrow{e}$ is $|\mathsf{E}|^{dmd(\Phi)}$, thus exponential in $|\Phi|$. □

A Hintikka set (see Definition 4.5) is a maximal subset of $Cl(\Phi)$ that is consistent with respect to propositional logic (points 2-4), common knowledge reflexivity (point 5), dynamic operators (point 6-7), executability of events (point 8-9) and postconditions (point 10).

Definition 4.5 A *Hintikka set* h over $Cl(\Phi)$ is a subset of $Cl(\Phi)$ that satisfies:

(1)	If $(\overrightarrow{e}, \varphi) \in h$	then	$in_{\overrightarrow{e}} \in h$;
(2)	$(\overrightarrow{e}, \varphi \wedge \psi) \in h$	iff	$(\overrightarrow{e}, \varphi) \in h$ and $(\overrightarrow{e}, \psi) \in h$;
(3)	$(\overrightarrow{e}, \varphi \vee \psi) \in h$	iff	$(\overrightarrow{e}, \varphi) \in h$ or $(\overrightarrow{e}, \psi) \in h$;
(4)	If $in_{\overrightarrow{e}} \in h$	then	$(\overrightarrow{e}, \varphi) \in h$ xor $(\overrightarrow{e}, \neg\varphi) \in h$[3];
(5)	If $(\overrightarrow{e}, C_G\varphi) \in h$	then	$(\overrightarrow{e}, \varphi) \in h$;
(6)	$(\overrightarrow{e}, \langle \mathsf{E}_0 \rangle \varphi) \in h$	iff	there exists $e \in \mathsf{E}_0$ s.t. $in_{\overrightarrow{e}::e} \in h$ and $(\overrightarrow{e}::e, \varphi) \in h$;[4]
(7)	$(\overrightarrow{e}, [\mathsf{E}_0]\varphi) \in h$	iff	for all $e \in \mathsf{E}_0$, we have $in_{\overrightarrow{e}::e} \in h$ implies $(\overrightarrow{e}::e, \varphi) \in h$;
(8)	$in_\epsilon \in h$;		
(9)	$in_{\overrightarrow{e}::e} \in h$	iff	$in_{\overrightarrow{e}} \in h$ and $(\overrightarrow{e}, pre(e)) \in h$;
(10)	$(\overrightarrow{e}::e, p) \in h$	iff	$(\overrightarrow{e}, post(e)(p)) \in h$.

Point (1) means that if a Hintikka set contains $(\overrightarrow{e}, \varphi)$, then it means that $\overrightarrow{e}$ should be executable (in the intuitive world represented by the Hintikka set). Point (4) means that Hintikka sets are consistent. Point (5) says that if φ is common knowledge then φ is true. Points (6) and (7) mimics the truth condition given in Definition 2.6. Point (8) means the empty sequence of events ϵ is always executable. Point (9) means that $\overrightarrow{e}::e$ is executable iff $\overrightarrow{e}$ is executable and the precondition of e holds after having executed $\overrightarrow{e}$. Point (10) means that the truth of atomic proposition p after a non-empty sequence $\overrightarrow{e}::e$ of events is given by the truth of its postcondition before the last event e.

Now, we define the following structure that takes care of the consistency of the box modalities K_a, C_G.

[4] Formula $\neg\varphi$ is the negation of φ in the following sense: the negative normal obtained by negating all connectives in φ.

[4] We explicitly mentioned $in_{\overrightarrow{e}::e} \in h$ for uniformity with the semantics. However, note that it is implied by point (1).

function isDELCK-sat?(Φ)
  Compute the Hintikka structure $\mathcal{H} := (H, (R_a)_{a \in Ag})$ for Φ
  repeat
    Remove any Hintikka set h from $\mathcal{H}$ if
    ($\hat{K}_a$) either there is $(\overrightarrow{e}, \hat{K}_a\psi) \in h$ but no $h' \in R_a(h)$ with $(\overrightarrow{e}', \psi) \in h$ with $\overrightarrow{e} \rightarrow^a \overrightarrow{e}'$ and $in_{\overrightarrow{e}'} \in h'$;
    ($\hat{C}_G$) or there is $(\overrightarrow{e}, \hat{C}_G\psi) \in h$ but no path $h = h_0 \rightarrow^{a_1} h_1 \dots h_k$ and no path $\overrightarrow{e} = \overrightarrow{e}^{(0)} \rightarrow^{a_1} \overrightarrow{e}^{(1)} \dots \rightarrow^{a_k} \overrightarrow{e}^{(k)}$ such that $(\overrightarrow{e}^{(k)}, \psi) \in h_k$ and $a_1, \dots, a_k \in G$ and $in_{\overrightarrow{e}^{(i)}} \in h_i$.
  until no more Hintikka sets are removed
  if there is still a Hintikka set in $\mathcal{H}$ containing (ϵ, Φ) **then accept**
  else reject
endFunction

Figure 4. Algorithm for the satisfiability problem of a **DELCK**-formula Φ.

Definition 4.6 The *Hintikka structure* for φ is $\mathcal{H} := (H, (R_a)_{a \in Ag})$ where:

- H is the set of all possible Hintikka sets over $Cl(\Phi)$;
- hR_ah' if the two following conditions holds:

(K_a) for all $(\overrightarrow{e}, K_a\varphi) \in h$ we have $(\overrightarrow{e}', \varphi) \in h'$ for all $\overrightarrow{e}'$ such that $\overrightarrow{e} \rightarrow^a \overrightarrow{e}'$ and $in_{\overrightarrow{e}'} \in h'$,

(C_G) for all $(\overrightarrow{e}, C_G\varphi) \in h$ we have $(\overrightarrow{e}', C_G\varphi) \in h'$ for all $\overrightarrow{e}'$ such that $\overrightarrow{e} \rightarrow^a \overrightarrow{e}'$ with $a \in G$ and $in_{\overrightarrow{e}'} \in h'$.

The size of the Hintikka structure is double-exponential in $|\varphi|$, since there are a double-exponential number of different Hintikka sets. We finish by giving the algorithm isDELCK-sat? (see Figure 4) whose **repeat...until** loop takes care of the consistency of diamond modalities, $\hat{K}_a, \hat{C}_a$. The algorithm starts with the full Hintikka structure. Points $(\hat{K}_a), (\hat{C}_a)$ remove worlds where $\hat{K}_a\psi$ and $\hat{C}_a\psi$ have no appropriate ψ-successor. We write $\overrightarrow{e} \rightarrow^a \overrightarrow{e}'$ if for all $(\overrightarrow{e}_i, \overrightarrow{e}'_i) \in R_a^{\mathcal{E}}$. Actually, the algorithm decides in double-exponential time whether a **DELCK**-formula is satisfiable (Propositions 4.7 and 4.8).

Proposition 4.7 *Algorithm isDELCK-sat? of Figure 4 runs in double-exponential time in* $|\Phi|$.

Proof The computation of $\mathcal{H}$ can be performed by brute-force: enumerate all subsets of $Cl(\Phi)$ and discard those which do not satisfy all conditions (1)-(10) of Definition 4.5. Compute R_a according to Definition 4.6. The loop is repeated at most the number of Hintikka sets in $\mathcal{H}$, that is $O(2^{2^{|\Phi|}})$ times, since at least one Hintikka set is removed or we exit the loop. Both tests $(\hat{K}_a)$ and $(\hat{C}_G)$ can be performed by depth-first search algorithm running in polynomial time in the size of the graph, that is of size double-exponential in $|\Phi|$. $\square$

Proposition 4.8 Φ *is* **DELCK**-*satisfiable iff isDELCK-sat? accepts* Φ.

Proof ($\Rightarrow$) Let $\mathcal{M}, w$ such that $\mathcal{M}, w \models \Phi$. Given a world u, we note $h(u)$

the Hintikka set obtained by taking $in_{\overrightarrow{e}}$ if $\overrightarrow{e}$ is executable in u and $(\overrightarrow{e}, \psi)$ if ψ holds in $u, \overrightarrow{e}$. We show that no Hintikka set $h(u)$ is removed from $\mathcal{H}$. In particular, $h(w)$ is not removed, and contains (ϵ, Φ) so the algorithm isDELCK-sat? accepts Φ.

($\Leftarrow$) Suppose isDELCK-sat? accepts Φ. We construct a model $\mathcal{M} = (W, (R_a)_{a \in Ag}, V)$ as follows:

- W is the set of Hintikka sets that remain in the structure at the end of the algorithm;
- R_a is the relation for agent a at the end of the algorithm;
- $V(h) = \{p \in AP \mid (\epsilon, p) \in h\}$.

The proof finishes by proving the following lemma:

Lemma 4.9 *(truth lemma) The properties $\mathcal{P}(in_{\overrightarrow{e}})$ and $\mathcal{P}((\overrightarrow{e}, \varphi))$ defined below hold:*

- *$\mathcal{P}(in_{\overrightarrow{e}})$: for all $h \in W$, $in_{\overrightarrow{e}} \in h$ iff $\overrightarrow{e}$ is executable in $\mathcal{M}, h$;*
- *$\mathcal{P}((\overrightarrow{e}, \varphi))$: for all $h \in W$, $(\overrightarrow{e}, \varphi) \in h$ iff $\mathcal{M} \otimes \mathcal{E}^{|\overrightarrow{e}|}, (h, \overrightarrow{e}) \models \varphi$.*

Proof The proof is performed by induction by assigning the following quantities: the quantity for $in_{\overrightarrow{e}}$ is $n|\mathcal{E}|$; the quantity for $(\overrightarrow{e}, \varphi)$ is $n|\mathcal{E}| + |\varphi|$ where n is the length of $\overrightarrow{e}$, $|\mathcal{E}$ and $|\varphi|$ are defined as in Definition 3.3, except that now we use the traditional clause $|C_G\varphi| := |\varphi| + 1$.

□

We conclude by applying the truth lemma (Lemma 4.9 to the Hintikka set h that contains (ϵ, Φ) and we obtain that $\mathcal{M}, h \models \Phi$. □

Remark 4.10 The 2EXPTIME upper bound also holds for the satisfiability problem of **DELCK** in **S5** Kripke models. We proceed as in [15] (p. 358). We add the following clauses to definition 4.5:

$$\text{(5')} \quad \text{If } (\overrightarrow{e}, K_a\varphi) \in h \quad \text{then} \quad (\overrightarrow{e}, \varphi) \in h;$$

We add the following clause in the definition of R_a in Definition 4.6:

$$\text{for all } \overrightarrow{e} \to^a \overrightarrow{e}', \text{ if } in_{\overrightarrow{e}} \in h \text{ and } in_{\overrightarrow{e}'} \in h' \text{ then } (\overrightarrow{e}, K_a\varphi) \in h \text{ iff } (\overrightarrow{e}', K_a\varphi) \in h'.$$

5 Lower bound of SAT

5.1 Reduction

The aim of this section is to prove the following theorem.

Theorem 5.1 *The satisfiability problem of* **DELCK** *is* 2-EXPTIME*-hard.*

Let us consider any 2-EXPTIME decision problem L. As AEXPSPACE = 2-EXPTIME [8], it is decided by an alternating Turing machine M that runs in exponential space. W.l.o.g we suppose that all executions halt [5] and no state

[5] If not, we add a double exponential counter to the machine and we abort the execution after a double exponential number of steps.

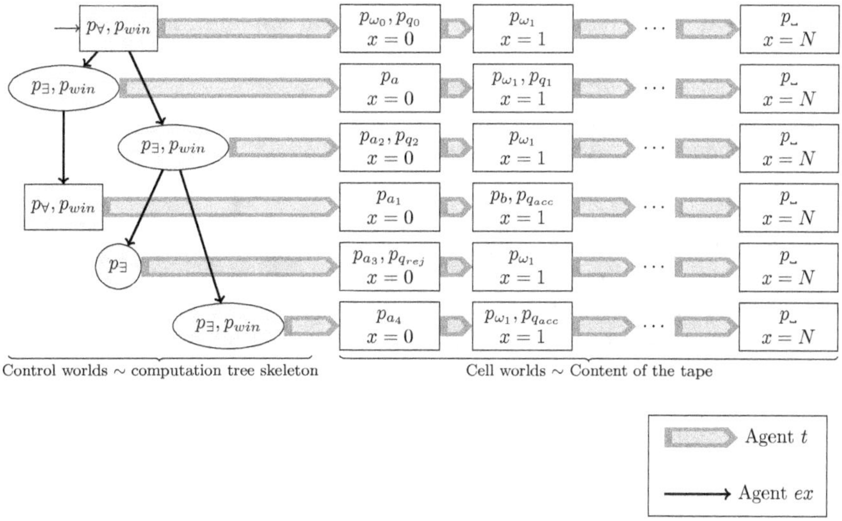

Figure 5. (Expected) Kripke model that represents the computation tree of M on the input instance ω.

is a negated state. We will define a polynomial reduction tr from L to the satisfiability problem of **DELCK**, that is tr will be computable in polynomial time, and ω is a positive instance of L if and only if $tr(\omega)$ is a satisfiable **DELCK**-formula.

The idea of $tr(\omega)$ is to enforce an expected form of a Kripke model as shown in Figure 5 that represents the computation tree of M starting with ω on the tape. The cursor of the machine remains in the N-first cell portion of the tape, where N is exponential in $|\omega|$. We define $N_0 = \log_2(N)$ for the rest of the section. N_0 is polynomial in $|\omega|$.

We introduce two agents: agent ex for the transitions in the computation tree and agent t for the linear structure of tapes. A configuration of the Turing machine is represented by a sequence of worlds linked by agent t: one so-called *control world* followed by *cell worlds*.

- The control world contains the type of the configuration: existential (resp. universal) if $p_\exists$ (resp. $p_\forall$) is true. A special atomic proposition p_{win} tags control worlds that correspond to winning configurations for player $\exists$.
- Cell worlds represent the cells of the tape and form a linear structure. They are indexed by x from $x = 0$ (left-most cell) to $x = N$ (right-most cell). In each cell world, p_a is true means that the corresponding cell contains letter $a \in \Gamma$. A proposition of the form p_q being true means that the cursor is at that cell and the current state is $q \in Q$.

Besides atomic propositions $p_\exists, p_\forall, p_a, a \in \Gamma$ and $p_q, q \in Q$, we also consider the list of atomic propositions for the bits of the cell index x: $x_1, \ldots, x_{N_0}$. We also consider another such list for another cell index v: $v_1, \ldots, v_{N_0}$. The index

v will be used to compare cell worlds of tapes of a configuration and a successor configuration different tapes during transitions.

The definition of $tr(\omega)$ needs multi-pointed event models $(\mathcal{E}^i, \mathsf{E}_0^i)$ given in Figure 6 non-deterministically and publicly choose the ith bit of value v. We also consider Boolean formulas $x \leq v$, $x = v$, $x = v - 1$ and finally $K_t x = x + 1$ (the value of $K_t x_0 \dots K_t x_{N_0}$ is equal to $x + 1$). We define the abbreviation $[\texttt{choosev}] = [\mathcal{E}^0, \mathsf{E}_0^0] \dots [\mathcal{E}^{N_0}, \mathsf{E}_0^{N_0}]$. Technically, it corresponds to non-deterministically choosing and publicly announcing a value for v.

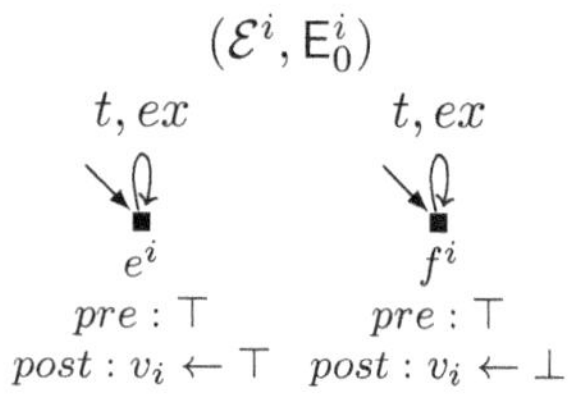

Figure 6: Multi-pointed event models $(\mathcal{E}^i, \mathsf{E}_0^i)$.

Definition 5.2 Formula $tr(\omega)$ is the conjunction of the formulas shown in Table 1.

In formulas of Table 1, common knowledge operators C_{ex} and $\hat{C}_{ex}$ are used to talk about any control world, while C_t and $\hat{C}_t$ talk about any cell world.

Importantly, notice that with **DELCK**, it is impossible to force the each world to have exactly one successor. Thus in general, the expected Kripke model is not as depicted in Figure 5. Instead, we ensure that cell worlds of depth k have the value $x = k$. We use formula (ix) that imposes the value of x to be the same in all successor cell worlds, and formula (x) saying that everywhere, $K_t x = x + 1$. Formula (xi) states that only one p_a is true in each cell world, formula (xii) states that only one p_q is true in some cell world, and formula (xiii) states that if p_q is true in a cell world, then no p'_q is true in all the t-successors. Formulas (xvii), (xviii) and (xix) define the initial tape. Transitions are ensured by formulas (xiv) to (xvi). These formulas automatically ensure that several cell worlds with the same index x have the same valuation over $x, p_a, a \in \Gamma, p_q, q \in Q$.

Formulas that handle transitions use integer v to pinpoint a cell index in the tape. It is used in formula (xiv) to tell that when the cursor is not in a cell world, then the letter remains the same during any transition. It is also used in formula (xv) to check the existence of all compatible transitions and in formula (xvi) to check that all successor control worlds and their tapes correspond to a transition.

Proposition 5.3 *$tr(\omega)$ is satisfiable if and only if ω is a positive instance of L.*

The lower bound given in Theorem 5.1 still holds for the variant of the satisfiability problem where we require the model to be **S5**, that is, epistemic relations, to be equivalence relations.

Valuations for control worlds

(i)	$C_{ex}(x=0)$	$x=0$ holds in all control worlds.
(ii)	$C_{ex}\ (p_\exists \leftrightarrow \bigvee_{q\mid g(q)=\exists} \hat{C}_t p_q)$ $\wedge\ (p_\forall \leftrightarrow \bigvee_{q\mid g(q)=\forall} \hat{C}_t p_q)$	$p_\forall$ and $p_\exists$ match the type of the state on the tape.
(iii)	$C_{ex}\left(\bigwedge_{a\in\Gamma} \neg p_a \wedge \bigwedge_{q\in Q} \neg p_q\right)$	Every p_a or p_q is false.

Winning condition

(iv)	$C_{ex}((\hat{C}_t p_{q_{acc}}) \to p_{win})$	If the current state is q_{acc} the world is marked as winning.
(v)	$C_{ex}((\hat{C}_t p_{q_{rej}}) \to \neg p_{win})$	If the current state is q_{rej} the world is marked as losing.
(vi)	$C_{ex}((p_\forall \wedge (C_t \neg p_{q_{acc}}))$ $\to (p_{win} \leftrightarrow K_{ex} p_{win}))$	If the current state is not q_{acc} and is universal, the world is marked as winning if all successor worlds are marked as winning.
(vii)	$C_{ex}((p_\exists \wedge (C_t \neg p_{q_{acc}}))$ $\to (p_{win} \leftrightarrow \hat{K}_{ex} p_{win}))$	If the current state is not q_{acc} and is existential, the world is marked as winning if one successor world is marked as winning.

Tape

(viii)	$C_{ex} K_t (x=0)$	The cell index of the left-most cell is 0.
(ix)	$C_{ex} K_t C_t \left(\bigwedge_{i=0}^{N} (K_t x_i \vee K_t \neg x_i)\right)$	On any tape world, the value of x is the same in all successors
(x)	$C_{ex} K_t C_t (K_t x = x+1)$	On any tape world, the value of x is incremented by 1 on all successors.
(xi)	$C_{ex} K_t C_t (\oplus_{a\in\Gamma} p_a)$	On any tape world, only one p_a is true and represent the current letter on the cell.
(xii)	$C_{ex} \hat{C}_t (\oplus_{q\in Q} p_q)$	On any tape, somewhere only one p_q is true
(xiii)	$C_{ex} C_t \bigwedge_{q\in Q} \left(p_q \to C_t \bigwedge_{q'\in Q} \neg p_{q'}\right)$	Anywhere, if p_q is true then no $p_{q'}$ is true anywhere on the rest of the tape.

Transitions

We define here $\varphi_{(q,a,q',b,d)} = C_t((x=v \to p_b \wedge \neg p_q) \wedge (x = v+d \to p_{q'}))$

(xiv)	$[\texttt{choosev}] C_{ex} \bigwedge_{a\in\Gamma} \hat{C}_t$ $\left(p_a \wedge \bigwedge_{q\in Q} \neg p_q \wedge x = v\right)$ $\to K_{ex} C_t (x = v \to p_a)$	On the tape, if no p_q is true and p_a is true, then at the same position on the successors' tapes, p_a is true.
(xv)	$\bigwedge_{(q,a,q',b,d)\in\delta} [\texttt{choosev}]$ $\Big(C_{ex}(\hat{C}_t(p_q \wedge p_a \wedge x = v)$ $\to \hat{K}_{ex} \varphi_{(q,a,q',b,d)})\Big)$	If there is a transition it must be present on the model.
(xvi)	$[\texttt{choosev}] \bigwedge_{a\in\Gamma} \bigwedge_{q\in Q} C_{ex}$ $\Big(\hat{C}_t(p_q \wedge p_a \wedge x = v)$ $\to K_{ex} \bigvee_{(q,a,q',b,d)\in\delta} \varphi_{(q,a,q',b,d)}\Big)$	In every world, any ex-successor must correspond to a transition.

Initial configuration

(xvii)	$\bigwedge_{i=0}^{\lvert\omega\rvert-1} C_t((x=i) \to p_{\omega(i)})$	The letters of the initial word are on the initial tape.
(xviii)	$C_t((x \geq \lvert\omega\rvert) \to p_{\_})$	Cells of index $\lvert\omega\rvert$ are blank.
(xix)	$K_t p_{q_0}$	Head in the left-most cell. Initially in the initial state.
(xx)	p_{win}	The initial control world is winning.

Table 1
Clauses of **DELCK**-formula $tr(\omega)$.

6 Succinct models for Dynamic Epistemic Logic

We define the succinct models for DEL, i.e. succinct Kripke models and event models. Such models were originally introduced in [9] and [10], but we propose a simplification of the definitions compared to their original definitions. In particular, the presentation of the new definitions of succinct Kripke models and succinct event models are neater: their sets of atomic propositions and not mixed as in [10].

6.1 Accessibility programs

Instead of describing the epistemic relations $R^a_{\mathcal{M}}$ and $R^{\mathcal{E}}_a$ respectively in Kripke models and event models in extension, we describe them in intention by using *accessibility programs.* Technically, we use Dynamic Logic with Propositional Assignments proposed by Herzig et al. ([3], [4],).

Definition 6.1 The syntax for accessibility programs is defined by the BNF $\pi ::= p{\leftarrow}\beta \mid \beta? \mid (\pi;\pi) \mid (\pi \cup \pi)$ where $p \in AP$, β is a Boolean formula.

Program $p{\leftarrow}\beta$ reads as "assign atomic proposition p to the truth value of β". Program $\beta?$ reads as "test β". Program $\pi_1;\pi_2$ reads as "execute π_1 then π_2". Program $(\pi_1 \cup \pi_2)$ reads as "either execute π_1 or π_2". We write $\mathtt{assign}(p_1, \ldots, p_n) = (p_1{\leftarrow}\bot \cup p_1{\leftarrow}\top); \ldots; (p_n{\leftarrow}\bot \cup p_n{\leftarrow}\top)$ for the program setting arbitrary values to $p_1, \ldots, p_n$.

Definition 6.2 The semantics of π is the binary relation over valuations defined by induction on π as follows, where $\mathtt{w}$ and $\mathtt{u}$ are valuations:

- $\mathtt{w} \xrightarrow{p\leftarrow\beta} \mathtt{u}$ if $(\mathtt{u} = \mathtt{w}\backslash\{p\}$ and $\mathtt{w} \not\models \beta)$ or $(\mathtt{u} = \mathtt{w} \cup \{p\}$ and $\mathtt{w} \models \beta)$;
- $\mathtt{w} \xrightarrow{\beta?} \mathtt{u}$ if $\mathtt{w} = \mathtt{u}$ and $\mathtt{w} \models \beta?$;
- $\mathtt{w} \xrightarrow{\pi_1;\pi_2} \mathtt{u}$ if there exists a valuation $\mathtt{v}$ such that $\mathtt{w} \xrightarrow{\pi_1} \mathtt{v}$ and $\mathtt{v} \xrightarrow{\pi_2} \mathtt{u}$;
- $\mathtt{w} \xrightarrow{\pi_1\cup\pi_2} \mathtt{u}$ if $\mathtt{w} \xrightarrow{\pi_1} \mathtt{u}$ or $\mathtt{w} \xrightarrow{\pi_2} \mathtt{u}$;

6.2 Succinct Kripke models

From now on, we suppose that we have a set AP to define the formulas.

Definition 6.3 A *succinct Kripke model* is a tuple $\mathfrak{M} = \langle AP_{\mathcal{M}}, \beta_{\mathcal{M}}, (\pi_a)_{a\in Ag}\rangle$ where $AP_{\mathcal{M}} \supseteq AP$ is a finite set of atomic propositions, $\beta_{\mathcal{M}}$ is a Boolean formula over $AP_{\mathcal{M}}$, and π_a is a program over $AP_{\mathcal{M}}$ for each agent a.

The Boolean formula β_M succinctly describes the set of epistemic states. Intuitively, each π_a succinctly describes the accessibility relation $\rightarrow_a$ for an agent a. A pointed succinct Kripke model is a pair $\mathfrak{M}, \mathtt{w}$ where $\mathfrak{M} = \langle AP_{\mathcal{M}}, \beta_{\mathcal{M}}, (\pi_a)_{a\in Ag}\rangle$ is a succinct Kripke model and $\mathtt{w}$ is a valuation satisfying $\beta_{\mathcal{M}}$.

Definition 6.4 Given a succinct Kripke model $\mathfrak{M} = \langle AP_{\mathcal{M}}, \beta_{\mathcal{M}}, (\pi_a)_{a\in Ag}\rangle$, the Kripke model represented by $\mathfrak{M}$, noted $\hat{\mathcal{M}}(\mathfrak{M})$ is the model $\mathcal{M} = (W, (R_a)_{a\in Ag}, V)$ where $W = \{\mathtt{w} \in \mathcal{V}(AP_{\mathcal{M}}) \mid \mathtt{w} \models \beta_{\mathcal{M}}\}$; $R_a = \left\{(\mathtt{w}, \mathtt{u}) \in W^2 \mid \mathtt{w} \xrightarrow{\pi_a} \mathtt{u}\right\}$; $V(\mathtt{w}) = \mathtt{w}$.

Example 6.5 In the muddy children example [19], each child does not know whether she is muddy or not, but knows the muddiness of the other children. If m_a is a proposition for "child a is muddy", a succinct Kripke model is $\mathfrak{M} = \langle AP_{\mathcal{M}}, \beta_{\mathcal{M}}, (\pi_a)_{a \in Ag} \rangle$ with $AP_{\mathcal{M}} = \{m_a, a \in Ag\}$, $\beta_{\mathcal{M}} = \top$ and $\pi_a = \mathtt{assign}(m_a)$. This representation is polynomial in the number of agents $|Ag|$ whereas the non-succinct Kripke model has exponential size in $|Ag|$.

Any Kripke model can be represented as a succinct Kripke model of polynomial size in the worst case. To do so, we define a succinct Kripke model $\mathfrak{M}_{\mathcal{M}}$ representing the Kripke model $\mathcal{M}$ with respect to a set of propositions AP.

Definition 6.6 Let $\mathcal{M} = (W, (R_a)_{a \in Ag}, V)$ be a Kripke model. We define the *succinct Kripke model* $\mathfrak{M}_{\mathcal{M}} = \langle AP_{\mathcal{M}}, \beta_{\mathcal{M}}, (\pi_a)_{a \in Ag} \rangle$ where: $AP_{\mathcal{M}} = AP \cup \{p_w \mid w \in W\}$; $\beta_{\mathcal{M}} = \exists!(\{p_w \mid w \in W\}) \wedge \bigwedge_{w \in W} p_w \rightarrow desc(V(w))$; $\pi_a = \bigcup_{wR_a u} p_w?; \mathtt{assign}(AP_{\mathcal{M}}); p_u?$.

Example 6.7 The Kripke model $\mathcal{M}$ from Figure 2 is modeled by the succinct Kripke model $\mathfrak{M}_{\mathcal{M}} = \langle AP_{\mathcal{M}}, \beta_{\mathcal{M}}, (\pi_a)_{a \in Ag} \rangle$ with $AP_{\mathcal{M}} = \{p, p_w, p_u\}$, $\beta_{\mathcal{M}} = \exists!(\{p_u, p_w\}) \wedge (p_w \rightarrow p) \wedge (p_u \rightarrow \neg p)$ and $\pi_a = \bigcup_{w_1, w_2 \in W} p_{w_1}?; \mathtt{assign}(AP_{\mathcal{M}}); p_{w_2}?$.

6.3 Succinct event models

We define succinct event models in the same spirit than succinct Kripke models.

Definition 6.8 A *succinct event model* is a tuple $\mathfrak{E} = \langle AP_{\mathcal{E}}, \chi_{\mathcal{E}}, (\pi_{a,\mathcal{E}})_{a \in Ag}, \mathsf{pre}, \mathsf{post} \rangle$ where $AP_{\mathcal{E}}$ is a set of atomic propositions disjoint from AP; $\chi_{\mathcal{E}}$ is a propositional formula over $AP_{\mathcal{E}}$ characterizing the set of events; $\pi_{a,\mathcal{E}}$ is a program over $AP_{\mathcal{E}}$ for all $a \in Ag$; pre is a propositional formula over $AP_{\mathcal{E}} \cup \mathcal{L}_{\mathbf{EL}}(AP)$ (meaning that any atom from $AP_{\mathcal{E}}$ cannot be under the scope of a K or a C operator); For all $p \in AP$, $\mathsf{post}(p)$ is a propositional formula over $AP_{\mathcal{E}} \cup \mathcal{L}_{\mathbf{EL}}(AP)$.

Definition 6.9 Given a succinct event model $\mathfrak{E} = \langle AP_{\mathcal{E}}, \chi_{\mathcal{E}}, (\pi_{a,\mathcal{E}})_{a \in Ag}, \mathsf{pre}, \mathsf{post} \rangle$, the event model represented by $\mathfrak{E}$, noted $\hat{\mathcal{E}}(\mathfrak{E})$ is the model $(\mathsf{E}, (R_a^{\mathcal{E}})_{a \in Ag}, pre, post)$ on AP where $\mathsf{E} = \{\mathrm{v_e} \in \mathcal{V}(AP_{\mathcal{E}}) \mid \mathrm{v_e} \models \chi\}$;$R_a^{\mathcal{E}} = \{(\mathrm{v_e}, \mathrm{v_e}') \mid \mathrm{v_e} \xrightarrow{\pi_{a,\mathcal{E}}} \mathrm{v_e}'\}$; $pre(\mathrm{v_e}) = \mathsf{pre} \wedge desc(\mathrm{v_e})$; $post(\mathrm{v_e}, p) = \mathsf{post}(p) \wedge desc(\mathrm{v_e})$.

Definition 6.10 Let $\mathcal{E} = (\mathsf{E}, (R_a^{\mathcal{E}})_{a \in Ag}, pre, post)$ be an event model on AP. We define the succinct event model $\mathfrak{E}_{\mathcal{E}} = \langle AP_{\mathcal{E}}, \chi_{\mathcal{E}}, (\pi_{a,\mathcal{E}})_{a \in Ag}, \mathsf{pre}, \mathsf{post} \rangle$ where $AP_{\mathcal{E}} = \{p_e \mid e \in \mathsf{E}\}$; $\chi_{\mathcal{E}} = \exists!(AP_{\mathcal{E}})$; $\pi_{a,\mathcal{E}} = \bigcup_{eR_a^{\mathcal{E}} f} p_e?; p_e \leftarrow \bot; p_f \leftarrow \top$; $\mathsf{pre} = \bigwedge_{e \in \mathsf{E}}(p_e \rightarrow pre(e))$; $\mathsf{post}(p) = \bigwedge_{e \in \mathsf{E}}(p_e \rightarrow post(e, p))$.

6.4 Complexity of decision problems

Naturally, the pointed Kripke model is replaced by a pointed succinct Kripke model in the model checking. Formulas contain dynamic modalities $\langle \mathfrak{E}, \beta_0 \rangle$ where $\mathfrak{E}, \beta_0$ is a pointed succinct event model, instead of $\langle \mathcal{E}, \mathsf{E}_0 \rangle$. This new language is called $\mathcal{L}^{suc}_{\mathbf{DELCK}}$. We translate a succinct formula $\varphi \in \mathcal{L}^{suc}_{\mathbf{DELCK}}$ into a formula $\tau(\varphi)$ as follows:

- $\tau(\langle \mathfrak{E}, \beta_0 \rangle \varphi) := \langle \hat{\mathcal{E}}(\mathfrak{E}), \{\mathrm{v_e} \in \mathcal{V}(AP_{\mathcal{E}}) \mid \mathrm{v_e} \models \beta_0\} \rangle \tau(\varphi)$.

Interestingly, the upper complexities of both the model checking problem and the satisfiability problem remain the same in the succinct case: the core reason is that $\underbrace{2^{poly(n)} \times \ldots 2^{poly(n)}}_{n \text{ times}} = 2^{poly(n)}$. Technical details are omitted due to space restriction.

7 Conclusion

Complexity results for dynamic epistemic logic with common knowledge was left open since the first complexity results in 2013 [1]. In this paper, we proved that the model checking of **DELCK** remains in PSPACE, even for succinct models. This result somehow justifies that BDD techniques are applicable for solving the model checking problem of **DELCK** in practice, as done in the tool DEMO [25].

We proved that the satisfiability problem of **DELCK** is 2-EXPTIME-hard. Actually, we only need trivial preconditions, Boolean postconditions and multi-pointed event models to obtain this lower bound. As a direct corollary, it implies that the satisfiability problem of the logic defined in [30] that contains knowledge, common knowledge operators, public assignments and non-deterministic $\cup$ over programs is also 2-EXPTIME-hard. The fall in PSPACE of the logic of public announcement and public assignment given in [28] is to due the absence of the common knowledge operator in their specification language. There is a long avenue of research to classify fragments of **DELCK** and evaluate the exact complexity of them. For instance, the exact of complexity of **DELCK** where event models are non-ontic (no postconditions) is an open question.

We also proved that the satisfiability problem of **DELCK** is in 2-EXPTIME. The proof technique is an adaptation of Pratt's technique for proving that **PDL** is in EXPTIME. Actually **DELCK** required such a deep machinery and more direct proofs (for instance via reduction axioms [26]) were not successful. We could infer a tableau method from our adaptation of Pratt's technique. We claim that one of the tableau rule requires an unbounded number of non-deterministic choices (actually exponential in the size of the input formula). Unfortunately, as far as we know, generic available tableau method provers, as Mettel2 [24] or Lotrec [11], only allows for a fixed number of non-deterministic choice in a given rule and do not provide any mechanism for allowing tableau rules with an unbounded number of non-deterministic choices.

Acknowledgments. We would like to thank Andreas Herzig, Renate A. Schmidt and Dmitry Tishkovsky for email discussions about tableau provers, and the anonymous reviewers for their useful comments.

References

[1] Aucher, G. and F. Schwarzentruber, *On the complexity of dynamic epistemic logic*, in: *Proceedings of the 14th Conference on Theoretical Aspects of Rationality and Knowledge (TARK 2013), Chennai, India*, 2013.
URL http://www.tark.org/proceedings/tark_jan7_13/p19-aucher.pdf

[2] Aumann, R. J., *Interactive epistemology I: Knowledge*, Int. J. Game Theory **28** (1999), pp. 263–300.
URL http://dx.doi.org/10.1007/s001820050111

[3] Balbiani, P., A. Herzig, F. Schwarzentruber and N. Troquard, *DL-PA and DCL-PC: Model checking and satisfiability problem are indeed in PSPACE*, CoRR **abs/1411.7825** (2014).
URL http://arxiv.org/abs/1411.7825

[4] Balbiani, P., A. Herzig and N. Troquard, *Dynamic logic of propositional assignments: A well-behaved variant of PDL*, in: *LICS*, 2013, pp. 143–152.

[5] Bolander, T. and M. B. Andersen, *Epistemic planning for single and multi-agent systems*, Journal of Applied Non-Classical Logics **21** (2011), pp. 9–34.
URL http://dx.doi.org/10.3166/jancl.21.9-34

[6] Bolander, T., M. H. Jensen and F. Schwarzentruber, *Complexity results in epistemic planning*, in: *Proceedings of the Twenty-Fourth International Joint Conference on Artificial Intelligence, IJCAI 2015, Buenos Aires, Argentina*, 2015, pp. 2791–2797.
URL http://ijcai.org/Abstract/15/395

[7] Chandra, A. K., D. Kozen and L. J. Stockmeyer, *Alternation*, J. ACM **28** (1981), pp. 114–133.
URL http://doi.acm.org/10.1145/322234.322243

[8] Chandra, A. K. and L. J. Stockmeyer, *Alternation*, in: *Foundations of Computer Science, 1976., 17th Annual Symposium on*, IEEE, 1976, pp. 98–108.

[9] Charrier, T. and F. Schwarzentruber, *Arbitrary public announcement logic with mental programs*, in: *Proceedings of the 2015 International Conference on Autonomous Agents and Multiagent Systems, AAMAS 2015, Istanbul, Turkey*, 2015, pp. 1471–1479.
URL http://dl.acm.org/citation.cfm?id=2773340

[10] Charrier, T. and F. Schwarzentruber, *A succinct language for dynamic epistemic logic*, in: *Proceedings of the 16th Conference on Autonomous Agents and MultiAgent Systems, AAMAS 2017, São Paulo, Brazil*, 2017, pp. 123–131.
URL http://dl.acm.org/citation.cfm?id=3091148

[11] del Cerro, L. F., D. Fauthoux, O. Gasquet, A. Herzig, D. Longin and F. Massacci, *Lotrec: The generic tableau prover for modal and description logics*, in: *Automated Reasoning, First International Joint Conference, IJCAR 2001, Siena, Italy, Proceedings*, 2001, pp. 453–458.
URL https://doi.org/10.1007/3-540-45744-5_38

[12] Engesser, T., T. Bolander, R. Mattmüller and B. Nebel, *Cooperative epistemic multi-agent planning with implicit coordination*, in: *Proceedings of the ICAPS-2015 Workshop on Distributed and Multi-Agent Planning (DMAP 2015)*, 2015.

[13] Gasquet, O., V. Goranko and F. Schwarzentruber, *Big brother logic: Visual-epistemic reasoning in stationary multi-agent systems*, Autonomous Agents and Multi-Agent Systems **30** (2016), pp. 793–825.
URL https://doi.org/10.1007/s10458-015-9306-4

[14] Halpern, J. Y. and Y. Moses, *Knowledge and common knowledge in a distributed environment*, J. ACM **37** (1990), pp. 549–587.
URL http://doi.acm.org/10.1145/79147.79161

[15] Halpern, J. Y. and Y. Moses, *A guide to completeness and complexity for modal logics of knowledge and belief*, Artif. Intell. **54** (1992), pp. 319–379.
URL https://doi.org/10.1016/0004-3702(92)90049-4

[16] Halpern, J. Y. and M. Y. Vardi, *Model checking vs. theorem proving: A manifesto*, in: *Proceedings of the 2nd International Conference on Principles of Knowledge Representation and Reasoning (KR'91). Cambridge, MA, USA*, 1991, pp. 325–334.

[17] Kozen, D., "Theory of Computation," Texts in Computer Science, Springer, 2006.
URL `https://doi.org/10.1007/1-84628-477-5`

[18] Ladner, R. E., *The computational complexity of provability in systems of modal propositional logic*, SIAM J. Comput. **6** (1977), pp. 467–480.
URL `https://doi.org/10.1137/0206033`

[19] McCarthy, J., *Formalization of two puzzles involving knowledge*, Unpublished note, Stanford University (1978).

[20] Meyer, J.-J. C. and W. Van Der Hoek, "Epistemic Logic for AI and Computer Science" **41**, Cambridge University Press, 1995.

[21] Papadimitriou, C. H., "Computational Complexity," Academic Internet Publ., 2007.

[22] Pratt, V. R., *A near-optimal method for reasoning about action*, J. Comput. Syst. Sci. **20** (1980), pp. 231–254.
URL `http://dx.doi.org/10.1016/0022-0000(80)90061-6`

[23] Schnoebelen, P., *The complexity of temporal logic model checking*, in: *Advances in Modal Logic 4, papers from the fourth conference on "Advances in Modal logic," held in Toulouse (France)*, 2002, pp. 393–436.

[24] Tishkovsky, D., R. A. Schmidt and M. Khodadadi, *The tableau prover generator mettel2*, in: *Logics in Artificial Intelligence - 13th European Conference, JELIA 2012, Toulouse, France*, 2012, pp. 492–495.
URL `https://doi.org/10.1007/978-3-642-33353-8_41`

[25] van Benthem, J., J. van Eijck, M. Gattinger and K. Su, *Symbolic model checking for dynamic epistemic logic*, in: *Logic, Rationality, and Interaction - 5th International Workshop, LORI 2015 Taipei, Taiwan*, 2015, pp. 366–378.
URL `https://doi.org/10.1007/978-3-662-48561-3_30`

[26] van Benthem, J., J. van Eijck and B. P. Kooi, *Logics of communication and change*, Inf. Comput. **204** (2006), pp. 1620–1662.
URL `http://dx.doi.org/10.1016/j.ic.2006.04.006`

[27] van de Pol, I., I. van Rooij and J. Szymanik, *Parameterized complexity results for a model of theory of mind based on dynamic epistemic logic*, in: *Proceedings Fifteenth Conference on Theoretical Aspects of Rationality and Knowledge, TARK 2015, Carnegie Mellon University, Pittsburgh, USA*, 2015, pp. 246–263.
URL `https://doi.org/10.4204/EPTCS.215.18`

[28] van Ditmarsch, H., A. Herzig and T. D. Lima, *Public announcements, public assignments and the complexity of their logic*, Journal of Applied Non-Classical Logics **22** (2012), pp. 249–273.
URL `https://doi.org/10.1080/11663081.2012.705964`

[29] Van Ditmarsch, H., W. van Der Hoek and B. Kooi, "Dynamic Epistemic Logic" **337**, Springer Science & Business Media, 2007.

[30] van Ditmarsch, H. P., W. van der Hoek and B. P. Kooi, *Dynamic epistemic logic with assignment*, in: *4th International Joint Conference on Autonomous Agents and Multiagent Systems (AAMAS 2005), Utrecht, The Netherlands*, 2005, pp. 141–148.
URL `http://doi.acm.org/10.1145/1082473.1082495`

Dependence Statements Are Strict Conditionals

Ivano Ciardelli

Munich Center for Mathematical Philosophy, Ludwig-Maximilians-Universität, Munich

Abstract

In this paper I discuss dependence statements like "*whether p determines whether q*". I propose to analyze such statements as involving a generalized strict conditional operator applied to two questions—a determining question and a determined one. The dependence statement is true or false at a world w according to whether, relative to the set of successors of w, every answer to the former yields an answer to the latter. This motivates an investigation of strict conditionals in the context of inquisitive logic. A sound and complete axiomatization of the resulting logic is established, both for the class of all Kripke frames, and for various notable frame classes.

Keywords: Dependency, inquisitive logic, strict conditional, questions

1 Introduction

Consider a process which may have different outcomes, leading to different truth values for the sentences p, q, r. I will be concerned with dependence statements like (1-a) and sentences that embed such statements, like (1-b-e). [1]

(1)
   a. Whether q is determined by whether p.
   b. Whether q is not determined by whether p.
   c. Whether q is determined by whether p, or it is determined by whether r.
   d. Alice knows that whether q is determined by whether p.
   e. Is it the case that whether q is determined by whether p?

The main question I will address is how statements like (1-a) are best formalized in a logical language so that we get a satisfactory analysis of these statements as well as of compound sentences like (1-b-e) in which these statements are embedded under connectives, modalities, and a polar question operator.

The kind of dependency that is involved in these statements is an interesting logical notion, which plays an important role in science. In recent years, this

[1] In agreement with the literature (e.g., [10]), I take the verb *determine* in (1) to mean *completely determine*, and to include the limit case in which the dependency is trivial (e.g., q is bound to be true regardless of p).

notion has come under attention in logic from the perspective of two historically independent but formally related traditions. The first tradition is associated with *dependence logic*. Originating with Väänänen's work on dependencies between bound variables in predicate logic [18], dependence logic evolved into a general theory of logical systems equipped with the tools to reason about dependency relations. In particular, the system of propositional dependence logic [21] and its extension with modal operators [19,9] are equipped with formulas expressing dependencies between truth-values of sentences. E.g., the formula $=(p,q)$ expresses the fact that the truth-value of q is functionally determined by the truth-value of p. In order to interpret such formulas, dependence logic uses a non-standard semantics in which formulas are interpreted not with respect to single possible worlds, but rather with respect to sets of worlds, called *teams*.

The second tradition is associated with *inquisitive logic*. Originating from a line of work that aimed at a uniform analysis of statements and questions [11,12,14,7,5], inquisitive logic evolved into a full-blown theory of logic in a context encompassing both kinds of sentences (for an overview, see [3]). In this tradition, too, formulas are interpreted not with respect to single possible worlds, representing states of affairs, but with respect to sets of worlds, which are viewed as states of information. The idea is that the meaning of a question is given not by specifying in what states of affairs it is true, but rather by specifying what information is needed to resolve it. As for statements, their meaning can be given by specifying what information is needed to establish them. Thus, in an information-based semantics statements and questions can be interpreted in terms of a single semantic relation, called *support*. A fundamental connection between questions and dependency was established in [2,4]: dependency is nothing but question entailment in context. E.g., the dependency expressed by (1-a) amounts to the fact that, relative to the set of possible outcomes, the question whether p entails the question whether q. Due to a systematic connection between contextual entailment and the implication connective, this entailment can be expressed within the language by means of the formula $?p \to ?q$, which is equivalent to the dependence logic formula $=(p,q)$. Various advantages of this modular analysis of dependencies are discussed in [2] and in Ch. 5 of [3].

Nevertheless, in this paper I argue that the formula $?p \to ?q$ (and, thus, also the equivalent dependence logic formula $=(p,q)$) is not the right way to formalize a dependence statement like (1-a). Rather, I propose that dependence statements are modal statements. The appropriate semantic structures to interpret such statements are Kripke models, where each possible world w is equipped with a corresponding set $R[w]$ of possibilities. A dependence statement is essentially a strict conditional, which is true or false at a world w depending on some global features of the set $R[w]$. It is, however, a strict conditional whose antecedent and consequent are questions. Thus, e.g., (1-a) should be formalized as the formula $?p \Rightarrow ?q$, which is true at w just in case the question $?p$ entails the question $?q$ relative to $R[w]$. Making this idea work requires extending the strict conditional operator to a logic equipped with questions. In this paper I define and investigate such an extension.

The paper is structured as follows: Section 2 provides some background on propositional inquisitive logic, InqB; Section 3 describes a number of problems that arise if we analyze dependence statements as InqB-implications; Section 4 shows that these problems disappear if dependence statements are analyzed as strict conditionals involving questions; Section 5 investigates the logic of strict implication in the inquisitive setting, leading up to a sound and complete axiomatization; Section 6 provides a comparison with a recent related proposal.

2 Propositional inquisitive logic

Propositional inquisitive logic, InqB, can be seen as a conservative extension of classical propositional logic with questions. The language $\mathcal{L}$ of the system is given by the following BNF, where p ranges over a set $\mathcal{P}$ of atomic sentences:

$$\varphi ::= p \mid \bot \mid \varphi \wedge \varphi \mid \varphi \rightarrow \varphi \mid \varphi \vee\!\!\!\vee \varphi$$

Formulas which do not contain occurrences of $\vee\!\!\!\vee$ are called *classical formulas.* The set $\mathcal{L}_c$ of classical formulas can be identified with the standard language of propositional logic. Negation and classical disjunction are defined by setting:

$$\neg\varphi := \varphi \rightarrow \bot \qquad \varphi \vee \psi := \neg(\neg\varphi \wedge \neg\psi)$$

The connective $\vee\!\!\!\vee$, called *inquisitive disjunction*, allows us to form questions. In particular, the formula $?p := p \vee\!\!\!\vee \neg p$ can be seen as a representation of the polar question *whether* p. In general, we set:

$$?\varphi := \varphi \vee\!\!\!\vee \neg\varphi$$

A model for InqB is a pair $M = \langle W, V \rangle$ consisting of a set W of possible worlds and a valuation function $V : W \times \mathcal{P} \rightarrow \{0, 1\}$. In order to interpret statements and questions in a uniform logical framework, InqB interprets formulas not in terms of truth relative to a possible world, but in terms of *support* relative to a set of worlds. Following a tradition that goes back to the work of Hintikka [13], such a set is referred to as an *information state*: the idea is that a set $s \subseteq W$ encodes a body of information which is compatible with the actual world being one of the worlds $w \in s$, and incompatible with it being one of the worlds $w \notin s$. The support relation for InqB is defined recursively as follows.

Definition 2.1 [Support for InqB]

- $M, s \models p$ iff $\forall w \in s : V(w, p) = 1$
- $M, s \models \bot$ iff $s = \emptyset$
- $M, s \models \varphi \wedge \psi$ iff $M, s \models \varphi$ and $M, s \models \psi$
- $M, s \models \varphi \vee\!\!\!\vee \psi$ iff $M, s \models \varphi$ or $M, s \models \psi$
- $M, s \models \varphi \rightarrow \psi$ iff $\forall t \subseteq s : M, t \models \varphi$ implies $M, t \models \psi$

The support relation has the following basic properties: the first says that support is preserved as information increases; the second says that every formula is supported by the inconsistent information state, $s = \emptyset$.

- Persistency: if $M, s \models \varphi$ and $t \subseteq s$, then $M, t \models \varphi$;
- Semantic *ex-falso*: $M, \emptyset \models \varphi$ for all $\varphi \in \mathcal{L}$.

The derived clauses for negation and classical disjunction can be given naturally in terms of the relation $s \between \varphi$ of *compatibility* between a state s and a formula φ, which holds in case s can be extended consistently to a state t that supports φ:

- $s \between \varphi$ iff there exists a non-empty $t \subseteq s$ such that $t \models \varphi$
- $s \models \neg\varphi$ iff it is not the case that $s \between \varphi$
- $s \models \varphi \vee \psi$ iff for all non-empty $t \subseteq s$, $t \between \varphi$ or $t \between \psi$

Truth at a world w is recovered as a special case of support, i.e., support at $\{w\}$.

Definition 2.2 [Truth]
We say that φ is true at world w in M, notation $M, w \models \varphi$, if $M, \{w\} \models \varphi$. The truth-set of φ in M is the set $|\varphi|_M = \{w \in M \mid M, w \models \varphi\}$.

For some formulas, support at a state s amounts to truth at each world $w \in s$. Formulas with this property are said to be *truth-conditional* (or *flat*, in the dependence logic literature). They can be viewed as corresponding to statements, whose semantics is completely determined by their truth-conditions.

Definition 2.3 [Truth-conditionality]
We say that φ is *truth-conditional* if for every model M and state s:
$M, s \models \varphi$ iff $\forall w \in s : M, w \models \varphi$

It is easy to verify by induction that all classical formulas are truth-conditional. Moreover, the truth-conditions that Definition 2.2 assigns to them are just the ones familiar from classical logic. Thus, for classical formulas, the support semantics given here is essentially equivalent to the standard truth-conditional semantics. In this sense, when restricted to statements, the above definition just provides an alternative semantics for classical propositional logic. The benefit of this alternative semantic setup is that it allows us to interpret not only statements, but also *questions*, which we take to be formulas of $\mathcal{L}$ which are not truth-conditional (on the relation between questions and truth-conditions, see Ch. 1 of [3]). For an example, take the formula $?p := p \mathbin{\vee\!\!\!\vee} \neg p$. We have:

$$M, s \models ?p \text{ iff } s \subseteq |p|_M \text{ or } s \subseteq |\neg p|_M$$

That is, $?p$ is supported at s just in case the information available in s settles whether or not p is true, i.e., resolves the question whether p. In general, if μ is a question, the relation $s \models \mu$ captures the fact that s contains enough information to resolve μ.

Now let us come back to the relation of dependency discussed in the beginning of the paper. Let s be an information state. It is natural to say that a question λ *determines* a question μ relative to s if in context of s, resolving λ implies resolving μ; that is, if any way of extending s to a stronger state $t \subseteq s$ that supports λ leads to a state that also supports μ. Generalizing to multiple determining questions, we say that a set of questions Λ determines a question

μ in s if extending s to a state $t \subseteq s$ that supports all questions in Λ leads to a state that also supports μ. This relation of dependency can be seen as the special case of a more general relation of contextual entailment (see [3,4]).

Definition 2.4 [Contextual entailment]
A set of formulas Φ entails a formula ψ relative to an information state s, notation $\Phi \models_s \psi$, if for all states $t \subseteq s$, if $t \models \varphi$ for all $\varphi \in \Phi$, then $t \models \psi$.

The following remark shows that, if Φ is a finite set $\{\varphi_1, \dots, \varphi_n\}$, the fact that $\varphi_1, \dots, \psi_n$ contextually entail ψ can be expressed by an implication.

Remark 2.5 $s \models \varphi_1 \wedge \dots \wedge \varphi_n \to \psi$ iff $\varphi_1, \dots, \varphi_n \models_s \psi$

In particular, since dependencies are special cases of contextual entailment, they are expressible within the language as implications between questions. To make this more concrete, let us look at one specific example.

Example 2.6 Suppose we want to express that, in the state s, whether q is determined by whether p. This means that in s, resolving the question $?p$ implies resolving the question $?q$. The relevant property of s can then be expressed by the implication $?p \to ?q$. Indeed, we have:

$$M, s \models ?p \to ?q \text{ iff } \exists f\colon \{0,1\} \to \{0,1\} \text{ s.t. } \forall w \in s : V(w,q) = f(V(w,p))$$

In words, s supports $?p \to ?q$ if and only if within s the truth-value of q is functionally determined by the truth-value of p.

This example generalizes straightforwardly. For any classical formula α, let us denote by $V(w,\alpha)$ the truth-value of α at w (1 if $w \in |\alpha|_M$ and 0 otherwise). Let $\alpha_1, \dots, \alpha_n, \beta$ be classical formulas: $?\alpha_1 \wedge \dots \wedge ?\alpha_n \to ?\beta$ expresses that, in s, whether β is true is determined by which of the α_is are true:

$$M, s \models ?\alpha_1 \wedge \dots \wedge ?\alpha_n \to ?\beta \text{ iff } \exists f : \{0,1\}^n \to \{0,1\} \text{ s.t. } \forall w \in s : V(w,\beta) = f(V(w,\alpha_1), \dots, V(w,\alpha_n))$$

3 Dependence statements aren't inquisitive conditionals

The discussion in the previous section illustrates how InqB provides a simple and elegant language to express dependencies. This might suggest that we can analyze dependence statements in InqB, translating, e.g., (1-a) as $?p \to ?q$. However, this analysis of (1-a) leads to a number of problems.

Consider first the truth-conditions of this sentence. Intuitively, (1-a) may well be false at a given possible world. E.g., it is intuitively false at the actual world, if we are talking about the rolling of a die, and p and q are taken to be, respectively, "the outcome is even" and "the outcome is prime". By contrast, $?p \to ?q$ is true at every world in every model: this is because relative to a singleton $\{w\}$, the truth-value of q is functionally determined by that of p in a trivial way. Thus, the truth-conditions of $?p \to ?q$ do not match those of (1-a).

Next, consider the interaction of dependence sentences with other operators. First take an embedding under negation. InqB has a negation operator $\neg$, which is used to formalize negative statements like *Alice is not home.* If we translate (1-a) as $?p \to ?q$, we'd want to translate (1-b) as $\neg(?p \to ?q)$. But this does not work: while (1-b) is a consistent statement, $\neg(?p \to ?q)$ is a contradiction.

One may try to solve this problem by changing the clause for negation, letting $s \models \neg p$ iff $s \not\models p$. However, this would not work as a general account of negation: in order to establish that *Alice is not home*, it is not enough that we have not established that she is, as this clause would have it; we must really exclude the possibility that she is home, as required by the InqB-negation. Thus, we would need two different negations, one to translate negations of ordinary statements, and the other to translate negations of dependence statements. This seems undesirable. While an extension of InqB with both negations has indeed been studied and axiomatized [15], the solution proposed here shows that it is possible to use a single negation in all cases. There is no need to stipulate a special meaning for negation when it applies to dependence statements.

Second, consider the case of a disjunctive dependence statement like (1-c). If we translate (1-a) as the formula $?p \to ?q$, we would want to translate (1-c) as the disjunction $(?p \to ?q) \vee (?r \to ?q)$. But again, this does not work well: while (1-c) may be true or false, the formula $(?p \to ?q) \vee (?r \to ?q)$ is a tautology.

Once more, one might want to blame this problem on the treatment of disjunction in InqB. Translating (1-c) by means of *inquisitive* disjunction, as $(?p \to ?q) \mathbin{\vee\!\!\!\vee} (?r \to ?q)$ fixes the problem, ensuring that (1-c) holds at a state only if one of its disjuncts does. But again, this would not work as a general account of disjunction: to establish that *Alice is at home or at school*, it is not necessary to establish in which of the two places she is, as would be required by treating disjunction as $\mathbin{\vee\!\!\!\vee}$. Thus, again, this strategy has the undesirable consequence that disjunctions in dependence statements need to be analyzed differently from disjunctions in other statements. By contrast, the proposal developed below shows that no departure from $\vee$ is needed to analyze (1-c). [2]

Next, consider the case of knowledge attributions, exemplified by (1-d). The epistemic extension of InqB, called *inquisitive epistemic logic* [8,1,3], comes with a simple analysis of the knowledge modality K_a. If $S_a \subseteq W \times W$ is a relation of epistemic accessibility for agent a, and $S_a[w] := \{v \in W \mid wS_av\}$, the truth-conditions for $K_a\varphi$ are as follows: [3]

$$M, w \models K_a\varphi \ \text{ iff } \ M, S_a[w] \models \varphi$$

This clause boils down to the standard epistemic logic clause when φ is truth-conditional, but it also allows us to deal with the case in which φ is a question.

---

[2] In dependence logic, disjunction is not defined via negation and conjunction, but taken as a primitive operator with the following support clause: $M, s \models \varphi \vee \psi$ iff $s = t_1 \cup t_2$ for some t_1, t_2 such that $M, t_1 \models \varphi$ and $M, t_2 \models \psi$. But this clause also renders $(?p \to ?q) \vee (?r \to ?q)$ a tautology. To see this, notice that any state s can be split into the states $s_q := s \cap |q|_M$ and $s_{\neg q} := s \cap |\neg q|_M$, and we have $M, s_q \models ?p \to ?q$ and $M, s_{\neg q} \models ?r \to ?q$. Thus, the problem with analyzing disjunctive dependence statements would not disappear if we treated disjunction by means of this clause. Thanks to an anonymous reviewer for pointing this out.

[3] In inquisitive epistemic logic, formulae $K_a\varphi$ are truth-conditional: the corresponding support clause just makes $K_a\varphi$ supported at s in case it is true at every world in s.

We can, e.g., analyze the statement *Alice knows whether* p directly as $K_a?p$, rather than having to paraphrase that statement as $K_a p \vee K_a \neg p$.

If we analyze (1-a) as $?p \to ?q$, we'd want to analyze (1-d) as $K_a(?p \to ?q)$. Once again, the result is not quite right. In inquisitive epistemic logic, this formula is equivalent to the following disjunction:

$$\bigvee_{\alpha \in S} K_a \alpha \qquad \text{where } S = \left\{ \begin{array}{ll} (p \to q) \wedge (\neg p \to q), & (p \to q) \wedge (\neg p \to \neg q) \\ (p \to \neg q) \wedge (\neg p \to q), & (p \to \neg q) \wedge (\neg p \to \neg q) \end{array} \right\}$$

This means that $K(?p \to ?q)$ is true only if Alice knows *how* the truth-value of q is determined by the one of p. But this is not required for the truth of (1-d): Alice may know that $?q$ is determined by $?p$ without knowing exactly how this dependency is realized; she might, e.g., be uncertain between two possibilities: (i) q is true if and only if p is; and (ii) q is true if and only if p is false.

Finally, consider the polar question (1-e) asking about the truth of (1-a). In inquisitive logic, if a statement is formalized by α, then the corresponding polar question is formalized by $?\alpha := \alpha \,\mathbb{V}\, \neg\alpha$. Thus, if we formalize (1-a) as $?p \to ?q$ we would want to formalize (1-e) as $?(?p \to ?q)$. However, since $\neg(?p \to ?q) \equiv \bot$, we have $?(?p \to ?q) = (?p \to ?q) \,\mathbb{V}\, \neg(?p \to ?q) \equiv ?p \to ?q$. Thus, the question in (1-e) would come out equivalent to the statement in (1-a)—clearly not the right result.

Summing up, then, in combination with the analysis of other logical items (negation, disjunction, knowledge, and the polar question operator) the assumption that (1-a) is translated as $?p \to ?q$ leads to blatantly wrong results.

4 Dependence statements are strict conditionals

All of the problems described in the previous section disappear if we formalize dependence statements not by means of the InqB-conditional $\to$, but rather by means of a generalized version of a strict conditional operator. To see how this solution works, the first step is to introduce such an operator into our logic. Thus, in this section we will work with the following language, denoted $\mathcal{L}^{\Rightarrow}$:[4]

$$\varphi ::= p \mid \bot \mid \varphi \wedge \varphi \mid \varphi \,\mathbb{V}\, \varphi \mid \varphi \to \varphi \mid \varphi \Rightarrow \varphi$$

The operators $\neg$, $\vee$, and ? are defined as in InqB; as before, we refer to $\mathbb{V}$-free formulas as *classical formulas*. With respect to standard logics equipped with a strict conditional operator, the novelty lies in the fact that this operator can be applied to questions. Thus, our language contains, e.g., formulas like $?p \Rightarrow ?q$. These are precisely the formulas that I propose to regard as formal counterparts of dependence statements.

[4] For simplicity, here I consider a single strict conditional operator $\Rightarrow$. One could implement the same ideas in a multi-modal language, equipped with a family $\{\Rightarrow_a \mid a \in \mathcal{A}\}$ of strict conditional operators, associated with corresponding accessibility relations $\{R_a \mid a \in \mathcal{A}\}$. The indices $a \in \mathcal{A}$ may be viewed as different agents endowed with different information, or as different processes associated with different sets of possible outcomes. In this setting, different dependencies will hold at a world w from the perspective of different $a \in \mathcal{A}$. The completeness results established below all extend straightforwardly to this case.

A strict implication is a modal operator. To interpret it, we need to equip our set of worlds with a binary relation R which captures relative possibility. This leads us naturally to work with the standard structure of a *Kripke model*, i.e., a triple $M = \langle W, R, V\rangle$ where W and V are as above, and $R \subseteq W \times W$.

Intuitively, wRv means that v is *possible* at w in the relevant sense. E.g., in our initial examples we were thinking of a process that may yield different outcomes. In this context, wRv will mean that v is a possible outcome of the process as it takes place at w. But this is just one of many possible interpretations. One could also, e.g., take wRv to mean that v is compatible with the information available to an agent at w; in this case, the relevant dependencies will be epistemic in nature. As far as I can see, there are just as many different flavors of dependency as there are flavors of modality.

The semantics for the system $\mathsf{Inq}^{\Rightarrow}$ based on the language $\mathcal{L}^{\Rightarrow}$ is given in terms of support clauses relative to an information state $s \subseteq W$ in a Kripke model. The clauses for atoms and connectives are the same as in Definition 2.1. The support clause for $\Rightarrow$ is as follows.

Definition 4.1 [Support for $\Rightarrow$] $M, s \models \varphi \Rightarrow \psi$ iff $\forall w \in s : \varphi \models_{R[w]} \psi$

This clause makes the formula $\varphi \Rightarrow \psi$ truth-conditional, with the following truth-conditions.

Definition 4.2 [Truth-conditions for $\Rightarrow$] $M, w \models \varphi \Rightarrow \psi$ iff $\varphi \models_{R[w]} \psi$

That is, $\varphi \Rightarrow \psi$ is true at w in case φ entails ψ relative to the information state $R[w]$. Making the contextual entailment relation explicit, this becomes:

- $M, w \models \varphi \Rightarrow \psi$ iff $\forall t \subseteq R[w] : M, t \models \varphi$ implies $M, t \models \psi$

To understand what this clause delivers, let us first look at the special case in which $\Rightarrow$ applies to two truth-conditional formulas. In this case, it is not hard to see that we recover the standard clause for the strict conditional operator.

Proposition 4.3 (Strict conditional recovered)
If α and β are truth-conditional, then $M, w \models \alpha \Rightarrow \beta$ iff $R[w] \cap |\alpha|_M \subseteq |\beta|_M$

It is also easy to check that all classical formulas are truth-conditional. This means that, when we restrict ourselves to the $\mathbin{\rotatebox[origin=c]{180}{$\wedge$}}$-free fragment of $\mathsf{Inq}^{\Rightarrow}$, what we have is just (a support-based implementation of) classical propositional logic augmented with a standard strict conditional operator.

The novel feature of $\mathsf{Inq}^{\Rightarrow}$ lies in the fact that our language contains not only statements, but also questions, and $\Rightarrow$ can be applied meaningfully to questions μ and ν to yield a modal statement $\mu \Rightarrow \nu$. What does this statement express? By Definition 4.2, $\mu \Rightarrow \nu$ is true at w just in case $\mu \models_{R[w]} \nu$. As we discussed in Section 2, the relation $\mu \models_{R[w]} \nu$ captures the fact that question μ determines question ν relative to the state $R[w]$. Thus, $\mu \Rightarrow \nu$ is a truth-conditional formula which is true or false at a world w according to whether μ determines ν in the associated set of worlds $R[w]$. This, I propose, is the right way to understand the semantics of dependence statements: a statement of the form *question μ determines question ν* should be rendered in our formal language not as the InqB-conditional $\mu \to \nu$, but as the strict conditional $\mu \Rightarrow \nu$. For instance, our dependence statement in (1-a) should be formalized as $?p \Rightarrow ?q$. This formula is true or false at w depending on whether the truth-value of q is functionally determined by the truth-value of p throughout the set $R[w]$.

$$M, w \models ?p \Rightarrow ?q \text{ iff } \exists f : \{0,1\} \to \{0,1\} \text{ s.t. } \forall v \in R[w] : V(v,q) = f(V(v,p))$$

Notice that $?p \Rightarrow ?q$ may well be false at a world w; so, the first problem we discussed in the previous section does not arise. Also, $?p \Rightarrow ?q$ is truth-conditional. Thus, dependence statements are translated by formulae that have the fundamental semantic property that we associate with statements, namely, truth-conditionality. This also ensures that embeddings under other operators work just as well for dependence statements as for ordinary statements.

Consider negation: unlike $\neg(?p \to ?q)$, the formula $\neg(?p \Rightarrow ?q)$ is consistent, and it is true at w just in case $?p \Rightarrow ?q$ is false, i.e., just in case $?p \not\models_{R[w]} ?q$. Thus, we get a good analysis of (1-b) as a negation of a dependence statement.

Consider disjunction: unlike $(?p \to ?q) \vee (?r \to ?q)$, the formula $(?p \Rightarrow ?q) \vee (?r \Rightarrow ?q)$ is not a tautology: it is a truth-conditional formula which is true at w just in case one of its disjuncts is true, i.e., in case $?p \models_{R[w]} ?q$ or $?r \models_{R[w]} ?q$. Thus we get a good analysis for (1-c) as a disjunction of dependence statements.

Next consider the embedding of a dependence statement $?p \Rightarrow ?q$ in the scope of a knowledge modality K_a. We get:

$$\begin{aligned} M, w \models K_a(?p \Rightarrow ?q) \quad &\text{iff} \quad M, S[w] \models ?p \Rightarrow ?q \\ &\text{iff} \quad \forall v \in S[w] : M, v \models ?p \Rightarrow ?q \\ &\text{iff} \quad \forall v \in S[w] : \exists f_v : \{0,1\} \to \{0,1\} \text{ such that} \\ &\qquad\qquad \forall u \in R[v] : V(u,q) = f_v(V(u,p)) \end{aligned}$$

Thus, $K_a(?p \Rightarrow ?q)$ is true at a world w in case $?p \Rightarrow ?q$ is true in all worlds compatible with the agent's knowledge at w. This means that for all worlds v compatible with the agent's knowledge we have a corresponding functional dependency that holds across $R[v]$. But different worlds $v, v' \in S[w]$ might be associated with different sets $R[v], R[v']$ of possibilities, in which the dependency of $?q$ on $?p$ might be realized by different functions $f_v, f_{v'}$. Although there might be a single dependence function f which works on $R[v]$ for all worlds $v \in S[w]$, there need not be. Thus, as we expect, the formula $K_a(?p \Rightarrow ?q)$ does not entail the following disjunction (although it is entailed by it):

$$\bigvee_{\alpha \in S} K_a \alpha \qquad \text{where } S = \left\{ \begin{array}{ll} (p \Rightarrow q) \wedge (\neg p \Rightarrow q), & (p \Rightarrow q) \wedge (\neg p \Rightarrow \neg q) \\ (p \Rightarrow \neg q) \wedge (\neg p \Rightarrow q), & (p \Rightarrow \neg q) \wedge (\neg p \Rightarrow \neg q) \end{array} \right\}$$

Thus, by combining the proposed analysis of dependence statements with the inquisitive analysis of the knowledge modality we obtain a good analysis of the knowledge attribution in (1-d)—one that predicts that knowing that a dependency holds does not require knowing how it is realized.

Finally, take the question (1-e): while $?(?p \to ?q)$ is equivalent to $?p \to ?q$, the formula $?(?p \Rightarrow ?q)$ is not equivalent to $?p \Rightarrow ?q$. The latter is a statement, while $?(?p \Rightarrow ?q)$ is a polar question that can be resolved in two different ways, namely, by establishing that $?p \Rightarrow ?q$ is true, or by establishing that it is false:

$$M, s \models ?(?p \Rightarrow ?q) \text{ iff } s \subseteq |?p \Rightarrow ?q|_M \text{ or } s \subseteq |\neg(?p \Rightarrow ?q)|_M$$

So, we get a good analysis of (1-e) as a polar question about the statement (1-a).

In conclusion, the proposed analysis, in addition to having a *prima facie* intuitive appeal, interacts well with the rest of the logical repertoire. When we analyze dependence statements as strict conditionals involving questions, the problems discussed in the previous section disappear. The fact that all these pieces fall naturally into place, I submit, provides a good case for the proposal, and motivates an investigation of the logic of strict conditionals in the inquisitive setting. To this I turn in the next section.

5 The inquisitive logic of strict conditionals

5.1 Declaratives and normal form

In inquisitive logic, entailment is defined naturally as preservation of support:

- $\Phi \models \psi$ iff $\forall M, s:$ if $M, s \models \varphi$ for all $\varphi \in \Phi$, then $M, s \models \psi$

Logical equivalence, denoted $\equiv$, is defined as support at the same states in all models, and coincides with mutual entailment.

Following a standard strategy in the study of inquisitive logics [1,2,3,4], we first isolate a fragment of $\mathcal{L}^{\Rightarrow}$ consisting only of truth-conditional formulas. The set of *declaratives*, $\mathcal{L}_!$, is given by the following BNF, where $\varphi \in \mathcal{L}^{\Rightarrow}$:

$$\alpha ::= p \mid \bot \mid \alpha \wedge \alpha \mid \varphi \to \alpha \mid \varphi \Rightarrow \varphi$$

In words, a formula α is a declarative iff the only occurrences of $\mathbin{\rotatebox[origin=c]{-90}{$\geqslant$}}$ in α, if any, are within the scope of a strict conditional or in the antecedent of $\to$. It is then straightforward to show the following.

Proposition 5.1 *All $\alpha \in \mathcal{L}_!$ are truth-conditional.*

As in InqB (see [3,4]), so also in $\mathsf{Inq}^{\Rightarrow}$ we can associate to each formula φ a set $\mathcal{R}(\varphi)$ of declaratives, called the resolutions of φ, which jointly characterize φ.

Definition 5.2 [Resolutions]

- $\mathcal{R}(\varphi) = \{\varphi\}$ if φ is an atom, $\bot$, or a strict conditional;
- $\mathcal{R}(\varphi \wedge \psi) = \{\alpha \wedge \beta \mid \alpha \in \mathcal{R}(\varphi) \text{ and } \beta \in \mathcal{R}(\psi)\}$
- $\mathcal{R}(\varphi \vee\!\!\!\vee \psi) = \mathcal{R}(\varphi) \cup \mathcal{R}(\psi)$
- $\mathcal{R}(\varphi \to \psi) = \{\bigwedge_{\alpha \in \mathcal{R}(\varphi)} (\alpha \to f(\alpha)) \mid f : \mathcal{R}(\varphi) \to \mathcal{R}(\psi)\}$

It is easy to see by induction that a declarative is the only resolution of itself.

Remark 5.3 For all $\alpha \in \mathcal{L}_!$, $\mathcal{R}(\alpha) = \{\alpha\}$.

On the other hand, if φ is a question then the elements of $\mathcal{R}(\varphi)$ can be thought of as syntactically generated answers to φ. For instance, $\mathcal{R}(?p) = \{p, \neg p\}$. An easy induction suffices to establish the following normal form result.

Definition 5.4 [Inquisitive normal form] For all $\varphi \in \mathcal{L}^{\Rightarrow}$, $\varphi \equiv \bigvee\!\!\!\!\bigvee \mathcal{R}(\varphi)$.

5.2 Proof system

In this section we describe a Hilbert-style proof-system for the logic $\mathsf{Inq}^{\Rightarrow}$. For convenience, we divide the axioms into a set of propositional axioms and a set of axioms which involve specifically the strict conditional operator.

The propositional axioms, inherited from the propositional logic InqB, are all instances of the following schemata, where $\varphi, \psi, \chi \in \mathcal{L}^{\Rightarrow}$ and $\alpha \in \mathcal{L}_!$:

(i) $\varphi \to (\psi \to \varphi)$

(ii) $(\varphi \to (\psi \to \chi)) \to ((\varphi \to \psi) \to (\varphi \to \chi))$

(iii) $\varphi \to (\psi \to \varphi \wedge \psi)$

(iv) $\varphi \wedge \psi \to \varphi, \quad \varphi \wedge \psi \to \psi$

(v) $\varphi \to \varphi \vee\!\!\!\vee \psi, \quad \psi \to \varphi \vee\!\!\!\vee \psi$

(vi) $(\varphi \to \chi) \to ((\psi \to \chi) \to (\varphi \vee\!\!\!\vee \psi \to \chi))$

(vii) $\bot \to \varphi$

(viii) $\neg\neg\alpha \to \alpha$

(ix) $(\alpha \to \varphi \vee\!\!\!\vee \psi) \to (\alpha \to \varphi) \vee\!\!\!\vee (\alpha \to \psi)$

Axioms (i)–(vii) are essentially the axioms of intuitionistic propositional logic with $\vee\!\!\!\vee$ in the role of intuitionistic disjunction (see [7] for discussion). Axiom (viii) captures the fact that declaratives are truth-conditional and, as a consequence, obey classical logic. Axiom (ix) captures the fact that, again due to truth-conditionality, declaratives distribute over inquisitive disjunctions. [5]

The axioms for the strict implication modality are all instances of the following schemata, where φ, ψ, χ stand for arbitrary formulas and α for a declarative:

(x) transitivity: $(\varphi \Rightarrow \psi) \to ((\psi \Rightarrow \chi) \to (\varphi \Rightarrow \chi))$

(xi) import-export: $(\varphi \wedge \psi \Rightarrow \chi) \leftrightarrow (\varphi \Rightarrow (\psi \to \chi))$

(xii) $\Rightarrow$-split: $(\alpha \Rightarrow \varphi \vee\!\!\!\vee \psi) \to (\alpha \Rightarrow \varphi) \vee (\alpha \Rightarrow \psi)$

The inference rules are modus ponens and conditional necessitation:

$$\frac{\varphi \to \psi \quad \varphi}{\psi}\ (\mathsf{MP}) \qquad\qquad \frac{\varphi \to \psi}{\varphi \Rightarrow \psi}\ (\mathsf{CN})$$

We write $\vdash \psi$ if the formula ψ is derivable in this system, $\varphi \dashv\vdash \psi$ if $\varphi \leftrightarrow \psi$ is derivable, and $\Phi \vdash \psi$ if $\varphi_1 \wedge \cdots \wedge \varphi_n \to \psi$ is derivable for some $\varphi_1, \ldots, \varphi_n \in \Phi$.

It is easy to check directly that all instances of the axioms are valid, and that the inference rules preserves validity, which implies that the proof system is sound. In the remainder of this section we show that it is also complete.

---

[5] The first axiomatization given for InqB [7] used slightly different axioms: double negation elimination was restricted to atoms, and instead of the $\vee\!\!\!\vee$-split axiom it included the Kreisel-Putnam axiom $(\neg\varphi \to \psi \vee\!\!\!\vee \chi) \to (\neg\varphi \to \psi) \vee\!\!\!\vee (\neg\varphi \to \chi)$ for arbitrary φ, ψ, χ. In the setting of $\mathsf{Inq}^{\Rightarrow}$, having double negation elimination for atoms only is not enough: we must minimally have it for strict conditionals as well. On the other hand, the choice between $\vee\!\!\!\vee$-split and the KP axiom is a matter of convenience. Using $\vee\!\!\!\vee$-split has an advantage in terms of generality, since this axiom also plays a role in non-classical versions of inquisitive logic [16,17,6].

As usual, completeness is established by constructing a canonical model. To show that the construction works, we first need a few results about our proof system. One thing that will play a crucial role is that the system has enough resources to justify the normal form result given by Proposition 5.4.

Lemma 5.5 $\varphi \dashv\vdash \bigvee\!\!\!\!\bigvee \mathcal{R}(\varphi)$

Proof. By induction on the structure of φ. The only step that requires some work is the one for $\varphi = (\psi \to \chi)$. But the proof has nothing to do with strict implication: it is essentially the same as for InqB (see Lemma 3.3.4 in [3]). □

The following two lemmata connecting provability of/from a formula to provability of/from its resolutions will play a crucial role in the completeness proof.

Lemma 5.6 *If* $\Gamma \subseteq \mathcal{L}_!$ *and* $\Gamma \vdash \varphi$*, then* $\Gamma \vdash \alpha$ *for some* $\alpha \in \mathcal{R}(\varphi)$

Proof. First we show the following claim:

$$\text{If } \vdash \varphi, \text{ then } \vdash \alpha \text{ for some } \alpha \in \mathcal{R}(\varphi) \tag{1}$$

We show (1) by induction on the length of the shortest proof of φ. If φ is provable with a proof of length 1, then φ is an axiom. Suppose φ is a propositional axiom: then it is straightforward to check, considering each case in turn, that a resolution of φ is a classical tautology. By way of example, suppose φ is an instance $\psi \to (\chi \to \psi)$ of axiom (i); the following tautology is a resolution of φ:

$$\bigwedge_{\alpha \in \mathcal{R}(\psi)} (\alpha \to \bigwedge_{\beta \in \mathcal{R}(\chi)} (\beta \to \alpha))$$

But in restriction to declaratives, our system contains a complete set of axioms for classical propositional logic, and so it proves all classical tautologies.

Next, suppose φ is one of the axioms for the strict conditional operator. All instances of such axioms are declaratives. Thus, by Remark 5.3, we have $\mathcal{R}(\varphi) = \{\varphi\}$; since φ is provable, obviously a resolution of φ is provable.

Now consider the inductive case. We have only two cases to consider:

(i) φ is obtained by modus ponens from formulas $\psi \to \varphi$ and ψ which are provable with shorter proofs. By induction hypothesis, the system proves a resolution γ of $\psi \to \varphi$ and a resolution β_0 of ψ. By definition of resolutions of an implication, γ has the form $\bigwedge_{\beta \in \mathcal{R}(\psi)} (\beta \to f(\beta))$ for some $f : \mathcal{R}(\psi) \to \mathcal{R}(\varphi)$. But then, clearly, $f(\beta_0)$ is a provable resolution of φ.

(ii) $\varphi = (\psi \Rightarrow \chi)$ is obtained by modal necessitation from a formula $\psi \to \chi$ that has a shorter proof. In this case, φ is a declarative; so, by Remark 5.3, $\mathcal{R}(\varphi) = \{\varphi\}$. Thus, our proof of φ is also a proof of a resolution of φ.

This completes the inductive proof of (1). To prove Lemma 5.6, suppose $\Gamma \subseteq \mathcal{L}_!$ and $\Gamma \vdash \varphi$. This means that $\vdash \gamma_1 \wedge \cdots \wedge \gamma_n \to \varphi$ for some $\gamma_1, \ldots, \gamma_n \in \Gamma$. Let $\gamma := \gamma_1 \wedge \cdots \wedge \gamma_n$: since γ is a declarative, by Remark 5.3 we have $\mathcal{R}(\gamma) = \{\gamma\}$. Thus, by definition of resolutions of an implication, we have:

$$\mathcal{R}(\gamma \to \varphi) = \{\gamma \to f(\gamma) \mid f : \{\gamma\} \to \mathcal{R}(\varphi)\} = \{\gamma \to \alpha \mid \alpha \in \mathcal{R}(\varphi)\}$$

Since $\vdash \gamma \to \varphi$, by (1) it follows that $\vdash \gamma \to \alpha$ for some resolution $\alpha \in \mathcal{R}(\varphi)$. Since obviously $\Gamma \vdash \gamma$, we can conclude that $\Gamma \vdash \alpha$. □

Lemma 5.7 *If $\Phi, \chi \nvdash \psi$, then $\Phi, \alpha \nvdash \psi$ for some $\alpha \in \mathcal{R}(\chi)$*

Proof. By contraposition, suppose $\Phi, \alpha \vdash \psi$ for all $\alpha \in \mathcal{R}(\chi)$. Using the axioms for $\mathbin{\backslash\!\!/}$ we infer that $\Phi, \mathbin{\backslash\!\!/}\mathcal{R}(\chi) \vdash \psi$. By Lemma 5.5 we have $\Phi, \chi \vdash \psi$.□

Using this lemma inductively, we obtain the following corollary (for the details, see the analogous result in the propositional setting, Lemma 3.3.7 in [3]).

Corollary 5.8 *If $\Phi \nvdash \psi$, then there is a set of declaratives $\Gamma \subseteq \mathcal{L}_!$ which contains a resolution of each $\varphi \in \Phi$ and such that $\Gamma \nvdash \psi$.*

In addition to these results, in the completeness proof we will use some provable features of strict conditionals, spelled out in the following lemmata.

Lemma 5.9 *If $\varphi' \vdash \varphi$ and $\psi \vdash \psi'$ then $\varphi \Rightarrow \psi \vdash \varphi' \Rightarrow \psi'$.*

Proof. From $\varphi' \vdash \varphi$ and $\psi \vdash \psi'$ we have $\vdash \varphi' \Rightarrow \varphi$ and $\vdash \psi \Rightarrow \psi'$ by conditional necessitation. By transitivity we get $\vdash (\varphi \Rightarrow \psi) \to (\varphi' \Rightarrow \psi')$. □

As an immediate corollary we get replacement of provably equivalent formulas.

Corollary 5.10 *If $\varphi \dashv\vdash \varphi'$ and $\psi \dashv\vdash \psi'$ then $\varphi \Rightarrow \psi \dashv\vdash \varphi' \Rightarrow \psi'$*

Another provable feature of $\Rightarrow$ concerns the behavior of $\mathbin{\backslash\!\!/}$ in antecedents.

Lemma 5.11 *For any $\varphi, \psi, \chi \in \mathcal{L}^{\Rightarrow}$, $(\varphi \mathbin{\backslash\!\!/} \psi \Rightarrow \chi) \dashv\vdash (\varphi \Rightarrow \chi) \wedge (\psi \Rightarrow \chi)$.*

Proof. By Corollary 5.10 and the import-export axiom we have that $\varphi \Rightarrow \chi \dashv\vdash (\top \wedge \varphi) \Rightarrow \chi \dashv\vdash \top \Rightarrow (\varphi \to \chi)$. Analogously we have $\psi \Rightarrow \chi \dashv\vdash \top \Rightarrow (\psi \to \chi)$. Thus, $(\varphi \Rightarrow \chi) \wedge (\psi \Rightarrow \psi) \dashv\vdash (\top \Rightarrow (\varphi \to \chi)) \wedge (\top \Rightarrow (\psi \to \chi))$.

We have $(\varphi \to \chi) \to ((\psi \to \chi) \to (\varphi \mathbin{\backslash\!\!/} \psi \to \chi))$ as an axiom. Using this, by conditional necessitation, transitivity, and import-export, we can show that $(\top \Rightarrow (\varphi \to \chi)) \wedge (\top \Rightarrow (\psi \to \chi)) \vdash \top \Rightarrow (\varphi \mathbin{\backslash\!\!/} \psi \to \chi)$. By import-export and Cor. 5.10, $\top \Rightarrow (\varphi \mathbin{\backslash\!\!/} \psi \to \chi) \dashv\vdash (\top \wedge (\varphi \mathbin{\backslash\!\!/} \psi)) \Rightarrow \chi \dashv\vdash \varphi \mathbin{\backslash\!\!/} \psi \Rightarrow \chi$. This shows the right-to-left direction. The converse direction follows from the provable monotonicity of $\Rightarrow$ on the left (Lemma 5.9) since $\varphi \vdash \varphi \mathbin{\backslash\!\!/} \psi$ and $\psi \vdash \varphi \mathbin{\backslash\!\!/} \psi$. □

Finally, the next lemma connects strict conditionals applying to arbitrary formulas to Boolean combinations of strict conditionals applying to declaratives.

Lemma 5.12 $\varphi \Rightarrow \psi \dashv\vdash \bigvee_{f:\mathcal{R}(\varphi)\to\mathcal{R}(\psi)} \bigwedge_{\alpha\in\mathcal{R}(\varphi)} (\alpha \Rightarrow f(\alpha))$

The proof of this lemma uses the following auxiliary result.

Lemma 5.13 *Let $\alpha, \beta, \gamma \in \mathcal{L}_!$. If $\alpha \vdash \gamma$ and $\beta \vdash \gamma$ then $\alpha \vee \beta \vdash \gamma$.*

Proof. This follows from the fact that, restricted to declaratives, our system includes a complete set of axioms for classical propositional logic. □

Proof of Lemma 5.12. Let $\alpha \in \mathcal{R}(\varphi)$. We have $\alpha \Rightarrow \psi \dashv\vdash \alpha \Rightarrow \mathbin{\backslash\!\!/}\mathcal{R}(\psi)$ by Lemma 5.5 and Cor. 5.10. We also have $\alpha \Rightarrow \mathbin{\backslash\!\!/}\mathcal{R}(\psi) \dashv\vdash \bigvee_{\beta\in\mathcal{R}(\psi)} (\alpha \Rightarrow \beta)$: the left-to-right direction uses the $\Rightarrow$-split axiom; the converse uses Lemma

5.13 once we notice that for each $\beta \in \mathcal{R}(\psi)$, $\alpha \Rightarrow \beta \vdash \alpha \Rightarrow \bigvee\!\!\!\!\bigvee \mathcal{R}(\psi)$ by Lemma 5.9. As a consequence, we have:

$$\bigwedge_{\alpha \in \mathcal{R}(\varphi)} (\alpha \Rightarrow \psi) \dashv\vdash \bigwedge_{\alpha \in \mathcal{R}(\varphi)} \bigvee_{\beta \in \mathcal{R}(\psi)} (\alpha \Rightarrow \beta) \tag{2}$$

By Lemma 5.11 we have $\bigwedge_{\alpha \in \mathcal{R}(\varphi)} (\alpha \Rightarrow \psi) \dashv\vdash \bigvee\!\!\!\!\bigvee \mathcal{R}(\varphi) \Rightarrow \psi$. By Lemma 5.5 and Corollary 5.10, the latter is provably equivalent to $\varphi \Rightarrow \psi$. Thus, the left-hand-side of (2) is provably equivalent to $\varphi \Rightarrow \psi$. As for the right hand side, recall that in restriction to declaratives, our system contains a complete set of axioms for classical propositional logic. In particular, we can distribute $\wedge$ over $\vee$. Using this we can show that the right-hand-side of (2) is provably equivalent to $\bigvee_{f:\mathcal{R}(\varphi)\to\mathcal{R}(\psi)} \bigwedge_{\alpha \in \mathcal{R}(\varphi)} (\alpha \Rightarrow f(\alpha))$. □

5.3 Completeness by canonical model

We call a set $\Gamma \subseteq \mathcal{L}_!$ a *declarative theory* if for all $\alpha \in \mathcal{L}_!$, $\Gamma \vdash \alpha$ only if $\alpha \in \Gamma$. We call Γ a *complete declarative theory* if it is a declarative theory and for every $\alpha \in \mathcal{L}_!$ it contains exactly one of α and $\neg\alpha$ (which implies consistency).

Definition 5.14 [Canonical model]
The canonical model for $\mathsf{Inq}^{\Rightarrow}$ is the model $M^c = \langle W^c, R^c, V^c \rangle$ where:

- W^c is the set of all complete declarative theories;
- $\Gamma R^c \Gamma'$ iff $\forall \alpha, \beta \in \mathcal{L}_! : (\alpha \Rightarrow \beta) \in \Gamma$ and $\alpha \in \Gamma'$ implies $\beta \in \Gamma'$
- $V^c(\Gamma, p) = 1$ iff $p \in \Gamma$

The following lemmata about complete declarative theories are mostly familiar from classical propositional and modal logic; I omit the straightforward proofs.

Lemma 5.15 *If $\Gamma \in W^c$ and $\alpha \vee \beta \in \Gamma$, then $\alpha \in \Gamma$ or $\beta \in \Gamma$.*

Lemma 5.16 *If $S \subseteq W^c$, then $\bigcap S$ is a declarative theory.*

Lemma 5.17 *If $\Delta \subseteq \mathcal{L}_!$ is consistent then $\Delta \subseteq \Gamma$ for some $\Gamma \in W^c$.*

The next two lemmata establish important connections between provability and the structure of the canonical model.

Lemma 5.18 *Let $S \subseteq W^c$. If $\bigcap S \nvdash \varphi \to \psi$, then for some subset $T \subseteq S$ we have $\bigcap T \vdash \varphi$ and $\bigcap T \nvdash \psi$.*

Proof. If $\bigcap S \nvdash \varphi \to \psi$, then $\bigcap S, \varphi \nvdash \psi$. By Lemma 5.7 there is some $\alpha \in \mathcal{R}(\varphi)$ such that $\bigcap S, \alpha \nvdash \psi$. Let $T := \{\Gamma \in S \mid \alpha \in \Gamma\}$. Since $\alpha \in \bigcap T$, by Lemma 5.5 we have $\bigcap T \vdash \varphi$. It remains to be seen that $\bigcap T \nvdash \psi$.

Towards a contradiction, suppose $\bigcap T \vdash \psi$. By Lemma 5.6 this implies $\bigcap T \vdash \beta$ for some $\beta \in \mathcal{R}(\psi)$. By Lemma 5.16 we have $\beta \in \bigcap T$. Thus, for all $\Gamma \in T$ we have $\beta \in \Gamma$, and so also $\alpha \to \beta \in \Gamma$. But now consider any $\Gamma \in S - T$. Since $\alpha \notin \Gamma$, by completeness we have $\neg\alpha \in \Gamma$ and so also $\alpha \to \beta \in \Gamma$. We have thus reached the conclusion that $\alpha \to \beta$ belongs to all $\Gamma \in S$, no matter whether $\Gamma \in T$ or $\Gamma \in S - T$. Thus, $\alpha \to \beta \in \bigcap S$. But this implies that $\bigcap S, \alpha \vdash \beta$, which by Lemma 5.5 gives $\bigcap S, \alpha \vdash \psi$, contrary to assumption. □

Lemma 5.19 *Let* $\Gamma \in W^c$. *If* $(\varphi \Rightarrow \psi) \notin \Gamma$ *then* $\bigcap R^c[\Gamma] \nvdash \varphi \to \psi$.

Proof. Suppose $(\varphi \Rightarrow \psi) \notin \Gamma$. By Equation (2) above, this means that Γ does not contain $\bigwedge_{\alpha\in\mathcal{R}(\varphi)} \bigvee_{\beta\in\mathcal{R}(\psi)}(\alpha \Rightarrow \beta)$. This means that there is some $\alpha_j \in \mathcal{R}(\varphi)$ such that for all $\beta \in \mathcal{R}(\psi)$, $\alpha_j \Rightarrow \beta \notin \Gamma$. Now let $\mathcal{R}(\psi) = \{\beta_1, \dots, \beta_n\}$. For $i \leq n$ set $\Delta_i := \{\gamma \in \mathcal{L}_! \mid (\alpha_j \Rightarrow \gamma) \in \Gamma\} \cup \{\neg\beta_i\}$. We claim that Δ_i is consistent. Towards a contradiction, suppose not. That means that there are $\gamma_1, \dots, \gamma_n$ such that $\alpha \Rightarrow \gamma_k \in \Gamma$ for each $k \leq n$ and $\gamma_1, \dots, \gamma_n, \neg\beta_i \vdash \bot$. Since $\neg\beta_i$ is a declarative and for declaratives we have all the axioms of classical propositional logic, we have $\gamma_1, \dots, \gamma_n \vdash \beta_i$. By conditional necessitation and transitivity we easily obtain $\alpha_j \Rightarrow \gamma_1, \dots, \alpha_j \Rightarrow \gamma_n \vdash \alpha_j \Rightarrow \beta_i$. Since all the formulas $\alpha_j \Rightarrow \gamma_i$ are in Γ, it follows that $\alpha_j \Rightarrow \beta_i \in \Gamma$, contrary to assumption.

Thus, Δ_i is indeed consistent. This means, by Lemma 5.17, that it can be extended to a $\Gamma_i \in W^c$. We claim that $\Gamma R^c \Gamma_i$. To see this, let $\gamma, \delta \in \mathcal{L}_!$ and suppose that $\gamma \Rightarrow \delta \in \Gamma$ and $\gamma \in \Gamma_i$. We need to show that $\delta \in \Gamma_i$.

By Lemma 5.9, $\gamma \Rightarrow \delta \vdash \alpha_j \wedge \gamma \Rightarrow \delta$. By the import-export axiom we have $\alpha_j \wedge \gamma \Rightarrow \delta \dashv\vdash \alpha_j \Rightarrow (\gamma \to \delta)$. Since $\gamma \Rightarrow \delta \in \Gamma$, also $\alpha_j \Rightarrow (\gamma \to \delta) \in \Gamma$, and so by definition of Δ_i we have $\gamma \to \delta \in \Delta_i \subseteq \Gamma_i$. Since Γ_i also contains γ, it follows that Γ_i contains δ, as we needed to show. Thus $\Gamma R^c \Gamma_i$.

Towards a contradiction, suppose that $\bigcap R^c[\Gamma] \vdash \varphi \to \psi$. By Lemma 5.6 and Lemma 5.15, for some $f : \mathcal{R}(\varphi) \to \mathcal{R}(\psi)$ we have $\bigcap R^c[\Gamma] \vdash \bigwedge_{\alpha\in\mathcal{R}(\varphi)}(\alpha \Rightarrow f(\alpha))$. In particular, $\bigcap R^c[\Gamma] \vdash \alpha_j \Rightarrow f(\alpha_j)$. Suppose $f(\alpha_j) = \beta_i$, so that $\bigcap R^c[\Gamma] \vdash \alpha_j \to \beta_i$. By Lemma 5.16 we obtain $(\alpha_j \to \beta_i) \in \bigcap R^c[\Gamma]$. Since $\Gamma_i \in R^c[\Gamma]$ we have $(\alpha_j \to \beta_i) \in \Gamma_i$. Since $\alpha_j \Rightarrow \alpha_j$ is provable, we have $(\alpha_j \Rightarrow \alpha_j) \in \Gamma$, and so $\alpha_j \in \Gamma_i$ by construction of Γ_i. Since Γ_i contains both $\alpha_j \to \beta_i$ and α_j, it follows that $\beta_i \in \Gamma_i$. But this is impossible, since by construction $\neg\beta_i \in \Gamma_i$ and Γ_i is consistent. Thus, $\bigcap R^c[\Gamma] \nvdash \varphi \to \psi$. □

With the help of these lemmata, the bridge between provability and semantics in the canonical model can be built. Usually, semantics is based on truth-conditions, and so the relevant bridge takes the form of a *truth lemma*, equating truth at a world with provability from it. In our setting, the fundamental semantic notion is not truth at a world, but support at an information state. Accordingly, the relevant bridge takes the form of a *support lemma*, which equates support at a state S in M^c with provability from the intersection of the theories in S (if $S = \emptyset$ we let $\bigcap S := \mathcal{L}_!$).

Lemma 5.20 (Support lemma)
For every $S \subseteq W^c$ *and* $\varphi \in \mathcal{L}^{\Rightarrow}$: $M^c, S \models \varphi$ *iff* $\bigcap S \vdash \varphi$.

Proof. By induction on the structure of φ. The only interesting cases are the inductive steps for the two implications $\to$ and $\Rightarrow$.

- $\varphi = (\psi \to \chi)$. Suppose $\bigcap S \vdash \psi \to \chi$. Take any $T \subseteq S$ with $M^c, T \models \psi$. By induction hypothesis we have $\bigcap T \vdash \psi$. Since $T \subseteq S$, $\bigcap T \supseteq \bigcap S$, so $\bigcap T \vdash \psi \to \chi$. By modus ponens it follows that $\bigcap T \vdash \chi$, which by induction hypothesis gives $M^c, T \models \chi$. This shows that $M^c, S \models \psi \to \chi$.

  For the converse direction, suppose $\bigcap S \nvdash \psi \to \chi$. By Lemma 5.18 there is a subset $T \subseteq S$ such that $\bigcap T \vdash \psi$ but $\bigcap T \nvdash \chi$. By induction hypothesis, this translates to $M^c, T \models \psi$ but $M^c, T \not\models \chi$, which implies $M^c, S \not\models \psi \to \chi$.

- $\varphi = (\psi \Rightarrow \chi)$. Suppose $\bigcap S \vdash \psi \Rightarrow \chi$. Since $\psi \Rightarrow \chi$ is a declarative, by Lemma 5.16 we have $(\psi \Rightarrow \chi) \in \bigcap S$. Now take any world $\Gamma \in S$ and take any $T \subseteq R[\Gamma]$ such that $M^c, T \models \psi$. We want to show that $M^c, T \models \chi$.

  Since $\Gamma \supseteq \bigcap S$, we have $\psi \Rightarrow \chi \in \Gamma$. Since Γ is a declarative theory, by Lemma 5.12, it must contain $\bigvee_{f:\mathcal{R}(\psi)\to\mathcal{R}(\chi)} \bigwedge_{\alpha\in\mathcal{R}(\psi)}(\alpha \Rightarrow f(\alpha))$. By Lemma 5.15 , Γ must then contain $\bigwedge_{\alpha\in\mathcal{R}(\psi)}(\alpha \Rightarrow f(\alpha))$ for some $f : \mathcal{R}(\psi) \to \mathcal{R}(\chi)$.

  Since $M^c, T \models \psi$, the induction hypothesis gives $\bigcap T \vdash \psi$. By Lemma 5.6 we have $\bigcap T \vdash \alpha_i$ for some $\alpha_i \in \mathcal{R}(\psi)$. By Lemma 5.16, this implies $\alpha_i \in \bigcap T$, and so $\alpha_i \in \Gamma'$ for all $\Gamma' \in T$. Now take any $\Gamma' \in T$: since $\Gamma R^c \Gamma'$, and since $\alpha_i \Rightarrow f(\alpha_i) \in \Gamma$ and $\alpha_i \in \Gamma'$, we must have $f(\alpha_i) \in \Gamma'$. This holds for all $\Gamma' \in T$, and so $\bigcap T \vdash f(\alpha_i)$. Since $f(\alpha_i) \in \mathcal{R}(\chi)$, it follows from Lemma 5.5 that $\bigcap T \vdash \chi$. By induction hypothesis, this gives $M^c, T \models \chi$.

  Summing up, we have shown that for all $\Gamma \in S$ and $T \subseteq R[\Gamma]$, $M^c, T \models \psi$ implies $M^c, T \models \chi$. This is just what is required to have $M^c, S \models \psi \Rightarrow \chi$.

  For the converse direction, suppose $\bigcap S \nvdash \psi \Rightarrow \chi$. Then $(\psi \Rightarrow \chi) \notin \Gamma$ for some $\Gamma \in S$. By Lemma 5.19 it follows that $\bigcap R^c[\Gamma] \nvdash \psi \to \chi$. Then by Lemma 5.18 there is some $T \subseteq R^c[\Gamma]$ with $\bigcap T \vdash \psi$ and $\bigcap T \nvdash \chi$. By the induction hypothesis, this translates to $M^c, T \models \varphi$ and $M^c, T \not\models \psi$. Thus, for some $\Gamma \in S$ and some $T \subseteq R^c[\Gamma]$ we have $M^c, \Gamma \models \psi$ and $M^c, \Gamma \not\models \chi$, which shows that $M^c, S \not\models \psi \Rightarrow \chi$. □

Finally, we can use the canonical model to prove our completeness result.[6]

Theorem 5.21 *For all* $\Phi \cup \{\psi\} \subseteq \mathcal{L}^{\Rightarrow}$, $\Phi \models \psi$ *iff* $\Phi \vdash \psi$.

Proof. Suppose $\Phi \nvdash \psi$. By Lemma 5.8, there is a set $\Delta \subseteq \mathcal{L}_!$ that contains a resolution of each $\varphi \in \Phi$ and such that $\Delta \nvdash \psi$. Let $S_\Delta := \{\Gamma \in W^c \mid \Delta \subseteq \Gamma\}$. We claim that (i) $M^c, S_\Delta \models \varphi$ for all $\varphi \in \Phi$, but (ii) $M^c, S_\Delta \not\models \psi$.

To show (i), take $\varphi \in \Phi$. For some $\alpha \in \mathcal{R}(\varphi)$ we have $\alpha \in \Delta \subseteq \bigcap S_\Delta$. The support lemma gives $M^c, S_\Delta \models \alpha$; by Proposition 5.4 we then get $M^c, S_\Delta \models \varphi$.

For (ii), suppose towards a contradiction that $M^c, S_\Delta \models \psi$. Proposition 5.4 implies that $M^c, S_\Delta \models \beta$ for some $\beta \in \mathcal{R}(\psi)$. By the support lemma we have $\bigcap S_\Delta \vdash \beta$, and so $\beta \in \Gamma$ for all $\Gamma \in S_\Delta$. This means that no $\Gamma \in W^c$ includes the set $\Delta \cup \{\neg\beta\}$. By Lemma 5.17, this is only possible if $\Delta \cup \{\neg\beta\} \vdash \bot$. From this we have $\Delta \vdash \neg\neg\beta$, and since for declaratives we have the double negation axiom, also $\Delta \vdash \beta$. By Lemma 5.5 it follows that $\Delta \vdash \psi$, contrary to assumption. □

5.4 Completeness for special frame classes

Particular interpretations of modal logic require restricting one's attention to special classes of Kripke frames, characterized by constraints on the relation R.

---

6 As an aside, notice that if we restrict our language and our axioms to classical formulas, this proof still works, yielding a completeness result for the logic of strict implication in a classical setting over all Kripke models. Interestingly, our axiomatization is quite different from the one of Veltman [20]. A closer look at the relation must be left for another occasion.

Three of the most important constraints are reflexivity (wRw for all w) transitivity (wRv & vRu implies wRu) and Euclideanness (wRv & wRu implies vRu). In this section I provide axioms for $\Rightarrow$ corresponding to these conditions, and show that augmenting our system for $\mathsf{Inq}^{\Rightarrow}$ with these axioms leads to a system which is sound and complete for the corresponding class of frames. The relevant axioms are the following ones, where $\varphi, \psi, \chi \in \mathcal{L}^{\Rightarrow}$ and $\alpha, \beta \in \mathcal{L}_!$.

- Reflexivity: $(\alpha \Rightarrow \beta) \to (\alpha \to \beta)$
- Transitivity: $(\varphi \Rightarrow \psi) \to (\chi \Rightarrow (\varphi \Rightarrow \psi))$
- Euclideanness: $\neg(\varphi \Rightarrow \psi) \to (\chi \Rightarrow \neg(\varphi \Rightarrow \psi))$

Theorem 5.22 *Adding one or more of the above axioms to our proof system results in a sound and complete system for the class of frames whose accessibility relation has the properties corresponding to those axioms.*

As usual, soundness is shown directly, verifying that all instances of the above axioms are valid over the corresponding class. For the completeness result, we construct a canonical model M^c_+ as above, but based on the extended proof system $\vdash_+$. Proceeding exactly as above we can show that if $\Phi \not\vdash_+ \psi$ then there is a state S in M^c_+ which supports all formulas in Φ but does not support ψ. Given this, we just need to show that the frame underlying the canonical model belongs to the relevant class. This is the role of the following lemma.

Lemma 5.23 (Canonicity) *If the proof system $\vdash_+$ includes one of the above axioms, then the canonical accessibility relation has the corresponding property.*

Proof.

- Suppose $\vdash_+$ contains all instances of the axiom for reflexivity. Consider a point Γ in the canonical model M^c_+. We want to show that $\Gamma R^c_+ \Gamma$. For this, take $\alpha, \beta \in \mathcal{L}_!$ and suppose $(\alpha \Rightarrow \beta) \in \Gamma$ and $\alpha \in \Gamma$: we need to show that $\beta \in \Gamma$. Since $(\alpha \Rightarrow \beta) \to (\alpha \to \beta)$ is an axiom, and since Γ contains $\alpha \Rightarrow \beta$ and is closed under deduction of declaratives in $\vdash_+$, we have $(\alpha \to \beta) \in \Gamma$. Since $\alpha \in \Gamma$, again by closure under deduction we have $\beta \in \Gamma$.
- Suppose $\vdash_+$ contains all instances of the axiom for transitivity. Suppose $\Gamma R^c_+ \Gamma'$ and $\Gamma' R^c_+ \Gamma''$. We want to show that $\Gamma R^c_+ \Gamma''$. For this, let $\alpha, \beta \in \mathcal{L}_!$ and suppose $(\alpha \Rightarrow \beta) \in \Gamma$ and $\alpha \in \Gamma''$: we need to show that $\beta \in \Gamma''$. Since $(\alpha \Rightarrow \beta) \to (\top \Rightarrow (\alpha \Rightarrow \beta))$ is an axiom and Γ is closed under deduction of declaratives in $\vdash_+$, we have $(\top \Rightarrow (\alpha \Rightarrow \beta)) \in \Gamma$. Since $\top \in \Gamma'$ and $\Gamma R^c_+ \Gamma'$, by definition of R^c_+ we have $(\alpha \Rightarrow \beta) \in \Gamma'$. Since $\alpha \in \Gamma''$ and $\Gamma' R^c_+ \Gamma''$, again by definition of R^c_+ we have $\beta \in \Gamma''$, as we wanted.
- Suppose $\vdash_+$ contains all instances of the axiom for Euclideanness. Suppose $\Gamma R^c_+ \Gamma'$ and $\Gamma R^c_+ \Gamma''$. We want to show that $\Gamma' R^c_+ \Gamma''$. For this, let $\alpha, \beta \in \mathcal{L}_!$ and suppose $(\alpha \Rightarrow \beta) \in \Gamma'$ and $\alpha \in \Gamma''$: we need to show that $\beta \in \Gamma''$. Towards a contradiction, suppose $\beta \notin \Gamma''$. Since $\Gamma R^c_+ \Gamma''$ and $\alpha \in \Gamma''$, by definition of R^c_+ it follows that $(\alpha \Rightarrow \beta) \notin \Gamma$. Since Γ is a complete theory of declaratives, $\neg(\alpha \Rightarrow \beta) \in \Gamma$. Since $\neg(\alpha \Rightarrow \beta) \to (\top \Rightarrow \neg(\alpha \Rightarrow \beta))$ is an axiom and Γ is closed under deduction of declaratives in $\vdash_+$, we have

$(\top \Rightarrow \neg(\alpha \Rightarrow \beta)) \in \Gamma$. Since $\top \in \Gamma'$ and $\Gamma R^c_+ \Gamma'$ we have $\neg(\alpha \Rightarrow \beta) \in \Gamma'$. But this is impossible, since Γ' is consistent and $(\alpha \Rightarrow \beta) \in \Gamma'$. □

6 Comparison with Goranko and Kuusisto's approach

In a recent paper [10], Goranko and Kuusisto (henceforth G&K) develop a logic of propositional dependency which shares some features of the present proposal. Indeed, the motivations for their work are similar to the ones that led us to investigate the system $\mathsf{Inq}^{\Rightarrow}$: e.g., G&K also aim at a system in which we can make good sense of negated and disjunctive dependence statements.

G&K's system is based on a modal language built up from atoms, connectives, and an operator D of flexible arity. Formulas are evaluated relative to a model $M = \langle W, V\rangle$ of the kind assumed above for InqB, and a world $w \in W$. Atoms and connectives are interpreted as usual; the clause for D is as follows:

$$M, w \models \mathsf{D}(\alpha_1, \dots, \alpha_n, \beta) \quad \text{iff} \quad \exists f : \{0,1\}^n \to \{0,1\} \text{ s.t. } \forall v \in W : \\ V(v, \beta) = f(V(v, \alpha_1), \dots, V(v, \alpha_n))$$

Now, with any model $M = \langle W, V\rangle$ we can associate a corresponding Kripke model $M^U = \langle W, U, V\rangle$, where U is the universal accessibility map, $U = W \times W$. Then, it is easy to see that we have the following equivalence.

$$M, w \models \mathsf{D}(\alpha_1, \dots, \alpha_n, \beta) \quad \text{iff} \quad M^U, w \models ?\alpha_1 \wedge \dots \wedge ?\alpha_n \Rightarrow ?\beta$$

This connection allows us to bring out precisely what the differences are between the proposal of G&K and the one I described in this paper.

Firstly, G&K interpret dependency by means of a universal accessibility relation, while I allow it to be interpreted by means of an arbitrary relation. For some applications, this is necessary: e.g., in a multi-agent epistemic setting, different agents will have access to different information, and so different epistemic dependencies will hold from the perspective of each agent (see footnote 4). Modeling such scenarios requires using non-universal accessibility relations.

However, the main difference between the two approaches lies in the way the dependency relation is construed. In G&K's system, it is construed as a relation between statements, while in the view I propose it is construed as a relation between questions. An advantage of G&K's approach is that the semantics of the system can be kept completely truth-conditional. No detour at the level of information states is necessary. This contrasts with the situation in $\mathsf{Inq}^{\Rightarrow}$: although a dependence statement $\varphi \Rightarrow \psi$ itself is truth-conditional, its truth-conditions depend crucially on the support conditions for φ and ψ.

On the other hand, the account of dependence statements provided by $\mathsf{Inq}^{\Rightarrow}$ is more general than the one given by G&K. This is because it can deal with dependencies between questions which are not necessarily *polar* questions of the form $?\alpha$ for some statement α. For a concrete example, consider a conditional question like *whether Sue will dance if Bob asks her.* This question is resolved in a state s if all the worlds in s in which Bob asks Sue to dance agree on whether she accepts. This is captured by formalizing this question as $a \to ?d$.

$$M, s \models a \to ?d \quad \text{iff} \quad M, s \cap |a|_M \models ?d$$

Notice that the conditional question does *not* correspond to the polar question $?(a \to d)$: establishing that Sue will not dance if Bob asks her $(a \to \neg d)$ is sufficient to resolve the conditional question, but it does not determine whether the conditional $a \to d$ is true or false, and thus it does not resolve $?(a \to d)$.

Now consider the following dependence statement, where the determined question is the conditional question we just examined.

(2) Whether Sue is in a good mood determines whether she will dance if Bob asks her.

This does not amount to a dependency between the truth-value of two statements; thus, it cannot be formalized by G&K's D operator. By contrast, in our setting (2) can be expressed straightforwardly, following our general strategy, as a strict conditional $?g \Rightarrow (a \to ?d)$ involving the polar question $?g$ (*whether Sue is in a good mood*) and the conditional question $a \to ?d$ (*whether she will dance if Bob asks her*). Spelling out the clauses, for this formula we have:

$$M, w \models ?g \Rightarrow (a \to ?d) \text{ iff } \exists f : \{0,1\} \to \{0,1\} \text{ such that } \forall v \in R[w] \cap |a|_M : V(v,d) = f(V(v,g))$$

Thus, (2) expresses the existence of a functional dependence of d on g not relative to the whole set $R[w]$ of accessible worlds, but only relative to a subset thereof, namely, the set of those accessible worlds where Bob asks Sue to dance. The system $\mathsf{Inq}^{\Rightarrow}$ gives us the means to express such more intricate dependence patterns, and to do so in a perspicuous, systematic, and compositional way.[7]

A further benefit of $\mathsf{Inq}^{\Rightarrow}$ is that it allows for a more fine-grained analysis of the logical operations involved in a dependence statement, which yields a better proof theory. Indeed, G&K's operator D translates to $\mathsf{Inq}^{\Rightarrow}$ as a rather complex combination of operators. Spelling out the abbreviation '?', the formula $\mathsf{D}(\alpha_1, \ldots, \alpha_n, \beta)$ corresponds to our $((\alpha_1 \vee\!\!\!\vee \neg\alpha_1) \wedge \cdots \wedge (\alpha_n \vee\!\!\!\vee \neg\alpha_n)) \Rightarrow (\beta \vee\!\!\!\vee \neg\beta)$. This shows that D bears the burden of performing many operations at once: negation, inquisitive disjunction, conjunction, and the strict conditional. As a consequence, this operator is not simple from a proof-theoretic point of view. For instance, G&K's axiomatization of it contains the following axiom:

$$\mathsf{D}(\alpha_1, \ldots, \alpha_n, \beta) \longleftrightarrow \bigvee_{\chi \in \mathrm{DNF}(\alpha_1, \ldots, \alpha_n)} ((\chi \leftrightarrow \beta) \wedge \mathsf{D}(\chi \leftrightarrow \beta))$$

where $\mathrm{DNF}(\alpha_1, \ldots, \alpha_n)$ is the set of disjunctions of formulas $\delta_1 \wedge \cdots \wedge \delta_n$ where $\delta_i \in \{\alpha_i, \neg\alpha_i\}$. Such an axiom is hardly the sort of characterization one would

[7] Indeed, notice that the translation of (2) as $?g \Rightarrow (a \to ?d)$ can be obtained systematically by applying the following translation rules, where $\rightsquigarrow$ abbreviates "translates to":

- if 'A' $\rightsquigarrow \alpha$ then 'whether A' $\rightsquigarrow ?\alpha$
- if 'A' $\rightsquigarrow \varphi$ and 'B' $\rightsquigarrow \psi$, then 'B if A' $\rightsquigarrow (\varphi \to \psi)$
- if 'A' $\rightsquigarrow \varphi$ and 'B' $\rightsquigarrow \psi$, then 'A determines B' $\rightsquigarrow (\varphi \Rightarrow \psi)$

Assuming that the syntactic parsing of (2) is "(whether (Sue is in a good mood)) determines ((whether (she will dance)) if (Bob asks her))", these rules deliver precisely $?g \Rightarrow (a \to ?d)$.

hope to have for the behavior of a primitive logical constant. By contrast, in $\mathsf{Inq}^{\Rightarrow}$ dependence statements are analyzed in terms of basic building blocks ($\bot$, $\to$, $\mathbin{\rotatebox[origin=c]{-90}{$\geqslant$}}$, $\wedge$, and $\Rightarrow$) each of which obeys simple and natural logical principles.

In spite of these differences in implementation, however, my proposal and the one of G&K converge on a fundamental conceptual point: dependence statements are modal statements, which are true or false at a world w according to whether a dependency holds in a set of worlds associated with w.

References

[1] Ciardelli, I., *Modalities in the realm of questions: Axiomatizing inquisitive epistemic logic*, in: R. Goré, B. Kooi and A. Kurucz, editors, *Advances in Modal Logic* (2014), pp. 94–113.

[2] Ciardelli, I., *Dependency as question entailment*, in: S. Abramsky, J. Kontinen, J. Väänänen and H. Vollmer, editors, *Dependence Logic: theory and applications*, Springer International Publishing Switzerland, 2016 pp. 129–181.

[3] Ciardelli, I., "Questions in Logic," Ph.D. thesis, ILLC, University of Amsterdam (2016).

[4] Ciardelli, I., *Questions as information types*, Synthese **195** (2018), pp. 321–365.

[5] Ciardelli, I., J. Groenendijk and F. Roelofsen, *Inquisitive semantics: A new notion of meaning*, Language and Linguistics Compass **7** (2013), pp. 459–476.

[6] Ciardelli, I., R. Iemhoff and F. Yang, *Questions and dependency in intuitionistic logic* (2018), manuscript, under review.

[7] Ciardelli, I. and F. Roelofsen, *Inquisitive logic*, Journal of Philosophical Logic **40** (2011), pp. 55–94.

[8] Ciardelli, I. and F. Roelofsen, *Inquisitive dynamic epistemic logic*, Synthese **192** (2015), pp. 1643–1687.

[9] Ebbing, J., L. Hella, A. Meier, J.-S. Müller, J. Virtema and H. Vollmer, *Extended modal dependence logic*, in: L. Libkin, U. Köhlenbach and R. de Queiroz, editors, *Proceedings of WoLLIC 2013*, Lecture Notes in Computer Science (2013), pp. 126–137.

[10] Goranko, V. and A. Kuusisto, *Logics for propositional determinacy and independence*, Review of Symbolic Logic (2018), DOI: 10.1007/s11229-016-1221-y.

[11] Groenendijk, J., *The logic of interrogation*, in: T. Matthews and D. Strolovitch, editors, *Semantics and Linguistic Theory* (1999), pp. 109–126.

[12] Groenendijk, J., *Inquisitive semantics: Two possibilities for disjunction*, in: P. Bosch, D. Gabelaia and J. Lang, editors, *Tbilisi Symposium on Language, Logic, and Computation*, Springer-Verlag, 2009 pp. 80–94.

[13] Hintikka, J., "Knowledge and Belief: An Introduction to the Logic of the Two Notions," Cornell University Press, 1962.

[14] Mascarenhas, S., *Inquisitive semantics and logic* (2009), Master Thesis, University of Amsterdam.

[15] Puncochar, V., *Weak negation in inquisitive semantics*, Journal of Logic, Language, and Information **24** (2015), pp. 323–355.

[16] Puncochar, V., *A generalization of inquisitive semantics*, Journal of Philosophical Logic **45** (2016), pp. 399–428.

[17] Puncochar, V., *Substructural inquisitive logics* (2018), manuscript, under review.

[18] Väänänen, J., "Dependence Logic: A New Approach to Independence Friendly Logic," Cambridge University Press, Cambridge, 2007.

[19] Väänänen, J., *Modal dependence logic.*, in: K. Apt and R. van Rooij, editors, *New Perspectives on Games and Interaction*, Amsterdam Univ. Press, 2008 pp. 237–254.

[20] Veltman, F., "Logics for Conditionals," Ph.D. thesis, University of Amsterdam (1985).

[21] Yang, F. and J. Väänänen, *Propositional logics of dependence*, Annals of Pure and Applied Logic **167** (2016), pp. 557–589.

One-Generated WS5-Algebras

Alex Citkin

Metropolitan Telecommunications
30 Upper Warren Way
Warren, NJ 07059

Abstract

We describe all finite one-generated **WS5**-algebras, and we describe and study the properties of free one-generated **WS5**-algebra. Using a splitting technique, we also prove that, in contrast to the variety of all **S5**-algebras, which is locally finite, and even though variety $\mathcal{M}$ of all **WS5**-algebras is finitely approximable, the variety $\mathcal{M}$ contains infinitely many non-finitely approximable subvarieties.

Keywords: Heyting algebra, Heyting algebra with additional operator,modal logic, weak logic WS5.

1 Introduction

In this paper, we study variety $\mathcal{M}$ of all **WS5**-algebras – the algebras that are models for modal logic WS5. Logic WS5 is known under different names for quite a long time[1]. It is a modal logic an assertoric part of which is the intuitionistic propositional logic, and modality is **S5**-type modality. This logic has attracted additional attention when it turned out that in Glivenko's theorem for intuitionistic modal logics, WS5 plays the same role as classical logic for intuitionistic logic (see [2]).

An algebraic semantic of WS5 are **WS5**–algebras. **WS5**-algebras are the Heyting algebras with an additional operator $\Box$ which satisfies the following condition:

$$(M_0)\ \Box \mathbf{1} \approx \mathbf{1};$$
$$(M_1)\ \Box x \to x \approx \mathbf{1};$$
$$(M_2)\ \Box(x \to y) \to (\Box x \to \Box y) \approx \mathbf{1};$$
$$(M_3)\ \Box x \to \Box\Box x \approx \mathbf{1};$$
$$(M_4)\ \neg\Box\neg\Box x \approx \Box x.$$

All **WS5**-algebras form a variety denoted by $\mathcal{M}$.

First, we recall all necessary facts about Heyting and **WS5**- algebras. Then, in Section 2, we describe all finite one-generated **WS5**-algebras, and using this description we prove that variety $\mathcal{M}$ of all **WS5**-algebras contains infinitely

[1] To be more precise, we use algebras that are models of logic L_4 from [10] – a unimodal logic on intuitionistic base.

many non-finitely approximable subvarieties. After this we turn to a description of the free one-generated algebra. In Section 4, we describe algebra $\mathbf{F}_{\mathcal{M}}(1)$ – the free one-generated algebra of $\mathcal{M}$, and we prove that $\mathbf{F}_{\mathcal{M}}(1)$ has a rather complex structure. Namely, $\mathbf{F}_{\mathcal{M}}(1)$ contains the infinite ascending and descending chains of open elements and the Heyting reduct of $\mathbf{F}_{\mathcal{M}}(1)$ is non-finitely generated as Heyting algebra.

1.1 Heyting Algebras

Algebra $\langle \mathsf{A}; \wedge, \vee, \rightarrow, \mathbf{0}, \mathbf{1}\rangle$, where $\langle \mathsf{A}; \wedge, \vee, \mathbf{0}, \mathbf{1}\rangle$ is a bounded distributive lattice, and $\rightarrow$ is a relative pseudo-complementation, is called a ***Heyting algebra***. We use the regular abbreviations: $\neg \mathsf{a} := \mathsf{a} \rightarrow \mathbf{0}$ and $\mathsf{a} \leq \mathsf{b} := \mathsf{a} \rightarrow \mathsf{b} = \mathbf{1}$. The variety of all Heyting algebras is denoted by $\mathcal{H}$, Heyting algebras are denoted by $\mathcal{A}, \mathcal{B}$, etc. (perhaps with indexes), and elements of Heyting algebras are denoted by $a, b, \ldots$ (perhaps with indexes).

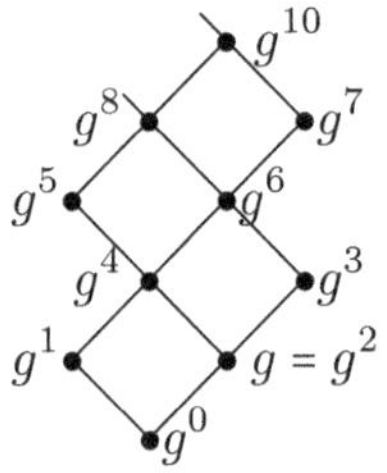

Fig.1.1. Degrees of an element.

For each element a of a Heyting algebra, we define the degrees of a as follows:

$$a^0 := \mathbf{0}, \quad a^1 := \neg a, \quad a^2 := a,$$
$$a^{2k+3} := a^{2k+1} \rightarrow a^{2k}, \quad a^{2k+4} := a^{2k+1} \vee a^{2k+2}$$

for all $k \geq 0$, and we let $a^\omega := \mathbf{1}$.

Let us recall that for each natural number m there exists a unique (up to isomorphism) one-generated Heyting m-element algebra which we denote by $\mathcal{Z}_m$. There is also a unique infinite one-generated Heyting algebra denoted by $\mathcal{Z}_\omega$. Thus, $\mathcal{Z}_2$ is a two-element Boolean algebra which we also denote by $\mathcal{B}_2$.

We will need the following simple observation.

Proposition 1.1 *Let $m > 1$ and element $g_m \in \mathcal{Z}_m$ be a generator of $\mathcal{Z}_m$. Then for every $k > 1$,*

(a) $g_m^m = \mathbf{1}$, and if $k < m$, then $g_m^k < \mathbf{1}$;

(b) if $k \geq m + 2$, then for every element $a \in \mathcal{Z}_m$, $a^k = \mathbf{1}$.

Proof. Proof can be done by a simple induction (or see Fig. 1.1). □

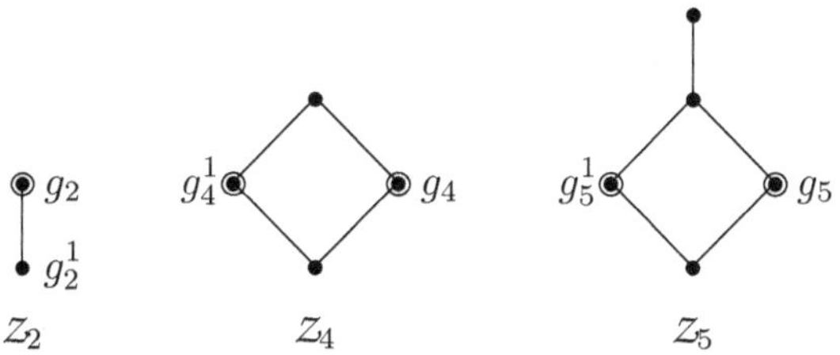

Fig. 1. One-generated Heyting algebras with distinct generators.

Let us observe that all one-generated Heyting algebras, except for Z_2, Z_4 and Z_5, have a unique generator, namely, $g = g^2$, while algebras Z_2, Z_4 and Z_5 have two distinct generators, namely $g = g^2$ and $\neg g = g^1$ (see Fig. 1). Let us note that only in these three one-generated Heyting algebras we have $g = \neg\neg g$, while in the rest of one-generated Heyting algebras $g \neq \neg\neg g$.

In the sequel, we use the notation $\mathbf{A}[\mathsf{a}]$ to underscore that we consider algebra $\mathbf{A}$ with generator a, for instance, $Z_4[g^1]$ means that we consider Z_4 as being generated by g^1.

Remark 1.2 It is worth noting that there is a significant difference between properties of generators of Z_2 and generators of Z_4 or Z_5: for Z_4 and Z_5 the map $\varphi : g \mapsto \neg g$ can be extended to automorphism, while for Z_2 it is not the case. In other words, for every pair of terms $t(x), r(x)$ for Z_4 and Z_5 we have

$$t(g) = r(g) \text{ if and only if } t(\neg g) = r(\neg g),$$

while for Z_2, the above does not hold: take $t(x) = \neg x$ and $r(x) = \mathbf{1}$.

1.2 WS5-Algebras

Algebra $\langle \mathsf{A}; \wedge, \vee, \rightarrow, \mathbf{0}, \mathbf{1}, \Box \rangle$ where $\langle \mathsf{A}; \wedge, \vee, \rightarrow, \mathbf{0}, \mathbf{1} \rangle$ is a Heyting algebra and $\Box$ satisfies identities (M_0) – (M_4), is called a ***WS5-algebra***. It is clear that the class of all **WS5**-algebras forms a variety which we denote by $\mathcal{M}$. All information about monadic Heyting algebras and, in particular about **WS5**-algebras that we use, can be found in [1]. To distinguish **WS5**-algebras from Heyting algebras, we denote **WS5**-algebras by $\mathbf{A}, \mathbf{B}, \mathbf{Z}$ perhaps with indexes, and elements of **WS5**-algebras are denoted by $\mathsf{a}, \mathsf{b}, \ldots$, perhaps with indexes.

An element a of a **WS5**-algebra $\mathbf{A}$ is called ***open*** if $\Box\mathsf{a} = \mathsf{a}$. Let us also observe that the following identities hold in $\mathcal{M}$ (see e.g. [10])

$$\begin{aligned} &\neg \Box x \approx \Box \neg \Box x; \\ &\Box x \rightarrow \Box y \approx \Box(\Box x \rightarrow \Box y); \\ &\Box(\Box x \vee y) \approx \Box x \vee \Box y. \end{aligned} \tag{1}$$

Hence, all open elements of **WS5**-algebra form a Boolean subalgebra. Thus, each **WS5**-algebra $\mathbf{A} := \langle \mathsf{A}; \wedge, \vee, \rightarrow, \mathbf{0}, \mathbf{1}, \Box \rangle$ can be viewed as a pair $\langle \mathcal{A}, \mathcal{B} \rangle$, where $\mathcal{A} = \langle \mathsf{A}; \wedge, \vee, \rightarrow, \mathbf{0}, \mathbf{1} \rangle$ is a Heyting algebra (a ***Heyting reduct*** of $\mathbf{A}$ – ***h-reduct*** for short), and $\mathcal{B}$ is the Boolean algebra of all open elements of $\mathcal{A}$.

1.3 WS5-Filters and Congruences

Definition 1.3 Recall (see e.g. [1]) that a subset $\mathsf{F} \subseteq \mathbf{A}$ is a **WS5-*filter*** of $\mathbf{A}$ if

(a) if $\mathsf{a}, \mathsf{b} \in \mathsf{F}$, then $\mathsf{a} \wedge \mathsf{b} \in \mathsf{F}$;

(b) if $\mathsf{a} \in \mathsf{F}$ and $\mathsf{a} \leq \mathsf{b}$, then $\mathsf{b} \in \mathsf{F}$;

(c) if $\mathsf{a} \in \mathsf{F}$, then $\Box\mathsf{a} \in \mathsf{F}$.

Since the meet of an arbitrary set of **WS5**-filters of a given **WS5**-algebra $\mathbf{A}$ forms a **WS5**–filter, for any subset of elements $\mathsf{D} \subseteq \mathbf{A}$ there is the least **WS5**-filter $[\mathsf{D})$ containing D. If $\mathsf{D} = \{\mathsf{f}\}$, i.e. if D consists of a single element f, **WS5**-filter $[\mathsf{D})$ is called ***principal***. A filter F is principal if and only if it has the smallest element.

There is a well known correspondence between congruences of $\mathbf{A}$ and **WS5**-filters of $\mathbf{A}$: if $\theta \in \mathsf{Con}(\mathbf{A})$, then the set $\mathsf{F}(\theta) := \{\mathsf{a} \in \mathbf{A} \,|\, \mathsf{a} \equiv \mathbf{1} \mod \theta\}$ is a **WS5**-filter of $\mathbf{A}$, and, given a **WS5**-filter F, the relation $\theta(\mathsf{F})$ defined by

$$\mathsf{a} \equiv \mathsf{b} \mod \theta(\mathsf{F}) \text{ if and only if } \mathsf{a} \leftrightarrow \mathsf{b} \in \mathsf{F}, \tag{2}$$

where $\mathsf{a} \leftrightarrow \mathsf{b} := (\mathsf{a} \to \mathsf{b}) \wedge (\mathsf{b} \to \mathsf{a})$, is a congruence of $\mathbf{A}$. And we write $\mathbf{A}/\mathsf{F}$ instead of $\mathbf{A}/\theta(\mathsf{F})$. Let us note that (2) establishes an isomorphism between lattice of filters of a algebra $\mathbf{A}$ and $\mathsf{Con}(\mathbf{A})$.

Because the set of all open elements of a **WS5**-algebra forms a Boolean algebra, it is not hard to demonstrate that a **WS5**-algebra is ***subdirectly irreducible*** (s.i. for short) if and only if it has exactly two open elements, namely $\mathbf{0}$ and $\mathbf{1}$, that is, s.i. **WS5**-algebra corresponds to a pair $\langle \mathcal{A}, \mathcal{B}_2 \rangle$, where $\mathcal{B}_2$ is the two-element Boolean algebra. Also, any nontrivial s.i. **WS5**-algebra $\mathbf{A}$ is ***simple***: $\mathbf{A}$ has precisely two congruences. By $\mathcal{M}_{fsi}$ we denotes the set of all finite s.i. **WS5**-algebras.

The following simple observation plays an important role in the sequel.

Proposition 1.4 *Let $\mathbf{A} = \prod_{i \in I} \mathbf{B}_i$ be a direct product of a family of nontrivial s.i.* **WS5***-algebras and $\mathsf{F} \subset \mathbf{A}$ be a principal filter. Then there is a subset $J \subseteq I$ such that*

$$\mathbf{A}/\theta(\mathsf{F}) \cong \prod_{j \in J} \mathbf{B}_j. \tag{3}$$

Proof. Let $\mathbf{A} = \prod_{i \in I} \mathbf{B}_i$ be a direct product of nontrivial s.i. **WS5**-algebra. Hence, all algebras $\mathbf{B}_i$ are simple. Thus, each $\mathbf{B}_i$ contains precisely two open elements: $\mathbf{0}$ and $\mathbf{1}$. Hence, open elements of $\mathbf{A}$ are precisely the elements each projection of which belongs to $\{\mathbf{0}, \mathbf{1}\}$.

Because filter F is principal, it has the smallest element f. By property (c) of the definition of **WS5**-filter, f is an open element, and hence, its every projection belongs to $\{\mathbf{0}, \mathbf{1}\}$. Because $\mathbf{A}/\theta(\mathsf{F})$ is nontrivial, $\mathsf{f} \neq \mathbf{0}$, therefore, element f has distinct from $\mathbf{0}$ projections. Let

$$J = \{i \in I \,|\, \pi_i(\mathsf{f}) = \mathbf{1}\}, \text{ where } \pi_i \text{ is the projection to } \mathbf{B}_i.$$

Then, from (2) for any pair of elements $\mathsf{a},\mathsf{b} \in \mathbf{A}$,

$$\mathsf{a} \equiv \mathsf{b} \mod \theta(\mathsf{F}) \text{ if and only if } \mathsf{a} \leftrightarrow \mathsf{b} \in \mathsf{F}.$$

That is,

$$\mathsf{a} \equiv \mathsf{b} \mod \theta(\mathsf{F}) \text{ if and only if } \pi_j(\mathsf{a}) = \pi_j(\mathsf{b}) \text{ for all } j \in J.$$

And this means that $\mathbf{A}/\theta(\mathsf{F}) \cong \prod_{j\in J} \mathbf{B}_j$. □

It is clear that all **WS5**-filters of any finite **WS5**-algebra are principal. Thus, the following holds.

Corollary 1.5 *Let* $\mathbf{A} = \prod_{i\in I} \mathbf{A}_i$ *be a finite* **WS5**-*algebra and all* $\mathbf{A}_i$ *be simple. If* $\mathbf{B}$ *is a nontrivial homomorphic image of* $\mathbf{A}$*, then* $\mathbf{B} \cong \prod_{i\in J} \mathbf{A}_j$ *for some* $J \subseteq I$.

1.4 Some Properties of Variety $\mathcal{M}$

Variety $\mathcal{M}$ is well-behaved. First, we note that every nontrivial s.i. algebra is simple, that is, variety $\mathcal{M}$ is ***semisimple*** and hence, each nontrivial **WS5**-algebra is a subdirect product of simple algebras.

Next, we recall (e.g. [5]) that a term $t(x,y,z)$ is a ***discriminator*** on algebra $\mathbf{A}$ if for any $\mathsf{a},\mathsf{b},\mathsf{c} \in \mathbf{A}$

$$t(\mathsf{a},\mathsf{b},\mathsf{c}) = \begin{cases} \mathsf{a} & \text{if } \mathsf{a} \neq \mathsf{b} \\ \mathsf{c} & \text{if } \mathsf{a} = \mathsf{b}; \end{cases}$$

and a variety $\mathcal{V}$ is ***discriminator*** if it is generated by a class of algebras having the same of discriminator discriminator term.

It is not hard to verify that term

$$t(x,y,z) = (z \wedge \Box((x \vee y) \to (x \wedge y))) \vee (x \wedge (\Box((x \vee y) \to (x \wedge y)) \to \mathbf{0})$$

is discriminator on s.i. **WS5**-algebras (cf. [11, p. 571]) and, hence, $\mathcal{M}$ is a discriminator variety.

Every discriminator variety is congruence-distributive and congruence-permutable (see e.g. [5, Theorem 9.4]). Hence, $\mathcal{M}$ is congruence-distributive and congruence-permutable.

We will use the following property of congruence-permutable varieties.

Proposition 1.6 *[5, Corollary 10.2] Let* $\mathbf{A}_1,\dots,\mathbf{A}_k$ *be simple algebras in a congruence-permutable variety* $\mathcal{V}$*. If* $\mathbf{B} \leq \mathbf{A}_1 \times \dots \times \mathbf{A}_n$ *is a subdirect product, then* $\mathbf{B} \cong \mathbf{A}_{i_1} \times \dots \times \mathbf{A}_{i_k}$ *for some* $\{i_i,\dots,i_k\} \subseteq \{1,\dots,n\}$.

Thus, in every discriminator variety $\mathcal{V}$ each finite algebra $\mathbf{A} \in \mathcal{V}$ is a direct product of simple algebras from $\mathcal{V}$. Moreover, Theorem 4.71 from [9] entails that each such decomposition contains the same number of factors.

Let us observe that for any **WS5**-algebra $\mathbf{A}$ and any elements $\mathsf{a},\mathsf{b},\mathsf{c},\mathsf{d} \in \mathbf{A}$

$$\mathsf{c} \equiv \mathsf{d} \mod \theta(\mathsf{a},\mathsf{b}) \text{ if and only if } \mathbf{A} \models (\Box(\mathsf{a} \leftrightarrow \mathsf{b}) \to \mathsf{c}) \approx (\Box(\mathsf{a} \leftrightarrow \mathsf{b}) \to \mathsf{d}),$$

where $\theta(\mathsf{a},\mathsf{b})$ is a congruence generated by elements a,b. Therefore (see [4]), variety $\mathcal{M}$ has equationally definable principal congruences (EDPC for short). We will also use that in each variety with EDPC (with finitely many fundamental operations), every nontrivial finite s.i. algebra is a splitting algebra ([3][Corollary 3.2], for Heyting algebras this property was observed in [7]), that is, for each nontrivial finite s.i. algebra $\mathbf{A}$ there is a term $s_{\mathbf{A}}$, such that for each algebra $\mathbf{B} \in \mathcal{M}$,

$$\begin{array}{l} \mathbf{B} \not\models s_{\mathbf{A}} \approx \mathbf{1} \text{ if and only if} \\ \mathbf{A} \text{ is embedded in some homomorphic image of } \mathbf{B}. \end{array} \tag{4}$$

If algebra $\mathbf{B}$ is s.i., and hence, it is simple and does not have nontrivial homomorphic images, we have

$$\mathbf{B} \not\models s_{\mathbf{A}} \approx \mathbf{1} \text{ if and only if } \mathbf{A} \text{ is embedded in } \mathbf{B}, \tag{5}$$

in particular, $\mathbf{A} \not\models s_{\mathbf{A}} \approx \mathbf{1}$.

2 One-generated WS5-algebras

It is clear that every one-generated **WS5**-algebra is a subdirect product of some one-generated s.i. algebras. This is why we start with describing all one-generated s.i. **WS5**-algebras.

2.1 Subdirectly Irreducible One-Generated WS5-algebras

Proposition 2.1 *An s.i.* **WS5***-algebra* $\mathbf{A} \in \mathcal{M}$ *is one-generated if and only if its h-reduct is a one-generated Heyting algebra.*

Proof. Suppose $\mathbf{A} \in \mathcal{V}$ is an s.i. one-generated **WS5**-algebra and g is a generator. Then for each element $\mathsf{a} \in \mathbf{A}$, there is a term $t(x)$ such that $\mathsf{a} = t(\mathsf{g})$. If $\Box r(x)$ is a subterm of $t(x)$, because $\mathbf{A}$ is s.i. and has just two open elements, either $\Box r(\mathsf{g}) = \mathbf{0}$, or $\Box r(\mathsf{g}) = \mathbf{1}$. In either case we can replace $r(x)$ respectively with $\mathbf{0}$ or $\mathbf{1}$, and obtain a new term $t'(x)$ such that $t'(\mathsf{g}) = \mathsf{a}$. In such a way we can reduce $t(x)$ to a Heyting term $t'(x)$ (which does not contain $\Box$) such that $t'(\mathsf{g}) = \mathsf{a}$.

Converse statement is trivial, because every generator of h-reduct is at the same time a generator of **WS5**-algebra. □

Thus, every nontrivial one-generated s.i. **WS5**-algebra is isomorphic to one of the following algebras: $\mathbf{Z}_k := \langle \mathcal{Z}_k, \mathcal{B}_2 \rangle, k = 2, 3, \ldots, \omega$.

Let us note that except for $\mathbf{Z}_2, \mathbf{Z}_4$ and $\mathbf{Z}_5$ all s.i. one-generated **WS5**-algebras have a unique generator.

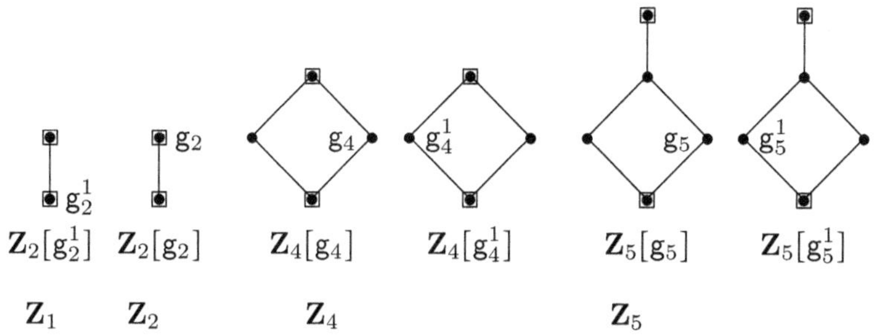

Fig. 2. Generators of $\mathbf{Z}_2, \mathbf{Z}_4$ and $\mathbf{Z}_5$.

WS5-algebras $\mathbf{Z}_2, \mathbf{Z}_4$ and $\mathbf{Z}_5$ have two distinct generators each, but the maps $\varphi : g_4 \mapsto g_4^1$ and $\psi : g_5 \mapsto g_5^1$ can be extended to automorphisms of respective algebras. Hence, if we consider algebras modulo automorphism, we can view $\mathbf{Z}_4$ and $\mathbf{Z}_5$ as having a unique generator. To simplify notation, we denote by $\mathbf{Z}_1$ an isomorphic copy of $\mathbf{Z}_2$ with generator $\mathbf{1}$, preserving $\mathbf{Z}_2$ for an isomorphic copy with the generator $\mathbf{0}$.

Let

$$\mathcal{Z} := \{\mathbf{Z}_k, k = 1, 2, 3, \dots\},$$

i.e. $\mathcal{Z}$ is the list of all finite s.i. one-generated algebras and all these algebras are simple. For each $\mathbf{Z}_i$, by g_i we denote the generator of $\mathbf{Z}_i$.

The following algebras will play the central role in our study of one-generated algebras. We let

$$\mathbf{P} := \prod_{i=1}^{\infty} \mathbf{Z}_i \tag{6}$$

and $\mathsf{g} \in \mathbf{P}$ be an element such that $\pi_i(\mathsf{g}) = \mathsf{g}_i$ for all $i > 0$ and we take

$$\mathbf{Z} \leq \mathbf{P} \text{ to be a subalgebra of } \mathbf{P} \text{ generated by element } \mathsf{g}. \tag{7}$$

Let us note that $\mathbf{Z}$ is a subdirect product of algebras $\mathbf{Z}_i, i > 0$, for element $\pi_i(\mathsf{g})$ generates whole factor $\mathbf{Z}_i$ for each $i > 0$.

We will need the following technical lemma.

Lemma 2.2 *Let $\mathsf{s}_m \in \mathbf{P}$ be an element such that $\pi_m(\mathsf{s}_m) = \mathbf{1}$ and $\pi_j(\mathsf{s}_m) = \mathbf{0}$ for all $j \neq m$. Then, $s_m(\mathsf{g}) \in \mathbf{Z}$ for every $m > 0$.*

Proof. We need to prove that, given element

$$\mathsf{g} = (\mathsf{g}_1, \mathsf{g}_2, \dots),$$

for each $m > 2$ there is a term $s_m(x)$ such that

$$s_m(\mathsf{g}) = \mathsf{s}_m = (\underbrace{\mathbf{0}, \dots, \mathbf{0}}_{m-1 \text{ times}}, \mathbf{1}, \mathbf{0}, \dots).$$

In other words, we need to prove that each element s_m belongs to the subalgebra of $\mathbf{Z}$ generated by element g.

Indeed, we know from Proposition 1.1(a) that $g_m^m = \mathbf{1}$ and $g_k^m < \mathbf{1}$ for all $m < k < \omega$. Thus,

$$\mathsf{g}^m = (\mathsf{g}_1^m, \mathsf{g}_2^m, \dots, \mathsf{g}_m^m, \mathsf{g}_{m+1}^m, \mathsf{g}_{m+2}^m, \dots).$$

and

$$\Box\mathsf{g}^m = (\Box\mathsf{g}_1^m, \Box\mathsf{g}_2^m, \dots, \Box\mathsf{g}_m^m, \Box\mathsf{g}_{m+1}^m, \Box\mathsf{g}_{m+2}^m, \dots) = (\Box\mathsf{g}_1^m, \Box\mathsf{g}_2^m, \dots, \mathbf{1}, \mathbf{0}, \mathbf{0}, \dots).$$

Hence,

$$\pi_m(\Box(\mathsf{g}^m)) = \mathbf{1} \text{ and } \pi_j(\Box(\mathsf{g}^m)) = \mathbf{0} \text{ for all } j > m.$$

Let

$$\tilde{\mathsf{s}}_i = \bigvee_{j=1}^{m} \Box(\mathsf{g}^j) = (\underbrace{1, \dots, 1}_{m \text{ times}}, 0, \dots).$$

Then, $\mathsf{s}_1 = \tilde{\mathsf{s}}_1 = \Box\mathsf{g}$ and $\mathsf{s}_m = \tilde{\mathsf{s}}_m \wedge \neg\tilde{\mathsf{s}}_{m-1}$ for all $m > 1$, and this observation the proof □

With each set $I \subseteq \{1, 2, \dots\}$ we associate a congruence $\theta(I)$ of $\mathbf{Z}$:

$$\mathsf{a} \equiv \mathsf{b} \mod \theta(I) \text{ if and only if } \pi_i(\mathsf{a}) = \pi_i(\mathsf{b}) \text{ for all } i \in I.$$

Corollary 2.3 *For any finite subset $\{i_1, \dots, i_n\} \subseteq \{1, 2, \dots\}$ and any elements $\mathsf{a}_i \in \mathbf{Z}_{i_j}, j = 1, \dots, n$, there is an element $\mathsf{a} \in \mathbf{Z}$ such that*

$$\pi_{i_j}(\mathsf{a}) = \mathsf{a}_{i_j}, \text{ for all } j = 1, \dots, n. \tag{8}$$

Proof. Recall that each algebra $\mathbf{Z}_{i_j}$ is generated by g_{i_j}. Hence, for some numbers $k_1, \dots, k_n$, we have $\mathsf{a}_{i_j} = \mathsf{g}_{i_j}^{k_j}$. Hence

$$(\mathsf{g}^{k_1} \wedge \mathsf{s}_{i_1}) \vee \dots \vee (\mathsf{g}^{k_n} \wedge \mathsf{s}_{i_n})$$

is a desired element of $\mathbf{Z}$. □

Corollary 2.4 *If $I \subseteq \{1, 2, \dots\}$ is a nonempty finite set of numbers, then direct product $\prod_{i \in I} \mathbf{Z}_i$ is (isomorphic to) $\mathbf{Z}/\theta(I)$.*

Proof. Suppose $I = \{i_1, \dots, i_n\}$. If $\mathsf{a} \in \mathbf{Z}$, by $\bar{\mathsf{a}}$ we denote $\theta(I)$-congruence class containing a. If $(\mathsf{a}_1, \dots, \mathsf{a}_n) \in \prod_{i \in I} \mathbf{Z}_i$, by virtue of Corollary 2.3, there is an element $\mathsf{a} \in \mathbf{Z}$ satisfying (8). Thus, the map $\phi : (\mathsf{a}_1, \dots, \mathsf{a}_n) \longrightarrow \bar{\mathsf{a}}$ is an isomorphism between $\mathbf{Z}/\theta(I)$ and $\prod_{i \in I} \mathbf{Z}_i$. □

2.2 Finite One-Generated WS5-algebras

Every finite one-generated **WS5**-algebra is a subdirect product of algebras from $\mathcal{Z}$. Hence, by virtue of Proposition 1.6, every finite one-generated **WS5**-algebra is a direct product of algebras from $\mathcal{Z}$. The following theorem gives a description of all finite one-generated **WS5**-algebras.

Theorem 2.5 *Finite* **WS5**-*algebra* $\mathbf{A}$ *is one-generated if and only if there is a subset* $I \subseteq \{1,2,,\dots\}$ *such that*

$$\mathbf{A} = \prod_{i \in I} \mathbf{Z}_i. \tag{9}$$

Proof. Suppose that $\mathbf{A}$ is finite one-generated **S5**-algebra. Then $\mathbf{A}$ is (isomorphic to) a direct product $\mathbf{B}_1 \times \dots \times \mathbf{B}_n$ of algebras from $\mathcal{Z}$. First, let us verify that for every $k > 2$, algebra $\mathbf{Z}_i$ cannot occur in this decomposition more than once.

For contradiction: assume that $\mathbf{Z}_k$ occurs in the decomposition twice. Then, $\mathbf{Z}_k^2$ is a homomorphic image of $\mathbf{A}$, and hence, $\mathbf{Z}_k^2$ is one-generated. Recall that for every $k \geq 1$, algebra $\mathbf{Z}_k$ has a unique (up to automorphism) generator [2]. Note also that any projection of a generator of algebra $\mathbf{Z}_k \times \mathbf{Z}_k$ is a generator of a respective factor. Thus, if g is a generator of $\mathbf{Z}_k \times \mathbf{Z}_k$, then $\mathsf{g} = (\mathsf{g}_k, \mathsf{g}_k)$, where g_k is a generator of $\mathbf{Z}_k$. It is not hard to see that element $(\mathsf{g}_k, \mathsf{g}_k)$ generates just a diagonal of $\mathbf{Z}_k \times \mathbf{Z}_k$ and not the whole algebra: for instance, element $(\mathbf{0}, \mathbf{1})$ does not belong to the diagonal of $\mathbf{Z}_k \times \mathbf{Z}_k$. Thus, we have arrived to contradiction.

Conversely, let $\{i_1, \dots, i_n\} \subseteq \{1,2,\dots\}$ and $\mathbf{A} = \mathbf{Z}_{i_1} \times \dots \times \mathbf{Z}_{i_n}$. Then, by Corollary 2.4, $\mathbf{A}$ is a homomorphic image of algebra $\mathbf{Z}$, and $\mathbf{A}$ is one-generated, for $\mathbf{Z}$ being one-generated. □

Let us underscore that in Theorem 2.5, I is a set, and therefore, the factors of each direct decomposition, except for $\mathbf{Z}_2$, are unique. Also, it is not hard to see that if $\mathbf{A} = \mathbf{A}_1 \times \dots \times \mathbf{A}_n$, where all algebras $\mathbf{A}_i$ are s.i., then open elements of $\mathbf{A}$ form a Boolean algebra of cardinality 2^n.

We say that a direct product of simple algebras is ***non-repetitive*** if all its factors are mutually non-isomorphic.

Thus, Theorem 2.5 asserts that every finite nontrivial one-generated **WS5**-algebra $\mathbf{A}$ is of one of the following types:

(a) $\mathbf{A}$;

(b) $\mathbf{Z}_2 \times \mathbf{A}$,

(c) $\mathbf{Z}_2 \times \mathbf{Z}_2 \times \mathbf{A}$,

where $\mathbf{A}$ is a non-repetitive product of algebras $\mathbf{Z}_i, i > 2$ and such a representation is unique modulo rearranging the factors.

Example 2.6 There are precisely four (up to isomorphism) distinct one-generated **WS5**-algebras of cardinality 12, namely

$$(\mathcal{Z}_2 \times \mathcal{Z}_6, \mathcal{B}_2^2), \quad (\mathcal{Z}_2 \times \mathcal{Z}_2 \times \mathcal{Z}_3, \mathcal{B}_2^3), \quad (\mathcal{Z}_3 \times \mathcal{Z}_4, \mathcal{B}_2^2), \quad (\mathcal{Z}_{12}, \mathcal{B}_2).$$

These algebras are not pairwise isomorphic because only $(\mathcal{Z}_2 \times \mathcal{Z}_2 \times \mathcal{Z}_3, \mathcal{B}_2^3))$ and $(\mathcal{Z}_3 \times \mathcal{Z}_4, \mathcal{B}_2^2)$ have isomorphic h-reducts, but the former algebra has eight open elements, while the latter – just four.

[2] This is where the distinction between $\mathbf{Z}_1$ and $\mathbf{Z}_2$ comes to play.

Example 2.7 There are precisely three (up to isomorphism) distinct one-generated **WS5**-algebras of cardinality 16, namely:

$$(\mathcal{Z}_2 \times \mathcal{Z}_8, \mathcal{B}_2^2), \quad (\mathcal{Z}_2 \times \mathcal{Z}_2 \times \mathcal{Z}_4, \mathcal{B}_2^3), \quad (\mathcal{Z}_{16}, \mathcal{B}_2).$$

Remark 2.8 In [6], Grigolia suggested an approach to description of a free monadic Heyting algebra with one free generator. From this description it does not immediately follow that a direct decomposition of a finite one-generated **WS5**-algebra, algebra $\mathbf{Z}_2$ can appear twice, while the rest of the factors are distinct modulo isomorphism.

3 Non-finitely approximable subvarieties of $\mathcal{M}$

If L is an extension of WS5, an ***assertoric fragment*** of L is an intermediate logic consisting of all assertoric formulas from L, that is, a logic consisting of all formulas of L without occurrences of $\Box$. The goal of this section is to show that there is an intermediate logic L enjoying the finite model property (f.m.p. for short) and such that it is an assertoric fragment of infinitely many extensions of WS5 without the f.m.p.

If $\mathcal{V}$ is a variety of **WS5**-algebras, then $\mathsf{L}(\mathcal{V})$ is an extension of WS5 consisting of all formulas valid in $\mathcal{V}$.

If $\mathcal{V}$ is a variety of **WS5**-algebras, by $\mathcal{V}^-$ we denote the variety of Heyting algebras generated by the Heyting reducts of all algebras from $\mathcal{V}$. Let us note that two logics $\mathsf{L}(\mathcal{V}_0)$ and $\mathsf{L}(\mathcal{V}_1)$ have the same assertoric fragment if and only iff $\mathcal{V}_0^- = \mathcal{V}_1^-$.

First, let us show that there are intermediate logics that are assertoric fragments of infinitely many extensions of WS5.

Proposition 3.1 *Let I be a set of natural numbers and let $\mathcal{V}_I$ be a variety generated by algebras $\{\mathbf{Z}_i, i \in I\}$ and $\mathbf{Z}_\omega$. Then, logics $\mathsf{L}(\mathcal{V}_I)$ and $\mathsf{L}(\mathcal{V}_\varnothing)$ have the same assertoric fragment.*

Proof. Indeed, it is clear that $\mathcal{V}_\varnothing \subseteq \mathcal{V}_I$, and hence, $\mathcal{V}_\varnothing^- \subseteq \mathcal{V}_I^-$. On the other hand, for each $i \in I$, $\mathcal{Z}_i$ is a homomorphic image of $\mathcal{Z}_\omega$, and hence, $\mathcal{V}_I^- \subseteq \mathcal{V}_\varnothing^-$. □

Let us recall that a variety $\mathcal{V}$ is said to be ***finitely approximable*** if $\mathcal{V}$ is generated by its finite algebras. If $\mathcal{V}$ is a finitely approximable variety, then each of its free algebras $\mathbf{F}_\mathcal{V}(n)$ is finitely approximable, that is, $\mathbf{F}_\mathcal{V}(n)$ is a subdirect product of the finite algebras (see [8, Chapter VI Theorem 5]). Variety $\mathcal{V}$ is finitely approximable if and only if $\mathsf{L}(\mathcal{V})$ enjoys the f.m.p.

Proposition 3.2 *The variety $\mathcal{V}_I^-$ from Proposition 3.1 is finitely approximable.*

Proof. The proof follows from the observation that variety $\mathcal{V}_\varnothing^-$ is generated by $\mathcal{Z}_\omega$ which is a subdirect product of finite algebras $\mathcal{Z}_i, i > 0$. And this means that variety $\mathcal{V}_\varnothing^-$ is generated by finite algebras, and thus, it is finitely approximable.□

A variety is ***locally finite*** if each of its finitely generated algebra is finite. A variety $\mathcal{V}$ is locally finite if and only if free algebra $\mathbf{F}(n)$ is finite for every finite n (see [5, Theorm 10.15]). It is clear that every locally finite variety is finitely approximable, and that every subvariety of locally finite variety is locally finite.

For instance, a variety of all **S5**-algebras is locally finite, while variety $\mathcal{M}$ of all **WS5**-algebras is not locally finite but it is nevertheless finitely approximable [1, Theorem 42].

For each $m > 6$, let $\mathcal{V}_m$ be a variety generated by algebras $\{\mathbf{Z}_k, 6 < k \le m\}$ and $\mathbf{Z}_\omega$. From Proposition 3.1 we know that $\mathcal{V}_m^- = \mathcal{V}_n^-$ for all $m, n > 6$.

Theorem 3.3 *For each $m > 6$, variety $\mathcal{V}_m$ is not finitely approximable.*

Proof. First, we observe that $\mathcal{V}_m$ contains the infinite one-generated algebra $\mathbf{Z}_\omega$. Hence, algebra $\mathbf{F}_{\mathcal{V}_m}(1)$ is infinite. Thus, to prove that $\mathcal{V}_m$ is non-finitely approximable, it suffices to demonstrate that $\mathbf{F}_{\mathcal{V}_m}(1)$ is not a subdirect product of finite algebras. Because $\mathbf{F}_{\mathcal{V}_m}(1)$ is one generated, every subdirect factor of $\mathbf{F}_{\mathcal{V}_m}(1)$ is one-generated too. Therefore, to prove that $\mathbf{F}_{\mathcal{V}_m}(1)$ is non-finitely approximable, it is enough to show that $\mathbf{F}_{\mathcal{V}_m}(1)$ does not belong to subvariety $\mathcal{V}_m^{(1)} \subseteq \mathcal{V}_m$ generated by all finite one-generated s.i. algebras from $\mathcal{V}_m$.

Next, we note that for any finite s.i. algebra $\mathbf{A} \in \mathcal{V}_m$, $\mathbf{A}$ is (isomorphic to) a subalgebra of one of the algebra $\mathbf{Z}_k$ for some $k \in [7, m] \cup \{\omega\}$.

Indeed, by (5), there is a term s such that $\mathbf{A} \not\models s \approx \mathbf{1}$ and for every $\mathbf{B} \in \mathcal{V}_m$, $\mathbf{B} \not\models s \approx \mathbf{1}$ entails that $\mathbf{A}$ is a subalgebra of $\mathbf{B}$. Because algebras $\mathbf{Z}_k$, where $k \in [7, m] \cup \{\omega\}$ generate variety $\mathcal{V}$, for some $k \in [7, m] \cup \{\omega\}$, identity $s \approx \mathbf{1}$ fails in $\mathbf{Z}_k$.

Also, the only proper subalgebras of algebra $\mathbf{Z}_k, k \in [7, m] \cup \{\omega\}$ are $\mathbf{Z}_2, \mathbf{Z}_3, \mathbf{Z}_5$. Indeed, each algebra $\mathbf{Z}_k, k \ge 7$ contains just three types of elements: if $\mathsf{a} \in \mathbf{Z}_k$, then

$$\neg\neg\mathsf{a} = \begin{cases} \mathsf{a} \\ \mathbf{1} \\ \text{neither } \mathsf{a}, \text{ nor } \mathbf{1} \end{cases}$$

If $\neg\neg\mathsf{a} = \mathsf{a}$ and $\mathsf{a} \in \{\mathbf{0}, \mathbf{1}\}$, then a generates (a subalgebra isomorphic to) $\mathbf{Z}_2$, otherwise, a generates $\mathbf{Z}_5$. If $\neg\neg\mathsf{a} = \mathbf{1}$, then a generates $\mathbf{Z}_3$. And, if $\neg\neg\mathsf{a} \ne \mathsf{a}$ and $\neg\neg\mathsf{a} \ne \mathbf{1}$, element a is the generator of $\mathbf{Z}_k$, and hence, a generates whole algebra $\mathbf{Z}_k$.

Thus, variety $\mathcal{V}_m$ contains finite s.i. one-generated algebras only from $\{\mathbf{Z}_k, k \in [7, m] \cup \{2, 3, 5\}\}$. That is, variety $\mathcal{V}_m^{(1)}$ is generated by finite set of finite algebras. Then, variety $\mathcal{V}_m^{(1)}$ is locally finite (see [5, Theorem 10.16]), and therefore, $\mathcal{V}_m^{(1)}$ does not contain infinite finitely-generated algebras. Thus, $\mathbf{F}_{\mathcal{V}_m}(1) \notin \mathcal{V}_m^{(1)}$. □

Corollary 3.4 *The logic of Heyting algebra $\mathcal{Z}_\omega$ enjoys the f.m.p. and it is an assertoric fragment of infinitely many extensions of* WS5 *without the f.m.p.*

4 Free WS5-Algebra on one free generator

The goal of this section is to prove that algebra $\mathbf{Z}$, introduced earlier (see (7)), is free in $\mathcal{M}$, and then to give an alternative intrinsic description of this algebra.

Theorem 4.1 *Algebra* $\mathbf{Z}$ *is freely generated in variety* $\mathcal{M}$ *by element* g, *that is* $\mathbf{Z} \cong \mathbf{F}_{\mathcal{M}}(1)$.

Proof. First, recall from [8, Section 12.2 Theorem 1] that for any variety $\mathcal{V}$, any algebra $\mathbf{A} \in \mathcal{V}$ generated by element $\mathsf{g} \in \mathbf{A}$, if for any identity $t(x) \approx r(x)$,

$$t(\mathsf{g}) = r(\mathsf{g}) \text{ yields } \mathcal{V} \models t(x) \approx r(x),$$

then g is a free generator.

To prove the contrapositive, we take any identity $t(x) \approx r(x)$, such that $\mathcal{M} \not\models t(x) \approx r(x)$. Because variety $\mathcal{M}$ is finitely approximable, there is a finite algebra $\mathbf{A} \in \mathcal{M}$ such that $\mathbf{A} \not\models t(x) \approx r(x)$, that is, for some $\mathsf{a} \in \mathbf{A}$, $t(\mathsf{a}) \neq r(\mathsf{a})$. We can safely assume that a generated $\mathbf{A}$.

By Birkhoff's theorem, algebra $\mathbf{A}$ is (isomorphic to) a subdirect product $\mathbf{A}_1 \times \cdots \times \mathbf{A}_n$ of s.i. **WS5**-algebras. Hence $\mathbf{A}_k \not\models t(x) \approx r(x)$ for some $1 \leq k \leq n$ and $t(\mathsf{a}_k) \neq r(\mathsf{a}_k)$, where $\mathsf{a}_k = \pi_k(\mathsf{a})$. Let us also note that element a_k generates $\mathbf{A}_k$, for element a generates $\mathbf{A}$ and $\mathbf{A}_k$ is a subdirect factor of $\mathbf{A}$.

Thus, algebra $\mathbf{A}_k$ is finite and s.i. and it is generated by element a_k. Hence, for some $m > 0$, $\mathbf{A}_k \cong \mathbf{Z}_m \in \mathcal{Z}$ and $t(\mathsf{g}_m) \neq r(\mathsf{g}_m)$. Recall that $\mathbf{Z}_m$ is a subdirect factor of algebra $\mathbf{Z}$ and that $\mathsf{g}_m = \pi_m(\mathsf{g})$, which entails $t(\mathsf{g}) \neq r(\mathsf{g})$ and this observation completes the proof. □

Let us give a criterion for an element of $\mathbf{P}$ to belong to $\mathbf{Z}$.

4.1 Leveled Elements

We regard elements of $\mathbf{P}$ as infinite vectors, and let $\pi_j, j > 0$ be a j-th component (j-th projection). For instance, if g is a generator defined by (7), then $\pi_j(\mathsf{g}) = \mathsf{g}_j$.

Definition 4.2 Let $k > 0$ and $m \in \{0, 1, \ldots, \omega\}$. An element $\mathsf{a} \in \mathbf{P}$ is called (k, m)-***leveled***, if for all $i \geq k$,

$$\pi_i(\mathsf{a}) = \mathsf{g}_i^m.$$

Thus, an element $\mathsf{a} \in \mathbf{P}$ is (k, m)-leveled if, starting from k-th component, each component of a is equal to the same degree of the respective generator, that is, a is of form

$$(\mathsf{a}_1, \ldots, \mathsf{a}_{k-1}, \mathsf{g}_k^m, \mathsf{g}_{k+1}^m, \ldots)$$

For instance, g is $(2,2)$-leveled, because each of its components, starting with $\pi_2(\mathsf{g})$, is the 2-nd degree of the respective generator (recall that $\mathsf{a}^2 = \mathsf{a}$).

Definition 4.3 An element $\mathsf{a} \in \mathbf{P}$ is ***leveled***, if it is (k, m)-leveled for some $k > 0$ and $m \in \{0, 1, \ldots, \omega\}$.

For instance, if a is a ***binary element***, that is each component of a is $\mathbf{0}$ or $\mathbf{1}$, then a is leveled if and only if it contains either a finite number of $\mathbf{0}$-components, or a finite number of $\mathbf{1}$-components.

It is obvious that if an element a is (k, m)-leveled, it is (k', m)-leveled for every $k' > k$. We will need a bit more stronger property of leveled elements.

Proposition 4.4 *Let* $\mathsf{a} \in \mathbf{P}$ *be a leveled element. Then for some* $k > 0$ *and* $m \in \{0, 1, \ldots, \omega\}$, *either element* a *is* (k, ω)*-leveled, or it is* (k, m)*-leveled and* $\pi_i(\mathsf{a}) < \mathbf{1}$ *for all* $i \geq k$.

Proof. Let $\mathsf{a} \in \mathbf{P}$ be a (k, m)-leveled element and $m < \omega$. Without loss of generality, we can assume that $k \geq m + 2$. Then, by Proposition 1.1(b), for all $j \geq k$, we have $\pi_j(\mathsf{a}) < \mathbf{1}$.. □

Proposition 4.5 *Let* $\mathbf{L} \subseteq \mathbf{P}$ *be a subset of all leveled elements of* $\mathbf{P}$. $\mathbf{L}$ *forms a subalgebra of* $\mathbf{P}$.

Proof. First, let us recall that, by the definition, $\mathsf{a}^0 = \mathbf{0}$ and $\mathsf{a}^\omega = \mathbf{1}$ for any element a. Hence, elements $\mathbf{0_P} = (\mathbf{0}, \mathbf{0}, \ldots)$ and $\mathbf{1_P} = (\mathbf{1}, \mathbf{1}, \ldots)$ are $(1, 0)$- and $(1, \omega)$-leveled. Thus, $\mathbf{0_P}, \mathbf{1_P} \in \mathbf{L}$, and we need to check only that $\mathbf{L}$ is closed under $\wedge, \vee, \rightarrow$ and $\Box$.

Let us start with $\Box$. Suppose that $\mathsf{a} \in \mathbf{L}$ and a is (k, m)-leveled. Then, by Proposition 4.4, we can assume that either $m = \omega$, or for each $j \geq k$, $\pi_j(\mathsf{a}) < \mathbf{1}$. Hence, either $\Box\mathsf{a}$ is (k, ω)-leveled, or $\Box\mathsf{a}$ is $(k, 0)$-leveled (because each $\mathbf{Z}_j$ is s.i., and therefore, $\Box\mathsf{b} = \mathbf{0}_{\mathbf{Z}_j}$ as long as $\mathsf{b} < \mathbf{1}_{\mathbf{Z}_j}$).

Now, suppose that $\mathsf{a}, \mathsf{b} \in \mathbf{L}$ are leveled elements. Without loss of generality we can assume that a is (k, m)-leveled and b is (k, n)-leveled, where $k > 1$ and $m, n \geq k + 2$. If $m = \omega$ or $n = \omega$, the statement is trivial.

Suppose that $m \neq \omega$, $n \neq \omega$ and $\circ \in \{\wedge, \vee, \rightarrow\}$. Then for each $j \geq k$ we have

$$\pi_j(\mathsf{a} \circ \mathsf{b}) = \mathsf{g}_j^m \circ \mathsf{g}_j^n,$$

and for some s, because g_j is a generator of Z_j, we have

$$\mathsf{g}_j^m \circ \mathsf{g}_j^n = \mathsf{g}_j^s.$$

Hence, $\mathsf{a} \circ \mathsf{b}$ is (k, s)-leveled, and this observation completes the proof. □

Now we are in a position to give a intrinsic description of $\mathbf{Z}$.

Theorem 4.6 *Algebra* $\mathbf{Z}$ *is a subalgebra of* $\mathbf{P}$ *consisting of all leveled elements.*

Proof. Let $\mathbf{L} \subseteq \mathbf{P}$ be a set of all leveled elements of $\mathbf{P}$. The generator g is leveled and, therefore $\mathsf{g} \in \mathbf{L}$. By Proposition 4.5, $\mathbf{L}$ is a subalgebra of $\mathbf{P}$, hence, $\mathbf{Z} \subseteq \mathbf{L}$.

To prove $\mathbf{L} \subseteq \mathbf{Z}$ it is sufficient to demonstrate that every leveled element can be expressed via g. More precisely, we will demonstrate that for every element $\mathsf{a} \in \mathbf{L}$ there is a term $t(x)$ such that $\mathsf{a} = t(\mathsf{g})$, where g is a generator of $\mathbf{Z}$.

Let us consider elements $\mathsf{s}_m, m = 1, 2, \ldots$ defined as in Lemma 2.2:

$$\pi_j(\mathsf{s}_m) = \begin{cases} \mathbf{1}, & \text{if } j = m; \\ \mathbf{0}, & \text{otherwise.} \end{cases}$$

It is clear that each element s_m is leveled, that is, $\mathsf{s}_m \in \mathbf{L}$ as well as $\mathsf{g} \in \mathbf{L}$. On the other hand, by virtue of Lemma 2.2, $\mathsf{s}_m \in \mathbf{Z}$ and $\mathsf{g} \in \mathbf{Z}$. Because $\mathbf{Z}$ is closed under WS5-operations, to prove $\mathbf{L} \subseteq \mathbf{Z}$ it is sufficient to show that each element $\mathsf{a} \in \mathbf{L}$ can be expressed via g and elements from $\{\mathsf{s}_k, k > 0\}$.

Let $\mathsf{a} \in \mathbf{L}$. Recall that $\mathsf{g} = (\mathsf{g}_1, \mathsf{g}_2, \dots)$ is a generator of $\mathbf{Z}$. Because a is leveled, a is (k, m)-leveled for some k and m. Hence, a is an element of form

$$\mathsf{a} = (\mathsf{g}_1^{m_1}, \dots, \mathsf{g}_{k-1}^{m_{k-1}}, \mathsf{g}_k^{m}, \mathsf{g}_{k+1}^{m}, \dots).$$

Let $\mathsf{s}_i \in \mathbf{P}, i > 0$, that is,

$$\mathsf{s}_i = (\underbrace{\mathbf{0}, \dots, \mathbf{0}}_{i-1 \text{ times}}, \mathbf{1}, \mathbf{0}, \dots).$$

Hence, for each $k > 0$,

$$\mathsf{s}_1 \vee \dots \vee \mathsf{s}_k = (\underbrace{\mathbf{1}, \dots, \mathbf{1}}_{k \text{ times}}, \mathbf{0}, \dots) \text{ and } \neg(\mathsf{s}_1 \vee \dots \vee \mathsf{s}_k) = (\underbrace{\mathbf{0}, \dots, \mathbf{0}}_{k \text{ times}}, \mathbf{1}, \dots).$$

Therefore we can express a in the following way:

$$\mathsf{a} = (\mathsf{s}_1 \wedge \mathsf{g}^{m_1}) \vee \dots \vee (\mathsf{s}_{k-1} \wedge \mathsf{g}^{m_{k-1}}) \vee (\neg(\mathsf{s}_1 \vee \dots \vee \mathsf{s}_{k-1}) \wedge \mathsf{g}^{m}). \tag{10}$$

And this completed the proof. □

4.2 Some Properties of $\mathbf{F}_{\mathcal{M}}(1)$

First, let as take a closer look at h-reduct of $\mathbf{F}_{\mathcal{M}}(1)$.

We say that a **WS5**-algebra $\mathbf{A}$ is ***finitely h-generated*** if h-reduct of $\mathbf{A}$ is finitely generated as Heyting algebra. In particular, any finite **WS5**-algebra is finitely h-generated.

Let us observe that (10) entails that elements g and $\mathsf{s}_i, i > 0$ generate h-reduct of $\mathbf{Z}$ as Heyting algebra. On the other hand, the following holds.

Corollary 4.7 $\mathbf{Z}$ *is non-finitely h-generated.*

Proof. Indeed, let us take any finite set of elements $\mathsf{a}_1, \dots, \mathsf{a}_n \in \mathbf{Z}$. Then, for every $0 < i \leq n$ every element a_i is (k_i, m_i)-leveled. Let $k = \max(k_i)$. Then every element a_i is (k, m_i)-leveled. Any Heyting operation over (k, m_i)-leveled elements yields a (k, m')-leveled element. So, for instance, element $\mathsf{s}_k \in \mathbf{Z}$ is not (k, m)-leveled for any m (it is $(k+1, 0)$-leveled) and, hence, it cannot by expressed via $\mathsf{a}_i, i \leq n$ and Heyting operations. Thus, elements $\mathsf{a}_1, \dots, \mathsf{a}_n$ do not generate h-reduct of $\mathbf{Z}$. □

Now, let us note that

$$\mathsf{s}_1 < \mathsf{s}_1 \vee \mathsf{s}_2 < \mathsf{s}_1 \vee \mathsf{s}_2 \vee \mathsf{s}_3 \dots, \text{ and subsequently, } \neg\mathsf{s}_1 > \neg(\mathsf{s}_1 \vee \mathsf{s}_2) > \neg(\mathsf{s}_1 \vee \mathsf{s}_2 \vee \mathsf{s}_3) \dots$$

Hence, the following holds.

Corollary 4.8 *Algebra* $\mathbf{F}_{\mathcal{M}}(1)$ *has infinite ascending and descending chains of open elements.*

An element $a \in \mathbf{A}$ is said to be an ***atom*** of $\mathbf{A}$ if $\mathbf{0} < a$ and there are no elements strongly between $\mathbf{0}$ and a. An algebra $\mathbf{A}$ is said to be ***atomic*** if there is a set $A \subseteq \mathbf{A}$ of atoms such that for every $b \in \mathbf{A}$ if $b > \mathbf{0}$, then $b \geq a$ for some $a \in A$.

Clearly, every finite algebra is atomic. Let us observe that algebra $\mathbf{Z}_2$ has the only atom: $\mathbf{1}$, while algebra $\mathbf{Z}_3$ has the only atom: its generator. Each algebra $\mathbf{Z}_m, m > 3$ has exactly two atoms: its generator g_m and $\neg g_m$. Hence, the following holds.

Corollary 4.9 (comp. [6, Theorem 5.2]) *Algebra* $\mathbf{F}_{\mathcal{M}}(1)$ *is atomic and has infinitely many atoms.*

Proof. The elements $s_1, s_2, s_3 \wedge g$ and $s_m \wedge g, s_m \wedge \neg g, m > 3$ form the complete list of atoms of $\mathbf{Z}$. It is not hard to see that for any element $a \in \mathbf{Z}$ (and even for every element from $\mathbf{P}$), if $a > 0$, there is an atom a' from the above list such that $a' \leq a$. □

Corollary 4.10 *Algebra* $\mathbf{F}_{\mathcal{M}}(1)$ *contains a single s.i. subalgebra, namely,* $\mathbf{Z}_2$.

Proof. Suppose that $\mathbf{A}$ is an s.i. subalgebra of $\mathbf{Z}$ and $\mathbf{A}$ has more than two elements. Let $a \in \mathbf{A}$ and $\mathbf{0} < a < \mathbf{1}$. There are just two possibilities: either $\neg a = \mathbf{0}$, or $\neg a > \mathbf{0}$.

Assume that $\neg a = \mathbf{0}$. Observe, that $\pi_1(a) \in \{\mathbf{0}, \mathbf{1}\}$ and, hence, $\neg a = \mathbf{0}$ entails $\pi_1(a) = \mathbf{1}$. Therefore, $\pi_1(\Box a) = \Box(\pi_1(a)) = \mathbf{1}$, that is, $\Box a \neq \mathbf{0}$, which is impossible, for $\mathbf{A}$ is s.i. and $a < \mathbf{1}$.

Assume that $\neg a > \mathbf{0}$. Let us note that $\neg a < \mathbf{1}$, for $a > \mathbf{0}$. Thus, we have $\mathbf{0} < a < \mathbf{1}$ and $\mathbf{0} < \neg a < \mathbf{1}$ and, hence, $\mathbf{0} \leq \Box a < \mathbf{1}$ and $\mathbf{0} \leq \Box\neg a < \mathbf{1}$. If we prove that $\Box a > \mathbf{0}$ or $\Box\neg a > \mathbf{0}$, we will be able to conclude that $\mathbf{A}$ is not s.i., because it contains more than two open elements.

Indeed, let us consider the first two projections of a. Note that $\pi_1(a), \pi_2(a) \in \{\mathbf{0}, \mathbf{1}\}$. Hence, there are just four distinct combinations of their values, so we have

$\Box(\pi_1(a), \pi_2(a), \dots)$	$\Box(\neg\pi_1(a), \neg\pi_2(a), \dots)$
$(\mathbf{0}, \mathbf{0}, \dots)$	$(\mathbf{1}, \mathbf{1}, \dots)$
$(\mathbf{0}, \mathbf{1}, \dots)$	$(\mathbf{1}, \mathbf{0}, \dots)$
$(\mathbf{1}, \mathbf{0}, \dots)$	$(\mathbf{0}, \mathbf{1}, \dots)$
$(\mathbf{1}, \mathbf{1}, \dots)$	$(\mathbf{0}, \mathbf{0}, \dots)$

It is clear that in any case either $\Box a > \mathbf{0}$, or $\Box\neg a > \mathbf{0}$. □

References

[1] Bezhanishvili, G., *Varieties of monadic Heyting algebras. I*, Studia Logica **61** (1998), pp. 367–402.
URL `http://dx.doi.org/10.1023/A:1005073905902`

[2] Bezhanishvili, G., *Glivenko type theorems for intuitionistic modal logics*, Studia Logica **67** (2001), pp. 89–109.
URL `http://dx.doi.org/10.1023/A:1010577628486`

[3] Blok, W. J. and D. Pigozzi, *On the structure of varieties with equationally definable principal congruences. I*, Algebra Universalis **15** (1982), pp. 195–227. URL `http://dx.doi.org/10.1007/BF02483723`

[4] Blok, W. J. and D. Pigozzi, *On the structure of varieties with equationally definable principal congruences. III*, Algebra Universalis **32** (1994), pp. 545–608. URL `http://dx.doi.org/10.1007/BF01195727`

[5] Burris, S. and H. P. Sankappanavar, "A Course in Universal Algebra," Graduate Texts in Mathematics **78**, Springer-Verlag, New York, 1981.

[6] Grigolia, R., *Free and projective Heyting and monadic Heyting algebras*, in: *Non-classical logics and their applications to fuzzy subsets (Linz, 1992)*, Theory Decis. Lib. Ser. B Math. Statist. Methods **32**, Kluwer Acad. Publ., Dordrecht, 1995 pp. 33–52.

[7] Jankov, V. A., *On the relation between deducibility in intuitionistic propositional calculus and finite implicative structures*, Dokl. Akad. Nauk SSSR **151** (1963), pp. 1293–1294, English translation in Sov. Math., Dokl. 4, 1203-1204 (1963).

[8] Mal'cev, A., "Algebraic Systems," Die Grundlehren der mathematischen Wissenschaften. Band 192. Berlin-Heidelberg-New York: Springer-Verlag; Berlin: Akademie-Verlag, 1973.

[9] McKenzie, R. N., G. F. McNulty and W. F. Taylor, "Algebras, Lattices, Varieties. Vol. I," The Wadsworth & Brooks/Cole Mathematics Series, Wadsworth & Brooks/Cole Advanced Books & Software, Monterey, CA, 1987.

[10] Ono, H., *On some intuitionistic modal logics*, Publ. Res. Inst. Math. Sci. **13** (1977/78), pp. 687–722.

[11] Werner, H., *Varieties generated by quasi-primal algebras have decidable theories* (1977), Colloq. Math. Soc. János Bolyai, Vol. 17, pp. 555–575.

Non-Normal Modal Logics: Bi-Neighbourhood Semantics and Its Labelled Calculi

Tiziano Dalmonte*, Nicola Olivetti*, Sara Negri⋆ [1]

*Aix Marseille Univ, Université de Toulon, CNRS, LIS, Marseille, France;
⋆ University of Helsinki, Finland

Abstract

The classical cube of non-normal modal logics is considered, and an alternative neighbourhood semantics is given in which worlds are equipped with sets of pairs of neighbourhoods. The intuition is that the two neighbourhoods of a pair provide independent positive and negative evidence (or support) for a formula. This bi-neighbourhood semantics is significant in particular for logics without the monotonicity property. It is shown that this semantics characterises the cube of non-normal modal logics and that there is a mutual correspondence between models in the standard and in the bi-neighbourhood semantics. On the basis of this alternative semantics, labelled sequent calculi are developed for all the logics of the classical cube. The calculi thus obtained are fully modular and have good structural properties, first of all, syntactic cut elimination. Moreover, they provide a decision procedure and an easy countermodel extraction, both in the bi-neighbourhood and in the standard semantics.

Keywords: Non-normal modal logics, (bi)-neighbourhood semantics, labelled sequent calculi.

1 Introduction

Non-normal modal logics are called in this way because they do not satisfy all the axioms and rules of the minimal normal modal logic K. They have been studied since the seminal work of C.I. Lewis, Scott, Lemmon, and Chellas (for an introduction see [1] and [5]) and can be seen as generalisations of standard modal logics. Non-normal modal logics have found an interest in several areas such as epistemic and deontic reasoning, reasoning about games, and reasoning about probabilistic notions such as 'truth in most of the cases'. In all these contexts the $\Box$ modality is better understood as non-normal. For instance, an epistemic interpretation of $\Box A$ as 'the agent knows A' for a non-omniscient agent would reject the rule of monotonicity (RM), that $A \to B$ implies $\Box A \to \Box B$, and possibly the rule of necessitation, the latter meaning in this case that the agent would know every logical validity. In deontic logic, where $\Box A$ is

[1] This work was partially supported by the Project TICAMORE ANR-16-CE91-0002-01, and by the Academy of Finland, research project no. 1308664.

interpreted as 'A is obligatory', some paradoxes like the 'gentle murder' can be avoided if $\Box$ is non-normal. Furthermore, if we interpret $\Box A$ as 'A is true in most of the cases' or 'A is highly probable', the modality $\Box$ will not be likely to satisfy axiom (C) $(\Box A \wedge \Box B) \rightarrow \Box(A \wedge B)$. Validity of this axiom would also fail in a game-theoretical interpretation, where $\Box A$ is interpreted as the agent's availability of a winning strategy to bring about A.

Non-normal modal logics have been studied essentially from a semantical point of view. The standard semantics (Chellas [1]) for these systems is defined in terms of *neighbourhood* models: these are possible world models, where each world w is equipped with a set of neighbourhoods $\mathcal{N}(w)$, each one of them being a set of worlds/states. The loose intuition is that each neighbourhood provides sufficient or relevant evidence to establish the truth of a formula of type $\Box A$. A formula $\Box A$ is forced by a world w if the truth-set of A belongs to $\mathcal{N}(w)$. By imposing further closure conditions on $\mathcal{N}(w)$, various non-normal modal logics can be obtained. The classical cube of non-normal modal logics is determined by considering any combination of the following three conditions: for any world w, (M) $\mathcal{N}(w)$ is closed under supersets, (C) it is closed under intersection, (N) it contains the whole set of possible worlds.

The study of proof systems for non-normal modal logics does not have a state of the art comparable with the one of proof systems for normal modal logics, for which there exist well-understood proof methods of many kinds.

There are several desiderata on proof systems:[2] they should be *standard*, that is, they should contain only a finite number of rules, each with a fixed number of premisses; logical operators should be dealt with dual rules (for the antecedent and the succedent) that introduce a *single* occurrence of a formula; the rules should be *analytic* and allow for a *syntactic* proof of cut elimination; the calculi should be *modular*, with stronger systems obtained simply by adding rules to a basic system; finally, they should provide a *decision procedure* (possibly of optimal complexity) whenever the logic is decidable, and from a failed proof it should be possible to extract directly a *countermodel* of the formula the validity of which is being checked.

Cut-free sequent calculi for non-normal modal logics have been studied by Lavendhomme and Lucas [9]; in their calculi, however, rules allow several formulas as principal and modularity does not obtain; further, a decision procedure is given but it is rather complicated in the non-monotonic case. Indrzejczak [6] has further developed the calculi by Lavendhomme and Lucas [9] extending them with standard axioms of normal modal logics (the non-normal counterpart of logics from **K** to **S5**). Gilbert and Maffezioli [3] investigate labelled calculi using three modalities, on the basis of the translation of non-normal modal logics into normal modal logics given by Gasquet and Herzig [2] and Kracht and Wolter [8]. As a bi-product of the general methodology employed, their calculi are also fully modular (i.e., modular with no exceptions)

[2] For general desiderata on proof systems see [7,15]; for modularity see [11], and for the extraction of countermodels from failed proof search see [12].

but computational issues are not discussed. Recently, Lellmann and Pimentel [10] have proposed linear *nested* sequent calculi for non-normal modal logics; their calculi are fully modular and allow for syntactic cut elimination, but it is not obvious how to get countermodels and a decision procedure out of them.

In this work, we propose labelled sequent calculi for the basic non-normal modal logics. Our calculi are based on *bi-neighbourhood* models, an alternative semantics more general than the standard one. Differently from the standard semantics, worlds in a bi-neighbourhood model are equipped with sets of *pairs* of neighbourhoods rather than single neighbourhoods. The intuition is that the two components of a pair provide independent positive and negative evidence (or support) for a proposition. Standard models correspond exactly to bi-neighbourhood models in which the two neighbourhoods of a pair are complement of each other. The bi-neighbourhood semantics is significant mostly for logics without the monotonicity property, as it collapses into the standard one in the monotonic case. We show *directly* that this semantics characterises non-normal modal logics, being sound and complete with respect to them. Moreover, each bi-neighbourhood model gives rise (in an effective way) to a standard model, providing thereby a mutual correspondence between models of the two kinds.

The new semantics is the starting point for developing labelled sequent calculi for non-normal modal logics. Our aim is to define calculi that satisfy all the above desiderata. The calculi presented in this work are standard (in the sense specified above) and are based on the same approach of Negri [13] of importing the semantics into the syntax by making use of labels; however, they differ significantly from those for non-monotonic systems. The main difference is that the calculi presented here make use of *pseudo-complement* neighbourhoods (corresponding to pairs in the bi-neighbourhood semantics) instead of the *covering* relation to express the inclusion of the truth-set of a formula in a neighbourhood. The new semantic element has the effect that the calculi presented here do not introduce relational formulas in the consequent of a sequent and thus avoid exponential branching in proof search. Departing from the standard neighbourhood semantics gives, as a further bonus, calculi that cover in a modular way the whole cube of non-normal modal logics.

We shall first present a version of the calculi with good proof-theoretical properties, the most important being syntactic cut elimination, from which syntactic completeness of the calculi follows. We then present a second version of the calculi with optimised rules for *closure under intersection.* We show that proof search in these calculi is always terminating, just by adopting a very simple strategy (with no additional mechanism needed). We then prove semantic completeness with respect to bi-neighbourhood models, whence also with respect to the standard semantics by virtue of the correspondence mentioned above. This means that from a failed proof search it is possible to extract directly a countermodel both in the bi-neighbourhood semantics and in the standard one. Since the models obtained in this way are finite, the semantic completeness proof provides in itself also a constructive proof of the finite model property. We finally give a syntactic proof of the fact that bi-neighbourhood semantics coincides with the standard semantics: if we force the two neighbourhoods of each pairs to be complements of each other, we do not get more provable formulas.

2 Non-normal modal logics and bi-neighbourhood semantics

In this section, we present the modal logic **E** and its extensions. We also present its standard semantics in terms of neighbourhood models and a more general semantics in terms of bi-neighbourhood models. We show that bi-neighbourhood semantics characterises logic **E** and its extensions and is equivalent to the standard neighbourhood semantics.

Let $\mathcal{L}$ be a propositional modal language based on countably many propositional variables, the Boolean connectives, and $\Box$. We use A, B, C and p, q as metavariables for arbitrary formulas and atoms of $\mathcal{L}$. $\Diamond A$ is an abbreviation for $\neg\Box\neg A$. Logic **E** is obtained by adding to classical propositional logic the rule of inference

$$\mathsf{RE}\ \frac{A \leftrightarrow B}{\Box A \leftrightarrow \Box B}$$

and can be extended further by choosing any combination of axioms M, C and N (below left), thus producing eight distinct logics. The resulting systems are denoted by $\mathbf{ES}_1\mathbf{...S}_n$, where $\mathrm{S}_i \in \{\mathrm{M,C,N}\}$ [3] (see the *classical cube* below on the right). We write $\mathbf{E}^*$ ($\mathbf{EM}^*$, $\mathbf{EC}^*$, $\mathbf{EN}^*$) to indicate any extension of **E** (**EM**, **EC**, **EN**) with some of these axioms and recall that the top extension coincides with **K**.

M	$\Box(A \wedge B) \to \Box A \wedge \Box B$
C	$\Box A \wedge \Box B \to \Box(A \wedge B)$
N	$\Box\top$

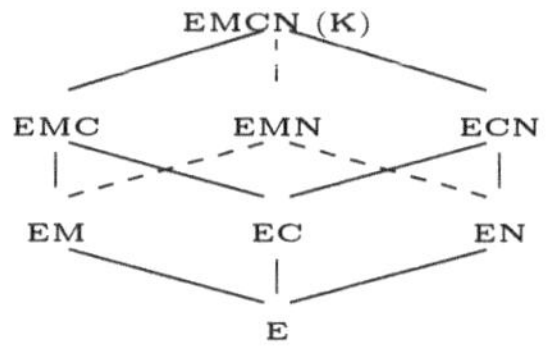

Definition 2.1 A *standard neighbourhood model* (just *standard model* in the following) is a triple $\mathcal{F} = \langle W, \mathcal{N}, V\rangle$, where W is a non-empty set, $\mathcal{N}$ is a function $W \longrightarrow \mathcal{PP}(W)$ and V is a valuation function for propositional variables of $\mathcal{L}$. A model is said to be *supplemented* if for all $\alpha, \beta \subseteq W$, $\alpha \in \mathcal{N}(w)$ and $\alpha \subseteq \beta$ implies $\beta \in \mathcal{N}(w)$; it is *closed under intersection* if $\alpha \in \mathcal{N}(w)$ and $\beta \in \mathcal{N}(w)$ implies $\alpha \cap \beta \in \mathcal{N}(w)$; and it *contains the unit* if for all $w \in W, W \in \mathcal{N}(w)$. The forcing relation $\mathcal{M}, w \models_{st} A$ is defined in the usual way for atomic formulas and Boolean connectives. For the modality we have $\mathcal{M}, w \models_{st} \Box A$ iff $[A]_{\mathcal{M}} \in \mathcal{N}(w)$, where $[A]_{\mathcal{M}}$ denotes the set $\{v \mid \mathcal{M}, v \models_{st} A\}$ of the worlds v that force A, also called the *truth set* of A.

[3] In the literature, in the presence of axiom M the letter E is sometimes omitted from the name of the systems, that are instead denoted by $\mathbf{MS}_1\mathbf{...S}_n$, where $\mathrm{S}_i \in \{\mathrm{C,N}\}$.

As a consequence of the above definition, we obtain the following truth condition for $\Diamond A$: $\mathcal{M}, w \models_{st} \Diamond A$ iff $[\neg A]_{\mathcal{M}} \notin \mathcal{N}(w)$.

Theorem 2.2 (Chellas [1]) *Logic* $\mathbf{E}(\mathbf{M}, \mathbf{C}, \mathbf{N})$ *is sound and complete with respect to standard models (which in addition are, respectively, supplemented, closed under intersection, or contain the unit).*

We now introduce a new semantics where pairs of neighbourhood are used to evaluate the truth of a modal formula.

Definition 2.3 A *bi-neighbourhood model* is a triple $\mathcal{M} = \langle W, \mathcal{N}, V\rangle$, where W is a non-empty set, V is a valuation function and $\mathcal{N}$ is a function that assigns to each world w a subset of $\mathcal{P}(W) \times \mathcal{P}(W)$ such that if $(\alpha, \beta) \in \mathcal{N}(w)$, then $\alpha \cap \beta = \emptyset$. Moreover, $\mathcal{M}$ is a N-model if for all $w \in W$, $(W, \emptyset) \in \mathcal{N}(w)$; it is a C-model if $(\alpha_1, \beta_1), (\alpha_2, \beta_2) \in \mathcal{N}(w)$ implies $(\alpha_1 \cap \alpha_2, \beta_1 \cup \beta_2) \in \mathcal{N}(w)$; and it is an M-model if for all $w \in W$, $(\alpha, \beta) \in \mathcal{N}(w)$ implies $\beta = \emptyset$.

The forcing relation $\mathcal{M}, w \models_{bi} A$ is defined as in Definition 2.1 except for the modality, for which the clause is as follows:

$$\mathcal{M}, w \models_{bi} \Box A \text{ iff } \begin{array}{l}\text{for some } (\alpha, \beta) \in \mathcal{N}(w) \text{ and all } v \in W,\\ v \in \alpha \text{ implies } \mathcal{M}, v \models_{bi} A, \text{ and } v \in \beta \text{ implies } \mathcal{M}, v \not\models_{bi} A.\end{array}$$

Observe that in case the considered model does not satisfy condition M (*i.e.* in the non-monotonic case), if α and β are complementary, this definition becomes equivalent to the standard one. From Definition 2.3 we obtain the following truth condition for $\Diamond A$: $\mathcal{M}, w \models_{bi} \Diamond A$ iff for all $(\alpha, \beta) \in \mathcal{N}(w)$, there is $v \in \alpha$ such that $\mathcal{M}, v \models_{bi} A$, or there is $u \in \beta$ such that $\mathcal{M}, u \not\models_{bi} A$. Notice also that bi-neghbourhood models satisfying condition M collapse into standard models, where $\Box$ coincides with the modality $\langle\ ]$ considered by Pacuit [14].

Theorem 2.4 *Logic* $\mathbf{E}\ (\mathbf{M}, \mathbf{C}, \mathbf{N})$ *is sound with respect to bi-neighbourhood (M,C,N-)models.*

Proof. It can be easily shown that each axiom is valid in the respective class of models and that all the rules preserve validity. □

Even if completeness of all logics $\mathbf{E}^*$ with respect to bi-neighbourhood models follows from Theorem 2.2 and the fact that standard models are particular cases of bi-neighbourhood models, it can be interesting to prove it directly by the canonical model construction. In the proof we do not consider the case of M as we saw it is standard. First of all, for any logic L based on the language $\mathcal{L}$ and for any set X of formulas of $\mathcal{L}$, we say that X is L-consistent if $X \not\vdash_{\mathsf{L}} \bot$, and that it is L-maximal consistent if it is L-consistent and for any formula $A \in \mathcal{L}$ such that $A \notin X$, $X \cup \{A\}$ is not L-consistent. We denote by Max_{L} the class of all L-maximal consistent sets of formulas of $\mathcal{L}$, and for any formula A we denote by $\uparrow A$ the set $\{Y \in Max_{\mathsf{L}} \mid A \in Y\}$. Before defining canonical models, we recall some basic properties of L-maximal consistent sets.

Lemma 2.5 *(a) Any* L*-consistent set of formulas* Γ *can be extended to an* L*-maximal consistent set. (b) If* $\Gamma \not\vdash_{\mathsf{L}} A$*, there is* X *in* Max_{L} *such that* $\Gamma \subseteq X$

and $X \notin\uparrow A$. *(c) If* $\nvdash_{\mathsf{L}} B \to A$, *there is* X *in* Max_{L} *such that* $X \in\uparrow B$ *and* $X \notin\uparrow A$.

Lemma 2.6 *Let* X *be an* L*-maximal consistent set. The usual properties of maximal consistent sets hold, in particular: (a) If* $\vdash_{\mathsf{L}} A$, *then* $A \in X$; *(b) if* $Y \vdash_{\mathsf{L}} A$ *and* $Y \subseteq X$, *then* $A \in X$; *(c) if* $\vdash_{\mathsf{L}} A \leftrightarrow B$ *and* $\Box A \in X$, *then* $\Box B \in X$; *(d)* $\uparrow(A \wedge B) =\uparrow A \cap \uparrow B$; *and (e)* $\uparrow(A \vee B) =\uparrow A \cup \uparrow B$.

Lemma 2.7 *Let the* canonical model $\mathcal{M}^c = \langle W^c, \mathcal{N}^c, V^c \rangle$ *for* L *be defined as follows:* $W^c = Max_{\mathsf{L}}$; *for any* $p \in \mathcal{L}$, $V^c(p) = \{X \in W^c \,|\, p \in X\}$; *for all* $X \in W$,

$$\mathcal{N}^c(X) = \{(\uparrow A, \uparrow\neg A) | \Box A \in X\}.$$

Then for any formula $B \in \mathcal{L}$ *we have* $\mathcal{M}^c, X \models B$ *iff* $B \in X$. *Moreover, (N) if* L *contains axiom* N, *then* $\mathcal{M}^c$ *is an N-model, and (C) if* L *contains axiom* C, *then* $\mathcal{M}^c$ *is a C-model.*

Proof. By induction on B. If B is p the claim holds by definition of V^c. If B is $\bot$, we have $\bot \notin X$ for every X, because X is consistent. If B is $C \circ D$, the proof is immediate by applying the inductive hypothesis and properties of maximal consistent sets. If B is $\Box C$: ($\Rightarrow$) Assume $\mathcal{M}^c, X \models \Box C$. Then for some $(\alpha, \beta) \in \mathcal{N}^c(X)$ and all $Y \in W^c$, $Y \in \alpha$ implies $\mathcal{M}^c, Y \models C$ and $Y \in \beta$ implies $\mathcal{M}^c, Y \not\models C$. By definition of $\mathcal{N}^c$, there is a formula D such that $\alpha =\uparrow D$, $\beta =\uparrow\neg D$ and $\Box D \in X$. Since by inductive hypothesis $[C]_{\mathcal{M}^c} =\uparrow C$, it holds that for all $Z \in W^c$, $Z \in\uparrow D$ implies $Z \in\uparrow C$ and $Z \in\uparrow C$ implies $Z \in\uparrow D$ (if $\mathcal{M}^c, Z \models C$, then $Z \notin\uparrow\neg D$, then $Z \in\uparrow D$); that is $\uparrow D =\uparrow C$. By the properties of maximal consistent sets, $\vdash_{\mathsf{L}} D \leftrightarrow C$. Since $\Box D \in X$, by Lemma 2.6 we have $\Box C \in X$. ($\Leftarrow$) Assume $\Box C \in X$. By definition, $(\uparrow C, \uparrow\neg C) \in \mathcal{N}^c(X)$. Since, by inductive hypothesis, $\uparrow C = [C]_{\mathcal{M}^c}$, we have that for all $Y \in W^c$, $Y \in\uparrow C$ implies $\mathcal{M}^c, Y \models C$ and $Y \in\uparrow\neg C$ implies $\mathcal{M}^c, Y \not\models C$ (because $Y \in\uparrow\neg C$ iff $Y \notin\uparrow C$). Thus $\mathcal{M}^c, X \models \Box C$.

(N) Since $\vdash_{\mathsf{L}} \Box\top$, for all $X \in W^c$ we have $\Box\top \in X$. Thus by definition, $(\uparrow\top, \uparrow\neg\top) \in \mathcal{N}^c(X)$, and by Lemma 2.6, $([\top]_{\mathcal{M}^c}, [\neg\top]_{\mathcal{M}^c}) = (W^c, \emptyset) \in \mathcal{N}^c(X)$.

(C) Assume $(\alpha_1, \beta_1), (\alpha_2, \beta_2) \in \mathcal{N}^c(X)$. Then, by definition, for some $C, D \in \mathcal{L}$, $\alpha_1 =\uparrow C$, $\beta_1 =\uparrow\neg C$, $\alpha_2 =\uparrow D$, $\beta_2 =\uparrow\neg D$ and $\Box C, \Box D \in X$. Thus by the properties of maximal consistent sets we have $\Box C \wedge \Box D \in X$ and, since X contains axiom C, also $\Box(C \wedge D) \in X$. Then $(\uparrow(C \wedge D), \uparrow\neg(C \wedge D)) \in \mathcal{N}^c(X)$, where $\uparrow(C \wedge D) = \alpha_1 \cap \alpha_2$ and $\uparrow\neg(C \wedge D) = \beta_1 \cup \beta_2$. □

Theorem 2.8 (Completeness of E*) *A formula* A *is a theorem of* $\mathbf{E}^*$ *if and only if is valid in the corresponding class of bi-neighbourhood models.*

We now show that from any bi-neighbourhood model we can build an equivalent standard model. As a matter of fact, we can relativise the construction and the equivalence to an arbitrary set of formulas $\mathcal{S}$ provided that it is closed under subformulas. In this way we have an effective procedure to transform a finite bi-neighbourhood model satisfying a given formula into a standard one satisfying the same formula. Because of the obvious equivalence of the two semantics in the monotonic case, the latter is not considered in the lemma.

Lemma 2.9 *Let $\mathcal{M} = \langle W, \mathcal{N}, V\rangle$ be a bi-neighbourhood model and $\mathcal{S}$ be a set of formulas of $\mathcal{L}$ closed under subformulas. We define the standard model $\mathcal{M}^{\mathcal{S}} = \langle W, \mathcal{N}^{\mathcal{S}}, V\rangle$ with the same W and V and by taking, for all $w \in W$,*

$$\mathcal{N}^{\mathcal{S}}(w) = \{[C]_{\mathcal{M}} \mid C \in \mathcal{S} \text{ and } \mathcal{M}, w \models \Box C\}.$$

Then for any formula $A \in \mathcal{S}$ and any world $w \in W$, $\mathcal{M}^{\mathcal{S}}, w \models A$ iff $\mathcal{M}, w \models A$. Moreover, (N) if $\top \in \mathcal{S}$ and $\mathcal{M}$ is a N-model, then $\mathcal{M}^{\mathcal{S}}$ contains the unit; and (C) if $\mathcal{S}$ is closed under conjunction and $\mathcal{M}$ is a C-model, then $\mathcal{M}^{\mathcal{S}}$ is closed under intersection.

Proof. By induction on the complexity of any formula B it can be easily shown that $[B]_{\mathcal{M}^{\mathcal{S}}} = [B]_{\mathcal{M}}$. Moreover it can be proved that (N) $\mathcal{M}^{\mathcal{S}}$ contains the unit whenever $\mathcal{M}$ is a N-model and $\top \in \mathcal{S}$ and (C) $\mathcal{M}^{\mathcal{S}}$ is closed under intersection whenever $\mathcal{M}$ is a C-model and $\mathcal{S}$ is closed under conjunction. □

Theorem 2.10 *A formula A is valid in bi-neighbourhood models if and only if it is valid in the standard models satisfying the corresponding model conditions (N, C and M).*

Proof. From right to left, the claim follows from Lemma 2.9. From left to right, observe that given a standard model $\mathcal{M}_{st}$, we obtain an equivalent bi-neighbourhood model $\mathcal{M}_{bi}$ by taking, for all $w \in W$, $\mathcal{N}_{bi}(w) = \{(\alpha, W \setminus \alpha) \mid \alpha \in \mathcal{N}_{st}(w)\}$. Moreover, $\mathcal{M}_{bi}$ is a N-model if $\mathcal{M}_{st}$ contains the unit, and $\mathcal{M}_{bi}$ is a C-model if $\mathcal{M}_{st}$ is closed under intersection. □

3 The calculi LSE*

In this section, we define our labelled calculi **LSE***. We first present their language and rules, then prove soundness with respect to bi-neighbourhood semantics and syntactic completeness.

Let $\mathbb{WL} = \{x, y, z, ...\}$ and $\mathbb{NL} = \{a, b, c, ...\}$ be two infinite sets, respectively of *world labels* and of *neighbourhood labels*. *Positive neighbourhood terms* (or just *terms*) are finite sets of neighbourhood labels, and are written $[a_1 \dots a_n]$. If t is a positive term, then $\overline{t}$ is a *negative* term. The term τ and its negative counterpart $\overline{\tau}$ are neighbourhood constants. If a (positive or negative) term contains exactly one label or it is τ or $\overline{\tau}$, then it is *atomic*, otherwise it is *complex*.

Intuitively, a positive complex term represents the intersection of its constituents, whereas a negative complex term represents the union of the negative counterparts of its constituents. Moreover, t and $\overline{t}$ are the two members of a pair of neighbourhoods in bi-neighbourhood models. Observe that the operation of overlining a term cannot be iterated: it can be applied only once for turning a positive term into a negative one. Two operations over terms are defined as follows: (a) Composition of positive terms:

$$[a_1 \dots a_n][b] = \begin{cases} [a_1 \dots a_n] & \text{if } b = a_i \text{ for some } i,\ 1 \leq i \leq n; \\ [a_1 \dots a_n b] & \text{otherwise.} \end{cases}$$

$$[a_1 \dots a_n][b_1 \dots b_m] = (...([a_1 \dots a_n][b_1])...[b_{m-1}])[b_m]$$

(b) Substitution of a positive term for a neighbourhood label inside a term:

$$[a](t/b) = \begin{cases} t & \text{if } b = a \\ [a] & \text{if } b \neq a \end{cases} \quad [a_1 \dots a_n](t/b) = [a_1](t/b)...[a_n](t/b) \quad \overline{s}(t/b) = \overline{s(t/b)}$$

Observe that these operations do not introduce multiple occurrences of the same label, thus their results are still neighbourhood terms. We write $\Gamma(t/a)$ to indicate that the substitution applies to all formulas in Γ. As immediate consequences of the definition we have: $\tau(t/a) = \tau$ and $\overline{\tau}(t/a) = \overline{\tau}$, $(sr)(t/a) = s(t/a)r(t/a)$, and $\overline{sr}(t/a) = \overline{s(t/a)r(t/a)}$.

Definition 3.1 The formulas of $\mathcal{L}_{\text{LS}}$ are of the following kinds:

$$\phi ::= x : A \mid x : t \mid x : \overline{t} \mid t : A \mid \overline{t} : A \mid t : x.$$

The semantic interpretation of formulas of $\mathcal{L}_{\text{LS}}$ is given in Definition 3.3. Intuitively, $x : A$ means that x forces A, $x : t$ (resp. $x : \overline{t}$) means that x is a world in neighbourhood t (resp. $\overline{t}$), $t : A$ (resp. $\overline{t} : A$) means that every world in t (resp. some world in $\overline{t}$) forces A, and $t : x$ means that the pair $(t, \overline{t})$ is a bi-neighbourhood of x.

We have chosen a polymorphic notation, in which the colon has a meaning that depends on the type of its arguments, because of its compactness. As we shall see the interpretation of a formula ϕ is uniquely determined.

Sequents are defined as usual as pairs $\Gamma \Rightarrow \Delta$ of finite multisets of formulas, however they must satisfy some restrictions in order to assure cut admissibility.

Definition 3.2 A sequent is a pair $\Gamma \Rightarrow \Delta$, where Γ and Δ are finite multisets of formulas of $\mathcal{L}_{\text{LS}}$, that respect the following conditions: (1) Δ contains only formulas of the kinds $x : A$, $t : A$ and $\overline{t} : A$ (whereas Γ may contain any formula of $\mathcal{L}_{\text{LS}}$); (2) If Γ is non-empty, then all world labels and all neighbourhood labels occurring in Δ occur also in Γ.[4] (3) If Γ is empty, then Δ contains only formulas of the kind $x : A$, and all these formulas are labelled by the same world label x. (4) If $x : t$ is in Γ, then there is a world label y such that $t : y$ is in Γ.

The calculi **LSE*** are defined by the rules in Figure 1. Observe that, in analogy with the calculi based on standard possible world semantics, the left-right rules are meaning conferring and directly derive from the semantic explanation of logical constants in terms of bi-neighbourhood semantics, whereas the rules that manipulate only labels provide modular extensions of the basic systems to yield all the systems of the modal cube.

In Figure 2, the derivations of rule RE and axioms M, N and C in the respective calculi will be shown (for RE we assume sequents $y : A \Rightarrow y : B$ and $y : B \Rightarrow y : A$ derivable for any label y). Observe that considering rule applications backwards, the restrictions on sequents of Definition 3.2 are necessarily satisfied: If the conclusion of an instance of a rule satisfies conditions (1)-(4), then its premisses also satisfy (1)-(4). On the other hand, if we consider forward applications of the rules, these must be obviously restricted in such

[4] A neighbourhood label a occurs in (or belongs to) a labelled formula ϕ (set of formulas, sequent) if there is a (positive or negative) term containing a in ϕ.

Initial sequents: $x : p, \Gamma \Rightarrow \Delta, x : p \qquad x : \bot, \Gamma \Rightarrow \Delta \qquad \Gamma \Rightarrow \Delta, x : \top$

Propositional rules: As for **G3K**.

$$\frac{x : t, x : A, t : A, \Gamma \Rightarrow \Delta}{x : t, t : A, \Gamma \Rightarrow \Delta}\ \mathrm{L}{\Vdash^{\forall}} \qquad \frac{x : t, \Gamma \Rightarrow \Delta, x : A}{\Gamma \Rightarrow \Delta, t : A}\ \mathrm{R}{\Vdash^{\forall}} \qquad \frac{x : \overline{t}, x : A, \Gamma \Rightarrow \Delta}{\overline{t} : A, \Gamma \Rightarrow \Delta}\ \mathrm{L}{\Vdash^{\exists}}$$

$$\frac{x : \overline{t}, \Gamma \Rightarrow \Delta, x : A, \overline{t} : A}{x : \overline{t}, \Gamma \Rightarrow \Delta, \overline{t} : A}\ \mathrm{R}{\Vdash^{\exists}} \qquad \frac{[a] : x, [a] : A, \Gamma \Rightarrow \Delta, \overline{[a]} : A}{x : \Box A, \Gamma \Rightarrow \Delta}\ \mathrm{L}\Box$$

$$\frac{t : x, \Gamma \Rightarrow \Delta, x : \Box A, t : A \qquad t : x, \overline{t} : A, \Gamma \Rightarrow \Delta, x : \Box A}{t : x, \Gamma \Rightarrow \Delta, x : \Box A}\ \mathrm{R}\Box$$

$$\frac{}{t : x, y : \overline{t}, \Gamma \Rightarrow \Delta}\ \mathrm{M} \qquad \frac{\tau : x, \Gamma \Rightarrow \Delta}{\Gamma \Rightarrow \Delta}\ \mathrm{N}\tau \qquad \frac{}{x : \overline{\tau}, \Gamma \Rightarrow \Delta}\ \mathrm{N}\overline{\tau}$$

$$\frac{ts : x, t : x, s : x, \Gamma \Rightarrow \Delta}{t : x, s : x, \Gamma \Rightarrow \Delta}\ \mathrm{C} \qquad \frac{x : t, x : s, x : ts, \Gamma \Rightarrow \Delta}{x : ts, \Gamma \Rightarrow \Delta}\ \mathrm{dec}$$

$$\frac{x : \overline{t}, x : \overline{ts}, \Gamma \Rightarrow \Delta \qquad x : \overline{s}, x : \overline{ts}, \Gamma \Rightarrow \Delta}{x : \overline{ts}, \Gamma \Rightarrow \Delta}\ \overline{\mathrm{dec}}$$

Application conditions:
x is fresh in $\mathrm{R}{\Vdash^{\forall}}$ and $\mathrm{L}{\Vdash^{\exists}}$, a is fresh in $\mathrm{L}\Box$, and x occurs in the conclusion of $\mathrm{N}\tau$.

Fig. 1. The calculi **LSE***.

a way that they satisfy (1)-(4). Notice also that if rule M is added to the basic calculus, our rules $\mathrm{L}\Box$ and $\mathrm{R}\Box$ become interderivable with the rules for monotonic $\Box$ given in [13]; the latter rules, rewritten with the present notation, are as follows:

$$\frac{[a] : x, [a] : A, \Gamma \Rightarrow \Delta}{x : \Box A, \Gamma \Rightarrow \Delta}\ \mathrm{L}\Box^{\mathrm{M}}\ (a \text{ fresh}) \qquad \frac{t : x, \Gamma \Rightarrow \Delta, x : \Box A, t : A}{t : x, \Gamma \Rightarrow \Delta, x : \Box A}\ \mathrm{R}\Box^{\mathrm{M}}$$

It can be shown that these calculi are sound with respect to bi-neighbourhood semantics. For this purpose, we need to introduce the notion of realisation.

Definition 3.3 Given a model $\mathcal{M} = \langle W, \mathcal{N}, V\rangle$, a *realisation* is a pair of functions (ρ, σ), where $\rho : \mathbb{WL} \longrightarrow W$, and $\sigma : \mathbb{NT} \longrightarrow \mathcal{P}(W)$ such that $\sigma(\tau) = W$, $\sigma(t) \cap \sigma(\overline{t}) = \emptyset$, $\sigma(ts) = \sigma(t) \cap \sigma(s)$ and $\sigma(\overline{ts}) = \sigma(\overline{t}) \cup \sigma(\overline{s})$. The relation $\mathcal{M} \models_{\rho,\sigma} \phi$ is defined by cases as follows:

$\mathcal{M} \models_{\rho,\sigma} x : t$ iff $\rho(x) \in \sigma(t)$, and $\mathcal{M} \models_{\rho,\sigma} x : \overline{t}$ iff $\rho(x) \in \sigma(\overline{t})$;
$\mathcal{M} \models_{\rho,\sigma} x : A$ iff $\mathcal{M}, \rho(x) \models A$;
$\mathcal{M} \models_{\rho,\sigma} t : A$ iff for all $w \in \sigma(t)$, $\mathcal{M}, w \models A$;
$\mathcal{M} \models_{\rho,\sigma} \overline{t} : A$ iff there is a $w \in \sigma(\overline{t})$ such that $\mathcal{M}, w \models A$;
$\mathcal{M} \models_{\rho,\sigma} t : x$ iff $(\sigma(t), \sigma(\overline{t})) \in \mathcal{N}(\rho(x))$.

Then given a sequent $\Gamma \Rightarrow \Delta$ we stipulate that $\mathcal{M} \models_{\rho,\sigma} \Gamma \Rightarrow \Delta$ iff whenever $\mathcal{M} \models_{\rho,\sigma} \phi$ for all formulas ϕ in Γ we also have $\mathcal{M} \models_{\rho,\sigma} \psi$ for a formula ψ in Δ. Moreover, $\Gamma \Rightarrow \Delta$ is valid in $\mathcal{M}$ iff for all realisations (ρ, σ) we have $\mathcal{M} \models_{\rho,\sigma} \Gamma \Rightarrow \Delta$, and it is valid in bi-neighbourhood (N,C,M)-models iff it is

(RE)
$$\dfrac{\dfrac{\dfrac{\dfrac{y:\overline{[a]},y:B,[a]:x,[a]:A \Rightarrow x:\Box B,\overline{[a]}:A,y:A}{y:\overline{[a]},y:B,[a]:x,[a]:A \Rightarrow x:\Box B,\overline{[a]}:A}\text{R}\Vdash\exists}{\overline{[a]}:B,[a]:x,[a]:A \Rightarrow x:\Box B,\overline{[a]}:A}\text{L}\Vdash\exists}}$$

$$\dfrac{\dfrac{\dfrac{\dfrac{y:A,y:[a],[a]:x,[a]:A \Rightarrow x:\Box B,\overline{[a]}:A,y:B}{y:[a],[a]:x,[a]:A \Rightarrow x:\Box B,\overline{[a]}:A,y:B}\text{L}\Vdash\forall}{[a]:x,[a]:A \Rightarrow x:\Box B,\overline{[a]}:A,[a]:B}\text{R}\Vdash\forall}{\dfrac{[a]:x,[a]:A \Rightarrow x:\Box B,\overline{[a]}:A}{x:\Box A \Rightarrow x:\Box B}\text{L}\Box}\text{R}\Box}$$

(M)
$$\dfrac{\dfrac{\dfrac{\dfrac{\dfrac{...,y:A,y:B,y:[a],[a]:A\wedge B \Rightarrow y:A,...}{...,y:A\wedge B,y:[a],[a]:A\wedge B \Rightarrow y:A,...}\wedge\text{L}}{...,y:[a],[a]:A\wedge B \Rightarrow y:A,...}\text{L}\Vdash\forall}{...,[a]:A\wedge B \Rightarrow [a]:A,...}\text{R}\Vdash\forall}\quad \dfrac{...,y:\overline{[a]},y:A,[a]:x \Rightarrow ...}{...,\overline{[a]}:A,[a]:x \Rightarrow ...}\text{L}\Vdash\exists}{\dfrac{[a]:x,[a]:A\wedge B \Rightarrow x:\Box A,\overline{[a]}:A\wedge B}{x:\Box(A\wedge B) \Rightarrow x:\Box A}\text{L}\Box}\text{M}$$

(N)
$$\dfrac{\dfrac{\dfrac{\tau:x,y:\tau \Rightarrow x:\Box\top,y:\top}{\tau:x \Rightarrow x:\Box\top,\tau:\top}\text{R}\Vdash\forall \quad \dfrac{\dfrac{\tau:x,y:\overline{\tau},y:\top \Rightarrow x:\Box\top}{\tau:x,\overline{\tau}:\top \Rightarrow x:\Box\top}\text{L}\Vdash\exists}{}}{\dfrac{\tau:x \Rightarrow x:\Box\top}{\Rightarrow x:\Box\top}\text{N}\tau}\text{R}\Box}\text{N}\overline{\tau}$$

(C)
$$\dfrac{\dfrac{...,y:\overline{[a]},y:A \Rightarrow \overline{[a]}:A,y:A...}{...,y:\overline{[a]},y:A \Rightarrow \overline{[a]}:A...}\text{R}\Vdash\exists \quad \dfrac{...,y:\overline{[b]},y:B,\Rightarrow \overline{[b]}:B,y:B,...}{...,y:\overline{[b]},y:B,\Rightarrow \overline{[b]}:B,...}\text{R}\Vdash\exists}{\dfrac{...,y:\overline{[a,b]},y:A,y:B \Rightarrow \overline{[a]}:A,\overline{[b]}:B,...}{\dfrac{...,y:\overline{[a,b]},y:A\wedge B \Rightarrow \overline{[a]}:A,\overline{[b]}:B,...}{...,\overline{[a,b]}:A\wedge B \Rightarrow \overline{[a]}:A,\overline{[b]}:B,...}\wedge\text{L}}\text{L}\Vdash\exists}\text{dec}$$

$$\dfrac{\text{branch left to the reader} \quad \dfrac{...,\overline{[a,b]}:A\wedge B \Rightarrow \overline{[a]}:A,\overline{[b]}:B,...}{\dfrac{[a,b]:x,[a]:x,[b]:x,[a]:A,[b]:B \Rightarrow x:\Box(A\wedge B),\overline{[a]}:A,\overline{[b]}:B}{\dfrac{[a]:x,[b]:x,[a]:A,[b]:B \Rightarrow x:\Box(A\wedge B),\overline{[a]}:A,\overline{[b]}:B}{x:\Box A,x:\Box B \Rightarrow x:\Box(A\wedge B)}\text{C}}\text{L}\Box^{(2)}}\text{R}\Box}$$

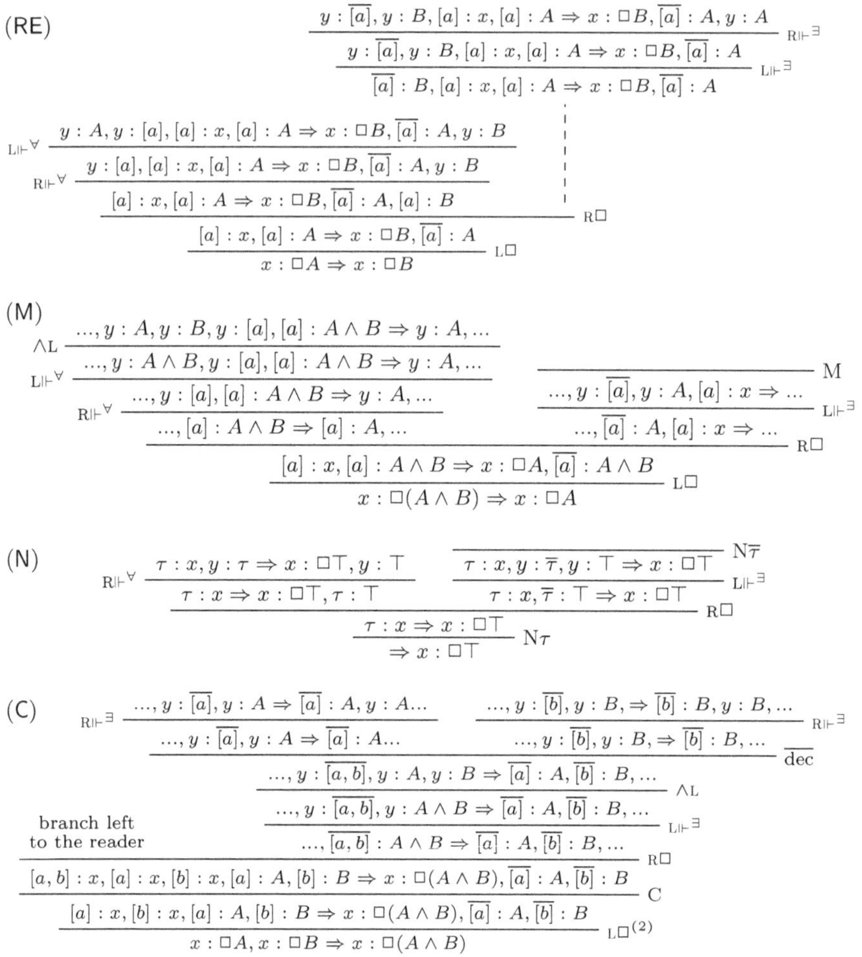

Fig. 2. Derivation of rule RE and axioms M, N and C in the respective calculi.

valid in every model $\mathcal{M}$ of the corresponding class.

By an easy induction on derivations we can prove the soundness of the calculi.

Theorem 3.4 *If a sequent $\Gamma \Rightarrow \Delta$ is derivable in $\mathbf{LSE}(\mathbf{N},\mathbf{C},\mathbf{M})$, then it is valid in the class of all bi-neighbourhood (N,C,M-)models.*

Observe that all rules are also sound in standard models in which $\bar{t}$ is interpreted as the real complement of t, with the exception of rule M which is incompatible with such an interpretation. In what follows, we prove the main structural properties of the calculus, most importantly admissibility of *cut*, from which we obtain the syntactic completeness of the calculus.

Proposition 3.5 *(a) Substitution of world labels and (b) substitution of pos-*

itive terms for neighbourhood labels are height-preserving admissible (hp-admissible) in **LSE***. Moreover, (c) the rules of left and right weakening are hp-admissible in* **LSE***; (d) all rules of* **LSE*** *are hp-invertible; and (e) the rules of left and right contraction are hp-admissible in* **LSE***.*

We aim to prove admissibility of the following cut rule:

$$\dfrac{\Gamma \Rightarrow \Delta, \phi \qquad \phi, \Gamma \Rightarrow \Delta}{\Gamma \Rightarrow \Delta}\ cut$$

where ϕ is any formula of $\mathcal{L}_{\text{LS}}$ that can occur on both sides of a sequent. Observe that any application of *cut* respects the restrictions on sequents of Definition 3.2. In order to prove admissibility of *cut* we need to define the weight of a labelled formula. Then by admissibility of *cut* it is easy to prove completeness of **LSE***.

Definition 3.6 The weight $w(\phi)$ of a formula ϕ of the form $x : A$, $t : A$ or $\overline{t} : A$ is the pair $\langle w(f(\phi)), w(l(\phi))\rangle$, where $f(\phi)$ and $l(\phi)$ are, respectively, the $\mathcal{L}$ formula A and the world label or neighbourhood term occurring in ϕ; $w(x) = 0$ and $w(t) = w(\overline{t}) = card(t)$, where $card(t)$ is the number of neighbourhood labels occurring in t; $w(p) = 1$, $w(A \circ B) = w(A) + w(B) + 1$, $w(\Box A) = w(A) + 1$. We consider weights of formulas lexicographically ordered.

Theorem 3.7 *Cut is admissible in* **LSE***.

Proof. By double induction, with primary induction on the weight of the cut formula and subinduction on the cut height. Observe that, because of Definition 3.2, cut formulas can be only of the kinds $x : A$, $t : A$ and $\overline{t} : A$. We only show some significant cases. (i) The last rule applied in the derivation of the left premiss of *cut* is N$\mathcal{T}$. The derivation on the left is converted into the one on the right (in this and in the other cases we implicitly use hp-admissibility of structural rules). Observe that the restrictions on sequents guarantee that in the right derivation the label condition on the application of N$\mathcal{T}$ is respected, *i.e.* it is not the case that ϕ contains the only occurrence of x.

$$cut\ \dfrac{\text{N}\mathcal{T}\ \dfrac{\mathcal{T} : x, \Gamma \Rightarrow \Delta, \phi}{\Gamma \Rightarrow \Delta, \phi} \qquad \phi, \Gamma \Rightarrow \Delta}{\Gamma \Rightarrow \Delta} \quad \leadsto \quad \dfrac{\dfrac{\mathcal{T} : x, \Gamma \Rightarrow \Delta, \phi \qquad \dfrac{\phi, \Gamma \Rightarrow \Delta}{\mathcal{T} : x, \phi, \Gamma \Rightarrow \Delta}\ wk}{\mathcal{T} : x, \Gamma \Rightarrow \Delta}\ cut}{\Gamma \Rightarrow \Delta}\ \text{N}\mathcal{T}$$

(ii) The cut formula is $x : \Box A$, principal in the last rule of the derivation of both premisses of *cut*:

$$\dfrac{\text{R}\Box\ \dfrac{\begin{array}{c} t : x, \Gamma \Rightarrow \Delta, x : \Box A, t : A \\ \vdots \\ t : x, \overline{t} : A, \Gamma \Rightarrow \Delta, x : \Box A \end{array}}{t : x, \Gamma \Rightarrow \Delta, x : \Box A} \qquad \dfrac{\dfrac{\mathcal{D}}{[a] : x, [a] : A, t : x, \Gamma \Rightarrow \Delta, \overline{[a]} : A}}{x : \Box A, t : x, \Gamma \Rightarrow \Delta}\ \text{L}\Box}{t : x, \Gamma \Rightarrow \Delta}\ cut$$

with a fresh in the application of L$\Box$. The derivation is converted into the following, with four applications of *cut*, each one having smaller height or a cut formula of smaller weight:

$$
\dfrac{\dfrac{t:x,\Gamma \Rightarrow \Delta, x:\Box A, t:A \qquad \dfrac{t:x, x:\Box A, \Gamma \Rightarrow \Delta}{t:x, x:\Box A, \Gamma \Rightarrow \Delta, t:A}\,wk}{t:x,\Gamma \Rightarrow \Delta, t:A}\,cut}{t:x,\Gamma \Rightarrow \Delta, t:A, \overline{t}:A}\,wk
$$

$$
\dfrac{\dfrac{\dfrac{\dfrac{\mathcal{D}(t/a)}{t:x, t:A, t:x, \Gamma \Rightarrow \Delta, \overline{t}:A}}{t:x, t:A, \Gamma \Rightarrow \Delta, \overline{t}:A}\,ctr \qquad \vdots}{t:x,\Gamma \Rightarrow \Delta, \overline{t}:A}\,cut \qquad \dfrac{t:x, \overline{t}:A, \Gamma \Rightarrow \Delta, x:\Box A \qquad \vdots\ \dfrac{t:x, x:\Box A, \Gamma \Rightarrow \Delta}{t:x, \overline{t}:A, x:\Box A, \Gamma \Rightarrow \Delta}\,wk}{t:x, \overline{t}:A, \Gamma \Rightarrow \Delta}\,cut}{t:x,\Gamma \Rightarrow \Delta}\,cut
$$

□

Theorem 3.8 *The calculus* **LSE*** *is complete with respect to the logic* **E***.

Proof. Straightforward by showing that any instance of the axioms and all the rules of **E*** are derivable in **LSE*** (cf. Figure 2), using *cut* when needed.□

4 The calculi $\mathbb{T}$LSE*

In this section, we present the calculi $\mathbb{T}$**LSE*** (where $\mathbb{T}$ stays for *terms*) which are refinements of the calculi **LSE*** for the cases in which complex terms are present. We show that these calculi are terminating and thereby provide a decision procedure for the respective logics, and we prove semantic completeness of the calculi with respect to bi-neighbourhood semantics. By simulating derivations in **LSE***, we also show that these calculi are syntactically complete, although, as explained below, a direct proof of cut elimination cannot be given, what justifies a separate presentation of the two calculi.

Observe that in **LSE*** it may happen that if the starting sequent contains n atomic terms $[a_1], ..., [a_n]$, a derivation branch - by application of rule C and repeated applications of dec - may take $\mathcal{O}(2^n)$ steps to generate a complex term t containing an arbitrary subset of $a_1, ..., a_n$. To prevent this situation we reformulate the rules for complex terms as follows:

Simplified rules for C:

$$
\dfrac{[a_1]:x, ..., [a_n]:x, [a_1 \dots a_n]:x, \Gamma \Rightarrow \Delta}{[a_1]:x, ..., [a_n]:x, \Gamma \Rightarrow \Delta}\ \mathrm{C}_{\mathbb{T}}
$$

$$
\dfrac{x:[a_1], ..., x:[a_n], \Gamma \Rightarrow \Delta}{x:[a_1 \dots a_n], \Gamma \Rightarrow \Delta}\ \mathrm{dec}_{\mathbb{T}}
\qquad
\dfrac{x:\overline{[a_1]}, \Gamma \Rightarrow \Delta \quad \dots \quad x:\overline{[a_n]}, \Gamma \Rightarrow \Delta}{x:\overline{[a_1 \dots a_n]}, \Gamma \Rightarrow \Delta}\ \overline{\mathrm{dec}_{\mathbb{T}}}
$$

Since these rules are easily derivable in **LSE***, it turns out that $\mathbb{T}$**LSE*** is sound. The rules for decomposition of terms are modified as follows: a complex term can be decomposed only into its atomic components and is not copied into the premiss; moreover by the simplified rule for C complex terms can be formed only by joining atomic terms. However, the calculi with the restricted rules are complete only with respect to sequents of a special form, as described in the next definition.

Definition 4.1 A sequent $\Gamma \Rightarrow \Delta$ of $\mathcal{L}_{\text{LS}}$ is *proper* if it satisfies all the following additional conditions: (1) If $t:A$ is in Γ, then t is atomic and different from τ; (2) $t:A$ is in Γ if and only if $\overline{t}:A$ is in Δ; (3) If $[a]$ occurs in $\Gamma \Rightarrow \Delta$, then there is exactly one formula A such that $[a] : A$ is in Γ; (4) If $[a_1 \dots a_n] : x$ is in Γ, then $[a_1] : x, ..., [a_n] : x$ are in Γ.

It follows from Definition 4.1 that if a formula $\overline{t} : A$ occurs in the right-hand side of a proper sequent $\Gamma \Rightarrow \Delta$, then $\overline{t}$ is atomic and different from $\overline{\tau}$, and $\overline{t} : A$ is the only formula of this kind labelled by $\overline{t}$ occurring in Δ. Trivially, since a sequent of the form $\Rightarrow x_0 : A$ is proper, restricting consideration to proper sequents is sufficient to prove the validity of any formula of $\mathbf{E}^*$.

It can be shown that the calculi $\mathbb{T}\mathbf{LSE}^*$ are syntactically complete as they can simulate $\mathbf{LSE}^*$ derivations restricted to proper sequents. As a preliminary condition, observe that any sequent occurring in a derivation of a proper sequent in $\mathbf{LSE}^*$ or $\mathbb{T}\mathbf{LSE}^*$ is proper, since whenever the conclusion of a rule of $\mathbf{LSE}^*$ or $\mathbb{T}\mathbf{LSE}^*$ is proper its premisses are also proper. The need of such an indirect proof is due to the fact that proper sequents are not preserved by substitution of neighbourhood terms, as it is needed for a direct proof of cut elimination. Although we do not have a syntactic proof of cut admissibility, we have a semantic proof of it: by the completeness of the calculi, the cut rule turns out to be admissible in each system.

By the restrictions of Definition 4.1 we obtain the following property, that will be needed in the proof of Theorem 4.11.

Proposition 4.2 *Every proper sequent of the form* $x : [a], x : \overline{[a]}, \Gamma \Rightarrow \Delta$ *is derivable in* $\mathbb{T}\mathbf{LSE}^*$.

Proof. Since $x : [a], x : \overline{[a]}, \Gamma \Rightarrow \Delta$ is proper, by definition there is a formula A such that $[a] : A$ is in Γ and $\overline{[a]} : A$ is in Δ. Then the sequent has the form $x : [a], x : \overline{[a]}, [a] : A, \Gamma' \Rightarrow \Delta', \overline{[a]} : A$ and is derivable as follows:

$$\dfrac{\dfrac{x : [a], x : \overline{[a]}, [a] : A, x : A, \Gamma' \Rightarrow \Delta', \overline{[a]} : A, x : A}{x : [a], x : \overline{[a]}, [a] : A, \Gamma' \Rightarrow \Delta', \overline{[a]} : A, x : A}\text{L}\Vdash^{\forall}}{x : [a], x : \overline{[a]}, [a] : A, \Gamma' \Rightarrow \Delta', \overline{[a]} : A}\text{R}\Vdash^{\exists}$$

□

The adequacy of rules $\mathrm{C}_{\mathbb{T}}$, $\mathrm{dec}_{\mathbb{T}}$ and $\overline{\mathrm{dec}_{\mathbb{T}}}$ is proved by the following proposition.

Proposition 4.3 (*a*) *Rules* $\mathrm{dec}_{\mathbb{T}}$ *and* $\overline{\mathrm{dec}_{\mathbb{T}}}$ *are invertible in* $\mathbb{T}\mathbf{LSE}^*$ *with respect to derivations of proper sequents.* (*b*) *Contraction is hp-admissible in* $\mathbb{T}\mathbf{LSE}^*$.

Theorem 4.4 *Any proper sequent derivable in* $\mathbf{LSE}^*$ *is derivable also in* $\mathbb{T}\mathbf{LSE}^*$*, whence the calculi* $\mathbb{T}\mathbf{LSE}^*$ *are complete for the corresponding logic.*

Proof. We just consider the most significant cases. If the last rule applied is C, then S has the form $t : x, s : x, \Gamma \Rightarrow \Delta$ and it was derived from the proper sequent $ts : x, t : x, s : x, \Gamma \Rightarrow \Delta$, that by inductive hypothesis is derivable in $\mathbb{T}\mathbf{LSE}^*$. Let t and s be the terms $[a_1 \dots a_n]$ and $[b_1 \dots b_m]$. Then ts is $[a_1 \dots a_n b_1 \dots b_m]$ (without possible repetitions). By definition of proper sequent, Γ contains $[a_i] : x$ and $[b_j] : x$ for all $1 \leq i \leq n$, $1 \leq j \leq m$. Then we can apply $\mathrm{C}_{\mathbb{T}}$ and obtain $t : x, s : x, \Gamma \Rightarrow \Delta$.

If the last rule applied is $\overline{\text{dec}}$, then S has the form $x : \overline{ts}, \Gamma \Rightarrow \Delta$ and it was derived from the proper sequents $x : \overline{t}, x : \overline{ts}, \Gamma \Rightarrow \Delta$ and $x : \overline{s}, x : \overline{ts}, \Gamma \Rightarrow \Delta$, that by inductive hypothesis are derivable in $\mathbb{T}$**LSE***. Let t and s be the terms $[a_1 \dots a_n]$ and $[b_1 \dots b_m]$. Then ts is $[a_1 \dots a_n b_1 \dots b_m]$ (without possible repetitions). Consider the first premiss, that is $x : \overline{[a_1 \dots a_n]}, x : \overline{[a_1 \dots a_n b_1 \dots b_m]}, \Gamma \Rightarrow \Delta$. By invertibility of $\overline{\text{dec}_\mathbb{T}}$ in $\mathbb{T}$**LSE***, $x : \overline{[a_i]}, x : \overline{[a_1 \dots a_n b_1 \dots b_m]}, \Gamma \Rightarrow \Delta$ is derivable for all $1 \leq i \leq n$. Again by invertibility of $\overline{\text{dec}_\mathbb{T}}$, $x : \overline{[a_i]}, x : \overline{[a_k]}, \Gamma \Rightarrow \Delta$ and $x : \overline{[a_i]}, x : \overline{[b_l]}, \Gamma \Rightarrow \Delta$ are derivable for all $1 \leq k \leq n$, $1 \leq l \leq m$. By applying the same procedure to the second premiss, we obtain that sequents $x : \overline{[b_j]}, x : \overline{[a_k]}, \Gamma \Rightarrow \Delta$ and $x : \overline{[b_j]}, x : \overline{[b_l]}, \Gamma \Rightarrow \Delta$ are derivable for all $1 \leq k \leq n$, $1 \leq j, l \leq m$. Now take all sequents $x : \overline{[a_i]}, x : \overline{[a_k]}, \Gamma \Rightarrow \Delta$ and $x : \overline{[b_j]}, x : \overline{[b_l]}, \Gamma \Rightarrow \Delta$ where $i = k$ and $j = l$. By contraction we obtain $x : \overline{[a_i]}, \Gamma \Rightarrow \Delta$ and $x : \overline{[b_j]}, \Gamma \Rightarrow \Delta$. Then by an application of $\overline{\text{dec}_\mathbb{T}}$ with all these sequents as premisses we derive $x : \overline{ts}, \Gamma \Rightarrow \Delta$. □

We now show that by adopting a simple strategy, proof search in $\mathbb{T}$**LSE*** always terminates in a finite number of steps, thereby providing a decision procedure for the corresponding logic. This is basically proved by showing that the set of labelled formulas which can occur in any sequent in any derivation branch is finite. In order to define the strategy, we introduce saturation conditions associated to the rules and the notion of saturated branch.

Definition 4.5 Let $\mathcal{B} = \{\Gamma_i \Rightarrow \Delta_i\}$ be a (finite or infinite) branch in a proof search in $\mathbb{T}$**LSE*** for $\Gamma \Rightarrow \Delta$. We define $\Gamma^* = \bigcup \Gamma_i$ and $\Delta^* = \bigcup \Delta_i$. The saturation conditions associated to each rule of $\mathbb{T}$**LSE*** are as follows: $(Init)$ for all i, there is no $x : p$ in $\Gamma_i \cap \Delta_i$; $x : \bot$ is not in Γ_i and $x : \top$ is not in Δ_i. Standard for propositional rules (omitted). (L$\Vdash^\forall$) If $t : A$ and $x : t$ are in Γ^*, then $x : A$ is in Γ^*. (R$\Vdash^\forall$) If $t : A$ is in Δ^*, then for a label x, $x : t$ is in Γ^* and $x : A$ is in Δ^*. (L$\Vdash^\exists$) If $\overline{t} : A$ is in Γ^*, then for a label x, $x : \overline{t}$ and $x : A$ are in Γ^*. (R$\Vdash^\exists$) If $\overline{t} : A$ is in Δ^* and $x : \overline{t}$ is in Γ^*, then $x : A$ is in Δ^*. (L$\Box$) If $x : \Box A$ is in Γ^*, then for a label a, $[a] : x$ and $[a] : A$ are in Γ^* and $\overline{[a]} : A$ is in Δ^*. (R$\Box$) If $x : \Box A$ is in Δ^* and $t : x$ is in Γ, then either $t : A$ is in Δ^* or $\overline{t} : A$ is in Γ^*. (Nτ) For every world label x occurring in $\Gamma^* \cup \Delta^*$, $\tau : x$ is in Γ^*. (N$\overline{\tau}$) $x : \overline{\tau}$ is not in Γ^*. (M) $t : x$ and $y : \overline{t}$ are not both in Γ^*. (C$_\mathbb{T}$) If $[a_1] : x$, ..., $[a_n] : x$ are in Γ^*, then $[a_1 \dots a_n] : x$ is in Γ^*. (dec$_\mathbb{T}$) If $x : [a_1 \dots a_n]$ is in Γ^*, then $x : [a_1]$, ..., $x : [a_n]$ are in Γ^*. ($\overline{\text{dec}_\mathbb{T}}$) If $x : \overline{[a_1 \dots a_n]}$ is in Γ^*, then $x : \overline{[a_1]}$ or, ..., or $x : \overline{[a_n]}$ is in Γ^*.

We say that $\mathcal{B}$ is saturated with respect to an application of a rule if the corresponding condition holds, and it is saturated with respect to $\mathbb{T}$**LSE*** if it is saturated with respect to all possible applications of any rule of $\mathbb{T}$**LSE***.

The *strategy* for constructing a root-first proof search tree in $\mathbb{T}$**LSE*** of the sequent $\Rightarrow x_0 : A$ obeys the following conditions: (i) No rule can be applied to an initial sequent; (ii) A specific application of a rule R to a formula ϕ (or to a pair of formulas ϕ and ψ) in a sequent $\Gamma_i \Rightarrow \Delta_i$ is not allowed if the branch from $\Rightarrow x_0 : A$ to $\Gamma_i \Rightarrow \Delta_i$ already fulfills the saturation condition for that application of R; (iii) If rules for N are present, as first step apply Nτ to x_0.

We now show that for each sequent $\Rightarrow x_0 : A$ this strategy produces either a proof of it or a finite tree in which all open branches are saturated.

Definition 4.6 Let $\mathcal{B}$ be a branch of a proof search in $\mathbb{T}\mathbf{LSE}^*$ for $\Rightarrow x_0 : A$, t a neighbourhood term and x, y world labels occurring in $\mathcal{B}$, and let $k(x) = min\{i \in \mathbb{N} \mid x \text{ is in } \Gamma_i\}$. The relations $\rightarrow_1 \subseteq \mathbb{WL} \times \mathbb{NT}$, $\rightarrow_2 \subseteq \mathbb{NT} \times \mathbb{WL}$, and $\rightarrow_w \subseteq \mathbb{WL} \times \mathbb{WL}$ are defined as follows:

$\rightarrow_1$) (i) $x \rightarrow_1 t$ if $t \neq \tau$ and $t : x$ is in Γ^*; (ii) $x_0 \rightarrow_1 \tau$;
(iii) $y \neq x_0$ implies $y \not\rightarrow_1 \tau$; and (iv) $x \rightarrow_1 \bar{t}$ if $x \rightarrow_1 t$.

$\rightarrow_2$) $t \rightarrow_2 x$ if for a $i \in \mathbb{N}$, $k(x) = i$ and $x : t$ is in Γ_i (for t positive or negative).

$\rightarrow_w$) $x \rightarrow_w y$ if for some (positive or negative) term t, $x \rightarrow_1 t$ and $t \rightarrow_2 y$.

Lemma 4.7 *Given a branch $\mathcal{B}$ in a proof search tree for $\Rightarrow x_0 : A$ built in accordance with the strategy we have that (a) the graph $\mathcal{T}_w$ determined by x_0 and the relation $\rightarrow_w$ is a tree with root x_0, and (b) all the world labels occurring in $\mathcal{B}$ are nodes of $\mathcal{T}_w$.*

Lemma 4.8 *Let for any world label x and any (positive or negative) term t, $md(x) = max\{md(A) \mid x : A \text{ is in } \Gamma^* \cup \Delta^*\}$ and $md(t) = max\{md(A) \mid t : A \text{ is in } \Gamma^* \cup \Delta^*\}$, where $md(A)$ is the modal degree of A defined in the standard way. Then for any x, y in $\mathcal{T}_w$ we have that $x \rightarrow_w y$ implies $md(y) < md(x)$.*

Proposition 4.9 *Given a branch $\mathcal{B}$ of a proof search for $\Rightarrow x_0 : A$, (a) any world label occurring in $\mathcal{B}$ generates at most finitely many terms, and (b) any term occurring in $\mathcal{B}$ generates at most finitely many world labels. Whence (c) $\mathcal{T}_w$ is finite.*

Proof. (a) Consider first atomic terms: A world label x generates an atomic term $[a]$ by an application of L$\Box$. By its saturation clause, L$\Box$ can be applied to each formula $x : \Box B$ at most once. Therefore the problem is reduced to counting how many different formulas $x : \Box B$ can occur in the branch. If x is x_0, *i.e.* the label occurring in the sequent $\Rightarrow x_0 : A$ at the root, then the number of these formulas is smaller than the length of A. If x is generated by a term t, then it is generated by an application of R$\Vdash^\forall$ with a formula $[b] : C$ in Δ^* principal in the rule application (or by an application of L$\Vdash^\exists$ with a formula $\overline{[b]} : C$ in Γ^* principal in the rule application). Thus all formulas $\Box B$ such that $x : \Box B$ is in the branch are subformulas of C or - if t is atomic and different from τ - subformulas of D, where D is the only formula such that $t : D$ is in Γ^* (or $\bar{t} : D$ is in Δ^*), whose existence is guaranteed by definition of proper sequents. For complex terms: If x generates n atomic (positive) terms, then - by means of $\mathrm{C}_{\mathbb{T}}$ - it generates at most $2^n - 1$ positive terms. Therefore the terms generated by x are in any case finitely many.

(b) A term t generates a world label y by an application of R$\Vdash^\forall$ or L$\Vdash^\exists$. By the saturation clauses of these rules, every expression $t : B$ produces at most one world label. Therefore the problem is reduced to counting how many different expressions $t : B$ can occur in the branch. First assume $t \neq \tau$ and t generated by x. Then the number of these expressions depends directly on the number of formulas $x : \Box B$ in Δ^*, which - as shown in point (a) - are finitely many. If $t = \tau$: By the properties of the calculus, if $\tau : B$ is in Δ^*, then B is a subformula of A, where A is the formula labelled by x_0 at the root. Thus the possible expressions $\tau : B$ in Δ^* are finitely many. Observe also that there is no $\overline{\tau} : B$ in Γ^*. In fact, by an application of $\mathrm{L}{\Vdash^{\exists}}$ this would give a formula $y : \overline{\tau}$ in Γ^*, against the saturation clause for $\mathrm{N}\overline{\tau}$.

(c) By the decrease in modal depth stated stated by Lemma 4.8 it follows that any branch of $\mathcal{T}_w$ has a finite length. Moreover, $\mathcal{T}_w$ is finitary: if $x \overset{w}{\to} y$, then by definition there is a term t such that $x \to_1 t \to_2 y$; but by points (a) and (b) x is related to finitely many terms and t is related to finitely many world labels. □

Theorem 4.10 *Any branch $\mathcal{B}$ of a proof search for $\Rightarrow x_0 : A$ built in accordance with the strategy is finite, therefore proof search for any sequent of the form $\Rightarrow x_0 : A$ always comes to an end after a finite number of steps. Furthermore, each branch is either closed or saturated.*

Proof. By Proposition 4.9, $\mathcal{B}$ contains finitely many world labels and neighbourhood terms. Moreover, by the properties of the calculus, in any formula $x : B$ (or $t : B$, $\overline{t} : B$) that can occur in $\mathcal{B}$, B is a subformula of A, where A is the formula labelled by x_0 in the root sequent. Therefore only a finite number of labelled formulas can occur in $\mathcal{B}$. Thus, since by the saturation conditions a rule is not applied more than once to the same labelled formula ϕ (or the same pair of formulas ϕ and ψ), there are always only finitely many possible rule applications. □

We now prove semantic completeness of the calculi. This result shows that given an unprovable formula we can extract a finite countermodel of it in the bi-neighbourhood semantics. Moreover, by Lemma 2.9 we can also get a standard countermodel. Observe that this result, combined with the soundness of $\mathbb{T}$**LSE***, provides a constructive proof of the finite model property both in the bi-neighbourhood and in the standard semantics.

Theorem 4.11 $\mathbb{T}$**LSE*** *is complete with respect to the corresponding class of bi-neighbourhood models.*

Proof. Given a saturated branch $\mathcal{B}$ in a proof search in $\mathbb{T}$**LSE*** for the proper sequent $\Gamma \Rightarrow \Delta$, we build a bi-neighbourhood countermodel $\mathcal{M}$ to $\Gamma \Rightarrow \Delta$ that makes all formulas in Γ^* true and all formulas in Δ^* false. Model $\mathcal{M} = \langle W, \mathcal{N}, V\rangle$ is defined as follows: $W = \{x \in \mathbb{WL} \mid x \text{ occurs in } \Gamma^* \cup \Delta^*\}$; $\alpha_{[a_1 \dots a_n]} = \{x \in W \mid \text{for all } 1 \leq i \leq n,\ x : [a_i] \text{ is in } \Gamma^*\}$; $\alpha_{\overline{[a_1 \dots a_n]}} = \{x \in W \mid \text{for some } 1 \leq i \leq n,\ x : \overline{[a_i]} \text{ is in } \Gamma^*\}$; $\alpha_\tau = W$; $\alpha_{\overline{\tau}} = \emptyset$; for any $x \in W$, $\mathcal{N}(x) = \{(\alpha_t, \alpha_{\overline{t}}) \mid t : x \text{ is in } \Gamma^*\}$; for any $p \in \mathcal{L}$, $V(p) = \{x \in W \mid x : p \text{ is in } \Gamma^*\}$. Then we define the realisation (ρ, σ) by choosing $\rho(x) = x$ for any world label x, and $\sigma(t) = \alpha_t$ for any positive or negative term t occurring in $\Gamma^* \cup \Delta^*$.

First of all observe that $\mathcal{M}$ and σ are well defined: By the definition of $\alpha_{[a_1 \dots a_n]}$ and $\alpha_{\overline{[a_1 \dots a_n]}}$ it follows immediately that $\sigma(ts) = \sigma(t) \cap \sigma(s)$ and $\sigma(\overline{ts}) = \sigma(\overline{t}) \cup \sigma(\overline{s})$. Moreover, $\sigma(t) \cap \sigma(\overline{t}) = \emptyset$. In fact, assume $t = [a_1 \dots a_n]$

and $\alpha_t \cap \alpha_{\overline{t}} \neq \emptyset$. By definition, for some $1 \leq i \leq n$ and some $y \in W$, $y : [a_i]$ and $y : \overline{[a_i]}$ are in Γ^*. Since such expressions are never deleted, this means that there is a sequent $\Gamma_j \Rightarrow \Delta_j$ in the brach $\mathcal{B}$ such that $y : [a_i]$ and $y : \overline{[a_i]}$ are in Γ_j. Then by Proposition 4.2, $\Gamma_j \Rightarrow \Delta_j$ is derivable, against the hypothesis that $\mathcal{B}$ is saturated. Finally, from this it follows that $(\alpha, \beta) \in \mathcal{N}(x)$ implies $\alpha \cap \beta = \emptyset$. By considering all possible cases, it is easy to prove by induction on the weight of ϕ that if ϕ is in Γ^*, then $\mathcal{M} \models_{\rho,\sigma} \phi$, and if ϕ is in Δ^*, then $\mathcal{M} \not\models_{\rho,\sigma} \phi$. Moreover, it can be shown that if $\mathbb{TLSE}^*$ contains the rules for C, then $\mathcal{M}$ is a C-model, if it contains the rules for N, then $\mathcal{M}$ is a N-model, and if it contains the rules for M, then $\mathcal{M}$ is a M-model. □

Example 4.12 Here is a failed derivation of an instance of axiom M in $\mathbb{TLSE}$:

$$\dfrac{\mathcal{A}: \text{closed} \qquad \dfrac{\dfrac{\dfrac{\mathcal{A}': \text{closed} \qquad \overset{\text{saturated branch } \mathcal{B}}{y : \overline{[a]}, y : p, [a] : x, [a] : p \wedge q \Rightarrow x : \Box p, \overline{[a]} : p \wedge q, y : q}}{y : \overline{[a]}, y : p, [a] : x, [a] : p \wedge q \Rightarrow x : \Box p, \overline{[a]} : p \wedge q, y : p \wedge q}\,\wedge\mathrm{L}}{y : \overline{[a]}, y : p, [a] : x, [a] : p \wedge q \Rightarrow x : \Box p, \overline{[a]} : p \wedge q}\,\mathrm{R}\Vdash^{\exists}}{[a] : x, [a] : p \wedge q, \overline{[a]} : p \Rightarrow x : \Box p, \overline{[a]} : p \wedge q}\,\mathrm{L}\Vdash^{\exists}}{\dfrac{[a] : x, [a] : p \wedge q \Rightarrow x : \Box p, \overline{[a]} : p \wedge q}{x : \Box(p \wedge q) \Rightarrow x : \Box p}\,\mathrm{L}\Box}\,\mathrm{R}\Box$$

The bi-neighbourhood model $\mathcal{M} = \langle W, \mathcal{N}, V \rangle$ defined directly from the saturated branch $\mathcal{B}$ is the following: $W = \{x, y\}$, $\mathcal{N}(x) = \{(\emptyset, \{y\})\}$, $\mathcal{N}(y) = \emptyset$, $V(p) = \{y\}$ and $V(q) = \emptyset$. Then we have $\mathcal{M}, x \not\models \Box p$ and, since $[p \wedge q]_{\mathcal{M}} = \emptyset$, we also have $\mathcal{M}, x \models \Box(p \wedge q)$, thus the sequent at the root is falsified.

If we now consider the set $\mathcal{S} = \{\Box(p \wedge q), \Box p, p \wedge q, p, q\}$ and we follow the definition in Lemma 2.9, we obtain the standard model $\mathcal{M}^{\mathcal{S}}$ in which $\mathcal{N}^{\mathcal{S}}(x) = \{[p \wedge q]_{\mathcal{M}}\} = \{\emptyset\}$ and $\mathcal{N}^{\mathcal{S}}(y) = \emptyset$. It is immediate to verify that also $\mathcal{M}^{\mathcal{S}}$ falsifies the sequent.

5 Proof-theoretic equivalence of the semantics

In the previous section, we have shown that $\mathbb{TLSE}^*$ is sound and complete with respect to bi-neighbourhood semantics, thus by virtue of Theorem 2.10 also with respect to the standard semantics. For the non-monotonic case we now give a proof-theoretical argument to show that the two semantics coincide (therefore we do not consider rule M in this section). More precisely, we show that interpreting the negative terms as true complements (as it happens in standard semantics) does not extend the set of provable formulas, whence the set of valid formulas. To this purpose we consider the following rule:

$$\dfrac{x : [a], \Gamma \Rightarrow \Delta \qquad x : \overline{[a]}, \Gamma \Rightarrow \Delta}{\Gamma \Rightarrow \Delta}\ \mathrm{cmp}\ (x, a \in \Gamma \cup \Delta)$$

and we show that it is admissible in $\mathbb{TLSE}^*$. Moreover, we also show easily that by using this rule we can directly build countermodels in the standard semantics. As before, the analysis is restricted to proper sequents. Observe that the application of cmp respects (backwards) the constraints of proper sequents.

Proposition 5.1 *Rule* cmp *is admissible in* $\mathbb{T}$**LSE*** *for derivations of proper sequents.*

Proof. First of all, by induction on the height of the derivations one can prove that (a) if $x : [a], [a] : B, x : B, \Gamma \Rightarrow \Delta$ is proper and derivable, then $[a] : B, x : B, \Gamma \Rightarrow \Delta$ is proper and derivable with a derivation of the same height; and (b) if $x : \overline{[a]}, \Gamma \Rightarrow \Delta, \overline{[a]} : B, x : B$ is proper and derivable and x is in Γ, then $\Gamma \Rightarrow \Delta, \overline{[a]} : B, x : B$ is proper and derivable with a derivation of the same height. Then by induction on the height of the application of cmp it is possible to show how to remove all its applications. We only show the most significant case, in which $x : [a]$ and $x : \overline{[a]}$ are both principal in the last rule of the derivation of the respective premisses. The only possibility is that the applied rules are $\text{L}\Vdash^{\forall}$ for the left premiss and $\text{R}\Vdash^{\exists}$ for the right premiss:

$$\dfrac{\dfrac{x : [a], x : B, [a] : B, \Gamma \Rightarrow \Delta, \overline{[a]} : C}{x : [a], [a] : B, \Gamma \Rightarrow \Delta, \overline{[a]} : C}\,\text{L}\Vdash^{\forall} \qquad \dfrac{x : \overline{[a]}, [a] : B, \Gamma \Rightarrow \Delta, \overline{[a]} : C, x : C}{x : \overline{[a]}, [a] : B, \Gamma \Rightarrow \Delta, \overline{[a]} : C}\,\text{R}\Vdash^{\exists}}{[a] : B, \Gamma \Rightarrow \Delta, \overline{[a]} : C}\,\text{cmp}$$

However, since $[a] : B, \Gamma \Rightarrow \Delta, \overline{[a]} : C$ is a proper sequent, we have that $B \equiv C$. Therefore the case under consideration is as follows:

$$\dfrac{\dfrac{x : [a], x : B, [a] : B, \Gamma \Rightarrow \Delta, \overline{[a]} : B}{x : [a], [a] : B, \Gamma \Rightarrow \Delta, \overline{[a]} : B}\,\text{L}\Vdash^{\forall} \qquad \dfrac{x : \overline{[a]}, [a] : B, \Gamma \Rightarrow \Delta, \overline{[a]} : B, x : B}{x : \overline{[a]}, [a] : B, \Gamma \Rightarrow \Delta, \overline{[a]} : B}\,\text{R}\Vdash^{\exists}}{[a] : B, \Gamma \Rightarrow \Delta, \overline{[a]} : B}\,\text{cmp}$$

where the premisses of $\text{L}\Vdash^{\forall}$ and $\text{R}\Vdash^{\exists}$ are proper. Then by (a) and (b) we have that also $x : B, [a] : B, \Gamma \Rightarrow \Delta, \overline{[a]} : B$, and $[a] : B, \Gamma \Rightarrow \Delta, \overline{[a]} : B, x : B$ are proper and are derivable with derivations of the same heights. Observe in particular that point (b) is here applicable because of the condition on the application of cmp and the definition of sequents, that guarantee that x is in Γ. By an application of *cut* to these sequents with $x : B$ as cut formula we then obtain $[a] : B, \Gamma \Rightarrow \Delta, \overline{[a]} : B$. □

Theorem 5.2 $\mathbb{T}$**LSE*** *is complete with respect to the corresponding class of standard models.*

Proof. Let $\mathcal{B}$ be a saturated branch in a proof search in $\mathbb{T}$**LSE*** for the proper sequent $\Gamma \Rightarrow \Delta$ satisfying also the saturation condition for rule cmp: If x and $[a]$ are in Γ^*, then $x : [a]$ is in Γ^* or $x : \overline{[a]}$ is in Γ^*. We then build a standard countermodel $\mathcal{M}$ to $\Gamma \Rightarrow \Delta$ that makes all formulas in Γ^* true and all formulas in Δ^* false. Let the realisation (ρ, σ) and the model $\mathcal{M}$ be defined as in Theorem 4.11 with the minor modification that for all $x \in W$, $\mathcal{N}(x) = \{\alpha_t \mid t : x \text{ is in } \Gamma^*\}$. We only need to prove that $\mathcal{M}$ is now a standard model, that is $\sigma(\overline{t}) = W \setminus \sigma(t)$. We already know that $\sigma(t) \cap \sigma(\overline{t}) = \emptyset$; we show that $\sigma(t) \cup \sigma(\overline{t}) = W$. If $t = \tau$, this holds by definition of $\sigma(\tau)$. Assume $t = [a_1 \dots a_n]$. By saturation of cmp, for all $1 \leq i \leq n$, $x : [a_i]$ or $x : \overline{[a_i]}$ is in Γ^*. If for some i, $x : \overline{[a_i]}$ is in Γ^*, then by definition $x \in \alpha_{\overline{t}}$. Otherwise $x : [a_i]$ is in Γ^* for all i, and by definition $x \in \alpha_t$. In addition observe also that by saturation of rules C_T and $N\tau$ we have that if $\mathbb{TLSE}^*$ contains the rules for C, then $\mathcal{M}$ is closed under intersection, and if $\mathbb{TLSE}^*$ contains the rules for N, then $\mathcal{M}$ contains the unit. □

Example 5.3 This example shows how to obtain directly a standard countermodel from a failed branch of a proof search in $\mathbb{TLSE}$ which is saturated also with respect to rule cmp. In the derivation below we extend the branch $\mathcal{B}$ of the proof search of Example 4.12 in order to get such a saturation.

$$
\begin{array}{c}
\text{saturated branch } \mathcal{C}_3 \\
x:\overline{[a]}, y:\overline{[a]}, y:p, [a]:x, [a]:p\wedge q \Rightarrow x:\Box p, \overline{[a]}:p\wedge q, y:q, x:p \\
\vdots \\
\text{saturated branch } \mathcal{C}_2 \\
\dfrac{\dfrac{x:\overline{[a]}, y:\overline{[a]}, y:p, [a]:x, [a]:p\wedge q \Rightarrow x:\Box p, \overline{[a]}:p\wedge q, y:q, x:q}{x:\overline{[a]}, y:\overline{[a]}, y:p, [a]:x, [a]:p\wedge q \Rightarrow x:\Box p, \overline{[a]}:p\wedge q, y:q, x:p\wedge q}\wedge R}{x:\overline{[a]}, y:\overline{[a]}, y:p, [a]:x, [a]:p\wedge q \Rightarrow x:\Box p, \overline{[a]}:p\wedge q, y:q} R\vdash\exists \\
\text{saturated branch } \mathcal{C}_1 \\
\wedge L \dfrac{x:[a], x:p, x:q, y:\overline{[a]}, y:p, [a]:x, [a]:p\wedge q \Rightarrow x:\Box p, \overline{[a]}:p\wedge q, y:q}{\dfrac{x:[a], x:p\wedge q, y:\overline{[a]}, y:p, [a]:x, [a]:p\wedge q \Rightarrow x:\Box p, \overline{[a]}:p\wedge q, y:q}{\dfrac{x:[a], y:\overline{[a]}, y:p, [a]:x, [a]:p\wedge q \Rightarrow x:\Box p, \overline{[a]}:p\wedge q, y:q}{y:\overline{[a]}, y:p, [a]:x, [a]:p\wedge q \Rightarrow x:\Box p, \overline{[a]}:p\wedge q, y:q} \text{cmp}}L\vdash^\forall}
\end{array}
$$

On the basis of the three open branches we define three models following the definition of Theorem 5.2. The branch $\mathcal{C}_1$ gives the model $\mathcal{M}_1 = \langle W, \mathcal{N}_1, V_1\rangle$, where $W = \{x, y\}$, $\mathcal{N}_1(x) = \{\{x\}\}$, $\mathcal{N}_1(y) = \emptyset$, $V_1(p) = \{x, y\}$ and $V_1(q) = \{x\}$. The branch $\mathcal{C}_2$ gives the model $\mathcal{M}_2 = \langle W, \mathcal{N}_2, V_2\rangle$, where $W = \{x, y\}$, $\mathcal{N}_2(x) = \{\emptyset\}$, $\mathcal{N}_2(y) = \emptyset$, $V_2(p) = \{y\}$, and $V_2(q) = \emptyset$. Finally, $\mathcal{C}_3$ gives the same model of $\mathcal{C}_2$. It is immediate to verify that they are countemodels to the sequent at the root. Observe that $\mathcal{M}_2$ is the model $\mathcal{M}^S$ of Example 4.12.

It is instructive to compare this example with the countermodels provided by the (rather complicated) decision procedure given by Lavendhomme and Lucas [9] (Example pp. 137-139). The first model they obtain is the following (after renaming variables): $\mathfrak{M} = \langle W, \mathcal{N}, V\rangle$ where $W = \{x, y\}$, $\mathcal{N}(x) = \{\{x\}\}$, $\mathcal{N}(y) = \{\{x, y\}, \{x\}\}$, $V(p) = \{x, y\}$ and $V(q) = \{x\}$. The second model is the same as $\mathfrak{M}$ except for $\mathcal{N}(y) = \{\{x\}\}$. Both models are very similar to our model $\mathcal{M}_1$, however $\mathcal{M}_1$ is simpler as $\mathcal{N}_1(y) = \emptyset$. This is essentially due to the fact that we do not need to saturate worlds with respect to boxed subformulas as in the procedure given in [9].

6 Conclusion

In this paper, we have proposed labelled calculi for the cube of basic non-normal modal logic. The calculi are based on bi-neighbourhood models, a variation of the standard neighbourhood models, where each world is equipped with a set of pairs of neighbourhoods. The two components of a pair provide separate positive and negative support for a formula. This semantics might be of independent interest, being perhaps more natural for logics without monotonicity. We have shown that this semantics characterises all non-normal modal logics and (in the non-monotonic case) a standard model can be directly built from a bi-neighbourhood one. The sequent calculi we propose are fully modular and standard. For logics containing axiom C we actually propose two versions of the calculi: the first allows a syntactic proof of cut admissibility, whereas the second handles a more restricted form of sequents and comprises more efficient rules for handling intersections of neighbourhoods. In any case, the calculi provide a decision procedure for the respective logics and they are semantically complete: from any failed derivation of a formula one can effectively (and easily) extract a countermodel, both a bi-neighbourhood and a standard one, of the formula. A number of issues deserve to be further investigated: first we aim to study how to get optimal decision procedures from the calculi. We then plan to study how our calculi are related to other proof systems known in the literature, in particular the calculi proposed in [9] and the structural calculi proposed recently in [10]. We also intend to extend our approach, both the bi-neighbourhood semantics and the calculi, to stronger non-normal modal logics determined by the analogous ones of the normal cube from **K** to **S5** and to logical systems below **E**. Finally, it might be useful to draw a detailed comparison between bi-neighbourhood semantics and *bi-lattice semantics* since there is a resemblance between the two and the latter has recently been provided with a display proof system in [4]. All these topics will be object of our future work.

References

[1] Chellas, B. F., "Modal Logic: An Introduction," Cambridge University Press, 1980.

[2] Gasquet, O. and A. Herzig, *From classical to normal modal logics*, in: *Proof theory of modal logic*, Springer, 1996 pp. 293–311.

[3] Gilbert, D. R. and P. Maffezioli, *Modular sequent calculi for classical modal logics*, Studia Logica **103** (2015), pp. 175–217.

[4] Greco, G., F. Liang, A. Palmigiano and U. Rivieccio, *Bilattice logic properly displayed*, Fuzzy Sets and Systems (2018).

[5] Hughes, G. and M. Cresswell, "A New Introduction to Modal Logic," Routledge, 1996.

[6] Indrzejczak, A., *Sequent calculi for monotonic modal logics*, Bulletin of the Section of logic **34** (2005), pp. 151–164.

[7] Indrzejczak, A., "Natural Deduction, Hybrid Systems and Modal Logics," Springer, 2010.

[8] Kracht, M. and F. Wolter, *Normal monomodal logics can simulate all others*, The Journal of Symbolic Logic **64** (1999), pp. 99–138.

[9] Lavendhomme, R. and T. Lucas, *Sequent calculi and decision procedures for weak modal systems*, Studia Logica **66** (2000), pp. 121–145.

[10] Lellmann, B. and E. Pimentel, *Proof search in nested sequent calculi*, in: *Logic for Programming, Artificial Intelligence, and Reasoning*, Springer, 2015, pp. 558–574.

[11] Negri, S., *Proof analysis in modal logic*, Journal of Philosophical Logic **34** (2005), pp. 507–544.

[12] Negri, S., *Proofs and countermodels in non-classical logics*, Logica Universalis **8** (2014), pp. 25–60.

[13] Negri, S., *Proof theory for non-normal modal logics: The neighbourhood formalism and basic results*, IFCoLog Journal of Logics and their Applications, Mints' memorial issue **4** (2017), pp. 1241–1286.

[14] Pacuit, E., "Neighborhood Semantics for Modal Logic," Springer, 2017.

[15] Wansing, H., *Sequent systems for modal logics*, in: *Handbook of philosophical logic*, Springer, 2002 pp. 61–145.

On the Complexity of Modal Separation Logics

Stéphane Demri

LSV, CNRS, ENS Paris-Saclay, Université Paris-Saclay, France

Raul Fervari

FAMAF, Universidad Nacional de Córdoba & CONICET, Argentina

Abstract

We introduce a modal separation logic MSL whose models are memory states from separation logic and the logical connectives include modal operators as well as separating conjunction and implication from separation logic. With such a combination of operators, some fragments of MSL can be seen as genuine modal logics whereas some others capture standard separation logics, leading to an original language to speak about memory states. We analyse the decidability status and the computational complexity of several fragments of MSL, leading to surprising results, obtained by designing proof methods that take into account the modal and separation features of MSL. For example, the satisfiability problem for the fragment of MSL with $\Diamond$, the inequality modality $\langle\neq\rangle$ and separating conjunction $*$ is shown TOWER-complete whereas the restriction either to $\Diamond$ and $*$ or to $\langle\neq\rangle$ and $*$ is only NP-complete.

Keywords: separation logics, relation-changing logics, satisfiability, model-checking, complexity, expressive power.

1 Introduction

Combining modalities and separating connectives. Separation logic is known as an assertion language to perform verification, by extending Hoare-Floyd logic in order to verify programs with mutable data structures [27,34]. Local reasoning is a key feature of separation logic and the separating conjunction $*$ allows us to state properties in disjoint parts of the memory. Moreover, the separating implication $-\!\!*$ asserts that whenever a fresh heap satisfies a property, its composition with the current heap satisfies another property. Hence, the separating connectives $*$ and $-\!\!*$ allow us to evaluate formulae in alternative models, which is a feature shared with many modal logics such as sabotage logics [39,29], logics of public announcements (see e.g., [30]), interval temporal logics [26] or relation-changing logics [4,1].

Many other examples of such logics can be found in the literature (see also [18]) but the modalities involved in such logics can be of a different nature. For instance, combinations of epistemic logics and abstract separation logics (such as variants of BI) can be found in [17,23]. Sometimes, the concept of separation is different and performed at a different level, for instance a simple separation logic is introduced in [25] in which separation is performed on valuations instead of being performed on heaps. A slightly different approach including description logics [5] was investigated in [24,15]. An interesting attempt to get a logic (namely CT^2) that captures both a very expressive description logic and a separation logic (the symbolic heap fragment) can be found in [15].

Our motivations. Most existing logics combining (epistemic, temporal, etc.) modalities and separating connectives are multi-dimensional logics and the modal dimension is often orthogonal with the separation dimension (see e.g. [10,17,23]), which allows to get proof methods combining adequately the modal part and the separation part. Our intention in this work is to introduce a modal separation logic whose models are Kripke-style structures that can be also viewed as memory states from separation logic, without being multi-dimensional. As a gain, it is possible to study the computational effects of the interaction between modalities and separating connectives but within a uniform framework and to push further the expressive power of the underlying modal logics as well as the expressive power of the underlying separation logics. Adding modalities to separation logics happens to be an original means to work on fragments of first-order separation logics. So, the logic MSL introduced herein can be understood as a *hybrid* separation logic, by analogy to hybrid versions of modal logics [7]. Note that a hybrid extension of Boolean BI is defined in [13], in which nominals are interpreted by heaps whereas herein, the nominals are interpreted by locations.

Our contributions. We introduce the logic MSL whose models are Kripke-style structures with domain $\mathbb{N}$ (understood as the set of locations) and the accessibility relation is finite and functional (understood as some heap $\mathfrak{h} : \mathbb{N} \rightarrow_{fin} \mathbb{N}$). In MSL, the modal connectives are $\Diamond$ and the inequality modality $\langle \neq \rangle$ [19] whereas the separating connectives are the separating conjunction $*$ and separating implication $-\!\!*$ (also known as the magic wand operator). These connectives allow to update dynamically the model under evaluation. Therefore, in MSL, $\Diamond$ provides a means to move within the model following the accessibility relation, $\langle \neq \rangle$ adds the possibility to jump to (almost) any location of the model, and the connectives $*$ and $-\!\!*$, removes or adds edges in the model respectively. The closest logic to MSL is probably the modal logic of heaps MLH [21] since they share the same class of frames. However, there are differences, notably MSL has propositional variables (unlike MLH whose atomic formulae are truth constants) and MSL does not contain the converse modality and the reflexive transitive closure modality. Moreover, MSL shares with some logics from [28,12] the feature of having propositional variables whose interpretation is unrestricted but in such logics, the propositional variables are interpreted as sets of memory states whereas in MSL, the variables are interpreted as sets of locations, as usual for modal logics.

- MSL restricted to $\Diamond$ and $*$, written MSL$(*, \Diamond)$, can be viewed as the minimal modal separation logic as it witnesses a simple interaction between $\Diamond$ and, on the other side $*$ and $\mathtt{emp}$ (formula stating that the heap domain is empty). By showing a small model property, we establish that the satisfiability problem for MSL$(*, \Diamond)$ is NP-complete. The same result is shown for MSL$(*, \langle\neq\rangle)$ by adapting arguments for the logic of elsewhere [36,20]. To obtain the NP upper bound, we need to show that underlying model-checking problems are in P, which requires a refined analysis as the model checking problem for propositional separation logic (even restricted to $*$) is already PSPACE-complete [14].
- As far as decidability is concerned, we show that the satisfiability problem for MSL$(*, \Diamond, \langle\neq\rangle)$ is decidable by translation into the weak monadic second-order theory of one unary function shown decidable in [33]. This extends the decidability proof of 1SL1$(*)$ from [11] as, now, propositional variables need to be taken into account. More surprisingly, even though both MSL$(*, \Diamond)$ and MSL$(*, \langle\neq\rangle)$ are NP-complete, we establish that the satisfiability problem for MSL$(*, \Diamond, \langle\neq\rangle)$ is TOWER-hard by reduction from the nonemptiness problem for star-free expressions [31,37,35]. To do so, we show an essential property: the formula $\exists\ \mathtt{x}, \mathtt{y}\ \mathtt{ls}(\mathtt{x}, \mathtt{y})$ from separation logic (see e.g. [6,16]) can be expressed in MSL$(*, \Diamond, \langle\neq\rangle)$, which allows us to encode finite words. The notion of TOWER-completeness is borrowed from [35].
- Using the fact that $\mathtt{ls}(\mathtt{x}, \mathtt{y})$ can be expressed in MSL$(*, \Diamond, \langle\neq\rangle)$ we also establish that MSL (i.e. MSL$(*, \Diamond, \langle\neq\rangle)$ augmented with the magic wand $-\!\!*$) admits an undecidable satisfiability problem by using the recent result from [22] about the undecidability of propositional separation logic (with $*$ and $-\!\!*$) augmented with the list segment predicate $\mathtt{ls}$.
- Along the paper, we also investigate variants of MSL (or some of its fragments) by slightly modifying the semantics or by adding other modal connectives. For instance, we provide a reduction from the satisfiability problem for MSL$(*, \Diamond)$ when the models are arbitrary countable Kripke-style models into global sabotage logic over general models [3].

2 Preliminaries

In this section we introduce the modal separation logic MSL, as well as several fragments that we briefly compare with propositional separation logic.

2.1 Modal separation logic MSL

Let PROP $= \{p_1, q_1, p_2, q_2, \ldots\}$ be a countably infinite set of propositional variables. Formulae for the logic MSL are defined by the grammar below:

$$\phi ::= p \mid \mathtt{emp} \mid \neg\phi \mid \phi \vee \phi \mid \Diamond\phi \mid \langle\neq\rangle\phi \mid \phi * \phi \mid \phi -\!\!* \phi,$$

where $p \in$ PROP. An MSL *model* is a tuple $\mathfrak{M} = \langle \mathbb{N}, \mathfrak{R}, \mathfrak{V} \rangle$ such that $\mathfrak{R} \subseteq \mathbb{N} \times \mathbb{N}$ is finite and functional, and $\mathfrak{V} : \text{PROP} \to \mathcal{P}(\mathbb{N})$. Since separation logics are interpreted on structures representing heaps, our formulas are interpreted on models where the accessibility relation is finite and functional. The models $\mathfrak{M}_1 = \langle \mathbb{N}, \mathfrak{R}_1, \mathfrak{V} \rangle$ and $\mathfrak{M}_2 = \langle \mathbb{N}, \mathfrak{R}_2, \mathfrak{V} \rangle$ are *disjoint* if $\mathfrak{R}_1 \cap \mathfrak{R}_2 = \emptyset$; when this holds, $\mathfrak{M}_1 \uplus \mathfrak{M}_2$ denotes the model corresponding to the disjoint union of $\mathfrak{M}_1$ and $\mathfrak{M}_2$, and $\mathfrak{M}_1 \subseteq \mathfrak{M}_2$ means that $\mathfrak{M}_1$ and $\mathfrak{M}_2$ have the same valuation and $\mathfrak{R}_1 \subseteq \mathfrak{R}_2$. Given $\mathfrak{M} = \langle \mathbb{N}, \mathfrak{R}, \mathfrak{V} \rangle$ and $\mathfrak{l} \in \mathbb{N}$, the satisfaction relation $\models$ is defined below (clauses for Boolean connectives are omitted):

$$\begin{array}{lll}
\mathfrak{M}, \mathfrak{l} \models p & \overset{\text{def}}{\Leftrightarrow} & \mathfrak{l} \in \mathfrak{V}(p) \\
\mathfrak{M}, \mathfrak{l} \models \texttt{emp} & \overset{\text{def}}{\Leftrightarrow} & \mathfrak{R} = \emptyset \\
\mathfrak{M}, \mathfrak{l} \models \Diamond\phi & \overset{\text{def}}{\Leftrightarrow} & \mathfrak{M}, \mathfrak{l}' \models \phi, \text{ for some } \mathfrak{l}' \in \mathbb{N} \text{ such that } (\mathfrak{l}, \mathfrak{l}') \in \mathfrak{R} \\
\mathfrak{M}, \mathfrak{l} \models \langle\neq\rangle\phi & \overset{\text{def}}{\Leftrightarrow} & \mathfrak{M}, \mathfrak{l}' \models \phi, \text{ for some } \mathfrak{l}' \in \mathbb{N} \text{ such that } \mathfrak{l}' \neq \mathfrak{l} \\
\mathfrak{M}, \mathfrak{l} \models \phi_1 * \phi_2 & \overset{\text{def}}{\Leftrightarrow} & \langle \mathbb{N}, \mathfrak{R}_1, \mathfrak{V} \rangle, \mathfrak{l} \models \phi_1 \text{ and } \langle \mathbb{N}, \mathfrak{R}_2, \mathfrak{V} \rangle, \mathfrak{l} \models \phi_2, \\
& & \text{for some partition } \{\mathfrak{R}_1, \mathfrak{R}_2\} \text{ of } \mathfrak{R} \\
\mathfrak{M}, \mathfrak{l} \models \phi_1 \mathrel{-\!\!*} \phi_2 & \overset{\text{def}}{\Leftrightarrow} & \text{for all } \mathfrak{M}' = \langle \mathbb{N}, \mathfrak{R}', \mathfrak{V} \rangle \text{ such that } \mathfrak{R} \cup \mathfrak{R}' \text{ is finite and} \\
& & \text{functional, and } \mathfrak{R} \cap \mathfrak{R}' = \emptyset, \\
& & \text{we have } \mathfrak{M}', \mathfrak{l} \models \phi_1 \text{ implies } \langle \mathbb{N}, \mathfrak{R} \cup \mathfrak{R}', \mathfrak{V} \rangle, \mathfrak{l} \models \phi_2.
\end{array}$$

The semantics for the modal operators and the separating connectives is the standard one, see e.g. [8,34]. Other standard connectives or formulae are used:

- $[\neq]\phi \overset{\text{def}}{=} \neg\langle\neq\rangle\neg\phi$ and $\Box\phi \overset{\text{def}}{=} \neg\Diamond\neg\phi$,
- $\langle \mathrm{U} \rangle\phi \overset{\text{def}}{=} \phi \vee \langle\neq\rangle\phi$ and $[\mathrm{U}]\phi \overset{\text{def}}{=} \neg\langle \mathrm{U} \rangle\neg\phi$,
- $\langle\,!\,\rangle\phi \overset{\text{def}}{=} \langle \mathrm{U} \rangle(\phi \wedge [\neq]\neg\phi)$ (unicity of the satisfaction of ϕ),
- the atomic formula $\texttt{size} = 1$ is a shortcut for $\neg\texttt{emp} \wedge \neg(\neg\texttt{emp} * \neg\texttt{emp})$.

The *satisfiability problem* for the logic MSL, takes as input a formula ϕ and asks whether there exist an MSL model $\mathfrak{M}$ and a location $\mathfrak{l}$ such that $\mathfrak{M}, \mathfrak{l} \models \phi$.

Not only our study includes MSL but above all, we also deal with fragments. For instance, the fragment with Boolean connectives and $\Diamond$ is the *basic modal logic* ML. Otherwise, as a convention, we always consider the Boolean part and the emptiness constant `emp`, and we put between parentheses the rest of (separating or modal) connectives we are considering. The main logics we consider are $\mathrm{MSL}(*, \Diamond)$, $\mathrm{MSL}(*, \langle\neq\rangle)$ and $\mathrm{MSL}(*, \Diamond, \langle\neq\rangle)$.

2.2 Nominals, program variables and separation logic in a nutshell

In all the fragments of MSL containing the inequality modality [19], it is known that nominals from hybrid logics [7] can be used since stating that p holds true in a unique location can be expressed by $\langle\,!\,\rangle p$. So, we can freely use nominals. Syntactically, nominals are taken from $\mathrm{PVAR} = \{\mathtt{x}, \mathtt{y}, \ldots\}$, that is actually also used as the set of *program variables* in separation logic (see below). Indeed, nominals and program variables are both interpreted by locations, as noticed in [24]. So, checking the satisfiability status of a formula ϕ containing $\mathtt{x}_1, \ldots, \mathtt{x}_n$ actually amounts to checking the satisfiability status of $(\bigwedge_{1 \leq i \leq n} \langle\,!\,\rangle \mathtt{x}_i) \wedge \phi$. A formula ϕ is said to be *global* iff its satisfaction does not depend on the location and we simply write $\mathfrak{M} \models \phi$ (instead of $\mathfrak{M}, \mathfrak{l} \models \phi$). Below, we show why these formulae are important to compare MSL with separation logics. Indeed, MSL behaves as a standard modal logic since the satisfaction relation has three arguments (a model, a location and a formula) but it can be also presented as a separation logic so that the satisfaction relation takes only two arguments, a model and a global formula. Let us briefly explain why separation logic can be viewed as a fragment of MSL. A *memory state* is a pair $(\mathfrak{s}, \mathfrak{h})$ such that $\mathfrak{s} : \mathrm{PVAR} \rightarrow \mathbb{N}$ (the *store*) and $\mathfrak{h} : \mathbb{N} \rightarrow_{\mathrm{fin}} \mathbb{N}$ is a partial function with finite domain (the *heap*). Models of the separation logic $\mathrm{SL}(*, \mathrel{-\!\!*})$ are memory states. When the respective domains of the heaps $\mathfrak{h}_1$ and $\mathfrak{h}_2$ are disjoint, we write $\mathfrak{h}_1 \uplus \mathfrak{h}_2$ to denote the heap corresponding to the disjoint union of $\mathfrak{h}_1$ and $\mathfrak{h}_2$. Formulae of $\mathrm{SL}(*, \mathrel{-\!\!*})$ are built from

$$\phi ::= \mathtt{x} = \mathtt{y} \mid \mathtt{x} \hookrightarrow \mathtt{y} \mid \mathtt{emp} \mid \neg\phi \mid \phi \wedge \phi \mid \phi * \phi \mid \phi \mathrel{-\!\!*} \phi,$$

where $\mathtt{x}, \mathtt{y} \in \mathrm{PVAR}$. The satisfaction relation $\models$ is defined as follows:

$$\begin{array}{lcl}
(\mathfrak{s}, \mathfrak{h}) \models \mathtt{x} = \mathtt{y} & \stackrel{\text{def}}{\Leftrightarrow} & \mathfrak{s}(\mathtt{x}) = \mathfrak{s}(\mathtt{y}) \\
(\mathfrak{s}, \mathfrak{h}) \models \mathtt{emp} & \stackrel{\text{def}}{\Leftrightarrow} & \mathrm{dom}(\mathfrak{h}) = \emptyset \\
(\mathfrak{s}, \mathfrak{h}) \models \mathtt{x} \hookrightarrow \mathtt{y} & \stackrel{\text{def}}{\Leftrightarrow} & \mathfrak{s}(\mathtt{x}) \in \mathrm{dom}(\mathfrak{h}) \text{ and } \mathfrak{h}(\mathfrak{s}(\mathtt{x})) = \mathfrak{s}(\mathtt{y}) \\
(\mathfrak{s}, \mathfrak{h}) \models \phi_1 * \phi_2 & \stackrel{\text{def}}{\Leftrightarrow} & \text{there are } \mathfrak{h}_1 \text{ and } \mathfrak{h}_2 \text{ such that } \mathfrak{h}_1 \uplus \mathfrak{h}_2 = \mathfrak{h}, \\
& & (\mathfrak{s}, \mathfrak{h}_1) \models \phi_1 \text{ and } (\mathfrak{s}, \mathfrak{h}_2) \models \phi_2 \\
(\mathfrak{s}, \mathfrak{h}) \models \phi_1 \mathrel{-\!\!*} \phi_2 & \stackrel{\text{def}}{\Leftrightarrow} & \text{for all } \mathfrak{h}_1, \text{ if } (\mathrm{dom}(\mathfrak{h}_1) \cap \mathrm{dom}(\mathfrak{h}) = \emptyset \text{ and} \\
& & (\mathfrak{s}, \mathfrak{h}_1) \models \phi_1), \text{ then } (\mathfrak{s}, \mathfrak{h} \uplus \mathfrak{h}_1) \models \phi_2.
\end{array}$$

Any memory state $(\mathfrak{s}, \mathfrak{h})$ can be viewed as the MSL model $\mathfrak{M} = \langle \mathbb{N}, \mathfrak{R}, \mathfrak{V} \rangle$ such that $\mathfrak{R} = \{(\mathfrak{l}, \mathfrak{h}(\mathfrak{l})) \mid \mathfrak{l} \in \mathrm{dom}(\mathfrak{h})\}$ and the restriction of $\mathfrak{V}$ to PVAR is equal to $\mathfrak{s}$. Actually, any formula ϕ of $\mathrm{SL}(*, \mathrel{-\!\!*})$ is satisfiable iff $t(\phi)$ is satisfiable in MSL where t is homomorphic for Boolean and separating connectives and,

$$t(\mathtt{x} = \mathtt{y}) \stackrel{\text{def}}{=} \langle \mathrm{U} \rangle (\mathtt{x} \wedge \mathtt{y}) \quad t(\mathtt{emp}) \stackrel{\text{def}}{=} \mathtt{emp} \quad t(\mathtt{x} \hookrightarrow \mathtt{y}) \stackrel{\text{def}}{=} \langle \mathrm{U} \rangle (\mathtt{x} \wedge \Diamond \mathtt{y}).$$

It is worth noting that each formula $t(\phi)$ is a global formula of MSL.

2.3 Alternative semantics

A *general model* $\mathfrak{M} = \langle \mathfrak{W}, \mathfrak{R}, \mathfrak{V} \rangle$ is such that $\mathfrak{W}$ is an arbitrary countable set, $\mathfrak{R} \subseteq \mathfrak{W} \times \mathfrak{W}$ and $\mathfrak{V} : \mathrm{PROP} \rightarrow \mathcal{P}(\mathfrak{W})$. This corresponds to standard (countable) Kripke structures with no frame condition. A *finite and functional model* $\mathfrak{M} = \langle \mathfrak{W}, \mathfrak{R}, \mathfrak{V} \rangle$ is such that $\mathfrak{W}$ is a finite set, $\mathfrak{R} \subseteq \mathfrak{W} \times \mathfrak{W}$ is functional and $\mathfrak{V}$ is a valuation. Without loss of generality, we assume $\mathfrak{W} \subseteq \mathbb{N}$. Each syntactic fragment $\mathfrak{L}$ of MSL gives rise to the logic $\mathfrak{L}^f$ (resp. $\mathfrak{L}^g$) where the models for $\mathfrak{L}^f$ are finite and functional models (resp. are general models). When $\mathfrak{L}$ includes $\mathrel{-\!\!*}$, the definition of $\models$ for $\mathfrak{L}^g$ is updated as follows:

$$\begin{array}{lcl}
\mathfrak{M}, \mathfrak{l} \models \phi_1 \mathrel{-\!\!*} \phi_2 & \stackrel{\text{def}}{\Leftrightarrow} & \text{for all } \mathfrak{M}' = \langle \mathfrak{W}, \mathfrak{R}', \mathfrak{V} \rangle \text{ such that } \mathfrak{R} \cap \mathfrak{R}' = \emptyset \\
& & \mathfrak{M}', \mathfrak{l} \models \phi_1 \text{ implies } \langle \mathfrak{W}, \mathfrak{R} \cup \mathfrak{R}', \mathfrak{V} \rangle, \mathfrak{l} \models \phi_2.
\end{array}$$

Note that the formula $(\top \mathrel{-\!\!*} \neg((\neg\mathtt{emp}) \mathrel{-\!\!*} \bot))$ is valid for MSL but not for MSL^f. The *model-checking problem for* MSL^f is defined in the usual way. As MSL can be viewed as a fragment of second-order logic (the second-order feature is needed to internalise the semantics of separating connectives), the model-checking problem for MSL^f is in PSPACE. More surprisingly, we show that the restriction to either $\mathrm{MSL}^f(*, \langle\neq\rangle)$ or $\mathrm{MSL}^f(*, \Diamond)$ is in P, whereas the restriction to $\mathrm{MSL}^f(*, \Diamond, \langle\neq\rangle)$ is already untractable.

Lemma 2.1 *The model-checking problem for* $\mathrm{MSL}^f(*, \Diamond, \langle\neq\rangle)$ *is* PSPACE-*hard.*

Proof Let $\mathcal{Q}_1\ p_1\ \cdots\ \mathcal{Q}_n\ p_n\ \phi$ be a QBF formula with $\{\mathcal{Q}_1, \ldots, \mathcal{Q}_n\} \subseteq \{\exists, \forall\}$ and ϕ is a propositional formula built over $\{p_1, \ldots, p_n\}$ and the Boolean connectives $\wedge$, $\vee$ and $\neg$ (only in front of atomic propositions). Satisfiability problem for QBF formulae is known to be PSPACE-complete [38].

In the reduction of $\varphi = \mathcal{Q}_1\ p_1\ \cdots\ \mathcal{Q}_n\ p_n\ \phi$, we introduce a finite and functional model $\mathfrak{M}_n = \langle \mathfrak{W}, \mathfrak{R}, \mathfrak{V}\rangle$ with $\mathfrak{W} = [0, 2n]$ such that $\mathcal{Q}_1\ p_1\ \cdots\ \mathcal{Q}_n\ p_n\ \phi$ is satisfiable iff $\mathfrak{M}_n, 0 \models t(\varphi)$, where $t(\cdot)$ is recursively defined below. The truth of the propositional variable p_i in QBF subformulae is encoded by the satisfaction of the formula $\langle\neq\rangle(p_i \wedge p_\top \wedge \Diamond\top)$ from $\mathrm{MSL}^f(*, \Diamond, \langle\neq\rangle)$.

First, let us complete the definition of $\mathfrak{M}_n$ over the propositional variables $\{p_\top, p_1, \ldots, p_n\}$.

$$\mathfrak{V}(p_i) \stackrel{\text{def}}{=} \{i, n+i\}, \text{ for all } i = 1, \ldots, n$$
$$\mathfrak{V}(p_\top) \stackrel{\text{def}}{=} [1, n]$$
$$\mathfrak{R} \stackrel{\text{def}}{=} \{(i, 0) \mid i \in [1, 2n]\}.$$

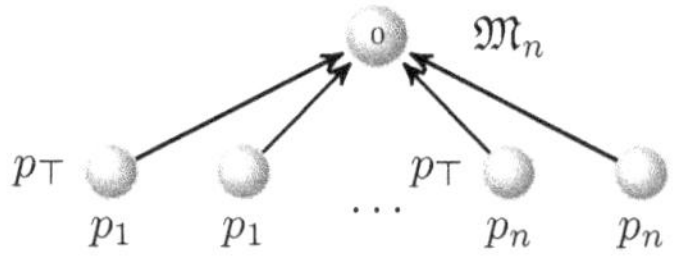

Let us define the map t as follows (homomorphic for Boolean connectives):

$$t(p_i) \stackrel{\text{def}}{=} \langle\neq\rangle(p_i \wedge p_\top \wedge \Diamond\top)$$
$$t(\exists\ p_i\ \psi) \stackrel{\text{def}}{=} (\mathtt{size} = 1 \wedge \langle\neq\rangle(p_i \wedge \Diamond\top)) * t(\psi)$$
$$t(\forall\ p_i\ \psi) \stackrel{\text{def}}{=} \neg((\mathtt{size} = 1 \wedge \langle\neq\rangle(p_i \wedge \Diamond\top)) * \neg t(\psi)).$$

For every $j \in [1, n+1]$, we write ϕ_j to denote the formula $\mathcal{Q}_j\ p_j\ \cdots\ \mathcal{Q}_n\ p_n\ \phi$. By definition, we have $\phi_1 = \mathcal{Q}_1\ p_1\ \cdots\ \mathcal{Q}_n\ p_n\ \phi$ and by convention $\phi_{n+1} = \phi$.

Given a model $\mathfrak{M} \subseteq \mathfrak{M}_n$ and a propositional valuation v, we write $\mathfrak{M} \approx_j v$ to denote the fact that:

- For all $i \in [1, j-1]$, exactly one location in $\{i, n+i\}$ has an outgoing edge.
- For all $i \in [j, n]$, all the locations in $\{i, n+i\}$ have an outgoing edge.
- For all $i \in [1, j-1]$, i has an outgoing edge iff $v(p_i) = \top$.

By induction on j, one can show that for all $j \in [1, n+1]$, if $\mathfrak{M} \approx_j v$, then $\mathfrak{M}, 0 \models t(\phi_j)$ iff $v \models \phi_j$. Details are omitted. So, as $\mathfrak{M}_n \approx_1 v$ for any v, we have $v \models \varphi$ iff $\mathfrak{M}_n, 0 \models t(\phi_1)$. As φ is a closed formula and $\phi_1 = \varphi$, φ is satisfiable iff $\mathfrak{M}_n, 0 \models t(\varphi)$. □

MSL can be seen as a logic with the ability to add or remove edges from the accessibility relation, closely related to relation-changing modal logics [1]. Below, we discuss the connections between MSL and the *global sabotage logic* $\mathrm{MSL}^g(\Diamond, \langle\mathsf{gsb}\rangle)$. Formulae of $\mathrm{MSL}^g(\Diamond, \langle\mathsf{gsb}\rangle)$ extends those of ML by adding the operator $\langle\mathsf{gsb}\rangle$ interpreted over general models $\mathfrak{M} = \langle \mathfrak{W}, \mathfrak{R}, \mathfrak{V}\rangle$ as:

$$\mathfrak{M}, \mathfrak{l} \models \langle \mathsf{gsb} \rangle \phi \stackrel{\text{def}}{\Leftrightarrow} \text{ for some } (\mathfrak{l}', \mathfrak{l}'') \in \mathfrak{R},\ \mathfrak{M}^{-}_{\mathfrak{l}',\mathfrak{l}''}, \mathfrak{l} \models \phi,$$

where $\mathfrak{M}^{-}_{\mathfrak{l}',\mathfrak{l}''} = \langle \mathfrak{W}, \mathfrak{R} \setminus \{(\mathfrak{l}', \mathfrak{l}'')\}, \mathfrak{V} \rangle$. $\mathrm{MSL}^g(\Diamond, \langle \mathsf{gsb} \rangle)$ can be encoded into $\mathrm{MSL}^g(*, \Diamond)$ by the translation t that is homomorphic for Boolean connectives and for $\Diamond$ and, $t(\langle \mathsf{gsb} \rangle \phi) \stackrel{\text{def}}{=} (\mathtt{size} = 1) * t(\phi)$. We have ϕ is satisfiable iff $t(\phi)$ is satisfiable for $\mathrm{MSL}^g(*, \Diamond)$. Similarly, $\mathrm{MSL}(\Diamond, \langle \mathsf{gsb} \rangle)$ is the variant of $\mathrm{MSL}^g(\Diamond, \langle \mathsf{gsb} \rangle)$ with MSL models.

3 Decision problems in TOWER

Below, we establish that the satisfiability problem for $\mathrm{MSL}(*, \Diamond, \langle \neq \rangle)$ is in TOWER [35], the class of problems of time complexity bounded by a tower of exponentials, whose height is an elementary function of the input. To do so, we design a reduction to the satisfiability problem for $\mathrm{MSL}^f(*, \Diamond, \langle \neq \rangle)$ and then we show that the satisfiability problem for $\mathrm{MSL}^f(*, \Diamond, \langle \neq \rangle)$ is in TOWER by translation into the weak MSO theory of one unary function. Notice that the difference between $\mathrm{MSL}(*, \Diamond, \langle \neq \rangle)$ and $\mathrm{MSL}^f(*, \Diamond, \langle \neq \rangle)$ is that models in $\mathrm{MSL}(*, \Diamond, \langle \neq \rangle)$ have finite relations over an infinite set of locations, while the set of locations in $\mathrm{MSL}^f(*, \Diamond, \langle \neq \rangle)$ models is finite. This proof is analogous to the decidability proof for 1SL1($*$) in [11] but our main technical task is to solve the satisfiability problem for $\mathrm{MSL}(*, \Diamond, \langle \neq \rangle)$ by using only propositional variables that hold true on a finite amount of locations. First, we show that locations satisfying the same propositional variables and with no successor satisfy the same formulae.

Lemma 3.1 *Let $p_1, \ldots, p_n$ be propositional variables, $\mathfrak{M} = \langle \mathbb{N}, \mathfrak{R}, \mathfrak{V} \rangle$ be a model and $\mathfrak{l} \neq \mathfrak{l}'$ be locations such that $\mathfrak{R}(\mathfrak{l}) = \mathfrak{R}(\mathfrak{l}') = \emptyset$ and, $\mathfrak{l}$ and $\mathfrak{l}'$ agree on $p_1, \ldots, p_n$. For all ϕ in $\mathrm{MSL}(*, \Diamond, \langle \neq \rangle)$ built over $p_1, \ldots, p_n$, $\mathfrak{M}, \mathfrak{l} \models \phi$ iff $\mathfrak{M}, \mathfrak{l}' \models \phi$.*

Let ϕ in $\mathrm{MSL}(*, \Diamond, \langle \neq \rangle)$ be built over $p_1, \ldots, p_n$. Let us define $T(\phi)$ as

$$T(\phi) \stackrel{\text{def}}{=} \phi \wedge \bigvee_{X \subseteq \{p_1, \ldots, p_n\}} \langle \mathrm{U} \rangle (\Box \perp \wedge \bigwedge_{p \in X} p \wedge \bigwedge_{p \notin X} \neg p \wedge \langle \neq \rangle (\Box \perp \wedge \bigwedge_{p \in X} p \wedge \bigwedge_{p \notin X} \neg p)).$$

When $\langle \neq \rangle$ is not present, the second conjunct can be removed (see Lemma 4.3, where we take $T(\phi) = \phi$). Such a conjunct states that there are two distinct locations with no successor that agree on propositional variables from X and it is needed since $\langle\, !\, \rangle p \wedge [\mathrm{U}]p$ is satisfiable for $\mathrm{MSL}^f(*, \langle \neq \rangle)$ but not for the logic $\mathrm{MSL}(*, \langle \neq \rangle)$.

Lemma 3.2 *ϕ is satisfiable in $\mathrm{MSL}(*, \Diamond, \langle \neq \rangle)$ iff $T(\phi)$ is satisfiable in $\mathrm{MSL}^f(*, \Diamond, \langle \neq \rangle)$.*

The complexity class TOWER has been introduced in [35] and sits between the class of elementary problems and the class of primitive recursive problems.

Theorem 3.3 *The satisfiability problem for $\mathrm{MSL}(*, \Diamond, \langle \neq \rangle)$ is in* TOWER.

By Lemma 3.2, there is a reduction from the satisfiability problem for $\mathrm{MSL}(*, \Diamond, \langle \neq \rangle)$ into the satisfiability problem for $\mathrm{MSL}^f(*, \Diamond, \langle \neq \rangle)$ that works in exponential time. There is also a (logspace) reduction from the satisfiability problem for $\mathrm{MSL}^f(*, \Diamond, \langle\neq\rangle)$ into the satisfiability problem for the weak MSO theory of one unary function whose structures are $\langle D, \mathfrak{f}, =\rangle$ where D is a countable domain, $\mathfrak{f}$ is a unary function ('weakness' refers to the fact that the monadic predicates are interpreted by *finite* sets). This theory is decidable, see e.g. [9, Corollary 7.2.11] and it can be shown in TOWER as it can be reduced to the satisfiability to the MSO theory of the infinite binary tree. In the proof of Theorem 3.3, the reduction from $\mathrm{MSL}^f(*, \Diamond, \langle\neq\rangle)$ simply internalises its semantics by using the (weak) second-order feature of the target logic.

In order to conclude this section, we consider the standard converse modality $\Diamond^{-1}$ (not originally in MSL), when it interacts with separating connectives. More precisely, $\mathfrak{M}, \mathfrak{l} \models \Diamond^{-1}\phi \overset{\text{def}}{\Leftrightarrow} \mathfrak{M}, \mathfrak{l}' \models \phi$, for some $\mathfrak{l}' \in \mathbb{N}$ such that $(\mathfrak{l}', \mathfrak{l}) \in \mathfrak{R}$. Although replacing $\Diamond$ by $\Diamond^{-1}$ does not sound as leading to a major variant, we will show that $\Diamond^{-1}$ already brings new difficulties.

Theorem 3.4 *The satisfiability problem for* $\mathrm{MSL}(*, \Diamond^{-1})$ *is* PSPACE*-hard as well as the model-checking problem for* $\mathrm{MSL}^f(*, \Diamond^{-1})$.

Note that $\mathrm{MSL}(*, \Diamond^{-1})$ contains $\mathrm{MSL}(\Diamond^{-1})$ that can be viewed as a slight variant of the modal logic K on finite trees, known to admit a PSPACE-complete satisfiability problem. So, PSPACE-hardness of the satisfiability problem for $\mathrm{MSL}(*, \Diamond^{-1})$ is quite expected.

By using the proof technique from the proof of Theorem 3.3, we can establish the result below where $\Diamond^{-1}$ is part of the modal operators.

Theorem 3.5 *The satisfiability problem for* $\mathrm{MSL}(*, \Diamond, \Diamond^{-1}, \langle\neq\rangle)$ *is in* TOWER.

As a conclusion, there is a huge gap for $\mathrm{MSL}(*, \Diamond^{-1})$ between the PSPACE-hardness for the satisfiability problem and the TOWER upper bound.

4 NP-complete fragments of MSL

In this section, we show that the satisfiability problems for $\mathrm{MSL}(*, \Diamond)$ and for $\mathrm{MSL}(*, \langle\neq\rangle)$ are NP-complete. In order to establish the NP upper bound, we reduce the problems to their variants with finite and functional models, we show a linear-size model property and finally, we prove that the model-checking problems are in P, dealing in each case with particular technical difficulties.

4.1 The minimal modal separation logic $\mathrm{MSL}(*, \Diamond)$

To show that $\mathrm{MSL}(*, \Diamond)$ has a linear-size model property (i.e., the cardinal of the relation can be bounded), we introduce an equivalence relation $\overset{s,n}{\sim}$ ($s \geq 0$ is a parameter about the number of edges and $n \geq 1$ is a parameter about the propositional variables) such that $\overset{s,n}{\sim}$-equivalent models satisfy the same formulae with less than s syntactic resources (to be defined) and built over $\{p_1, \ldots, p_n\}$. First, we need to explain how to decompose models with respect to the parameters s and n and, the relation $\overset{s,n}{\sim}$ is defined by using such a decomposition. As $\mathfrak{R}$ is functional, what matters is the structure of $\mathfrak{R}$ reduced

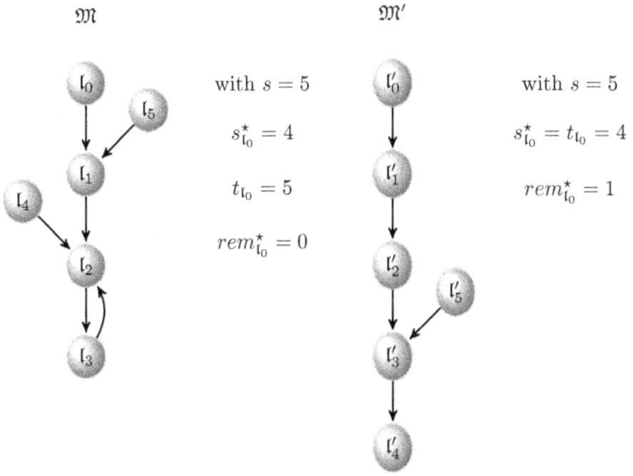

Figure 1. Decomposition.

to at most the s first steps from a given location as well as the total number of edges, counting up to s. Below, we show that this abstraction is correct with respect to the expressive power of MSL$(*, \Diamond)$, see e.g. Lemma 4.2. Let $\mathfrak{M} = \langle \mathbb{N}, \mathfrak{R}, \mathfrak{V} \rangle$ be a model, $\mathfrak{l} \in \mathbb{N}$ and $s \geq 0$, we define $\mathfrak{W}_{\mathfrak{l},s}$ and $\mathfrak{R}_{\mathfrak{l},s}$ as follows.

- $\mathfrak{W}_{\mathfrak{l},s} \stackrel{\text{def}}{=} \{(i, \mathfrak{l}_i) \mid i \in [0, s],\ \exists\ \mathfrak{l}_0, \ldots, \mathfrak{l}_i,\ \mathfrak{l} = \mathfrak{l}_0 \mathfrak{R} \mathfrak{l}_1 \cdots \mathfrak{R} \mathfrak{l}_{i-1} \mathfrak{R} \mathfrak{l}_i\}$. We also write $t_{\mathfrak{l}} = \max\{i \mid (i, \mathfrak{l}_i) \in \mathfrak{W}_{\mathfrak{l},s}\}$ (so $t_{\mathfrak{l}} \leq s$).
- $\mathfrak{R}_{\mathfrak{l},s} \stackrel{\text{def}}{=} \{(\mathfrak{l}_i, \mathfrak{l}_{i+1}) \mid i \in [0, t_{\mathfrak{l}} - 1] \text{ and } (i, \mathfrak{l}_i), (i+1, \mathfrak{l}_{i+1}) \in \mathfrak{W}_{\mathfrak{l},s}\}$. We also write $s^\star_{\mathfrak{l}} \stackrel{\text{def}}{=} \mathsf{card}(\mathfrak{R}_{\mathfrak{l},s})$ and $rem^\star_{\mathfrak{l}} = \min(s - \mathsf{card}(\mathfrak{R}_{\mathfrak{l},s}), \mathsf{card}(\mathfrak{R} \setminus \mathfrak{R}_{\mathfrak{l},s}))$. So, $s^\star_{\mathfrak{l}} \leq t_{\mathfrak{l}} \leq s$ and $s^\star_{\mathfrak{l}} + rem^\star_{\mathfrak{l}} \leq s$.

Let $\mathfrak{M}$, $\mathfrak{M}'$ be models, $\mathfrak{l}, \mathfrak{l}' \in \mathbb{N}$ and $s \geq 0$, $n \geq 1$ such that $\mathfrak{W}_{\mathfrak{l},s}$ and $\mathfrak{R}_{\mathfrak{l},s}$ are defined as above and $\mathfrak{W}'_{\mathfrak{l}',s}$ and $\mathfrak{R}'_{\mathfrak{l}',s}$ are related to $\mathfrak{M}'$, $\mathfrak{l}'$ and s. Let us define the relation $\overset{s,n}{\sim}$: $\mathfrak{M}, \mathfrak{l} \overset{s,n}{\sim} \mathfrak{M}', \mathfrak{l}' \overset{\text{def}}{\Leftrightarrow}$ the conditions below are satisfied:

- We have $t_{\mathfrak{l}} = t_{\mathfrak{l}'} (\stackrel{\text{def}}{=} t)$. Say, $\mathfrak{W}_{\mathfrak{l},s} = \{(0, \mathfrak{l}_0), \ldots, (t, \mathfrak{l}_t)\}$ and $\mathfrak{W}'_{\mathfrak{l}',s} = \{(0, \mathfrak{l}'_0), \ldots, (t, \mathfrak{l}'_t)\}$.
- For all $i \in [0, t]$, $\mathfrak{l}_i$ in $\mathfrak{M}$ and $\mathfrak{l}'_i$ in $\mathfrak{M}'$ agree on $\{p_1, \ldots, p_n\} \subset$ PROP.
- For all $i, j \in [0, t-1]$, we have $\mathfrak{l}_i = \mathfrak{l}_j$ iff $\mathfrak{l}'_i = \mathfrak{l}'_j$. Hence, $s^\star_{\mathfrak{l}} = s^\star_{\mathfrak{l}'} (\stackrel{\text{def}}{=} s^\star)$.
- We have $rem^\star_{\mathfrak{l}} = rem^\star_{\mathfrak{l}'} (\stackrel{\text{def}}{=} rem^\star)$.

The binary relation $\overset{s,n}{\sim}$ is an equivalence relation. In Figure 1, $\mathfrak{M}, \mathfrak{l}_0 \overset{4,n}{\sim} \mathfrak{M}', \mathfrak{l}'_0$ (assuming that $\mathfrak{l}_i$ and $\mathfrak{l}'_i$ agree on $\{p_1, \ldots, p_n\}$ for every $i \in [0, 3]$, and, $\mathfrak{l}_2/\mathfrak{l}'_2$ and $\mathfrak{l}'_4$ agree too). By contrast, $\mathfrak{M}, \mathfrak{l}_0 \overset{5,n}{\sim} \mathfrak{M}', \mathfrak{l}'_0$ does not hold. Lemma 4.1 below is essential to justify that $\overset{s,n}{\sim}$ behaves properly with disjoint unions of models. Its proof is tedious as numerous cases are needed.

Lemma 4.1 *Let $s, s_1, s_2 \geq 1$ with $s = s_1 + s_2$, $\mathfrak{M}, \mathfrak{l} \overset{s,n}{\sim} \mathfrak{M}', \mathfrak{l}'$ and $\mathfrak{M}_1, \mathfrak{M}_2$ be models such that $\mathfrak{M} = \mathfrak{M}_1 \uplus \mathfrak{M}_2$. There are models $\mathfrak{M}'_1$ and $\mathfrak{M}'_2$ such that $\mathfrak{M}' = \mathfrak{M}'_1 \uplus \mathfrak{M}'_2$, $\mathfrak{M}_1, \mathfrak{l} \overset{s_1,n}{\sim} \mathfrak{M}'_1, \mathfrak{l}'$ and $\mathfrak{M}_2, \mathfrak{l} \overset{s_2,n}{\sim} \mathfrak{M}'_2, \mathfrak{l}'$.*

Given a formula ϕ in MSL$(*, \Diamond)$, let us define its *esize* (written esize(ϕ)):

- $\text{esize}(p) \stackrel{\text{def}}{=} \text{esize}(\texttt{emp}) \stackrel{\text{def}}{=} 1$, $\text{esize}(\neg\phi) \stackrel{\text{def}}{=} \text{esize}(\phi)$, $\text{esize}(\Diamond\phi) \stackrel{\text{def}}{=} 1 + \text{esize}(\phi)$,
- $\text{esize}(\phi \wedge \psi) \stackrel{\text{def}}{=} \max(\text{esize}(\phi), \text{esize}(\psi))$, $\text{esize}(\phi * \psi) \stackrel{\text{def}}{=} \text{esize}(\phi) + \text{esize}(\psi)$.

Note that esize(ϕ) is greater than the modal degree of ϕ, and approximatively, esize(ϕ) provides an upper bound on the number of edges that need to be considered in a model for ϕ (so it will play the role of the value s). For technical reasons, we have assumed that esize$(p) = 1$, so that esize$(\phi) \geq 1$ for any ϕ.

Lemma 4.2 *Let $s, n \geq 1$. For all formulae ϕ in* MSL$(*, \Diamond)$ *with* esize$(\phi) \leq s$ *and built over $p_1, \ldots, p_n$, we have $\mathfrak{M}, \mathfrak{l} \stackrel{s,n}{\sim} \mathfrak{M}', \mathfrak{l}'$ implies $\mathfrak{M}, \mathfrak{l} \models \phi$ iff $\mathfrak{M}', \mathfrak{l}' \models \phi$.*

The following quantitative result is crucial to get the NP upper bound.

Lemma 4.3 *Let ϕ be a formula in* MSL$(*, \Diamond)$*. ϕ is satisfiable iff ϕ is satisfiable in a finite and functional model with* $\mathsf{card}(\mathfrak{R}) \leq \text{esize}(\phi)$.

It remains to show that the model-checking problem for MSL$^f(*, \Diamond)$ is in P. The main difficulty rests on the fact that evaluating an $*$-formula may require to consider an exponential number of pairs of disjoint submodels. Fortunately, only a polynomial amount of disjoint unions are shown relevant. At the beginning of this section, we defined a decomposition of any MSL model based on the parameter $s \geq 0$. Such a decomposition was useful to show Lemma 4.3. A similar decomposition can be done with finite and functional models. More precisely, let $\mathfrak{M} = \langle \mathfrak{W}, \mathfrak{R}, \mathfrak{V} \rangle$ be a finite and functional model and $\mathfrak{l} \in \mathfrak{W}$. One can easily define the set $\mathfrak{W}_{\mathfrak{l},s}$, the relation $\mathfrak{R}_{\mathfrak{l},s}$ and the values $t_{\mathfrak{l}}$, $s^{\star}_{\mathfrak{l}}$ and $rem^{\star}_{\mathfrak{l}}$. Consequently, an equivalence relation $\stackrel{s,n}{\sim}$ can be also defined on finite and functional pointed models leading to a natural variant of Lemma 4.2 involving finite and functional models instead of MSL models.

In order to check whether $\mathfrak{M}, \mathfrak{l} \models \phi$ holds, we start by building a submodel $\mathfrak{M}' = \langle \mathfrak{W}, \mathfrak{R}', \mathfrak{V} \rangle \subseteq \mathfrak{M}$ with $\mathsf{card}(\mathfrak{R}') \leq \text{esize}(\phi)$ and check whether $\mathfrak{M}', \mathfrak{l} \models \phi$ holds. The submodel $\mathfrak{M}'$ can be built in polynomial time in the size of $\mathfrak{M}$ and in s. In forthcoming Algorithm 1, instead of working with models, we operate with slightly more abstract structures. An *abstract frame up to s* is a pair $\mathcal{F} = ((\mathfrak{l}_0, \ldots, \mathfrak{l}_t), r)$ where $r \geq 0$, $(\mathfrak{l}_0, \ldots, \mathfrak{l}_t) \in \mathbb{N}^+$ (standing for locations linked by edges) and the conditions below hold:

(truncation) $t^* + r \leq s$ and $t \leq s$ with $t^* = \mathsf{card}(\{(\mathfrak{l}_i, \mathfrak{l}_{i+1}) \mid i \in [0, t-1]\})$.
(maximality) $t < s$ implies there is no $i < t$ such that $\mathfrak{l}_i = \mathfrak{l}_t$.
(functionality) for all $i < j < t$, we have $\mathfrak{l}_i = \mathfrak{l}_j$ implies $t = s$ and $\mathfrak{l}_{i+1} = \mathfrak{l}_{j+1}$.

Given a finite and functional model $\mathfrak{M} = \langle \mathfrak{W}, \mathfrak{R}, \mathfrak{V} \rangle$, $\mathfrak{l} \in \mathfrak{W}$, and $s \geq 0$, we write abst$(\mathfrak{M}, \mathfrak{l}, s)$ to denote the abstraction $((\mathfrak{l}_0, \ldots, \mathfrak{l}_t), r)$ with $\{(0, \mathfrak{l}_0), \ldots, (t, \mathfrak{l}_t)\} = \mathfrak{W}_{\mathfrak{l},s}$ and $r = rem^{\star}_{\mathfrak{l}}$. An abstract frame is not explicitly equipped with a propositional valuation but in forthcoming Algorithm 1, we manipulate such structures as the associated propositional valuation will be systematically the one induced by the valuation of the input model. Let shrink$(\mathfrak{M}, \mathfrak{l}, s)$ be the finite and functional model $\mathfrak{M}' = \langle \mathfrak{W}, \mathfrak{R}', \mathfrak{V} \rangle$ such that $\mathfrak{R}' \stackrel{\text{def}}{=} \{(\mathfrak{l}_i, \mathfrak{l}_{i+1}) \mid i \in [0, t-1]\} \cup \{(\mathfrak{n}_1, \mathfrak{n}'_1), \ldots, (\mathfrak{n}_r, \mathfrak{n}'_r)\}$, where

$\{(\mathfrak{n}_1, \mathfrak{n}'_1), \ldots, (\mathfrak{n}_r, \mathfrak{n}'_r)\}$ is a set of r egdes in $\mathfrak{R} \setminus \mathfrak{R}_{\mathfrak{l},s}$ and the locations $\mathfrak{n}_1, \ldots, \mathfrak{n}_r$ are minimal. Lemma 4.4 below justifies the correctness of the abstraction.

Lemma 4.4 *Let* $s \geq 0$, $\mathfrak{M} = \langle \mathfrak{W}, \mathfrak{R}, \mathfrak{V} \rangle$ *be finite and functional and* $\mathfrak{l} \in \mathfrak{W}$ *with* $\mathfrak{M}' = \mathrm{shrink}(\mathfrak{M}, \mathfrak{l}, s)$. *Then* $\mathfrak{M}, \mathfrak{l} \overset{s,n}{\sim} \mathfrak{M}', \mathfrak{l}$ *and* $\mathrm{abst}(\mathfrak{M}, \mathfrak{l}, s) = \mathrm{abst}(\mathfrak{M}', \mathfrak{l}, s)$.

Let us define a notion of disjoint union between abstract frames to mimic the disjoint union of models. Let $s = s_1 + s_2$, $s, s_1, s_2 \geq 1$, $\mathcal{F} = ((\mathfrak{l}_0, \ldots, \mathfrak{l}_t), r)$ be an abstract frame up to s, $\mathcal{F}_i = ((\mathfrak{l}^i_0, \ldots, \mathfrak{l}^i_{t_i}), r^i)$ be an abstract frame up to s_i, with $i \in \{1, 2\}$. We write $\mathcal{F} = \mathcal{F}_1 \uplus \mathcal{F}_2 \overset{\text{def}}{\Leftrightarrow}$ (i)–(v) below hold ($i \in \{1, 2\}$):

(i) $\max(t_1, t_2) \leq t$, $t_1 \times t_2 = 0$ and, if $t > 0$ then $t_1 + t_2 > 0$.
(ii) $(\mathfrak{l}^i_0, \ldots, \mathfrak{l}^i_{t_i}) = (\mathfrak{l}_0, \ldots, \mathfrak{l}_{t_i})$.
(iii) $0 < t_i < \min(s_i, t)$ implies $r^{3-i} > 0$.
(iv) $0 < t_i$ implies $r^1 + r^2 \leq r + t^* - t^*_i$.
(v) $0 < t_i$ and $r^1 + r^2 < r + t^* - t^*_i$ imply $r^i = s_i - t^*_i$ or $r^{3-i} = s_{3-i}$.

Though (i)–(v) sound reasonable at first glance, the best way to understand what is really needed, is by proving Lemma 4.5 and Lemma 4.6. Similarly, given abstract frames $\mathcal{F}_1 = ((\mathfrak{l}^1_0, \ldots, \mathfrak{l}^1_{t_1}), r^1)$ up to s_1 and $\mathcal{F}_2 = ((\mathfrak{l}^2_0, \ldots, \mathfrak{l}^2_{t_2}), r^2)$ up to s_2 with $s_1 \leq s_2$, we write $\mathcal{F}_1 \subseteq \mathcal{F}_2$ whenever $(\mathfrak{l}^1_0, \ldots, \mathfrak{l}^1_{t_1})$ is a factor of $(\mathfrak{l}^2_0, \ldots, \mathfrak{l}^2_{t_2})$ and, $r^1 + t^*_1 \leq r^2 + t^*_2$.

Algorithm 1 below operates with abstract frames and its correctness is partly based on forthcoming Lemma 4.5 and Lemma 4.6. For instance, Lemma 4.5 can be understood as a correctness result: disjoint unions of models lead to the satisfaction of the conditions (i)-(v) at the level of abstract frames.

Lemma 4.5 *Let* $s = s_1 + s_2$ *with* $s, s_1, s_2 \geq 1$. *Let* $\mathfrak{M}$, $\mathfrak{M}_1$ *and* $\mathfrak{M}_2$ *be finite and functional models such that* $\mathfrak{M} = \mathfrak{M}_1 \uplus \mathfrak{M}_2$. *For all* $\mathfrak{l} \in \mathfrak{W}$, *we have* $\mathrm{abst}(\mathfrak{M}, \mathfrak{l}, s) = \mathrm{abst}(\mathfrak{M}_1, \mathfrak{l}, s_1) \uplus \mathrm{abst}(\mathfrak{M}_2, \mathfrak{l}, s_2)$.

By contrast, Lemma 4.6 below can be understood as a completeness result: the satisfaction of (i)-(v) can always be mimicked at the level of models.

Lemma 4.6 *Let* $s = s_1 + s_2$ *with* $s, s_1, s_2 \geq 1$. *Let* $\mathfrak{M}$ *be finite and functional,* $\mathcal{F}_i$ *be an abstract frame up to* s_i *(*$i \in \{1, 2\}$*) such that* $\mathrm{abst}(\mathfrak{M}, \mathfrak{l}, s) = \mathcal{F}_1 \uplus \mathcal{F}_2$. *There are* $\mathfrak{M}_1$ *and* $\mathfrak{M}_2$ *such that* $\mathfrak{M} = \mathfrak{M}_1 \uplus \mathfrak{M}_2$, $\mathcal{F}_i = \mathrm{abst}(\mathfrak{M}_i, \mathfrak{l}, s_i)$ *(*$i \in \{1, 2\}$*).*

What is essential is the fact that the number of non-equivalent decompositions is polynomial and not exponential in the size of $\mathcal{F}$, which is a serious guarantee to obtain a model checking algorithm running in polynomial time.

Lemma 4.7 *Let* $s = s_1 + s_2$ *with* $s, s_1, s_2 \geq 1$, $\mathcal{F} = ((\mathfrak{l}_0, \ldots, \mathfrak{l}_t), r)$ *be an abstract frame up to* s. *We have* $\mathsf{card}(\{(\mathcal{F}_1, \mathcal{F}_2) \mid \mathcal{F} = \mathcal{F}_1 \uplus \mathcal{F}_2,\ \mathcal{F}_i \text{ up to } s_i\}) \leq 2(s+1)(s_1+1)(s_2+1)$.

Algorithm 1 below uses first principles of dynamic programming as well as the map $\mathrm{shrink}(\mathcal{F}, s)$ defined as follows (abstract version of the shrink construction on models): $\mathrm{shrink}(((\mathfrak{l}_0, \ldots, \mathfrak{l}_t), r), s) \overset{\text{def}}{=} ((\mathfrak{l}_0, \ldots, \mathfrak{l}_{t'}), r')$ with

- $t' = \min(s, t)$,

- $t'_* = \mathsf{card}(\{\mathfrak{l}_1, \ldots, \mathfrak{l}_{t'}\})$ and,
- $r' = \min(s - t'_*, r + (\mathsf{card}(\{\mathfrak{l}_1, \ldots, \mathfrak{l}_t\}) - t'_*))$.

One can show that $\mathrm{shrink}(((\mathfrak{l}_0, \ldots, \mathfrak{l}_t), r), s) \subseteq ((\mathfrak{l}_0, \ldots, \mathfrak{l}_t), r)$ and $\mathrm{shrink}(((\mathfrak{l}_0, \ldots, \mathfrak{l}_t), r), s)$ is an abstract frame up to s. Algorithm 1 only computes values for $T(\mathrm{shrink}(\mathcal{F}, \mathrm{esize}(\psi_k)), k)$ as it would be time-consuming (and useless) to compute all the values $T(\mathcal{F}, k)$. This is enforced by the values in the **for** loops and by line 2. The map $\mathrm{shrink}(\cdot, \cdot)$ is also further needed for conjunctions as the measure $\mathrm{esize}(\cdot)$ involves a maximum for conjunctions.

Algorithm 1 Model Checking $\mathrm{MSL}^f(*, \Diamond)$

In: A finite and functional model $\mathfrak{M} = \langle \mathfrak{W}, \mathfrak{R}, \mathfrak{V} \rangle$, a location $\mathfrak{l} \in \mathfrak{W}$, an $\mathrm{MSL}(*, \Diamond)$ formula ϕ
Out: Return 1 iff $\mathfrak{M}, \mathfrak{l} \models \phi$.
1: **function** $\mathrm{MC}(\mathfrak{M}, \mathfrak{l}, \phi)$
2: $((\mathfrak{l}_0, \ldots, \mathfrak{l}_L), R) := \mathrm{abst}(\mathfrak{M}, \mathfrak{l}, \mathrm{esize}(\phi))$ $\triangleright$ $\mathsf{card}(\{\mathfrak{l}_1, \ldots, \mathfrak{l}_L\}) + R \leq \mathrm{esize}(\phi)$
3: $\psi_1, \ldots, \psi_M$ subformulae of ϕ in increasing size $\triangleright$ $\psi_M = \phi$
4: **for** $k \leftarrow 1$ **to** M **do**
5: **for** $j \leftarrow L$ **downto** 0 **do**
6: **for** $len \leftarrow 0$ **to** $\max\{len' \in [0, L-j] \mid \mathsf{card}(\{\mathfrak{l}_j, \ldots, \mathfrak{l}_{j+len'}\}) \leq \mathrm{esize}(\psi_k)\}$ **do**
7: **for** $r \leftarrow 0$ **to** $\max\{r' \in [0, R] \mid \mathsf{card}(\{\mathfrak{l}_j, \ldots, \mathfrak{l}_{j+len}\}) + r' \leq \mathrm{esize}(\psi_k)\}$ **do**
8: $\mathcal{F} := ((\mathfrak{l}_j, \ldots, \mathfrak{l}_{j+len}), r)$ $\triangleright$ $\mathcal{F} = \mathrm{shrink}(\mathcal{F}, \mathrm{esize}(\psi_k))$
9: **case** ψ_k **of**
10: $\mathtt{emp}$: $T(\mathcal{F}, k) := 1$ if ($len = 0$ and $r = 0$), otherwise 0.
11: p: $T(\mathcal{F}, k) := 1$ if $\mathfrak{l}_j \in \mathfrak{V}(p)$, otherwise 0.
12: $\neg\psi_{k'}$: $T(\mathcal{F}, k) := 1 - T(\mathcal{F}, k')$ $\triangleright$ $k' < k$
13: $\psi_{k_1} \wedge \psi_{k_2}$: $\triangleright$ $k_1, k_2 < k$
14: $T(\mathcal{F}, k) := \min(T(\mathrm{shrink}(\mathcal{F}, \mathrm{esize}(\psi_{k_1})), k_1), T(\mathrm{shrink}(\mathcal{F}, \mathrm{esize}(\psi_{k_2})), k_2))$
15: $\Diamond\psi_{k'}$: if ($len > 0$), $\mathcal{F}' := \mathrm{shrink}(((\mathfrak{l}_{j+1}, \ldots, \mathfrak{l}_{j+len}), r), \mathrm{esize}(\psi_{k'}))$ $\triangleright$ $k' < k$
16: $T(\mathcal{F}, k) := 1$ if ($len > 0$) and $T(\mathcal{F}', k') = 1$, otherwise 0.
17: $\psi_{k_1} * \psi_{k_2}$: $\triangleright$ $k_1, k_2 < k$
18: $s_1 := \mathrm{esize}(\psi_{k_1})$; $s_2 := \mathrm{esize}(\psi_{k_2})$ $\triangleright$ $\mathrm{esize}(\psi_k) = s_1 + s_2$
19: $T(\mathcal{F}, k) := \max\{\min(T(\mathcal{F}_1, k_1), T(\mathcal{F}_2, k_2)) \mid \mathcal{F} = \mathcal{F}_1 \uplus \mathcal{F}_2,\ \mathcal{F}_i \text{ up to } s_i\}$
20: **end case**
21: **return** $T(((\mathfrak{l}_0, \ldots, \mathfrak{l}_L), R), M)$

Due to the organisation of the **for** loops, each time the algorithm computes $T(\mathcal{F}, k)$, it requires values of the form $T(\mathcal{F}', k')$, always with $\mathcal{F}' \subseteq \mathcal{F}$ and $k' < k$, so the algorithm is properly defined. The algorithm runs in polynomial time thanks to Lemma 4.7. The following lemma establishes that the algorithm is correct and explains what is the intention behind computing the values $T(\mathcal{F}, k)$.

Lemma 4.8 *For all $k \in [1, M]$, for all abstract frames $\mathcal{F} = ((\mathfrak{l}, \ldots), R')$ up to $\mathrm{esize}(\psi_k)$ with $\mathcal{F} \subseteq ((\mathfrak{l}_0, \ldots, \mathfrak{l}_L), R)$, when the model-checking algorithm ends, $T(\mathcal{F}, k) = 1$ iff for all finite and functional submodels $\mathfrak{M}' \subseteq \mathfrak{M}$ such that $\mathrm{abst}(\mathfrak{M}', \mathfrak{l}, \mathrm{esize}(\psi_k)) = \mathcal{F}$, we have $\mathfrak{M}', \mathfrak{l} \models \psi_k$.*

So the model checking problem for $\mathrm{MSL}^f(*, \Diamond)$ is in P, and we can conclude.

Theorem 4.9 *The satisfiability problem for $\mathrm{MSL}(*, \Diamond)$ is NP-complete.*

From Section 2, we recall that $\mathrm{MSL}(\Diamond, \langle \mathsf{gsb} \rangle)$ is defined as a fragment of $\mathrm{MSL}(*, \Diamond)$ with the translation $t(\langle \mathsf{gsb} \rangle \phi) \stackrel{\text{def}}{=} (\mathtt{size} = 1) * t(\phi)$ (global sabotage modal operator). As a corollary of Theorem 4.9, we obtain the result below.

Corollary 4.10 *The satisfiability problem of* $\mathrm{MSL}(\Diamond, \langle \mathsf{gsb} \rangle)$ *is* NP*-complete.*

4.2 The fragment $\mathrm{MSL}(*, \langle \neq \rangle)$

We also establish that the satisfiability problem for $\mathrm{MSL}(*, \langle \neq \rangle)$ is NP-complete and its model-checking problem is in P. To do so, we reduce the problems from $\mathrm{MSL}(*, \langle \neq \rangle)$ to $\mathrm{MSL}^f(*, \langle \neq \rangle)$ and we show a small model property. Given ϕ in $\mathrm{MSL}(*, \langle \neq \rangle)$, let us define its $*$-*weight* $w_*(\phi)$ as follows:

- $w_*(p) \stackrel{\mathrm{def}}{=} 0$, $w_*(\mathsf{emp}) \stackrel{\mathrm{def}}{=} 1$, $w_*(\neg\phi) \stackrel{\mathrm{def}}{=} w_*(\langle \neq \rangle \phi) \stackrel{\mathrm{def}}{=} w_*(\phi)$,
- $w_*(\phi \wedge \psi) \stackrel{\mathrm{def}}{=} \max(w_*(\phi), w_*(\psi))$, $w_*(\phi * \psi) \stackrel{\mathrm{def}}{=} w_*(\phi) + w_*(\psi)$.

Lemma 4.11 *Let* $\alpha \geq 0$ *and* $\mathfrak{M} = \langle \mathbb{N}, \mathfrak{R}, \mathfrak{V} \rangle$ *and* $\mathfrak{M}' = \langle \mathbb{N}, \mathfrak{R}', \mathfrak{V} \rangle$ *be* MSL *models such that* $\min(\mathsf{card}(\mathfrak{R}), \alpha) = \min(\mathsf{card}(\mathfrak{R}'), \alpha)$. *Then, for all locations* $\mathfrak{l}$ *and formulae* ϕ *in* $\mathrm{MSL}(*, \langle \neq \rangle)$ *such that* $w_*(\phi) \leq \alpha$, *we have* $\mathfrak{M}, \mathfrak{l} \models \phi$ *iff* $\mathfrak{M}', \mathfrak{l} \models \phi$.

As a corollary, if ϕ in $\mathrm{MSL}(*, \langle \neq \rangle)$ is satisfiable, then it has a model with at most $w_*(\phi)$ edges. Let us refine this. Let $\mathfrak{M} = \langle \mathbb{N}, \mathfrak{R}, \mathfrak{V} \rangle$ be an MSL model with $\mathsf{card}(\mathfrak{R}) = \beta$, $\mathfrak{l} \in \mathbb{N}$ and ϕ be in $\mathrm{MSL}(*, \langle \neq \rangle)$ such that $\mathfrak{M}, \mathfrak{l} \models \phi$. Let $\psi_1, \ldots, \psi_N$ be the subformulae of ϕ such that $\langle \neq \rangle \psi_1, \ldots, \langle \neq \rangle \psi_N$ are the only subformulae of ϕ whose outermost connective is $\langle \neq \rangle$. For all $i \in [1, N]$ and all $\beta' \in [0, \beta]$, we define at most *two* locations $\mathfrak{l}_1^{i,\beta'}$ and $\mathfrak{l}_2^{i,\beta'}$ as follows.

- Given $\mathfrak{R}' \subseteq \mathfrak{R}$ with $\mathsf{card}(\mathfrak{R}') = \beta'$, we have $\langle \mathbb{N}, \mathfrak{R}', \mathfrak{V} \rangle, \mathfrak{l}_1^{i,\beta'} \models \psi_i$ and $\langle \mathbb{N}, \mathfrak{R}', \mathfrak{V} \rangle, \mathfrak{l}_2^{i,\beta'} \models \psi_i$. By Lemma 4.11, this definition makes sense as two models with the same valuation and with the same cardinal of the relation satisfy the same formulae.
- If possible we require that $\mathfrak{l}_1^{i,\beta'}$ and $\mathfrak{l}_2^{i,\beta'}$ are distinct, otherwise if there is only one location satisfying ψ_i in $\langle \mathbb{N}, \mathfrak{R}', \mathfrak{V} \rangle$, we require $\mathfrak{l}_1^{i,\beta'} = \mathfrak{l}_2^{i,\beta'}$.
- If no location satisfies ψ_i in $\langle \mathbb{N}, \mathfrak{R}', \mathfrak{V} \rangle$, then by default $\mathfrak{l}_1^{i,\beta'} = \mathfrak{l}_2^{i,\beta'} = \mathfrak{l}$.

Let $\mathfrak{W} \stackrel{\mathrm{def}}{=} \{\mathfrak{l}\} \cup \{\mathfrak{l}_j^{i,\beta'} \mid j \in \{1,2\},\ i \in [1,N],\ \beta' \in [0,\beta]\}$.

Lemma 4.12 *We have* $\langle \mathfrak{W}, \mathfrak{R}, \mathfrak{V} \rangle, \mathfrak{l} \models \phi$.

So, $\mathrm{MSL}(*, \langle \neq \rangle)$ satisfies a small model property.

Corollary 4.13 *Let* ϕ *be a formula in* $\mathrm{MSL}(*, \langle \neq \rangle)$. ϕ *is satisfiable iff* ϕ *is* $\mathrm{MSL}^f(*, \langle \neq \rangle)$ *satisfiable in a model with* $\mathsf{card}(\mathfrak{W}) \leq 1 + 2|\phi| \times w_*(\phi)$.

It remains to characterise the complexity of the model-checking problem for $\mathrm{MSL}^f(*, \langle \neq \rangle)$.

Lemma 4.14 *The model-checking problem for* $\mathrm{MSL}^f(*, \langle \neq \rangle)$ *is in* P.

Proof Let $\mathfrak{M} = \langle \mathfrak{W}, \mathfrak{R}, \mathfrak{V} \rangle$ be a finite and functional model, $\mathfrak{l} \in \mathfrak{W}$, and ϕ be a formula in $\mathrm{MSL}(*, \langle \neq \rangle)$. Let $\psi_1, \ldots, \psi_M$ be the subformulae of ϕ ordered in increasing size. We assume $\mathfrak{W} = [0, K]$ for some $K \geq 0$, $\mathfrak{l} = 0$ and $\mathsf{card}(\mathfrak{R}) = \beta$. In order to determine whether $\mathfrak{M}, \mathfrak{l} \models \phi$, we use a labelling algorithm and we complete a table $T(i, j, k)$ with $i \in [0, K]$, $j \in [0, \beta]$ and $k \in [1, M]$ that takes the value 1 iff $\langle \mathfrak{W}, \mathfrak{R}', \mathfrak{V} \rangle, i \models \psi_k$ with $\mathsf{card}(\mathfrak{R}') = j$ (dynamic programming is used here as usual). The polynomial-time upper bound is mainly due to the fact (see Lemma 4.12) that what matters in a partition $\{\mathfrak{R}'_1, \mathfrak{R}'_2\}$ of $\mathfrak{R}' \subseteq \mathfrak{R}$ is the respective cardinalities of $\mathfrak{R}'_1$ and $\mathfrak{R}'_2$.

Algorithm 2 Model Checking $\mathrm{MSL}^f(*, \langle\neq\rangle)$

In: A finite and functional model $\mathfrak{M} = \langle [0, K], \mathfrak{R}, \mathfrak{V} \rangle$, $K \geq 0$, an $\mathrm{MSL}(*, \langle\neq\rangle)$ formula ϕ
Out: Return 1 iff $\mathfrak{M}, 0 \models \phi$.

function $\mathrm{MC}(\mathfrak{M}, \mathfrak{l}, \phi)$
  $\psi_1, \ldots, \psi_M$ subformulae of ϕ in increasing size ▷ $\psi_M = \phi$
  $\beta := \mathsf{card}(\mathfrak{R})$
  for $j \leftarrow 0$ **to** β **do**
    for $k \leftarrow 0$ **to** M **do**
      for $i \leftarrow 0$ **to** K **do**
        case ψ_k **of**
        emp: $T(i, j, k) := 1$ if $(j = 0)$, otherwise 0
        p: $T(i, j, k) := 1$ if $i \in \mathfrak{V}(p)$, otherwise 0
        $\neg\psi_{k'}$: $T(i, j, k) := 1 - T(i, j, k')$ ▷ $k' < k$
        $\psi_{k_1} \wedge \psi_{k_2}$: $T(i, j, k) := \min(T(i, j, k_1), T(i, j, k_2))$ ▷ $k_1, k_2 < k$
        $\langle\neq\rangle\psi_{k'}$: ▷ $k' < k$
          $T(i, j, k) := \max(T(1, j, k'), \ldots, T(i-1, j, k'), T(i+1, j, k'), \ldots, T(K, j, k'))$
        $\psi_{k_1} * \psi_{k_2}$: ▷ $k_1, k_2 < k$
          $T(i, j, k) := \max\{\min(T(i, I, k'_1), T(i, J, k'_2)) \mid I + J = j \text{ and } I, J \geq 0\}$
        end case
  return $T(0, \beta, M)$

It is worth noting that computing $T(i, j, k)$ always requires values $T(i', j', k')$ that have already got a value and the whole procedure requires polynomial-time in $\beta + M + K$. The correctness of $T(i, j, k) = 1$ iff $\langle \mathfrak{W}, \mathfrak{R}', \mathfrak{V} \rangle, i \models \psi_k$ with $\mathsf{card}(\mathfrak{R}') = j$ is then by an easy verification. The satisfaction of $\mathfrak{M}, \mathfrak{l} \models \phi$ is therefore stored in $T(0, \beta, M)$. □

Again, we are able to establish an NP upper bound.

Theorem 4.15 *The satisfiability problem for* $\mathrm{MSL}(*, \langle\neq\rangle)$ *is* NP*-complete.*

5 $\mathrm{MSL}(*, \Diamond, \langle\neq\rangle)$: a TOWER-complete fragment of MSL

In this section, we show that the satisfiability problem for $\mathrm{MSL}(*, \Diamond, \langle\neq\rangle)$ is TOWER-complete. The upper bound is from Section 3 whereas the proof for TOWER-hardness consists of two parts. First, we show that there is a formula in $\mathrm{MSL}(*, \Diamond, \langle\neq\rangle)$ that characterises the linear structures. Then, we reduce the nonemptiness problem for star-free expressions into the satisfiability problem.

5.1 Encoding linear structures

The goal of this section is to design a global formula in $\mathrm{MSL}(*, \Diamond, \langle\neq\rangle)$, namely $\phi_{\exists 1\mathrm{s}}$, such that for all models $\mathfrak{M}$, we have $\mathfrak{M} \models \phi_{\exists 1\mathrm{s}}$ iff either $\mathfrak{R}$ is empty or $\mathfrak{R} = \{(\mathfrak{l}_0, \mathfrak{l}_1), \ldots, (\mathfrak{l}_{n-1}, \mathfrak{l}_n)\}$ for some $n \geq 1$ such that for all $i \neq j \in [0, n]$, we have $\mathfrak{l}_i \neq \mathfrak{l}_j$. In that case, we say that $\mathfrak{M}$ is *linear*. Given a finite set $X \subseteq \mathrm{PROP}$, the relation $\{(\mathfrak{l}_0, \mathfrak{l}_1), \ldots, (\mathfrak{l}_{n-1}, \mathfrak{l}_n)\}$ encodes the finite word $b_1 \cdots b_n$ where each letter b_j is equal to $\{p \in X \mid \mathfrak{l}_j \in \mathfrak{V}(p)\}$ (the labelling of the location $\mathfrak{l}_0$ is irrelevant for the encoding). When $\mathfrak{R}$ is empty, the pair $\mathfrak{M}, \mathfrak{l}$ encodes the empty string.

Note that $\phi_{\exists\mathtt{ls}}$ shall be free of propositional variables, which is not so surprising as it expresses a property about the structure of the model. This corresponds to the natural counterpart of the *list segment predicate* $\mathtt{ls}(\mathtt{x},\mathtt{y})$ in separation logic, defined as follows:

$$(\mathfrak{s},\mathfrak{h}) \models \mathtt{ls}(\mathtt{x},\mathtt{y}) \quad \overset{\text{def}}{\Leftrightarrow} \quad \begin{array}{l} \text{either } (\mathrm{dom}(\mathfrak{h}) = \emptyset \text{ and } \mathfrak{s}(\mathtt{x}) = \mathfrak{s}(\mathtt{y})) \text{ or} \\ \mathfrak{h} = \{\mathfrak{l}_0 \mapsto \mathfrak{l}_1, \mathfrak{l}_1 \mapsto \mathfrak{l}_2, \ldots, \mathfrak{l}_{n-1} \mapsto \mathfrak{l}_n\} \text{ with } n \geq 1, \\ \mathfrak{l}_0 = \mathfrak{s}(\mathtt{x}),\ \mathfrak{l}_n = \mathfrak{s}(\mathtt{y}) \text{ and for all } i \neq j \in [0,n],\ \mathfrak{l}_i \neq \mathfrak{l}_j. \end{array}$$

So, the formula $\phi_{\exists\mathtt{ls}}$ expresses a property that corresponds to $\exists\ \mathtt{x},\mathtt{y}\ \mathtt{ls}(\mathtt{x},\mathtt{y})$ from (first-order) separation logic.

Given an MSL model $\mathfrak{M} = \langle \mathbb{N}, \mathfrak{R}, \mathfrak{V} \rangle$, let us introduce a few notions that are helpful to build the formula $\phi_{\exists\mathtt{ls}}$. As $\mathrm{MSL}(*, \Diamond, \langle\neq\rangle)$ does not include $\Diamond^{-1}$ and $\langle\star\rangle$ (unlike MLH [21]), we need to characterise linear structures by combining intricate properties. By way of example, stating that each location has at most one predecessor can be easily expressed with $[\mathrm{U}](\neg(\Diamond^{-1}\top * \Diamond^{-1}\top))$, but, obviously, this formula does not belong to $\mathrm{MSL}(*, \Diamond, \langle\neq\rangle)$.

A *loop* in $\mathfrak{M}$ is a sequence of locations $(\mathfrak{l}_0, \ldots, \mathfrak{l}_n)$ for some $n \geq 1$ such that $\mathfrak{l}_0 = \mathfrak{l}_n$ and for all $i \in [0, n-1]$, $(\mathfrak{l}_i, \mathfrak{l}_{i+1}) \in \mathfrak{R}$. $\mathfrak{M}$ *has at most one maximally connected component* (MCC) whenever for all $\mathfrak{l}, \mathfrak{l}'$ such that $\mathfrak{R}(\mathfrak{l})$ and $\mathfrak{R}(\mathfrak{l}')$ are non-empty, there is $\mathfrak{l}^+$ such that $(\mathfrak{l}, \mathfrak{l}^+) \in \mathfrak{R}^+$ and $(\mathfrak{l}', \mathfrak{l}^+) \in \mathfrak{R}^+$, where $\mathfrak{R}^+$ is the transitive closure of $\mathfrak{R}$. A location $\mathfrak{l}$ is a *leaf* in $\mathfrak{M}$ if $\mathfrak{R}(\mathfrak{l}) \neq \emptyset$ and $\mathfrak{R}^{-1}(\mathfrak{l}) = \emptyset$, and $\mathfrak{l}$ is a *pre-root* if $\mathfrak{R}(\mathfrak{l}) = \{\mathfrak{l}'\}$ for some $\mathfrak{l}'$ and $\mathfrak{R}(\mathfrak{l}') = \emptyset$. In Figure 2 we illustrate these concepts. This terminology making reference to trees is best understood if we think the definitions with respect to $\mathfrak{R}^{-1}$.

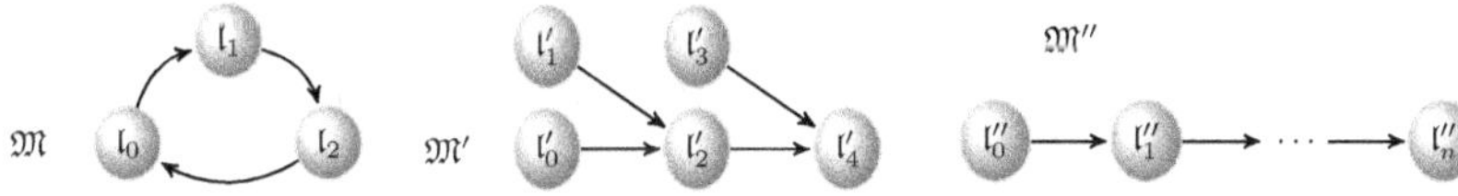

Figure 2. $\mathfrak{M}$ is a MCC and a loop, with no leaves and no pre-roots; $\mathfrak{M}'$ is a MCC with three leaves ($\mathfrak{l}'_0$, $\mathfrak{l}'_1$ and $\mathfrak{l}'_3$) and two pre-roots ($\mathfrak{l}'_2$ and $\mathfrak{l}'_3$); $\mathfrak{M}''$ is linear.

Obviously, if $\mathfrak{M}$ is linear, then it is loop-free, it has at most one MCC and has a unique leaf in case $\mathfrak{M}$ is non-empty. The result below states the converse, and below we explain how to express all these properties.

Lemma 5.1 *Let $\mathfrak{M}$ be an* MSL *model with a non-empty relation. $\mathfrak{M}$ is linear iff $\mathfrak{M}$ is loop-free and has a unique leaf.*

Let us introduce the global formula $\mathsf{Loop} \overset{\text{def}}{=} \top * (([\mathrm{U}]\Box\Diamond\top) \wedge \neg\mathtt{emp})$.

Lemma 5.2 *Let $\mathfrak{M}$ be an* MSL *model. $\mathfrak{M} \models \mathsf{Loop}$ iff $\mathfrak{M}$ has at least one loop.*

Let us consider the formulae below (whose semantics is given in Lemma 5.3).

$$\begin{array}{ll} \mathsf{PRoot} \overset{\text{def}}{=} \Diamond\Box\bot; & \mathsf{UniqTreePRoot} \overset{\text{def}}{=} \neg\mathsf{Loop} \wedge ((\neg(\neg\mathtt{emp} * \neg\mathtt{emp})) \vee \langle\,!\,\rangle\mathsf{PRoot}) \\ \mathsf{Leaf} \overset{\text{def}}{=} & (\Diamond\top \wedge \mathtt{size} = 1) \vee \\ & (\Diamond\top \wedge \neg\mathsf{PRoot} \wedge ((\mathtt{size} = 1 \wedge \Diamond\top) * \mathsf{UniqTreePRoot})). \end{array}$$

Lemma 5.3 *Let $\mathfrak{M} = \langle \mathbb{N}, \mathfrak{R}, \mathfrak{V} \rangle$ be a model and $\mathfrak{l} \in \mathbb{N}$.*

(I) $\mathfrak{M}, \mathfrak{l} \models \mathsf{PRoot}$ *iff* $\mathfrak{l}$ *is a pre-root.*

(II) $\mathfrak{M}, \mathfrak{l} \models \mathsf{UniqTreePRoot}$ *iff* $\mathfrak{M}$ *is loop-free and either* $\mathfrak{R}$ *is empty or (*$\mathfrak{M}$ *has at most one MCC and a unique pre-root).*

(III) Assuming that $\mathfrak{M} \models \mathsf{UniqTreePRoot}$*, we have* $\mathfrak{M}, \mathfrak{l} \models \mathsf{Leaf}$ *iff* $\mathfrak{l}$ *is a leaf.*

The proof is rather tedious and is intrinsically related to the definition of the formulae. Let $\phi_{\exists \mathtt{ls}}$ be $\mathtt{emp} \vee (\mathsf{UniqTreePRoot} \wedge \langle\, ! \,\rangle \mathsf{Leaf})$. By combination of the previous lemmas and using that if $\mathfrak{M}$ is linear and non-empty, then $\mathfrak{M}$ has at most one MCC and a unique pre-root, we get the result below.

Theorem 5.4 *Let $\mathfrak{M} = \langle \mathbb{N}, \mathfrak{R}, \mathfrak{V} \rangle$ be a model. $\mathfrak{M} \models \phi_{\exists \mathtt{ls}}$ iff $\mathfrak{M}$ is linear.*

The formula $\mathtt{ls}(\mathtt{x}, \mathtt{y})$ can be therefore encoded by the formula below:

$$\phi_{\mathtt{ls}(\mathtt{x},\mathtt{y})} \stackrel{\text{def}}{=} \phi_{\exists \mathtt{ls}} \wedge ((\mathtt{emp} \wedge \langle \mathrm{U} \rangle (\mathtt{x} \wedge \mathtt{y})) \vee (\langle \mathrm{U} \rangle (\mathtt{x} \wedge \mathsf{Leaf}) \wedge \langle \mathrm{U} \rangle (\mathsf{PRoot} \wedge \Diamond \mathtt{y}))).$$

5.2 The reduction

In this section, we show that the satisfiability problem for $\mathrm{MSL}(*, \Diamond, \langle \neq \rangle)$ is TOWER-hard by reduction from the nonemptiness problem for star-free expressions [31,35]. The proof takes advantage of Theorem 5.4 to encode finite words and separating conjunction will be helpful to encode concatenation, whereas complement and union operators in the star-free expressions are taken care by negation and disjunction, respectively. Our proof is reminiscent to developments from [21, Section 3] as it is essential to be able to encode finite words. Instead of reducing the satisfiability problem for Propositional Interval Temporal Logic [32] as done in [21, Section 3], we define a reduction from the nonemptiness problem for star-free expressions. A *star-free expression* e over some alphabet Σ is defined by

$$e ::= a \mid \varepsilon \mid e \cup e \mid ee \mid \sim e,$$

where $a \in \Sigma$ and ε denotes the empty string. Star-free expressions e are interpreted by languages $\mathrm{L}(e) \subseteq \Sigma^*$ as follows:

- $\mathrm{L}(a) \stackrel{\text{def}}{=} \{a\}$ for all $a \in \Sigma$; $\mathrm{L}(\varepsilon) \stackrel{\text{def}}{=} \{\varepsilon\}$; $\mathrm{L}(\sim e) \stackrel{\text{def}}{=} \Sigma^* \setminus \mathrm{L}(e)$;
- $\mathrm{L}(e \cup e') \stackrel{\text{def}}{=} \mathrm{L}(e) \cup \mathrm{L}(e')$; $\mathrm{L}(ee') \stackrel{\text{def}}{=} \{\mathfrak{w}\mathfrak{w}' \in \Sigma^* \mid \mathfrak{w} \in \mathrm{L}(e), \mathfrak{w}' \in \mathrm{L}(e')\}$.

The *nonemptiness problem* consists in checking whether $\mathrm{L}(e) \neq \emptyset$. The problem is shown decidable with a non elementary procedure in [31,37] and refined to TOWER-completeness in [35].

Given a finite alphabet $\Sigma = \{a_1, \ldots, a_\alpha\}$, we use the models encoding finite words thanks to the formula $\phi_{\exists \mathtt{ls}}$ and furthermore, we require that $[\mathrm{U}] \bigvee_i \mathsf{a}_i$ where $\mathsf{a}_i \stackrel{\text{def}}{=} p_i \wedge \bigwedge_{j \neq i} \neg p_j$. So, for every $\mathfrak{w} \in \Sigma^*$, there is a pair $\mathfrak{M}, \mathfrak{l}$ encoding $\mathfrak{w}$. We define a relation $\rhd$ that establishes this correspondence: $\mathfrak{w} \rhd \mathfrak{M}, \mathfrak{l} \stackrel{\text{def}}{\Leftrightarrow}$ $\mathfrak{M}$ is linear and

- If $\mathfrak{w} = \varepsilon$, then $\mathfrak{M}$ has an empty accessibility relation and $\mathfrak{l}$ is arbitrary.
- If $\mathfrak{w} = a_{i_1} \cdots a_{i_n}$ $(n \geq 1)$, then $\mathfrak{M}$ has n edges and $\mathfrak{l}$ is the unique leaf. With $\mathfrak{R} = \{(\mathfrak{l}_0, \mathfrak{l}_1), \ldots, (\mathfrak{l}_{n-1}, \mathfrak{l}_n)\}$, for all $k \in [1, \alpha]$, $\mathfrak{V}(p_k) = \{\mathfrak{l}_j \mid j \geq 1,\ i_j = k\}$.

The correspondence between finite words in Σ^* and pairs $\mathfrak{M}, \mathfrak{l}$ satisfies a nice property as far as splitting a word into two disjoint subwords is concerned.

Lemma 5.5 *Let $\mathfrak{w} \rhd \mathfrak{M}, \mathfrak{l}$ with $\mathfrak{w} = \mathfrak{w}_1\mathfrak{w}_2 \in \Sigma^*$. There exist linear models $\mathfrak{M}_1$ and $\mathfrak{M}_2$ and $\mathfrak{l}'$ such that $\mathfrak{M} = \mathfrak{M}_1 \uplus \mathfrak{M}_2$, $\mathfrak{w}_1 \rhd \mathfrak{M}_1, \mathfrak{l}$ and $\mathfrak{w}_2 \rhd \mathfrak{M}_2, \mathfrak{l}'$.*

Another technical lemma is needed for the proof of Lemma 5.7.

Lemma 5.6 *Let $\mathfrak{w} \rhd \mathfrak{M}, \mathfrak{l}$ with $\mathfrak{M} = \mathfrak{M}_1 \uplus \mathfrak{M}_2$ and, $\mathfrak{M}_1$ and $\mathfrak{M}_2$ are linear. There are $\mathfrak{w}_1, \mathfrak{w}_2 \in \Sigma^*$ and $\mathfrak{l}' \in \mathbb{N}$ such that $\mathfrak{w} = \mathfrak{w}_1\mathfrak{w}_2$, $\mathfrak{w}_1 \rhd \mathfrak{M}_1, \mathfrak{l}$ and $\mathfrak{w}_2 \rhd \mathfrak{M}_2, \mathfrak{l}'$.*

Each expression e is translated as

$$T(e) \stackrel{\text{def}}{=} ([\mathrm{U}] \textstyle\bigvee_i \mathsf{a}_i) \wedge \phi_{\exists \mathtt{ls}} \wedge (\mathsf{emp} \wedge t(e)) \vee (\neg \mathsf{emp} \wedge \mathsf{Leaf} \wedge t(e)),$$

where $t(\cdot)$ is recursively defined. The four disjuncts in $t(e_1e_2)$ below correspond to cases depending on the emptiness of subwords.

$$\begin{array}{llll} t(\varepsilon) & \stackrel{\text{def}}{=} \mathsf{emp} & t(a_i) & \stackrel{\text{def}}{=} (\Diamond \mathsf{a}_i) \wedge \mathtt{size} = 1 \\ t({\sim} e) & \stackrel{\text{def}}{=} \neg t(e) & t(e_1 \cup e_2) & \stackrel{\text{def}}{=} t(e_1) \vee t(e_2) \\ t(e_1e_2) & \stackrel{\text{def}}{=} \psi_1 \vee \psi_2 \vee \psi_3 \vee \psi_4 & & \end{array}$$

$$\psi_1 \stackrel{\text{def}}{=} \mathsf{emp} \wedge t(e_1) \wedge t(e_2) \quad \psi_2 \stackrel{\text{def}}{=} (t(e_1) \wedge \mathsf{emp}) * t(e_2) \quad \psi_3 \stackrel{\text{def}}{=} t(e_1) * (t(e_2) \wedge \mathsf{emp})$$

$$\psi_4 \stackrel{\text{def}}{=} (\phi_{\exists \mathtt{ls}} \wedge \neg \mathsf{emp} \wedge t(e_1)) * (\phi_{\exists \mathtt{ls}} \wedge \neg \mathsf{emp} \wedge \langle \mathrm{U} \rangle (\mathsf{Leaf} \wedge t(e_2))).$$

In ψ_4, to evaluate $t(e_2)$, we move to the unique leaf of the linear structure.

Lemma 5.7 *Let $\mathfrak{w} \in \Sigma^*$, and $\mathfrak{M}$ be a linear model such that $\mathfrak{w} \rhd \mathfrak{M}, \mathfrak{l}$. For every star-free expression e, we have $\mathfrak{w} \in \mathrm{L}(e)$ iff $\mathfrak{M}, \mathfrak{l} \models t(e)$.*

As a consequence,

Lemma 5.8 *Given $\alpha \geq 1$, $\Sigma = \{a_1, \ldots, a_\alpha\}$ and a star-free expression e built on Σ, $\mathrm{L}(e) \neq \emptyset$ iff the formula $T(e)$ is* MSL$(*, \Diamond, \langle \neq \rangle)$ *satisfiable.*

Finally, we get the TOWER-completeness.

Theorem 5.9 *The satisfiability problem for* MSL$(*, \Diamond, \langle \neq \rangle)$ *is* TOWER-*complete.*

6 When the magic wand strikes back

In this short section, we show that the satisfiability problem for MSL is actually undecidable by taking advantage of previous results. All the previous complexity results, deal with fragments that are $-\!\!*$-free. It is well-known that adding the separating connective $-\!\!*$ can dramatically augment the expressive power or the complexity, see e.g. [11]. Below, the expressive strength of $-\!\!*$ is again illustrated, via a reduction from propositional separation logic augmented with the list segment predicate `ls` [22]. By contrast, it is known that the modal logic for heaps MLH restricted to $*$ is decidable [21], but it is open whether the addition of $-\!\!*$ leads to undecidability.

First, note that the interval temporal logic with the operators C, D and T over the class of finite strict orders (equivalently, one may consider only the finite intervals of $\mathbb{N}$) is shown to admit an undecidable satisfiability problem in [26] and to be non recursively enumerable. By contrast, the version of the logic in which the propositional valuation of an interval only depends on the first value of the interval (the locality condition) is decidable as satisfiability can be reduced to the satisfiability problem for first-order logic over $\langle \mathbb{N}, \leq, +1 \rangle$. As we have seen in the paper, the formula $\phi_{\exists \mathtt{ls}}$ can enforce a linear structure whose labelling depends on the first location (corresponding to the locality condition) but it is unclear how to reduce the undecidable version to MSL, even though there is a clear correspondence between the chop operator C and $*$, and between the operators D and T, and $-\!\!*$. Instead, our undecidability proof for (full) MSL is by reducing the satisfiability problem for $\mathrm{SL}(*, -\!\!*, \mathtt{ls})$, recently shown undecidable in [22].

Notice that for the translation of $\mathrm{SL}(*, -\!\!*, \mathtt{ls})$ formulae, the most complex part is the encoding of the atomic formulae $\mathtt{ls}(\mathtt{x}, \mathtt{y})$. However, all this work has already been done in Section 5 when we encode linear structures with $\mathrm{MSL}(*, \Diamond, \langle \neq \rangle)$. Then, what gives us undecidability is essentially the inclusion of the $-\!\!*$ operator.

Let us define the translation $t(\cdot)$ from $\mathrm{SL}(*, -\!\!*, \mathtt{ls})$ into MSL formulas, which is homomorphic for Boolean and separation connectives, and

$$t(\mathtt{emp}) \stackrel{\text{def}}{=} \mathtt{emp} \quad t(\mathtt{x} = \mathtt{y}) \stackrel{\text{def}}{=} \langle \mathrm{U} \rangle (\mathtt{x} \wedge \mathtt{y}) \quad t(\mathtt{x} \hookrightarrow \mathtt{y}) \stackrel{\text{def}}{=} \langle \mathrm{U} \rangle (\mathtt{x} \wedge \Diamond \mathtt{y})$$

$$t(\mathtt{ls}(\mathtt{x}, \mathtt{y})) \stackrel{\text{def}}{=} \phi_{\exists \mathtt{ls}} \wedge ((\mathtt{emp} \wedge \langle \mathrm{U} \rangle (\mathtt{x} \wedge \mathtt{y})) \vee (\langle \mathrm{U} \rangle (\mathtt{x} \wedge \mathsf{Leaf}) \wedge \langle \mathrm{U} \rangle (\mathsf{PRoot} \wedge \Diamond \mathtt{y}))),$$

where $\mathtt{x}, \mathtt{y}$ are nominals and $\phi_{\exists \mathtt{ls}}$ defined as in Section 5.1. We get the result below.

Lemma 6.1 *Let ϕ be an $\mathrm{SL}(*, -\!\!*, \mathtt{ls})$ formula. ϕ is satisfiable iff $t(\phi)$ is satisfiable in* MSL*.*

As the satisfiability problem for $\mathrm{SL}(*, -\!\!*, \mathtt{ls})$ is recently shown undecidable [22], we get the following result.

Theorem 6.2 *The satisfiability problem for* MSL *is undecidable.*

Another consequence is the non-finite axiomatisability of MSL, which is inherited from $\mathrm{SL}(*, -\!\!*, \mathtt{ls})$. As a corollary, the modal logic for heaps MLH (including $-\!\!*$) augmented with propositional variables is undecidable [21] as MSL is one of its fragments.

Furthermore, the satisfiability problem of $\mathrm{MSL}^g(\Diamond, \langle \mathsf{gsb} \rangle)$ is undecidable [2]. Therefore, when considering general models, the minimal modal separation logic $\mathrm{MSL}^g(*, \Diamond)$ is also undecidable (use a map t such that $t(\langle \mathsf{gsb} \rangle \phi) \stackrel{\text{def}}{=} (\mathtt{size} = 1) * t(\phi)$).

7 Conclusion

We have introduced the logic MSL and studied several of its fragments. For $\mathrm{MSL}(*, \Diamond)$, we proved that the satisfiability problem is NP-complete whereas the model-checking problem is in P. A similar complexity characterisation is provided for $\mathrm{MSL}(*, \langle \neq \rangle)$. Surprisingly, we have shown that the satisfiability problem for $\mathrm{MSL}(*, \Diamond, \langle \neq \rangle)$ is TOWER-complete. A key element of our TOWER-hardness proof is the ability to express the property $\exists\ \mathtt{x}, \mathtt{y}\ \mathtt{ls}(\mathtt{x}, \mathtt{y})$ from separation logic. Hence, we are able to show that MSL admits an undecidable satisfiability problem. Along the paper, we also investigated variants of MSL (or some of its fragments) by slightly modifying the semantics or by adding other modal connectives. For instance, we have proved that the satisfiability problem for $\mathrm{MSL}(\Diamond, \langle \mathsf{gsb} \rangle)$ is (only) NP-complete.

Most of the results are summarised in the table below.

	Model checking (with finite models)	Satisfiability
$\mathrm{MSL}(*, \Diamond)$, $\mathrm{MSL}(*, \langle \neq \rangle)$	P	NP-complete
$\mathrm{MSL}(*, \Diamond, \langle \neq \rangle)$	PSPACE-complete	TOWER-complete
MSL	PSPACE-complete	Undecidable
$\mathrm{MSL}(*, \Diamond^{-1})$	PSPACE-complete	PSPACE-hard, in TOWER
$\mathrm{MSL}(\Diamond, \langle \mathsf{gsb} \rangle)$	P	NP-complete

Understanding the effects of the interactions between modal operators and separating connectives is still to be strengthened and many interesting problems are left open. By way of example, we have shown that the satisfiability problem for $\mathrm{MSL}(*, \Diamond^{-1})$ is PSPACE-hard and in TOWER but a complexity characterisation is not yet known. Similarly, the satisfiability problem for $\mathrm{MSL}(*, \Diamond, \langle \neq \rangle)$ is shown TOWER-complete, what about its slight variant $\mathrm{MSL}(*, \Diamond, \langle \mathrm{U} \rangle)$? The decidability status of MSL^f and MLH [21] is also open. Finally, the design of proof systems for modal separation logics remains a challenging question.

Ackowledgements. We would like to thank Alessio Mansutti (LSV, France) for helpful suggestions and enlightning discussions. This work was partially supported by ANPCyT-PICTs-2016-0215, SeCyT-UNC, and the Laboratoire International Associé INFINIS.

References

[1] Areces, C., R. Fervari and G. Hoffmann, *Relation-changing modal operators*, IGPL **23** (2015), pp. 601–627.

[2] Areces, C., R. Fervari, G. Hoffmann and M. Martel, *Satisfiability for relation-changing logics*, JLC (2018), accepted, subject to minor revisions.

[3] Areces, C., R. Fervari, G. Hoffmann and M. Martel, *Undecidability of relation-changing modal logics*, in: *Dynamic Logic. New Trends and Applications - First International Workshop, DALI 2017, Brasilia, Brazil*, LNCS **10669** (2018), pp. 1–16.

[4] Aucher, G., P. Balbiani, L. Fariñas del Cerro and A. Herzig, *Global and local graph modifiers*, ENTCS **231** (2009), pp. 293–307.

[5] Baader, F., I. Horrocks, C. Lutz and U. Sattler, "An Introduction to Description Logic," CUP, 2017.

[6] Berdine, J., C. Calcagno and P. O'Hearn, *A decidable fragment of separation logic*, in: *FST&TCS'04*, LNCS **3328** (2004), pp. 97–109.

[7] Blackburn, P., *Representation, reasoning, and relational structures: A hybrid logic manifesto*, IGPL **8** (2000), pp. 339–365.

[8] Blackburn, P., M. de Rijke and Y. Venema, "Modal Logic," CUP, 2001.

[9] Börger, E., E. Grädel and Y. Gurevich, "The Classical Decision Problem," Perspectives in Mathematical Logic, Springer, 1997.

[10] Brochenin, R., S. Demri and E. Lozes, *Reasoning about sequences of memory states*, APAL **161** (2009), pp. 305–323.

[11] Brochenin, R., S. Demri and E. Lozes, *On the almighty wand*, IC **211** (2012), pp. 106–137.

[12] Brotherston, J. and M. Kanovich, *Undecidability of propositional separation logic and its neighbours*, JACM **61** (2014).

[13] Brotherston, J. and J. Villard, *Parametric completeness for separation theories*, in: *POPL'14* (2014), pp. 453–464.

[14] Calcagno, C., P. O'Hearn and H. Yang, *Computability and complexity results for a spatial assertion language for data structures*, in: *FSTTCS'01*, LNCS **2245** (2001), pp. 108–119.

[15] Calvanese, D., T. Kotek, M. Simkus, H. Veith and F. Zuleger, *Shape and content - A database-theoretic perspective on the analysis of data structures*, in: *IFM'14*, LNCS **8739** (2014), pp. 3–17.

[16] Cook, B., C. Haase, J. Ouaknine, M. Parkinson and J. Worrell, *Tractable reasoning in a fragment of separation logic*, in: *CONCUR'11*, LNCS **6901** (2011), pp. 235–249.

[17] Courtault, J.-R. and D. Galmiche, *A modal BI logic for dynamic resource properties*, in: *LFCS'13*, LNCS **7734** (2013), pp. 134–148.

[18] Dawar, A., P. Gardner and G. Ghelli, *Expressiveness and complexity of graph logic*, IC **205** (2007), pp. 263–310.

[19] de Rijke, M., *The modal logic of inequality*, JSL **57** (1992), pp. 566–584.

[20] Demri, S., *A simple tableau system for the logic of elsewhere*, in: *TABLEAUX'96*, LNAI **1071** (1996), pp. 177–192.

[21] Demri, S. and M. Deters, *Two-variable separation logic and its inner circle*, ACM ToCL **2** (2015).

[22] Demri, S., E. Lozes and A. Mansutti, *The effects of adding reachability predicates in propositional separation logic*, in: *FOSSACS'18*, LNCS **10803** (2018), pp. 476–493.

[23] Galmiche, D., P. Kimmel and D. Pym, *A substructural epistemic resource logic*, in: *ICLA'17*, LNCS **10119** (2017), pp. 106–122.

[24] Georgieva, L. and P. Maier, *Description logics for shape analysis*, in: *SEFM'05* (2005), pp. 321–331.

[25] Herzig, A., *A simple separation logic*, in: *WoLLIC'13*, LNCS **8071** (2013), pp. 168–178.

[26] Hodkinson, I., A. Montanari and G. Sciavicco, *Non-finite axiomatizability and undecidability of interval temporal logics with C, D, and T*, in: *CSL'08*, LNCS **5213** (2008), pp. 308–322.

[27] Ishtiaq, S. and P. O'Hearn, *BI as an assertion language for mutable data structures*, in: *POPL'01* (2001), pp. 14–26.

[28] Larchey-Wendling, D. and D. Galmiche, *Nondeterministic phase semantics and the undecidability of Boolean BI*, ACM ToCL **14** (2013).

[29] Löding, C. and P. Rohde, *Model checking and satisfiability for sabotage modal logic*, in: *FST&TCS'03*, LNCS **2914** (2003), pp. 302–313.

[30] Lutz, C., *Complexity and succinctness of public announcement logic*, in: *AAMAS'06* (2006), pp. 137–143.

[31] Meyer, A. and L. Stockmeyer, *Word problems requiring exponential time*, in: *STOC'73* (1973), pp. 1–9.

[32] Moszkowski, B., *Reasoning about digital circuits*, Technical Report STAN-CS-83-970, Dept. of Computer Science, Stanford University, Stanford, CA (1983).

[33] Rabin, M., *Decidability of second-order theories and automata on infinite trees*, Transactions of the American Mathematical Society **41** (1969), pp. 1–35.

[34] Reynolds, J., *Separation logic: A logic for shared mutable data structures*, in: *LICS'02* (2002), pp. 55–74.

[35] Schmitz, S., *Complexity hierarchies beyond Elementary*, ACM Transactions on Computation Theory **8** (2016), pp. 3:1–3:36.

[36] Segerberg, K., *A note on the logic of elsewhere*, Theoria **47** (1981), pp. 183–187.

[37] Stockmeyer, L., "The Complexity of Decision Problems in Automata Theory and Logic," Ph.D. thesis, Department of Electrical Engineering, MIT (1974).

[38] Stockmeyer, L., *The polynomial-time hierarchy*, TCS **3** (1977), pp. 1–21.

[39] van Benthem, J., *An Essay on sabotage and obstruction*, in: *Mechanizing Mathematical Reasoning, Essays in Honor of Jörg Siekmann on the Occasion of his 69th Birthday* (2005), pp. 268–276.

An Intuitionistic Axiomatization of 'Eventually'

Martín Diéguez [1]

Lab-STICC
CERV, Ecole Nationale d'Ingénieurs de Brest
Brest, France

David Fernández-Duque

Department of Mathematics
Ghent University
Ghent, Belgium

Abstract

The language of linear temporal logic can be interpreted over a class of structures called *expanding posets*. This gives rise to the intuitionistic temporal logic $\mathsf{ITL}^{\mathsf{e}}$, recently shown to be decidable by Boudou and the authors. In this article we completely axiomatize the 'henceforth'-free fragment of this logic.

Keywords: axiomatization, intuitionistic modal logic, completeness, temporal logic.

1 Introduction

Intuitionistic logic is the basis for constructive reasoning and temporal logics are an important tool for reasoning about dynamic processes. One would expect that a combination of the two would yield a powerful framework in which to model phenomena involving both computation and time, an idea explored by Davies [6] and Maier [25]. This is not the only potential application of such a logic: in view of the topological interpretation of the intuitionistic implication, one may instead use it to model *space* and time [15]. This makes it important to study these logics, which in particular did not previously enjoy a complete axiomatization in the presence of 'infinitary' tenses. Our goal in this paper is to present such an axiomatization for 'next' and 'eventually'.

1.1 State-of-the-art

There are several (poly)modal logics which may be used to model time, and some have already been studied in an intuitionistic setting, e.g. tense logics by Davoren [7] and propositional dynamic logic with iteration by Nishimura [27]. Here we are specifically concerned with intuitionistic analogues of discrete-time

[1] Martín Diéguez is funded by the ANR-12-ASTR-0020 Project STRATEGIC.

linear temporal logic. Versions of such a logic in finite time have been studied by Kojima and Igarashi [21] and Kamide and Wansing [20]. Nevertheless, logics over infinite time have proven to be rather difficult to understand, in no small part due to their similarity to intuitionistic modal logics such as IS4, whose decidablitiy is still an open problem [28].

In recent times, Balbiani, Boudou and the authors have made some advances in this direction, showing that the intermediate logic of temporal here-and-there is decidable and enjoys a natural axiomatization [3] and identifying two conservative temporal extensions of intuitionistic logic, denoted $\mathsf{ITL^e}$ and $\mathsf{ITL^p}$ (see §2.1). These logics are based on the temporal language with $\bigcirc$ ('next'), $\Diamond$ ('eventually') and $\Box$ ('henceforth'); note that unlike in the classical case, the latter two are not inter-definable [2]. Both logics are given semantically and interpreted over the class of *dynamic posets,* structures of the form $\mathcal{F} = (W, \preccurlyeq, S)$ where $\preccurlyeq$ is a partial order on W used to interpret implication and $S \colon W \to W$ is used to interpret tenses. If $w \preccurlyeq v$ implies that $S(w) \preccurlyeq S(v)$ we say that $\mathcal{F}$ is an *expanding poset;* $\mathsf{ITL^e}$ is then defined to be the set of valid formulas for the class of expanding posets, while $\mathsf{ITL^p}$ is the logic of *persistent posets,* where $\mathcal{F}$ has the additional *backward confluence* condition stating that if $v \succcurlyeq S(w)$, then there is $u \succcurlyeq w$ such that $S(u) = v$.

Unlike $\mathsf{ITL^e}$, the logic $\mathsf{ITL^p}$ satisfies the familiar Fischer Servi axioms [16]; nevertheless, $\mathsf{ITL^e}$ has some technical advantages. We have shown that $\mathsf{ITL^e}$ has the small model property while $\mathsf{ITL^p}$ does not [4]; this implies that $\mathsf{ITL^e}$ is decidable. It is currently unknown if $\mathsf{ITL^p}$ is even axiomatizable, and in fact its modal cousin $\mathsf{LTL} \times \mathsf{S4}$ is not computably enumerable [17]. On the other hand, while $\mathsf{ITL^e}$ is axiomatizable in principle, the decision procedure we currently know uses model-theoretic techniques and does not suggest a natural axiomatization.

In [5] we laid the groundwork for an axiomatic approach to intuitionistic temporal logics, identifying a family of natural axiom systems that were sound for different classes of structures, including a 'minimal' logic ITL^0 based on a standard axiomatization for LTL. There we consider a wider class of models based on topological semantics and show that ITL^0 is sound for these semantics, while

(a) $\Box(p \vee q) \to \Diamond p \vee \Box q$ (b) $\Box(\bigcirc p \to p) \wedge \Box(p \vee q) \to p \vee \Box q$

are Kripke-, but not topologically, valid, from which it follows that these principles are not derivable in ITL^0.

On the other hand, it is also shown in [5] that for $\varphi \in \mathcal{L}_\Diamond$, the following are equivalent:

(i) φ is topologically valid,

(ii) φ is valid over the class of expanding posets,

(iii) φ is valid over the class of finite quasimodels.

Quasimodels are discussed in §3 and are the basis of the completeness for dynamic topological logic presented in [13], which works for topological, but not Kripke, semantics. This suggests that similar techniques could be employed to give a completeness proof for a natural logic over the $\Box$-free fragment, but not necessarily over the full temporal language; in fact, we do not currently have a useful notion of quasimodel in the presence of $\Box$. Moreover, (a) and (b) are not valid in most intuitionistic modal logics, and there is little reason at this point to suspect that no other independent validities are yet to be discovered. For this reason, in this manuscript we restrict our attention to the $\Box$-free fragment of the temporal language, which we denote $\mathcal{L}_\Diamond$, and we will work with the logic $\mathsf{ITL}^0_\Diamond$, a $\Box$-free version of ITL^0.

1.2 Our main result

The goal of this article is to prove that $\mathsf{ITL}^0_\Diamond$ is complete for the class of expanding posets (Theorem 7.5). The completeness proof follows the general scheme of that for linear temporal logic [24]: a set of 'local states', which we will call *moments,* is defined, where a moment is a representation of a potential point in a model (or, in our case, a quasimodel). To each moment w one then assigns a characteristic formula $\chi(w)$ in such a way that $\chi(w)$ is consistent if and only if w can be included in a model, from which completeness can readily be deduced.

In the LTL setting, a moment is simply a maximal consistent subset of a suitable finite set Σ of formulas. For us a moment is instead a finite labelled tree, and the formula $\chi(w)$ must characterize w up to *simulation;* for this reason we will henceforth write $\mathrm{Sim}(w)$ instead of $\chi(w)$. The required formulas $\mathrm{Sim}(w)$ can readily be constructed in $\mathcal{L}_\Diamond$ (Proposition 5.3).

Note that it is *failure* of $\mathrm{Sim}(w)$ that characterizes the property of simulating w, hence the *possible* states will be those moments w such that $\mathrm{Sim}(w)$ is unprovable. The set of possible moments will form a quasimodel falsifying a given unprovable formula φ (Corollary 7.4), from which it follows that such a φ is falsified on some model as well (Theorem 3.7). Thus any unprovable formula is falsifiable, and Theorem 7.5 follows.

Layout

Section 2 introduces the syntax and semantics of $\mathsf{ITL}^{\mathsf{e}}$, and Section 3 discusses labelled structures, which generalize both models and quasimodels. Section 4 discusses the canonical model, which properly speaking is a deterministic weak quasimodel. Section 5 reviews simulations and dynamic simulations, including their definability in the intuitionistic language. Section 6 constructs the initial quasimodel and establishes its basic properties, but the fact that it is actually a quasimodel is proven only in Section 7 where it is shown that the quasimodel is *ω-sensible,* i.e. it satisfies the required condition to interpret $\Diamond$. The completeness of $\mathsf{ITL}^0_\Diamond$ follows from this.

Appendix A gives an explicit construction of simulation formulas and Appendix B reviews the construction of the initial weak quasimodel from [15].

2 Syntax and semantics

Fix a countably infinite set $\mathbb{P}$ of 'propositional variables'. The language $\mathcal{L}$ of intuitionistic (linear) temporal logic ITL is given by the grammar

$$\varphi, \psi ::= \ p \mid \bot \mid \varphi \wedge \psi \mid \varphi \vee \psi \mid \varphi \to \psi \mid \bigcirc\varphi \mid \Diamond\varphi \mid \Box\varphi,$$

where $p \in \mathbb{P}$. As usual, we use $\neg\varphi$ as a shorthand for $\varphi \to \bot$ and $\varphi \leftrightarrow \psi$ as a shorthand for $(\varphi \to \psi) \wedge (\psi \to \varphi)$. We read $\bigcirc$ as 'next', $\Diamond$ as 'eventually', and $\Box$ as 'henceforth'. Given any formula φ, we denote the set of subformulas of φ by $\mathrm{sub}(\varphi)$. We will work mainly in the language $\mathcal{L}_\Diamond$, defined as the sublanguage of $\mathcal{L}$ without the modality $\Box$, although the full language will be discussed occasionally.

2.1 Semantics

Formulas of $\mathcal{L}$ are interpreted over expanding posets. An *expanding poset* is a tuple $\mathcal{D} = (|\mathcal{D}|, \preccurlyeq_\mathcal{D}, S_\mathcal{D})$, where $|\mathcal{D}|$ is a non-empty set of moments, $\preccurlyeq_\mathcal{D}$ is a partial order over $|\mathcal{D}|$, and $S_\mathcal{D}$ is a function from $|\mathcal{D}|$ to $|\mathcal{D}|$ satisfying the *forward confluence* condition that for all $w, v \in |\mathcal{D}|$, if $w \preccurlyeq_\mathcal{D} v$ then $S_\mathcal{D}(w) \preccurlyeq S_\mathcal{D}(v)$. We will omit the subindices in $\preccurlyeq_\mathcal{D}$, $S_\mathcal{D}$ when $\mathcal{D}$ is clear from context and write $v \prec w$ if $v \preccurlyeq w$ and $v \neq w$. An *intuitionistic dynamic model,* or simply *model,* is defined to be a tuple $\mathcal{M} = (|\mathcal{M}|, \preccurlyeq_\mathcal{M}, S_\mathcal{M}, V_\mathcal{M})$ consisting of an expanding poset equipped with a valuation function $V_\mathcal{M}$ from $|\mathcal{M}|$ to sets of propositional variables that is $\preccurlyeq$-monotone in the sense that for all $w, v \in |\mathcal{M}|$, if $w \preccurlyeq v$ then $V_\mathcal{M}(w) \subseteq V_\mathcal{M}(v)$. In the standard way, we define $S^0_\mathcal{M}(w) = w$ and $S^{k+1}_\mathcal{M}(w) = S\left(S^k_\mathcal{M}(w)\right)$. Then we define the satisfaction relation $\models$ inductively by:

(i) $\mathcal{M}, w \models p$ if $p \in V_\mathcal{M}(w)$;

(ii) $\mathcal{M}, w \not\models \bot$;

(iii) $\mathcal{M}, w \models \varphi \wedge \psi$ if $\mathcal{M}, w \models \varphi$ and $\mathcal{M}, w \models \psi$;

(iv) $\mathcal{M}, w \models \varphi \vee \psi$ if $\mathcal{M}, w \models \varphi$ or $\mathcal{M}, w \models \psi$;

(v) $\mathcal{M}, w \models \bigcirc\varphi$ if $\mathcal{M}, S_\mathcal{M}(w) \models \varphi$;

(vi) $\mathcal{M}, w \models \varphi \to \psi$ if $\forall v \succcurlyeq w$, if $\mathcal{M}, v \models \varphi$ then $\mathcal{M}, v \models \psi$;

(vii) $\mathcal{M}, w \models \Diamond\varphi$ if there exists k such that $\mathcal{M}, S^k_\mathcal{M}(w) \models \varphi$;

(viii) $\mathcal{M}, w \models \Box\varphi$ if for all k, $\mathcal{M}, S^k_\mathcal{M}(w) \models \varphi$.

As usual, a formula φ is *valid over a class of models* Ω if, for every world w of every model $\mathcal{M} \in \Omega$, $\mathcal{M}, w \models \varphi$. The set of valid formulas over an arbitrary expanding poset will be called $\mathsf{ITL^e}$, or *expanding intuitionistic temporal logic;* the terminology was coined in [4] and is a reference to the closely-related *expanding products* of modal logics [17]. The main result of [4] is the following.

Theorem 2.1 $\mathsf{ITL^e}$ *is decidable.*

Nevertheless, Theorem 2.1 is proved using purely model-theoretic techniques that do not suggest an axiomatization in an obvious way. In [5] we introduced the axiomatic system ITL^0, inspired by standard axiomatizations for LTL. As we will see, adapting this system to $\mathcal{L}_\Diamond$ yields a sound and complete deductive calculus for the class of expanding posets.

2.2 The axiomatization

Our axiomatization obtained from propositional intuitionistic logic [26] by adding standard axioms and inference rules of LTL [24], although modified to use $\Diamond$ instead of $\Box$. To be precise, the logic $\mathsf{ITL}^0_\Diamond$ is the least set of $\mathcal{L}_\Diamond$-formulas closed under the following axiom schemes and rules:

(A1) All intuitionistic tautologies

(A2) $\neg \bigcirc \bot$

(A3) $\bigcirc\varphi \wedge \bigcirc\psi \to \bigcirc(\varphi \wedge \psi)$

(A4) $\bigcirc(\varphi \vee \psi) \to \bigcirc\varphi \vee \bigcirc\psi$

(A5) $\bigcirc(\varphi \to \psi) \to (\bigcirc\varphi \to \bigcirc\psi)$

(A6) $\varphi \vee \bigcirc\Diamond\varphi \to \Diamond\varphi$

(R1) $\dfrac{\varphi \quad \varphi \to \psi}{\psi}$

(R2) $\dfrac{\varphi}{\bigcirc\varphi}$

(R3) $\dfrac{\varphi \to \psi}{\Diamond\varphi \to \Diamond\psi}$

(R4) $\dfrac{\bigcirc\varphi \to \varphi}{\Diamond\varphi \to \varphi}$

The axioms (A2)-(A5) are standard for a functional modality. Axiom (A6) is the dual of $\Box\varphi \to \varphi \wedge \bigcirc\Box\varphi$. The rule (R3) replaces the dual K-axiom $\Box(\varphi \to \psi) \to (\Diamond\varphi \to \Diamond\psi)$, while (R4) is dual to the induction rule $\frac{\varphi \to \bigcirc\varphi}{\varphi \to \Box\varphi}$. As we show next, we can also derive the converses of some of these axioms. Below, for a set of formulas Γ we define $\bigcirc\Gamma = \{\bigcirc\varphi : \varphi \in \Gamma\}$, and empty conjunctions and disjunctions are defined by $\bigwedge \varnothing = \top$ and $\bigvee \varnothing = \bot$.

Lemma 2.2 *Let $\varphi \in \mathcal{L}_\Diamond$ and $\Gamma \subseteq \mathcal{L}_\Diamond$ be finite. Then, the following are derivable in $\mathsf{ITL}^0_\Diamond$:*

(i) $\bigcirc\bigwedge\Gamma \leftrightarrow \bigwedge\bigcirc\Gamma$

(ii) $\bigcirc\bigvee\Gamma \leftrightarrow \bigvee\bigcirc\Gamma$

(iii) $\Diamond\varphi \to \varphi \vee \bigcirc\Diamond\varphi$.

Proof. For the first two claims, one direction is obtained from repeated use of axioms (A3) or (A4) and the other is proven using (R2) and (A5); note that the second claim requires (A2) to treat the case when $\Gamma = \varnothing$. Details are left to the reader.

For the third claim, reasoning within $\mathsf{ITL}^0_\Diamond$, note that $\varphi \to \Diamond\varphi$ holds by (A6) and propositional reasoning, hence $\bigcirc\varphi \to \bigcirc\Diamond\varphi$ by (R2), (A5) and (R1). In a similar way, $\bigcirc\Diamond\varphi \to \Diamond\varphi$ holds by (A6) and propositional reasoning, so $\bigcirc\bigcirc\Diamond\varphi \to \bigcirc\Diamond\varphi$ does by (R2), (A5) and (R1). Hence, $\bigcirc\varphi \vee \bigcirc\bigcirc\Diamond\varphi \to \bigcirc\Diamond\varphi$ holds. Using (A4) and some propositional reasoning we obtain $\bigcirc(\varphi \vee \bigcirc\Diamond\varphi) \to \varphi \vee \bigcirc\Diamond\varphi$. But then, by (R4), $\Diamond(\varphi \vee \bigcirc\Diamond\varphi) \to \varphi \vee \bigcirc\Diamond\varphi$; since $\Diamond\varphi \to \Diamond(\varphi \vee \bigcirc\Diamond\varphi)$ can be proven using (R3), we obtain $\Diamond\varphi \to \varphi \vee \bigcirc\Diamond\varphi$, as needed. $\Box$

For purposes of this discussion, a *logic* may be any set $\Lambda \subseteq \mathcal{L}$, and we may write $\Lambda \vdash \varphi$ instead of $\varphi \in \Lambda$. Then, Λ is *sound* for a class of structures Ω if, whenever $\Lambda \vdash \varphi$, it follows that $\Omega \models \varphi$. The following is essentially proven in [5]:

Theorem 2.3 $\mathsf{ITL}^0_\Diamond$ *is sound for the class of expanding posets.*

Note however that a few of the axioms and rules have been modified to fall within $\mathcal{L}_\Diamond$, but these modifications are innocuous and their correctness may be readily checked by the reader. We remark that, in contrast to Lemma 2.2, $(\bigcirc p \to \bigcirc q) \to \bigcirc(p \to q)$ is not valid [2], hence by Theorem 2.3, it is not derivable.

3 Labelled structures

The central ingredient of our completeness proof is given by non-deterministic *quasimodels,* introduced by Fernández-Duque in the context of dynamic topological logic [10] and later adapted to intuitionistic temporal logic [15].

3.1 Two-sided types

Quasimodels are structures whose worlds are labelled by types, as defined below. More specifically, following [5], our quasimodels will be based on two-sided types.

Definition 3.1 Let $\Sigma \subseteq \mathcal{L}_\Diamond$ be closed under subformulas and $\Phi^-, \Phi^+ \subseteq \Sigma$. We say that the pair $\Phi = (\Phi^-; \Phi^+)$ is a *two-sided Σ-type* if:

(a) $\Phi^- \cap \Phi^+ = \varnothing$,

(b) $\Phi^- \cup \Phi^+ = \Sigma$,

(c) $\bot \notin \Phi^+$,

(d) if $\varphi \wedge \psi \in \Sigma$, then $\varphi \wedge \psi \in \Phi^+$ if and only if $\varphi, \psi \in \Phi^+$,

(e) if $\varphi \vee \psi \in \Sigma$, then $\varphi \vee \psi \in \Phi^+$ if and only if $\varphi \in \Phi^+$ or $\psi \in \Phi^+$,

(f) if $\varphi \to \psi \in \Phi^+$, then either $\varphi \in \Phi^-$ or $\psi \in \Phi^+$, and

(g) if $\Diamond\varphi \in \Phi^-$ then $\varphi \in \Phi^-$.

The set of two-sided Σ-types will be denoted T_Σ.

We will write $\Phi \preccurlyeq_T \Psi$ if $\Phi^+ \subseteq \Psi^+$ (or, equivalently, if $\Psi^- \subseteq \Phi^-$). If $\Sigma \subseteq \Delta$ are both closed under subformulas, $\Phi \in T_\Sigma$ and $\Psi \in T_\Delta$, we will write $\Phi \subseteq_T \Psi$ if $\Phi^- \subseteq \Psi^-$ and $\Phi^+ \subseteq \Psi^+$.

Often (but not always) we will want Σ to be finite, in which case given $\Delta \subseteq \mathcal{L}_\Diamond$ we write $\Sigma \Subset \Delta$ if Σ is finite and closed under subformulas. It is not hard to check that $\preccurlyeq_T$ is a partial order on T_Σ. Whenever Ξ is an expression denoting a two-sided type, we write Ξ^- and Ξ^+ to denote its components. Elements of $T_{\mathcal{L}_\Diamond}$ are *full types.* Note that Fernández-Duque [15] uses one-sided types, but it is readily checked that a one-sided Σ-type Φ as defined there can be regarded as a two-sided type Ψ by setting $\Psi^+ = \Phi$ and $\Psi^- = \Sigma\backslash\Phi$. Henceforth we will refer to two-sided types simply as *types.*

3.2 Quasimodels

Next we will define quasimodels; these are similar to models, except that valuations are replaced with a labelling function ℓ. We first define the more basic notion of *Σ-labelled frame.*

Definition 3.2 Let $\Sigma \subseteq \mathcal{L}_\Diamond$ be closed under subformulas. A *Σ-labelled frame* is a triple $\mathcal{F} = (|\mathcal{F}|, \preccurlyeq_\mathcal{F}, \ell_\mathcal{F})$, where $\preccurlyeq_\mathcal{F}$ is a partial order on $|\mathcal{F}|$ and $\ell_\mathcal{F} \colon |\mathcal{F}| \to T_\Sigma$ is such that

(a) whenever $w \preccurlyeq_\mathcal{F} v$ it follows that $\ell_\mathcal{F}(w) \preccurlyeq_T \ell_\mathcal{F}(v)$, and

(b) whenever $\varphi \to \psi \in \ell^-_\mathcal{F}(w)$, there is $v \succcurlyeq_\mathcal{F} w$ such that $\varphi \in \ell^+_\mathcal{F}(v)$ and $\psi \in \ell^-_\mathcal{F}(v)$.

We say that $\mathcal{F}$ *falsifies* $\varphi \in \mathcal{L}_\Diamond$ if $\varphi \in \ell^-(w)$ for some $w \in W$.

As before, we may omit the subindexes in $\preccurlyeq_\mathcal{F}$, $S_\mathcal{F}$ and $\ell_\mathcal{F}$ when $\mathcal{F}$ is clear from context. Labelled frames model only the intuitionistic aspect of the logic. For the temporal dimension, let us define a new relation over types.

Definition 3.3 Let $\Sigma \subseteq \mathcal{L}_\Diamond$ be closed under subformulas. We define a relation $S_T \subseteq T_\Sigma \times T_\Sigma$ by $\Phi \; S_T \; \Psi$ iff for all $\varphi \in \mathcal{L}$:

(a) if $\bigcirc\varphi \in \Phi^+$ then $\varphi \in \Psi^+$,

(b) if $\bigcirc\varphi \in \Phi^-$ then $\varphi \in \Psi^-$,

(c) if $\Diamond\varphi \in \Phi^+$ and $\varphi \in \Phi^-$ then $\Diamond\varphi \in \Psi^+$, and

(d) if $\Diamond\varphi \in \Phi^-$, then $\Diamond\varphi \in \Psi^-$.

Quasimodels are then defined as labelled frames with a suitable binary relation.

Definition 3.4 Given $\Sigma \subseteq \mathcal{L}_\Diamond$ closed under subformulas, a *Σ-quasimodel* is a tuple $\mathcal{Q} = (|\mathcal{Q}|, \preccurlyeq_\mathcal{Q}, S_\mathcal{Q}, \ell_\mathcal{Q})$ where $(|\mathcal{Q}|, \preccurlyeq_\mathcal{Q}, \ell_\mathcal{Q})$ is a labelled frame and $S_\mathcal{Q}$ is a binary relation over $|\mathcal{Q}|$ that is

(i) *serial:* for all $w \in |\mathcal{Q}|$ there is $v \in |\mathcal{Q}|$ such that $w \; S_\mathcal{Q} \; v$;

(ii) *forward-confluent:* if $w \preccurlyeq_\mathcal{Q} w'$ and $w \; S_\mathcal{Q} \; v$, there is v' such that $v \preccurlyeq_\mathcal{Q} v'$ and $w' \; S_\mathcal{Q} \; v'$;

(iii) *sensible:* if $w \; S_\mathcal{Q} \; v$ then $\ell_\mathcal{Q}(w) \; S_T \; \ell_\mathcal{Q}(v)$, and

(iv) *ω-sensible:* whenever $\Diamond\varphi \in \ell^+_\mathcal{Q}(w)$, there are $n \geqslant 0$ and v such that $w \; S^n_\mathcal{Q} \; v$ and $\varphi \in \ell^+_\mathcal{Q}(v)$.

A forward-confluent, sensible Σ-labelled frame is a *weak Σ-quasimodel,* and if $S_\mathcal{Q}$ is a function we say that $\mathcal{Q}$ is *deterministic.*

We may write *quasimodel* instead of Σ-quasimodel when Σ is clear from context, and *full quasimodel* instead of *$\mathcal{L}_\Diamond$-quasimodel.* Similar conventions apply to labelled structures, weak quasimodels, etc.

Definition 3.5 Let $\mathcal{Q}$ be a weak quasimodel and let U be such that $U \subseteq |\mathcal{Q}|$.

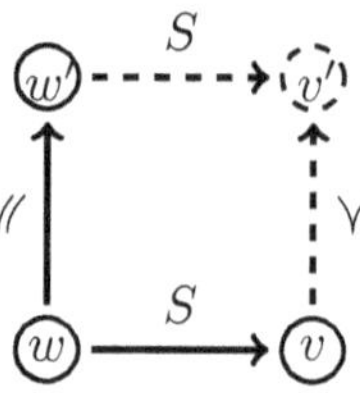

Fig. 1. If S is forward-confluent, then the above diagram can always be completed.

The restriction of $\mathcal{Q}$ with respect to U is defined to be the structure

$$\mathcal{Q} \upharpoonright U = (|\mathcal{Q} \upharpoonright U|, \preccurlyeq_{\mathcal{Q}\upharpoonright U}, S_{\mathcal{Q}\upharpoonright U}, \ell_{\mathcal{Q}\upharpoonright U}),$$

where:

(i) $|\mathcal{Q} \upharpoonright U| = U$;

(ii) $\preccurlyeq_{\mathcal{Q}\upharpoonright U} = \preccurlyeq_{\mathcal{Q}} \cap (U \times U)$;

(iii) $S_{\mathcal{Q}\upharpoonright U} = S_{\mathcal{Q}} \cap (U \times U)$;

(iv) $\ell_{\mathcal{Q}\upharpoonright U} = \ell_{\mathcal{Q}} \cap (U \times T_\Sigma)$.

Lemma 3.6 *If $\mathcal{Q}$ is a weak quasimodel, $U \subseteq |\mathcal{Q}|$ is upward closed and $S_{\mathcal{Q}} \upharpoonright U$ is serial and ω-sensible, then $\mathcal{Q} \upharpoonright U$ is a quasimodel.*

Proof. We must show that $\mathcal{Q}$ satisfies all properties of Definition 3.4. First we check that

$$(U, \preccurlyeq_{\mathcal{Q}\upharpoonright U}, \ell_{\mathcal{Q}\upharpoonright U})$$

is a labelled frame. The relation $\preccurlyeq_{\mathcal{Q}\upharpoonright U}$ is a partial order, since restrictions of partial orders are partial orders. Similarly, if $x \preccurlyeq_{\mathcal{Q}\upharpoonright U} y$ it follows that $x \preccurlyeq_{\mathcal{Q}} y$, so that from the definition of $\ell_{\mathcal{Q}\upharpoonright U}$ it is easy to deduce that $\ell_{\mathcal{Q}\upharpoonright U}(x) \preccurlyeq_T \ell_{\mathcal{Q}\upharpoonright U}(y)$.

To check that condition (b) holds, let us take $x \in U$ and a formula $\varphi \to \psi \in \ell^-_{\mathcal{Q}\upharpoonright U}(x)$. By definition, $\varphi \to \psi \in \ell^-_{\mathcal{Q}}(x)$ so there exists $y \in |\mathcal{Q}|$ such that $x \preccurlyeq_{\mathcal{Q}} y$, $\varphi \in \ell^+_{\mathcal{Q}}(y)$ and $\psi \in \ell^-_{\mathcal{Q}}(y)$. Since U is upward closed then $y \in U$ and, by definition, $x \preccurlyeq_{\mathcal{Q}\upharpoonright U} y$, $\varphi \in \ell^+_{\mathcal{Q}\upharpoonright U}(y)$ and $\psi \in \ell^-_{\mathcal{Q}\upharpoonright U}(y)$, as needed.

Now we check that the relation $S_{\mathcal{Q}\upharpoonright U}$ satisfies (i)-(iv). Note that $S_{\mathcal{Q}\upharpoonright U}$ is serial and ω-sensible by assumption and it is clearly sensible as $S_{\mathcal{Q}}$ was already sensible, so it remains to see that $S_{\mathcal{Q}\upharpoonright U}$ is forward-confluent. Take $x, y, z \in U$ such that $x \preccurlyeq_{\mathcal{Q}\upharpoonright U} y$ and $x \; S_{\mathcal{Q}\upharpoonright U} \; z$. By definition $x \preccurlyeq_{\mathcal{Q}} y$ and $x \; S_{\mathcal{Q}} \; z$. Since $S_{\mathcal{Q}}$ is confluent, there exists $t \in |\mathcal{Q}|$ such that $z \preccurlyeq_{\mathcal{Q}} t$ and $y \; S_{\mathcal{Q}} \; t$. Since U is upward closed $t \in U$ and, by definition, $y \; S_{\mathcal{Q}\upharpoonright U} \; t$ and $z \preccurlyeq_{\mathcal{Q}\upharpoonright U} t$. □

The following result of [5] will be crucial for our completeness proof.

Theorem 3.7 *A formula $\varphi \in \mathcal{L}_\Diamond$ is falsifiable over the class of expanding posets if and only if it is falsifiable over the class of finite,* sub(φ)*-quasimodels.*

As usual, if φ is not derivable, we wish to produce an expanding poset where φ is falsified, but in view of Theorem 3.7, it suffices to falsify φ on a quasimodel. This is convenient, as quasimodels are much easier to construct than models.

4 The canonical model

The standard canonical model for $\mathsf{ITL}^0_\Diamond$ is only a full, weak, deterministic quasimodel rather than a proper model. Nevertheless, it will be a useful ingredient in our completeness proof. Since we are working over an intuitionistic logic, the role of maximal consistent sets will be played by prime types, which we define below; recall that *full types* are elements of $T_{\mathcal{L}_\Diamond}$.

Definition 4.1 Given two sets of formulas Γ and Δ, we say that Δ is a consequence of Γ (denoted by $\Gamma \vdash \Delta$) if there exist finite $\Gamma' \subseteq \Gamma$ and $\Delta' \subseteq \Delta$ such that $\mathsf{ITL}^0_\Diamond \vdash \bigwedge \Gamma' \to \bigvee \Delta'$.

We say that a pair of sets $\Phi = (\Phi^-, \Phi^+)$ is *full* if $\Phi^- \cup \Phi^+ = \mathcal{L}_\Diamond$, and *consistent* if $\Phi^+ \nvdash \Phi^-$. A full, consistent type is a *prime type*. The set of prime types will be denoted T_∞.

Note that we are using the standard interpretation of $\Gamma \vdash \Delta$ in Gentzen-style calculi. When working within a turnstyle, we will follow the usual proof-theoretic conventions of writing Γ, Δ instead of $\Gamma \cup \Delta$ and φ instead of $\{\varphi\}$. Observe that there is no clash in terminology regarding the use of the word *type:*

Lemma 4.2 *If Φ is a prime type then Φ is an $\mathcal{L}_\Diamond$-type.*

Proof. Let Φ be a prime type; we must check that Φ satisfies all conditions of Definition 3.1. Condition (b) holds by assumption, and conditions (a) and (c) follow from the consistency of Φ.

The proofs of the other conditions are all similar to each other. For example, for (f), suppose that $\varphi \to \psi \in \Phi^+$ and $\varphi \notin \Phi$. Since Φ is full, it follows that $\varphi \in \Phi^+$. But $\big(\varphi \wedge (\varphi \to \psi)\big) \to \psi$ is an intuitionistic tautology, so using the fact that Φ is consistent we see that $\psi \notin \Psi^-$, which once again using condition (b) gives us $\psi \in \Phi^+$. For condition (g) we use (A6): if $\Diamond\varphi \in \Phi^-$ and $\varphi \in \Phi^+$ we would have that Φ is inconsistent, hence $\varphi \in \Phi^-$. The rest of the conditions are left to the reader. □

As with maximal consistent sets, prime types satisfy a Lindenbaum property.

Lemma 4.3 (Lindenbaum Lemma) *Let Γ and Δ be sets of formulas. If $\Gamma \nvdash \Delta$ then there exists a prime type Φ such that $\Gamma \subseteq \Phi^+$ and $\Delta \subseteq \Phi^-$.*

Proof. The proof is standard, but we provide a sketch. Let $\varphi \in \mathcal{L}_\Diamond$. Note that either $\Gamma, \varphi \nvdash \Delta$ or $\Gamma \nvdash \Delta, \varphi$, for otherwise by a cut rule (which is intuitionistically derivable) we would have $\Gamma \vdash \Delta$. Thus we can add φ to $\Gamma \cup \Delta$, and by repeating this process for each element of $\mathcal{L}_\Diamond$ (or using Zorn's lemma) we can find suitable Φ. □

Given a set A, let $\mathbb{I}_A$ denote the identity function on A. The canonical model $\mathcal{M}_c$ is then defined as the labelled structure

$$\mathcal{M}_c = (|\mathcal{M}_c|, \preccurlyeq_c, S_c, \ell_c) \stackrel{\text{def}}{=} (T_{\mathcal{L}_\Diamond}, \preccurlyeq_T, S_T, \mathbb{I}_{T_\infty}) \upharpoonright T_\infty;$$

in other words, $\mathcal{M}_c$ is the set of prime types with the usual ordering and successor relations. Note that ℓ_c is just the identity (i.e., $\ell_c(\Phi) = \Phi$). We will usually omit writing ℓ_c, as it has no effect on its argument.

Next we show that $\mathcal{M}_c$ is a full, weak, deterministic quasimodel. For this, we must prove that it has all properties required by Definition 3.4.

Lemma 4.4 *$\mathcal{M}_c$ is a labelled frame.*

Proof. We know that $\preccurlyeq_T$ is a partial order and restrictions of partial orders are partial orders, so $\preccurlyeq_c$ is a partial order. Moreover, ℓ_c is the identity, so $\Phi \preccurlyeq_c \Psi$ implies $\ell_c(\Phi) \preccurlyeq_T \ell_c(\Psi)$.

Now let $\Phi \in |\mathcal{M}_c|$ and assume that $\varphi \to \psi \in \Phi^-$. Note that $\Phi^+, \varphi \nvdash \psi$, for otherwise by intuitionistic reasoning we would have $\Phi^+ \vdash \varphi \to \psi$, which is impossible if Φ is a prime type. By Lemma 4.3, there is a prime type Ψ with $\Phi^+ \cup \{\varphi\} \subseteq \Psi^+$ and $\psi \in \Psi^-$. It follows that $\Phi \preccurlyeq_c \Psi$, $\varphi \in \Psi^+$ and $\psi \in \Psi^-$, as needed. □

Lemma 4.5 *S_c is a forward-confluent function.*

Proof. For a set $\Gamma \subseteq \mathcal{L}_\Diamond$, recall that we have defined $\bigcirc\Gamma = \{\bigcirc\varphi : \varphi \in \Gamma\}$. It will be convenient to introduce the notation

$$\ominus\Gamma = \{\varphi : \bigcirc\varphi \in \Gamma\}.$$

With this, we show that S_c is functional and forward-confluent.

FUNCTIONALITY. We claim that for all $\Phi, \Psi \in |\mathcal{M}_c|$,

$$\Phi \; S_c \; \Psi \text{ if and only if } \Psi = (\ominus\Phi^-, \ominus\Phi^+). \tag{1}$$

We must check that $\Psi \in |\mathcal{M}_c|$. To see that Ψ is full, let $\varphi \in \mathcal{L}_\Diamond$ be so that $\varphi \notin \Psi^-$. It follows that $\bigcirc\varphi \notin \Phi^-$, but Φ is full, so $\bigcirc\varphi \in \Phi^+$ and thus $\varphi \in \Psi^+$. Since φ was arbitrary, $\Psi^- \cup \Psi^+ = \mathcal{L}_\Diamond$.

Next we check that Ψ is consistent. If not, let $\Gamma \subseteq \Psi^+$ and $\Delta \subseteq \Psi^-$ be finite and such that $\bigwedge\Gamma \to \bigvee\Delta$ is derivable. Using (R2) and (A5) we see that $\bigcirc\bigwedge\Gamma \to \bigcirc\bigvee\Delta$ is derivable, which in view of Lemma 2.2 implies that $\bigwedge\bigcirc\Gamma \to \bigvee\bigcirc\Delta$ is derivable as well. But $\bigcirc\Gamma \subseteq \Phi^+$ and $\bigcirc\Delta \subseteq \Phi^-$, contradicting the fact that Φ is consistent.

Thus $\Psi \in |\mathcal{M}_c|$, and $\Phi \; S_c \; \Psi$ holds provided that $\Phi \; S_T \; \Psi$. It is clear that clauses (a) and (b) of Definition 3.3 hold. If $\Diamond\varphi \in \Phi^+$ and $\varphi \notin \Phi^+$, it follows that $\varphi \in \Phi^-$. By Lemma 2.2 $\Diamond\varphi \to \varphi \vee \bigcirc\Diamond\varphi$ is derivable, so we cannot have that $\bigcirc\Diamond\varphi \in \Phi^-$ and hence $\bigcirc\Diamond\varphi \in \Phi^+$, so that $\Diamond\varphi \in \Psi^+$. Similarly, if $\Diamond\varphi \in \Phi^-$ we have that $\bigcirc\Diamond\varphi \in \Phi^-$, for otherwise we obtain a contradiction from (A6). Therefore, $\Diamond\varphi \in \Psi^-$ as well.

To check that Ψ is unique, suppose that $\Theta \in |\mathcal{M}_c|$ is such that $\Phi \; S_c \; \Theta$. Then if $\varphi \in \Psi^+$ it follows from (1) that $\bigcirc\varphi \in \Phi^+$ and hence $\varphi \in \Theta^+$; by the same argument, if $\varphi \in \Psi^-$ it follows that $\varphi \in \Theta^-$, and hence $\Theta = \Psi$.

FORWARD CONFLUENCE: Now that we have shown that S_c is a function, we may treat it as such. Suppose that $\Phi \preccurlyeq_c \Psi$; we must check that $S_c(\Phi) \preccurlyeq_c S_c(\Psi)$.

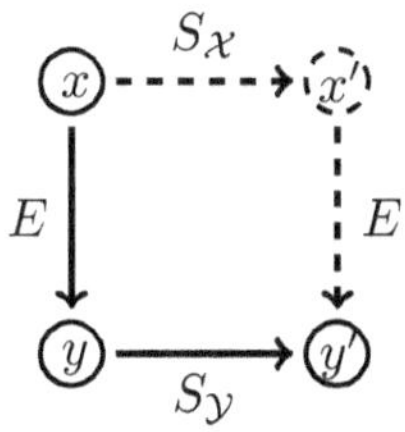

Fig. 2. If $E \subseteq |\mathcal{X}| \times |\mathcal{Y}|$ is a dynamical simulation, this diagram can always be completed.

Let $\varphi \in S_c^+(\Phi)$. Using (1), we have that $\bigcirc\varphi \in \Phi^+$, hence $\bigcirc\varphi \in \Psi^+$ and thus $\varphi \in S_c(\Psi^+)$. Since $\varphi \in S_c(\Phi)$ was arbitrary we obtain $S_c^+(\Phi) \preccurlyeq_c S_c^+(\Psi)$, as needed. □

From all these lemmas we conclude the following result.

Proposition 4.6 *The canonical model is a deterministic weak quasimodel.*

Proof. In view of Definition 3.4, we need (i) $(|\mathcal{M}_c|, \preccurlyeq_c, \ell_c)$ to be a labelled frame, (ii) S_c to be a sensible forward-confluent function, and (iii) ℓ_c to have $T_{\mathcal{L}_\Diamond}$ as its codomain. The first item is Lemma 4.4. That S_c is a forward-confluent function is Lemma 4.5, and it is sensible since $\Phi\ S_c\ \Psi$ precisely when $\Phi\ S_T\ \Psi$. Finally, if $\Phi \in |\mathcal{M}_c|$ then $\ell_c(\Phi) = \Phi$, which is an element of $T_{\mathcal{L}_\Diamond}$ by Lemma 4.2. □

5 Simulations

Simulations are relations between worlds in labelled spaces, and give rise to the appropriate notion of 'substructure' for modal and intuitionistic logics. We have used them to prove that a topological intuitionistic temporal logic has the finite quasimodel property [15], and they will also be useful for our completeness proof. Below, recall that $\Phi \subseteq_T \Psi$ means that $\Phi^- \subseteq \Psi^-$ and $\Phi^+ \subseteq \Psi^+$.

Definition 5.1 *Let $\Sigma \subseteq \Delta \subseteq \mathcal{L}_\Diamond$ be closed under subformulas, $\mathcal{X}$ be a Σ-labelled frame and $\mathcal{Y}$ be Δ-labelled. A forward-confluent relation $E \subseteq |\mathcal{X}| \times |\mathcal{Y}|$ is a* simulation *if, whenever $x\ E\ y$, $\ell_{\mathcal{X}}(x) \subseteq_T \ell_{\mathcal{Y}}(y)$. If there exists a simulation E such that $x\ E\ y$, we write $(\mathcal{X}, x) \rightharpoonup (\mathcal{Y}, y)$.*

The relation E is a dynamic simulation *between $\mathcal{X}$ and $\mathcal{Y}$ if $S_{\mathcal{Y}}E \subseteq ES_{\mathcal{X}}$.*

The following is proven in [15]. While the details of the construction given there are not important for our current purposes, the interested reader may find an overview in Appendix B. Below, recall that $\Sigma \Subset \mathcal{L}_\Diamond$ means that Σ is finite and closed under subformulas.

Theorem 5.2 *Given $\Sigma \Subset \mathcal{L}_\Diamond$, there exists a finite weak quasimodel $\mathcal{I}_\Sigma$ such that if $\mathcal{A}$ is any deterministic weak quasimodel then $\rightharpoonup \subseteq |\mathcal{I}_\Sigma| \times |\mathcal{A}|$ is a surjective dynamic simulation.*

Points of $\mathcal{I}_\Sigma$ are called *moments*. One can think of $\mathcal{I}_\Sigma$ as a finite initial structure over the category of labelled weak quasimodels. Next, we will internalize the notion of simulating elements of $\mathcal{I}_\Sigma$ into the temporal language. This is achieved by the formulas $\mathrm{Sim}(w)$ given by the next proposition.

Proposition 5.3 *Given $\Sigma \Subset \mathcal{L}_\Diamond$ and a finite Σ-labelled frame $\mathcal{W}$, there exist formulas $(\mathrm{Sim}(w))_{w\in|\mathcal{W}|}$ such that for any fully labelled frame $\mathcal{X}$, $w \in |\mathcal{W}|$ and $x \in |\mathcal{X}|$, $\mathrm{Sim}(w) \in \ell^-(x)$ if and only if there is $y \succcurlyeq x$ such that $(\mathcal{W}, w) \rightleftharpoons (\mathcal{X}, y)$.*

Proof. An explicit construction is given in Appendix A. □

The next proposition allows us to emulate model-theoretic reasoning within $\mathcal{L}_\Diamond$.

Proposition 5.4 *Fix $\Sigma \Subset \mathcal{L}_\Diamond$ and let $\mathcal{I} = \mathcal{I}_\Sigma$, $w \in |\mathcal{I}|$ and $\psi \in \Sigma$.*

(i) *If $\psi \in \ell^-(w)$, then $\vdash \psi \to \mathrm{Sim}(w)$.*

(ii) *If $\psi \in \ell^+(w)$, then $\vdash \big(\psi \to \mathrm{Sim}(w)\big) \to \mathrm{Sim}(w)$.*

(iii) *If $w \preccurlyeq v$, then $\vdash \mathrm{Sim}(v) \to \mathrm{Sim}(w)$.*

(iv) $\vdash \displaystyle\bigwedge_{\psi\in\ell^-_{\mathcal{I}}(w)} \mathrm{Sim}(w) \to \psi.$

(v) $\vdash \bigcirc \displaystyle\bigwedge_{wS_{\mathcal{I}}v} \mathrm{Sim}(v) \to \mathrm{Sim}(w).$

Proof.

(i) First assume that $\psi \in \ell^-(w)$, and toward a contradiction that $\nvdash \psi \to \mathrm{Sim}(w)$. By the Lindenbaum lemma there is $\Gamma \in |\mathcal{M}_c|$ such that $\psi \to \mathrm{Sim}(w) \in \Gamma^-$. Thus for some $\Theta \succcurlyeq_c \Gamma$ we have that $\psi \in \Theta^+$ and $\mathrm{Sim}(w) \in \Theta^-$. But then by Proposition 5.3 we have that $(\mathcal{W}, w) \rightleftharpoons (\mathcal{M}_c, \Delta)$ for some $\Delta \succcurlyeq_c \Theta$, so that $\psi \in \Delta^-$, and by upwards persistence $\psi \in \Theta^-$, contradicting the consistency of Θ.

(ii) If $\psi \in \ell^+(w)$, we proceed similarly. Assume toward a contradiction that $\nvdash \big(\psi \to \mathrm{Sim}(w)\big) \to \mathrm{Sim}(w)$. Then, reasoning as above there is $\Theta \in |\mathcal{M}_c|$ such that $\psi \to \mathrm{Sim}(w) \in \Theta^+$ and $\mathrm{Sim}(w) \in \Theta^-$. From Proposition 5.3 we see that there is $\Delta \succcurlyeq_c \Theta$ such that $(\mathcal{W}, w) \rightleftharpoons (\mathcal{M}_c, \Delta)$, so that $\psi \in \Delta^+$ and, once again by Proposition 5.3, $\mathrm{Sim}(w) \in \Delta^-$. It follows that $\psi \to \mathrm{Sim}(w) \notin \Delta^+$; but in view of upward persistence, this contradicts that $\psi \to \mathrm{Sim}(w) \in \Theta^+$.

(iii) Suppose that $v \succcurlyeq w$. Reasoning as above, it suffices to show that if $\Gamma \in |\mathcal{M}_c|$ is such that $\mathrm{Sim}(w) \in \Gamma^-$, then also $\mathrm{Sim}(v) \in \Gamma^-$. But if $\mathrm{Sim}(w) \in \Gamma^-$, there is $\Theta \succcurlyeq_c \Gamma$ such that $(\mathcal{I}, w) \rightleftharpoons (\mathcal{M}_c, \Theta)$. By forward confluence $(\mathcal{I}, v) \rightleftharpoons (\mathcal{M}_c, \Delta)$ for some $\Delta \succcurlyeq_c \Theta$. Thus by Proposition 5.3, $\mathrm{Sim}(v) \in \Delta^-$ and by upwards persistence $\mathrm{Sim}(v) \in \Gamma^-$. Since $\Gamma \in |\mathcal{M}_c|$ was arbitrary, the claim follows.

(iv) We prove that if $\Gamma \in |\mathcal{M}_c|$ is such that

$$\bigwedge_{\psi\in\ell^-(w)} \mathrm{Sim}(w) \in \Gamma^+, \tag{2}$$

then $\psi \in \Gamma^+$. If (2) holds then by Theorem 5.2, there is $w \in |\mathcal{I}|$ with $(\mathcal{I}, w) \rightleftharpoons$

$(\mathcal{M}_c, \Gamma)$. By Proposition 5.3, $\mathrm{Sim}(w) \in \Gamma^-$, hence it follows from (2) that $\psi \notin \ell^-(w)$; but w is Σ-typed and $\psi \in \Sigma$, so $\psi \in \ell^+(w)$ and thus $\psi \in \Gamma^+$, as required.

(v) Suppose that $\Gamma \in |\mathcal{M}_c|$ is such that

$$\bigcirc \bigwedge_{w S_{\mathcal{I}} v} \mathrm{Sim}(v) \in \Gamma^+, \tag{3}$$

and assume toward a contradiction that $\mathrm{Sim}(w) \in \Gamma^-$. By Proposition 5.3 $(\mathcal{I}, w) \rightleftharpoons (\mathcal{M}_c, \Delta)$ for some $\Delta \succcurlyeq_c \Gamma$. Since $\rightleftharpoons$ is a dynamic simulation, it follows that there is $v \in |\mathcal{I}|$ with $w \; S_{\mathcal{I}} \; v$ and $(\mathcal{I}, v) \rightleftharpoons (\mathcal{M}_c, S_c(\Delta))$, so that $\mathrm{Sim}(v) \in (S_c(\Delta))^-$. It follows that $\bigcirc\mathrm{Sim}(v) \in \Gamma^-$, since S_c is sensible and Γ is full. But $\Delta \succcurlyeq_c \Gamma$, so that $\bigcirc\mathrm{Sim}(v) \in S_c^-(\Gamma)$ as well, contradicting (3). □

6 The initial quasimodel

We are now ready to define our initial quasimodel. Given a finite set Σ of formulas, we will define a quasimodel $\mathcal{J}_\Sigma$ falsifying all unprovable Σ-types. This quasimodel is a substructure of $\mathcal{I}_\Sigma$, containing only moments which are *possible* in the following sense.

Definition 6.1 Fix $\Sigma \Subset \mathcal{L}_\Diamond$. We say that a moment $w \in |\mathcal{I}_\Sigma|$ is *possible* if $\nvdash \mathrm{Sim}(w)$, and denote the set of possible Σ-moments by J_Σ.

With this we are ready to define our initial structure, which as we will see later is indeed a quasimodel.

Definition 6.2 Given $\Sigma \Subset \mathcal{L}_\Diamond$, we define the *initial structure* for Σ by $\mathcal{J}_\Sigma = \mathcal{I}_\Sigma \upharpoonright J_\Sigma$.

Our strategy from here on will be to show that canonical structures are indeed quasimodels; once we establish this, completeness of $\mathsf{ITL}^0_\Diamond$ is an easy consequence. The most involved step will be showing that the successor relation on $\mathcal{J}_\Sigma$ is ω-sensible, but we begin with some simpler properties.

Lemma 6.3 *Let Σ be a finite set of formulas, $\mathcal{I} = \mathcal{I}_\Sigma$ and $\mathcal{J} = \mathcal{J}_\Sigma$. Then, $|\mathcal{J}|$ is an upward-closed subset of $|\mathcal{I}|$ and $S_{\mathcal{J}}$ is serial.*

Proof. To check that $|\mathcal{J}|$ is upward closed, let $w \in |\mathcal{J}|$ and suppose $v \succcurlyeq w$. Now, by Proposition 5.4.iii, we have that

$$\vdash \mathrm{Sim}(v) \to \mathrm{Sim}(w);$$

hence if w is possible, so is v.

To see that $S_{\mathcal{J}}$ is serial, observe that by Proposition 5.4.v, if $w \in |\mathcal{J}| \subseteq |\mathcal{I}|$,

$$\vdash \bigcirc \bigwedge_{w S_{\mathcal{I}} v} \mathrm{Sim}(v) \to \mathrm{Sim}(w).$$

Since w is possible, it follows that for some v with $w \; S_{\mathcal{I}} \; v$, v is possible as well, for otherwise $\bigcirc \bigwedge_{w S_{\mathcal{I}} v} \mathrm{Sim}(v)$ would be equivalent to $\bigcirc\top$, allowing us to deduce $\mathrm{Sim}(w)$. But then $v \in |\mathcal{J}|$, as needed. □

7 ω-Sensibility

In this section we will show that $S_{\mathcal{J}}$ is ω-sensible, the most difficult step in proving that $\mathcal{J} = \mathcal{J}_\Sigma$ is a quasimodel. In other words, we must show that, given $w \in |\mathcal{J}_\Sigma|$ and $\Diamond\psi \in \ell^+(w)$, there is a finite path

$$w = w_0 \; S_{\mathcal{J}} \; w_1 \; S_{\mathcal{J}} \ldots S_{\mathcal{J}} \; w_n,$$

where $\psi \in \ell^+(w_n)$ and $w_i \in |\mathcal{J}_\Sigma|$ for all $i \leqslant n$.

Definition 7.1 Let $\Sigma \Subset \mathcal{L}_\Diamond$ and $w, v \in |\mathcal{J}_\Sigma|$. Say that v is *reachable* from w if there is a finite path

$$\vec{u} = (u_0, ..., u_n)$$

of possible moments with $u_0 = w$, $u_n = v$, and $u_i \; S_{\mathcal{J}} \; u_{i+1}$ for all $i < n$. We denote the set of all possible moments that are reachable from w by $R(w)$.

Lemma 7.2 *If* $\Sigma \Subset \mathcal{L}_\Diamond$ *and* $w \in |\mathcal{J}_\Sigma|$ *then*

$$\vdash \bigcirc \bigwedge_{v \in R(w)} \mathrm{Sim}(v) \to \bigwedge_{v \in R(w)} \mathrm{Sim}(v).$$

Proof. Let $\mathcal{I} = \mathcal{I}_\Sigma$. By Proposition 5.4.v we have that, for all $v \in R(w)$,

$$\vdash \bigcirc \bigwedge_{v S_{\mathcal{I}} u} \mathrm{Sim}(u) \to \mathrm{Sim}(v).$$

Now, if $u \notin |\mathcal{J}_\Sigma|$, then $\vdash \mathrm{Sim}(u)$, hence by (R2) $\vdash \bigcirc\mathrm{Sim}(u)$, and we can remove $\mathrm{Sim}(u)$ from the conjunction using Lemma 2.2 and propositional reasoning. Since $v \in R(w)$ was arbitrary, this shows that

$$\vdash \bigcirc \bigwedge_{v \in R(w)} \mathrm{Sim}(v) \to \bigwedge_{v \in R(w)} \mathrm{Sim}(v).$$

□

From this we obtain the following, which evidently implies ω-sensibility:

Proposition 7.3 *If* $w \in |\mathcal{J}_\Sigma|$ *and* $\Diamond\psi \in \ell^+(w)$, *then there is* $v \in R(w)$ *such that* $\psi \in \ell^+(v)$.

Proof. Towards a contradiction, assume that $w \in J_\Sigma$ and $\Diamond\psi \in \ell^+(w)$ but, for all $v \in R(w)$, $\psi \in \ell^-(w)$.

By Lemma 7.2,

$$\vdash \bigcirc \bigwedge_{v \in R(w)} \mathrm{Sim}(v) \to \bigwedge_{v \in R(w)} \mathrm{Sim}(v).$$

By the $\Diamond$-induction rule (R4),

$$\vdash \Diamond \bigwedge_{v \in R(w)} \mathrm{Sim}(v) \to \bigwedge_{v \in R(w)} \mathrm{Sim}(v);$$

in particular,

$$\vdash \Diamond \bigwedge_{v \in R(w)} \mathrm{Sim}(v) \to \mathrm{Sim}(w). \tag{4}$$

Now let $v \in R(w)$. By Proposition 5.4.i and the assumption that $\psi \in \ell^-(v)$ we have that

$$\vdash \psi \to \mathrm{Sim}(v),$$

and since v was arbitrary,

$$\vdash \psi \to \bigwedge_{v \in R(w)} \mathrm{Sim}(v).$$

Using distributivity (R3) we further have that

$$\vdash \Diamond\psi \to \Diamond \bigwedge_{v \in R(w)} \mathrm{Sim}(v).$$

This, along with (4), shows that

$$\vdash \Diamond\psi \to \mathrm{Sim}(w);$$

however, by Proposition 5.4.ii and our assumption that $\Diamond\psi \in \ell^+(w)$ we have that

$$\vdash \big(\Diamond\psi \to \mathrm{Sim}(w)\big) \to \mathrm{Sim}(w),$$

hence by modus ponens we obtain $\vdash \mathrm{Sim}(w)$, which contradicts the assumption that $w \in J_\Sigma$. We conclude that there can be no such w. □

Corollary 7.4 *Given any finite set Σ of formulas, $\mathcal{J}_\Sigma$ is a quasimodel.*

Proof. Let $\mathcal{J} = \mathcal{J}_\Sigma$. By Lemma 6.3, $|\mathcal{J}|$ is upwards closed in $|\mathcal{I}_\Sigma|$ and $S_\mathcal{J}$ is serial, while by Proposition 7.3, $S_\mathcal{J}$ is ω-sensible. It follows from Lemma 3.6 that $\mathcal{J}$ is a quasimodel. □

We are now ready to prove that $\mathsf{ITL}^0_\Diamond$ is complete.

Theorem 7.5 *If $\varphi \in \mathcal{L}_\Diamond$ is valid over the class of expanding posets, then $\mathsf{ITL}^0_\Diamond \vdash \varphi$.*

Proof. We prove the contrapositive. Suppose φ is an unprovable formula and let

$$W = \left\{ w \in \mathcal{I}_{\mathrm{sub}(\varphi)} : \varphi \in \ell^-(w) \right\}.$$

Then, by Proposition 5.4.iv we have that

$$\vdash \bigwedge_{w \in W} \mathrm{Sim}(w) \to \varphi;$$

since φ is unprovable, it follows that some $w^* \in W$ is possible and hence $w^* \in J_{\mathrm{sub}(\varphi)}$. By Corollary 7.4, $\mathcal{J}_{\mathrm{sub}(\varphi)}$ is a quasimodel, so that by Theorem 3.7, φ is falsifiable in some expanding poset. □

Concluding remarks

We have provided a sound and complete axiomatization for the $\Box$-free fragment of the expanding intuitionistic temporal logic $\mathsf{ITL}^{\mathsf{e}}$. With this we may develop syntactic techniques to decide validity over the class of expanding posets, complementing the semantic methods presented by Boudou and the authors [4] and possibly leading to an elementary decision procedure.

Many questions remain open in this direction, perhaps most notably an extension to the full language with $\Box$. This is likely to be a much more challenging problem, as the language with 'henceforth' can distinguish between Kripke and topological models and hence methods based on non-deterministic quasimodels do not seem feasible. Along these lines, one can consider the LTL connectives *until* and *release*. In an upcoming contribution, we study the decidability of the full LTL language with respect to the $\mathsf{ITL}^{\mathsf{e}}$ semantics. It is likely that the techniques presented here can be extended to handle the 'until' operator, which can be used to define $\Diamond$; on the other hand, 'release' can be used to define $\Box$, and axiomatizing it should be at least as difficult as axiomatizing the logic with 'henceforth'.

The question of axiomatizing $\mathsf{ITL}^{\mathsf{p}}$ (with persistent domains) is also of interest, but here it is possible that the logic is not even axiomatizable in principle. It may be that methods from products of modal logics [23] can be employed here; for example, one can reduce tiling or related problems to show that certain products such as $\mathsf{LTL} \times \mathsf{S4}$ are not computably enumerable. However, even if such a reduction is possible, working over the more limited intuitionistic language poses an additional challenge. Even computational lower bounds for these logics are not yet available, aside from the trivial PSPACE bound obtained from the purely propositional fragment.

Appendix

A Simulation formulas

In this appendix, we show that there exist $\mathcal{L}_\Diamond$ formulas defining points in finite frames up to simulability, i.e. that if $\mathcal{W}$ is a finite frame and $w \in |\mathcal{W}|$, there exists a formula $\mathrm{Sim}(w)$ such that for all labelled frames $\mathcal{M}$ and all $x \in |\mathcal{M}|$, $\mathcal{M}, x \models x$ if and only if $(\mathcal{W}, w) \rightleftharpoons (\mathcal{M}, x)$. In contrast, such formulas do not exist in the classical modal language for finite $\mathsf{S4}$ models [11], but they *can* be constructed using a polyadic 'tangled' modality. This tangled modalilty was proven to be expressively equivalent to the μ-calculus over transitive frames by Dawar and Otto [8], and later axiomatized for several classes of models by Fernández-Duque [12] and Goldblatt and Hodkinson [18,19].

Simulation formulas were used in [13] to provide a sound and complete axiomatization of *dynamic topological logic* [1,22], a classical tri-modal system closely related to $\mathsf{ITL}^{\mathsf{e}}$, where the intuitionistic implication is replaced by an $\mathsf{S4}$ modality. One can use the fact that simulability is not definable over the modal language to prove that the natural axiomatization suggested by Kremer and Mints [22] of dynamic topological logic was incomplete for its topological, let alone its Kripke, semantics [14].

While simulability is not modally definable, it is definable over the language of intuitionistic logic, as finite frames [9] (and hence models) are already definable up to simulation in the intuitionistic language. This may be surprising, as the intuitionistic language is less expressive than the modal language; however, intuitionistic models are posets rather than arbitrary preorders, and this allows us to define simulability formulas by recusion on $\prec$.

Definition A.1 *Fix* $\Sigma \Subset \mathcal{L}_\Diamond$ *and let* $\mathcal{W}$ *be a finite* Σ*-labeled frame. Given* $w \in |\mathcal{W}|$*, we define a formula* $\mathrm{Sim}(w)$ *by backwards induction on* $\preccurlyeq\ = \ \preccurlyeq_{\mathcal{W}}$ *by*

$$\mathrm{Sim}(w) = \bigwedge \ell^+(w) \to \bigvee \ell^-(w) \vee \bigvee_{v \succ w} \mathrm{Sim}(v).$$

Proposition A.2 *Given* $\Sigma \Subset \Delta \subseteq \mathcal{L}_\Diamond$*, a finite* Σ*-labelled frame* $\mathcal{W}$*, a* Δ*-labelled frame* $\mathcal{X}$ *and* $w \in |\mathcal{W}|$*,* $x \in |\mathcal{X}|$*:*

(i) *if* $\mathrm{Sim}(w) \in \ell^-_{\mathcal{X}}(x)$ *then there is* $y \succcurlyeq x$ *such that* $(\mathcal{W}, w) \rightleftharpoons (\mathcal{X}, y)$*, and*

(ii) *if there is* $y \succcurlyeq x$ *such that* $(\mathcal{W}, w) \rightleftharpoons (\mathcal{X}, y)$ *then* $\mathrm{Sim}(w) \notin \ell^+_{\mathcal{X}}(x)$*.*

Proof. Each claim is proved by backward induction on $\preccurlyeq$.

(i) Let us first consider the base case, when there is no $v \succ w$. Assume that $\mathrm{Sim}(w) \in \ell^-(x)$. From the definition of labelled frame $\bigwedge \ell^+_{\mathcal{W}}(w) \in \ell^+_{\mathcal{X}}(y)$ and $\bigvee \ell^-_{\mathcal{W}}(w) \in \ell^-_{\mathcal{X}}(y)$ for some $y \succcurlyeq x$. From the definition of type it follows that $\ell^+_{\mathcal{W}}(w) \subseteq \ell^+_{\mathcal{X}}(y)$ and $\ell^-_{\mathcal{W}}(w) \subseteq \ell^-_{\mathcal{X}}(y)$, so that $\ell_{\mathcal{W}}(w) \subseteq_T \ell_{\mathcal{X}}(y)$. It follows that $E \stackrel{\text{def}}{=} \{(w, y)\}$ is a simulation, so $(\mathcal{W}, w) \rightleftharpoons (\mathcal{X}, y)$.

For the inductive step, let us assume that the lemma is proved for all $v \succ w$. Assume that $\mathrm{Sim}(w) \in \ell^-_{\mathcal{X}}(x)$. From Condition (b) it follows that $\bigwedge \ell^+_{\mathcal{W}}(w) \in \ell^+_{\mathcal{X}}(y)$, $\bigvee \ell^-_{\mathcal{W}}(w) \in \ell^-_{\mathcal{X}}(y)$ and $\bigvee_{v \prec w} \mathrm{Sim}(v) \in \ell^-_{\mathcal{X}}(y)$ for some $y \succcurlyeq x$. By following a similar reasoning as in the base case we can conclude that $\ell_{\mathcal{W}}(w) \subseteq \ell_{\mathcal{X}}(y)$, and moreover, that $\mathrm{Sim}(v) \in \ell^-_{\mathcal{X}}(y)$ for all $v \succ w$. By induction hypothesis we conclude that for all $v \succ w$, there exists a simulation E_v such that $v \ E_v \ z_v$ for some $z_v \succcurlyeq y$. Let

$$E \stackrel{\text{def}}{=} \{(w, y)\} \cup \bigcup_{v \succ w} E_v \ .$$

The reader may check that E is a simulation and that $w \ E \ y \succcurlyeq x$, so that $(\mathcal{W}, w) \rightleftharpoons (\mathcal{X}, y)$, as needed.

(ii) For the base case, assume that $(\mathcal{W}, w) \rightleftharpoons (\mathcal{X}, y)$ for some $y \succcurlyeq x$, so there exists a simulation E such that $w \ E \ y$. It follows that $\ell^+_{\mathcal{W}}(w) \subseteq \ell^+_{\mathcal{X}}(y)$ and $\ell^-_{\mathcal{W}}(w) \subseteq \ell^-_{\mathcal{X}}(y)$. From conditions (d) and (e) of the definition of type (Definition 3.1), it follows that $\bigwedge \ell^+_{\mathcal{W}}(w) \notin \ell^-_{\mathcal{X}}(y)$ and $\bigvee \ell^-_{\mathcal{W}}(w) \notin \ell^+_{\mathcal{X}}(y)$. But then, condition (f) gives us $\mathrm{Sim}(w) \notin \ell^+_{\mathcal{X}}(y)$, so $\mathrm{Sim}(w) \notin \ell^+_{\mathcal{X}}(x)$.

For the inductive step, by the same reasoning as in the base case it follows that $\bigwedge \ell^+_{\mathcal{W}}(w) \notin \ell^-_{\mathcal{X}}(y)$ and $\bigvee \ell^-_{\mathcal{W}}(w) \notin \ell^+_{\mathcal{X}}(y)$. Now, let v be such that $v \succ w$. Since E is forward confluent then $v \ E \ z_v$ for some $z_v \succcurlyeq y$. By induction hypothesis, $\mathrm{Sim}(v) \notin \ell^+(z_v)$, so $\mathrm{Sim}(v) \notin \ell^+(y)$. Since v was arbitrary we conclude that $\bigvee_{v>w} \mathrm{Sim}(v) \notin \ell^+(y)$. Finally, from condition (f) of Definition 3.1 and the fact that $y \preccurlyeq x$ we get that $\mathrm{Sim}(w) \notin \ell^+(x)$. □

B The finite initial frame

In this appendix we review the construction of the structure $\mathcal{I}_\Sigma$ of Theorem 5.2. The worlds of this structure are called *irreducible* Σ*-moments*. The intuition is that a Σ-moment represents all the information that holds at the same 'moment of time'. Recall that we write $\Sigma \Subset \mathcal{L}_\Diamond$ if $\Sigma \subseteq \mathcal{L}_\Diamond$ is finite and closed under subformulas. We omit all proofs, which can be found in [15].

Definition B.1 *Let $\Sigma \Subset \mathcal{L}_\Diamond$. A* Σ-moment *is a Σ-labelled space $\boldsymbol{w}$ such that $(|\boldsymbol{w}|, \preccurlyeq_{\boldsymbol{w}})$ is a finite tree with unique root $r_{\boldsymbol{w}}$.*

Note that moments can be arbitrarily large. In order to obtain a finite structure we will restrict the set of moments to those that are, in a sense, no bigger than they need to be. To be precise, we want them to be minimal with respect to $\trianglelefteq$, which we define below.

Definition B.2 *Let $\Sigma \Subset \mathcal{L}_\Diamond$ and $\boldsymbol{w}, \boldsymbol{v}$ be Σ-moments. We write*

(i) *$\boldsymbol{w} \sqsubseteq \boldsymbol{v}$ if $|\boldsymbol{w}| \subseteq |\boldsymbol{v}|$, $\preccurlyeq_{\boldsymbol{w}} = \preccurlyeq_{\boldsymbol{v}} \upharpoonright |\boldsymbol{w}|$, and $\ell_{\boldsymbol{w}} = \ell_{\boldsymbol{v}} \upharpoonright |\boldsymbol{w}|$;*

(ii) *$\boldsymbol{w} \trianglelefteq \boldsymbol{v}$ if if $\boldsymbol{w} \sqsubseteq \boldsymbol{v}$ and there is a forward confluent, surjective function $\pi\colon |\boldsymbol{v}| \to |\boldsymbol{w}|$ such that $\ell_{\boldsymbol{v}}(v) = \ell_{\boldsymbol{w}}(\pi(v))$ for all $v \in |\boldsymbol{v}|$ and $\pi^2 = \pi$. We say that $\boldsymbol{w}$ is a* reduct *of $\boldsymbol{v}$ and π is a* reduction.

Note that the condition $\pi^2 = \pi$ is equivalent to requiring $\pi(w) = w$ whenever $w \in |\boldsymbol{w}|$. Irreducible moments are the minimal moments under $\trianglelefteq$.

Definition B.3 *Let $\Sigma \Subset \mathcal{L}_\Diamond$. A Σ-moment $\boldsymbol{w}$ is* irreducible *if whenever $\boldsymbol{w} \trianglelefteq \boldsymbol{v}$, it follows that $\boldsymbol{w} = \boldsymbol{v}$. The set of irreducible moments is denoted I_Σ.*

To view I_Σ as a labeled frame, we need to equip it with a suitable partial order.

Definition B.4 *Let $\boldsymbol{w} \in I_\Sigma$. For $w \in |\boldsymbol{w}|$, let $\boldsymbol{w}[w] = \boldsymbol{w} \upharpoonright \uparrow w$, i.e.,*

$$\boldsymbol{w}[w] = (\uparrow w\ ,\ \preccurlyeq_{\boldsymbol{w}} \upharpoonright \uparrow w\ ,\ \ell_{\boldsymbol{w}} \upharpoonright \uparrow w\ ).$$

We write $\boldsymbol{v} \preccurlyeq \boldsymbol{w}$ if $\boldsymbol{v} = \boldsymbol{w}[w]$ for some $w \in |\boldsymbol{w}|$.

It is shown in [15] that if $\boldsymbol{w}$ is irreducible and $\boldsymbol{v} \preccurlyeq \boldsymbol{w}$, $\boldsymbol{v}$ is irreducible as well. To obtain a weak quasimodel, it remains to define a sensible relation on I_Σ.

Definition B.5 *If $\Sigma \Subset \mathcal{L}_\Diamond$ and $\boldsymbol{w}, \boldsymbol{v} \in I_\Sigma$, we define $\boldsymbol{v} \mapsto \boldsymbol{w}$ if there exists a sensible, forward-confluent relation $S \subseteq |\boldsymbol{v}| \times |\boldsymbol{w}|$ such that $r_{\boldsymbol{v}}\ S\ r_{\boldsymbol{w}}$.*

We are now ready to define our initial weak quasimodel.

Definition B.6 *Given $\Sigma \Subset \mathcal{L}_\Diamond$, we define $\mathcal{I} = \mathcal{I}_\Sigma$ to be the structure $(|\mathcal{I}|, \preccurlyeq_{\mathcal{I}}, S_{\mathcal{I}}, \ell_{\mathcal{I}})$, where $|\mathcal{I}| = I_\Sigma$, $\boldsymbol{v} \preccurlyeq_{\mathcal{I}} \boldsymbol{w}$ if and only if $\boldsymbol{v} \succcurlyeq \boldsymbol{w}$, $\boldsymbol{w}\ S_{\mathcal{I}}\ \boldsymbol{v}$ if and only if $\boldsymbol{w} \mapsto \boldsymbol{v}$, and $\ell_{\mathcal{I}}(\boldsymbol{w}) = \ell_{\boldsymbol{w}}(r_{\boldsymbol{w}})$.*

Note that in this construction, the moments accessible from $\boldsymbol{w}$ are *smaller* than $\boldsymbol{w}$, and thus we use the reverse partial order to interpret implication. The structure $\mathcal{I}_\Sigma$ is always finite, a fact that is used in an essential way in our completeness proof. Below, 2^n_m denotes the superexponential function.

Theorem B.7 *Let* $\Sigma \Subset \mathcal{L}_\Diamond$ *and let* $s = \#\Sigma$. *Then,* $\mathcal{I}_\Sigma$ *is a weak* Σ*-quasimodel and* $\#I_\Sigma \leqslant 2^{s^2+s}_{s+1}$. *Moreover, if* $\Sigma \Subset \mathcal{L}_\Diamond$ *and* $\mathcal{A}$ *is any deterministic weak quasimodel then* $\rightleftharpoons \subseteq I_\Sigma \times |\mathcal{A}|$ *is a surjective dynamic simulation.*

In fact, the claim proven in Fernández-Duque [15] is more general in that $\mathcal{A}$ may belong to a wider class of topological weak qusimodels, but this special case is sufficient for our purposes.

References

[1] Artëmov, S. N., J. M. Davoren and A. Nerode, *Modal logics and topological semantics for hybrid systems*, Technical Report MSI 97-05 (1997).

[2] Balbiani, P., J. Boudou, M. Diéguez and D. Fernández-Duque, *Bisimulations for intuitionistic temporal logics*, arXiv **1803.05078** (2018).

[3] Balbiani, P. and M. Diéguez, *Temporal here and there*, in: M. Loizos and A. Kakas, editors, *Logics in Artificial Intelligence* (2016), pp. 81–96.

[4] Boudou, J., M. Diéguez and D. Fernández-Duque, *A decidable intuitionistic temporal logic*, in: *26th EACSL Annual Conference on Computer Science Logic (CSL)*, 2017, pp. 14:1–14:17.

[5] Boudou, J., M. Diéguez, D. Fernández-Duque and F. Romero, *Axiomatic systems and topological semantics for intuitionistic temporal logic*, arXiv **1803.05077** (2018).

[6] Davies, R., *A temporal-logic approach to binding-time analysis*, in: *Proceedings, 11th Annual IEEE Symposium on Logic in Computer Science, New Brunswick, New Jersey, USA*, 1996, pp. 184–195.

[7] Davoren, J. M., *On intuitionistic modal and tense logics and their classical companion logics: Topological semantics and bisimulations*, Annals of Pure and Applied Logic **161** (2009), pp. 349–367.

[8] Dawar, A. and M. Otto, *Modal characterisation theorems over special classes of frames*, Annals of Pure and Applied Logic **161** (2009), pp. 1–42, extended journal version LICS 2005 paper.

[9] de Jongh, D. and F. Yang, *Jankov's theorems for intermediate logics in the setting of universal models*, in: *Logic, Language, and Computation - 8th International Tbilisi Symposium on Logic, Language, and Computation, TbiLLC 2009, Bakuriani, Georgia. Revised Selected Papers*, 2009, pp. 53–76.

[10] Fernández-Duque, D., *Non-deterministic semantics for dynamic topological logic*, Annals of Pure and Applied Logic **157** (2009), pp. 110–121.

[11] Fernández-Duque, D., *On the modal definability of simulability by finite transitive models*, Studia Logica **98** (2011), pp. 347–373.

[12] Fernández-Duque, D., *Tangled modal logic for spatial reasoning*, in: T. Walsh, editor, *Proceedings of IJCAI*, 2011, pp. 857–862.

[13] Fernández-Duque, D., *A sound and complete axiomatization for dynamic topological logic*, Journal of Symbolic Logic **77** (2012), pp. 947–969.

[14] Fernández-Duque, D., *Non-finite axiomatizability of dynamic topological logic*, ACM Transactions on Computational Logic **15** (2014), pp. 4:1–4:18.

[15] Fernández-Duque, D., *The intuitionistic temporal logic of dynamical systems*, arXiv **1611.06929** (2016).

[16] Fischer Servi, G., *Axiomatisations for some intuitionistic modal logics*, in: *Rendiconti del Seminario Matematico* (1984), pp. 179–194.

[17] Gabelaia, D., A. Kurucz, F. Wolter and M. Zakharyaschev, *Non-primitive recursive decidability of products of modal logics with expanding domains*, Annals of Pure and Applied Logic **142** (2006), pp. 245–268.

[18] Goldblatt, R. and I. Hodkinson, *Spatial logic of tangled closure operators and modal mu-calculus*, Annals of Pure and Applied Logic **168** (2017), pp. 1032–1090.

[19] Goldblatt, R. and I. M. Hodkinson, *The finite model property for logics with the tangle modality*, Studia Logica **106** (2018), pp. 131–166.

[20] Kamide, N. and H. Wansing, *Combining linear-time temporal logic with constructiveness and paraconsistency*, Journal of Applied Logic **8** (2010), pp. 33–61.

[21] Kojima, K. and A. Igarashi, *Constructive linear-time temporal logic: Proof systems and Kripke semantics*, Information and Computation **209** (2011), pp. 1491–1503.

[22] Kremer, P. and G. Mints, *Dynamic topological logic*, Annals of Pure and Applied Logic **131** (2005), pp. 133–158.

[23] Kurucz, A., F. Wolter, M. Zakharyaschev and D. M. Gabbay, "Many-Dimensional Modal Logics: Theory and Applications", Volume 148 (Studies in Logic and the Foundations of Mathematics), North Holland, 2003.

[24] Lichtenstein, O. and A. Pnueli, *Propositional temporal logics: Decidability and completeness*, Logic Jounal of the IGPL **8** (2000), pp. 55–85.

[25] Maier, P., *Intuitionistic LTL and a new characterization of safety and liveness*, in: J. Marcinkowski and A. Tarlecki, editors, *18th EACSL Annual Conference on Computer Science Logic (CSL)* (2004), pp. 295–309.

[26] Mints, G., "A Short Introduction to Intuitionistic Logic," University Series in Mathematics, Springer, 2000.

[27] Nishimura, H., *Semantical analysis of constructive PDL*, Publications of the Research Institute for Mathematical Sciences, Kyoto University **18** (1982), pp. 427–438.

[28] Simpson, A. K., "The Proof Theory and Semantics of Intuitionistic Modal Logic," Ph.D. thesis, University of Edinburgh, UK (1994).

On the Logics with Propositional Quantifiers Extending S5Π

Yifeng Ding [1]

Group in Logic and the Methodology of Science
University of California, Berkeley

Abstract

Scroggs's theorem on the extensions of S5 is an early landmark in the modern mathematical studies of modal logics. From it, we know that the lattice of normal extensions of S5 is isomorphic to the inverse order of the natural numbers with infinity and that all extensions of S5 are in fact normal. In this paper, we consider extending Scroggs's theorem to modal logics with propositional quantifiers governed by the axioms and rules analogous to the usual ones for ordinary quantifiers. We call them Π-logics. Taking S5Π, the smallest normal Π-logic extending S5, as the natural counterpart to S5 in Scroggs's theorem, we show that all normal Π-logics extending S5Π are complete with respect to their complete simple S5 algebras, that they form a lattice that is isomorphic to the lattice of the open sets of the disjoint union of two copies of the one-point compactification of $\mathbb{N}$, that they have arbitrarily high Turing-degrees, and that there are non-normal Π-logics extending S5Π.

Keywords: Propositional quantifiers, Scroggs's theorem, lattice of modal logics, algebraic semantics.

1 Introduction

In this paper, we study the modal logics with propositional quantifiers extending the well-studied modal logic S5. Modal logics with propositional quantifiers have been of considerable interest to many modal logicians since their appearances in Fine's dissertation [9] and an early paper by Bull [6]. However, much of the interest is devoted to a few particular systems (e.g., [19,18,4,2,5]) and the expressive power under Kripke semantics (e.g., [7,21,20,16,11,1,3]), and there is an obvious lack of general study of classes of such logics. An exemplary early general study of propositional modal logics is found in Scroggs's famous 1959 paper [22], and it is our intention here to extend it to modal logics with propositional quantifiers.

To this end, we must first define, in general, what is a modal logic with propositional quantifiers. Since we consider here only logics with one modal operator, the language $\mathcal{L}\Pi$ defined below suffices.

[1] Email: yf.ding@berkeley.edu

Definition 1.1 Let $\mathcal{L}\Pi$ be the language with the following grammar

$$\varphi ::= p \mid \top \mid \neg\varphi \mid (\varphi \wedge \varphi) \mid \Box\varphi \mid \forall p\varphi$$

where $p \in \mathsf{Prop}$, a countably infinite set of propositional *variables.*[2] Other Boolean connectives, $\bot$, and $\Diamond$ are defined as usual.

As is common in the general study of modal logics, we take a modal logic with propositional quantifiers to be a set of formulas satisfying certain closure conditions, which represent the necessary axioms and rules for connectives with fixed meaning. There are many readings of the propositionally quantified sentence $\forall p\varphi$, which result in different axioms and semantics (see [10] for example), but here we take the most straightforward reading: "no matter what proposition p expresses, φ." From a purely logical point of view, this reading should warrant the following widely accepted principles, which we call the Π-*principles*:

- All instances of the universal distribution axiom schema: $\forall p(\varphi \to \psi) \to (\forall p\varphi \to \forall p\psi)$.
- All instances of the universal instantiation axiom schema: $\forall p\varphi \to \varphi^p_\psi$ where ψ is substitutable for p in φ, and φ^p_ψ is the result of this substitution.
- All instances of the vacuous quantification axiom schema: $\varphi \to \forall p\varphi$ where p is not free in φ.
- Universalization rule: if φ is derivable, then $\forall p\varphi$ is derivable.

Then the modal logics with propositional quantifiers, which we call Π-logics in accordance with [6] and most recently [15], can now be defined.

Definition 1.2 A Π-*logic* is a set Λ of formulas in $\mathcal{L}\Pi$ such that Λ contains all instances of propositional tautologies and axioms in the Π-principles, and is closed under *modus ponens* and the only rule, universalization, in the Π-principles.

A *normal* Π-logic Λ is a Π-logic that contains the K axiom and is further closed under necessitation: if $\varphi \in \Lambda$, then $\Box\varphi \in \Lambda$.

For any normal modal logic L in the usual basic modal language, let $\mathsf{L}\Pi$ be the smallest (in terms of inclusion) normal Π-logic containing L.

Then, for example, S5Π is the smallest normal Π-logic extending S5, and KΠ is the smallest normal Π-logic extending K, which is just the smallest normal Π-logic.

Following Scroggs, we address the following questions in this paper regarding the Π-logics extending S5Π, the set of which we call NextΠ(S5Π).

General completeness of logics in NextΠ(S5Π) It is well known that S5Π is incomplete with respect to its Kripke frames if propositional quantifiers can vary the valuation of propositional variables to any set of worlds. This was

[2] In contrast, $\top$ is a propositional *constant.* Later we will have another propositional constant.

observed by Fine already in [9] and is in stark contrast to the situation without propositional quantifiers: as is shown by Scroggs, all modal logics in the basic modal language extending S5 are complete with their finite Kripke frames with a totally connected relation. However, Scroggs's proof is algebraic in spirit, and indeed, an algebraic semantics for $\mathcal{L}\Pi$ based on modal algebras is more natural for the normal Π-logics, given our straightforward reading of $\forall p\varphi$. Algebraically, $\forall p\varphi$ is interpreted as the meet (greatest lower bound) of all possible semantic values of φ when we only vary the valuation of p. In short, $\forall p\varphi$ expresses an arbitrary meet. Dually, $\exists p\varphi$ expresses an arbitrary join. For this to work, however, we need the modal algebras to be complete in the sense that for any set of elements in the algebra, the meet and join of this set exist. We will show that all logics in $\mathsf{Next}\Pi(\mathsf{S5}\Pi)$ are complete with respect to their complete simple S5 algebras, to be defined later.

The lattice structure of $\mathsf{Next}\Pi(\mathsf{S5}\Pi)$ From the general completeness for logics in $\mathsf{Next}\Pi(\mathsf{S5}\Pi)$, the lattice structure of $\mathsf{Next}\Pi(\mathsf{S5}\Pi)$ can be reduced to the lattice structure of classes of algebras defined by logics in $\mathsf{Next}\Pi(\mathsf{S5}\Pi)$. We will show from this that the logics in $\mathsf{Next}\Pi(\mathsf{S5}\Pi)$ correspond to the closed sets of a Stone space $\mathcal{S}$, which is homeomorphic to the disjoint union of two copies of the one-point compactification of $\mathbb{N}$ with the natural order topology. Then the lattice $\langle \mathsf{Next}\Pi(\mathsf{S5}\Pi), \subseteq \rangle$ is isomorphic to the lattice of open sets of $\mathcal{S}$ ordered by inclusion.

The computability of logics in $\mathsf{Next}\Pi(\mathsf{S5}\Pi)$ From the correspondence between the logics and the closed sets, we also obtain that there are logics in $\mathsf{Next}\Pi(\mathsf{S5}\Pi)$ of arbitrarily high Turing-degree. While it is known that many natural modal logics with propositional quantifiers are of very high complexity [10,16], this shows that we may still need to face the problem even above S5.

The non-normal Π-logics extending $\mathsf{S5}\Pi$ We will also show that there are many non-normal Π-logics extending $\mathsf{S5}\Pi$, contrary to the situation in the basic modal language, where all modal logics extending S5 are normal. However, we leave a complete study of the non-normal Π-logics extending $\mathsf{S5}\Pi$ to future work.

The plan to address these questions is as follows. In § 2, we present the semantics for $\mathcal{L}\Pi$ and collect the necessary results already appearing in [9] and more recently in [15]. In § 3, we show that, in terms of validity or theoremhood, every formula in $\mathcal{L}\Pi$ is equivalent to a Boolean combination of a few simple formulas. This serves as a good preparation for § 4, where we construct a topological space $\mathcal{S}$ based on all complete simple S5 algebras, which encodes what classes of algebras are definable in terms of validity by $\mathcal{L}\Pi$. Crucially, $\mathcal{S}$ is a Stone space. In § 5, we prove all the main results, which make essential use of the fact that $\mathcal{S}$ is a Stone space and, in particular, that $\mathcal{S}$ is compact. This allows us to prove completeness without using the usual Lindenbaum algebra and quotient construction, though we need to rely on the already proven completeness of $\mathsf{S5}\Pi$. Finally, we conclude with related open problems in § 6.

2 Preliminaries

Recall that a modal algebra is a pair $\langle B, \Box\rangle$ where B is a Boolean algebra and $\Box$ is a unary operator on B satisfying $\Box 1 = 1$ and $\Box(a \wedge b) = \Box a \wedge \Box b$ for any $a, b \in B$. In most cases, we will conflate the notation of an algebra and its carrier set, and we will take $\neg, \wedge, \vee, \Box$ to be the complement, meet, join, and modal operators in modal algebra, despite that they are also in our formal language $\mathcal{L}\Pi$. The usual abbreviations also apply to operations on modal algebras, including $\Diamond a := \neg\Box\neg a$ for all $a \in B$. When confusion may arise, we will use $\neg_B, \wedge_B, \vee_B, \Box_B$ for the operators in a modal algebra B. A modal algebra B is *complete* when its Boolean part is a complete Boolean algebra. Then the semantics for $\mathcal{L}\Pi$ can be defined as follows.

Definition 2.1 For any modal algebra B, a *valuation* V on B is a function from Prop to B. When B is complete, any such valuation can then be extended to a $\mathcal{L}\Pi$-*valuation* $\widehat{V}$ from $\mathcal{L}\Pi$ to B defined recursively by:

(i) $\widehat{V}(p) = V(p)$ for all $p \in \mathsf{Prop}$;

(ii) $\widehat{V}(\top) = 1$; $\widehat{V}(\neg\varphi) = \neg\widehat{V}(\varphi)$; $\widehat{V}(\varphi \wedge \psi) = \widehat{V}(\varphi) \wedge \widehat{V}(\psi)$; $\widehat{V}(\Box\varphi) = \Box\widehat{V}(\varphi)$;

(iii) $\widehat{V}(\forall p\varphi) = \bigwedge\{\widehat{V'}(\varphi) \mid V' : \mathsf{Prop} \to B, V' \sim_p V\}$, where we define $V' \sim_p V$ by $V'(q) = V(q)$ for any $q \in \mathsf{Prop} \setminus \{p\}$.

A formula $\phi \in \mathcal{L}\Pi$ is *valid* on a complete modal algebra B, written as $B \vDash \phi$, if for all valuations V on B, $\widehat{V}(\phi) = 1$.

Since we are only interested in Π-logics extending S5Π, we only need modal algebras validating S5. In fact, we only need a very special class of such modal algebras called simple S5 algebras.

Definition 2.2 A *simple S5 algebra* is pair $\langle B, \Box\rangle$ where B is a non-trivial Boolean algebra and $\Box$ is the unary function on B defined for $a \in B$ by

$$\Box a = \begin{cases} 1 & \text{if } a = 1 \\ 0 & \text{otherwise.} \end{cases} \quad \text{Then } \Diamond a = \begin{cases} 1 & \text{if } a \neq 0 \\ 0 & \text{otherwise.} \end{cases}$$

Let us denote the class of all simple S5 algebras by $\mathsf{sS5A}$ and the class of all complete simple S5 algebras by $\mathsf{csS5A}$.

modal algebras validating S5 are also known as *monadic algebra* (see [14,13]). However, in the context of monadic algebras, $\Diamond$ and $\Box$ operators are usually denoted by $\exists$ and $\forall$, which we need for propositional quantifiers. We also remark that our simple S5 algebras are indeed simple in its general algebraic sense: they have no non-trivial congruence relation. These algebras are also known as *Henle algebras*.

To formulate completeness with respect to $\mathsf{csS5A}$, it is natural to use the following Galois connection:

Definition 2.3 For any class $\mathsf{C} \subseteq \mathsf{csS5A}$, define $\mathrm{Log}(\mathsf{C}) = \{\varphi \in \mathcal{L}\Pi \mid \forall B \in \mathsf{C}, B \vDash \varphi\}$. We also write $\mathrm{Log}(\{B\})$ as simply $\mathrm{Log}(B)$ for any $B \in \mathsf{csS5A}$.

Conversely, for any set of formulas $\Gamma \subseteq \mathcal{L}\Pi$, define $\mathrm{Alg}(\Gamma) = \{B \in \mathsf{csS5A} \mid \forall\varphi \in \Gamma, B \vDash \varphi\}$. Similarly, $\mathrm{Alg}(\varphi)$ abbreviates $\mathrm{Alg}(\{\varphi\})$.

This finishes the semantics for $\mathcal{L}\Pi$, and now we march into expanding $\mathcal{L}\Pi$, as Fine did in [9], to $\mathcal{L}\Pi\mathrm{Mg}$. This is instrumental for formulating the quantifier elimination on which completeness for $\mathsf{S5\Pi}$ alone in [9,15] depends, and all our new results will also need it. In the following, let $\mathbb{N}_+$ be the set of positive natural numbers, and $\mathbb{N}^\infty$ be the set of natural numbers plus an infinite element ∞. Also, we will use $\mathbb{N}_+^\infty$, which has ∞ but not 0.

Definition 2.4 ([9]) Define $\mathcal{L}\Pi\mathrm{Mg}$ by extending the grammar for $\mathcal{L}\Pi$ with a propositional *constant* g (not in Prop) and countably many new unary operators $\{\mathrm{M}_i \mid i \in \mathbb{N}_+\}$. Then, define $\mathcal{L}\mathrm{Mg}$ as the quantifier free fragment of $\mathcal{L}\Pi\mathrm{Mg}$, which has the following grammar:

$$\varphi ::= p \mid \top \mid \mathrm{g} \mid \Box\varphi \mid \mathrm{M}_i\varphi \mid \neg\varphi \mid (\varphi \wedge \varphi)$$

with $p \in \mathsf{Prop}$.

For future convenience, we refer to the elements in $\mathsf{Prop} \cup \{\top, g\}$ in general as propositional *letters*, and we define $\mathrm{md}(\varphi)$ to be the *modal depth* of φ defined as usual, with M_i's and $\Box$ all treated as modal operators, $\mathrm{free}(\varphi)$ to be the set of free propositional variables in φ, and the *quantificational depth* of φ the maximal length of any chain of nested quantifers in φ, analogous to the usual definition in first-order logics.

Let us also define as in [9] for every $\alpha \in \mathcal{L}\Pi\mathrm{Mg}$ an important formula $\mathrm{atom}(\alpha)$:

$$\mathrm{atom}(\alpha) := \Diamond\alpha \wedge \forall q(\Box(q \to \alpha) \vee \Box(q \to \neg\alpha)) \tag{1}$$

where $q \in \mathsf{Prop}$ does not occur in α. To fix this choice, we assume that there is an enumeration of Prop fixed from the outset. Then whenever we need fresh propositional letters in a definition, the definition picks out the first available propositional variable.

Here g is intended to express the proposition that some atomic proposition is true, and $\mathrm{M}_i\varphi$ the proposition that φ is entailed by at least i many atomic propositions. Hence, g should be evaluated to the join of the atoms in a modal algebra. But this requires that the join exists. Let us call a modal algebra *separable* if the join of its atoms exists. Then we can give the semantics for $\mathcal{L}\mathrm{Mg}$ and $\mathcal{L}\Pi\mathrm{Mg}$ on appropriate modal algebras.

Definition 2.5 For any separable modal algebra B, define g (or g_B when ambiguity arises) as the join of all atoms of B, and M_i an operator on B as follows:

$$M_i a = \begin{cases} 1 & \text{if there are at least } i \text{ distinct atoms below } a \\ 0 & \text{otherwise} \end{cases}$$

for $i \in \mathbb{N}_+$.

Then, any valuation V on B can be recursively extended to an $\mathcal{L}$Mg-*valuation* $\widehat{V}$ from $\mathcal{L}$Mg to B by the same clauses for Boolean connectives and $\Box$ as in Definition 2.1, plus the following two clauses:

(i) $\widehat{V}(\mathrm{g}) = g_B$

(ii) $\widehat{V}(\mathrm{M}_i\varphi) = M_i\widehat{V}(\varphi)$.

If B is actually complete, define the $\mathcal{L}\Pi$Mg-*valuation* extending $\widehat{V}$ by combining the clauses above and in Definition 2.1. It is not hard to see that the $\mathcal{L}$Mg-valuation, $\mathcal{L}\Pi$-valuation, and $\mathcal{L}\Pi$Mg-valuation extending V are compatible. Hence by $\widehat{V}$, we always mean the defined valuation with the maximal domain. This will be either an $\mathcal{L}$Mg-valuation or an $\mathcal{L}\Pi$Mg-valuation, depending on whether the codomain of V is merely separable or is complete. We also extend the definition of validity and also the Alg operator to formulas in $\mathcal{L}\Pi$Mg in the obvious way.

Regarding atom(α), it is intended to express the proposition that α expresses an atomic proposition. Its definition does not always achieve this intended meaning, but assuming that it is interpreted on complete simple S5 algebras, this definition indeed singles out atoms in simple S5 algebras. The proof of this can be found in [15].

Lemma 2.6 *For any $\alpha \in \mathcal{L}\Pi$Mg, any complete simple S5 algebra B, and any valuation V on B, we have*

$$\widehat{V}(\mathrm{atom}(\alpha)) = \begin{cases} 1 & \textit{if } \widehat{V}(\alpha) \textit{ is an atom in } B \\ 0 & \textit{otherwise.} \end{cases}$$

Now to the logics in the language $\mathcal{L}\Pi$Mg. They are obtained by adding two axiom schemata to define the new operators M_i's and g by formulas in $\mathcal{L}\Pi$.

Definition 2.7 For any normal Π-logic Λ, define ΛMg as the smallest normal Π-logic (with formula variables in the schemata and rules of the definition now ranging over $\mathcal{L}\Pi$Mg) that contains the following two axiom schemata for each $n \in \mathbb{N}_+$:

$$\mathrm{M}_n\phi \leftrightarrow \exists q_1 \cdots \exists q_n\Big(\bigwedge_{1\leqslant i<j\leqslant n} \Box(q_i \to \neg q_j) \wedge \bigwedge_{1\leqslant i\leqslant n} (\mathrm{atom}(q_i) \wedge \Box(q_i \to \varphi))\Big) \quad (\mathsf{M})$$

$$\mathrm{g} \leftrightarrow \exists q(q \wedge \mathrm{atom}(q)) \quad (\mathsf{g})$$

where $q_1, \cdots q_n \in \mathsf{Prop}$ do not occur in φ, and $q \in \mathsf{Prop}$.

With the help of Lemma 2.6, it is not hard to directly observe that both (M) and (g) are sound. In fact, we have the following theorem, mostly by Fine and Holliday, on which our new results depend.

Lemma 2.8 S5ΠMg *is a conservative extension of* S5Π. *Namely,* S5ΠMg $\cap$ $\mathcal{L}\Pi$ = S5Π. *Also,* S5ΠMg *is sound and complete with respect to* csS5A.

Proof. For any $\varphi \in \mathcal{L}\Pi$, if $\varphi \in \mathsf{S5\Pi Mg}$, then we can replace all M_i's and g in its derivation by their definitions in Definition 2.7. The resulting derivation is in $\mathsf{S5\Pi}$. Hence $\varphi \in \mathsf{S5\Pi}$. This shows that $\mathsf{S5\Pi Mg} \cap \mathcal{L}\Pi \subseteq \mathsf{S5\Pi}$. The other direction is trivial. This is observed and first used by Fine in [9].

It is first shown algebraically in [15] that $\mathsf{S5\Pi Mg}$ is sound and that $\mathsf{S5\Pi}$ is sound and complete with respect to $\mathsf{csS5A}$. Now for any $\varphi \in \mathcal{L}\Pi\mathrm{Mg}$ that is valid in $\mathsf{csS5A}$, we can first replace all M_i's and g in φ by their definitions to obtain ψ. Then $\varphi \leftrightarrow \psi \in \mathsf{S5\Pi Mg}$. We know that $\mathsf{S5\Pi Mg}$ is sound on $\mathsf{csS5A}$. So ψ is valid. Then, since $\psi \in \mathcal{L}\Pi$, $\psi \in \mathsf{S5\Pi}$, which means $\psi \in \mathsf{S5\Pi Mg}$. By *modus ponens*, $\varphi \in \mathsf{S5\Pi Mg}$. □

The proof of the completeness of $\mathsf{S5\Pi}$ in [15] relies on a fairly intricate quantifier elimination in $\mathsf{S5\Pi Mg}$ found first by Fine in [9], which says that for any $\varphi \in \mathcal{L}\Pi$, there is a formula $\psi \in \mathcal{L}\mathrm{Mg}$ such that $\phi \leftrightarrow \psi \in \mathsf{S5\Pi Mg}$. We will also make use of this technical result. In fact, ψ can be chosen from a much smaller fragment of $\mathcal{L}\mathrm{Mg}$. Following Fine, we call them model descriptions and define them now.

Definition 2.9 For any $\varphi \in \mathcal{L}\Pi\mathrm{Mg}$, first define the following abbreviations:

$$\mathrm{Q}_0 := \neg\mathrm{M}_1\varphi; \quad \mathrm{Q}_i\varphi := \mathrm{M}_i\varphi \wedge \mathrm{M}_{i+1}\varphi, i \in \mathbb{N}_+; \quad \mathrm{N}\varphi := \Diamond(\neg\mathrm{g} \wedge \varphi).$$

For any finite subset $P \subseteq \mathsf{Prop}$, a *state description* s over P is a conjunction of literals from P in which every $p \in P$ occurs. We follow the convention that an empty conjunction is $\top$ and an empty disjunction is $\bot$. Let 2^P be the set of all state descriptions over P. Then, a *model description of degree* n *over* P is a conjunction of:

(i) either g or $\neg$g;

(ii) a state description $a \in 2^P$;

(iii) for each $s \in 2^P$, either $\mathrm{M}_n s$ or some $\mathrm{Q}_n s$ for some $i < n$;

(iv) for each $s \in 2^P$, either $\mathrm{N}s$ or $\mathrm{N}\neg s$.

Lemma 2.10 ([9], § 4.2) *For any* $\varphi \in \mathcal{L}\Pi$, *there exists a* $qf(\varphi) \in \mathcal{L}\mathrm{Mg}$ *such that* $\varphi \leftrightarrow qf(\varphi) \in \mathsf{S5\Pi Mg}$. *Moreover,* $qf(\varphi)$ *is a disjunction of model descriptions over* $\mathrm{free}(\varphi)$ *of degree* 2^n *where* n *is the quantification degree of* φ.

For the construction of qf, the reader can also see the appendix of [15].

3 Semantical and syntactical reduction

In this section, we show that for any $\varphi \in \mathcal{L}\Pi$, any complete simple S5 algebra A, and any $\Lambda \in \mathsf{Next\Pi(S5\Pi)}$, we can construct a formula, which we call $basic(\varphi)$, such that:

- $\varphi \in \Lambda$ iff $basic(\varphi) \in \Lambda\mathsf{Mg}$;
- $A \models \varphi$ iff $A \models basic(\varphi)$;
- $basic(\varphi)$ is a Boolean combination of $\Diamond\neg g$ and $\mathrm{M}_i\top$ for $i \in \mathbb{N}_+$.

To facilitate the proof, let us first define a number of useful fragments of $\mathcal{L}\Pi$.

Definition 3.1 Recall that $\mathcal{L}$Mg is the quantifier free fragment of $\mathcal{L}\Pi$Mg. Now, Define the following propositional-variable-free fragments of $\mathcal{L}$Mg where i ranges in $\mathbb{N}_+$:

$$\begin{aligned} \mathcal{S}\mathrm{Mg} \ni \varphi &::= \top \mid \mathrm{g} \mid \neg\varphi \mid (\varphi \wedge \varphi) \mid \Box\varphi \mid \mathrm{M}_i\varphi \\ \mathcal{S}_{\leqslant 1}\mathrm{Mg} \ni \varphi &::= \top \mid \mathrm{g} \mid \Diamond\neg\mathrm{g} \mid \mathrm{M}_i\top \mid \neg\varphi \mid (\varphi \wedge \varphi) \\ \mathcal{S}\mathrm{Basic} \ni \varphi &::= \top \mid \Diamond\neg\mathrm{g} \mid \mathrm{M}_i\top \mid \neg\varphi \mid (\varphi \wedge \varphi)\ . \end{aligned}$$

The $\mathcal{S}$ instead of $\mathcal{L}$ in their names means "$\mathcal{S}$entence." It is not hard to see that $\mathcal{S}$Mg collects all propositional-variable-free formulas in $\mathcal{L}$Mg and that $\mathcal{S}_{\leqslant 1}$Mg collects some formulas with modal depth at most 1 in $\mathcal{S}$Mg, which are enough for our purposes.

For any $\varphi \in \mathcal{L}\Pi$, we will construct $basic(\varphi)$ as the following with u and $comp$ to be defined:

$$basic(\varphi) = comp(\Box qf(u(\varphi))).$$

Here, $u(\varphi)$ is the universal closure of φ, which is defined as $\forall p_1 \forall p_2 \cdots \forall p_n \varphi$ where $p_1, p_2, \cdots, p_n$ enumerate the free propositional letters in φ. And recall that qf returns the result of quantifier elimination. Since $u(\varphi)$ has no free propositional variable, according to Lemma 2.10, $qf(u(\varphi)) \in \mathcal{S}$Mg is a disjunction of model descriptions of some finite degree over $\varnothing$. From Definition 2.9, we can see that all model descriptions of degree n over $\varnothing$ are of the form

$$\pm g \wedge \mathrm{M}_i\top \wedge \neg\mathrm{M}_{i+1}\top \wedge \pm\Diamond\neg\mathrm{g} \quad \text{or} \quad \pm\, g \wedge \mathrm{M}_n\top \wedge \pm\Diamond\neg\mathrm{g}$$

where $i < n$, $\pm$ stands for $\neg$ or nothing, and $\mathrm{M}_0\top$ for $\top$. In short, $qf(u(\varphi))$ is a Boolean combination of g, $\mathrm{M}_i\top$ for $i \in \mathbb{N}_+$, and $\Diamond\neg g$, and hence is in $\mathcal{S}_{\leqslant 1}$Mg.

Now we construct $comp$ as a function that simplifies a boxed modal description over $\varnothing$ to a formula in $\mathcal{S}$Basic in a provably equivalent way.

Lemma 3.2 *For any ψ a disjunction of model descriptions of degree n over $\varnothing$, there is a formula $comp(\Box\psi) \in \mathcal{S}$Basic such that $\Box\psi \leftrightarrow comp(\Box\psi) \in$ S5ΠMg.*

Proof. Let pos be the number of model descriptions in ψ where g appears positively and neg the number of model descriptions in ψ where g appears negatively. For any $1 \leqslant i \leqslant pos$, let α_i be the result of deleting the conjunct g in the ith model description in ψ where g appears positively, and similarly define β_i for $1 \leqslant i \leqslant neg$, where we need to delete the $\neg$g conjunct.

Let $\alpha = \bigvee_{1 \leqslant i \leqslant pos} \alpha_i$ and $\beta = \bigvee_{1 \leqslant i \leqslant neg} \beta_i$, which are now Boolean combinations of $\mathrm{M}_i\top$'s and $\Diamond\neg$g. Then obviously $\psi \leftrightarrow ((\mathrm{g} \wedge \alpha) \vee (\neg\mathrm{g} \wedge \beta))$ and $\Box\psi \leftrightarrow \Box((\mathrm{g} \wedge \alpha) \vee (\neg\mathrm{g} \wedge \beta))$ are in S5ΠMg using propositional tautologies and normality. Let us write for any $\varphi_1, \varphi_2 \in \mathcal{L}\Pi$Mg, $\varphi_1 \equiv_{\mathsf{S5\Pi Mg}} \varphi_2$ iff

$\varphi_1 \leftrightarrow \varphi_2 \in \mathsf{S5\Pi Mg}$. Then we have

$$\begin{aligned}
\Box\psi &\equiv_{\mathsf{S5\Pi Mg}} \Box((\mathrm{g} \wedge \alpha) \vee (\neg\mathrm{g} \wedge \beta)) &(2)\\
&\equiv_{\mathsf{S5\Pi Mg}} \Box((\mathrm{g} \vee \neg\mathrm{g}) \wedge (\mathrm{g} \vee \beta) \wedge (\neg\mathrm{g} \vee \alpha) \wedge (\alpha \vee \beta)) &(3)\\
&\equiv_{\mathsf{S5\Pi Mg}} \Box(\mathrm{g} \vee \neg\mathrm{g}) \wedge \Box(\mathrm{g} \vee \beta) \wedge \Box(\neg\mathrm{g} \vee \alpha) \wedge \Box(\alpha \vee \beta) &(4)\\
&\equiv_{\mathsf{S5\Pi Mg}} \Box(\mathrm{g} \vee \Box\beta) \wedge \Box(\neg\mathrm{g} \vee \Box\alpha) \wedge \Box(\Box\alpha \vee \Box\beta) &(5)\\
&\equiv_{\mathsf{S5\Pi Mg}} (\Box\mathrm{g} \vee \Box\beta) \wedge (\Box\neg\mathrm{g} \vee \Box\alpha) \wedge (\Box\alpha \vee \Box\beta) &(6)\\
&\equiv_{\mathsf{S5\Pi Mg}} (\Box\mathrm{g} \vee \beta) \wedge (\Box\neg\mathrm{g} \vee \alpha) \wedge (\alpha \vee \beta) &(7)\\
&\equiv_{\mathsf{S5\Pi Mg}} (\neg\Diamond\neg\mathrm{g} \vee \beta) \wedge (\neg\mathrm{M}_1\top \vee \alpha) \wedge (\alpha \vee \beta). &(8)
\end{aligned}$$

In the above chain of provable equivalences, (5), (7), and (8) require more explanation. Note that in $\mathsf{S5\Pi Mg}$, we have $\Box\forall p\varphi \equiv_{\mathsf{S5\Pi Mg}} \forall p\Box\varphi$, and dually, $\Diamond\exists p\varphi \equiv_{\mathsf{S5\Pi Mg}} \exists p\Diamond\varphi$. With all the axioms in $\mathsf{S5}$, and also the axiom (M) defined in Definition 2.7, $\mathrm{M}_i\top \equiv_{\mathsf{S5\Pi Mg}} \Box\mathrm{M}_i\top$. Then, $\alpha \equiv_{\mathsf{S5\Pi Mg}} \Box\alpha$ and $\beta \equiv_{\mathsf{S5\Pi Mg}} \Box\beta$, since α and β are Boolean combinations of $\mathrm{M}_i\top$ and $\Diamond\neg g$. Thus we have (5) and (7). For (8), some manipulation of axioms (g) and (M) gives us $\Box\neg g \equiv_{\mathsf{S5\Pi Mg}} \neg\mathrm{M}_1\top$. The rest of the equivalences are standard $\mathsf{S5}$ reasoning. We can then define $basic(\Box\psi)$ to be the right hand side in (8), which is provably equivalent to $\Box\psi$ and is in $\mathcal{S}$Basic. □

Now we show that for any $\varphi \in \mathcal{L}\Pi$, $basic(\varphi)$ has the required properties.

Lemma 3.3 *For any $\varphi \in \mathcal{L}\Pi$ and $\Lambda \in \mathsf{Next\Pi(S5\Pi)}$, $\varphi \in \Lambda$ iff $basic(\varphi) \in \Lambda\mathsf{Mg}$.*

Proof. Since $\Lambda\mathsf{Mg}$ is a conservative extension of Λ (by Lemma 2.8), $\varphi \in \Lambda$ iff $\varphi \subset \Lambda\mathsf{Mg}$. Using universalization and also universal instantiation, $\varphi \in \Lambda\mathsf{Mg}$ iff $u(\varphi) \in \Lambda\mathsf{Mg}$. Since $\Lambda \in \mathsf{Next\Pi(S5\Pi)}$, $\Lambda\mathsf{Mg}$ extends $\mathsf{S5\Pi Mg}$. Together with the quantifier elimination result in Lemma 2.10, $qf(u(\varphi)) \leftrightarrow u(\varphi) \in \Lambda\mathsf{Mg}$. Thus $u(\varphi) \in \Lambda\mathsf{Mg}$ iff $qf(u(\varphi)) \in \Lambda\mathsf{Mg}$. By necessitation and also the T axiom derivable in $\mathsf{S5}$, $qf(u(\varphi)) \in \Lambda\mathsf{Mg}$ iff $\Box qf(u(\varphi)) \in \Lambda\mathsf{Mg}$. Finally, due to Lemma 3.2 and the fact that $qf(u(\varphi))$ is indeed a disjunction of model descriptions of some finite degree over $\varnothing$, we have $basic(\varphi) = comp(\Box qf(u(\varphi))) \leftrightarrow \Box qf(u(\varphi)) \in \Lambda\mathsf{Mg}$. Thus $basic(\varphi) \in \Lambda\mathsf{Mg}$ iff $\Box qf(u(\varphi)) \in \Lambda\mathsf{Mg}$. Connecting all the equivalences, $\varphi \in \Lambda$ iff $basic(\varphi) \in \Lambda\mathsf{Mg}$. □

On the semantical side, we first make an easy observation.

Lemma 3.4 *For any complete (resp. separable) modal algebra B and any $\varphi \in \mathcal{L}\Pi\mathrm{Mg}$ (resp. $\mathcal{L}\mathrm{Mg}$) such that $\mathrm{free}(\varphi) = \varnothing$, for any two valuations V and V' on B, $\widehat{V}(\varphi) = \widehat{V'}(\varphi)$.*

Due to this observation, we define for any separable modal algebra B, a fixed trivial valuation V_B which maps every $p \in \mathsf{Prop}$ to 1_B. Then $B \models \varphi$ iff $\widehat{V_B}(\varphi) = 1_B$ for any $\varphi \in \mathcal{L}\mathrm{Mg}$ (or $\mathcal{L}\Pi\mathrm{Mg}$ when B is complete) such that $\mathrm{free}(\varphi) = \varnothing$.

Then we can prove the semantical requirement for *basic*.

Lemma 3.5 *For any $\varphi \in \mathcal{L}\Pi$ and any complete simple S5 algebra B, $B \models \varphi$ iff $B \models basic(\varphi)$.*

Proof. First consider the following chain of equivalences:

$$
\begin{array}{llr}
B \vDash \varphi & \text{iff } B \vDash u(\varphi) & \text{by the definition of validity} \\
& \text{iff } \widehat{V_B}(u(\varphi)) = 1 & \text{by the definition of validity} \\
& \text{iff } \widehat{V_B}(qf(u(\varphi))) = 1 & B \text{ validates } \mathsf{S5\Pi Mg}, \text{ and} \\
& & u(\varphi) \leftrightarrow qf(u(\varphi)) \in \mathsf{S5\Pi Mg} \\
& \text{iff } \widehat{V_B}(\Box(qf(u(\varphi)))) = 1 & \Box 1 = 1 \text{ and } \Box a \neq 1 \text{ if } a \neq 1 \\
& \text{iff } \widehat{V_B}(comp(\Box(qf(u(\varphi))))) = 1 & B \text{ validates } \mathsf{S5\Pi Mg}, \text{ and} \\
& & \text{Lemma 3.2}
\end{array}
$$

Note that $basic(\varphi)$ must have no free propositional variable, because $basic(\varphi) = comp(\Box qf(u(\varphi)))$, $\text{free}(u(\varphi)) = \varnothing$, and neither qf nor $comp$ introduces new free variables. Then by the observation in the previous lemma, $B \vDash basic(\varphi)$ iff $\widehat{V_B}(basic(\varphi)) = 1$. Connecting all the equivalences, $B \vDash \varphi$ iff $B \vDash basic(\varphi)$. □

4 Types and type space

In the last section, we have shown that many formulas are equivalent in terms of validity or theoremhood. In this section, we do the same to the algebras: many algebras are equivalent in terms of the formulas in $\mathcal{L}\Pi$ they validate. This equivalence relation can in fact bring csS5A from a class to a countable set, which we will call the type space. Then, to study the classes of algebras definable by formulas in $\mathcal{L}\Pi$, we can just study the sets of types of the algebras in those classes. This in turn gives us a topology on the type space. Now we start with the definition of the types, which is in fact a much simplified version of the famous Tarski invariant for Boolean algebras (see § 5.5 in [8]), due to the completenss of the algebras we are interested in.

Definition 4.1 For any complete simple S5 algebra B, its *type* $t(B)$ is a pair $\langle t_0(B), t_1(B)\rangle$ where

$$
t_0(B) = \begin{cases} 1 & \text{if } g \neq 1 \\ 0 & \text{if } g = 1, \end{cases} \quad t_1(B) = \begin{cases} i \in \mathbb{N} & \text{if } B \text{ has exactly } i \text{ atoms} \\ \infty & \text{if } B \text{ has infinitely many atoms.} \end{cases} \tag{9}
$$

Recall that g of B is the join of its atoms. Hence, t_0 says whether this algebra contains an atomless part, and t_1 counts the atoms it has. Let S be the set of all types of complete simple S5 algebras, the type space.

Proposition 4.2 $S = (\{0,1\} \times \mathbb{N}^\infty) \backslash \{\langle 0,0\rangle\}$.

Proof. Apparently, $S \subseteq \{0,1\} \times \mathbb{N}^\infty$ as any type is a pair $\langle t_0, t_1\rangle$ where the first component can only be 0 or 1 and the second component can only be a natural number of ∞. Also, if the type of a complete simple S5 algebra A is $\langle 0,0\rangle$, then A has no atom and also no atomless part, which means A is trivial and thus not a complete simple S5 algebra in our Definition 2.5. So we have shown the inclusion of left to right. Now to the other direction. The right-hand-side can be decomposed into three parts:

- $\langle 0, n\rangle$ for $n \in \mathbb{N}_+$. Types of this form can be realized by the Boolean algebra B_n of the powerset of a set of n elements, with $\Box$ defined as in Definition 2.5.
- $\langle 1, 0\rangle$. To realize this type, take the countable free Boolean algebra B. It is well known that B is atomless, but not complete. However, we can take the MacNeille completion B^+ of B, a complete Boolean algebra (unique up to isomorphism) such that B embeds into and that every element of B^+ is a join of images of elements of B. For a construction of this B^+, see [12], Chap. 25. Then B^+ is complete and atomless, as if there is an atom, it must be the image of an atom of B, but B is atomless. Now turn B^+ to a simple S5 algebra by defining $\Box$ as in Definition 2.5. Then B^+ has type $\langle 1, 0\rangle$.
- $\langle 1, n\rangle$ for $n \in \mathbb{N}_+$. Consider the product of B^+ and B_n with $\Box$ again defined as above. It is not hard to see that it has an atomless part: g of this algebra is $\langle 1, 0\rangle$. It also has n many atoms, listed by $\langle 0, a\rangle$ where a range over atoms in B_n. Thus this simple S5 algebra has type $\langle 1, n\rangle$.

Hence we realized all types in the right-hand-side. So the inclusion from right to left is also shown. □

Now we define the equivalences between algebras. Then we will show that types capture this equivalence relation.

Definition 4.3 For any two complete simple S5 algebras A, B, and any $\mathcal{L} \in \{\mathcal{L}\Pi, \mathcal{S}\text{Basic}\}$ we say $A \equiv_{\mathcal{L}} B$ if for any sentence $\phi \in \mathcal{L}$, $A \vDash \varphi$ iff $B \vDash \varphi$.

Lemma 4.4 *For any two complete simple S5 algebras A, B, $A \equiv_{\mathcal{L}\Pi} B$ iff $A \equiv_{\mathcal{S}\text{Basic}} B$.*

Proof. Immediate from Lemma 3.5. □

Lemma 4.5 *For any two complete simple S5 algebras A, B, $t(A) = t(B)$ iff $A \equiv_{\mathcal{S}\text{Basic}} B$. Hence, together with Lemma 4.4, $t(A) = t(B)$ iff $A \equiv_{\mathcal{L}\Pi} B$.*

Proof. Recall that for all $\varphi \in \mathcal{S}\text{Basic}$ and any complete simple S5 algebra A, $A \vDash \varphi$ iff $\widehat{V_A}(\varphi) = 1$, because $\text{free}(\varphi) = \varnothing$. Also notice that for $\varphi = \Diamond\neg g$ or $\text{M}_i\top$ for any $i \in \mathbb{N}_+$, $\widehat{V_A}(\varphi)$ is either 0 or 1. Since $\mathcal{S}\text{Basic}$ consists of all and only the Boolean combinations of these formulas, in fact for any $\varphi \in \mathcal{S}\text{Basic}$, $\widehat{V_A}(\varphi) \in \{0, 1\}$, and in other words, $A \vDash \varphi$ or $A \vDash \neg\varphi$.

Now suppose $t(A) = t(B)$. Then, conflating 1_A and 1_B, and also 0_A and 0_B, we can easily verify that $\widehat{V_A}(\Diamond\neg \text{g}) = \widehat{V_B}(\Diamond\neg \text{g})$ and that for all $i \in \mathbb{N}_+$, $\widehat{V_A}(\text{M}_i\top) = \widehat{V_B}(\text{M}_i\top)$. Then a simple induction propagates these equalities to all $\varphi \in \mathcal{S}\text{Basic}$. Thus we see that if $t(A) = t(B)$, then $A \equiv_{\mathcal{S}\text{Basic}} B$.

On the other hand, if $t(A) \neq t(B)$, then there are two cases:

- $t_1(A) \neq t_1(B)$. In this case, $\Diamond\neg\text{g}$ distinguishes the two algebras.
- $t_2(A) \neq t_2(B)$. Let n be the smaller number among them. Then $n \in \mathbb{N}$, $n + 1 \in \mathbb{N}_+$, and $\neg\text{M}_{n+1}$ distinguishes the two algebras.

Hence, if $t(A) \neq t(B)$, then $A \not\equiv_{\mathcal{S}\text{Basic}} B$. □

Due to this lemma, the function t can be seen as the quotient map from csS5A to csS5A$/\equiv_{\mathcal{L}\Pi}$. This means that the Galois connection between csS5A and $\mathcal{L}\Pi$ by Alg and Log can be reduced to the following Galois connection between S and $\mathcal{L}\Pi$.

Definition 4.6 For any type $s \in S$ and any $\varphi \in \mathcal{L}\Pi$, let us write $s \vDash \varphi$ just in case for any $A \in$ csS5A such that $t(A) = s$, $A \vDash \varphi$.

Then define Type(Γ) for every $\Gamma \subseteq \mathcal{L}\Pi$ as $\{s \in S \mid \forall \varphi \in \Gamma, s \vDash \varphi\}$, with Type($\varphi$) again abbreviating Type($\{\varphi\}$). On the other direction, define Log(T) for any subset T of S as $\{\varphi \in \mathcal{L}\Pi \mid \forall s \in T, s \vDash \varphi\}$, with Log($s$) abbreviating Log($\{s\}$) for any $s \in S$ as well.

Then, we can collect the following easy but useful observations.

Lemma 4.7 *For any* $\Gamma \subseteq \mathcal{L}\Pi$, $\varphi, \psi \in \mathcal{S}$Basic*:*

- $\mathrm{Alg}(\Gamma) = t^{-1}(\mathrm{Type}(\Gamma))$, $t(\mathrm{Alg}(\Gamma)) = \mathrm{Type}(\Gamma)$, *and then* $\mathrm{Log}(\mathrm{Type}(\Gamma)) = \mathrm{Log}(\mathrm{Alg}(\Gamma))$*;*
- $\mathrm{Type}(\Gamma) = \bigcap\{\mathrm{Type}(\varphi) \mid \varphi \in \Gamma\}$*;*
- $\mathrm{Type}(\neg\varphi) = S \setminus \mathrm{Type}(\varphi)$, $\mathrm{Type}(\varphi \wedge \psi) = \mathrm{Type}(\varphi) \cap \mathrm{Type}(\psi)$.

Using this lemma, we can study the following topology that will be important to us for both general completeness and the lattice structure of NextΠ(S5Π).

Definition 4.8 Let $\mathcal{S}$ be the topological space with the type space S as the underlying set and $\{\mathrm{Type}(\varphi) \mid \varphi \in \mathcal{L}\Pi\}$ as basic opens.

Lemma 4.9 *$\mathcal{S}$ is a Stone space, homeomorphic to the disjoint union of two copies of the one-point compactification of $\mathbb{N}$ with the usual order topology.*

Proof. From Lemma 3.5, the basic opens of $\mathcal{S}$ are just sets in $\{\mathrm{Type}(\varphi) \mid \varphi \in \mathcal{S}\mathrm{Basic}\}$. By the third bullet in Lemma 4.7, we know that $\{\mathrm{Type}(\varphi) \mid \varphi \in \mathcal{S}\mathrm{Basic}\}$ is a field of sets on S. Thus $\mathcal{S}$ is zero-dimensional. To see that $\mathcal{S}$ is Hausdorff, take two different $s_1, s_2 \in S$. Recall that S is t(csS5A). So we can find two complete simple S5 algebras B_1 and B_2 such that $t(B_1) = s_1$ and $t(B_2) = s_2$. Then, by Lemma 3.5, $B_1 \not\equiv_{\mathcal{S}\mathrm{Basic}} B_2$. So we can find a formula $\varphi \in \mathcal{S}$Basic such Type(φ) that separates B_1 and B_2. So $\mathcal{S}$ is Hausdorff.

To show that $\mathcal{S}$ is compact, we need a more detailed analysis of $\{\mathrm{Type}(\varphi) \mid \varphi \in \mathcal{S}\mathrm{Basic}\}$. First, note that $\mathrm{Type}(\Diamond\neg g) = \{1\} \times \mathbb{N}^\infty$, $\mathrm{Type}(\neg\Diamond\neg g) = \{0\} \times \mathbb{N}^\infty_+$, and they partition S into two parts. Let us name them by S_1 and S_0 respectively. Hence $\mathcal{S}$ is the disjoint union of $\mathcal{S}_1$ and $\mathcal{S}_0$ defined as the subspaces of S on S_1 and S_0 respectively. So we only need to show that they are both compact. On $\mathcal{S}_1$, the basic clopens are now Boolean combinations of $\mathrm{Type}(\mathrm{M}_i)$ for $i \in \mathbb{N}_+$ and $\mathrm{Type}(\Diamond\neg g)$, all restricted to S_1. But $\mathrm{Type}(\Diamond\neg g) = S_1$. Then the clopens are actually the field of sets on S_1 generated by $\{\mathrm{Type}(\mathrm{M}_n) \cap S_1 \mid n \in \mathbb{N}\} = \{\{\langle 1, i\rangle \mid n \leqslant i \leqslant \infty\} \mid n \in \mathbb{N}\}$. Hence it is not hard to see that $\mathcal{S}_1$ is just (homeomorphic to) the one-point compactification of the order topology on $\mathbb{N}$. The situation for $\mathcal{S}_0$ is almost the same, except that the space is on $\mathbb{N}^\infty_+$. But it is still homeomorphic to the one-point compactification of $\mathbb{N}$. □

5 Main results

Now we are prepared to prove the main results regarding Π-logics extending S5Π. Let us start with the general completeness.

Theorem 5.1 *For any* $\Lambda \in \mathsf{Next\Pi(S5\Pi)}$, $\Lambda = \mathrm{Log}(\mathrm{Alg}(\Lambda))$.

Proof. As is shown in Lemma 4.7, $\mathrm{Log}(\mathrm{Alg}(\Lambda)) = \mathrm{Log}(\mathrm{Type}(\Lambda))$. Also, it is trivial that $\Lambda \subseteq \mathrm{Log}(\mathrm{Type}(\Lambda))$. Hence we just need to show that, for any $\varphi \in \mathcal{L}\Pi$, if $\varphi \in \mathrm{Log}(\mathrm{Type}(\Lambda))$, then $\varphi \in \Lambda$.

Let us assume the antecedent. Then for any $s \in \mathrm{Type}(\Lambda)$, $s \vDash \varphi$. In other words, $\mathrm{Type}(\varphi) \supseteq \mathrm{Type}(\Lambda)$. As we observed in Lemma 4.7, $\mathrm{Type}(\Lambda) = \bigcap\{\mathrm{Type}(\psi) \mid \psi \in \Lambda\}$. By Lemma 3.5, $\mathrm{Type}(\psi) = \mathrm{Type}(basic(\psi))$. Note also that $\psi \in \Lambda$ iff $basic(\psi) \in \Lambda\mathsf{Mg}$, which is shown in Lemma 3.3. Thus the set $\{\mathrm{Type}(\psi) \mid \psi \in \Lambda\} \subseteq \{\mathrm{Type}(\psi) \mid \psi \in \Lambda\mathsf{Mg} \cap \mathcal{S}\mathrm{Basic}\}$. On the other hand, for any $\psi \in \Lambda\mathsf{Mg} \cap \mathcal{S}\mathrm{Basic}$, using the axioms defining M_i and g, there is a $\psi' \in \mathcal{L}\Pi$ such that $\psi \leftrightarrow \psi' \in \Lambda\mathsf{Mg}$. This means that ψ' is in $\Lambda\mathsf{Mg}$, hence also in Λ, and that $\mathrm{Type}(\psi) = \mathrm{Type}(\psi')$, using Lemma 2.8. Hence $\{\mathrm{Type}(\psi) \mid \psi \in \Lambda\} = \{\mathrm{Type}(\psi) \mid \psi \in \Lambda\mathsf{Mg} \cap \mathcal{S}\mathrm{Basic}\}$, which we now call F.

Now this is a filter of basic clopens in $\mathcal{S}$ for the following reasons.

- For any $X, Y \in F$, we can find $\alpha, \beta \in \Lambda\mathsf{Mg} \cap \mathcal{S}\mathrm{Basic}$ such that $X = \mathrm{Type}(\alpha)$ and $Y = \mathrm{Type}(\beta)$. Now $\alpha \wedge \beta \in \Lambda\mathsf{Mg} \cap \mathcal{S}\mathrm{Basic}$, since $\Lambda\mathsf{Mg}$ has all propositional tautologies and *modus ponens*. Hence $X \cap Y = \mathrm{Type}(\alpha) \cap \mathrm{Type}(\beta) = \mathrm{Type}(\alpha \wedge \beta) \in F$.
- Recall that the basic clopens in $\mathcal{S}$ are just $\{\mathrm{Type}(\beta) \mid \beta \in \mathcal{S}\mathrm{Basic}\}$. For any $X \in F$ and any basic clopen Y such that $X \subseteq Y$, we first find $\alpha \in \Lambda\Pi\mathsf{Mg} \cap \mathcal{S}\mathrm{Basic}$ and $\beta \in \mathcal{S}\mathrm{Basic}$ such that $X = \mathrm{Type}(\alpha)$ and $Y = \mathrm{Type}(\beta)$. Then note that $\mathrm{Type}(\alpha \to \beta) = (S \setminus X) \cup Y = S$, since $X \subseteq Y$. Then by the completeness of $\mathsf{S5\Pi Mg}$ (Lemma 2.8), $\alpha \to \beta \in \mathsf{S5\Pi Mg}$. Then by *modus ponens* in $\Lambda\mathsf{Mg}$, which extends $\mathsf{S5\Pi Mg}$ as Λ extends $\mathsf{S5}$, $\beta \in \Lambda\mathsf{Mg}$. Remember that $\beta \in \mathcal{S}\mathrm{Basic}$. Hence $\beta \in \Lambda\mathsf{Mg} \cap \mathcal{S}\mathrm{Basic}$ and $Y = \mathrm{Type}(\beta) \in F$.

Thus we have $\mathrm{Type}(\Lambda) = \cap F$, a filter of basic clopens in $\mathcal{S}$, and we assumed that $\mathrm{Type}(\varphi) \supseteq \mathrm{Type}(\Lambda)$. Take $basic(\varphi)$. We have $\mathrm{Type}(basic(\varphi)) = \mathrm{Type}(\varphi)$ and that it is basic clopen in $\mathcal{S}$. We have shown that $\mathcal{S}$ is a Stone space in Lemma 4.9. Hence by compactness, there is actually an element $Z \in F$ such that $\mathrm{Type}(\varphi) \subseteq Z$. By the definition of F, we can find a $\psi \in \Lambda\mathsf{Mg} \cap \mathcal{S}\mathrm{Basic}$ such that $Z = \mathrm{Type}(\psi)$. Then $\mathrm{Type}(\psi \to basic(\varphi)) = S$. By completeness again, we have $\psi \to basic(\varphi) \in \mathsf{S5\Pi Mg}$ and thus also $\Lambda\mathsf{Mg}$. Then, since ψ is taken in $\Lambda\mathsf{Mg}$, $basic(\varphi) \in \Lambda\mathsf{Mg}$. By Lemma 3.3, $\varphi \in \Lambda$. This finishes the completeness of Λ. □

Then we describe the lattice structure of $\mathsf{Next\Pi(S5\Pi)}$.

Theorem 5.2 *The lattice* $\langle \mathsf{Next\Pi(S5\Pi)}, \subseteq \rangle$ *is isomorphic to the lattice of the open sets of* $\mathcal{S}$. *The isomorphism is* $\Lambda \mapsto S \setminus \mathrm{Type}(\Lambda)$, *or in the other direction,* $X \mapsto \mathrm{Log}(S \setminus X)$.

Proof. It is shown in the proof of Theorem 5.1 that for any $\Lambda \in \mathsf{Next\Pi(S5\Pi)}$, Type($\Lambda$) is the intersection of a filter of the basic opens of $\mathcal{S}$. By the basic theory of Stone spaces, this means that Type(Λ) is always a closed set in $\mathcal{S}$. Also, for any $\Lambda_1, \Lambda_2 \in \mathsf{Next\Pi(S5\Pi)}$, obviously $\Lambda_1 \subseteq \Lambda_2$ iff Type(Λ_1) $\supseteq$ Type(Λ_2). This means that, if we can establish that for every closed set $X \in S$, there is a $\Lambda \in \mathsf{Next\Pi(S5\Pi)}$ such that $X = \text{Type}(\Lambda)$, then the lattice structure of $\mathsf{Next\Pi(S5\Pi)}$ is precisely the inverse lattice of the closed sets of $\mathcal{S}$, or just the lattice of the open sets of $\mathcal{S}$, and the isomorphism will be given by $\Lambda \mapsto S \setminus \text{Type}(\Lambda)$.

Now take an arbitrary closed set X in $\mathcal{S}$. Then $\text{Log}(X) \in \mathsf{Next\Pi(S5\Pi)}$ as it is the set of formulas in $\mathcal{L}\Pi$ valid on a class of complete simple S5 algebras. Then what remains to be shown is that $\text{Type}(\text{Log}(X)) = X$. Again, the direction $X \subseteq \text{Type}(\text{Log}(X))$ is trivial. Now take an arbitrary type $s \in S \setminus X$. Then we just need to show that $s \notin \text{Type}(\text{Log}(X))$. Since $\mathcal{S}$ is a Stone space, X is closed, and $s \notin X$, we know that s and X can be separated by a basic clopen. Then, we can find a $\varphi \in \mathcal{S}\text{Basic}$ such that $X \subseteq \text{Type}(\varphi)$ but $s \notin \text{Type}(\varphi)$. But then, $\varphi \in \text{Log}(X)$. Since $\text{Type}(\text{Log}(X)) = \bigcap\{\text{Type}(\psi) \mid \psi \in \text{Log}(X)\}$, we see that $s \notin \text{Type}(\text{Log}(X))$. This finishes the proof. □

Since we have shown in the process of proving Theorem 4.9 that $\mathcal{S}$ is isomorphic to the disjoint union of two copies of the one-point compactification of $\mathbb{N}$, we have the following corollary.

Corollary 5.3 *The lattice* $\langle \mathsf{Next\Pi(S5\Pi)}, \subseteq \rangle$ *is isomorphic to the lattice of open sets of the disjoint union of two copies of the one-point compactification of* $\mathbb{N}$, *which is further isomorphic to the lattice of filters of the direct product of two copies of the field of finite and cofinite sets in* $\mathbb{N}$.

Another corollary of this characterization of all logics in $\mathsf{Next\Pi(S5\Pi)}$ is that, in terms of computability, there are arbitrarily complex logics (coded as sets of natural numbers in some natural way). More precisely, for any $X \subseteq \mathbb{N}$, there is a $\Lambda \in \mathsf{Next\Pi(S5\Pi)}$ such that X and Λ are Turing-equivalent.

Theorem 5.4 *For any* $X \subseteq \mathbb{N}$, $\text{Log}(\{1\} \times (X \cup \{\infty\}))$ *is Turing-equivalent to* X.

Proof. It is not hard to see that $\{1\} \times (X \cup \{\infty\})$ is a closed set in $\mathcal{S}$. Let us name $\{1\} \times (X \cup \{\infty\})$ by C, and $\{1\} \times \mathbb{N}^{\infty}$ by S_1 as we did before. Then $C = \text{Type}(\text{Log}(C))$, and thus for any $\varphi \in \mathcal{L}\Pi$, $\varphi \in \text{Log}(C)$ iff $\text{Type}(\varphi) \supseteq C$.

To reduce $\text{Log}(C)$ to X, the core idea is that to decide $\text{Type}(\varphi) \supseteq C$, we only need to see whether $\text{Type}(\varphi) \cap S_1 \supseteq C$, as $C \subseteq S_1$. Also $\text{Type}(\varphi) \cap S_1 = \text{Type}(\mathit{basic}(\varphi)) \cap S_1$, which is a finite union of intervals with finite left end points in S_1, with these end points readily computable from φ. Then, when X, and hence C, is given by an oracle, to decide whether $\varphi \in \text{Log}(C)$, we just need to do the following:

- If there is no cofinite interval, then return "no". This is correct because only a cofinite interval contains $\langle 1, \infty \rangle$, which is in C by definition.
- Otherwise, for each $s \in S_1$ that is not in $\text{Type}(\varphi)$, of which there are only finitely many, check whether $s \in C$. If the oracle for C ever returns "yes",

then return "no", as now $\mathrm{Type}(\varphi)$ is not a superset of C. Otherwise, $S_1 \cap \mathrm{Type}(\varphi) \supseteq C$, and $\varphi \in \Lambda$.

When X, and hence C, is finite, then we can have an algorithm that directly checks whether for each $s \in C$, s is also in $\mathrm{Type}(\varphi)$. In sum, $Log(C)$ can be reduced to X.

On the other hand, suppose $\mathrm{Log}(C)$ is given by an oracle. Then to compute whether $n \in X$ for some $n \in \mathbb{N}$, we only need to use the formula $\varphi_n = \neg \mathrm{M}_{n-1} \vee \mathrm{M}_{n+1}$, with M_{-1} and M_0 here defined as $\top$. Note that $\mathrm{Type}(\varphi_n) = S_1 \setminus \{\langle 1, n\rangle\}$. This means that:

- If $\varphi_n \in \mathrm{Log}(C)$, then $S_1 \setminus \{\langle 1, n\rangle\} \supseteq C$, and hence $\langle 1, n\rangle \notin C$ and $n \notin X$.
- If $\varphi_n \notin \mathrm{Log}(C)$, then $S_1 \setminus \{\langle 1, n\rangle\} \not\supseteq C$. Then $\langle 1, n\rangle \in C$ and $n \in X$.

Thus we only need to use the oracle to decide whether $\varphi_n \in \mathrm{Log}(C)$ and then return the opposite answer. □

Regarding the non-normal Π-logics extending S5Π, we limit ourselves in this paper to merely point out that there are many such logics. Algebraically, non-normal modal logics come from *matrices* (see §1.5 of [17]), which are algebras of propositions with a set of designated truth values. To exhibit a non-normal Π-logic extending S5Π, we can use just one particular structure. Let B be the complete simple S5 algebra whose Boolean part is the direct product of the powerset algebra of $\mathbb{N}$ and the MacNeille completion of the free Boolean algebra with countably many generators. Note that $t(B) = \langle 1, \infty\rangle$. Now consider the following set:

$$\Lambda = \{\varphi \in \mathcal{L}\Pi \mid \forall V : \mathsf{Prop} \to B, \widehat{V}(\varphi) \geqslant g\}.$$

It is not hard to see that $\Lambda \supseteq \mathrm{Log}(B)$, as the latter collects formulas whose valuation stay at 1, hence necessarily above g. Also, Λ is a Π-logic. In particular, universalization is valid because if φ only evaluates to elements above g, then $\forall\varphi$ evaluates to the meet of those elements above g, which must stay above g. Moreover, $\exists q(q \wedge \mathrm{atom}(q)) \in \Lambda$, as this formula evaluates precisely to g. However, $\Box\exists q(q \wedge \mathrm{atom}(q))$ is not in Λ, since $\Box g$ is $\bot$, because $g \neq 1$ in B: there is an atomless part in B. This means we obtained a non-normal Π-logic extending a normal Π-logic $\mathrm{Log}(B)$ which has no proper while consistent normal extension: the only closed proper subset of $\{B\}$ in $\mathcal{S}$ is $\varnothing$. Obviously, for any complete simple S5 algebra B that has both a non-trivial atomless part and a non-trivial atomic part, we can obtain a non-normal Π-logic in the same fashion. We could also use the requirement that $\widehat{V}(\varphi) \geqslant \neg g$, which will result in non-normal Π-logics including $\neg g$ but not $\Box\neg g$.

6 Conclusion

In this paper, we investigated Π-logics extending S5Π. In particular, we see that complete simple S5 algebras are semantically adequate for all normal Π-logics extending S5Π, that the lattice of these normal Π-logics are isomorphic to the lattice of the open sets of the type space $\mathcal{S}$ that is homeomorphic to the disjoint union of two copies of the one-point compactification of $\mathbb{N}$, that they can have arbitrarily high Turing-degree, and that they do not exhaust all the Π-logics extending $\mathsf{S5\Pi}$ as there are non-normal ones.

A major unresolved problem though, is the characterization of all Π-logics, instead of only the normal ones, extending $\mathsf{S5\Pi}$. We conjecture that a similar strategy can be used, though we need to be more careful about the choice of types. With an informative characterization, we may also be able to find a simple syntactical condition for a Π-logic extending $\mathsf{S5\Pi}$ to be normal and describe how the normal ones are distributed in the lattice of all Π-logics extending $\mathsf{S5\Pi}$.

Finally, we ask whether there is a way to prove all the results, especially the completeness of $\mathsf{S5\Pi}$ and stronger Π-logics, without using explicitly quantifier elimination. That this is important is because for many modal logics L, there is little hope that one can obtain a manageable quantifier elimination for $\mathsf{L\Pi}$. Hence, we need some technique that can be more easily generalized.

References

[1] Antonelli, G. A. and R. H. Thomason, *Representability in second-order propositional poly-modal logic*, The Journal of Symbolic Logic **67** (2002), pp. 1039–1054.

[2] Belardinelli, F. and W. Van Der Hoek, *Epistemic quantified boolean logic: Expressiveness and completeness results*, in: *24th International Joint Conferences on Artificial Intelligence (IJCAI 2015)*, 2015, pp. 2748–2754.

[3] Belardinelli, F. and W. Van Der Hoek, *A semantical analysis of second-order propositional modal logic*, in: *Thirtieth AAAI Conference on Artificial Intelligence*, 2016.

[4] Belardinelli, F., H. van Ditmarsch and W. van der Hoek, *Second-order propositional announcement logic*, in: *Proceedings of the 2016 International Conference on Autonomous Agents & Multiagent Systems*, International Foundation for Autonomous Agents and Multiagent Systems, 2016, pp. 635–643.

[5] Besnard, P., J.-M. Guinnebault and E. Mayer, *Propositional quantification for conditional logic*, in: *Qualitative and Quantitative Practical Reasoning*, Springer, 1997 pp. 183–197.

[6] Bull, R., *On modal logic with propositional quantifiers*, The Journal of Symbolic Logic **34** (1969), pp. 257–263.

[7] ten Cate, B., *Expressivity of second order propositional modal logic*, Journal of Philosophical Logic **35** (2006), pp. 209–223.

[8] Chang, C. C. and H. J. Keisler, "Model Theory," Elsevier, 1990.

[9] Fine, K., "For Some Proposition and So Many Possible Worlds," Ph.D. thesis, University of Warwick (1969).
URL `http://wrap.warwick.ac.uk/72219/`

[10] Fine, K., *Propositional quantifiers in modal logic*, Theoria **36** (1970), pp. 336–346.

[11] Ghilardi, S. and M. Zawadowski, *Undefinability of propositional quantifiers in the modal system S4*, Studia Logica **55** (1995), pp. 259–271.

[12] Givant, S. and P. Halmos, "Introduction to Boolean Algebras," Springer Science & Business Media, 2008.

[13] Halmos, P. R., *Algebraic Logic. I. Monadic boolean algebras*, Compositio Math **12** (1955), p. 7.

[14] Halmos, P. R., "Algebraic Logic," Courier Dover Publications, 2016.

[15] Holliday, W. H., *A note on algebraic semantics for S5 with propositional quantifiers*, Notre Dame Journal of Formal Logic (Forthcoming).
URL `https://escholarship.org/uc/item/303338xr`

[16] Kaminski, M. and M. Tiomkin, *The expressive power of second-order propositional modal logic*, Notre Dame Journal of Formal Logic **37** (1996), pp. 35–43.

[17] Kracht, M., "Tools and Techniques in Modal Logic," Elsevier Amsterdam, 1999.

[18] Kremer, P., *Propositional quantification in the topological semantics for S4*, Notre Dame Journal of Formal Logic **38** (1997), pp. 295–313.
[19] Kuhn, S., *A simple embedding of T into double S5*, Notre Dame Journal of Formal Logic **45** (2004), pp. 13–18.
[20] Kuusisto, A., *A modal perspective on monadic second-order alternation hierarchies*, Advances in Modal Logic **2008** (2008), pp. 231–247.
[21] Kuusisto, A., *Second-order propositional modal logic and monadic alternation hierarchies*, Annals of Pure and Applied Logic **166** (2015), pp. 1–28.
[22] Scroggs, S. J., *Extensions of the Lewis system S5*, The Journal of Symbolic Logic **16** (1951), pp. 112–120.

Chain-Monadic Second Order Logic over Regular Automatic Trees and Epistemic Planning Synthesis

Gaëtan Douéneau-Tabot [1]

ENS Paris-Saclay, Université Paris-Saclay

Sophie Pinchinat [2]

IRISA, Université de Rennes

François Schwarzentruber [3]

IRISA, ENS Rennes

Abstract

We consider infinite relational structures that have a finite presentation by means of a finite tuple of finite-state automata, and already known as *automatic structures*. While it is well established that model checking against first-order logic is decidable over automatic structures, we show how this seminal result can be adapted for a restricted class of automatic structures called *regular automatic trees* and an extension of the logic based on *chain-MSO*, and written cMSO (where MSO stands for monadic second-order logic). The logic cMSO, as chain-MSO, is interpreted over trees, and its second-order quantifiers range over subsets of branches. In the setting of regular automatic trees, we relate cMSO and logics of knowledge and time, among which the branching epistemic linear-time mu-calculus $\mathsf{BL}^{\mathsf{lin}}_{\mu}\mathrm{K}$. We finally apply our results to dynamic epistemic logic and its related epistemic planning problem: when restricting to event models with propositional preconditions and postconditions, the relational structures arising from epistemic planning problems turn out to be regular automatic trees. This already established latter central property allows us to derive the (already known) decidability of the epistemic planning problem as a mere corollary, but also to enlarge the class of decidable epistemic planning problems to goals expressed in $\mathsf{BL}^{\mathsf{lin}}_{\mu}\mathrm{K}$, with an effective way of computing the set of all successful (possibly infinite) plans.

Keywords: Automatic structures, chain Monadic Second-Order logic, Epistemic planning synthesis.

---

[1] gaetan.doueneau@ens-paris-saclay.fr

[2] sophie.pinchinat@irisa.fr

[3] francois.schwarzentruber@ens-rennes.fr

1 Introduction

In artificial intelligence, classical planning consists in generating a finite sequence of actions for achieving a given goal. The so-called *epistemic planning* [6] generalizes classical planning with epistemic goal (agent a knows that...) and complex actions (public announcements, private announcements, etc.). Epistemic planning is based on Dynamic epistemic logic ([2], [25]) and is undecidable in the general case [7].

Actions are represented by event models[4] that are Kripke structures of events, instead of possible worlds, equipped with a precondition and a postcondition. When event preconditions and postconditions are propositional (Boolean) only (intuitively, announcements are Boolean formulas and uncertainty is only about the content of the messages and physical changes), epistemic planning has been shown to be decidable ([29],[15]).

Regrettably, epistemic planning as considered so far concerns finite plans, and is in essence a reachability problem. However, real-life applications may challenge infinite plans. We give three examples of planning goals that require an infinite plan.

(i) safety properties, e.g. 'invariantly, an intruder a does not know the location of the piece of jewelry more than 3 consecutive steps'.

(ii) recurring bounded properties, e.g. 'all drones know that the region is safe every 20 steps';

(iii) strategic reasoning, e.g. 'with the current plan, the drone a never knows the region is safe but every 10 steps, there is a(nother) plan to let the drone a eventually know the region is safe'.

Nevertheless, infinite plans have been considered in [15] in the DEL (Dynamic Epistemic Logic) setting. In this approach, *DEL structures*, those arising from iterating ad infinitum the triggering of events, are seen as infinite relational structures. Maubert has shown that, when all event preconditions and postconditions are propositional, DEL structures are *automatic structures* [5,18]. Automatic structures are relational structures which have a presentation by means of a finite family of finite-state automata. As noticed by Maubert, the decidability result of epistemic planning (for epistemic goals) relies on the fact that model checking against first-order logic is decidable over automatic structures: in a DEL structure, the skeleton is a tree whose nodes (or equivalently finite branches) are finite plans, i.e. sequences of triggered events, and this tree structure is augmented with "transverse" binary relations between nodes in possibly different branches. These extra relations denote epistemic relations between histories. In DEL structures, the existence of a plan reduces to the existence of a point in this automatic (tree) structure where the epistemic planning goal (hence expressible in FO) holds. After all, the whole epistemic planning problem can be expressed in FO whose outermost quantifier is ex-

[4] also called sometimes "action models" in the literature.

istential. Additionally, automata constructions that answer the query of an FO-formula on an automatic structure allow to synthesize an automaton that recognizes the set of all (finite) plans achieving the goal. Regarding infinite plans, Maubert relies on the involved theory of *uniform strategies* he has developed [15]. This theory enables him to deal with epistemic planning instances whose goals are expressed in the branching-time epistemic logic CTL*K, the extension of temporal logic CTL* [12] with epistemic modalities. The setting does lead to synthesizing infinite plans, but the synthesis relies on a bottom-up technique over the goal formula that prevents the setting from being extended to logics featuring fix-points; while statements (ii) and (iii) above cannot be expressed in CTL*K, a well-tuned logic with fix-point or monadic with second-order quantifiers would suffice.

In an attempt to enlarge the specification language for epistemic planning goals, that goes beyond CTL*K while remaining decidable, one should observe that using the full *monadic second-order logic* (MSO) is hopeless: by [22], model checking against MSO over the binary tree equipped with the transverse binary relation "equal level" is undecidable. Relation "equal level" is an instance of an agent epistemic relation where the agent does not observe anything from the current finite sequence of triggered events but its length. However, there are variants of MSO, where second-order quantifications are constrained enough to yield a decidable model checking problem over infinite trees of the kind of DEL structures. For example, relying on [12], and as depicted in Figure 1, there exist mainly three interpretations of the second-order quantifiers, yielding full MSO, path-MSO, and chain-MSO. The full MSO logic can be interpreted over arbitrary relational structures and the second-order quantifiers range over all subsets of the structure domain. On the contrary, the path-MSO logic is interpreted on (possibly infinite) trees and the quantifiers range over paths/branches [5] of the tree. Finally, chain-MSO, also interpreted on trees only, requires the second-order quantifier to range over *chains*, that are subsets of paths [12]. It is known that the expressive power of chain-MSO strictly subsumes the one of path-MSO, the former allowing to describe arbitrary ω-regular properties of branches, while the latter captures only star-free properties, see [12].

In this paper, we extend chain-MSO of [21,12] with relations in the tree, e.g. interpreting epistemic modalities, making statements (ii) and (iii) expressible. This extension is written cMSO. We introduce *regular automatic trees*, a strict but large subclass of automatic structures that encompass DEL structures with propositional events. We then show that model checking against cMSO is decidable over *regular automatic trees*. Our decidability result is strongly inspired from the proof technique in [23, Th. 5.2] to show that chain-MSO with the "equal level" predicate is decidable over n-ary trees. Our proof yields automata constructions that can be exploited to achieve the synthesis of plans in epistemic planning.

[5] The difference between paths and branches is irrelevant by [12].

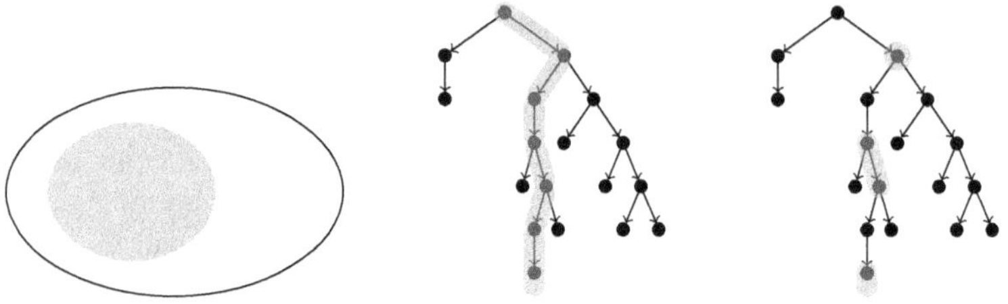

(a) MSO
quantification over any subset.

(b) path-MSO
quantification over any path in a tree.

(c): cMSO
quantification over any chain in a tree.

Figure 1. Different restrictions of second-order quantifications.

Noticeably, the logic cMSO captures the linear-time mu-calculus ([4], [26]) enriched with path quantifiers and epistemic modalities, here written $\mathrm{BL}^{\mathsf{lin}}_{\mu}\mathrm{K}$ (the acronym stands for '**B**ranching **L**ogic of the **lin**ear-time μ-calculus with **K**nowledge). In essence, the logic $\mathrm{BL}^{\mathsf{lin}}_{\mu}\mathrm{K}$ is an epistemic extension of the logic ECTL*, for "extended CTL*", of [12], but in a mu-calculus style rather than with the use of automata modalities, in the sense of [28]. Since CTL*K can easily be embedded into $\mathrm{BL}^{\mathsf{lin}}_{\mu}\mathrm{K}$ (the hard-coded linear-time temporal operators of CTL* are based on fix-points expressible in the linear-time mu-calculus), our result significantly exceeds the one of [15] whose proof technique does not allow for considering arbitrary fix-point formulas.

The presented contribution is derived from the preliminary work available in [10]. In Section 2, we recall the notion of automatic structures and results on model checking against FO-formulas and MSO-formulas. In Section 3, we present the logic cMSO, as well as the subclass of so-called *regular automatic trees*. We prove that model checking against cMSO is decidable over this subclass (Theorem 3.9). In Section 4, we compare cMSO with some logics of knowledge and time, namely CTLK, CTL*K and $\mathrm{BL}^{\mathsf{lin}}_{\mu}\mathrm{K}$, with increasing expressive power. Section 5 is dedicated to the application of our results to the *generalized epistemic planning problem* (Definition 5.7) where goals are arbitrary $\mathrm{BL}^{\mathsf{lin}}_{\mu}\mathrm{K}$-statements and instances of the planning domain rely on DEL specifications with propositional events.

2 Automatic structures and decidable theories

In this section we recall seminal results on automatic structures: namely that first-order logic FO is decidable over these structures while monadic second-order logic MSO is not. The interested reader may refer to [5,18] for further details.

2.1 Structures and logics

Logics FO and MSO are interpreted over relation structures.

Definition 2.1 A *relational structure* is a structure of the form $\mathcal{A} = \langle D, R_1 \dots R_p \rangle$ where D is a non-empty set called the *domain*; $R_1 \dots R_p$ are relations over D of arity $r_1 \dots r_p$, respectively; namely $R_i \subseteq D^{r_i}$.

The set of symbols $\{R_1 \dots R_p\}$ is called the *signature* of $\mathcal{A}$. We take the convention to write $R_i(d_1, \dots, d_{r_i})$ for $(d_1, \dots, d_{r_i}) \in R_i$. We distinguish the particular class of structures that are infinite trees of fixed finite degree n, called *n-ary trees*, that are central in our contribution. These structures represent computation trees (with a root of address ε, i-successor relations, and prefix binary relation between addresses) augmented with additional relations, such as epistemic relations.

Definition 2.2 Given a finite set $E := \{1, \dots, n\}$ of directions, we call *n-ary tree* a structure $\mathcal{T} = \langle D, r, S_1, \dots, S_n, R_1, \dots, R_p \rangle$ where:

- D is a prefix-closed subset of E^* (the addresses of the nodes);
- r is a unary relation that holds only for ε, the root address of the tree;
- $S_e(x, xe)$ whenever $xe \in D$;
- $R_1, \dots, R_p$ are additional relations on D.

In the particular case of a tree, elements of D are *nodes*.

Notice that in Definition 2.2 of a tree no condition was imposed on the relations $R_1 \dots R_p$. They may therefore correspond to transversal links between nodes just as epistemic relations would do, e.g. the "equal level" binary relation.

Example 2.3 The structure $\mathcal{T}_2 = \langle E^*, S_1, S_2 \rangle$ is a 2-ary tree called the *infinite full binary tree*. Also the structure $\mathcal{T}_2^{\mathsf{el}} = \langle E^*, S_1, S_2, \mathsf{el} \rangle$ obtained by augmenting $\mathcal{T}_2$ with the binary relation "at equal level in the tree" ($\mathsf{el}(x, y)$ holds if, and only if, $|x| = |y|$) is another example of a 2-ary tree.

First-order logic (FO) over relational structures $\mathcal{A} = \langle D, R_1 \dots R_p \rangle$ concerns formulas that conform to the following syntax: $\varphi ::= R_i(x_1 \dots x_{r_i}) \mid \neg\varphi \mid (\varphi \wedge \varphi) \mid \exists x \varphi$, where x, x_1, x_{r_i} are first-order variables whose interpretation ranges over the domain of relational structures.

When turning to the more expressive monadic second-order logic (MSO), one allows for second-order variables that range over subsets of the domain of relational structures. Formally, the syntax of MSO is given by: $\varphi ::= R_i(x_1 \dots x_{r_i}) \mid x \in X \mid \neg\varphi \mid (\varphi \wedge \varphi) \mid \exists x \varphi \mid \exists X \varphi$, where x is a first-order variable and X is a *second-order* variable whose interpretation ranges over subsets of the domain. In the following, we let $\mathcal{V}_1$ and $\mathcal{V}_2$ denote respectively the set of first-order variables and second-order variables.

As usual, in a formula, a variable $x \in \mathcal{V}_1$ (resp. $X \in \mathcal{V}_2$) is *free* if it is not under the scope of a quantifier, namely some Qx (resp. QX) with $Q \in \{\forall, \exists\}$. A formula is *closed* if it contains no free variables. An *assignment* in domain D is a function σ that maps a variable of $\mathcal{V}_1$ onto an element of D and a variable of $\mathcal{V}_2$ onto a subset of D. Given an assignment σ in D, a variable $x \in \mathcal{V}_1$ and $d \in D$, we let $\sigma[x \mapsto d]$ be the assignment on D that coincides with σ but maps x onto d. Similarly, we define $\sigma[X \mapsto D']$ where $D' \subseteq D$ for the second-order variables.

Due to limited space, we do not provide the semantics of MSO, see for example [11].

2.2 Automatic presentations

We now describe how some relational structures can be encoded using formal languages, following the presentation of [18] . By *alphabet* we mean a finite set of symbols, named *letters*. A *word* over Σ is a finite sequence of letters. We denote by Σ^* the set of such sequences. For $u \in \Sigma^*$ and $n \geq 0$, let $u[n]$ be the $n+1$-th letter of u (when defined). We assume familiarity with the basic definitions of automata theory and the properties of regular languages.

Let $\Box$ be a fresh padding symbol. If Σ is an alphabet, we define $\Sigma_\Box$ as $\Sigma \uplus \{\Box\}$. The following definition describes an encoding for k-tuples of words as a word on the product alphabet $\Sigma_\Box^n$; since words of the tuple may have different length, the padding symbol $\Box$ is used to align the words.

Definition 2.4 If $u, v \in \Sigma^*$, their *convolution* $u \otimes v$ is the word of $(\Sigma_\Box \times \Sigma_\Box)^*$ of length $\max(|u|, |v|)$ such that $(u \otimes v)[n] = (u[n], v[n])$ if $n < \min(|u|, |v|)$; $(u \otimes v)[n] = (u[n], \Box)$ if $|v| \leq n < |u|$; and $(u \otimes v)[n] = (\Box, v[n])$ if $|u| \leq n < |v|$. Convolution is defined similarly for k-tuples of words.

Example 2.5 The convolution of $\mathsf{aaba} \otimes \mathsf{b} \otimes \mathsf{ba}$ over alphabet $\Sigma = \{\mathsf{a}, \mathsf{b}\}$ is the four-letter word $\begin{pmatrix}\mathsf{a}\\\mathsf{b}\\\mathsf{b}\end{pmatrix}\begin{pmatrix}\mathsf{a}\\\Box\\\mathsf{a}\end{pmatrix}\begin{pmatrix}\mathsf{b}\\\Box\\\Box\end{pmatrix}\begin{pmatrix}\mathsf{a}\\\Box\\\Box\end{pmatrix}$ over alphabet $\Sigma_\Box^3$.

Definition 2.6 Let $\mathcal{A} = \langle D, R_1 \dots R_p\rangle$ be a relational structure. An *automatic presentation* of A is a tuple of regular languages $(L_D, L_1, \dots, L_n)$ meeting the following conditions: (1) there exists a one-to-one *encoding function* $\mathtt{enc} : D \to L_D$, and (2) for all R_i (arity r_i), $L_i = \{u_1 \otimes \cdots \otimes u_{r_i} \mid \forall j, u_j \in L_D \text{ and } (\mathtt{enc}^{-1}(u_1), \dots, \mathtt{enc}^{-1}(u_{r_i})) \in R_i\}$.

The alphabet Σ of L_D is called the *encoding alphabet*. The inverse $\mathtt{enc}^{-1}$ of the encoding function is the *decoding function*. We write $\mathtt{enc}(D)$ for L_D, and more generally, $\mathtt{enc}(R)$ for the set $\{\mathtt{enc}(d_1) \otimes \dots \otimes \mathtt{enc}(d_r) \mid (d_1, \dots, d_r) \in R\}$, for an arbitrary relation $R \subseteq D^r$.

A structure is *automatic* if it has (at least) an automatic presentation. A presentation can effectively be described by a tuple of finite automata $(\mathcal{M}_D, \mathcal{M}_1, \dots, \mathcal{M}_p)$ recognizing $(L_D, L_1, \dots, L_n)$.

Remark 2.7 We may assume that equality is among the relations R_i, represented by the regular language $\{u \otimes u \mid u \in L_D\}$. In the literature, the standard definition of automatic presentations allows an element to have several encodings, whenever equality can be presented by some regular language. However, both definitions enable to present the same structures [5].

Example 2.8 (i) Finite structures are automatic, since finite languages are regular.

(ii) $\langle \mathbb{N}, \leq\rangle$ is automatic, with an automatic presentation over the unary alphabet $\{1\}$ by letting $\mathtt{enc}(n) = 1^n$. Automaton $\mathcal{M}_\leq$ of Figure 2 verifies there are less 1's in the first component than in the second component.

(iii) Both trees $\mathcal{T}_2$ and $\mathcal{T}_2^{\mathsf{el}}$ are automatic.

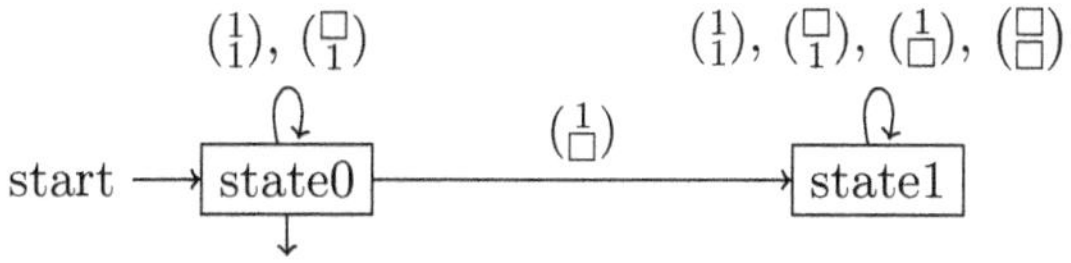

Figure 2. The finite-state automaton $\mathcal{M}_{\leq}$ of Example 2.8 (ii).

We recall the known fundamental theorem of [18, Th. 3.1] regarding the model checking problem against FO over automatic structures.

Definition 2.9 Model checking against FO

Input : A presentation $(\mathcal{M}_D, \mathcal{M}_1, \ldots, \mathcal{M}_p)$ of a relational structure $\mathcal{A}$ and a closed FO-formula φ over the corresponding signature.

Output : *Yes* if $\mathcal{A} \models \varphi$, *no* otherwise.

Theorem 2.10 *[18, Th. 3.1] Model checking against FO is decidable.*

The main ingredient of the proof of Theorem 2.10 relies on the construction of Proposition 2.11 which heavily relies on the closure of regular languages by intersection, negation and projection over components.

Let $\varphi(x_1 \ldots x_n) \in \mathsf{FO}$ where $x_1 \ldots x_n \in \mathcal{V}_1$ are the only free variables, and let $\mathcal{A}$ be a relational structure. We write $\varphi^{\mathcal{A}} := \{(d_1 \ldots d_n) \in D^n \mid \mathcal{A}, [x_i \mapsto d_i]_{1 \leq i \leq n} \models \varphi[x_1 \ldots x_n]\}$ for the set of tuples that satisfy φ.

Proposition 2.11 *There is an algorithm that given an automatic presentation of a relational structure $\mathcal{A}$ with encoding function* enc*, and a first-order formula $\varphi(x_1, \ldots, x_n)$, outputs an automaton that recognizes* $\mathsf{enc}(\varphi^{\mathcal{A}})$.

Theorem 2.10 can sometimes be extended to MSO. For example, there are very specific cases where MSO can be decided, among which the typical example of the automatic structure $(\mathbb{N}, <)$.

Proposition 2.12 *[3] The MSO-theory of a structure having an automatic presentation based on a unary encoding alphabet is decidable.*

Also, by Rabin's Theorem [17], MSO is decidable over the full binary tree with signature r, S_1, S_2. However, MSO becomes undecidable over the full binary relation augmented with the "equal level" relation el (see Example 2.3), although the obtained structure is automatic.

Theorem 2.13 *[22] Model checking against MSO over the binary tree $\mathcal{T}_2^{\mathsf{el}}$ is undecidable.*

We now focus on a subclass of automatic structures where the undecidability frontier can be pushed back.

3 Model checking against cMSO over RA trees

We consider the variant of MSO stemming from "chain MSO", written cMSO in this paper, which contrary to FO and MSO, can only be interpreted over infinite trees. The second-order quantifiers range over *chains*, namely subsets of nodes along a branch, as depicted in Figure 1(c). The logic cMSO has already been defined in [21,12,23] for n-ary trees and with a signature restricted to the successor relations $S_1, \dots, S_n$; the main results concern the decidability of the model checking of chain-MSO over the full binary tree, and its expressivity. In our setting, the signature of cMSO can be arbitrary, so that properties involving e.g. knowledge and time can be expressed.

We also exhibit a subclass of automatic structures called *regular automatic trees* that are automatic trees where the encoding of a node is its address in the tree. This subclass is large and encompasses infinite models arising from the unfolding of finite-state systems but also in dynamic epistemic logic. We establish that model checking against cMSO is decidable over regular automatic trees (Theorem 3.9).

3.1 The logic cMSO over trees

We recall a tree structure $\mathcal{T} = \langle D, r, S_1, \dots, S_n, R_1, \dots, R_p \rangle$ over a finite set $E = \{1, \dots, n\}$ of directions (Definition 2.2), where D is a prefix-closed subset of E^* describing the nodes through their addresses in the tree. In the following, we let S^* be the reflexive and transitive closure of the *generalized successor relation* $S := \bigcup_{i=1}^{n} S_i$.

Definition 3.1 In a tree structure $\mathcal{T}$ of domain $D \subseteq E^*$, a subset $C \subseteq D$ is a *chain* if it is totally ordered with respect to S^*, i.e. for all $d, d' \in C$, either $S^*(d, d')$ or $S^*(d', d)$ holds.
We denote by $Ch(\mathcal{T})$ the set of chains of the tree $\mathcal{T}$.

Example 3.2 In the full binary tree $\mathcal{T}_2$, the set $\{1^{2n} \mid n \in \mathbb{N}\}$ is a chain, whereas $\{1, 2\}$ is not.

The notion of chains enables one to consider an interpretation of MSO formulas in trees, named cMSO, where the second-order quantifiers are restricted to chains.

Definition 3.3 The logic cMSO is interpreted in a tree $\mathcal{T} = \langle D, r, S_1, \dots, S_n, R_1, \dots, R_p \rangle$ with an assignment σ in D for free variables as follows.

$\mathcal{T}, \sigma \models R_i(x_1 \dots x_{r_i})$	iff $(\sigma(x_1) \dots \sigma(x_{r_i})) \in R_i$;
$\mathcal{T}, \sigma \models x \in X$	iff $\sigma(x) \in \sigma(X)$;
$\mathcal{T}, \sigma \models \neg\varphi$	iff $\mathcal{T}, \sigma \not\models \varphi$;
$\mathcal{T}, \sigma \models (\varphi \wedge \psi)$	iff $\mathcal{T}, \sigma \models \varphi$ and $\mathcal{T}, \sigma \models \psi$;
$\mathcal{T}, \sigma \models \exists x \varphi$	iff there exists $d \in D$ such that $\mathcal{T}, \sigma[x \mapsto d] \models \varphi$;
$\mathcal{T}, \sigma \models \exists X \varphi$	iff there exists $C \in Ch(\mathcal{T})$ s. th. $\mathcal{T}, \sigma[x \mapsto C] \models \varphi$.

For a closed formula φ, we simply write $\mathcal{T} \models \varphi$ without specifying the assignment.

The logic cMSO is close to the logic path-MSO [21,12] which restricts second-order quantification to maximal branches only. Actually the latter is a subsystem of the former: the cMSO-formula $\mathsf{pathfrom}[X, x_0] := x_0 \in X \wedge \forall x \{x \in X \rightarrow [(\exists y S(x,y) \rightarrow \exists y(S(x,y) \wedge y \in X)) \wedge \neg S(x, x_0)]\}$ expresses that chain X is a maximal path starting at note x_0, which corresponds to the very quantifier in path-MSO. This translation has already been noticed many times in the literature.

3.2 Regular automatic trees

We now turn to a particular class of trees, called *regular automatic trees*, which are automatic trees for the most intuitive encoding of nodes, that is by their address in the tree. Since the domain is required to be regular, such trees arise from the unfolding of some finite-state transition systems, hence the terminology of *regular trees*.

Definition 3.4 A tree $\mathcal{T} = \langle D, r, S_1, \ldots, S_n, R_1, \ldots, R_p \rangle$ on the set of directions $E = \{1, \ldots, n\}$ is a *regular automatic (RA) tree* if the identity function $\mathtt{enc} : D \rightarrow E^*$ describes an automatic presentation of $\mathcal{T}$ over the encoding alphabet E. This presentation is called the *canonical presentation* of $\mathcal{T}$.

Since the encoding function of a RA tree is the identity function, the regularity of $S_1, \ldots, S_n$ and $\preceq$ directly follows from that of D. Thus, a tree is a regular automatic tree if, and only if, $D \subseteq E^*$ is regular, and for all R_i (of arity r_i), $L_i = \{d_1 \otimes \cdots \otimes d_{r_i} \mid (d_1, \ldots, d_{r_i}) \in R_i\}$ is regular.

Example 3.5 The tree $\mathcal{T}_2^{\mathsf{el}}$ of Example 2.3 is a regular automatic tree.

As stated in the following proposition, RA trees form a strict subclass of automatic trees.

Proposition 3.6 *There are automatic trees that are not RA trees.*

Proof

Let $E = \{1, 2\}$ we consider the binary tree $\mathcal{T}^{\text{auto}}_{\text{notreg}} = \langle D, r, S_1, S_2, \mathsf{el} \rangle$ of Figure 3(a) where $D = \{1^m 2^k \mid 0 \leq k \leq m\}$ and el is the "equal level" relation of Example 2.3.

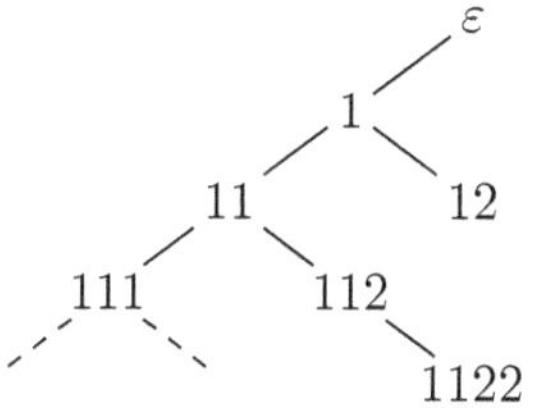

(a) The tree $\mathcal{T}^{\text{auto}}_{\text{notreg}}$.

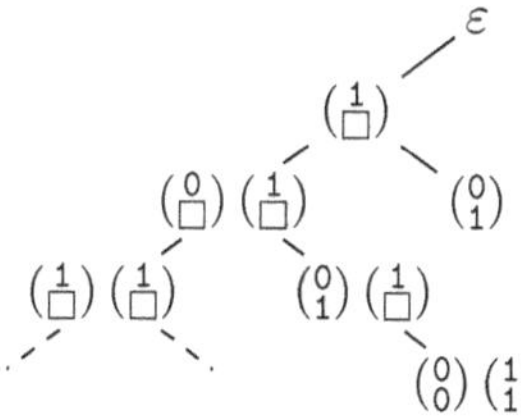

(b) Encodings of nodes of $\mathcal{T}^{\text{auto}}_{\text{notreg}}$.

Figure 3. An automatic tree that is not regular.

As D is not a regular language, $\mathcal{T}^{\text{auto}}_{\text{notreg}}$ is not a RA tree, but it is automatic: consider the encoding alphabet $\Sigma = (\{\mathtt{0}, \mathtt{1}\}_\Box)^2$ and let $\mathtt{enc}(1^m 2^k) = \mathtt{bin}(m) \otimes$

$\mathtt{bin}(k)$ be the encoding function, as presented in Figure 3(b), where $\mathtt{bin}(m)$ (resp. $\mathtt{bin}(k)$) the binary representation of m (resp. k) with least significant digit first. For example, $\mathtt{enc}(112) = \mathtt{enc}(1^2 2^1) = \mathtt{bin}(2) \otimes \mathtt{bin}(1) = \binom{0}{1}\binom{1}{\square}$. The reader should easily get convinced that $\mathtt{enc}(D)$, $\mathtt{enc}(S_1)$, $\mathtt{enc}(S_2)$ and $\mathtt{enc}(\mathsf{el})$ are regular languages. □

When restricting to RA trees, some usual relations over trees can be easily captured by automata over the canonical encoding of nodes. We recall the following additional binary relations over trees.

- The reflexive and transitive closure S^* of the generalized successor binary relation S;
- The binary relation $(d, d') \in \preccurlyeq$ whenever node d is not deeper than node d' in $\mathcal{T}$;
- The binary relation el, where $(d, d') \in \mathsf{el}$ whenever nodes d and d' are at the same depth in the tree.
- The binary equality relation $=$.

Lemma 3.7 *Let $\langle D, r, S_1, \ldots, S_n, R_1, \ldots, R_p\rangle$ be a RA tree. Then the relational structure $\langle D, r, S_1, \ldots, S_n, R_1 \ldots R_p, S^*, \preccurlyeq, \mathsf{el}, =\rangle$ is also an RA tree.*

Restricting to the class of RA trees allows for decidability results of model checking against logics that go beyond FO.

3.3 Model checking against cMSO

We now describe the main result of this section, as stated by Theorem 3.9, where the model checking problem against cMSO is as follows.

Definition 3.8 MODEL CHECKING AGAINST cMSO

Input : $\langle \mathcal{M}_D, \mathcal{M}_r, (\mathcal{M}'_i)_{1\leq i\leq n}, (\mathcal{M}_i)_{1\leq i\leq p}\rangle$ the canonical presentation of a RA tree $\mathcal{T}$ and a closed cMSO-formula φ over the corresponding signature.

Output : *Yes* if $\mathcal{T} \models \varphi$, *no* otherwise.

Theorem 3.9 can be read as a variant of Theorem 2.10 where, on the one hand, the logic FO is extended, and, on the other hand, the class of automatic structures is restricted. It can also be read as a generalization of [23, Th. 5.2].

Theorem 3.9 *Model checking against cMSO is decidable over RA trees.*

The rest of this section is dedicated to the proof of Theorem 3.9. This proof goes over the proof made in [23, Th. 5.2] to generalize the result to arbitrary RA trees.

Before starting the proof, we may assume without loss of generality, that all the variables occurring in the formulas are second-order variables. This means to add an extra unary predicate Sing() which conveys that a set is a singleton, and to add inclusion formulas of the form $X \subseteq Y$ so as to capture previous formulas of the form $x \in X$. The reader may refer to [23] for this routine transformation. The syntax then becomes as follows.

$$\begin{array}{l}\varphi ::= \mathsf{Sing}(X) \mid X \subseteq Y \mid R_i(X_1 \dots X_{r_i}) \mid \neg\varphi \mid (\varphi \wedge \varphi) \mid \exists X \varphi, \\ \text{where } X, Y, X_1, \dots \in \mathcal{V}_2 \end{array} \quad (1)$$

The new syntax requires to adjust the semantics of the formulas, and in particular of those of the form $R_i(X_1, \dots, X_{r_i})$, by imposing that each X_j is interpreted as a singleton, this is routine. The logic resulting from (1) is expressively equivalent to cMSO to the extent that a singleton is a chain. Therefore, in the following, we still use cMSO when using the syntax of (1).

A key idea to develop automata construction to model check against cMSO is that a set of addresses is a chain if, and only if, it is contained in the set of all prefixes of some infinite word. Given a chain $C \subseteq E^*$, we let $\mathsf{Branches}(C) := \bigcap\{uE^\omega \mid u \in C\}$ be the set of infinite words whose set of prefixes contains C. In particular, $\mathsf{Branches}(C)$ is a singleton if, and only if, C is infinite.

For $v \in \mathsf{Branches}(C)$, we write $\mathsf{mbranch}(v, C)$ for the infinite word over $E \times \{0,1\}$ whose first component is v and whose second component is marked by 1 if the current prefix of v belongs to C, and by 0 otherwise. Formally, $\mathsf{mbranch}(v, C) = m[0]m[1]m[2]\dots$ with $m[i] = (v[i], b)$ where $b = 1$ if $v[0]v[1]\dots v[i] \in C$, and $b = 0$ otherwise.

The words $\mathsf{mbranch}(v, C)$ are used to encode the chain C. We will overload the encoding identity function enc of RA trees to denote the encoding of chains. We letting $\mathsf{enc}(C) := \{\mathsf{mbranch}(v, C) \mid v \in \mathsf{Branches}(C)\}$. Also, we extend the notation to $\mathsf{enc}(C_1, \dots, C_m)$ that contains infinite words over alphabet $(E \times \{0,1\})^m$ by letting $\mathsf{enc}(C_1, \dots, C_m) := \mathsf{enc}(C_1) \otimes \dots \otimes \mathsf{enc}(C_m)$ where $\otimes$ is the convolution extended to sets of infinite words in a natural manner. For $m = 0$, $\mathsf{enc}()$ is the singleton containing the unique infinite word over the singleton alphabet $(\Sigma \times \{0,1\})^0$.

We now provide an automata-theoretic approach for model checking against cMSO formulas. We first establish the existence of a Büchi automaton $\mathcal{C}_m$ that verifies if an infinite word over alphabet $(E \times \{0,1\})^m$ denotes a m-tuple of chains, as stated by Lemma 3.10.

Lemma 3.10 *One can effectively construct a Büchi automaton $\mathcal{C}_m$ that recognizes the encoding of m-tuples of chains, namely the language* $\bigcup_{C_1,\dots,C_m \in Ch(\mathcal{T})} \mathsf{enc}(C_1, \dots, C_m)$.

The automaton $\mathcal{C}_m$ of Lemma 3.10 runs m copies of the domain automaton $\mathcal{M}_D$. Each copy reads one of the m infinite words over E extracted from the infinite input word over $(E \times \{0,1\})^m$. Automaton $\mathcal{C}_m$ rejects if one of these copies rejects.

Let $\mathcal{T} = \langle D, r, S_1, \dots, S_n, R_1, \dots, R_p\rangle$ be a RA tree over the set of directions $E = \{1, \dots, n\}$ with $T = \langle \mathcal{M}_D, \mathcal{M}_r, (\mathcal{M}'_i)_{1 \le i \le n}, (\mathcal{M}_i)_{1 \le i \le p}\rangle$ its canonical presentation, and let φ be a cMSO-formula[6], with free variables $X_1, \dots, X_m$.

[6] in the language defined by the grammar (1)

We define the set

$$\mathtt{enc}(\varphi^{\mathcal{T}}) := \bigcup_{\substack{C_1,\ldots,C_m \in Ch(\mathcal{T}) \\ \mathcal{T},[X_i \to C_i]_{1\leq i\leq m} \models \varphi}} \mathtt{enc}(C_1,\ldots,C_m) \tag{2}$$

Proposition 3.11 *Given φ a cMSO-formula*[7] *with free variables $X_1, \ldots, X_m$, one can effectively construct a Büchi automaton $\mathcal{B}_{\varphi,\mathcal{T}}$ that recognizes $\mathtt{enc}(\varphi^{\mathcal{T}})$.*

We can now conclude the proof of Theorem 3.9. Observe that model checking against formula φ over the RA tree $\mathcal{T}$ amounts to deciding whether the language of the Büchi automaton $\mathcal{B}_{\varphi,\mathcal{T}}$ of Proposition 3.11 is non-empty. Indeed, this language is not empty if, and only if, there is some assignment σ of the second order-variables $X_1, \ldots, X_m$ such that $\mathcal{T}, \sigma \models \varphi$. In particular, if φ is closed ($m = 0$), by convention we have:

- If $\mathcal{T} \not\models \varphi$, the union in Equation (2) is empty as it ranges over an empty set;
- If $\mathcal{T} \models \varphi$, the union in Equation (2) is not empty as it is equal to $\mathtt{enc}()$.

Notice that Proposition 2.12 from [3] regarding the decidablity of the full MSO-theory (and not only cMSO) for automatic presentations based on a singleton alphabet is now a direct corollary of Theorem 3.9 since every set of words over a singleton alphabet is a chain.

From Theorem 3.9 on the decidability of cMSO over RA trees, and because binary relations $S^*, \preccurlyeq, \mathsf{el}, =$ come for free with automata constructions (Lemma 3.7), we have the following.

Corollary 3.12 *Model checking against cMSO$[r, S_1, \ldots, S_n, R_1, \ldots, R_p, S^*, \preccurlyeq, \mathsf{el}, =]$ is decidable over a RA trees.*

Notice that if we restrict to trees with only relations $S_1 \ldots S_n$, model checking against the full logic MSO becomes decidable (Rabin's Theorem [17]). However, considering extra relations (e.g. "equal level") makes the model checking against MSO undecidable over RA trees, as we already saw in Theorem 2.13 of Section 2, which is not the case for cMSO by Corollary 3.12.

Also, since every path quantifier in a path-MSO-formula can be translated in cMSO, and since considering extra arbitrary relations in trees, such as epistemic relations, does not harm this translation, we obtain the following result, which to our knowledge has not been considered in the literature.

Corollary 3.13 *Given a RA tree $\mathcal{T} = \langle D, r, S_1, \ldots, S_n, R_1, \ldots, R_p \rangle$ and a close path-MSO-formula φ over the signature of $\mathcal{T}$, it can be decided if $\mathcal{T} \models \varphi$.*

We now turn to logical formalisms that combine knowledge and time in the spirit of [13].

[7] with syntax of (1)

4 Logics of knowledge and time over RA trees

We now exploit the peculiarities of RA trees to consider extensions of several classic formalisms and study their expressivity. We first recall our criterion for expressivity.

Definition 4.1 A logic $\mathcal{L}$ is *embedded into* a logic $\mathcal{L}'$, whenever there is an effective translation that maps every $\mathcal{L}$-formula onto an equivalent $\mathcal{L}'$-formula, namely for every regular automatic tree $\mathcal{T}$, we have: $\mathcal{T} \models \varphi$ if, and only if, $\mathcal{T} \models \varphi'$.

Notice that when $\mathcal{L}$ is embedded into $\mathcal{L}'$, the decidability of the model checking against $\mathcal{L}'$ entails the decidability of the model checking against $\mathcal{L}$.

In tree structures, *time* is captured by the generalized successor binary relation $S := \bigcup_{i=1}^{n} S_i$, as in computation trees for temporal logics such as CTL [8], CTL^* [12], etc. Depending on their arities, the other relations $R_1, \ldots, R_p$ have different roles. Unary relations label nodes with atomic propositions that range over some fixed set $AP = \{p, q, \ldots\}$. Binary relations, written $K_1 \ldots K_m$ as the epistemic modalities in multi-agent epistemic logic, will play the role of *knowledge* modalities as in classic epistemic logic. On this basis, we shortly write cMSOK for the logic $\mathsf{cMSO}[r, S_1, \ldots, S_n, (p)_{p\in AP}, K_1 \ldots K_m]$.

The ability to quantify over chains allows us to capture properties along branches of the trees that are naturally stated in linear-time mu-calculus. Recall that MSO along linear orders, hence branches, is expressively equivalent to Büchi automata that capture all ω-regular properties, as shown by [24], and so does the linear-time mu-calculus [9]. We therefore introduce the logic $\mathrm{BL}_\mu^{\mathsf{lin}}\mathrm{K}$ that is based on the linear-time mu-calculus [4,26], equipped with knowledge modalities but also with second order quantifications over branches of the tree, just as CTL^* extends LTL to the branching-time semantics.

The logic $\mathrm{BL}_\mu^{\mathsf{lin}}\mathrm{K}$ is called the *branching epistemic linear-time mu-calculus*; in the spirit of CTL^*, it relies on two kinds of formulas: *state formulas* (Φ) and *path formulas* (φ).

Definition 4.2 The syntax of $\mathrm{BL}_\mu^{\mathsf{lin}}\mathrm{K}$ is as follows.

- State formulas: $\Phi ::= p \,|\, K_i\Phi \,|\, \neg\Phi \,|\, (\Phi \wedge \Phi) \,|\, \mathsf{E}\varphi$ where $i \in \{1, \ldots, m\}$.
- Path formulas: $\varphi ::= Z \,|\, \Phi \,|\, \neg\varphi \,|\, (\varphi \wedge \varphi) \,|\, \mathsf{X}\varphi \,|\, \mu Z.\ \varphi[Z]$ where φ is closed in $\mathsf{E}\varphi$, $Z \in \mathcal{V}$ is under the scope of an even number of negations in $\varphi[Z]$.

State formula $K_i\Phi$ is read as 'Agent i knows Φ' and state formula $\mathsf{E}\varphi$ is read as 'there is a path starting from the current state that satisfies φ'. Path formula $\mathsf{X}\varphi$ is read as 'φ holds in the next state'. Path formula $\mu Z.\ \varphi[Z]$ is the linear mu-calculus fix-point construction. State formula $\mathsf{A}\varphi$ is an abbreviation of $\neg\mathsf{E}\neg\varphi$ and is read as 'φ holds in all paths starting from the current state'. All formulas of $\mathrm{BL}_\mu^{\mathsf{lin}}\mathrm{K}$ are interpreted in a tree $\mathcal{T} = \langle D, r, S_1, \ldots, S_n, K_1 \ldots K_m, (p)_{p\in AP}\rangle$. State formulas are interpreted on nodes, while path formulas are interpreted on branches.

- State formulas:

$\mathcal{T}, d \models p$	iff	$d \in p$;
$\mathcal{T}, d \models K_i\Phi$	iff	for all $d' \in D$ s.t. $(d, d') \in K_i$ we have $\mathcal{T}, d' \models \Phi$;
$\mathcal{T}, d \models \neg\Phi$	iff	$\mathcal{T}, d \not\models \Phi$;
$\mathcal{T}, d \models (\Phi \wedge \Psi)$	iff	$\mathcal{T}, d \models \Phi$ and $\mathcal{T}, d \models \Psi$;
$\mathcal{T}, d \models \mathsf{E}\varphi$	iff	there exists a maximal chain C with least element d (i.e. a branch starting from d) such that $\mathcal{T}, C \models \varphi$;

- Path formulas: we add a valuation $\sigma : \mathcal{V} \to 2^{\mathbb{N}}$ for the free second-order variables in the formulas, and write σ^{+i} for the valuation $\sigma^{+i}(Z) = \{n \mid n + i \in \sigma(Z)\}$.

$\mathcal{T}, \pi, \sigma \models Z$	iff	$0 \in \sigma(Z)$
$\mathcal{T}, \pi, \sigma \models \Phi$	iff	$\mathcal{T}, \pi(0) \models \Phi$;
$\mathcal{T}, \pi, \sigma \models (\varphi \wedge \psi)$	iff	$\mathcal{T}, \pi, \sigma \models \varphi$ and $\mathcal{T}, \pi, \sigma \models \psi$;
$\mathcal{T}, \pi, \sigma \models \neg\varphi$	iff	$\mathcal{T}, \pi, \sigma \not\models \varphi$;
$\mathcal{T}, \pi, \sigma \models \mathsf{X}\varphi$	iff	$\mathcal{T}, \pi(1)\pi(2)\ldots, \sigma^{+1} \models \varphi$;
$\mathcal{T}, \pi, \sigma \models \mu Z.\ \varphi[Z]$	iff	0 is in the least-fix point of $[\![\varphi[Z]]\!]^{\sigma}_{\pi}$

$$\text{where } \begin{cases} [\![\varphi[Z]]\!]^{\sigma}_{\pi} : 2^{\mathbb{N}} \to 2^{\mathbb{N}} \\ \qquad P \mapsto \{i \in \mathbb{N} \mid \pi(i)\pi(i+1)\ldots, (\sigma[Z \mapsto P])^{+i} \models \varphi[Z]\}. \end{cases}$$

A tree (resp. a forest) satisfies a state formula if it satisfies it from its root node (resp. from all its distinguished root nodes).

Notice that the least fix-point always exists. Indeed, the function $[\![\varphi[Z]]\!]^{\sigma}_{\pi}$ is monotone (recall that we have required that every variable to occur under the scope of an even number of negations) and the Knaster-Tarski Theorem [20] applies in the complete lattice $(2^{\mathbb{N}}, \subseteq)$. Besides, we can alternatively define the least fix-point of $[\![\varphi[Z]]\!]^{\sigma}_{\pi}$ as $\bigcap\{F \subseteq 2^{\mathbb{N}} \mid [\![\varphi[Z]]\!]^{\sigma}_{\pi}(F) \subseteq F\}$.

Example 4.3
- $\mathcal{T}, \pi \models \mu Z.\ p$ if, and only if, $\mathcal{T}, \pi(0) \models p$.
- $\mathcal{T}, \pi \models \mu Z.\ (\mathsf{X}Z \vee p)$ if, and only if, there exists a node $\pi(i)$ along π such that $\mathcal{T}, \pi(i) \models p$.
- $\mathcal{T}, \pi \models \mu Z.\ (\psi \vee (\varphi \wedge \mathsf{X}Z))$ if, and only if, there exists a node $\pi(j)$ such that $\mathcal{T}, \pi(j) \models \psi$ and $\mathcal{T}, \pi(i) \models \varphi$ for every $0 \leq i < j$.

By means of fix-points in $\mathrm{BL}^{\mathsf{lin}}_{\mu}\mathrm{K}$, we capture the linear-time logic LTL [16], and since we also have the existential path quantifier E, we can embed the full logic CTL* of [12] in $\mathrm{BL}^{\mathsf{lin}}_{\mu}\mathrm{K}$.

Now the epistemic modalities K_i allow us to capture CTL*K, the logic CTL* equipped with knowledge modalities. More precisely, we introduce the following macro for 'φ until ψ': $\varphi\mathsf{U}\psi$ is an abbreviation of formula $\mu Z.(\psi \vee (\varphi \wedge \mathsf{X}Z))$.

Definition 4.4 CTL*K is the syntactic fragment of $\mathrm{BL}^{\mathsf{lin}}_{\mu}\mathrm{K}$ defined by:

- State formulas: $\Phi ::= \Phi \wedge \Phi \mid \neg\Phi \mid \mathsf{E}\varphi \mid K_i\Phi \mid p$ where $i \in \{1, \ldots, m\}$.
- Path formulas: $\varphi ::= \Phi \mid \neg\varphi \mid \varphi \wedge \varphi \mid \mathsf{X}\varphi \mid \varphi\mathsf{U}\varphi$.

The syntactic fragment CTLK of CTL*K has only state formulas:

$$\Phi ::= \Phi \wedge \Phi \mid \neg\Phi \mid \mathsf{EX}\varphi \mid \mathsf{E}(\Phi\mathsf{U}\Phi) \mid \mathsf{A}(\Phi\mathsf{U}\Phi) \mid K_i\Phi \mid p$$

Proposition 4.5 *CTLK is embedded into* $\mathsf{FO}[K_1 \dots K_m, (p)_{p \in \mathsf{AP}}, S, S^*, \preccurlyeq, =]$.

We end this section with the main Theorem 4.6 that states the decidability of model checking against $\mathrm{BL}_\mu^{\mathrm{lin}}$Kover RA trees. First of all, it should be clear that the following holds.

Theorem 4.6 *$BL_\mu^{lin}K$ is embedded into cMSOK.*

Corollary 4.7 *Model checking against $BL_\mu^{lin}K$ is decidable over automatic regular trees.*

Figure 4 is a recap of the expressivity results we have obtained (arrows resulting from transitivity are omitted), where an arrow $\mathcal{L} \to \mathcal{L}'$ means that $\mathcal{L}$ is embedded into $\mathcal{L}'$.

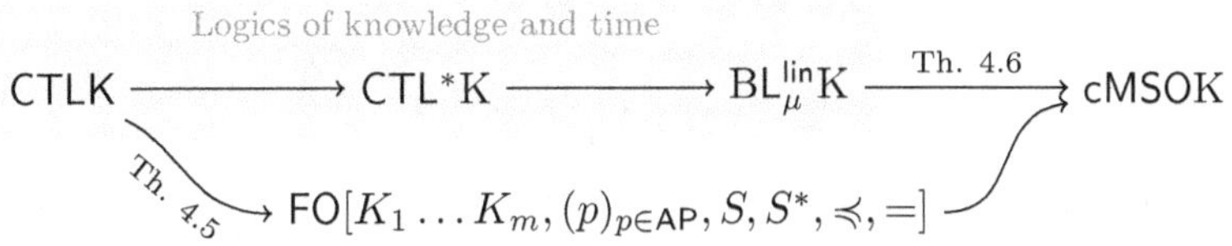

Figure 4. Embeddings between the various logics.

5 Application to epistemic planning synthesis

As announced, we show that Theorem 3.9 together with Proposition 3.11 have an interesting impact in the domain of epistemic planning. The epistemic planning setting as introduced by [6] relies on Dynamic epistemic logic, that we recall here. Next we explain how the original problem can be generalized to attain more expressive planning goals, such as statements (i)–(iii) of the introduction, while maintaining the decidability frontier for free. Our results generalize the ones obtained in [15,1], but make great use of the already established property [15, Lemma 22, p. 109] that, under the assumption that preconditions and postconditions of events are propositional, the relational structure that contains all plan candidates is automatic, and actually it is a RA tree (see Theorem 5.5).

5.1 Preliminaries on Dynamic epistemic logic

As earlier, AP denotes the set of atomic propositions with typical elements $p, q, \dots$, and Ag is finite set of agents with typical elements $a, b, \dots$. The propositional language is denoted by $\mathscr{L}_{Prop}$ and the language of multi-agent epistemic modal logic is denoted by $\mathscr{L}_{EL}$.

Definition 5.1 A Kripke model $\mathcal{M} = (W, (K_a)_{a \in Ag}, V)$ is defined by a non-empty set W of epistemic worlds, epistemic relations $(K_a)_{a \in Ag} \subseteq W \times W$ and a valuation function $V : W \to 2^{AP}$.

A pair $(\mathcal{M}, w)$ is called a *pointed epistemic model.*

Figure 5 bottom left shows a pointed epistemic model.

The dynamic of the system is captured by *event models.* An event model is like a Kripke model but possible worlds are instead events equipped with a

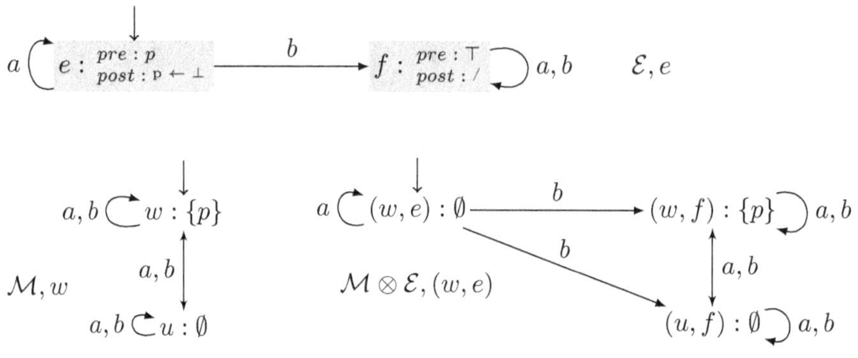

Figure 5. Example of a DEL product.

precondition and a postcondition. We intentionally denote the set of events E with the same notation that the set of directions in trees (Definition 2.2), as they will denote directions in the DEL structure (see Definition 5.4).

Definition 5.2 An *event model* $\mathcal{E} = (E, (K_a^{\mathcal{E}})_{a \in Ag}, pre, post)$ is defined by a non-empty set of *events* E, epistemic relations $(K_a^{\mathcal{E}})_{a \in Ag} \subseteq E \times E$, a precondition function $pre : E \to \mathscr{L}_{EL}$ and a postcondition function $post : E \times AP \to \mathscr{L}_{Prop}$. When $pre : E \to \mathscr{L}_{Prop}$, the event model $\mathcal{E}$ is *propositional.*

A pair $(\mathcal{E}, e)$ is called a *pointed event model*, where e represents the actual event.

Definition 5.3 Let $\mathcal{M} = (W, (K_a)_{a \in Ag}, V)$ be a Kripke model. Let $\mathcal{E} = (E, (K_a^{\mathcal{E}})_{a \in Ag}, pre, post)$ be an event model. The *product* of $\mathcal{M}$ and $\mathcal{E}$ is $\mathcal{M} \otimes \mathcal{E} = (W', (K_a)', V')$ where:

- $W' = \{(w, e) \in W \times E \mid \mathcal{M}, w \models pre(e)\}$;
- $((w, e), (w', e')) \in K'_a$ iff $(w, w') \in K_a$ and $(e, e') \in K_a^{\mathcal{E}}$;
- $V'((w, e)) = \{p \in AP \mid \mathcal{M}, w \models post(e, p)\}$.

Figure 5 top shows an example of a pointed event model with two events e and f. The actual event is e but agent b imagines event f as the sole possible event. Also, the product is given in bottom right.

5.2 DEL structures

In this subsection, following [15], we incorporate the initial epistemic model $\mathcal{M}$ and the infinitely many the products $\mathcal{M} \otimes \mathcal{E}^n := \mathcal{M} \otimes \underbrace{\mathcal{E} \otimes \cdots \otimes \mathcal{E}}_{n \text{ times}}$ into a single structure, called a *DEL structure*, denoted by $\mathcal{M}\mathcal{E}^*$. Worlds of $\mathcal{M} \otimes \mathcal{E}^n$, called *histories*, are naturally denoted by words of the form $we_1 \dots e_n$ where w is a world of $\mathcal{M}$ and $e_1, \dots, e_n$ are events of $\mathcal{E}$. E.g. the world $((w, e_1), e_2)$ is denoted we_1e_2. The pair $(\mathcal{M}, \mathcal{E})$ is called the *DEL presentation* of $\mathcal{M}\mathcal{E}^*$. A DEL presentation $(\mathcal{M}, \mathcal{E})$ is *propositional* if $\mathcal{E}$ is propositional.

Definition 5.4 [15, , Def. 54, p. 105] The *DEL structure* denoted by $(\mathcal{M}, \mathcal{E})$ is the structure $\mathcal{ME}^* = (\mathcal{H}, (S_e)_{e\in E}, (K_a)_{a\in Ag}, (p)_{p\in AP})$, where $\mathcal{H}$ is the disjoint union of the sets of worlds of $\mathcal{M} \otimes \mathcal{E}^n$ – namely, the *histories*; $(h, h') \in S_e$ whenever $h' = he$ for all events e; $(h, h') \in K_a$ whenever h and h' are worlds of $\mathcal{M} \otimes \mathcal{E}^n$, for some n, and $(h, h') \in K'_a$ where K'_a is the epistemic relation for agent a in $\mathcal{M} \otimes \mathcal{E}^n$; and finally, $h \in p$ whenever $p \in V(h)$ in $\mathcal{M} \otimes \mathcal{E}^n$.

From the proof of [15, Lemma 22, p. 109], one can easily show that:

Theorem 5.5 *Propositional DEL structures $\mathcal{ME}^*$ are finite sets (forests) of RA trees, and their presentation is effectively computable from $(\mathcal{M}, \mathcal{E})$.*

We now turn to the epistemic planning problems.

5.3 Generalized epistemic planning and plan synthesis

The *epistemic planning problem*, as originally stated by [6] is defined as follows.

Definition 5.6 Epistemic planning problem

Input: a pointed epistemic model $\mathcal{M}, w$, an event model $\mathcal{E}$, an $\mathscr{L}_{EL}$-formula ψ;

Output: Yes, if there is a sequence of events $e_1, \ldots, e_n$ in $\mathcal{E}$ such that $\mathcal{ME}^n, we_1 \ldots e_n \models \psi$.

We generalize this epistemic planning problem by considering a model checking problem against $\mathrm{BL}^{\mathsf{lin}}_{\mu}\mathrm{K}$ over DEL structures.

Definition 5.7 Generalized Epistemic planning problem

Input: a pointed epistemic model $\mathcal{M}, w$, an event model $\mathcal{E}$, an $\mathrm{BL}^{\mathsf{lin}}_{\mu}\mathrm{K}$-formula φ;

Output: Yes, if $\mathcal{ME}^* \models \varphi$.

The problem of Definition 5.7 is indeed a generalization of epistemic planning problem of Definition 5.6 in the sense that the latter can be reduced to the former by model checking the formula $\varphi := \mathsf{EF}\psi$, where ψ is the planning goal.

The following example shows the relevance of generalized epistemic planning.

Example 5.8 We give $\mathrm{BL}^{\mathsf{lin}}_{\mu}\mathrm{K}$-formulas φ for the three statements (i)–(iii) of the introduction.

(i) 'invariantly, an intruder a does not know the location of the piece of jewelry more than 3 consecutive steps' can be expressed as $\mathsf{EG}(\Theta \rightarrow (\mathsf{X}\neg\Theta \vee \mathsf{X}^2\neg\Theta \vee \mathsf{X}^3\neg\Theta)$ where $\Theta := \bigvee_{\ell \in Loc} K_a \mathrm{pieceOfJewelleryIn}(\ell)$.

(ii) 'all drones know that the region is safe every 20 steps' can be expressed as $\mathsf{E}\nu Z.(\bigwedge_{a\in Ag} K_a \mathrm{regionSafe} \wedge \mathsf{X}^{20} Z)$.

(iii) 'with the current plan, the drone a never knows the region is safe but every 10 steps, there is a(nother) plan to let the drone a eventually know the region is safe' can be expressed as $\mathsf{EG}[\neg K_a \mathrm{regionSafe} \wedge \nu Z.(\mathsf{EF} K_a \mathrm{regionSafe} \wedge \mathsf{X}^{10} Z)]$.

The following result subsumes the decidability result for the epistemic planning problem for propositional DEL presentations ([29],[15]), and is a mere corollary of Theorems 5.5, 4.6 and 3.9.

Theorem 5.9 *The generalized epistemic planning problem is decidable for propositional DEL presentations.*

In other terms, in the literature, decidability was established for reachability goals and Theorem 5.9 says that the problem remains decidable for goals that are arbitrary $\mathrm{BL}^{\mathsf{lin}}_{\mu}\mathrm{K}$-formulas φ. However, when considering planning, only formulas φ of the form $\mathsf{E}\varphi'$ are relevant. For such formulas, we can effectively build an automaton that recognizes exactly all plans achieving φ'. This automaton arises from the automata constructions for $\mathrm{BL}^{\mathsf{lin}}_{\mu}\mathrm{K}$-formulas, made possible by the automata constructions for cMSO-formulas, and the fact that every $\mathrm{BL}^{\mathsf{lin}}_{\mu}\mathrm{K}$-formula can be effectively translated into a cMSO-formula. The algorithm to obtain this "plan automaton" is as follows.

(i) Compute the cMSOK-formula χ equivalent to φ' (Th. 4.6). By the translation of $\mathrm{BL}^{\mathsf{lin}}_{\mu}\mathrm{K}$ into cMSO, formula χ has a single free second-order variable X, interpreted as a path, i.e. a plan;

(ii) Compute $\mathcal{B}_{\chi}$ (Proposition 3.11) which accepts all paths achieving goal χ, or equivalently φ'.

Actually, automaton $\mathcal{B}_{\chi}$ reads infinite words over alphabet $E \times \{0, 1\}$, where the second component of the letters is always equal to 1, since the specified chain is required to be a path. Projecting a word accepted by $\mathcal{B}_{\chi}$ onto the first components of its letters provides a plans.

6 Discussion

We have adapted the seminal result of the decidability of the FO-theory of every automatic structure by augmenting the logic up to cMSO, a logic relying on chain-MSO of [12,21], but by restricting to the subclass of RA trees. A nice application of this result is the decidability of the generalized epistemic planning problem for propositional DEL presentations.

Regarding the computational complexity of our algorithms, it should be observed that the automata construction to model check against cMSO (Proposition 3.11) is non elementary in the number of alternations between existential and universal quantifiers in the formula (each underlying negation yields an exponentiation blow-up for the complementation). Still, it would be relevant to investigate in practice, if the automata have particular shapes that do not reach this non-elementary worst-case upper-bound complexity.

Compared to the work of Maubert [15] with CTL*K, we offer the entire logic cMSO (or $\mathrm{BL}^{\mathsf{lin}}_{\mu}\mathrm{K}$) that strictly subsumes CTL*K, and yet allowing to solve the epistemic planning problems over the same class of planning domains (i.e. propositional DEL presentations). However, it should be noticed that [15] also considers a different problem called the *epistemic protocol synthesis problem*, which differs from the generalized epistemic planning problem: in the epistemic protocol synthesis problem, one has to prune the tree structure so that the remaining satisfies a CTL^*K-goal. This subject is out of the scope of this paper, but suggests some comments.

The existence of a pruning, i.e. a sub-tree, that satisfies the protocol goal, requires the use of second-order quantifiers ranging over arbitrary subsets of nodes, that is the full logic MSO which cannot be model checked (Theorem 2.13). It is a open question whetherthere is a logic in between cMSO and full MSO to solve the epistemic protocol synthesis problem with $\mathrm{BL}_\mu^{\mathsf{lin}}\mathrm{K}$ specifications. The existence of a pruning of the tree that satisfies the protocol goal, requires the use of second-order quantifiers ranging over arbitrary subsets of nodes, that is the full logic MSO which cannot be model checked (Theorem 2.13). It is an open question whether or not there is room for a logic in between cMSO and full MSO to solve the epistemic protocol synthesis problem with $\mathrm{BL}_\mu^{\mathsf{lin}}\mathrm{K}$ specifications. Also, the synthesis technique deployed by [15] actually applies to a family of tree structures that is larger than the one of RA trees. At the moment, the technique does not adapt to logics which involve arbitrary fix-points. It is an open question whether or not the work of [15] can be faithfully extended to deal the entire alternation-free fragment of $\mathrm{BL}_\mu^{\mathsf{lin}}\mathrm{K}$.

Acknowledgments We thank the anonymous reviewers for their careful reading of our manuscript and their many insightful comments and suggestions.

References

[1] Aucher, G., B. Maubert and S. Pinchinat, *Automata techniques for epistemic protocol synthesis*, in: *Proceedings 2nd International Workshop on Strategic Reasoning, Grenoble, France*, 2014, pp. 97–103.

[2] Baltag, A., L. S. Moss and S. Solecki, *The logic of public announcements and common knowledge and private suspicions*, in: *Proceedings of the 7th Conference on Theoretical Aspects of Rationality and Knowledge (TARK-98), Evanston, IL, USA*, 1998, pp. 43–56.

[3] Barany, V., "Automatic Presentations of Infinite Structures." Ph.D. thesis, RWTH Aachen University (2007).

[4] Barringer, H., R. Kuiper and A. Pnueli, *A really abstract concurrent model and its temporal logic*, in: *Conference Record of the Thirteenth Annual ACM Symposium on Principles of Programming Languages, St. Petersburg Beach, Florida, USA*, 1986, pp. 173–183.

[5] Blumensath, A. and E. Grädel, *Automatic structures*, in: *15th Annual IEEE Symposium on Logic in Computer Science, Santa Barbara, California, USA*, 2000, pp. 51–62.

[6] Bolander, T., *A gentle introduction to epistemic planning: The DEL approach*, in: *Proceedings of the Ninth Workshop on Methods for Modalities*, M4M@ICLA 2017, Indian Institute of Technology, Kanpur, 2017, pp. 1–22.

[7] Bolander, T. and M. B. Andersen, *Epistemic planning for single and multi-agent systems*, Journal of Applied Non-Classical Logics **21** (2011), pp. 9–34.

[8] Clarke, E. M. and E. A. Emerson, *Design and synthesis of synchronization skeletons using branching time temporal logic*, in: *Workshop on Logic of Programs*, Springer, 1981, pp. 52–71.

[9] Dam, M., *Fixed points of Büchi automata*, in: *International Conference on Foundations of Software Technology and Theoretical Computer Science*, Springer, 1992, pp. 39–50.

[10] Douéneau, G., *Sur les propriétés régulières des arbres* (2015), internship report, IRISA Rennes.

[11] Enderton, H., "A Mathematical Introduction to Logic," Academic press, 2001.

[12] Hafer, T. and W. Thomas, *Computation tree logic CTL* and path quantifiers in the monadic theory of the binary tree*, in: *Automata, Languages and Programming, 14th International Colloquium, ICALP87, Karlsruhe, Germany, Proceedings*, 1987, pp. 269–279.

[13] Halpern, J. Y. and M. Y. Vardi, *The complexity of reasoning about knowledge and time. I. Lower bounds*, J. Comput. System Sci. **38** (1989), pp. 195–237, 18th Annual ACM Symposium on Theory of Computing (Berkeley, CA, 1986).

[14] Kozen, D., *Results on the propositional μ-calculus*, Theoretical Computer Science **27** (1983), pp. 333–354.

[15] Maubert, B., "Logical Foundations of Games with Imperfect Information: Uniform strategies. (Fondations logiques des jeux à information imparfaite: Stratégies uniformes)," Ph.D. thesis, University of Rennes 1, France (2014).

[16] Pnueli, A., *The temporal logic of programs*, in: *Foundations of Computer Science, 1977., 18th Annual Symposium on*, IEEE, 1977, pp. 46–57.

[17] Rabin, M. O., *Decidability of second-order theories and automata on infinite trees*, Transactions of the American Mathematical Society **141** (1969), pp. 1–35.

[18] Rubin, S., *Automata presenting structures: A survey of the finite string case*, Bulletin of Symbolic Logic **14** (2008), pp. 169–209.

[19] Schwarzentruber, F., *Hintikka's world: agents with higher-order knowledge (demo)*, in: *Proceedings of the 27th International Joint Conference on Artificial Intelligence (IJCAI) and the 23rd European Conference on Artificial Intelligence (ECAI)*, Stockholm, 2018.

[20] Tarski, A., *A lattice-theoretical fixpoint theorem and its applications*, Pacific Journal of Mathematics **5** (1955), pp. 285–309.

[21] Thomas, W., *On chain logic, path logic, and first-order logic over infinite trees*, in: *Proceedings of the Symposium on Logic in Computer Science (LICS '87), Ithaca, New York, USA*, 1987, pp. 245–256.

[22] Thomas, W., *Automata on infinite objects*, in: *Handbook of formal languages*, Volume B (1990), pp. 133–191.

[23] Thomas, W., *Infinite trees and automaton-definable relations over ω-words*, Theoretical Computer Science **103** (1992), pp. 143–159.

[24] Thomas, W., *Languages, automata, and logic*, in: *Handbook of Formal Languages*, Springer, 1997 pp. 389–455.

[25] Van Ditmarsch, H., W. van Der Hoek and B. Kooi, "Dynamic Epistemic Logic," Springer Science & Business Media, 2007.

[26] Vardi, M. Y., *A temporal fixpoint calculus*, in: *Proceedings of the 15th ACM SIGPLAN-SIGACT symposium on Principles of programming languages*, ACM, 1988, pp. 250–259.

[27] Vardi, M. Y., *The Büchi complementation saga*, in: *Annual Symposium on Theoretical Aspects of Computer Science*, Springer, 2007, pp. 12–22.

[28] Wolper, P., *Temporal logic can be more expressive*, Information and Control **56** (1983), pp. 72–99.

[29] Yu, Q., X. Wen and Y. Liu, *Multi-agent epistemic explanatory diagnosis via reasoning about actions*, in: *IJCAI 2013, Proceedings of the 23rd International Joint Conference on Artificial Intelligence*, 2013, pp. 1183–1190.

Cut-Free Sequent Calculi for Products and Relativised Products of Modal Logics

Birgit Elbl [1]

UniBw München
85577 Neubiberg, Germany

Abstract

Using structured labels, we define one-sided, contraction-free and cut-free sequent calculi for a class of products and relativised products of modal logics, including some examples which can not be characterised by a finite Hilbert-style axiomatisation. To this end, we introduce the product of labelled calculi. For the special case of products with S5, the method yields two different systems, one based on equivalence frames the other on universal frames, and we show how to translate derivations from one to the other and vice versa. In these calculi, all logical rules are height-preserving invertible. Furthermore, we prove that contraction and cut are admissible.

Keywords: proof theory, cut-free sequent calculus, labelled deduction, products of modal logics, relativised product.

1 Introduction

Building products of structures is a very natural standard construction in mathematics. The product formation of Kripke structures is just a special case. The *product* of n modal *logics* $\mathcal{L}_1, \ldots, \mathcal{L}_n$ is defined to be the set of modal formulas that hold in all products of frames $\mathfrak{F}_1, \ldots, \mathfrak{F}_n$ where each $\mathfrak{F}_i$ is a frame for $\mathcal{L}_i$. As a combination method for modal logics, products have been introduced in [17,18,3]. *Relativised products* are determined by classes of subframes of product frames. A detailed discussion of these notions and major results can be found in [2] and [8]. In particular, axiomatisations for interesting classes are known but there are also very natural logics as the n-dimensional products $\mathbf{K}^n$ and $\mathbf{S5}^n$, $n \geq 3$, which are not finitely axiomatisable (see [7] for $\mathbf{K}^n$, [6,9] for further results).

Labelled calculi provide a general method for constructing *sequent systems* for modal logics, see [11,4]. In that approach, the basic judgements are labelled formulas $x\colon A$ or relational atoms xRy where x, y are labels that denote worlds in a Kripke frame. The statement $x : A$ could be read as "A holds at world x", the expression xRy stands for "y is reachable from x". The different systems

[1] Birgit.Elbl@unibw.de

share a set of logical rules, while each of them includes a set of *frame rules* that reflect the special conditions for the Kripke frames in question.

Since then, several extensions of the language of judgements have been considered. To capture logics based on neighbourhood semantics, labelled calculi have been defined in [13,5,12] that use a multi-sorted labelled language and judgements $a \Vdash^{\exists} A$ and $a \Vdash^{\forall} A$ where the first assertion means that formula A holds at some world x in a and the second stands for "A holds at all worlds x in a". For the logic of subset spaces, a labelled sequent calculus has been presented in [1]. One characteristic of subset spaces is that formulas are evaluated with respect to pairs of *points* x and *opens* u so that $x \in u$, and this is used in the design of the calculus. The language uses two sorts of labels, labelled formulas have the form $(x, u) : A$ where x, u are labels of sort 1 or 2 respectively, and a third type of judgement is introduced referring to the status of the pair (x, u) as a world. The ingredients of the latter approach are used and generalised here to deal with products and relativised products.

In order to obtain *calculi for products*, we define the *product of calculi*. Throughout we work with one-sided sequent systems in the style of the GS-calculi in [16], but extended by labels. The calculi to be combined are cut-free unimodal systems with pairwise distinct modalities. Usually they are formulated with a corresponding symbol for the accessibility relation but we allow the exception of the extremely simple system for S5 which is based on universal frames. The language of the compound system contains the modalities and relational symbols of the components plus a world-predicate. The labels attached to formulas are tuples of simple labels. The logical rules are adapted to this situation and extra frame rules can be added to determine the type of relativisation.

Arbitrary relativised products, expanding products, and non-relativised products serve as examples for instantiations of the general scheme. For the non-relativised case, an equivalent simplified variant is also presented. To illustrate how to work with the calculi, we consider the alternative ways to describe an S5-component, either based on equivalence frames or based on universal frames. In our setting, the equivalence of the two approaches takes the form of comparing the sets of derivable formulas of the corresponding calculi, and we define the supporting translations.

The product calculi are contraction-free and cut-free. We prove that all logical rules are height-preserving invertible and that contraction and cut are admissible. To round off the theory, we sketch a proof of completeness.

2 Preliminaries

2.1 One-sided labelled sequent calculi

We presuppose a fixed set PV of propositional letters. Modal *formulas* are built from the elements of PV using propositional connectives and modalities. In the multi-modal case we use modalities $\Box_1, \ldots, \Box_n, \Diamond_1, \ldots, \Diamond_n$, in the unimodal case also $\Box, \Diamond$. As usual, $\Box_i$ and $\Diamond_i$ are dual to each other. We will present a calculus in the Schütte-Tait style, similar to the GS-calculi in [16].

$$(\text{ax})\ \frac{}{\Gamma, x\colon P, x\colon \neg P} \qquad (\wedge)\ \frac{\Gamma, x\colon A \qquad \Gamma, x\colon B}{\Gamma, x\colon A \wedge B} \qquad (\vee)\ \frac{\Gamma, x\colon A, x\colon B}{\Gamma, x\colon A \vee B}$$

$$(\Box)\ \frac{\Gamma, x\overline{R}y, y\colon A}{\Gamma, x\colon \Box A}\ !(y) \qquad (\Diamond)\ \frac{\Gamma, x\overline{R}y, x\colon \Diamond A, y\colon A}{\Gamma, x\overline{R}y, x\colon \Diamond A}$$

Table 1
The system **GS3K**

Hence we work with formulas in *negation normal form.* This means that our formulas contain no implication and the only negated subformulas are negated propositional variables. Propositional variables P are also called *positive literals*, while the corresponding $\neg P$ is a *negative literal.* Negation *for non-atoms* is given by

$$\neg\Box_i A :\equiv \Diamond_i \neg A \quad \neg\Diamond_i A :\equiv \Box_i \neg A \quad \neg\Box A :\equiv \Diamond\neg A \quad \neg\Diamond A :\equiv \Box\neg A$$

$$\neg\neg P :\equiv P \quad \neg(A \wedge B) :\equiv \neg A \vee \neg B \quad \neg(A \vee B) :\equiv \neg A \wedge \neg B$$

We use $A \to B$ as a shorthand for $\neg A \vee B$. In two-sided sequent derivations of formulas in negation normal form, the left rules for logical symbols will never be used. In the one-sided style, these rules are removed completely, and negative literals are kept in the sole multiset corresponding to the right part of two-sided sequents. Thus the number of rules is reduced drastically, and what's more, the redundancy caused by pairs of dual rules (e.g. right $\wedge$ / left $\vee$) is removed.

Labelled formulas have the form $x : A$ where A is a modal formula and x is taken from a fixed set of labels. As weakening and contraction shall be absorbed into the logical rules, we start with the propositional, cut-free part of the calculus **GS3** in [16]. Prefixing formulas with labels, we obtain the non-modal part of the calculus **GS3K** in Table 1. Originally, Negri's labelled calculi are written in a two-sided style. Then sequents would have the form $\Gamma \Rightarrow \Delta$ where Γ, Δ are multisets of labelled formulas, and relational atoms have the form xRy. Logical axioms for relational atoms are not necessary. If they are not included, the atoms xRy occur on the left of the sequent arrow only. A two-sided sequent $A_1, \ldots, A_m \Rightarrow B_1, \ldots, B_n$ corresponds to the multiset $\neg A_1, \ldots, \neg A_m, B_1, \ldots, B_n$ in the one-sided style. With this transformation, the relational atoms would occur *negated* only in the one-sided system. We rather introduce relational symbols $\overline{R}$ or $\overline{R}_i$ for the complement relation right from the beginning and avoid negation. So relational atoms in the one-sided style have the form $x\overline{R}y$ or $x\overline{R}_iy$, *judgements* are relational atoms or labelled formulas, and a sequent is a multiset of judgements. Transferring Negri's modal rules to the one-sided style, we obtain the system **GS3K** in Table 1. Here $!(y)$ abbreviates the usual eigenvariable condition that y does not occur in the conclusion.

Adding relational rules, one can construct labelled calculi for a large variety of modal systems. The general method presented in [11] applies to normal modal logics which are characterised by universal axioms or, more generally, geometric implications as frame conditions. The two-sided versions of the resulting systems are studied in detail in [11]. As an illustration of the one-sided

$$(\text{ax})\ \frac{}{\Gamma, x\colon P, x\colon \neg P} \qquad (\wedge)\ \frac{\Gamma, x\colon A \qquad \Gamma, x\colon B}{\Gamma, x\colon A \wedge B} \qquad (\vee)\ \frac{\Gamma, x\colon A, x\colon B}{\Gamma, x\colon A \vee B}$$

$$(\Box)\ \frac{\Gamma, y\colon A}{\Gamma, x\colon \Box A}\ !(\mathrm{y}) \qquad (\Diamond)\ \frac{\Gamma, x\colon \Diamond A, y\colon A}{\Gamma, x\colon \Diamond A}$$

Table 2
The system **GS3S5u**

analogues, we present the rules obtained by direct transformation of the properties reflexivity, symmetry, and transitivity:

$$\frac{\Gamma, x\overline{R}x}{\Gamma}\ (\text{ref}) \qquad \frac{\Gamma, x\overline{R}z, x\overline{R}y, y\overline{R}z}{\Gamma, x\overline{R}y, y\overline{R}z}\ (\text{trans}) \qquad \frac{\Gamma, y\overline{R}x, x\overline{R}y}{\Gamma, x\overline{R}y}\ (\text{symm})$$

Note however that in some cases further rules have to be added to satisfy the *closure condition*. In the one-sided style, the geometric rule scheme has the form

$$\frac{\Gamma, U_1, \ldots, U_k, V_1^1, \ldots, V_{l_1}^1 \qquad \ldots \qquad \Gamma, U_1, \ldots, U_k, V_1^m, \ldots, V_{l_m}^m}{\Gamma, U_1, \ldots, U_k}\ !(\bar{x})$$

where the U_i and V_j^i are relational atoms and are called the *principal atoms* of that rule. The closure condition now refers to inferences obtained from instances of this scheme by contracting several occurrences of principal formulas of the conclusion and the corresponding occurrences in the premisses into one. It postulates that these should also be inferences of the system. The case that R is the universal relation could be formalised by a rule which allows to remove any atom $x\overline{R}y$. In that case, however, it would be more natural to remove relational atoms completely and modify the modal rules accordingly. The resulting system **GS3S5u** is shown in Table 2. These one-sided versions have already been used in [1] as the starting point for the development of a calculus. Here, they will be combined to product calculi in Sec. 3.

2.2 Products and relativised products of modal logics

The binary product combines two frames for uni- or multi-modal logics. More generally, one can consider the higher dimensional product of n frames. To simplify notation, we discuss the products of unimodal frames only. As a shorthand, we use $m..n$ for $\{m, \ldots, n\}$ where $m, n \in \mathbb{N}$. For $m > n$, $m..n$ denotes the empty set. Given $n \geq 2$, the *product* of frames $\mathfrak{F}_i = (\mathcal{W}_i, \mathcal{R}_i)$, $i \in 1..n$, is the n-frame

$$\mathfrak{F}_1 \times \ldots \times \mathfrak{F}_n = (\mathcal{W}_1 \times \ldots \times \mathcal{W}_n, \hat{\mathcal{R}}_1, \ldots, \hat{\mathcal{R}}_n)$$

where, for each $i \in 1..n$, the binary relation $\hat{\mathcal{R}}_i$ on $\mathcal{W}_1 \times \ldots \times \mathcal{W}_n$ is given by

$$(u_1, \ldots, u_n)\hat{\mathcal{R}}_i(v_1, \ldots, v_n) \text{ iff } u_i\mathcal{R}_i v_i \text{ and } u_j = v_j, \text{ for } j \neq i.$$

Given n modal logics $\mathcal{L}_i$ formulated in languages that have no modal operator in common, the *product of* $\mathcal{L}_1, \ldots, \mathcal{L}_n$ is the modal logic determined by the class of product frames $\mathcal{F}_1 \times \ldots \times \mathcal{F}_n$ where the $\mathcal{F}_i$ are frames for $\mathcal{L}_i$. For example, $\mathbf{K}^n = \underbrace{\mathbf{K} \times \ldots \times \mathbf{K}}_{n}$ is the logic determined by all n-dimensional product frames, and $\mathbf{S5}^n$ is the logic determined by product frames $\mathcal{F}_1 \times \ldots \times \mathcal{F}_n$ where all $\mathcal{F}_i$ are equivalence frames. The class of products of universal frames $(W_i, W_i \times W_i)$ also determines the logic $\mathbf{S5}^n$. For the axiomatisations of products, the following *commutator axioms* are used: [2]

$$\begin{array}{ll} com^l{}_{i,j} :\equiv \Diamond_j \Diamond_i P \to \Diamond_i \Diamond_j P & com^r{}_{i,j} :\equiv \Diamond_i \Diamond_j P \to \Diamond_j \Diamond_i P (\equiv com^l{}_{j,i}) \\ com_{i,j} :\equiv com^l{}_{i,j} \wedge com^r{}_{i,j} & chr_{i,j} :\equiv \Diamond_i \Box_j P \to \Box_j \Diamond_i P \end{array}$$

The *commutator* $[\mathcal{L}_1, \ldots, \mathcal{L}_n]$ of modal logics $\mathcal{L}_i$, $i = 1, \ldots, n$, is the smallest n-modal logic containing all the $\mathcal{L}_i$ and the axioms $com_{i,j}$ and $chr_{i,j}$ for all $i, j \in 1..n$, $i \neq j$. It is easy to see that the axioms $com_{i,j}$ and $chr_{i,j}$ hold in all product frames. So the product of Kripke complete modal logics is an extension of the commutator. If the converse is also true, $\mathcal{L}_1, \ldots, \mathcal{L}_n$ is called *product-matching.* If $\mathcal{L}_1, \mathcal{L}_2$ are Kripke complete and Horn axiomatisable logics then $\mathcal{L}_1, \mathcal{L}_2$ is product-matching (see [8], Th. 21, and [2], Ch. 5 for a proof). For $n \geq 3$, any n-modal logic $\mathcal{L}$ such that $\mathbf{K}^n \subseteq \mathcal{L} \subseteq \mathbf{S5}^n$ is not finitely axiomatisable (see [8], Th. 25, and [6]). Now let $n \geq 2$, $\mathcal{K}$ be a class of subframes of n-ary product frames, and $\mathcal{L}_1, \ldots, \mathcal{L}_n$ be Kripke complete unimodal logics formulated in languages that have no modal operator in common. The $\mathcal{K}$-*relativised product* $(\mathcal{L}_1 \times \ldots \times \mathcal{L}_n)^{\mathcal{K}}$ is the logic determined by the class of all subframes $\mathcal{G}$ of product frames $\mathcal{F}_1 \times \ldots \times \mathcal{F}_n$ where each $\mathcal{F}_i$ is a frame for $\mathcal{L}_i$, and $\mathcal{G} \in \mathcal{K}$. Note that the usual product is the special case where $\mathcal{K}$ contains all products of frames for $\mathcal{L}_i$. Let SF_n denote the class of all subframes of n-ary product frames. The SF_n-relativised products are also called *arbitrarily relativised products.* For $N \subseteq 1..n$, an n-ary N-*expanding relativised product frame* (see [10]) is a subframe $\mathcal{G} = (W, \ldots)$ of a product frame $\mathcal{F} = \mathcal{F}_1 \times \ldots \times \mathcal{F}_n$, of frames $\mathcal{F}_i = (W_i, R_i)$, $i \in 1..n$, where $(u_1, \ldots, u_{j-1}, v, u_{j+1}, \ldots, u_n) \in W$ for all $j \in N$, $(u_1, \ldots, u_n) \in W$, $v \in W_j$ satisfying $u_j R_j v$. Let EX_n^N be the class of all n-ary N-expanding relativised product frames. These determine the n-ary N-*expanding relativised product.* Axiomatisations for some expanding products can be found in [10]. An overview of results concerning products and other methods of combination can be found in [8].

3 Products of calculi

3.1 Language and rules

Components of the construction. We will combine n labelled calculi $C_1, \ldots, C_n$ and extend the result by further relational rules. We confine ourselves to unimodal logics and assume that the i-th calculus uses modalities $\Box_i, \Diamond_i$ where $\Box_1, \ldots, \Box_n, \Diamond_1, \ldots, \Diamond_n$ are pairwise distinct. We presuppose pairwise distinct relational symbols $R_1, \ldots, R_n$ and assume that the calculus C_i for logic $\mathcal{L}_i$ is formulated using symbol R_i. The case of **GS3S5u**, however,

[2] Here $\equiv$ stands for syntactic identity.

shall be included. To simplify notation, we assume that for some $r \in 0..n$ the first r calculi use the relation symbol R_i, whereas the remaining ones are copies of **GS3S5u**. The given calculi should follow the style in Sec. 2.1. In particular, the logical rules are the ones presented there, and the relational rules follow the given rule scheme and satisfy the closure condition. For the remaining section, we let $n \in \mathbb{N}_+$, $r \in 0..n$, and fix calculi $C_1, \ldots, C_n$ as just described. We use L_i for the corresponding infinite set of labels and add a further infinite set L of labels for the product calculus. For our convenience, we assume that the $L, L_1, \ldots, L_n$ are pairwise disjoint. This is not strictly necessary, as we can deduce the type of every occurrence of a label in a judgement, but it helps to shorten conditions like "x does not occur *as an L_i-label* in the sequent Γ" and similar. To deal with relativisations, the language will in general be extended by a unary predicate symbol $\overline{W}$ which is applied to labels in L. The symbol $\overline{W}$ stands for the complement of the relation "is-a-world" and is used to formalise conditions of the form "if α is a world then ...". Properties of $\overline{W}, \overline{R}_1, \ldots \overline{R}_r$ can be described by further relational rules, see below.

The structure of labels. To deal with an n-dimensional product logic, we choose $L = L_1 \times \ldots \times L_n$. In order to obtain a compact and uniform notation for the calculus, we define the *access* and *update* operations by $(x_1, \ldots, x_n)(i) := x_i$ and $(x_1, \ldots, x_n)[i \leftarrow y] := (x_1, \ldots, x_{i-1}, y, x_{i+1}, \ldots, x_n)$ for $(x_1, \ldots, x_n) \in L$, $i \in 1..n$ and $y \in L_i$. Remember that the elements of L_i are the labels of calculus C_i, typically primitive symbols. The labels $\alpha \in L$, however, are compound and can be considered as terms built from label "variables" $x_1, \ldots, x_n$ in $\bigcup_{i=1}^n L_i$. The notion *occurs in* and the operation *substitution* refer to this feature of L-labels. Apart from that, the system does not rely on the tuple notation but only on the fact that access and update satisfy the following conditions:

- $\alpha[i \leftarrow x](i) = x$ for all $\alpha \in L, i \in 1..n, x \in L_i$
- $\alpha[i \leftarrow x](j) = \alpha(j)$ for all $\alpha \in L, i, j \in 1..n, x \in L_i$ so that $i \neq j$
- $\forall i \in \{1, \ldots, n\}(\alpha(i) = \beta(i)) \rightarrow \alpha = \beta$ for all $\alpha, \beta \in L$

We use $x, x_1, x', y, z, \ldots$ for elements of $\bigcup_{i=1}^n L_i$ and $\alpha, \alpha_1, \alpha', \beta, \gamma, \ldots$ for elements of L. Let $x_1, \ldots, x_k$ be pairwise distinct labels in $\bigcup_{i=1}^n L_i$ and $i_1, \ldots, i_k \in 1..n$ so that $x_\nu \in L_{i_\nu}$ for all $\nu \in 1..k$. A finite set $\theta = \{y_1/x_1, \ldots, y_k/x_k\}$ where $y_\nu \in L_{i_\nu}$ for every $\nu \in 1..k$ is called *finite label substitution.* (We do not exclude $y_\nu \equiv x_\nu$.) For any expression E, $E\theta$ is the result of substituting simultaneously every occurrence of x_ν by y_ν, $\nu = 1, \ldots, k$. We abbreviate $\alpha[i \leftarrow x][j \leftarrow y]$ by $\alpha[i, j \leftarrow x, y]$ if i, j are distinct.

Judgements. *Labelled formulas* now have the form $\alpha : A$ where $\alpha \in L$ and A is a modal formula. *Judgements* are labelled formulas or relational atoms of the form $x\overline{R}_i y$ where $x, y \in L_i$ or $\overline{W}\alpha$ where $\alpha \in L$. The first kind of relational atoms is called R-relational. A *sequent* Γ is a finite multiset of judgements. The notion *occurs in* and the substitution operation are extended to judgements and sequents in the straightforward way. Due to our choice of label sets, the condition that the atom $\overline{W}\alpha$ or some labelled formula $\alpha : A$ occurs in some sequent Γ can be abbreviated to "α occurs in Γ" for $\alpha \in L$.

Definition 3.1 Let $\mathcal{F}_1, \ldots, \mathcal{F}_n$ be frames, $\mathcal{F}_i = (\mathcal{W}_i, \mathcal{R}_i)$ for all $i \in 1..n$, and $\mathcal{F} = (\mathcal{W}, \hat{\mathcal{R}}_1, \ldots, \hat{\mathcal{R}}_n)$ be a subframe of their product. Furthermore let $\mathcal{V} : \mathcal{W} \to \mathrm{PV} \to \mathbb{B}$ be a valuation, and $\mathcal{M} = (\mathcal{F}, \mathcal{V})$. For every $w \in \mathcal{W}$, we write $w \models_{\mathcal{M}} A$ for "A is true at the world w of $\mathcal{M}$". Let $\ell_i : L_i \to \mathcal{W}_i$, $i = 1, \ldots, n$, be mappings, set $\bar{\ell} = \ell_1, \ldots, \ell_n$, and define $\bar{\ell}(\alpha) = (\ell_1(\alpha(1)), \ldots, \ell_n(\alpha(n)))$ for every $\alpha \in L$. Then, based on the validity of formulas, we define the validity of judgements and sequents as follows:

$$
\begin{aligned}
(\mathcal{M}, \bar{\ell}) &\models \overline{W}\alpha &&\iff \bar{\ell}(\alpha) \notin \mathcal{W} \\
(\mathcal{M}, \bar{\ell}) &\models \alpha : A &&\iff \text{if } \bar{\ell}(\alpha) \in \mathcal{W} \text{ then } \bar{\ell}(\alpha) \models_{\mathcal{M}} A \\
(\mathcal{M}, \bar{\ell}) &\models x\overline{R}_i y &&\iff (\ell_i(x), \ell_i(y)) \notin \mathcal{R}_i \text{ for } i \in 1..n \\
(\mathcal{M}, \bar{\ell}) &\models \Gamma &&\iff (\mathcal{M}, \bar{\ell}) \models J \text{ for some judgement } J \text{ in } \Gamma
\end{aligned}
$$

As we use it to formalise conditions on worlds, it is most natural to introduce a corresponding predicate and write $\overline{W}\alpha$ but $\alpha : \bot$ where $\bot$ is any contradictory proposition has the same interpretation.

Relational rules. Let T_i stand for the relational rules of calculus C_i, $i \in 1..r$. These shall be included, where now the context Γ consists of judgements in the extended language. Furthermore, we want to add rules that refer to $\overline{W}$ and possibly several $\overline{R}_i$. It is required that the additional rules have the form

$$
\frac{\Gamma, U_1, \ldots, U_k, V_1^1, \ldots, V_{l_1}^1 \qquad \cdots \qquad \Gamma, U_1, \ldots, U_k, V_1^m, \ldots, V_{l_m}^m}{\Gamma, U_1, \ldots, U_k} \; !(\bar{x})?(\bar{\alpha})
$$

where $\bar{x}, \bar{\alpha}$ are lists of labels, the U_i are R-relational and V_j^i arbitrary relational atoms. These atoms are called *principal*. For the general results below, we need no further restriction on the V_ν^μ. The intended application, however, is describing relativised products. Typically, the condition on subframes consists in postulating that some tuples are worlds if certain preconditions are fulfilled. Then the V_ν^μ would simply be of the form $\overline{W}\beta$. Furthermore, none of the systems discussed in 3.2 makes use of eigenvariables. It is understood that every instance of the rule scheme that satisfies the side conditions is an accepted inference. The $!(\bar{x})$ abbreviates, as usual, the eigenvariable condition for the label $\bar{x} = x_1, \ldots, x_k$, while $?(\bar{\alpha})$ stands for the additional requirement that the elements of $\bar{\alpha} = \alpha_1, \ldots, \alpha_\ell$ do occur in the conclusion. The closure condition now refers to inferences obtained from instances of this rule scheme, in which several $U_{j_1}, \ldots, U_{j_o}$ become identical by substitution. We postulate that the result of contracting these into one, and also contracting the corresponding occurrences in the premisses into one, is an inference of the system. Then the calculi described below depend on

- $n \in \mathbb{N}_+$, the dimension of the product,
- some $r \in \mathbb{N}$ so that $0 \leq r \leq n$ which determines the number of calculi that are no copies of **GS3S5u**,

$$(\mathrm{ax})\ \frac{}{\Gamma, \alpha\colon P, \alpha\colon \neg P} \qquad (\vee)\ \frac{\Gamma, \alpha\colon A, \alpha\colon B}{\Gamma, \alpha\colon A \vee B} \qquad (\wedge)\ \frac{\Gamma, \alpha\colon A \qquad \Gamma, \alpha\colon B}{\Gamma, \alpha\colon A \wedge B}$$

For all $i \in 1..r$:

$$(\Box_i)\ \frac{\Gamma, \overline{W}\alpha, \alpha(i)\overline{R}_i x, \alpha[i \leftarrow x]\colon A}{\Gamma, \alpha\colon \Box_i A}\ !(x)$$

For all $i \in (r+1)..n$:

$$(\Box_i)\ \frac{\Gamma, \overline{W}\alpha, \alpha[i \leftarrow x]\colon A}{\Gamma, \alpha\colon \Box_i A}\ !(x)$$

For all $i \in 1..r$:

$$(\Diamond_i)\ \frac{\Gamma, \alpha(i)\overline{R}_i x, \alpha\colon \Diamond_i A, \alpha[i \leftarrow x]\colon A}{\Gamma, \alpha(i)\overline{R}_i x, \alpha\colon \Diamond_i A}\ ?(\alpha[i \leftarrow x])$$

For all $i \in (r+1)..n$:

$$(\Diamond_i)\ \frac{\Gamma, \alpha\colon \Diamond_i A, \alpha[i \leftarrow x]\colon A}{\Gamma, \alpha\colon \Diamond_i A}\ ?(\alpha[i \leftarrow x])$$

Table 3
The logical rules of $\mathbf{LS}^{n,r}_{W\times}(T)$

- the decision whether to include $\overline{W}$ or not, and
- some set $T \supseteq \bigcup_{i=1}^{r} T_i$ of relational rules as described above.

The resulting labelled system will be denoted by $\mathbf{LS}^{n,r}_{W\times}(T)$ or $\mathbf{LS}^{n,r}_{\times}(T)$ respectively. Slightly misusing notation, as the choice of T depends on the language, we write $\mathbf{LS}^{n,r}_{(W)\times}(T)$ to talk about both variants.

Logical rules. The logical rules of $\mathbf{LS}^{n,r}_{W\times}(T)$ are given in Table 3. The relational part T may introduce further axioms, in which case we refer to the axiom scheme presented in Tab. 3 as *logical axioms.* The side condition in $(\Diamond_i)$ is necessary for soundness, in contrast to the extra $\overline{W}\alpha$ in the premiss of rule $(\Box_i)$. If desired, one could safely add also the variant of the $\Box_i$-rule without $\overline{W}\alpha$. The version above, however, enables us to contract some $\overline{W}\alpha$ into the newly built $\alpha : \Box_i A$, and we will use this possibility below. Note that every relational atom $x\overline{R}_i y$ and every label $\alpha \in L$ in the conclusion of a logical rule occurs also in all premisses.

The world predicate $\overline{W}$ combined with the side conditions "?" are introduced for relativisations. For the non-relativised products, we present the simplified variants $\mathbf{LS}^{n,r}_{\times}(T)$ obtained by removing them. They are given as follows: The rules (ax), $(\wedge)$ and $(\vee)$ are the same as in $\mathbf{LS}^{n,r}_{W\times}(T)$, except that now the context does not contain atoms $\overline{W}\alpha$. The rules for the modalities are given in Tab. 4. We write $\mathbf{LS}^{n,r}_{(W)\times}(T) \vdash \Gamma$ if Γ is derivable in $\mathbf{LS}^{n,r}_{W\times}(T)$, furthermore we use $\vdash^k$ if there is a derivation of height $\leq k$. A modal *formula* A is derivable if the sequent $\alpha : A$ is derivable for some $\alpha \in L$.

Lemma 3.2 (Soundness) *If $\mathbf{LS}^{n,r}_{W\times}(T) \vdash \Gamma$ then $(\mathcal{M}, \bar{\ell}) \models \Gamma$ for all models $\mathcal{M} = (\mathcal{F}, \mathcal{V})$ based on subframes $\mathcal{F}$ of product frames that satisfy all frame conditions determined by T and all appropriate assignment functions $\bar{\ell}$.*

Proof By induction on the height of the derivation. Let $\mathcal{F}_i = (\mathcal{W}_i, \mathcal{R}_i)$ for all

For all $i \in 1..r$:

$$(\Box_i)\ \frac{\Gamma, \alpha(i)\overline{R}_i x, \alpha[i \leftarrow x]\colon A}{\Gamma, \alpha\colon \Box_i A}\ !(x)$$

$$(\Diamond_i)\ \frac{\Gamma, \alpha(i)\overline{R}_i x, \alpha\colon \Diamond_i A, \alpha[i \leftarrow x]\colon A}{\Gamma, \alpha(i)\overline{R}_i x, \alpha\colon \Diamond_i A}$$

For all $i \in (r+1)..n$:

$$(\Box_i)\ \frac{\Gamma, \alpha[i \leftarrow x]\colon A}{\Gamma, \alpha\colon \Box_i A}\ !(x)$$

$$(\Diamond_i)\ \frac{\Gamma, \alpha\colon \Diamond_i A, \alpha[i \leftarrow x]\colon A}{\Gamma, \alpha\colon \Diamond_i A}$$

Table 4
The rules for the modalities in $\mathbf{LS}^{n,r}_{\times}(T)$

$i \in 1..n$, $\mathcal{F} = (\mathcal{W}, \ldots)$ and $\bar{\ell} = \ell_1, \ldots, \ell_n$. We consider the case $\Diamond_i$, $i \in 1..r$. Let $\Gamma = \Gamma', \alpha(i)\overline{R}_i x, \alpha : \Diamond_i A$. By IH $(\mathcal{M}, \bar{\ell}) \models \Gamma, \alpha[i \leftarrow x] : A$. Obviously, Γ holds if $(\ell_i(\alpha(i)), \ell_i(x)) \notin \mathcal{R}_i$. As $\alpha[i \leftarrow x]$ occurs in Γ, that sequent holds if $\bar{\ell}(\alpha[i \leftarrow x]) \notin \mathcal{W}$. Otherwise the validity of $\alpha[i \leftarrow x] : A$ implies validity of $\alpha : \Diamond_i A$, hence again validity of Γ. $\Box$

3.2 Instantiations of the scheme

In this section we consider instantiations of the general scheme to deal with products and several kinds of relativised products. Let T_i again stand for the relational rules of calculus C_i and $T' := \bigcup_{i=1}^{r} T_i$.

Arbitrary and expanding relativisations. To deal with relativisations, we need the version $\mathbf{LS}^{n,r}_{W\times}(T)$ where the language includes the predicate symbol $\overline{W}$ and the rules refer to it. For arbitrarily relativised products, we add no rule for $\overline{W}$ in the calculus. For the expanding relativisations, we introduce the following relational axioms:

$$(i\text{EX})\ \frac{\Gamma, \alpha(i)\overline{R}_i y, \overline{W}\alpha[i \leftarrow y]}{\Gamma, \alpha(i)\overline{R}_i y}\ ?(\alpha) \qquad (i\text{EXu})\ \frac{\Gamma, \overline{W}\alpha[i \leftarrow y]}{\Gamma}\ ?(\alpha)$$

For $N \subseteq 1..n$, let $T^{N\text{-EX}}$ be obtained from T' by adding the rules (iEX) for $i \in N \cap 1..r$ and (iEXu) for $i \in N \cap (r+1)..n$. To illustrate the application of these frame rules we present derivations of the axioms of the expanding commutator.

Lemma 3.3 *Let $N \subseteq 1..n$. The formulas $com^l_{i,j}$ and $chr_{i,j}$ are derivable in $\mathbf{LS}^{n,r}_{W\times}(T^{N-EX})$ if $i \in N$.*

Proof Let $i \in N$. Let P be a propositional symbol, $\alpha \in L$, and $L_i \ni x \neq \alpha(i)$, $L_j \ni y \neq \alpha(j)$ for some $j \in 1..n$, $i \neq j$. We present the derivations for the case that $i, j \leq r$. As abbreviation, let $\Gamma := \overline{W}\alpha, \alpha(j)\overline{R}_j y, \overline{W}\alpha[j \leftarrow y], \alpha(i)\overline{R}_i x$. Note that due to ($i$EX) we can infer Γ, Δ from $\Gamma, \Delta, \overline{W}\alpha[i \leftarrow x]$. So we get the following derivation in $\mathbf{LS}^{n,r}_{W\times}(T^{N\text{-EX}})$:

$$\dfrac{\dfrac{\dfrac{\dfrac{\dfrac{\dfrac{\Gamma, \overline{W}\alpha[i\leftarrow x], \alpha[i,j\leftarrow x,y] : \neg P, \alpha : \Diamond_i\Diamond_j P, \alpha[i\leftarrow x] : \Diamond_j P, \alpha[i,j\leftarrow x,y] : P}{\Gamma, \overline{W}\alpha[i\leftarrow x], \alpha[i,j\leftarrow x,y] : \neg P, \alpha : \Diamond_i\Diamond_j P, \alpha[i\leftarrow x] : \Diamond_j P}(\Diamond_j)}{\Gamma, \overline{W}\alpha[i\leftarrow x], \alpha : \Diamond_i\Diamond_j P, \alpha[i,j\leftarrow x,y] : \neg P}(\Diamond_i)}{\Gamma, \alpha : \Diamond_i\Diamond_j P, \alpha[i,j\leftarrow x,y] : \neg P}(i\mathrm{EX})}{\overline{W}\alpha, \alpha(j)\overline{R}_j y, \alpha : \Diamond_i\Diamond_j P, \alpha[j\leftarrow y] : \Box_i\neg P}(\Box_i)}{\alpha : \Box_j\Box_i\neg P, \alpha : \Diamond_i\Diamond_j P}(\Box_j)}{\alpha : \Box_j\Box_i\neg P \vee \Diamond_i\Diamond_j P}(\vee)$$

The derivation of $\mathrm{chr}_{i,j}$ can be found in the appendix. The remaining cases are similar. □

Note that $\mathrm{com}^l_{j,i}$ is also derivable if $i > r$ but not in general for $i \leq r$.

Non-relativised products. As the product is the special case of the relativised product where every tuple is a world, one can use $\mathbf{LS}^{n,r}_{W\times}(T^{\mathrm{all}})$ where T^{all} is obtained from T' by adding the rule:

$$(\text{all-}W)\ \dfrac{\Gamma, \overline{W}\alpha}{\Gamma}$$

This allows for a unified treatment of relativised and non-relativised products. In the latter case however, the symbol $\overline{W}$ is dispensable. The modal rules for the simplified version without $\overline{W}$ were given in Table 4. In Sec. 4 we show that the two systems for non-relativised products are equivalent. In $\mathbf{LS}^{n,r}_{\times}(T')$, all axioms $\mathrm{com}^l_{i,j}$, $\mathrm{com}^r_{i,j}$ and $\mathrm{chr}_{i,j}$, $i \neq j$, are derivable. To see this, reinspect the derivations in 3.3 and simplify them. Moreover, it is straightforward how to transform derivations in C_i in derivations in $\mathbf{LS}^{n,r}_{\times}(T')$ that prove the same modal formula. For conservativity of the extension, we remark that the derivations in $\mathbf{LS}^{n,r}_{\times}(T')$ have the following properties:

Lemma 3.4 *Let Γ be a sequent and $I \subseteq 1..n$ so that $i \in I$ for all modalities $\Box_i, \Diamond_i$ in Γ. Consider a derivation d of Γ in $\mathbf{LS}^{n,r}_{(W)\times}(T')$. Then for every labelled formula $\beta : B$ in d, B is a subformula of some formula in Γ and for all $j \in 1..n$, $j \notin I$, there is some labelled formula $\alpha : A$ in Γ for which $\beta(j) = \alpha(j)$.*

Proof Inspection of the rules of the calculus. □

Due to this fact, every derivation in $\mathbf{LS}^{n,r}_{\times}(T')$ of some $\alpha : A$ where A contains only $\Diamond_i, \Box_i$ can be transformed in a straightforward way into a derivation in C_i with endsequent $\alpha(i) : A$.

3.3 Basic proof-theoretic properties

For the remainder of this section, we assume a fixed set $T \supseteq \bigcup_{i=1}^{r} T_i$ of rules to be given which satisfies the conditions of Sec. 3.1.

Lemma 3.5 *The following holds for $\mathbf{LS}^{n,r}_{(W)\times}(T)$:*

(i) (renaming) *Let d be a derivation with endsequent Γ, $i \in 1..n$, and $x, y \in L_i$ so that y does not occur in d. Then replacing every occurrence of x in d by y yields a derivation of the endsequent $\Gamma\{y/x\}$.*

(ii) (label substitution) *Let θ be a finite label substitution. Then* $\vdash^k \Gamma$ *implies* $\vdash^k \Gamma\theta$.

(iii) (weakening) $\vdash^k \Gamma \Longrightarrow \vdash^k \Gamma, J$ *for every judgement* J

Proof Straightforward induction on the height of the given derivation. We use Fact (i) in the proofs of (ii),(iii) to avoid a clash with eigenvariables. □

Due to (renaming), we can always assume that the eigenvariables in a given derivation d are only used as eigenvariable for one particular inference and do not occur in any further judgement, sequent or given derivation other than d. We make use of this fact in many of the proofs below, without mentioning it explicitly.

Lemma 3.6 (Relational contraction) *Let Γ be a sequent, $x, y \in L_i$, and $\alpha \in L$.*

(i) (R-contraction) $\vdash^k \Gamma, x\overline{R}_i y, x\overline{R}_i y \Longrightarrow \vdash^k \Gamma, x\overline{R}_i y$

(ii) (W-contraction) $\vdash^k \Gamma, \overline{W}\alpha \Longrightarrow \vdash^k \Gamma$ *if α occurs in Γ.*

Proof By induction on k, using the closure condition for relational rules. □

Theorem 3.7 (Invertibility of logical rules) *The following holds for the calculus $\boldsymbol{LS}^{n,r}_{(W)\times}(T)$:*

(i) ($\wedge$-inversion) $\vdash^k \Gamma, \alpha\colon A \wedge B$ *implies* $\vdash^k \Gamma, \alpha\colon A$ *and* $\vdash^k \Gamma, \alpha\colon B$

(ii) ($\vee$-inversion) $\vdash^k \Gamma, \alpha\colon A \vee B$ *implies* $\vdash^k \Gamma, \alpha\colon A, \alpha\colon B$

(iii) ($\Box_i$-inversion) $\vdash^k \Gamma, \alpha\colon \Box_i A$ *implies* $\vdash^k \Gamma, \alpha(i)\overline{R}_i x, (\overline{W}\alpha,)\alpha[i \leftarrow x]\colon A$ *for every $x \in L_i$ if $i \in 1..r$, and* $\vdash^k \Gamma, (\overline{W}\alpha,)\alpha[i \leftarrow x] : A$ *if $i \in (r+1)..n$.*

Proof By induction on the height of the derivation. Note that in the W-version the sequent(s) after inversion contain every $\beta \in L$ that is in the sequent before inversion. So context conditions $?(\beta)$ are not destroyed. □

Inversion is helpful in proofs of non-derivability. For example, we can show the non-derivability of the formulas $\mathrm{com}^{\ell}_{i,j}$, $i \neq j$, $i, j \in 1..r$, in $\mathbf{LS}^{n,r}_{W\times}(\bigcup_{i=1}^{r} T_i)$ as follows: Suppose we had a derivation of $\alpha : \Box_j\Box_i\neg P \vee \Diamond_i\Diamond_j P$. Then by (inversion) and (weakening) we obtain a derivation of

$$\begin{aligned}&\overline{W}\alpha, \overline{W}\alpha[j \leftarrow y], \alpha(j)\overline{R}_j y, \alpha(i)\overline{R}_i x,\\ &\alpha[i, j \leftarrow x, y] : \neg P, \alpha : \Diamond_i\Diamond_j P, \alpha : \Diamond_j P, \alpha : P, \alpha[j \leftarrow y] : P\end{aligned}$$

for fresh $x \in L_i, y \in L_j$. As the rules in $\bigcup_{i=1}^{r} T_i$, read from bottom to top, add only R-relational atoms, every sequent in that derivation satisfies the side condition $?(\beta)$ for the same expressions β. Hence every ($\Diamond_j$)- or ($\Diamond_i$)-rule, also read from bottom to top, adds no new labelled formulas. As a consequence, the topmost sequent contains the same labelled formulas as the endsequent, and hence it is no axiom.

Moreover, inversion is crucial to the proof of the admissibility of contraction.

Theorem 3.8 (Admissibility of contraction) *Let Γ be a sequent, A a formula and $\alpha \in L$. If $\boldsymbol{LS}^{n,r}_{(W)\times}(T) \vdash^k \Gamma, \alpha\colon A, \alpha\colon A$ then $\boldsymbol{LS}^{n,r}_{(W)\times}(T) \vdash^k \Gamma, \alpha\colon A$.*

Proof By induction on the height of the derivation. In the case that one of the distinguished occurrences of $\alpha : A$ is constructed, we use inversion combined with the IH. Furthermore, in the case of $(\Box_i)$, (R-contraction) (and (W-contraction)) is used. □

4 Comparison of the variations

Again, we fix $n \in \mathbb{N}_+$, $r \in 0..n$, and calculi $C_1, \dots, C_n$ as described in Sec. 2.1, let T_i, $i \in 1..r$, stand for the relational rules in C_i and $T' := \bigcup_{i=1}^{n} T_i$. Furthermore, let T^{all} be obtained from T' by adding the rule (all-W). This rule formalises the fact that every tuple is a world, and can be used to deal with the product as a special case of relativised products. For the non-relativised case however, we also introduced the simplification $\mathbf{LS}^{n,r}_{\times}(T')$ without $\overline{W}$-predicate.

Lemma 4.1 *Let Γ be a sequent not containing $\overline{W}$. Then Γ is derivable in $\boldsymbol{LS}^{n,r}_{W\times}(T^{all})$ if and only if it is derivable in $\boldsymbol{LS}^{n,r}_{\times}(T')$.*

Proof Consider first a $\mathbf{LS}^{n,r}_{W\times}(T^{\text{all}})$-derivation. Removing all instances of the rule (all-W) and all occurrences of some $\overline{W}\beta$ yields a derivation in $\mathbf{LS}^{n,r}_{\times}(T')$. Now we turn to the converse direction and proceed by induction on the height of the given derivation. We present the constructions for the cases of the modalities for $i \in 1..r$.

$(\Box_i)$: Suppose $\Gamma', \alpha(i)\overline{R}_i x, \alpha[i \leftarrow x] : A$ is the endsequent of the derivation obtained by applying the IH. Using (weakening), we can add $\overline{W}\alpha$ to this, and subsequently apply $(\Box_i)$ to deduce $\Gamma', \alpha : \Box_i A$.

$(\Diamond_i)$: Suppose $\Gamma', \alpha : \Diamond_i A, \alpha(i)\overline{R}_i x, \alpha[i \leftarrow x] : A$ is the endsequent of the derivation obtained by applying the IH. Using (weakening), we add $\overline{W}\alpha[i \leftarrow x]$ to this, and, as $\alpha[i \leftarrow x]$ occurs in the result, apply $(\Diamond_i)$ to deduce the sequent $\Gamma', \alpha : \Diamond_i A, \alpha(i)\overline{R}_i x, \overline{W}\alpha[i \leftarrow x]$. Finally, $\overline{W}\alpha[i \leftarrow x]$ is removed by (all-W). □

As a consequence, the sets of derivable formulas in $\mathbf{LS}^{n,r}_{W\times}(T^{\text{all}})$ and $\mathbf{LS}^{n,r}_{\times}(T')$ coincide.

The second type of variation discussed here concerns the formalisation of the S5 components. As **GS3S5u** offers a simplified version without relational symbol for the logic S5, we admitted that system as component calculus. In the definition of the product however, it is stressed that in general one has to consider *all* frames for the logics. So alternatively, we use systems based on the rules for equivalence. For $i \in (r+1)..n$, let T_i consist of the rules (ref), (symm), (trans) formulated for symbol $\overline{R}_i$. Furthermore let $T'_s := T' \cup \bigcup_{i=r+1}^{s} T_i$ for $s \in r..n$. Then the last $n - r$ components of $\mathbf{LS}^{n,n}_{(W)\times}(T'_n)$ are also systems for S5, the logical rules for all components follow the same pattern, and we do not have to distinguish the cases $i \leq r$ and $i > r$ when we *argue about* the product calculus. On the other hand, the components based on **GS3S5u**, which are used in $\mathbf{LS}^{n,r}_{(W)\times}(T')$, are obtained from $\mathbf{LS}^{n,n}_{(W)\times}(T'_n)$ by removing the relational symbols $\overline{R}_i$ for $i \in (r+1)..n$ and the corresponding rules. They are simpler and hence preferable when we *work with* the calculus. For the comparison of these systems, we first prove the following auxiliary lemma.

Lemma 4.2 *Let $s \in (r+1)..n$ and Γ be a sequent so that $\boldsymbol{LS}^{n,s-1}_{\times}(T'_{s-1}) \vdash^k \Gamma$. Then there is a $\boldsymbol{LS}^{n,s-1}_{\times}(T'_{s-1})$-derivation with endsequent Γ and height $\leq k$ so that, for every instance 5*

$$(\star\star) \qquad \frac{\Gamma', \alpha : \Diamond_s A, \alpha[s \leftarrow x] : A}{\Gamma', \alpha : \Diamond_s A}$$

of $(\Diamond_s)$, the label x does occur in the conclusion.

Proof By induction on k. Consider the case in which the last inference has the form $(\star\star)$ where x does *not* occur in the conclusion. By (substitution) we obtain a derivation of height $\leq k-1$ where the endsequent is $\Gamma', \alpha : \Diamond_s A, \alpha : A$. By IH we can assume that the $\Diamond_s$-inferences in that derivation are as requested. Now we apply $(\Diamond_s)$ to build the derivation of Γ. The remaining cases are trivial. □

Definition 4.3 Let $s \in 1..n$, $L', L'' \subseteq L_s$, and Γ be a sequent. Then define

$$\begin{array}{lcl} L_s(\Gamma) & = & \{x \in L_s \mid x \text{ occurs in } \Gamma\} \\ L'\overline{R}_s L'' & = & \text{the least multiset that contains all relational} \\ & & \text{atoms } x'\overline{R}_s x'' \text{ where } x' \in L' \text{ and } x'' \in L'' \end{array}$$

Lemma 4.4 *Let $s \in (r+1)..n$ and Γ be a sequent. Then $\boldsymbol{LS}^{n,s-1}_{(W)\times}(T'_{s-1}) \vdash \Gamma$ implies $\boldsymbol{LS}^{n,s}_{(W)\times}(T'_s) \vdash \Gamma, L_s(\Gamma)\overline{R}_s L_s(\Gamma)$.*

Proof First, we consider the calculus $\mathbf{LS}^{n,s-1}_{\times}(T'_{s-1})$. By Lem. 4.2 we can assume that, for every application $(\star\star)$ of a $(\Diamond_s)$-rule in the given derivation, the variable x does occur in the conclusion. Now we proceed by induction on the height of such a derivation. Let Γ be the endsequent and $\Pi = L_s(\Gamma)\overline{R}_s L_s(\Gamma)$. We distinguish cases according to the rule applied last. We present the cases of the modalities $(\Diamond_s)$ and $(\Box_s)$.

$(\Diamond_s)$: Assume that the last inference has the form $(\star\star)$. As x occurs in the conclusion, we have $L_s(\Gamma) = L_s(\Gamma', \alpha : \Diamond_s A, \alpha[s \leftarrow x] : A)$. By IH the sequent $\Gamma', \alpha : \Diamond_s A, \alpha[s \leftarrow x] : A, \Pi$ is derivable in $\mathbf{LS}^{n,s}_{\times}(T'_s)$. As $\alpha(s)\overline{R}_s x$ occurs in Π, we can deduce $\Gamma', \alpha : \Diamond_s A, \Pi$.

$(\Box_s)$: Assume the last inference is:

$$(\Box_s)\ \frac{\Gamma', \alpha[s \leftarrow x] : A}{\Gamma', \alpha : \Box_s A}\ !(x)$$

Let $\Pi' = L_s(\Gamma')\overline{R}_s L_s(\Gamma')$. By IH the sequent

$$\Gamma', \alpha[s \leftarrow x] : A, \Pi', \{x\}\overline{R}_s L_s(\Gamma'), L_s(\Gamma')\overline{R}_s\{x\}, x\overline{R}_s x$$

is derivable in $\mathbf{LS}^{n,s}_{\times}(T'_s)$. Now we apply (ref) to remove $x\overline{R}_s x$ and (symm) to remove all elements of $\{x\}\overline{R}_s L_s(\Gamma')$.

Subcase 1: $\alpha(s) \in L_s(\Gamma')$. Then, for every $y \in L_s(\Gamma') \setminus \{\alpha(s)\}$, the atom $y\overline{R}_s\alpha(s)$ is in Π', and we can remove $y\overline{R}_s x$ using (trans). Now we have a

deduction of $\Gamma', \alpha[s \leftarrow x] : A, \Pi', \alpha(s)\overline{R}_s x$, and the only occurrences of x in this sequent are those explicitly mentioned. So we can apply $(\Box_s)$ to deduce Γ, Π'.
Subcase 2: $\alpha(s) \notin L_s(\Gamma')$. Then

$$\Pi = \Pi', L_s(\Gamma')\overline{R}_s\{\alpha(s)\}, \{\alpha(s)\}\overline{R}_s L_s(\Gamma'), \alpha(s)\overline{R}_s\alpha(s)$$

and we use (weakening) to obtain a derivation of:

$$\Gamma', \alpha[s \leftarrow x] : A, \Pi, \alpha(s)\overline{R}_s x, L_s(\Gamma')\overline{R}_s\{x\}$$

As, for every $y \in L_s(\Gamma')$, the sequent Π contains $y\overline{R}_s\alpha(s)$, we can apply (trans) to remove $y\overline{R}_s x$. Subsequently, we use $(\Box_s)$ to derive $\Gamma', \alpha : \Box_s A, \Pi$.

The case $\mathbf{LS}^{n,s-1}_{W\times}(T'_{s-1})$ is even simpler. The side condition for $(\Diamond_s)$ ensures that the substituted label x occurs in the conclusion. Moreover, for every instance of a $(\Box_s)$ with conclusion $\Gamma', \alpha : \Box_s A$, we know that α occurs in the premiss, hence we proceed as in the first subcase above. $\Box$

Theorem 4.5 *Let $s \in (r+1)..n$, A be a modal formula and $\alpha \in L$. Then $\mathbf{LS}^{n,s-1}_{(W)\times}(T'_{s-1}) \vdash \alpha : A \iff \mathbf{LS}^{n,s}_{(W)\times}(T'_s) \vdash \alpha : A$.*

Proof Removing all relational atoms $x\overline{R}_s y$ and all instances of rules in $T'_s \setminus T'_{s-1}$ in a $\mathbf{LS}^{n,s}_{(W)\times}(T'_s)$-derivation, we obtain a derivation in $\mathbf{LS}^{n,s-1}_{(W)\times}(T'_{s-1})$. For the converse, we consider a derivation of $\alpha : A$ in $\mathbf{LS}^{n,s-1}_{(W)\times}(T'_{s-1})$ and apply Lem. 4.4 to obtain a derivation of $\alpha(s)\overline{R}_s\alpha(s), \alpha : A$ in $\mathbf{LS}^{n,s}_{(W)\times}(T'_s)$. Using (ref), we can deduce $\alpha : A$. $\Box$

Corollary 4.6 *Let A be a formula and $\alpha \in L$. Then $\mathbf{LS}^{n,r}_{(W)\times}(T') \vdash \alpha : A \iff \mathbf{LS}^{n,n}_{(W)\times}(T'_n) \vdash \alpha : A$.*

5 Admissibility of cut

Next we turn to the cut rule. As we have shown in Sec. 4 that the variants for non-relativised products as well as the systems where $r < n$ are equivalent to some $\mathbf{LS}^{n,n}_{W\times}(\hat{T})$, we consider only the latter type of calculus. Note that the sound version of cut is

$$(\text{cut}) \frac{\Gamma, \alpha : A \qquad \Pi, \alpha : \neg A}{\Gamma, \Pi} ?(\alpha)$$

with the side condition $?(\alpha)$. If both $\alpha : A$ and $\alpha : \neg A$ are true in a model $\mathcal{M}$ w.r.t. the labelling $\bar{\ell}$ then $\bar{\ell}(\alpha)$ is no world in $\mathcal{M}$. If α occurs in Γ, Π then this implies that Γ, Π holds.

Theorem 5.1 (Admissibility of cut) *If $\mathbf{LS}^{n,n}_{W\times}(T) \vdash^k \Gamma, \alpha : A$ and also $\mathbf{LS}^{n,n}_{W\times}(T) \vdash^m \Pi, \alpha : \neg A$ where α occurs in Γ, Π, then $\mathbf{LS}^{n,n}_{W\times}(T) \vdash \Gamma, \Pi$*

Proof By induction on A, side induction on $k + m$.
Case 1: $\alpha : A$ is not constructed in the last inference of the first derivation or $\alpha : \neg A$ is not constructed in the last inference of the second derivation. If that derivation consists of an axiom only, then Γ or Π is an axiom, hence Γ, Π

is. Otherwise, we apply the side induction hypothesis, and deduce Γ, Π in one step. That this is possible depends on the following facts:

- The side IH is applicable: If $\alpha : A$ is not principal in the last inference and α occurs in Γ, then α occurs also in the part Γ' of the premiss $\Gamma', \alpha : A$ of that last inference, similar for the second derivation.
- The potential context conditions of the last inference are not destroyed: Renaming eigenvariables first takes care of the eigenvariable condition. If we consider the first derivation and α does not occur in the part Γ' of the premiss $\Gamma', \alpha : A$, then it occurs in Π, as we postulated that it occurs in Γ, Π. A similar argument applies to the second derivation.

Case 2: Both $\alpha : A$ and $\alpha : \neg A$ are principal in the last inference. If A is a positive or negative literal then $\alpha : \neg A$ must be a labelled formula in Γ and $\alpha : A$ must be a labelled formula in Π. Hence Γ, Π is an axiom in this case. If the principal symbols in A and $\neg A$ are $\vee, \wedge$, we just have to apply the IH and (contraction). We present the more involved case that their principal symbols are modalities. W.l.o.g. the principal symbol in A is some $\Box_i$. Then the principal symbol in $\neg A$ is $\Diamond_i$. Let B be a formula and $x, y \in L_i$ so that the premiss of the first derivation is $\Gamma, \overline{W}\alpha, \alpha(i)\overline{R}_i x, \alpha[i \leftarrow x] : B$ and the premiss of the second is $\Pi, \alpha : \Diamond_i B, \alpha[i \leftarrow y] : \neg B$. (Note that $\alpha(i)\overline{R}_i y$ is an element of Π). Using (substitution) we obtain a derivation of $\Gamma, \overline{W}\alpha, \alpha(i)\overline{R}_i y, \alpha[i \leftarrow y] : B$. Combining the first derivation with the immediate subderivation of the second and applying SIH, we get a derivation of $\Gamma, \Pi, \alpha[i \leftarrow y] : \neg B$. Now either $y = \alpha(i)$, in which case $\alpha[i \leftarrow y] = \alpha$ and occurs in Γ, Π, or $\alpha[i \leftarrow y]$ must occur in Π, as the side condition $?(\alpha[i \leftarrow y])$ was satisfied for the last inference in the second derivation. In both cases we can apply IH to obtain a derivation of $\Gamma, \overline{W}\alpha, \alpha(i)\overline{R}_i y, \Gamma, \Pi$, and applications of (contraction), (R-contraction) and (W-contraction) complete the proof in this case. □

6 Completeness

In the cases where an axiomatisation of the (relativised) product is known, this can be used to show the completeness of the product calculi. As (cut) is admissible in the product calculi $\mathbf{LS}^{n,n}_{W\times}(T)$, the set of derivable formulas is closed under modus ponens. Obviously, it is also closed under necessitation. So completeness follows, if the axioms are derivable. In particular, for product-matching logics, the derivability of the commutator axioms yields completeness. A similar remark applies to the cases where the expanding commutator and the e-commutator coincide, and to the cases where the arbitrarily relativised product is indeed the fusion of the logics. This argument, however, does not cover, for example, the two most simple systems $\mathbf{LS}^{n,n}_{\times}(\emptyset)$ and $\mathbf{LS}^{n,0}_{\times}(\emptyset)$ which are candidates for the logics $\mathbf{K}^n$ and $\mathbf{S5}^n$. Hence we conclude with presenting a general, direct proof of completeness in the style of [15,14]. In the sequel, we sketch the argument and transfer more details to the appendix.

We consider a possibly infinite proof-search tree $\mathfrak{T}$ for a sequent Γ which is constructed by repeatedly extending a finite deduction tree at its leaves.

In every step, a rule of the system is applied bottom-up. If the result $\mathfrak{T}$ of this process is a finite tree where all leaves carry axioms, we have found a derivation of Γ. The construction is organised in such a manner that we can define a countermodel in the remaining cases. If $\mathfrak{T}$ contains a non-axiom leaf where no reduction step is applicable, we choose such a node, and let $\pi = N_1, \ldots, N_p$ denote the path from the root to that leaf. Otherwise the tree $\mathfrak{T}$ is a finitely-branching, infinite tree. Then let $\pi = N_1, \ldots, N_\nu, \ldots$ be an infinite maximal path starting at the root. Let Δ be the union of all judgements at the nodes N_ν of π. Define $\mathcal{W}_i := \{x \in L_i \mid x \text{ occurs in } \Delta\}$ for $i \in 1..n$, $\mathcal{R}_i := \{(x,y) \in \mathcal{W}_i \mid x\overline{R}_i y \text{ occurs in } \Delta\}$ for $i = 1..r$, $\mathcal{R}_i := \mathcal{W}_i \times \mathcal{W}_i$ for $i \in (r+1)..n$, and $\mathcal{F}_i := (\mathcal{W}_i, \mathcal{R}_i)$ for $i \in 1..n$. In case of $\mathbf{LS}^{n,r}_{W\times}(T)$, we let $\mathcal{W} := \{\alpha \in L \mid \alpha \text{ occurs in } \Delta\}$ otherwise $\mathcal{W} := \mathcal{W}_1 \times \ldots \times \mathcal{W}_n$. In case of the non-relativised product, we let $\mathcal{F}$ denote the product frame of the $\mathcal{F}_i$, otherwise the subframe of the product determined by $\mathcal{W}$. Define $\mathcal{V} : \mathcal{W} \to \mathrm{PV} \to \mathbb{B}$ by

$$\mathcal{V}(\alpha)(P) := \begin{cases} \mathbf{t} \text{ if } \alpha : \neg P \text{ occurs on } \pi \\ \mathbf{f} \text{ otherwise} \end{cases}$$

Note that $\mathcal{V}(\alpha)(P) = \mathbf{f}$ if $\alpha : P$ occurs on π, as $\alpha : P$ and $\alpha : \neg P$ can not occur both on π. Now consider the model $\mathcal{M} := (\mathcal{F}, \mathcal{V})$, let $\ell_i(x) = x$ for all $x \in \mathcal{W}_i$ and $\bar{\ell} = \ell_1, \ldots, \ell_n$. (For y not in π, $\ell_i(y)$ is irrelevant.) For this model, $(\mathcal{M}, \bar{\ell}) \not\models \alpha : A$ for all $\alpha : A$ in Δ can be shown by induction on A. This completes the proof of the following theorem:

Theorem 6.1 *Let A be a formula which is valid in all (subframes of) frames that are products of frames for C_i. Then $\boldsymbol{LS}^{n,r}_{(W)\times}(T) \vdash \alpha : A$.*

7 Conclusion

We have developed a general strategy for building *products* of labelled calculi. This way we obtain systems for a large class of products and relativised products of modal logics. In particular, we obtain rather handy systems for $\mathbf{K}^n$ and $\mathbf{S5}^n$. For product matching logics, these calculi offer an alternative to the axiomatisation as commutators. Note that the regular rule scheme is sufficient for the product if it is sufficient for the components. In contrast to the one-dimensional system obtained by transforming the frame conditions for the commutator into rules, the construction of the product calculus does not introduce rules with eigenvariables. We considered arbitrary relativisations, expanding relativisations and non-relativised products, and we presented proof-theoretic arguments for some basic facts. The theory of combinations of modal logics, however, comprises many results that were not touched here. Product calculi could be used to add proof-theoretic arguments to the picture but this is still up to further work.

Appendix

A Derivations of the commutator axioms

Derivation of $\mathrm{chr}_{i,j}$ in $\mathbf{LS}^{n,r}_{W\times}(T^{N\text{-EX}})$: Let $\Delta := \overline{W}\alpha, \alpha(i)\overline{R}_i x, \overline{W}\alpha, \alpha(j)\overline{R}_j y$ and $\Delta' := \Delta, \overline{W}\alpha[i,j\leftarrow x,y], \alpha[i\leftarrow x] : \Diamond_j\neg P$.

$$
\dfrac{\dfrac{\dfrac{\dfrac{\dfrac{\dfrac{\Delta', \alpha[j\leftarrow y] : \Diamond_i P, \alpha[i,j\leftarrow x,y] : \neg P, \alpha[i,j\leftarrow x,y] : P}{\Delta, \overline{W}\alpha[i,j\leftarrow x,y], \alpha[i\leftarrow x] : \Diamond_j\neg P, \alpha[j\leftarrow y] : \Diamond_i P, \alpha[i,j\leftarrow x,y] : \neg P}(\Diamond_i)}{\Delta, \overline{W}\alpha[i,j\leftarrow x,y], \alpha[i\leftarrow x] : \Diamond_j\neg P, \alpha[j\leftarrow y] : \Diamond_i P}(\Diamond_j)}{\Delta, \alpha[i\leftarrow x] : \Diamond_j\neg P, \alpha[j\leftarrow y] : \Diamond_i P}(i\mathrm{EX})}{\overline{W}\alpha, \alpha(i)\overline{R}_i x, \alpha[i\leftarrow x] : \Diamond_j\neg P, \alpha : \Box_j\Diamond_i P}(\Box_j)}{\alpha : \Box_i\Diamond_j\neg P, \alpha : \Box_j\Diamond_i P}(\Box_i)}{\alpha : \Box_i\Diamond_j\neg P \vee \Box_j\Diamond_i P}(\vee)
$$

Derivation of $\mathrm{com}^l_{i,j}$ in $\mathbf{LS}^{n,r}_{\times}(T)$ if $i,j \in 1..r$, $i \neq j$ where $x \in L_i$ and $y \in L_j$ are fresh:

$$
\dfrac{\dfrac{\dfrac{\dfrac{\dfrac{\alpha(j)\overline{R}_j y, \alpha(i)\overline{R}_i x, \alpha[i,j\leftarrow x,y] : \neg P, \alpha : \Diamond_i\Diamond_j P, \alpha[i\leftarrow x] : \Diamond_j P, \alpha[i,j\leftarrow x,y] : P}{\alpha(j)\overline{R}_j y, \alpha(i)\overline{R}_i x, \alpha[i,j\leftarrow x,y] : \neg P, \alpha : \Diamond_i\Diamond_j P, \alpha[i\leftarrow x] : \Diamond_j P}(\Diamond_j)}{\alpha(j)\overline{R}_j y, \alpha(i)\overline{R}_i x, \alpha : \Diamond_i\Diamond_j P, \alpha[i,j\leftarrow x,y] : \neg P}(\Diamond_i)}{\alpha(j)\overline{R}_j y, \alpha : \Diamond_i\Diamond_j P, \alpha[j\leftarrow y] : \Box_i\neg P}(\Box_i)}{\alpha : \Box_j\Box_i\neg P, \alpha : \Diamond_i\Diamond_j P}(\Box_j)}{\alpha : \Box_j\Box_i\neg P \vee \Diamond_i\Diamond_j P}(\vee)
$$

Derivation of $\mathrm{chr}_{i,j}$ in $\mathbf{LS}^{n,r}_{\times}(T)$ if $i,j \in 1..r$, $i \neq j$ where $x \in L_i$ and $y \in L_j$ are fresh:

$$
\dfrac{\dfrac{\dfrac{\dfrac{\dfrac{\alpha(i)\overline{R}_i x, \alpha(j)\overline{R}_j y, \alpha[i\leftarrow x] : \Diamond_j\neg P, \alpha[j\leftarrow y] : \Diamond_i P, \alpha[i,j\leftarrow x,y] : \neg P, \alpha[i,j\leftarrow x,y] : P}{\alpha(i)\overline{R}_i x, \alpha(j)\overline{R}_j y, \alpha[i\leftarrow x] : \Diamond_j\neg P, \alpha[j\leftarrow y] : \Diamond_i P, \alpha[i,j\leftarrow x,y] : \neg P}(\Diamond_i)}{\alpha(i)\overline{R}_i x, \alpha(j)\overline{R}_j y, \alpha[i\leftarrow x] : \Diamond_j\neg P, \alpha[j\leftarrow y] : \Diamond_i P}(\Diamond_j)}{\alpha(i)\overline{R}_i x, \alpha[i\leftarrow x] : \Diamond_j\neg P, \alpha : \Box_j\Diamond_i P}(\Box_j)}{\alpha : \Box_i\Diamond_j\neg P, \alpha : \Box_j\Diamond_i P}(\Box_i)}{\alpha : \Box_i\Diamond_j\neg P \vee \Box_j\Diamond_i P}(\vee)
$$

B Proof of Completeness

We construct a possibly infinite proof-search tree $\mathfrak{T}$ for a sequent Γ by repeatedly extending a finite deduction tree at its leaves. In every step, a rule of the system is applied bottom-up. First, we add some details concerning the single steps:

(i) In case of $(\vee)$ or $(\wedge)$, the expansion is completely determined. Note that applying $(\wedge)$ leads to a branching of the tree, duplicating the remaining applicable steps at this node. We can improve the procedure by observing that a reduction of $\alpha : A \wedge B$ is redundant if $\alpha : A$ or $\alpha : B$ already occurs on the path to the current node, and exclude redundant reductions. A similar remark applies to $\alpha : A \vee B$ if both $\alpha : A$ and $\alpha : B$ occur on the path to the node.

(ii) On reducing a $\Box_i$-formula, we have to choose an eigenvariable. We stipulate that this should be a label which is fresh with respect to the construction so far, and assume some arbitrary but fixed function which gives us the next fresh label of the requested type. A reduction of $\alpha : \Box_i A$ is redundant if some $\alpha[i \leftarrow x] : A$ does already occur on the path to the current node.

(iii) In the premiss of $(\Diamond_i)$, the label x which is substituted is not uniquely determined. Here we decide that, if $(\Diamond_i)$ is chosen, a finite sequence of $(\Diamond_i)$-steps should be performed, corresponding to different choices of x:
   (a) In the calculus $\mathbf{LS}^{n,r}_{W\times}(T)$, the side condition $?(\alpha[i \leftarrow x])$ takes care that there are only finitely many alternatives. We postulate that all of them should be considered.
   (b) In the calculus $\mathbf{LS}^{n,r}_{\times}(T)$, the label x could be any element of L_i. We decide that every label x which occurs on the path to the actual node should be considered. As the principal formula $\alpha : \Diamond_i A$ is repeated in the premiss, it stays reducible and will be reconsidered, as soon as new labels have been introduced.

   We apply all those corresponding steps, for which the formula $\alpha[i \leftarrow x]$ does not already occur on the path to the node.

(iv) Relational rules are treated similarly. In general they may combine the aspects discussed above: branching of the tree, choosing an eigenvariable, considering several instances at once. These should be dealt with as in the case of logical rules. A bottom-up-application of a relational rule is redundant if, for some of its premisses, all atoms which would be introduced by this step are already present on the path to the current node.

We work on all leaves in parallel. The process is nevertheless non-deterministic, as we do not fix how to choose the steps. We just give some rules. Observe that, if a possible reduction is not chosen, the relevant formulas for this step are copied to all premisses. We call these copies and the copies thereof upwards on the path *descendants*. The postulates are as follows:
Termination: Nodes that carry axioms are not extended.
Continuation: If a node does not carry an axiom and at least one non-redundant step is possible, then one of these steps is chosen and performed.
Fairness: For all nodes N, non-redundant reduction steps at this node and the corresponding relevant formulas, as well as paths π in the final tree starting at N, we have the following: If none of the descendants of the relevant formulas are chosen on that path, although the corresponding reduction stays non-redundant, then the path ends with a leaf carrying an axiom.

Fairness can be achieved by assigning at every stage higher priority to those reductions already enabled to those constructed by choosing an alternative step.

Now consider the result $\mathfrak{T}$ of this process. If this a finite tree where all leaves carry axioms, we have found a derivation of Γ. If it contains a non-axiom leaf where no reduction step is applicable, we choose such a node, and let $\pi = N_1, \ldots, N_p$ denote the path from the root to that leaf. If none of these is true, then the tree $\mathfrak{T}$ is a finitely-branching, infinite tree. Then let $\pi = N_1, \ldots, N_\nu, \ldots$ be an infinite maximal path starting at the root. Let Δ be the union of all judgements at the nodes N_ν. Based on Δ, we define a countermodel.

Let $\mathcal{W}_i := \{x \in L_i \mid x \text{ occurs in } \Delta\}$ for $i \in 1..n$, $\mathcal{R}_i := \{(x, y) \in \mathcal{W}_i \mid x\overline{R}_i y \text{ occurs in } \Delta\}$ for $i = 1..r$, $\mathcal{R}_i := \mathcal{W}_i \times \mathcal{W}_i$ for $i \in (r+1)..n$, and $\mathcal{F}_i := (\mathcal{W}_i, \mathcal{R}_i)$ for $i \in 1..n$. In case of $\mathbf{LS}^{n,r}_{W\times}(T)$, let $\mathcal{W} := \{\alpha \in L \mid \alpha \text{ occurs in } \Delta\}$ otherwise $\mathcal{W} := \mathcal{W}_1 \times \ldots \times \mathcal{W}_n$. Observe that the conditions given by the rules in T are satisfied. To see this, consider a frame condition

$$\forall\bar{y}\forall\bar{z}(\bigwedge_{\nu=1}^{k} U_\nu \wedge W\alpha_1 \wedge \ldots \wedge W\alpha_\ell \to \exists\bar{x} \bigvee_{\mu=1}^{m} \bigwedge_{\rho=1}^{l_m} V_\nu^\mu)$$

and the corresponding frame rule

$$\frac{\Gamma, U_1, \ldots, U_k, V_1^1, \ldots, V_{l_1}^1 \qquad \ldots \qquad \Gamma, U_1, \ldots, U_k, V_1^m, \ldots, V_{l_m}^m}{\Gamma, U_1, \ldots, U_k} \; !(\bar{x})?(\bar{\alpha})$$

Here $\bar{y}$ is the list of labels in $U_1, \ldots, U_k$ and $\bar{z}$ is the list of labels in $V_1^1, \ldots, V_{l_m}^m$ that are neither eigenvariables nor occur in $\bar{y}$. (We assume $\bar{x}, \bar{y}, \bar{z}$ to be pairwise distinct.) Let θ be a finite label substitution, replacing the labels $\bar{y}, \bar{z}$ by appropriate labels in $\bigcup_{i=1}^{n} \mathcal{W}_i$ so that its application turns $\forall\bar{y}\forall\bar{z}(\bigwedge_{\nu=1}^{k} U_\nu \wedge W\alpha_1 \wedge \ldots \wedge W\alpha_\ell)$ into a true assertion where we assume the interpretation given by $(\mathcal{W}, \mathcal{R}_1, \ldots, \mathcal{R}_n)$. This means that $\alpha_1\theta, \ldots, \alpha_\ell\theta$ as well as $U_1\theta, \ldots, U_k\theta$ occur at some nodes of π. Furthermore, the labels $z_\rho\theta$ occur in Δ. As relational atoms in the conclusion of an inference are also contained in all premisses, there must be some node N of π where *all* these atoms are present and all $z_\rho\theta$ occur on the path from the root to N. At this point, the application of the relational rule would be possible. As π does not contain an axiom, the reduction of the descendants must either become redundant or actually be performed. In both cases, for some $\mu \in 1..m$, all atoms $V_1^\mu\theta, \ldots, V_{l_\mu}^\mu\theta$ occur on the path π. The verification of the frame conditions of C_i is similar but simpler. Consequently, the frames $(\mathcal{W}_i, \mathcal{R}_i)$ are frames for C_i. In case of the non-relativised product, we let $\mathcal{F}$ denote their product frame, otherwise the subframe of the product determined by $\mathcal{W}$. We define $\mathcal{V} : \mathcal{W} \to \mathrm{PV} \to \mathbb{B}$ by

$$\mathcal{V}(\alpha)(P) := \begin{cases} \mathbf{t} \text{ if } \alpha : \neg P \text{ occurs on } \pi \\ \mathbf{f} \text{ otherwise} \end{cases}$$

Note that $\mathcal{V}(\alpha)(P) = \mathbf{f}$ if $\alpha : P$ occurs on π, as $\alpha : P$ and $\alpha : \neg P$ can not occur both on π.

We consider the model $\mathcal{M} := (\mathcal{F}, \mathcal{V})$, let $\ell_i(x) = x$ for all $x \in \mathcal{W}_i$, and $\bar{\ell} = \ell_1, \ldots, \ell_n$. (For y not in π, $\ell_i(y)$ is irrelevant.) For this model, $(\mathcal{M}, \bar{\ell}) \not\models \alpha : A$ for all $\alpha : A$ in Δ , which completes the proof of the theorem, and can be shown by induction on A. We present the case $A \equiv \Diamond_i B$ in the version with $\overline{W}$. Let $x \in L_i$ so that $\alpha[i \leftarrow x]$ and $\alpha(i)\overline{R}_i x$ occur on π. Then at some node N of π, there is a sequent Γ' containing $\alpha : \Diamond_i B$ so that $\alpha[i \leftarrow x]$ and $\alpha(i)\overline{R}_i x$ occur in Γ'. Then either the corresponding reduction is performed on the path or it becomes redundant. In both cases, we know that $\alpha[i \leftarrow x] : B$ occurs on π. Application of the IH yields $(\mathcal{M}, \bar{\ell}) \not\models \alpha[i \leftarrow x] : B$. The remaining cases are similar.

References

[1] Elbl, B., *A cut-free sequent calculus for the logic of subset spaces*, in: L. Beklemishev, S. Demri and A. Máté, editors, *Advances in Modal Logic, AiML 11*, Advances in Modal Logic **11**, 2016, pp. 268–287.

[2] Gabbay, D., A. Kurucz, F. Wolter and M. Zakharyaschev, "Many-Dimensional Modal Logics: Theory and Applications," Number 148 in Studies in Logic, Elsevier, 2003.

[3] Gabbay, D. and V. Shehtman, *Products of modal logics. part I.*, Journal of the IGPL **6** (1998), pp. 73–146.

[4] Garg, D., V. Genovese and S. Negri, *Countermodels from sequent calculi in multi-modal logics*, in: *Proceedings of the 2012 27th Annual IEEE/ACM Symposium on Logic in Computer Science*, LICS '12 (2012), pp. 315–324.

[5] Girlando, M., S. Negri, N. Olivetti and V. Risch, *The logic of conditional beliefs: Neighbourhood semantics and sequent calculus*, in: L. Beklemishev, S. Demri and A. Máté, editors, *Advances in Modal Logic, AiML 11*, Advances in Modal Logic **11**, 2016, pp. 322–341.

[6] Hirsch, R., I. Hodkinson and A. Kurucz, *On modal logics between* $\boldsymbol{K} \times \boldsymbol{K} \times \boldsymbol{K}$ *and* $\boldsymbol{S5} \times \boldsymbol{S5} \times \boldsymbol{S5}$, Journal of Symbolic Logic **67** (2002), pp. 221–234.

[7] Kurucz, A., *On axiomatising products of Kripke frames*, Journal of Symbolic Logic **65** (2000), pp. 923–945.

[8] Kurucz, A., *Combining modal logics*, in: P. Blackburn, J. van Benthem and F. Wolter, editors, *Handbook of Modal Logic*, Studies in Logic and Practical Reasoning, Elsevier, 2007 pp. 869–924.

[9] Kurucz, A. and S. Marcelino, *Non-finitely axiomatisable two-dimensional modal logics*, Journal of Symbolic Logic **77** (2012), pp. 970–986.

[10] Kurucz, A. and M. Zakharyaschev, *A note on relativised products of modal logics.*, in: P. Balbiani, N.-Y. Suzuki, F. Wolter and M. Zakharyaschev, editors, *Advances in Modal Logic*, Advances in Modal Logic **4** (2003), pp. 221–242.

[11] Negri, S., *Proof analysis in modal logic*, Journal of Philosophical Logic **34** (2005), pp. 507–544.

[12] Negri, S., *Proof theory for non-normal modal logics: The neighbourhood formalism and basic results*, IFCoLog Journal of Logics and their Applications (2017).

[13] Negri, S. and N. Olivetti, *A sequent calculus for preferential conditional logic based on neighbourhood semantics*, in: *Automated Reasoning with Analytic Tableaux and Related Methods*, LNCS **9323**, 2015, pp. 115–134.

[14] Schütte, K., *Ein System des verknüpfenden Schließens*, Archiv für mathematische Logik und Grundlagenforschung (1956).

[15] Schütte, K., "Proof Theory," Springer, 1977.

[16] Schwichtenberg, H. and A. Troelstra, "Basic Proof Theory," Cambridge University Press, 1996.

[17] Segerberg, K., *Two-dimensional modal logic*, Journal of Philosophical Logic **2** (1973), pp. 77–96.

[18] Shehtman, V., *Two-dimensional modal logics*, Mathematical Notices of the USSR Academy of Sciences **23** (1978), pp. 417–424, (Translated from Russian).

Ruitenburg's Theorem via Duality and Bounded Bisimulations

Silvio Ghilardi

Università degli Studi di Milano, Italy

Luigi Santocanale [1]

LIS, CNRS UMR 7020, Aix-Marseille Université

Abstract

For a given intuitionistic propositional formula A and a propositional variable x occurring in it, define the infinite sequence of formulae $\{ A^i \}_{i \geq 1}$ by letting A^1 be A and A^{i+1} be $A(A^i/x)$. Ruitenburg's Theorem [8] says that the sequence $\{ A^i \}_{i \geq 1}$ (modulo logical equivalence) is ultimately periodic with period 2, i.e. there is $N \geq 0$ such that $A^{N+2} \leftrightarrow A^N$ is provable in intuitionistic propositional calculus. We give a semantic proof of this theorem, using duality techniques and bounded bisimulations ranks.

Keywords: Ruitenburg's Theorem, Sheaf Duality, Bounded Bisimulations.

1 Introduction

Let us call an infinite sequence

$$a_1, a_2, \ldots, a_i, \ldots$$

ultimately periodic iff there are N and k such that for all $s_1, s_2 \geq N$, we have that $s_1 \equiv s_2 \mod k$ implies $a_{s_1} = a_{s_2}$. If (N, k) is the smallest (in the lexicographic sense) pair for which this happens, we say that N is an *index* and k a *period* for the ultimately periodic sequence $\{ a_i \}_i$. Thus, for instance, an ultimately periodic sequence with index N and period 2 looks as follows

$$a_1, \ldots, a_N, a_{N+1}, a_N, a_{N+1}, \ldots$$

A typical example of an ultimately periodic sequence is the sequence of the iterations $\{ f^i \}_i$ of an endo-function f of a finite set. Whenever infinitary data are involved, ultimate periodicity comes often as a surprise.

Ruitenburg's Theorem is in fact a surprising result stating the following: take a formula $A(x, \underline{y})$ of intuitionistic propositional calculus (IPC) (by the

[1] Partially supported by the "LIA LYSM AMU CNRS ECM INdAM"

notation $A(x, \underline{y})$ we mean that the only propositional letters occurring in A are among $x, \underline{y}$ - with $\underline{y}$ being, say, the tuple $y_1, \ldots, y_n$) and consider the sequence $\{ A^i(x, \underline{y}) \}_{i \geq 1}$ so defined:

$$A^1 :\equiv A, \quad \ldots, \quad A^{i+1} :\equiv A(A^i/x, \underline{y}) \tag{1}$$

where the slash means substitution; then, *taking equivalence classes under provable bi-implication in* (*IPC*), *the sequence* $\{ [A^i(x, \underline{y})] \}_{i \geq 1}$ *is ultimately periodic with period 2.* The latter means that there is N such that

$$\vdash_{IPC} A^{N+2} \leftrightarrow A^N \quad . \tag{2}$$

An interesting consequence of this result is that *least (and greatest) fixpoints of monotonic formulae are definable in* (*IPC*) [7,6,4]: this is because the sequence (1) becomes increasing when evaluated on $\bot/x$ (if A is monotonic in x), so that the period is decreased to 1. Thus the index of the sequence becomes a finite upper bound for the fixpoints approximation convergence.

Ruitenburg's Theorem was shown in [8] via a, rather involved, purely syntactic proof. The proof has been recently formalized inside the proof assistant COQ by T. Litak (see `https://git8.cs.fau.de/redmine/projects/ruitenburg1984`). In this paper we supply a semantic proof, using duality and bounded bisimulation machinery.

Bounded bisimulations are a standard tool in non classical logics [2] which is used in order to characterize satisfiability of bounded depth formulae and hence definable classes of models: examples of the use of bounded bisimulations include for instance [9], [5], [10], [3].

Duality has a long tradition in algebraic logic (see e.g. [1] for the Heyting algebras case): many phenomena look more transparent whenever they are analyzed in the dual categories, especially whenever dualities can convert coproducts and colimits constructions into more familiar 'honest' products and limits constructions. The duality we use here is taken from [5] and has a mixed geometric/combinatorial nature. In fact, the geometric environment shows *how to find* relevant mathematical structures (products, equalizers, images,...) using their standard definitions in sheaves and presheaves; on the other hand, the combinatorial aspects show that such constructions *are definable*, thus meaningful from the logical side. In this sense, notice that we work with finitely presented algebras, and our combinatoric ingredients (Ehrenfeucht-Fraissé games, etc.) replace the topological ingredients which are common in the algebraic logic literature (working with arbitrary algebras instead).

The paper is organized as follows. In Section 2 we show how to formulate Ruitenburg's Theorem in algebraic terms and how to prove it via duality in the easy case of classical logic (where index is always 1). This Section supplies the methodology we shall follow in the whole paper. After introducing the required duality ingredients for finitely presented Heyting algebras (this is done in Section 3 - the material of this Section is taken from [5]), we show how to extend the basic argument of Section 2 to finite Kripke models in Section 4.

This extension does not directly give Ruitenburg's Theorem, because it supplies a bound for the indexes of our sequences which is dependent on the poset a given model is based on. This bound is made uniform in Section 6 (using the ranks machinery introduced in Section 5), thus finally reaching our goal.

2 The Case of Classical Logic

We explain our methodology in the much easier case of classical logic. In classical propositional calculus (CPC), Ruitenburg's Theorem holds with index 1 and period 2, namely given a formula $A(x, \underline{y})$, we need to prove that

$$\vdash_{CPC} A^3 \leftrightarrow A \tag{3}$$

holds (here A^3 is defined like in (1)).

2.1 The algebraic reformulation

First, we transform the above statement (3) into an algebraic statement concerning free Boolean algebras. We let $\mathcal{F}_B(\underline{z})$ be the free Boolean algebra over the finite set $\underline{z}$. Recall that $\mathcal{F}_B(\underline{z})$ is the Lindenbaum-Tarski algebra of classical propositional calculus restricted to a language having just the $\underline{z}$ as propositional variables.

Similarly, morphisms $\mu : \mathcal{F}_B(x_1, \ldots, x_n) \longrightarrow \mathcal{F}_B(\underline{z})$ bijectively correspond to n-tuples of equivalence classes of formulae $A_1(\underline{z}), \ldots, A_n(\underline{z})$ in $\mathcal{F}_B(\underline{z})$: the map μ corresponding to the tuple $A_1(\underline{z}), \ldots, A_n(\underline{z})$ associates with the equivalence class of $B(x_1, \ldots, x_n)$ in $\mathcal{F}_B(x_1, \ldots, x_n)$ the equivalence class of $B(A_1/x_1, \ldots, A_n/x_n)$ in $\mathcal{F}_B(\underline{z})$.

Composition is substitution, in the sense that if $\mu : \mathcal{F}_B(x_1, \ldots, x_n) \longrightarrow \mathcal{F}_B(\underline{z})$ is induced, as above, by $A_1(\underline{z}), \ldots, A_n(\underline{z})$ and if $\nu : \mathcal{F}_B(y_1, \ldots, y_m) \longrightarrow \mathcal{F}_B(x_1, \ldots, x_n)$ is induced by $C_1(x_1, \ldots, x_n), \ldots, C_m(x_1, \ldots, x_n)$, then the composite map $\mu \circ \nu : \mathcal{F}_B(y_1, \ldots, y_m) \longrightarrow \mathcal{F}_B(\underline{z})$ is induced by the m-tuple $C_1(A_1/x_1, \ldots, A_n/x_n), \ldots, C_m(A_1/x_1, \ldots, A_n/x_n)$.

How to translate the statement (3) in this setting? Let $\underline{y}$ be $y_1, \ldots, y_n$; we can consider the map $\mu_A : \mathcal{F}_B(x, y_1, \ldots, y_n) \longrightarrow \mathcal{F}_B(x, y_1, \ldots, y_n)$ induced by the $n+1$-tuple of formulae $A, y_1, \ldots, y_n$; then, taking in mind that in Lindenbaum algebras identity is modulo provable equivalence, the statement (3) is equivalent to

$$\mu_A^3 = \mu_A \quad . \tag{4}$$

This raises the question: which endomorphisms of $\mathcal{F}_B(x, \underline{y})$ are of the kind μ_A for some $A(x, \underline{y})$? The answer is simple: consider the 'inclusion' map ι of $\mathcal{F}_B(\underline{y})$ into $\mathcal{F}_B(x, \underline{y})$ (this is the map induced by the n-tuple $y_1, \ldots, y_n$): the maps $\mu : \mathcal{F}_B(x, \underline{y}) \longrightarrow \mathcal{F}_B(x, \underline{y})$ that are of the kind μ_A are precisely the maps μ such that $\mu \circ \iota = \iota$, i.e. those for which the triangle

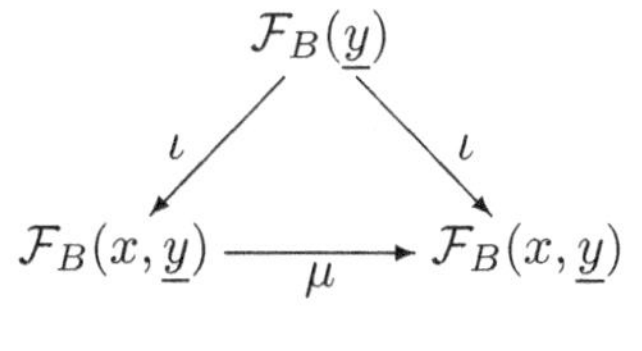

commutes.

It is worth making a little step further: since the free algebra functor preserves coproducts, we have that $\mathcal{F}_B(x, \underline{y})$ is the coproduct of $\mathcal{F}_B(\underline{y})$ with $\mathcal{F}_B(x)$ - the latter being the free algebra on one generator. In general, let us denote by $\mathcal{A}[x]$ the coproduct of the Boolean algebra $\mathcal{A}$ with the free algebra on one generator (let us call $\mathcal{A}[x]$ the *algebra of polynomials* over $\mathcal{A}$).

A slight generalization of statement (4) now reads as follows:

- let $\mathcal{A}$ be a finitely presented Boolean algebra[2] and let the map $\mu : \mathcal{A}[x] \longrightarrow \mathcal{A}[x]$ commute with the coproduct injection $\iota : \mathcal{A} \longrightarrow \mathcal{A}[x]$

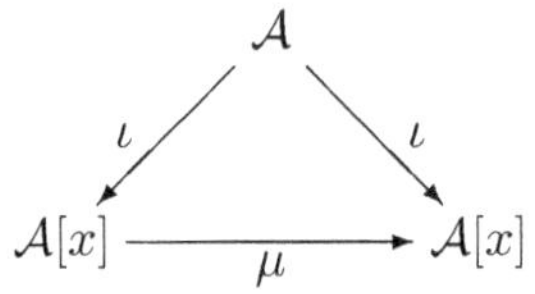

Then we have

$$\mu^3 = \mu \quad . \tag{5}$$

2.2 Duality

The gain we achieved with statement (5) is that the latter is a purely categorical statement, so that we can re-interpret it in dual categories. In fact, a good duality may turn coproducts into products and make our statement easier - if not trivial at all.

Finitely presented Boolean algebras are dual to finite sets; the duality functor maps coproducts into products and the free Boolean algebra on one generator to the two-elements set $\mathbf{2} = \{0, 1\}$ (which, by chance is also a subobject classifier for finite sets). Thus statement (5) now becomes

- let T be a finite set and let the function $f : T \times \mathbf{2} \longrightarrow T \times \mathbf{2}$ commute with the product projection $\pi_0 : T \times \mathbf{2} \longrightarrow T$

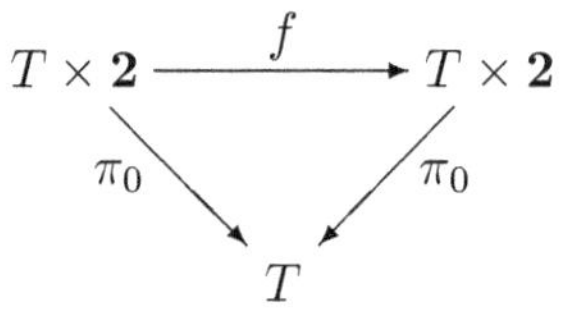

Then we have

$$f^3 = f \quad . \tag{6}$$

In this final form, statement (6) is now just a trivial exercise, which is solved as follows. Notice first that f can be decomposed as $\langle \pi_0, \chi_S \rangle$ (incidentally, χ_S is the characteristic function of some $S \subseteq T \times \mathbf{2}$). Now, if $f(a, b) = (a, b)$ we trivially have also $f^3(a, b) = f(a, b)$; suppose then $f(a, b) = (a, b') \neq (a, b)$. If

---

[2] Recall that an algebra is finitely presented iff it is isomorphic to the quotient of a finitely generated free algebra by a finitely generated congruence. In the case of Boolean algebra 'finitely presented' is the same as 'finite', but it is not anymore like that in the case of Heyting algebras.

$f(a,b') = (a,b')$, then $f^3(a,b) = f(a,b) = (a,b')$, otherwise $f(a,b') = (a,b)$ (there are only two available values for b!) and even in this case $f^3(a,b) = f(a,b)$.

Let us illustrate these cases by thinking of the action of f on $A \times \mathbf{2}$ as one-letter deterministic automaton:

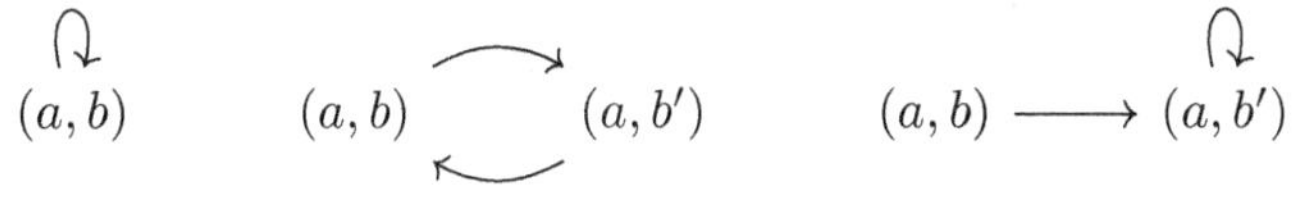

This means that on each irreducible component of the action the pairs index/period are among $(0,1)$, $(0,2)$, $(1,1)$. Out of these pairs we can compute the global index/period of f by means of a max/lcm formula: $(1,2) = (\max\{0,0,1\}, \mathrm{lcm}\{1,2\})$.

3 Duality for Heyting Algebras

In this Section we supply definitions, notation and statements from [5] concerning duality for finitely-presented Heyting algebras. Proofs of the facts stated in this section can all be found in [5, Chapter 4].

A partially ordered set (poset, for short) is a set endowed with a reflexive, transitive, antisymmetric relation (to be always denoted with $\leq$). A poset P is *rooted* if it has a greatest element, that we shall denote by $\rho(P)$. If a finite poset L is fixed, we call an *L-evaluation* or simply an *evaluation* a pair $\langle P, u\rangle$, where P is a rooted finite poset and $u : P \longrightarrow L$ is an order-preserving map.

Evaluations *restrictions* are introduced as follows. If $\langle P, u\rangle$ is an L-evaluation and if $p \in P$, then we shall denote by u_p the L-evaluation, $\langle \downarrow p, u \circ i\rangle$, where $\downarrow p = \{ p' \in P \mid p' \leq p \}$ and $i : \downarrow p \subseteq P$ is the inclusion map; briefly, u_p is the restriction of u to the downset generated by p.

Evaluations have a strict relationship with finite Kripke models: we show in detail the connection. If $\langle L, \leq\rangle$ is $\langle \mathcal{P}(\underline{x}), \supseteq\rangle$ (where $\underline{x} = x_1, \ldots, x_n$ is a finite list of propositional letters), then an L-evaluation $u : P \longrightarrow L$ is called a *Kripke model* for the propositional intuitionistic language built up from $\underline{x}$.[3] Given such a Kripke model u and an IPC formula $A(\underline{x})$, the *forcing* relation $u \models A$ is inductively defined as follows:

$$\begin{array}{ll} u \models x_i & \text{iff } x_i \in u(\rho(P)) \\ u \not\models \bot & \\ u \models A_1 \wedge A_2 & \text{iff } (u \models A_1 \text{ and } u \models A_2) \\ u \models A_1 \vee A_2 & \text{iff } (u \models A_1 \text{ or } u \models A_2) \\ u \models A_1 \rightarrow A_2 & \text{iff } \forall q\ (u_q \models A_1 \Rightarrow u_q \models A_2)\ . \end{array}$$

We define for every $n \in \omega$ and for every pair of L-evaluations u and v, the notions of being *n-equivalent* (written $u \sim_n v$). We also define, for two L-evaluations u, v, the notion of being *infinitely equivalent* (written $u \sim_\infty v$).

[3] According to our conventions, we have that (for $p, q \in P$) if $p \leq q$ then $u(p) \supseteq u(q)$, that is we use $\leq$ where standard literature uses $\geq$.

Let $u : P \longrightarrow L$ and $v : Q \longrightarrow L$ be two L-evaluations. The *game* we are interested in has two players, Player 1 and Player 2. Player 1 can choose either a point in P or a point in Q and Player 2 must answer by choosing a point in the other poset; the only rule of the game is that, if $\langle p \in P, q \in Q \rangle$ is the last move played so far, then in the successive move the two players can only choose points $\langle p', q' \rangle$ such that $p' \leq p$ and $q' \leq q$. If $\langle p_1, q_1 \rangle, \ldots, \langle p_i, q_i \rangle, \ldots$ are the points chosen in the game, Player 2 wins iff for every $i = 1, 2, \ldots,$ we have that $u(p_i) = v(q_i)$. We say that

- $u \sim_\infty v$ iff *Player 2 has a winning strategy* in the above game with infinitely many moves;
- $u \sim_n v$ (for $n > 0$) iff *Player 2 has a winning strategy* in the above game with n moves, i.e. he has a winning strategy provided we stipulate that the game terminates after n moves;
- $u \sim_0 v$ iff $u(\rho(P)) = v(\rho(Q))$ (recall that $\rho(P), \rho(Q)$ denote the roots of P, Q).

Notice that $u \sim_n v$ always implies $u \sim_0 v$, by the fact that L-evaluations are order-preserving. We shall use the notation $[v]_n$ for the equivalence class of an L-valuation v via the equivalence relation $\sim_n$.

The following Proposition states a basic fact (keeping the above definition for $\sim_0$ as base case for recursion, the Proposition also supplies an alternative recursive definition for $\sim_n$):

Proposition 3.1 *Given two L-evaluations $u : P \longrightarrow L, v : Q \longrightarrow L$, and $n > 0$, we have that $u \sim_{n+1} v$ iff $\forall p \in P\, \exists q \in Q\ (u_p \sim_n v_q)$ and vice versa.*

It can be shown that in case $L = \mathcal{P}(x_1, \ldots, x_n)$ (i.e. when L-evaluations are just ordinary finite Kripke models over the language built up from the propositional variables $x_1, \ldots, x_n$), two evaluations are $\sim_\infty$-equivalent (resp. $\sim_n$-equivalent) iff they force the same formulas (resp. the same formulas up to implicational degree n). This can be explained in a formal way as follows. For an IPC formula $A(\underline{x})$, define the *implicational degree* $d(A)$ as follows:

(i) $d(\bot) = d(x_i) = 0$, for $x_i \in \underline{x}$;

(ii) $d(A_1 * A_2) = max[d(A_1), d(A_2)]$, for $* = \wedge, \vee$;

(iii) $d(A_1 \to A_2) = max[d(A_1), d(A_2)] + 1$.

Then one can prove [10] that: (1) $u \sim_\infty v$ holds precisely iff $(u \models A \Leftrightarrow v \models A)$ holds for all formulae $A(\underline{x})$; (2) for all n, $u \sim_n v$ holds precisely iff $(u \models A \Leftrightarrow v \models A)$ holds for all formulae $A(\underline{x})$ with $d(A) \leq n$.[4]

The above discussion motivates a sort of identification of formulae with sets of evaluations closed under restrictions and under $\sim_n$ for some n. Thus, *bounded bisimulations* (this is the way the relations $\sim_n$ are sometimes called) supply the combinatorial ingredients for our duality; for the picture to be complete,

[4] For the statement (1) to be true, it is essential our evaluations to be defined over *finite* posets.

however, we also need a geometric environment, which we introduce using presheaves.

Recall that posets can be viewed as topological spaces whose open subsets are the downward closed subsets. If P, Q are posets, then a map $f : Q \longrightarrow P$ is continuous iff it is order-preserving and open in the topological sense iff it satisfies the following condition forall $q \in Q, p \in P$

$$p \leq f(q) \Rightarrow \exists q' \in Q \; (q' \leq q \;\&\; f(q') = p) \;\; .$$

We shall consider continuous open maps between finite posets and say that rest that $f : Q \longrightarrow P$ is *open* iff it is order-preserving and moreover open in the topological sense. [5]

Let $\mathbf{P_0}$ be the category of finite rooted posets and open maps between them; a *presheaf* over $\mathbf{P_0}$ is a contravariant functor from $\mathbf{P_0}$ to the category of sets and functions, that is, a functor $H : \mathbf{P}_0^{op} \longrightarrow \mathbf{Set}$. Let us recall what this means: a functor $H : \mathbf{P}_0^{op} \longrightarrow \mathbf{Set}$ associates to each finite rooted poset P a set $H(P)$; if $f : Q \longrightarrow P$ is an open map, then we are also given a function $H(f) : H(P) \longrightarrow H(Q)$; moreover, identities are sent to identities, while composition is reversed, $H(g \circ f) = H(f) \circ H(g)$.

Our presheaves form a category whose objects are presheaves over $\mathbf{P_0}$ and whose maps are natural transformations; recall that a natural transformation $\psi : H \longrightarrow H'$ is a collection of maps $\psi_P : H(P) \longrightarrow H'(P)$ (indexed by the objects of $\mathbf{P_0}$) such that for every map $f : Q \longrightarrow P$ in $\mathbf{P_0}$, we have $H'(f) \circ \psi_P = \psi_Q \circ H(f)$. Throughout the paper, we shall usually omit the subscript P when referring to the P-component ψ_P of a natural transformation ψ.

The basic example of presheaf we need in the paper is described as follows. Let L be a finite poset and let h_L be the contravariant functor so defined:

- for a finite poset P, $h_L(P)$ is the set of all L-evaluations;
- for an open map $f : Q \longrightarrow P$, $h_L(f)$ takes $v : P \longrightarrow L$ to $v \circ f : Q \longrightarrow L$.

The presheaf h_L is actually a sheaf (for the canonical Grothendieck topology over $\mathbf{P_0}$); we won't need this fact, [6] but we nevertheless call h_L the *sheaf of L-evaluations* (presheaves of the kind h_L, for some L, are called *evaluation sheaves*).

Notice the following fact: if $\psi : h_L \longrightarrow h_{L'}$ is a natural transformation, $v \in h_L(P)$ and $p \in P$, then $\psi(v_p) = (\psi(v))_p$ (this is due to the fact that the inclusion $\downarrow p \subseteq P$ is an open map, hence an arrow in $\mathbf{P_0}$); thus, we shall feel free to use the (non-ambiguous) notation $\psi(v)_p$ to denote $\psi(v_p) = (\psi(v))_p$.

The notion of *bounded bisimulation index* (*b-index*, for short) [7] takes to-

[5] Open surjective maps are called p-morphisms in the standard non classical logics terminology.

[6] The sheaf structure becomes essential for instance when one has to compute images - images are the categorical counterparts of second order quantifiers, see [5].

[7] This is called 'index' tout court in [5]; here we used the word 'index' for a different notion, since Section 1.

gether structural and combinatorial aspects. We say that a natural transformation $\psi : h_L \longrightarrow h_{L'}$ *has b-index* n iff for every $v : P \longrightarrow L$ and $v' : P' \longrightarrow L$, we have that $v \sim_n v'$ implies $\psi(v) \sim_0 \psi(v')$.

The following Proposition lists basic facts about b-indexes (in particular, it ensures that natural transformations having a b-index do compose):

Proposition 3.2 *Let $\psi : h_L \longrightarrow h_{L'}$ have b-index n; then it has also b-index m for every $m \geq n$. Moreover, for every $k \geq 0$, for every $v : P \longrightarrow L$ and $v' : P' \longrightarrow L$, we have that $v \sim_{n+k} v'$ implies $\psi(v) \sim_k \psi(v')$.*

We are now ready to state duality theorems. As it is evident from the discussion in Section 2, it is sufficient to state a duality for the category of finitely generated free Heyting algebras; although it would not be difficult to give a duality for finitely presented Heyting algebras, we just state a duality for the intermediate category of Heyting algebras freely generated by a finite bounded distributive lattice (this is quite simple to state and is sufficient for proving Ruitenburg's Theorem).

Theorem 3.3 *The category of Heyting algebras freely generated by a finite bounded distributive lattice is dual to the subcategory of presheaves over $\mathbf{P_0}$ having as objects the evaluations sheaves and as arrows the natural transformations having a b-index.*

We leave for an extended version of this paper a proof of the above Theorem (such a proof is contained in [5] and does not play a role in the sequel), however we give some hints on how to reconstruct it. Say that a sub-presheaf S of h_L is *definable* if, for some $n \geq 0$, $v \in S(P)$ and $v \sim_n u$ imply $v \in S(Q)$ (P, Q are the domains of v, u respectively). Such a sub-presheaf corresponds to the set of finite models of a propositional formula. It turns out that a natural transformation f has a b-index iff the inverse image along f of a definable sub-presheaf is definable: *precisely such maps are the duals of substitutions.*

It is important to notice that in the subcategory mentioned in the above Theorem 3.3, products are computed as in the category of presheaves. This means that they are computed pointwise, like in the category of sets: in other words, we have that $(h_L \times h_{L'})(P) = h_L(P) \times h_{L'}(P)$ and $(h_L \times h_{L'})(f) = h_L(f) \times h_{L'}(f)$, for all P and f. Notice moreover that $h_{L\times L'}(P) \simeq h_L(P) \times h_{L'}(P)$, so we have $h_{L\times L'} \simeq h_L \times h_{L'}$; in addition, the two product projections have b-index 0. The situation strongly contrasts with other kind of dualities, see [1] for example, for which products are difficult to compute. The ease by which products are computed might be seen as the principal reason for tackling a proof of Ruitenburg's Theorem by means of sheaf duality.

As a final information, we need to identify the dual of the free Heyting algebra on one generator:

Proposition 3.4 *The dual of the free Heyting algebra on one generator is $h_{\mathbf{2}}$, where $\mathbf{2}$ is the two-element poset $\{0, 1\}$ with $1 \leq 0$.*

4 Indexes and Periods over Finite Models

Taking into consideration the algebraic reformulation from Section 2 and the information from the previous section, we can prove Ruitenburg's Theorem for (IPC) by showing that *all natural transformations from $h_L \times h_{\mathbf{2}}$ into itself, commuting over the first projection π_0 and having a b-index, are ultimately periodic with period 2.* Spelling this out, this means the following. Fix a finite poset L and a natural transformation $\psi : h_L \times h_{\mathbf{2}} \longrightarrow h_L \times h_{\mathbf{2}}$ having a b-index such that the diagram

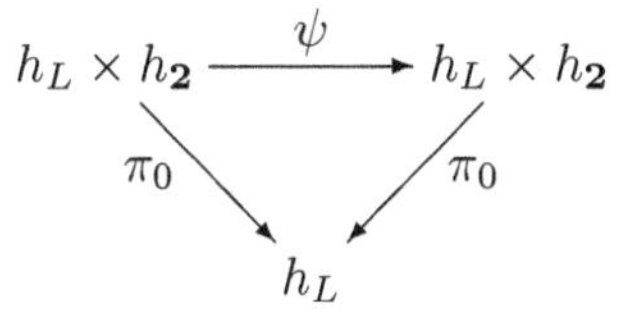

commutes; we have to find an N such that $\psi^{N+2} = \psi^N$, according to the dual reformulation of (2).

From the commutativity of the above triangle, we can decompose ψ as $\psi = \langle \pi_0, \chi \rangle$, where both $\pi_0 : h_L \times h_{\mathbf{2}} \longrightarrow h_L$ and $\chi : h_L \times h_{\mathbf{2}} \longrightarrow h_{\mathbf{2}}$ have a b-index; we assume that $n \geq 1$ is a b-index for both of them. Incidentally, since projections have b-index 0, we can take n to be a b-index of χ. *We let such $\psi = \langle \pi_0, \chi \rangle$ and n be fixed for the rest of the paper.*

Notice that for $(v, u) \in h_L(P) \times h_{\mathbf{2}}(P)$, we have

$$\psi^k(v, u) = (v, u_k)$$

where we put

$$u_0 := u \quad \text{and} \quad u_{k+1} := \chi(v, u_k)\,. \tag{7}$$

Since P and L are finite, it is clear that the sequence $\{\, \psi^k(v, u) \mid k \geq 0 \,\}$ (and obviously also the sequence $\{\, u_k \mid k \geq 0 \,\}$) must become ultimately periodic.

We show in this section that, for each finite poset P and for each $(v, u) \in h_L(P)$, the period of the sequence $\{\, \psi^k(v, u) \mid k \geq 0 \,\}$ has 2 as an upper bound, whereas the index of $\{\, \psi^k(v, u) \mid k \geq 0 \,\}$ can be bounded by the maximum length of the chains in the finite poset P (in the next section, we shall bound such an index independently on P, thus proving Ruitenburg's Theorem).

Call $(v, u) \in h_L(P) \times h_{\mathbf{2}}(P)$ *2-periodic* (or just *periodic*[8]) iff we have $\psi^2(v, u) = (v, u)$; a point $q \in P$ is similarly said periodic in (v, u) iff $(v, u)_q$ is periodic. We shall only say that p is periodic if an evaluation is given and understood from the context. We call a point *non-periodic* if it is not periodic (w.r.t. a given evaluation).

Lemma 4.1 *Let $(v, u) \in h_L(P) \times h_{\mathbf{2}}(P)$ and $p \in P$ be such that all $q \in P$, $q < p$, are periodic. Then either $(v, u)_p$ is periodic or $\psi(v, u)_p$ is periodic. Moreover, if $(v, u)_p$ is non-periodic and $u_0(p) = u(p) = 1$, then $u_1(p) = \chi(u, v)(p) = 0$.*

[8] From now on, 'periodic' will mean '2-periodic', i.e. 'periodic with period 2'.

Proof. We work by induction on the height of p (i.e. on the maximum $\leq$-chain starting with p in P). If the height of p is 1, then the argument is the same as in the classical logic case (see Section 2).

If the height is greater than one, then we need a simple combinatorial check about the possible cases that might arise. Recalling the above definition (7) of the **2**-evaluations u_n, the induction hypothesis implies that there is M big enough so that so for all $k \geq M$ and $q < p$, $(u_{k+2})_q = (u_k)_q$.

Let $\Downarrow p = \{ q \in P \mid q < p \}$. We shall represent $(u_k)_p$ as a pair $\binom{a_k}{x_k}$, where $a_k = u_k(p)$ and x_k is the restriction of $(u_k)_p$ to $\Downarrow p$.

Let us start by considering a first repeat (i, j) of the sequence $\{ a_{M+k} \}_{k \geq 0}$ - that is i is the smallest i such that there is $j > 0$ such that $a_{M+i+j} = a_{M+i}$ and j is the smallest such j. Since the a_{M+n} can only take value 0 or 1, we must have $i + j \leq 2$. We show that the sequence $\{ (u_{M+k})_p \}_{k \geq 0}$ has first repeat taken from

$$(0, 1), (0, 2), (1, 1), (1, 2) .$$

This shall imply in the first two cases that $(v, u)_p$ is periodic or, in the last two cases, that $\psi(v, u)_p$ is periodic. To our goal, let $x = x_M$ and $y = x_{M+1}$ (recall that we do now know whether $x = y$).

Notice that, if $j = 2$, then $i = 0$ and a first repeat for $\{ (u_k)_p \}_{k \geq M}$, is $(0, 2)$, as in the diagram below

$$(u_M)_p = \begin{pmatrix} a \\ x \end{pmatrix}, \; (u_{M+1})_p = \begin{pmatrix} b \\ y \end{pmatrix}, \; (u_{M+2})_p = \begin{pmatrix} a \\ x \end{pmatrix}.$$

Therefore, let us assume $j = 1$ (so $i \in \{ 0, 1 \}$). Consider firstly $i = 0$:

$$(u_M)_p = \begin{pmatrix} a \\ x \end{pmatrix}, \; (u_{M+1})_p = \begin{pmatrix} a \\ y \end{pmatrix}, \; (u_{M+2})_p = \begin{pmatrix} c \\ x \end{pmatrix}, \; (u_{M+3})_p = \begin{pmatrix} d \\ y \end{pmatrix}.$$

If $x = y$, then we have a repeat at $(0, 1)$. Also, if $a = 1$, then the mappings x and y are uniformly 1,[9] so again $x = y$ and $(0, 1)$ is a repeat.

So let us assume $x \neq y$ and $a = 0$. If $c = a$, then we have the repeat $(0, 2)$ as above. Otherwise $c = 1$, so $x = 1$. We cannot have $d = 1$, otherwise $1 = x = y$. Thus $d = 0 = a$, and the repeat is $(1, 2)$.

Finally, consider $i = 1$ (so $a \neq b$ and $j = 1$):

$$(u_M)_p = \begin{pmatrix} a \\ x \end{pmatrix}, \; (u_{M+1})_p = \begin{pmatrix} b \\ y \end{pmatrix}, \; (u_{M+2})_p = \begin{pmatrix} b \\ x \end{pmatrix}, \; (u_{M+3})_p = \begin{pmatrix} d \\ y \end{pmatrix}.$$

We have two subcases: $b = 1$ and $b = 0$. If $b = 1$, then $a = 0$ and $x = 1 = y$: we have a repeat at $(1, 1)$.

In the last subcase, we have $b = 0$, $a = 1$ and now if $d = 0$ we have a repeat at $(1, 2)$ and if $d = 1$ we have a repeat $(1, 1)$ (because $d = a = 1$ implies $y = 1$ and $x = 1$).

The last statement of the Lemma is also obvious in view of the fact that if $a = b = 1$, then $x = y = 1$, so p is periodic. □

[9] Recall that our evaluations are order-preserving maps and we have $1 \leq 0$ in **2**.

Corollary 4.2 *Let N_P be the height of P; then $\psi^{N_P}(v,u)$ is periodic for all $(v,u) \in h_L(P) \times h_2(P)$.*

Proof. An easy induction on N_P, based on the previous Lemma. □

5 Ranks

Ranks (already introduced in [2]) are a powerful tool suggested by bounded bisimulations; in our context the useful notion of rank is given below. Recall that $\psi = \langle \pi_0, \chi \rangle$ and that $n \geq 1$ is a b-index for ψ and χ.

Let $(v,u) \in h_L(P) \times h_2(P)$ be given. The *type* of a periodic point $p \in P$ is the pair of equivalence classes

$$\langle [(v_p, u_p)]_{n-1}, [\psi(v_p, u_p)]_{n-1} \rangle. \tag{8}$$

The *rank* of a point p (that we shall denote by $rk(p)$) is the cardinality of the set of distinct types of the periodic points $q \leq p$. Since $\sim_{n-1}$ is an equivalence relation with finitely many equivalence classes, the rank cannot exceed a positive number $R(L,n)$ (that can be computed in function of L,n).

Clearly we have $rk(p) \geq rk(q)$ in case $p \geq q$. Notice that an application of ψ does not decrease the rank of a point: this is because the pairs (8) coming from a periodic point just get swapped after applying ψ. A non-periodic point $p \in P$ has *minimal rank* iff we have $rk(p) = rk(q)$ for all non-periodic $q \leq p$.

Lemma 5.1 *Let $p \in P$ be a non-periodic point of minimal rank in $(v,u) \in h_L(P) \times h_2(P)$; suppose also that (v,u) is constant on the set of all non-periodic points in $\downarrow p$. Then we have $\psi^m(v,u)_{q_0} \sim_n \psi^m(v,u)_{q_1}$ for all $m \geq 0$ and for all non-periodic points $q_0, q_1 \leq p$.*

Proof. We let Π be the set of periodic points of (v,u) that are in $\downarrow p$ and let Π^c be $(\downarrow p) \setminus \Pi$. Let us first observe that for every $r \in \Pi^c$, we have

$$\begin{aligned}\{\langle [(v_s,u_s)]_{n-1}, [\psi(v_s,u_s)]_{n-1} \rangle \mid s \leq r,\ s \text{ is periodic}\} \\ = \{\langle [(v_s,u_s)]_{n-1}, [\psi(v_s,u_s)]_{n-1} \rangle \mid s \leq p,\ s \text{ is periodic}\}\end{aligned}$$

(indeed the inclusion $\subseteq$ is because $r \leq p$ and the inclusion $\supseteq$ is by the minimality of the rank of p). Saying this in words, we have that "for every periodic $s \leq p$ there is a periodic $s' \leq r$ such that $(v_s,u_s) \sim_{n-1} (v_{s'},u_{s'})$ and $\psi(v_s,u_s) \sim_{n-1} \psi(v_{s'},u_{s'})$"; also (by the definition of 2-periodicity), "for all $m \geq 0$, for every periodic $s \leq p$ there is a periodic $s' \leq r$ such that $\psi^m(v_s,u_s) \sim_{n-1} \psi^m(v_{s'},u_{s'})$". By letting both q_0,q_1 playing the role of r, we get:

Fact. For every $m \geq 0$, for every $q_0, q_1 \in \Pi^c$, for every periodic $s \leq q_0$ there is a periodic $s' \leq q_0$ such that $\psi^m(v_s,u_s) \sim_{n-1} \psi^m(v_{s'},u_{s'})$ (and vice versa).

We now prove the statement of the theorem by induction on m; take two points $q_0, q_1 \in \Pi^c$.

For $m = 0$, $(v,u)_{q_0} \sim_n (v,u)_{q_1}$ is established as follows: as long as Player 1 plays in Π^c, we know (v,u) is constant so that Player 2 can answer with an identical move still staying within Π^c; as soon as he plays in Π, Player 2 uses the above Fact to win the game.

The inductive case $\psi^{m+1}(v,u)_{q_0} \sim_n \psi^{m+1}(v,u)_{q_1}$ is proved in the same way, using the Fact (which holds for the integer $m+1$) and observing that ψ^{m+1} is constant on Π^c. The latter statement can be verified as follows: by the induction hypothesis we have $\psi^m(v,u)_q \sim_n \psi^m(v,u)_{q'}$, so we derive from Proposition 3.2 $\psi^{m+1}(v,u)_q \sim_0 \psi^{m+1}(v,u)_{q'}$, for all $q, q' \in \Pi^c$; that is, ψ^{m+1} is constant on Π^c. □

6 Ruitenburg's Theorem

We can finally prove:

Theorem 6.1 (Ruitenburg's Theorem for IPC) *There is $N \geq 1$ such that we have $\psi^{N+2} = \psi^N$.*

Proof. Let L be a finite poset and let $R := R(L,n)$ be the maximum rank for n, L (see the previous section). Below, for $e \in L$, we let $|e|$ be the height of e in L, i.e. the maximum size of chains in L whose maximum element is e; we let also $|L|$ be the maximum size of a chain in L. We make an induction on natural numbers $l \geq 1$ and show the following: *(for each $l \geq 1$) there is $N(l)$ such that for every (v,u) and $p \in dom(v,u)$ such that $l \geq |v(p)|$, we have that $\psi^{N(l)}(v_p, u_p)$ is periodic.* Once this is proved, the statement of the Theorem shall be proved with $N = N(|L|)$.[10]

If $l = 1$, it is easily seen that we can put $N(l) = 1$ (this case is essentially the classical logic case).

Pick a p with $|v(p)| = l > 1$; let N_0 be the maximum of the values $N(l_0)$ for $l_0 < l$:[11] we show that we can take $N(l)$ to be $N_0 + 2R$.

Firstly, let $(v, u_0) := \psi^{N_0}(v,u)$ so all q with $|v(q)| < l$ are periodic in (v, u_0). After such iterations, suppose that p is not yet periodic in (v, u_0). We let r be the minimum rank of points $q \leq p$ which are not periodic (all such points q must be such that $v(q) = v(p)$); we show that after *two iterations* of ψ, all points $p_0 \leq p$ having rank r become periodic or increase their rank, thus causing the overall minimum rank below p to increase: this means that after at most $2(R-r) \leq 2R$ iterations of ψ, all points below p (p itself included!) become periodic (otherwise said, we take $R - r$ as the secondary parameter of our double induction).

Pick $p_0 \leq p$ having minimal rank r; thus we have that all $q \leq p_0$ in (v, u_0) are now either periodic or have the same rank and the same v-value as p_0 (by

[10] It will turn out that $N(l)$ is $2R(l-1)+1$.

[11] It is easily seen that we indeed have $N_0 = N(l-1)$.

the choice of N_0 above). Let us divide the points of $\downarrow p_0$ into four subsets:

$$\begin{array}{ll} E_{per} & := \{\, q \mid q \text{ is periodic} \,\} \\ E_0 & := \{\, q \mid q \notin E_{per} \;\&\; \forall q' \leq q \; (q' \notin E_{per} \Rightarrow u_0(q') = 0) \,\} \\ E_1 & := \{\, q \mid q \notin E_{per} \;\&\; \forall q' \leq q \; (q' \notin E_{per} \Rightarrow u_0(q') = 1) \,\} \\ E_{01} & := \{\, q \mid q' \notin E_{per} \cup E_1 \cup E_0 \,\}\,. \end{array}$$

Let us define a *frontier point* to be a non-periodic point $f \leq p$ such that all $q < f$ are periodic. Notice that, since all the elements strictly below a frontier point f are periodic, such an f belongs to E_i, where $i = u_0(f)$. Therefore, by Lemma 4.1, all frontier points become periodic after applying ψ. Take a point $q \in E_i$ and a frontier point f below it; since q also has minimal rank and the hypotheses of Lemma 5.1 are satisfied for $(v, u)_q$, we have in particular that $\psi^m(v, u_0)_{q'} = \psi^m(v, u_0)_f$ for all $m \geq 0$ and all non-periodic $q' \leq q$, and hence $\psi(v, u_0)_q$ is periodic too.

Thus, if we apply ψ, we have that in $(v, u_1) := \psi(v, u_0)$ all points in $E_{per} \cup E_0 \cup E_1$ become periodic, together with possibly some points in E_{01}. The latter points get in any case u_1-value equal to 0. This can be seen as follows. If any such point gets u_1-value equal to 1, then all points below it get the same u_1-value. Yet, by definition, these points are above some frontier point in E_1 and frontier points in E_1 get u_1-value 0 by the second statement of Lemma 4.1.

If $p_0 \in E_0$ has become periodic, we are done; we are also done if the rank of p_0 increases, because this is precisely what we want. If p_0 has not become periodic and its rank has not increased, then now all the non-periodic points below p_0 in (v, u_1) have u_1-value 0 (by the previous remark) and have the same rank as p_0. Thus, they are the set E_0 computed in (v, u_1) (instead of in (v, u_0)) and we know by the same considerations as above that it is sufficient to apply ψ once more to make them periodic. □

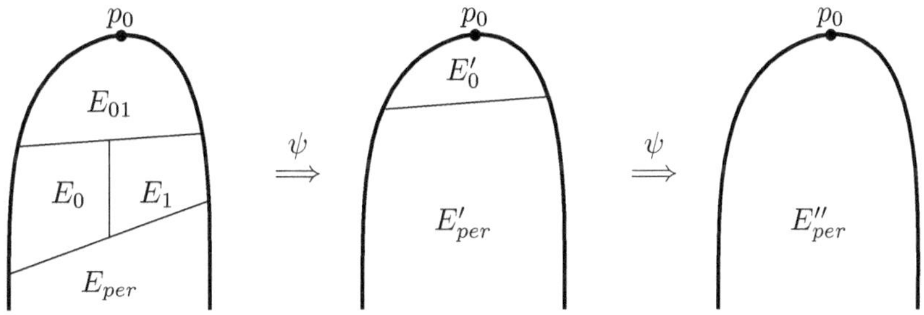

Notice that some crucial arguments used in the above proof (starting from the induction on $|e|$ itself) make essential use of the fact that evaluations are order-preserving, so such arguments are not suitable for modal logics.

The above proof of Theorem 6.1 gives a bound for N which is not optimal, when compared with the bound obtained via syntactic means in [8] (the syntactic computations in [4] for fixpoints convergence are also better). Thus refining indexes of ultimate periodicity of our sequences within semantic arguments remains as an open question.

References

[1] Esakia, L., *Topological Kripke models*, Soviet Math. Dokl. **15** (1974), pp. 147–151.
[2] Fine, K., *Logics containing* K4. *II*, J. Symbolic Logic **50** (1985), pp. 619–651.
[3] Ghilardi, S., *Best solving modal equations*, Ann. Pure Appl. Logic **102** (2000), pp. 183–198.
[4] Ghilardi, S., M. J. Gouveia and L. Santocanale, *Fixed-point elimination in the intuitionistic propositional calculus*, in: *Foundations of Software Science and Computation Structures, FOSSACS 2016, Proceedings*, 2016, pp. 126–141.
[5] Ghilardi, S. and M. Zawadowski, "Sheaves, Games, and Model Completions: A Categorical Approach to Nonclassical Propositional Logics," Springer Publishing Company, Incorporated, 2011, 1st edition.
[6] Mardaev, S., *Definable fixed points in modal and temporal logics: A survey*, Journal of Applied Non-Classical Logics **17** (2007), pp. 317–346.
[7] Mardaev, S. I., *Least fixed points in Grzegorczyk's Logic and in the intuitionistic propositional logic*, Algebra and Logic **32** (1993), pp. 279–288.
[8] Ruitenburg, W., *On the period of sequences* $(a^n(p))$ *in intuitionistic propositional calculus*, The Journal of Symbolic Logic **49** (1984), pp. 892–899.
[9] Shavrukov, V. Y., *Subalgebras of diagonalizable algebras of theories containing arithmetic*, Dissertationes Math. (Rozprawy Mat.) **323** (1993), p. 82.
[10] Visser, A., *Uniform interpolation and layered bisimulation*, in: *Gödel '96 (Brno, 1996)*, Lecture Notes Logic **6**, Springer, Berlin, 1996 pp. 139–164.

Counterfactual Logic: Labelled and Internal Calculi, Two Sides of the Same Coin?

Marianna Girlando**, Sara Negri*, Nicola Olivetti* [1]

*Aix Marseille Univ, Université de Toulon, CNRS, LIS, Marseille, France;
* University of Helsinki, Finland

Abstract

Lewis' Logic $\mathbb{V}$ is the fundamental logic of counterfactuals. Its proof theory is here investigated by means of two sequent calculi based on the connective of comparative plausibility. First, a labelled calculus is defined on the basis of Lewis' sphere semantics. This calculus has good structural properties and provides a decision procedure for the logic. An internal calculus, recently introduced, is then considered. In this calculus, each sequent in a derivation can be interpreted directly as a formula of $\mathbb{V}$. In spite of the fundamental difference between the two calculi, a mutual correspondence between them can be established in a constructive way. In one direction, it is shown that any derivation of the internal calculus can be translated into a derivation in the labelled calculus. The opposite direction is considerably more difficult, as the labelled calculus comprises rules which cannot be encoded by purely logical rules. However, by restricting to derivations in normal form, derivations in the labelled calculus can be mapped into derivations in the internal calculus. On a general level, these results aim to contribute to the understanding of the relations between labelled and internal proof systems for logics belonging to the realm of modal logic and its extensions, a topic still relatively unexplored.

Keywords: Counterfactual logic, sequent calculus, labelled sequent calculi, translation between calculi.

1 Introduction

In 1973 David Lewis presented, in a short but dense monograph, a family of logics for counterfactual reasoning. Lewis denoted counterfactual implication with the modal connective $A > B$, the meaning of which is "if A had been the case, then B would have been the case". A counterfactual implication $A > B$ is true if B is true in any state of affairs differing minimally from the actual one and in which A is true. Standard possible worlds semantics had to be generalized to capture the counterfactual conditional; to this aim Lewis proposed *sphere semantics*, a new semantics of topological flavour, which

[1] This work was partially supported by the Project TICAMORE ANR-16-CE91-0002-01 and by the Academy of Finland, research project no. 1308664.

later inspired a wealth of new endeavours in the field of modal logic and its applications [19].

Lewis' development was essentially axiomatic, and it took one decade before the first (quite complex) Gentzen-style proof system for the logic of counterfactuals was proposed [11]; the calculus is non-standard, as it contains infinitely many rules. The meaning explanation of the counterfactual conditional is not truth-functional in the conventional sense; for this reason, one cannot find natural deduction or sequent calculus rules as one does for the classical connectives. Labelled sequent calculi offer an answer to the problem of developing a proof-theoretic semantics for logics based on possible worlds semantics [17] and have been shown indeed apt to capture any modal logic characterized by first-order conditions on their Kripke frames [1]. Recently, the labelled approach has been extended to deal also with neighbourhood semantics [13,14]. For the logic of counterfactuals, labelled sequent calculi have been proposed both for a generalized relational semantics, based on ternary accessibility relations [16] and for preferential conditional logic based on a broader version of neighbourhood semantics [15].

Parallel to these developments, alternative inference styles, not directly referring to formalized semantics in their rules, have been developed by a number of authors. In such calculi, rather than adding labels for elements of the characteristic semantic structures, one enriches the structure of sequents with new structural connectives (cf. [11,3], and more recently [9,18,7,6]). These calculi are usually referred to as "internal," since their sequents can be directly interpreted in the language of the corresponding logic.

Natural questions arise on the relationships between the two different inference styles of labelled and internal sequent calculi: What is the expressive power of internal calculi in relation to labelled calculi? Can one find an equally general way of generating internal calculi?

A summary of the relationships between various calculi for normal propositional modal logics was presented in [20, pp. 116, 206] together with a conjecture of a general interpretability of tree hypersequents into labelled calculi. This conjecture has been confirmed by in [8] through an embedding of nested sequent calculi and tree-hypersequent calculi into a suitable subclass of labelled calculi, those in which the relational structure forms a tree. A similar result has been proved for nested sequent calculi and labelled tableaux [2].

The question we address in this article is whether we can establish a translation between the labelled and the internal calculi for the logics of counterfactuals. As is often the case in logic, the problem is better tackled from the analysis of a significant case study: thus, we shall focus on logic $\mathbb{V}$, the most basic system among the counterfactual logics presented by Lewis.

Following Lewis, in Section 2 we define the language of $\mathbb{V}$ taking as primitive the comparative plausibility operator $A \preccurlyeq B$, meaning "A is at least as plausible as B", instead of the counterfactual conditional. The two connectives are interdefinable, but the former has a simpler explanation: $A \preccurlyeq B$ is true at x if every sphere of x that meets B also meets A. Since this condition is universal, it is readily translated into left and right labelled rules. Then, we add rules for "meet", the existential forcing relation, and rules for the propositional base as well as rules for spheres, all these latter unchanged with respect to the calculus presented in [15]. The resulting calculus, **G3V**, is a new and non-trivial proof system for $\mathbb{V}$ (Section 3). The calculus has all the typical structural properties of the G3-family of sequent calculi (hence the name) and enjoys a simple completeness proof via a proof-or-countermodel construction. This latter yields the finite model property, and thus an effective decision procedure, for the logic.

In Section 4, the internal sequent calculus $\mathcal{I}^{\mathsf{i}}_{\mathbb{V}}$, introduced in [5], is recalled. Differently from the labelled calculus, each $\mathcal{I}^{\mathsf{i}}_{\mathbb{V}}$ sequent can be directly interpreted as a formula of the language. The translation from the internal to the labelled calculus (Section 5) can be directly specified by adding to the labelled calculus a few admissible rules. The converse direction (Section 6) is far from immediate. At each step of inference, labelled calculi display many relational formulas that cannot be directly translated into the language of $\mathbb{V}$. This overload has somehow to be disciplined to transform a labelled derivation into an internal one. The core of the procedure lies in the identification of a "normal form" for **G3V** derivations, basically corresponding to the requirement that the relational structure of a derivation forms a tree, and in the Jump lemma (Lemma 6.5), that allows to focus on particular subsets of a labelled sequent, namely, to formulas labelled with the same world label. We shall show that, thanks to these restriction, we are able to translate **G3V** derivations into $\mathcal{I}^{\mathsf{i}}_{\mathbb{V}}$ derivations.

Both directions of the translation are defined by means of an inductive procedure. The present paper should give enough support to the claim that, at least the present case study, internal and labelled sequent calculi can be considered as two sides of the same coin.

2 Logic $\mathbb{V}$

The language of $\mathbb{V}$ is defined as: $P \mid \bot \mid \neg A \mid A \vee A \mid A \wedge A \mid A \supset A \mid A \preccurlyeq A$ where $\preccurlyeq$ is the comparative plausibility operator. A formula $A \preccurlyeq B$ means "A is at least as plausible as B". The counterfactual conditional operator $>$ can be defined in terms of $\preccurlyeq$: $A > B \;\equiv\; (\bot \preccurlyeq A) \vee \neg((A \wedge \neg B) \preccurlyeq (A \wedge B))$[2]. Thus, for a counterfactual conditional $A > B$ to be true, it must be that either A is impossible, or that $A \wedge \neg B$ is less plausible than $A \wedge B$.

An axiomatization of $\mathbb{V}$ is given extending classical propositional logic with the following axioms and inference rules:

$$(\text{CPR})\ \frac{\vdash B \supset A}{\vdash A \preccurlyeq B} \qquad (\text{CPA})\ (A \preccurlyeq A \vee B) \vee (B \preccurlyeq A \vee B)$$

$$(\text{TR})\ (A \preccurlyeq B) \wedge (B \preccurlyeq C) \supset (A \preccurlyeq C) \qquad (\text{CO})\ (A \preccurlyeq B) \vee (B \preccurlyeq A)$$

Following [10], we define a semantics for $\mathbb{V}$ in terms of sphere models.

---

[2] And, vice versa, the comparative plausibility operator can be defined in terms of the counterfactual conditional: $A \preccurlyeq B \;\equiv\; ((A \vee B) > \bot) \vee \neg((A \vee B) > \neg A)$.

Definition 2.1 A *sphere model* is a triple $\mathcal{M} = \langle W, S, [\![\ ]\!]\rangle$ where W is a non-empty set of possible words; S is a function $S : W \to \mathcal{P}(\mathcal{P}(W))$, and $[\![\ ]\!] : Atm \to \mathcal{P}(W)$ is the propositional evaluation. We assume S to satisfy the properties of *Non-emptiness*: $\forall \alpha \in S(x) \,.\, \alpha \neq \emptyset$ and of *Nesting*: $\forall \alpha, \beta \in S(x) \,.\, \alpha \subseteq \beta \ \vee \ \beta \subseteq \alpha$.

We write $w \Vdash A$ to denote truth of formula A at world w. Truth conditions for Boolean combinations of formulas are the standard ones. As for comparative plausibility, a formula $A \preccurlyeq B$ is true at a world x if for all the spheres $\alpha \in S(x)$, if α contains a world that satisfies B then it must also contain a world that satisfies A. Using the existential forcing relation from [15] $\alpha \Vdash^{\exists} A$ iff $\exists y \in \alpha.\ y \Vdash A$ the truth condition for comparative plausibility[3] can be formally stated as follows: $x \Vdash A \preccurlyeq B$ iff $\forall \alpha \in S(x).\ \alpha \Vdash^{\exists} B \to \alpha \Vdash^{\exists} A$.

A formula A is *valid in a sphere model* $\mathcal{M}$ if for all $w \in W$, $w \Vdash A$. We say that A is *valid* if A is valid in every sphere model.

3 The labelled calculus G3V

The calculus **G3V** (Figure 1) displays two sorts of labels: $x, y, z, \dots$ for worlds, and $a, b, c, \dots$ for spheres. All formulas occurring in a sequent are labelled: an expression $x : A$ means that world x forces A and $a \Vdash^{\exists} A$ means that sphere a contains a world that forces A. The propositional rules can be found in [12].

Initial sequents $\quad x : p, \Gamma \Rightarrow \Delta, x : p \qquad x : \bot, \Gamma \Rightarrow \Delta$

Rules for local forcing

$$\frac{x \in a, x : A, \Gamma \Rightarrow \Delta}{a \Vdash^{\exists} A, \Gamma \Rightarrow \Delta}\ L\Vdash^{\exists}\ (x \text{ fresh}) \qquad \frac{x \in a, \Gamma \Rightarrow \Delta, x : A, a \Vdash^{\exists} A}{x \in a, \Gamma \Rightarrow \Delta, a \Vdash^{\exists} A}\ R\Vdash^{\exists}$$

Propositional rules: rules of **G3K**

Rules for comparative plausibility

$$\frac{a \Vdash^{\exists} B, a \in S(x), \Gamma \Rightarrow \Delta, a \Vdash^{\exists} A}{\Gamma \Rightarrow \Delta, x : A \preccurlyeq B}\ R\preccurlyeq\ (a \text{ fresh})$$

$$\frac{a \in S(x), x : A \preccurlyeq B, \Gamma \Rightarrow \Delta, a \Vdash^{\exists} B \quad a \Vdash^{\exists} A, a \in S(x), x : A \preccurlyeq B, \Gamma \Rightarrow \Delta}{a \in S(x), x : A \preccurlyeq B, \Gamma \Rightarrow \Delta}\ L\preccurlyeq$$

Rules for inclusion and nesting

$$\frac{x \in a, a \subseteq b, x \in b, \Gamma \Rightarrow \Delta}{x \in a, a \subseteq b, \Gamma \Rightarrow \Delta}\ L\subseteq$$

$$\frac{a \subseteq b, a \in S(x), b \in S(x), \Gamma \Rightarrow \Delta \qquad b \subseteq a, a \in S(x), b \in S(x), \Gamma \Rightarrow \Delta}{a \in S(x), b \in S(x), \Gamma \Rightarrow \Delta}\ \mathsf{Nes}$$

Fig. 1. Rules of **G3V**

[3] Truth condition for the (defined) conditional operator is the following: $x \Vdash A > B$ iff $\forall \alpha \in S(x).\ \alpha \nVdash^{\exists} A$ *or* $\exists \beta \in S(x).\ \beta \Vdash^{\exists} A \ \&\ \beta \Vdash^{\forall} A \supset B$, for $\alpha \Vdash^{\forall} A$ iff $\forall y \in \alpha.\ y \Vdash A$.

Theorem 3.1 (Soundness) *If a sequent* $\Gamma \Rightarrow \Delta$ *is derivable in* **G3V**, *then it is valid in all sphere models.*

Proof. As in [15], we need to define the notion of *realization*, which interprets labelled sequents in sphere semantics; soundness is then proved by straightforward induction on the height of the derivation. Given a sphere model $\mathcal{M} = \langle W, I, [\![\ ]\!] \rangle$, a set S of world labels, and a set N of sphere labels, an SN*-realization* over $\mathcal{M}$ is a pair of functions (ρ, σ) such that $\rho : S \to W$ is a function that assigns to each world label $x \in S$ an element $\rho(x) \in W$, and $\sigma : N \to \mathcal{P}(W)$ is a function that assigns to each sphere label $a \in N$ a sphere $\sigma(a) \in S(w)$, for $w \in W$. *Satisfiability* of a formula $\mathcal{F}$ under an SN-realization is defined by cases as follows: $\mathcal{M} \vDash_{\rho,\sigma} a \in S(x)$ if $\sigma(a) \in S(\rho(x))$; $\mathcal{M} \vDash_{\rho,\sigma} a \subseteq b$ if $\sigma(a) \subseteq \sigma(b)$; $\mathcal{M} \vDash_{\rho,\sigma} x \in a$ if $\rho(x) \in \sigma(a)$; $\mathcal{M} \vDash_{\rho,\sigma} x : P$ if $\rho(x) \in [\![P]\!]$, for P atomic[4]; $\mathcal{M} \vDash_{\rho,\sigma} a \Vdash^{\exists} A$ if $\sigma(a) \Vdash^{\exists} A$; $\mathcal{M} \vDash_{\rho,\sigma} x : A \preccurlyeq B$ if for all $\alpha \in S(\rho(x))$, if $\alpha \Vdash^{\exists} B$ then $\alpha \Vdash^{\exists} A$. A sequent $\Gamma \Rightarrow \Delta$ is *valid in* $\mathcal{M}$ *under the* (ρ, σ) *realization* iff $\mathcal{M} \nvDash_{\rho,\sigma} F$ for all $F \in \Gamma$ or $\mathcal{M} \vDash_{\rho,\sigma} G$ for some $G \in \Delta$. A sequent is *valid in* $\mathcal{M}$ if it is valid under any (ρ, σ) realization. □

The calculus enjoys admissibility of all the structural rules. For admissibility of weakening, invertibility of all the rules and admissibility of contraction, the proof is an extension of the proofs in [15]. In order to prove the admissibility of cut, we need a substitution lemma, spelled out as in [15] with the addition of the clause for comparative plausibility, namely $x : A \preccurlyeq B[x/y] \equiv y : A \preccurlyeq B$. Then, we need a suitable definition of weight of a labelled formula.

Definition 3.2 Given a labelled formula $\mathcal{F}$, we define the pure part $p(\mathcal{F})$ and the label part $l(\mathcal{F})$ of $\mathcal{F}$ as follows: $p(x : A) = p(a \Vdash^{\exists} A) = A$; $l(x : A) = x$; $l(a \Vdash^{\exists} A) = a$. The *weight* of a labelled formula is defined as an ordered pair $\langle w(p(\mathcal{F}), w(l(\mathcal{F})) \rangle$ where

- for all world labels. x, $w(x) = 0$; for all sphere labels a, $w(a) = 1$;
- $w(P) = w(\bot) = 1$; $w(A \circ B) = w(A) + w(B) + 1$ for $\circ$ conjunction, disjunction or implication; $w(A \preccurlyeq B) = w(A) + w(B) + 2$.

Theorem 3.3 *The rule of cut is admissible in* **G3V**.

Proof. 3.3 Primary induction on the weight of formulas and secondary induction on the sum of heights of derivations of the premisses. The proof is by cases, according to the last rules applied in the derivations of the premisses of Cut. We only show the case in which both occurrences of the cut formula are principal and derived by $\mathsf{R}\preccurlyeq$ and $\mathsf{L}\preccurlyeq$ respectively.

$$\dfrac{\dfrac{\overset{(1)}{b \in S(x), b \Vdash^{\exists} B, \Gamma \Rightarrow \Delta, b \Vdash^{\exists} A}}{\Gamma \Rightarrow \Delta, x : A \preccurlyeq B}\,\mathsf{R}\preccurlyeq \qquad \dfrac{(2) \qquad (3)}{a \in S(x), x : A \preccurlyeq B, \Gamma' \Rightarrow \Delta'}\,\mathsf{L}\preccurlyeq}{\Gamma, a \in S(x), \Gamma' \Rightarrow \Delta, \Delta'}\,Cut$$

[4] This definition can be extended in the standard way to the propositional formulas of the language.

where $(2) = a \in S(x), x : A \preccurlyeq B, \Gamma' \Rightarrow \Delta', a \Vdash^{\exists} B$ and $(3) = a \Vdash^{\exists} A, a \in S(x), x : A \preccurlyeq B, \Gamma' \Rightarrow \Delta'$ are the left and right premisses of $\mathsf{L}\preccurlyeq$. In (1), substitute variable b with variable a. The derivation is transformed as follows:

$$\dfrac{\dfrac{\overset{(2)}{a \in S(x), x : A \preccurlyeq B, \Gamma' \Rightarrow \Delta', a \Vdash^{\exists} B} \quad \dfrac{(1)[a/b] \qquad (3)}{\Gamma, a \Vdash^{\exists} B, a \in S(x), \Gamma' \Rightarrow \Delta, \Delta'}\,Cut}{\Gamma, a \in S(x), a \in S(x), \Gamma', \Gamma' \Rightarrow \Delta, \Delta', \Delta'}\,Cut}{\Gamma, a \in S(x), \Gamma' \Rightarrow \Delta, \Delta'}\,\mathsf{Ctr}$$

where the upper occurrence of cut has smaller sum of heights than the original cut, and the lower occurrence of cut is applied to formulas of smaller weight than the cut formula. □

The following rule is needed to define the translation in Section 5. The proof is by easy induction on the height of the derivation.

Lemma 3.4 *Rule* Mon∃ *is admissible.*

$$\dfrac{b \subseteq a, \Gamma \Rightarrow \Delta, a \Vdash^{\exists} A, b \Vdash^{\exists} A}{b \subseteq a, \Gamma \Rightarrow \Delta, a \Vdash^{\exists} A}\,\mathsf{Mon}\exists$$

Remark 3.5 Notational convention: given multisets of formulas $\Gamma = \{A_1, \ldots, A_m\}$ and $\Sigma = \{S_1, \ldots, S_k\}$, we shall write $x : \Gamma$ and $a \Vdash^{\exists} \Sigma$ as abbreviations for $x : A_1, \ldots, x : A_m$ and $a \Vdash^{\exists} S_1, \ldots, a \Vdash^{\exists} S_k$ respectively.

Theorem 3.6 (Syntactic completeness) *If a formula A is valid in $\mathbb{V}$, sequent $\Rightarrow x : A$ is derivable in* **G3V**.

Proof. Using completeness for the axiomatic system, it suffices to show that the axioms and inference rules of $\mathbb{V}$ are derivable in **G3V**. By means of example, derivation of axiom (CO) in which we have omitted the right premiss of Nes.

$$\dfrac{\dfrac{\dfrac{\dfrac{\dfrac{\dfrac{\dfrac{\overset{\nabla}{a \subseteq b, a \in S(x), b \in S(x), y \in a, y \in b, y : B, b \Vdash^{\exists} A \Rightarrow a \Vdash^{\exists} A, y : B, b \Vdash^{\exists} B}}{a \subseteq b, a \in S(x), b \in S(x), y \in a, y \in b, y : B, b \Vdash^{\exists} A \Rightarrow a \Vdash^{\exists} A, b \Vdash^{\exists} B}\,\mathsf{R}\Vdash^{\exists}}{a \subseteq b, a \in S(x), b \in S(x), y \in a, y : B, b \Vdash^{\exists} A \Rightarrow a \Vdash^{\exists} A, b \Vdash^{\exists} B}\,\mathsf{L}\subseteq}{a \subseteq b, a \in S(x), b \in S(x), a \Vdash^{\exists} B, b \Vdash^{\exists} A \Rightarrow a \Vdash^{\exists} A, b \Vdash^{\exists} B}\,\mathsf{L}\Vdash^{\exists}}{a \in S(x), b \in S(x), a \Vdash^{\exists} B, b \Vdash^{\exists} A \Rightarrow a \Vdash^{\exists} A, b \Vdash^{\exists} B}\,\mathsf{Nes}}{a \in S(x), a \Vdash^{\exists} B \Rightarrow a \Vdash^{\exists} A, x : B \preccurlyeq A}\,\mathsf{R}\preccurlyeq}{\Rightarrow x : A \preccurlyeq B, x : B \preccurlyeq A}\,\mathsf{R}\preccurlyeq}{\Rightarrow x : A \preccurlyeq B \vee B \preccurlyeq A}\,R\vee$$

□

Bottom-up proof search in **G3V** comes to an end with the adoption the following strategy: In any derivation branch (i) no rule can be applied to an initial sequent and (ii) no redundant application of any rule is allowed, where the standard notion of *redundant application* of a rule is defined as follows. Let $\mathcal{B} = S_0, S_1, \ldots$ be a derivation branch with S_i sequent $\Gamma_i \Rightarrow \Delta_i$ for $i = 1, 2, \ldots$. An application of a **G3V** rule (R) to S_i is *redundant* if whenever the branch up to S_i contains the conclusion of that application of (R) then it also contains at least one premiss of that application of (R). We say that a branch

$\mathcal{B} = S_0, S_1, \ldots$ to be *saturated* if every **G3V** rule is redundant, and $\mathcal{B}$ to be *open* if it does not contain an initial sequent.

To check the validity of a formula A, we try to build a derivation with root $\Rightarrow x : A$ according to the strategy. We show that the process of building such a derivation always comes to an end in a finite number of steps. We actually prove the result for a more general case: the one in which the sequent at the root of the derivation is *simple.* This notion will be useful in the following sections.

Definition 3.7 A sequent $\Gamma \Rightarrow \Delta$ is *simple* if (i) it contains only one world label x, (ii) if Γ is non-empty, then x occurs in Γ, (iii) for all neighbourhood labels a occurring in $\Gamma \Rightarrow \Delta$, $a \in S(x)$ occurs in Γ and (iv) $x \in a$ does not occur in $\Gamma \Rightarrow \Delta$.

Definition 3.8 Given a branch $\mathcal{B} = S_0, S_1, \ldots$ where $S_i = \Gamma_i \Rightarrow \Delta_i$ for $i = 1, 2, \ldots$ and let $\Pi_{\mathcal{B}} = \bigcup_i \Gamma_i$. We define the following relations:

- $x \rightarrow_{\Pi_{\mathcal{B}}} a$ if $a \in S(x)$ occurs in $\Pi_{\mathcal{B}}$;
- $a \rightarrow_{\Pi_{\mathcal{B}}} y$ if for some $S_i = \Gamma_i \Rightarrow \Delta_i$, $y \in a$ occurs in Γ_i and y does not occur in any S_j with $j < i$;
- $x \rightarrow_{\Pi_{\mathcal{B}}} y$ if there exists an a such that $x \rightarrow_{\Pi_{\mathcal{B}}} a$ and $a \rightarrow_{\mathcal{B}} y$;
- $x \rightarrow^*_{\Pi_{\mathcal{B}}} y$ is the transitive closure of $x \rightarrow_{\Pi_{\mathcal{B}}} y$.

Lemma 3.9 *Let $\mathcal{B} = S_0, S_1, \ldots$ be any branch of a derivation of a simple sequent $\Gamma \Rightarrow \Delta$ where x_0 is the only world label appearing in S_0. Then: (a) for every label x occurring in any sequent of $\mathcal{B}$, it holds that $x_0 \rightarrow^*_{\Pi_{\mathcal{B}}} x$; (b) the relation $\rightarrow^*_{\Pi_{\mathcal{B}}}$ forms a tree $\mathcal{T}_{x_0}$ with root x_0; (c) if $x \rightarrow^*_{\Pi_{\mathcal{B}}} y$ then $m(y) < m(x)$, where for a world label u, $m(u)$ is the maximal modal degree of all formulas C such that $u : C$ occurs in any sequent of $\mathcal{B}$; (d) If $\mathcal{B}$ is built according to the strategy, then for every x the set $\{x \mid x \rightarrow_{\Pi_{\mathcal{B}}} a\}$ is finite, and for every a the set $\{y \mid a \rightarrow_{\mathcal{B}} y\}$ is finite, whence the tree $\mathcal{T}_{x_0}$ is finite*[5] .

The proof of this lemma relies on the fact that world and sphere labels are introduced analysing only *once* (by the irredundancy restriction) the subformulas of the sequent at the root.

Theorem 3.10 (Termination) *Let $\Gamma \Rightarrow \Delta$ be a simple sequent. Proof search for $\Gamma \Rightarrow \Delta$ always terminates.*

Proof. We prove that any derivation of $\Gamma \Rightarrow \Delta$ built according to the strategy is finite. Let $\mathcal{B} = S_0, S_1, S_k, S_{k+1} \ldots$ be any branch with $S_0 = \Gamma \Rightarrow \Delta$; by Lemma 3.9 the set of world labels and the set of sphere labels occurring in $\mathcal{B}$ are both finite. Since all pure formulas occurring in $\mathcal{B}$ are subformulas of the root sequent $\Gamma \Rightarrow \Delta$, also the set of labelled formulas that may occur in the *whole* $\mathcal{B}$ is finite. Whence by the strategy (no redundant application of the rules), any sequent S_i is finite and the branch $\mathcal{B} = S_0, S_1, S_k, S_{k+1} \ldots$ must

[5] As a matter of fact it can be proved that the sets $\{x \mid x \rightarrow_{\Pi_{\mathcal{B}}} a\}$, $\{y \mid a \rightarrow_{\mathcal{B}} y\}$ and the tree $\mathcal{T}_{x_0}$ are not only finite, but bounded in size by some function of the size of the sequent $\Gamma \Rightarrow \Delta$ at the root.

come to an end after a finite number steps, that is it must be $\mathcal{B} = S_0, \ldots, S_k$, where either S_k is an initial sequent or the whole $\mathcal{B}$ is saturated. We have shown that every derivation of $\Gamma \Rightarrow \Delta$ is finite. □

Termination yields a decision procedure: to check provability of formula A, build a proof search tree $\mathcal{D}$ with root $\Rightarrow x : A$. By the previous theorem, $\mathcal{D}$ is finite: either every branch of $\mathcal{D}$ terminates with an initial sequent, and $\mathcal{D}$ is a derivation of A, or $\mathcal{D}$ contains an open saturated branch. In the former case A is provable; in the latter case it is not, and it is possible to extract a countermodel of A from the open branch. Thus we can give an alternative (semantic) completeness proof for **G3V**. Observe that the following theorem combined with soundness of **G3V** provides a constructive proof of the *finite model property* of $\mathbb{V}$.

Theorem 3.11 (Semantic completeness) *If a formula A is not derivable in* **G3V**, *there is a* finite *sphere model $\mathcal{M}$ such that A is not valid in $\mathcal{M}$.*

Proof. By Theorem 3.10 any derivation $\mathcal{D}$ with root S_0 sequent $\Rightarrow x_0 : A$ contains *a finite open branch* $S_0, \ldots, S_k$, with S_i sequent $\Gamma_i \Rightarrow \Delta_i$. Define a model $\mathcal{M} = (W, S, [\![\ ]\!])$, where $W = \{x \mid x \text{ occurs in } \bigcup_i^k \Gamma_i\}$. Given a sphere label a, we define a sphere $\alpha_a = \{y \in W \mid y \in a \text{ occurs in } \bigcup_i^k \Gamma_i\}$, and $S(x) = \{\alpha_a \mid a \in S(x) \text{ occurs in } \bigcup_i^k \Gamma_i\}$. For any atom P, $[\![P]\!] = \{x \in W \mid x : P \text{ occurs in } \bigcup_i^k \Gamma_i\}$. We consider the SN-realisation (ρ, σ) where $\rho(x) = x$ and $\sigma(a) = \alpha_a$. For all $i = 0, \ldots, k$ and for any formula $\mathcal{F}$ we show that if $\mathcal{F}$ occurs in Γ_i then $\mathcal{M} \vDash_{\rho,\sigma} \mathcal{F}$ and if $\mathcal{F}$ occurs in Δ_i, then $\mathcal{M} \nvDash_{\rho,\sigma} \mathcal{F}$. In particular for labelled formulas $\mathcal{F}$ of the form $x : B$ and $a \Vdash^{\exists} B$ we proceed by induction of on the weight $w(\mathcal{F})$. Since $x_0 : A$ occurs in Δ_0 A is not valid in $\mathcal{M}$. □

4 Internal sequent calculus $\mathcal{I}_{\mathbb{V}}^{\mathsf{i}}$

The sequent calculus $\mathcal{I}_{\mathbb{V}}^{\mathsf{i}}$ for $\mathbb{V}$ was proposed in [5]; we recall here the basic notions. Sequents of $\mathcal{I}_{\mathbb{V}}^{\mathsf{i}}$ are composed of formulas and *blocks*, where for formulas $A_1, ..., A_n, B$, a block is a syntactic structure $[A_1, .., A_n \lhd B]$, representing the disjunction $(A_1 \preccurlyeq B) \vee \cdots \vee (A_n \preccurlyeq B)$.

Definition 4.1 A *block* is an expression of the form $[\Sigma \lhd B]$, where Σ is a multiset of formulas and B is a formula. A *sequent* is an expression of the form $\Gamma \Rightarrow \Delta$, where Γ is a multiset of formulas and Δ is a multiset of formulas and blocks.
The *formula interpretation* of a sequent is given by:

$$\iota(\Gamma \Rightarrow \Delta', [\Sigma_1 \lhd B_1], \ldots, [\Sigma_n \lhd B_n]) := \bigwedge \Gamma \rightarrow \bigvee \Delta' \vee \bigvee_{1 \leq i \leq n} \bigvee_{A \in \Sigma_i} (A \preccurlyeq B_i)$$

We write $[\Theta, \Sigma \lhd B]$ for $[(\Theta, \Sigma) \lhd B]$, with Θ, Σ denoting multiset union.

Rules of $\mathcal{I}_{\mathbb{V}}^{\mathsf{i}}$ are shown in Figure 2. Proofs of admissibility of weakening and contraction, invertibility of all rules and admissibility of cut can be found in [5], as well as proof of admissibility of the following rule (needed in Section 6).

Initial sequents $p, \Gamma \Rightarrow \Delta, p \qquad \Gamma, \bot \Rightarrow \Delta$

Propositional rules (standard)

Rules for comparative plausibility

$$\frac{\Gamma \Rightarrow \Delta, [A \lhd B]}{\Gamma \Rightarrow \Delta, A \preccurlyeq B} \ \mathsf{R}{\preccurlyeq}^{\mathsf{i}}$$

$$\frac{\Gamma, A \preccurlyeq B \Rightarrow \Delta, [B, \Sigma \lhd C] \quad \Gamma, A \preccurlyeq B \Rightarrow \Delta, [\Sigma \lhd A], [\Sigma \lhd C]}{\Gamma, A \preccurlyeq B \Rightarrow \Delta, [\Sigma \lhd C]} \ \mathsf{L}{\preccurlyeq}^{\mathsf{i}}$$

Rules for blocks

$$\frac{\Gamma \Rightarrow \Delta, [\Sigma_1, \Sigma_2 \lhd A], [\Sigma_2 \lhd B] \quad \Gamma \Rightarrow \Delta, [\Sigma_1 \lhd A], [\Sigma_1, \Sigma_2 \lhd B]}{\Gamma \Rightarrow \Delta, [\Sigma_1 \lhd A], [\Sigma_2 \lhd B]} \ \mathsf{Com}^{\mathsf{i}}$$

$$\frac{A \Rightarrow \Sigma}{\Gamma \Rightarrow \Delta, [\Sigma \lhd A]} \ \mathsf{Jump}$$

Fig. 2. Rules of $\mathcal{I}^{\mathsf{i}}_{\mathbb{V}}$

Lemma 4.2 *Weakening inside blocks is admissible in* $\mathcal{I}^{\mathsf{i}}_{\mathbb{V}}$.

$$\frac{\Gamma \Rightarrow \Delta, [\Sigma \lhd C]}{\Gamma \Rightarrow \Delta, [A, \Sigma \lhd C]} \ \mathsf{Wk_B}$$

Example 4.3 Derivation in $\mathcal{I}^{\mathsf{i}}_{\mathbb{V}}$ of axiom (CO):

$$\dfrac{\dfrac{\dfrac{\dfrac{\dfrac{\overset{\triangledown}{B \Rightarrow A, B}}{\Rightarrow [A, B \lhd B], [B \lhd A]} \ \mathsf{Jump} \qquad \dfrac{\overset{\triangledown}{A \Rightarrow A, B}}{\Rightarrow [A \lhd B], [A, B \lhd A]} \ \mathsf{Jump}}{\Rightarrow [A \lhd B], [B \lhd A]} \ \mathsf{Com}^{\mathsf{i}}}{\Rightarrow [A \lhd B], B \preccurlyeq A} \ \mathsf{R}{\preccurlyeq}^{\mathsf{i}}}{\Rightarrow A \preccurlyeq B, B \preccurlyeq A} \ \mathsf{R}{\preccurlyeq}^{\mathsf{i}}}{\Rightarrow (A \preccurlyeq B) \vee (B \preccurlyeq A)} \ \vee_R$$

5 From the internal to the labelled calculus

In this section and in the next one we shall present a mutual translation between calculi $\mathcal{I}^{\mathsf{i}}_{\mathbb{V}}$ and **G3V**. Since we aim at translating derivations, we introduce a notation to represent derivations, applicable to both calculi.

Definition 5.1 Let INIT be a **G3V** / $\mathcal{I}^{\mathsf{i}}_{\mathbb{V}}$ initial sequent, and let SEQ denote a **G3V** / $\mathcal{I}^{\mathsf{i}}_{\mathbb{V}}$ sequent $\Gamma \Rightarrow \Delta$. Let R be a **G3V** / $\mathcal{I}^{\mathsf{i}}_{\mathbb{V}}$ rule. A derivation is the following object, where (1) and (2) are sequents:

$$\mathcal{D} : \ \overset{\triangledown}{\text{INIT}} \ ; \ \dfrac{\overset{\mathcal{D}_1}{(1)}}{\text{SEQ}} \ R \ ; \ \dfrac{\overset{\mathcal{D}_1}{(1)} \quad \overset{\mathcal{D}_2}{(2)}}{\text{SEQ}} \ R$$

In this section, we show how $\mathcal{I}^{\mathsf{i}}_{\mathbb{V}}$ derivations can be translated into derivations in **G3V**. We first define t, translation for sequents; then, we specify a function taking as argument $\mathcal{I}^{\mathsf{i}}_{\mathbb{V}}$ derivations and producing in output derivations in **G3V**

+ Wk + Ctr + $\mathsf{Mon}\exists$, and prove that the the translation specified by the function is correct.

Definition 5.2 Given a world label x, a list of countably many sphere labels $\bar{a} = a_1\, a_2 \ldots a_n$ and multisets of formulas $\Gamma, \Delta, \Sigma_1, \ldots, \Sigma_n$, define:

- $t(\Gamma)^x := x : F_1, ..., x : F_k$
- $t(\Gamma \Rightarrow \Delta, [\Sigma_1 \lhd B_1], ..., [\Sigma_n \lhd B_n])^{x,\bar{a}} := a_1 \in S(x), ..., a_n \in S(x),\ a_1 \Vdash^{\exists} B_1, ..., a_n \Vdash^{\exists} B_n, t(\Gamma)^x \Rightarrow t(\Delta)^x, a_1 \Vdash^{\exists} \Sigma_1, ..., a_n \Vdash^{\exists} \Sigma_n$

The translation takes as parameter one world label, x, and sphere labels $\bar{a}$: the idea is that for each block $[\Sigma_i \lhd B_i]$ we introduce a new sphere label a_i such that $a_i \in S(x)$, and formulas $a_i \Vdash^{\exists} B_i$ in the antecedent and $a_i \Vdash^{\exists} \Sigma_i$ in the consequent. These formulas correspond to the semantic condition for a block i.e., a disjunction of $\preccurlyeq$ formulas in sphere models.

We now describe function $\{\ \}^{x,\bar{a}}$ that takes as input a $\mathcal{I}^{\mathsf{i}}_{\mathbb{V}}$ derivation $\mathcal{D}$ and produces as output a **G3V** derivation $\{\mathcal{D}\}^{x,\bar{a}}$. The parameters of the function are the labels x and $\bar{a}$; these are the world and sphere labels used to translate the root sequent of $\mathcal{D}$. For $\bar{a} = (a_1 \ldots a_n)$, we write $\bar{a}b$ to denote the list $(a_1 \ldots a_n\ b)$. The function for propositional rules is immediate: from a translation of the premiss(es) derive a translation of the conclusion applying the corresponding **G3V** rule. For R rule of a calculus, we denote by $R(n)$ n applications of R.

(init) $$\left\{\overset{\nabla}{\text{INIT}}\right\}^{x,\bar{a}} \rightsquigarrow \overset{\nabla}{t(\text{INIT})^{x,\bar{a}}}$$

($\mathsf{R}\preccurlyeq^{\mathsf{i}}$) $$\left\{\dfrac{\overset{\mathcal{D}_1}{\Gamma \Rightarrow \Delta, [A \lhd B]}}{\Gamma \Rightarrow \Delta, A \preccurlyeq B}\ \mathsf{R}\preccurlyeq^{\mathsf{i}}\right\}^{x,\bar{a}} \rightsquigarrow \dfrac{\overset{\{\mathcal{D}_1\}^{x,\bar{a}\,b}}{t(\Gamma \Rightarrow \Delta, [A \lhd B])^{x,\bar{a}\,b}}}{t(\Gamma \Rightarrow \Delta, A \preccurlyeq B)^{x,\bar{a}}}\ \mathsf{R}\preccurlyeq$$

($\mathsf{L}\preccurlyeq^{\mathsf{i}}$) $$\left\{\dfrac{\overset{\mathcal{D}_1}{A \preccurlyeq B, \Gamma \Rightarrow \Delta\, [\Sigma, B \lhd C]} \quad \overset{\mathcal{D}_2}{A \preccurlyeq B, \Gamma \Rightarrow \Delta, [\Sigma \lhd A], [\Sigma \lhd C]}}{A \preccurlyeq B, \Gamma \Rightarrow \Delta, [\Sigma \lhd A]}\ \mathsf{L}\preccurlyeq^{\mathsf{i}}\right\}^{x,\bar{a}\,b} \rightsquigarrow$$

$$\dfrac{\overset{\{\mathcal{D}_1\}^{x,\bar{a}\,b}}{t(A \preccurlyeq B, \Gamma \Rightarrow \Delta\, [\Sigma, B \lhd C])^{x,\bar{a}\,b}} \quad \dfrac{\overset{\{\mathcal{D}_2\}^{x,\bar{a}\,b\,c\,[c/b]}}{t(A \preccurlyeq B, \Gamma \Rightarrow \Delta, [\Sigma \lhd A], [\Sigma \lhd C])^{x,\bar{a}\,b\,c\,[c/b]}}}{b \in S(x), b \Vdash^{\exists} A, b \Vdash^{\exists} C, t(\Gamma)^{x,\bar{a}} \Rightarrow t(\Delta)^{x,\bar{a}}, b \Vdash^{\exists} \Sigma}\ \mathsf{Ctr}}{t(A \preccurlyeq B, \Gamma \Rightarrow \Delta, [\Sigma \lhd A])^{x,\bar{a}\,b}}\ \mathsf{L}\preccurlyeq$$

($\mathsf{Com}^{\mathsf{i}}$) $$\left\{\dfrac{\overset{\mathcal{D}_1}{\Gamma \Rightarrow \Delta, [\Sigma_1, \Sigma_2 \lhd A], [\Sigma_2 \lhd B]} \quad \overset{\mathcal{D}_2}{\Gamma \Rightarrow \Delta, [\Sigma \lhd A], [\Sigma, \Pi \lhd B]}}{\Gamma \Rightarrow \Delta, [\Sigma_1 \lhd A], [\Sigma_1 \lhd B]}\ \mathsf{Com}^{\mathsf{i}}\right\}^{x,\bar{a}\,b\,c} \rightsquigarrow$$

$$\dfrac{\dfrac{\dfrac{\overset{\{\mathcal{D}_1\}^{x,\bar{a}\,b\,c}}{t(\Gamma \Rightarrow \Delta, [\Sigma_1, \Sigma_2 \lhd A], [\Sigma_2 \lhd B])^{x,\bar{a}\,b\,c}}}{b \subseteq c, b \Vdash^{\exists} A, c \Vdash^{\exists} B, t(\Gamma)^{x,\bar{a}} \Rightarrow t(\Delta)^{x,\bar{a}}, b \Vdash^{\exists} \Sigma_1, b \Vdash^{\exists} \Sigma_2, c \Vdash^{\exists} \Sigma_2}\ \mathsf{Wk}}{b \subseteq c, b \Vdash^{\exists} A, c \Vdash^{\exists} B, t(\Gamma)^{x,\bar{a}} \Rightarrow t(\Delta)^{x,\bar{a}}, b \Vdash^{\exists} \Sigma_1, c \Vdash^{\exists} \Sigma_2}\ \mathsf{Mon}\exists \quad \dfrac{\dfrac{\overset{\{\mathcal{D}_2\}^{x,\bar{a}\,b\,c}}{t(2)^{x,\bar{a}\,b\,c}}}{(2')}\ \mathsf{Wk}}{(2'')}\ \mathsf{Mon}\exists}{t(\Gamma \Rightarrow \Delta, [\Sigma_1 \lhd A], [\Sigma_1 \lhd B])^{x,\bar{a}\,b\,c}}\ \mathsf{Nes}$$

(2) $\Gamma \Rightarrow \Delta, [\Sigma \lhd A], [\Sigma, \Pi \lhd B]$;
(2′) $b \subseteq c, b \Vdash^{\exists} A, c \Vdash^{\exists} B, t(\Gamma)^{x,\bar{a}} \Rightarrow t(\Delta)^{x,\bar{a}}, b \Vdash^{\exists} \Sigma_1, c \Vdash^{\exists} \Sigma_1, c \Vdash^{\exists} \Sigma_2$;
(2″) $c \subseteq b, b \Vdash^{\exists} A, c \Vdash^{\exists} B, t(\Gamma)^{x,\bar{a}} \Rightarrow t(\Delta)^{x,\bar{a}}, b \Vdash^{\exists} \Sigma_1, c \Vdash^{\exists} \Sigma_2$

(Jump)

$$\left\{\frac{\begin{array}{c}\mathcal{D}_1\\ x:\Sigma\Rightarrow x:A\end{array}}{\Gamma\Rightarrow\Delta,[\Sigma\lhd A]}\ \mathsf{Jump}\right\}^{x,\bar{a}\,b} \rightsquigarrow \cfrac{\cfrac{\cfrac{\begin{array}{c}t\{\mathcal{D}_1\}^{x\,[x/y]}\\ t(x:\Sigma\Rightarrow x:A)^{x\,[x/y]}\end{array}}{y\in b, b\in S(x), y:A, t(\Gamma)^{x,\bar{a}}\Rightarrow t(\Delta)^{x,\bar{a}}, y:\Sigma, b\Vdash^{\exists}\Sigma}\ Wk}{y\in b, b\in S(x), y:A, t(\Gamma)^{x,\bar{a}}\Rightarrow t(\Delta)^{x,\bar{a}}, b\Vdash^{\exists}\Sigma}\ \mathsf{R}\Vdash^{\exists}(n)}{t(\Gamma\Rightarrow\Delta,[\Sigma\lhd A])^{x,\bar{a}\,b}}\ \mathsf{L}\Vdash^{\exists}$$

Theorem 5.3 *Let $\mathcal{D}$ be a $\mathcal{I}^{\mathsf{i}}_{\mathbb{V}}$ derivation of $\Gamma \Rightarrow \Delta$. Then $\{\mathcal{D}\}^{x,\bar{a}}$ is a derivation of $t(\Gamma \Rightarrow \Delta)^{x,\bar{a}}$ in* **G3V**.

Proof. By induction on the height h of the derivation of the sequent. If $h = 0$, $\Gamma \Rightarrow \Delta$ is a $\mathcal{I}^{\mathsf{i}}_{\mathbb{V}}$ initial sequent, and $t(\Gamma \Rightarrow \Delta)^{x,\bar{a}}$ is a **G3V** initial sequent. If $h > 0$, $\Gamma \Rightarrow \Delta$ must have been derived applying a rule of $\mathcal{I}^{\mathsf{i}}_{\mathbb{V}}$. All cases easily follow applying the clauses of the procedure described above.

[$\mathsf{R} \preccurlyeq^{\mathsf{i}}$] The translation of the premiss of $\mathsf{R} \preccurlyeq^{\mathsf{i}}$ is the **G3V** sequent $t(\Gamma \Rightarrow \Delta, A \preccurlyeq B)^{x,\bar{a}} = b \in S(x), b \Vdash^{\exists} B, t(\Gamma)^{x,\bar{a}} \Rightarrow t(\Delta)^{x,\bar{a}}, b \Vdash^{\exists} A$. Applying $\mathsf{R} \preccurlyeq$ we obtain the translation of the conclusion: $t(\Gamma \Rightarrow \Delta, A \preccurlyeq B)^{x,\bar{a}} = t(\Gamma)^{x,\bar{a}} \Rightarrow t(\Delta)^{x,\bar{a}}, x : A \preccurlyeq B$.

[$\mathsf{L} \preccurlyeq^{\mathsf{i}}$] The translations of the premisses are the **G3V** sequents: $t(A \preccurlyeq B, \Gamma \Rightarrow \Delta, [\Sigma, B \lhd C])^{x,\bar{a}\,b} = b \in S(x), b \Vdash^{\exists} C, x : A \preccurlyeq B, t(\Gamma)^{x,\bar{a}} \Rightarrow t(\Delta)^{x,\bar{a}}, b \Vdash^{\exists} \Sigma, b \Vdash^{\exists} B$; $t(A \preccurlyeq B, \Gamma \Rightarrow \Delta, [\Sigma \lhd A], [\Sigma \lhd C])^{x,\bar{a}\,b\,c} = b \in S(x), c \in S(x), b \Vdash^{\exists} A, c \Vdash^{\exists} C, t(\Gamma)^{x,\bar{a}} \Rightarrow t(\Delta)^{x,\bar{a}}, b \Vdash^{\exists} \Sigma, c \Vdash^{\exists} \Sigma$. We substitute the sphere label c with b in the second sequent, obtaining $b \in S(x), b \in S(x), b \Vdash^{\exists} A, c \Vdash^{\exists} C, t(\Gamma)^{x,\bar{a}} \Rightarrow t(\Delta)^{x,\bar{a}}, b \Vdash^{\exists} \Sigma, b \Vdash^{\exists} \Sigma$. After application of contraction, application of $\mathsf{L} \preccurlyeq$ to this sequent and to the translation of the first premiss yields sequent $b \in S(x), b \Vdash^{\exists} C, x : A \preccurlyeq B, t(\Gamma)^{x,\bar{a}} \Rightarrow t(\Delta)^{x,\bar{a}}, b \Vdash^{\exists} \Sigma$. This sequent is the translation of the $\mathcal{I}^{\mathsf{i}}_{\mathbb{V}}$ sequent $A \preccurlyeq B, \Gamma \Rightarrow \Delta, [\Sigma \lhd C]$, with parameters $x, \bar{a}\,b$.

[$\mathsf{Com}^{\mathsf{i}}$] The translations of the premisses are the **G3V** sequents: $t(\Gamma \Rightarrow \Delta, [\Sigma_1, \Sigma_2 \lhd A], [\Sigma_2 \lhd B])^{x,\bar{a}\,b\,c} = b \in S(x), c \in S(x), b \Vdash^{\exists} A, c \Vdash^{\exists} B, t(\Gamma)^{x,\bar{a}} \Rightarrow t(\Delta)^{x,\bar{a}}, b \Vdash^{\exists} \Sigma_1, b \Vdash^{\exists} \Sigma_2, c \Vdash^{\exists} \Sigma_2$; $t(\Gamma \Rightarrow \Delta, [\Sigma_1 \lhd A], [\Sigma_1, \Sigma_2 \lhd B])^{x,\bar{a}\,b\,c} = b \in S(x), c \in S(x), b \Vdash^{\exists} A, c \Vdash^{\exists} B, t(\Gamma)^{x,\bar{a}} \Rightarrow t(\Delta)^{x,\bar{a}}, b \Vdash^{\exists} \Sigma_1, c \Vdash^{\exists} \Sigma_2, c \Vdash^{\exists} \Sigma_2$. We add by weakening $b \subseteq c$ to the translation of the first premiss and $c \subseteq b$ to the translation of the second, and apply rule $\mathsf{Mon}\exists$ to both. A final application of Nes yields the desired sequent: $t(\Gamma \Rightarrow \Delta, [\Sigma_1 \lhd A], [\Sigma_2 \lhd b])^{x,\bar{a}\,b\,c} = b \in S(x), c \in S(x), b \Vdash^{\exists} A, c \Vdash^{\exists} B, t(\Gamma)^{x,\bar{a}} \Rightarrow t(\Delta)^{x,\bar{a}}, b \Vdash^{\exists} \Sigma_1, c \Vdash^{\exists} \Sigma_2$.

[Jump] The translation of the premiss of Jump is the sequent $t(x : \Sigma \Rightarrow x : A)^x = x : \Sigma \Rightarrow x : A$. We substitute x with a fresh world label y, and apply the transformations described above; we obtain the sequent $b \in S(x), b \Vdash^{\exists} A, t(\Gamma)^{x,\bar{a}} \Rightarrow t(\Delta)^{x,\bar{a}}, b \Vdash^{\exists} \Sigma$, which is the translation of the sequent $\Gamma \Rightarrow \Delta, [\Sigma \lhd A]$, the conclusion of Jump, with parameters $x, \bar{a}\,b$.

□

Example 5.4 This **G3V** derivation is obtained translating the $\mathcal{I}^{\mathsf{i}}_{\mathbb{V}}$ derivation of Example 4.3. In the application of Nes only the left premiss is shown.

$$\begin{array}{c}
\nabla \\
y : B \Rightarrow y : A, y : B \\
\hline
a \in S(x), b \in S(x), y \in a, y \in b, y : B, b \Vdash^{\exists} A \Rightarrow a \Vdash^{\exists} A, a \Vdash^{\exists} B, b \Vdash^{\exists} B, y : A, y : B \quad \mathsf{Wk} \\
\hline
a \in S(x), b \in S(x), y \in a, y \in b, y : B, b \Vdash^{\exists} A \Rightarrow a \Vdash^{\exists} A, a \Vdash^{\exists} B, b \Vdash^{\exists} B \quad \mathsf{R}\Vdash^{\exists}(2) \\
\hline
a \in S(x), b \in S(x), a \Vdash^{\exists} B, b \Vdash^{\exists} A \Rightarrow a \Vdash^{\exists} A, a \Vdash^{\exists} B, b \Vdash^{\exists} B \quad \mathsf{L}\Vdash^{\exists} \\
\hline
a \subseteq b, a \in S(x), b \in S(x), a \Vdash^{\exists} B, b \Vdash^{\exists} A \Rightarrow a \Vdash^{\exists} A, a \Vdash^{\exists} B, b \Vdash^{\exists} B \quad \mathsf{Wk} \\
\hline
a \subseteq b, a \in S(x), b \in S(x), a \Vdash^{\exists} B, b \Vdash^{\exists} A \Rightarrow a \Vdash^{\exists} A, b \Vdash^{\exists} B \quad \mathsf{Mon}\exists \\
\hline
a \in S(x), b \in S(x), a \Vdash^{\exists} B, b \Vdash^{\exists} A \Rightarrow a \Vdash^{\exists} A, b \Vdash^{\exists} B \quad \mathsf{Nes} \\
\hline
a \in S(x), a \Vdash^{\exists} B \Rightarrow x : B \preccurlyeq A, a \Vdash^{\exists} A \quad \mathsf{R}\preccurlyeq^{\mathsf{i}} \\
\hline
\Rightarrow x : A \preccurlyeq B, x : B \preccurlyeq A \quad \mathsf{R}\preccurlyeq^{\mathsf{i}} \\
\hline
\Rightarrow x : A \preccurlyeq B \vee B \preccurlyeq A \quad R\vee
\end{array}$$

6 From the labelled to the internal calculus

The inverse translation takes care of translating **G3V** derivations into $\mathcal{I}^{\mathsf{i}}_{\mathbb{V}}$ derivations. With respect to the previous translation, we are faced with an additional difficulty: there are **G3V** derivable sequents that cannot be translated into $\mathcal{I}^{\mathsf{i}}_{\mathbb{V}}$ sequents or, equivalently, there are *more* **G3V** derivable sequents than $\mathcal{I}^{\mathsf{i}}_{\mathbb{V}}$ derivable sequents. For this reason, proving that if $t(\Gamma \Rightarrow \Delta)^x$ is derivable in **G3V** then $\Gamma \Rightarrow \Delta$ is derivable in $\mathcal{I}^{\mathsf{i}}_{\mathbb{V}}$ would not work: in the **G3V** derivation of $t(\Gamma \Rightarrow \Delta)^x$ there could occur some sequents that are not in the range of the translation t.

Thus, we need a more complex proof strategy. After defining a translation s for sequents, we shall introduce the notion of normal form derivations in **G3V**: the idea is that we cannot translate *any* derivation, but only those constructed following a certain order of application of the rules. We will prove that any derivation in **G3V** can be transformed into a normal form derivation. Then, we shall prove the fundamental Jump lemma, which allows us to "skip" the sequents that we cannot translate in the translation of a derivation. Finally, we shall define a function to translate derivations from **G3V** to $\mathcal{I}^{\mathsf{i}}_{\mathbb{V}}$.

Definition 6.1 Let $\Gamma \Rightarrow \Delta$ be a sequent of the form

$$\mathcal{R}^{a_1}, \ldots, \mathcal{R}^{a_n}, a_1 \in S(x), ..., a_n \in S(x), a_1 \Vdash^{\exists} A_1, ..., a_n \Vdash^{\exists} A_n, x : \Gamma^P \Rightarrow x : \Delta^P, a_1 \Vdash^{\exists} \Sigma_1, ..., a_n \Vdash^{\exists} \Sigma_n$$

where: *a*) each $\mathcal{R}^{a_i}$ contains zero or more inclusions $a_i \subseteq a_j$ for $1 \leqslant i \leqslant j \leqslant n$; *b*) Γ^P and Δ^P are composed only of propositional and $\preccurlyeq$ formulas; *c*) for each a_i, there is exactly one formula $a_i \Vdash^{\exists} A_i$ in the antecedent, and at least one formula $a_i \Vdash^{\exists} B_i$ in the consequent. The translation s takes as parameter a world label x, label of Γ^P and Δ^P, and is defined as

$$s(\Gamma \Rightarrow \Delta)^x := \Gamma^P \Rightarrow \Delta^P, \Pi$$

where: Γ^P is obtained from $x : \Gamma^P$ by removing the label x, Δ^P is obtained from $x : \Delta^P$ by removing the label x, and Π contains n blocks $[\{\Sigma_1\} \lhd A_1], ..., [\{\Sigma_n\} \lhd A_n]$, where each $\{\Sigma_i\}$ is the multiset union $\{\Sigma_i\} = \Sigma_i \cup \bigcup \{\Sigma_j \mid a_i \subseteq a_j \textit{ occurs in } \mathcal{R}^{a_i}\}$.

For instance, consider the translation s of the following sequent:

$$s(a_1 \subseteq a_2, a_2 \subseteq a_3, a_1 \subseteq a_3, a_1 \in S(x), a_2 \in S(x), a_3 \in S(x), a_1 \Vdash^{\exists} A_1, a_2 \Vdash^{\exists} A_2, a_3 \Vdash^{\exists} A_3, x : \Gamma \Rightarrow x : \Delta, a_1 \Vdash^{\exists} \Sigma_1, a_2 \Vdash^{\exists} \Sigma_2, a_3 \Vdash^{\exists} \Sigma_3)^x := \\ := s(\Gamma)^x \Rightarrow s(\Delta)^x, [\Sigma_1, \Sigma_2, \Sigma_3 \lhd A_1], [\Sigma_2, \Sigma_3 \lhd A_2], [\Sigma_3 \lhd A_3]$$

Intuitively, s re-assembles the blocks from formulas labelled with the same sphere label. Furthermore, for each inclusion $a_i \subseteq a_j$ we add to the corresponding block also formulas Σ_j such that $a_j \Vdash^{\exists} \Sigma_j$ occurs in the consequent of the labelled sequent. Thus, each block in the internal calculus consists of $\preccurlyeq$-formulas relative to some sphere i.e., labelled with the same sphere label in **G3V**.

We now introduce the notion of normal form derivations and state the Jump lemma.

Definition 6.2 Given a world label x and a sequent $\Gamma \Rightarrow \Delta$, the sequent is *saturated with respect to variable* x (is x-saturated) if: *a*) if $x : A$ belongs to $\Gamma \cup \Delta$, then A is atomic or $A \equiv B \preccurlyeq C$ and $x : B \preccurlyeq C$ does not belong to Δ; *b*) if $x : A \preccurlyeq B$ and $a \in S(x)$ occur in Γ, either $a \Vdash^{\exists} A$ occurs in Δ or $a \Vdash^{\exists} B$ occurs in Γ; *c*) if $a \in S(x)$ and $b \in S(x)$ occur in Γ, either $a \subseteq b$ or $b \subseteq a$ occurs in Γ. A sequent is *x-hypersaturated with respect to variable x* if, for all $a \in S(x)$, the following hold: *a*) for each $a \in S(x)$, no formulas $a \Vdash^{\exists} B$ occurs in Γ; *b*) for each $a \in S(x)$, $y \in a$ occurring in Γ and for each $a \Vdash^{\exists} B$ occurring in Δ, there is a formula $y : B$ occurring in Δ; *c*) for each $a \in S(x)$, $b \in S(x)$, $a \subseteq b$ and $y \in a$ occurring in Γ, there is a formula $y \in b$ occurring in Γ. Given a branch $\mathcal{B}$ of a derivation of $\Gamma \Rightarrow \Delta$, we say that $\mathcal{B}$ is in normal form with respect to x if from the root sequent $\Gamma \Rightarrow \Delta$ upwards the following holds: first all propositional and $\preccurlyeq$ rules are applied, until an x-saturated sequent is reached, and then rules $\mathsf{R} \Vdash^{\exists}$, $\mathsf{L} \Vdash^{\exists}$ and $\mathsf{L} \subseteq$ are applied to the x-saturated sequent, until a sequent which is x-hypersaturated is reached. We say that a derivation of $\Gamma \Rightarrow \Delta$ is *in normal form* with respect to x if all its branches are in normal form. [6]

Lemma 6.3 *Given a* **G3V** *derivable sequent* $\Gamma \Rightarrow \Delta$ *which is the result of a translation* t *and a variable* x *occurring in it, we can transform any derivation of* $\Gamma \Rightarrow \Delta$ *into a derivation in normal form with respect to variable* x.

Proof. Induction on the height of the derivation of $\Gamma \Rightarrow \Delta$. Let $\rightarrow^*_{\Pi_{\mathcal{B}}}$ be the relation between world labels occurring in the union of all antecedents of a branch, as in Definition 3.8. If the sequent is an axiom, we are done. If the height of the derivation is greater than zero, we proceed by cases: if there are no labels y different from x such that $x \rightarrow^*_{\Pi_{\mathcal{B}}} y$, then x is the only label in the branch. The derivation of $\Gamma \Rightarrow \Delta$ will use only propositional rules; thus, the branch is in normal form with respect to x. If there is some label y such that

[6] Recall Definition 3.7; if a **G3V** derivable sequent $\Gamma \Rightarrow \Delta$ is the result of a translation t (i.e. there exists a $\mathcal{I}^{i}_{\mathbb{V}}$ sequent $\Gamma^I \Rightarrow \Delta^I$ such that $t(\Gamma^I \Rightarrow \Delta^I)^{x,\bar{a}} = \Gamma \Rightarrow \Delta$), the sequent is *simple*, and the labels occurring in its derivation form a tree according to the relation $\rightarrow^*_{\Pi_{\mathcal{B}}}$ (Lemma 3.9). This result is not unexpected: as in the case of [8], we are able to translate only tree-form sequents.

$x \rightarrow^*_{\Pi_\mathcal{B}} y$, transform each branch of the derivation of $\Gamma \Rightarrow \Delta$ as follows. Sequent $\Gamma \Rightarrow \Delta$ contains at most one world label and possibly some sphere labels $a, b \ldots$ such that $a \in S(x), b \in S(x) \ldots$. Labels $y \in a$ might be introduced only by $\mathsf{L} \Vdash^\exists$. If some rules are applied to formulas $a \Vdash^\exists A$ or to formulas $y \in a, a \subseteq b$ or to formulas $y : A$, when there are still some rules (non-redundantly) applicable to formulas $x : A$ or $a \in S(x), b \in S(x)$, apply *first* the rules for $x : A$ and $a \in S(x), b \in S(x)$, until the x-saturated sequent is reached. Similarly, if some rules are applied to a formula $y : A$ when there are still some rules which can be (non-redundantly) applied to formulas $a \Vdash^\exists A$ or to formulas $y \in a, a \subseteq b$, apply these latter rules until an x-hypersaturated sequent is reached, before proceeding to apply rules for $y : A$. In both cases, permuting the rules in the derivation does not represent a problem: rules applicable to $x : A$ and to $y : A$ involve different active formulas. As for rules applicable to $a \Vdash^\exists A$ with $a \in S(x)$, observe that the normal form "respects" the order in which labels are generated in the tree. For instance, the normal form with respect to x requires that rule $\mathsf{R} \preccurlyeq$, generating spheres $a \in S(x)$, has to be applied to $x : C \preccurlyeq D$ *before* rules $\mathsf{R} \Vdash^\exists$ or $\mathsf{L} \Vdash^\exists$ might be applied to $b \Vdash^\exists C$ or $b \Vdash^\exists D$. To obtain a normal form derivation with respect to x, we have to apply the procedure to all branches; we might also have to add some rules to obtain the x-saturated and x-hypersaturated sequents. □

Definition 6.4 We give here a simplified version of Definition 3.8. Given a multiset of labelled formulas Π, define: *a)* $x \rightarrow_\Pi a$ if $a \in S(x)$ occurs in Π; *b)* $a \rightarrow_\Pi y$ if $y \in a$ occurs in Π; *c)* $x \rightarrow_\Pi y$ if there exists an a such that $x \rightarrow_\Pi a$ and $a \rightarrow_\Pi y$ occur in Π. Let $W_\Pi(x)$ be the reflexive and transitive closure of $x \rightarrow_\Pi y$: $W_\Pi(x) = \{y \mid x \rightarrow^*_\Pi y\}$. Let $N_\Pi(x) = \{b \mid \exists u \,.\, x \rightarrow^*_\Pi u \textit{ and } u \rightarrow_\Pi b\}$. These sets represents respectively the set of world labels accessible from a world label x, and the set of sphere labels accessible from a world label x occurring in Π. Define Σ^Π_x as the union of the sets:

$$\begin{aligned}
\Sigma^\Pi_x \;=\;& \{u : F \mid u : F \textit{ occurs in } \Sigma \textit{ and } u \in W_\Pi(x)\} \cup \\
& \cup \{a \Vdash^\exists B \mid a \Vdash^\exists B \textit{ occurs in } \Sigma \textit{ and } a \in N_\Pi(x)\} \cup \\
& \cup \{b \in S(y) \mid b \in S(y) \textit{ occurrs in } \Pi,\ b \in N_\Pi(x) \textit{ and } y \in W_\Pi(x)\} \cup \\
& \cup \{a \subseteq b \mid a \subseteq b \textit{ occurs in } \Pi \textit{ and } a, b \in N_\Pi(x)\} \cup \\
& \cup \{z \in a \mid z \in a \textit{ occurs in } \Pi \textit{ and } a \in N_\Pi(x)\}.
\end{aligned}$$

Lemma 6.5 (Jump lemma) *Let $\Gamma \Rightarrow \Delta$ be a derivable* **G3V** *sequent. If the labels occurring in $W_\Gamma(x)$ have a tree structure, for each label x occurring in the sequent, it holds that either 1) sequent $\Gamma^\Gamma_x \Rightarrow \Delta^\Gamma_x$ or 2) sequent $\Gamma - \Gamma^\Gamma_x \Rightarrow \Delta - \Delta^\Gamma_x$ is derivable, with the same derivation height.*

Proof. To simplify the notation, we write Γ^* for Γ^Γ_x and Δ^* for Δ^Γ_x respectively. The proof is by induction on the height of the derivation, and by distinction of cases. If $\Gamma \Rightarrow \Delta$ is an initial sequent, it has the form $u : P, \Gamma' \Rightarrow \Delta', u : P$. If $u \in W^\Gamma_x$, we have that $u : P, \Gamma^* \Rightarrow \Delta^*, u : P$ is and initial sequent, hence derivable, and we are in case 1. If $u \notin W^\Gamma_x$, then $u : P \in \Gamma - \Gamma^\Gamma_x$, and we obtain case 2. For the propositional rules, we show only the case of $L \rightarrow$.

$$\frac{\Gamma \Rightarrow \Delta, u : B \quad u : A, \Gamma \Rightarrow \Delta}{u : A \to B, \Gamma \Rightarrow \Delta} \; L\to$$

Suppose $u \in W_x^\Gamma$. We have to show that either 1) the sequent $u : A \to B, \Gamma^* \Rightarrow \Delta^*$ is derivable, or that 2) sequent $\Gamma - \Gamma^* \Rightarrow \Delta - \Delta^*$ is derivable. By inductive hypothesis applied to both premisses, we have that either a) $\Gamma^* \Rightarrow \Delta^*, u : B$ is derivable or b) $\Gamma - \Gamma^* \Rightarrow \Delta - \Delta^*$ is derivable, and that either c) $u : A, \Gamma^* \Rightarrow \Delta^*$ is derivable or d) $\Gamma - \Gamma^* \Rightarrow \Delta - \Delta^*$ is derivable. If *a*) and *c*) are derivable, we apply $L \to$ and obtain the derivable sequent $u : A \to B, \Gamma^* \Rightarrow \Delta^*$ (which is case 1). If *a*) and *d*) are derivable, *d*) is already the sequent corresponding to case 2 of the statement; the same holds if *b*) and *c*) are derivable, and if *b*) and *d*) are derivable. If $u \notin W_x^\Gamma$, we want to show that either 1) $\Gamma^* \Rightarrow \Delta^*$ id derivable or that 2) $\Gamma - \Gamma^*, u : A \to B \Rightarrow \Delta - \Delta^*$ is derivable. Again, by inductive hypothesis we have that either *a*) $\Gamma - \Gamma^*, u : B \Rightarrow \Delta - \Delta^*$ or *b*) $\Gamma^* \Rightarrow \Delta^*$ and either *c*) $\Gamma - \Gamma^* \Rightarrow \Delta - \Delta^*, u : A$ or *d*) $\Gamma^* \Rightarrow \Delta^*$ are derivable. If *a*) and *c*) are derivable, we obtain case 2; otherwise we are in case 1.
If $\Gamma \Rightarrow \Delta$ has been derived by $\mathsf{L} \Vdash^{\exists}$, we have:

$$\frac{\Gamma, y \in a, y : B \Rightarrow \Delta}{\Gamma, a \Vdash^{\exists} B \Rightarrow \Delta} \; \mathsf{L}\Vdash^{\exists}$$

where y does not occur in Γ and Δ. If $a \in N_\Gamma(x)$ then $y \in W_\Gamma(x)$; by inductive hypothesis, we have that either *a*) $\Gamma^*, y : B, y \in a \Rightarrow \Delta^*$ is derivable, or *b*) $\Gamma - \Gamma^* \Rightarrow \Delta - \Delta*$ is derivable. In the former case, a step of $\mathsf{L} \Vdash^{\exists}$ gives that $\Gamma^*, a \Vdash^{\exists} B \Rightarrow \Delta^*$ is derivable. If *b*) is derivable, we already have our desired sequent (case 2). If $a \notin N_\Gamma(x)$, then $y \notin W_\Gamma(x)$. By inductive hypothesis, either $\Gamma^* \Rightarrow \Delta^*$ is derivable, and we are done, or $\Gamma - \Gamma^*, y : B, y \in a \Rightarrow \Delta - \Delta^*$ is derivable. Apply $\mathsf{L} \Vdash^{\exists}$ to obtain the sequent $\Gamma - \Gamma^*, a \Vdash^{\exists} B \Rightarrow \Delta - \Delta^*$.
For the remaining cases: $\mathsf{R} \preccurlyeq$ is similar to $\mathsf{L} \Vdash^{\exists}$. Rules Nes and $\mathsf{L} \preccurlyeq$ are similar to $L \to$. $\mathsf{R} \Vdash^{\exists}$ and $\mathsf{L} \subseteq$ are immediate, since they do not introduce new labels and have just one premiss. □

Example 6.6 Suppose that the following sequent is derivable.

$$a \in S(x), b \in S(x), y \in a, y : B, z \in a, z : C \Rightarrow a \Vdash^{\exists} A, b \Vdash^{\exists} B, z : B, y : A$$

Consider label z: $N_z^\Gamma = \varnothing$ and $W_z^\Gamma = \{z\}$. Thus, Γ_z^Γ coincides with $z : C$, and Δ_z^Γ coincides with $z : B$, and either $z : C \Rightarrow z : A$ is derivable, or the rest of the sequent is derivable, namely $a \in S(x), b \in S(x), y \in a, y : B, z \in a \Rightarrow a \Vdash^{\exists} A, y : A, b \Vdash^{\exists} B$.

Lemma 6.7 *If a sequent* $a \Vdash^{\exists} B, a \Vdash^{\exists} C, \Gamma \Rightarrow \Delta$ *is derivable in* **G3V**, *then either* $a \Vdash^{\exists} B, \Gamma \Rightarrow \Delta$ *or* $a \Vdash^{\exists} C, \Gamma \Rightarrow \Delta$ *are derivable in* **G3V** *with same derivation height.*

Proof. By induction on the height of the derivation. The only relevant case is $R = \mathsf{L} \Vdash^{\exists}$, applied to one of the formulas $a \Vdash^{\exists} A$ or $a \Vdash^{\exists} B$.

$$\frac{y \in a, y : A, a \Vdash^{\exists} B, \Gamma \Rightarrow \Delta}{a \Vdash^{\exists} A, a \Vdash^{\exists} B, \Gamma \Rightarrow \Delta} \; \mathsf{L}\Vdash^{\exists}$$

Apply the Jump lemma to the premiss, obtaining that either 1) $y : A \Rightarrow$ is derivable, or 2) $y \in a, a \Vdash^{\exists} B, \Gamma \Rightarrow \Delta$ is derivable. In the former case, apply weakening and $\mathsf{L} \Vdash^{\exists}$ to obtain a derivation of $a \Vdash^{\exists} A, \Gamma \Rightarrow \Delta$. In the latter case, by invertibility of $\mathsf{L} \Vdash^{\exists}$ we have that sequent $y \in a, w \in a, w : B, \Gamma \Rightarrow \Delta$ is derivable, for some $w \notin \Gamma, \Delta$. We substitute variable y with variable w. The substitution does not affect other formulas than $y \in a$, since $y, w \notin \Gamma, \Delta$. Contraction and $\mathsf{L} \Vdash^{\exists}$ give the sequent $a \Vdash^{\exists} B, \Gamma \Rightarrow \Delta$. □

In order to define a translation $[\![\ ]\!]^x$ for **G3V** normal form derivations we need to define a sub-translation $[\ ]^x$, that takes care of the translation of a derivation from the root sequent up to the x-saturated sequents. Theorem 6.9 will take care of translating the upper part of the normal form derivations: from x-saturated sequents to x-hypersaturated sequents, making an essential use of the Jump lemma.

The translation $[\ ]^x$ takes as parameter a world label x, the one used to translate the root sequent. For $\mathsf{L} \preccurlyeq$ and Nes we explicitly define sets of inclusions that might occur in the sequent: $\mathcal{R}^a = \{a \subseteq c_1, \dots, a \subseteq c_n\}$; $\mathcal{R}^b = \{b \subseteq d_1, \dots, b \subseteq d_k\}$. We will also use the corresponding mulisets of formulas: $\{\Omega\} = \{c \Vdash^{\exists} \Omega \,|\, c \Vdash^{\exists} \Omega \textit{ occurs in } \Delta \textit{ and } a \subseteq c \textit{ occurs in } \mathcal{R}^a\}$ and $\{\Xi\} = \{d \Vdash^{\exists} \Xi \,|\, d \Vdash^{\exists} \Xi \textit{ occurs in } \Delta \textit{ and } b \subseteq d \textit{ occurs in } \mathcal{R}^b\}$.

(init)
$$\left[\overset{\nabla}{\mathrm{INIT}}\right]^x \rightsquigarrow \overset{\nabla}{s(\mathrm{INIT})^x}$$

$(\mathsf{R} \preccurlyeq)$
$$\left[\dfrac{\begin{array}{c}\mathcal{D}_1\\ a \in S(x), a \Vdash^{\exists} B, \Gamma \Rightarrow \Delta, a \Vdash^{\exists} A\end{array}}{\Gamma \Rightarrow \Delta, x : A \preccurlyeq B}\,\mathsf{R}\!\preccurlyeq\right]^x \rightsquigarrow \dfrac{\begin{array}{c}[\mathcal{D}_1]^x\\ s(a \in S(x), a \Vdash^{\exists} B, \Gamma \Rightarrow \Delta, a \Vdash^{\exists} A)^x\end{array}}{s(\Gamma \Rightarrow \Delta, x : A \preccurlyeq B)^x}\,\mathsf{R}\!\preccurlyeq^{\mathsf{i}}$$

$(\mathsf{L} \preccurlyeq)$
$$\left[\dfrac{\begin{array}{c}\mathcal{D}_1\\ (1)\end{array} \quad \begin{array}{c}\mathcal{D}_2\\ (2)\end{array}}{\mathsf{Conc}}\,\mathsf{L}\!\preccurlyeq\right]^x \rightsquigarrow \dfrac{\begin{array}{c}[\mathcal{D}_1]^x\\ s(1)^x\end{array} \quad \dfrac{\begin{array}{c}[\mathcal{D}^-]^x\\ s(S_1)^x\end{array}}{Q}\,\mathsf{Wk}}{s(\mathsf{Conc})^x}\,\mathsf{L}\!\preccurlyeq^{\mathsf{i}}$$

$(1) = a \in S(x), a \Vdash^{\exists} C, x : A \preccurlyeq B, \mathcal{R}^a, \Gamma \Rightarrow \Delta, a \Vdash^{\exists} \Sigma, a \Vdash^{\exists} B$
$(2) = a \in S(x), a \Vdash^{\exists} A, a \Vdash^{\exists} C, x : A \preccurlyeq B, \Gamma \Rightarrow \Delta, a \Vdash^{\exists} \Sigma$
$\mathsf{Conc} = a \in S(x), a \Vdash^{\exists} C, x : A \preccurlyeq B, \mathcal{R}^a, \Gamma \Rightarrow \Delta, a \Vdash^{\exists} \Sigma$
$S_1 = a \in S(x), a \Vdash^{\exists} A, x : A \preccurlyeq B, \mathcal{R}^a, \Gamma \Rightarrow \Delta, a \Vdash^{\exists} \Sigma$ (from (2) by Lemma 6.7)
$Q = A \preccurlyeq B, s(\Gamma)^x \Rightarrow s(\Delta)^x, [\Sigma, \{\Omega\} \lhd C], [\Sigma, \{\Omega\} \lhd A]$

(Nes)
$$\left[\dfrac{\begin{array}{c}\mathcal{D}_1\\ (1)\end{array} \quad \begin{array}{c}\mathcal{D}_2\\ (2)\end{array}}{\mathsf{Conc}}\,\mathsf{Nes}\right]^x \rightsquigarrow \dfrac{\dfrac{\begin{array}{c}[\mathcal{D}_1]^x\\ s(1)^x\end{array}}{P}\,\mathsf{Wk_B} \quad \dfrac{\begin{array}{c}[\mathcal{D}_2]^x\\ s(2)^x\end{array}}{Q}\,\mathsf{Wk_B}}{s(\mathsf{Conc})^x}\,\mathsf{Com}^{\mathsf{i}}$$

The underlined formulas are added by $\mathsf{Wk_B}$.
$(1) = a \subseteq b, a \in S(x), b \in S(x), \mathcal{R}^a, \mathcal{R}^b, a \Vdash^{\exists} A, a \Vdash^{\exists} B, \Gamma \Rightarrow \Delta, a \Vdash^{\exists} \Sigma, b \Vdash^{\exists} \Pi$
$(2) = b \subseteq a, a \in S(x), b \in S(x), \mathcal{R}^a, \mathcal{R}^b, a \Vdash^{\exists} A, a \Vdash^{\exists} B, \Gamma \Rightarrow \Delta, a \Vdash^{\exists} \Sigma, b \Vdash^{\exists} \Pi$
$\mathsf{Conc} = a \in S(x), b \in S(x), \mathcal{R}^a, \mathcal{R}^b, a \Vdash^{\exists} A, a \Vdash^{\exists} B, \Gamma \Rightarrow \Delta, a \Vdash^{\exists} \Sigma, b \Vdash^{\exists} \Pi$
$\mathrm{P} = s(\Gamma)^x \Rightarrow s(\Delta)^x, \left[\Sigma, \Pi, \{\Omega\}, \underline{\{\Xi\}} \lhd A\right], [\Pi, \{\Xi\} \lhd B]$
$\mathrm{Q} = s(\Gamma)^x \Rightarrow s(\Delta)^x, [\Sigma, \{\Omega\} \lhd A], \left[\Sigma, \Pi, \underline{\{\Omega\}}, \{\Xi\} \lhd B\right]$

Lemma 6.8 *Let $\mathcal{D}^S$ be a* **G3V** *derivation of $\Gamma \Rightarrow \Delta$ from x-saturated sequents $\Gamma_1^S \Rightarrow \Delta_1^S, \dots, \Gamma_n^S \Rightarrow \Delta_n^S$; then $[\mathcal{D}^S]^x$ is a derivation of $s(\Gamma \Rightarrow \Delta)^x$ in $\mathcal{I}_{\mathbb{V}}^{\mathsf{i}}$ from sequents $s(\Gamma_1^{S^-} \Rightarrow \Delta_1^{S^-})^x, \dots, s(\Gamma_n^{S^-} \Rightarrow \Delta_n^{S^-})^x$, where for each $\Gamma_i^{S^-} \Rightarrow \Delta_i^{S^-}$ it* *holds that* $\Gamma_i^{S^-} \cup \Delta_i^{S^-} \subseteq \Gamma_i^S \cup \Delta_i^S$.

Proof. By distinction of cases, and by induction on the height of the derivation. If $h = 0$, $\Gamma \Rightarrow \Delta$ is a **G3V** initial sequent, and its translation $s(\Gamma \Rightarrow \Delta)$ is a $\mathcal{I}_{\mathbb{V}}^{\mathsf{i}}$ initial sequent. The propositional cases are obtained applying the corresponding $\mathcal{I}_{\mathbb{V}}^{\mathsf{i}}$ rule to the translation(s) of the premiss(es).
[$\mathsf{R} \preccurlyeq$] Translation of the premiss: $s(a \in S(x), a \Vdash^{\exists} B, \Gamma \Rightarrow \Delta, a \Vdash^{\exists} A)^x = s(\Gamma)^x \Rightarrow s(\Delta)^x, [A \lhd B]$; translation of the conclusion: $s(\Gamma \Rightarrow \Delta, x : A \preccurlyeq B)^x = s(\Gamma)^x \Rightarrow s(\Delta)^x, A \preccurlyeq B$.
[$\mathsf{L} \preccurlyeq$] $s(1)^x = A \preccurlyeq B, s(\Gamma)^x \Rightarrow s(\Delta)^x, [\Sigma, B, \{\Omega\} \lhd C]$. The right premiss (2) cannot be translated, since it features in the antecedent *two* formulas with the same sphere label: $a \Vdash^{\exists} A$ and $a \Vdash^{\exists} C$. By Lemma 6.7 we have that we have that either sequent $S_1 = a \in S(x), a \Vdash^{\exists} A, x : A \preccurlyeq B, \Gamma \Rightarrow \Delta, a \Vdash^{\exists} \Sigma$ or $S_2 = a \in S(x), a \Vdash^{\exists} C, x : A \preccurlyeq B, \Gamma \Rightarrow \Delta, a \Vdash^{\exists} \Sigma$ are derivable (possibly both). However, sequent S_2 is the same sequent as the conclusion of $\mathsf{L} \preccurlyeq$; thus, if we replace the right premiss with this sequent, the application of $\mathsf{L} \preccurlyeq$ would be useless, and we can ignore the case. Thus, replace the right premiss with S_1. Let $\mathcal{D}^-$ be the derivation for S_1: for Lemma 6.7 the derivation has height less or equal than $\mathcal{D}_2$, derivation of (2). Moreover, observe that $\mathcal{D}^-$ is a *subderivation* of $\mathcal{D}_2$: it displays the same formulas (and the same rules) except for formula $a \Vdash^{\exists} A$ (and the rules applied to it). Let $\Gamma_1^{S^-} \Rightarrow \Delta_1^{S^-}, \dots, \Gamma_k^{S^-} \Rightarrow \Delta_k^{S^-}$ be the x-saturated sequents from which $\mathcal{D}^-$ is derived. Each of these sequent is composed of *less* formulas than the x- saturated sequents $\Gamma_1^S \Rightarrow \Delta_1^S, \dots, \Gamma_k^S \Rightarrow \Delta_k^S$ from which S_2 was derived. This is the reason why we translate x-saturated sequents which are composed of not exactly the same formulas of the original x-saturated sequents, but of a subset of them. Apply the translation to S_1: $s(S_1)^x = A \preccurlyeq B, s(\Gamma)^x \Rightarrow s(\Delta)^x, [\Sigma, \{\Omega\} \lhd C]$. Add the missing block to $s(S_1)^x$ by weakening. Application of $\mathsf{L} \preccurlyeq^{\mathsf{i}}$ yields the translation of the conclusion of $\mathsf{L} \preccurlyeq$: $s(\mathsf{Conc})^x = A \preccurlyeq B, s(\Gamma)^x \Rightarrow s(\Delta)^x, [\Sigma, \{\Omega\} \lhd C]$.
[Nes] Sequent $s(1)^x$ is $s(\Gamma)^x \Rightarrow s(\Delta)^x, [\Sigma, \Pi, \{\Omega\} \lhd A], [\Pi, \{\Xi\} \lhd B]$; sequent $s(2)^x$ is $s(\Gamma)^x \Rightarrow s(\Delta)^x, [\Sigma, \{\Omega\} \lhd A], [\Sigma, \Pi, \{\Xi\} \lhd B]$; sequent $s(\mathsf{Conc})^x$ is $s(\Gamma)^x \Rightarrow s(\Delta)^x, [\Sigma, \{\Omega\} \lhd A], [\Pi, \{\Xi\} \lhd B]$.
Sets $\{\Omega\}$ and $\{\Xi\}$ account for the formulas to be added inside blocks, in correspondence with inclusions in $\mathcal{R}^a$ and $\mathcal{R}^b$ (see Definition 6.1). If both sets $\mathcal{R}^a$ and $\mathcal{R}^b$ are empty, $\{\Omega\}$ and $\{\Xi\}$ are empty there is no needed to apply $\mathsf{Wk_B}$ to either of the premisses. If $\mathcal{R}^a$ is not empty, $\{\Omega\}$ is not empty; we need to apply $\mathsf{Wk_B}$ to the second block of $s(2)^x$; Similarly, if $\mathcal{R}^b$ is not empty, $\{\Xi\}$ is not empty; we need to apply $\mathsf{Wk_B}$ to the first block of $s(1)^x$. If both $\{\Omega\}$ and $\{\Xi\}$ are not empty, combine the two above strategies. □

We are finally ready to define the full translation for derivations.

Theorem 6.9 *Let $\mathcal{D}$ be a* **G3V** *derivation of* $\Gamma \Rightarrow \Delta$ *in normal form with respect to some* x *in* $\Gamma \cup \Delta$. *Then* $[\![\mathcal{D}]\!]^x$ *is a* $\mathcal{I}_{\mathbb{V}}^{\mathsf{i}}$ *derivation of* $s(\Gamma \Rightarrow \Delta)^x$.

Proof. By induction on the height of the derivation. Since $\mathcal{D}$ is in normal form with respect to x, it will contain a subderivation $\mathcal{D}^S$ of $\Gamma \Rightarrow \Delta$ from x-saturated sequents $\Gamma_1^S \Rightarrow \Delta_1^S, \dots, \Gamma_n^S \Rightarrow \Delta_n^S$. Apply translation $[\ ]^x$ to $\mathcal{D}^S$, and obtain a derivation of $s(\Gamma \Rightarrow \Delta)^x$ from $s(\Gamma_1^{S^-} \Rightarrow \Delta_1^{S^-})^x, \dots, s(\Gamma_n^{S^-} \Rightarrow \Delta_n^{S^-})^x$ (Lemma 6.8). Each x-saturated sequent $\Gamma_i^S \Rightarrow \Delta_i^S$ has the form[7]: $\mathcal{R}^{a_1}, \dots, \mathcal{R}^{a_n}, a_1 \in S(x), \dots, a_n \in S(x), a_1 \Vdash^\exists A_1, \dots, a_n \Vdash^\exists A_n, x : \Gamma^P \Rightarrow$ $\Rightarrow x : \Delta^P, a_1 \Vdash^\exists \Sigma_1, \dots, a_n \Vdash^\exists \Sigma_n$. Its translation according to s and x is: $s(\Gamma_i^S \Rightarrow \Delta_i^S)^x = \Gamma^P \Rightarrow \Delta^P, [\{\Sigma_1\} \lhd A_1], \dots, [\{\Sigma_n\} \lhd A_n]$. For each $\Gamma_i^S \Rightarrow \Delta_i^S$, apply the following transformation: go up in the derivation until the x-hypersaturated sequent $\Gamma_i^H \Rightarrow \Delta_i^H$ is reached. The sequent will have the following form: $\mathcal{R}^{a_1}, \dots, \mathcal{R}^{a_n}, a_1 \in S(x), \dots, a_n \in S(x), \{y_1 \in a_1\}, \dots, \{y_n \in a_n\}, y_1 : A_1, \dots, y_n : A_n, x : \Gamma^P \Rightarrow x : \Delta^P, y_1 : \{\Sigma_1\}, \dots y_n : \{\Sigma_n\}, a_1 \Vdash^\exists \Sigma_1, \dots, a_n \Vdash^\exists \Sigma_n$ where $\{y_i \in a_i\}$ is a shorthand for $y_i \in a_i \cup \{y_i \in a_j \mid a_i \subseteq a_j \in \mathcal{R}^{a_i} \textit{ and } y \in a_i\}$. By Jump lemma either $y_1 : A_1 \Rightarrow y_1 : \{\Sigma_1\}$ is derivable, or the following sequent is derivable: $\mathcal{R}^{a_1}, \dots, \mathcal{R}^{a_n}, a_1 \in S(x), \dots, a_n \in S(x), \{y_1 \in a_1\}, \dots, \{y_n \in a_n\}, y_2 : A_2 \dots y_n : A_n, x : \Gamma^P \Rightarrow x : \Delta^P, y_2 : \{\Sigma_2\}, \dots y_n : \{\Sigma_n\}, a_1 \Vdash^\exists \Sigma_1, \dots, a_n \Vdash^\exists \Sigma_n$.

If $y_1 : A_1 \Rightarrow y_1 : \{\Sigma_1\}$ is not derivable, apply the Jump lemma to the above sequent, and iterate the procedure until a derivable sequent $y_i : A_i \Rightarrow y_i : \{\Sigma_i\}$ is found, for some $1 \leqslant i \leqslant n$. The existence of such a derivable sequent is guaranteed by the Jump lemma, and by the fact that a) the x-hypersaturated sequent is derivable and b) the x-hypersaturated sequent is not derivable in virtue of the part $x : \Gamma^P \Rightarrow x : \Delta^P$. If this was the case, the proof search would have stopped way before, since only propositional rules would have been applied in the derivation.

Suppose $y_i : A_i \Rightarrow y_i : \{\Sigma_i\}$ is derivable; $s(y_i : A_i \Rightarrow y_i : \{\Sigma_i\})^y = A_i \Rightarrow \{\Sigma_i\}$. Application of Jump to this sequent yields the translation of the x-saturated sequent $s(\Gamma^S \Rightarrow \Delta^S)^x = \Gamma^P \Rightarrow \Delta^P, [\{\Sigma_1\} \lhd A_1], \dots, [\{\Sigma_i\} \lhd A_i] \dots, [\{\Sigma_n\} \lhd A_n]$. Then, $[\![\ ]\!]^{y_i}$ has to be recursively invoked to translate, with variable y_i as a parameter, the derivation of sequent $y_i : A_i \Rightarrow y_i : \{\Sigma_i\}$; of smaller height than derivation of $\Gamma \Rightarrow \Delta$.

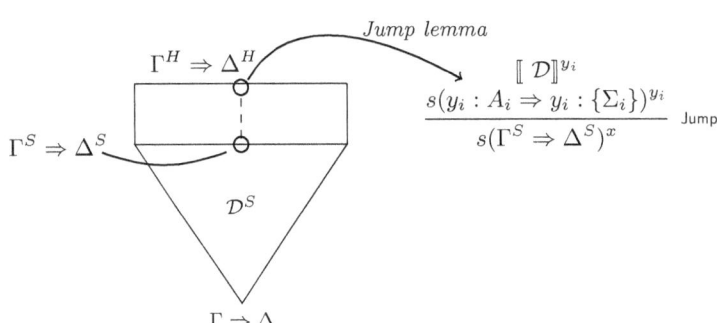

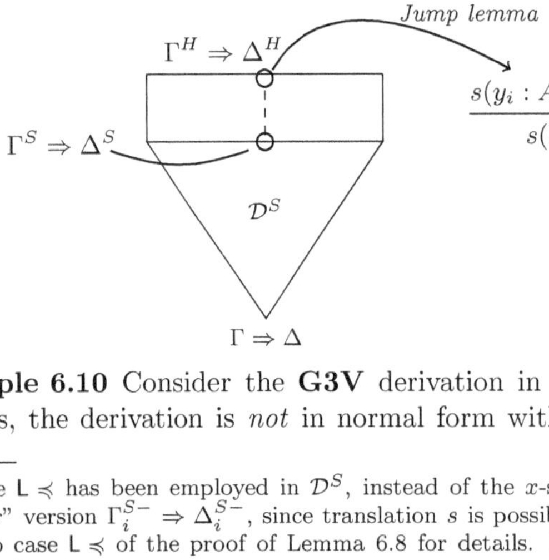

$$\dfrac{\dfrac{[\![\mathcal{D}]\!]^{y_i}}{s(y_i : A_i \Rightarrow y_i : \{\Sigma_i\})^{y_i}}}{s(\Gamma^S \Rightarrow \Delta^S)^x}\ \textsf{Jump}$$

□

Example 6.10 Consider the **G3V** derivation in the proof of Theorem 3.6. As it is, the derivation is *not* in normal form with respect to x: we have to

[7] If rule $\mathsf{L} \preccurlyeq$ has been employed in $\mathcal{D}^S$, instead of the x-saturated sequent we choose its "smaller" version $\Gamma_i^{S-} \Rightarrow \Delta_i^{S-}$, since translation s is possibly not applicable to $\Gamma_i^S \Rightarrow \Delta_i^S$. Refer to case $\mathsf{L} \preccurlyeq$ of the proof of Lemma 6.8 for details. In any case, the proof strategy remains the same.

saturate its upper part with respect to rules $\mathsf{L} \Vdash^{\exists}$, $\mathsf{R} \Vdash^{\exists}$ and $\mathsf{L} \subseteq$, to the effect that the two upper sequents become two x-hypersaturated sequents. Then, the part of the derivation up to the x-saturated sequents (in this case: premisses of Nesting) is translated employing $[\ ]^x$. Then, we apply $[\![\ ]\!]^x$: consider the left premiss of Nes (x-saturated), and go up to the x-hypersaturated sequent $y \in a, y \in b, z \in a, a \subseteq b, y : B, z : A \Rightarrow y : A, z : B, a \Vdash^{\exists} A, b \Vdash^{\exists} B$. From Lemma 6.5, we have that either a) $y : B \Rightarrow y : A, y : B$ is derivable, or b) $y \in a, y \in b, z \in a, z : A \Rightarrow z : B, a \Vdash^{\exists} A, b \Vdash^{\exists} B$ is derivable. Sequent a) is derivable. Thus, we translate sequent $y : B \Rightarrow y : A, y : B$ in the internal sequent calculus, obtaining $B \Rightarrow A, B$. An application of Jump allows us to obtain the sequent which is the leftmost application of $\mathsf{Com}^{\mathsf{i}}$. A similar reasoning is applied to translate the right premiss of Nes.

$$
\dfrac{\dfrac{\dfrac{\dfrac{\dfrac{\overset{\nabla}{B \Rightarrow A, B}}{\Rightarrow [A, B \lhd B], [B \lhd A]}\ \mathsf{Jump} \qquad \Rightarrow [A \lhd B], [A, B \lhd A]}{\Rightarrow [A \lhd B], [B \lhd A]}\ \mathsf{Com}^{\mathsf{i}}}{\Rightarrow [A \lhd B], B \preccurlyeq A}\ \mathsf{R}\preccurlyeq}{\Rightarrow A \preccurlyeq B, B \preccurlyeq A}\ \mathsf{R}\preccurlyeq}{\Rightarrow A \preccurlyeq B \vee B \preccurlyeq A}\ L\vee
$$

7 Conclusions

In this paper we defined **G3V**, a proof-theoretically well-behaved labelled calculus that provides an effective decision procedure for logic $\mathbb{V}$. Then, we have considered the calculus $\mathcal{I}^{\mathsf{i}}_{\mathbb{V}}$, the only internal and standard calculus known the logic [5]. We have shown that it is possible to translate directly derivations of the internal calculus $\mathcal{I}^{\mathsf{i}}_{\mathbb{V}}$ into derivations of the labelled calculus **G3V**. The opposite mapping is considerably more complex: we are able to translate derivations of the labelled calculus into derivations of the internal calculus provided (i) they satisfy a kind of normal form, and (ii) the relation between labels is essentially tree-like. It is worth noticing that this latter requirement is analogous to the tree-like restriction needed for mapping labelled calculi for standard modal logic [8] into nested sequent ones.

The present results are the first attempt to relate two basically different types of calculi for logics well beyond standard modal logics; despite their syntactic difference the two calculi are intrinsically related.

Many issues deserve to be further investigated. First, we aim at analysing the computational cost of the translation, namely what is the size of translated derivations with respect to the size of the input ones. Then, we can use the mapping to transfer results and properties of one calculus to the other: in one direction, syntactic cut-elimination and countermodel extraction, relatively easy to prove in the labelled calculus can be inherited in the internal calculus, for which these results are more difficult to prove. In the opposite direction, complexity bound and interpolation should be provable directly for the internal calculus, similarly to [9]. These results could be transferred to the labelled calculus, for which they are presently not known. Furthermore, since the mappings are given by functional procedures, we are interested in implementing an automated translation between derivations. Finally, the present results concern only logic $\mathbb{V}$: they could be extended to the other logics of the Lewis' cube for which internal calculi exist [6,4].

References

[1] Dyckhoff, R. and S. Negri, *Geometrisation of first-order logic*, The Bulletin of Symbolic Logic **21** (2015), pp. 123–163.

[2] Fitting, M., *Prefixed tableaus and nested sequents*, Annals of Pure and Applied Logic **163** (2012), pp. 291–313.

[3] Gent, I. P., *A sequent or tableaux-style system for Lewis's counterfactual logic* $\mathbb{VC}$, Notre Dame Journal of Formal Logic **33** (1992), pp. 369–382.

[4] Girlando, M., B. Lellmann and N. Olivetti, *Hypersequent calculus for the logic of conditional belief: Preliminary results*, Accepted for publication in EICNCL Floc Workshop Proceedings (2018).

[5] Girlando, M., B. Lellmann, N. Olivetti and G. L. Pozzato, *Standard sequent calculi for Lewis' logics of counterfactuals*, in: L. Michael and A. Kaks, editors, *European Conference on Logics in Artificial Intelligence*, Springer, 2016, pp. 272–287.

[6] Girlando, M., B. Lellmann, N. Olivetti and G. L. Pozzato, *Hypersequent calculi for Lewis conditional logics with uniformity and reflexivity*, in: R. A. Schmidt and C. Nalon, editors, *International Conference on Automated Reasoning with Analytic Tableaux and Related Methods*, Springer, 2017, pp. 131–148.

[7] Girlando, M., S. Negri, N. Olivetti and V. Risch, *The logic of conditional belief: Neighbourhood semantics and sequent calculus*, in: *Advances in Modal Logic*, 11, 2016, pp. 322–341.

[8] Goré, R. and R. Ramanayake, *Labelled tree sequents, tree hypersequents and nested (deep) sequents*, Advances in Modal Logic **9** (2012), pp. 279–299.

[9] Lellmann, B. and D. Pattinson, *Sequent systems for Lewis' conditional logics*, in: L. F. del Cerro, A. Herzig and J. Mengin, editors, *JELIA 2012*, LNAI **7519**, Springer-Verlag Berlin Heidelberg, 2012 pp. 320–332.

[10] Lewis, D. K., "Counterfactuals," Blackwell, 1973.

[11] de Swart, H. C., *A Gentzen-or Beth-type system, a practical decision procedure and a constructive completeness proof for the counterfactual logics VC and VCS*, The Journal of Symbolic Logic **48** (1983), pp. 1–20.

[12] Negri, S., *Proof analysis in modal logic*, Journal of Philosophical Logic **34** (2005), p. 507.

[13] Negri, S., *Non-normal modal logics: A challenge to proof theory*, The Logica Yearbook (2017), pp. 125–140.

[14] Negri, S., *Proof theory for non-normal modal logics: The neighbourhood formalism and basic results*, IFCoLog Journal of Logics and their Applications **4** (2017), pp. 1241–1286.

[15] Negri, S. and N. Olivetti, *A sequent calculus for preferential conditional logic based on neighbourhood semantics*, in: H. De Nivelle, editor, *International Conference on Automated Reasoning with Analytic Tableaux and Related Methods*, Springer-Verlag, 2015, pp. 115–134.

[16] Negri, S. and G. Sbardolini, *Proof analysis for Lewis counterfactuals*, The Review of Symbolic Logic **9** (2016), pp. 44–75.

[17] Negri, S. and J. von Plato, *Meaning in use*, in: H. Wansing, editor, *Dag Prawitz on Proofs and Meaning*, Springer, 2015 pp. 239–257.

[18] Olivetti, N. and G. Pozzato, *A standard internal calculus for Lewis counterfactual logics*, in: *International Conference on Automated Reasoning with Analytic Tableaux and Related Methods 9323*, Springer, 2015, pp. 270–286.

[19] Pacuit, E., "Neighbourhood Semantics for Modal Logics," Springer, 2017.

[20] Poggiolesi, F., "Gentzen Calculi for Modal Propositional Logic," 32, Springer Science & Business Media, 2010.

The Bimodal Logic of Commuting Difference Operators Is Decidable

Christopher Hampson

King's College London
Department of Informatics
30 Aldwych, London, WC2B 4BG

Abstract

In this paper, we show that the bimodal logic of two commuting difference operators [**Diff**, **Diff**] is decidable (in N2EXPTIME), despite lacking the finite model property and thereby remaining impervious to standard filtration techniques. The proof of decidability involves an exponential-time reduction from the satisfiability problem of the commutator [**Diff**, **Diff**] to that of the product logic **Diff** × **Diff**, via an intermediate quasimodel construction. To the best of the author's knowledge, this marks the first example of a (non-trivial) reduction between a commutator $[L_h, L_v]$ and its respective product logic $L_h \times L_v$, where the two logics do not coincide. By adapting this same technique, we are able to establish the finite model property for [**S5**, **Diff**], without recourse to standard filtration techniques that break down for logics that are not Horn-axiomatizable, such as that of the difference operator.

Keywords: Two-dimensional modal logics, difference operator, product logic, commutator, decidable, fmp

1 Introduction

The logic of the difference operator **Diff**, first described by von Wright [23] as the logic of 'elsewhere', is the set of all propositional unimodal formulas that are valid in all *difference frames* $\mathfrak{F} = (W, \neq)$, in which each possible world is accessible from every other *distinct* possible world. In isolation, the logic of the difference operator has been extensively studied [5,6,21], and shares connections with *nominals* [9] as well as *graded modalities* [1], each of which finds applications in description logics [2].

Segerberg [21] proved that **Diff** is sound and complete with respect to the class of all frames that are symmetric and *weakly-transitive* [1]:

$$\forall x \forall y \forall z \big(xRy \wedge yRz \rightarrow (x = z \vee xRz)\big)$$

Consequently, **Diff** can be axiomatized as the smallest normal modal logic containing the formulas:

[1] Segerberg refers to this property as *alio-transitivity*.

$$(B) := p \to \Box\Diamond p \qquad \text{and} \qquad (w4) := \Diamond\Diamond p \to (p \vee \Diamond p),$$

corresponding to symmetry and weak-transitivity, respectively. It is a routine exercise to show that, like the above axiomatization, any axiomatization for **Diff** must contain some formulas that are not equivalent to any universal Horn-formula [4, Proposition 6.2.2]; that is to say **Diff** is not *Horn-axiomatizable*.

As a unimodal logic, **Diff** shares many similarities with the logic **S5** of all equivalence relations: the two logics are finitely axiomatisable (with **S5** being axiomatized over **Diff** with the addition of the axiom $(T) := \Box p \to p$), both logics have NP-complete satisfiability problems (the latter can be polynomially reducible to the former), and the structure of their frames is quite similar (with the class of frames **Diff** extending the class of frames for **S5** by allowing frames with some irreflexive points). However, these similarities do not extend to their bimodal counterparts, where their interactions with other logics often differ considerably [12,13].

In what follows, we will be interested in combining unimodal logics, expressed over the bimodal language $\mathcal{ML}_2$ whose formulas are given by the following grammar:

$$\varphi ::= p_i \mid \neg\varphi \mid (\varphi_1 \wedge \varphi_2) \mid \Diamond_h\varphi \mid \Diamond_v\varphi$$

where $p_i \in \mathsf{Prop}$ ranges over a countably infinite set of propositional variables. The other Boolean connectives and $\Box_j\varphi$ are defined in the usual way with the addition of $\Box_j^+\varphi := \varphi \wedge \Box_j\varphi$ and $\Diamond_j^+\varphi := \varphi \vee \Diamond_j\varphi$, for $j = h, v$.

We define the size of an $\mathcal{ML}_2$-formula to be it's length $\|\varphi\|$, taken to be the number of symbols it comprises, and note that $|\mathsf{sub}(\varphi)| \leq \|\varphi\|$, where $\mathsf{sub}(\varphi)$ denotes the set of all subformulas of φ. Formulas of $\mathcal{ML}_2$ are interpreted over Kripke models $\mathfrak{M} = (\mathfrak{F}, \mathfrak{V})$, in which $\mathfrak{F} = (W, R_h, R_v)$ is a *bimodal Kripke frame*, where $R_h, R_v \subseteq W^2$ are binary relations on W, and $\mathfrak{V} : \mathsf{Prop} \to 2^W$ is a propositional valuation. Satisfiability is defined in the usual way with $\Diamond_j\varphi$ being interpreted over R_j, for $j = h, v$. For convenience, we also write $R_j^+ := R_j \cup \{(w, w) : w \in W\}$ so that $\Diamond_j^+\varphi$ can be interpreted over R_j^+, for $j = h, v$.

The product construction, first investigated by Segerberg [20] and later extended by Shehtman [22], provides a natural semantic way of combining unimodal logics and has been extensively studied since its inception [8,7,16]. Given two unimodal Kripke frames $\mathfrak{F}_h = (W_h, R_h)$ and $\mathfrak{F}_v = (W_v, R_v)$, we define their *product frame* to be the bimodal Kripke frame $\mathfrak{F}_h \times \mathfrak{F}_v = (W_h \times W_v, \overline{R}_h, \overline{R}_v)$, where $W_h \times W_v$ is the Cartesian product of W_h and W_v, and where $\overline{R}_h$ and $\overline{R}_v$ act component-wise on $W_h \times W_v$, such that:

$$(u, v)\overline{R}_h(u', v') \iff uR_hu' \text{ and } v = v',$$

$$(u, v)\overline{R}_v(u', v') \iff u = u' \text{ and } vR_vv',$$

for all $(u, v), (u', v') \in W_h \times W_v$.

The *product logic* $L_h \times L_v$ of two unimodal logic is characterised by all those formulas that are valid in every product frame $\mathfrak{F}_h \times \mathfrak{F}_v$, in which $\mathfrak{F}_j$ is a frame for L_j, for $j = h, v$.

Among those formulas that common to all product logics are those of the two constituent logics L_h and L_v (rewritten with the appropriate operators: $\Diamond_h$ and $\Diamond_v$, respectively), together with the formulas $\Diamond_v\Diamond_h p \to \Diamond_h\Diamond_v p$, $\Diamond_h\Diamond_v p \to \Diamond_v\Diamond_h p$, and $\Diamond_h\Box_v p \to \Box_v\Diamond_h p$, corresponding to the following frame properties inherent to the structure of all product frames:

- *Left-commutativity:* $\forall x\forall y\forall z\big(xR_v y \wedge yR_h z \to \exists u(xR_h u \wedge uR_v z)\big)$
- *Right-commutativity:* $\forall x\forall y\forall z\big(xR_h y \wedge yR_v z \to \exists u(xR_v u \wedge uR_h z)\big)$
- *Church-Rosser property:* $\forall x\forall y\forall z\big(xR_h y \wedge xR_v z \to \exists u(yR_v u \wedge zR_h u)\big)$.

We define the *commutator* of L_h and L_v, denoted $[L_h, L_v]$, to be the smallest modal logic axiomatized by these formulas. It follows that $[L_h, L_v] \subseteq L_h \times L_v$ for any choice of L_h and L_v. Furthermore, in the case where both L_h and L_v are Horn-axiomatizable, the two logics are known to coincide; in which case we say that L_h and L_v are *product matching.*

Theorem 1.1 (Gabbay–Shehtman [8, Theorem 7.12]) *Let L_h and L_v be any two Kripke complete, Horn-axiomatizable unimodal logics. Then L_h and L_v are product-matching.*

However, in general, the product logic may admit many more formulas than its commutator sublogic. Indeed, it is straightforward to see that $[\mathbf{Diff}, \mathbf{Diff}] \neq \mathbf{Diff} \times \mathbf{Diff}$, as can be evidenced by the following frame $\mathfrak{F} = (\{a_1, a_2, a_3, b_1, b_2\}, R_h, R_v)$, where

$$R_h = \{(u, v) : u \neq v\} \qquad \text{and} \qquad R_v = \{(a_i, a_j), (b_i, b_j) : i \neq j\}.$$

Despite being a frame for $[\mathbf{Diff}, \mathbf{Diff}]$, it is straightforward to check that $\mathfrak{F}$ is not the p-morphic image of *any* product frame for $\mathbf{Diff} \times \mathbf{Diff}$. Hence, the Jankov-Fine frame formula (see [3]) for $\mathfrak{F}$ is therefore an example of a formula that is satisfiable with respect to $[\mathbf{Diff}, \mathbf{Diff}]$ but not with respect to $\mathbf{Diff} \times \mathbf{Diff}$.

Indeed, in a forthcoming paper [14], it is shown that every bimodal logic between $\mathbf{K} \times \mathbf{Diff}$ and $\mathbf{S5} \times \mathbf{Diff}$ cannot be axiomatized using only finitely many variables and, thus, there are infinitely many logics separating $[\mathbf{Diff}, \mathbf{Diff}]$ from $\mathbf{Diff} \times \mathbf{Diff}$.

Theorem 1.2 (Hampson et al. [14]) *Let L be any Kripke complete bimodal logic such that $\mathbf{K} \times \mathbf{Diff} \subseteq L \subseteq \mathbf{S5} \times \mathbf{Diff}$. Then L cannot be axiomatized using only finitely many propositional variables.*

Consequently, although the satisfiability problem for $\mathbf{Diff} \times \mathbf{Diff}$ can be reduced to that of the decidable ([18]) two-variable fragment of first-order logic with counting quantifiers [2], we cannot appeal to this result directly in order to establish the decidability of $[\mathbf{Diff}, \mathbf{Diff}]$.

---

[2] By exploiting the ability to express 'elsewhere' with $\exists^{\neq} x\varphi := \exists_{>1} x\varphi \vee (\neg\varphi \wedge \exists_{>0} x\varphi)$.

Furthermore, as demonstrated below, the commutator [**Diff**, **Diff**] lacks the finite model property and so is impervious to standard filtration techniques that are commonly employed to establish decidability of commutators [8].

2 Main Results

2.1 [Diff, Diff] lacks the finite model property

In this section, we first establish that the logic of two commuting difference operators lacks the finite model property, thereby necessitating the need for the alternative approach taken in this paper. To this end, let φ_∞ be the conjunction of the following formulas:

$$\Diamond_h \Diamond_v (p \wedge \neg q \wedge \Box_h \neg p \wedge \Box_v \neg q), \tag{1}$$

$$\Box_h \Diamond_v (p \wedge \neg q \wedge \Box_h \neg p), \tag{2}$$

$$\Box_v \Diamond_h (q \wedge \neg p \wedge \Box_v \neg q), \tag{3}$$

where $p, q \in \mathsf{Prop}$ are propositional variables.

Lemma 2.1 *Let $\mathfrak{F} = (W, R_h, R_v)$ be any bimodal Kripke frame such that:*

(i) *R_h and R_v commute,*

(ii) *R_h and R_v are both* weakly-Euclidean*:*

$$\forall x \forall y \forall z \ \big(xR_j y \wedge xR_j z \to (y = z \vee yR_j z)\big), \qquad \textit{for } j = h, v.$$

If φ_∞ is satisfiable in $\mathfrak{F}$ then $\mathfrak{F}$ must be infinite.

Proof. Let $\mathfrak{F} = (W, R_h, R_v)$ be as described and suppose that $\mathfrak{M}, r \models \varphi_\infty$, for some model $\mathfrak{M} = (\mathfrak{F}, \mathfrak{V})$ based on $\mathfrak{F}$, with $r \in W$. We define, inductively, four infinite sequences:

$$\langle x_k \in W : k < \omega \rangle, \qquad \langle y_k \in W : k < \omega \rangle,$$

$$\langle u_k \in W : k < \omega \rangle, \quad \text{and} \quad \langle v_k \in W : k < \omega \rangle,$$

such that $\mathfrak{M}, u_0 \models \Box_v \neg q$, and for all $k < \omega$:

(inf1) $rR_h x_k$ and $rR_v y_k$,

(inf2) $x_k R_v u_k$ and $x_{k+1} R_v v_k$,

(inf3) $y_k R_h v_k$ and $y_k R_h u_k$,

(inf4) $\mathfrak{M}, u_k \models p \wedge \neg q \wedge \Box_h \neg p$,

(inf5) $\mathfrak{M}, v_k \models q \wedge \neg p \wedge \Box_v \neg q$.

At this stage we do not assume that all the points are distinct from one another.

Firstly, by (1), there is some $x_0, u_0 \in W$ such that $rR_h x_0$, $x_0 R_v u_0$ and $\mathfrak{M}, u_0 \models p \wedge \neg q \wedge \Box_h \neg p \wedge \Box_v \neg q$. Then by (i), there is some $y_0 \in W$ such that $rR_v y_0$ and $y_0 R_h u_0$. Whence, it follows from (3) that there is some $v_0 \in W$ such that $y_0 R_h v_0$ and $\mathfrak{M}, v_0 \models q \wedge \neg p \wedge \Box_v \neg q$.

Now, suppose we have already defined x_k, y_k, u_k, v_k, for some $k < \omega$. By **(inf1)** and **(inf3)**, we have that $rR_v y_k$ and $y_k R_h v_k$. Then, by (i), there is some $x_{k+1} \in W$ such that $rR_h x_{k+1}$ and $x_{k+1} R_v v_k$. Whence, it follows from (2) that there is some $u_{k+1} \in W$ such that $x_{k+1} R_v u_{k+1}$ and $\mathfrak{M}, u_{k+1} \models p \wedge \neg q \wedge \Box_h \neg p$.

Now, by (i), there is some $y_{k+1} \in W$ such that $rR_v y_{k+1}$ and $y_{k+1} R_h u_{k+1}$. Whence, it follows from the formula given in (3) that there is some $v_{k+1} \in W$ such that $y_{k+1} R_h v_{k+1}$ and $\mathfrak{M}, v_{k+1} \models q \wedge \neg p \wedge \Box_v \neg q$. Hence, by induction on the length, we may extend each of the four sequences indefinitely, as illustrated in Figure 1.

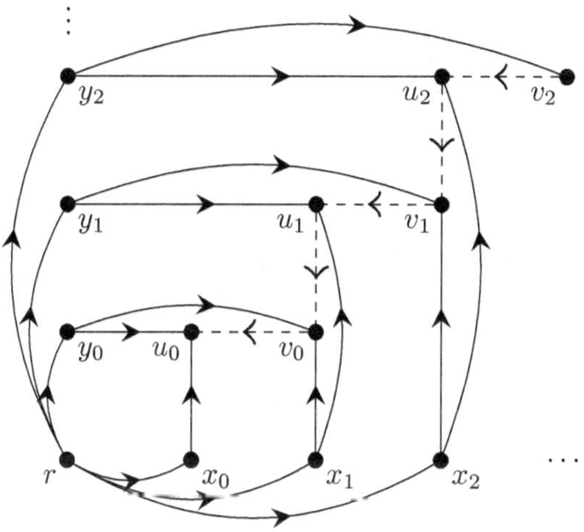

Fig. 1. Illustration of the model generated by φ_∞.

To show that each of the u_k are distinct, consider the following sequence of formulas $\langle \psi_k : k < \omega \rangle$, defined inductively, by taking $\psi_0 := \Box_v \neg q$, and

$$\psi_{k+1} \;:=\; \Diamond_v\big(q \wedge \Diamond_h(p \wedge \psi_k)\big),$$

for $k < \omega$. We claim that, for all $k < \omega$:

$$\mathfrak{M}, u_k \models \psi_k \qquad \text{and} \qquad \mathfrak{M}, u_\ell \not\models \psi_k \quad \text{for } \ell < k. \tag{4}$$

Indeed, it is immediate from the definition of u_0 that $\mathfrak{M}, u_0 \models \psi_0$. So suppose that $\mathfrak{M}, u_k \models \psi_k$, for some $k < \omega$, and that $\mathfrak{M}, u_\ell, \not\models \psi_k$ for all $\ell < k$. By **(inf4)**–**(inf5)**, we must have that $u_i \neq v_j$, for all $i, j < \omega$. Therefore, it follows from **(inf2)**–**(inf3)** and (ii), that $v_k R_h u_k$ and $u_{k+1} R_v v_k$. Whence, by **(inf4)**–**(inf5)**, we deduce that $\mathfrak{M}, u_{k+1} \models \psi_{k+1}$.

Now suppose, to the contrary, that $\mathfrak{M}, u_\ell \models \psi_{k+1}$ for some $\ell \leq k$. Then there is some $u, v \in W$ such that $u_\ell R_v v$, $vR_h u$, $\mathfrak{M}, v \models q$, and $\mathfrak{M}, u \models p \wedge \psi_k$. It then follows from **(inf2)**–**(inf5)** and (ii) that $v = v_{\ell-1}$ and $u = u_{\ell-1}$. This is to say that $\mathfrak{M}, u_{\ell-1} \models \psi_k$, where $\ell - 1 < k$, contrary to our induction hypothesis. Hence, $\mathfrak{M}, u_\ell \not\models \psi_{k+1}$, for all $\ell < k+1$.

It then follows from (4) that each of the $u_k \in W$ must be distinct. Being such, we must have that $\mathfrak{F}$ is infinite, as required. □

It is straightforward to check that every frame for $[\mathbf{Diff}, \mathbf{Diff}]$ satisfied conditions (i)–(ii) of Lemma 2.1. Furthermore, we note that φ_∞ is satisfiable with respect to $[\mathbf{Diff}, \mathbf{Diff}]$, as evidenced by the model $\mathfrak{M} = (\mathfrak{F}, \mathfrak{V})$, where $\mathfrak{F} = (\omega, \neq) \times (\omega, \neq)$ and $\mathfrak{V}$ is such that

$$\mathfrak{V}(p) = \{(k,k) : 0 < k < \omega\} \qquad \text{and} \qquad \mathfrak{V}(q) = \{(k+1,k) : 0 < k < \omega$$

for which it is straightforward to check that $\mathfrak{M}, (0,0) \models \varphi_\infty$. Hence, it follows that $[\mathbf{Diff}, \mathbf{Diff}]$ does not possess the finite model property and, as such, does not admit filtration.

Theorem 2.2 *The bimodal logic of two commuting difference operators* $[\mathbf{Diff}, \mathbf{Diff}]$ *does not possess the finite model property.*

It is known that $\mathbf{Diff} \times \mathbf{Diff}$ lacks the finite *product* model property, as it can be viewed as a syntactic variant of the two-variable fragment of first-order logic with counting quantifiers $\{\exists_{>0}, \exists_{>1}\}$, which is known to lack the fmp [10]. However, since $\mathfrak{F}_\infty$ is the product of two difference frames, as a further corollary of Lemma 2.1, we have the stronger result that φ_∞ is satisfiable with respect to $\mathbf{Diff} \times \mathbf{Diff}$ but cannot be satisfied in *any* finite frame for $\mathbf{Diff} \times \mathbf{Diff}$; product or otherwise!

Corollary 2.3 *The product logic* $\mathbf{Diff} \times \mathbf{Diff}$ *does not posses the* (abstract) *finite model property.*

2.2 Quasimodels for $[\mathbf{Diff}, \mathbf{Diff}]$

Despite the lack of any finite model property for $[\mathbf{Diff}, \mathbf{Diff}]$, we are still able to obtain a N2ExpTime upper-bound on the complexity of its satisfiability problem by a exponential-time reduction to that of $\mathbf{Diff} \times \mathbf{Diff}$, whose NExpTime-completeness follows from that of the two-variable fragment of first-order logic with counting quantifiers [18,19].

Theorem 2.4 *The satisfiability problem for the bimodal logic of two commuting difference operators* $[\mathbf{Diff}, \mathbf{Diff}]$ *is decidable in* N2ExpTime.

To facilitate this reduction, we employ a variation of the quasimodel technique [24,7] as an intermediary stage in the reduction. By limiting the size of the constituent *quasistates* to 'small' states, we incur at most an exponential increase in the complexity.

Definition 2.5 (Types and Quasistates) Let $\varphi \in \mathcal{ML}_2$ be an arbitrary bimodal formula of size n, and define a *type* for φ to be any Boolean-saturated subset $t \subseteq \mathsf{sub}(\varphi)$, which is to say that:

(tp1) $\neg\psi \in t$ if and only if $\psi \notin t$, for all $\neg\psi \in \mathsf{sub}(\varphi)$,

(tp2) $(\psi_1 \wedge \psi_2) \in t$ if and only if $\psi_1 \in t$ and $\psi_2 \in t$, for all $(\psi_1 \wedge \psi_2) \in \mathsf{sub}(\varphi)$.

Note that the set of all types for φ can be constructing in time that is at most exponential in the size of $\mathsf{sub}(\varphi)$.

A *quasistate* for φ is a tuple (T, S_h, S_v) such that:

(qs1) T is a non-empty set of types for φ, and $S_h, S_v \subseteq T^2$ are binary relations on T.

(qs2) For all $t, t' \in T$, if $t \neq t'$ then tS_ht' and tS_vt',

(qs3) *(internal coherence)* For all $t \in T$ and $\Diamond_j\psi \in \mathsf{sub}(\varphi)$,

$$\exists t' \in T;\ tS_jt' \text{ and } \psi \in t' \qquad \Longrightarrow \qquad \Diamond_j\psi \in t,$$

for $j = h, v$,

Let $\mathfrak{Q}$ denote the set of all possible quasistates for φ, of which there can be at most finitely many; albeit of the order of 2^{2^n}, where $n = |\mathsf{sub}(\varphi)|$.

Given a quasistate $q = (T, S_h, S_v) \in \mathfrak{Q}$, we say that $\Diamond_i\psi \in \mathsf{sub}(\varphi)$ is a *defect* of q if there is some $t \in T$ such that $\Diamond_j\psi \in t$, and there is no $t' \in T$ such that tS_jt' and $\psi \in T$. Let $D_q \subseteq \mathsf{sub}(\varphi)$ denote the set of all defects of q. For convenience, we write $\psi \in \bigcup q$ if there is some $t \in T$ such that $\psi \in t$, and $\psi \in \bigcap q$ if $\psi \in t$ for all $t \in T$.

Definition 2.6 (Quasimodels) Given a set of quasistates $\mathfrak{S} \subseteq \mathfrak{Q}$, we define a $\mathfrak{S}$-*quasimodel* for φ to be a triple $(X, Y, \boldsymbol{q})$ such that

(qm1) X and Y are non-empty sets, and $\boldsymbol{q}$ is a function associating each pair $(x, y) \in X \times Y$ with a quasistate $\boldsymbol{q}(x, y) = (T^{x,y}, S_h^{x,y}, S_v^{x,y}) \in \mathfrak{S}$,

(qm2) There is some $x_0 \in X$ and $y_0 \in Y$ such that $\varphi \in \bigcup \boldsymbol{q}(x_0, y_0)$,

(qm3) *($\Diamond_h$-coherence)* For all $x \in X$, $y \in Y$ and $\Diamond_h\psi \in \mathsf{sub}(\varphi)$,

$$\exists x' \in X;\ x \neq x' \text{ and } \psi \in \bigcup \boldsymbol{q}(x', y) \qquad \Longrightarrow \qquad \Diamond_h\psi \in \bigcap \boldsymbol{q}(x, y),$$

(qm4) *($\Diamond_v$-coherence)* For all $x \in X$, $y \in Y$ and $\Diamond_v\psi \in \mathsf{sub}(\varphi)$,

$$\exists y' \in Y;\ y \neq y' \text{ and } \psi \in \bigcup \boldsymbol{q}(x, y') \qquad \Longrightarrow \qquad \Diamond_v\psi \in \bigcap \boldsymbol{q}(x, y),$$

(qm5) *($\Diamond_h$-saturation)* For all $x \in X$, $y \in Y$ and $\Diamond_h\psi \in \mathsf{sub}(\varphi)$,

$$\Diamond_h\psi \in D_{\boldsymbol{q}(x,y)} \qquad \Longrightarrow \qquad \exists x' \in X;\ x \neq x' \text{ and } \psi \in \bigcup \boldsymbol{q}(x', y),$$

(qm6) *($\Diamond_v$-saturation)* For all $x \in X$, $y \in Y$ and $\Diamond_v\psi \in \mathsf{sub}(\varphi)$,

$$\Diamond_v\psi \in D_{\boldsymbol{q}(x,y)} \qquad \Longrightarrow \qquad \exists y' \in Y;\ y \neq y' \text{ and } \psi \in \bigcup \boldsymbol{q}(x, y'),$$

(qm7) For all $x, x' \in X$ and $y, y' \in Y$,

$$h\text{-}\mathsf{size}(\boldsymbol{q}(x, y)) = 1 \qquad \Longrightarrow \qquad h\text{-}\mathsf{size}(\boldsymbol{q}(x, y')) = 1,$$

$$v\text{-}\mathsf{size}(\boldsymbol{q}(x, y)) = 1 \qquad \Longrightarrow \qquad v\text{-}\mathsf{size}(\boldsymbol{q}(x', y)) = 1,$$

where j-size$(\boldsymbol{q}(x,y)) = |T^{x,y}| + \left|\{t \in T^{x,y} : tS_j^{x,y}t\}\right|$ is a measure of the horizontal and vertical 'size' of each quasistate, for $j = h, v$ respectively.

We first show that by taking $\mathfrak{S} = \mathfrak{Q}$ to be the set of all possible quasistates for φ, our $\mathfrak{Q}$-quasimodels adequately capture the notion of satisfiability with respect to $[\mathbf{Diff}, \mathbf{Diff}]$.

Lemma 2.7 *A formula φ is satisfiable with respect to $[\mathbf{Diff}, \mathbf{Diff}]$ if and only if there is a $\mathfrak{Q}$-quasimodel for φ, where $\mathfrak{Q}$ is the set of all quasistates for φ.*

Proof.

($\Rightarrow$) Suppose that $\mathfrak{M}, r \models \varphi$ for some model $\mathfrak{M} = (\mathfrak{F}, \mathfrak{V})$, where $\mathfrak{F} = (W, R_h, R_v)$ is a frame for $[\mathbf{Diff}, \mathbf{Diff}]$. We define an equivalence relation on W by taking

$$u \sim v \qquad \Longleftrightarrow \qquad uR_h^+ v \quad \text{and} \quad uR_v^+ v$$

for all $u, v \in W$. Now let X and Y be defined by taking

$$X = \{[u] : rR_h^+ u\} \qquad \text{and} \qquad Y = \{[v] : rR_v^+ v\},$$

where $[u]$ denotes the $\sim$-equivalence class containing $u \in W$. We define an intermediary function $h : X \times Y \to W/\!\sim$ as follows: For each $[u] \in X$ and $[v] \in Y$, we have that $rR_h^+ u$ and $rR_v^+ v$. Hence, by the Church-Rosser property, there is some $w \in W$ such that $uR_v^+ w$ and $vR_h^+ w$. Moreover, for all $w' \in W$ such that $uR_v^+ w'$ and $vR_h^+ w'$ we must have that $wR_h^+ w'$ and $wR_v^+ w'$, since both R_h^+ and R_v^+ are equivalence relations, which is to say that $w \sim w'$. Hence, we may uniquely define $h([u],[v]) = [w]$ to be the equivalence class containing w.

With every $w \in W$ we associate a type

$$\boldsymbol{t}(w) = \{\psi \in \mathsf{sub}\,(\varphi) : \mathfrak{M}, w \models \psi\},$$

and for each $(x,y) \in X \times Y$, we may associate the quasistate $\boldsymbol{q}(x,y) = (T^{x,y}, S_h^{x,y}, S_v^{x,y})$ by taking:

- $T^{x,y} = \{\boldsymbol{t}(w) : w \in h(x,y)\}$, and
- $\boldsymbol{t}(u) S_i^{x,y} \boldsymbol{t}(v)$ if and only if there is some $u', v' \in h(x,y)$ such that $\boldsymbol{t}(u) = \boldsymbol{t}(u')$, $\boldsymbol{t}(v) = \boldsymbol{t}(v')$ and $u' R_i v'$, for $i = h, v$.

It is straightforward to verify that $(X, Y, \boldsymbol{q})$ is a $\mathfrak{Q}$-quasimodel for φ (the details for which can be found in the Appendix.

($\Leftarrow$) Conversely, suppose that $(X, Y, \boldsymbol{q})$ is a $\mathfrak{Q}$-quasimodel for φ. We define a model $\mathfrak{M} = (\mathfrak{F}, \mathfrak{V})$, where $\mathfrak{F} = (W, R_h, R_v)$, by taking

$$W = \{(x,y,t) : x \in X,\ y \in Y \text{ and } t \in T^{x,y}\},$$

where $\boldsymbol{q}(x,y) = (T^{x,y}, S_h^{x,y}, S_v^{x,y}) \in \mathfrak{Q}$, and defining $R_h, R_v \subseteq W \times W$, such that

$$(x,y,t) R_h (x',y',t') \quad \Longleftrightarrow \quad y = y' \text{ and } \left(x \neq x' \text{ or } tS_h^{x,y}t'\right),$$

$$(x,y,t) R_v (x',y',t') \quad \Longleftrightarrow \quad x = x' \text{ and } \left(y \neq y' \text{ or } tS_v^{x,y}t'\right)$$

for all $(x, y, t), (x', y', t') \in W$.

Since $S_h^{x,y}$ and $S_v^{x,y}$ are such that $tS_i^{x,y}t'$ for all $t \neq t'$, it is straightforward to verify that R_h and R_v commute and are both symmetric and weakly-transitive, which is to say that $\mathfrak{F}$ is a frame for $[\mathbf{Diff}, \mathbf{Diff}]$.
Finally, for each propositional variable $p \in \mathsf{sub}(\varphi)$, take

$$\mathfrak{V}(p) = \{(x, y, t) \in W \ : \ p \in t\}.$$

It remains to show that $\mathfrak{M}$ is a model for φ.

We claim that for all $(x, y, t) \in W$ and $\psi \in \mathsf{sub}(\varphi)$,

$$\mathfrak{M}, (x, y, t) \models \psi \qquad \Longleftrightarrow \qquad \psi \in t. \tag{I.H.}$$

The cases where ψ is a propositional variable or a Boolean combination of smaller formulas are trivial and follow immediately from the definitions.
So suppose that $\mathfrak{M}, (x, y, t) \models \Diamond_h \alpha$, for some $\alpha \in \mathsf{sub}(\varphi)$. It follows that there is some $(x', y', t') \in W$ such that $(x, y, t) R_h (x', y', t')$ and $\mathfrak{M}, (x', y', t') \models \alpha$. By the induction hypothesis (I.H.) we find that $\alpha \in t'$. Furthermore, by the definition of R_h we have that $y = y'$ and either $x \neq x'$ or else $x = x'$ and $tS_h^{x,y}t'$.

- Suppose that $x \neq x'$, then by **(qm3)** we have that $\Diamond_h \alpha \in \bigcap \boldsymbol{q}(x, y)$ and therefore $\Diamond_h \alpha \in t$, by definition.
- Otherwise, we must have that $tS_h^{x,y}t'$ and so it follows from **(qs3)** that $\Diamond_h \alpha \in t$.

Conversely, suppose that $\Diamond_h \alpha \in t$. Then we have two cases to consider, depending on whether or not $\Diamond_h \alpha$ is a defect of $\boldsymbol{q}(x, y)$:

- If $\Diamond_h \alpha$ is not a defect of $\boldsymbol{q}(x, y)$, then by definition there is some $t' \in \boldsymbol{q}(x, y)$ such that $tS_h^{x,y}t'$ and $\alpha \in t'$. By the induction hypothesis we have that $\mathfrak{M}, (x, y, t') \models \alpha$. Moreover, by definition, we have that $(x, y, t) R_h (x, y, t')$ and so $\mathfrak{M}, (x, y, t) \models \Diamond_h \alpha$, as required.
- If $\Diamond_h \alpha$ is a defect of $\boldsymbol{q}(x, y)$, then by **(qm5)** there is some $x' \in X$ such that $x \neq x'$ and $\alpha \in \bigcup \boldsymbol{q}(x', y)$, which is to say that $\alpha \in t'$ for some $t' \in \boldsymbol{q}(x', y)$. Again, by the induction hypothesis, we have that $\mathfrak{M}, (x', y, t') \models \alpha$. Furthermore, by definition, we have that $(x, y, t) R_h (x', y, t')$ and so $\mathfrak{M}, (x, y, t) \models \Diamond_h \alpha$, as required.

The case where ψ is of the form $\Diamond_v \alpha$, for some $\alpha \in \mathsf{sub}(\varphi)$, is analogous. Hence we have that $\mathfrak{M}, (x, y, t) \models \psi$ if and only if $\psi \in t$, for all $(x, y, t) \in W$ and $\psi \in \mathsf{sub}(\varphi)$. In particular, it follows from **(qm2)** that there is some $(x_0, y_0, t_0) \in W$ such that $\mathfrak{M}, (x_0, y_0, t_0) \models \varphi$, as required.

□

Thus we have reduced the problem of deciding whether φ is satisfiable with respect to $[\mathbf{Diff}, \mathbf{Diff}]$ to that of checking whether φ has a suitable quasimodel. This exercise is fruitless, however, unless we have some means by which we can effectively search for quasimodels, which may still be infinite!

Fortunately, owing to the rigid grid-like structure of our quasimodels, we may further reduce the problem of checking whether φ has a quasimodel to that of satisfiability with respect to $\mathbf{Diff} \times \mathbf{Diff}$, whose satisfiability problem is known to be NEXPTIME-complete [18]. This approach is similar to the one described in [16, Lemma 32] in which the problem of identifying $(\mathbf{K4.3} \times \mathbf{K})$-quasimodels is reduced to that of satisfiability for some monadic second-order formula qm^m, defined therein.

First, given a set of quasistates $\mathfrak{S} \subseteq \mathfrak{Q}$, we associate with each quasistate $q \in \mathfrak{S}$ some propositional variable $\mathsf{QS}_q \in \mathsf{Prop}$. We then define, for each $\psi \in \mathsf{sub}(\varphi)$, the following abbreviations

$$\mathsf{SOME}^{\mathfrak{S}}_{\psi} := \bigvee \{\mathsf{QS}_q : q \in \mathfrak{S} \text{ and } \psi \in \bigcup q\}, \tag{5}$$

$$\mathsf{ALL}^{\mathfrak{S}}_{\psi} := \bigvee \{\mathsf{QS}_q : q \in \mathfrak{S} \text{ and } \psi \in \bigcap q\}, \tag{6}$$

$$\mathsf{DEFECT}^{\mathfrak{S}}_{\psi} := \bigvee \{\mathsf{QS}_q : q \in \mathfrak{S} \text{ and } \psi \in D_q\}. \tag{7}$$

Furthermore, we define

$$\mathsf{SIZE}^{\mathfrak{S}}_{j} := \bigvee \{\mathsf{QS}_q : \boldsymbol{q} \in \mathfrak{S} \text{ and } j\text{-}\mathsf{size}(q) = 1\}, \tag{8}$$

for $j = h, v$.

Take $\mathfrak{S}$-qm_{φ} to be the conjunction of the following formulas:

$$\boxplus \bigvee_{q \in \mathfrak{S}} \mathsf{QS}_q \wedge \boxplus \bigwedge_{\substack{q,q' \in \mathfrak{S} \\ q \neq q'}} \neg(\mathsf{QS}_q \wedge \mathsf{QS}_{q'}) \wedge \diamondplus \mathsf{SOME}^{\mathfrak{S}}_{\varphi}, \tag{9}$$

$$\boxplus \bigwedge_{\Diamond_j \psi \in \mathsf{sub}(\varphi)} (\Diamond_i \mathsf{SOME}^{\mathfrak{S}}_{\psi} \to \mathsf{ALL}^{\mathfrak{S}}_{\Diamond_j \psi}), \qquad \text{for } j = h, v, \tag{10}$$

$$\boxplus \bigwedge_{\Diamond_j \psi \in \mathsf{sub}(\varphi)} (\mathsf{DEFECT}^{\mathfrak{S}}_{\Diamond_j \psi} \to \Diamond_j \mathsf{SOME}^{\mathfrak{S}}_{\psi}), \qquad \text{for } j = h, v, \tag{11}$$

$$\boxplus (\mathsf{SIZE}^{\mathfrak{S}}_{h} \to \Box_v \mathsf{SIZE}^{\mathfrak{S}}_{h}) \quad \wedge \quad \boxplus (\mathsf{SIZE}^{\mathfrak{S}}_{v} \to \Box_h \mathsf{SIZE}^{\mathfrak{S}}_{v}), \tag{12}$$

where $\boxplus\varphi := \Box^+_h \Box^+_v \varphi$ and $\diamondplus\varphi := \Diamond^+_h \Diamond^+_v \varphi$. With these formulas, we are able to establish an equivalence between the existence of a $\mathfrak{S}$-quasimodel for φ and the satisfiability of $\mathfrak{S}$-qm_{φ} with respect to $\mathbf{Diff} \times \mathbf{Diff}$, as is demonstrated below in Lemma 2.8.

Lemma 2.8 *The formula* $\mathfrak{S}$-qm_{φ} *is satisfiable with respect to* $\mathbf{Diff} \times \mathbf{Diff}$ *if and only if* φ *has an* $\mathfrak{S}$*-quasimodel, for all* $\mathfrak{S} \subseteq \mathfrak{Q}$.

Proof.

($\Rightarrow$) Suppose that $\mathfrak{M}, (r_h, r_v) \models \mathfrak{S}$-$\mathsf{qm}_{\varphi}$, for some product model $\mathfrak{M} = (\mathfrak{F}_h \times \mathfrak{F}_v, \mathfrak{V})$, where $\mathfrak{F}_j = (W_j, \neq)$ is a difference frame, for $j = h, v$. We define a quasimodel $(X, Y, \boldsymbol{q})$ by taking

$$X = W_h, \qquad Y = W_v, \qquad \text{and} \qquad \boldsymbol{q}(x, y) = q \iff (x, y) \in \mathfrak{V}(\mathsf{QS}_q),$$

for all quasistates $q \in \mathfrak{S}$.

By (9), we can be assured that $\boldsymbol{q}(x, y)$ is well-defined, for all $(x, y) \in X \times Y$, and that there is some $x_0 \in X$, $y_0 \in Y$ such that $\varphi \in \bigcup \boldsymbol{q}(x_0, y_0)$, as required for condition **(qm2)**. Conditions **(qm3)**–**(qm4)** are satisfied by (10), while conditions **(qm5)**–**(qm6)** are satisfied by (11). Finally, (12) ensures that condition **(qm7)** is satisfied. Hence $(X, Y, \boldsymbol{q})$ is an appropriate $\mathfrak{S}$-quasimodel for φ, as required.

($\Leftarrow$) Conversely, suppose that $(X, Y, \boldsymbol{q})$ is an $\mathfrak{S}$-quasimodel for φ. Let $\mathfrak{F}_h = (X, \neq)$ and $\mathfrak{F}_v = (Y, \neq)$ be difference frames on X and Y, respectively, and define a new model $\mathfrak{M} = (\mathfrak{F}_h \times \mathfrak{F}_v, \mathfrak{V})$ over $\mathfrak{F}_h \times \mathfrak{F}_v$, by taking

$$(x, y) \in \mathfrak{V}(\mathsf{QS}_q) \qquad \Longleftrightarrow \qquad \boldsymbol{q}(x, y) = q,$$

for all $x \in X$, $y \in Y$, and all quasistates $q \in \mathfrak{S}$. The following are then immediate consequences of the definitions:

$$\begin{aligned}
\mathfrak{M}, (x, y) \models \mathsf{SOME}^{\mathfrak{S}}_{\psi} \quad &\Longleftrightarrow \quad \psi \in \textstyle\bigcup \boldsymbol{q}(x, y), \\
\mathfrak{M}, (x, y) \models \mathsf{ALL}^{\mathfrak{S}}_{\psi} \quad &\Longleftrightarrow \quad \psi \in \textstyle\bigcap \boldsymbol{q}(x, y), \\
\mathfrak{M}, (x, y) \models \mathsf{DEFECT}^{\mathfrak{S}}_{\psi} \quad &\Longleftrightarrow \quad \psi \in D_{\boldsymbol{q}(x,y)}, \\
\mathfrak{M}, (x, y) \models \mathsf{SIZE}^{\mathfrak{S}}_{h} \quad &\Longleftrightarrow \quad h\text{-}\mathsf{size}(\boldsymbol{q}(x, y)) = 1, \\
\mathfrak{M}, (x, y) \models \mathsf{SIZE}^{\mathfrak{S}}_{v} \quad &\Longleftrightarrow \quad v\text{-}\mathsf{size}(\boldsymbol{q}(x, y)) = 1.
\end{aligned}$$

It is then straightforward to check that each of the conjuncts (9)–(12) reflect the conditions **(qm2)**–**(qm8)** on $(X, Y, \boldsymbol{q})$ being a quasimodel for φ. Hence we must have that $\mathfrak{S}\text{-}\mathsf{qm}_{\varphi}$ is satisfiable with respect to **Diff** $\times$ **Diff**, as required.

□

Hence, it follows from Lemmas 2.7–2.8 that φ is satisfiable with respect to [**Diff**, **Diff**] if and only if $\mathfrak{Q}\text{-}\mathsf{qm}_{\varphi}$ is satisfiable with respect to **Diff** $\times$ **Diff**, where $\mathfrak{Q}$ is the set of all quasistates. Since the satisfiability problem for the product logic **Diff** $\times$ **Diff** is decidable, so too must be the satisfiability problem for the commutator [**Diff**, **Diff**]. However, the size of $\mathfrak{Q}\text{-}\mathsf{qm}_{\varphi}$ is *doubly-exponential* in the size of $\mathsf{sub}(\varphi)$. Together with the optimal NExpTime upper-bound on the satisfiability problem for **Diff** $\times$ **Diff**, this would provide only a N3ExpTime upper-bound on the satisfiability problem for [**Diff**, **Diff**].

2.3 Small Quasimodels for [Diff, Diff]

To redress the issue highlighted above, we can restrict our attention to a much smaller set of quasistates $\mathfrak{Q}_{\mathsf{sm}} \subseteq \mathfrak{Q}$ that is at most (singly-) exponential in the size of φ, thereby reducing the upper-bound on the satisfiability problem for [**Diff**, **Diff**] from N3ExpTime to N2ExpTime.

Definition 2.9 (Small Quasistates) A quasistate $\boldsymbol{q} = (T, S_h, S_v)$ for φ is said to be *small* if it satisfies the condition that:

(sm1) $|T| \leq 2n^2$ is at most quadratic in $n = |\mathsf{sub}(\varphi)|$.

Let $\mathfrak{Q}_{\mathsf{sm}} \subseteq \mathfrak{Q}$ denote the set of all small quasistates for φ.

Note that $|\mathfrak{Q}_{\mathsf{sm}}|$ is a most exponential in the size of φ, since there can be at most 2^n types for φ, and at most $\sum_{k=1}^{N}(2^n)^k \leq N \cdot 4^{n^3}$ possible candidates for T, where $N = 2n^2$ is the maximum size of T. Furthermore, there are at most $2^{|T|} \leq 4^{n^2}$ candidates for S_h and S_v, since, by **(qs2)**, they are both completely defined by their reflexive elements. Since the set of all types can be constructed in exponential-time, so too can be the set of all small quasistates.

Lemma 2.10 *φ has a $\mathfrak{Q}$-quasimodel if and only if φ also has a $\mathfrak{Q}_{\mathsf{sm}}$-quasimodel, comprising only* small *quasistates.*

Proof. The right-to-left direction is trivial, since $\mathfrak{Q}_{\mathsf{sm}} \subseteq \mathfrak{Q}$, and so every $\mathfrak{Q}_{\mathsf{sm}}$-quasimodel is also a $\mathfrak{Q}$-quasimodel. For the converse, it is sufficient to show that every quasistate $q \in \mathfrak{Q}$ can be replaced with small quasistate $q' \in \mathfrak{Q}_{\mathsf{sm}}$ such that

$$\bigcup q \;=\; \bigcup q', \qquad \bigcap q \;=\; \bigcap q', \qquad \text{and} \qquad D_q = D_{q'}$$

since each of the conditions **(qm1)**–**(qm7)** makes reference only to these properties of its constituent quasistates.

To this end, let $q = (T, S_h, S_v)$ be an arbitrary quasistate, and for each $\psi \in \bigcup q$, fix some $t_\psi \in T$ such that $\psi \in t_\psi$. Let $T_0 \subseteq T$ be the subset comprising all such types. For each $t \in T_0$ and $\Diamond_j\alpha \in t$ such that $\Diamond_j\alpha \notin D_q$, fix some $s_{(t,\alpha)} \in T$ such that $tS_j s_{(t,\alpha)}$ and $\alpha \in s_{(t,\alpha)}$, and take T_1 to be the set of all such types. We may then define $q' = (T', S'_h, S'_v)$ by taking

$$T' \;:=\; T_0 \cup T_1 \qquad \text{and} \qquad tS'_jt' \quad\Longleftrightarrow\quad tS_jt'$$

for all $t, t' \in T'$ and $j = h, v$. It clear that q' is quasistate for φ and that, by construction $\bigcap q = \bigcap q'$ and $\bigcup q = \bigcup q'$. Furthermore, it is clear from the construction that $|T| \leq (n + n^2) \leq 2n^2$, and so $q' \in \mathfrak{Q}_{\mathsf{sm}}$. All that remains is to show that $D_q = D_{q'}$.

- It is straightforward to verify that $D_q \subseteq D_{q'}$, since T' is a subset of T and so cannot provide any remedies to any of the defects of q. For the other direction, suppose that $\Diamond_j\alpha \notin D_q$, and suppose that $t \in T'$ is such that $\Diamond_j\alpha \in t$. If $t \in T_0$ then by construction there is some $s_{(t,\alpha)} \in T_1 \subseteq T'$ such that $tS_j s_{(t,\alpha)}$ and $\alpha \in s_{(t,\alpha)}$, which is to say that $\Diamond_j\alpha \notin D_{q'}$. On the other hand, if $t \in T_1$ then there is some $t' \in T_0$ such that $t'S_jt$. Moreover, since $\Diamond_j\alpha \notin D_q$, there must be some $t'' \in T$ such that tS_jt'' and $\alpha \in t''$. By **(qs2)**, either $t'' = t' \in T'$ or $t''S_jt$. In the latter case, by **(qs3)**, we must have that $\Diamond_j\alpha \in t''$ and so, by construction, there is some $s_{(t'',\alpha)} \in T_1 \subseteq T'$ such that $\alpha \in s_{(t'',\alpha)}$. In both cases, there is some $s \in T'$ such that tS_js and $\alpha \in s$, which is to say that $\Diamond_j\alpha \notin D_{q'}$, as required.

Hence, it follows that every $\mathfrak{Q}$-quasimodel for φ can be transformed into a $\mathfrak{Q}_{\mathsf{sm}}$-quasimodel for φ, in which each quasistate is small. $\square$

It then follows from Lemmas 2.7, 2.8 and 2.10 that φ is satisfiable with respect to $[\mathbf{Diff}, \mathbf{Diff}]$ if and only if $\mathfrak{Q}_{\mathsf{sm}}\text{-}\mathsf{qm}_\varphi$ is satisfiable with respect to $\mathbf{Diff} \times \mathbf{Diff}$, where $\mathfrak{Q}_{\mathsf{sm}}$ is the set of all small quasistates for φ. Furthermore, since the size of $\mathfrak{Q}_{\mathsf{sm}}\text{-}\mathsf{qm}_\varphi$ is at most exponential in the size of $\mathsf{sub}(\varphi)$ and can be constructed in exponential-time, the above reduction from commutator $[\mathbf{Diff}, \mathbf{Diff}]$ to its product $\mathbf{Diff} \times \mathbf{Diff}$ incurs, at most, an *exponential* increase in complexity. As the satisfiability problem for $\mathbf{Diff} \times \mathbf{Diff}$ can be decided in NEXPTIME, so it follows that the satisfiability problem for $[\mathbf{Diff}, \mathbf{Diff}]$ can be decided in N2EXPTIME, thereby completing the proof of Theorem 2.4.

It is worth noting that this is in-line with what is typically achieved from standard filtration techniques which place double-exponential bounds on the size of the filtered models (see, for example, Gabbay et al. [7, Theorem 5.27]).

2.4 The Finite Model Property of $[\mathbf{S5}, \mathbf{Diff}]$

One immediate consequence of Theorem 2.4 is that the the satisfiability problem for $[\mathbf{S5}, \mathbf{Diff}]$ is also decidable, since it can be identified with a term-definable fragment of $[\mathbf{Diff}, \mathbf{Diff}]$, by rewriting $\Diamond_h \psi := \psi \vee \Diamond_h \psi$.

However, in this case we are able to prove a stronger result, by appealing to the fact that, unlike $\mathbf{Diff} \times \mathbf{Diff}$, the product $\mathbf{S5}$ and $\mathbf{Diff}$ possess the exponential *product* fmp [11], which is to say that every formula φ that is satisfiable with respect to $\mathbf{S5} \times \mathbf{Diff}$ can be satisfied in a *product* model for $\mathbf{S5} \times \mathbf{Diff}$ that is at most exponential in the size of φ. This affords us the possibility of adapting the above strategy by finitizing the resulting product model and thereby placing an upper-bound on the size of the satisfying models for $[\mathbf{S5}, \mathbf{Diff}]$, despite the lack of a known method of filtration.

Theorem 2.11 *The commutator* $[\mathbf{S5}, \mathbf{Diff}]$ *has the doubly-exponential sized finite model property.*

Proof. Let φ be an $\mathcal{ML}_2$ formula, and define $\mathfrak{Q}^{\bullet}_{\mathsf{sm}} \subseteq \mathfrak{Q}$ to be the set of all small *horizontally reflexive* quasistates, in which $tS_h t$, for all $t \in T$. We claim that φ is satisfiable with respect to $[\mathbf{S5}, \mathbf{Diff}]$ if and only if there is an $\mathfrak{Q}^{\bullet}_{\mathsf{sm}}$-quasimodel for φ (the proof is analogous to that of Lemmas 2.7).

Since every type belonging to a horizontally reflexive quasistate is reflexive, if $\Diamond_h \psi \in D_{\boldsymbol{q}(x,y)}$ then we must necessarily have that $\psi \notin \bigcup \boldsymbol{q}(x, y)$. Hence, for $\mathfrak{Q}^{\bullet}_{\mathsf{sm}}$-quasimodels, condition **(qm5)** of Definition 2.6 is equivalent to:

(qm5′) For all $x \in X$, $y \in Y$ and $\Diamond_h \psi \in \mathsf{sub}(\varphi)$,

$$\Diamond_h \psi \in D_{\boldsymbol{q}(x,y)} \qquad \Longrightarrow \qquad \exists x' \in X;\ \psi \in \bigcup \boldsymbol{q}(x', y)$$

From here it is not difficult to adapt the proof of Lemma 2.8, to show that φ has a $\mathfrak{Q}^{\bullet}_{\mathsf{sm}}$-quasimodel if and only if $\mathfrak{Q}^{\bullet}_{\mathsf{sm}}\text{-}\mathsf{qm}_\varphi$ is satisfiable with respect to $\mathbf{S5} \times \mathbf{Diff}$. Moreover, the size of the resulting quasimodel is at most (singly-) exponential in the size of the satisfying model.

Hence, if φ is satisfiable with respect to $[\mathbf{S5}, \mathbf{Diff}]$ then $\mathfrak{Q}^{\bullet}_{\mathsf{sm}}\text{-}\mathsf{qm}_\varphi$ is satisfiable with respect to $\mathbf{S5} \times \mathbf{Diff}$. Furthermore, since $\mathbf{S5} \times \mathbf{Diff}$ has the exponential *product* fmp [11], $\mathfrak{Q}^{\bullet}_{\mathsf{sm}}\text{-}\mathsf{qm}_\varphi$ can be satisfied in a product model that is at most exponential in the size of $\mathfrak{Q}^\bullet_{\mathsf{sm}}$-$\mathsf{qm}_\varphi$. This, in turn, can be converted back to a model for φ that is at most exponential in the size of $\mathfrak{Q}^\bullet_{\mathsf{sm}}$-$\mathsf{qm}_\varphi$ and at most *doubly-exponential* in the size of φ, as required. □

Again, it is worth noting that in [8], this same upper-bound is achieved through a method of filtration for a large collection of commutators of the form $[L, \mathbf{S5}]$, where L can be axiomatized exclusively with proposition-free formulas, and formulas from among $\{p \to \Box\Diamond p, \Diamond^k p \to \Diamond p : k > 0\}$. However, neither this nor any other method of filtration is known for the case where L is the logic of the difference operator.

3 Discussion

This paper provides a first glance into the behaviour of some commutators of modal logics which are neither product matching nor are both Horn-axiomatizable. We conclude with a discussion of some open problems and directions for future work:

- The satisfiability problem for $[\mathbf{Diff}, \mathbf{Diff}]$ is known to be NExpTime-hard [17], and so it remains open as to where lies the precise complexity? Is it possible to improve upon the N2ExpTime upper-bound on the complexity of the satisfiability problem for $[\mathbf{Diff}, \mathbf{Diff}]$, or is it, perhaps, possible to exploit the infinite 'grid'-like structure of Lemma 2.1 to encode some N2ExpTime-hard problem?
- A natural generalization of the logic of the difference operator is provided by Jansana's [15] family of logics **Kn.4B**, for $n > 1$, axiomatized by the following formulas:

$$(n.4) := [n]p \to [n+1]p \qquad \text{and} \qquad (n.B) := p \to [n]\langle n\rangle p,$$

  where $\langle 0\rangle p := p$, $\langle n\rangle p := \Diamond^n p \vee \langle n-1\rangle p$, and $[n]\varphi := \neg\langle n\rangle\neg\varphi$. These logics are characterised by the class of frames in which every possible world is reachable from every other in fewer than n transitions; in particular, we have that $\mathbf{Diff} = \mathbf{K1.4B}$. Can the above techniques be adapted to construct a reduction between $[\mathbf{Kn.4B}, \mathbf{Km.4B}]$ and $\mathbf{Kn.4B} \times \mathbf{Km.4B}$, for $n, m \geq 1$?
- Finally, if the techniques employed here could be suitably extended, this would serve to limit the search for any examples of Kripke complete modal logics L_h and L_v such that one and only one of the logics $[L_h, L_v]$ and $L_h \times L_v$ is decidable; a question posed in [7].

Acknowledgements I would like thank Agi Kurucz for many fruitful discussions, upon which this work is based, as well as the anonymous reviewers for their insightful comments and suggestions.

A Appendix

Claim A.1 *The triple $(X, Y, \boldsymbol{q})$ of Lemma 2.7 is a $\mathfrak{Q}$-quasimodel for φ.*

Proof. First, we must verify that each $\boldsymbol{q}(x,y) \in \mathfrak{Q}$ is indeed a quasistate for φ:

- Clearly $T^{x,y}$ is a non-empty set of types, since $h(x,y)$ is non-empty by construction, as required for **(qs1)**.
- Suppose that $\boldsymbol{t}(u), \boldsymbol{t}(v) \in T^{x,y}$ are such that $\boldsymbol{t}(u) \neq \boldsymbol{t}(v)$, for some $u, v \in h(x,y)$. In particular, we have that $uR_h v$ and $uR_v v$. It then follows immediately from the definition that $\boldsymbol{t}(u) S_j^{x,y} \boldsymbol{t}(v)$ for $j = h, v$, as required for **(qs2)**.
- Finally, suppose that $\boldsymbol{t}(u), \boldsymbol{t}(v) \in T^{x,y}$ and $\Diamond_j \psi \in \mathsf{sub}(\varphi)$ are such that $\boldsymbol{t}(u) S_j^{x,y} \boldsymbol{t}(v)$ and $\psi \in \boldsymbol{t}(v)$. It follows by definition that there is some $u', v' \in h(x,y)$ such that $\boldsymbol{t}(u) = \boldsymbol{t}(u')$, $\boldsymbol{t}(v) = \boldsymbol{t}(v')$ and $u' R_j v'$. Hence we have that $\alpha \in \boldsymbol{t}(v')$ and consequently, that $\Diamond_j \psi \in \boldsymbol{t}(u') = \boldsymbol{t}(u)$, as required for **(qs3)**.

Next, we must check that $(X, Y, \boldsymbol{q})$ satisfies all the conditions **(qm1)**–**(qm7)** to be a suitable $\mathfrak{Q}$-quasimodel for φ:

- By definition, $\mathfrak{M}, r \models \varphi$, and by definition we have that $[r] \in X$ and $[r] \in Y$. Therefore, we may take $x_0 = y_0 = r \in h([r],[r])$ such that $\varphi \in \boldsymbol{t}(r)$ and hence, by construction, $\varphi \in \bigcup \boldsymbol{q}(x_0, y_0)$, as required for **(qm2)**.
- For **(qm3)**, suppose that $x, x' \in X$, $y \in Y$ and $\Diamond_h \psi \in \mathsf{sub}(\varphi)$ are such that $x \neq x'$ and $\psi \in \bigcup \boldsymbol{q}(x', y)$, which is to say that $\psi \in \boldsymbol{t}(w')$ for some $w' \in h(x', y)$. By construction we have that $h(x', y) \neq h(x, y)$, since R_h^+ is an equivalence relation. It then follows, again from the fact that R_h^+ is an equivalence relation, that $wR_h w'$ for all $w \in h(x,y)$ and $w' \in h(x', y)$. Hence, we have that $\Diamond_h \psi \in \boldsymbol{t}(w)$ for all $w \in h(x,y)$, which is to say that $\Diamond_h \psi \in \bigcap \boldsymbol{q}(x,y)$, as required. Condition **(qm4)** is analogous.
- For **(qm5)**, suppose that $x \in X$, $y \in Y$ and $\Diamond_h \psi \in D_{\boldsymbol{q}(x,y)}$. By definition, there is some $u \in h(x,y)$ such that $\Diamond_h \psi \in \boldsymbol{t}(u)$ and there is no $v \in h(x,y)$ such that $\boldsymbol{t}(u) S_h^{x,y} \boldsymbol{t}(v)$ and $\psi \in \boldsymbol{t}(v)$. However, since $\mathfrak{M}, u \models \Diamond_h \psi$, there must be some $v' \in W$ such that $uR_h v'$ and $\psi \in \boldsymbol{t}(v')$. It then follows that there is some $x' \in X$ such that $x \neq x'$ and $v' \in h(x', y)$. Hence we have that $\psi \in \bigcup \boldsymbol{q}(x', y)$, as required. Condition **(qm6)** is analogous.
- For **(qm7)**, suppose that $x \in X$ and $y, y' \in Y$ are such that $y \neq y'$ and $h\text{-}\mathsf{size}(\boldsymbol{q}(x,y)) > 1$ then by definition there are some $t, t' \in T^{x,y}$ such that $tS_h^{x,y} t'$ (note that t and t' may or may not be identical). Hence there are some $u, v \in h(x,y)$ such that $\boldsymbol{t}(u) = t$, $\boldsymbol{t}(v) = t$ and $uR_h v$. Let $u' \in h(x, y')$ then since $y \neq y'$ and R_v^+ is an equivalence relation, we have that $uR_v u'$. Hence, by the Church-Rosser property, there is some $v' \in W$ such that $u' R_h v'$ and $vR_v v'$. Moreover, since R_v^+ is an equivalence relation, we have that $uR_v^+ v'$, and thus $v' \in h(x, y')$. Hence there are $\boldsymbol{t}(u'), \boldsymbol{t}(v') \in T^{x,y'}$ such that $\boldsymbol{t}(u') S_h^{x,y'} \boldsymbol{t}(v')$. It then follows that $h\text{-}\mathsf{size}(\boldsymbol{q}(x, y')) > 1$, as required. The case for $v\text{-}\mathsf{size}(\boldsymbol{q}(x,y))$ is analogous.

Hence it follows that $(X, Y, \boldsymbol{q})$ is a $\mathfrak{Q}$-quasimodel for φ, as required. □

References

[1] Areces, C., G. Hoffmann and A. Denis, *Modal logics with counting*, in: A. Dawar and R. de Queiroz, editors, *Logic, Language, Information and Computation*, Lecture Notes in Artificial Intelligence **6188**, Springer, 2010 pp. 98–109.

[2] Baader, F., D. Calvanese, D. L. McGuinness, D. Nardi and P. F. Patel-Schneider, "The Description Logic Handbook: Theory, Implementation, and Applications," Cambridge University Press, New York, NY, USA, 2003.

[3] Blackburn, P., M. de Rijke and Y. Venema, "Modal Logic," Cambridge University Press, New York, NY, USA, 2001.

[4] Chang, C. C. and H. J. Keisler, "Model Theory," Studies in Logic and the Foundations of Mathematics **73**, North-Holland Publishing Company, 1990, 3rd edition.

[5] de Rijke, M., *The modal logic of inequality*, The Journal of Symbolic Logic **57** (1992), pp. 566–584.

[6] de Rijke, M., "Extending Modal Logic," Ph.D. thesis, Institute of Logic, Language and Computation, University van Amsterdam (1993).

[7] Gabbay, D. M., A. Kurucz, F. Wolter and M. Zakharyaschev, "Many-Dimensional Modal Logics: Theory and Applications," Studies in Logic and the Foundations of Mathematics **148**, Elsevier, 2003.

[8] Gabbay, D. M. and V. Shehtman, *Products of modal logics. Part I*, Logic Journal of the IGPL **6** (1998), pp. 73–146.

[9] Gargov, G. and V. Goranko, *Modal logic with names*, Journal of Philosophical Logic **22** (1993), pp. 607–636.

[10] Grädel, E., M. Otto and E. Rosen, *Two-variable logic with counting is decidable*, in: *Proceedings of the 12th Annual IEEE Symposium on Logic in Computer Science (LICS'97)*, IEEE, 1997, pp. 306–317.

[11] Hampson, C., *Decidable first-order modal logics with counting quantifiers*, Advances in Modal Logic **11** (2016), pp. 382–400.

[12] Hampson, C. and A. Kurucz, *On modal products with the logic of 'elsewhere'*, Advances in Modal Logic **9** (2012), pp. 339–347.

[13] Hampson, C. and A. Kurucz, *Undecidable propositional bimodal logics and one-variable first-order linear temporal logics with counting*, ACM Transactions on Computational Logic (TOCL) **16** (2015).

[14] Hampson, C., A. Kurucz, S. Kikot and S. Marcelino, *Non-finitely axiomatisable 2D modal product logics with infinite canonical axiomatisations*, (to appear).

[15] Jansana, R., *Some logics related to von Wright's logic of place*, Notre Dame Journal of Formal Logic **35** (1994), pp. 88–98.

[16] Kurucz, A., *Combining modal logics*, in: P. Blackburn, J. van Benthem and F. Wolter, editors, *Handbook of Modal Logic*, Studies in Logic and Practical Reasoning **3**, Elsevier, 2007 pp. 869–924.

[17] Marx, M., *Complexity of products of modal logics*, Journal of Logic and Computation **9** (1999), pp. 197–214.

[18] Pacholski, L., W. Szwast and L. Tendera, *Complexity of two-variable logic with counting*, in: *Proceedings of the 12th Annual IEEE Symposium on Logic in Computer Science (LICS'97)*, IEEE, 1997, pp. 318–327.

[19] Pratt-Hartmann, I., *Complexity of the two-variable fragment with counting quantifiers*, Journal of Logic, Language and Information **14** (2005), pp. 369–395.

[20] Segerberg, K., *Two-dimensional modal logic*, Journal of Philosophical Logic **2** (1973), pp. 77–96.

[21] Segerberg, K., *A note on the logic of elsewhere*, Theoria **46** (1980), pp. 183–187.

[22] Shehtman, V., *Two-dimensional modal logics*, Mathematical Notes of the USSR Academy of Sciences **23** (1978), pp. 417–424, (Translated from Russian).

[23] von Wright, G. H., *A modal logic of place*, in: E. Sosa, editor, *The philosophy of Nicolas Rescher*, Dordrecht, 1979 pp. 65–73.

[24] Wolter, F., *The product of converse* **PDL** *and polymodal* **K**, Journal of Logic and Computation **10** (2000), pp. 223–251.

Relational Semantics for the Turing Schmerl Calculus

Eduardo Hermo Reyes [1]

University of Barcelona
Department of Philosophy

Joost J. Joosten [2]

University of Barcelona
Department of Philosophy

Abstract

In [13] the authors introduced the propositional modal logic **TSC** (which stands for Turing Schmerl Calculus) which adequately describes the provable interrelations between different kinds of Turing progressions. The current paper defines a model $\mathcal{J}$ which is proven to be a universal model for **TSC**. The model $\mathcal{J}$ is a slight modification of the intensively studied $\mathcal{I}$: Ignatiev's universal model for the closed fragment of Gödel Löb's polymodal provability logic **GLP**.

Keywords: Provability logic, strictly positive fragments, Turing progressions, universal model.

1 Introduction

Turing progressions arise by iteratedly adding consistency statements to a base theory. Different notions of consistency give rise to different Turing progressions. In [13], the authors introduced the system **TSC** (sometimes denoted by Cyrillic letter **Tse** and Latin **C**) that generates exactly all relations that hold between these different Turing progressions given a particular set of natural consistency notions. The system was proven to be arithmetically sound and complete for a natural interpretation, named the *Formalized Turing progressions* (FTP) interpretation. A brief overview of this work can be found in Section 2.1 together with Theorem 3.8.

In this paper we discuss relational semantics of **TSC** by considering a small modification on Ignatiev's frame, which is a universal frame for the variable-free fragment of Japaridze's provability logic **GLP**.

[1] ehermo.reyes@ub.edu

[2] jjoosten@ub.edu

2 Strictly positive signature

TSC is built-up from a positive propositional modal signature using *ordinal modalities.* Let Λ be a fixed recursive ordinal throughout the paper with some properties as specified in Remark 3.4. By ordinal modalities we denote modalities of the form $\langle n^\alpha \rangle$ where $\alpha \in \Lambda$ and $n \in \omega$ (named *exponent* and *base*, respectively). The set of formulas in this language is defined as follows:

Definition 2.1 By $\mathbb{F}$ we denote the smallest set such that:

i) $\top \in \mathbb{F}$;
ii) If $\varphi, \psi \in \mathbb{F} \Rightarrow (\varphi \wedge \psi) \in \mathbb{F}$;
iii) if $\varphi \in \mathbb{F}$, $n < \omega$ and $\alpha < \Lambda \Rightarrow \langle n^\alpha \rangle \varphi \in \mathbb{F}$.

For any formula ψ in this signature, we define the set of base elements occurring in ψ. That is:

Definition 2.2 The set of base elements occurring in any modality of a formula $\psi \in \mathbb{F}$ is denoted by $\mathsf{N\text{-}mod}(\psi)$. We recursively define $\mathsf{N\text{-}mod}$ as follows:

i) $\mathsf{N\text{-}mod}(\top) = \emptyset$;
ii) $\mathsf{N\text{-}mod}(\varphi \wedge \psi) = \mathsf{N\text{-}mod}(\varphi) \cup \mathsf{N\text{-}mod}(\psi)$;
iii) $\mathsf{N\text{-}mod}(\langle n^\alpha \rangle \psi) = \{n\} \cup \mathsf{N\text{-}mod}(\psi)$.

2.1 (FTP) interpretation

In [13], the authors introduced an arithmetical interpretation in which modal formulas are intended to be read as *Turing progressions*; hierarchies of theories that arise by transfinitely iterating n-consistency statements. These progressions can be defined according to the following conditions below:

T1. $(T)_n^0 := T$ where T is an initial or base theory;

T2. $(T)_n^{\alpha+1} := (T)_n^\alpha \cup \{\mathrm{Con}_n((T)_n^\alpha)\}$;

T3. $(T)_n^\lambda := \bigcup_{\beta<\lambda} (T)_n^\beta$, for λ a limit ordinal below Λ.

However, conditions T1-T3 can be reformulated by the unique following clause:

$$(T)_n^\alpha := T \cup \{\mathrm{Con}_n((T)_n^\beta) : \beta < \alpha\} \quad \text{for } \alpha < \Lambda,\ n < \omega.$$

This presentation, known as *Smooth Turing progressions*, was studied by Beklemishev in among others [5] and [1].

Given such a family of theories $(T)_n^\alpha$ we have that they can be represented within $\mathbf{EA}^+$ through some arithmetical formula numerating their axioms. Here, $\mathbf{EA}^+$ is Robinson's arithmetic $\mathbf{Q}$ together with induction for bounded formulas.

Suppose we are given some elementary well-ordering $(D, \prec)$. Consider the elementary formula $\tau_n^{\sigma(z)}(x, y)$ where x is a variable for an ordinal $\alpha \in D$, y stands for the coding of some arithmetical formula and $\sigma(z)$ is an elementary formula enumerating the axioms of some base theory. Hence, roughly speaking, the formula tells us that the formula coded by y is an axiom of $(T)_n^\alpha$ where the initial theory is numerated by the elementary formula $\sigma(z)$ and $\mathbf{EA}^+$ is numerated by $\epsilon(x)$.

We say that $\tau_n^{\sigma(z)}(\alpha, x)$ enumerates the α-th theory of a progression based on iteration of consistency along $(D, \prec)$ with base $\sigma(z)$ if:

$$\mathbf{EA}^+ \vdash \tau_n^{\sigma(z)}(\alpha, x) \leftrightarrow ((\epsilon(x) \vee \sigma(x)) \vee \exists \beta\, (\prec(\beta, \alpha) \wedge x = \ulcorner \mathrm{Con}_n(\tau_n^{\sigma(z)}(\dot{\beta}, y)\urcorner)).$$

The existence of such $\tau_n^{\sigma(z)}(\alpha, x)$ is guaranteed by the fixed point theorem.

Let us introduce now the arithmetical interpretation of our modal formulae in terms of the τ-formulae. Let $\mathcal{L}_{\mathbb{N}}$ denote the set of formulas in the usual language of arithmetic.

Definition 2.3 An arithmetical interpretation is a map $* : \mathbb{F} \longrightarrow \mathcal{L}_{\mathbb{N}}$ inductively defined as follows:

(i) $(\top)^*(x) = \epsilon(x)$;
(ii) $(\varphi \wedge \psi)^*(x) = (\varphi)^*(x) \vee (\psi)^*(x)$
(iii) $(\langle n^\alpha \rangle \varphi)^*(x) = \tau_n^{\varphi^*(y)}(\alpha, x)$.

Since $\mathbb{F}$ has no propositional variables, we can identify a modal formula with its arithmetical interpretation unambiguously. Moreover, for the sake of clarity, and since we are working in the close fragment, we will use the following notation: given $\varphi \in \mathbb{F}$ by Th_φ we denote Th_σ where $\varphi^*(x) = \sigma(x)$, following Definition 2.3. If $\varphi^*(x) = \epsilon(x)$ we use just $\mathbf{EA}^+$ instead of Th_ϵ.

3 The logic TSC

In this section we introduce the logic **TSC** whose main goal is to express valid relations that hold between the corresponding Turing progressions. For this purpose we shall consider a kind of special formulas named *monomial normal forms* which are used in the axiomatization of the calculus **TSC**.

Monomial normal forms are conjunctions of monomials with an additional condition on the occurring exponents. In order to formulate this condition we first need to define the *hyper-exponential* as studied in [9].

Definition 3.1 For every $n \in \omega$ the *hyper-exponential* functions $e^n : \mathrm{On} \to \mathrm{On}$ are recursively defined as follows: e^0 is the identity function, $e^1 : \alpha \mapsto -1 + \omega^\alpha$ and $e^{n+m} = e^n \circ e^m$.

We will use e to denote e^1. Note that for α not equal to zero we have that $e(\alpha)$ coincides with the regular ordinal exponentiation with base ω; that is, $\alpha \mapsto \omega^\alpha$. However, it turns out that hyper-exponentials have the nicer algebraic properties in the context of provability logics.

The next definition may seem a bit ad-hoc in a purely syntactical setting so we provide some minimal motivation. Monomials are terms of the form $\langle n^\alpha \rangle \top$, for $n < \omega$ and $\alpha < \Lambda$. The simplest form of stating information about Turing progressions will be by means of a conjunction of monomials. However, the arithmetical behavior of monomials tells us that monomials will imply other monomials. In our normal form, we wish to only include those monomials that add new information to the entire expression giving rise to the technical Condition c below.

Definition 3.2 The set of formulas in *monomial normal form*, MNF, is inductively defined as follows:

i) $\top \in \mathsf{MNF}$;

ii) $\langle n^{\alpha} \rangle \top \in \mathsf{MNF}$, for any $n < \omega$ and $\alpha < \Lambda$;

iii) if *a)* $\langle n_0^{\alpha_0} \rangle \top \wedge \ldots \wedge \langle n_k^{\alpha_k} \rangle \top \in \mathsf{MNF}$;

b) $n < n_0$;

c) α of the form $e^{n_0 - n}(\alpha_0) \cdot (2 + \delta)$ for some $\delta < \Lambda$,

then $\langle n^{\alpha} \rangle \top \wedge \langle n_0^{\alpha_0} \rangle \top \wedge \ldots \wedge \langle n_k^{\alpha_k} \rangle \top \in \mathsf{MNF}$.

The derivable objects of **TSC** are *sequents* i.e. expressions of the form $\varphi \vdash \psi$ where $\varphi, \psi \in \mathbb{F}$. We will use the following notation: by $\varphi \equiv \psi$ we will denote that both $\varphi \vdash \psi$ and $\psi \vdash \varphi$ are derivable. Also, by convention we take that for any n, $\langle n^0 \rangle \varphi$ is just φ.

Definition 3.3 **TSC** is given by the following set of axioms and rules:

Axioms:

(i) $\varphi \vdash \varphi, \quad \varphi \vdash \top$;

(ii) $\varphi \wedge \psi \vdash \varphi, \quad \varphi \wedge \psi \vdash \psi$;

(iii) Monotonicity axioms: $\langle n^{\alpha} \rangle \varphi \vdash \langle n^{\beta} \rangle \varphi$, for $\beta < \alpha$;

(iv) Co-additivity axioms: $\langle n^{\beta+\alpha} \rangle \varphi \equiv \langle n^{\alpha} \rangle \langle n^{\beta} \rangle \varphi$;

(v) Reduction axioms: $\langle (n+m)^{\alpha} \rangle \varphi \vdash \langle n^{e^m(\alpha)} \rangle \varphi$;

(vi) Schmerl axioms:

$$\langle n^{\alpha} \rangle \big(\langle n_0^{\alpha_0} \rangle \top \wedge \psi \big) \equiv \langle n^{e^{n_0 - n}(\alpha_0) \cdot (1+\alpha)} \rangle \top \wedge \langle n_0^{\alpha_0} \rangle \top \wedge \psi$$

for $n < n_0$ and $\langle n_0^{\alpha_0} \rangle \top \wedge \psi \in \mathsf{MNF}$.

Rules:

(i) If $\varphi \vdash \psi$ and $\varphi \vdash \chi$, then $\varphi \vdash \psi \wedge \chi$;

(ii) If $\varphi \vdash \psi$ and $\psi \vdash \chi$, then $\varphi \vdash \chi$;

(iii) If $\varphi \vdash \psi$, then $\langle n^{\alpha} \rangle \varphi \vdash \langle n^{\alpha} \rangle \psi$;

(iv) If $\varphi \vdash \psi$, then $\langle n^{\alpha} \rangle \varphi \wedge \langle m^{\beta+1} \rangle \psi \vdash \langle n^{\alpha} \rangle \big(\varphi \wedge \langle m^{\beta+1} \rangle \psi \big)$ for $n > m$.

It is worth mentioning the special character of Axioms (v) and (vi) since both axioms are modal formulations of principles related to Schmerl's fine structure theorem, also known as *Schmerl's formulas* (see [16] and [3]).

Remark 3.4 As we see in the axioms of our logic, they only make sense if the ordinals occuring in them are available. Recall that Λ is fixed to be a recursive ordinal all through the paper. Moreover, some usable closure conditions on Λ naturally suggest themselves. Since it suffices to require that for $n < \omega$ that $\alpha, \beta < \Lambda \ \Rightarrow \ \alpha + e^n(\beta) < \Lambda$, we shall for the remainder assume that Λ is an ε-number, that is, a positive fixpoint of e whence $e(\Lambda) = \Lambda = \omega^{\Lambda}$.

In [13], the authors proved that for any formula φ, there is a unique equivalent ψ in monomial normal form.

Theorem 3.5 *For every formula* φ *there is a unique* $\psi \in \mathsf{MNF}$ *such that* $\varphi \equiv \psi$.

In virtue of the Reduction axioms, a formula $\psi \in \mathsf{MNF}$ may bear implicit information on monomials $\langle n^{\alpha} \rangle \top$ for $n \notin \mathsf{N\text{-}mod}(\psi)$. The next definition is made to retrieve this information.

Definition 3.6 Let $\psi := \langle n_0^{\alpha_0} \rangle \top \wedge \ldots \wedge \langle n_k^{\alpha_k} \rangle \top \in \mathsf{MNF}$. By $\pi_{n_i}(\psi)$ we denote the corresponding exponent α_i. Moreover, for $m \notin \mathsf{N\text{-}mod}(\psi)$, with $n_k > m$, $\pi_m(\psi)$ is set to be $e\big(\pi_{m+1}(\psi)\big)$ and for $m' > n_k$, $\pi_{m'}(\psi)$ is defined to be 0.

The following theorems are proven in [13]. The first one provides a succinct derivability condition between monomial normal forms while the second one establishes the soundness and completeness of the system with respect to the (FTP) interpretation:

Theorem 3.7 *For any* $\psi_0, \ \psi_1 \in \mathsf{MNF}$, *where* $\psi_0 := \langle n_0^{\alpha_0} \rangle \top \wedge \ldots \wedge \langle n_k^{\alpha_k} \rangle \top$ *and* $\psi_1 := \langle m_0^{\beta_0} \rangle \top \wedge \ldots \wedge \langle m_j^{\beta_j} \rangle \top$. *We have that* $\psi_0 \vdash \psi_1$ *iff for any* $n < \omega$, $\pi_n(\psi_0) > \pi_n(\psi_1)$.

Theorem 3.8 *For any* $\varphi, \ \psi \in \mathbb{F}$,

$$\varphi \vdash \psi \Longleftrightarrow \boldsymbol{EA}^{+} \vdash \forall x \ \big(\Box_{\mathsf{Th}_{\psi}}(x) \to \Box_{\mathsf{Th}_{\varphi}}(x)\big).$$

4 A variation on Ignatiev's Frame

The purpose of this section is to define a modal model $\mathcal{J}$ which is universal for our logic. That is, any derivable sequent will hold everywhere in the model whereas any non-derivable sequent will be refuted somewhere in the model.

The model will be based on special sequences of ordinals. In order to define them, we need the following central definition.

Definition 4.1 We define *ordinal logarithm* as $\ell(0) := 0$ and $\ell(\alpha + \omega^{\beta}) := \beta$.

With this last definition we are now ready to introduce the set of worlds of our frame.

Definition 4.2 By Ig^{ω} we denote the set of *ℓ-sequences* or *Ignatiev sequences*. That is, the set of sequences $x := \langle x_0, x_1, x_2, \ldots \rangle$ where for $i < \omega$, $x_{i+1} \leq \ell(x_i)$.

Given a ℓ-sequence x, if all but finitely many of its elements are zero, we will write $\langle x_0, \ldots, x_n, \mathbf{0} \rangle$ to denote such ℓ-sequence or even simply $\langle x_0, \ldots, x_n \rangle$ whenever $x_{n+1} = 0$.

Next, we can define our frame, which is a variation of Ignatiev's frame.

Definition 4.3 $\mathcal{J}_\Lambda := \langle I, \{R_n\}_{n<\omega} \rangle$ is defined as follows:

$$I := \{x \in \mathrm{Ig}^\omega : x_i < \Lambda \text{ for } i < \omega\}$$

and

$$xR_ny :\Leftrightarrow (\forall m \leq n\ x_m > y_m \ \wedge \ \forall i > n\ x_i \geq y_i).$$

Since Λ is a fixed ordinal along the paper, from now on we suppress the subindex Λ.

The observations collected in the next lemma all have elementary proofs. Basically, the lemma confirms that the R_n are good to model provability logic and respect the increasing strength of the provability predicates $[n]$.

Lemma 4.4

(i) *Each R_n for $n \in \omega$ is* transitive*: $xR_ny \wedge yR_nz \Rightarrow xR_nz$;*

(ii) *Each R_n for $n \in \omega$ is* Noetherian*: each non-empty $X \subseteq I$ has an R_n-maximal element $y \in X$, i.e., $\forall x{\in}X\ \neg yR_nx$;*

(iii) *The relations R_n are* monotone *in n in the sense that: $xR_ny \Rightarrow xR_my$ whenever $n > m$.*

Note that Item (ii) is equivalent to stating that there are no infinite ascending R_n chains. In other words, the converse of R_n is well-founded.

We define the auxiliary relations R_n^α for any $n < \omega$ and $\alpha < \Lambda$. The idea is that the R_n^α will model the $\langle n^\alpha \rangle$ modality.

Definition 4.5 Given $x, y \in I$ and R_n on I, we recursively define $xR_n^\alpha y$ as follows:

(i) $xR_n^0y \ :\Leftrightarrow \ x = y$;

(ii) $xR_n^{1+\alpha}y \ :\Leftrightarrow \ \forall \beta{<}1{+}\alpha\ \exists z\ \big(xR_nz \ \wedge \ zR_n^\beta y\big)$.

Let us introduce some simple observations about the R_n^α relations.

Proposition 4.6 *Given $x, y \in I$, $n < \omega$ and $\alpha < \Lambda$:*

$$xR_n^{\alpha+1}y \ \Leftrightarrow \ \exists z\ \big(xR_nz \ \wedge \ zR_n^\alpha y\big).$$

Proof We make a case distinction on α with $\alpha = 0$ being trivial. For $\alpha > 0$, we have that $\alpha = 1 + \gamma$ for some $\gamma \leq \alpha$. With the help of this fact, we can reason as follows:

$$\begin{aligned} xR_n^{\alpha+1}y \ &\Leftrightarrow \ xR_n^{1+\gamma+1}y; \\ &\Leftrightarrow \ \forall \beta < 1 + \delta + 1\, \exists z\ \big(xR_nz \ \wedge \ zR_n^\beta y\big); \\ &\Rightarrow \ \exists z\big(xR_nz \ \wedge \ zR_n^\alpha y\big), \text{ in particular.} \end{aligned}$$

Thus, $xR_n^{\alpha+1}y \Rightarrow \exists z(xR_n z \wedge zR_n^{\alpha}y)$. For right-to-left implication we proceed analogously:

$$\begin{aligned}\exists z(xR_n z \wedge zR_n^{\alpha}y) &\Leftrightarrow \exists z\ (xR_n z \wedge zR_n^{1+\gamma}y);\\ &\Leftrightarrow \exists z\ (xR_n z \wedge zR_n^{1+\gamma}y) \wedge\\ &\qquad \forall\beta < 1+\gamma\ \exists z'\ (zR_n z' \wedge z'R_n^{\beta}y);\\ &\Rightarrow \forall\beta' < 1+\gamma+1\ \exists u\ (xR_n u \wedge uR_n^{\beta'}y);\\ &\Leftrightarrow xR_n^{1+\gamma+1}y;\\ &\Leftrightarrow xR_n^{\alpha+1}y.\end{aligned}$$

□

Proposition 4.7 *Let $x, y \in I$, $n < \omega$ and $\lambda < \Lambda$ such that $\lambda \in Lim$:*

$$xR_n^{\lambda}y \Leftrightarrow \forall\beta < \lambda\ xR_n^{1+\beta}y.$$

Proof For left-to-right implication, notice that if $xR_n^{\lambda}y$ then by definition, we have that $\forall\beta < \lambda\ \exists u\ (xR_n u \wedge uR_n^{\beta}y)$. Therefore, in particular, we obtain that $\forall\beta < \lambda\ \exists u\ (xR_n u \wedge uR_n^{1+\beta}y)$ thus by transitivity, $\forall\beta < \lambda\ \ xR_n^{1+\beta}y$. For the other direction, if $\forall\beta < \lambda\ xR_n^{1+\beta}y$, then in particular, $\forall\beta < \lambda\ xR_n^{\beta+1}y$ and then, by Proposition 4.6, $\forall\beta < \lambda\ \exists u\ (xR_n u \wedge uR_n^{\beta}y)$, that is, $xR_n^{\lambda}y$. □

It is easy to see that for example $\langle\omega, \mathbf{0}\rangle R_0^n\langle m, \mathbf{0}\rangle$ for each $n, m \in \omega$, so that also $\langle\omega, \mathbf{0}\rangle R_0^{\omega}\langle m, \mathbf{0}\rangle$ for each $m \in \omega$. Clearly, we do not have $\langle\omega, \mathbf{0}\rangle R_0^{\omega+1}\langle m, \mathbf{0}\rangle$ for any $m \in \omega$ but we do have $\langle\omega+1, \mathbf{0}\rangle R_0^{\omega+1}\langle m, \mathbf{0}\rangle$ for all $m \in \omega$.

We also note that the dual definition $x\overline{R}_n^0 y :\Leftrightarrow x = y$; and $x\overline{R}_n^{1+\alpha} y :\Leftrightarrow \forall\,\beta{<}1{+}\alpha\ \exists z\ (x\overline{R}_n^{\beta} z \wedge z\overline{R}_n y)$ does not make much sense on our frames. For example we could have $\langle\omega, \mathbf{0}\rangle\overline{R}_0^{\alpha}\langle 0, \mathbf{0}\rangle$ for any ordinal $\alpha > 0$.

With the the auxiliary relations R_n^{α}, we give the following definition for a formula φ being true in a point x of $\mathcal{J}$.

Definition 4.8 Let $x \in I$ and $\varphi \in \mathbb{F}$. By $x \Vdash \varphi$ we denote the validity of φ in x that is recursively defined as follows:

- $x \Vdash \top$ for all $x \in I$;
- $x \Vdash \varphi \wedge \psi$ iff $x \Vdash \varphi$ and $x \Vdash \psi$;
- $x \Vdash \langle n^{\alpha}\rangle\varphi$ iff there is $y \in I$, $xR_n^{\alpha}y$ and $y \Vdash \varphi$.

Here are some easy observations on the R_n^{α} relations which among others tell us that all the R_n^{α} serve the purpose of a provability predicate for any $n \in \omega$ and $\alpha < \Lambda$.

Lemma 4.9

(i) *Each $R_n^{1+\alpha}$ for $n \in \omega$ and α an ordinal is* transitive*: $xR_n^{1+\alpha}y \wedge yR_n^{1+\alpha}z \Rightarrow xR_n^{1+\alpha}z$;*

Figure 1. A fragment of our frame $\mathcal{J}$. The dashed arrows represent R_0 relations, while the continuous arrows represent R_1 relations.

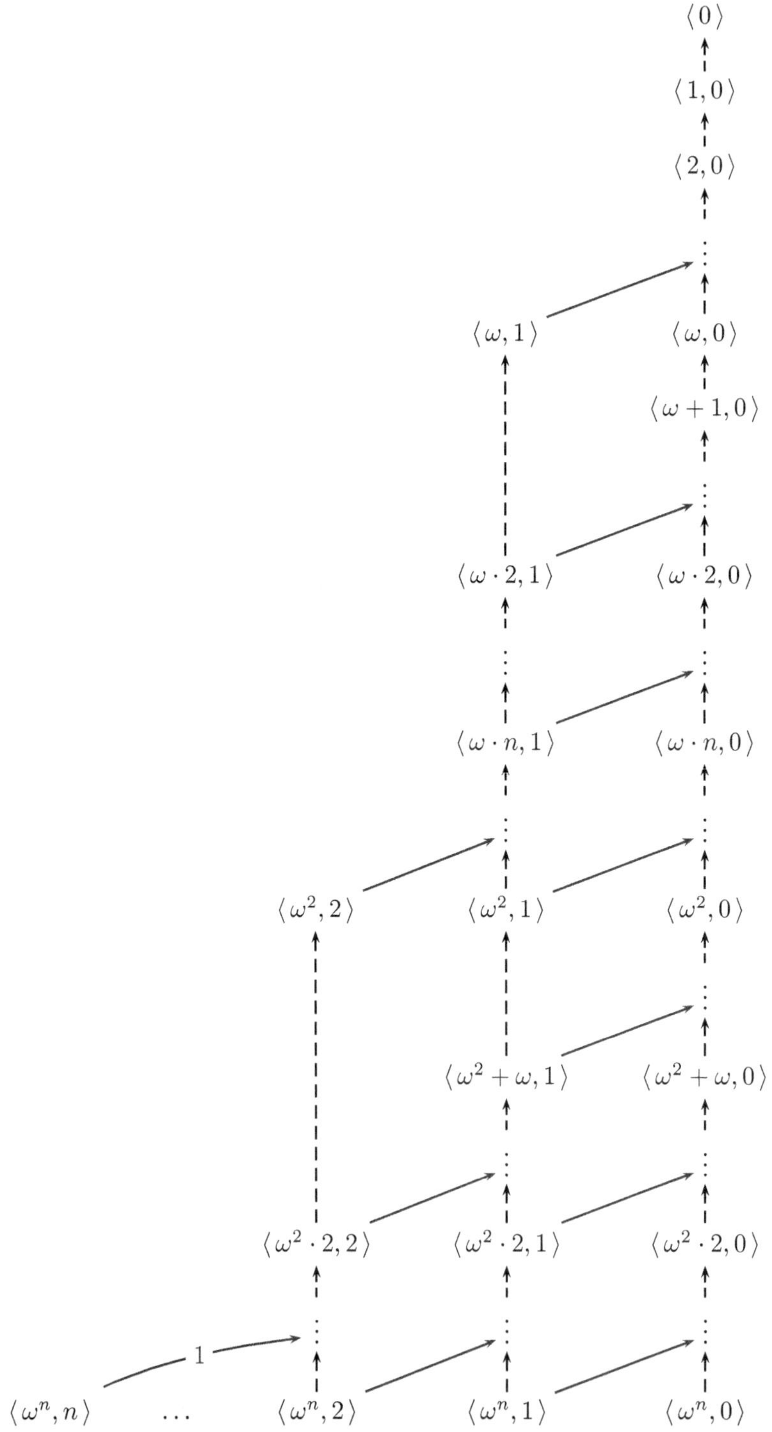

(ii) *Each* $R_n^{1+\alpha}$ *for* $n \in \omega$ *and* α *an ordinal is* Noetherian*: each non-empty* $X \subseteq I$ *has an* $R_n^{1+\alpha}$*-maximal element* $y \in X$*, i.e.,* $\forall x{\in}X\ \neg y R_n^{1+\alpha} x$*;*

(iii) *The relations* $R_n^{1+\alpha}$ *are* monotone *in* n *in the sense that:* $xR_n^{1+\alpha}y \Rightarrow xR_m^{1+\alpha}y$ *whenever* $n > m$*;*

(iv) *The relations* $R_n^{1+\alpha}$ *are* monotone *in* $1+\alpha$ *in the sense that:* $xR_n^{1+\alpha}y \Rightarrow xR_n^{1+\beta}y$ *whenever* $1+\beta < 1+\alpha$*.*

Proof The first three items follow directly from Lemma 4.4 by an easy transfinite induction. The last item is also easy. □

5 A characterization for transfinite accessibility

The intuitive idea behind the $xR_n^{\alpha}y$ assertion, is that this tells us that there exists a chain of 'length' α of R_n steps leading from the point x up to the point y. The following useful lemma tries to capture this intuition.

Lemma 5.1 *For* $x, y \in I$ *and* $n < \omega$ *we have that the following are equivalent*

(i) $xR_n^{1+\alpha}y$

(ii) *For each* $\beta < 1+\alpha$ *there exists a collection* $\{x^{\gamma}\}_{\gamma<\beta}$ *so that*
  (a) xR_nx^{γ} *for any* $\gamma < \beta$*,*
  (b) $x^0 = y$ *and,*
  (c) *for any* $\gamma' < \gamma < \beta$ *we have* $x^{\gamma}R_nx^{\gamma'}$*.*

Proof By induction on α. □

We shall now provide a characterization of the $R_n^{1+\alpha}$ relations. To this end, let us for convenience define

$$xR_{-1}^{\zeta}y \quad :\Leftrightarrow \quad \forall n{>}0\ \ x_n \geq y_n.$$

With this notation the following theorem makes sense.

Theorem 5.2 *For* $x, y \in I$ *and* $n < \omega$ *we have that the following are equivalent*

(i) $xR_n^{1+\alpha}y$*;*

(ii) $x_n \geq y_n + \big(1 + e(y_{n+1})\big) \cdot (1+\alpha)$ *and* $xR_{n-1}^{e(1+\alpha)}y$*;*

(iii)
$$\begin{array}{ll} x_n \geq y_n + \big(1 + e(y_{n+1})\big) \cdot (1+\alpha) & \textit{and,} \\ x_m > y_m \ \textit{for}\ m < n & \textit{and,} \\ x_m \geq y_m \ \textit{for}\ m > n. & \end{array}$$

We dedicate the remainder of this section to proving this theorem and move there through a series of lemmas. The first lemma in this series is pretty obvious. It tells us that if we can move from x to y in α many steps, then the distance between x_n and y_n must allow α many steps; That is, they lie at least α apart.

Lemma 5.3 *For* $x, y \in I$ *and* $n < \omega$ *and any ordinal* $\alpha < \Lambda$*, if* $xR_n^{\alpha}y$ *then* $x_n \geq y_n + \alpha$*.*

Proof By an easy induction on α. □

However, how many R_n steps one can make is not entirely determined by the n coordinates of the points. For example, there is just a single R_0 step from the point $\langle \omega \cdot 2, 1 \rangle$ to the point $\langle \omega, 1 \rangle$ whereas these points lie ω apart on the '0 coordinate'. The following lemma tells us how for R_n steps, the n-th coordinates are affected by the values of the $n+1$-th coordinate.

Lemma 5.4 *For $x, y \in I$ and $n < \omega$ with $xR_n^{1+\alpha}y$, we have*

$$x_n \geq y_n + e(y_{n+1}) \cdot (1 + \alpha).$$

In order to give a smooth presentation of this proof, we first give two simple technical lemmas with useful observations on the ordinals and ordinal functions involved.

Lemma 5.5 *For α, β and γ ordinals we have*

(i) $\ell(\beta) \geq 1 + \alpha \iff \beta \in e(1+\alpha) \cdot (1 + \mathsf{On})$,

(ii) *If* $(1+\alpha) < \beta$ *and* $\gamma \in e(\beta) \cdot (1 + \mathsf{On})$, *then* $\gamma \in e(1+\alpha) \cdot (1 + \mathsf{On})$,

(iii) $e(\beta + (1+\alpha)) = e(\beta) \cdot e(1+\alpha)$,

(iv) *For α a limit ordinal, we have that*

$$xR_n^\alpha y \iff \forall 1{+}\beta{<}\alpha\, \exists z\, (xR_n z \wedge zR_n^{1+\beta} y).$$

Proof The first two items then can easily be seen by using a Cantor Normal Form expression with base ω. For Item (i), we use the fact that $\beta \in \mathrm{Lim}$ together with that if $\ell(\beta) \geq 1 + \alpha$, then $\beta \geq e(\ell(\beta)) \geq e(1+\alpha)$. For Items (ii) and (iii) we use that $e(1+\omega) = \omega^{1+\omega} = \omega^1 \cdot \omega^\omega$. The last item follows from Definition 4.5 together with the fact that $1 + \alpha = \alpha \in \mathrm{Lim}$. □

Lemma 5.6 *For $x, y \in I$ and $n < \omega$, $xR_n y \Longrightarrow x_n \geq y_n + e(x_{n+1})$.*

Proof We make a case distinction on x_n. If $x_n \in \mathrm{Succ}$ then is trivial since $e(x_{n+1}) = 0$. If $x_n \in \mathrm{Lim}$, and furthermore, x_n is an additively indecomposable limit ordinal, it follows from the fact that $x_n > y_n$ and $x_n \geq e(x_{n+1})$. Otherwise, we can rewrite x_n as $\alpha + e(\beta)$ for some $\beta \geq x_{n+1}$, and y_n as $\delta + \omega^\gamma$. If $y_n \leq \alpha$ then clearly $x_n \geq y_n + e(x_{n+1})$. If $\alpha = \delta$ and $\gamma < \beta$, then notice that $\omega^\gamma + e(\beta) = e(\beta)$ Thus, we have that $\alpha + e(\beta) = \delta + \omega^\gamma + e(\beta) \geq y_n + e(x_{n+1})$. □

With these technical lemmas at hand we can now prove Lemma 5.4.

Proof By induction on α. For $\alpha := 0$, we check that $x_n \geq y_n + e(y_{n+1})$. Note that since $xR_n y$ then $x_n \geq y_n + e(x_{n+1})$ and $x_{n+1} \geq y_{n+1}$, then $x_n \geq y_n + e(y_{n+1})$. For $\alpha := \beta + 1$, if $xR_n^{1+\beta+1}y$ then there is $z \in I$ such that $xR_n z$ and $zR_n^{1+\beta}y$. Thus, we have the following:

(i) $x_n \geq z_n + e(z_{n+1})$;

(ii) $z_n \geq y_n + e(y_{n+1}) \cdot (1 + \beta)$.

Therefore, $x_n \geq y_n + e(y_{n+1}) \cdot (1+\beta) + e(z_{n+1})$. Since $e(z_{n+1}) \geq e(y_{n+1})$ then $x_n \geq y_n + e(y_{n+1}) \cdot (1+\beta) + e(y_{n+1})$ i.e. $x_n \geq y_n + e(y_{n+1}) \cdot (1+\beta+1)$. For $\alpha \in \mathrm{Lim}$, notice that by IH, we have that $x_n \geq y_n + e(y_{n+1}) \cdot (1+\delta)$ for $\delta < \alpha$. Thus, $x_n \geq y_n + e(y_{n+1}) \cdot (1+\alpha)$. □

Combining Lemma 5.4 and Lemma 5.3 we get the following.

Corollary 5.7 *For $x, y \in I$ and $n < \omega$ we have that*

$$xR_n^{1+\alpha} y \Rightarrow x_n \geq y_n + \big(1 + e(y_{n+1})\big) \cdot (1+\alpha).$$

This corollary takes care of part of the implication from Item (i) to Item (ii) in Theorem 5.2. We will now focus on the implication from Item (iii) to Item (i) but before we do so, we first formulate a simple yet useful lemma.

Lemma 5.8 *For $x, y \in I$, if $xR_{m+1}y$, then $x_m \geq y_m + e(x_{m+1})$.*

Proof Since R_{m+1} is contained in R_m, if $xR_{m+1}y$ then $xR_m y$ and thus by Lemma 5.6, $x_m \geq y_m + e(x_{m+1})$. □

With this technical lemma we can obtain the next step in the direction from Item (iii) to Item (i) in Theorem 5.2.

Lemma 5.9 *For $x, y \in I$ and $n < \omega$ we have that if*

$$\begin{array}{ll} x_n \geq y_n + \big(1 + e(y_{n+1})\big) \cdot (1+\alpha) & \textit{and,} \\ x_m > y_m \textit{ for } m < n & \textit{and,} \\ x_m \geq y_m \textit{ for } m > n. & \end{array}$$

then

$$xR_n^{1+\alpha} y.$$

Proof We use Lemma 5.1 whence are done if we can find for each $\beta < 1+\alpha$ there exists a collection $\{x^\gamma\}_{\gamma<\beta}$ so that

(i) $xR_n x^\gamma$ for any $\gamma < \beta$,

(ii) $x_0 = y$ and,

(iii) for any $\gamma' < \gamma < \beta$ we have $x^\gamma R_n x^{\gamma'}$.

We define x^γ uniformly as follows. We define $x^0 := y$ and

$$x_m^{1+\gamma} \;:=\; \begin{cases} y_m & \text{in case } m > n, \\ y_m + \big(1 + e(y_{n+1})\big) \cdot (1+\gamma) & \text{in case } m = n, \\ y_m + e(y_{m+1}) & \text{in case } m < n. \end{cases}$$

We make a collection of simple observations:

i Each x^γ is an element of I for any $\gamma < \alpha$ since $x_{m+1}^\gamma \leq \ell(x_m^\gamma)$ for any m;

ii We now see that $xR_n x^\gamma$ for each $\gamma < \alpha$. For $m > n$ we obviously have that $x_m \geq x_m^\gamma$ and also $x_n > x_n^\gamma$ is clear. By induction we see that $x_m > x_m^\gamma$

using Lemma 5.8 and the fact that e is a strictly monotonously growing ordinal function;

iii $x_0 = y$ by definition;

iv By strict monotonicity of e, we see that for any $\gamma' < \gamma < \alpha$ we have $x^{\gamma} R_n x^{\gamma'}$.

□

We are now ready to prove Theorem 5.2.

Proof From Item (ii) to Item (iii) is easy and from Item (iii) to Item (i) is Lemma 5.9 so we focus on the remaining implication.

As mentioned before, half of the implication from Item (i) to Item (ii) follows from Corollary 5.7 so that it remains to show that $xR_n^{1+\alpha}y \Rightarrow xR_{n-1}^{e(1+\alpha)}y$. For $n = 0$ this is trivial and in case $n \neq 0$ we reason as follows.

Since $xR_n^{1+\alpha}y$ we get in particular that $x_n \geq y_n + 1 + \alpha$. Thus, by Lemma 5.8 we see

$$x_{n-1} \geq y_{n-1} + e(x_n) \geq y_{n-1} + e(y_n + 1 + \alpha).$$

Now using the fact (Lemma 5.5) that $e(y_n + 1 + \alpha) = e(y_n) \cdot e(1 + \alpha)$ we see, making a case distinction whether $y_n = 0$ or not and using that $e(1 + \alpha)$ is a limit ordinal, that

$$x_{n-1} \geq y_{n-1} + (1 + e(y_n) \cdot (1 + e(1 + \alpha)).$$

The result now follows from an application of Lemma 5.9. □

6 Definable sets

In this section we shall define a translation between formulas in MNF and Ignatiev sequences with finite support as well as a way of characterizing subsets of I. Moreover, we shall see how some of these subsets of I can be related to the extensions of formulas.

Definition 6.1 Let $\psi := \langle n_0^{\alpha_0} \rangle \top \wedge \ldots \wedge \langle n_k^{\alpha_k} \rangle \top \in \mathsf{MNF}$. By x_ψ we denote the sequence $\langle \pi_i(\psi) \rangle_{i<\omega}$.

In virtue of Definition 3.6, we can observe that for $\psi \in \mathsf{MNF}$, we have that $x_\psi \in \mathrm{Ig}^\omega$. Furthermore, we shall see that x_φ is the "first" point in I where φ holds. First we can make some simple observations.

Lemma 6.2

(i) *For any* $x \in I$, $x \Vdash \langle n^\alpha \rangle \top$ *iff* $x_n \geq \alpha$;

(ii) *For any* $\psi \in \mathsf{MNF}$, $x_\psi \Vdash \psi$.

Proof The second item follows from the first one and Definition 6.1. For the right-to-left implication of the first item, assume $x_n \geq \alpha > 0$. Therefore, for $i < n$, we have that $x_i > 0$ and for $i' > n$, $x_{i'} \geq 0$. Thus, by Theorem 5.2, $xR_n^\alpha \langle 0 \rangle$ and so $x \Vdash \langle n^\alpha \rangle \top$. For the other direction, assume $x \Vdash \langle n^\alpha \rangle \top$

for $\alpha > 0$. Hence, there is $y \in I$ such that $xR_n^\alpha y$ and $y \Vdash \top$. By Theorem 5.2, $x_n \geq y_n + (1 + e(y_{n+1})) \cdot \alpha$ and so, $x_n \geq \alpha$. The case $\alpha = 0$ is straightforward.□

The following two definitions introduce the extension of Ignatiev sequences and the extension of formulas, respectively.

Definition 6.3 Given $x \in I$, by $[\![x]\!]$ we denote the set of ℓ-sequences which are coordinate-wise at least as big as x. That is, we define $[\![x]\!] := \{y \in I : y_i \geq x_i \text{ for every } i < \omega\}$.

Definition 6.4 Let $\varphi \in \mathbb{F}$. By $[\![\varphi]\!]$ we denote the set of worlds where φ holds i.e. $[\![\varphi]\!] = \{x \in I : x \Vdash \varphi\}$.

The following lemma relates definitions 6.3 and 6.4.

Lemma 6.5 *For any $\varphi \in \mathbb{F}$, there is $x := \langle x_0, \ldots, x_k, 0\rangle \in I$ such that $[\![\varphi]\!] = [\![x]\!]$.*

Proof The proof goes by induction on φ. The base case is trivial. For **the conjunctive case**, let $\varphi = \psi \wedge \chi$. By the I.H. we have that there are y, $z \in I$ such that $[\![\psi]\!] = [\![y]\!]$ and $[\![\chi]\!] = [\![z]\!]$. Moreover, by the I.H. we also have that $y := \langle y_0, \ldots, y_j, 0\rangle$ and $z := \langle z_0, \ldots, z_i, 0\rangle$. Let n be the index of the rightmost non-zero component. Hence we can define x as follows:

- $x_i = \max(y_i, z_i)$ for $i \geq n$;
- $x_i = \min\{\delta : \delta \geq \max(y_i, z_i) \ \&\ l(\delta) \geq x_{i+1}\}$ for $i < n$.

We can easily check that $x \in I$. Next, we check that for any $x' \in I$, we have that $x' \Vdash \psi \wedge \chi$ iff $x' \in [\![x]\!]$. For right-to-left implication, consider $x' \in [\![x]\!]$. Thus, for $k < \omega$, we have that both $x'_k \geq x_k \geq y_k$ and $x'_k \geq x_k \geq z_k$. Thus, $x' \in [\![y]\!] \cap [\![z]\!]$ and so by the I.H. $x' \Vdash \psi \wedge \chi$. For the other direction, consider $x' \in I$ such that $x' \Vdash \psi \wedge \chi$. Clearly, for $i > n$, we have that $x'_i \geq x_i$. We check by induction on k that $x'_{n-k} \geq x_{n-k}$. For the base case, since $x' \Vdash \psi \wedge \chi$, then by the I.H. $x' \in [\![y]\!] \cap [\![z]\!]$ and so $x'_n \geq y_n$ and $x'_n \geq z_n$. Thus, $x'_n \geq \max(y_n, z_n) = x_n$. For the inductive step, by definition of Ignatiev sequences together with the I.H., we have that $l(x'_{n-(k+1)}) \geq x'_{n-k} \geq x_{n-k}$ and since $x' \Vdash \psi \wedge \chi$, then $x'_{n-(k+1)} \geq \max(y_{n-(k+1)}, z_{n-(k+1)})$. Therefore, being $x_{n-(k+1)}$ the minimal ordinal satisfying both conditions, we can conclude that $x'_{n-(k+1)} \geq x_{n-(k+1)}$. Hence, $[\![\psi \wedge \chi]\!] = [\![x]\!]$.

For **the modality case**, let $\varphi := \langle n^\alpha \rangle \psi$ with $\alpha > 0$. Thus, by the I.H. there is $y \in I$ such that $[\![\psi]\!] = [\![y]\!]$ and $y := \langle y_0, \ldots, y_j, 0\rangle$. We can define x as follows:

- $x_i = y_i$ for $i > n$;
- $x_n = y_n + (1 + e(y_{n+1})) \cdot \alpha$;
- $x_i = \min\{\delta : \delta \geq y_i \ \&\ l(\delta) \geq x_{i+1}\}$ for $i < n$.

As in the previous case, we can easily check that $x \in I$. We claim that $[\![x]\!] = [\![\langle n^\alpha \rangle \psi]\!]$. Let $x' \in [\![x]\!]$. By Theorem 5.2 we can see that $xR_n^\alpha y$. Hence, since $x'_i \geq x_i$ for $i < \omega$, $x' R_n^\alpha y$ and so $x' \Vdash \langle n^\alpha \rangle \psi$. For the other inclusion, consider

$x' \in I$ such that $x' \Vdash \langle n^\alpha \rangle \psi$. By the I.H. and Theorem 5.2, we can easily check that for $i > n$, we have that $x'_i \geq x_i$. For $i \leq n$, we proceed by an easy induction on k to see that $z_{n-k} \geq x_{n-k}$. The base case follows directly from Theorem 5.2. For the inductive step, by definition of Ignatiev sequences together with the I.H., we have that $l(x'_{n-(k+1)}) \geq x'_{n-k} \geq x_{n-k}$. Since $x' \Vdash \langle n^\alpha \rangle \psi$, then there is $z \in I$ such that $xR^\alpha_n z$ and $z \Vdash \psi$. Thus, by the I.H., $z \in [\![y]\!]$, and so we have that $x'_{n-(k+1)} > z_{n-(k+1)} \geq y_{n-(k+1)}$. Therefore, we get that $l(x'_{n-(k+1)}) \geq x_{n-k}$ and $x'_{n-(k+1)} > y_{n-(k+1)}$. Thus, since $x_{n-(k+1)}$ is the least ordinal satisfying both conditions, we have that $x'_{n-(k+1)} \geq x_{n-(k+1)}$. □

7 Soundness

To prove the soundness of **TSC**, let us begin by semantically define the entailment between our modal formulas.

Definition 7.1 For any formulas $\varphi, \psi \in \mathbb{F}$, we write $\varphi \models \psi$ iff for all $x \in I$, if $x \Vdash \varphi$ then $x \Vdash \psi$. Analogously, we write $\varphi \equiv_{\mathcal{J}} \psi$ iff for any $x \in I$, we have that $x \Vdash \varphi$ iff $x \Vdash \psi$.

With our notion of semantical entailment we can formulate our soundness theorem.

Theorem 7.2 (Soundness) *For any formulas $\varphi, \psi \in \mathbb{F}$, if $\varphi \vdash \psi$ then $\varphi \models \psi$.*

Proof By induction on the length of a **TSC** proof of $\varphi \vdash \psi$. It is easy to see that the first three rules preserve validity. With respect to the axioms, the first two axioms are easily seen to be valid. The the correctness of reduction axiom is given by Theorem 5.2. The remaining axioms and rules are separately proven to be sound in the remainder of this section. □

We start by proving the soundness of co-additivity axiom i.e.

$$\langle n^\alpha \rangle \langle n^\beta \rangle \varphi \equiv_{\mathcal{I}} \langle n^{\beta+\alpha} \rangle \varphi.$$

Proposition 7.3 *For any $x, z \in I$, $n < \omega$ and $\alpha, \beta < \Lambda$,*

$$\exists y \in I \ \left(xR^\alpha_n y \text{ and } yR^\beta_n z\right) \iff xR^{\beta+\alpha}_n z.$$

Proof We proceed by transfinite induction on α with the base case being trivial. For $\alpha \in \text{Succ}$, let $\alpha := \delta + 1$ for some δ. Therefore:

$$\begin{aligned}
xR^\alpha_n y \text{ and } yR^\beta_n z \ &\Leftrightarrow \ xR^{\delta+1}_n y \text{ and } yR^\beta_n z; \\
&\Leftrightarrow \ \exists u \ \left(xR_n u \ \wedge \ uR^\delta_n y \ \wedge \ yR^\beta_n z\right); \\
&\Leftrightarrow \ \exists u \ \left(xR_n u \ \wedge \ uR^{\beta+\delta}_n z\right), \text{ by the I.H. }; \\
&\Leftrightarrow \ xR^{\beta+\delta+1}_n z; \\
&\Leftrightarrow \ xR^{\beta+\alpha}_n z.
\end{aligned}$$

For $\alpha \in \text{Lim}$, we have that $xR_n^\alpha y$ and $yR_n^\beta z \Leftrightarrow \forall \delta < \alpha \left(xR_n^{1+\delta} y \wedge yR_n^\beta z\right)$ by Proposition 4.7. By the I.H. we obtain $\forall \delta < \alpha\ xR_n^{\beta+1+\delta} z$ and so $xR_n^{\beta+\alpha} z$. □

With this last result, we get the co-additivity of the R_n^α relations. This together with Definition 4.8 gives us the following corollary.

Corollary 7.4 *The co-additivity axiom is sound.*

Proof By Definition 4.8, $x \Vdash \langle n^\alpha \rangle \langle n^\beta \rangle \varphi$ iff there are $y, z \in I$ such that $xR_n^\alpha y$, $yR_n^\beta z$ and $z \Vdash \varphi$. Thus, by Proposition 7.3, $x \Vdash \langle n^\alpha \rangle \langle n^\beta \rangle \varphi$ iff $xR_n^{\beta+\alpha} z$ and $z \Vdash \varphi$ i.e. $x \Vdash \langle n^{\beta+\alpha} \rangle \varphi$. □

Proposition 7.5 *The monotonicity axiom is sound, that is:*

$$\langle n^\alpha \rangle \varphi \models \langle n^\beta \rangle \varphi$$

for $\beta < \alpha$.

Proof With the help Lemma 4.9, Item (iv), we have that if $x \Vdash \langle n^\alpha \rangle \varphi$ then $x \Vdash \langle n^\beta \rangle \varphi$ for β, $0 < \beta < \alpha$. We check that if $x \Vdash \langle n^1 \rangle \varphi$ then $x \Vdash \varphi$ by induction on φ.

The Base and the conjunctive cases are straightforward, so we consider $\varphi := \langle m^\delta \rangle \psi$ and assume $x \Vdash \langle n^1 \rangle \langle m^\delta \rangle \psi$. We make the following case distinction:

- If $n = m$, then by soundness of co-additivity axiom together with Lemma 4.9, Item (iv) we have that $x \Vdash \langle m^\delta \rangle \psi$;
- If $n > m$, then by monotonicity property of $R_n^{1+\alpha}$ together with soundness of co-additivity axiom and Lemma 4.9, Item (iv) we have that $x \Vdash \langle m^\delta \rangle \psi$;
- If $n < m$, then there are $y, z \in I$ such that $x\, R_n\, y\, R_m^\delta\, z$ and $z \Vdash \psi$. Thus, we can easily check that $x\, R_m^\delta\, z$, and so $x \Vdash \langle m^\delta \rangle \psi$.

□

The following proposition establishes the correction of the Schmerl axiom by using the translation between formulas in monomial normal form and Ignatiev sequences.

Proposition 7.6 *The Schmerl axiom is sound i.e.*

$$\langle n^\alpha \rangle (\langle n_0^{\alpha_0} \rangle \top \wedge \psi) \equiv_{\mathcal{J}} \langle n^{e^{n_0-n}(\alpha_0)\cdot(1+\alpha)} \rangle \top \wedge \langle n_0^{\alpha_0} \rangle \top \wedge \psi$$

for $n < n_0$ *and* $\langle n_0^{\alpha_0} \rangle \top \wedge \psi \in \mathsf{MNF}$.

Proof For the left-to-right direction, assume $x \Vdash \langle n^\alpha \rangle (\langle n_0^{\alpha_0} \rangle \top \wedge \psi)$. Thus, by soundness of monotonicity axiom, we have that $x \Vdash \langle n_0^{\alpha_0} \rangle \top \wedge \psi$. Therefore, we only need to check that $x \Vdash \langle n^{e^{n_0-n}(\alpha_0)\cdot(1+\alpha)} \rangle \top$. Notice that $x \Vdash \langle n^\alpha \rangle \langle n_0^{\alpha_0} \rangle \top$ and so there are $y, z \in I$ such that $xR_n^\alpha y R_{n_0}^{\alpha_0} z$. By Theorem 5.2 we have that

$$x_n \geq y_n + (1 + e(y_{n+1})) \cdot \alpha. \tag{1}$$

Also notice that since $yR_{n_0}^{\alpha_0} z$ then $yR_n^{e^{n_0-n}(\alpha_0)} z$ and $yR_{n+1}^{e^{n_0-n+1}(\alpha_0)} z$. Hence by Theorem 5.2 $y_n \geq e^{n_0-n}(\alpha_0)$ and $y_{n+1} \geq e^{n_0-n+1}(\alpha_0)$. Combining this with 1 we get that $x_n \geq e^{n_0-n}(\alpha_0) + \big(1 + e(e^{n_0-n+1}(\alpha_0))\big) \cdot \alpha = e^{n_0-n}(\alpha_0) \cdot (1+\alpha)$. Thus, in particular, we have that $xR_n^{e^{n_0-n}(\alpha_0)\cdot(1+\alpha)}\langle 0 \rangle$ and so, $x \Vdash \langle n^{e^{n_0-n}(\alpha_0)\cdot(1+\alpha)} \rangle \top$.

For the other direction, assume $x \Vdash \langle n^{e^{n_0-n}(\alpha_0)\cdot(1+\alpha)} \rangle \top \wedge \langle n_0^{\alpha_0} \rangle \top \wedge \psi$. Hence, $x \Vdash \langle n^{e^{n_0-n}(\alpha_0)\cdot(1+\alpha)} \rangle \top$ and so, by Lemma 6.2, Item (i), $x_n \geq e^{n_0-n}(\alpha_0) \cdot (1+\alpha) = e^{n_0-n}(\alpha_0) + (1+e^{n_0-n}(\alpha_0)) \cdot \alpha$. Since $\langle n_0^{\alpha_0} \rangle \top \wedge \psi \in \mathsf{MNF}$ consider $y_{\langle n_0^{\alpha_0} \rangle \top \wedge \psi}$. Notice that $\pi_n(\langle n_0^{\alpha_0} \rangle \top \wedge \psi) = e^{n_0-n}(\alpha_0)$, thus by Defintion 6.1 and Theorem 5.2 we can easly check that $xR_n^{\alpha} y_{\langle n_0^{\alpha_0} \rangle \top \wedge \psi}$ and by Lemma 6.2, Item (i), $y_{\langle n_0^{\alpha_0} \rangle \top \wedge \psi} \Vdash \langle n_0^{\alpha_0} \rangle \top \wedge \psi$. Therefore, $x \Vdash \langle n^{\alpha} \rangle (\langle n_0^{\alpha_0} \rangle \top \wedge \psi)$.□

Lastly, we check the soundness of Rule (iv) by applying the relation between definable sets and the extension of Ignatiev sequences proved in Lemma 6.5. This next result concludes the soundness proof of **TSC**.

Proposition 7.7 *If $\varphi \models \psi$ then, for $m < n$:*

$$\langle n^{\alpha} \rangle \varphi \wedge \langle m^{\beta+1} \rangle \psi \models \langle n^{\alpha} \rangle (\varphi \wedge \langle m^{\beta+1} \rangle \psi).$$

Proof Assume $\varphi \models \psi$ and let $x \in I$ such that $x \Vdash \langle n^{\alpha} \rangle \varphi \wedge \langle m^{\beta+1} \rangle \psi$. Since $\varphi \models \psi$, by Lemma 6.5, there are $y, z \in I$ such that $[\![y]\!] = [\![\varphi]\!] \subseteq [\![\psi]\!] = [\![z]\!]$. Let $y', z' \in I$ such that $[\![y']\!] = [\![\langle n^{\alpha} \rangle \varphi]\!]$ and $[\![z']\!] = [\![\langle m^{\beta+1} \rangle \psi]\!]$, and $w \in I$ such that $[\![w]\!] = [\![\varphi \wedge \langle m^{\beta+1} \rangle \psi]\!]$. Since $y \in [\![z]\!]$, we know that $w_i = y_i$ for $i > m$. For the remaining components, we have that:

- $w_m = \max\big(y_m, z'_m\big)$;
- $w_i = \min\{\delta : \delta \geq \max(y_i, z'_i)\ \&\ l(\delta) \geq w_{i+1}\}$ for $i < m$.

On the other hand, since $x \Vdash \langle n^{\alpha} \rangle \varphi \wedge \langle m^{\beta+1} \rangle \psi$, we have the following:

- $x_i \geq y_i$ for $i > n$;
- $x_n \geq y'_n$;
- $x_i \geq \min\{\delta : \delta \geq y'_i\ \&\ l(\delta) \geq x_{i+1}\}$ for $i,\ m < i < n$;
- $x_i \geq \min\{\delta : \delta \geq \max(y'_i, z'_i)\ \&\ l(\delta) \geq x_{i+1}\}$ for $i \leq m$.

It remains to be checked that $xR_n^{\alpha}w$. Clearly, $x_i \geq w_i$ for $i > n$. Also, since $w_n = y_n$, $w_{n+1} = y_{n+1}$ and $x_n \geq y'_n = y_n + \big(1 + e(y_{n+1})\big) \cdot \alpha$ we have that $x_n \geq w_n + \big(1 + e(w_{n+1})\big) \cdot \alpha$. Thus, we need to see that $x_i > w_i$ for $i < n$. For $i,\ m < i < n$, we can easily check that $y'_i > y_i = w_i$, and so $x_i > w_i$. For $i \leq m$, we show by induction on k that $x_{m-k} > w_{m-k}$. For the base case, we can have that $x_{m+1} > w_{m+1}$. Also we can observe that $\max(y'_m, z'_m) \geq \max(y_m, z'_m)$. Therefore $x_m > w_m$. For the inductive step, by the I.H. we have that $x_{m-k} > w_{m-k}$. Again, $\max(y'_{m-(k+1)}, z'_{m-(k+1)}) \geq \max(y_{m-(k+1)}, z'_{m-(k+1)})$, and so $x_{m-(k+1)} > w_{m-(k+1)}$. Hence, in virtue of Theorem 5.2 we get that $xR_n^{\alpha}w$, that is, $x \Vdash \langle n^{\alpha} \rangle (\varphi \wedge \langle m^{\beta+1} \rangle \psi)$. □

Although it is not needed later in this paper, we find it useful to observe that for any $x = \langle x_0, \ldots, x_k, 0 \rangle \in I$ there is $\psi \in \mathsf{MNF}$ so that $[\![x]\!] = [\![\psi]\!] = [\![x_\psi]\!]$. Having finite support is essential since e.g. the Ignatiev sequence $\langle \varepsilon_0, \varepsilon_0, \ldots \rangle \in I$ is not modally definable. To this regard, in [12] it is shown that a universal model for **TSC** can be built by just considering Ignatiev sequences with finite support.

8 Completeness

To establish the completeness of our system, first we need the following proposition that characterizes the non-derivability between formulas in monomial normal form.

Proposition 8.1 *Given φ, $\psi \in \mathsf{MNF}$, if $\varphi \nvdash \psi$ then there is $m_I \in \mathsf{N\text{-}mod}(\psi)$ such that $\pi_{m_I}(\varphi) < \pi_{m_I}(\psi)$.*

Proof This follows directly from Theorem 3.7 and Definition 3.6. □

Now we are ready to prove the completeness of **TSC**.

Theorem 8.2 (Completeness) *Given formulas φ, $\psi \in \mathbb{F}$, if $\varphi \models \psi$, then $\varphi \vdash \psi$.*

Proof By Theorem 3.5, w.l.o.g. let φ, $\psi \in \mathsf{MNF}$ such that $\varphi := \langle n_0^{\alpha_0} \rangle \top \wedge \ldots \wedge \langle n_k^{\alpha_k} \rangle \top$ and $\psi := \langle m_0^{\beta_0} \rangle \top \wedge \ldots \wedge \langle m_j^{\beta_j} \rangle \top$. Reasoning by contraposition, suppose $\varphi \nvdash \psi$. Therefore, by Proposition 8.1, we can conclude that for some $m_I \in \mathsf{N\text{-}mod}(\psi)$, we have that $\pi_{m_I}(\varphi) < \pi_{m_I}(\psi)$. Thus, consider the Ignatiev sequence x_φ. By Lemma 6.2, Item (ii), $x_\varphi \Vdash \varphi$ but $x_\varphi \nVdash \langle m_I^{\beta_I} \rangle \top$. Hence, $x_\varphi \nVdash \psi$ and so $\varphi \not\models \psi$. □

9 A calculus without normal forms

In this section we shall introduce a presentation of **TSC** that makes no use of formulas in monomial normal form. To this purpose, we shall introduce the notions of *increasing worms* and *m-β-ordinals*, and replace Schmerl's axiom by a new principle that establishes the derivability between these new formulas. The system obtained from this replacement is named **TSC***.

Definition 9.1 The set of *increasing worms*, denoted by IW is inductively defined as follows:

i) $\top \in \mathsf{IW}$;

ii) $\langle n^\alpha \rangle \top \in \mathsf{IW}$ for any $n < \omega$ and α, $0 < \alpha < \Lambda$;

iii) if $\langle n^\alpha \rangle A \in \mathsf{IW}$ and $m < n$, then $\langle m^\beta \rangle \langle n^\alpha \rangle A \in \mathsf{IW}$.

Definition 9.2 Let $\langle n^\alpha \rangle A \in \mathsf{IW}$, $m < n$ and $\beta < \Lambda$. By $o_m^\beta(\langle n^\alpha \rangle A)$ we denote the *m-β-ordinal* of $\langle n^\alpha \rangle A$, that is recursively defined as follows:

i) $o_m^\beta(\langle n^\alpha \rangle \top) = e^{n-m}(\alpha) \cdot (1 + \beta)$;

ii) $o_m^\beta(\langle n^\alpha \rangle A) = e^{n-m}(o_n^\alpha(A)) \cdot (1 + \beta)$.

For any $m < \omega$ and $\beta < \Lambda$, we set $o_m^\beta(\top)$ to be zero.

By **TSC*** we denote the system obtained by substituting in **TSC** the Schmerl axiom by the following principle:

$$\text{IW axioms: } \langle n^{\alpha} \rangle A \equiv \langle n^{o_n^{\alpha}(A)} \rangle \top \ \wedge \ A$$

for $\langle n^{\alpha} \rangle A \in \mathsf{IW}$.

9.1 MNF's and IW's

To prove the equivalence between both systems, first we shall see how formulas in monomial normal form and increasing worms are related. Therefore, in the following lemmata we state how every formula in monomial normal form is equivalent to an increasing worm, modulo **TSC***, and likewise, that every increasing worm is **TSC**-equivalent to a formula in monomial normal form.

From now on we will use the following notation: given $\varphi,\ \psi \in \mathbb{F}$, we write $\varphi \vdash_{\mathbf{TSC}} \psi$ ($\varphi \vdash_{\mathbf{TSC}^*} \psi$) to denote that the sequent $\varphi \vdash \psi$ is derivable in **TSC** (**TSC***). Analogously, we use $\varphi \equiv_{\mathbf{TSC}} \psi$ ($\varphi \equiv_{\mathbf{TSC}^*} \psi$) to denote that both $\varphi \vdash \psi$ and $\psi \vdash \varphi$ are derivable in **TSC** (**TSC***).

Lemma 9.3 *For every* $\psi := \langle n_0^{\alpha_0} \rangle \top \wedge \ldots \wedge \langle n_k^{\alpha_k} \rangle \top \in \mathsf{MNF}$ *there is an* $A \in \mathsf{IW}$ *such that:*

(i) $A \equiv_{\mathbf{TSC}^*} \psi$;

(ii) $A := \langle n_0^{\beta_0} \rangle \ldots \langle n_k^{\beta_k} \rangle \top$ *where:*
   (a) $\alpha_k = \beta_k$ *and*
   (b) $\alpha_i = o_{n_i}^{\beta_i}(\langle n_{i+1}^{\beta_{i+1}} \rangle \ldots \langle n_k^{\beta_k} \rangle \top)$ *for* $i,\ 0 \leq i < k$.

Proof By induction on k. The base case is trivial and the inductive case follows from the I.H. and the IW axiom. □

Lemma 9.4 *For any* $A := \langle n_0^{\beta_0} \rangle \ldots \langle n_k^{\beta_k} \rangle \top \in \mathsf{IW}$ *there is a unique* $\psi \in \mathsf{MNF}$ *such that:*

(i) $\psi \equiv_{\mathbf{TSC}} A$;

(ii) $\psi := \langle n_0^{\alpha_0} \rangle \top \wedge \ldots \wedge \langle n_k^{\alpha_k} \rangle \top$ *where:*
   (a) $\beta_k = \alpha_k$ *and*
   (b) $\alpha_i = o_{n_i}^{\beta_i}(\langle n_{i+1}^{\beta_{i+1}} \rangle \ldots \langle n_k^{\beta_k} \rangle \top)$ *for* $i,\ 0 \leq i < k$.

Proof By induction k. The base case is straightforward. The inductive step follows from the Schmerl axiom together with the I.H. □

9.2 Equivalence between TSC* and TSC

With these results established in the previous subsection, we are ready to prove the equivalence of both systems by checking the interderivability of Schmerl and IW axioms.

Proposition 9.5 *The Schmerl axiom is derivable in* ***TSC**** *i.e.*

$$\langle n^{\alpha} \rangle (\langle n_0^{\alpha_0} \rangle \top \ \wedge \ \psi) \equiv_{\mathbf{TSC}^*} \langle n^{e^{n_0 - n}(\alpha_0) \cdot (1+\alpha)} \rangle \top \wedge \langle n_0^{\alpha_0} \rangle \top \ \wedge \ \psi$$

for $n < n_0$ *and* $\langle n_0^{\alpha_0} \rangle \top \wedge \psi \in \mathsf{MNF}$.

Proof By Lemma 9.3 we have that:

$$\langle n^\alpha \rangle (\langle n_0^{\alpha_0} \rangle \top \wedge \psi) \equiv_{\mathbf{TSC}^*} \langle n^\alpha \rangle \langle n_0^{\beta_0} \rangle A$$

with $\langle n_0^{\alpha^0} \rangle \top \wedge \psi \equiv_{\mathbf{TSC}^*} \langle n_0^{\beta_0} \rangle A \in \mathsf{IW}$. Thus, by the IW axiom, we get that

$$\langle n^\alpha \rangle (\langle n_0^{\alpha_0} \rangle \top \wedge \psi) \equiv_{\mathbf{TSC}^*} \langle n^{o_n^\alpha (\langle n_0^{\beta_0} \rangle A)} \rangle \top \wedge \langle n_0^{\beta_0} \rangle A,$$

and so

$$\langle n^\alpha \rangle (\langle n_0^{\alpha_0} \rangle \top \wedge \psi) \equiv_{\mathbf{TSC}^*} \langle n^{o_n^\alpha (\langle n_0^{\beta_0} \rangle A)} \rangle \top \wedge \langle n_0^{\alpha_0} \rangle \top \wedge \psi.$$

Since $o_n^\alpha(\langle n_0^{\beta_0} \rangle A) = e^{n_0 - n}(o_{n_0}^{\beta_0}(A)) \cdot (1+\alpha)$ and by Lemma 9.3, $o_{n_0}^{\beta_0}(A) = \alpha_0$, we can conclude that $o_n^\alpha(\langle n_0^{\beta_0} \rangle A) = e^{n_0 - n}(\alpha_0) \cdot (1+\alpha)$ and therefore:

$$\langle n^\alpha \rangle (\langle n_0^{\alpha_0} \rangle \top \wedge \psi) \equiv_{\mathbf{TSC}^*} \langle n^{e^{n_0 - n}(\alpha_0)\cdot(1+\alpha)} \rangle \top \wedge \langle n_0^{\alpha_0} \rangle \top \wedge \psi.$$

□

Proposition 9.6 *The IW axiom is derivable in* ***TSC*** *i.e.*

$$\langle n^\alpha \rangle A \equiv_{\boldsymbol{TSC}} \langle n^{o_n^\alpha (A)} \rangle \top \wedge A$$

for $\langle n^\alpha \rangle A \in \mathsf{IW}$.

Proof By induction on A with base case being trivial. For the inductive step, let $A := \langle n_0^{\beta_0} \rangle A'$. By Lemma 9.4, we have that

$$\langle n^\alpha \rangle A \equiv_{\mathbf{TSC}} \langle n^\alpha \rangle (\langle n_0^{o_{n_0}^{\beta_0}(A')} \rangle \top \wedge \psi)$$

for $\langle n_0^{\beta_0} \rangle A' \equiv_{\mathbf{TSC}} \langle n_0^{o_{n_0}^{\beta_0}(A')} \rangle \top \wedge \psi \in \mathsf{MNF}$, and so by Schmerl's axiom:

$$\langle n^\alpha \rangle A \equiv_{\mathbf{TSC}} \langle n^{e^{n_0 - n}(o_{n_0}^{\beta_0}(A'))\cdot(1+\alpha)} \rangle \wedge \langle n_0^{o_{n_0}^{\beta_0}(A')} \rangle \top \wedge \psi.$$

Thus,

$$\langle n^\alpha \rangle A \equiv_{\mathbf{TSC}} \langle n^{o_n^\alpha (\langle n_0^{\beta_0} \rangle A')} \rangle \wedge \langle n_0^{o_{n_0}^{\beta_0}(A')} \rangle \top \wedge \psi,$$

that is, $\langle n^\alpha \rangle A \equiv_{\mathbf{TSC}} \langle n^{o_n^\alpha (\langle n_0^{\beta_0} \rangle A')} \rangle \wedge A$. □

Corollary 9.7 *For any* $\varphi, \psi \in \mathbb{F}$,

$$\varphi \vdash_{\boldsymbol{TSC}} \psi \iff \varphi \vdash_{\boldsymbol{TSC}^*} \psi.$$

Proof By induction on the length of the proof. It follows immediately from Propositions 9.5 and 9.6. □

References

[1] Beklemishev, L. D., *Provability logics for natural Turing progressions of arithmetical theories*, Studia Logica **50** (1991), pp. 109–128.

[2] Beklemishev, L. D., *Iterated local reflection vs iterated consistency*, Annals of Pure and Applied Logic **75** (1995), pp. 25–48.

[3] Beklemishev, L. D., *Proof-theoretic analysis by iterated reflection*, Archive for Mathematical Logic **42** (2003), pp. 515–552.

[4] Beklemishev, L. D., *Provability algebras and proof-theoretic ordinals, I*, Annals of Pure and Applied Logic **128** (2004), pp. 103–124.

[5] Beklemishev, L. D., *Reflection principles and provability algebras in formal arithmetic*, Uspekhi Matematicheskikh Nauk **60** (2005), pp. 3–78, in Russian. English translation in: Russian Mathematical Surveys, 60(2): 197–268, 2005.

[6] Beklemishev, L. D., *Calibrating provability logic*, *Advances in Modal Logic*, **9** (2012), pp. 89–94.

[7] Beklemishev, L. D., *Positive provability logic for uniform reflection principles*, ArXiv: **1304.4396 [math.LO]** (2013).

[8] Dashkov, E. V., *On the positive fragment of the polymodal provability logic GLP*, Mathematical Notes **91** (2012), pp. 318–333.

[9] Fernández-Duque, D. and J. J. Joosten, *Hyperations, Veblen progressions and transfinite iteration of ordinal functions*, Annals of Pure and Applied Logic **164** (2013), pp. 785–801.

[10] Gödel, K., *Über formal unentscheidbare Sätze der Principia Mathematica und verwandter Systeme, I*, Monatshefte für Mathematik und Physik **38** (1931), pp. 173–198.

[11] Hájek, P. and P. Pudlák, "Metamathematics of First Order Arithmetic," Springer-Verlag, Berlin, Heidelberg, New York, 1993.

[12] Hermo-Reyes, E., *A finitely supported frame for the Turing Schmerl calculus*, ArXiv **1804.10452 [math.LO]** (2018).

[13] Hermo-Reyes, E. and J. J. Joosten, *The logic of Turing progressions*, ArXiv **1604.08705 [math.LO]** (2017).

[14] Japaridze, G., *The polymodal provability logic*, in: *Intensional Logics and Logical Structure of Theories: Materials from the Fourth Soviet-Finnish Symposium on Logic*, Metsniereba, Tbilisi, 1988 (in Russian).

[15] Joosten, J. J., *Turing-Taylor expansions for arithmetic theories*, Studia Logica **104** (2016), pp. 1225–1243.

[16] Schmerl, U. R., *A fine structure generated by reflection formulas over primitive recursive arithmetic*, in: *Logic Colloquium '78 (Mons, 1978)*, Stud. Logic Foundations Math. **97**, North-Holland, Amsterdam, 1979 pp. 335–350.

[17] Turing, A., *Systems of logics based on ordinals*, Proceedings of the London Mathematical Society **45** (1939), pp. 161–228.

The Temporal Logic of Two-Dimensional Minkowski Spacetime with Slower-Than-Light Accessibility Is Decidable

Robin Hirsch [1] Brett McLean [2]

Department of Computer Science
University College London
Gower Street, London WC1E 6BT

Abstract

We work primarily with the Kripke frame consisting of two-dimensional Minkowski spacetime with the irreflexive accessibility relation 'can reach with a slower-than-light signal'. We show that in the basic temporal language, the set of validities over this frame is decidable. We then refine this to PSPACE-complete. In both cases the same result for the corresponding reflexive frame follows immediately. With a little more work we obtain PSPACE-completeness for the validities of the Halpern–Shoham logic of intervals on the real line with two different combinations of modalities.

Keywords: temporal logic, basic temporal language, Minkowski spacetime, frame validity, Halpern–Shoham logic.

1 Introduction

Minkowski spacetime refers to the flat spacetime of special relativity, where in any inertial coordinates $(\boldsymbol{r}, ct)$, light travels in straight lines at a 45° angle to the time axis. To view this as a Kripke frame, there are at least four natural accessibility relations to chose from: reflexive and irreflexive versions of 'can reach with a lightspeed-or-slower signal' and 'can reach with a slower-than-light signal'. See below for the formal definitions in the two-dimensional case.

For the basic *modal* language, Goldblatt found that, regardless of the number of spatial dimensions, both reflexive choices produce the logic **S4.2** (reflexivity, transitivity, confluence) [5]. Shapirovsky and Shehtman proved the irreflexive slower-than-light logic is **OI.2**—transitivity, seriality, confluence, *two-density* (see Section 2)—again regardless of dimension [15]. Both **S4.2** and **OI.2** are PSPACE-complete [13]. In contrast, Shapirovsky has shown that with irreflexive *exactly lightspeed* accessibility, validity is undecidable [14].

[1] r.hirsch@ucl.ac.uk

[2] b.mclean@cs.ucl.ac.uk

For the basic *temporal* language, the problems of axiomatising and determining the complexity of the validities of these frames had all been open for decades—Shehtman recommended an investigation of the two-dimensional case (one time and one space dimension) in the concluding remarks of [16]. In the two-dimensional case, starting with coordinates (r, ct), we may rotate the axes through 45° to get coordinates (x, y), where $x = \frac{1}{\sqrt{2}}(ct + r)$, $y = \frac{1}{\sqrt{2}}(ct - r)$. With these coordinates the reflexive lightspeed-or-slower relation $\leq$, and the irreflexive slower-than-lightspeed relation $\prec$, are given by

$$(x, y) \leq (x', y') \iff x \leq x' \wedge y \leq y', \quad (x, y) \prec (x', y') \iff x < x' \wedge y < y \quad (1)$$

where $<, \leq$ on the right are the usual irreflexive/reflexive orderings of the reals, and $<, \preceq$ are obtained from $\leq, \prec$ by deleting/adding the identity, respectively.

Recently, Hirsch and Reynolds managed to show that for this two-dimensional case, with either reflexive or irreflexive lightspeed-or-slower accessibility, the validity problem is PSPACE-complete [7]. However, they were unable to obtain decidability/complexity results for slower-than-light accessibility. Indeed decidability appears in item (2) in the list of open problems at the end of their paper. In this paper we solve Hirsch and Reynolds' problem, eventually proving the following.

Theorem 6.1 *On the frame consisting of two-dimensional Minkowski spacetime equipped with the irreflexive slower-than-light accessibility relation, the set of validities of the basic temporal language is* PSPACE*-complete. The same is true with reflexive slower-than-light accessibility.*

The proof of Theorem 6.1 follows that of [7] very closely. The only additional insight needed is that all pertinent information about the behaviour of a valuation on a light-line can be captured in a finite way—see Definition 2.3—, despite all of a light-line's points being mutually inaccessible.

The proof of the main result is structured as follows.

(i) Given a fixed formula ϕ whose satisfiability is to be determined, we define the maximal consistent sets of subformulas/negated subformulas of ϕ, where consistency is with respect to the class of all temporal frames.

(ii) We define *surrectangles* by recording a maximal consistent set at each point of a rectangle, together with some information about the maximal consistent sets holding near but beyond the boundaries of the rectangle.

(iii) Starting in Section 3, we define *biboundaries*, also based on the maximal consistent sets. Biboundaries have finite specifications and the intuition for them is as a record of the information contained near the boundary of a surrectangle, plus a little from its interior. Indeed we define the biboundary ∂^s determined by a given surrectangle s.

(iv) We define three operations on biboundaries: *joins*, *limits*, and *shuffles*, and we define the set of *fabricated biboundaries* to be those biboundaries formed by iterating these operations, starting from certain basic biboundaries. This is all computable, and the iterative procedure must terminate, for the biboundaries are finite in number.

(v) We show that every fabricated biboundary is given by ∂^s for some surrectangle s, by describing a recursive construction of s. (Section 4)

(vi) Conversely, we show that for every surrectangle s, the biboundary ∂^s is fabricated. (Section 5)

(vii) We deduce the decidability of the validity problem over our frame.

(viii) In Section 6, we give a procedure for deciding if a biboundary is fabricated using polynomial space, and also note the satisfiability task is PSPACE-hard. Consequently our result that validity is decidable is refined to validity being PSPACE-complete.

In [6], Halpern and Shoham introduced a family of modal logics in which the entities under discussion are intervals. These logics have subsequently come to be highly influential, and their axiomatisability, decidability, and complexity extensively studied [19,2,9,4,11,10,3]. In Section 7, we describe how the proof of the complexity of validity for two-dimensional Minkowski spacetime can be adapted to prove PSPACE-completeness of the temporal logic of intervals where the accessibility relation is `overlaps` $\cup$ `meets` $\cup$ `before`, or its reflexive closure.

2 Preliminaries and surrectangles

We often, but not exclusively, follow the terminology and notation of [7]. We take as primitive the propositional connectives $\neg$ and $\vee$, and modal operators $\mathbf{F}$ and $\mathbf{P}$; the usual abbreviations apply. The semantics is the usual semantics on temporal frames—Kripke frames for which the accessibility relation for $\mathbf{P}$ is the converse of that for $\mathbf{F}$. An example of a formula that is valid over two-dimensional slower-than-light frames but not over irreflexive lightspeed-or-slower frames is the two-density formula

$$(\mathbf{F}p \wedge \mathbf{F}q) \to \mathbf{F}(\mathbf{F}p \wedge \mathbf{F}q)$$

asserting that if x precedes both z_1 and z_2, then there exists y that is between x and z_1 *and* between x and z_2.

Throughout, ϕ is a fixed formula whose satisfiability is to be determined. The **closure** $\mathrm{Cl}(\phi)$ of ϕ is the set of all subformulas and negated subformulas of ϕ. A **maximal consistent set** is a subset of $\mathrm{Cl}(\phi)$ that is satisfiable in some temporal frame and is maximal with respect to the inclusion ordering, subject to the satisfiability constraint. We denote the set of maximal consistent sets by MCS. It is well known that satisfiability in a temporal frame is decidable, indeed PSPACE-complete [17]. Hence the set of all maximal consistent sets of ϕ is a (total) computable function of ϕ. The relation on MCS given by

$$\begin{aligned} m \lesssim n \iff \quad & \forall \mathbf{F}\psi \in \mathrm{Cl}(\phi)\ ((\psi \in n \to \mathbf{F}\psi \in m) \wedge (\mathbf{F}\psi \in n \to \mathbf{F}\psi \in m)) \\ & \wedge \forall \mathbf{P}\psi \in \mathrm{Cl}(\phi)\ ((\psi \in m \to \mathbf{P}\psi \in n) \wedge (\mathbf{P}\psi \in m \to \mathbf{P}\psi \in n)) \end{aligned}$$

is transitive, and so defines a preorder on reflexive elements. We call a $\lesssim$-equivalence class a **cluster** and write $\leq$ for the partial order on clusters induced by $\lesssim$. The notation $<$ means $\leq$ but not equal. We extend these notations to compare a cluster c and a (not necessarily $\lesssim$-reflexive) maximal consistent set m as follows. Write $m \leq c$ if for all $n \in c$ we have $m \lesssim n$, and write $m < c$ if for all $n \in c$ we have $(m \lesssim n \wedge m \neq n)$. Similarly for $c \leq m$ and $c < m$.

Definition 2.1 A formula of the form $\mathbf{F}\psi$ is a **future defect** of a maximal consistent set m if $\mathbf{F}\psi \in m$. A future defect of a set S of maximal consistent sets is a future defect of any member of S *unless* we explicitly identify S as a cluster. The formula $\mathbf{F}\psi$ is a future defect of a cluster c if $\mathbf{F}\psi$ is contained in some $m \in c$, but, for all $n \in c$, we have $\psi \notin n$. A future defect $\mathbf{F}\psi$ is **passed up** to a set S of maximal consistent sets if either ψ or $\mathbf{F}\psi$ belongs to some $m \in S$. A **past defect** is defined similarly.

Recall from (1) the ordering $\prec$ and the associated 'can reach with a slower-than-light signal' Kripke frame $(\mathbb{R}^2, \prec)$. Subsets of $\mathbb{R}^2$ inherit the same ordering. The notation $\uparrow \boldsymbol{x}$ denotes the set $\{\boldsymbol{y} \mid \boldsymbol{x} \prec \boldsymbol{y}\}$. (The set $\boldsymbol{y}$ is drawn from should be clear.) We define the operations $\vee$ and $\wedge$ by

$$(x, y) \vee (x', y') = (\max\{x, x'\}, \max\{y, y'\}),$$
$$(x, y) \wedge (x', y') = (\min\{x, x'\}, \min\{y, y'\}).$$

(According to lightspeed-or-slower accessibility $\vee$ is join and $\wedge$ is meet). We also define the partial order $\lhd$ on $\mathbb{R}^2$ by $(x, y) \lhd (x', y') \iff x \leq x', y \geq y'$, and $(x, y) \neq (x', y')$. When this holds we say that (x, y) is 'northwest' of (x', y').

The **diagonal dual** of a condition c on a frame/model whose domain is a subset of $\mathbb{R}^2$ is the condition that c holds on the frame/model obtained by swapping the x- and y-axes (which will not affect accessibility). The **temporal dual** is the result of reversing the x-axis and reversing the y-axis (which reverses accessibility), and also swapping $\mathbf{F}$ and $\mathbf{P}$ in formulas. When we say 'all duals' we mean the diagonal dual, the temporal dual, and the diagonal temporal dual.

Suppose we have a preorder-preserving map f from a subset of $\mathbb{R}^2$ to MCS and that f is defined on $\mathcal{U} \cap \uparrow \boldsymbol{x}$ for some $\mathcal{U}$ an open neighbourhood (in $\mathbb{R}^2$) of $\boldsymbol{x}$. Then it is easy to see there is a unique cluster c such that there exists a neighbourhood $\mathcal{U}'$ of $\boldsymbol{x}$ such that f only takes values in c on $\mathcal{U}' \cap \uparrow \boldsymbol{x}$. We denote this cluster by $f^+(\boldsymbol{x})$. The analogous value for $\downarrow \boldsymbol{x}$ we denote $f^-(\boldsymbol{x})$.

Let $(x, y), (x', y)$ be two distinct points on a horizontal line segment in the domain of the preorder-preserving map f. Then $(x, y) \not\prec (x', y)$ so we do not know the relation between $f(x, y)$ and $f(x', y)$. However, it is easy to see that $x \leq x' \implies f^+(x, y) \lesssim f^+(x', y)$. Similarly for f^-. If f^+ is constantly equal to c on an open line segment l and defined at the left end of l then f^+ also equals c there.[3] Similarly for f^- with right ends. All diagonal duals of statements in this paragraph (that is, statements for vertical lines) hold similarly.

The following concept will be used in definitions that follow, and informs the way we think of a bi-trace (Definition 2.3) as specifying behaviour *on* a horizontal or vertical line segment.

[3] For line segments, 'open' and 'closed' have their usual meanings of 'excludes end points' and 'includes end points', respectively.

Definition 2.2 Let c^- and c^+ be clusters. The **interpolant** of c^- and c^+ is the set of all $m \in \mathsf{MCS}$ with $c^- \leq m \leq c^+$ such that all future defects of m are passed up to c^+, and all past defects of m are passed down to c^-.

Definition 2.3 A **bi-trace** (of length n) is two sequences $c_0^+ \leq \cdots \leq c_n^+$ and $c_0^- \leq \cdots \leq c_n^-$ of clusters, and one sequence $b_1, \ldots, b_n \in \mathsf{MCS}$, with the following constraints.

- For each $i \leq n$ we have $c_i^- \leq c_i^+$, and the interpolant of c_i^- and c_i^+ is nonempty.
- For each $i < n$ either $c_i^- < c_{i+1}^-$ or $c_i^+ < c_{i+1}^+$.
- For each $i < n$ we have $c_i^- \leq b_{i+1} \leq c_{i+1}^+$.
- For each $i < n$ all future defects of b_{i+1} are passed up to c_{i+1}^+ and all past defects passed down to c_i^-.

Formally, we consider a bi-trace to be the interleaving of its three sequences, with the advantage that we can use the notation $(c_0^-, c_0^+, b_1, c_1^-, \ldots, c_n^+)$ to indicate a bi-trace. We call the c_i^+'s the **upper clusters**, the c_i^-'s the **lower clusters**, c_0^+ and c_0^- the **initial clusters**, and c_n^+ and c_n^- the **final clusters**. Each pair c_i^-, c_i^+ is a **cluster pair** of the bi-trace, and each b_i a **transition value**. The top part of Figure 1 suggests how to visualise a bi-trace. There are only finitely many bi-traces (because of the second condition in their definition).

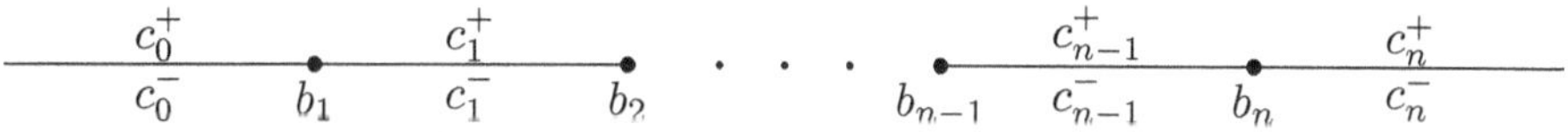

Figure 1. A bi-trace

Definition 2.4 Let $t_1 = (c_0^-, c_0^+, b_1, \ldots, c_n^+)$ and $t_2 = (e_0^-, e_0^+, d_1, \ldots, e_k^+)$ be bi-traces, and $a \in \mathsf{MCS}$. Then $t_1 + a + t_2$ is defined and equal to $(c_0^-, c_0^+, b_1, \ldots, c_n^+, a, e_0^-, e_0^+, d_1, \ldots, e_k^+)$ if this is a bi-trace. It is also defined if the final clusters of t_1 equal the initial clusters of t_2, and a is in the interpolant of c_n^- and c_n^+, in which case it equals $(c_0^-, c_0^+, b_1, \ldots, c_n^+, d_1, \ldots, e_k^+)$.

We now define surrectangles. Intuitively, the domain of a surrectangle is a rectangle plus infinitesimally more beyond any closed edges of the rectangle (a 'surreal rectangle'), and a surrectangle records a valuation on this domain.

Definition 2.5 A **rectangle** is a product of two intervals of $\mathbb{R}$ (with unbounded and single-point intervals both allowed); it is **degenerate** if either interval is a single point, otherwise it is **nondegenerate**. An **edge** of a rectangle R is an edge of the closure of R in $\mathbb{R}^2$ and is not considered to include its end points; a **closed edge** of R is an edge of R contained in R. An **upper edge** of R is either a horizontal edge with maximal vertical component, or a vertical edge with maximal horizontal component, a lower edge is defined dually. A rectangle is **open/closed** if it is open/closed in $\mathbb{R}^2$. The notation $[\boldsymbol{b}, \boldsymbol{t}]$, for points $\boldsymbol{b} = (b_1, b_2)$ and $\boldsymbol{t} = (t_1, t_2)$, signifies the closed rectangle $\{(x, y) \in \mathbb{R}^2 \mid b_1 \leq x \leq t_1 \text{ and } b_2 \leq y \leq t_2\}$.

A **surrectangle** consists of the following data.

(1) A preorder-preserving map $f\colon (R, \prec) \to (\mathsf{MCS}, \lesssim)$, for some *nondegenerate* rectangle R. We call R the **rectangle** and f the **core map** of the surrectangle.

(2) For each closed horizontal upper edge e of R with endpoints (x_0, y) and (x', y),[4] a finite sequence $c_0^+ \leq \cdots \leq c_n^+$ of clusters and a sequence $(x_1, y), \ldots, (x_n, y) \in e$, with $x_0 < \cdots < x_n < x_{n+1}$, defining x_{n+1} to be x'. Similar finite sequences for any closed vertical and/or lower edges.[5]

And is required to satisfy the following constraints.

(3) For each closed horizontal upper edge as above, f^- is constant on each open line segment $((x_i, y), (x_{i+1}, y))$ (let this constant cluster be c_i^-), and $(c_0^-, c_0^+, b_1, c_1^-, \ldots, c_n^-, c_n^+)$ forms a bi-trace, where b_i is defined to be $f(x_i, y)$, for each i. Also, all duals of this constraint. (Figure 2 suggests how to visualise a closed edge.)

(4) For any point $\boldsymbol{x} \in R$, if $\mathbf{F}\psi \in f(\boldsymbol{x})$ either
- **resolved internally:** there is $\boldsymbol{y} \in \uparrow \boldsymbol{x}$ such that $\psi \in f(\boldsymbol{y})$,
- **passed upwards:** R has a boundary point $\boldsymbol{y}$ either due north or due east of $\boldsymbol{x}$ such that $\mathbf{F}\psi$ is passed up to $f(\boldsymbol{y})$.

(5) The temporal dual of (4) holds.

In (2), we call $c_0^+, \ldots, c_n^+$ the **supplementary clusters** of e, we call $(x_1, y), \ldots, (x_n, y)$ the **transition points** of e, and we call $f(x_1, y), \ldots, f(x_n, y)$ the **transition values**. Note that the clusters c_i^- are determined by f^- in (3), and for edges not contained in R (for example if the rectangle is unbounded in the corresponding direction) supplementary clusters and transition points are not defined.

Let $\boldsymbol{b}$ and $\boldsymbol{t}$ be respectively the lower-left and upper-right corners of R. Then $f^+(\boldsymbol{b})$ and $f^-(\boldsymbol{t})$ are necessarily defined. The **height** of the surrectangle is the maximum possible length of a chain of clusters (not necessarily in the image of f) from its **lower cluster** $f^+(\boldsymbol{b})$ to its **upper cluster** $f^-(\boldsymbol{t})$. Surrectangles also inherit descriptions such as open/closed from their underlying rectangle.

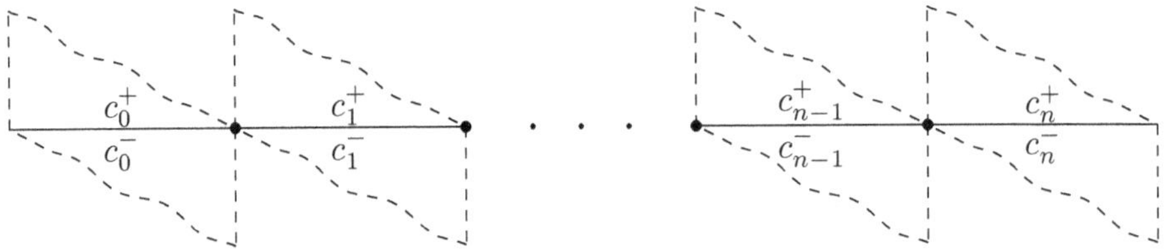

Figure 2. A closed edge of a surrectangle

[4] Here, it could be that $x_0 = -\infty$ and/or $x' = \infty$; this does not present any problems.

[5] For lower edges, these supplementary clusters are the ones denoted c_i^-.

3 Biboundaries

Definition 3.1 A **biboundary** is a partial map ∂ on $\{-, +, b, t, l, r, N, S, E, W\}$. It must be defined on $-$ and $+$, and be cluster-valued there. If ∂ is defined on N, S, E, or W, it is bi-trace-valued there, and if defined on b, t, l, or r, it is MCS-valued there. It is defined on b if and only if it is defined on S and W; similarly for t and N, E, for l and N, W, and for r and S, E. The following conditions must also be satisfied.

(i) If $\partial(b)$ is defined, then $\partial(b) \leq \partial(-)$ and every future defect of $\partial(b)$ is passed up to $\partial(-)$.

(ii) If $\partial(W)$ is defined, then the initial upper cluster of $\partial(W)$ equals $\partial(-)$.

(iii) If $\partial(l)$ is defined, then it is less than or equal to the initial upper cluster of $\partial(N)$, and every future defect of $\partial(l)$ is passed up to that cluster.

(iv) Every future defect of $\partial(+)$ is passed up to the interpolant of the final clusters of $\partial(N)$, to the interpolant of the final clusters of $\partial(E)$, or to $\partial(t)$, with ∂ defined in the places appropriate to the case.

(v) All duals of (i)–(iv) hold (with the evident meaning of duals).

Since there are only finitely many maximal consistent sets, clusters, and bi-traces, there are only finitely many biboundaries. A biboundary is **closed** if it is defined on N, S, E, and W (hence also on b, l, r, and t).

Let s be a surrectangle. The biboundary ∂^s determined by ∂ is defined in the obvious way.

We now define three types of operations on biboundaries: joins, limits, and shuffles. See Figure 3 for visual representations of limits and shuffles.

Definition 3.2 A biboundary ∂ is the **vertical join** of biboundaries ∂_1 and ∂_2, written $\partial_1 \oplus_- \partial_2$, if

- $\partial_1(N)$ and $\partial_2(S)$ are both defined and are equal,
- either $\partial_1(W)$, $\partial_2(W)$, and $\partial(W)$ are all defined, $\partial_1(l) = \partial_2(b)$, and $\partial(W) = \partial_1(W) + \partial_1(l) + \partial_2(W)$, or $\partial_1(W), \partial_2(W)$ and $\partial(W)$ are all undefined; similarly for E,
- ∂ agrees with ∂_1 on b, S, r, and $-$, and with ∂_2 on l, N, t, and $+$.

The diagonal-dual concept is a **horizontal join**, written $\partial_1 \oplus_| \partial_2$.[6]

Definition 3.3 A biboundary ∂^* is the **southeastern limit** of a biboundary ∂_0 using biboundaries $\partial_1, \partial_2, \partial_3$ if

- $\partial_0 = (\partial_2 \oplus_| \partial_3) \oplus_- (\partial_0 \oplus_| \partial_1)$,
- the lower cluster of $\partial_1(E)$ is constantly $\partial_0(+)$,
- the upper cluster of $\partial_2(S)$ is constantly $\partial_0(-)$,
- ∂^* agrees with ∂_0 over $\{-, +, l, W, N\}$,

[6] The $-$ and $|$ subscripts indicate the orientation of the shared edge.

- $\partial^*(S)$, if defined, is a bi-trace where the upper cluster is constantly $\partial_0(-)$,
- $\partial^*(E)$, if defined, is a bi-trace where the lower cluster is constantly $\partial_0(+)$.

A **northwestern limit** is defined dually.

If Δ is a set of biboundaries and there are $\partial_0, \partial_1, \partial_2, \partial_3 \in \Delta$ such that ∂^* is the southeastern limit of ∂_0 using $\partial_1, \partial_2, \partial_3$, then ∂^* is a southeastern limit over Δ. Northwestern limits are dual. We say that ∂^* is a limit over Δ if it is either a southeastern or northwestern limit over Δ.

Definition 3.4 Let Δ be a collection of *closed* biboundaries. The biboundary ∂' is a **shuffle** of Δ if there is a *nonempty* set $M \subseteq \mathsf{MCS}$ such that

(i) if $\partial'(W)$ is defined, then all upper clusters of $\partial'(W)$ equal $\partial'(-)$,

(ii) every future defect $\mathbf{F}\psi$ of $\partial'(-)$ is passed up to some $m \in M$, or there is a $\partial \in \Delta$ such that $\mathbf{F}\psi$ is passed up to either $\partial(b), \partial(l), \partial(r)$, the interpolant of some cluster pair of $\partial(W)$ or $\partial(S)$, or some transition value of $\partial(W)$ or $\partial(S)$,

(iii) for all $\partial \in \Delta$, we have $\partial(t) \leq \partial'(+)$, all future defects of $\partial(t)$ are passed up to $\partial'(+)$, and all upper clusters of $\partial(N)$ equal $\partial'(+)$,

(iv) for all $m \in M$, we have $m \leq \partial'(+)$, and all future defects of every m are passed up to $\partial'(+)$,

(v) all duals of (i), (ii), (iii), and (iv) hold.

Now we are ready to define the biboundaries that we proceed to show are precisely those obtained from surrectangles.

Definition 3.5 A **ground fabricated biboundary** is a biboundary ∂ such that

(i) $\partial(-) = \partial(+)$,

(ii) if $\partial(N)$ is defined, then all lower clusters of $\partial(N)$ equal $\partial(+)$,

(iii) all duals of (ii) hold.

A **fabricated biboundary** is either a ground fabricated biboundary, or a biboundary obtained recursively as the join, limit, or shuffle of fabricated biboundaries.

4 From fabricated biboundaries to surrectangles

In this section we show that every fabricated biboundary is the biboundary obtained from some surrectangle, by describing how to construct such a surrectangle from a given biboundary. We use the recursive structure of fabricated biboundaries as given by their definition.

When we say a function f fills X **densely** with M we mean that $f(\boldsymbol{x}) \in M$ for all $\boldsymbol{x} \in X$, and for each $m \in M$ the set $f^{-1}(m)$ is dense in X. It is clear that if $\mathcal{U}$ is an open subset of $\mathbb{R}^2$ and c is a cluster, then there exists f that fills $\mathcal{U}$ densely with c (and this remains true when restrictions are placed on the behaviour of f outside of $\mathcal{U}$). Similarly when $\mathcal{U}$ is an open line segment.

Further, if for a biboundary ∂ we have a surrectangle satisfying $\partial^s = \partial$, we may assume that for every closed edge e of s and each associated cluster pair c_i^-, c_i^+ between transition points $\boldsymbol{x}_i, \boldsymbol{x}_{i+1}$, the core map of s fills $(\boldsymbol{x}_i, \boldsymbol{x}_{i+1})$ densely with the interpolant of c_i^- and c_i^+. Hence if surrectangles s_1 and s_2 have a common edge e, on which we obtain the same bi-trace and transition points from both surrectangles, then we may assume s_1 and s_2 agree on e.

Lemma 4.1 *Let s be a surrectangle with core map f and let g_1 and g_2 be order-preserving bijections $\mathbb{R} \to \mathbb{R}$. Then there is a surrectangle s' with core map $(x, y) \mapsto f(g_1(x), g_2(y))$, the same supplementary clusters and transition values as s, and a transition point $(g_1^{-1}(x), g_2^{-1}(y))$ for every transition point (x, y) of s. Moreover, s' yields the same biboundary as s.*

The proof of Lemma 4.1 is routine, and omitted. The proofs of the following four lemmas may be found in the appendix. Figure 3 illustrates the proofs of the last two.

Lemma 4.2 (ground biboundaries) *Let ∂ be a ground fabricated biboundary. Then there exists a surrectangle s such that $\partial^s = \partial$.*

Lemma 4.3 (joins) *Let ∂_1 and ∂_2 be biboundaries such that the vertical join $\partial_1 \oplus_- \partial_2$ exists, and suppose there exist surrectangles s_1 and s_2 with $\partial^{s_1} = \partial_1$ and $\partial^{s_2} = \partial_2$. Then there exists a surrectangle s such that $\partial^s = \partial_1 \oplus_- \partial_2$.*

Similarly for horizontal joins.

Lemma 4.4 (limits) *Let ∂^* be a southeastern limit of ∂_0 using $\partial_1, \partial_2, \partial_3$, and suppose there are surrectangles s_0, s_1, s_2, s_3 such that $\partial^{s_i} = \partial_i$, for $i < 4$. Then there exists a surrectangle s^* such that $\partial^{s^*} = \partial^*$. Similarly for northwestern limits.*

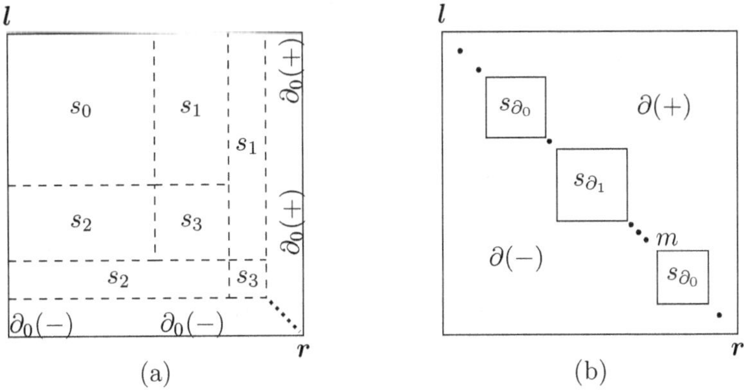

Figure 3. (a) A surrectangle for a southeastern limit of ∂_0 using $\partial_1, \partial_2, \partial_3$, and (b) a surrectangle for a shuffle of $\partial_0, \partial_1, \ldots$

Lemma 4.5 (shuffles) *Let ∂' be a shuffle of Δ, and suppose that for all $\partial \in \Delta$ there is a surrectangle s_∂ with $\partial^{s_\partial} = \partial$. Then there is a surrectangle s' with $\partial^{s'} = \partial'$.*

By Lemma 4.2, the set Δ of biboundaries that can be obtained from surrectangles contains the ground fabricated biboundaries. By Lemma 4.3, Lemma 4.4, and Lemma 4.5, it is closed under joins, limits, and shuffles. Hence Δ contains all fabricated biboundaries. This proves, as promised, the following.

Lemma 4.6 *Let ∂ be a fabricated biboundary. Then there exists a surrectangle s such that $\partial^s = \partial$.*

5 From surrectangles to fabricated biboundaries

In this section we show that every biboundary obtained from a surrectangle is a fabricated biboundary (Lemma 5.6). We do this by induction on the height of the surrectangle.

If a surrectangle s has height 0, then the upper and lower clusters of ∂^s are equal, and from that it is easy to see that all conditions for ∂^s to be a ground fabricated biboundary are satisfied.

In the remainder of this section, we assume that for every surrectangle s of height no greater than N, the biboundary ∂^s is fabricated, and aim to prove the statement for all surrectangles of height $N+1$. First, note that by Lemma 4.1, we only need prove the result for surrectangles with bounded domains. So henceforth we take bounded domains as an assumption.

Let s be a surrectangle with domain R, core map f, northwest corner $\boldsymbol{l}$, southeast corner $\boldsymbol{r}$, lower cluster c^-, and upper cluster c^+. Let $I^- = f^{-1}(c^-)$ be the subset of R that maps to the lower cluster, and $I^+ = f^{-1}(c^+)$. Define Γ^- to be the *closure* (in $\mathbb{R}^2$) of the intersection of the boundary of I^- with the interior of R. Define Γ^+ similarly. Elementary topology shows that Γ^- and Γ^+ are homeomorphic to closed (proper) line segments, meeting the boundary of R only at their endpoints, and linearly ordered by $\lhd$; see [7, Lemmas 2.10 and 2.11] for details. The notation $s\restriction_{[\boldsymbol{x},\boldsymbol{y}]}$ signifies the surrectangle formed in the obvious way by restriction of s to $[\boldsymbol{x},\boldsymbol{y}] \cap R$ (assuming this is nondegenerate). It is straightforward to check that $s\restriction_{[\boldsymbol{x},\boldsymbol{y}]}$ is indeed a surrectangle.

As before, proofs of lemmas are contained in the appendix.

Lemma 5.1 *If Γ^- and Γ^+ are disjoint then ∂^s is fabricated.*

If Γ^- and Γ^+ are *not* disjoint, define a binary relation $\equiv$ over $R \setminus (\downarrow I^- \cup \uparrow I^+)$ by letting $\boldsymbol{x} \equiv \boldsymbol{y} \iff$ for *all* nondegenerate rectangles $[\boldsymbol{w},\boldsymbol{z}] \subseteq [\boldsymbol{x} \wedge \boldsymbol{y}, \boldsymbol{x} \vee \boldsymbol{y}]$ the biboundary $\partial^{s\restriction[\boldsymbol{w},\boldsymbol{z}]}$ is fabricated—clearly reflexive and symmetric, and also transitive (use a join of four biboundaries and the induction hypothesis), so an equivalence relation. By using joins and the induction hypothesis if necessary, we may assume $\boldsymbol{l}, \boldsymbol{r} \in R \setminus (\downarrow I^- \cup \uparrow I^+)$. We aim to show that $\boldsymbol{l} \equiv \boldsymbol{r}$.

Lemma 5.2 *Let $\boldsymbol{x}_0 \lhd \boldsymbol{x}_1 \lhd \ldots \in R \setminus (\downarrow I^- \cup \uparrow I^+)$ be an infinite sequence converging to $\boldsymbol{x}$. If for all i we have $\boldsymbol{y} \equiv \boldsymbol{x}_i$ then $\boldsymbol{y} \equiv \boldsymbol{x}$.*

Let $P = \{\boldsymbol{l}, \boldsymbol{r}\} \cup (\Gamma^- \cap \Gamma^+)$—a closed set, inheriting a linear order $\lhd$ from Γ^-, and a subset of $R \setminus (\downarrow I^- \cup \uparrow I^+)$. Define a binary relation $\approx$ over P as the smallest equivalence relation such that

(a) $\boldsymbol{p} \approx \boldsymbol{q}$ whenever $\boldsymbol{p}$ is an immediate $\lhd$-successor of $\boldsymbol{q}$,

(b) $\boldsymbol{p} \approx \boldsymbol{q}$ whenever $[\boldsymbol{p} \wedge \boldsymbol{q}, \boldsymbol{p} \vee \boldsymbol{q}]$ is degenerate,

(c) all equivalence classes are topologically closed (in $\mathbb{R}^2$, equivalently, in P).

Lemma 5.3 *$\boldsymbol{p} \approx \boldsymbol{q}$ implies $\boldsymbol{p} \equiv \boldsymbol{q}$.*

For any $\approx$-equivalence class E, let $l(E), r(E)$ be the extreme points with respect to $\lhd$ (equal if E is a singleton). As E is closed, $l(E), r(E) \in E$. Write $R(E)$ for the closed rectangle $[l(E) \wedge r(E), l(E) \vee r(E)]$ (a singleton if and only if E is a singleton). Since $\boldsymbol{p} \approx \boldsymbol{q}$ we know $l(E) \equiv r(E)$.

Lemma 5.4 *Either all elements of P are $\approx$-equivalent, or there are uncountably many singleton equivalence classes of $\approx$.*

Let $\boldsymbol{x} = r(E)$, where E is the $\approx$-equivalence class of $\boldsymbol{l}$, and let $\boldsymbol{y} = l(E')$ where E' is the $\approx$-equivalence class of $\boldsymbol{r}$. Since $\boldsymbol{x}$ is the most southeastern point in E, and has no immediate $\lhd$-successor in P, there is no point of Γ^- due south of $\boldsymbol{x}$, similarly no point of Γ^+ due east of $\boldsymbol{x}$. Dual conditions hold for $\boldsymbol{y}$. It follows that f^+ is constantly c^- on the south and west edges of $[\boldsymbol{x} \wedge \boldsymbol{y}, \boldsymbol{x} \vee \boldsymbol{y}]$, and f^- is constantly c^+ on the north and east edges.

Since $\boldsymbol{l} \approx \boldsymbol{x}$ and $\boldsymbol{y} \approx \boldsymbol{r}$ we know that $\boldsymbol{l} \equiv \boldsymbol{x}$ and $\boldsymbol{y} \equiv \boldsymbol{r}$. The next lemma shows that $\boldsymbol{x} \equiv \boldsymbol{y}$. We omit the proof, which is to check each of the conditions of Definition 3.4.

Lemma 5.5 *Suppose the upper cluster on the south and west edges of s is constantly c^-, and the lower cluster on the north and east edges of s is constantly c^+. Then ∂^s is a shuffle of $\partial^{s \restriction R(E)}$ where E ranges over $\approx$-equivalence classes.*

Using joins and the induction hypothesis, this proves $\boldsymbol{l} \equiv \boldsymbol{r}$. Hence we have obtained our goal.

Lemma 5.6 *Let s be any surrectangle. The biboundary ∂^s is fabricated.*

Combining Lemmas 4.6 and 5.6 we have the following.

Lemma 5.7 *A biboundary ∂ is of the form ∂^s for some surrectangle s if and only if ∂ is a fabricated biboundary.*

Now we are in a position to prove our first main result.

Theorem 5.8 *It is decidable whether a formula of the basic temporal language is valid on the frame consisting of two-dimensional Minkowski spacetime equipped with the irreflexive slower-than-light accessibility relation. The same is true with reflexive slower-than-light accessibility.*

Proof Decidability of validity is equivalent to decidability of satisfiability; we prove the latter.

We first show that satisfiability of ϕ on $(\mathbb{R}^2, \prec)$ is equivalent to the existence of an open surrectangle having some point assigned a maximal consistent set containing ϕ. An open surrectangle consists only of its core map f. Given a valuation v on $(\mathbb{R}^2, \prec)$ and a point $\boldsymbol{x}$ at which ϕ holds, define f by $f(\boldsymbol{y}) = \{\psi \in \mathrm{Cl}(\phi) \mid (\mathbb{R}^2, \prec), v, \boldsymbol{y} \models \psi\}$, and then f will be a surrectangle on $\mathbb{R}^2$ with $\phi \in f(\boldsymbol{x})$. Conversely, given such an f, for any propositional variable

p if p appears in ϕ, define $v(p) = \{\boldsymbol{y} \mid p \in f(\boldsymbol{y})\}$, and otherwise define $v(p)$ arbitrarily. Then ϕ holds at $\boldsymbol{x}$ under valuation v.

Let $NE(x, y) = \{(x', y') \in \mathbb{R}^2 \mid x \leq x' \text{ and } y \leq y'\}$ and define $NW(\boldsymbol{x})$, $SE(\boldsymbol{x})$, and $SW(\boldsymbol{x})$ similarly, in the evident way. The existence of an open surrectangle having some point $\boldsymbol{x}$ assigned a maximal consistent set containing ϕ is equivalent to the existence of four surrectangles with domains $NE(\boldsymbol{x})$, $NW(\boldsymbol{x})$, $SE(\boldsymbol{x})$, and $SW(\boldsymbol{x})$, agreeing on their shared edges and at $\boldsymbol{x}$, and with ϕ contained in the maximal consistent set assigned to $\boldsymbol{x}$. By Lemma 5.7, this is in turn equivalent to the existence of four fabricated biboundaries of the appropriate types that match up in the appropriate way.

Since satisfiability of ϕ is equivalent to the existence of a set of four fabricated biboundaries with properties that are easily checked, and the set of all fabricated biboundaries is finite and computable, the satisfiability problem is decidable. This completes the proof of the irreflexive case.

The reflexive case follows by the reduction given by recursively replacing subformulas of ϕ of the form $\mathbf{F}\varphi$ with $\varphi \wedge \mathbf{F}\varphi$, and similarly for $\mathbf{P}$. □

6 A PSPACE procedure for fabricated biboundaries

In this section, we refine the decidability results of Theorem 5.8 to show the validity problems are PSPACE-complete.

Theorem 6.1 *On the frame consisting of two-dimensional Minkowski spacetime equipped with the irreflexive slower-than-light accessibility relation, the set of validities of the basic temporal language is* PSPACE*-complete. The same is true with reflexive slower-than-light accessibility.*

As mentioned in the introduction, for the reflexive frame, the validities of the purely *modal* fragment of the basic temporal language form **S4.2**, and for the irreflexive frame, **OI.2**. These are both known to be PSPACE-complete [13], so the validity problems for the *entire* basic temporal language are PSPACE-hard.

We provide a nondeterministic polynomial space algorithm for satisfiability, for the irreflexive frame. Hence validity is in coNPSPACE. By the Immerman–Szelepcsényi theorem [8,18], coNPSPACE = NPSPACE, and by Savitch's theorem [12], NPSPACE = PSPACE, giving the result. The reflexive case follows by the same reduction as before.

Throughout this section, let the length of ϕ be n. By structural induction, the length of a formula bounds the number of its subformulas, so the cardinality of $\mathrm{Cl}(\phi)$ is linear in n. Hence any maximal consistent set—a subset of $\mathrm{Cl}(\phi)$—can be stored using a linear number of bits, and $|\mathsf{MCS}|$ is at most exponential in n. All pertinent information about any cluster c can also be stored in a linear number of bits, for we only need record $\{\psi \in \mathrm{Cl}(\phi) \mid \exists m \in c : \psi \in m\}$. The maximal length of a chain of distinct clusters or irreflexive members of MCS is also linear in n. Hence any biboundary can be stored using a quadratic number of bits, and the number of biboundaries is exponential in n.

Having nondeterministically chosen a bit string representing some $m \subseteq$

$\mathrm{Cl}(\phi)$—a putative maximal consistent set—the conjunct over all formulas in m is quadratic in n, so its satisfiability can be determined in polynomial space [17]. Hence we can determine if a chosen bit string represents a maximal consistent set using polynomial space.

We can similarly nondeterministically 'guess' sets of the form $\{\psi \in \mathrm{Cl}(\phi) \mid \exists m \in c : \psi \in m\}$ for some cluster c, using polynomial space. To do this, first guess a maximal consistent set m and check it is reflexive; keep m in memory. Then one-by-one for each bit string representing some $m' \subseteq \mathrm{Cl}(\phi)$, check $m' \in \mathsf{MCS}$ and that $m \lesssim m' \lesssim m$. If so, add all elements of m' to an ongoing collection of formulas, discarding m' after each iteration.

From the preceding discussion, it is clear we can also determine if a chosen string of bits represents a *biboundary* using polynomial space.

The algorithm for satisfiability of ϕ first nondeterministically chooses four bit strings and checks they represent four compatible biboundaries, with ϕ at the appropriate corners—performed using polynomial space. Then for each of these in turn, it is checked whether the biboundary is fabricated. The remainder of the proof is devoted to showing that this check can be performed using polynomial space. The procedure to do this is shown in Algorithm 1.

Algorithm 1 Nondeterministic procedure to decide whether ∂ is fabricated

```
procedure FABRICATED(∂) choose either
  option 0
    check ∂ is ground;
  option 1
    choose ∂1, ∂2; check they are biboundaries
    check their join (choose some direction) is ∂; release ∂
    check FABRICATED(∂1); tail-call FABRICATED(∂2)
  option 2
    choose ∂1, ∂2, ∂3, ∂4; check they are biboundaries
    check ∂ is the limit (choose direction) of ∂1 using ∂2, ∂3, ∂4; release ∂
    check FABRICATED(∂2), FABRICATED(∂3); release ∂2, ∂3; check FABRIC-
ATED(∂4); tail-call FABRICATED(∂1)
  option 3
    choose k ∈ {0, 1, ..., n}, ∂1, ..., ∂k; check they are biboundaries
    choose m1, ..., mn; check they are maximal consistent sets
    check ∂ is the shuffle of ∂1, ..., ∂k using m1, ..., mn; release ∂
    for i = 1, ..., k do
      check FABRICATED(∂i)
    end for
end procedure
```

We assume that during execution of Algorithm 1, the formula ϕ is a global constant (and therefore is n too). All choices are made nondeterministically. At a **check** the algorithm fails if the check fails, otherwise it proceeds. A **tail-call** uses tail recursion. The algorithm succeeds if it terminates without failure.

Clearly Algorithm 1 can only succeed for a biboundary ∂ if ∂ is fabricated. Let $S(N)$ be the amount of space required by Algorithm 1 to succeed for any fabricated biboundary of height N. We prove that $S(N) = \mathcal{O}((N+1)n^3)$ (as a function of N and n) by induction on N. Hence Algorithm 1 requires space $\mathcal{O}(n^4)$ to check any biboundary.

If ∂ is a fabricated biboundary of height 0, then it is ground, so success can be achieved by taking the first branch, requiring space $\mathcal{O}(n^2)$.

For the inductive step, we show that $S(N+1) \leq S(N) + \mathcal{O}(n^3)$. The proof has a similar structure to the proof in the previous section. Let ∂ be a fabricated biboundary of height $N+1$, and s be a surrectangle such that $\partial^s = \partial$. Define Γ^-, Γ^+, $\boldsymbol{l}$, and $\boldsymbol{r}$ as in the previous section. Let δ be the space required to store any biboundary. Thus δ is a quadratic function of n, independent of N.

As always, proofs of lemmas are relegated to the appendix.

Lemma 6.2 *If Γ^- and Γ^+ are disjoint, the space required is bounded by $S(N)+\delta$.*

Lemma 6.3 *If Γ^- and Γ^+ intersect only on the boundary of s, the space required is bounded by $S(N) + 3\delta$.*

As in the previous section, let $P = \{\boldsymbol{l}, \boldsymbol{r}\} \cup (\Gamma^- \cap \Gamma^+)$. Additionally, let $\Delta = \{\partial^{s\restriction_{[\boldsymbol{x}\wedge\boldsymbol{y},\boldsymbol{x}\vee\boldsymbol{y}]}} \mid \boldsymbol{x}, \boldsymbol{y} \in P$, and $\boldsymbol{y}$ is an immediate $\lhd$-successor of $\boldsymbol{x}\}$.

Lemma 6.4 *If the biboundary of s is a shuffle of a subset Δ' of Δ, the space required is bounded by $S(N) + 3\delta + n\delta$.*

If $\boldsymbol{l}$ and $\boldsymbol{r}$ are each the limit of elements of P in the interior of s, then the biboundary of s is the shuffle of Δ', using M, where $\Delta' = \{\partial^{s\restriction_{[\boldsymbol{x}\wedge\boldsymbol{y},\boldsymbol{x}\vee\boldsymbol{y}]}} \mid \boldsymbol{x}, \boldsymbol{y} \in P \setminus \{\boldsymbol{l}, \boldsymbol{r}\}$, and $\boldsymbol{y}$ is the successor of $\boldsymbol{x}\}$ and $M = P \setminus \{\boldsymbol{l}, \boldsymbol{r}\}$.

Now suppose only one of $\boldsymbol{l}$, $\boldsymbol{r}$ is a limit of elements of P in the interior of s. Say $\boldsymbol{l}$ is such a limit, but $\boldsymbol{r}$ is not. If $\boldsymbol{r}$ has a direct predecessor $\boldsymbol{r}'$, then $\boldsymbol{r}'$ is in the interior of s. If there are points of P (strictly) due north of $\boldsymbol{r}$, let the northmost one be $\boldsymbol{w}$. Then either $\boldsymbol{w}$ is a limit, and using a join we can reduced to the case of the previous paragraph, or there is a point $\boldsymbol{r}'$ either an immediate predecessor or due west of $\boldsymbol{w}$. In the second case, $\boldsymbol{r}'$ is in the interior of $\boldsymbol{s}$. Similarly if there are points of P due west of $\boldsymbol{r}$. In each case we obtain an $\boldsymbol{r}'$ in the interior of P and the biboundary ∂_{SE} of $s\restriction_{[\boldsymbol{r}\wedge\boldsymbol{r}',\boldsymbol{r}\vee\boldsymbol{r}']}$ can be checked in $S(N) + 3\delta$, by Lemma 6.3.

We are going to obtain the biboundary of s as the join of four biboundaries. In the southeast corner is ∂_{SE}. If the biboundary of $s\restriction_{[\boldsymbol{l}\wedge\boldsymbol{r}',\boldsymbol{l}\vee\boldsymbol{r}']}$ is of reduced height then we are done. Otherwise, the other three biboundaries we use are modifications of restrictions. For the northwest corner, take the biboundary of $s\restriction_{[\boldsymbol{l}\wedge\boldsymbol{r}',\boldsymbol{l}\vee\boldsymbol{r}']}$. Modify its south edge to the bi-trace (c^-, c^-) and its east edge to (c^+, c^+). This is still a biboundary; call it ∂_{NW}. For the northeast corner, take the biboundary of $s\restriction_{[\boldsymbol{r}',\boldsymbol{t}]}$ (which is ground). Modify its west edge to the bi-trace (c^+, c^+). This is still a biboundary (and still ground). Similarly for the southwest corner.

The biboundary of s is the join of these four biboundaries. It only remains to show ∂_{NW} is fabricated, and that this can be checked efficiently. As ∂_{NW} is of full height, its upper cluster is c^+ and lower c^-. By appealing to the facts that s is a surrectangle and $s\restriction_{[l\wedge r', l\vee r']}$ is also a surrectangle, we can see that ∂_{NW} is the shuffle of Δ', using M, where $\Delta' = \{\partial^{s\restriction_{[x\wedge y, x\vee y]}} \mid \boldsymbol{x}, \boldsymbol{y} \in (P\backslash\{\boldsymbol{l}, \boldsymbol{r}\}) \cap [\boldsymbol{l}\wedge \boldsymbol{r}', \boldsymbol{l}\vee \boldsymbol{r}']$, and $\boldsymbol{y}$ is the successor of $\boldsymbol{x}\}$, and $M = (P\backslash\{\boldsymbol{l}, \boldsymbol{r}\}) \cap [\boldsymbol{l}\wedge \boldsymbol{r}', \boldsymbol{l}\vee \boldsymbol{r}']$. The set Δ' is a subset of Δ, so by Lemma 6.4, the biboundary ∂_{NW} can be checked using no more than $S(N) + 3\delta + n\delta$ space. Hence ∂ can been checked within $S(N) + \mathcal{O}(n^3)$, as promised.

The case where neither $\boldsymbol{l}$ nor $\boldsymbol{r}$ are limits is similar.

7 Halpern–Shoham logic

In this section we explain how the PSPACE procedure for the temporal validities of $(\mathbb{R}^2, \prec)$ can be modified to procedures performing the same function for certain Halpern–Shoham logics of intervals on the real line.

In Halpern–Shoham logic, intervals are identified with pairs (x, y) of points, with either $x < y$ (**strict interval semantics**) or $x \leq y$ (**non-strict semantics**).[7] There are thirteen different atomic relations that may hold between two strict intervals. Following Allen [1], we call these `equals`, `before`, `after`, `during`, `contains`, `starts`, `started_by`, `finishes`, `finished_by`, `meets`, `met_by`, `overlaps`, and `overlapped_by`.[8] Modalities that may be included in a Halpern–Shoham logic are any corresponding to a relation given by the union of some of these thirteen.

Let $H_<$ be the open half-plane $\{(x, y) \in \mathbb{R}^2 \mid x < y\}$. The frame $(H_<, \prec)$ is precisely the frame of strict intervals of $\mathbb{R}$ with the relation `overlaps` $\cup$ `meets` $\cup$ `before`. Hence the temporal logic of $(H, \prec)$ is the strict Halpern–Shoham logic of $\mathbb{R}$ with two modalities corresponding to this relation and its converse. As in [5], it can be shown that an arbitrary finitely generated directed partial order is a p-morphic image of the reflexive closure of $(H_<, \prec)$. Hence an arbitrary purely modal formula (not using the past modality) is in **S4.2** if and only if it is valid for the reflexive closure of $(H_<, \prec)$, and hence the validity of temporal formulas over this frame is PSPACE-hard. The PSPACE-hardness of irreflexive $(H_<, \prec)$ itself, follows.

A **surtriangle** is similar to a surrectangle, but the domain of the core map f is either $\{(x, y) \in H_< \mid a \leq x,\ y \leq b\}$, for some $a < b$, or a similar set where either $a \leq x$ is replaced by $a < x$, or $y \leq b$ is replaced by $y < b$, or both. A finite sequence of upper supplementary clusters and transition points are defined along $\{(x, b) \in H_< \mid a < x < b\}$ if contained in the domain of f, and similarly for $\{(a, y) \in H_< \mid a < y < b\}$ with lower supplementary clusters.

[7] Thus there is no distinction between open, half-open, and closed intervals, and unbounded intervals are not present.

[8] When point-intervals are present, pairs of intervals of the form $((x, x), (x, y))$ for $x < y$ are considered by Halpern and Shoham to stand in the relation `starts` (and not in `meets`). Similarly for `started_by`, `finishes`, and `finished_by`.

A biboundary is called **triangular** if its domain is disjoint from $\{b, t, r, S, E\}$. A surtriangle determines a triangular biboundary in the obvious way. If τ_1 is a triangular biboundary containing its northern edge, τ_2 a triangular biboundary containing its western edge, and ∂ a (rectangular) biboundary closed on its southern and eastern edges, such that $\tau_1(N) = \partial(S)$ and $\partial(W) = \tau_2(E)$, then the join $J(\tau_1, \partial, \tau_2)$ is the triangular biboundary formed by joining the three parts together. (The inconsequential southeast corner of ∂ is discarded.) A triangular biboundary is **fabricated** if it is either ground, or the join of a ground triangular biboundary, a fabricated rectangular biboundary, and a triangular biboundary of strictly smaller depth. The proofs of the following lemmas and theorem are similar to the proofs of Lemma 5.7, Theorems 5.8, and Theorem 6.1.

Lemma 7.1 *A triangular biboundary is fabricated if and only if it is the triangular biboundary of some surtriangle.*

Lemma 7.2 *There is a* PSPACE *algorithm to determine whether a triangular biboundary is fabricated.*

Theorem 7.3 *Let $H_< = \{(x, y) \in \mathbb{R}^2 \mid x < y\}$, and let R be either the relation given by $(x, y)R(x', y') \iff x < x'$ and $y < y'$ (so $R =$ `overlaps` $\cup$ `meets` $\cup$ `before`) or the reflexive closure of this relation (so $R =$ `equals` $\cup$ `overlaps` $\cup$ `meets` $\cup$ `before`). Then the temporal logic of $(H_<, R)$ is* PSPACE*-complete.*

Appendix

Proof of Lemma 4.2 (ground biboundaries) Let R be the appropriate rectangle (given ∂) with $(0, 1) \times (0, 1) \subseteq R \subseteq [0, 1] \times [0, 1]$. Define $f \colon R \to \mathsf{MCS}$ as filling the interior of R densely with the cluster $\partial(+)$. Define f on any corners of R in the evident way.

If ∂ is defined on N, with $\partial(N) = (c_0^-, c_0^+, b_1, \ldots, c_n^+)$ say, then define $f(\frac{i}{n+1}, 1) = b_i$ for each i, and fill each line segment $((\frac{i}{n+1}, 1), (\frac{i+1}{n+1}, 1))$ densely with the interpolant of c_i^- and c_i^+. Define the supplementary clusters for this edge to be $c_0^+, \ldots, c_n^+$, and the sequence of transition points to be $(\frac{1}{n+1}, 1), \ldots, (\frac{n}{n+1}, 1)$. Similarly if ∂ is defined on S, E, or W.

The conditions on biboundaries in general, together with those for ground fabricated ones in particular, ensure our construction satisfies all conditions necessary to be a surrectangle.

Proof of Lemma 4.3 (joins) We know the bi-trace obtained from the northern edge of s_1 equals that obtained from the southern edge of s_2, and in particular these edges have the same number of transition points. Then by Lemma 4.1, we may assume these two edges and their cluster pairs, transition points, transition values, and values of f at any end-points coincide. As we remarked in the opening of Section 4, this is sufficient to assume s_1 and s_2 agree on the common edge.

We define s with domain the union of the domains of s_1 and s_2 in the obvious way. Then it is straightforward to check that s forms a surrectangle and that $\partial^s = \partial_1 \oplus_- \partial_2$.

The proof for horizontal joins is completely analogous.

Proof of Lemma 4.4 (limits) Since ∂^* is a southeastern limit of ∂_0 using $\partial_1, \partial_2, \partial_3$, the vertical join of ∂_2, ∂_0 is defined, and similarly for other joins. Figure 3(a) illustrates how a surrectangle for ∂^* can be constructed using transformed copies of the surrectangles s_0, s_1, s_2, s_3. The domain of the surrectangle is a rectangle with interior $(0,1) \times (0,1)$. The upper-left quadrant $(0, \frac{1}{2}) \times (\frac{1}{2}, 1)$ together with boundaries dictated by ∂_0 is a copy of s_0. Each rectangle $(0, 1-\frac{1}{2^k}) \times (\frac{1}{2^{k+1}}, \frac{1}{2^k})$, plus appropriate boundaries, is a copy of s_2, for $k \geq 1$. Use Lemma 4.1 to ensure agreement along common boundaries. Transformed copies of s_1 are used dually. The square $[1-\frac{1}{2^k}, 1-\frac{1}{2^{k+1}}] \times [\frac{1}{2^{k+1}}, \frac{1}{2^k}]$ is a copy of s_3, for $k \geq 1$. This covers the interior of the unit square, and northern, western edges, and northwest corner, if included in the domain of ∂^*. Use ∂^* to define the corners b, r, t if included in the domain of ∂^*. Use the interpolant of the lower and upper clusters of $\partial^*(S)$ densely along the southern edge of the unit square, if S is included in the domain of ∂^*, similarly for the eastern edge.

The proof for northwestern limits is completely analogous.

Proof of Lemma 4.5 (shuffles) The closure of the rectangle of s' will be $[0,1] \times [0,1]$. Let $M \subseteq \mathsf{MCS}$ be such that the conditions in Definition 3.4, the definition of a shuffle, are satisfied. Let d be the open diagonal line-segment $((0,1),(1,0))$. We describe how to construct s' in stages, as illustrated in Figure 3(b). First fill the area of $(0,1) \times (0,1)$ below d densely with $\partial'(-)$, and fill the area above this diagonal densely with $\partial'(+)$. Fill d with any element of M.

For any edge that ∂' indicates should be closed in s', assign the appropriate sequence of supplementary clusters, and evenly spaced transition points. Assign the appropriate transition values at the transition points, and between transition points fill the edge densely with the appropriate interpolant. If ∂' indicates any corners are required, assign them the appropriate maximal consistent set.

Next, we successively modify the interior of the construction. Each point will be updated at most once, so this process has a well-defined limit. We maintain a finite set S of disjoint open subsegments of d, initialised to $\{d\}$. At a later stage, pick $d' \in S$ and $b \in \Delta \cup M$. If $b \in \Delta$, reassign the closed rectangle whose diagonal is the central third of d', using the surrectangle s_b. Otherwise, reassign the midpoint $\boldsymbol{m}$ of d' with the maximal consistent set b. In either case, the segment d' is replaced by two in S—in the first case the open initial and final thirds of d', in the second, the two halves of $d' \setminus \{\boldsymbol{m}\}$. Schedule the choices of d' and b so that for every segment d' occurring in this construction, for every $b \in \Delta \cup M$, a choice d^*, b is eventually selected, where d^* is a subsegment of d'.

The limit of this process gives a well-defined map f from the appropriate subset of $[0,1] \times [0,1]$ to MCS, and supplementary clusters where appropriate. We argue that together these form a surrectangle, s'.

The constraints placed on ∂' by the ordering conditions in the definition of a biboundary, together with the ordering conditions in the definition of a shuffle, ensure that f is preorder preserving. Our construction has the property that every point below d that did not undergo reassignment has an open subset of points in its future that also did not undergo reassignment, *and* has a copy of each s_∂, and each $m \in M$ in its future. Noting this, it is straightforward to check that all future defects are either resolved internally or passed upwards. The other conditions in the definition of a surrectangle are also straightforward to check.

Clearly the biboundary of s' is ∂', as required.

Proof of Lemma 5.1 If Γ^- and Γ^+ are disjoint, then by the boundedness of R, and since they are closed, they are bounded away from each other, with a bound $\varepsilon > 0$ say. Then we may divide R into a *finite* grid of rectangles each with a diagonal shorter than ε. The restriction of s to such a rectangle has height N or less, so yields a fabricated biboundary. Then ∂^s is a join of such biboundaries, and is therefore itself fabricated.

Proof of Lemma 5.2 Using the inductive hypothesis, we may assume that the biboundary of $s\restriction_{[\boldsymbol{y}\wedge\boldsymbol{x},\boldsymbol{y}\vee\boldsymbol{x}]}$ is of height $N+1$, that is, both Γ^- and Γ^+ intersect the interior of $[\boldsymbol{x}\wedge\boldsymbol{y},\boldsymbol{x}\vee\boldsymbol{y}]$. As each $\boldsymbol{x}_i$ is in $R\setminus(\downarrow I^- \cup \uparrow I^+)$, we know $s\restriction_{[\boldsymbol{x}_i,\boldsymbol{x}\vee\boldsymbol{y}]}$ (if $[\boldsymbol{x}_i,\boldsymbol{x}\vee\boldsymbol{y}]$ is nondegenerate) has height at most N, so its biboundary is fabricated, similarly for $s\restriction_{[\boldsymbol{x}\wedge\boldsymbol{y},\boldsymbol{x}_i]}$. Hence, using joins, it suffices to prove the biboundary of $s\restriction_{[\boldsymbol{x}_i\wedge\boldsymbol{x},\boldsymbol{x}_i\vee\boldsymbol{x}]}$ is fabricated, for some i. Then it follows using Lemma 5.1 and the inductive hypothesis that we may assume Γ^- and Γ^+ meet at $\boldsymbol{x}$, that no point due west of $\boldsymbol{x}$ is in Γ^-, and that no point due north of $\boldsymbol{x}$ is in Γ^+. Together, these assumptions imply the upper cluster on the southern edge of $s\restriction_{[\boldsymbol{y}\wedge\boldsymbol{x},\boldsymbol{y}\vee\boldsymbol{x}]}$ is constantly c^-, and the lower cluster on the eastern edge is constantly c^+.

Since there are only finitely many biboundaries, by taking a subsequence, we can assume the biboundary of $s\restriction_{[\boldsymbol{y}\wedge\boldsymbol{x}_i,\boldsymbol{y}\vee\boldsymbol{x}_i]}$ is constant, and by our previous assumptions we may also assume that $\partial^{s\restriction_{[\boldsymbol{x}_0,\boldsymbol{y}\vee\boldsymbol{x}_1]}}$ has constant lower cluster $\partial^s(+)$ on its eastern edge, and $\partial^{s\restriction_{[\boldsymbol{y}\wedge\boldsymbol{x}_1,\boldsymbol{x}_0]}}$ has constant upper cluster $\partial^s(-)$ on its southern edge. Then the biboundary of $s\restriction_{[\boldsymbol{y}\wedge\boldsymbol{x},\boldsymbol{y}\vee\boldsymbol{x}]}$ is a southeastern limit of the biboundary of $s\restriction_{[\boldsymbol{y}\wedge\boldsymbol{x}_0,\boldsymbol{y}\vee\boldsymbol{x}_0]}$ using the biboundaries of $s\restriction_{[\boldsymbol{x}_0,\boldsymbol{y}\vee\boldsymbol{x}_1]}$, $s\restriction_{[\boldsymbol{y}\wedge\boldsymbol{x}_1,\boldsymbol{x}_0]}$, and $s\restriction_{[\boldsymbol{x}_0\wedge\boldsymbol{x}_1,\boldsymbol{x}_0\vee\boldsymbol{x}_1]}$. All of these are fabricated, by the hypothesis that $\boldsymbol{y}\equiv\boldsymbol{x}_1$. Hence the biboundary of $s\restriction_{[\boldsymbol{y}\wedge\boldsymbol{x},\boldsymbol{y}\vee\boldsymbol{x}]}$ is fabricated. Furthermore the biboundary of the restriction of s to any nondegenerate $[\boldsymbol{w},\boldsymbol{z}] \subsetneq [\boldsymbol{y}\wedge\boldsymbol{x},\boldsymbol{y}\vee\boldsymbol{x}]$ is clearly fabricated, so $\boldsymbol{y}\equiv\boldsymbol{x}$.

Proof of Lemma 5.3 It is clear that if $[\boldsymbol{p}\wedge\boldsymbol{q},\boldsymbol{p}\vee\boldsymbol{q}]$ is degenerate then $\boldsymbol{p}\equiv\boldsymbol{q}$. If $\boldsymbol{p}$ is an immediate $\lhd$-successor of $\boldsymbol{q}$ pick $\boldsymbol{y}\lhd\boldsymbol{x}_0\lhd\boldsymbol{x}_1\lhd\dots$ in the interior of $[\boldsymbol{p}\wedge\boldsymbol{q},\boldsymbol{p}\vee\boldsymbol{q}]$, converging to $\boldsymbol{p}$. Since $\boldsymbol{y},\boldsymbol{x}_i$ are in the interior of the rectangle, the upper and lower clusters of $s\restriction_{[\boldsymbol{y}\wedge\boldsymbol{x}_i,\boldsymbol{y}\vee\boldsymbol{x}_i]}$ are bounded away, hence the biboundary is fabricated. By Lemma 5.2 the biboundary of $s\restriction_{[\boldsymbol{y}\wedge\boldsymbol{p},\boldsymbol{y}\vee\boldsymbol{p}]}$ is fabricated. Similarly, by considering northwestern limits, the biboundary of $s\restriction_{[\boldsymbol{p}\wedge\boldsymbol{q},\boldsymbol{p}\vee\boldsymbol{q}]}$ is fabricated and $\boldsymbol{p}\equiv\boldsymbol{q}$. By Lemma 5.2 and its diagonal dual, the $\equiv$-equivalence classes are topologically closed. The lemma follows.

Proof of Lemma 5.4 The upper boundary Γ^- of I^- is homeomorphic to the closed unit interval $[0,1]$, and such a homeomorphism f will map $R(E) \cap \Gamma^-$ to an interval, for any $\approx$-equivalence class E. The set I^- is the disjoint union of the closed sets $R(E) \cap \Gamma^-$ as E ranges over $\approx$-equivalence classes. Hence $\{f(R(E) \cap \Gamma^-) \mid E \in P/{\approx}\}$ is a partition of $[0,1]$ into closed intervals. It follows that if there is more than one $\approx$-equivalence class, then uncountably many of these closed intervals are singletons; see [7, Lemma 2.9]. For each singleton interval, the corresponding $\approx$-equivalence class must itself be a singleton.

Proof of Lemma 6.2 If Γ^- and Γ^+ are disjoint, they are bounded apart, by ε say. Then either Γ^- is bounded away from north edge of s by $\frac{\varepsilon}{\sqrt{2}}$, or Γ^+ is bounded away from the *west* edge by $\frac{\varepsilon}{\sqrt{2}}$. Without loss of generality, assume the latter. Let s_1 be the restriction of s to the region at a distance no more than $\frac{\varepsilon}{\sqrt{2}}$ from the west edge of s, and s_2 be the restriction of s to the region *at least* $\frac{\varepsilon}{\sqrt{2}}$ from the west edge. Then $\partial = \partial^{s_1} \oplus_| \partial^{s_2}$, and ∂^{s_1} has height no greater than N. Hence it is possible to choose option 1 with ∂^{s_1}, ∂^{s_2}, and check ∂_1 is fabricated using no more than $S(N) + \delta$ space. The tail-recursive call is to the biboundary of s_2, which is smaller than s in one dimension by $\frac{\varepsilon}{\sqrt{2}}$ and—if still of height $N+1$—has upper and lower clusters still bounded apart by ε. Hence iterating the described choice scheme requires space $S(N)+\delta$ each iteration, and must eventually result in a reduction of the height of the second biboundary s_2 in the join, after which the recursive call to s_2 can succeed with $S(N)$ space. The maximum space required is $S(N) + \delta$, as claimed.

Proof of Lemma 6.3 If Γ^- and Γ^+ intersect only at *one* point on the boundary of s, by Lemma 6.2, we may assume (without exceeding our space allowance) it is either $\boldsymbol{l}$ or $\boldsymbol{r}$—without loss of generality, assume $\boldsymbol{r}$. Then for some $\boldsymbol{x}$ and $\boldsymbol{y}$ in the interior of s, the biboundary ∂ is the limit of $\partial^{s\restriction[\boldsymbol{l}\wedge\boldsymbol{x},\boldsymbol{l}\vee\boldsymbol{x}]}$ using $\partial^{s\restriction[\boldsymbol{x},\boldsymbol{l}\vee\boldsymbol{y}]}$, $\partial^{s\restriction[\boldsymbol{l}\wedge\boldsymbol{y},\boldsymbol{x}]}$, and $\partial^{s\restriction[\boldsymbol{x}\wedge\boldsymbol{y},\boldsymbol{x}\vee\boldsymbol{y}]}$, and both $\partial^{s\restriction[\boldsymbol{x},\boldsymbol{l}\vee\boldsymbol{y}]}$ and $\partial^{s\restriction[\boldsymbol{l}\wedge\boldsymbol{y},\boldsymbol{x}]}$ are of height no greater than N. Whilst checking $\partial^{s\restriction[\boldsymbol{x},\boldsymbol{l}\vee\boldsymbol{y}]}$ and $\partial^{s\restriction[\boldsymbol{l}\wedge\boldsymbol{y},\boldsymbol{x}]}$, at most $S(N)$ space is needed, plus 3δ to store the other three biboundaries. Whilst checking $\partial^{s\restriction[\boldsymbol{l}\wedge\boldsymbol{y},\boldsymbol{x}]}$, by Lemma 6.2, at most $S(N) + \delta$ is needed, plus δ to store $\partial^{s\restriction[\boldsymbol{l}\wedge\boldsymbol{x},\boldsymbol{l}\vee\boldsymbol{x}]}$. Lastly, $\partial^{s\restriction[\boldsymbol{l}\wedge\boldsymbol{x},\boldsymbol{l}\vee\boldsymbol{x}]}$ is checked, and $S(N)+\delta$ is sufficient by Lemma 6.2. The maximum space needed is bounded by $S(N) + 3\delta$.

If Γ^- and Γ^+ intersect at *two* points on the boundary of s, then using the same choice scheme, $\Gamma^- \cap [\boldsymbol{l}\wedge\boldsymbol{x}, \boldsymbol{l}\vee\boldsymbol{x}]$ and $\Gamma^+ \cap [\boldsymbol{l}\wedge\boldsymbol{x}, \boldsymbol{l}\vee\boldsymbol{x}]$ will intersect only on the boundary of $s\restriction_{[\boldsymbol{l}\wedge\boldsymbol{x},\boldsymbol{l}\vee\boldsymbol{x}]}$, and at at most one point. Then by the previous case, checking the biboundary of $s\restriction_{[\boldsymbol{l}\wedge\boldsymbol{x},\boldsymbol{l}\vee\boldsymbol{x}]}$ can be done with $S(N)+3\delta$ space. As this is done last, the bound $S(N) + 3\delta$ remains intact.

Proof of Lemma 6.4 If a biboundary ∂' is a shuffle over Δ' using $M' \subseteq \mathsf{MCS}$ then it is the shuffle of any subset Δ'' of Δ' using any subset M'' of M', so long as any future defect of $\partial'(-)$ is passed up somehow to Δ'' or M'', and any past defect of $\partial'(+)$ is passed down to Δ'' or M''. There are at most n defects, past or future, so any shuffle is the shuffle of at most n biboundaries, using at most n auxiliary maximal consistent sets. Hence by choosing option 3, Algorithm 1 can succeed in checking ∂'. Since each of the up to n subchecks requires no more than $S(N) + 3\delta$ space, checking ∂' can be accomplished using $S(N) + 3\delta + n\delta$ space.

References

[1] Allen, J. F., *Maintaining knowledge about temporal intervals*, Communications of the ACM **26** (1983), pp. 832–843.

[2] Bresolin, D., D. Della Monica, V. Goranko, A. Montanari and G. Sciavicco, *Decidable and undecidable fragments of Halpern and Shoham's interval temporal logic: Towards a complete classification*, in: I. Cervesato, H. Veith and A. Voronkov, editors, *Proceedings of 15th Conference on Logic for Programming, Artificial Intelligence, and Reasoning* (2008), pp. 590–604.

[3] Bresolin, D., A. Kurucz, E. Muñoz Velasco, V. Ryzhikov, G. Sciavicco and M. Zakharyaschev, *Horn fragments of the Halpern–Shoham interval temporal logic*, ACM Transactions on Computational Logic **18** (2017), pp. 22:1–22:39.

[4] Bresolin, D., A. Montanari, P. Sala and G. Sciavicco, *What's decidable about Halpern and Shoham's interval logic? The maximal fragment ABBL*, in: *Pproceedings of 26th IEEE Symposium on Logic in Computer Science*, 2011, pp. 387–396.

[5] Goldblatt, R., *Diodorean modality in Minkowski spacetime*, Studia Logica **39** (1980), pp. 219–236.

[6] Halpern, J. Y. and Y. Shoham, *A propositional modal logic of time intervals*, Journal of the ACM **38** (1991), pp. 935–962.

[7] Hirsch, R. and M. Reynolds, *The temporal logic of two dimensional Minkowski spacetime is decidable*, The Journal of Symbolic Logic (2018), (in press), arXiv:1507.04903.

[8] Immerman, N., *Nondeterministic space is closed under complementation*, SIAM Journal on Computing **17**, pp. 935–938.

[9] Marcinkowski, J. and J. Michaliszyn, *The ultimate undecidability result for the Halpern–Shoham logic*, in: *Proceedings of 26th IEEE Symposium on Logic in Computer Science*, 2011, pp. 377–386.

[10] Marcinkowski, J. and J. Michaliszyn, *The undecidability of the logic of subintervals*, Fundamenta Informaticae **XX** (2013), pp. 1–25.

[11] Montanari, A. and P. Sala, *An optimal tableau system for the logic of temporal neighborhood over the reals*, in: *Proceedings of 19th Symposium on Temporal Representation and Reasoning*, 2012, pp. 39–46.

[12] Savitch, W. J., *Relationships between nondeterministic and deterministic tape complexities*, Journal of Computer and System Sciences **4** (1970), pp. 177–192.

[13] Shapirovsky, I., *On PSPACE-decidability in transitive modal logics*, in: R. Schmidt, I. Pratt-Hartmann, M. Reynolds and H. Wansing, editors, *Advances in Modal Logic* (2005), pp. 269–287.

[14] Shapirovsky, I., *Simulation of two dimensions in unimodal logics*, in: P. Balbiani, N.-Y. Suzuki, F. Wolter and M. Zakharyaschev, editors, *Advances in Modal Logic* (2010), pp. 371–391.

[15] Shapirovsky, I. and V. B. Shehtman, *Chronological future modality in Minkowski spacetime*, in: F. Wolter, H. Wansing, M. de Rijke and M. Zakharyaschev, editors, *Advances in Modal Logic* (2003), pp. 437–460.

[16] Shehtman, V. B., *Modal logics of domains on the real plane*, Studia Logica **42** (1983), pp. 63–80.

[17] Spaan, E., *The complexity of propositional tense logics*, in: M. de Rijke, editor, *Diamonds and Defaults: Studies in Pure and Applied Intensional Logic*, Springer, 1993 pp. 287–307.

[18] Szelepcsényi, R., *The method of forced enumeration for nondeterministic automata*, Acta Informatica **26** (1988), pp. 279–284.

[19] Venema, Y., *Expressiveness and completeness of an interval tense logic*, Notre Dame Journal of Formal Logic **31** (1990), pp. 529–547.

One Modal Logic to Rule Them All?

Wesley H. Holliday

University of California, Berkeley

Tadeusz Litak

Friedrich-Alexander-Universität Erlangen-Nürnberg

Abstract

In this paper, we introduce an extension of the modal language with what we call the *global quantificational modality* $[\forall p]$. In essence, this modality combines the propositional quantifier $\forall p$ with the global modality A: $[\forall p]$ plays the same role as the compound modality $\forall p \mathsf{A}$. Unlike the propositional quantifier by itself, the global quantificational modality can be straightforwardly interpreted in any Boolean Algebra Expansion (BAE). We present a logic GQM for this language and prove that it is complete with respect to the intended algebraic semantics. This logic enables a conceptual shift, as what have traditionally been called different "modal logics" now become $[\forall p]$-universal theories over the base logic GQM: instead of defining a new logic with an axiom schema such as $\Box\varphi \to \Box\Box\varphi$, one reasons in GQM about what follows from the globally quantified formula $[\forall p](\Box p \to \Box\Box p)$.

Keywords: global quantificational modalities, propositional quantifiers, Boolean algebra expansions, Boolean algebras with operators, modal consequence.

1 Introduction

In this paper, we investigate the effect of extending modal syntax with the *global quantificational modality* $[\forall p]$. This proposal arose from our efforts to tackle two foundational problems in modal logic, which under careful inspection turn out to be related to each other.

1.1 The proliferation of modal "logics"

According to a standard view, there is a striking contrast between first-order logic and modal logic: the former leads to a single system of First-Order Logic, while the latter leads to a vast landscape of different logics. In some contexts this has prompted the "suggestion... that the great proliferation of modal logics is an epidemy from which modal logic ought to be cured" [8, p. 25].[1] One could object that first-order logic is not so monolithic, given the choice between classical, intuitionistic, superintuitionistic, or substructural bases. But even in

[1] The authors of [8] attribute this suggestion to others rather than endorsing it themselves.

the classical context, there are objections to the claimed contrast, now coming from the modal side. As van Benthem [4] writes:

> [T]hese systems are not "different modal logics", but different special theories of particular kinds of accessibility relation. We do not speak of "different first-order logics" when we vary the underlying model class. There is no good reason for that here, either. (p. 93)

Yet in modal logic there remains a distinction between theories and logics: the set of formulas satisfied at all points (or a point) in a model counts as a *theory*, but not a logic, while the set of formulas validated at all points (or a point) of a frame counts as a *logic*. Here we will not address the question in the philosophy of logic about what should count as a "logic". Instead, we answer the question: is there a mathematically appealing way in which what are traditionally called "modal logics" are special theories relative to one logical system?

1.2 The riddle of propositional quantification

The other problem we are concerned with is that of conservatively handling propositional quantifiers. Historically, propositional quantifiers were considered in modal logic from the very beginning. Most of the literature quotes references such as Kripke [23], Bull [7], Fine [13], and Kaplan [21]. In fact, however, propositional quantifiers were already present not only in Ruth Barcan Marcus's post-war papers [3], but even in a chapter about the "existence postulate" by C. I. Lewis in his famous 1932 monograph with Langford [24, § VI.6]. Lewis's postulate is classically equivalent to $\exists p(\Diamond p \wedge \Diamond\neg p)$, and he insisted that "it is only through such principles that the outlines of a logical system can be positively delineated" [24, p. 181].

The problem, however, is that the addition of propositional quantifiers is not necessarily conservative for a given logic. It appears most natural, for example, to interpret them using infinite meets and joins in (algebras dual to) a suitable semantics (see § 3). Unfortunately, there are logics that are not even weakly complete with respect to lattice-complete algebras [25,27,26,29,33]. Moreover, even for standard logics such algebras might well validate undesirable quantified principles, as shown by "Kaplan's paradox" [22] for possible world semantics (see § 4). Logics with propositional quantifiers also tend to display very bad computational behavior over the dual algebras of Kripke frames: even for logics as strong as S4.2, propositional quantification over Kripke frames produces a system as complex as full second-order logic [13,20].

1.3 Our proposal

In this paper, we intend to solve both problems by showing, on the one hand, how "different modal logics" can indeed be seen as different theories over a single base logic, and on the other hand, how each and every modal logic can be conservatively extended with a form of propositional quantification. This is made possible by extending the language with the *global quantificational modality* $[\forall p]$, which combines the propositional quantifier $\forall p$ with the global modality A. Semi-formally, one can introduce it as $[\forall p]\varphi := \forall p \mathsf{A}\varphi$. One can then think of the global modality as definable by taking a fresh variable in $[\forall p]$ and introduce further global quantificational modalities (GQMs) as follows:

$$\langle\exists p\rangle\varphi := \neg[\forall p]\neg\varphi = \exists p\mathsf{E}\varphi$$
$$[\exists p]\varphi := \langle\exists p\rangle\mathsf{A}\varphi = \exists p\mathsf{A}\varphi$$
$$\langle\forall p\rangle\varphi := \neg[\exists p]\neg\varphi = \forall p\mathsf{E}\varphi.$$

This language of global quantificational modalities, formally defined in § 2, will be our object of study. In § 3, we show that in a global sense, this language is as expressive as the standard language of second-order propositional modal logic over the lattice-complete algebras used to interpret that language; and in § 4, we show that the flexibility to interpret our language in incomplete algebras provides a response to "Kaplan's paradox" for possible world semantics. In § 5, we introduce our logic GQM, and in § 6 we show that GQM solves the twin problems of proliferation (§ 1.1) and nonconservativity (§ 1.2). Toward proving the completeness of GQM with respect to its intended semantics, we establish prenex normal form results in § 7 and mutual translations with the first-order theory of "discriminator BAEs" in § 8. The storyline culminates with the completeness theorem at the end of § 8. We conclude in § 9.

Several proofs are deferred to appendices. A number of proofs involving syntactic derivations are given in an extended technical report online [17]. An open-source git repository containing formalizations of our proofs in Coq by Michael Sammler is available at https://gitlab.cs.fau.de/lo22tobe/GQM-Coq.

2 Language and semantics

In § 1.3, we introduced the idea of $[\forall p]$ in terms of the propositional quantifier and the global modality. Officially, we take $[\forall p]$ as primitive.

Fix a countably infinite set Prop of propositional variables and define:

$$\mathcal{L}_{\text{GQM}} \qquad \varphi ::= p \mid \neg\varphi \mid (\varphi \wedge \varphi) \mid \Box\varphi \mid [\forall p]\varphi,$$

where $p \in \mathsf{Prop}$. We treat $\vee$, $\rightarrow$, $\leftrightarrow$, and $\Diamond$ as abbreviations as usual and define:

- $\mathsf{A}\varphi := [\forall r]\varphi$ for an $r \in \mathsf{Prop}$ not free (in the usual sense) in φ; $\mathsf{E}\varphi := \neg\mathsf{A}\neg\varphi$;
- $\langle\exists p\rangle\varphi := \neg[\forall p]\neg\varphi$, $[\exists p]\varphi := \langle\exists p\rangle\mathsf{A}\varphi$, and $\langle\forall p\rangle\varphi := [\forall p]\mathsf{E}\varphi$.
- $\bot := (p \wedge \neg p)$ and $\top := \neg\bot$ for some $p \in \mathsf{Prop}$; [2]
- for each GQM $\{Qp\} \in \{[\forall p], [\exists p], \langle\exists p\rangle, \langle\forall p\rangle\}$, its dual $\overline{\{Qp\}}$ is defined in the obvious way, i.e., $[\forall p]$ is dual to $\langle\exists p\rangle$, and $[\exists p]$ is dual to $\langle\forall p\rangle$;
- for $* \in \{\wedge, \vee\}$, let G_* be A if $* = \wedge$ and E otherwise, and let us use plain G to stand for A or E (uniformly in a formula) in results that hold for both;
- for any formulas φ, ψ and propositional variable p, φ^p_ψ is the result of substituting ψ for all free occurrences of p in φ.

[2] Note that since $\bot$ can be defined as $[\forall p]p$, another elegant choice would be to have $\rightarrow$ as the *only* Boolean primitive.

Let $\mathcal{L}_{\Box}$ ($\mathcal{L}_{\Box\mathsf{A}}$) be the set of GQM formulas in which no global quantificational modalities (no global quantificational modalities other than A and E) appear.

Remark 2.1 The use of a single unary $\Box$ is for simplicity only. What follows could instead be developed in a polymodal language with polyadic modalities.

We now introduce the intended algebraic semantics for $\mathcal{L}_{\mathrm{GQM}}$.

Definition 2.2 A *Boolean algebra expansion* (BAE) is a tuple $\mathfrak{A} = \langle A, \neg, \wedge, \bot, \top, \Box\rangle$ where $\langle A, \neg, \wedge, \bot, \top\rangle$ is a Boolean algebra and $\Box : A \to A$.

Definition 2.3

(i) A $\mathcal{C}$-BAE (resp. $\mathcal{A}$-BAE) is a BAE whose Boolean reduct is complete (resp. atomic).

(ii) A BAO is a BAE with a *normal* $\Box$, i.e., $\Box$ distributes over all finite meets.

(iii) A $\mathcal{V}$-BAO is a BAO in which $\Box$ distributes over all existing meets.

We may concatenate '$\mathcal{C}$', '$\mathcal{A}$', and '$\mathcal{V}$' to indicate multiple properties; e.g., a $\mathcal{CA}$-BAE is a BAE whose Boolean reduct is both complete and atomic. This is a convention used in our earlier papers [27,26,29,16,15].

Definition 2.4 A *valuation* on a BAE $\mathfrak{A}$ is a function $\theta : \mathsf{Prop} \to \mathfrak{A}$ that extends to a function $\tilde{\theta} : \mathcal{L}_{\mathrm{GQM}} \to \mathfrak{A}$ as follows:

$$\tilde{\theta}(p) := \theta(p) \qquad \tilde{\theta}(\neg\varphi) := \neg\tilde{\theta}(\varphi)$$
$$\tilde{\theta}(\varphi \wedge \psi) := \tilde{\theta}(\varphi) \wedge \tilde{\theta}(\psi) \qquad \tilde{\theta}(\Box\varphi) := \Box\tilde{\theta}(\varphi)$$
$$\tilde{\theta}([\forall p]\varphi) := \begin{cases} \top & \text{if } \tilde{\gamma}(\varphi) = \top \text{ for all valuations } \gamma \sim_p \theta \\ \bot & \text{otherwise} \end{cases}$$

where $\gamma \sim_p \theta$ iff γ and θ disagree at most at p.

A formula φ is *valid in* $\mathfrak{A}$ iff for every valuation θ on $\mathfrak{A}$, $\tilde{\theta}(\varphi) = \top$. Let $\vDash_{\mathsf{GQM}} \varphi$ iff φ is valid in all BAEs, in which case φ is simply *valid.*

Lemma 2.5 *For any valuation θ on a* BAE $\mathfrak{A}$*:*

$$\tilde{\theta}(\mathsf{A}\varphi) = \begin{cases} \top & \text{if } \tilde{\theta}(\varphi) = \top \\ \bot & \text{otherwise} \end{cases} \qquad \tilde{\theta}(\mathsf{E}\varphi) = \begin{cases} \top & \text{if } \tilde{\theta}(\varphi) \neq \bot \\ \bot & \text{otherwise} \end{cases}$$
$$\tilde{\theta}(\langle\exists p\rangle\varphi) = \begin{cases} \top & \text{if } \exists\gamma \sim_p \theta.\tilde{\gamma}(\varphi) \neq \bot \\ \bot & \text{otherwise} \end{cases} \qquad \tilde{\theta}([\exists p]\varphi) = \begin{cases} \top & \text{if } \exists\gamma \sim_p \theta.\tilde{\gamma}(\varphi) = \top \\ \bot & \text{otherwise} \end{cases}$$
$$\tilde{\theta}(\langle\forall p\rangle\varphi) = \begin{cases} \top & \text{if } \forall\gamma \sim_p \theta.\tilde{\gamma}(\varphi) \neq \bot \\ \bot & \text{otherwise} \end{cases}.$$

Several definitions of semantic consequence are available, but as our default we pick the algebraic analogue of global model consequence [5, § 1.5].

Definition 2.6 Given $\Gamma \cup \{\varphi\} \subseteq \mathcal{L}_{\mathrm{GQM}}$, let $\Gamma \vDash^{\mathsf{A}}_{\mathrm{GQM}} \varphi$ iff for any BAE $\mathfrak{A}$ and $\theta : \mathsf{Prop} \to \mathfrak{A}$, if $\tilde{\theta}(\gamma) = \top$ for each $\gamma \in \Gamma$, then $\tilde{\theta}(\varphi) = \top$.

One of our main goals is to find a proof system complete with respect to $\vDash^{\mathsf{A}}_{\mathrm{GQM}}$. For this relation we have the following semantic deduction theorem.

Lemma 2.7 *For any formulas* $\varphi_1,\ldots,\varphi_n,\psi \in \mathcal{L}_{\mathsf{GQM}}$, $\{\varphi_1,\ldots,\varphi_n\} \vDash^{\mathsf{A}}_{\mathrm{GQM}} \psi$ *iff* $\vDash_{\mathsf{GQM}} \mathsf{A}(\varphi_1 \wedge \cdots \wedge \varphi_n) \to \mathsf{A}\psi$.

Proof. Immediate from Definition 2.6 and Lemma 2.5. □

We also distinguish two senses in which formulas may be equivalent.

Definition 2.8 For any $\varphi,\psi \in \mathcal{L}_{\mathrm{GQM}}$ and class $\mathcal{K}$ of BAEs:

(i) φ and ψ are *equivalent over* $\mathcal{K}$ iff for every $\mathfrak{A} \in \mathcal{K}$ and valuation θ on $\mathfrak{A}$, $\tilde{\theta}(\varphi) = \tilde{\theta}(\psi)$ (or equivalently, $\varphi \leftrightarrow \psi$ is valid in $\mathfrak{A}$);

(ii) φ and ψ are *globally equivalent over* $\mathcal{K}$ iff for every $\mathfrak{A} \in \mathcal{K}$ and valuation θ on $\mathfrak{A}$, $\tilde{\theta}(\varphi) = \top$ iff $\tilde{\theta}(\psi) = \top$ (or equivalently, $\mathsf{A}\varphi \leftrightarrow \mathsf{A}\psi$ is valid in $\mathfrak{A}$).

(iii) φ and ψ are *equivalent* (resp. *globally equivalent*) iff they are equivalent (resp. globally equivalent) over the class of all BAEs.

Remark 2.9 Since $\mathcal{L}_{\mathrm{GQM}}$ can be interpreted in arbitrary BAEs, it can be interpreted in any frames that give rise to BAEs, e.g.: Kripke frames (corresponding to $\mathcal{CAV}$-BAOs); relational possibility frames [15] (corresponding to $\mathcal{CV}$-BAOs); neighborhood frames (corresponding to $\mathcal{CA}$-BAEs); neighborhood possibility frames [15] (corresponding to $\mathcal{C}$-BAEs); discrete general frames [10] (corresponding to $\mathcal{AV}$-BAOs); discrete general neighborhood frames (corresponding to $\mathcal{A}$-BAEs); general neighborhood frames (corresponding to BAEs).

Remark 2.10 Given that the "predicate lifting" approach in coalgebraic logic reduces any set-based coalgebra to a neigborhood frame, it would be interesting to investigate coalgebraic applications of GQMs.

3 Reduction of SOPML to GQM

The standard language $\mathcal{L}_{\mathrm{SOPML}}$ of second-order propositional modal logic replaces $[\forall p]$ by the propositional quantifier $\forall p$. At first one might expect that the implicit global modality in $[\forall p]$ reduces the expressivity of $\mathcal{L}_{\mathrm{GQM}}$ relative to the language $\mathcal{L}_{\mathrm{SOPML}_{\mathsf{A}}}$ of second-order propositional modal logic plus the global modality. In fact, we will show that every $\mathrm{SOPML}_{\mathsf{A}}$ formula is globally equivalent to a GQM formula over standard semantics. For convenience, in this section we regard GQM formulas as $\mathrm{SOPML}_{\mathsf{A}}$ formulas with $[\forall p]$ as $\forall p\mathsf{A}$.

First, let us recall the algebraic semantics for $\mathcal{L}_{\mathrm{SOPML}_{\mathsf{A}}}$ that interprets $\forall p$ using the meets in a $\mathcal{C}$-BAE, as in, e.g., [16].

Definition 3.1 We extend a valuation θ on a $\mathcal{C}$-BAE $\mathfrak{A}$ to a valuation $\tilde{\theta}$: $\mathcal{L}_{\mathrm{SOPML}_{\mathsf{A}}} \to \mathfrak{A}$ using the clauses for $\neg$, $\wedge$, and $\Box$ from Definition 2.4 plus:

$$\tilde{\theta}(\forall p\varphi) = \bigwedge\{\tilde{\gamma}(\varphi) \mid \gamma \sim_p \theta\} \qquad \tilde{\theta}(\mathsf{A}\varphi) = \begin{cases} \top & \text{if } \tilde{\theta}(\varphi) = \top \\ \bot & \text{otherwise.} \end{cases}$$

Dually, $\exists p\varphi$ is interpreted using the join. The definitions of local and global equivalence from Definition 2.8 transfer in the obvious way to $\mathcal{L}_{\mathrm{SOPML}_{\mathsf{A}}}$.

We will reduce $\mathcal{L}_{\mathrm{SOPML}_{\mathsf{A}}}$ to $\mathcal{L}_{\mathrm{GQM}}$ over $\mathcal{C}$-BAEs using a prenex form result. In [9] it was shown that over $\mathcal{CAV}$-BAOs, every $\mathcal{L}_{\mathrm{SOPML}}$ formula is equivalent to a prenex one, i.e., a formula of the form $Q_1 p_1 \dots Q_n p_n \varphi$ where $Q_i \in \{\forall, \exists\}$ and φ is quantifier-free. In fact, the following more general result holds.

Proposition 3.2

(i) *Over* $\mathcal{CV}$-BAOs, *every* SOPML *formula is equivalent to a prenex* SOPML *formula.*

(ii) *Over* $\mathcal{C}$-BAEs, *every* SOPML$_{\mathsf{A}}$ *formula is equivalent to a prenex* SOPML$_{\mathsf{A}}$ *formula.*

Proof. The proof of part (i) is the same as in [9, Prop. 3] except that we give a different argument for pulling the quantifier out of $\Diamond\forall p\varphi$, which does not assume $\mathcal{A}$. For consistency with [9], we work with $\Diamond$, though everything dualizes easily. For any normal $\Box$, we have the following equivalence:

$$\Diamond\psi \Leftrightarrow \exists q(\Diamond q \wedge \Box(q \to \psi))$$

for a variable q that does not occur in ψ. Thus, we have the following equivalences where $q \neq p$ and q does not occur in φ:

$$\begin{aligned}
\Diamond\forall p\varphi &\Leftrightarrow \exists q(\Diamond q \wedge \Box(q \to \forall p\varphi)) \text{ setting } \psi := \forall p\varphi\\
&\Leftrightarrow \exists q(\Diamond q \wedge \Box\forall p(q \to \varphi)) \text{ because } q \neq p\\
&\Leftrightarrow \exists q(\Diamond q \wedge \forall p\Box(q \to \varphi)) \text{ by } \mathcal{V} \text{ for } \Box\\
&\Leftrightarrow \exists q\forall p(\Diamond q \wedge \Box(q \to \varphi)) \text{ because } q \neq p.
\end{aligned}$$

For part (ii), we use the fact that A distributes over arbitrary meets, so the reasoning for part (i) shows that $\mathsf{E}\forall p\varphi$ is equivalent to $\exists q\forall p(\mathsf{E}q \wedge \mathsf{A}(q \to \varphi))$, which implies ($\star$): $\mathsf{A}\exists p\psi$ is equivalent to $\forall q\exists p(\mathsf{E}q \to \mathsf{E}(q \wedge \psi))$.

Now for any $\Box$ in a $\mathcal{C}$-BAE, we have the following equivalence:

$$\Diamond\psi \Leftrightarrow \exists q(\Diamond q \wedge \mathsf{A}(q \leftrightarrow \psi))$$

for a variable q that does not occur in ψ. Thus, we have the following equivalences where $q \neq p$ and q does not occur in φ:

$$\begin{aligned}
\Diamond\forall p\varphi &\Leftrightarrow \exists q(\Diamond q \wedge \mathsf{A}(q \leftrightarrow \forall p\varphi)) \text{ setting } \psi := \forall p\varphi\\
&\Leftrightarrow \exists q(\Diamond q \wedge \mathsf{A}(\forall p(q \to \varphi) \wedge \exists p(\varphi \to q))) \text{ because } q \neq p\\
&\Leftrightarrow \exists q(\Diamond q \wedge \mathsf{A}(\forall p(q \to \varphi) \wedge \exists r(\varphi^p_r \to q))) \text{ for a fresh } r\\
&\Leftrightarrow \exists q(\Diamond q \wedge \mathsf{A}\forall p\exists r\underbrace{((q \to \varphi) \wedge (\varphi^p_r \to q))}_{\alpha}) \text{ because } r \neq p\\
&\Leftrightarrow \exists q(\Diamond q \wedge \forall p\mathsf{A}\exists r\alpha) \text{ by } \mathcal{V} \text{ for } \mathsf{A}\\
&\Leftrightarrow \exists q(\Diamond q \wedge \forall p\forall q'\exists r(\mathsf{E}q' \to \mathsf{E}(q' \wedge \alpha))) \text{ by } (\star) \text{ where } q' \text{ is fresh}\\
&\Leftrightarrow \exists q\forall p\forall q'\exists r(\Diamond q \wedge (\mathsf{E}q' \to \mathsf{E}(q' \wedge \alpha))) \text{ because } q \neq p, q \neq q', q \neq r.
\end{aligned}$$

The rest of the proof is as in [9]. □

Proposition 3.3 *If* α *is a prenex* SOPML$_{\mathsf{A}}$ *formula, then* $\mathsf{A}\alpha$ *is equivalent over* $\mathcal{C}$-BAEs *to a* GQM *formula.*

Proof. Proved in Appendix A as Proposition A.2. □

Theorem 3.4 *Every* SOPML$_\mathsf{A}$ *formula is globally equivalent over* $\mathcal{C}$-BAEs *to a* GQM *formula.*

Proof. By Proposition 3.2.(ii), φ is globally equivalent over $\mathcal{C}$-BAEs to a prenex SOPML$_\mathsf{A}$ formula ψ. Then since ψ is globally equivalent to $\mathsf{A}\psi$, it follows by Proposition 3.3 that ψ is globally equivalent to a GQM formula. □

Theorem 3.5 *The set of* GQM *formulas valid over any class of* $\mathcal{C}$-BAEs *containing the class of* $\mathcal{CAV}$-BAOs *validating* S4.2 *is not recursively enumerable.*

Proof. [Sketch] Fine [13, Prop. 6] (cf. [20]) showed that the set of SOPML sentences valid in $\mathcal{CAV}$-BAOs validating S4.2 is not recursively enumerable. The property of a BAE being an $\mathcal{AV}$-BAO is expressible in $\mathcal{L}_{\mathrm{GQM}}$; we leave this to the reader as an exercise (cf. § 8, [18, § 9]). Let $\chi_{\mathcal{AV}}$ be the corresponding sentence. The validity of an SOPML sentence φ over S4.2 $\mathcal{CAV}$-BAOs is equivalent to the validity of the GQM sentence $(\chi_{\mathcal{AV}} \wedge [\forall]\mathsf{S4.2}) \to \varphi^*$ over $\mathcal{C}$-BAEs, where φ^* is obtained from $\mathsf{A}\varphi$ by Theorem 3.4 and $[\forall]\mathsf{S4.2}$ is the GQM statement of the S4.2 axioms. Thus, the existence of a semi-decision procedure contradicting the statement would yield a semi-decision procedure contradicting [13,20]. □

Another route would be to use results of Thomason [32]. Both [32] and [20] deal with a stronger property: reducibility of full second-order consequence. We postpone the details to a sequel paper.

4 Interlude: "Kaplan's paradox"

In a festschrift for Ruth Barcan Marcus, Kaplan [22] posed a problem for possible world semantics involving propositional quantification. In brief, Kaplan claimed that the following should be consistent for a non-monotonic $\Box$:

- $\kappa := \forall p \mathsf{E} \forall q(\Box q \leftrightarrow \mathsf{A}(p \leftrightarrow q))$.

For example, if $\Box$ means "it is entertained at time t that...", then κ says that for all propositions p, it could have been that p was the unique proposition entertained at time t. Kaplan argued that *logic* should not rule this out. Yet he noted that κ is unsatisfiable in possible world semantics with $\forall$ quantifying over the powerset. For the truth of κ would yield an injection from the powerset of the set of worlds to the set of worlds. In fact, as Yifeng Ding (p.c.) observed, it is unsatisfiable in any $\mathcal{C}$-BAE. The truth of κ would yield (a) an injection from the algebra to an antichain of elements. But since the algebra is complete, (b) every subset of the antichain has a join, and all such joins are distinct. Together (a) and (b) contradict Cantor's theorem.

We will show there is a GQM formula φ, regarded as an SOPML$_\mathsf{A}$ formula as in § 3, such that (i) in any logic that derives some plausible equivalences, φ is provably equivalent to (the A-necessitation of) Kaplan's formula, so intuitively the truth of φ implies the truth of Kaplan's formula, and (ii) φ can be made true in an incomplete BAE. First, in any modal logic with propositional quantifiers in which the equivalences in the proof of Proposition 3.2.(ii) are provable, (the A-necessitation of) Kaplan's formula is provably equivalent to

- $\mathsf{A}\forall p\exists r(\mathsf{E}r \wedge \forall q\mathsf{A}(r \to (\Box q \leftrightarrow \mathsf{A}(p \leftrightarrow q))))$. [3]

Then using Barcan's equivalence $\mathsf{A}\forall p\psi \leftrightarrow \forall p\mathsf{A}\psi$ and S5 reasoning with E and A, the preceding formula is provably equivalent to

- $\forall p\mathsf{A}\exists r\mathsf{A}(\mathsf{E}r \wedge \forall q\mathsf{A}(r \to (\Box q \leftrightarrow \mathsf{A}(p \leftrightarrow q))))$,

which becomes the GQM formula

- $[\forall p][\exists r](\mathsf{E}r \wedge [\forall q](r \to (\Box q \leftrightarrow \mathsf{A}(p \leftrightarrow q))))$.

Now this formula can be made true in a BAE. Pick any infinite set X, and let $\mathfrak{A}$ be the Boolean algebra of its finite and cofinite subsets. Clearly, not only is X identifiable with the set $At(\mathfrak{A})$ of atoms of $\mathfrak{A}$, but also there is an injective (and hence bijective) map $\Box : \mathfrak{A} \to At(\mathfrak{A})$. It is easy to see that the formula above evaluates to $\top$ in $\mathfrak{A}$ with this interpretation of $\Box$. Thus, according to the logic GQM, the GQM translation of Kaplan's formula is consistent. [4]

5 Axiomatization

Let us now turn to formulating a complete proof system for $\mathcal{L}_{\mathrm{GQM}}$.

Definition 5.1 The logic GQM is the smallest set of formulas containing the following axioms and closed under the following rules.

propositional axioms

- all classical propositional tautologies.

axioms for $[\forall p]$

- distribution: $[\forall p](\varphi \to \psi) \to ([\forall p]\varphi \to [\forall p]\psi)$;
- instantiation: $[\forall p]\varphi \to \varphi^p_\psi$ where ψ is substitutable for p in φ; [5]
- global instantiation: $[\forall p]\varphi \to [\forall r]\varphi^p_\psi$ where ψ is substitutable for p in φ and r is not free in φ^p_ψ;
- quantificational 5 axiom: $\neg[\forall p]\varphi \to [\forall r]\neg[\forall p]\varphi$ where r is not free in $[\forall p]\varphi$.

axioms linking $[\forall p]$ and $\Box$

- $\Box$-congruence: $[\forall p](\varphi \leftrightarrow \psi) \to (\Box\varphi \leftrightarrow \Box\psi)$.

rules

- modus ponens: if $\vdash_{\mathsf{GQM}} \varphi$ and $\vdash_{\mathsf{GQM}} \varphi \to \psi$, then $\vdash_{\mathsf{GQM}} \psi$;
- $[\forall p]$-necessitation: if $\vdash_{\mathsf{GQM}} \varphi$, then $\vdash_{\mathsf{GQM}} [\forall p]\varphi$;
- universal generalization: if $\vdash_{\mathsf{GQM}} \alpha \to [\forall p]\varphi$ and q is not free in α, then

---

3 Of course the quantifier $\forall q$ can be pulled to the front for a prenex form, but here we opt for better human readability.

4 After submitting this paper, we were informed by John Hawthorne of the paper [2] in which the finite-cofinite algebra has also been used in response to Kaplan's paradox.

5 The definition of ψ being substitutable for p in φ is the obvious analogue of the definition of a term t being substitutable for a variable x in a first-order formula [12, p. 113]: no propositional variable in ψ should be captured by a quantifier in φ upon substituting ψ for p.

$\vdash_{\mathsf{GQM}} \; \alpha \to \; [\forall q][\forall p]\varphi.$

Here '$\vdash_{\mathsf{GQM}} \varphi$' means $\varphi \in \mathsf{GQM}$. We write '$\vdash \varphi$' when no confusion will arise.

Let us now record some useful theorems and metatheorems.

Lemma 5.2 *If q is substitutable for p in φ, and q is not free in φ, then* $\vdash [\forall p]\varphi \leftrightarrow [\forall q]\varphi^p_q$.

Proof. See the extended technical report [17]. □

Lemma 5.3 *If* $\vdash \varphi \to \psi$, *then* $\vdash \{Qp\}\varphi \to \{Qp\}\psi$.

Proof. See the extended technical report [17]. □

Lemma 5.4

(i) $\vdash \mathsf{A}(\varphi \to \psi) \to (\mathsf{A}\varphi \to \mathsf{A}\psi)$;

(ii) $\vdash \mathsf{G}_*(\varphi * \psi) \leftrightarrow (\mathsf{G}_*\varphi * \mathsf{G}_*\psi)$;

(iii) *if* $\vdash \varphi \to \psi$, *then* $\vdash \mathsf{G}\varphi \to \mathsf{G}\psi$;

(iv) $\vdash \mathsf{A}\varphi \to \varphi$;

(v) $\vdash \varphi \to \mathsf{E}\varphi$;

(vi) $\vdash \mathsf{E}\varphi \leftrightarrow \mathsf{AE}\varphi$;

(vii) $\vdash \mathsf{EA}\varphi \leftrightarrow \mathsf{A}\varphi$;

(viii) $\vdash \mathsf{GG}\varphi \leftrightarrow \mathsf{G}\varphi$;

(ix) $\vdash \{Qp\}\mathsf{A}\psi \leftrightarrow [Qp]\psi$;

(x) $\vdash \{Qp\}\mathsf{E}\psi \leftrightarrow \langle Qp\rangle\psi$;

(xi) $\vdash \{Qp\}\psi \leftrightarrow \mathsf{A}\{Qp\}\psi$;

(xii) $\vdash \{Qp\}\psi \leftrightarrow \mathsf{E}\{Qp\}\psi$.

Proof. See the extended technical report [17]. □

Let us now introduce a relation of syntactic consequence. In the following definition, Γ may be regarded as a set of globally true premises.

Definition 5.5 Given $\Gamma \cup \{\varphi\} \subseteq \mathcal{L}_{\mathrm{GQM}}$, let $\Gamma \vdash^{\mathsf{A}}_{\mathsf{GQM}} \varphi$ iff φ belongs to the smallest set Λ of GQM formulas that includes $\Gamma \cup \mathsf{GQM}$ and is closed under modus ponens and A-necessitation: if $\psi \in \Lambda$, then $\mathsf{A}\psi \in \Lambda$.

Now we obtain a syntactic deduction theorem parallel to Lemma 2.7.

Lemma 5.6 *For any formulas* $\varphi_1, \ldots, \varphi_n, \psi \in \mathcal{L}_{\mathsf{GQM}}$, $\{\varphi_1, \ldots, \varphi_n\} \vdash^{\mathsf{A}}_{\mathsf{GQM}} \psi$ *iff* $\vdash_{\mathsf{GQM}} \mathsf{A}(\varphi_1 \wedge \cdots \wedge \varphi_n) \to \mathsf{A}\psi$.

Proof. By application of Lemma 5.4. □

By design, $\vdash^{\mathsf{A}}_{\mathsf{GQM}}$ is sound with respect to $\vDash^{\mathsf{A}}_{\mathrm{GQM}}$.

Lemma 5.7 *For* $\Gamma \cup \{\varphi\} \subseteq \mathcal{L}_{\mathrm{GQM}}$, $\Gamma \vdash^{\mathsf{A}}_{\mathsf{GQM}} \varphi$ *implies* $\Gamma \vDash^{\mathsf{A}}_{\mathrm{GQM}} \varphi$.

Proof. Straightforward induction. □

It follows from Lemma 5.7 and the example in § 4 (or Theorem 3.5) that GQM is incomplete with respect to validity over the class of $\mathcal{C}$-BAEs. However, we will see in § 8 that GQM is complete with respect to validity over all BAEs.

6 Conservativity and modal logics as GQM theories

Before proving completeness, we show that GQM solves our two problems from § 1: the proliferation problem and the nonconservativity problem.

A *congruential modal logic* is a set $\mathsf{L} \subseteq \mathcal{L}_\Box$ containing all propositional tautologies and closed under uniform substitution, modus ponens, and the rule that if $\varphi \leftrightarrow \psi \in \mathsf{L}$, then $\Box\varphi \leftrightarrow \Box\psi \in \mathsf{L}$. Let GQM-L be the smallest set of formulas that includes $\mathsf{GQM} \cup \mathsf{L}$ and is closed under all three rules of GQM.

Proposition 6.1 (Conservativity) *For any* $\varphi \in \mathcal{L}_\Box$, $\varphi \in \mathsf{GQM\text{-}L}$ *iff* $\varphi \in \mathsf{L}$.

Proof. The Lindenbaum-Tarski algebra for L is a BAE in which every $\varphi \in \mathsf{GQM\text{-}L}$ is valid and in which any $\mathcal{L}_\Box$ formula not in L can be refuted. □

A set $\Sigma \subseteq \mathcal{L}_\Box$ *axiomatizes* a congruential modal logic L iff L is the smallest congruential modal logic such that $\Sigma \subseteq \mathsf{L}$.

Theorem 6.2 *If* Σ *axiomatizes* L, *then we have the following equivalence:* $\varphi \in \mathsf{L}$ *iff there are* $\psi_1, \ldots, \psi_n \in \Sigma$ *such that* $\vdash_{\mathsf{GQM}} [\forall \boldsymbol{p}](\psi_1 \wedge \cdots \wedge \psi_n) \to \varphi$, *where* $\boldsymbol{p}$ *is the tuple of variables occurring in* $\psi_1, \ldots, \psi_n$.

Proof. From right to left, we have:

$\vdash_{\mathsf{GQM}} [\forall \boldsymbol{p}](\psi_1 \wedge \cdots \wedge \psi_n) \to \varphi$
$\Rightarrow \varphi \in \mathsf{GQM\text{-}L}$ by $[\forall]$-necessitation to $\psi_1 \wedge \cdots \wedge \psi_n \in \mathsf{L}$ and modus ponens
$\Rightarrow \varphi \in \mathsf{L}$ by Proposition 6.1.

From left to right, the proof proceeds by induction on the length of derivations. Details are in the extended technical report [17]. Also see Remark 8.7.□

We can easily rephrase Theorem 6.2 in the language of "theories."

Definition 6.3 A $\vdash_{\mathsf{GQM}}$-*theory* is a set of GQM formulas that includes GQM and is closed under modus ponens.

Corollary 6.4 *If* $\Sigma \subseteq \mathcal{L}_\Box$ *axiomatizes a congruential modal logic* L, *then we have the following equivalence:* $\varphi \in \mathsf{L}$ *iff* φ *belongs to the smallest* $\vdash_{\mathsf{GQM}}$-*theory that includes* $[\forall]\Sigma = \{[\forall \boldsymbol{p}]\varphi \mid \varphi \in \Sigma$ *and* $\boldsymbol{p}$ *are the variables in* $\varphi\}$.

Given this reduction of modal logics to $\vdash_{\mathsf{GQM}}$-theories, we have the following.

Corollary 6.5 GQM *theoremhood is undecidable.*

Proof. In light of Theorem 6.2, a decision procedure for GQM would yield a decision procedure for every finitely axiomatizable modal logic. But there are undecidable logics with finite axiomatizations [11, § 16.4]. □

Theorem 3.5 showed that the set of GQM formulas valid over $\mathcal{C}$-BAEs is not recursively enumerable. Our completeness result in Theorem 8.6 will yield that the situation is better over general algebraic semantics.

7 Prenex forms

Our path to completeness begins with suitable normal and prenex forms.

7.1 Weak prenex forms

Definition 7.1

(i) A *nontrivial weak prenex* (NWP) *formula* is a formula of the form $\{\!\!\{Qp\}\!\!\}\,\varphi$, where $\{\!\!\{Qp\}\!\!\}$ is a nonempty sequence of GQMs and φ is a $\mathcal{L}_{\Box\mathsf{A}}$-formula.

(ii) A *normal clause* is a disjunction each disjunct of which is either (a) a literal, (b) of the form $\Box\psi$ or $\Diamond\psi$ with ψ quantifier free, or (c) a formula in NWP form.

(iii) A *conjunctive normal form weak prenex* (CNFWP) *formula* is a conjunction of normal clauses.

The following is a key lemma for the purposes of showing that formulas can be transformed into equivalent CNFWP formulas.

Lemma 7.2

(i) $\vdash (\mathsf{G}_{*}\alpha * \{Qp\}\beta) \leftrightarrow \{Qp\}(\mathsf{G}_{*}\alpha * \beta)$ *where p is not free in α;*

(ii) $\vdash \mathsf{A}(\varphi \vee \{\!\!\{Qp\}\!\!\}\psi) \leftrightarrow (\mathsf{A}\varphi \vee \{\!\!\{Qp\}\!\!\}\psi)$;

(iii) $\vdash \Box(\alpha \wedge (\varphi \vee \{Qp\}\psi)) \leftrightarrow ((\{Qp\}\psi \wedge \Box\alpha) \vee (\neg\{Qp\}\psi \wedge \Box(\alpha \wedge \varphi)))$.

Proof. See the extended technical report [17]. □

Theorem 7.3 *For every $\varphi \in \mathcal{L}_{\mathrm{GQM}}$:*

(i) *φ is provably equivalent to a CNFWP formula;*

(ii) *$\mathsf{A}\varphi$ is provably equivalent to an NWP formula.*

Proof. Proved in Appendix B as Theorem B.2. □

7.2 Pure weak prenex forms

The following special case of NWP form will be essential in relating $\mathcal{L}_{\mathrm{GQM}}$ to the first-order language in § 8.

Definition 7.4 A formula is in *pure weak prenex form* (PWP) iff it is of the form $\{\!\!\{Qp\}\!\!\}\mathsf{G}\varphi$ where $\{\!\!\{Qp\}\!\!\}$ is a sequence of $[\forall p_i]$ and $\langle\exists p_i\rangle$ GQMs only, G is either A or E, and φ is a $\mathcal{L}_{\Box\mathsf{A}}$-formula.

Theorem 7.5 *Every NWP formula is provably equivalent to a PWP formula.*

Proof. By induction on the length of the quantifier prefix. Assuming φ is a PWP formula, we must show that $\{Qp\}\varphi$ is equivalent to a PWP formula. If $\{Qp\} \in \{[\forall p], \langle\exists p\rangle\}$, there is nothing to do. Case 1: $\{Qp\} := \langle\forall p\rangle$. By Lemma 5.4.(ix)-(x), where r is not free in φ, $\langle\forall p\rangle\varphi$ is equivalent to $[\forall p]\langle\exists r\rangle\varphi$, which is a PWP formula. Case 2: $\{Qp\} := [\exists p]$. By Lemma 5.4.(ix)-(x), where r is not free in φ, $[\exists p]\varphi$ is equivalent to $\langle\exists p\rangle[\forall r]\varphi$, which is a PWP formula. □

8 Completeness via FO-theory of discriminator BAEs

Using the prenex results of § 7, we will now prove the completeness of GQM via mutual translations with the first-order theory of discriminator BAEs.

Definition 8.1 A *Boolean algebra expansion with a discriminator* ($\mathrm{BAE}_{\mathsf{A}}$) is a tuple $\mathfrak{A} = \langle A, \neg, \wedge, \bot, \top, \Box, \mathsf{A}\rangle$ where $\langle A, \neg, \wedge, \bot, \top, \Box\rangle$ is a BAE and A is the dual form of the *unary discriminator term* [19], i.e., an algebraic counterpart of the global modality: $\mathsf{A}a = \top$ if $a = \top$, and $\mathsf{A}a = \bot$ otherwise.

Let $\mathrm{FO}_{\mathrm{BAE}_{\mathsf{A}}}$ (resp. $\mathrm{FO}_{\mathrm{BAE}}$) be the set of first-order formulas in the $\mathrm{BAE}_{\mathsf{A}}$ (resp. BAE) signature (recycling Prop for our set of first-order variables).

The class of all $\mathrm{BAE}_{\mathsf{A}}$s is elementary, although it is not exactly a variety (an equationally definable class): rather, it is the class of all *simple* members of the corresponding variety [19, Thm. 3]. BAEs and $\mathrm{BAE}_{\mathsf{A}}$s are in 1-1 correspondence: $\mathrm{BAE}_{\mathsf{A}}$s have BAEs as reducts; every BAE $\mathfrak{A}$ can be trivially extended to a $\mathrm{BAE}_{\mathsf{A}}$ $\mathfrak{A}_{\mathsf{A}}$; and both operations are mutual inverses.

In a similar way, we can assign to every formula of $\mathrm{FO}_{\mathrm{BAE}_{\mathsf{A}}}$ a formula equivalent to a PWP formula (where $\sim$ and $\&$ are the negation and conjunction connectives in the first-order language, whereas $\neg$ and $\wedge$ in the first-order language are function symbols for the Boolean algebraic operations):

$$(\varphi \approx \psi)_* := \mathsf{A}(\varphi \leftrightarrow \psi) \qquad (\sim\alpha)_* := \neg(\alpha)_*$$
$$(\alpha \,\&\, \beta)_* := ((\alpha)_* \wedge (\beta)_*) \qquad (\forall p\alpha)_* := [\forall p](\alpha)_*.$$

Note that the *terms* in the $\mathrm{FO}_{\mathrm{BAE}_{\mathsf{A}}}$ formula become formulas of $\mathcal{L}_{\mathrm{GQM}}$, with the Boolean function symbols becoming propositional connectives.

In the reverse direction, define for each PWP formula:

$$(\mathsf{A}\varphi)^* := \varphi \approx \top \qquad (\mathsf{E}\varphi)^* := \varphi \not\approx \bot$$
$$([\forall p]\varphi)^* := \forall p(\varphi)^* \qquad (\langle\exists p\rangle\varphi)^* := \exists p(\varphi)^*.$$

Any A or E GQMs inside φ become function symbols in the $\mathrm{FO}_{\mathrm{BAE}_{\mathsf{A}}}$ translation.

Lemma 8.2 *For any nontrivial*[6] BAE $\mathfrak{A}$, $\theta : \mathsf{Prop} \to \mathfrak{A}$, *and* $\alpha \in \mathrm{FO}_{\mathrm{BAE}_{\mathsf{A}}}$*:*

$$\mathfrak{A}, \theta \vDash \alpha \text{ iff } \tilde{\theta}((\alpha)_*) = \top \quad \text{and} \quad \mathfrak{A}, \theta \nvDash \alpha \text{ iff } \tilde{\theta}((\alpha)_*) = \bot.$$

Proof. By induction on the complexity of α. The atomic case follows directly from properties of the connective $\leftrightarrow$, Lemma 2.5, and the fact that in a nontrivial Boolean algebra, $\top$ and $\bot$ are distinct. The Boolean cases follow from the first-order satisfaction definition, the inductive hypothesis, and the algebraic behavior of $\top$ and $\bot$. The GQM case is by Definition 2.4. □

Corollary 8.3 *For any* $\Delta \cup \{\alpha\} \subseteq \mathrm{FO}_{\mathrm{BAE}_{\mathsf{A}}}$, $\Delta \vDash_{\mathrm{FO}_{\mathrm{BAE}_{\mathsf{A}}}} \alpha$ *iff* $(\Delta)_* \vDash^{\mathsf{A}}_{\mathrm{GQM}} (\alpha)_*$.

Proof. Immediate from Lemma 8.2 and the definitions of consequence. □

Theorem 8.4

(i) *For any PWP formula* $\varphi \in \mathcal{L}_{\mathrm{GQM}}$, $\varphi \vdash^{\mathsf{A}}_{\mathsf{GQM}} ((\varphi)^*)_*$ *and* $((\varphi)^*)_* \vdash^{\mathsf{A}}_{\mathsf{GQM}} \varphi$.

(ii) *For any* $\Delta \cup \{\alpha\} \subseteq \mathrm{FO}_{\mathrm{BAE}_{\mathsf{A}}}$, $\Delta \vdash_{\mathrm{FO}_{\mathrm{BAE}_{\mathsf{A}}}} \alpha$ *implies* $(\Delta)_* \vdash^{\mathsf{A}}_{\mathsf{GQM}} (\alpha)_*$.

(iii) *For any* $\Delta \cup \{\alpha\} \subseteq \mathrm{FO}_{\mathrm{BAE}_{\mathsf{A}}}$, $\Delta \vdash_{\mathrm{FO}_{\mathrm{BAE}_{\mathsf{A}}}} \alpha$ *iff* $(\Delta)_* \vdash^{\mathsf{A}}_{\mathsf{GQM}} (\alpha)_*$.

[6] By a nontrivial BAE, we mean a BAE in which $\top \neq \bot$.

Proof. For part (i), given $\langle\exists p\rangle\psi := \neg[\forall p]\neg\psi$, we have (for a fresh q):

$$(\{\!\!\{\boldsymbol{Qp}\}\!\!\}\mathsf{A}\varphi)^* = \boldsymbol{Qp}\forall q(\varphi \approx \top) \qquad ([\boldsymbol{Qp}]\mathsf{E}\varphi)^* = \boldsymbol{Qp}\exists q(\varphi \not\approx \bot)$$
$$((\{\!\!\{\boldsymbol{Qp}\}\!\!\}\mathsf{A}\varphi)^*)_* = \{\!\!\{\boldsymbol{Qp}\}\!\!\}\mathsf{A}(\varphi \leftrightarrow \top) \quad ((\{\!\!\{\boldsymbol{Qp}\}\!\!\}\mathsf{E}\varphi)^*)_* = \{\!\!\{\boldsymbol{Qp}\}\!\!\}\mathsf{E}\neg(\varphi \leftrightarrow \bot).$$

It is an easy exercise using Lemmas 5.3 and 5.4 to show that $\{\!\!\{\boldsymbol{Qp}\}\!\!\}\mathsf{A}\varphi$ is GQM-equivalent to $\{\!\!\{\boldsymbol{Qp}\}\!\!\}\mathsf{A}(\varphi \leftrightarrow \top)$ and $\{\!\!\{\boldsymbol{Qp}\}\!\!\}\mathsf{E}\varphi$ to $\{\!\!\{\boldsymbol{Qp}\}\!\!\}\mathsf{E}\neg(\varphi \leftrightarrow \bot)$.

For part (ii), see Appendix C. Part (iii) is obtained from (ii) by noting that the opposite direction follows from the soundness of GQM (Lemma 5.7), Corollary 8.3, and the completeness of $\mathrm{FO}_{\mathrm{BAE}_\mathsf{A}}$. □

An astute reader will note here that even though $(\cdot)_*$ and $(\cdot)^*$ are mutual inverses up to equivalence, the matrix of $((\alpha)_*)^*$ consists of a single equation or its negation, including for those α whose matrix is a nontrivial conjunction of disjunctions. This is in keeping with general discriminator theory [34].

Corollary 8.5

(i) *For any* $\Delta \cup \{\alpha\} \subseteq \mathrm{FO}_{\mathrm{BAE}_\mathsf{A}}$, $\Delta \vDash_{\mathrm{FO}_{\mathrm{BAE}_\mathsf{A}}} \alpha$ *iff* $(\Delta)_* \vdash^{\mathsf{A}}_{\mathsf{GQM}} (\alpha)_*$.

(ii) *For any set of PWP fomulas* $\Gamma \cup \{\varphi\} \subseteq \mathcal{L}_{\mathrm{GQM}}$, $\Gamma \vdash^{\mathsf{A}}_{\mathsf{GQM}} \varphi$ *iff* $(\Gamma)^* \vDash_{\mathrm{FO}_{\mathrm{BAE}_\mathsf{A}}} (\varphi)^*$.

Proof. For part (i), we proceed as follows:

$$\begin{aligned}\Delta \vDash_{\mathrm{FO}_{\mathrm{BAE}_\mathsf{A}}} \alpha &\Leftrightarrow \Delta \vdash_{\mathrm{FO}_{\mathrm{BAE}_\mathsf{A}}} \alpha && \text{by completeness of } \mathrm{FO}_{\mathrm{BAE}_\mathsf{A}}\\ &\Leftrightarrow (\Delta)_* \vdash^{\mathsf{A}}_{\mathsf{GQM}} (\varphi)_* && \text{by Theorem 8.4.(iii).}\end{aligned}$$

For part (ii), we have:

$$\begin{aligned}(\Gamma)^* \vDash_{\mathrm{FO}_{\mathrm{BAE}_\mathsf{A}}} (\varphi)^* &\Leftrightarrow ((\Gamma)^*)_* \vdash^{\mathsf{A}}_{\mathsf{GQM}} ((\varphi)^*)_* && \text{by part (i)}\\ &\Leftrightarrow \Gamma \vdash^{\mathsf{A}}_{\mathsf{GQM}} \varphi && \text{by Theorem 8.4.(i).}\end{aligned}$$ □

Theorem 8.6 (Completeness) *For any* $\Gamma \cup \{\varphi\} \subseteq \mathcal{L}_{\mathrm{GQM}}$,

$$\Gamma \vdash^{\mathsf{A}}_{\mathsf{GQM}} \varphi \textit{ iff } \Gamma \vDash^{\mathsf{A}}_{\mathrm{GQM}} \varphi.$$

Proof. First, as far as the consequence relation $\vdash^{\mathsf{A}}_{\mathsf{GQM}}$ is concerned, we can prefix all formulas in $\Gamma \cup \{\varphi\}$ by A (by Lemma 5.6) and then transform them into equivalent PWP formulas (Theorems 7.3.(ii) and 7.5). Corollary 8.5 established that $\Gamma \vdash^{\mathsf{A}}_{\mathsf{GQM}} \varphi$ iff $(\Gamma)^* \vDash_{\mathrm{FO}_{\mathrm{BAE}_\mathsf{A}}} (\varphi)^*$. By Corollary 8.3, this is equivalent to $((\Gamma)^*)_* \vDash_{\mathrm{FO}_{\mathrm{BAE}_\mathsf{A}}} ((\varphi)^*)_*$. The result then follows by Theorem 8.4.(i). □

Remark 8.7 The use of Theorem 8.4 in this section should be compared with [6, Thm. 3.7] and [28, Lem. 19]. Our success in establishing the equivalence between the (global) GQM-consequence relation and that of $\mathrm{FO}_{\mathrm{BAE}}$ means that we can internalize the metatheory of modal logics concisely and in a generic way using "bridge theorems" of abstract algebraic logic [6,1,14]. For lack of space, we are not pursuing this option further in this paper, but a good illustration of how GQM can be used in such a generalization can be found in our recent paper [18, § 9], which in fact led us to the invention of this formalism. Bases of admissible rules (see, e.g., [31]) seem to provide another promising candidate. Details and more examples will be provided in a sequel paper.

9 Conclusions

We have seen that GQM provides the sought-after way of viewing "modal logics" as theories relative to one logical system, while also offering a generic and conservative way to enrich any modal logic with propositional quantifiers. This study led us to new perspectives on the first-order correspondence language for BAEs (§ 8) on the one hand and SOPML on the other hand (§ 3). In the first case, the equivalence between $\mathrm{FO}_{\mathrm{BAE}_{\mathsf{A}}}$-consequence and $\vdash^{\mathsf{A}}_{\mathsf{GQM}}$-consequence illustrates a curious use of techniques from abstract algebraic logic (cf. Remark 8.7) beyond their usual scope. In the second case, we were led to new prenex normal form results. We also believe that focusing on the syntax of GQM and its algebraic semantics can lead to a clarification of philosophical problems concerning propositional quantification, such as Kaplan's paradox (§ 4).

Along the way, a number of issues have been postponed to a follow-up paper. In particular, we mentioned that over dual, set-based semantics GQM-consequence may be intractable, as it is over Kripke frames. On the other hand, given that modal logics can be identified with (fragments of) universal GQM-theories, the existence of a rich modal completeness apparatus indicates that for suitable fragments of GQM and formulas of specific syntactic shapes, developing GQM model theory is not hopeless and may yield additional insights in modal logic. This will be a subject of future investigation, as will the systematic internalization of "bridge theorems" mentioned in Remark 8.7.

Another issue we have not touched on is that of Gentzen-style proof theory for (well-behaved fragments of) GQM. Since the difficulty of developing Gentzen systems for many modal logics was behind the idea "that the great proliferation of modal logics is an epidemy from which modal logic ought to be cured" [8, p. 25], it would be of interest to see if GQM could help here as well.

A further intriguing possibility is that of weakening the classical base of GQM to an intuitionistic one. Just as modal logics are $[\forall]$-universal theories in classical GQM, intermediate (modal) logics could be $[\forall]$-universal theories in intuitionistic GQM. There is a connection here with the origins of modal logic: not only was C. I. Lewis a proponent of propositional quantification in modal logic, as well as perhaps the earliest opponent of modal proliferation, but also he seemed interested in the idea of strict implication on an intuitionistic base (see [30] for discussion). It is argued in [30] that moving Lewis's strict implication to an intuitionistic base is indeed a conceptually fruitful step. The enrichment of that system with GQM may be the ultimate Lewisian logic.

One need not stop at intuitionistic logic. With the power of the global quantificational modality, there is the possibility that even vaster swaths of "logics" could become special theories over a generalized version of GQM. Alternatively, the classical base of GQM could be retained, while the connectives of different "logics" are treated as modal operators in BAEs, whose behavior is governed by [∀]-universal GQM formulas. Under one of these approaches, a version of GQM could bring us closer to the idea of "one logic to rule them all."

Acknowledgement

We thank Johan van Benthem, Yifeng Ding, Peter Fritz, Lloyd Humberstone, Michael Sammler, and the referees for AiML for their helpful feedback.

Appendix

A Proof of Proposition 3.3

To prove Proposition 3.3, it is convenient to have another equivalence for $\mathsf{A}\exists p\psi$ (other than ($\star$) in the proof of Proposition 3.2) in part (ii) of the following.

Lemma A.1 *The following are valid in all $\mathcal{C}$-BAEs:*

(i) $\mathsf{A}\forall p\psi \leftrightarrow \forall p\mathsf{A}\psi$;

(ii) $\mathsf{A}\exists p\psi \leftrightarrow \forall q\mathsf{A}(\mathsf{E}q \to \exists r\mathsf{A}(\mathsf{E}r \wedge (r \to q) \wedge \exists p\mathsf{A}(r \to \psi)))$ *where q and r do not occur in ψ.*

Proof. Part (i) again follows from the distribution of A over arbitrary meets.

Part (ii) follows from the Boolean algebraic fact that for any $\mathcal{C}$-BA $\mathfrak{A}$ and $Y \subseteq \mathfrak{A}$, we have $\bigvee Y = \top$ (take $Y = \{\tilde{\gamma}(\psi) \mid \gamma \sim_p \theta\}$) iff for all $x \in \mathfrak{A}$ (take x as the semantic value of q), if $x \neq 0$, then there exists a $y \in Y$ such that $x \wedge y \neq 0$, which is equivalent to there being a $z \in \mathfrak{A}$ (take z to be the semantic value of r) such that $z \neq 0$, $z \leq x$, and z is under some element of Y. □

Proposition A.2 *If α is a prenex SOPML$_{\mathsf{A}}$ formula, then $\mathsf{A}\alpha$ is equivalent over $\mathcal{C}$-BAEs to a GQM formula.*

Proof. We continue to regard GQM formulas as SOPML$_{\mathsf{A}}$ formulas as in § 3.

The proof is by induction on the number of quantifiers in the prenex formula $\alpha := Q_1p_1 \dots Q_np_n\chi$. Let $\alpha' := Q_2p_2 \dots Q_np_n\chi$.

Case 1: $Q_1 = \forall$. Then by Lemma A.1.(i), $\mathsf{A}\alpha$ is equivalent to $\forall p_1\mathsf{A}\alpha'$, which is equivalent to $\forall p_1\mathsf{AA}\alpha'$. Since α' has fewer quantifiers than α, by the inductive hypothesis $\mathsf{A}\alpha'$ is equivalent to a GQM formula β. Hence $\forall p_1\mathsf{AA}\alpha'$ is equivalent to the GQM formula $\forall p_1\mathsf{A}\beta$.

Case 2: $Q_1 = \exists$. Then by Lemma A.1.(ii), $\mathsf{A}\alpha$ is equivalent to

(1) $\forall q\mathsf{A}(\mathsf{E}q \to \exists r\mathsf{A}(\mathsf{E}r \wedge (r \to q) \wedge \exists p_1\mathsf{A}(r \to \alpha')))$

and hence to

(2) $\forall q\mathsf{A}(\mathsf{E}q \to \exists r\mathsf{A}(\mathsf{E}r \wedge (r \to q) \wedge \exists p_1\mathsf{AA}(r \to \alpha')))$.

Since r is not among $p_2, \dots, p_n$, (2) is equivalent to

(3) $\forall q\mathsf{A}(\mathsf{E}q \to \exists r\mathsf{A}(\mathsf{E}r \wedge (r \to q) \wedge \exists p_1\mathsf{AA}Q_2p_2 \dots Q_np_n(r \to \chi)))$.

Since $Q_2p_2 \dots Q_np_n(r \to \chi)$ has fewer quantifiers than α, by the inductive hypothesis $\mathsf{A}Q_2p_2 \dots Q_np_n(r \to \chi)$ is equivalent to a GQM formula γ. Hence (3) is equivalent to the GQM formula

(4) $\forall q\mathsf{A}(\mathsf{E}q \to \exists r\mathsf{A}(\mathsf{E}r \wedge (r \to q) \wedge \exists p_1\mathsf{A}\gamma))$. □

B Proof of Theorem 7.3

In order to prove Theorem 7.3 (Theorem B.2), we first need the following lemma (in the proof, 'PL' stands for propositional logic).

Lemma B.1

(i) *If* $\alpha_1, \ldots, \alpha_m$ *are each NWP formulas, then* $\alpha_1 * \cdots * \alpha_m$ *is provably equivalent to* $\mathsf{G}_*(\alpha_1 * \cdots * \alpha_m)$.

(ii) *If* $\alpha_1, \ldots, \alpha_n$ *are each NWP formulas, then* $\alpha_1 * \cdots * \alpha_n$ *is provably equivalent to an NWP formula.*

(iii) *If* α *is a normal clause, then* $\mathsf{A}\alpha$ *is provably equivalent to an NWP formula.*

(iv) *If* φ *is a CNFWP formula, then* $\mathsf{A}\varphi$ *is provably equivalent to an NWP formula.*

Proof. (i) We have:

(1) $\vdash (\alpha_1 * \cdots * \alpha_n) \leftrightarrow (\mathsf{G}_*\alpha_1 * \cdots * \mathsf{G}_*\alpha_n)$ by Lemma 5.4.(xi)-(xii) since each α_i is an NWP formula

(2) $\vdash (\mathsf{G}_*\alpha_1 * \cdots * \mathsf{G}_*\alpha_n) \leftrightarrow \mathsf{G}_*(\alpha_1 * \cdots * \alpha_n)$ by Lemma 5.4.(ii)

(3) $\vdash (\alpha_1 * \cdots * \alpha_n) \leftrightarrow \mathsf{G}_*(\alpha_1 * \cdots * \alpha_n)$ from (1) and (2) by PL.

(ii) By Lemma 5.2, we may assume without loss of generality that (a) no propositional variable occurs both free and bound in $\alpha_1, \ldots, \alpha_n$. The proof is by induction on the number of nonvacuous GQMs (i.e., GQMs binding variables, unlike A and E) occurring in $\alpha_1, \ldots, \alpha_n$. First, by Lemma 5.4.(xi)-(xii), we may replace each α_i with an equivalent NWP formula α'_i containing no more GQMs and in which no vacuous GQM occurs before a nonvacuous GQM. Thus, if no α'_i begins with a nonvacuous GQM, then $\alpha'_1 * \cdots * \alpha'_n$ is already an NWP formula, so we are done. Now suppose that some α'_i, say α'_n, is of the form $\{Qp\}\varphi$ where $\{Qp\}$ is nonvacuous. Since $\alpha'_1, \ldots, \alpha'_{n-1}$ are each NWP formulas, $\alpha := (\alpha'_1 * \cdots * \alpha'_{n-1})$ is equivalent to $\mathsf{G}_*\alpha$ by part (i). Then we have:

(4) $\vdash \alpha \leftrightarrow \mathsf{G}_*\alpha$

(5) $\vdash (\alpha'_1 * \cdots * \alpha'_n) \leftrightarrow (\mathsf{G}_*\alpha * \{Qp\}\varphi)$ by (4) and PL

(6) $\vdash (\mathsf{G}_*\alpha * \{Qp\}\varphi) \leftrightarrow (\mathsf{G}_*\alpha * \{Qp\}\{Qr\}\varphi)$ where r is not free in φ or α, by Lemma 5.4.(ix)-(x)

(7) $\vdash (\mathsf{G}_*\alpha * \{Qp\}\{\forall r\}\varphi) \leftrightarrow \{Qp\}(\mathsf{G}_*\alpha * \{Qr\}\varphi)$ by Lemma 7.2.(i), since p is not free in α by (a) above

(8) $\vdash \{Qp\}(\mathsf{G}_*\alpha * \{\forall r\}\varphi) \leftrightarrow \{Qp\}(\alpha * \{Qr\}\varphi)$ by (4), PL, and Lemma 5.3

(9) $\vdash (\alpha'_1 * \cdots * \alpha'_n) \leftrightarrow \{Qp\}(\alpha'_1 * \cdots * \alpha'_{n-1} * \{Qr\}\varphi)$ by (5)–(8) by PL.

Since $\alpha'_1, \ldots, \alpha'_{n-1}, \{Qr\}\varphi$ are each NWP formulas, and there is one fewer nonvacuous GQM in $\alpha'_1, \ldots, \alpha'_{n-1}, \{Qr\}\varphi$ than in $\alpha'_1, \ldots, \alpha'_n$, the inductive hypothesis implies that $\alpha'_1 * \cdots * \alpha'_{n-1} * \{Qr\}\varphi$ is equivalent to an NWP formula. It follows by Lemma 5.3 that $\{Qp\}(\alpha'_1 * \cdots * \alpha'_{n-1} * \{Qr\}\varphi)$ is equivalent to an NWP formula, so by (9), $\alpha'_1 * \cdots * \alpha'_n$ is equivalent to an NWP formula, which means that $\alpha_1 * \cdots * \alpha_n$ is equivalent to an NWP formula.

(iii) The proof is by induction on the number of disjuncts in a normal clause. Suppose α is $\beta_1 \vee \cdots \vee \beta_m$. If no β_k is an NWP formula, then $\mathsf{A}\alpha$ is already an NWP formula. Suppose $\beta_m := \{\!\!\{ Qp \}\!\!\}\gamma$ is an NWP formula, and let $\beta := \beta_1 \vee \cdots \vee \beta_{m-1}$. Then β is a normal clause, so by the inductive hypothesis, $\mathsf{A}\beta$ is equivalent to an NWP formula δ. Now we have:

(10) $\vdash \alpha \leftrightarrow (\beta \vee \{\!\!\{ Qp \}\!\!\}\gamma)$ by our assumption of what α is

(11) $\vdash \mathsf{A}\alpha \leftrightarrow \mathsf{A}(\beta \vee \{\!\!\{ Qp \}\!\!\}\gamma)$ from (10) by Lemma 5.4.(iii)

(12) $\vdash \mathsf{A}(\beta \vee \{\!\!\{ Qp \}\!\!\}\gamma) \leftrightarrow (\mathsf{A}\beta \vee \{\!\!\{ Qp \}\!\!\}\gamma)$ by Lemma 7.2.(ii)

(13) $\vdash \mathsf{A}\beta \leftrightarrow \delta$ by the inductive hypothesis

(14) $\vdash (\mathsf{A}\beta \vee \{\!\!\{ Qp \}\!\!\}\gamma) \leftrightarrow (\delta \vee \{\!\!\{ Qp \}\!\!\}\gamma)$ from (12) and (13) by PL

(15) $\vdash \mathsf{A}\alpha \leftrightarrow (\delta \vee \{\!\!\{ Qp \}\!\!\}\gamma)$ from (11), (12), and (14) by PL.

Since δ is an NWP formula, $\delta \vee \{\!\!\{ Qp \}\!\!\}\gamma$ is a disjunction of NWP formulas. Thus, by part (ii), $\delta \vee \{\!\!\{ Qp \}\!\!\}\gamma$ is equivalent to an NWP formula, and hence by (15), $\mathsf{A}\alpha$ is equivalent to an NWP formula.

(iv) Suppose φ is a CNFWP formula $\alpha_1 \wedge \cdots \wedge \alpha_n$. Hence $\mathsf{A}\varphi$ is equivalent to $\mathsf{A}\alpha_1 \wedge \cdots \wedge \mathsf{A}\alpha_n$ by Lemma 5.4.(ii). Since each α_i is a normal clause, part (iii) implies that each $\mathsf{A}\alpha_i$ is equivalent to an NWP formula χ_i. Hence $\mathsf{A}\alpha_1 \wedge \cdots \wedge \mathsf{A}\alpha_n$ is equivalent to $\chi_1 \wedge \cdots \wedge \chi_n$, which by part (ii) is equivalent to an NWP formula. Thus, φ is equivalent to an NWP formula. □

Theorem B.2 *For every $\varphi \in \mathcal{L}_{\mathrm{GQM}}$:*

(i) *φ is provably equivalent to a CNFWP formula;*

(ii) *$\mathsf{A}\varphi$ is provably equivalent to an NWP formula.*

Proof. We prove part (i) by induction on φ. The base case for propositional variables is immediate. Suppose φ is $\neg\psi$. By the inductive hypothesis, ψ is equivalent to a CNFWP formula. One then uses de Morgan and distributive laws to show that $\neg\psi$ is also equivalent to a CNFWP formula. Suppose φ is $\psi_1 \wedge \psi_2$. By the inductive hypothesis, ψ_1 and ψ_2 are both equivalent to CNFWP formulas ψ_1' and ψ_2'. Then $\psi_1 \wedge \psi_2$ is equivalent to the CNFWP formula $\psi_1' \wedge \psi_2'$.

Suppose φ is $[\forall p]\psi$. By the inductive hypothesis, ψ is equivalent to a CNFWP formula χ, which implies that $\mathsf{A}\psi$ is equivalent to $\mathsf{A}\chi$ by Lemma 5.3. Hence $[\forall p]\psi$, which is equivalent to $[\forall p]\mathsf{A}\psi$ by Lemma 5.4.(ix), is equivalent to $[\forall p]\mathsf{A}\chi$ by Lemma 5.3. By Lemma B.1.(iv), $\mathsf{A}\chi$ is equivalent to an NWP formula, from which it follows by Lemma 5.3 that $[\forall p]\mathsf{A}\chi$ is equivalent to an NWP formula. Such a formula is in CNFWP.

Suppose φ is $\Box\psi$. By the inductive hypothesis, ψ is equivalent to a CNFWP formula $\alpha_1 \wedge \cdots \wedge \alpha_n$. We will prove that for any CNFWP formula $\sigma_1 \wedge \cdots \wedge \sigma_k$, $\Box(\sigma_1 \wedge \cdots \wedge \sigma_k)$ is equivalent to a formula in CNFWP, by induction on the number of GQMs occurring in $\sigma_1 \wedge \cdots \wedge \sigma_k$. If no σ_i contains a disjunct in NWP, then no σ_i contains a GQM, which means $\Box(\sigma_1 \wedge \cdots \wedge \sigma_k)$ is already in CNFWP. So suppose that some σ_i, say σ_k, contains as a disjunct an NWP formula $\{Qp\}\gamma$. Hence σ_k is equivalent to $\beta \vee \{Qp\}\gamma$ for a normal clause β. Let $\sigma := \sigma_1 \wedge \cdots \wedge \sigma_{k-1}$. Thus, $\Box(\sigma_1 \wedge \cdots \wedge \sigma_k)$ is equivalent to $\Box(\sigma \wedge (\beta \vee \{Qp\}\gamma))$, which by Lemma 7.2.(iii) is equivalent to

(1) $(\{Qp\}\gamma \wedge \Box\sigma) \vee (\neg\{Qp\}\gamma \wedge \Box(\sigma \wedge \beta))$.

Now σ and $\sigma \wedge \beta$ are CNFWP formulas containing fewer GQMs than $\sigma_1 \wedge \cdots \wedge \sigma_k$. Hence by the inductive hypothesis, there are CNFWP formulas χ_1 and χ_2 such that $\Box\sigma$ is equivalent to χ_1 and $\Box(\sigma \wedge \beta)$ is equivalent to χ_2. Thus, $\Box(\sigma_1 \wedge \cdots \wedge \sigma_k)$ is equivalent to

(2) $(\{Qp\}\gamma \wedge \chi_1) \vee (\neg\{Qp\}\gamma \wedge \chi_2)$.

Since $\{Qp\}\gamma$ is an NWP formula and χ_1 and χ_2 are CNFWP formulas, (2) can be transformed into an equivalent CNFWP using distributive laws and the fact that $\neg\{Qp\}\gamma$ is equivalent to the NWP formula $\overline{\{Qp\}}\neg\gamma$.

Part (ii) follows from part (i) and Lemma B.1.(iv). □

C Proof of Theorem 8.4.(ii)

In order to prove Theorem 8.4.(ii), we first recall the needed first-order apparatus, for which we follow Enderton [12]. A *generalization* of a first-order formula φ is any formula of the form $\forall p_1 \ldots \forall p_n \varphi$ for $n \geq 0$. Enderton takes as axioms all generalizations of the following:

- all substitution instances of propositional tautologies;
- $\forall p\varphi \to \varphi^p_t$ where the term t is substitutable for p in φ;
- $\forall p(\varphi \to \psi) \to (\forall p\varphi \to \forall p\psi)$;
- $\varphi \to \forall p\varphi$ where p does not occur free in φ;
- $p \approx p$, and $p \approx q \to (\varphi \to \varphi')$ where φ is atomic and φ' is obtained from φ by replacing p in zero or more places by q.

In addition, we add all generalizations of the following axioms for the elementary theory of nontrivial discriminator BAEs:

- first-order axioms of Boolean algebras;
- $\forall p((p \approx \top \,\&\, \mathsf{A}p \approx \top)\,\mathrm{OR}\,(p \not\approx \top \,\&\, \mathsf{A}p \approx \bot))$ and $\top \not\approx \bot$.

Let $\Gamma \vdash_{\mathrm{FO}_{\mathrm{BAE}_\mathsf{A}}} \varphi$ iff φ belongs to the smallest set of $\mathrm{FO}_{\mathrm{BAE}_\mathsf{A}}$ formulas that includes all the axioms above, is closed under modus ponens, and includes Γ.

Lemma C.1 *For any $\varphi \in \mathrm{FO}_{\mathrm{BAE}_\mathsf{A}}$, term t and variable p:*

(i) *if p is not free in φ, then p is not free in $(\varphi)_*$;*

(ii) *if t is substitutable for p in φ, then t is substitutable for p in $(\varphi)_*$;*

(iii) $(\varphi^p_t)_* = ((\varphi)_*)^p_t$.

Lemma C.2 *For every $\varphi \in \mathrm{FO}_{\mathrm{BAE}_\mathsf{A}}$, $\vdash_{\mathsf{GQM}} (\varphi)_* \leftrightarrow \mathsf{A}(\varphi)_*$.*

Proof. A straightforward induction using Lemma 5.4. □

We are now ready to prove Theorem 8.4.(ii).

Proof. The proof is by induction on the length of $\vdash_{\mathrm{FOBAE}_{\mathsf{A}}}$ proofs. We first check that the translation of each axiom is a theorem of GQM. Since GQM has the $[\forall]$-necessitation rule that if $\vdash_{\mathsf{GQM}} \varphi$, then $\vdash_{\mathsf{GQM}} [\forall p]\varphi$, it suffices to check that each of the ungeneralized axioms translates to a theorem of GQM:

- The translation of any propositional tautology is clearly also a propositional tautology.
- By Lemma C.1.(iii), $(\forall p\varphi \to \varphi^p_t)_* = [\forall p](\varphi)_* \to ((\varphi)_*)^p_t$, which by Lemma C.1.(ii) is an instance of instantiation.
- $(\forall p(\varphi \to \psi) \to (\forall p\varphi \to \forall p\psi))_* = [\forall p]((\varphi)_* \to (\psi)_*) \to ([\forall p](\varphi)_* \to [\forall p](\psi)_*)$, which is an instance of distribution.
- $(\varphi \to \forall p\varphi)_* = (\varphi)_* \to [\forall p](\varphi)_*$, and we have $\vdash (\varphi)_* \to \mathsf{A}(\varphi)_*$ by Lemma C.2 and hence $\vdash (\varphi)_* \to [\forall p](\varphi)_*$ by Lemma 5.2 since p is not free in $(\varphi)_*$ (by Lemma C.1).
- $(p \approx p)_* := \mathsf{A}(p \leftrightarrow p)$, which is obtained from the tautology $p \leftrightarrow p$ by $[\forall]$-necessitation.
- $(p \approx q \to (\varphi \to \varphi'))_* = \mathsf{A}(p \leftrightarrow q) \to ((\varphi)_* \to ((\varphi)_*)')$ where $((\varphi)_*)'$ is obtained from $(\varphi)_*$ by replacing the appropriate occurrences of p by q. Proving that $\vdash_{\mathsf{GQM}} \mathsf{A}(p \leftrightarrow q) \to ((\varphi)_* \to ((\varphi)_*)')$ is routine.
- The translation of any axiom of Boolean algebra is clearly derivable in GQM using PL and $[\forall]$-necessitation.
- $(\forall p((p \approx \top \,\&\, \mathsf{A}p \approx \top)\ \mathrm{OR}\ (p \not\approx \top \,\&\, \mathsf{A}p \approx \bot)))_*$ is

$$[\forall p]\big((\mathsf{A}(p \leftrightarrow \top) \wedge \mathsf{A}(\mathsf{A}p \leftrightarrow \top)) \vee (\neg\mathsf{A}(p \leftrightarrow \top) \wedge \mathsf{A}(\mathsf{A}p \leftrightarrow \bot))\big),$$

  which is straightforward to derive using Lemma 5.4 and $[\forall]$-necessitation.
- $(\top \neq \bot)_* = \neg\mathsf{A}(\top \leftrightarrow \bot)$, which is derivable by instantiation and PL.

Finally, any application of modus ponens for $\vdash_{\mathrm{FOBAE}_{\mathsf{A}}}$ can be matched—using the inductive hypothesis—by an application of modus ponens for $\vdash^{\mathsf{A}}_{\mathsf{GQM}}$. $\Box$

References

[1] Andréka, H., I. Németi and I. Sain, *Algebraic logic*, in: D. M. Gabbay and F. Guenthner, editors, *Handbook of Philosophical Logic*, Kluwer, Dordrecht, 2001, 2nd edition pp. 133–249.

[2] Bacon, A., J. Hawthorne and G. Uzquiano, *Higher-order free logic and the Prior-Kaplan paradox*, Canadian Journal of Philosophy **46** (2016), pp. 493–541.

[3] Barcan, R. C., *The identity of individuals in a strict functional calculus of second order*, The Journal of Symbolic Logic **12** (1947), pp. 12–15.

[4] van Benthem, J., "Modal Logic for Open Minds," CSLI Publications, Stanford, 2010.

[5] Blackburn, P., M. de Rijke and Y. Venema, "Modal Logic," Cambridge Tracts in Theoretical Computer Science **53**, Cambridge University Press, Cambridge, 2001.

[6] Blok, W. J. and B. Jónsson, *Equivalence of consequence operations*, Studia Logica **83** (2006), pp. 91–110.

[7] Bull, R. A., *On modal logic with propositional quantifiers*, The Journal of Symbolic Logic **34** (1969), pp. 257–263.

[8] Bull, R. A. and K. Segerberg, *Basic modal logic*, in: D. M. Gabbay and F. Guenthner, editors, *Handbook of Philosophical Logic, Vol. 3*, Kluwer Academic Publishers, Dordrecht, 2001, 2nd edition pp. 1–82.

[9] ten Cate, B., *Expressivity of second order propositional modal logic*, Journal of Philosophical Logic **35** (2006), pp. 209–223.

[10] ten Cate, B. and T. Litak, *The importance of being discrete*, Technical Report PP-2007-39, Institute for Logic, Language and Computation, University of Amsterdam (2007).

[11] Chagrov, A. V. and M. Zakharyaschev, "Modal Logic," Oxford Logic Guides, Clarendon Press, Oxford, 1997.

[12] Enderton, H. B., "A Mathematical Introduction to Logic," Harcourt Academic Press, 2001, 2nd edition.

[13] Fine, K., *Propositional quantifiers in modal logic*, Theoria **36** (1970), pp. 336–346.

[14] Font, J. M., R. Jansana and D. Pigozzi, *A survey of abstract algebraic logic*, Studia Logica **74** (2003), pp. 13–97.

[15] Holliday, W. H., *Possibility frames and forcing for modal logic (February 2018)* (2018), UC Berkeley Working Paper in Logic and the Methodology of Science. URL `https://escholarship.org/uc/item/0tm6b30q`

[16] Holliday, W. H., *A note on algebraic semantics for S5 with propositional quantifiers*, Notre Dame Journal of Formal Logic (Forthcoming). URL `https://escholarship.org/uc/item/303338xr`

[17] Holliday, W. H. and T. Litak, *One modal logic to rule them all? (extended technical report)* (2018), UC Berkeley Working Paper in Logic and the Methodology of Science. URL `https://escholarship.org/uc/item/07v9360j`

[18] Holliday, W. H. and T. Litak, *Complete additivity and modal incompleteness*, Review of Symbolic Logic (Forthcoming), URL `http://www.escholarship.org/uc/item/8pp4d94t`.

[19] Jipsen, P., *Discriminator varieties of Boolean algebras with residuated operators*, in: *Algebraic Methods in Logic and in Computer Science*, Banach Center Publications **28**, Institute of Mathematics, Polish Academy of Sciences, Warszawa, 1993 pp. 239–252.

[20] Kaminski, M. and M. Tiomkin, *The expressive power of second-order propositional modal logic*, Notre Dame Journal of Formal Logic **37** (1996), pp. 35–43.

[21] Kaplan, D., *S5 with quantifiable propositional variables*, The Journal of Symbolic Logic **35** (1970), p. 355.

[22] Kaplan, D., *A problem in possible world semantics*, in: W. Sinnott-Armstrong, D. Raffman and N. Asher, editors, *Modality, Morality, and Belief: Essays in Honor of Ruth Barcan Marcus*, Cambridge University Press, Cambridge, 1995 pp. 41–52.

[23] Kripke, S., *A completeness theorem in modal logic*, Journal of Symbolic Logic **24** (1959), pp. 1–14.

[24] Lewis, C. and C. Langford, "Symbolic Logic," The Century Company, New York, 1932.

[25] Litak, T., *Modal incompleteness revisited*, Studia Logica **76** (2004), pp. 329–342.

[26] Litak, T., "An Algebraic Approach to Incompleteness in Modal Logic," Ph.D. thesis, Japan Advanced Institute of Science and Technology (2005).

[27] Litak, T., *On notions of completeness weaker than Kripke completeness*, in: R. Schmidt, I. Pratt-Hartmann, M. Reynolds and H. Wansing, editors, *Advances in Modal Logic, Vol. 5*, College Publications, London, 2005 pp. 149–169.

[28] Litak, T., *Isomorphism via translation*, in: G. Governatori, I. M. Hodkinson and Y. Venema, editors, *Advances in Modal Logic, Vol. 6*, College Publications, London, 2006 pp. 333–351.

[29] Litak, T., *Stability of the Blok theorem*, Algebra Universalis **58** (2008), pp. 385–411.

[30] Litak, T. and A. Visser, *Lewis meets Brouwer: Constructive strict implication*, Indagationes Mathematicae **29** (2018), pp. 36–90, a special issue on "L.E.J. Brouwer, fifty years later". URL `https://arxiv.org/abs/1708.02143`.

[31] Rybakov, V. V., "Admissibility of Logical Inference Rules," Elsevier, Amsterdam, 1997.

[32] Thomason, S. K., *Reduction of second-order logic to modal logic*, Zeitschrift für mathematische Logik und Grundlagen der Mathematik **21** (1975), pp. 107–114.

[33] Vosmaer, J., *A new proof of an old incompleteness theorem*, Bulletin of the Section of Logic **39** (2010), pp. 199–204.

[34] Werner, H., "Discriminator Algebras," Akademie Verlag, Berlin, 1978.

Cut-Free Modal Theory of Definite Descriptions

Andrzej Indrzejczak [1]

Department of Logic, University of Łódź
Lindleya 3/5, 90–131 Łódź
e-mail:andrzej.indrzejczak@filozof.uni.lodz.pl

Abstract

We present a standard sequent calculus for first-order modal logic with definite descriptions. It is equivalent to Garson's system which is a generalization and simplification of the approach originally introduced in Q3 system of Thomason. This particular theory of definite descriptions is based on free logic with identity and existence predicate where both rigid and nonrigid terms are present. We show that, despite of the complexities unavoidable for any characterization of definite descriptions, it is possible to provide a structural proof theoretic analysis of such theory. In particular, cut elimination theorem for this sequent calculus is proved in a constructive manner. We briefly consider some possible extensions of this calculus. Finally, some other approaches to modal description theories, due to Goldblatt, and to Fitting and Mendelsohn, are also discussed from the standpoint of structural proof theory.

Keywords: first-order modal logic, free logic, definite descriptions, sequent calculus, cut elimination.

1 Introduction

The aim of this paper is to present a cut-free sequent calculus for first-order modal logic with definite descriptions. It seems that a satisfactory structural proof theory for such logics is not yet developed. We mean by that a formalization provided in terms of sequent calculi enabling analysis of proofs. Roughly speaking, in order to allow such an analysis, suitable sequent calculus must be defined in terms of rules having analytical character and admitting cut elimination. These requirements will be discussed in more detail in section 4.

The first problem requiring a clarification is which logic should be taken into account. There is a variety of first-order modal logics based on different assumptions concerning such questions as existence or denotation [2] and we can hardly say that some of them are treated as commonly acceptable. The same

---

[1] The results reported in this paper are supported by the National Science Centre, Poland (grant number: DEC-2017/25/B/HS1/01268).

[2] For a survey see e.g. Garson [10] or Fitting and Mendelsohn [7].

may be said about theories of definite descriptions. Since the publication of B. Russell's famous paper "On Denoting" [28] several theories were formulated but neither can be claimed to be a definitive solution to the problem of descriptions. In particular, a treatment of improper descriptions which fail to designate a unique object leads to significant differences between several approaches. Many researchers dealing with the problem of definite descriptions follow the Russellian route and eliminate them in favour of ordinary first-order logic with identity but such reductionist approach has serious disadvantages. However, there is an older tradition, starting with Frege [8], [9], in which definite descriptions are treated as genuine terms and a fixed denotation is assigned to all improper descriptions. This account was formally developed by Kalish and Montague [21] but it has also some disadvantages. It seems that a detailed treatment of definite descriptions requires richer resources beyond those offered by classical logic. In fact, a construction of a satisfactory theory of definite descriptions was one of the aims of developing free logics, as reported by Bencivenga [2]. Lambert [24] shows also that free logics offer an useful setting for comparison of Russellian and Fregean approaches to definite descriptions.

Modal logics and semantics of possible worlds provide even better framework for construction of such a theory. A good witness to this claim is a detailed study of first-order modal logics with complex terms of different kinds, including definite descriptions, developed by Fitting and Mendelsohn [7]. It is probably the most subtle theory of definite descriptions which is rich enough for expressing differences between terms that designate existent and nonexistent object, and terms that do not (and even cannot) designate. As such it certainly deserves attention but it is difficult to provide a suitable sequent formalization of it. We will comment on these problems in the last section.

Another formalization of modal logic with definite descriptions, also discussed in the last section, is due to Goldblatt [12]. His approach does not require introduction of some extra machinery beyond standard apparatus. It is in fact not difficult to provide adequate sequent calculus for it but the problem of proving cut elimination is open.

It seems that for the aims of proof theoretic analysis, the approach presented by Garson [11] is a better option. He provided elegant, relatively simple, yet well justified treatment of definite descriptions on the basis of some variant of free logic. It is a slightly strenghtened version of Lambert's system [23] of minimal free description theory MFD in the language with modalities. A strenghtening is due not only to the addition of modalities but also to the addition of a rule specifying the relationship between rigid and nonrigid terms. The first version of such logic was developed by Thomason [30] under the name Q3 but with some unecessary complications. Garson provides much simpler formalization of this system in terms of natural deduction, and this formulation will be our basic point of reference. In what follows we present a sequent calculus formalization of Garson's system and prove cut elimination theorem for it. Hence "Free" in the title is purposely ambigous in the sense of being cut-free formalization of free modal logic.

2 Garson's System !S

The system is formulated in the standard predicate language with identity and existence predicate and with iota-operator forming definite descriptions from formulae of the language. More precisely, we will use the following categories of expressions denoted by the following symbols:

- denumerably infinite set of bound variables $VAR = \{x, y, z, ...\}$
- denumerably infinite set of free variables (rigid names) $CON = \{a, b, c, ...\}$
- denumerably infinite set of predicate symbols $PRED = \{A, B, C, ...\}$
- connectives: $\neg, \wedge, \vee, \rightarrow, \leftrightarrow, \Box$
- predicates of identity and existence: $=$, E
- (free) quantifiers: $\forall, \exists$
- iota-operator: $\imath$

In general we will use the same symbols in the metalanguage but with aditional metavariables φ, ψ, χ used for any formulae and $\Gamma, \Delta, \Pi, \Sigma$ for their multisets. A definition of a term and formula is standard; note however, that we do not admit formulae containing $x, y, ...$ not bound by quantifiers or iota-operator. Accordingly, the category of terms covers free variables and descriptions which will be written as $\imath x\varphi$ where φ is a formula in the scope of iota-operator. Metavariables $t, t_1, ...$ will be applied for any terms, including descriptions. Moreover, we will use a metavariable d for denoting any definite description if its structure is not essential. $\varphi[x/t]$ is officially used for the operation of correct substitution of a term t for x. However, to simplify matters, we will be also using freely in proof schemata a notation $\varphi(x), \varphi(a), \varphi(t)$. In particular, $\varphi(x)$ will be used to denote that φ (being a scope of some operator which binds x) contains at least one occurrence of free x, whereas $\varphi(a)$ or $\varphi(t)$ will denote the result of substitution.

Note that to simplify matters, and following Garson's policy, we do not introduce function symbols and we regard only elements of CON as rigid, and definite descriptions as nonrigid terms. It is possible to divide all terms into rigid and nonrigid, then to subdivide both classes into simple (names) and complex terms and the latter into descriptions and functional terms. It seems that such syntactic extensions require only additional notational complications (two-sorted language) and no substantial changes into presented systems are necessary. However, we will see that admitting rigid definite descriptions or universal instantiation on nonrigid terms may lead to troubles in proving cut elimination for sequent calculus formulation. We will comment on this problem in the last section while discussing Goldblatt's approach.

Garson presents his system as Jaśkowski-style natural deduction. We omit propositional details of his system and briefly recall only his rules for quantifiers, identity and descriptions:

($\forall E$) $\forall x\varphi \vdash Ea \rightarrow \varphi[x/a]$, where a is any (rigid) constant.
($\forall I$) $Ea \rightarrow \varphi[x/a] \vdash \forall x\varphi$, where a is neither in active assumptions nor in

φ.

$(=I)\ \vdash t = t$
$(=E)\ \varphi[x/t_1], t_1 = t_2, \vdash\ \varphi[x/t_2]$
$(\exists i)\quad a \neq d\ \vdash\ \bot$, where a is neither in active assumptions nor in d.
$(=\Box)\ \varphi\ \vdash\ \Box\varphi$, where φ is $a = b$ or $a \neq b$
$(\imath E)\ Ea, a = \imath x\varphi(x) \vdash \varphi(a) \wedge \forall x(\varphi(x) \to x = a)$
$(\imath I)\ Ea, \varphi(a) \wedge \forall x(\varphi(x) \to x = a) \vdash a = \imath x\varphi(x)$

Note that in the rules for quantifiers and boxed identity only rigid terms are allowed to instantiate variables, whereas in other identity rules all terms may appear. $(\exists i)$ is a special rule which guarantees that all descriptions have some denotation although not necessarily in the actual world. This rule provides a form of "rigidification" of nonrigid terms.

In what follows we will use two equivalent rules for definite descriptions of the form:

$Ea, a = \imath x\varphi(x) \vdash \forall x(\varphi(x) \leftrightarrow x = a)$
$Ea, \forall x(\varphi(x) \leftrightarrow x = a) \vdash a = \imath x\varphi(x)$

Although on the ground of free logic $\forall x(\varphi(x) \leftrightarrow x = a)$ is not equivalent to $\varphi(a) \wedge \forall x(\varphi(x) \to x = a)$, in the presence of Ea they are equivalent since $\varphi(a) \wedge \forall x(\varphi(x) \to x = a)$ implies $\forall x(\varphi(x) \leftrightarrow x = a)$ and the latter with Ea implies $\varphi(a) \wedge \forall x(\varphi(x) \to x = a)$.

Garson's system !S (where S is the name of suitable propositional modal logic) is adequate with respect to relational semantics with varying domains (of objects), actualist quantification and both rigid and nonrigid terms. Semantical characterisation will not be used in the remaining sections. However, for better understanding of the meaning of his system's principles and intuitions behind them, we will recall briefly the notion of a model for this logic. Our characterisation is a slightly modified version of Garson's semantics but giving equivalent results. In particular, for easier comparison with more standard semantics of first order languages we admit variables $x, y, \ldots$ as occurring free as well, and our free variables $a, b, \ldots$ treat as individual rigid constants. It does not make any essential differences with Garson's version.

A model is any structure $\mathfrak{M} = \langle \mathcal{W}, \mathcal{R}, D, d, I_w \rangle$, where $\mathcal{W}, \mathcal{R}$ is a standard modal frame, D is a nonempty domain, $d : W \longrightarrow \mathcal{P}(D)$ is a function which assigns a set of (existent) objects to every world, and I_w is a family of world's relative functions of interpretation for predicate symbols, defined as follows:

$I_w(P^n) \subseteq D^n$, for every n-argument predicate and world.

An assignment a is defined in a standard way as $a : VAR \cup CON \longrightarrow D$, similarly for the notion of x-variant. Interpretation $I^a_w(t)$ of a term t in w under an assignment a is just $a(t)$ for elements of VAR and CON. Now, I^a_w for definite descriptions is defined in terms of satisfaction relation, so we recall it first (essential clauses only):

$$
\begin{array}{lll}
\mathfrak{M}, a, w \models P^n(t_1, ..., t_n) & \text{iff} & \langle I^a_w(t_1), ..., I^a_w(t_n)\rangle \in I_w(P^n) \\
\mathfrak{M}, a, w \models t_1 = t_2 & \text{iff} & I^a_w(t_1) = I^a_w(t_2) \\
\mathfrak{M}, a, w \models Et & \text{iff} & I^a_w(t) \in d(w) \\
\mathfrak{M}, a, w \models \neg\varphi & \text{iff} & \mathfrak{M}, a, w \not\models \varphi \\
\mathfrak{M}, a, w \models \varphi \to \psi & \text{iff} & \mathfrak{M}, a, w \not\models \varphi \text{ or } \mathfrak{M}, a, w \models \psi \\
\mathfrak{M}, a, w \models \Box\varphi & \text{iff} & \mathfrak{M}, a, w' \models \varphi \text{ for any } w' \text{ such that } \mathcal{R}ww' \\
\mathfrak{M}, a, w \models \forall x\varphi & \text{iff} & \mathfrak{M}, a^x_o, w \models \varphi \text{ for all } o \in d(w) \\
\mathfrak{M}, a, w \models \exists x\varphi & \text{iff} & \mathfrak{M}, a^x_o, w \models \varphi \text{ for some } o \in d(w)
\end{array}
$$

Now, for any definite description: If there is a unique $o \in d(w)$ such that $\mathfrak{M}, a^x_o, w \models \varphi$, then $I^a_w(\imath x\varphi) = o$; otherwise $I^a_w(\imath x\varphi) \notin d(w)$.

Definitions of truth in a model, satisfiability, validity and entailment are standard. Note that we obtain different normal modal logics by restricting $\mathcal{R}$ suitably. Thus for T-modality which we have chosen as a fixed representative, $\mathcal{R}$ must be reflexive. We omit the details of adequacy proof and direct a reader to Garson [11]. One should note that in this semantics improper descriptions are explained as having a nonexistent designatum in respective world. It means that every description has a designatum but not in the sense of Fregean theory of the chosen object where all improper descriptions have a unique designatum. Improper descriptions just have designates somewhere. Such an approach is also in contrast to Fitting and Mendelsohn's solution where one can treat as proper description a term which designates in other world and improper descriptions are terms that do not designate at all. But, as Garson pointed out, this question may be treated as a way of interpretation of worlds in a model rather than an issue requiring a technical regulation in the semantics.

3 Sequent System SC!S

We will use a version of Gentzen's LK calculus but with sequents built not from finite lists but from multisets of formulae and with all rules multiplicative (i.e. context-free in case of many-premiss rules). Since modal details are not essential we just fix rules adequate for logic T, hence the concrete sequent calculus specified below should be named SC!T. Clearly, one can use rules characterising other modal logics, weaker (like K) or stronger (like S4 – as in Thomason's Q3). Note however, that if we want to have our system cut-free, modal rules should be taken only from the, rather modest, list of those systems where cut elimination holds[3].

In what follows, for rules with more than two premisses we will use Γ, Δ in conclusions always to denote multiset unions of $\Gamma_1, ..., \Gamma_n, \Delta_1, ..., \Delta_n$ occurring in premisses. The system consists of the following rules:

$$(AX)\quad \varphi \Rightarrow \varphi \qquad\qquad (Cut)\quad \frac{\Gamma \Rightarrow \Delta, \varphi \qquad \varphi, \Pi \Rightarrow \Sigma}{\Gamma, \Pi \Rightarrow \Delta, \Sigma}$$

[3] For a survey see e.g. Fitting [6], Goré [13], Indrzejczak [15], Poggiolesi [27] or Wansing [31], [32].

$(W\Rightarrow)\ \dfrac{\Gamma\Rightarrow\Delta}{\varphi,\Gamma\Rightarrow\Delta}$ $(\Rightarrow W)\ \dfrac{\Gamma\Rightarrow\Delta}{\Gamma\Rightarrow\Delta,\varphi}$

$(C\Rightarrow)\ \dfrac{\varphi,\varphi,\Gamma\Rightarrow\Delta}{\varphi,\Gamma\Rightarrow\Delta}$ $(\Rightarrow C)\ \dfrac{\Gamma\Rightarrow\Delta,\varphi,\varphi}{\Gamma\Rightarrow\Delta,\varphi}$

$(\neg\Rightarrow)\ \dfrac{\Gamma\Rightarrow\Delta,\varphi}{\neg\varphi,\Gamma\Rightarrow\Delta}$ $(\Rightarrow\neg)\ \dfrac{\varphi,\Gamma\Rightarrow\Delta}{\Gamma\Rightarrow\Delta,\neg\varphi}$

$(\wedge\Rightarrow)\ \dfrac{\varphi,\psi,\Gamma\Rightarrow\Delta}{\varphi\wedge\psi,\Gamma\Rightarrow\Delta}$ $(\Rightarrow\wedge)\ \dfrac{\Gamma\Rightarrow\Delta,\varphi \quad \Pi\Rightarrow\Sigma,\psi}{\Gamma,\Pi\Rightarrow\Delta,\Sigma,\varphi\wedge\psi}$

$(\vee\Rightarrow)\ \dfrac{\varphi,\Gamma\Rightarrow\Delta \quad \psi,\Pi\Rightarrow\Sigma}{\varphi\vee\psi,\Gamma,\Pi\Rightarrow\Delta,\Sigma}$ $(\Rightarrow\vee)\ \dfrac{\Gamma\Rightarrow\Delta,\varphi,\psi}{\Gamma\Rightarrow\Delta,\varphi\vee\psi}$

$(\rightarrow\Rightarrow)\ \dfrac{\Gamma\Rightarrow\Delta,\varphi \quad \psi,\Pi\Rightarrow\Sigma}{\varphi\rightarrow\psi,\Gamma,\Pi\Rightarrow\Delta,\Sigma}$ $(\Rightarrow\rightarrow)\ \dfrac{\varphi,\Gamma\Rightarrow\Delta,\psi}{\Gamma\Rightarrow\Delta,\varphi\rightarrow\psi}$

$(\leftrightarrow\Rightarrow)\ \dfrac{\Gamma\Rightarrow\Delta,\varphi,\psi \quad \varphi,\psi,\Pi\Rightarrow\Sigma}{\varphi\leftrightarrow\psi,\Gamma,\Pi\Rightarrow\Delta,\Sigma}$ $(\Rightarrow\leftrightarrow)\ \dfrac{\varphi,\Gamma\Rightarrow\Delta,\psi \quad \psi,\Pi\Rightarrow\Sigma,\varphi}{\Gamma,\Pi\Rightarrow\Delta,\Sigma,\varphi\leftrightarrow\psi}$

$(\Box\Rightarrow)\ \dfrac{\varphi,\Gamma\Rightarrow\Delta}{\Box\varphi,\Gamma\Rightarrow\Delta}$ $(\Rightarrow\Box)\ \dfrac{\Gamma\Rightarrow\varphi}{\Box\Gamma\Rightarrow\Box\varphi}$

$(\forall\Rightarrow)\ \dfrac{\Gamma\Rightarrow\Delta,Ea \quad \varphi[x/a],\Pi\Rightarrow\Sigma}{\forall x\varphi,\Gamma,\Pi\Rightarrow\Delta,\Sigma}$ $(\Rightarrow\forall)^1\ \dfrac{Ea,\Gamma\Rightarrow\Delta,\varphi[x/a]}{\Gamma\Rightarrow\Delta,\forall x\varphi}$

$(\exists\Rightarrow)^1\ \dfrac{Ea,\varphi[x/a],\Gamma\Rightarrow\Delta}{\exists x\varphi,\Gamma\Rightarrow\Delta}$ $(\Rightarrow\exists)\ \dfrac{\Gamma\Rightarrow\Delta,Ea \quad \Pi\Rightarrow\Sigma,\varphi[x/a]}{\Gamma,\Pi\Rightarrow\Delta,\Sigma,\exists x\varphi}$

1. where a is not in Γ,Δ and φ.

$(=\Rightarrow)\ \dfrac{t=t,\Gamma\Rightarrow\Delta}{\Gamma\Rightarrow\Delta}$ $(=d\Rightarrow)^2\ \dfrac{a=d,\Gamma\Rightarrow\Delta}{\Gamma\Rightarrow\Delta}$

2. where a is not in Γ,Δ,d.

$$(\Rightarrow=)^3\ \dfrac{\Gamma_1\Rightarrow\Delta_1,\varphi[x/t_1] \qquad \Gamma_2\Rightarrow\Delta_2,t_1=t_2 \qquad \varphi[x/t_2],\Gamma_3\Rightarrow\Delta_3}{\Gamma\Rightarrow\Delta}$$

3. where φ is atomic, t_1,t_2 are any terms.

$(=\Box)\ \dfrac{\Gamma\Rightarrow\Delta,a=b \quad \Box a=b,\Pi\Rightarrow\Sigma}{\Gamma,\Pi\Rightarrow\Delta,\Sigma}$ $(\neq\Box)\ \dfrac{a=b,\Gamma\Rightarrow\Delta \quad \Box a\neq b,\Pi\Rightarrow\Sigma}{\Gamma,\Pi\Rightarrow\Delta,\Sigma}$

$$(\Rightarrow\imath)^4\ \dfrac{\Gamma_1\Rightarrow\Delta_1,Ea \qquad Eb,\varphi[x/b],\Gamma_2\Rightarrow\Delta_2,a=b \qquad Eb,a=b,\Gamma_3\Rightarrow\Delta_3,\varphi[x/b]}{\Gamma\Rightarrow\Delta,a=\imath x\varphi}$$

4. where b is not in Γ,Δ,φ

$$(\imath\Rightarrow)\ \dfrac{\Gamma_1\Rightarrow\Delta_1,Ea \quad \Gamma_2\Rightarrow\Delta_2,Eb \qquad \Gamma_3\Rightarrow\Delta_3,\varphi[x/b],a=b \qquad \varphi[x/b],a=b,\Gamma_4\Rightarrow\Delta_4}{a=\imath x\varphi,\Gamma\Rightarrow\Delta}$$

A definition of a proof is standard, as well as definitions of principal, side and parametric formulae in rule's applications. It is easy to demonstrate the soundness of this calculus but we'll rather prove it indirectly by showing that it is not stronger than Garson's system (Theorem 2 in the Appendix). As for completeness one may suspect that the presented system is too weak. Thomason [30] introduced a generalised versions of $(\forall I)$ and $(\exists i)$ in order to prove completeness of Q3. Such rules are also used by Goldblatt [12] under the name template rules. Garson [11] avoids rules of this kind because they are derivable in his natural deduction system due to the presence of modal subproofs. In the standard sequent calculus it is not possible to represent modal nesting mechanism involved here, so we cannot derive such rules. Nevertheless, they are admissible and we can add to SC!S suitable counterparts of such rules of the form:

$$(T \Rightarrow \forall)\quad \frac{\Rightarrow \varphi_1 \to \Box(\varphi_2 \to \dots \Box(\varphi_n \to (Ea \to \psi[x/a])\dots)}{\Rightarrow \varphi_1 \to \Box(\varphi_2 \to \dots \Box(\varphi_n \to \forall x\psi)\dots)}$$

$$(T = d \Rightarrow)\quad \frac{\Rightarrow \varphi_1 \to \Box(\varphi_2 \to \dots \Box(\varphi_n \to a \neq d)\dots)}{\Rightarrow \varphi_1 \to \Box(\varphi_2 \to \dots \Box(\varphi_n \to \bot)\dots)}$$

where $n > 1$ (for $n = 1$ both rules are derivable), a is not in any φ_i, ψ, d.

However, such rules are necessary only if we prove completeness by means of canonical models and using Thomason's strategy of saturation. For sequent calculus without these rules (and without cut) we can adapt completeness proof provided by Garson for his tableau system which works without the need of using template rules. Thus we conclude that the system is equivalent to Garson's !S (!T in particular) and admits cut elimination; proofs of both results are in the Appendix. Note also that even the version with added template rules admits cut elimination.

Below, in order to see how the system works we will provide some examples of proofs. For better readability we underline side-formulae of all rule-applications.

$$(\Rightarrow=)\ \dfrac{Ea \Rightarrow \underline{Ea} \qquad a = t \Rightarrow \underline{a = t} \qquad \underline{Et} \Rightarrow Et}{(\exists\Rightarrow)\ \dfrac{\underline{Ea}, \underline{a = t} \Rightarrow Et}{\exists x(x = t) \Rightarrow Et}}$$

This proof works for any term, rigid or nonrigid, but for the converse we must provide two different proofs. In case of rigid terms it is trivial:

$$\dfrac{Ea \Rightarrow \underline{Ea} \qquad \dfrac{a = a \Rightarrow a = a}{\Rightarrow \underline{a = a}}\ (=\Rightarrow)}{Ea \Rightarrow \exists x(x = a)}\ (\Rightarrow \exists)$$

However, for descriptions we must apply $(= d \Rightarrow)$:

$$
(=d\Rightarrow)\,\dfrac{(C\Rightarrow)\,\dfrac{(\Rightarrow\exists)\,\dfrac{(\Rightarrow=)\,\dfrac{Ed\Rightarrow\underline{Ed}\qquad a=d\Rightarrow\underline{a=d}\qquad \underline{Ea}\Rightarrow Ea}{a=d,Ed\Rightarrow\underline{Ea}}\qquad a=d\Rightarrow\underline{a=d}}{\underline{a=d},a=d,Ed\Rightarrow\exists x(x=d)}}{\underline{a=d},Ed\Rightarrow\exists x(x=d)}}{Ed\Rightarrow\exists x(x=d)}
$$

The next proof of $E\imath xAx\Rightarrow A\imath xAx$ is much more involved and the converse is not provable. First we construct a proof:

$$
(W\Rightarrow)\,\dfrac{(\Rightarrow=)\,\dfrac{Aa\Rightarrow\underline{Aa}\qquad a=\imath xAx\Rightarrow\underline{a=\imath xAx}\qquad \underline{A\imath xAx}\Rightarrow A\imath xAx}{Aa,a=\imath xAx\Rightarrow A\imath xAx}}{\underline{a=b},\underline{Aa},a=\imath xAx\Rightarrow A\imath xAx}
$$

let S_1 denote the last sequent (the root) of this proof-tree; it is then used to obtain:

$$
(\imath\Rightarrow)\,\dfrac{Ea\Rightarrow\underline{Ea}\quad Eb\Rightarrow\underline{Eb}\qquad (\Rightarrow W)\,\dfrac{a=b\Rightarrow a=b}{a=b\Rightarrow\underline{a=b},\underline{Aa}}\qquad S_1}{b=\imath xAx,Ea,\underline{Eb},a=b,a=\imath xAx\Rightarrow A\imath xAx}
$$

again the last sequent S_2 of the above is used to obtain:

$$
(C\Rightarrow)\,\dfrac{(\Rightarrow=)\,\dfrac{Ea\Rightarrow\underline{Ea}\qquad a=b\Rightarrow\underline{a=b}\qquad S_2}{\underline{Ea},\underline{a=b},b=\imath xAx,Ea,a=b,a=\imath xAx\Rightarrow A\imath xAx}}{b=\imath xAx,Ea,\underline{a=b},a=\imath xAx\Rightarrow A\imath xAx}
$$

similarly (the root) S_3 is applied in:

$$
(=d\Rightarrow)\,\dfrac{(C\Rightarrow)\,\dfrac{(\Rightarrow=)\,\dfrac{b=\imath xAx\Rightarrow\underline{b=\imath xAx}\qquad a=\imath xAx\Rightarrow\underline{a=\imath xAx}\qquad S_3}{\underline{b=\imath xAx},\underline{a=\imath xAx},b=\imath xAx,Ea,a=\imath xAx\Rightarrow A\imath xAx}}{\underline{b=\imath xAx},Ea,a=\imath xAx\Rightarrow A\imath xAx}}{\underline{Ea},a=\imath xAx\Rightarrow A\imath xAx}
$$

and S_4 is eventually used in:

$$
(=d\Rightarrow)\,\dfrac{(C\Rightarrow)\,\dfrac{(\Rightarrow=)\,\dfrac{a=\imath xAx\Rightarrow\underline{a=\imath xAx}\qquad E\imath xAx\Rightarrow\underline{E\imath xAx}\qquad S_4}{\underline{a=\imath xAx},E\imath xAx,a=\imath xAx\Rightarrow A\imath xAx}}{E\imath xAx,\underline{a=\imath xAx}\Rightarrow A\imath xAx}}{E\imath xAx\Rightarrow A\imath xAx}
$$

Notice that a and b were new in both applications of $(=d\Rightarrow)$ and that all applications of $(\Rightarrow=)$ were on atomic formulae, as required.

We finish this section with an interesting and very useful result. It shows that in the present setting E cannot be in general defined in terms of identity.

Theorem 3.1 (i) $\vdash\Gamma\Rightarrow\Delta,Et$ *iff* $\vdash\Gamma\Rightarrow\Delta,\exists xx=t$

(ii) *If* $\vdash Et,\Gamma\Rightarrow\Delta$, *then* $\vdash\exists xx=t,\Gamma\Rightarrow\Delta$

(iii) *If* $\vdash\exists xx=t,\Gamma\Rightarrow\Delta$, *then* $\vdash Et,\Gamma\Rightarrow\Delta$, *provided t is nonrigid.*

Proof. By induction on the height of proofs. The basis of the equivalence and both implications hold by the provability of sequents establishing equivalence of Et and $\exists xx=t$ stated above. So we need to prove only the inductive steps.

In case of the left-right direction of the first equivalence, and of the second item, a proof is trivial since Et may occur only as a parametric formula or introduced by weakening and, due to context insensitivity of almost all rules, we can safely replace it with $\exists xx = t$. Note that neither $(\Rightarrow \Box)$, nor any rule calling for fresh constant might make any harm. The right-left direction of the first item, and the third item require additional work since we must take into account also cases when $\exists xx = t$ is the principal formula of the last applied rule. In the first case we have:

$$(\Rightarrow \exists)\ \frac{\Gamma_1 \Rightarrow \Delta_1, Ea \qquad \Gamma_2 \Rightarrow \Delta_2, a = t}{\Gamma \Rightarrow \Delta, \exists xx = t}$$

Now, from both premisses together with $Et \Rightarrow Et$ we obtain by $(\Rightarrow=)$ $\Gamma \Rightarrow \Delta, Et$.

In the second case we have:

$$(\exists \Rightarrow)\ \frac{Ea, a = t, \Gamma \Rightarrow \Delta}{\exists xx = t, \Gamma \Rightarrow \Delta}$$

with a fresh, and we proceed as follows:

$$(\Rightarrow=)\ \cfrac{a = t \Rightarrow a = t \qquad Et \Rightarrow Et \qquad Ea, a = t, \Gamma \Rightarrow \Delta}{(= d \Rightarrow)\ \cfrac{(C \Rightarrow)\ \cfrac{a = t, a = t, Et, \Gamma \Rightarrow \Delta}{a = t, Et, \Gamma \Rightarrow \Delta}}{Et, \Gamma \Rightarrow \Delta}}$$

a is again fresh. Notice that the application of $(= d \Rightarrow)$ was necessary here which explains the proviso in the last item.

□

4 Comments on Rules

We finish the presentation of the sequent calculus for !S with some remarks concerning the shape of rules and possible extensions of the system. Both the selection of these particular rules for identity and the shape of rules for iota-operator were dictated by the need of proving cut elimination. The problems of possible applications for proof-search was not our concern here; we only pause to mention that standard tableau system may be easily obtained on the basis of SC!S rules. The remarks below help to make clear if they offer some advantages in actual proof construction.

The rules for iota-operator were constructed by decomposition of Garson's natural deduction rules in such a way as to obtain a well-behaved pair of rules. By this we mean some requirements put on the rules of standard sequent calculus analysed in Wansing [31] and Poggiolesi [27] like symmetry, explicitness and separation. Rules which exhibit all these properties jointly, may be called canonical, after Avron and Lev [1]. The rules for boolean connectives and quantifiers in SC!S provide a clear example of canonical rules, whereas $(\Rightarrow \Box)$ fails to be separated and explicit. The rules for iota-operator are not canonical either, although they satisfy conditions of explicitness and symmetry. They are not fully separated since in the principal formulae identity is present in addition to iota-operator, and identity is treated here as a logical constant.

From the point of view of cut elimination proof, more important is the fact that both rules for description satisfy a property of reductivity. It was stated in general form for rules of hypersequent calculi in Metcalfe, Olivetti and Gabbay [25] and we may roughly define it as follows: A pair of introduction rules $(\Rightarrow \star)$, $(\star \Rightarrow)$ for a constant $\star$ is reductive if an application of cut on cut formulae introduced by these rules may be replaced by the series of cuts made on less complex formulae, in particular on their subformulae[4]. Reductivity permits induction on the complexity of cut formula in the course of proving cut elimination (see the proof in the Appendix as an exemplification). Hence it is important that all rules for connectives (including necessity) and for free quantifiers, as well as both rules for descriptions, satisfy this property.

In case of the rules for identity the situation is worse but it is not a fault of this system only. In fact, it is possible to demonstrate that identity cannot be formalised by means of canonical rules (see Indrzejczak [20]). As for the rules applied in SC!S we can observe that they are in a (reversed) sense symmetric, but they are not explicit and, in case of $(= \Box)$ and $(\neq \Box)$, also not separated. The problem of their reductivity simply does not arise since they are not introduction rules and, in fact, it is an advantage here allowing cut elimination. Still, $(\Rightarrow =)$, $(= \Box)$ and $(\neq \Box)$ may seem like some restricted forms of cut in disguise. It is true to some extent but it is a reasonable price for possibility of proving general cut elimination. One may ask however, if some other choices do not provide better solution. Suppose we will use additional axiomatic sequents of the form $\Rightarrow t = t$, $\Rightarrow a = d$ (with a not in d) and $t_1 = t_2, \varphi[x/t_1] \Rightarrow \varphi[x/t_2]$. In such case it is not possible to eliminate cuts with at least one premiss being of such form and having definite description as one of the arguments of identity, while the other premiss having this identity is deduced by one of the rules for $\imath$. Similar problem arises for sequents of the form $a = b \Rightarrow \Box a = b$, $a \neq b \Rightarrow \Box a \neq b$, although here it is generated not by identities with descriptions but by $(\Rightarrow \Box)$. The problem is due to the fact that $(\Rightarrow \Box)$ is context sensitive and, in general, not permutable with other rules, so after reduction of the height we cannot apply the rule with the same result. The same problem is encountered if we use rules of the form:

$$(\Rightarrow = \Box)\ \frac{\Gamma \Rightarrow \Delta, a = b}{\Gamma \Rightarrow \Delta, \Box a = b} \qquad (\Rightarrow \neq \Box)\ \frac{\Gamma \Rightarrow \Delta, a \neq b}{\Gamma \Rightarrow \Delta, \Box a \neq b}$$

Notice however that if instead of our $(\Rightarrow \Box)$ we will use a rule for transitive logic like S4, of the form:

$$\frac{\Box\Gamma \Rightarrow \varphi}{\Box\Gamma \Rightarrow \Box\varphi}$$

[4] Again, one can refer here to Avron and Lev [1], where the criterion of coherency of canonical rules yields the same result.

the above rules will work. We decided however to choose rules which are insensitive to the changes in background modal rules and work uniformly with any modal logic for which cut-free sequent calculus exists.

One could also think about using some other rules for expressing Leibniz Law instead of our $(\Rightarrow=)$. For instance, Negri and von Plato's rule seems to be a reasonable option:

$$\frac{t_1 = t_2, \varphi[x/t_1], \varphi[x/t_2], \Gamma \Rightarrow \Delta}{t_1 = t_2, \varphi[x/t_1], \Gamma \Rightarrow \Delta}$$

But such choice also does not work if $t_1 = t_2$ is a cut formula with description. Consider a situation where it is principal in both premisses but in the left premiss introduced by $(\Rightarrow \imath)$; in such cases there is no possibility of replacing this cut with cuts made on premisses of these two rule's applications. Similarly, if we introduce a rule used in Indrzejczak [19] for the formalization of Fregean description theory:

$$(\Rightarrow=')\ \frac{\Gamma \Rightarrow \Delta, \varphi[x/t_1] \qquad \Pi \Rightarrow \Sigma, t_1 = t_2}{\Gamma, \Pi \Rightarrow \Delta, \Sigma, \varphi[x/t_2]}$$

This time we have a problem if $\varphi[x/t_2]$ is an identity statement with description and cut formula in the right premiss is deduced by $(\imath \Rightarrow)$. The same remarks apply to other possible rules – in fact there are four more (see p.79 of Indrzejczak [15] or [20]). In general, in cases where cut-formula is an identity with description, principal in both premisses, and in one premiss introduced by the rule for identity whereas in the other by the rule for $\imath$, there is no possibility of reduction of the height or complexity of cut formula.

The most important disadvantage is that we must sacrifice subformula property in the strict form. In proofs there may appear not only subformulae of formulae from the root but also atomic formulae as well as boxed identities and negated identities. But note that this is only a little more (boxed identities and their negations) than in case of sequent calculi for several axiomatic theories provided by Negri and von Plato [26].

5 Extensions and Alternatives

SC!T may be easily extended in at least three ways: by enriching the language, by strenghtening the theory of descriptions, by strengthening the background modal logic. The last option is obvious. One can easily provide modal rules for most known normal modal logic but not many of them admit cut elimination. Standard sequent calculi are rather weak tool in this respect and to provide cut-free characterisation of such logics like B or S5, not to mention bimodal temporal logics, one must use non-standard, generalised framework like hypersequent calculi or nested calculi (see the references listed in footnote 3.). On the other hand, we can develop in such a way a suitable theory of description also in the setting of weaker modal logics, like regular, monotonic or even congruent. Cut-free characterisation of many such logics was provided in Indrzejczak [14] and [16], and may be lifted from propositional level to first-order level with descriptions.

As for the first option, it is not problematic to add lambda operator in such a way as to cover Garson's system λS which permits expression of scope differences like de re/de dicto distinction. In Garson [11] it is formalised by means of one axiom:

$$\lambda x\varphi(x)(a) \leftrightarrow \varphi(a)$$

where $\lambda x\varphi(x)$ is a predicate abstracted from a formula φ. It is rather restricted use of lambda operator but still permitting important extension of expressive power. In the setting of sequent calculus it is enough to add two rules:

$$(\Rightarrow \lambda)\ \frac{\Gamma \Rightarrow \Delta, \varphi(a)}{\Gamma \Rightarrow \Delta, \lambda x\varphi(x)(a)} \qquad (\lambda \Rightarrow)\ \frac{\varphi(a), \Gamma \Rightarrow \Delta}{\lambda x\varphi(x)(a), \Gamma \Rightarrow \Delta}$$

It is obvious that their addition cannot spoil the proof of cut elimination since they are also reductive.

Finally, we will take a look at possible strengthenings of this system and some alternatives. If we do not consider elements connected with the treatment of modalities and rigid/nonrigid term distinctions but focus only on the rules for descriptions it is obvious that Thomason's and Garson's theory of descriptions is essentially the minimal free description theory MFD of Lambert [23], [24]. We have a strengthening of the latter not only in the sense that it is developed on the basis of modal logic since it is a conservative extension. More important is the addition of a rule $(\exists i)$ which in our system is covered by $(= d \Rightarrow)$. Yet this may be still considered to be rather weak theory of definite descriptions in the sense that it is rather concerned with proper definite descriptions. There are some costs of that, e.g. the law of extensionality of the form: $\forall x(\varphi \leftrightarrow \psi) \rightarrow \imath x\varphi = \imath x\psi$ which is technically useful, cannot be proved although we can prove its weaker version: $E\imath x\varphi \rightarrow (\forall x(\varphi \leftrightarrow \psi) \rightarrow \imath x\varphi = \imath x\psi)$. In the setting of classical logic, Fregean approach with the chosen object being denotation of all improper descriptions provides a solution to this problem. It may be criticised as artificial (Garson's theory does not equate all improper descriptions) but technically it seems to be more plausible. Fregean approach was developed formally by Kalish and Montague [21] and recently also received a cut-free sequent calculus in Indrzejczak [19]. A counterpart of this approach in the setting of free logic is even easier to formulate and was provided by Scott [29]. It is enough to add an axiom $\neg Ed \rightarrow d = \imath x x \neq x$. Since a converse of this implication also holds for Scott's logic we can express it in the setting of SC by means of two rules:

$$(\Rightarrow id)\ \frac{Ed, \Gamma \Rightarrow \Delta,}{\Gamma \Rightarrow \Delta, d = \imath x(x \neq x)} \qquad (id \Rightarrow)\ \frac{\Gamma \Rightarrow \Delta, Ed}{d = \imath x(x \neq x), \Gamma \Rightarrow \Delta}$$

Again it is clear that our proof of cut elimination still holds for such extension. One can prove in this logic a full version of the law of extensionality for descriptions. Lambert [24] considered also other extensions of MFD, weaker than Scott's logic, and some of them may be also expressed quite easily in our framework by means of reasonably simple rules. We illustrate the issue with two examples. The logic FD1 is an extension of MFD by means of the addition of cancellation law: $t = (\imath x\varphi = t)$, An effect of this axiom is directly obtained by means of the rule similar to $(=\Rightarrow)$ and $(= d \Rightarrow)$ but with such identity in the antecedent of the premiss. If we are interested just in having extensionality principle for descriptions we can add it as an axiom to MFD. This way we obtain the system FDV due to van Fraasen. The same effect can be obtained in SC by the addition of the rule:

$$(\Rightarrow d_1 = d_2)\ \frac{\varphi[x/a], \Gamma \Rightarrow \Delta, \psi[x/a] \qquad \psi[x/a], \Pi \Rightarrow \Sigma, \varphi[x/a]}{\Gamma, \Pi \Rightarrow \Delta, \Sigma, \imath x\varphi = \imath x\psi}$$

where a is not in $\varphi, \psi, \Gamma, \Delta, \Pi, \Sigma$.

The proof of cut elimination presented in the Appendix works for all such extensions of SC!S.

There are at least two approaches to definite descriptions in modal logic framework which are significantly different from the hierarchy of theories developed in free logic setting; one is due to Goldblatt [12], and the other to Fitting and Mendelsohn [7]. The former uses two-sorted language to make a distinction between rigid and nonrigid terms. In many respects the strategy of expressing relations between both kinds of terms is similar to the approach of Thomason/Garson and a rule which is a counterpart of $(\exists i)$ is applied. However, there are also strong differences. Goldblatt's theory is based on a semantics where individual variables range not over objects from D but over substances which are defined as partial functions from W to D. One of the consequences of this choice is invalidation of reflexivity of identity; a weaker axiom $t = t' \to t = t$ is postulated instead. Note also that $t = t$ is equivalent in his system to Et, hence this axiom may be expressed as $t = t' \to Et$. Although semantical machinery of Goldblatt's approach is a bit more complicated, on the syntactical side it has no serious consequences. The only axiom for definite descriptions[5] has the form: $\imath x\varphi = t \leftrightarrow Et \wedge \forall x(\varphi \leftrightarrow x = t)$, where t is rigid and does not have free x. Note that t may be also a rigid description. Goldblatt shows that in his approach descriptions cannot be eliminated in the Russelian way; it would be possible only for rigid descriptions and on the condition that some chosen nonexistent object would be added in the spirit of Fregean approach.

We can easily obtain an equivalent formalization of Goldblatt's axiom in standard sequent calculus by means of the following rules:

$$(\Rightarrow \imath)^1\ \frac{\Gamma_1 \Rightarrow \Delta_1, Et \qquad Ea, \varphi[x/a], \Gamma_2 \Rightarrow \Delta_2, t = a \qquad Ea, t = a, \Gamma_3 \Rightarrow \Delta_3, \varphi[x/a]}{\Gamma \Rightarrow \Delta, t = \imath x\varphi}$$

[5] Strictly speaking it works for modal logics characterised by Kripkean models. Golblatt considers also logics characterised by non-Kripkean models, where truth conditions for quantifiers are weaker, and such logics require also a template rule for description.

1. where a is not in Γ, Δ, φ

$$(\imath \Rightarrow)\ \frac{Et, \Gamma_1 \Rightarrow \Delta_1, Ea \qquad Et, \Gamma_2 \Rightarrow \Delta_2, \varphi[x/a], t = a \qquad Et, \varphi[x/a], t = a, \Gamma_3 \Rightarrow \Delta_3}{t = \imath x\varphi, \Gamma \Rightarrow \Delta}$$

However, one should remember that also rules for identity and quantifiers need to be changed in order to comply with Goldblatt's system. For completeness we need template rules like $(T \Rightarrow \forall)$ and $(T = d \Rightarrow)$. We do not develop this system here for the lack of space; only suitable rules for definite descriptions were displayed above just for comparison with Garson's approach.

Both rules for Goldblatt's theory of descriptions are also reductive, so we can expect that cut elimination holds in the way demonstrated in the Appendix. There is one serious difficulty however. In Goldblatt's system rules for quantifiers work not only for rigid but also for nonrigid terms. Moreover, descriptions may be also rigid. It may seem an advantage, in comparison to Garson's system, but in fact it is not, since rules for quantifiers are not reductive. We mean here the fact that in the course of the application of $(\Rightarrow \forall)$ a variable of suitable sort may be instantiated with definite description of arbitrary complexity, so the instance of substituted formula is not less complex and we cannot obtain reduction of the complexity of cut formula. To avoid the problem we should have the instantiation of quantifiers restricted to rigid terms and only nonrigid definite descriptions as in Garson's system. Another possibility would be to apply the solution from Indrzejczak [19] where all terms have the same complexity measure but this solution does not work either. After changing the definition of complexity in such a way the above rules for descriptions do not allow for reduction of cut-degree (see Appendix) since in their premisses a term is unpacked and occurs as a formula which is at least as complex as the identity with description in the conclusion. It was not a problem for Fregean system from [19] since all rules for definite descriptions introduce them only to succedents and the situation with cut on such formulae as principal simply does not arise. Summing up, in contrast to Garson's approach, a version of description theory present in Goldblatt is harder for proving cut elimination theorem, and for the time being we are unable to provide a solution to this problem.

Finally, there is an alternative approach to first-order modal logics with descriptions provided by Fitting and Mendelsohn [7]. As we mentioned in the Introduction, their theory of definite descriptions is perhaps a subtler solution since it does not equate existence and designation and makes distinctions between different kinds of improper descriptions. It is again significantly different than any theory belonging to Lambert hierarchy since, for example $t = t$ is also not a thesis in general but holds only for designating terms. In this respect, their theory is similar to Goldblatt's approach but, in contrast, it is too rich on the syntactical level to be dealt with by means of standard resources which are used in this research. In particular, tableau systems for these logics are based on the complex machinery of labels attached not only to formulae but also to nonrigid terms to fix their denotations in possible worlds.

In fact, insufficiency of the standard apparatus for formalization of Fitting and Mendelsohn's theory is more connected with the way in which nonrigid terms in general are dealt with, than with the specific features of definite descriptions in their theory. In short, nonrigid terms are "rigidified" by means of labels, and lambda operator is used just to permit predication on nonrigid terms, in contrast to Garson's restricted solution.

It seems that the framework of hybrid logics is promising here. One could possibly extend the approach to hybrid first-order modal logic provided by Blackburn and Marx [3] to cover this theory of descriptions, and apply cut-free sequent calculi for hybrid logics developed by Braüner [4] or Indrzejczak [18]. But this is a future project; for the time being we are concerned with the abilities of standard sequent calculi with no labels or nominals.

A Appendix

We will show that for every $\Gamma \vdash \varphi$ derivable in !S we can provide a proof of $\Gamma \Rightarrow \varphi$ in SC!S and conversely. Since for the propositional part as well as for (free) quantifiers and identity rules the equivalence is clear we restrict the consideration to rules for descriptions.

Theorem A.1 *If* $\Gamma \vdash_{!S} \varphi$, *then* $\vdash_{SC!S} \Gamma \Rightarrow \varphi$.

Proof. It is sufficient to show that $(\exists i)$ is a derivable rule in SC!S and that sequents corresponding to both rules for $\imath$ are provable. As for the first:

$$\dfrac{\dfrac{a = d \Rightarrow \neg\neg a = d \qquad \dfrac{\Gamma \Rightarrow \neg a = d}{\neg\neg a = d, \Gamma \Rightarrow}(\neg\Rightarrow)}{a = d, \Gamma \Rightarrow}(Cut)}{\Gamma \Rightarrow}(= d \Rightarrow)$$

The sequents corresponding to Garson's rules are:

$Ea, a = \imath x\varphi(x) \Rightarrow \forall x(\varphi(x) \leftrightarrow x = a)$
$Ea, \forall x(\varphi(x) \leftrightarrow x = a) \Rightarrow a = \imath x\varphi(x)$

We can prove the first one in the following way. First, using $(\imath \Rightarrow)$ twice, we construct:

$$\dfrac{Ea \Rightarrow Ea \qquad Eb \Rightarrow Eb \qquad \varphi(b) \Rightarrow \varphi(b), a = b \qquad \varphi(b), a = b \Rightarrow a = b}{Ea, Eb, a = \imath x\varphi(x), \varphi(b) \Rightarrow a = b}$$

and

$$\dfrac{Ea \Rightarrow Ea \qquad Eb \Rightarrow Eb \qquad a = b \Rightarrow \varphi(b), a = b \qquad \varphi(b), a = b \Rightarrow \varphi(b)}{Ea, Eb, a = \imath x\varphi(x), a = b \Rightarrow \varphi(b)}$$

Both, by the application of $(\Rightarrow\leftrightarrow)$, contractions and $(\Rightarrow \forall)$ yield:

$$\dfrac{Ea, Eb, a = \imath x\varphi(x) \Rightarrow \varphi(b) \leftrightarrow a = b}{Ea, a = \imath x\varphi(x) \Rightarrow \forall x(\varphi(x) \leftrightarrow a = x)}$$

where b is new in the next to last sequent.

For the second sequent we prove first:

$$\dfrac{Eb \Rightarrow Eb \qquad \dfrac{\varphi(b), a=b \Rightarrow a=b \qquad \varphi(b) \Rightarrow \varphi(b), a=b}{\varphi(b) \leftrightarrow a=b, \varphi(b) \Rightarrow a=b}(\leftrightarrow\Rightarrow)}{\forall x(\varphi(x) \leftrightarrow a=x), Eb, \varphi(b) \Rightarrow a=b}(\forall\Rightarrow)$$

and

$$\dfrac{Eb \Rightarrow Eb \qquad \dfrac{\varphi(b), a=b \Rightarrow \varphi(b) \qquad a=b \Rightarrow \varphi(b), a=b}{\varphi(b) \leftrightarrow a=b, a=b \Rightarrow \varphi(b)}(\leftrightarrow\Rightarrow)}{\forall x(\varphi(x) \leftrightarrow a=x), Eb, a=b \Rightarrow \varphi(b)}(\forall\Rightarrow)$$

Let S_1 and S_2 denote the roots ot the above proof-trees. Then we obtain:

$$\dfrac{\dfrac{Ea \Rightarrow Ea \qquad S_1 \qquad S_2}{Ea, \forall x(\varphi(x) \leftrightarrow a=x), \forall x(\varphi(x) \leftrightarrow a=x) \Rightarrow a=\imath x\varphi(x)}(\Rightarrow \imath)}{Ea, \forall x(\varphi(x) \leftrightarrow a=x) \Rightarrow a=\imath x\varphi(x)}(C\Rightarrow)$$

where b is new. □

Theorem A.2 *If* $\vdash_{SC!S} \Gamma \Rightarrow \Delta$, *then* $\Gamma \vdash_{!S} \bigvee\Delta$, *where* $\bigvee\Delta$ *is a disjunction of elements of* Δ.

Proof. Since one can simulate in Garson's ND system !S all the rules of SC!S we can use the reduct of the latter without the rules for description and $(= d \Rightarrow)$ but with an analog of $(\exists i)$ and two additional axiomatic sequents corresponding to rules:

$Ea, a = \imath x\varphi(x) \Rightarrow \forall x(\varphi(x) \leftrightarrow x = a)$
$Ea, \forall x(\varphi(x) \leftrightarrow x = a) \Rightarrow a = \imath x\varphi(x)$

Call this calculus SC!S*. It is obviously equivalent to !S so it will be enough to demonstrate that the three rules in question are derivable in SC!S*. Derivability of $(= d \Rightarrow)$ is trivial by $(\Rightarrow \neg)$ and $(\exists i)$ but for the remaining rules proofs are more involved. Let us start with $(\Rightarrow \imath)$. On the basis of the first sequent and the first premiss we obtain:

$$\dfrac{\Gamma_1 \Rightarrow \Delta_1, Ea \qquad Ea, a=\imath x\varphi(x) \Rightarrow \forall x(\varphi(x) \leftrightarrow x=a)}{a=\imath x\varphi(x), \Gamma_1 \Rightarrow \Delta_1, \forall x(\varphi(x) \leftrightarrow x=a)}(Cut)$$

Three remaining premisses yield:

$$\dfrac{\Gamma_2 \Rightarrow \Delta_2, Eb \qquad \dfrac{\Gamma_3 \Rightarrow \Delta_3, \varphi(b), a=b \qquad \varphi(b), a=b, \Gamma_4 \Rightarrow \Delta_4}{\varphi(b) \leftrightarrow a=b, \Gamma_3, \Gamma_4 \Rightarrow \Delta_3, \Delta_4}(\leftrightarrow\Rightarrow)}{\forall x(\varphi(x) \leftrightarrow a=x), \Gamma_2, \Gamma_3, \Gamma_4 \Rightarrow \Delta_2, \Delta_3, \Delta_4}(\forall\Rightarrow)$$

By cut on these two root sequents we obtain $a = \imath x\varphi(x), \Gamma \Rightarrow \Delta$.

For the second rule, from the first premiss and the second sequent we obtain:

$$\dfrac{\Gamma_1 \Rightarrow \Delta_1, Ea \qquad Ea, \forall x(\varphi(x) \leftrightarrow a=x) \Rightarrow a=\imath x\varphi(x)}{\forall x(\varphi(x) \leftrightarrow a=x), \Gamma_1 \Rightarrow \Delta_1, a=\imath x\varphi(x)}(Cut)$$

its root S in combination with the remaining premisses yields:

$$
\dfrac{(C\Rightarrow)\dfrac{(\Rightarrow\leftrightarrow)\dfrac{Eb, \varphi(b), \Gamma_2 \Rightarrow \Delta_2, a = b \qquad Eb, a = b, \Gamma_3 \Rightarrow \Delta_3, \varphi(b)}{Eb, Eb, \Gamma_2, \Gamma_3 \Rightarrow \Delta_2, \Delta_3, \varphi(b) \leftrightarrow a = b}}{(\Rightarrow\forall)\dfrac{Eb, \Gamma_2, \Gamma_3 \Rightarrow \Delta_2, \Delta_3, \varphi(b) \leftrightarrow a = b}{\Gamma_2, \Gamma_3 \Rightarrow \Delta_2, \Delta_3, \forall x(\varphi(x) \leftrightarrow a = x)}} \qquad S}{\Gamma \Rightarrow \Delta, a = \imath x\varphi(x)}\ (Cut)
$$

where b is new. □

As a preliminary step for proving cut elimination we need:

Lemma A.3 (Substitution) *If* $\vdash_k \Gamma \Rightarrow \Delta$, *then* $\vdash_k (\Gamma \Rightarrow \Delta)[a/t]$.

Proof. By induction on the height of a proof. It is straightforward but tedious exercise. Note that we provided not sheer admissibility but height-preserving admissibility. □

Moreover, we assume that all proofs satisfy the condition of regularity – every constant which is fresh by side condition on the respective rule must be fresh in the entire proof, not only on the branch where the application of this rule takes place. Clearly, every proof may be systematically transformed into regular proof by Substitution lemma.

Let us define the notions of cut-degree and proof-degree:

(i) Cut-degree is the complexity of cut-formula φ, i.e. the number of connectives and operators occurring in φ and is denoted as $d\varphi$;

(ii) Proof-degree ($d\mathcal{D}$) is the maximal cut-degree in $\mathcal{D}$.

The proof of cut elimination theorem is based on two lemmata which make a reduction first on the right and then on the left premiss of cut. The general strategy of proof was originally developed for hypersequent calculi by Metcalfe, Olivetti and Gabbay [25] and later extensively used in this framework (see e.g. Ciabattoni, Metcalfe, Montagna [5], Indrzejczak [17], Kurokawa [22]). However it is also applicable to standard sequent calculi (see Indrzejczak [18], [20]) and allows for elegant proof which helps to avoid many complexities inherent in other methods of proving cut elimination. Here are the key lemmata:

Lemma A.4 (Right reduction) *Let* $\mathcal{D}_1 \vdash \Gamma \Rightarrow \Delta, \varphi$ *and* $\mathcal{D}_2 \vdash \varphi^k, \Pi \Rightarrow \Sigma$ *with* $d\mathcal{D}_1, d\mathcal{D}_2 < d\varphi$, *and* φ *principal in* $\Gamma \Rightarrow \Delta, \varphi$, *then we can construct a proof* $\mathcal{D}$ *such that* $\mathcal{D} \vdash \Gamma^k, \Pi \Rightarrow \Delta^k, \Sigma$ *and* $d\mathcal{D} < d\varphi$.

Lemma A.5 (Left reduction) *Let* $\mathcal{D}_1 \vdash \Gamma \Rightarrow \Delta, \varphi^k$ *and* $\mathcal{D}_2 \vdash \varphi, \Pi \Rightarrow \Sigma$ *with* $d\mathcal{D}_1, d\mathcal{D}_2 < d\varphi$, *then we can construct a proof* $\mathcal{D}$ *such that* $\mathcal{D} \vdash \Gamma, \Pi^k \Rightarrow \Delta, \Sigma^k$ *and* $d\mathcal{D} < d\varphi$.

Proof. For lemma 4 by induction on the height of $\mathcal{D}_2$. The basis is trivial. Induction step requires consideration of all cases of possible derivation of $\varphi^k, \Pi \Rightarrow \Sigma$ and the role of cut-formula in the transition. In cases where all occurrences of φ are parametric we simply apply the induction hypotheses to premisses of $\varphi^k, \Pi \Rightarrow \Sigma$ and then apply to them respective rule – it is essentially due to the context independence of almost all rules and regularity of proofs which prevents violation of side conditions. If one of the occurrences of φ in the premiss(es) is a side formula of the last rule we must additionaly apply weakening to restore the lacking formula before the application of a rule. Note also that the situation with $(\Rightarrow \Box)$ as the last applied rule is not of this case since φ, being some $\Box\psi$, is not present in the premiss (ψ is).

In cases where one occurrence of φ in $\varphi^k, \Pi \Rightarrow \Sigma$ is principal we make use of the fact that φ in the left premiss is principal too (note that for C and W it is trivial). We analyse two cases.

Case of $\forall x\varphi(x)$:

$$\dfrac{\dfrac{Ea, \Gamma \Rightarrow \Delta, \varphi(a)}{\Gamma \Rightarrow \Delta, \forall x\varphi(x)} \qquad \dfrac{\forall x\varphi(x)^i, \Pi_1 \Rightarrow \Sigma_1, Eb \qquad \varphi(b), \forall x\varphi(x)^j, \Pi_2 \Rightarrow \Sigma_2}{\forall x\varphi(x)^k, \Pi \Rightarrow \Sigma}}{\Gamma^k, \Pi \Rightarrow \Delta^k, \Sigma}$$

where $k = i + j + 1$ and a is fresh, hence by Substitution Lemma we have:

$Eb, \Gamma \Rightarrow \Delta, \varphi(b)$

By the induction hypothesis we have:

$\Gamma^i, \Pi_1 \Rightarrow \Delta^i, \Sigma_1, Eb$
$\varphi(b), \Gamma^j, \Pi_2 \Rightarrow \Delta^j, \Sigma_2$

Now we can build a proof:

$$\dfrac{\Gamma^i, \Pi_1 \Rightarrow \Delta^i, \Sigma_1, Eb \qquad \dfrac{Eb, \Gamma \Rightarrow \Delta, \varphi(b) \qquad \varphi(b), \Gamma^j, \Pi_2 \Rightarrow \Delta^j, \Sigma_2}{Eb, \Gamma^{j+1}, \Pi_2 \Rightarrow \Delta^{j+1}, \Sigma_2}}{\Gamma^k, \Pi \Rightarrow \Delta^k, \Sigma}$$

Case of $a = \imath x\varphi(x)$:

In the right premiss we have $a = \imath x\varphi(x)^k, \Pi \Rightarrow \Sigma$ deduced from:

$a = \imath x\varphi(x)^i, \Pi_1 \Rightarrow \Sigma_1, Ea$
$a = \imath x\varphi(x)^j, \Pi_2 \Rightarrow \Sigma_2, Eb$
$a = \imath x\varphi(x)^n, \Pi_3 \Rightarrow \Sigma_3, \varphi(b), a = b$
$a = \imath x\varphi(x)^m, \varphi(b), a = b, \Pi_4 \Rightarrow \Sigma_4$

where $k = i + j + n + m + 1$ and by the induction hypothesis we obtain:

(a) $\Gamma^i, \Pi_1 \Rightarrow \Delta^i, \Sigma_1, Ea$
(b) $\Gamma^j, \Pi_2 \Rightarrow \Delta^j, \Sigma_2, Eb$
(c) $\Gamma^n, \Pi_3 \Rightarrow \Delta^n, \Sigma_3, \varphi(b), a = b$
(d) $a = b, \varphi(b), \Gamma^m, \Pi_4 \Rightarrow \Delta^m, \Sigma_4$

In the left premiss we have:

$$\dfrac{\Gamma_1 \Rightarrow \Delta_1, Ea \qquad Ec, \varphi(c), \Gamma_2 \Rightarrow \Delta_2, a = c \qquad Ec, a = c, \Gamma_3 \Rightarrow \Delta_3, \varphi(c)}{\Gamma \Rightarrow \Delta, a = \imath x\varphi(x)}$$

where c is not in Γ, Δ, φ hence by Substitution Lemma we obtain:

(e) $Eb, \varphi(b), \Gamma_2 \Rightarrow \Delta_2, a = b$
(f) $Eb, a = b, \Gamma_3 \Rightarrow \Delta_3, \varphi(b)$

These sequents may be combined, by cuts and contractions, in the following way:

$$\dfrac{\dfrac{\Gamma^n, \Pi_3 \Rightarrow \Delta^n, \Sigma_3, \varphi(b), a = b \qquad \dfrac{\Gamma^j, \Pi_2 \Rightarrow \Delta^j, \Sigma_2, Eb \qquad Eb, a = b, \Gamma_3 \Rightarrow \Delta_3, \varphi(b)}{a = b, \Gamma^j, \Pi_2, \Gamma_3 \Rightarrow \Delta^j, \Sigma_2, \Delta_3, \varphi(b)}}{\Gamma^n, \Pi_3, \Gamma^j, \Pi_2, \Gamma_3 \Rightarrow \Delta^n, \Sigma_3, \Delta^j, \Sigma_2, \Delta_3, \varphi(b), \varphi(b)}}{\Gamma^n, \Pi_3, \Gamma^j, \Pi_2, \Gamma_3 \Rightarrow \Delta^n, \Sigma_3, \Delta^j, \Sigma_2, \Delta_3, \varphi(b)}$$

and

$$\dfrac{\dfrac{\dfrac{\Gamma^j, \Pi_2 \Rightarrow \Delta^j, \Sigma_2, Eb \qquad Eb, \varphi(b), \Gamma_2 \Rightarrow \Delta_2, a = b}{\varphi(b), \Gamma^j, \Pi_2, \Gamma_2 \Rightarrow \Delta^j, \Sigma_2, \Delta_2, a = b} \qquad a = b, \varphi(b), \Gamma^m, \Pi_4 \Rightarrow \Delta^m, \Sigma_4}{\varphi(b), \varphi(b), \Gamma^m, \Pi_4, \Gamma^j, \Pi_2, \Gamma_2 \Rightarrow \Delta^m, \Sigma_4, \Delta^j, \Sigma_2, \Delta_2}}{\varphi(b), \Gamma^m, \Pi_4, \Gamma^j, \Pi_2, \Gamma_2 \Rightarrow \Delta^m, \Sigma_4, \Delta^j, \Sigma_2, \Delta_2}$$

By cut on the last two sequents and several contractions we obtain $\Gamma^k, \Pi \Rightarrow \Delta^k, \Sigma$. Note that all cuts are of lower degree hence we are done. □

The proof of the Left Reduction Lemma is similar but on the height of $\mathcal{D}_1$. The only difference is that now we do not assume that cut-formula in the right premiss is principal. Therefore, when cut-formula is principal in the left premiss we apply first the induction hypothesis and next the rule in question to side-formulae. The new proof of the left premiss satisfies the assumption of the Right Reduction Lemma, so we can safely apply it and, possibly after some applications of structural rules, obtain the result. If the last rule was ($\Rightarrow=$) or ($= \Box$) and one of the active formulae from succedent was involved, then we obtain the result by the induction hypothesis from this single premiss, like in case of contraction. Hence no essentially new cases appear and we can skip detailed analysis. Eventually, on the basis of the Left Reduction Lemma we obtain cut elimination by successive decreasing of cut degree in the input proof. Therefore:

Theorem A.6 (Cut Elimination) *If* $\Gamma \Rightarrow \Delta$ *is provable, then it is provable without applications of* (Cut).

Note that this result applies also to SC!S in extended form, i.e. with template rules ($T \Rightarrow \forall$) and ($T = d \Rightarrow$). Since the only possible rule applied to their premiss is ($\Rightarrow\rightarrow$) both cases are treated as cases of implication and no special operations are needed.

References

[1] Avron, A. and I. Lev, *Canonical propositional Gentzen-type systems*, in: *Proceedings of IJCAR'01, vol. 2083*, LNCS, 2001 pp. 529–543.

[2] Bencivenga, E., *Free logics*, in: *Handbook of Philosophical Logic, vol III*, Reidel Publishing Company, Dordrecht, 1986 pp. 373–426.
[3] Blackburn, P. and M. Marx, *Tableaux for quantified hybrid logic*, in: *Tableaux 2002, LNAI 2381*, Springer, 2002 pp. 38–52.
[4] Braüner, T., "Hybrid Logic and its Proof-Theory," Springer, 2011.
[5] Ciabattoni, A., G. Metcalfe and F. Montagna, *Algebraic and proof-theoretic characterizations of truth stressers for MTL and its extensions*, Fuzzy Sets and Systems **161/3** (2010), pp. 369–389.
[6] Fitting, M., "Proof Methods for Modal Logic," Kluwer, Dordrecht, 1984.
[7] Fitting, M. and R. L. Mendelsohn, "First-Order Modal Logic," Kluwer, Dordrecht, 1998.
[8] Frege, G., *Über Sinn und Bedeutung*, Zeitschrift für Philosophie und Philosophische Kritik **100** (1892), pp. 25–50.
[9] Frege, G., "Grundgesetze der Arithmmetic I," Jena, 1893.
[10] Garson, J. M., *Quantification in modal logic*, in: *Handbook of Philosophical Logic, vol II*, Reidel Publishing Company, Dordrecht, 1984 pp. 249–307.
[11] Garson, J. M., "Modal Logic for Philosophers," Cambridge University Press, Cambridge, 2006.
[12] Goldblatt, R., "Quantifiers, Propositions and Identity," Cambridge University Press, Cambridge, 2011.
[13] Goré, R., *Tableau methods for modal and temporal logics*, *Handbook of Tableau Methods*, Kluwer Academic Publishers, Dordrecht, 1999 pp. 297–396.
[14] Indrzejczak, A., *Sequent calculi for monotonic modal logic*, Bulletin of the Section of Logic **34/3** (2005), pp. 151–164.
[15] Indrzejczak, A., "Natural Deduction, Hybrid Systems and Modal Logics," Springer, 2010.
[16] Indrzejczak, A., *Admissibility of cut in congruent modal logics*, Logic and Logical Philosophy **20/3** (2011), pp. 189–203.
[17] Indrzejczak, A., *Eliminability of cut in hypersequent calculi for some modal logics of linear frames*, Information Processing Letters **115/2** (2015), pp. 75–81.
[18] Indrzejczak, A., *Simple cut elimination proof for hybrid logic*, Logic and Logical Philosophy **25/2** (2016), pp. 129–141.
[19] Indrzejczak, A., *Fregean description theory in proof-theoretical setting*, Logic and Logical Philosophy **online first** (2018), DOI:10.12775/LLP.2018.008.
[20] Indrzejczak, A., *Rule-maker theorem and its applications*, submitted (2018).
[21] Kalish, D. and R. Montague, *Remarks on descriptions and natural deduction*, Archiv. für Mathematische Logik und Grundlagen Forschung **3** (1957), pp. 50–73.
[22] Kurokawa, H., *Hypersequent calculi for modal logics extending S4*, in: *New Frontiers in Artificial Intelligence (2013)*, Springer, 2014 pp. 51–68.
[23] Lambert, K., *Notes on E!: A theory of descriptions*, Philosophical Studies **13** (1962), pp. 51–59.
[24] Lambert, K., *Free logic and definite descriptions*, in: *New Essays in Free Logic*, Springer, 2001 pp. 37–48.
[25] Metcalfe, G., N. Olivetti and D. Gabbay, "Proof Theory for Fuzzy Logics," Springer, 2008.
[26] Negri, S. and J. von Plato, "Structural Proof Theory," Cambridge University Press, 2001.
[27] Poggiolesi, F., "Gentzen Calculi for Modal Propositional Logics," Springer, 2011.
[28] Russell, B., *On denoting*, Mind **14** (1905), pp. 479–493.
[29] Scott, D., *Existence and description in formal logic*, in: *Bertrand Russell: Philosopher of the Century*, London, 1967 pp. 181–200.
[30] Thomason, R., *Some completeness results for modal predicate calculi*, in: *Philosophical Problems in Logic*, Reidel Publishing Company, Dordrecht, 1970 pp. 56–76.
[31] Wansing, H., "Displaying Modal Logics," Kluwer Academic Publishers, Dordrecht, 1999.
[32] Wansing, H., *Sequent systems for modal logics*, in: F. G. D. Gabbay, editor, *Handbook of Philosophical Logic*, 2002 pp. 89–133.

A Logic for Temporal Conditionals and a Solution to the Sea Battle Puzzle

Fengkui Ju [1]
School of Philosophy, Beijing Normal University

Gianluca Grilletti [2]
Institute for Logic, Language and Computation, University of Amsterdam

Valentin Goranko [3]
Department of Philosophy, Stockholm University
and Department of Mathematics, University of Johannesburg (visiting professorship)

Abstract

Temporal reasoning with conditionals is more complex than both classical temporal reasoning and reasoning with timeless conditionals, and can lead to some rather counter-intuitive conclusions. For instance, Aristotle's famous "Sea Battle Tomorrow" puzzle leads to a fatalistic conclusion: whether there will be a sea battle tomorrow or not, but that is necessarily the case now. We propose a branching-time logic LTC to formalise reasoning about temporal conditionals and provide that logic with adequate formal semantics. The logic LTC extends the Nexttime fragment of CTL^*, with operators for model updates, restricting the domain to only future moments where antecedent is still possible to satisfy. We provide formal semantics for these operators that implements the restrictor interpretation of antecedents of temporalized conditionals, by suitably restricting the domain of discourse. As a motivating example, we demonstrate that a naturally formalised in our logic version of the 'Sea Battle' argument renders it unsound, thereby providing a solution to the problem with fatalist conclusion that it entails, because its underlying reasoning per cases argument no longer applies when these cases are treated not as material implications but as temporal conditionals. On the technical side, we analyze the semantics of LTC and provide a series of reductions of LTC-formulae, first recursively eliminating the dynamic update operators and then the path quantifiers in such formulae. Using these reductions we obtain a sound and complete axiomatization for LTC, and reduce its decision problem to that of the modal logic KD.

Keywords: conditionals, temporal settings, restrictors, the Sea Battle Puzzle

---

[1] Email: fengkui.ju@bnu.edu.cn. Supported by the National Social Science Foundation of China (No. 12CZX053) and the Major Program of the National Social Science Foundation of China (NO. 17ZDA026).

[2] Email: grilletti.gianluca@gmail.com. Supported by the European Research Council (ERC) under the European Union's Horizon 2020 research and innovation programme (No. 680220).

[3] Email: valentin.goranko@philosophy.su.se. Partly supported by research grant 2015-04388 of the Swedish Research Council.

1 Introduction

Temporal conditionals The original philosophical motivation for the present work is rooted in the concept of *conditional* (see e.g. [4], [1]). Classical logic only deals with the simplest kind of conditionals, viz. material implications. However, the logical laws for material implications sometimes lead to counter-intuitive, or just plainly inconsistent conclusions when applied to other types of conditionals. Here is a simple example from [19]. There are two marbles in a black box and we know that one of them is blue and the other is red. Someone picks a marble from the box but we do not see which one. Intuitively the sentence "*the picked marble must be blue or the picked marble must be red*" is false. However, each of the following sentences are true: "*the picked marble is not blue or is not red. if the picked marble is not blue then it must be red. if the picked marble is not red then it must be blue.*" These sentences seem to imply that the picked marble must be blue or the picked marble must be red.

Some conditionals involve temporality. Here is an example from chess: "*if you attack the opponent's queen with your knight in the next move, you will eventually lose the game.*" We call such conditionals *temporal conditionals*. Various logical problems are concerned with such conditionals and a typical one is the *Sea Battle Puzzle*. This puzzle goes back to Aristotle and is also closely related to the Master Argument of Diodorus Cronus. Briefly, it goes as follows[4]: either there will be a sea battle tomorrow or not; if there is a sea battle tomorrow, it is necessarily so; if there is no sea battle tomorrow, it is necessarily so. Thus, either there will necessarily be a sea battle tomorrow or there will necessarily be no sea battle tomorrow. The conclusion of this argument seems fatalistic and unacceptable for many, but its premises seem mostly fine.

Restrictors Several works, including [8], [18], [7] and [20] have proposed treatment of the semantics of conditionals based on the concept of *restrictor*, which can be traced back to Ramsey's Test [16]. The restrictor-based view interprets conditionals differently from material implication. By this view, the conditional 'if ϕ then ψ' is not a connective relating two sentences. Instead, the if-clause, 'if ϕ', is a device for restricting discourse domains that are collections of possibilities. Thus, 'if ϕ then ψ' is true with respect to a domain iff ψ is true with respect to the domain restricted by 'if ϕ'.

Reading 'if ϕ' as a restrictor can resolve some logical problems involving conditionals. The two puzzles mentioned above have the same inference pattern, that is, reasoning per cases: "ϕ_1 or ϕ_2; if ϕ_1 then ψ_1; if ϕ_2 then ψ_2; therefore, ψ_1 or ψ_2". As pointed out in [2], [6] and [10], reasoning per cases is not generally sound under the restrictor reading of conditionals. Here is why. Assume that 'ϕ_1 or ϕ_2', 'if ϕ_1 then ψ_1' and 'if ϕ_2 then ψ_2' are true with respect to a discourse

[4] Aristotle does not present the puzzle clearly and there are different versions of it in the literature. We use the one from [5].

domain Δ. Let Δ^{ϕ_1} and Δ^{ϕ_2} be the respective results of restricting Δ with 'if ϕ_1' and 'if ϕ_2'. What 'if ϕ_1 then ψ_1' and 'if ϕ_2 then ψ_2' say is just that ψ_1 is true with respect to Δ^{ϕ_1} and ψ_2 is true with respect to Δ^{ϕ_2}. But if neither the truth of ψ_1 nor the truth of ψ_2 is upward monotonic relative to discourse domains, then it is possible that neither ψ_1 nor ψ_2 is true with respect to Δ.

Under the restrictor view of conditionals, the two puzzles are not puzzling any more. The discourse domain of the marble puzzle consists of epistemic possibilities concerning the color of the picked marble. *Must* is not a upward monotonic notion relative to classes of epistemic possibilities. So the argument in this puzzle is not sound. The discourse domain of the Sea Battle Puzzle consists of possible futures and *necessity* is not a upward monotonic notion relative to classes of possible futures. Then the argument in this puzzle is not sound either.

Absolute and relative necessity Temporal restrictors shrink discourse domains which consist of *possible* futures. We argue that there are two senses of possibility: *ontological* and *genuine* possibility.

In reality we make decisions to do something or not to do something. So we will not do whatever we are able to do. Call the class of things we are able to do *the ability domain*. Call the ability domain excluding the things we decide to refrain from doing *the intension domain*. A future is ontologically possible if we can realize it by doing things in the ability domain. A future is genuinely possible if we can realize it by doing things in the intension domain.

Here is an example. There will be an exam tomorrow morning and there are two things for a student to do this evening: preparing for the exam and watching a football match. Doing the former will enable the student to pass the exam but doing the latter will not. The student decides to watch the match. In this case, the future where the student passes the exam is just ontologically possible but not genuinely possible.

Respectively, we can distinguish two senses of necessity: *absolute* and *relative* necessity. A proposition is absolutely necessary if it is the case for every ontologically possible future. A proposition is relatively necessary if it is the case for every genuinely possible future.

The second/third premise of the Sea Battle Puzzle, "*if there is a/no sea battle tomorrow, it is necessarily so*", can be called *the principle of necessity of truth*. As there are two senses of necessity, there are two principles of necessity of truth.

We argue that the principle of absolute necessity of truth does not hold. Assume that a fleet admiral is able to do two things: a and b. Doing a will cause a sea battle tomorrow but doing b will not. He decides to do a. In this case, it is plausible to say that there will be a sea battle tomorrow. But it is wrong to say that there will be a sea battle tomorrow no matter what the admiral will do in the absolute sense.

We think, however, that the principle of relative necessity of truth does hold. Let Δ be the class of all the choices for which the agent is open. Assume that it is not relatively necessary that ϕ will be the case. Then there is a b in Δ such that doing b will make ϕ false. In this situation, it is strange to say that ϕ will be the case, as the agent might do b. So if ϕ will be the case, it is relatively necessary that ϕ will be the case.

The first premise of the Sea Battle Puzzle, "*either there will be a sea battle tomorrow or not*", can be called *the principle of excluded future middle.* We think that this principle is commonly accepted, at least in classical (non-intuitionistic) reasoning. So all the three premises of the puzzle hold if necessity is understood in the relative sense.

However, the conclusion of this puzzle is problematic for both senses of necessity. Again, suppose that doing a will cause a sea battle tomorrow but doing b will not. Assume that the admiral has not made the decision to do a or b yet. Then it is wrong to say that there will necessarily be a sea battle tomorrow, and also wrong to say that there will necessarily be no sea battle tomorrow. So the conclusion of the puzzle is false.

Thus, the Sea Battle Puzzle is only logically problematic for the case of relative necessity, but not (necessarily) for the case of absolute necessity.

Our technical proposal and contributions In this paper we introduce a new system of extended temporal logic LTC to formalise temporal conditionals. This logic can express relative necessity. We will show that the three premises of the Sea Battle Puzzle are valid in this logic but its conclusion is not, thus indicating that the reasoning behind it is not sound with respect to this logic. The logic LTC extends the Nexttime fragment of CTL^* with *operators for model updates*, restricting the temporal domain to those future moments where antecedent is still possible to satisfy. As an illustrating example, we demonstrate that a suitably formalised in our logic version of the 'Sea Battle' argument is not sound, thereby providing a solution to the problem with the fatalist conclusion that it entails. As discussed earlier, the technical core of the proposed solution is that the underlying reasoning per cases argument no longer applies when these cases are treated not as material implications but as temporal conditionals. On the technical side, we analyze the semantics of LTC and provide a series of reductions of LTC-formulae, first recursively eliminating the dynamic operators and then the path quantifiers in such formulae. Using these reductions we obtain a sound and complete axiomatization for LTC and also eventually reduce its decision problem to that of the modal logic KD.

Thus, we consider the contributions of this work to be two-fold: to develop a logical system for formalization and analysis of temporal conditionals, and to apply it to the clarification and solution of various philosophical logical problems that they present.

Structure of the paper In Section 2 we introduce the logic of temporal conditionals LTC and provide and discuss its formal semantics. In Section 3 we show that the update operator in LTC properly captures temporal conditionals. Section 4 proposes and discusses our formal solution to the Sea Battle Puzzle, based on its formalisation in LTC. In Section 5 we prove that, by means of effective translations, the model update operator can be eliminated and then the path quantifier and temporal operator in *state formulas* can be replaced by the classical box modality. Using these results, in Section 6 we establish a sound and complete axiomatization of LTC and show its decidability. We conclude with brief remarks in Section 7.

2 The logic for temporal conditionals LTC

We assume that the reader has basic familiarity with modal and branching-time temporal logics, e.g. within [3, Chapters 4, 5, 7.1].

Let Φ_0 be a fixed countable set of atomic propositions and let p range over it. The language Φ_{LTC} of LTC involves Φ_0, the propositional connectives $\top, \neg, \wedge$ and the modalities $\mathbf{A}, \mathrm{X}$ and $[\cdot]$, where $\cdot$ stands for a formula (see further). Φ_{LTC} and $\Phi_{\mathrm{X}[\cdot]}$, a fragment of Φ_{LTC}, are defined as follows:

$$\Phi_{\mathrm{X}[\cdot]} : \psi ::= p \mid \top \mid \neg\psi \mid (\psi \wedge \psi) \mid \mathrm{X}\psi \mid [\psi]\psi$$
$$\Phi_{\mathsf{LTC}} : \phi ::= p \mid \top \mid \neg\phi \mid (\phi \wedge \phi) \mid \mathrm{X}\phi \mid \mathbf{A}\phi \mid [\psi]\phi$$

The other propositional connectives: $\perp$ (falsum), $\vee, \rightarrow, \leftrightarrow$ are defined in the usual way. We also define $\mathbf{E}\phi := \neg\mathbf{A}\neg\phi$.

We can naturally distinguish state and path formulae of LTC (just like in the logic CTL*), defined by mutual induction as follows:

$$\text{State formulae } \Phi^{s}_{\mathsf{LTC}} : \quad \phi ::= p \mid \top \mid \neg\phi \mid (\phi \wedge \phi) \mid \mathbf{A}\theta \mid [\psi]\phi, \text{ where } \psi \in \Phi_{\mathsf{PC}}$$
$$\text{Path formulae } \Phi^{p}_{\mathsf{LTC}} : \quad \theta ::= \phi \mid \neg\theta \mid (\theta \wedge \theta) \mid \mathrm{X}\theta \mid [\psi]\theta, \text{ where } \psi \in \Phi_{\mathrm{X}[\cdot]}$$

We define the abbreviation $\Box := \mathbf{A}\mathrm{X}$. What follow are four fragments of Φ_{LTC} which will be used for technical purposes later.

$$\Phi_{\mathsf{PC}} : \psi ::= p \mid \top \mid \neg\psi \mid (\psi \wedge \psi)$$
$$\Phi_{\mathrm{X}} : \psi ::= p \mid \top \mid \neg\psi \mid (\psi \wedge \psi) \mid \mathrm{X}\psi$$
$$\Phi_{\mathbf{A}\mathrm{X}} : \psi ::= p \mid \top \mid \neg\psi \mid (\psi \wedge \psi) \mid \mathrm{X}\psi \mid \mathbf{A}\phi$$
$$\Phi_{\Box} : \psi ::= p \mid \top \mid \neg\psi \mid (\psi \wedge \psi) \mid \Box\psi$$

Let us give some intuition on the logical operators in LTC. Informally, $\mathrm{X}\phi$ means that ϕ will be the case in the next moment. $\mathbf{A}\phi$ means that no matter how the agent will act in the future, ϕ is the case now, that is, ϕ is *necessary*. $\mathbf{A}$ can be viewed as a universal quantifier over possible futures. Later we will see that possible futures in the semantic setting of LTC are *genuinely possible* futures in the intuitive sense. So $\mathbf{A}\phi$ indicates the relative necessity. Respectively, $\mathbf{E}\phi$ states that the agent has a way to act in the future so that ϕ is the case now, that is, ϕ is *possible* [5].

---

[5] It seems strange to say that the agent has a way to act in the future so that ϕ is the case now. Actually this is fine, as whether a sentence involving future is true or not now might be dependent on how the agent will act in the future. Indeed, the truth of "there will be a sea battle tomorrow" depends on what the admirals of the fighting fleets will decide tonight.

Intuitively, $[\psi]\phi$ indicates that given ψ, ϕ is the case. Assume that one's decision to make ψ true will always make ψ true[6]. Then another intended understanding of $[\psi]\phi$ is as follows: the occurrence of ψ in $[\psi]$ represents the action of deciding to make ψ true and $[\psi]\phi$ is read as that ϕ is the case after the agent decides to make ψ true. This will be seen more clearly from the formal semantics.

Now, some technical terminology and notation. Let W be a nonempty set of states and let R be a binary relation on it. A (finite or infinite) sequence $w_0 \dots w_n(\dots)$ of states is called an *R-sequence* if $w_0 R \dots R w_n(\dots)$. Note that w is an R-sequence, for any $w \in W$. Next, (W, R) is a *tree* if there is a state $r \in W$, called the *root*, such that for any w, then there is a unique R-sequence starting with r and ending with w. It is easy to see that if there is a root, then it is unique and R is irreflexive. Further, R is *serial* if, for any $w \in W$ there is a $u \in W$ such that Rwu. We say that a tree (W, R) is serial if R is.

A serial tree can be understood as a time structure encoding an agent's actions (the transitions) and states in time (the nodes). A branching in the tree is interpreted as a situation where the agent can choose between different possible actions. The seriality corresponds to the fact that the agent can always perform an action at any given time.

Fix a serial tree (W, R). A finite R-sequence $w_0 \dots w_n$ starting at the root is called a *history* of w_n. For any states w and u, u is a *historical* state of w if there is a R-sequence $u_0 \dots u_n$ such that $0 < n$, $u_0 = u$ and $u_n = w$. If u is a historical state of w, we also say that w is a *future* state of u. Note that no state in a tree can be a historical or future state of itself.

An infinite R-sequence is called a *path*. A path starting at the root is a *timeline*. A path $\pi = w_0 w_1 \dots$ *passes through* a state x if $x = w_i$ for some i. For any path π, we use $\pi(i)$ to denote the $i+1$-th element of π, ${}^i\pi$ the prefix of π to the $i+1$-th element, and π^i the suffix of π from the $i+1$-th element. For example, if $\pi = w_0 w_1 \dots$, then $\pi(2) = w_2$, ${}^2\pi = w_0 w_1 w_2$ and $\pi^2 = w_2 w_3 \dots$. For any history $w_0 \dots w_n$ and path $u_0 u_1 \dots$, if $w_n = u_0$, let $w_0 \dots w_n \otimes u_0 \dots$ denote the timeline $w_0 \dots w_n u_1 \dots$.

$\mathfrak{M} = (W, R, r, V)$ is a (LTC-)*model* if (W, R) is a serial tree with r as the root and V is a valuation from Φ_0 to 2^W. Figure 1 illustrates a model.

Now we will define by mutual induction two important semantic concepts: "*ϕ being true at a state $\pi(i)$ relative to a timeline π in a model $\mathfrak{M}$*", denoted $\mathfrak{M}, \pi, i \Vdash \phi$ (Definition 2.1), and "*the result of updating $\mathfrak{M}$ at w with ϕ*", denoted $\mathfrak{M}^{\phi}_{w}$ (Definition 2.2). The mutual induction is needed because $\mathfrak{M}, \pi, i \Vdash [\phi]\psi$ is defined in terms of $\mathfrak{M}^{\phi}_{\pi(i)}$ which, in turn, is defined in terms of $\mathfrak{M}, \pi, i \Vdash \phi$.

6 In this paper we maintain the assumption that making a decision to do something will always result in this thing being done.

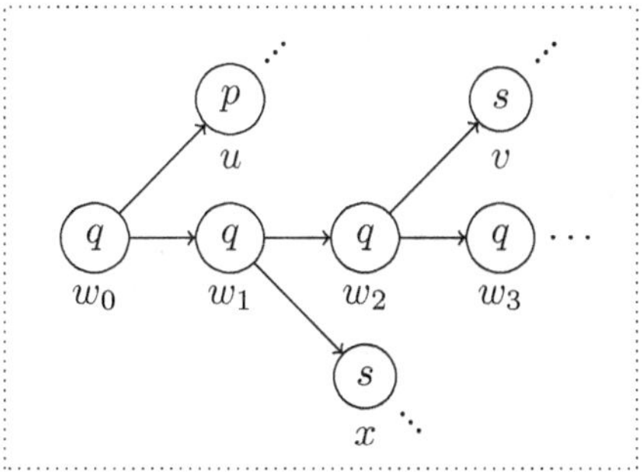

Fig. 1. A fragment of a LTC-model.

Definition 2.1 [Semantics]

$$\begin{array}{lcl}
\mathfrak{M},\pi,i \Vdash p & \Leftrightarrow & \pi(i) \in V(p) \\
\mathfrak{M},\pi,i \Vdash \top & & \\
\mathfrak{M},\pi,i \Vdash \neg\phi & \Leftrightarrow & \text{not } \mathfrak{M},\pi,i \Vdash \phi \\
\mathfrak{M},\pi,i \Vdash \phi \wedge \psi & \Leftrightarrow & \mathfrak{M},\pi,i \Vdash \phi \text{ and } \mathfrak{M},\pi,i \Vdash \psi \\
\mathfrak{M},\pi,i \Vdash \mathrm{X}\phi & \Leftrightarrow & \mathfrak{M},\pi,i+1 \Vdash \phi \\
\mathfrak{M},\pi,i \Vdash \mathbf{A}\phi & \Leftrightarrow & \text{for any path } \rho \text{ starting at } \pi(i),\ \mathfrak{M},{}^{i}\pi \otimes \rho, i \Vdash \phi \\
\mathfrak{M},\pi,i \Vdash [\phi]\psi & \Leftrightarrow & \mathfrak{M}^{\phi}_{\pi(i)},\pi,i \Vdash \psi \text{ if } (\mathfrak{M}^{\phi}_{\pi(i)},\pi) \text{ is well-defined (see below)}
\end{array}$$

(Note that $\mathfrak{M},\pi,i \Vdash [\phi]\psi$ holds vacuously if $(\mathfrak{M}^{\phi}_{\pi(i)},\pi)$ is not well-defined.)

The truth condition of $\mathbf{A}\phi$ at $\mathfrak{M},\pi,i$ can also be equivalently stated as follows: $\mathfrak{M},\tau,i \Vdash \phi$ for any timeline τ passing through $\pi(i)$. If π is a timeline in $\mathfrak{M}$ and $\pi(i) = w$, we will sometimes write $\mathfrak{M},\pi,w \Vdash \phi$ instead of $\mathfrak{M},\pi,i \Vdash \phi$. We say that ϕ is *achievable* at w in $\mathfrak{M}$ if $\mathfrak{M},\pi,w \Vdash \mathbf{E}\phi$, i.e. $\mathfrak{M},\pi,w \Vdash \phi$ for some timeline π containing w.

Note that the truth of a state formula at a state, relative to a timeline, is not dependent on the timeline, but this is not so for path formulae. Sometimes, if ϕ is a state formula, we write $\mathfrak{M},w \Vdash \phi$ without specifying a timeline.

Definition 2.2 [Model updates and well-definedness] Suppose ϕ is achievable at w in $\mathfrak{M}$. We define the set $X^{\phi}_{w} \subseteq W$ as follows: for any $x \in W$, $x \in X^{\phi}_{w}$ iff

(i) x is a future state of w, and
(ii) there is no timeline ρ passing through w and x such that $\mathfrak{M},\rho,w \Vdash \phi$.

Now, we define $\mathfrak{M}^{\phi}_{w} = (W - X^{\phi}_{w}, R', r, V')$ as the *restriction* of $\mathfrak{M}$ to $W - X^{\phi}_{w}$, that is, $R' = R \cap (W - X^{\phi}_{w})^2$ and $V'(p) = V(p) \cap (W - X^{\phi}_{w})$ for each $p \in \Phi_0$.

If ϕ is not achievable at w, we declare $\mathfrak{M}^{\phi}_{w}$ undefined because in this case all successors of w are in X^{ϕ}_{w} and therefore R' is not serial.

If π is a timeline in $\mathfrak{M}$, such that $\mathfrak{M}^{\phi}_{\pi(i)}$ is defined, and π belongs to $\mathfrak{M}^{\phi}_{\pi(i)}$, then we say that $(\mathfrak{M}^{\phi}_{\pi(i)},\pi)$ is *well-defined.*

Suppose that ϕ is achievable at w. It can be easily verified that $r \notin X^{\phi}_{w}$ and X^{ϕ}_{w} is closed under R. So $(W - X^{\phi}_{w}, R')$ is still a tree with r as the root. It can also be verified that R' is serial when $\mathfrak{M}^{\phi}_{w}$ is defined. Then $\mathfrak{M}^{\phi}_{w}$ is a LTC-model. Later, we will see that a timeline π in $\mathfrak{M}$ is a timeline in $\mathfrak{M}^{\phi}_{\pi(i)}$ iff $\mathfrak{M}, \pi, i \Vdash \phi$.

Updating $\mathfrak{M}$ with ϕ at w means to shrink $\mathfrak{M}$ by removing the states in X^{ϕ}_{w}. The set X^{ϕ}_{w} can be understood as follows. Assume that the agent is at w and decides to make ϕ true. After the decision is made, some future states are not possible anymore. A state becomes impossible if, when the agent travels in time to it, it would no longer be possible to make ϕ true at w, no matter what happens afterwards. X^{ϕ}_{w} is the collection of these states. Figure 2 illustrates how a formula updates a model.

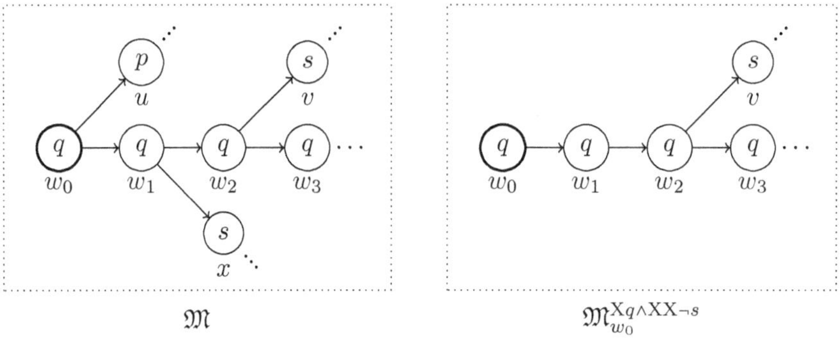

Fig. 2. This figure illustrates how a model is updated and reduced by a formula. The model on the right is the result of updating the one on the left at w_0 with $Xq \wedge XX\neg s$.

A formula ϕ of LTC is *valid* (resp., *satisfiable*) if $\mathfrak{M}, \pi, i \Vdash \phi$ for any (resp., for some) $\mathfrak{M}$, π and i. Note that a path formula ϕ is valid (resp., satisfiable) iff the state formula $\mathbf{A}\phi$ is valid (resp., $\mathbf{E}\phi$ is satisfiable). A set Γ of formulae *entails* a formula ϕ, denoted $\Gamma \vDash \phi$, if for any $\mathfrak{M}$, π and i, if $\mathfrak{M}, \pi, i \Vdash \psi$ for each $\psi \in \Gamma$, then $\mathfrak{M}, \pi, i \Vdash \phi$. We write $\psi_1, \dots, \psi_n \vDash \phi$ for $\{\psi_1, \dots, \psi_n\} \vDash \phi$. As usual, we also say that ϕ and ψ are *equivalent*, denoted $\psi \equiv \phi$, if $\psi \vDash \phi$ and $\phi \vDash \psi$. Note that, as expected, $\equiv$ is a congruence with respect to all operators in the language of LTC, incl. $[\cdot]$: if $\phi, \psi \in \Phi_{X[\cdot]}$ and $\phi \equiv \psi$, then $[\phi]\chi \equiv [\psi]\chi$ for any $\chi \in \Phi_{\mathsf{LTC}}$, and if $\chi, \theta \in \Phi_{\mathsf{LTC}}$ and $\chi \equiv \theta$, then $[\phi]\chi \equiv [\phi]\theta$ for any $\phi \in \Phi_{X[\cdot]}$.

3 Restrictors work as intended

The update with ϕ at a state in a model restricts the class of possible futures starting at that state. The following theorem indicates that it restricts the class of possible futures exactly as we wish: it excludes those possible futures which do not satisfy ϕ.

For any ϕ in Φ_{LTC}, we define *the temporal depth* ϕ^{td} (intuitively, ϕ^{td} indicates how far in the future ϕ can see) and *the combined depth* ϕ^{cd} of ϕ, as follows:

$p^{td} = 0$	$p^{cd} = 0$
$\top^{td} = 0$	$\top^{cd} = 0$
$(\neg\phi)^{td} = \phi^{td}$	$(\neg\phi)^{cd} = \phi^{cd}$
$(\phi \wedge \psi)^{td} = \max\{\phi^{td}, \psi^{td}\}$	$(\phi \wedge \psi)^{cd} = \max\{\phi^{cd}, \psi^{cd}\}$
$(\mathrm{X}\phi)^{td} = \phi^{td} + 1$	$(\mathrm{X}\phi)^{cd} = \phi^{cd} + 1$
$(\mathbf{A}\phi)^{td} = \phi^{td}$	$(\mathbf{A}\phi)^{cd} = \phi^{cd} + 1$
$([\phi]\psi)^{td} = \max\{\phi^{td}, \psi^{td}\}$	$([\phi]\psi)^{cd} = \max\{\phi^{cd}, \psi^{cd}\}$

Lemma 3.1 *Let ϕ be a formula in $\Phi_{\mathrm{X}[\cdot]}$, $\mathfrak{M}$ be a model and w be a state in it.*

(a) For any natural number $n \geq \phi^{td}$ and any pair of timelines π and τ of $\mathfrak{M}$ passing through w and sharing the same n states after w, it holds that $\mathfrak{M}, \pi, w \Vdash \phi$ iff $\mathfrak{M}, \tau, w \Vdash \phi$.

(b) For any timeline π of $\mathfrak{M}$ passing through w, $(\mathfrak{M}^{\phi}_{w}, \pi)$ is well-defined iff $\mathfrak{M}, \pi, w \Vdash \phi$.

Proof. We will prove both claims simultaneously by induction on ϕ. The base cases $\phi = p$ and $\phi = \top$ are straightforward. Suppose that both claims hold for all subformulae of ϕ. We spell out here only the two non-trivial cases.

Case $\phi = \mathrm{X}\psi$.

(a) Let $n \geq (\mathrm{X}\psi)^{td}$ and π and τ be two timelines of $\mathfrak{M}$ passing through w and sharing the same n states after w. Let $\pi(i) = \tau(i) = w$. We have the following equivalences:

$$
\begin{array}{ll}
& \mathfrak{M}, \pi, i \Vdash \mathrm{X}\psi \\
\Leftrightarrow & \mathfrak{M}, \pi, i+1 \Vdash \psi \\
{}^{*}\Leftrightarrow & \mathfrak{M}, \tau, i+1 \Vdash \psi \\
\Leftrightarrow & \mathfrak{M}, \tau, i \Vdash \mathrm{X}\psi
\end{array}
$$

The equivalence marked by * holds by the inductive hypothesis for (a) applied to ψ, as π and τ share the same $n-1$ states after $\pi(i+1)$ and $\psi^{td} \leq n-1$.

(b) Suppose π is a timeline in $\mathfrak{M}$ passing through w and $\pi(i) = w$. Note that $(\mathfrak{M}^{\mathrm{X}\psi}_{w}, \pi)$ is well-defined iff $\mathfrak{M}^{\mathrm{X}\psi}_{w}$ is defined and π is a path in it. Clearly if $\mathfrak{M}, \pi, w \Vdash \mathrm{X}\psi$ then $\mathfrak{M}^{\mathrm{X}\phi}_{w}$ is defined. To show that $(\mathfrak{M}^{\mathrm{X}\psi}_{w}, \pi)$ is well-defined iff $\mathfrak{M}, \pi, w \Vdash \mathrm{X}\psi$, it now suffices to show that, assuming $\mathfrak{M}^{\mathrm{X}\phi}_{w}$ is defined, π is a path in $\mathfrak{M}^{\mathrm{X}\psi}_{w}$ iff $\mathfrak{M}, \pi, w \Vdash \mathrm{X}\psi$. Assume $\mathfrak{M}^{\mathrm{X}\psi}_{w}$ is defined. Then:

$$
\begin{array}{ll}
& \pi \text{ is a path in } \mathfrak{M}^{\mathrm{X}\psi}_{w} \\
\Leftrightarrow & \text{for any } u \text{ of } \pi, \text{ if } u \text{ is a future state of } w, \text{ then there is a timeline} \\
& \tau \text{ passing through } u \text{ such that } \mathfrak{M}, \tau, i \Vdash \mathrm{X}\psi, \text{ i.e. } \mathfrak{M}, \tau, i+1 \Vdash \psi \\
{}^{*}\Leftrightarrow & \mathfrak{M}, \pi, i+1 \Vdash \psi \\
\Leftrightarrow & \mathfrak{M}, \pi, i \Vdash \mathrm{X}\psi
\end{array}
$$

Here is why the marked equivalence holds. The direction from the right to the left is clear. Assume the left. Let n be a natural number such that $\psi^{td} \leq n$. Let $u = \pi(i+1+n)$. Then there is a timeline τ passing through u such that $\mathfrak{M}, \tau, i+1 \Vdash \psi$. Note that π and τ share the same n states after $\pi(i+1)$. By the inductive hypothesis for (a) applied to ψ, we have $\mathfrak{M}, \pi, i+1 \Vdash \psi$.

Case $\phi = [\psi]\chi$.

(a) Let $n \geq ([\psi]\chi)^{td}$ and π and τ be two timelines of $\mathfrak{M}$ passing through w and sharing the same n states after w. We have the following equivalences:

$$\begin{array}{rl} & \mathfrak{M}, \pi, w \Vdash [\psi]\chi \\ \Leftrightarrow & (\mathfrak{M}^{\psi}_{w}, \pi) \text{ is not well-defined or } \mathfrak{M}^{\psi}_{w}, \pi, w \Vdash \chi \\ {}^{*}\Leftrightarrow & \mathfrak{M}, \pi, w \nVdash \psi \text{ or } \mathfrak{M}^{\psi}_{w}, \pi, w \Vdash \chi \\ {}^{*}\Leftrightarrow & \mathfrak{M}, \tau, w \nVdash \psi \text{ or } \mathfrak{M}^{\psi}_{w}, \tau, w \Vdash \chi \\ {}^{*}\Leftrightarrow & (\mathfrak{M}^{\psi}_{w}, \tau) \text{ is not well-defined or } \mathfrak{M}^{\psi}_{w}, \tau, w \Vdash \chi \\ \Leftrightarrow & \mathfrak{M}, \tau, w \Vdash [\psi]\chi \end{array}$$

The first and third equivalence marked by * holds by the inductive hypothesis for (b) applied to ψ. The second equivalence marked by * holds by the inductive hypothesis for (a) applied to ψ, as $\psi^{td} \leq n$.

(b) Let π be a timeline in $\mathfrak{M}$ passing through w and $\pi(i) = w$. Again, note that $(\mathfrak{M}^{[\psi]\chi}_{w}, \pi)$ is well-defined iff $\mathfrak{M}^{[\psi]\chi}_{w}$ is defined and π is a path in it. Also note that if $\mathfrak{M}, \pi, w \Vdash [\psi]\chi$, then $\mathfrak{M}^{[\psi]\chi}_{w}$ is defined. To show that $(\mathfrak{M}^{[\psi]\chi}_{w}, \pi)$ is well-defined iff $\mathfrak{M}, \pi, w \Vdash [\psi]\chi$, it now suffices to show that, assuming $\mathfrak{M}^{[\psi]\chi}_{w}$ is defined, π is a path in $\mathfrak{M}^{[\psi]\chi}_{w}$ iff $\mathfrak{M}, \pi, w \Vdash [\psi]\chi$. Assume that $\mathfrak{M}^{[\psi]\chi}_{w}$ is defined. Then:

$$\begin{array}{rl} & \pi \text{ is a path in } \mathfrak{M}^{[\psi]\chi}_{w} \\ \Leftrightarrow & \text{for any } u \text{ of } \pi, \text{ if } u \text{ is a future state of } w, \text{ then there is a timeline} \\ & \tau \text{ passing through } u \text{ such that } \mathfrak{M}, \tau, w \Vdash [\psi]\chi \\ \Leftrightarrow & \text{for any } u \text{ of } \pi, \text{ if } u \text{ is a future state of } w, \text{ then there is a timeline} \\ & \tau \text{ passing through } u \text{ such that if } (\mathfrak{M}^{\psi}_{w}, \tau) \text{ is well-defined, then} \\ & \mathfrak{M}^{\psi}_{w}, \tau, w \Vdash \chi \\ {}^{*}\Leftrightarrow & \text{for any } u \text{ of } \pi, \text{ if } u \text{ is a future state of } w, \text{ then there is a timeline} \\ & \tau \text{ passing through } u \text{ such that if } \mathfrak{M}, \tau, w \Vdash \psi, \text{ then } \mathfrak{M}^{\psi}_{w}, \tau, w \Vdash \chi \\ {}^{*}\Leftrightarrow & \text{if } \mathfrak{M}, \pi, w \Vdash \psi, \text{ then } \mathfrak{M}^{\psi}_{w}, \pi, w \Vdash \chi \\ \Leftrightarrow & \text{if } (\mathfrak{M}^{\psi}_{w}, \pi) \text{ is well-defined, then } \mathfrak{M}^{\psi}_{w}, \pi, w \Vdash \chi \\ \Leftrightarrow & \mathfrak{M}, \pi, w \Vdash \chi \end{array}$$

The first equivalence marked by * holds by the inductive hypothesis for (b) applied to ψ. Here is why the second *-marked equivalence holds. The direction from the right to the left is clear. Assume the left. Let n be a natural number such that $\psi^{td}, \chi^{td} \leq n$. Let $u = \pi(i+n)$. Then there is a timeline τ passing through u such that if $\mathfrak{M}, \tau, w \Vdash \psi$, then $\mathfrak{M}^{\psi}_{w}, \tau, w \Vdash \chi$. Note that π and τ share the same n elements after w. By the inductive hypothesis for (a) applied to ψ and χ, we have that if $\mathfrak{M}, \pi, w \Vdash \psi$, then $\mathfrak{M}^{\psi}_{w}, \pi, w \Vdash \chi$. □

The next corollary follows immediately from Lemma 3.1.

Corollary 3.2 *Let $\mathfrak{M}$ be a model, w a state and ϕ a formula of $\Phi_{X[\cdot]}$ achievable at w (and therefore $\mathfrak{M}^{\phi}_{w}$ is defined). Then any timeline π in $\mathfrak{M}$ passing through w is also a timeline in $\mathfrak{M}^{\phi}_{w}$ iff $\mathfrak{M}, \pi, w \Vdash \phi$.*

The following lemma states that $\Phi_{X[\cdot]}$ can be effectively reduced to Φ_X.

Lemma 3.3 *For any ϕ in $\Phi_{X[\cdot]}$, there is a ψ in Φ_X equivalent to ϕ.*

Proof. We first show that for any $\psi \in \Phi_X$, $[\phi]\psi$ is equivalent to $\phi \to \psi$. Note that the truth of the formulae in Φ_X at a state relative to a path in a model is completely determined by the information of the state and the path. So for any $\psi \in \Phi_X$, $\mathfrak{M}, \pi, w \Vdash \psi$ iff $\mathfrak{M}^{\phi}_{w}, \pi, w \Vdash \psi$. Let ψ be in Φ_X. Then: $\mathfrak{M}, \pi, w \Vdash [\phi]\psi$ $\Leftrightarrow$ if $\mathfrak{M}, \pi, w \Vdash \phi$ then $\mathfrak{M}^{\phi}_{w}, \pi, w \Vdash \psi$ $\Leftrightarrow$ if $\mathfrak{M}, \pi, w \Vdash \phi$ then $\mathfrak{M}, \pi, w \Vdash \psi$ $\Leftrightarrow$ $\mathfrak{M}, \pi, w \Vdash \phi \to \psi$. For any χ in $\Phi_{X[\cdot]}$, by use of the previous result, we can eliminate step-by-step the update operator $[\cdot]$ in χ. □

Next theorem points out some connections between $[\phi]\psi$ and $\phi \to \psi$.

Theorem 3.4 *$[\phi]\psi$ collapses to $\phi \to \psi$ if either of the following holds:*

(a) ϕ is a state formula;
(b) ψ is in $\Phi_{X[\cdot]}$.

Proof. (a) Suppose ϕ is a state formula. Let $\mathfrak{M}$ be a model and w a state in it. Then ϕ is either true or false at w. Suppose first that ϕ is true at w. Then the update with ϕ at w does not change $\mathfrak{M}$. Then for any timeline π passing through w, $[\phi]\psi$ is true at w relative to π iff ψ – which is now equivalent to $\phi \to \psi$ – is true at w relative to π. Now, suppose ϕ is false at w. Then the update with ϕ at w fails, so both $[\phi]\psi$ and $\phi \to \psi$ are trivially true at w relative to any timeline.

(b) Suppose ψ is in $\Phi_{X[\cdot]}$. By Lemma 3.3, there is a $\psi' \in \Phi_X$ equivalent to ψ. Then $[\phi]\psi$ is equivalent to $[\phi]\psi'$. By the proof for Lemma 3.3, $[\phi]\psi'$ is equivalent to $\phi \to \psi'$. □

Note that $[\phi]\psi$ does not generally collapse to $\phi \to \psi$ if ϕ is not a state formula and ψ contains the operator **A**. A typical example is $[\mathrm{X}s]\mathbf{A}\mathrm{X}s$: e.g., if there is a sea battle tomorrow, it is necessarily so.

4 Our solution to the Sea Battle Puzzle

We first show that the principle of relative necessity of truth, formalised in the logic as $[\phi]\mathbf{A}\phi$, is valid.

Lemma 4.1 *Let $(\mathfrak{M}^{\phi}_{w}, \pi)$ be well-defined. For any ψ in $\Phi_{X[\cdot]}$, $\mathfrak{M}^{\phi}_{w}, \pi, w \Vdash \psi$ iff $\mathfrak{M}, \pi, w \Vdash \psi$.*

Proof. By Lemma 3.3, $\Phi_{X[\cdot]}$ can be reduced to Φ_X. As mentioned in the proof for Lemma 3.3, the truth of the formulae in Φ_X at a state relative to a timeline in a model is completely determined by the information of the state and the timeline. Then the claim follows. □

Theorem 4.2 *$[\phi]\mathbf{A}\phi$ is valid, where $\phi \in \Phi_{X[\cdot]}$.*

Proof. Suppose $\mathfrak{M}, \pi, w \nVdash [\phi]\mathbf{A}\phi$. This means that $(\mathfrak{M}^{\phi}_{w}, \pi)$ is well-defined but $\mathfrak{M}^{\phi}_{w}, \pi, w \nVdash \mathbf{A}\phi$. The latter condition entails the existence of a timeline τ in $\mathfrak{M}^{\phi}_{w}$ passing through w such that $\mathfrak{M}^{\phi}_{w}, \tau, w \nVdash \phi$. By Lemma 4.1, $\mathfrak{M}, \tau, w \nVdash \phi$. By Corollary 3.2, τ is not a timeline in $\mathfrak{M}^{\phi}_{w}$ – a contradiction. □

Let s denote the proposition "there is a sea battle". The Sea Battle Puzzle can now be formalised as

$$\mathrm{X}s \vee \mathrm{X}\neg s, [\mathrm{X}s]\mathbf{A}\mathrm{X}s, [\mathrm{X}\neg s]\mathbf{A}\mathrm{X}\neg s \vDash \mathbf{A}\mathrm{X}s \vee \mathbf{A}\mathrm{X}\neg s.$$

It is easy to show that $\mathrm{X}\phi \vee \mathrm{X}\neg\phi$ is valid. The validity of the other two premises follows from Theorem 4.2. On the other hand, it is also easy to see that $\mathbf{A}\mathrm{X}s \vee \mathbf{A}\mathrm{X}\neg s$ is not valid. Therefore, the logical argument behind the Puzzle of Sea Battle is not sound.

Remark 4.3 There have been various proposed solutions to the Sea Battle Puzzle in the literature, including Łukasiewicz's three-valued logic [12], Prior's Peircean temporal logic [15], Prior's Ockhamist temporal logic [15], the true futurist theory [13], the supervaluationist theory [17] and the relativist theory [9]. These solutions accept the validity of the argument and argue that not all its premises are true. They focus on the following issue: how do we ascribe truth values to statements such as "there will be a sea battle tomorrow"? These statements are called in the literature *future contingents*, as they are about the future but do not have an absolute sense. In the first two solutions mentioned above, the principle of excluded future middle fails. In others, the principle of necessity of truth fails. Except for Łukasiewicz's three-valued logic, the other solutions use branching time models. We refer to [14] for detailed comparison between these solutions.

5 Expressivity of LTC

In this section, we show two results. Firstly, the update operator $[\phi]$ can be eliminated from any formula in Φ_{LTC}. So Φ_{LTC} can be reduced to $\Phi_{\mathrm{X}\mathbf{A}}$. Secondly, when restricted to state formulas, $\Phi_{\mathrm{X}\mathbf{A}}$ are equally expressive with $\Phi_{\Box}$. Later we will see that the two results enable us to obtain a complete axiomatization for LTC.

To eliminate $[\phi]$ from $[\phi]\psi$, the strategy is to 'massage' $[\phi]$ into ψ, deeper and deeper, by applying some valid reduction principles, until it meets atomic propositions and dissolves. The only difficulty arises when $[\phi]$ meets the operator X. To handle this, some preliminary treatment of ϕ is needed.

Recall that Φ_{PC} is the set of purely propositional formulae.

Lemma 5.1 *For any ϕ in Φ_{X}, there are $\psi_1, \ldots, \psi_n$ in Φ_{PC} and $\chi_1, \ldots, \chi_n$ in Φ_{X} such that ϕ is equivalent to $(\psi_1 \vee \mathrm{X}\chi_1) \wedge \cdots \wedge (\psi_n \vee \mathrm{X}\chi_n)$.*

Proof. The claim follows easily by firstly transforming ϕ to a CNF, where all subformulae beginning with X are treated as literals, and then applying the valid equivalences $\neg\mathrm{X}\phi \equiv \mathrm{X}\neg\phi$ and $\mathrm{X}\phi \vee \mathrm{X}\psi \equiv \mathrm{X}(\phi \vee \psi)$. □

Lemma 5.2 *For all $\phi, \psi \in \Phi_{\mathrm{X}[\cdot]}$ and $\chi \in \Phi_{\mathsf{LTC}}$, $[\phi \wedge \psi]\chi \leftrightarrow [\phi][\psi]\chi$ is valid.*

Proof. Firstly, we show that $\mathfrak{M}_w^{\phi\wedge\psi}$ is defined iff $(\mathfrak{M}_w^{\phi})_w^{\psi}$ is defined.

Assume that $\mathfrak{M}_w^{\phi\wedge\psi}$ is defined, which implies $\mathfrak{M}, \pi, w \Vdash \phi \wedge \psi$ for some timeline π passing through w. Then $\mathfrak{M}, \pi, w \Vdash \phi$ and $\mathfrak{M}, \pi, w \Vdash \psi$. By Lemma 3.1, $\mathfrak{M}_w^{\phi}$ is defined and π is a timeline in $\mathfrak{M}_w^{\phi}$. By Lemma 4.1, $\mathfrak{M}_w^{\phi}, \pi, w \Vdash \psi$ and thus also $(\mathfrak{M}_w^{\phi})_w^{\psi}$ is defined.

Now, assume that $(\mathfrak{M}^\phi_w)^\psi_w$ is defined, and thus $\mathfrak{M}^\phi_w, \pi, w \Vdash \psi$ for some timeline π in $\mathfrak{M}^\phi_w$ passing through w. Moreover, by Theorem 4.2, $\mathfrak{M}, \pi, w \Vdash [\phi]\mathbf{A}\phi$. As $(\mathfrak{M}^\phi_w, \pi)$ is well-defined, $\mathfrak{M}^\phi_w, \pi, w \Vdash \mathbf{A}\phi$, hence $\mathfrak{M}^\phi_w, \pi, w \Vdash \phi$. Thus, $\mathfrak{M}^\phi_w, \pi, w \Vdash \phi \wedge \psi$ and consequently, by Lemma 4.1, $\mathfrak{M}, \pi, w \Vdash \phi \wedge \psi$. So, $\mathfrak{M}^{\phi\wedge\psi}_w$ is defined, as wanted.

Secondly, we show $\mathfrak{M}^{\phi\wedge\psi}_w = (\mathfrak{M}^\phi_w)^\psi_w$. It suffices to show that $\mathfrak{M}^{\phi\wedge\psi}_w$ and $(\mathfrak{M}^\phi_w)^\psi_w$ have the same timelines passing through w. By Lemma 3.1 and Lemma 4.1, the following holds:

π is a timeline in $\mathfrak{M}^{\phi\wedge\psi}_w$ through w
$\Leftrightarrow$ $\mathfrak{M}, \pi, w \Vdash \phi \wedge \psi$
$\Leftrightarrow$ π is a timeline in $\mathfrak{M}^\phi_w$ through w and $\mathfrak{M}, \pi, w \Vdash \psi$
$\Leftrightarrow$ π is a timeline in $\mathfrak{M}^\phi_w$ through w and $\mathfrak{M}^\phi_w, \pi, w \Vdash \psi$
$\Leftrightarrow$ π is a timeline in $(\mathfrak{M}^\phi_w)^\psi_w$ through w

Consequently,

$\mathfrak{M}, \pi, w \Vdash [\phi \wedge \psi]\chi$
$\Leftrightarrow$ if $(\mathfrak{M}^{\phi\wedge\psi}_w, \pi)$ is well-defined, then $\mathfrak{M}^{\phi\wedge\psi}_w, \pi, w \Vdash \chi$
$\Leftrightarrow$ if $\mathfrak{M}^{\phi\wedge\psi}_w$ is defined and π is a timeline in $\mathfrak{M}^{\phi\wedge\psi}_w$, then $\mathfrak{M}^{\phi\wedge\psi}_w, \pi, w \Vdash \chi$
$\Leftrightarrow$ if $(\mathfrak{M}^\phi_w)^\psi_w$ is defined and π is a timeline in $(\mathfrak{M}^\phi_w)^\psi_w$, then $(\mathfrak{M}^\phi_w)^\psi_w, \pi, w \Vdash \chi$
$\Leftrightarrow$ if $\mathfrak{M}^\phi_w$ is defined, π is a timeline in $\mathfrak{M}^\phi_w$, $(\mathfrak{M}^\phi_w)^\psi_w$ is defined and π is a timeline in $(\mathfrak{M}^\phi_w)^\psi_w$, then $(\mathfrak{M}^\phi_w)^\psi_w, \pi, w \Vdash \chi$
$\Leftrightarrow$ if $(\mathfrak{M}^\phi_w, \pi)$ is well-defined and $((\mathfrak{M}^\phi_w)^\psi_w, \pi)$ is well-defined, then $(\mathfrak{M}^\phi_w)^\psi_w, \pi, w \Vdash \chi$
$\Leftrightarrow$ if $(\mathfrak{M}^\phi_w, \pi)$ is well-defined, then if $((\mathfrak{M}^\phi_w)^\psi_w, \pi)$ is well-defined, then $(\mathfrak{M}^\phi_w)^\psi_w, \pi, w \Vdash \chi$
$\Leftrightarrow$ if $(\mathfrak{M}^\phi_w, \pi)$ is well-defined, then $\mathfrak{M}^\phi_w, \pi, w \Vdash [\psi]\chi$
$\Leftrightarrow$ $\mathfrak{M}, \pi, w \Vdash [\phi][\psi]\chi$

□

In the sequel, $[\phi \wedge \psi]\chi \leftrightarrow [\phi][\psi]\chi$ is called the axiom $\mathbf{Ax}_{[\wedge]}$.

Lemma 5.3 *Let $\phi \in \Phi_{\mathrm{X}[\cdot]}$ and $\mathfrak{M}, \pi, i \Vdash \mathrm{X}\phi$. Then the generated submodels of $\mathfrak{M}^{\mathrm{X}\phi}_{\pi(i)}$ and $\mathfrak{M}^\phi_{\pi(i+1)}$ at $\pi(i+1)$ coincide.*

Proof. As $\mathfrak{M}, \pi, i \Vdash \mathrm{X}\phi$, $\mathfrak{M}^{\mathrm{X}\phi}_{\pi(i)}$ is defined and $\mathfrak{M}, \pi, i+1 \Vdash \phi$. Then $\mathfrak{M}^\phi_{\pi(i+1)}$ is defined. By Corollary 3.2, for every timeline τ passing through $\pi(i+1)$, we have the following equivalences: τ is a timeline in $\mathfrak{M}^{X\phi}_{\pi(i)} \Leftrightarrow \mathfrak{M}, \tau, i \Vdash \mathrm{X}\phi \Leftrightarrow \mathfrak{M}, \tau, i+1 \Vdash \phi \Leftrightarrow \tau$ is a timeline in $\mathfrak{M}^\phi_{\pi(i+1)}$. From this the claim follows easily.□

Lemma 5.4 *The following equivalences are valid, where* $\phi \in \Phi_{\mathsf{PC}}$ *and* $\mathrm{X}\psi \in \Phi_{\mathrm{X}}$*:*

$$\begin{array}{rl}\mathbf{Ax}_{[\cdot]p}: & [\phi \vee \mathrm{X}\psi]p \leftrightarrow ((\phi \vee \mathrm{X}\psi) \to p)\\ \mathbf{Ax}_{[\cdot]\top}: & [\phi \vee \mathrm{X}\psi]\top \leftrightarrow \top\\ \mathbf{Ax}_{[\cdot]\neg}: & [\phi \vee \mathrm{X}\psi]\neg\chi \leftrightarrow ((\phi \vee \mathrm{X}\psi) \to \neg[\phi \vee \mathrm{X}\psi]\chi)\\ \mathbf{Ax}_{[\cdot]\wedge}: & [\phi \vee \mathrm{X}\psi](\chi \wedge \xi) \leftrightarrow ([\phi \vee \mathrm{X}\psi]\chi \wedge [\phi \vee \mathrm{X}\psi]\xi)\\ \mathbf{Ax}_{[\cdot]\mathrm{X}}: & [\phi \vee \mathrm{X}\psi]\mathrm{X}\chi \leftrightarrow ((\phi \to \mathrm{X}\chi) \wedge (\neg\phi \to \mathrm{X}[\psi]\chi))\\ \mathbf{Ax}_{[\cdot]\mathbf{A}}: & [\phi \vee \mathrm{X}\psi]\mathbf{A}\chi \leftrightarrow ((\phi \vee \mathrm{X}\psi) \to \mathbf{A}[\phi \vee \mathrm{X}\psi]\chi)\end{array}$$

Proof. The first 4 equivalences are straightforward.

$\mathbf{Ax}_{[\cdot]\mathrm{X}}$. Assume $\mathfrak{M}, \pi, i \nVdash [\phi \vee \mathrm{X}\psi]\mathrm{X}\chi$. Then $(\mathfrak{M}^{\phi\vee\mathrm{X}\psi}_{\pi(i)}, \pi)$ is well-defined and $\mathfrak{M}^{\phi\vee\mathrm{X}\psi}_{\pi(i)}, \pi, i \nVdash \mathrm{X}\chi$. By Lemma 3.1, $\mathfrak{M}, \pi, i \Vdash \phi \vee \mathrm{X}\psi$. Assume $\mathfrak{M}, \pi, i \Vdash \phi$. As ϕ is in Φ_{PC}, $\mathfrak{M}^{\phi\vee\mathrm{X}\psi}_{\pi(i)} = \mathfrak{M}$. Then $\mathfrak{M}, \pi, i \nVdash \mathrm{X}\chi$. Then $\mathfrak{M}, \pi, i \nVdash \phi \to \mathrm{X}\chi$. Assume $\mathfrak{M}, \pi, i \Vdash \neg\phi$. As ϕ is in Φ_{PC}, $\mathfrak{M}^{\phi\vee\mathrm{X}\psi}_{\pi(i)} = \mathfrak{M}^{\mathrm{X}\psi}_{\pi(i)}$. Then $\mathfrak{M}^{\mathrm{X}\psi}_{\pi(i)}, \pi, i \nVdash \mathrm{X}\chi$. Then $\mathfrak{M}^{\mathrm{X}\psi}_{\pi(i)}, \pi, i{+}1 \nVdash \chi$. By Lemma 5.3, $\mathfrak{M}^{\psi}_{\pi(i+1)}, \pi, i{+}1 \nVdash \chi$. Then $\mathfrak{M}, \pi, i{+}1 \nVdash [\psi]\chi$. Hence $\mathfrak{M}, \pi, i \nVdash \mathrm{X}[\psi]\chi$. Therefore, $\mathfrak{M}, \pi, i \nVdash \neg\phi \to \mathrm{X}[\psi]\chi$.

Assume $\mathfrak{M}, \pi, i \nVdash \phi \to \mathrm{X}\chi$. Then $\mathfrak{M}, \pi, i \Vdash \phi$ and $\mathfrak{M}, \pi, i \nVdash \mathrm{X}\chi$. As ϕ is in Φ_{PC}, $\mathfrak{M}^{\phi\vee\mathrm{X}\psi}_{\pi(i)} = \mathfrak{M}$. Then $\mathfrak{M}^{\phi\vee\mathrm{X}\psi}_{\pi(i)}, \pi, i \nVdash \mathrm{X}\chi$. Then $\mathfrak{M}, \pi, i \nVdash [\phi \vee \mathrm{X}\psi]\mathrm{X}\chi$. Assume $\mathfrak{M}, \pi, i \nVdash \neg\phi \to \mathrm{X}[\psi]\chi$. Then $\mathfrak{M}, \pi, i \Vdash \neg\phi$ and $\mathfrak{M}, \pi, i \nVdash \mathrm{X}[\psi]\chi$. As ϕ is in Φ_{PC}, $\mathfrak{M}^{\phi\vee\mathrm{X}\psi}_{\pi(i)} = \mathfrak{M}^{\mathrm{X}\psi}_{\pi(i)}$. Then $\mathfrak{M}, \pi, i{+}1 \nVdash [\psi]\chi$. Then $\mathfrak{M}^{\psi}_{\pi(i+1)}, \pi, i{+}1 \nVdash \chi$. By Lemma 5.3, $\mathfrak{M}^{\mathrm{X}\psi}_{\pi(i)}, \pi, i{+}1 \nVdash \chi$. Then $\mathfrak{M}^{\mathrm{X}\psi}_{\pi(i)}, \pi, i \nVdash \mathrm{X}\chi$. Hence $\mathfrak{M}^{\phi\vee\mathrm{X}\psi}_{\pi(i)}, \pi, i \nVdash \mathrm{X}\chi$. Therefore, $\mathfrak{M}, \pi, i \nVdash [\phi \vee \mathrm{X}\psi]\mathrm{X}\chi$.

$\mathbf{Ax}_{[\cdot]\mathbf{A}}$. Let $\alpha = \phi \vee \mathrm{X}\psi$. Assume $\mathfrak{M}, \pi, i \nVdash \alpha \to \mathbf{A}[\alpha]\chi$. Then $\mathfrak{M}, \pi, i \Vdash \alpha$ but $\mathfrak{M}, \pi, i \nVdash \mathbf{A}[\alpha]\chi$. By Lemma 3.1, π is in $\mathfrak{M}^{\alpha}_{\pi(i)}$. Then there is a timeline τ passing through $\pi(i)$ such that $\mathfrak{M}, \tau, i \nVdash [\alpha]\chi$. Then $\mathfrak{M}^{\alpha}_{\pi(i)}, \tau, i \nVdash \chi$. Then $\mathfrak{M}^{\alpha}_{\pi(i)}, \tau, i \nVdash \mathbf{A}\chi$. Then $\mathfrak{M}^{\alpha}_{\pi(i)}, \pi, i \nVdash \mathbf{A}\chi$. Then $\mathfrak{M}, \pi, i \nVdash [\alpha]\mathbf{A}\chi$.

Assume $\mathfrak{M}, \pi, i \nVdash [\alpha]\mathbf{A}\chi$. Then $(\mathfrak{M}^{\alpha}_{\pi(i)}, \pi)$ is well-defined and $\mathfrak{M}^{\alpha}_{\pi(i)}, \pi, i \nVdash \mathbf{A}\chi$. By Lemma 3.1, $\mathfrak{M}, \pi, i \Vdash \alpha$. Then there is a timeline τ passing through $\pi(i)$ in $\mathfrak{M}^{\alpha}_{\pi(i)}$ such that $\mathfrak{M}^{\alpha}_{\pi(i)}, \tau, i \nVdash \chi$. Then $\mathfrak{M}, \tau, i \nVdash [\alpha]\chi$. Then $\mathfrak{M}, \pi, i \nVdash \mathbf{A}[\alpha]\chi$. Hence $\mathfrak{M}, \pi, i \nVdash \alpha \to \mathbf{A}[\alpha]\chi$. □

Recall that $\Phi\mathrm{X}\mathbf{A}$ denotes the language generated from Φ_0 under X and **A**.

Theorem 5.5 *There is an effective translation* $\mathbf{t} : \Phi_{\mathsf{LTC}} \to \Phi_{\mathrm{X}\mathbf{A}}$ *preserving formulae up to equivalence, i.e. such that* $\phi \equiv \mathbf{t}(\phi)$ *for every* $\phi \in \Phi_{\mathsf{LTC}}$.

Proof. Define a language $\Phi_{\mathrm{X}\mathbf{A}[\vee]}$ as follows, where $\eta \in \Phi_{\mathsf{PC}}$ and $\chi \in \Phi_{\mathrm{X}}$:

$$\phi ::= p \mid \top \mid \neg\phi \mid (\phi \wedge \phi) \mid \mathrm{X}\phi \mid \mathbf{A}\phi \mid [\eta \vee \mathrm{X}\chi]\phi$$

There is a translation $f : \Phi_{\mathsf{LTC}} \to \Phi_{\mathrm{X}\mathbf{A}[\vee]}$ preserving formulae up to equivalence. The reason is as follows. Take a formula $[\phi]\psi$ in Φ_{LTC}. By Lemma 3.3, ϕ is equivalent to some ϕ' in Φ_{X}. Then $[\phi]\psi$ is equivalent to $[\phi']\psi$. By Lemma 5.1, ϕ' is equivalent to a conjunction $\phi'_1 \wedge \cdots \wedge \phi'_n$ for some $\phi'_i = \eta_i \vee \mathrm{X}\chi_i$, where $\eta_i \in \Phi_{\mathsf{PC}}$ and $\chi_i \in \Phi_{\mathrm{X}}$, for $i = 1, ..., n$. Then $[\phi']\psi$ is equivalent to $[\phi'_1 \wedge \cdots \wedge \phi'_n]\psi$. By Lemma 5.2, $[\phi'_1 \wedge \cdots \wedge \phi'_n]\psi$ is equivalent to $[\phi'_1] \ldots [\phi'_n]\psi$.

Define a translation $g : \Phi_{\mathbf{XA}[\vee]} \to \Phi_{\mathbf{XA}}$ in the following way:

(i) $p^g = p$
(ii) $\top^g = \top$
(iii) $(\neg\psi)^g = \neg\psi^g$
(iv) $(\psi \wedge \chi)^g = \psi^g \wedge \chi^g$
(v) $(\mathrm{X}\psi)^g = \mathrm{X}\psi^g$
(vi) $(\mathbf{A}\psi)^g = \mathbf{A}\psi^g$
(vii) (a) $([\eta \vee \mathrm{X}\chi]p)^g = ((\eta \vee \mathrm{X}\chi) \to p)^g$
(b) $([\eta \vee \mathrm{X}\chi]\top)^g = \top^g$
(c) $([\eta \vee \mathrm{X}\chi]\neg\psi)^g = ((\eta \vee \mathrm{X}\chi) \to \neg[\eta \vee \mathrm{X}\chi]\psi)^g$
(d) $([\eta \vee \mathrm{X}\chi](\psi \wedge \xi)^g = ([\eta \vee \mathrm{X}\chi]\psi \wedge [\eta \vee \mathrm{X}\chi]\xi)^g$
(e) $([\eta \vee \mathrm{X}\chi]\mathrm{X}\psi)^g = ((\eta \to \mathrm{X}\psi) \vee (\neg\eta \to \mathrm{X}[\chi]\psi))^g$
(f) $([\eta \vee \mathrm{X}\chi]\mathbf{A}\psi)^g = ((\eta \vee \mathrm{X}\chi) \to \mathbf{A}[\eta \vee \mathrm{X}\chi]\psi)^g$
(g) $([\eta \vee \mathrm{X}\chi][\alpha \vee \mathrm{X}\beta]\psi)^g = ([\eta \vee \mathrm{X}\chi](([\alpha \vee \mathrm{X}\beta]\psi)^g)^g$

Now we show by three layers of induction that for any $\phi \in \Phi_{\mathbf{XA}[\vee]}$, $\phi^g \in \Phi_{\mathbf{XA}}$ and $\phi^g \equiv \phi$. We consider only the former result; with Lemma 5.4, the argument for the latter result is similar. We put the first layer of induction on ϕ. The only non-trivial case is $\phi = ([\eta \vee \mathrm{X}\chi]\gamma$. Then we put the second layer of induction on γ. We can easily go through all the subcases except the one $\gamma = [\alpha \vee \mathrm{X}\beta]\psi$. Assume $\gamma = [\alpha \vee \mathrm{X}\beta]\psi$. By the inductive hypothesis for the second layer of induction, $([\alpha \vee \mathrm{X}\beta]\psi)^g \in \Phi_{\mathbf{XA}}$. Then by the third layer of induction on $([\alpha \vee \mathrm{X}\beta]\psi)^g$, we can get $([\eta \vee \mathrm{X}\chi]\psi)^g \in \Phi_{\mathbf{XA}}$.

Now, the translation $\mathbf{t} : \Phi_{\mathsf{LTC}} \to \Phi_{\mathbf{XA}}$ defined as $g \circ f$ satisfies the claim. □

Example 5.6 This example demonstrates how the translation $\mathbf{t}$ is computed in the non-trivial case of $\mathbf{t}([\phi]\psi)$. It also illustrates the exponential blow-up that the translation causes to the length of the input formula in this case.

$$
\begin{aligned}
& [(p \vee \mathrm{X}q) \wedge (r \vee \mathrm{X}s)]\mathrm{X}t \\
\equiv\ & [p \vee \mathrm{X}q][r \vee \mathrm{X}s]\mathrm{X}t \\
\equiv\ & [p \vee \mathrm{X}q]\big((r \to \mathrm{X}t) \wedge (\neg r \to \mathrm{X}[s]t)\big) \\
\equiv\ & [p \vee \mathrm{X}q]\Big((r \to \mathrm{X}t) \wedge \big(\neg r \to \mathrm{X}(s \to t)\big)\Big) \\
\equiv\ & [p \vee \mathrm{X}q](r \to \mathrm{X}t) \wedge [p \vee \mathrm{X}q]\big(\neg r \to \mathrm{X}(s \to t)\big) \\
\equiv\ & ([p \vee \mathrm{X}q]r \to [p \vee \mathrm{X}q]\mathrm{X}t) \wedge \big([p \vee \mathrm{X}q]\neg r \to [p \vee \mathrm{X}q]\mathrm{X}(s \to t)\big) \\
\equiv\ & \Big(\big((p \vee \mathrm{X}q) \to r\big) \to [p \vee \mathrm{X}q]\mathrm{X}t\Big) \wedge \Big(\big((p \vee \mathrm{X}q) \to \neg r\big) \to [p \vee \mathrm{X}q]\mathrm{X}(s \to t)\Big) \\
\equiv\ & \left(\big((p \vee \mathrm{X}q) \to r\big) \to \big((p \to \mathrm{X}t) \wedge (\neg p \to \mathrm{X}[q]t)\big)\right) \wedge \\
& \left(\big((p \vee \mathrm{X}q) \to \neg r\big) \to \Big(\big(p \to \mathrm{X}(s \to t)\big) \wedge \big(\neg p \to \mathrm{X}[q](s \to t)\big)\Big)\right) \\
\equiv\ & \left(\big((p \vee \mathrm{X}q) \to r\big) \to \Big((p \to \mathrm{X}t) \wedge \big(\neg p \to \mathrm{X}(q \to t)\big)\Big)\right) \wedge
\end{aligned}
$$

$$\Big(\big((p \vee Xq) \to \neg r\big) \to \Big(\big(p \to X(s \to t)\big) \wedge \big(\neg p \to X(q \to (s \to t))\big)\Big)\Big)$$

Lemma 5.7 *Let η be a state formula and χ a formula in $\Phi_{\mathbf{XA}}$. Then the following equivalences are valid:*

(i) $\mathbf{A}(\eta \vee \chi) \leftrightarrow (\eta \vee \mathbf{A}\chi)$
(ii) $\mathbf{A}XX\chi \leftrightarrow \mathbf{A}X\mathbf{A}X\chi$

Theorem 5.8 *There is an effective translation* **u** *from the class of state formulas of $\Phi_{\mathbf{XA}}$ to $\Phi_{\Box}$ preserving formulae up to equivalence.*

Proof. (Sketch.) Induction on the combined depth ϕ^{cd} of ϕ. When $\phi^{cd} = 0$ then ϕ is already in $\Phi_{\Box}$. Suppose $\phi^{cd} = k + 1$ and the claim holds for all state formulae θ such that $\theta^{cd} \leq k$.

Now, the formula ϕ is a boolean combination of propositional formulae and formulae of the type $\mathbf{A}\psi$. In the latter, ψ is equivalent to a CNF, with disjunctions of the type $\delta = \alpha \vee \mathbf{E}\theta \vee \mathbf{A}\theta_1 \vee \ldots \vee \mathbf{A}\theta_k \vee X\chi$, where $\alpha \in \Phi_{\mathsf{PC}}$. Since each of $\alpha, \mathbf{E}\theta, \mathbf{A}\theta_1, ..., \mathbf{A}\theta_k$ is a state formula of combined depth $\leq k$, by using the inductive hypothesis we can assume that each of these is already replaced by an equivalent formula from $\Phi_{\Box}$. Then $\mathbf{A}\psi$ is equivalent to a conjunction of state formulae of the type $\mathbf{A}\delta = \mathbf{A}(\alpha \vee \mathbf{E}\theta \vee \mathbf{A}\theta_1 \vee \ldots \vee \mathbf{A}\theta_k \vee X\chi)$, which, by Lemma 5.7, is equivalent to $\alpha \vee \mathbf{E}\theta \vee \mathbf{A}\theta_1 \vee \ldots \vee \mathbf{A}\theta_k \vee \mathbf{A}X\chi$. Note that $\chi^{cd} \leq k-1$. After transforming, likewise, χ to a CNF, distributing $\mathbf{A}$ over the conjunctions in it and pulling out the state subformulae from these disjunctions outside $\mathbf{A}$, we reduce $\mathbf{A}X\chi$ above to an equivalent boolean combination of formulae from $\Phi_{\Box}$ and formulae of the type $\mathbf{A}XX\gamma$. The latter is equivalent to $\mathbf{A}X\mathbf{A}X\gamma$ by Lemma 5.7. Now, since $\gamma^{cd} \leq k - 2$ then $(\mathbf{A}X\gamma)^{cd} \leq k$, we can use the inductive hypothesis to replace $\mathbf{A}X\gamma$ with an equivalent formula from $\Phi_{\Box}$. By doing that for all such subformulae $\mathbf{A}XX\gamma$ we obtain an equivalent from $\Phi_{\Box}$ for $\mathbf{A}X\chi$, and therefore can define such an equivalent $\mathbf{u}(\mathbf{A}\delta)$ for $\mathbf{A}\delta$. Finally, we define $\mathbf{u}(\phi)$ to be the conjunction of all these $\mathbf{u}(\mathbf{A}\delta)$. □

6 Complete axiomatization of LTC

Here we present a sound and complete axiomatic system $\mathsf{AxSys}_{\mathsf{LTC}}$ for the logic LTC, consisting of the following groups of axioms schemes and inference rules:

I. Any complete system of axioms for PC.

II. The usual axiom schemes for X:

$\mathbf{Ax}_{\mathsf{K(X)}}$: $X(\phi \to \psi) \to (X\phi \to X\psi)$

$\mathbf{Ax}_{\mathsf{D(X)}}$: $\neg X\bot$

$\mathbf{Ax}_{\mathsf{Fun(X)}}$: $\neg X\phi \to X\neg\phi$

II. The S5 axioms for $\mathbf{A}$, plus the following:

$\mathbf{Ax}_{\mathsf{St}(\mathbf{A})}$: $\phi \to \mathbf{A}\phi$, for any state formula ϕ[7].

[7] It suffices to state that axiom for atomic propositions only, and the rest would be derivable,

$\mathbf{Ax}_{\mathbf{A}XX}$: $\mathbf{A}XX\phi \to \mathbf{A}X\mathbf{A}X\phi$

IV. Axiom schemes for $[\cdot]$: the valid schemes $\mathbf{Ax}_{[\wedge]}$ from Lemma 5.2 and $\mathbf{Ax}_{[\cdot]p}$, $\mathbf{Ax}_{[\cdot]\top}$, $\mathbf{Ax}_{[\cdot]\neg}$, $\mathbf{Ax}_{[\cdot]\wedge}$, $\mathbf{Ax}_{[\cdot]X}$, $\mathbf{Ax}_{[\cdot]\mathbf{A}}$ from Lemma 5.4.

V. Inference rules: Modus Ponens (MP) and the necessitation rules:

$$\mathsf{Nec}_{\mathbf{A}}: \frac{\vdash \phi}{\vdash \mathbf{A}\phi}, \quad \mathsf{Nec}_{X}: \frac{\vdash \phi}{\vdash X\phi}, \quad \mathsf{Nec}_{[\cdot]}: \frac{\vdash \phi}{\vdash [\psi]\phi}, \quad \mathsf{Equiv}_{[\cdot]}: \frac{\vdash \phi \leftrightarrow \psi}{\vdash [\phi]\chi \leftrightarrow [\psi]\chi}$$

The subsystem $\mathsf{AxSys}_{\mathbf{A}X}$ consists of axioms I, II, III, and the rules in V.

Recall that $\Box\phi$ denotes $\mathbf{A}X\phi$ and $\Phi_{\Box}$ is the respective fragment of $\Phi_{X\mathbf{A}}$ built from Φ_0 only by using the propositional connectives and $\Box$. We define the subsystem $\mathsf{AxSys}_{\Box}$, consisting of the KD axioms for $\Box$:

$\mathbf{Ax}_{\mathsf{K}(\Box)}$: $\Box(\phi \to \psi) \to (\Box\phi \to \Box\psi)$,

$\mathbf{Ax}_{\mathsf{D}(\Box)}$: $\neg \Box \bot$,

plus the rules MP and $\mathsf{Nec}_{\Box}: \frac{\vdash\phi}{\vdash\Box\phi}$.

Note that:

(i) The axioms and rules of $\mathsf{AxSys}_{\Box}$ are derivable in $\mathsf{AxSys}_{\mathbf{A}X}$.
(ii) $\mathsf{AxSys}_{\Box}$, being canonical, is sound and complete for the validities in $\Phi_{\Box}$.

Now, we are going to establish a series of equivalent reductions of formulae of LTC, all derivable in $\mathsf{AxSys}_{\mathsf{LTC}}$, which will enable us to eventually reduce the proof of completeness of $\mathsf{AxSys}_{\mathsf{LTC}}$ to the completeness of the much simpler logic $\mathsf{AxSys}_{\Box}$. We begin with some useful derivations, used further in the proofs.

Lemma 6.1 *The following are derivable in* $\mathsf{AxSys}_{\mathsf{LTC}}$*:*

(i) $\vdash_{\mathsf{LTC}} [\bot]\phi \leftrightarrow \top$
(ii) $\vdash_{\mathsf{LTC}} [\phi]\psi \leftrightarrow (\phi \to \psi)$*, where* ϕ *is in* Φ_{PC}

Proof. (i) By induction on ϕ. The only non-trivial case is $\phi = X\psi$. Since $\vdash_{\mathsf{LTC}} \bot \leftrightarrow (\bot \vee X\bot)$, we obtain $\vdash_{\mathsf{LTC}} [\bot]X\psi \leftrightarrow [\bot \vee X\bot]X\psi$ by $\mathsf{Equiv}_{[\cdot]}$. Then, using $\mathbf{Ax}_{[\cdot]X}$ we derive $\vdash_{\mathsf{LTC}} [\bot \vee X\bot]X\psi \leftrightarrow ((\bot \to X\psi) \wedge (\neg\bot \to X[\bot]\psi))$. By the inductive hypothesis, $\vdash_{\mathsf{LTC}} [\bot]\psi \leftrightarrow \top$. Then $\vdash_{\mathsf{LTC}} [\bot \vee X\bot]X\psi \leftrightarrow ((\bot \to X\psi) \wedge (\neg\bot \to X\top))$. Therefore, $\vdash_{\mathsf{LTC}} [\bot \vee X\bot]X\psi \leftrightarrow ((\bot \to X\psi) \wedge (\neg\bot \to \top))$. Hence, $\vdash_{\mathsf{LTC}} [\bot \vee X\bot]X\psi \leftrightarrow \top$. Therefore, $\vdash_{\mathsf{LTC}} [\bot]X\psi \leftrightarrow \top$.

(ii) By induction on ψ. The only non-trivial cases are $\psi = X\chi$ and $\psi = \mathbf{A}\chi$.

Case $\psi = X\chi$. As $\vdash_{\mathsf{LTC}} \phi \leftrightarrow (\phi \vee X\bot)$, we derive $\vdash_{\mathsf{LTC}} [\phi]X\chi \leftrightarrow [\phi \vee X\bot]X\chi$ by $\mathsf{Equiv}_{[\cdot]}$. Then, $\vdash_{\mathsf{LTC}} [\phi \vee X\bot]X\chi \leftrightarrow ((\phi \to X\chi) \wedge (\neg\phi \to X[\bot]\chi))$, by $\mathbf{Ax}_{[\cdot]X}$. By claim (i), $\vdash_{\mathsf{LTC}} [\bot]\phi \leftrightarrow \top$. Hence, $\vdash_{\mathsf{LTC}} [\phi \vee X\bot]X\chi \leftrightarrow ((\phi \to X\chi) \wedge (\neg\phi \to X\top))$. Then $\vdash_{\mathsf{LTC}} [\phi \vee X\bot]X\chi \leftrightarrow ((\phi \to X\chi) \wedge (\neg\phi \to \top))$, so $\vdash_{\mathsf{LTC}} [\phi \vee X\bot]X\chi \leftrightarrow (\phi \to X\chi)$.

Case $\psi = \mathbf{A}\chi$. By $\mathbf{Ax}_{[\cdot]\mathbf{A}}$ we have $\vdash_{\mathsf{LTC}} [\phi]\mathbf{A}\chi \leftrightarrow (\phi \to \mathbf{A}[\phi]\chi)$. By the inductive hypothesis, $\vdash_{\mathsf{LTC}} [\phi]\chi \leftrightarrow (\phi \to \chi)$. Then $\vdash_{\mathsf{LTC}} [\phi]\mathbf{A}\chi \leftrightarrow (\phi \to$

---

but we prefer to streamline the deductive system.

$\mathbf{A}(\phi \to \chi))$. Note that $\phi \in \Phi_{\mathsf{PC}}$. Then $\vdash_{\mathsf{LTC}} \mathbf{A}(\phi \to \chi) \leftrightarrow (\phi \to \mathbf{A}\chi)$, hence $\vdash_{\mathsf{LTC}} [\phi]\mathbf{A}\chi \leftrightarrow (\phi \to \mathbf{A}\chi)$. □

Lemma 6.2 *For every* LTC*-formula* ϕ, $\vdash_{\mathsf{AxSys_{LTC}}} \mathbf{t}(\phi) \leftrightarrow \phi$.

Proof. By induction on ϕ, by applying the axioms for $[\cdot]$, the derivations in Lemma 6.1, and the inference rules of $\mathsf{AxSys_{LTC}}$, one can formalise the proof of Theorem 5.5 in $\mathsf{AxSys_{LTC}}$. The details are routine and we leave them out. □

Lemma 6.3 *For every* LTC*-formula* ϕ, *if* ϕ *is* $\mathsf{AxSys_{LTC}}$*-consistent then* $\mathbf{t}(\phi)$ *is* $\mathsf{AxSys_{\mathbf{AX}}}$*-consistent.*

Proof. Equivalently (since $\mathbf{t}(\neg\phi) = \neg\mathbf{t}(\phi)$), we have to prove that for every LTC-formula ϕ, if $\vdash_{\mathsf{AxSys_{\mathbf{AX}}}} \mathbf{t}(\phi)$ then $\vdash_{\mathsf{AxSys_{LTC}}} \phi$. Indeed, take a derivation $\vdash_{\mathsf{AxSys_{\mathbf{AX}}}} \mathbf{t}(\phi)$. It is also a derivation in $\mathsf{AxSys_{LTC}}$. Append to it the derivation $\vdash_{\mathsf{AxSys_{LTC}}} \mathbf{t}(\phi) \leftrightarrow \phi$ from Lemma 6.2, to obtain $\vdash_{\mathsf{AxSys_{LTC}}} \mathbf{t}(\phi) \to \phi$. Then, by MP we obtain $\vdash_{\mathsf{AxSys_{LTC}}} \phi$. □

Lemma 6.4 *Every state formula* $\psi \in \Phi_{\mathbf{XA}}$ *is provably equivalent in* $\mathsf{AxSys_{\mathbf{AX}}}$ *to* $\mathbf{u}(\psi) \in \Phi_{\Box}$.

Proof. The equivalences used in defining the function $\mathbf{u}$ in the proof for Theorem 5.8, including those in Lemma 5.7, are all derivable in $\mathsf{AxSys_{\mathbf{AX}}}$. □

Lemma 6.5 *For every formula* $\psi \in \Phi_{\mathbf{XA}}$ *there is an effectively computable formula* $\mathbf{v}(\psi) \in \Phi_{\Box}$ *such that* ψ *and* $\mathbf{v}(\psi)$ *are equally satisfiable and equally consistent in* $\mathsf{AxSys_{\mathbf{AX}}}$.

Proof. (Sketch.) Here 'consistency' will mean 'consistency in $\mathsf{AxSys_{\mathbf{AX}}}$'. We will prove the claim for consistency; the proof for satisfiability is fully analogous.

The definition of $\mathbf{v}(\psi)$ extends that of $\mathbf{u}(\phi)$ for state formulae ϕ in Lemma 6.4 as follows. The formula ψ is provably in $\mathsf{AxSys_{\mathbf{AX}}}$ equivalent to some, hereafter fixed, formula ψ' in DNF, where every disjunct is of the type $\delta = \theta \wedge \mathrm{X}\chi$, for some state formula θ (note that χ may also be $\top$). Then, consistency of ψ is equivalent (provably in $\mathsf{AxSys_{\mathbf{AX}}}$) to consistency of some of these disjuncts. Note that $\theta \wedge \mathrm{X}\chi$ is consistent iff $\theta \wedge \mathbf{E}\mathrm{X}\chi$ is so. Indeed, $\vdash_{\mathsf{AxSys_{\mathbf{AX}}}} \theta \to \neg\mathrm{X}\chi$ iff $\vdash_{\mathsf{AxSys_{\mathbf{AX}}}} \mathbf{A}(\theta \to \neg\mathrm{X}\chi)$ iff $\vdash_{\mathsf{AxSys_{\mathbf{AX}}}} \theta \to \mathbf{A}\neg\mathrm{X}\chi$. Lastly, $\theta \wedge \mathbf{E}\mathrm{X}\chi$ is consistent iff $\mathbf{u}(\theta) \wedge \mathbf{u}(\mathbf{E}\mathrm{X}\chi)$ is so. Now, if ψ is consistent we define $\mathbf{v}(\psi)$ to be $\mathbf{u}(\theta) \wedge \mathbf{u}(\mathbf{E}\mathrm{X}\chi)$ for the first disjunct δ in ψ' that is consistent, else we define $\mathbf{v}(\psi)$ to be $\bot$. The claim of the lemma follows by construction. □

Theorem 6.6 $\mathsf{AxSys_{LTC}}$ *is sound and complete for the logic* LTC.

Proof. The soundness of the axioms for $[\cdot]$ follows from Lemmas 5.2, and 5.4. The soundness of $\mathrm{Ax}_{\mathbf{AXX}}$ is by Lemma 5.7. All others axioms are well-known and proving their soundness, as well as that of the inference rules, is routine.

For the completeness, we need to show that every $\mathsf{AxSys_{LTC}}$-consistent formula ϕ is satisfiable in a LTC-model. We prove that by a series of reductions, as follows. Let ϕ $\mathsf{AxSys_{LTC}}$-consistent. Then, by Lemma 6.3, $\mathbf{t}(\phi)$ is $\mathsf{AxSys_{\mathbf{AX}}}$-consistent. Next, by Lemma 6.5, $\mathbf{v}(\mathbf{t}(\phi))$ is $\mathsf{AxSys_{\mathbf{AX}}}$-consistent, and therefore $\mathsf{AxSys_{\Box}}$-consistent. Since $\mathsf{AxSys_{\Box}}$ is complete for the logic KD, it follows that

$\mathbf{v}(\mathbf{t}(\phi))$ is satisfiable, hence $\mathbf{t}(\phi)$ is satisfiable by Lemma 6.5, and therefore ϕ is satisfiable, too, by Theorem 5.5. □

Theorem 6.7 *The satisfiability problem for* LTC *is decidable, in* ExpSpace.

Proof. The translations and reductions defined here effectively reduce the satisfiability problem for LTC to that of KD, which is PSpace-complete (see e.g. [11]). However, as illustrated by Example 5.6, the translation $\mathbf{t}$ can cause an exponential blow-up to the length of the formula, hence the complexity upper bound that we provide based on the results obtained here is ExpSpace. However, the number of different subformulae need not grow exponentially, so it is conceivable that a more refined, on-the-fly application of the translation $\mathbf{t}$, or an alternative decision method may bring that complexity down. We leave the question of establishing the precise complexity of LTC open. □

7 Concluding remarks

In this work we have proposed the logic LTC for formalising reasoning with and about temporal conditionals, for which we establish a series of technical results, eventually leading to a sound and complete axiomatization and a decision procedure. We show how formalization on LTC can provide a solution to the Sea Battle Puzzle, by showing that the underlying reasoning per cases argument is not sound in that logic.

The present work suggests a variety of interesting follow-up developments:

- On the conceptual side, we believe that the logic LTC or suitable variations of it can be applied to resolve other philosophical problems arising from reasoning about temporal conditionals. We also intend to adapt that logic for reasoning about temporal counterfactuals, as well as to related type of reasoning, that often occurs in the theory of extensive forms, about the players' behaviour on and off the equilibrium path, non-credible threats, etc., leading, inter alia, to the more refined notion of subgame-perfect equilibrium and myopic equilibrium.
- On the technical side, we note that the Nexttime fragment of the logic CTL*, which is precisely the fragment $\Phi_{\mathbf{XA}}$ of LTC, seems not to have been studied so far. The present work provides, inter alia, a new complete axiomatization result for it. We intend to extend the present results to extensions of LTC, also involving long-term temporal operators, by relating them with more expressive, yet still manageable fragments of CTL*.
- In particular, an extension of the logic LTC with past and future temporal modalities can likewise resolve the problem raised by the Diodorean Master Argument, by rendering that argument unsound with respect to the resulting semantics.

Acknowledgment

Thanks go to Maria Aloni, Johan van Benthem, Alessandra Marra, Floris Roelofsen, Frank Veltman, and the audience of seminars or workshops at Delft University of Technology, Beijing Normal University and Tsinghua University, for their useful comments and suggestions. Valentin Goranko is grateful to the Department of Philosophy of Beijing Normal University for supporting his visit there, during which part of the present work was done. Thanks are also due to the three anonymous referees for helpful comments and some corrections.

References

[1] Arlo-Costa, H. and P. Egré, *The logic of conditionals*, in: E. Zalta, editor, *The Stanford Encyclopedia of Philosophy*, Metaphysics Research Lab, Stanford University, 2007.

[2] Cantwell, J., *Changing the modal context*, Theoria **74** (2010), pp. 331–351.

[3] Demri, S., V. Goranko and M. Lange, "Temporal Logics in Computer Science," Cambridge Tracts in Theoretical Computer Science **58**, Cambridge University Press, 2016.

[4] Edgington, D., *Indicative conditionals*, in: E. Zalta, editor, *The Stanford Encyclopedia of Philosophy*, Metaphysics Research Lab, Stanford University, 2014.

[5] Garrett, B., "What Is This Thing Called Metaphysics?" Taylor & Francis, 2017.

[6] Kolodny, N. and J. MacFarlane, *Ifs and oughts*, The Journal of Philosophy **107** (2010), pp. 115–143.

[7] Kratzer, A., *Conditionals*, Chicago Linguistics Society **22** (1986), pp. 1–15.

[8] Lewis, D., *Adverbs of quantification*, in: E. Keenan, editor, *Formal Semantics of Natural Language*, Cambridge University Press, 1975 pp. 178–188.

[9] MacFarlane, J., *Future contingents and relative truth*, The Philosophical Quarterly **53** (2003), pp. 321–336.

[10] Marra, A., *When is reasoning by cases valid? Some remarks on persistence and information-sensitivity*, Manuscript.

[11] Marx, M., *Complexity of modal logic*, in: P. Blackburn, J. van Benthem and F. Wolter, editors, *Handbook of Modal Logic*, Elsevier, 2007 pp. 139–179.

[12] McCall, S., editor, "Polish Logic 1920-1939," Oxford, 1967.

[13] Øhrstrøm, P., *Problems regarding the future operator in an indeterministic tense logic*, Danish Yearbook of Philosophy **18** (1981), pp. 81–95.

[14] Øhrstrøm, P. and F. Hasle, *Future contingents*, in: E. Zalta, editor, *The Stanford Encyclopedia of Philosophy*, Metaphysics Research Lab, Stanford University, 2015.

[15] Prior, A., "Past, Present and Future," Oxford, 1967.

[16] Ramsey, F., *General propositions and causality*, in: D. Mellor, editor, *Philosophical Papers*, Cambridge University Press, 1990 pp. 145–163.

[17] Thomason, R., *Indeterminist time and truth value gaps*, Theoria **36** (1970), pp. 264–281.

[18] van Benthem, J., *Foundations of conditional logic*, Journal of Philosophical Logic **13** (1984), pp. 303–349.

[19] Veltman, F., "Logics for Conditionals," Ph.D. thesis, University of Amsterdam (1985).

[20] Veltman, F., *Defaults in update semantics*, Journal of Philosophical Logic **25** (1996), pp. 221–261.

On Strictly Positive Modal Logics with S4.3 Frames

Stanislav Kikot

Department of Computer Science and Information Systems
Birkbeck, University of London, U.K.

Agi Kurucz

Department of Informatics
King's College London, U.K.

Frank Wolter

Department of Computer Science
University of Liverpool, U.K.

Michael Zakharyaschev

Department of Computer Science and Information Systems
Birkbeck, University of London, U.K.

Abstract

We investigate the lattice of strictly positive (SP) modal logics that contain the SP-fragment of the propositional modal logic S4.3 of linear quasi-orders. We are interested in Kripke (in)completeness of these logics, their computational complexity, as well as the definability of Kripke frames by means of SP-implications. We compare the lattice of these SP-logics with the lattice of normal modal logics above S4.3. We also consider global consequence relations for SP-logics, focusing on definability and Kripke completeness.

Keywords: strictly positive modal logic, meet-semilattices with monotone operators, Kripke completeness, definability, decidability.

1 Introduction

The lattice NExtS4.3 of propositional normal modal logics containing S4.3 is a rare example of a non-trivial class of 'well-behaved' modal logics. Indeed, all of them are finitely axiomatisable, have the finite model property [6,10], and are decidable in coNP [22]. Although Fine [10] complained that "the full lattice [of these logics] is one of great complexity", its structure is perfectly understandable (though difficult to depict) if one recalls the fact [7] that every $L \in$ NExtS4.3 can be axiomatised by Yankov (aka characteristic or frame) formulas for finite linear quasi-ordered frames.

In this paper, our concern is the fragment of the modal language that comprises implications $\sigma \to \tau$, where σ and τ are *strictly positive* modal formulas [3,23] constructed from variables using $\wedge$, $\Diamond$, and the constant $\top$. We call such implications *SP-implications*. A natural algebraic semantics for SP-implications is given by meet-semilattices with monotone operators (SLOs, for short) [21,2,16]. We denote the corresponding syntactic consequence relation by $\vdash_{\mathsf{SLO}}$ and call any $\vdash_{\mathsf{SLO}}$-closed set of SP-implications an *SP-logic*. The lattice of varieties of SLOs validating the SP-implicational S4 axioms $p \to \Diamond p$ and $\Diamond\Diamond p \to \Diamond p$ (*closure* SLOs) was studied by Jackson [15] (see also [9]). Jackson showed, in particular, that the SP-fragment P_{S4} of S4 is *complex* [12] in the sense that each closure SLO can be embedded into the full complex algebra of some Kripke frame for S4. It follows that the SP-logic axiomatised by $\{p \to \Diamond p, \Diamond\Diamond p \to \Diamond p\}$ is (*Kripke*) *complete* in the sense that its SP-implicational $\vdash_{\mathsf{SLO}}$-consequences coincide with those over Kripke frames, that is, with P_{S4}. On the other hand, it is shown in [16] that the SP-extension of P_{S4} with

$$\Diamond(p \wedge q) \wedge \Diamond(p \wedge r) \to \Diamond(p \wedge \Diamond q \wedge \Diamond r) \qquad (wcon)$$

is not complex; yet, it is complete and axiomatises the SP-fragment $P_{\mathsf{S4.3}}$ of S4.3. Both P_{S4} and $P_{\mathsf{S4.3}}$ are decidable in polynomial time [1,21,16].

Our aim here is to investigate SP-logics above $P_{\mathsf{S4.3}}$—focusing on completeness, definability (of Kripke frames) and computational complexity—and compare them with classical modal logics above S4.3. Our first important observation is that, unlike classical modal formulas, a randomly chosen SP-implication will most probably axiomatise an *incomplete* SP-extension of $P_{\mathsf{S4.3}}$. Such are, for example, $\Diamond p \to \Diamond q$ and $\Diamond p \wedge \Diamond q \to \Diamond(p \wedge q)$ [16]. In Section 3, we construct infinite sequences of SP-logics sharing the same two-point frames. But the main result of this paper is that we do identify *all complete* SP-logics above $P_{\mathsf{S4.3}}$.

Over S4.3 frames, modal formulas define exactly those frame classes that are closed under cofinal subframes [10,24]. On the other hand, over P_{S4}, SP-implications are capable of defining only FO-definable classes closed under subframes [16] (which implies that frames for any proper extension of $P_{\mathsf{S4.3}}$ are of bounded depth). In Section 4, we show that *all* FO-definable subframe-closed classes of S4.3-frames are SP-*definable*. We also give *finite axiomatisations* for the SP-logics of these SP-definable classes by means of SP-analogues of subframe formulas [11,24], and prove that each of the resulting SP-logics is decidable in *polynomial time*.

In Section 5, we investigate the lattice $\mathrm{Ext}^{+}P_{\mathsf{S4.3}}$ of SP-logics containing $P_{\mathsf{S4.3}}$ by comparing it with the lattice NExtS4.3. Following [4], we consider two maps: one associates with every modal logic L its SP-fragment $\pi(L)$, the other one associates with every SP-logic P the modal logic $\mu(P) = \mathsf{S4.3} \oplus P$. We give a frame-theoretic characterisation of π and show that each $\pi^{-1}(P)$ has a greatest element. On the other hand, as follows from Section 3, $\mu^{-1}(L)$ often contains infinitely many incomplete logics.

In Section 6, we consider rules $\varrho = \frac{\iota_1,\dots,\iota_n}{\iota}$ of SP-implications. In modal logics above S4, ϱ can be expressed by the (non-SP) formula $\Box(\iota_1 \wedge \cdots \wedge \iota_n) \to \iota$. We show that a class of S4.3-frames is modally definable iff it is definable by SP-rules. In general, an SP-logic is complete with respect to SP-rules iff it is complex [16]. We show that the only SP-logics in $\mathrm{Ext}^+ P_{\mathsf{S4.3}}$ complete with respect to SP-rules are P_{S5} and the logic of the singleton cluster. These are also the only complete SP-logics for which deciding valid SP-rules is in polynomial time [17]. We make a step towards axiomatising SP-rules by giving axiomatisations for the SP-rule logics of n-element clusters, $n > 1$.

2 Preliminaries

We assume that the reader is familiar with basic notions of modal logic and Kripke semantics [7]. In particular, $\mathfrak{M}, w \models \varphi$ means that φ holds at world w in Kripke model $\mathfrak{M}$, and $\mathfrak{F} \models \varphi$ says that φ is valid in Kripke frame $\mathfrak{F}$. We write $\mathcal{C} \models \varphi$, for a class $\mathcal{C}$ of frames, if φ is valid in every $\mathfrak{F} \in \mathcal{C}$.

We denote SP-implications by $\iota = (\sigma \to \tau)$ and refer to σ, τ as *terms*. By regarding ι as the equality $\sigma \wedge \tau = \sigma$, we can naturally evaluate it in SLOs, i.e., structures $\mathfrak{A} = (A, \wedge, \top, \Diamond)$, where $(A, \wedge, \top)$ is a semilattice with top element $\top$ and $\Diamond(a \wedge b) \wedge \Diamond b = \Diamond(a \wedge b)$, for any $a, b \in A$. We write $\mathfrak{A} \models \iota$ to say that $\sigma \wedge \tau = \sigma$ holds in $\mathfrak{A}$ under any valuation. For a set $P \cup \{\iota\}$ of SP-implications, we write $P \models_{\mathsf{SLO}} \iota$ if $\mathfrak{A} \models \iota$ for every SLO $\mathfrak{A}$ validating all implications in P. We denote by $\boldsymbol{P}[\mathfrak{A}]$ the set of SP-implications ι for which $\mathfrak{A} \models \iota$.

Since equational consequence can be characterised syntactically by Birkhoff's equational calculus [5,13], it is readily seen that

$$P \models_{\mathsf{SLO}} \iota \quad \text{iff} \quad P \vdash_{\mathsf{SLO}} \iota,$$

where $P \vdash_{\mathsf{SLO}} \iota$ means that there is a sequence (derivation) $\iota_0, \dots, \iota_n = \iota$, with each ι_i being a substitution instance of a member in P or one of the axioms

$$p \to p, \qquad p \to \top, \qquad p \wedge q \to q \wedge p, \qquad p \wedge q \to p,$$

or obtained from earlier members of the sequence using one of the rules

$$\frac{\sigma \to \tau \quad \tau \to \varrho}{\sigma \to \varrho}, \qquad \frac{\sigma \to \tau \quad \sigma \to \varrho}{\sigma \to \tau \wedge \varrho}, \qquad \frac{\sigma \to \tau}{\Diamond\sigma \to \Diamond\tau}$$

(see also the Reflection Calculus RC of [2,8]). We say that SP-implications ι and ι' are *P-equivalent*, if $P \cup \{\iota\} \vdash_{\mathsf{SLO}} \iota'$ and $P \cup \{\iota'\} \vdash_{\mathsf{SLO}} \iota$. We say that terms σ and τ are *P-equivalent* if $P \vdash_{\mathsf{SLO}} \sigma \to \tau$ and $P \vdash_{\mathsf{SLO}} \tau \to \sigma$.

For any class $\mathcal{C}$ of Kripke frames, the set of modal formulas (with full Booleans) validated by the frames in $\mathcal{C}$ is called the *modal logic* of $\mathcal{C}$ and denoted by $\boldsymbol{L}[\mathcal{C}]$; the restriction of $\boldsymbol{L}[\mathcal{C}]$ to SP-implications is called the *SP-logic* of $\mathcal{C}$ and denoted by $\boldsymbol{P}[\mathcal{C}]$. For finite $\mathcal{C} = \{\mathfrak{F}_1, \dots, \mathfrak{F}_n\}$, we write $\boldsymbol{L}[\mathfrak{F}_1 \dots, \mathfrak{F}_n]$ and $\boldsymbol{P}[\mathfrak{F}_1, \dots, \mathfrak{F}_n]$, respectively. For any set Φ of modal formulas, $\mathsf{Fr}(\Phi)$ is the class of frames validating Φ. Every Kripke frame $\mathfrak{F} = (W, R)$ gives rise to a SLO $\mathfrak{F}^\star = (2^W, \cap, W, \Diamond^+)$ where $\Diamond^+ X = \{w \in W \mid R(w,v) \text{ for some } v \in X\}$, for $X \subseteq W$ (that is, $\mathfrak{F}^\star$ is the $(\wedge, \Diamond, \top)$-type reduct of the *full complex algebra of* $\mathfrak{F}$ [12]). As Kripke models over $\mathfrak{F}$ and valuations in the algebra $\mathfrak{F}^\star$ are the same thing, we always have $\boldsymbol{P}[\mathfrak{F}] = \boldsymbol{P}[\mathfrak{F}^\star]$.

S4.3 is the modal logic whose rooted Kripke frames are linear quasi-orders $\mathfrak{F} = (W, R)$, i.e., R is a reflexive and transitive relation on W with xRy or yRx, for any $x, y \in W$. From now on, we refer to rooted frames for S4.3 as simply *frames.* A *cluster* in $\mathfrak{F}$ is any set of the form $C(x) = \{y \in W \mid xRy \;\&\; yRx\}$. If $|W| \leq \omega$ and the number of clusters in $\mathfrak{F}$ is finite, we write $\mathfrak{F} = (m_1, \dots, m_n)$, $1 \leq m_i \leq \omega$, to say that the ith cluster in $\mathfrak{F}$ (starting from the root) has m_i points. In this case, we also say that $\mathfrak{F}$ is of *depth* n and write $d(\mathfrak{F}) = n$. A linear order with n points is denoted by $\mathfrak{L}_n$.

We call frame $\mathfrak{F} = (W, R)$ a *subframe* of a frame $\mathfrak{F}' = (W', R')$ and write $\mathfrak{F} \subseteq \mathfrak{F}'$ if $W \subseteq W'$ and R is the restriction of R' to W. A subframe $\mathfrak{F}$ of $\mathfrak{F}'$ is *proper* if W is a proper subset of W'. We say that $\mathfrak{F}$ is a *cofinal subframe* of $\mathfrak{F}'$ and write $\mathfrak{F} \Subset \mathfrak{F}'$ if $\mathfrak{F} \subseteq \mathfrak{F}'$ and, for any $x \in W$ and $y \in W'$ with $xR'y$, there is $z \in W$ with $yR'z$. (As we only deal with frames for S4.3, $\mathfrak{F} \Subset \mathfrak{F}'$ iff $\mathfrak{F}$ is a p-morphic image of $\mathfrak{F}'$.)

The normal extension L of S4.3 with a set Φ of modal formulas is denoted by $L = \mathsf{S4.3} \oplus \Phi$; NExtS4.3 $= \{\mathsf{S4.3} \oplus \Phi \mid \Phi$ is a set of modal formulas$\}$. The minimal SP-logic P containing $P_{\mathsf{S4.3}}$ and a set Φ of SP-implications is denoted by $P = P_{\mathsf{S4.3}} + \Phi$; Ext$^+ P_{\mathsf{S4.3}} = \{P_{\mathsf{S4.3}} + \Phi \mid \Phi$ is a set of SP-implications$\}$.

Given P and ι, we write $P \models_{\mathsf{Kr}} \iota$ iff $\iota \in \boldsymbol{P}[\mathsf{Fr}(P)]$. It is easy to see that

$$P \vdash_{\mathsf{SLO}} \iota \text{ always implies } P \models_{\mathsf{Kr}} \iota. \tag{1}$$

We call P *complete* if the converse also holds for every ι. As shown in [16], $P_{\mathsf{S4.3}}$ is complete and $\mathsf{Fr}(P_{\mathsf{S4.3}}) = \mathsf{Fr}(\mathsf{S4.3})$.

Throughout (see Figs. 1, 3 and 6), we describe examples of SLOs for $P_{\mathsf{S4.3}}$ by Hasse diagrams where $\circ$ vertices represent (closed) elements a with $\Diamond a = a$ and $\bullet$ vertices represent elements a with $\Diamond a > a$. Note that, for each $\bullet$ element a, $\Diamond a$ is the (unique) smallest $\circ$ element b with $b > a$.

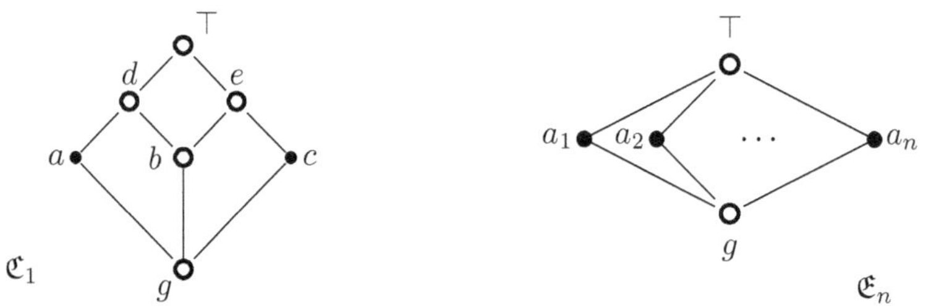

Fig. 1. SLOs for $P_{\mathsf{S4.3}}$.

3 Incomplete SP-logics

Axiomatising by SP-implications the SP-logics of even very simple frame classes turns out to be a challenging problem because many natural and innocuously looking axiom-candidates will most probably give incomplete SP-logics. For instance, let

$$\sigma_n(p,q) \;=\; \underbrace{\Diamond\big(p \wedge \Diamond(q \wedge \Diamond(p \wedge \dots)\dots\big)}_{\Diamond \text{ used } n \text{ times, } 1 \le n < \omega}.$$

Consider the following SP-implications (the reader may find it useful to compare them with the SP-implications defined in the next section):

$$\begin{aligned}
\boldsymbol{\iota}^2_{fun} &= \Diamond(p \wedge q) \wedge \Diamond(p \wedge r) \wedge \Diamond(q \wedge r) \to \Diamond(p \wedge q \wedge r), && (2)\\
\boldsymbol{\epsilon}_1 &= \Diamond p \wedge \Diamond q \to \Diamond(p \wedge \Diamond q), && (3)\\
\boldsymbol{\epsilon}_2 &= \Diamond(p \wedge \Diamond q) \wedge \Diamond(p \wedge \Diamond r) \wedge \Diamond(q \wedge \Diamond r) \to \Diamond\big(p \wedge \Diamond(q \wedge \Diamond r)\big), && (4)\\
\boldsymbol{\varphi} &= \Diamond(p \wedge \Diamond q) \wedge \Diamond(p \wedge \Diamond r) \wedge \Diamond(q \wedge r) \to \Diamond\big(p \wedge \Diamond(q \wedge r)\big), && (5)\\
\boldsymbol{\beta}_n &= \sigma_n(p,q) \wedge \sigma_n(q,p) \;\to\; \Diamond(p \wedge q), && (6)\\
\boldsymbol{\gamma}_n &= \sigma_{n+1}(p,q) \;\to\; \Diamond(p \wedge q), &&\\
\boldsymbol{\delta}_n &= \sigma_{n+1}(p,q) \;\to\; \sigma_{n+1}(q,p). &&
\end{aligned}$$

Figure 2 describes inclusions between the SP-logics above $P^2_{fun} = P_{\mathsf{S4.3}} + \boldsymbol{\iota}^2_{fun}$ axiomatised by these SP-implications. Using results from [15] and Claim 3.1 below, it is not hard to see that all of the depicted inclusions are proper.

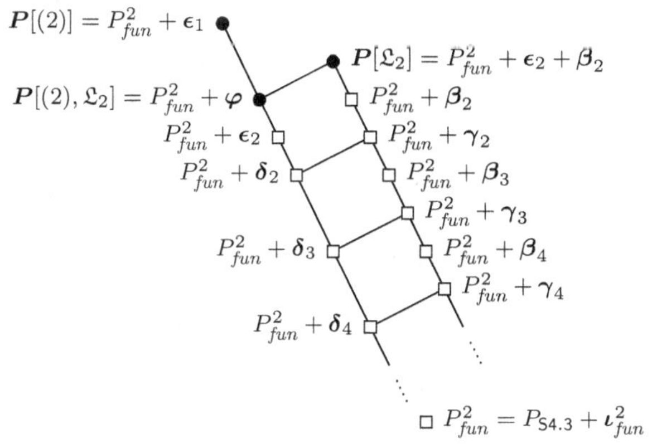

Fig. 2. Complete and incomplete SP-logics above $P^2_{fun} = P_{\mathsf{S4.3}} + \boldsymbol{\iota}^2_{fun}$.

The only complete SP-logics in the picture are shown by • (that they are axiomatised by the indicated SP-implications follows from [15]; see also Theorem 4.4 below). On the other hand, it is not hard to check that, for each SP-logic P in Fig. 2, if P is below $\boldsymbol{P}[(2), \mathfrak{L}_2]$ then the rooted frames for P are $\mathfrak{L}_1, (2), \mathfrak{L}_2$, and if P is below $\boldsymbol{P}[\mathfrak{L}_2]$ but not below $\boldsymbol{P}[(2), \mathfrak{L}_2]$ then the rooted frames for P are $\mathfrak{L}_1, \mathfrak{L}_2$. Therefore, all SP-logics shown by □ in the picture are incomplete.

Claim 3.1 *(i)* $P^2_{\mathit{fun}} + \epsilon_2 \not\models_{\mathsf{SLO}} \varphi$ *and* $P^2_{\mathit{fun}} + \epsilon_2 \not\models_{\mathsf{SLO}} \boldsymbol{\beta}_2$, *(ii)* $P^2_{\mathit{fun}} + \boldsymbol{\beta}_2 \not\models_{\mathsf{SLO}} \epsilon_2$, *(iii)* $P^2_{\mathit{fun}} + \gamma_n \not\models_{\mathsf{SLO}} \boldsymbol{\beta}_n$, *(iv)* $P^2_{\mathit{fun}} + \boldsymbol{\beta}_{n+1} \not\models_{\mathsf{SLO}} \boldsymbol{\delta}_n$, *(v)* $P^2_{\mathit{fun}} + \boldsymbol{\delta}_n \not\models_{\mathsf{SLO}} \boldsymbol{\beta}_{n+1}$.

Proof. We use the SLOs in Fig. 3. Claim *(i)* can be shown by $\mathfrak{A}_1$, *(ii)* by $\mathfrak{A}_2$, *(iii)* by $\mathfrak{C}_n$, *(iv)* by $\mathfrak{B}_n$, and *(v)* by $\mathfrak{D}_n$. □

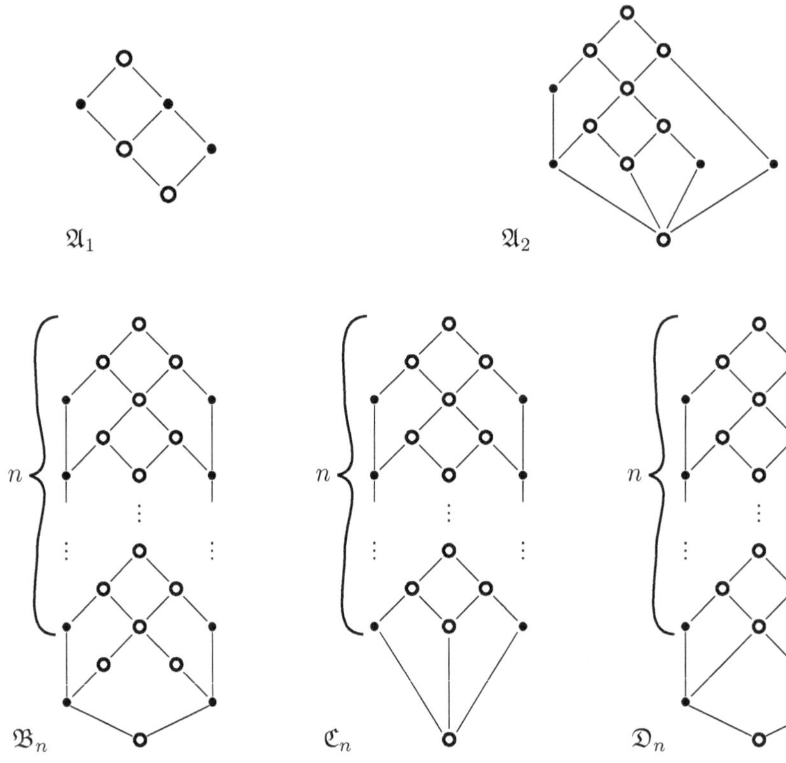

Fig. 3. SLOs validating P^2_{fun}.

The frames for the incomplete SP-logics above have at most 2 points. Similar incomplete SP-logics can clearly be defined for more complex frames, which makes the lattice $\mathrm{Ext}^+ P_{\mathsf{S4.3}}$ much more involved compared to NExtS4.3.

4 Axiomatising and defining frame classes by SP-implications

As any SP-implication ι is a Sahlqvist formula, $\mathsf{Fr}(\iota)$ is FO-definable [20]. As shown in [16], the class of *reflexive and transitive* frames validating ι is closed under subframes. Using the results from [11,24] on FO-definable subframe logics, we obtain the following theorem (which can be readily proved directly):

Theorem 4.1 *For every* $\iota \notin P_{\mathsf{S4.3}}$, *there exists* $n < \omega$ *such that* $\mathsf{Fr}(\iota)$ *is closed under subframes and any frame in* $\mathsf{Fr}(\iota)$ *is of depth* $< n$. *Thus, for any SP-logic* $P \in \mathrm{Ext}^+ P_{\mathsf{S4.3}}$, *either* $P = P_{\mathsf{S4.3}}$ *or* $\mathsf{Fr}(P)$ *is of bounded depth.*

Our aim now is to axiomatise the SP-logic of every FO-definable subframe-closed class of S4.3-frames. To this end, recall from [11,24] that, for any finite frame $\mathfrak{F}$, there are modal formulas $\alpha(\mathfrak{F})$ and $\alpha^\sharp(\mathfrak{F})$, called the *subframe* and *cofinal subframe formulas for* $\mathfrak{F}$, respectively, such that, for any frame $\mathfrak{G}$,

$$\mathfrak{G} \not\models \alpha(\mathfrak{F}) \quad \text{iff} \quad \mathfrak{F} \subseteq \mathfrak{G}, \qquad\qquad \mathfrak{G} \not\models \alpha^\sharp(\mathfrak{F}) \quad \text{iff} \quad \mathfrak{F} \Subset \mathfrak{G}. \tag{7}$$

In fact, every $L \in \mathrm{NExt}\mathsf{S4.3}$ can be represented in the form

$$L = \mathsf{S4.3} \oplus \alpha^\sharp(\mathfrak{F}_1) \oplus \cdots \oplus \alpha^\sharp(\mathfrak{F}_m),$$

for some $\mathfrak{F}_i$. If $\mathcal{C}$ is a nonempty FO-definable class of frames closed under subframes and $N < \omega$ is the maximal depth of frames in $\mathcal{C}$, then there exists a finite set $F_{\mathcal{C}}$ of frames of depth $\leq N$ such that

$$\boldsymbol{L}[\mathcal{C}] = \mathsf{S4.3} \oplus \alpha(\mathfrak{L}_{N+1}) \oplus \{\alpha(\mathfrak{F}) \mid \mathfrak{F} \in F_{\mathcal{C}}\}. \tag{8}$$

Note that $\mathsf{S4.3} \oplus \alpha(\mathfrak{F}) = \mathsf{S4.3} \oplus \alpha^\sharp(\mathfrak{F}) \oplus \alpha^\sharp(\mathfrak{F}^\circ)$, where $\mathfrak{F}^\circ$ is obtained from $\mathfrak{F}$ by adding a single-point cluster on top of $\mathfrak{F}$, and that $\mathsf{S4.3}\oplus\alpha(\mathfrak{L}_n) = \mathsf{S4.3}\oplus\alpha^\sharp(\mathfrak{L}_n)$.

In the remainder of this section we (i) construct SP-analogues $\boldsymbol{\kappa}^N(\mathfrak{F})$ of the subframe formulas $\alpha(\mathfrak{F})$ such that $\mathfrak{G} \models \boldsymbol{\kappa}^N(\mathfrak{F})$ iff $d(\mathfrak{G}) \leq N$ and $\mathfrak{G} \models \alpha(\mathfrak{F})$; and (ii) prove that, using some of the $\boldsymbol{\kappa}^N(\mathfrak{F})$ formulas, one can axiomatise all but two complete SP-logics properly extending $P_{\mathsf{S4.3}}$.

Let $\mathfrak{F} = (n_1, \ldots, n_f)$ be a finite frame with $f \leq N < \omega$. We begin by defining an SP-implication $\boldsymbol{\iota}^N(\mathfrak{F})$ as follows. First, we take the terms

$$\tau(\mathfrak{F}) \;=\; \Diamond(\bigwedge P_1 \wedge \Diamond(\bigwedge P_2 \wedge \Diamond(\ldots \Diamond \bigwedge P_f)\ldots)), \tag{9}$$

where the P_i are pairwise disjoint sets of variables with $|P_i| = n_i$, for $i \leq f$. Next, denote by $\Sigma^N(\mathfrak{F})$ the set of all terms of the form

$$\Diamond(\bigwedge Q_1 \wedge \Diamond(\bigwedge Q_2 \wedge \Diamond(\ldots \Diamond \bigwedge Q_N)\ldots)),$$

where there exist $1 = x_1 < x_2 < \cdots < x_{f+1} = N + 1$ such that:

$$\text{for any } i \text{ and } j, \text{ if } x_i \leq j < x_{i+1} \text{ then } Q_j \subseteq P_i \text{ and } |Q_j| \geq |P_i| - 1, \tag{10}$$

$$\text{there is } i \text{ such that } |Q_j| = |P_i| - 1 \text{ for any } j \text{ with } x_i \leq j < x_{i+1}. \tag{11}$$

(Note that $\bigwedge \emptyset = \top$, so when some $Q_j = \emptyset$, the corresponding term in $\Sigma^N(\mathfrak{F})$ is P_{S4}-equivalent to a term of modal depth $< N$.) It is not hard to see that $|\Sigma^N(\mathfrak{F})| \leq (\max_i |P_i| + 1)^N$, and so

$$|\Sigma^N(\mathfrak{F})| \text{ is polynomial in } |\mathfrak{F}|. \tag{12}$$

Finally, we set

$$\boldsymbol{\iota}^N(\mathfrak{F}) \;=\; \bigwedge_{\sigma \in \Sigma^N(\mathfrak{F})} \sigma \;\to\; \tau(\mathfrak{F}). \tag{13}$$

For example, $\iota^1((n))$ is

$$\bigwedge_{\substack{Q\subseteq\{p_1,\dots,p_n\}\\|Q|=n-1}} \Diamond\bigwedge Q \;\to\; \Diamond(p_1\wedge\dots\wedge p_n)$$

defining $(n-1)$-functionality (in particular, $\iota^1((3))$ is ι^2_{fun} in (2)). For every $m \geq n$, the SP-implication $\iota^m(\mathfrak{L}_n)$ is (P_{S4}-equivalent to)

$$\bigwedge_{i=1}^{n} \Diamond\Big[p_1\wedge\Diamond\Big(p_2\wedge\Diamond\big[\dots\Diamond(p_{i-1}\wedge\Diamond[p_{i+1}\wedge\Diamond(\dots\Diamond p_n)\dots])\dots\big]\Big)\Big] \to \Diamond\Big(p_1\wedge\Diamond(p_2\wedge\Diamond(\dots\Diamond p_n)\dots)\Big)$$

defining the property of having depth $< n$. In particular, $\iota^n(\mathfrak{L}_n)$ is ϵ_{n-1} in (3)–(4), for $n = 2, 3$; $\iota^2((1,2))$ is φ in (5), and $\iota^2((2))$ is β_2 in (6).

In order to define $\kappa^N(\mathfrak{F})$, we require the following:

Claim 4.2 [15, Lemma 7.7] *For any finite set Φ of SP-implications, there is a single SP-implication ι_Φ such that $P_{\mathsf{S4}} + \Phi = P_{\mathsf{S4}} + \iota_\Phi$.*

Now, we set

$$\kappa^N(\mathfrak{F}) \;=\; \iota_{\{\iota^{N+1}(\mathfrak{L}_{N+1}),\iota^N(\mathfrak{F})\}}.$$

Observe that $P_{\mathsf{S4}} + \iota^{n+1}(\mathfrak{L}_{n+1}) + \iota^n(\mathfrak{L}_n) = P_{\mathsf{S4}} + \iota^n(\mathfrak{L}_n)$, and so $\kappa^n(\mathfrak{L}_n)$ is P_{S4}-equivalent to $\iota^n(\mathfrak{L}_n)$.

Theorem 4.3 *For any finite* S4.3*-frame $\mathfrak{F}$, any N with $d(\mathfrak{F}) \leq N < \omega$, and any* S4.3*-frame $\mathfrak{G}$, we have $\mathfrak{G} \models \kappa^N(\mathfrak{F})$ iff $d(\mathfrak{G}) \leq N$ and $\mathfrak{G} \models \alpha(\mathfrak{F})$.*

Proof. By (7), it suffices to show that (i) if $\mathfrak{F} \subseteq \mathfrak{G}$ then $\mathfrak{G} \not\models \iota^N(\mathfrak{F})$; and (ii) if $d(\mathfrak{G}) \leq N$ and $\mathfrak{F} \not\subseteq \mathfrak{G}$ then $\mathfrak{G} \models \iota^N(\mathfrak{F})$.

(i) Let $\mathfrak{F} = (n_1, \dots, n_f)$ and $\mathfrak{F} \subseteq \mathfrak{G}$. Then there exist clusters $(\zeta_1), \dots, (\zeta_f)$ in $\mathfrak{G}$ such that $(\zeta_1, \dots, \zeta_f)$ is a subframe of $\mathfrak{G}$ and $\zeta_i \geq n_i$, for $i \leq f$. Take the term $\tau(\mathfrak{F})$ from (9). We define a model $\mathfrak{M}$ based on $\mathfrak{G}$ in the following way. For each $i \leq f$, we

- take n_i distinct points from (ζ_i) and let each of them validate a different $n_i - 1$-element subset Q of P_i;
- take some point from (ζ_i) and let it validate $\bigcup_{j=1,j\neq i}^{f} P_j$.

All other points in $\mathfrak{G}$ validate no variables. Take any point r in the root cluster of $\mathfrak{G}$. Since for every $i \leq f$ we have $\mathfrak{M}, x \not\models \bigwedge P_i$ for any $x \in (\zeta_i)$, we clearly have $\mathfrak{M}, r \not\models \tau(\mathfrak{F})$. On the other hand, take some $\sigma \in \Sigma^N(\mathfrak{F})$ of the form $\Diamond(\bigwedge Q_1 \wedge \Diamond(\bigwedge Q_2 \wedge \Diamond(\dots\Diamond\bigwedge Q_N)\dots))$. By (11), there is $i \leq f$ such that $|Q_j| = n_i - 1$ for all those Q_j for which $Q_j \subseteq P_i$. Therefore, by (10) for every $j \leq N$ we have $\mathfrak{M}, x_j \models \bigwedge Q_j$ for some $x_j \in (\zeta_i)$, and so $\mathfrak{M}, r \models \sigma$ as required.

(ii) is proved by induction on $d(\mathfrak{F})$. To begin with, we show that if $(n) \not\subseteq \mathfrak{G}$ for some $n < \omega$, then $\mathfrak{G} \models \iota^N((n))$ for any $N \geq d(\mathfrak{G})$. To this end, suppose $\tau((n)) = \Diamond\bigwedge P_1$ for some set P_1 of variables with $|P_1| = n$. Let $\mathfrak{M}$ be a model based on $\mathfrak{G} = (\zeta_1, \ldots, \zeta_g)$ such that the left-hand side $\bigwedge \Sigma^N((n))$ of $\iota^N((n))$ holds at some point r in $\mathfrak{M}$. (We may assume that $r \in (\zeta_1)$.) We claim that

$$\text{there is } j^* \leq g \text{ such that, for every } Q \subseteq P_1,\ |Q| = n-1, \text{ we have } \mathfrak{M}, y_Q \models \bigwedge Q \text{ for some } y_Q \in (\zeta_{j^*}). \quad (14)$$

Indeed, suppose otherwise. Then, for every $j \leq g$, there is an $n-1$-element subset Q_j of P_1 such that $\mathfrak{M}, y \not\models \bigwedge Q_j$ for any $y \in (\zeta_j)$. Take the term

$$\delta = \Diamond(\bigwedge Q_1 \wedge \Diamond(\ldots \Diamond \bigwedge Q_g) \ldots).$$

By the choice of the Q_j, it is easy to see that $\mathfrak{M}, r \not\models \delta$. On the other hand, as $g \leq N$, there is $\sigma \in \Sigma^N((n))$ with $P_{\mathsf{S4}} \vdash_{\mathsf{SLO}} \sigma \to \delta$, and so $\mathfrak{M}, r \models \delta$, which is a contradiction proving (14). Now as $(n) \not\leq \mathfrak{G}$, it follows that $\zeta_{j^*} < n$. Therefore, by the pigeonhole principle and (14), there is $x \in (\zeta_{j^*})$ with $\mathfrak{M}, x \models \bigwedge P_1$, and so $\mathfrak{M}, r \models \tau((n))$, as required.

Now let $d(\mathfrak{F}) > 1$ and suppose inductively that, for all $\mathfrak{F}'$, if $d(\mathfrak{F}') < d(\mathfrak{F})$ then $\mathfrak{G} \models \iota^N(\mathfrak{F}')$ for all N and $\mathfrak{G}$ such that $N \geq \max\big(d(\mathfrak{G}), d(\mathfrak{F}')\big)$ and $\mathfrak{F}' \not\leq \mathfrak{G}$. Take $\mathfrak{F} = (n_1, \ldots, n_f)$, $\mathfrak{G} = (\zeta_1, \ldots, \zeta_g)$, $N \geq \max(f, g)$ and $\mathfrak{F} \not\leq \mathfrak{G}$. Take $\iota^N(\mathfrak{F})$ from (13), and suppose $\mathfrak{M}$ is a model based on $\mathfrak{G}$ such that

$$\mathfrak{M}, r \models \bigwedge_{\sigma \in \Sigma^N(\mathfrak{F})} \sigma, \quad \text{for some } r \in (\zeta_1). \quad (15)$$

Since $P_{\mathsf{S4}} \vdash_{\mathsf{SLO}} \sigma \to \Diamond \bigwedge P_f$ for some $\sigma \in \Sigma^N(\mathfrak{F})$, by (15), there is x with $\mathfrak{M}, x \models \bigwedge P_f$. Let

$$m = \max\{j \mid \mathfrak{M}, x \models \bigwedge P_f \text{ for some } x \in (\zeta_j)\} \quad (16)$$

and let $\mathfrak{M}^-$ be the restriction of $\mathfrak{M}$ to $(\zeta_1, \ldots, \zeta_m)$. Let $\mathfrak{F}^- = (n_1, \ldots, n_{f-1})$. Then $\tau(\mathfrak{F}^-) = \Diamond(\bigwedge P_1 \wedge \Diamond(\bigwedge P_2 \wedge \Diamond(\ldots \Diamond \bigwedge P_{f-1}) \ldots))$. We show that

$$\mathfrak{M}^-, r \models \tau(\mathfrak{F}^-), \quad (17)$$

from which $\mathfrak{M}, r \models \tau(\mathfrak{F})$ would clearly follow by (16).

To begin with, if $m = g$ then $\mathfrak{M}^- = \mathfrak{M}$. As $f - 1 \leq N$, there is $\sigma \in \Sigma^N(\mathfrak{F})$ such that $P_{\mathsf{S4}} \vdash_{\mathsf{SLO}} \sigma \to \tau(\mathfrak{F}^-)$, and so we have $\mathfrak{M}^-, r \models \tau(\mathfrak{F}^-)$ by (15). Now suppose that $m < g$. Two cases are possible:

Case 1: $\mathfrak{F}^- \not\leq (\zeta_1, \ldots, \zeta_m)$. Let $k = \max(m, f-1)$. By IH, we then have

$$(\zeta_1, \ldots, \zeta_m) \models \iota^k(\mathfrak{F}^-). \quad (18)$$

Also, as $m < g \leq N$ and $f - 1 \leq N - 1 < N$, we have $k < N$. Thus, for every $\delta \in \Sigma^k(\mathfrak{F}^-)$ of the form $\Diamond(\bigwedge Q_1 \wedge \Diamond(\bigwedge Q_2 \wedge \Diamond(\ldots \Diamond \bigwedge Q_k) \ldots))$, there is some $\sigma \in \Sigma^N(\mathfrak{F})$ with $P_{\mathsf{S4}} \vdash_{\mathsf{SLO}} \sigma \to \delta^+$ for the term

$$\delta^+ = \Diamond\Big(\bigwedge Q_1 \wedge \Diamond(\ldots \Diamond(\bigwedge Q_k \wedge \Diamond(\bigwedge P_f)))\Big).$$

Thus, $\mathfrak{M}, r \models \delta^+$ by (15), and so $\mathfrak{M}^-, r \models \delta$ follows by (16). By (18), we have $\mathfrak{M}^-, r \models \tau(\mathfrak{F}^-)$, and so (17) holds as required.

Case 2: $\mathfrak{F}^- \subseteq (\zeta_1, \dots, \zeta_m)$. Then $f-1 \leq m$ and $(f-1)+(g-m) \leq g \leq N$. Thus, for any sequence $\bar{Q} = (Q_{m+1}, \dots, Q_g)$ of $(n_f - 1)$-element subsets of P_f, there is $\sigma \in \Sigma^N(\mathfrak{F})$ such that

$$P_{\mathsf{S4}} \vdash_{\mathsf{SLO}} \sigma \to \Diamond(\bigwedge P_1 \wedge \Diamond(\dots \Diamond(\bigwedge P_{f-1} \wedge \delta_{\bar{Q}}) \dots))$$

for the term

$$\delta_{\bar{Q}} \;=\; \Diamond(\bigwedge Q_{m+1} \wedge \Diamond(\dots \Diamond \bigwedge Q_g) \dots).$$

So we have $\mathfrak{M}, r \models \Diamond(\bigwedge P_1 \wedge \Diamond(\dots \Diamond(\bigwedge P_{f-1} \wedge \delta_{\bar{Q}}) \dots))$ by (15). Now suppose (17) does not hold. Then, for every such $\bar{Q}$, we have $\mathfrak{M}, x_{\bar{Q}} \models \delta_{\bar{Q}}$ for some $x_{\bar{Q}} \in (\zeta_{m+1})$. Thus, it is easy to see that there is j^* with $m < j^* \leq g$ such that, for each of the n_f-many $(n_f - 1)$-element subsets Q of P_f, we have $\mathfrak{M}, y_Q \models \bigwedge Q$ for some $y_Q \in (\zeta_{j^*})$. (Otherwise, for every j with $m < j \leq g$, take some Q_j such that $\mathfrak{M}, y \not\models \bigwedge Q_j$ for any $y \in (\zeta_j)$, and then consider the sequence $\bar{Q} = (Q_{m+1}, \dots, Q_g)$.) As $\mathfrak{F}^- \subseteq (\zeta_1, \dots, \zeta_m)$ and $\mathfrak{F} \not\subseteq \mathfrak{G}$, it follows that $\zeta_{j^*} < n_f$. Therefore, by the pigeonhole principle $\mathfrak{M}, x \models \bigwedge P_f$ for some $x \in (\zeta_{j^*})$, contrary to $j^* > m$ and (16), proving that (17) holds as required. □

We now prove our main theorem on definability, completeness (axiomatisability) and computational complexity. We show that, using some of the $\boldsymbol{\kappa}^N(\mathfrak{F})$ formulas, we can axiomatise all but two complete SP-logics properly extending $P_{\mathsf{S4.3}}$. We begin with the two exceptions. First, it is straightforward to see that $\boldsymbol{P}[\emptyset] = P_{\mathsf{S4.3}} + (p \to q)$. Second, it is shown in [16, Theorem 29] that $\boldsymbol{P}[\mathcal{C}_{id}] = P_{\mathsf{S4.3}} + (\Diamond p \to p)$, where

$$\mathcal{C}_{id} = \{(W, R) \mid \forall x, y \in W \,(R(x,y) \leftrightarrow x = y)\}.$$

Now, suppose that $\mathcal{C}$ is a nonempty FO-definable class of S4.3-frames that is closed under subframes and different from $\mathsf{Fr}(P_{\mathsf{S4.3}})$, and let $N < \omega$ be the maximal depth of frames in $\mathcal{C}$. Take some finite set $F_{\mathcal{C}}$ of depth $\leq N$ frames such that (8) holds. This $F_{\mathcal{C}}$ consists of 'forbidden subframes' in the sense that, for any frame $\mathfrak{G}$ of depth $\leq N$,

$$\mathfrak{G} \in \mathcal{C} \quad \text{iff} \quad \mathfrak{F} \not\subseteq \mathfrak{G} \text{ for any } \mathfrak{F} \in F_{\mathcal{C}}. \tag{19}$$

Note that such an $F_{\mathcal{C}}$ is not unique. In (8), this is not a problem since $P_{\mathsf{S4}} + \alpha(\mathfrak{F}_1) \models_{\mathsf{Kr}} \alpha(\mathfrak{F}_2)$ whenever $\mathfrak{F}_1 \subseteq \mathfrak{F}_2$ by (7). However, it is not always the case that $P_{\mathsf{S4.3}} + \boldsymbol{\kappa}^N(\mathfrak{F}_1) \vdash_{\mathsf{SLO}} \boldsymbol{\kappa}^N(\mathfrak{F}_2)$ whenever $\mathfrak{F}_1 \subseteq \mathfrak{F}_2$. (For example, it follows from Claim 3.1 that $P_{\mathsf{S4.3}} + \boldsymbol{\kappa}^2((2)) \not\vdash_{\mathsf{SLO}} \boldsymbol{\kappa}^2((1,2))$.) Therefore, in what follows, we assume that $F_{\mathcal{C}}$ has the following closure property:

$$\begin{array}{l}\text{For every } \mathfrak{G} \text{ with } d(\mathfrak{G}) \leq N,\\ \qquad \text{if } \mathfrak{G} \notin \mathcal{C} \text{ then there is } \mathfrak{F} \in F_{\mathcal{C}} \text{ such that } \mathfrak{F} \subseteq \mathfrak{G} \text{ and } d(\mathfrak{F}) = d(\mathfrak{G}).\end{array} \tag{20}$$

Now take the minimal set $F_\mathcal{C}$ of depth $\leq N$ frames such that (19) and (20) hold. It is easy to see that $F_\mathcal{C}$ is always finite and unique (up to isomorphism of its frames). (For example, if $\mathcal{C}$ consists of all frames isomorphic to some subframe of $(2,1)$, then $F_\mathcal{C} = \{(1,2),(3),(3,1)\}$.) Let

$$P_\mathcal{C} = \begin{cases} P_{\mathsf{S4.3}} + \{\boldsymbol{\kappa}^N(\mathfrak{F}) \mid \mathfrak{F} \in F_\mathcal{C}\}, & \text{if } F_\mathcal{C} \neq \emptyset, \\ P_{\mathsf{S4.3}} + \boldsymbol{\kappa}^{N+1}(\mathfrak{L}_{N+1}), & \text{otherwise.} \end{cases}$$

Theorem 4.4 *Let $\mathcal{C}$ be any FO-definable class of* S4.3*-frames that is closed under subframes and different from $\emptyset$, $\mathcal{C}_{id}$ and $\mathsf{Fr}(P_{\mathsf{S4.3}})$. Then the following hold:*

(i) $\mathcal{C} = \mathsf{Fr}(P_\mathcal{C})$.

(ii) $P_\mathcal{C}$ is Kripke complete, and so $\boldsymbol{P}[\mathcal{C}] = P_\mathcal{C}$.

(iii) $P_\mathcal{C}$ is decidable in PTIME.

Proof. *(i)* is a straightforward consequence of Claim 4.2 and Theorem 4.3.

(ii) By an *$F_\mathcal{C}$-normal form* we mean any term of the following form: $\Diamond(\bigwedge P^1 \wedge \Diamond(\bigwedge P^2 \wedge \Diamond(\dots \Diamond \bigwedge P^k)\dots))$, where $k \leq N$ and each P^i is a finite nonempty set of variables for which there is no $\mathfrak{F} \in F_\mathcal{C}$ such that $\mathfrak{F} \subseteq (|P^1|, |P^2|, \dots, |P^k|)$. We do not require that the P^i are disjoint.

Claim **4.4.1** *For every term σ, there is a set N_σ of $F_\mathcal{C}$-normal forms such that $|N_\sigma|$ is polynomial in the size of σ, and $\Diamond\sigma$ is $P_\mathcal{C}$-equivalent to $\bigwedge_{\varrho \in N_\sigma} \varrho$.*

Proof. We proceed via a series of steps (a)–(c).

(a) As shown in [16, Claim 48.1]), $\Diamond\sigma$ is $P_{\mathsf{S4.3}}$-equivalent to a conjunction of terms of the form $\Diamond(\bigwedge Q^1 \wedge \Diamond(\bigwedge Q^2 \wedge \Diamond(\dots \Diamond \bigwedge Q^k)\dots))$, where $k < \omega$ and each Q^i is a finite set of variables. Each such term describes a full linear branch in the 'term tree' of $\Diamond\sigma$, and so

(p1) each k is polynomial in the size $|\sigma|$ of σ;

(p2) each $|Q^i|$ of each term is polynomial in $|\sigma|$; and

(p3) the overall number of terms we obtain this way is polynomial in $|\sigma|$.

(b) By $\boldsymbol{\iota}^{N+1}(\mathfrak{L}_{N+1})$, we can take $k \leq N$ in (a). By (12) and **(p1)**–**(p3)**, the number of terms we thus obtain is polynomial in $|\sigma|$, so we are done if $F_\mathcal{C} = \emptyset$.

(c) Let $F_\mathcal{C} \neq \emptyset$. We claim that each term in (b) is $P_{\mathsf{S4}} + \{\boldsymbol{\iota}^N(\mathfrak{F}) \mid \mathfrak{F} \in F_\mathcal{C}\}$-equivalent to a conjunction of $F_\mathcal{C}$-normal forms. Indeed, suppose

$$\chi = \Diamond(\bigwedge Q^1 \wedge \Diamond(\bigwedge Q^2 \wedge \Diamond(\dots \Diamond \bigwedge Q^k)\dots))$$

is a term in (b), for some $k \leq N$ and finite sets Q^i of variables. By P_{S4}, we may assume that each Q^i is nonempty. If such a term is not an $F_\mathcal{C}$-normal form, then $\mathfrak{F} \subseteq (|Q^1|, |Q^2|, \dots, |Q^k|)$ for some $\mathfrak{F} \in F_\mathcal{C}$. By (19) and (20), we may assume that $d(\mathfrak{F}) = k$. Let $\mathfrak{F} = (n_1, \dots, n_k)$ and $n_i \leq |Q^i|$ for $i \leq k$. For $i \leq k$, let $P_i = \{p^i_1, \dots, p^i_{n_i}\}$ be the variables in $\boldsymbol{\iota}^N(\mathfrak{F})$ and $Q^i = \{q^i_1, \dots, q^i_{|Q_i|}\}$. We define a substitution S for the variables in $\bigcup_{i \leq k} P_i$ as follows: for any $i \leq k$ and any $j \le n_i - 1$, we substitute q^i_j for p^i_j, and $q^i_{n_i} \wedge \dots \wedge q^i_{|Q_i|}$ for $p^i_{n_i}$. Then the S-instance of $\tau(\mathfrak{F})$ is χ. It is easy to see that $P_{\mathsf{S4}} \vdash_{\mathsf{SLO}} \tau(\mathfrak{F}) \to \bigwedge \Sigma^N(\mathfrak{F})$ always holds, and so $P_{\mathsf{S4}} \vdash_{\mathsf{SLO}} \chi \to S(\sigma)$ for the S-instance $S(\sigma)$ of every $\sigma \in \Sigma^N(\mathfrak{F})$. On the other hand, we have $\iota^N(\mathfrak{F}) \vdash_{\mathsf{SLO}} \bigwedge_{\sigma \in \Sigma^N(\mathfrak{F})} S(\sigma) \to \chi$. By (10) and (11), each $S(\sigma)$ is of the form $\Diamond(\bigwedge R^1 \wedge \Diamond(\bigwedge R^2 \wedge \Diamond(\dots \Diamond \bigwedge R^N)\dots))$ such that

(n1) for every $i \le N$, there is some $j_i \le k$ with $R^i \subseteq Q^{j_i}$;

(n2) there is $j \le k$ such that $|R^i| < |Q^j|$, for every i with $j_i = j$.

Take those $R^{i_1}, \dots, R^{i_\ell}$ that are nonempty. Then $S(\sigma)$ is P_{S4}-equivalent to

$$\vartheta = \Diamond\Big(\bigwedge R^{i_1} \wedge \Diamond(\bigwedge R^{i_2} \wedge \Diamond(\dots \Diamond \bigwedge R^{i_\ell})\dots)\Big).$$

If such a ϑ is not an $F_{\mathcal{C}}$-normal form, then again there is some $\mathfrak{F}' \in F_{\mathcal{C}}$ such that $d(\mathfrak{F}') = \ell$ and $\mathfrak{F}' \subseteq \big(|R^{i_1}|, |R^{i_2}|, \dots, |R^{i_\ell}|\big)$. Thus, we can continue by 'applying' $\iota^N(\mathfrak{F}')$ to ϑ. By **(n1)** and **(n2)**, sooner or later the procedure stops and we obtain a set of $F_{\mathcal{C}}$-normal forms. In fact, by (12) and **(p1)**–**(p3)**, both the number of required steps and the number of terms we obtain in each step are polynomial in $|\sigma|$. □

For a set Σ of terms, let $var(\Sigma)$ be the set of propositional variables in Σ. We write $var(\sigma)$ for $var(\{\sigma\})$.

CLAIM **4.4.2** *For every finite set $\Sigma \cup \{\xi\}$ of $F_{\mathcal{C}}$-normal forms such that $var(\Sigma) \subseteq var(\xi)$, if $P_{\mathsf{S4}} \not\vdash_{\mathsf{SLO}} \bigwedge \Sigma \to \xi$ then $\mathfrak{G} \not\models \bigwedge \Sigma \to \xi$, for some $\mathfrak{G} \in \mathcal{C}$.*

Proof. Suppose $\xi = \Diamond(\bigwedge P^1 \wedge \Diamond(\bigwedge P^2 \wedge \Diamond(\dots \Diamond \bigwedge P^k)\dots))$ where $k \le N$, and each P^i is a finite set of variables such that there is no $\mathfrak{F} \in F_{\mathcal{C}}$ with $\mathfrak{F} \subseteq \big(|P^1|, |P^2|, \dots, |P^k|\big)$. Then $\big(|P^1|, |P^2|, \dots, |P^k|\big) \in \mathcal{C}$ by (19). By P_{S4}, we may assume that neither $P^i \subseteq P^{i+1}$ nor $P^{i+1} \subseteq P^i$ for any $i \le k$. Let $\mathfrak{M}$ be the following model over $\big(|P^1|, |P^2|, \dots, |P^k|\big)$: for every $i \le k$, let each point in $\big(|P^i|\big)$ validate $\bigcup_{j=1}^k P^j \setminus \{p\}$ for each different $p \in P^i$. Then clearly $\mathfrak{M}, r \not\models \xi$ for any $r \in \big(|P^i|\big)$. On the other hand, take any $\chi \in \Sigma$. As $var(\chi) \subseteq var(\xi)$, χ is of the form $\Diamond(\bigwedge R^1 \wedge \Diamond(\bigwedge R^2 \wedge \Diamond(\dots \Diamond \bigwedge R^\ell)\dots))$ such that $\ell \le N$ and $R^i \subseteq \bigcup_{j=1}^k P^j$ for $i \le \ell$. As $P_{\mathsf{S4}} \not\vdash_{\mathsf{SLO}} \chi \to \xi$, there is no subsequence $(R^{i_1}, \dots, R^{i_k})$ of $(R^1, \dots, R^\ell)$ with $P^j \subseteq R^{i_j}$ for $j \le k$. So it is not hard to check that $\mathfrak{M}, r \models \chi$. Therefore, $\mathfrak{M}, r \models \bigwedge \Sigma$, and so $\big(|P^1|, |P^2|, \dots, |P^k|\big) \not\models \bigwedge \Sigma \to \xi$. □

CLAIM **4.4.3** [15, Lemma 5.1] *Suppose σ and τ are terms.*

(a) $\sigma \to \tau$ is $\emptyset$-equivalent to $\sigma_\tau[\top] \to \tau$, where $\sigma_\tau[\top]$ is obtained from σ by substituting $\top$ for all variables not in $var(\tau)$.

(b) If $P_{\mathsf{S4}} + (\sigma \to \tau) \not\vdash_{\mathsf{SLO}} \Diamond p \to p$, then $\sigma \to \tau$ is P_{S4}-equivalent to $\Diamond\sigma \to \Diamond\tau$.

Proof. *(a)* is a straightforward.

(*b*) We clearly have $\sigma \to \tau \vdash_{\mathsf{SLO}} \Diamond\sigma \to \Diamond\tau$. For the other direction, suppose

$$\sigma = \bigwedge_{p \in P_\sigma} p \wedge \bigwedge_{\chi \in T_\sigma} \Diamond\chi \qquad \text{and} \qquad \tau = \bigwedge_{p \in P_\tau} p \wedge \bigwedge_{\vartheta \in T_\tau} \Diamond\vartheta,$$

for some sets P_σ, P_τ of variables, and sets T_σ, T_τ of terms. We claim that

$$P_\tau \subseteq P_\sigma. \tag{21}$$

Suppose otherwise. In this case, we take some $p \in P_\tau \setminus P_\sigma$, and let σ^- be obtained from σ by replacing all variables different from p with $\top$. Then we have $P_{\mathsf{S4}} + (\sigma \to \tau) \vdash_{\mathsf{SLO}} \sigma^- \to p$. As $P_{\mathsf{S4}} + (\sigma \to \tau) \not\vdash_{\mathsf{SLO}} \top \to p$, we may assume that $var(\tau) \subseteq var(\sigma)$, and so $p \in var(\sigma)$. It follows that $P_{\mathsf{S4}} \vdash_{\mathsf{SLO}} \Diamond p \to \sigma^-$, and so $P_{\mathsf{S4}} + (\sigma \to \tau) \vdash_{\mathsf{SLO}} \Diamond p \to p$, which is a contradiction proving (21). As we clearly have $P_{\mathsf{S4}} \vdash_{\mathsf{SLO}} \sigma \to \bigwedge P_\sigma \wedge \Diamond\sigma$, and $P_{\mathsf{S4}} \vdash_{\mathsf{SLO}} \Diamond\tau \to \bigwedge_{\vartheta \in T_\tau} \Diamond\vartheta$, by (21) we obtain $P_{\mathsf{S4}} + (\Diamond\sigma \to \Diamond\tau) \vdash_{\mathsf{SLO}} \sigma \to \tau$, as required. □

Now the proof of (*ii*) can be completed as follows. Suppose $P_{\mathcal{C}} \models_{\mathsf{Kr}} \sigma \to \tau$. As $\mathcal{C}$ is different from $\emptyset$ and $\mathcal{C}_{id}$, it follows that $P_{\mathsf{S4}} + (\sigma \to \tau) \not\vdash_{\mathsf{SLO}} \Diamond p \to p$, and so $\sigma \to \tau$ is P_{S4}-equivalent to $\Diamond\sigma \to \Diamond\tau$ by Claim 4.4.3 (b). Thus, by Claim 4.4.1 and (1), we have

$$P_{\mathcal{C}} \models_{\mathsf{Kr}} \bigwedge_{\chi \in N_\sigma} \chi \to \bigwedge_{\vartheta \in N_\tau} \vartheta,$$

and so $P_{\mathcal{C}} \models_{\mathsf{Kr}} \bigwedge_{\chi \in N_\sigma} \chi \to \vartheta$, for every $\vartheta \in N_\tau$. Take any $\vartheta \in N_\tau$. It is straightforward to see that by substituting $\top$ for some variables in an $F_{\mathcal{C}}$-normal form we obtain a term that is P_{S4}-equivalent to an $F_{\mathcal{C}}$-normal form. So by (*i*) and Claims 4.4.3 (a), 4.4.2, we have $P_{\mathsf{S4}} \vdash_{\mathsf{SLO}} \bigwedge_{\chi \in N_\sigma} \chi \to \vartheta$. Therefore,

$$P_{\mathsf{S4}} \vdash_{\mathsf{SLO}} \bigwedge_{\chi \in N_\sigma} \chi \to \bigwedge_{\vartheta \in N_\tau} \vartheta,$$

and so $P_{\mathcal{C}} \vdash_{\mathsf{SLO}} \sigma \to \tau$ follows by Claim 4.4.1.

(*iii*) follows from Claims 4.4.1 and 4.4.2, and the tractability of P_{S4}. □

It is to be noted that Theorem 4.4 does not hold for $\mathcal{C} = \mathcal{C}_{id}$. In this case $N = 1$ and $F_{\mathcal{C}_{id}} = \{(2)\}$. It is easy to see that

$$P_{\mathsf{S4.3}} + \boldsymbol{\kappa}^1((2)) = P_{\mathsf{S4.3}} + \boldsymbol{\iota}^1((2)) = P_{\mathsf{S4.3}} + \big(\Diamond p \wedge \Diamond q \to \Diamond(p \wedge q)\big).$$

As shown in [16, Theorem 29], this SP-logic is incomplete.

Note also that the axiomatisations given by Theorem 4.4 are not necessarily independent and often can be simplified. For instance, it is not hard to show the following:

- $P_{\mathsf{S4}} + \boldsymbol{\kappa}^N(\mathfrak{F}_1) \vdash_{\mathsf{SLO}} \boldsymbol{\kappa}^N(\mathfrak{F}_2)$ whenever $\mathfrak{F}_1 \subseteq \mathfrak{F}_2$ and $d(\mathfrak{F}_1) = d(\mathfrak{F}_2) \leq N$.
- $P_{\mathsf{S4.3}} + \boldsymbol{\kappa}^N(\mathfrak{F}) \vdash_{\mathsf{SLO}} \boldsymbol{\kappa}^N(\mathfrak{F}^\circ)$ whenever $d(\mathfrak{F}) < N$, where $\mathfrak{F}^\circ$ is obtained from $\mathfrak{F}$ by adding a single-point cluster on top of $\mathfrak{F}$.

5 Modal logics vs. SP-logics

The set NExtS4.3 ordered by $\subseteq$ forms a distributive lattice, a pseudo-Boolean algebra, to be more precise [19]. On the other hand, as follows from [15, Proposition 5.11], the lattice $\mathrm{Ext}^+P_{\mathsf{S4.3}}$ is not even modular because it contains the pentagon N_5 lattice; see Fig. 2.

Following [4], we compare these two lattices of logics using the map $\pi\colon \mathrm{NExtS4.3} \to \mathrm{Ext}^+P_{\mathsf{S4.3}}$, which gives the *SP-fragment* $\pi(L)$ of any modal logic L, and the map $\mu\colon \mathrm{Ext}^+P_{\mathsf{S4.3}} \to \mathrm{NExtS4.3}$, which gives the *modal extension* $\mu(P) = \mathsf{S4.3} \oplus P$ of any SP-logic P. We call L a *modal companion* of $\pi(L)$, and P an *SP-companion* of $\mu(P)$. It was observed in [4] that μ and π are monotone and form a Galois connection: $\mu(P) \subseteq L$ iff $P \subseteq \pi(L)$. In this section, we obtain answers to some open questions from [4] in the context of NExtS4.3 and $\mathrm{Ext}^+P_{\mathsf{S4.3}}$. First, we give a characterisation of those SP-logics that have modal companions (which actually holds for all multi-modal SP-logics).

Theorem 5.1 *An SP-logic P has a modal companion iff P is complete.*

Proof. ($\Rightarrow$) If $\pi(L) = P$ and $\iota \notin P$, then $\iota \notin \mathsf{S4.3} \oplus P$, and so, by Sahlqvist completeness, there is a Kripke frame for P refuting ι, as required. ($\Leftarrow$) It is readily seen that, if P is complete, then $\mathsf{S4.3} \oplus P$ is its modal companion. □

Next, we give a frame-theoretic characterisation of modal companions of any complete SP-logic in $\mathrm{Ext}^+P_{\mathsf{S4.3}}$, which consists of two parts. The first part describes the set $\pi^{-1}(P_{\mathsf{S4.3}})$ of modal companions of $P_{\mathsf{S4.3}}$ and shows that it comprises exactly the logics in NExtS4.3 that have frames of unbounded depth. We say that such logics are of *slice* ω.

Theorem 5.2 *For any modal logic $L \in$ NExtS4.3, we have $\pi(L) = P_{\mathsf{S4.3}}$ iff $L \subseteq \mathsf{Grz.3} = \mathsf{S4.3} \oplus \alpha((2))$. Thus, Grz.3 is the greatest companion of $P_{\mathsf{S4.3}}$.*

Proof. ($\Rightarrow$) If $L \not\subseteq \mathsf{Grz.3}$, then $\alpha(\mathfrak{L}_n) \in L$, for some $n < \omega$. By Theorem 4.3, it follows that $\boldsymbol{\kappa}^n(\mathfrak{L}_n) \in L$, contrary to $\pi(L) = P_{\mathsf{S4.3}}$. ($\Leftarrow$) Suppose there is $\iota \in \pi(L) \setminus P_{\mathsf{S4.3}}$. By Theorem 4.1, there is $n < \omega$ such that $\mathfrak{L}_n \not\models \iota$, which is impossible since $\mathfrak{L}_n \models \mathsf{Grz.3}$. □

In other words, $\pi^{-1}(P_{\mathsf{S4.3}})$ comprises S4.3 and all those (infinitely many) logics in NExtS4.3 whose classes of frames are *not FO-definable*. By a standard Löwenheim–Skolem–Tarski argument, if $\mathcal{C}$ is FO-definable, then both $\boldsymbol{L}[\mathcal{C}]$ and $\boldsymbol{P}[\mathcal{C}]$ are determined by the countable frames in $\mathcal{C}$. So the remaining logics can be classified according to the depth of their countable frames. (This classification was suggested in [14,18].) We say that a modal logic L is *of slice n* ($0 < n < \omega$) if $\mathfrak{L}_n \models L$ but $\mathfrak{L}_{n+1} \not\models L$. Within NExtS4.3, the logics L of slice n form the infinite interval

$$\mathsf{S4.3} \oplus \alpha(\mathfrak{L}_{n+1}) = \boldsymbol{L}[(\underbrace{\omega, \dots, \omega}_{n})] \subseteq L \subseteq \boldsymbol{L}[\mathfrak{L}_n] = \mathsf{S4.3} \oplus \alpha(\mathfrak{L}_{n+1}) \oplus \alpha((2)).$$

In particular, the logics L of slice 1 form the (infinite) interval $\mathsf{S5} \subseteq L \subseteq \boldsymbol{L}[\mathfrak{L}_1]$. (It is shown in [18] that all logics of finite slices above S4 are locally finite.)

We now characterise the modal companions of complete SP-logics properly extending $P_{\mathsf{S4.3}}$. Given the class $\mathsf{Fr}^\omega(P)$ of all countable rooted frames for P, denote by $\downarrow\mathsf{Fr}^\omega(P)$ its smallest subclass whose closure under subframes gives $\mathsf{Fr}^\omega(P)$. (It is not hard to see that $\downarrow\mathsf{Fr}^\omega(P)$ always exists.)

Theorem 5.3 *For any complete SP-logic* $P \supsetneq P_{\mathsf{S4.3}}$,

$$\pi^{-1}(P) = \{L \in \mathrm{NExt}\mathsf{S4.3} \mid \downarrow\mathsf{Fr}^\omega(P) \subseteq \mathsf{Fr}^\omega(L) \subseteq \mathsf{Fr}^\omega(P)\}.$$

Thus, $\boldsymbol{L}[\downarrow\mathsf{Fr}^\omega(P)]$ *is the greatest modal companion of* P.

Proof. Suppose $\downarrow\mathsf{Fr}^\omega(P) \subseteq \mathsf{Fr}^\omega(L) \subseteq \mathsf{Fr}^\omega(P)$ and show that $\pi(L) = P$. If $\iota \in L$, then $\mathsf{Fr}^\omega(L) \models \iota$, and so $\downarrow\mathsf{Fr}^\omega(P) \models \iota$. By Theorem 4.1, it follows that $\mathsf{Fr}^\omega(P) \models \iota$ and $\iota \in P$. The implication $\iota \in P \Rightarrow \iota \in L$ is trivial.

Next, we have to prove that every modal companion L of P belongs to the specified interval. It suffices to show that if $L = \mathsf{S4.3} \oplus P \oplus \alpha^\sharp(\mathfrak{F})$ is a modal companion of P, then $\downarrow\mathsf{Fr}^\omega(P) \models \alpha^\sharp(\mathfrak{F})$. Suppose otherwise and take some $\mathfrak{G} \in \downarrow\mathsf{Fr}^\omega(P)$ such that $\mathfrak{F} \Subset \mathfrak{G}$. By the construction of $\downarrow\mathsf{Fr}^\omega(P)$, we can always find a *finite* frame $\mathfrak{G}'$ such that $\mathfrak{F} \Subset \mathfrak{G}' \subseteq \mathfrak{G}$ and $\mathfrak{G}' \not\subseteq \mathfrak{H}$, for any $\mathfrak{H}$ in $\downarrow\mathsf{Fr}^\omega(P)$ different from $\mathfrak{G}$. But then $\alpha(\mathfrak{G}') \in L$, and so $\boldsymbol{\kappa}^n(\mathfrak{G}') \in P$ by Theorem 4.3, where n is the slice of P, which is impossible because $\mathfrak{G}' \in \mathsf{Fr}^\omega(P)$. □

Intuitively, $\boldsymbol{L}[\downarrow\mathsf{Fr}^\omega(P)]$ saturates $\mathsf{S4.3} \oplus P$ with all those formulas $\alpha^\sharp(\mathfrak{F})$ that do not derive any $\alpha(\mathfrak{G}) \notin \mathsf{S4.3} \oplus P$. We illustrate this by a few examples.

Example 5.4 (1) The SP-logic $\boldsymbol{P}[(\omega,\omega)] = P_{\mathsf{S4.3}} + \boldsymbol{\kappa}^3(\mathfrak{L}_3)$ has only one modal companion $\mathsf{S4.3} \oplus \alpha(\mathfrak{L}_3)$ since $(\omega,\omega) \not\models \alpha^\sharp(\mathfrak{F})$, for any $\mathfrak{F}$ of depth ≤ 2.

(2) The SP-logic $\boldsymbol{P}[(2,1)] = P_{\mathsf{S4.3}} + \boldsymbol{\kappa}^2((1,2)) + \boldsymbol{\kappa}^2((3))$ has two modal companions: $\mathsf{S4.3} \oplus \alpha(\mathfrak{L}_3) \oplus \alpha\big((1,2)\big) \oplus \alpha\big((3)\big) = \boldsymbol{L}[(2),(2,1)]$ and its extension with $\alpha^\sharp\big((2)\big)$, i.e., $\boldsymbol{L}[(2,1)]$. Note that $\mathsf{S4} \oplus \alpha^\sharp\big((2)\big) = \mathsf{S4} \oplus \Box\Diamond p \to \Diamond\Box p = \mathsf{S4.1}$.

(3) The companions L of $\boldsymbol{P}[(\omega,1)] = P_{\mathsf{S4.3}} + \boldsymbol{\kappa}^2((1,2))$ form the infinite interval

$$\boldsymbol{L}[(\omega),(\omega,1)] = \mathsf{S4.3} \oplus \alpha(\mathfrak{L}_3) \oplus \alpha\big((1,2)\big) \subseteq L \subseteq \mathsf{S4.3} \oplus \alpha(\mathfrak{L}_3) \oplus \alpha^\sharp\big((2)\big) = \boldsymbol{L}[(\omega,1)].$$

As follows from our results above, π maps NExt$\mathsf{S4.3}$ onto the complete SP-logics in $\mathrm{Ext}^+P_{\mathsf{S4.3}}$. On the other hand, for a complete SP-logic P, there may exist Kripke incomplete logics P' such that $\mathsf{Fr}(P) = \mathsf{Fr}(P')$. All these logics form the set $\mu^{-1}(\mathsf{S4.3} \oplus P)$. Thus, for every $L \in$ NExt$\mathsf{S4.3}$, the set $\mu^{-1}(L)$ has $\pi(L)$ as its greatest element; see Fig. 4. We do not know whether $\mu^{-1}(L)$ always has a least element, whether it has non-finitely axiomatisable SP-logics, and whether there are a continuum of them.

As we saw above, any SP-logic different from $P_{\mathsf{S4.3}}$ belongs to some finite slice. In Fig. 5, we show slice 1 and a part of slice 2 (to minimise clutter, we only give the frames or SLOs determining modal and SP-logics; as before, □ indicates incomplete SP-logics). The structure of slice 1 was detailed in [15]. As shown in [16], it has two incomplete SP-logics: $\boldsymbol{P}[\mathfrak{S}]$ and $\boldsymbol{P}[\mathfrak{C}_1]$, where $\mathfrak{S}$ is

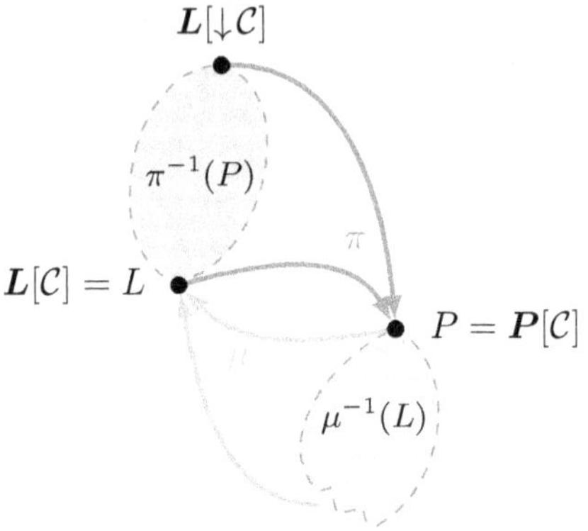

Fig. 4. Maps π and μ.

a SLO with two elements $a \leq \top$ such that $\Diamond a = \Diamond\top = \top$, and $\mathfrak{E}_1$ is in Fig. 1. Note that $\mu\big(\boldsymbol{P}[\mathfrak{S}]\big)$ is inconsistent, while $\mu^{-1}\big(\boldsymbol{L}[(1)]\big) = \big\{\boldsymbol{P}[(1)], \boldsymbol{P}[\mathfrak{E}_1]\big\}$. Slice 2 is much more involved. Its sublattice of Kripke complete logics comprises the SP-logics of the form $\boldsymbol{P}[\mathcal{C}]$, where $\mathcal{C}$ is a finite set of frames of depth ≤ 2 at least one of which is of depth 2. As shown in Section 3, modal logics $\boldsymbol{L}[\mathcal{C}]$ of slice 2 typically have infinitely many SP-companions in $\mu^{-1}(L)$; see Fig. 2. On the other hand, Example 5.4 shows SP-logics with multiple modal companions.

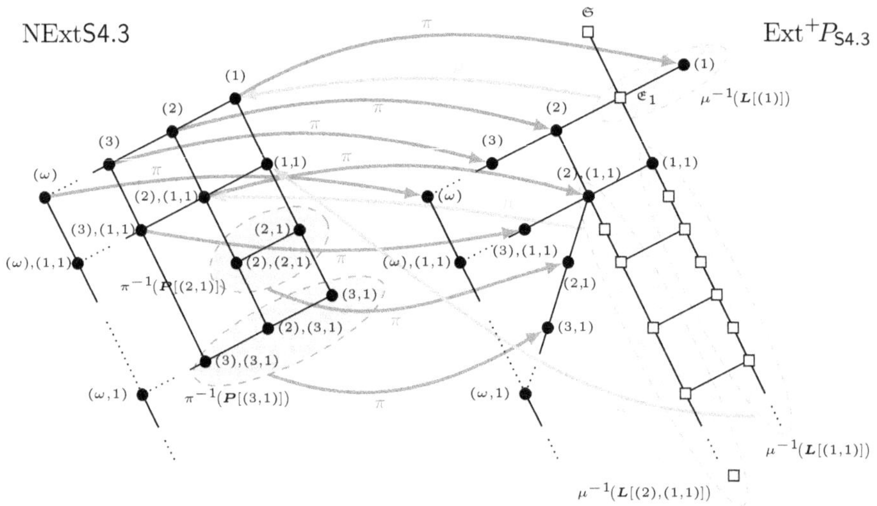

Fig. 5. Modal and SP-logics of slices 1 and 2.

6 SP-rules

An SP-rule, ϱ, takes the form $\frac{\iota_1,\dots,\iota_n}{\iota}$, where $\iota_1, \dots, \iota_n, \iota$ are SP-implications. We say that ϱ is *valid* in a SLO $\mathfrak{A}$ and write $\mathfrak{A} \models \varrho$ if $\mathfrak{A}$ satisfies the *quasi-*

equation $(\iota_1^* \wedge \ldots \wedge \iota_n^*) \to \iota^*$, where $(\sigma \to \tau)^* = (\sigma \wedge \tau = \sigma)$. We identify the rule $\frac{\emptyset}{\iota}$ with ι. Given a set R of SP-rules and an SP-rule ϱ, we set $R \models_{\mathsf{SLO}} \varrho$ if $\mathfrak{A} \models \varrho$ for any SLO $\mathfrak{A}$ validating every rule in R. We denote by $\boldsymbol{R}[\mathfrak{A}]$ the set of SP-rules that are valid in $\mathfrak{A}$ and, for a frame $\mathfrak{F}$, set $\boldsymbol{R}[\mathfrak{F}] = \boldsymbol{R}[\mathfrak{F}^*]$. For a class $\mathcal{C}$ of SLOs (or frames), we define the *SP-rule logic* $\boldsymbol{R}[\mathcal{C}]$ of $\mathcal{C}$ as the intersection of all $\boldsymbol{R}[\mathfrak{A}]$ with $\mathfrak{A} \in \mathcal{C}$. Clearly, there is a one-to-one correspondence between SP-rule logics and quasi-varieties of SLOs. The minimal SP-rule logic R containing $P_{\mathsf{S4.3}}$ and a set Φ of SP-rules is denoted by $P_{\mathsf{S4.3}}+\Phi$. We call R *globally complete* in case $\varrho \in R$ iff $\varrho \in \boldsymbol{R}[\mathfrak{F}]$ for any frame $\mathfrak{F}$ validating all rules in R. If Φ is a set of SP-implications, then $P_{\mathsf{S4.3}} + \Phi$ is complete whenever it is globally complete. However, the converse only holds for two non-trivial SP-logics containing $P_{\mathsf{S4.3}}$:

Theorem 6.1 *A non-trivial* $P \in \mathrm{Ext}^+ P_{\mathsf{S4.3}}$ *is globally complete iff* P *is complex iff* $P \in \{P_{\mathsf{S5}}, \boldsymbol{P}[(1)]\}$.

Proof. The first equivalence is shown in [16]. For the second one, it is known from [16] that P_{S5} and $\boldsymbol{P}[(1)]$ are complex. Conversely, let $P \notin \{P_{\mathsf{S5}}, \boldsymbol{P}[(1)]\}$. Suppose first that $(1,1) \in \mathsf{Fr}(P)$. Then the SLO $\mathfrak{C}_1$ in Fig. 1 validates P but there is no frame $\mathfrak{F} \in \mathsf{Fr}(P_{\mathsf{S4.3}})$ such that $\mathfrak{C}_1$ is embeddable into $\mathfrak{F}^*$. Thus, P is not complex. Now suppose $(1,1) \notin \mathsf{Fr}(P)$. Then $P \supset P_{\mathsf{S5}}$ and there exists a minimal $n > 2$ such that $(n) \notin \mathsf{Fr}(P)$. Then the SLO $\mathfrak{E}_n$ in Fig. 1 validates P but there is no frame $\mathfrak{F} \in \mathsf{Fr}(P)$ such that $\mathfrak{E}_n$ is embeddable into $\mathfrak{F}^*$. □

It follows that the axiomatisations of SP-logics given above do not provide axiomatisations of the corresponding SP-rule logics, except for two cases. This is in sharp contrast to normal modal logics containing S4.3 where the introduction of rules does not extend the expressive power of formulas as any rule $\varrho = \frac{\iota_1,\ldots,\iota_n}{\iota}$ can be equivalently expressed by the formula $\Box(\iota_1 \wedge \cdots \wedge \iota_n) \to \iota$. We next observe that SP-rules have sufficient expressive power to define any modally definable class of S4.3-frames:

Theorem 6.2 *For every finite* S4.3*-frame* $\mathfrak{F}$*, there is an SP-rule* $\varrho(\mathfrak{F})$ *such that, for any* S4.3*-frame* $\mathfrak{G}$*, we have* $\mathfrak{G} \not\models \varrho(\mathfrak{F})$ *iff* $\mathfrak{F}$ *is a p-morphic image of* $\mathfrak{G}$.

Proof. Let $\mathfrak{F} = (W, R)$ be a finite S4.3-frame with root r. For each $x \in W$, take a variable p_x and the following set $\Delta_{\mathfrak{F}}$ of SP-implications: $p_x \to \Diamond p_y$ if $R(x,y)$, $p_x \wedge \Diamond p_y \to q$ if $\neg R(x,y)$, $p_x \wedge p_y \to q$ for $x \neq y$, $\top \to \Diamond p_y$ for y in the final cluster of $\mathfrak{F}$. Let

$$\varrho(\mathfrak{F}) = \frac{\Delta_{\mathfrak{F}}}{p_r \to \Diamond q}.$$

It is straightforward to show that $\varrho(\mathfrak{F})$ is as required. □

Corollary 6.3 *A class of frames is modally definable iff it is SP-rule definable.*

We now return to the axiomatisation problem for SP-rule logics. The SLOs in Fig. 1 show that the rules from Theorem 6.2 do not axiomatise globally complete SP-rule logics. Here, we give an axiomatisation of $\boldsymbol{R}[(2)]$, which can

Fig. 6. The SLOs $\mathfrak{A}_3$ and $\mathfrak{A}_4$.

be easily generalised to any $\boldsymbol{R}[(n)]$ with $n > 2$. Consider the rules

$$\varrho^1 = \frac{(a \to \Diamond c_i), \quad (a \land \Diamond(c_i \land c_j) \to b), \ 1 \leq i \neq j \leq 3}{a \to b}$$

$$\varrho^2 = \frac{(c_1 \to b), (c_2 \to b), (a \to \Diamond c_1), (a \to \Diamond c_2), (a \land \Diamond(c_1 \land c_2) \to b)}{a \to b}$$

$$\varrho^3 = \frac{(a \to \Diamond c_1), (a \to \Diamond c_2), (c_1 \to b), (a \land \Diamond(c_1 \land c_2) \to b), (a \land \Diamond(a \land c_2) \to b)}{a \to b}$$

It is easy to see that a cluster (n) validates all the ϱ^i iff $n \leq 2$. Thus, restricted to the set of clusters, each rule ϱ^i defines the intended class of frames. Let $R_{(2)} = P_{\mathsf{S5}} + \{\varrho^1, \varrho^2, \varrho^3\}$.

Theorem 6.4 *(i)* $R_{(2)} = \boldsymbol{R}[(2)]$. *(ii)* $P_{\mathsf{S5}} + \Phi \neq \boldsymbol{R}[(2)]$, *for any proper subset* Φ *of* $\{\varrho^1, \varrho^2, \varrho^3\}$.

Proof. For the proof of (ii), take $\mathfrak{C}_3$ in Fig. 1, and $\mathfrak{A}_3$ and $\mathfrak{A}_4$ in Fig. 6. Then $\mathfrak{C}_3, \mathfrak{A}_3, \mathfrak{A}_4$ all validate P_{S5} and

- $\mathfrak{C}_3 \not\models \varrho^1$, but $\mathfrak{C}_3 \models \varrho^2, \varrho^3$;
- $\mathfrak{A}_3 \not\models \varrho^2$, but $\mathfrak{A}_3 \models \varrho^1, \varrho^3$;
- $\mathfrak{A}_4 \not\models \varrho^3$, but $\mathfrak{A}_4 \models \varrho^1, \varrho^2$.

For (i), it suffices to provide an embedding of any SLO $\mathfrak{A}$ with $\mathfrak{A} \models R_{(2)}$ into $\mathfrak{F}^*$, for a union $\mathfrak{F}$ of two-element clusters. Recall that a *filter* F in $\mathfrak{A}$ is a subset of A containing $\top$ and such that $a \in F$ and $a \leq b$ imply $b \in F$ and $a, b \in F$ imply $a \land b \in F$. For a filter F in $\mathfrak{A}$, we set $\Diamond F = \{\Diamond a \mid a \in F\}$ and define $\mathfrak{F} = (W, R)$ using a set $\mathcal{X}$ of pairs of filters in $\mathfrak{A}$. Let $(F_1, F_2) \in \mathcal{X}$ if F_1, F_2 are filters in $\mathfrak{A}$ such that

- $\Diamond F_1 \subseteq F_2$ and $\Diamond F_2 \subseteq F_1$;
- if $F_1 \subseteq \Diamond F'$ or $F_2 \subseteq \Diamond F'$ for a filter F', then $F_1 \supseteq F'$ or $F_2 \supseteq F'$.

For any $w = (F_1, F_2) \in \mathcal{X}$, take fresh 1_w and 2_w and set

$$W = \{1_w, 2_w \mid w \in \mathcal{X}\}, \quad R = \{(1_w, 2_w), (1_w, 1_w), (2_w, 2_w), (2_w, 1_w) \mid w \in \mathcal{X}\}.$$

By definition, $\mathfrak{F} = (W, R)$ is a union of two-element clusters. We show that $f(a) = \{1_{F_1,F_2} \mid a \in F_1\} \cup \{2_{F_1,F_2} \mid a \in F_2\}$ is an embedding of $\mathfrak{A}$ into $\mathfrak{F}^*$. We first show that if $a \neq b$, then $f(a) \neq f(b)$. We may assume that $a \not\leq b$. Let F_0 be a maximal filter containing a such that $b \notin F_0$. We show that there exists a pair $(F_1, F_2) \in \mathcal{X}$ with $a \in F_1$ and $b \notin F_1$.

Let $\mathcal{M}$ be the set of all maximal filters G in $\mathfrak{A}$ such that $\Diamond G \subseteq F_0$ and $\Diamond F_0 \subseteq G$. Observe that there exists a filter $G \in \mathcal{M}$ containing a set $X \subseteq A$ iff $\Diamond(a_1 \wedge \cdots \wedge a_n) \in F_0$ for all $a_1, \ldots, a_n \in X$. It follows that, for every filter F' in $\mathfrak{A}$ with $F_0 \subseteq \Diamond F'$, there exists $G \in \mathcal{M}$ with $F' \subseteq G$. We now make a case distinction according to the cardinality of $\mathcal{M}$.

- $|\mathcal{M}| = 1$. Let $\mathcal{M} = \{G\}$. Then $G \supseteq F_0$ and $(F_0, G) \in \mathcal{X}$. We have $a \in F_0$ and $b \notin F_0$, as required.
- $|\mathcal{M}| = 2$. Let $\mathcal{M} = \{G_0, G_1\}$. We distinguish between two cases:
  Case 1: $G_0, G_1 \supseteq F_0$. Since neither $G_0 \subseteq G_1$ nor $G_1 \subseteq G_0$, we obtain $G_0 \neq F_0$ and $G_1 \neq F_0$. Thus, $b \in G_0 \cap G_1$. Then we find $c_1' \in G_0$ and $c_2' \in G_1$ and $a' \in F_0$ such that $a' \wedge \Diamond(c_1' \wedge c_2') \leq b$. Then, using the condition that filters are closed under $\wedge$, we find $c_1 \in G_0$, $c_2 \in G_1$, and $a \in F_0$ such that $c_1 \leq b$, $c_2 \leq b$, $a \leq \Diamond c_1$, $a \leq \Diamond c_2$, and $a \wedge \Diamond(c_1 \wedge c_2) \leq b$. By rule ϱ^2, $a \leq b$, and we have derived a contradiction as $b \in F_0$ follows.
  Case 2: $G_1 \not\supseteq F_0$. Then $G_0 \supseteq F_0$ as there exists at least one filter $G \in \mathcal{M}$ with $G \supseteq F_0$. Assume first that $G_0 \neq F_0$. Then $b \in G_0$. Similarly to Case 1 we thus find $c_1 \in G_0$, $c_2 \in G_1$, and $a \in F_0$ such that $a \leq \Diamond c_1$, $a \leq \Diamond c_2$, $c_1 \leq b$, $a \wedge \Diamond(c_1 \wedge c_2) \leq b$, and $a \wedge \Diamond(a \wedge c_2) \leq b$. But then, by rule ϱ^3, $a \leq b$, and we have derived a contradiction. Assume now that $G_0 = F_0$. Then $(F_0, G_1) \in \mathcal{X}$. We are done as $a \in F_0$ and $b \notin F_0$.
- $|\mathcal{M}| \geq 3$. Let $G_0, G_1, G_2 \in \mathcal{M}$. Then we find $a \in F_0$, $c_1 \in G_0$, $c_2 \in G_1$, and $c_3 \in G_2$ such that $a \leq \Diamond c_i$ for $1 \leq i \leq 3$ and $a \wedge \Diamond(c_i \wedge c_j) \leq b$ for $1 \leq i \neq j \leq 3$. But then $a \leq b$, by rule ϱ^1, and we have derived a contradiction.

We next show that $f(\Diamond a) = \Diamond f(a)$ for all $a \in A$. Suppose first $1_{F_1,F_2} \in \Diamond f(a)$. Then $a \in F_1$ or $a \in F_2$. Then $\Diamond a \in F_1$. Then $1_{F_1,F_2} \in f(\Diamond a)$, as required. Conversely, suppose $1_{F_1,F_2} \in f(\Diamond a)$. Then $\Diamond a \in F_1$. Then there exists a filter F' with $a \in F'$ and $F_1 \subseteq \Diamond F'$ and $F' \subseteq \Diamond F_1$. By definition, $F_1 \supseteq F'$ or $F_2 \supseteq F'$. Thus $1_{F_1,F_2} \in \Diamond f(a)$. Finally, $f(a_1 \wedge a_2) = f(a_1) \cap f(a_2)$ can we proved in a straightforward way. □

The computational complexity of deciding $\boldsymbol{R}[\mathcal{C}]$ has been analysed in [17]. In contrast to SP-implications, deciding SP-rules is often CONP-hard. In fact, if $\mathcal{C}$ is a nonempty class of S4.3-frames of the form $\mathsf{Fr}(P)$ for some SP-logic $P \in \mathrm{Ext}^+ P_{\mathsf{S4.3}}$, then $\boldsymbol{R}[\mathcal{C}]$ is in PTIME iff $\mathcal{C}$ is the class of all clusters or a singleton cluster. Otherwise, $\boldsymbol{R}[\mathcal{C}]$ is CONP-complete.

Acknowledgements. We are grateful to the anonymous reviewers for spotting several mistakes in the preliminary version of this paper. Thanks are also due to Marcel Jackson for his inspiring paper [15] and to Yoshihito Tanaka without whom this work would never have been done.

References

[1] Baader, F., S. Brandt and C. Lutz, *Pushing the $\mathcal{EL}$ envelope*, in: L.P. Kaelbling and A. Saffiotti, editors, *Proc. of the 19th International Joint Conference on Artificial Intelligence (IJCAI-2005)* (2005), pp. 364–369.

[2] Beklemishev, L., *Calibrating provability logic: From modal logic to reflection calculus*, in: T. Bolander, T. Braüner, S. Ghilardi and L. Moss, editors, *Advances in Modal Logic, vol. 9*, College Publications, 2012 pp. 89–94.

[3] Beklemishev, L., *Positive provability logic for uniform reflection principles*, Annals of Pure and Applied Logic **165** (2014), pp. 82–105.

[4] Beklemishev, L., *A note on strictly positive logics and word rewriting systems*, in: S. Odintsov, editor, *Larisa Maksimova on Implication, Interpolation, and Definability*, Outstanding Contributions to Logic **15**, Springer, 2018 pp. 61–70.

[5] Birkhoff, G., *On the structure of abstract algebras*, Proc. Cambridge Phil. Soc. **31** (1935), pp. 433–454.

[6] Bull, R., *That all normal extensions of S4.3 have the finite model property*, Zeitschrift für Mathematische Logik und Grundlagen der Mathematik **12** (1966), pp. 341–344.

[7] Chagrov, A. and M. Zakharyaschev, "Modal Logic," Oxford Logic Guides **35**, Clarendon Press, Oxford, 1997.

[8] Dashkov, E., *On the positive fragment of the polymodal provability logic* GLP, Mathematical Notes **91** (2012), pp. 318–333.

[9] Davey, B., M. Jackson, J. Pitkethly and M. Talukder, *Natural dualities for three classes of relational structures*, Algebra Universalis (2007), pp. 1–22.

[10] Fine, K., *The logics containing S*4.3, Zeitschrift für Mathematische Logik und Grundlagen der Mathematik **17** (1971), pp. 371–376.

[11] Fine, K., *Logics containing K4, part II*, Journal of Symbolic Logic **50** (1985), pp. 619–651.

[12] Goldblatt, R., *Varieties of complex algebras*, Annals of Pure and Applied Logic **44** (1989), pp. 173–242.

[13] Grätzer, G., "Universal Algebra," Springer, 1979, 2nd edition.

[14] Hosoi, T., *On intermediate logics*, Journal of the Faculty of Science, University of Tokyo **14** (1967), pp. 293–312.

[15] Jackson, M., *Semilattices with closure*, Algebra Universalis **52** (2004), pp. 1–37.

[16] Kikot, S., A. Kurucz, Y. Tanaka, F. Wolter and M. Zakharyaschev, *Kripke completeness of strictly positive modal logics over meet-semilattices with operators*, CoRR **abs/1708.03403** (2017). URL `http://arxiv.org/abs/1708.03403`

[17] Kurucz, A., F. Wolter and M. Zakharyaschev, *Islands of tractability for relational constraints: Towards dichotomy results for the description logic $\mathcal{EL}$*, in: L. Beklemishev, V. Goranko and V. Shehtman, editors, *Advances in Modal Logic, vol. 8* (2010), pp. 271–291.

[18] Maksimova, L., *Modal logics of finite slices*, Algebra and Logic **14** (1975), pp. 188–197.

[19] Maksimova, L. and V. Rybakov, *Lattices of modal logics*, Algebra and Logic **13** (1974), pp. 105–122.

[20] Sahlqvist, H., *Completeness and correspondence in the first and second order semantics for modal logic*, in: S. Kanger, editor, *Procs. of the 3rd Scandinavian Logic Symposium*, North-Holland, 1975 pp. 110–143.

[21] Sofronie-Stokkermans, V., *Locality and subsumption testing in $\mathcal{EL}$ and some of its extensions*, in: C. Areces and R. Goldblatt, editors, *Advances in Modal Logic, vol. 7* (2008), pp. 315–339.

[22] Spaan, E., "Complexity of Modal Logics," Ph.D. thesis, Department of Mathematics and Computer Science, University of Amsterdam (1993).

[23] Svyatlovskiy, M., *Axiomatization and polynomial solvability of strictly positive fragments of certain modal logics*, Mathematical Notes, **103** (2018), pp. 952–967.

[24] Zakharyaschev, M., *Canonical formulas for K4. Part II: Cofinal subframe logics*, Journal of Symbolic Logic **61** (1996), pp. 421–449.

Normal Extensions of KTB of Codimension 3

James Koussas

Department of Mathematics and Statistics
La Trobe University
Melbourne, Australia

Tomasz Kowalski

Department of Mathematics and Statistics
La Trobe University
Melbourne, Australia

Yutaka Miyazaki

Osaka University of Economics and Law
Osaka, Japan

Michael Stevens

Research School of Information Sciences and Engineering
Australian National University
Canberra, Australia

Abstract

It is known that in the lattice of normal extensions of the logic **KTB** there are unique logics of codimensions 1 and 2, namely, the logic of a single reflexive point, and the logic of the total relation on two points. A natural question arises about the cardinality of the set of normal extensions of **KTB** of codimension 3. Generalising two finite examples found by a computer search, we construct an uncountable family of (countable) graphs, and prove that certain frames based on these produce a continuum of normal extensions of **KTB** of codimension 3. We use algebraic methods, which in this case turn out to be better suited to the task than frame-theoretic ones.

Keywords: Normal extensions, KTB-algebras, Subvarieties

1 Introduction

The Kripke semantics of **KTB** is the class of reflexive and symmetric frames, that is, frames whose accessibility relation is a *tolerance*. Since irreflexivity is not modally definable, it can be argued that **KTB** is *the* logic of simple graphs. Yet **KTB** is much less investigated that its transitive cousins, and in fact certain tools working very well for transitive logics (for example, canonical formulas) have no **KTB** counterparts working nearly as well. Among the articles dealing specifically with **KTB** and its extensions, Kripke incompleteness in various guises was investigated in [15] and [6], interpolation in [7] and [9], normal forms in [16], and splittings in [17], [10] and [8]. In the present article we focus on the upper part of the lattice of normal (axiomatic) extensions of **KTB**, or viewed dually, the lower part of the lattice of subvarieties of the corresponding variety of modal algebras.

The article is centred around a single construction, so it is structured rather simply: in the present section we give necessary preliminaries, in Section 2 we outline the history of the problem, in Section 3 we present the main construction and in Section 4 we draw the conclusion that there are uncountably many extensions of **KTB** of codimension 3.

Although we will use algebraic methods, we wish to move rather freely between graphs, frames and algebras. To make these transitions smooth we now establish a few conventions, the general principle behind them being that italic capitals stand for graphs, blackboard bold capitals for Kripke frames, and boldface capitals for algebras. With every simple graph $G = \langle V; E \rangle$, finite or infinite, we associate a Kripke frame $\mathbb{G}$ with the same universe and the reflexive closure of E as the accessibility relation. For example, $\mathbb{K}_i$ will be a looped version of K_i, the complete graph on i vertices. Thus, $\mathbb{K}_1$ is a single reflexive point, and $\mathbb{K}_2$ a two-element cluster. We will refer to these frames simply as graphs, unless the context calls for disambiguation. For a graph $\mathbb{G}$, we will write $\mathsf{Cm}(\mathbb{G})$, to denote its complex algebra. The figure below illustrates our conventions.

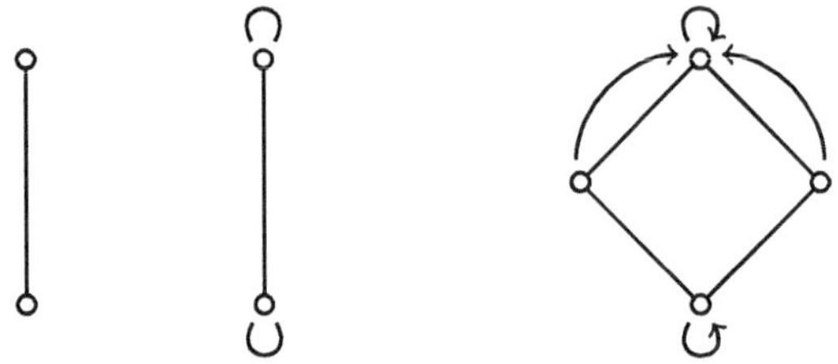

Fig. 1. Diagrams of K_2, $\mathbb{K}_2$ and $\mathsf{Cm}(\mathbb{K}_2)$.

If $\mathbb{G}$ is infinite, $\mathsf{Cm}(\mathbb{G})$ will typically be too big for our purposes, but certain special subalgebras of $\mathsf{Cm}(\mathbb{G})$ will play a critical role. These algebras are mathematically the same as general (descriptive) frames over $\mathbb{G}$, so the machinery of bounded morphisms reduces in these cases to verifying whether the identity map is one. The identity map is of course frame-theoretically invisible, so all that remains is algebra. This is essentially why algebraic methods are better suited to the task.

We assume familiarity with the basics of universal algebra and model theory. To be more precise, ultraproducts and Łoś Theorem, Jónsson's Lemma for congruence-distributive varieties, and some consequences of the congruence extension property will suffice. All of these concepts are covered in [1] and [3]. Our algebraic notation is standard: we use upright I, H, S, P, and $\mathrm{P_U}$ for the usual class operators of taking isomorphic copies, homomorphic images, subalgebras, direct products and ultraproducts, respectively. We also write $\mathrm{Si}(\mathcal{C})$ for the class of subdirectly irreducible algebras in $\mathcal{C}$. The variety generated by a class of algebras $\mathcal{C}$ we denote by $\mathrm{Var}(\mathcal{C})$, so Var is a shorthand for HSP. When we deal with Boolean algebras of sets, we use the standard set theoretical $\cup$ and $\cap$, and we write $\sim X$ instead of $\neg X$ for the complement of X.

1.1 KTB-algebras

A **KTB**-*algebra* is an algebraic structure $\mathbf{A} = \langle A; \vee, \wedge, \neg, \Diamond, 0, 1\rangle$ such that $\langle A; \vee, \wedge, \neg, 0, 1\rangle$ is a Boolean algebra, and $\Diamond$ a unary operation satisfying the following conditions:

(i) $\Diamond 0 = 0$,

(ii) $\Diamond(x \vee y) = \Diamond x \vee \Diamond y$,

(iii) $x \leqslant \Diamond x$,

(iv) $x \leqslant \Box\Diamond x$,

where $\Box$, as usual, stands for $\neg\Diamond\neg$ The last two conditions can be rendered as identities and so the class of **KTB**-algebras is a variety, which we will denote by $\mathcal{B}$. The inequality (iv) is also equivalent to:

$$x \wedge \Diamond y = 0 \iff \Diamond x \wedge y = 0.$$

Therefore, $\Diamond$ is a *self-conjugate operator* in the sense of [4], [5] and so $\mathcal{B}$ is a variety of self-conjugate Boolean Algebras with Operators (BAOs). Incidentally, the equational axiomatisation above is equivalent to the quasiequational one below:

(1) $x \leqslant y \Longrightarrow \Diamond x \leqslant \Diamond y$,

(2) $x \leqslant \Diamond x$,

(3) $x \leqslant \Box\Diamond x$.

For completeness, we include the following well known propositions (see [4], [13], [1] and [3] for proofs and useful exercises). The first two deal with **KTB**-algebras, and the third one recalls some crucial facts from universal algebra.

Proposition 1.1 *For any graph $G = \langle V; E\rangle$, the algebra $\mathsf{Cm}(\mathbb{G})$ is a* **KTB**-*algebra. The class of all such algebras generates the variety $\mathcal{B}$.*

Proposition 1.2 *The variety $\mathcal{B}$ is congruence distributive and has the congruence extension property.*

Proposition 1.3 *Let $\mathcal{V}$ be a variety of algebras, and $\mathcal{C}$ a subclass of $\mathcal{V}$.*

(i) *If $\mathcal{V}$ has the congruence extension property, $\mathbf{A}$ is a simple algebra in $\mathcal{V}$ and $\mathbf{B} \in \mathrm{IS}(\mathbf{A})$, then $\mathbf{B}$ is simple.*

(ii) *If $\mathcal{V}$ has the congruence extension property, then $\mathrm{HS}(\mathcal{C}) = \mathrm{SH}(\mathcal{C})$.*

(iii) *If* $\mathcal{V}$ *is congruence distributive, then* $\mathrm{Si}(\mathrm{Var}(\mathcal{C})) = \mathrm{Si}(\mathrm{HSP_U}(\mathcal{C}))$.

(iv) *We have* $\mathcal{V} = \mathrm{Var}(\mathrm{Si}(\mathcal{V}))$.

As usual, we define the term operations $\Diamond^n$, one for each n, recursively, putting $\Diamond^0 x = x$ and $\Diamond^{n+1}x = \Diamond\Diamond^n x$.

Definition 1.4 Let $\mathbf{B} = \langle B; \vee, \wedge, \neg, \Diamond, 0, 1\rangle \in \mathcal{B}$. Then the map $\gamma\colon B \to B$ given by $\gamma(x) = \Box\Diamond x$ is a closure operator on $\mathbf{B}$, which we call the *natural closure operator* on $\mathbf{B}$.

The following properties of natural closure operators will be useful.

Lemma 1.5 *Let* $\mathbf{B} = \langle B; \vee, \wedge, \neg, \Diamond, 0, 1\rangle \in \mathcal{B}$ *and let* γ *denote the natural closure operator on* $\mathbf{B}$.

(i) *If* $x \in B$ *is* γ*-closed, then* $\neg x = \Diamond\neg\Diamond x$ *and* $\Diamond\neg x = \Diamond^2\neg\Diamond x$.

(ii) *If* $x \in B$, *then* $\Diamond\gamma(x) = \Diamond x$.

Proof. Let $x \in B$. If x is γ-closed, then $x = \Box\Diamond x$, thus $\neg x = \Diamond\neg\Diamond x$ and so $\Diamond\neg x = \Diamond^2\neg\Diamond x$, hence (i) holds.

As γ is a closure operator, we have $x \leqslant \gamma(x)$, hence $\Diamond x \leqslant \Diamond\gamma(x)$. Similarly, $\neg\Diamond x \leqslant \gamma(\neg\Diamond x) = \Box\Diamond\neg\Diamond x = \neg\Diamond\Box\Diamond x = \neg\Diamond\gamma(x)$, so $\Diamond\gamma(x) \leqslant \Diamond x$. Thus, $\Diamond\gamma(x) = \Diamond x$, hence (ii) holds. □

Lemma 1.6 *Let* $\mathbf{B} = \langle B; \vee, \wedge, \neg, \Diamond, 0, 1\rangle \in \mathcal{B}$ *and let* γ *be the natural closure operator of* $\mathbf{B}$. *If* $\mathbf{B} \models \exists x\colon x \neq 0 \ \&\ \Diamond x \neq 1$ *and* $\mathbf{B} \models \forall x\colon x \neq 0 \to \Diamond^n x = 1$, *for some* $n \in \omega \setminus \{0\}$, *then there is a* γ*-closed* $y \in B$ *with* $\Diamond y \neq 1$ *and* $\Diamond^2 y = 1$.

Proof. Let x be a witness of $\exists x\colon x \neq 0 \ \&\ \Diamond x \neq 1$ in $\mathbf{B}$. By assumption, $\mathbf{B} \models \forall x\colon x \neq 0 \to \Diamond^n x = 1$, so we must have $\Diamond^n x = 1$. Hence, there is some $m \in \{1, \ldots, n-1\}$ with $\Diamond^m x \neq 1$ and $\Diamond^{m+1}x = 1$. By Lemma 1.5(ii), $\Diamond\gamma(\Diamond^{m-1}x) = \Diamond^m x \neq 1$ and $\Diamond^{m+1}x = 1$. Since $\gamma(\Diamond^{m-1}x)$ is γ-closed, putting $y = \gamma(\Diamond^{m-1}(x))$, we get a γ-closed $y \in B$ with $\Diamond y \neq 1$ and $\Diamond^2 y = 1$, as required. □

2 The history of the problem

A logic L is said to have codimension n, in some lattice Λ of logics, if there exists a descending chain $L_0 \succ \cdots \succ L_n$ of logics from Λ, such that L_0 is inconsistent, $L_n = L$, and L_{i-1} covers L_i for each $i \in \{0, \ldots, n\}$. Lattices of nonclassical logics are typically very complicated, so looking at logics of small codimensions is one way of analysing these lattices. In particular, finding the smallest n for which there are uncountably many logics of codimension n in Λ indicates at which level the lattice gets really badly complicated.

Let NExt(**KTB**) stand for the lattice of normal extensions of **KTB**, where we identify logics with their sets of theorems. We intend to show that for $\Lambda = \mathrm{NExt}(\mathbf{KTB})$ the smallest such n is 3.

Remark 2.1 If we identified logics with their *consequence operations*, rather than their sets of theorems, NExt(**KTB**) would be the the lattice of normal *axiomatic* extensions of **KTB**. Let us call the lattice of all normal extensions of

KTB, whether axiomatic or not, CNExt(**KTB**). Then NExt(**KTB**) is a subposet of CNExt(**KTB**). However, the codimension of a logic $L \in$ NExt(**KTB**) can be smaller in NExt(**KTB**) than the codimension of L in CNExt(**KTB**). It follows from results of Blanco, Campercholi and Vaggione (see Theorem 1 in [2]) that for any logic $L \in$ NExt(**KTB**) of codimension at least 2, NExt(L) is *strictly* contained in CNExt(L).

Let Subv($\mathcal{B}$) stand for the lattice of subvarieties of $\mathcal{B}$. Then, the usual dual isomorphism between NExt(**KTB**) and Subv($\mathcal{B}$) holds, and therefore logics of codimension n in NExt(**KTB**) correspond to varieties of height n in Subv($\mathcal{B}$). The next theorem gives a complete picture of Subv($\mathcal{B}$) up to height 2, and therefore, dually, of NExt(**KTB**) down to codimension 2. The second statement in the theorem is due to the third author (see [17]).

Theorem 2.2 *The lattice* Subv($\mathcal{B}$) *has exactly one atom, namely* Var($\mathsf{Cm}(\mathbb{K}_1)$). *This atom in turn has exactly one cover, namely* Var($\mathsf{Cm}(\mathbb{K}_2)$).

A natural question then arises about the cardinality of the "set" of varieties covering Var($\mathsf{Cm}(\mathbb{K}_2)$). It is easy to show that this "set" is infinite: countably many varieties covering Var($\mathsf{Cm}(\mathbb{K}_2)$) were constructed by the second and fourth author in an unpublished note [12], using certain finite graphs. But finite graphs clearly could not suffice for a construction of uncountably many varieties covering Var($\mathsf{Cm}(\mathbb{K}_2)$). A construction of an appropriate uncountable family of countably infinite graphs began by finding two finite ones, called below $\mathbb{G}_1$ and $\mathbb{G}_2$:

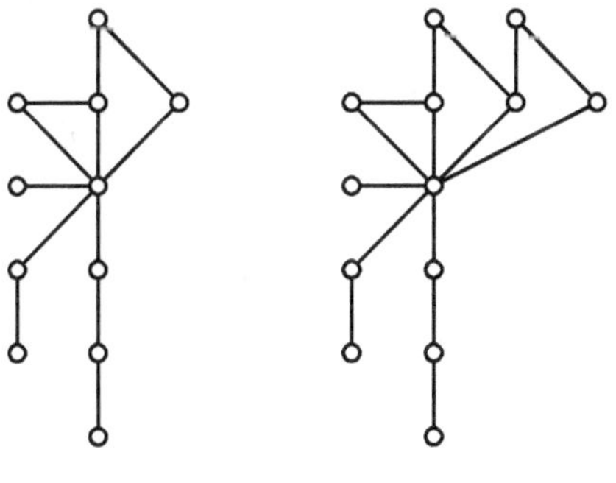

Fig. 2. Graph drawings of $\mathbb{G}_1$ and $\mathbb{G}_2$ (with loops omitted).

These were found by the second and fourth authors through a computer search, performed with the help of Brendan McKay's `nauty` (see [14]). All non-isomorphic graphs with up to 13 vertices were generated, and checked for the property of not admitting any bounded morphism, except the identity map, the constant map onto $\mathbb{K}_1$, and a bounded morphism onto $\mathbb{K}_2$. By finiteness, this is sufficient (and also necessary) for the logic of such a graph $\mathbb{G}$ to be of codimension 3, or, equivalently, for Var($\mathsf{Cm}(\mathbb{G})$) to be a cover of Var($\mathsf{Cm}(\mathbb{K}_2)$).

Two of these graphs are depicted in Figure 2. They were the only ones that revealed a workable family resemblance to one another. They were also so different from the finite graphs considered in [12] as to be completely unexpected to the finders. Verifying by hand that the bounded morphism condition mentioned above indeed holds, is tedious but not difficult, and so it was proved that $\mathrm{Var}(\mathsf{Cm}(\mathbb{G}_1))$ and $\mathrm{Var}(\mathsf{Cm}(\mathbb{G}_2))$ indeed cover $\mathrm{Var}(\mathsf{Cm}(\mathbb{K}_2))$, confirming the computer-assisted finding.

Extending the zigzaging pattern infinitely to the right is then a no-brainer, and a suitable twisting of the zigzag produces an uncountable family of pairwise non-isomorphic graphs. The next step is to take certain subalgebras of the complex algebras of these infinite graphs (unlike in the finite case, the full complex algebras may not do), and prove that the varieties they generate are pairwise distinct and cover $\mathrm{Var}(\mathsf{Cm}(\mathbb{K}_2))$. The three last authors did produce a rough approximation to a proof, which was convincing enough (for them) to announce the result (see [11]). However, the full proof was never published, and in fact it did not exist, as the details were never satisfactorily verified. The three authors dispersed around the globe and the proof was left unfinished. It took about 10 years, and the first author, to produce a complete proof. We are going to present it now.

3 Construction

Before we begin, we make one more remark on the methods. The construction presented below may at first glance suggest that the reasoning about ultrapowers, which will play an important part in the proofs, is not necessary, because everything that could go wrong in an ultrapower already goes wrong in the original algebra. Were it so, the proofs could be greatly simplified, but unfortunately the first glance is misleading. There exists an infinite **KTB**-algebra $\mathbf{A}$ such that $\mathrm{HS}(\mathbf{A})$ does not contain $\mathsf{Cm}(\mathbb{K}_3)$, but $\mathrm{HSP_U}(\mathbf{A})$ does, so $\mathbf{A}$ does not generate a cover of $\mathrm{Var}(\mathbb{K}_2)$. Considering ultrapowers is necessary, at least in principle.

Now, for the construction. Firstly, we will need the following Lemma, which is an easy consequence of Proposition 1.3(iii).

Lemma 3.1 *We have* $\mathrm{Si}(\mathrm{Var}(\mathsf{Cm}(\mathbb{K}_2))) = \mathrm{I}(\{\mathsf{Cm}(\mathbb{K}_1), \mathsf{Cm}(\mathbb{K}_2)\})$.

Next, we state a sufficient set of conditions for an algebra in $\mathcal{B}$ to generate a variety of height 3.

Lemma 3.2 *Let* $\mathbf{A} \in \mathcal{B}$ *and assume that* $\mathbf{A}$ *has the following properties:*

(i) $\mathbf{A}$ *is infinite;*

(ii) $\mathsf{Cm}(\mathbb{K}_2) \in \mathrm{IS}(\mathbf{A})$*;*

(iii) *every member of* $\mathrm{P_U}(\mathbf{A})$ *is simple;*

(iv) *for all* $\mathbf{B} \in \mathrm{ISP_U}(\mathbf{A})$, *we have* $\mathbf{B} \cong \mathsf{Cm}(\mathbb{K}_1)$, $\mathbf{B} \cong \mathsf{Cm}(\mathbb{K}_2)$ *or* $\mathbf{A} \in \mathrm{IS}(\mathbf{B})$.

Then $\mathrm{Var}(\mathbf{A})$ *is of height 3.*

Proof. Based on (iii), $\mathrm{HP_U}(\mathbf{A}) = \mathrm{I}(\{\mathbf{T}\} \cup \mathrm{P_U}(\mathbf{A}))$, for some trivial $\mathbf{T} \in \mathcal{B}$. So, by Proposition 1.3, $\mathrm{Si}(\mathrm{Var}(\mathbf{A})) = \mathrm{Si}(\mathrm{HSP_U}(\mathbf{A})) = \mathrm{Si}(\mathrm{SHP_U}(\mathbf{A})) = \mathrm{ISP_U}(\mathbf{A})$. Clearly, $\mathbf{A} \in \mathrm{ISP_U}(\mathbf{A})$, so (i), (ii) and Lemma 3.1 tell us that $\mathrm{Var}(\mathbf{A})$ properly extends $\mathrm{Var}(\mathsf{Cm}(\mathbb{K}_2))$. Let $\mathcal{V}$ be a variety with $\mathrm{Var}(\mathsf{Cm}(\mathbb{K}_2)) \subseteq \mathcal{V} \subseteq \mathrm{Var}(\mathbf{A})$. Since $\mathcal{V} \subseteq \mathrm{Var}(\mathbf{A})$, we have $\mathrm{Si}(\mathcal{V}) \subseteq \mathrm{Si}(\mathrm{Var}(\mathbf{A})) = \mathrm{ISP_U}(\mathbf{A})$. Combining this with (iv) and Lemma 3.1, we find that $\mathrm{Si}(\mathcal{V}) \subseteq \mathrm{Si}(\mathsf{Cm}(\mathbb{K}_2))$ or $\mathbf{A} \in \mathrm{IS}(\mathcal{V}) = \mathcal{V}$. So, by Proposition 1.3, we must have $\mathcal{V} = \mathrm{Var}(\mathsf{Cm}(\mathbb{K}_2))$ or $\mathcal{V} = \mathrm{Var}(\mathbf{A})$. Hence, $\mathrm{Var}(\mathbf{A})$ covers $\mathrm{Var}(\mathsf{Cm}(\mathbb{K}_2))$, so $\mathrm{Var}(\mathbf{A})$ has height 3, as claimed. □

Our construction of a continuum of subvarieties of $\mathcal{B}$ of height 3 begins with the following definition.

Definition 3.3 Let $\mathbb{E}$ denote the set of positive even numbers, let $A = \{a\}$, $B = \{b_1, b_2, b_3\}$, $C = \{c_1, c_2\}$, $D = \{d\}$, $U = \{u_i \mid i \in \omega \setminus \{0\}\}$ and $L = \{\ell_i \mid i \in \omega\}$ be pairwise disjoint, and assume that $u_i \neq u_j$, $\ell_i \neq \ell_j$, $b_i \neq b_j$ and $c_i \neq c_j$ whenever $i \neq j$. Now, define $U_i := \{u_i\}$, for all $i \in \omega \setminus \{0\}$, $L_i := \{\ell_i\}$, for all $i \in \omega$, $B_i := \{b_i\}$, for all $i \in \{1,2,3\}$, $C_i := \{c_i\}$, for all $i \in \{1,2\}$, and $P := \{b_1, c_1, d\}$. For each $N \subseteq \mathbb{E}$, let $\mathbb{F}_N$ be the graph $\langle W; R_N \rangle$, where $W := A \cup B \cup C \cup D \cup U \cup L$ and R_N is the relation defined by

$$x \mathrel{R_N} y \iff x = y \text{ or } \{x, y\} = \begin{cases} \{a, b_i\}, \text{ for some } i \in \{1,2,3\}, \\ \{b_i, c_i\}, \text{ for some } i \in \{1,2\}, \\ \{c_1, d\}, \\ \{\ell_0, \ell_1\}, \\ \{a, \ell_i\}, \text{ for some } i \in \omega, \\ \{\ell_i, u_i\}, \text{ for some } i \in \omega \setminus \{0\}, \\ \{\ell_i, u_{i-1}\}, \text{ for some } i \in \mathbb{E}, \\ \{\ell_i, u_{i+1}\}, \text{ for some } i \in N \text{ or} \\ \{\ell_{i+1}, u_i\}, \text{ for some } i \in \mathbb{E} \setminus N. \end{cases}$$

As usual with graphs, a picture is worth a thousand words. Certainly it is worth all the words of the definition above. Here it is.

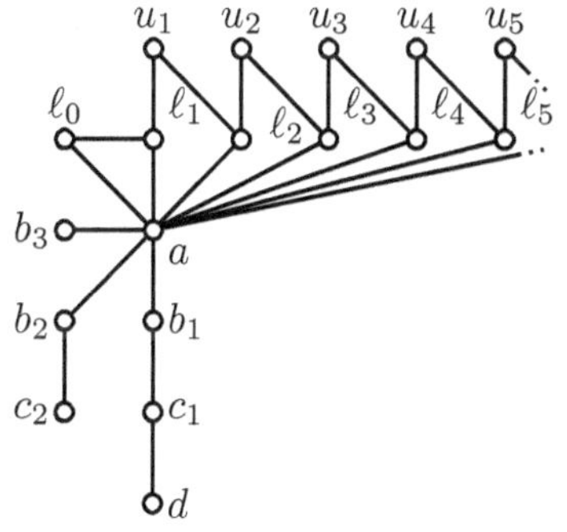

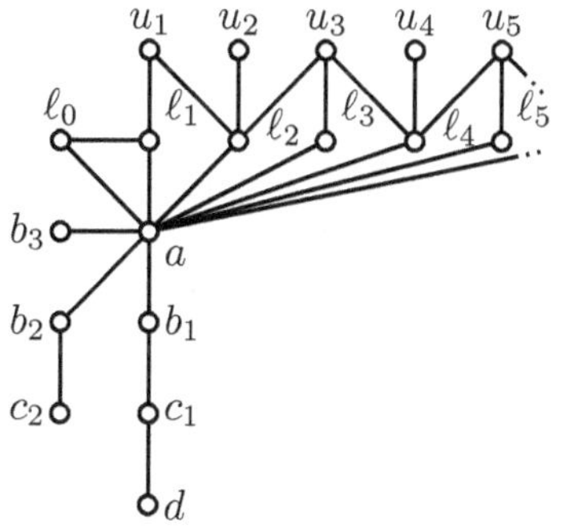

Fig. 3. Graph drawings of (finite sections of) $\mathbb{F}_\varnothing$ and $\mathbb{F}_{\{2,4\}}$ (with loops omitted).

Accordingly, in the proofs, we will frequently refer to Fig. 3, as well as to Fig. 4 below, rather than to Definition 3.3. Next, we define the algebras essential to our construction. The notation is as in Definition 3.3.

Definition 3.4 For each $N \subseteq \mathbb{E}$, let $\mathbf{D}_N$ be the subalgebra of $\mathsf{Cm}(\mathbb{F}_N)$ generated by D and let D_N be the universe of $\mathbf{D}_N$.

From now on, we will use $\Diamond_N$ to stand for R_N^{-1}, and we will omit the subscript N if there is no danger of confusion.

Lemma 3.5 *Let $N \subseteq \mathbb{E}$. Then $A, B_1, B_3, C_1, D, L_i, U_j, B_2 \cup L, C_2 \cup U \in D_N$, for all $i \in \omega$ and all $j \in \omega \setminus \{0\}$.*

Proof. By definition, $D \in D_N$. So, based on Fig. 3, $C_1 = \Diamond D \cap {\sim} D \in D_N$. Similarly, $B_1 = \Diamond C_1 \cap {\sim}\Diamond D \in D_N$ and $C_2 \cup U = {\sim}\Diamond^4 D \in D_N$, hence we have $A = \Diamond B_1 \cap {\sim}\Diamond C_1 \in D_N$ and $B_3 = {\sim}(\Diamond^2(C_2 \cup U) \cup \Diamond^2 D) \in D_N$. From this, it follows that $L_0 = {\sim}(B_3 \cup \Diamond(C_2 \cup U) \cup \Diamond^3 D) \in D_N$, so we have $B_2 \cup L = (\Diamond(C_2 \cup U) \cup L_0) \cap {\sim}(C_2 \cup U) \in D_N$. Similarly, we must have $L_1 = \Diamond L_0 \cap {\sim}(A \cup L_0) \in D_N$, which implies that $U_1 = \Diamond L_1 \cap {\sim}\Diamond A \in D_N$.

It remains to establish that $L_i, U_j \in D_N$, for all $i \in \omega$ and all $j \in \omega \setminus \{0\}$; we proceed by induction. Assume that $L_i, U_i \in D_N$, for some odd $i \in \omega$.

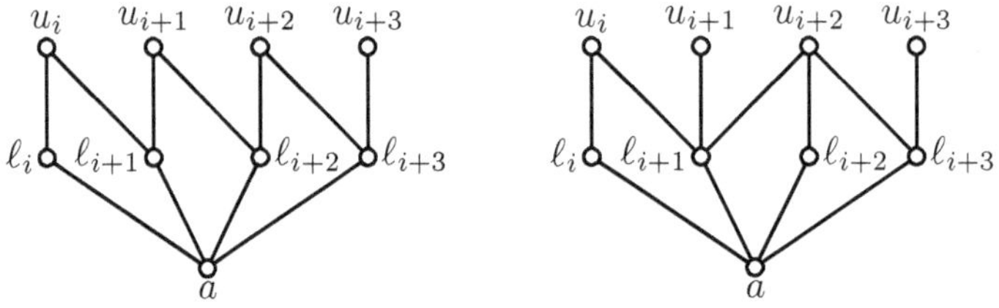

Fig. 4. Graph drawings for Lemma 3.5.

Firstly, assume that $i+1 \notin N$. By Fig. 4, $L_{i+1} = \Diamond U_i \cap {\sim}\Diamond L_i \in D_N$, which implies that $U_{i+1} = \Diamond L_{i+1} \cap {\sim}(\Diamond A \cup U_i) \in D_N$. This implies that $L_{i+2} = \Diamond U_{i+1} \cap {\sim}\Diamond L_{i+1} \in D_N$, so $U_{i+2} = \Diamond L_{i+2} \cap {\sim}(\Diamond A \cup U_{i+1}) \in D_N$. Thus, $L_{i+1}, L_{i+2}, U_{i+1}, U_{i+2} \in D_N$ if $i+1 \notin N$.

Next, assume that $i+1 \in N$. From Fig. 4, $L_{i+1} = \Diamond U_i \cap {\sim}\Diamond L_i \in D_N$, so we have $U_{i+1} \cup U_{i+2} = \Diamond L_{i+1} \cap {\sim}(\Diamond A \cup U_i) \in D_N$. Using these results, we find that we must have $L_{i+2} \cup L_{i+3} = \Diamond(U_{i+1} \cup U_{i+2}) \cap {\sim}\Diamond L_{i+1} \in D_N$. From this, it follows that $U_{i+2} = (U_{i+1} \cup U_{i+2}) \cap \Diamond(L_{i+2} \cup L_{i+3}) \in D_N$, which implies that $U_{i+1} = (U_{i+1} \cup U_{i+2}) \cap {\sim}U_{i+2} \in D_N$. From these results, $X := \Diamond(L_{i+2} \cup L_{i+3}) \cap {\sim}(\Diamond A \cup U_{i+2}) \in D_N$. Based on Fig 4, we must have $u_{i+3} \in X$ and $a, \ell_{i+2}, u_{i+2} \notin X$, hence $\ell_{i+2} \notin \Diamond X$ and $\ell_{i+3} \in \Diamond X$. Thus, $L_{i+2} = (L_{i+2} \cup L_{i+3}) \cap {\sim}\Diamond X \in D_N$, so $L_{i+1}, L_{i+2}, U_{i+1}, U_{i+2} \in D_N$ if $i+1 \in N$.

In every case, we have $L_{i+1}, L_{i+2}, U_{i+1}, U_{i+2} \in D_N$. Hence, by induction, $U_i, L_j \in D_N$, for all $i \in \omega$ and all $j \in \omega \setminus \{0\}$, so we are done. □

Corollary 3.6 *Let $N \subseteq \mathbb{E}$. Then the algebra $\mathbf{D}_N$ is infinite.*

Lemma 3.7 *Let $N \subseteq \mathbb{E}$. Then $\mathsf{Cm}(\mathbb{K}_2) \in \mathrm{IS}(\mathbf{D}_N)$.*

Proof. From Lemma 3.5, it follows that $X := B \cup D \cup L \in D_N$. Based on Fig. 1 and Fig. 4, the subalgebra of $\mathbf{D}_N$ generated by X is isomorphic to $\mathsf{Cm}(\mathbb{K}_2)$, hence $\mathsf{Cm}(\mathbb{K}_2) \in \mathrm{IS}(\mathbf{D}_N)$, as claimed. □

Based on Fig 4, if $N \subseteq \mathbb{E}$, then every vertex other than d is joined to a by a path of length of at most 2 in $\mathbb{F}_N$, so $\mathbf{D}_N \models \forall x\colon x \neq 0 \to \Diamond^5 x = 1$. The following Lemma is an easy consequence of this observation and Łoś's Theorem.

Lemma 3.8 *Let $N \subseteq \mathbb{E}$. Then every member of $\mathrm{P_U}(\mathbf{D}_N)$ is simple.*

Lemma 3.9 *Let $N \subseteq \mathbb{E}$, let F be an ultrafilter over a set I and let $\mathbf{S}$ be a subalgebra of $\mathbf{D}_N^I/F$. If $\mathbf{S} \models \forall x\colon x \neq 0 \to \Diamond x = 1$, then $\mathbf{S} \cong \mathsf{Cm}(\mathbb{K}_1)$ or $\mathbf{S} \cong \mathsf{Cm}(\mathbb{K}_2)$.*

Proof. Let S be the universe of $\mathbf{S}$ and define a map $\bar{X}\colon I \to D_N$ by $i \mapsto X$, for each $X \in D_N$. Suppose, for a contradiction, that there exist $X, Y \in D_N^I$ with $X/F, Y/F \in S \setminus \{\bar{\varnothing}/F, \bar{W}/F\}$, $X/F \neq Y/F$ and $X/F \neq \neg Y/F$. Clearly, we must have $\{i \in I \mid d \in X(i)\} \cup \{i \in I \mid d \in {\sim}X(i)\} = I \in F$, which implies that $\{i \in I \mid d \in X(i)\} \in F$ or $\{i \in I \mid d \in {\sim}X(i)\} \in F$. Similarly, $\{i \in I \mid d \in Y(i)\} \in F$ or $\{i \in I \mid d \in {\sim}Y(i)\} \in F$. Without loss of generality, we can assume that both $\{i \in I \mid d \in X(i)\} \in F$ and $\{i \in I \mid d \in Y(i)\} \in F$, since we can interchange X with $\neg X$ and Y with $\neg Y$ (if necessary).

Clearly, $\neg X/F \neq \bar{\varnothing}/F$ and $\neg Y/F \neq \bar{\varnothing}/F$, so $\Diamond\neg X/F = \bar{W}/F = \Diamond\neg Y/F$, since $\mathbf{S} \models \forall x\colon x \neq 0 \to \Diamond x = 1$. We have $\{i \in I \mid d \in X(i)\} \in F$ and $\{i \in I \mid d \in Y(i)\} \in F$, so $\{i \in I \mid d \notin {\sim}X(i)\} \in F$ and $\{i \in I \mid d \notin {\sim}Y(i)\} \in F$. By Fig. 3, $\{i \in I \mid c_1 \in {\sim}X(i)\} \in F$ and $\{i \in I \mid c_1 \in {\sim}Y(i)\} \in F$. Thus, $\{i \in I \mid c_1, d \notin X(i) \cap {\sim}Y(i)\} \in F$ and $\{i \in I \mid c_1, d \notin {\sim}X(i) \cap Y(i)\} \in F$. By Fig. 3, $\{i \in I \mid d \notin \Diamond(X \wedge \neg Y)(i)\} \in F$ and $\{i \in I \mid d \notin \Diamond(\neg X \wedge Y)(i)\} \in F$, hence $\Diamond(X/F \wedge \neg Y/F) \neq \bar{W}/F$ and $\Diamond(\neg X/F \wedge Y/F) \neq \bar{W}/F$. Since $X/F \neq Y/F$ and $X/F \neq \neg Y/F$, it follows that $X/F \wedge \neg Y/F \neq \bar{\varnothing}/F$ or $\neg X/F \wedge Y/F \neq \varnothing/F$, so this contradicts the fact that $\mathbf{S} \models \forall x\colon x \neq 0 \to \Diamond x = 1$. Thus, we must have $|S| \leqslant 4$. Since $\mathbf{S} \models \forall x\colon x \neq 0 \to \Diamond x = 1$ and $\mathbf{D}_N$ has no trivial subalgebras, this implies that $\mathbf{S} \cong \mathsf{Cm}(\mathbb{K}_1)$ or $\mathbf{S} \cong \mathsf{Cm}(\mathbb{K}_2)$, as claimed. □

Lemma 3.10 *Let $N \subseteq \mathbb{E}$, let γ be the natural closure operator of $\mathbf{D}_N$ and let X be a γ-closed element of D_N with $\Diamond X \neq W$ and $\Diamond^2 X = W$. Then $\Diamond{\sim}X \neq W$.*

Proof. Firstly, assume that $a \in X$. Based on Fig. 3, we have $a, b_3 \in \Diamond X$, hence $a, b_3 \notin {\sim}\Diamond X$. So, by Lemma 1.5(i), $b_3 \notin \Diamond{\sim}\Diamond X = {\sim}X$, hence $a, b_3 \in X$. By Fig. 3, $b_3 \notin \Diamond{\sim}X$, which implies that $\Diamond{\sim}X \neq W$ if $a \in X$.

Now, assume that $a \notin X$. We claim that $a \in \Diamond X$; suppose that $a \notin \Diamond X$. By Fig. 3, we have $b_3 \notin X$, hence $a, b_3 \notin X$ and $a \notin \Diamond X$. Thus, $b_3 \notin \Diamond X$, which contradicts the fact that $\Diamond^2 X = W$. It follows that $a \in \Diamond X$, as claimed. By Fig. 3, we must have $b_2 \in X$ or $c_2 \in X$, as $a \notin X$ and $\Diamond^2 X = W$. Hence, $a, b_2, c_2 \in \Diamond X$, so by Lemma 1.5(i), we have $c_2 \notin \Diamond^2{\sim}\Diamond X = \Diamond{\sim}X$. From this, it follows that $\Diamond{\sim}X \neq W$ if $a \notin X$, so $\Diamond{\sim}X \neq W$, as claimed. □

Lemma 3.11 *Let $N \subseteq \mathbb{E}$, let F be an ultrafilter over a set I, let $\mathbf{S}$ be a subalgebra of $\mathbf{D}_N^I/F$, let S be the universe of $\mathbf{S}$, let $\bar{X}\colon I \to D_N^I$ be defined by $i \mapsto X$, for each $X \in D_N$, let γ be the natural closure operator of $\mathbf{S}$ and let $X \in D_N^I$ with $X/F \in S$ and $X/F \neq \bar{\varnothing}/F$.*

(i) *If $\{i \in I \mid P \cap X(i) = \varnothing\} \in F$, then $\bar{D}/F \in S$.*

(ii) *If* $\{i \in I \mid P \subseteq \Diamond X(i)\} \in F$ *and* $\Diamond X/F \neq \bar{W}/F$, *then* $\bar{D} \in S$.

(iii) *If* $\{i \in I \mid P \subseteq \Diamond{\sim}X(i)\} \in F$, $\Diamond X/F \neq \bar{W}/F$, $\Diamond^2 X/F = \bar{W}/F$ *and* X/F *is* γ*-closed, then* $\bar{D}/F \in S$.

Proof. Assume that $\{i \in I \mid P \cap X(i) = \varnothing\} \in F$. By Fig. 3, if $Y \in D_N \setminus \{\varnothing\}$ with $Y \cap P = \varnothing$, then $\Diamond^2 Y = {\sim}D$, $\Diamond^3 Y = {\sim}D$ or $\Diamond^4 Y = {\sim}D$. Thus, $\{i \in I \mid D = {\sim}\Diamond^2 X(i)\} \cup \{i \in I \mid D = {\sim}\Diamond^3 X(i)\} \cup \{i \in I \mid D = {\sim}\Diamond^4 X(i)\} = I \in F$, so we must have $\{i \in I \mid D = {\sim}\Diamond^2 X(i)\} \in F$, $\{i \in I \mid D = {\sim}\Diamond^3 X(i)\} \in F$ or $\{i \in I \mid D = {\sim}\Diamond^4 X(i)\} \in F$. Clearly, this implies that $\neg\Diamond^2 X/F = \bar{D}/F$, $\neg\Diamond^3 X/F = \bar{D}/F$ or $\neg\Diamond^4 X/F = \bar{D}/F$, so (i) holds.

Now, to prove (ii), assume that we have $\{i \in I \mid P \subseteq \Diamond X(i)\} \in F$ and $\Diamond X/F \neq \bar{W}/F$. Then $\{i \in I \mid P \cap {\sim}\Diamond X(i) = \varnothing\} \in F$, $\neg\Diamond X/F \neq \bar{\varnothing}/F$ and $\neg\Diamond X/F \in S$. By the previous result, $\bar{D}/F \in S$, so (ii) holds.

To prove (iii), assume that $\{i \in I \mid P \subseteq \Diamond{\sim}X(i)\} \in F$, $\Diamond X/F \neq \bar{W}/F$, $\Diamond^2 X/F = \bar{W}/F$ and X/F is γ-closed. From Lemma 3.10 and Łoś's Theorem, it follows that $\Diamond\neg X/F \neq \bar{W}/F$. So, based on the previous result, $\bar{D}/F \in S$. Thus, the three required results hold. □

Lemma 3.12 *Let* $N \subseteq \mathbb{E}$, *let* F *be an ultrafilter over a set* I *and let* $\mathbf{S}$ *be a subalgebra of* $\mathbf{D}_N^I/F$. *If* $\mathbf{S} \models \exists x\colon x \neq 0 \;\&\; \Diamond x \neq 1$, *then* $\mathbf{D}_N \in \mathrm{IS}(\mathbf{S})$.

Proof. Let S be the universe of $\mathbf{S}$, let γ be the natural closure operator of $\mathbf{S}$ and define a map $\bar{X}\colon I \to D_N$ by $i \mapsto X$, for each $X \in D_N$. Since D generates $\mathbf{D}_N$ and the natural diagonal map embeds $\mathbf{D}_N$ into $\mathbf{D}_N^I/F$, it will be enough to show that $\bar{D}/F \in S$.

By Lemma 1.6, there is some $X \in D_N^I$ such that $X/F \in S$, $\Diamond X/F \neq \bar{W}/F$, $\Diamond^2 X/F = \bar{W}/F$ and X/F is γ-closed, as $\mathbf{S} \models \exists x\colon x \neq 0 \;\&\; \Diamond x \neq 1$ and $\mathbf{S} \models \forall x\colon x \neq 0 \to \Diamond^5 x = 1$. If $Y \subseteq W$, we either have $c_2 \in Y$ or $c_2 \in {\sim}Y$, so by Fig. 3, we must have $P \subseteq \Diamond Y$ or $P \subseteq \Diamond{\sim}Y$ if $Y \subseteq W$. From this, it follows that $\{i \in I \mid P \subseteq \Diamond X(i)\} \cup \{i \in I \mid P \subseteq \Diamond{\sim}X(i)\} = I \in F$, so $\{i \in I \mid P \subseteq \Diamond X(i)\} \in F$ or $\{i \in I \mid P \subseteq \Diamond{\sim}X(i)\} \in F$. By Lemma 3.11, $\bar{D}/F \in S$ and we are done. □

Lemma 3.13 *Let* $N \subseteq \mathbb{E}$. *Then* $\mathrm{Var}(\mathbf{D}_N)$ *is of height 3.*

Proof. By Corollary 3.6 and Lemma 3.7, $\mathbf{D}_N$ is infinite and $\mathsf{Cm}(\mathbb{K}_2) \in \mathrm{IS}(\mathbf{D}_N)$. By Lemma 3.8, each element of $\mathrm{P_U}(\mathbf{D}_N)$ is simple. By Lemmas 3.9 and 3.12, we must have $\mathbf{B} \cong \mathsf{Cm}(\mathbb{K}_1)$, $\mathbf{B} \cong \mathsf{Cm}(\mathbb{K}_2)$ or $\mathbf{D}_N \in \mathrm{IS}(\mathbf{B})$ if $\mathbf{B} \in \mathrm{ISP_U}(\mathbf{D_N})$. So, by Lemma 3.2, $\mathrm{Var}(\mathbf{D}_N)$ has height 3, as claimed. □

Now it remains to show that for distinct $N, M \subseteq \mathbb{E}$, the varieties $\mathrm{Var}(\mathbf{D}_N)$ and $\mathrm{Var}(\mathbf{D}_M)$ are distinct.

Lemma 3.14 *Let* $N \subseteq \mathbb{E}$ *and let* $X \in D_N \setminus \{\varnothing\}$ *with* $\Diamond_N^4 X \neq W$. *Then* $X = D$ *or* $\Diamond_N^4 X = {\sim}D$.

Proof. Based on Fig. 3, if $a \in \Diamond X$ or $c_2 \in X$, then we must have $\Diamond^4 X = W$, hence $X \subseteq U \cup C_2 \cup D$. By Fig. 3, $\Diamond^4 D = W \setminus (C_2 \cup U) \neq W$. Similarly, $\Diamond^4 X = W$ if $d \in X$ and $X \cap (C_2 \cup U) \neq \varnothing$, and $\Diamond^4 X = {\sim}D$ if $X \subseteq C_2 \cup U$. Since $\Diamond^4 X \neq W$, we must have $X = D$ or $\Diamond^4 X = {\sim}D$, as claimed. □

Lemma 3.15 *Let $M, N \subseteq \mathbb{E}$ and let $u\colon \mathbf{D}_M \to \mathbf{D}_N$ be an embedding. Then $u(D) = D$.*

Proof. Suppose, for a contradiction, that $u(D) \neq D$. Since u is an embedding, $u(D) \neq \varnothing$. Based on Fig. 3, $\Diamond^4_N u(D) = u(\Diamond^4_M D) = u(W \setminus (C_2 \cup U)) \neq W$, since u is an embedding. So, by Lemma 3.14, we must have $\Diamond^4_N u(D) = {\sim}D$. By Lemma 3.5, we have $U_1 \in D_M$ and $(C_1 \cup U) \setminus U_1 = (C_1 \cup U) \cap {\sim}U_1 \in D_M$. Now,

$$u(U_1) \cup u((C_2 \cup U) \setminus U_1) = u(C_2 \cup U) = u({\sim}\Diamond^4_M D) = {\sim}\Diamond^4_N u(D) = D,$$

hence we must have $u(U_1) = D = u((C_2 \cup U) \setminus U)$, $u(U_1) = \varnothing = u(\varnothing)$ or $u((C_2 \cup U) \setminus U_1) = \varnothing = u(\varnothing)$, which contradicts the fact that u is an embedding. Thus, $u(D) = D$, as claimed. □

Lemma 3.16 *Let $M, N \subseteq \mathbb{E}$ with $M \neq N$. Then* $\mathrm{Var}(\mathbf{D}_M) \neq \mathrm{Var}(\mathbf{D}_N)$.

Proof. Suppose, for a contradiction, that we have $\mathrm{Var}(\mathbf{D}_M) = \mathrm{Var}(\mathbf{D}_N)$. By Lemmas 3.8 and 3.12, there are embeddings $u\colon \mathbf{D}_M \to \mathbf{D}_N$ and $v\colon \mathbf{D}_N \to \mathbf{D}_M$. As $M \neq N$, we have $(M \setminus N) \cup (N \setminus M) \neq \varnothing$. Let $i := \min((M \setminus N) \cup (N \setminus M))$. Without loss of generality, we can assume that $i \in M$, since we can interchange M with N (if necessary). From the proof of Lemma 3.5, there are unary terms t_A, t_{L_i} and $t_{U_{i-1}}$ with $t_A^{\mathbf{D}_M}(D) = A = t_A^{\mathbf{D}_N}(D)$, $t_{L_i}^{\mathbf{D}_M}(D) = L_i = t_{L_i}^{\mathbf{D}_N}(D)$ and $t_{U_{i-1}}^{\mathbf{D}_M}(D) = U_{i-1} = t_{U_{i-1}}^{\mathbf{D}_N}(D)$, since i is the minimum of $(M \setminus N) \cup (N \setminus M)$. Now, let $t(x)$ be the unary term defined by

$$t(x) := \Diamond t_{L_i}(x) \wedge \neg(t_A(x) \vee t_{L_i}(x) \vee t_{U_{i-1}}(x)).$$

Based on Fig. 3 and Fig. 4, we have $t^{\mathbf{D}_M}(D) = U_i$ and $t^{\mathbf{D}_N}(D) = U_i \cup U_{i+1}$. Using Lemma 3.14, we find that

$$v(U_i) \cup v(U_{i+1}) = v(t^{\mathbf{D}_N}(D)) = t^{\mathbf{D}_M}(v(D)) = t^{\mathbf{D}_M}(D) = U_i,$$

so we have $v(U_i) = U_i = v(U_{i+1})$, $v(U_i) = \varnothing = v(\varnothing)$ or $v(U_{i+1}) = \varnothing = v(\varnothing)$. This contradicts the injectivity of v, so $\mathrm{Var}(\mathbf{D}_M) \neq \mathrm{Var}(\mathbf{D}_N)$, as claimed. □

4 Conclusion

We have constructed a continuum of subvarieties of $\mathcal{B}$ of height 3. Our main result follows immediately.

Theorem 4.1 *The class of normal axiomatic extensions of* **KTB** *of codimension 3 is of size continuum.*

It will be of interest to see what our result implies about subquasivarieties of $\mathcal{B}$ of small height, or, equivalently, about logics in CNExt(**KTB**) of small codimension (see Remark 2.1). However, from Blanco, Campercholi and Vaggione [2] it follows that even the lattice of subquasivarieties of $\mathrm{Var}(\mathsf{Cm}(\mathbb{K}_2))$ is not a chain, so the lattice of subquasivarieties of $\mathrm{Var}(\mathbf{D}_N)$ may be already quite complex, in particular, it may be of height strictly greater than 3.

References

[1] Bergman, C., "Universal Algebra. Fundamentals and Selected Topics," Pure and Applied Mathematics **301**, CRC Press, Boca Raton, 2012.

[2] Blanco, J., M. Campercholi and D. Vaggione, *The subquasivariety lattice of a discriminator variety*, Adv. Math. **159** (2001), pp. 18–50.

[3] Burris, S. and H. P. Sankappanavar, "A Course in Universal Algebra," Graduate Texts in Mathematics **78**, Springer-Verlag, New York-Berlin, 1981.

[4] Jónsson, B. and A. Tarski, *Boolean algebras with operators. I*, Amer. J. Math. **73** (1951), pp. 891–939.

[5] Jónsson, B. and A. Tarski, *Boolean algebras with operators. II*, Amer. J. Math. **74** (1952), pp. 127–162.

[6] Kostrzycka, Z., *On non-compact logics in NEXT(KTB)*, Math. Log. Q. **54** (2008), pp. 617–624.

[7] Kostrzycka, Z., *On interpolation and Halldén-completeness in NEXT(KTB)*, Bull. Sect. Logic Univ. Łódź **41** (2012), pp. 23–32.

[8] Kostrzycka, Z., *All splitting logics in the lattice* $NEXT(KTB.3'A)$, Sci. Issues Jan Długosz Univ. Czest. Math. **21** (2016), pp. 31–61.

[9] Kostrzycka, Z., *Interpolation in normal extensions of the Brouwer logic*, Bull. Sect. Logic Univ. Łódź **45** (2016), pp. 171–184.

[10] Kowalski, T. and Y. Miyazaki, *All splitting logics in the lattice NExt(KTB)*, in: *Towards Mathematical Philosophy*, Trends Log. Stud. Log. Libr. **28**, Springer, Dordrecht, 2009 pp. 53–67.

[11] Kowalski, T., Y. Miyazaki and M. Stevens, *Quasi-p morphisms and small varieties of KTB-algebras*, `http://people.maths.ox.ac.uk/~hap/tancl07/tancl07-kowalski.pdf`, TACL'07, Oxford, 2–5 August 2007.

[12] Kowalski, T. and M. Stevens, *Minimal varieties of KTB-algebras* (2005), unpublished note.

[13] Kracht, M., "Tools and Techniques in Modal Logic," Studies in Logic and the Foundations of Mathematics **142**, North Holland Publishing Co., Amsterdam, 1999.

[14] McKay, B. and A. Piperno, *nauty and Traces*, `https://users.cecs.anu.edu.au/~bdm/nauty/`.

[15] Miyazaki, Y., *Kripke incomplete logics containing KTB*, Studia Logica **85** (2007), pp. 303–317.

[16] Miyazaki, Y., *Normal forms for modal logics KB and KTB*, Bull. Sect. Logic Univ. Łódź **36** (2007), pp. 183–193.

[17] Miyazaki, Y., *A splitting logic in NExt(KTB)*, Studia Logica **85** (2007), pp. 381–394.

Coherence in Modal Logic

Tomasz Kowalski

Department of Mathematics and Statistics
La Trobe University, Melbourne, Australia
T.Kowalski@latrobe.edu.au

George Metcalfe [1]

Mathematical Institute
University of Bern, Switzerland
george.metcalfe@math.unibe.ch

Abstract

A variety is said to be coherent if the finitely generated subalgebras of its finitely presented members are also finitely presented. In [19] it was shown that coherence forms a key ingredient of the uniform deductive interpolation property for equational consequence in a variety, and a general criterion was given for the failure of coherence (and hence uniform deductive interpolation) in varieties of algebras with a term-definable semilattice reduct. In this paper, a more general criterion is obtained and used to prove the failure of coherence and uniform deductive interpolation for a broad family of modal logics, including K, KT, K4, and S4.

Keywords: Modal Logic, Coherence, Uniform Interpolation, Model Completion, Free Algebras, Compact Congruences.

1 Introduction

A variety — equivalently, a class of algebras defined by equations — is said to be *coherent* if every finitely generated subalgebra of a finitely presented member of the variety is again finitely presented. The notion of coherence originated in sheaf theory and has been studied quite widely in algebra, mostly in connection with groups, rings, modules, monoids, and lattices (see, e.g., [7,3,27,14]). It has also been considered from a more general model-theoretic perspective by Wheeler [31,32], who proved, among other things, that coherence of a variety is implied by, and in conjunction with amalgamation and a further property implies, the existence of a model completion for its first-order theory.

[1] The second author acknowledges support from Swiss National Science Foundation grant 200021_165850 and the EU Horizon 2020 research and innovation programme under the Marie Skłodowska-Curie grant agreement No 689176.

In [19] it was shown that there exists a close relationship between coherence and the logical notion of *uniform interpolation*. Following [29], right uniform deductive interpolation is defined for equational consequence in a variety $\mathcal{V}$ as an extension of deductive interpolation (studied in, e.g., [16,25,21,4,23]) and related to properties of compact congruences on the free algebras of $\mathcal{V}$. It was proved in [19] that right uniform deductive interpolation is equivalent to the conjunction of deductive interpolation and coherence. Since deductive interpolation is equivalent to amalgamation in the presence of the congruence extension property (see, e.g., [23]), coherence provides the key extra ingredient for an algebraic characterization of right uniform deductive interpolation.

In this paper, we investigate coherence in *modal logic*: more precisely, we establish the failure of this property — and hence also the failure of uniform deductive interpolation — for broad families of normal modal logics.

Following Pitts' seminal proof of uniform interpolation for intuitionistic propositional logic IPC [26], uniform interpolation was established also for a number of important modal logics. In particular, semantic proofs making use of bisimulation quantifiers were given by Visser in [30] for the basic normal modal logic K (see also [13]), Grzegorczyk logic S4Grz, and Gödel-Löb logic GL (first proved by Shavrukov [28]), and syntactic Pitts-style proofs were given by Bílková in [1] for K and KT. Relationships between uniform interpolation and bisimulation quantifiers for the modal μ-calculus and other fixpoint modal logics have been studied in some depth in [6,5,22].

Crucially for the topic of this paper, however, the above-mentioned proofs establish an "implication-based" uniform interpolation property, that for IPC, GL, and S4Grz implies the "consequence-based" uniform deductive interpolation property studied in [29,19] (and implicitly in [13]), but not for K or KT. Indeed, it is proved in [19] that any coherent variety of modal algebras must have equationally definable principal congruences (EDPC) or, equivalently, that the corresponding modal logic must be weakly transitive, i.e., admit as a theorem $\boxdot^n x \to \boxdot^{n+1} x$ (where $\boxdot x := x \wedge \Box x$) for some $n \in \mathbb{N}$. Since this is not the case for K or KT, these logics are not coherent and do not admit uniform deductive interpolation. In the case of K, the failure of uniform deductive interpolation was already observed (at least implicitly) in [11], where it was shown that the description logic $\mathcal{ALC}$, a notational variant of multi-modal K, does not have this property; note also that in [20] an algorithmic characterization was given of the formulas of this logic that do admit deductive uniform interpolants.

Failure of coherence for a family of non-weakly transitive modal logics (also substructural logics, bi-intuitionistic logic, and lattices) was established in [19] via a general criterion stating that in a coherent variety $\mathcal{V}$ of algebras with a term-definable semilattice reduct, any increasing and monotone term satisfying a certain fixpoint embedding condition in $\mathcal{V}$ admits a fixpoint obtained by iterating the term finitely many times. It was left open in [19], however, as to whether this criterion could be used to show also the failure of coherence for weakly transitive modal logics. Ghilardi and Zawadowski proved in [12] that S4 does not admit uniform interpolation, and in [13] gave a description of the axiomatic extensions of this logic admitting a model completion. In this paper, we obtain similar results using a more general criterion for the failure of coherence that allows extra variables in terms satisfying the fixpoint embedding condition in some variety of modal algebras.

In Section 2, we provide the required algebraic background and recall the relationship between uniform deductive interpolation and coherence established in [19]. In Section 3, we then state and prove our new criterion for the failure of coherence. Finally, in Section 4, we apply the criterion with carefully chosen terms to obtain failures of coherence for broad families of non-weakly transitive and weakly transitive modal logics, including K, KT, K4, and S4. Failure of coherence implies failure of uniform deductive interpolation and absence of a model completion, so our results overlap with those obtained by Ghilardi and Zawadowski in [12,13]. However, our proofs are arguably simpler, since they require only finding a suitable term; moreover, the method is not confined to modal logics but can be applied to many other families of non-classical logics and varieties of algebras.

2 Uniform Deductive Interpolation and Coherence

In this section, we recall a general account of uniform deductive interpolation for varieties of algebras that was first presented in [29], and a relationship, obtained in [19], between right uniform deduction interpolation and the model-theoretic notion of coherence. For logics admitting a variety as an equivalent algebraic semantics (in particular, normal modal logics), these notions and results can also be easily translated into a logical setting.

Let us fix an algebraic signature $\mathcal{L}$ with at least one constant symbol[2] and a variety of $\mathcal{L}$-algebras $\mathcal{V}$. Given any set of variables $\overline{x}$, denote by $\mathbf{Tm}(\overline{x})$ the *$\mathcal{L}$-term algebra over $\overline{x}$* and by $\mathbf{F}(\overline{x})$ the *free algebra of $\mathcal{V}$ over $\overline{x}$*, which may be defined as the quotient of $\mathbf{Tm}(\overline{x})$ by the congruence $\Theta_{\mathcal{V}}$ defined by $s\,\Theta_{\mathcal{V}}\,t \;:\Leftrightarrow\; \mathcal{V} \models s \approx t$. We write $t(\overline{x})$, $\varepsilon(\overline{x})$, or $\Sigma(\overline{x})$ to denote that the variables of an $\mathcal{L}$-term t, $\mathcal{L}$-equation ε, or set of $\mathcal{L}$-equations Σ, respectively, are included in $\overline{x}$. We also use t to denote the corresponding $\Theta_{\mathcal{V}}$ equivalence class of t in $\mathbf{F}(\overline{x})$, relying on context to avoid ambiguity; similarly, we use ε and Σ to denote corresponding pairs of elements and sets of pairs of elements, respectively, of $\mathbf{F}(\overline{x})$. We assume throughout the paper that $\overline{x}$, $\overline{y}$, etc. denote disjoint sets, and that $\overline{x}, \overline{y}$ denotes their disjoint union.

Consequence in $\mathcal{V}$ is defined as follows. For a set of $\mathcal{L}$-equations $\Sigma \cup \{\varepsilon\}$ containing exactly the variables in the set $\overline{x}$, let

$$\Sigma \models_{\mathcal{V}} \varepsilon \;:\Longleftrightarrow\; \text{for every } \mathbf{A} \in \mathcal{V} \text{ and homomorphism } e\colon \mathbf{Tm}(\overline{x}) \to \mathbf{A},$$
$$\Sigma \subseteq \ker(e) \implies \varepsilon \in \ker(e).$$

For a set of $\mathcal{L}$-equations $\Sigma \cup \Delta$, we write $\Sigma \models_{\mathcal{V}} \Delta$ if $\Sigma \models_{\mathcal{V}} \varepsilon$ for all $\varepsilon \in \Delta$.

[2] The restriction to one constant symbol is not necessary but simplifies certain aspects of the presentation.

We say that $\mathcal{V}$ admits *deductive interpolation* if for any sets $\overline{x}, \overline{y}, \overline{z}$ and set of equations $\Sigma(\overline{x}, \overline{y}) \cup \{\varepsilon(\overline{y}, \overline{z})\}$ satisfying $\Sigma \models_{\mathcal{V}} \varepsilon$, there exists a set of equations $\Pi(\overline{y})$ satisfying

$$\Sigma \models_{\mathcal{V}} \Pi \quad \text{and} \quad \Pi \models_{\mathcal{V}} \varepsilon.$$

Deductive interpolation and its relationships with other logical and algebraic properties have been investigated by many authors (see, e.g., [16,25,21,4,23]). In particular, it is well known (see, e.g., [23]) that if $\mathcal{V}$ has the amalgamation property, then it admits deductive interpolation and, conversely, if $\mathcal{V}$ admits deductive interpolation and has the congruence extension property, then it has the amalgamation property.

Observe now (or consult [29]*Proposition 2.10) that $\mathcal{V}$ admits deductive interpolation if and only if for any finite sets $\overline{x}, \overline{y}$ and finite set of equations $\Sigma(\overline{x}, \overline{y})$, there exists a set of equations $\Pi(\overline{y})$ such that for any equation $\varepsilon(\overline{y}, \overline{z})$,

$$\Sigma \models_{\mathcal{V}} \varepsilon \iff \Pi \models_{\mathcal{V}} \varepsilon.$$

Following [29], we say that $\mathcal{V}$ admits *right uniform deductive interpolation* if $\Pi(\overline{y})$ in the preceding condition is required to be finite. [3] It is then easily proved (see [29]*Proposition 3.5) that $\mathcal{V}$ admits right uniform deductive interpolation if and only if

(i) $\mathcal{V}$ admits deductive interpolation;

(ii) for any finite sets $\overline{x}, \overline{y}$ and finite set of equations $\Sigma(\overline{x}, \overline{y})$, there exists a set of equations $\Pi(\overline{y})$ satisfying for any equation $\varepsilon(\overline{y})$,

$$\Sigma \models_{\mathcal{V}} \varepsilon \iff \Pi \models_{\mathcal{V}} \varepsilon.$$

These notions may also be expressed in terms of congruences on the finitely generated free algebras of $\mathcal{V}$. Denote the congruence lattice of an algebra $\mathbf{A}$ by $\mathrm{Con}\,\mathbf{A}$ and the join-semilattice of compact (equivalently, finitely generated) congruences on $\mathbf{A}$ by $\mathrm{KCon}\,\mathbf{A}$, and write $\mathrm{Cg}_{\mathbf{A}}(S)$ to denote the congruence on $\mathbf{A}$ generated by $S \subseteq A^2$. Recall also (see [23]*Lemma 2) that for any sets of equations $\Sigma(\overline{x}), \Delta(\overline{x})$,

$$\Sigma \models_{\mathcal{V}} \Delta \iff \mathrm{Cg}_{\mathbf{F}(\overline{x})}(\Delta) \subseteq \mathrm{Cg}_{\mathbf{F}(\overline{x})}(\Sigma).$$

Observe next that the natural inclusion map $i\colon \mathbf{F}(\overline{y}) \to \mathbf{F}(\overline{x}, \overline{y})$ "lifts" to the adjoint pair of maps

$$i^*\colon \mathrm{Con}\,\mathbf{F}(\overline{y}) \to \mathrm{Con}\,\mathbf{F}(\overline{x}, \overline{y});\quad \Theta \mapsto \mathrm{Cg}_{\mathbf{F}(\overline{x},\overline{y})}(i[\Theta])$$
$$i^{-1}\colon \mathrm{Con}\,\mathbf{F}(\overline{x}, \overline{y}) \to \mathrm{Con}\,\mathbf{F}(\overline{y});\quad \Psi \mapsto i^{-1}[\Psi] = \Psi \cap \mathrm{F}(\overline{y})^2.$$

[3] Note that a similar definition can be given for left uniform deductive interpolation (see [29]), but in this paper, we focus only on failures of right uniform deductive interpolation, indeed only on cases where coherence fails.

It is then straightforward to show that $\mathcal{V}$ admits deductive interpolation if and only if for any finite sets $\overline{x}, \overline{y}, \overline{z}$, the following diagram commutes:

$$\begin{array}{ccc} \mathrm{Con}\,\mathbf{F}(\overline{x},\overline{y}) & \xrightarrow{\;i^{-1}\;} & \mathrm{Con}\,\mathbf{F}(\overline{y}) \\ {\scriptstyle j^*}\big\downarrow & & \big\downarrow{\scriptstyle l^*} \\ \mathrm{Con}\,\mathbf{F}(\overline{x},\overline{y},\overline{z}) & \xrightarrow[\;k^{-1}\;]{} & \mathrm{Con}\,\mathbf{F}(\overline{y},\overline{z}) \end{array}$$

where i, j, k, and l are the inclusion maps between the corresponding finitely generated free algebras.

Given any natural inclusion map $i\colon \mathbf{F}(\overline{y}) \to \mathbf{F}(\overline{x},\overline{y})$, the map i^* will always restrict to a map $i^*|_{\mathrm{KCon}\,\mathbf{F}(\overline{y})}\colon \mathrm{KCon}\,\mathbf{F}(\overline{y}) \to \mathrm{KCon}\,\mathbf{F}(\overline{x},\overline{y})$. On the other hand, i^{-1} restricts to $i^{-1}|_{\mathrm{KCon}\,\mathbf{F}(\overline{x},\overline{y})}\colon \mathrm{KCon}\,\mathbf{F}(\overline{x},\overline{y}) \to \mathrm{KCon}\,\mathbf{F}(\overline{y})$, yielding the right adjoint of $i^*|_{\mathrm{KCon}\,\mathbf{F}(\overline{y})}$, if and only if i preserves compact congruences; that is, if and only if for any compact congruence Ψ on $\mathbf{F}(\overline{x},\overline{y})$, also $\Psi \cap \mathrm{F}(\overline{y})^2$ is compact. It is not hard to see that this is exactly the case when condition (ii) for right uniform deductive interpolation is satisfied. Moreover, it was shown in [19] that this property is equivalent to the property of coherence, studied from a more general model-theoretic perspective by Wheeler in [31,32]. Recall that an algebra $\mathbf{A} \in \mathcal{V}$ is *finitely presented* (in $\mathcal{V}$) if it isomorphic to $\mathbf{F}(\overline{x})/\Theta$ for some finite set $\overline{x}$ and $\Theta \in \mathrm{KCon}\,\mathbf{F}(\overline{x})$.

Theorem 2.1 ([19]*Theorem 2.3) *The following are equivalent:*

(1) *For any finite sets $\overline{x}, \overline{y}$ and any finite set of equations $\Sigma(\overline{x},\overline{y})$, there exists a finite set of equations $\Pi(\overline{y})$ such that for any equation $\varepsilon(\overline{y})$,*

$$\Sigma \models_{\mathcal{V}} \varepsilon \iff \Pi \models_{\mathcal{V}} \varepsilon.$$

(2) *For any finite sets $\overline{x}, \overline{y}$ and any compact congruence Θ on $\mathbf{F}(\overline{x},\overline{y})$, the congruence $\Theta \cap F(\overline{y})^2$ on $\mathbf{F}(\overline{y})$ is compact.*

(3) *$\mathcal{V}$ is* coherent*; that is, all finitely generated subalgebras of finitely presented members of $\mathcal{V}$ are finitely presented.*[4]

It follows that $\mathcal{V}$ admits right uniform deductive interpolation if and only if $\mathcal{V}$ is coherent and admits deductive interpolation. As mentioned already in the introduction, it was proved by Wheeler in [31] that the coherence of $\mathcal{V}$ is implied by (and, in conjunction with amalgamation and another property, implies) the existence of a model completion for the first-order theory of $\mathcal{V}$. Hence establishing the failure of coherence for a variety yields also the failure of uniform deductive interpolation and lack of a model completion. Examples of coherent varieties include abelian groups and any locally finite variety (since

[4] Note that, by our earlier assumption, coherence is defined here only for varieties in a signature $\mathcal{L}$ that contains at least one constant symbol; this restriction is not essential, but allows for a neater presentation.

in these varieties, finitely generated algebras are finitely presented), lattice-ordered abelian groups and MV-algebras (see [29]*Example 3.7), and Heyting algebras (the critical part of Pitts' theorem [26]). Note, however, that the variety of groups is not coherent, since every finitely generated recursively presented group embeds into a finitely presented group [15].

3 A General Criterion

In this section, we show that in any coherent variety $\mathcal{V}$ of algebras with a term-definable join-semilattice reduct, each term $t(x, \bar{u})$ that is increasing and monotone, and satisfies a fixpoint embedding condition (with respect to x), has a fixpoint obtained by iterating t finitely many times in the first argument. This result provides a general criterion for establishing the failure of coherence and therefore also the failure of right uniform deductive interpolation and the lack of a model completion. It generalizes a similar result obtained in [19] by allowing extra variables $\bar{u}$ in the term $t(x, \bar{u})$; in Section 4 we make good use of this flexibility in dealing with weakly transitive modal logics.

Given any term $t(x, \bar{u})$, define inductively

$$t^0(x, \bar{u}) := x \quad \text{and} \quad t^{k+1}(x, \bar{u}) := t(t^k(x, \bar{u}), \bar{u}) \ \text{ for } k \in \mathbb{N}.$$

For any algebra $\mathbf{A}$ and term $s(x_1, \ldots, x_n)$, the term function $s^{\mathbf{A}} \colon A^n \to A$ is defined inductively in the usual way; for convenience, however, we often omit the superscript $^{\mathbf{A}}$ when referring to such functions.

Theorem 3.1 *Let $\mathcal{V}$ be a coherent variety of $\mathcal{L}$-algebras with a term-definable join-semilattice reduct and a term $t(x, \bar{u})$ satisfying*

$$\mathcal{V} \models x \leq t(x, \bar{u}) \quad \textit{and} \quad \mathcal{V} \models x \leq y \ \Rightarrow \ t(x, \bar{u}) \leq t(y, \bar{u}).$$

Suppose also that $\mathcal{V}$ satisfies the following fixpoint embedding condition *with respect to $t(x, \bar{u})$:*

(FE) *For any finitely generated $\mathbf{A} \in \mathcal{V}$ and $a, \bar{b} \in A$, there exists an algebra $\mathbf{B} \in \mathcal{V}$ such that $\mathbf{A}$ is a subalgebra of $\mathbf{B}$ and the join $\bigvee_{k \in \mathbb{N}} t^k(a, \bar{b})$ exists in $\mathbf{B}$ and satisfies*

$$\bigvee_{k \in \mathbb{N}} t^k(a, \bar{b}) = t(\bigvee_{k \in \mathbb{N}} t^k(a, \bar{b}), \bar{b}).$$

Then $\mathcal{V} \models t^n(x, \bar{u}) \approx t^{n+1}(x, \bar{u})$ for some $n \in \mathbb{N}$.

Proof. Let $\mathcal{V}$ and $t(x, \bar{u})$ be as in the statement of the theorem. Note first that the fact that t is increasing and monotone easily implies that for any $n \in \mathbb{N}$,

$$\mathcal{V} \models t^n(x, \bar{u}) \leq t^{n+1}(x, \bar{u}).$$

To establish the converse inequality for some $n \in \mathbb{N}$, we define sets of equations

$$\Sigma = \{y \leq x, x \leq z, x \approx t(x, \bar{u})\} \quad \text{and} \quad \Pi = \{t^k(y, \bar{u}) \leq z \mid k \in \mathbb{N}\}$$

and prove that for any equation $\varepsilon(y, z, \bar{u})$,

$$(\star) \qquad \Sigma \models_{\mathcal{V}} \varepsilon(y, z, \bar{u}) \iff \Pi \models_{\mathcal{V}} \varepsilon(y, z, \bar{u}).$$

For the right-to-left direction, it suffices to observe that $\Sigma \models_{\mathcal{V}} t^k(y, \bar{u}) \leq z$ for each $k \in \mathbb{N}$. For the left-to-right direction, suppose contrapositively that $\Pi \not\models_{\mathcal{V}} \varepsilon(y, z, \bar{u})$. Since only finitely many variables occur in Π, there exist a finitely generated $\mathbf{A} \in \mathcal{V}$ and a homomorphism $e\colon \mathbf{Tm}(y, z, \bar{u}) \to \mathbf{A}$ such that $\Pi \subseteq \ker(e)$, but $\varepsilon \notin \ker(e)$. Let $a = e(y)$, $\bar{b} = e(\bar{u})$. By assumption, $\mathbf{A}$ is a subalgebra of some $\mathbf{B} \in \mathcal{V}$ such that $\bigvee_{k \in \mathbb{N}} t^k(a, \bar{b})$ exists in $\mathbf{B}$ and

$$\bigvee_{k \in \mathbb{N}} t^k(a, \bar{b}) = t(\bigvee_{k \in \mathbb{N}} t^k(a, \bar{b}), \bar{b}).$$

Since x does not appear in $\Pi \cup \{\varepsilon\}$, we may extend e to a homomorphism $e\colon \mathbf{Tm}(x, y, z, \bar{u}) \to \mathbf{B}$ by defining

$$e(x) = \bigvee_{k \in \mathbb{N}} t^k(a, \bar{b}).$$

But $t^k(a, \bar{b}) \leq e(z)$ for each $k \in \mathbb{N}$, so clearly $e(y) \leq e(x) \leq e(z)$. Moreover, by the fixpoint embedding condition,

$$e(x) = \bigvee_{k \in \mathbb{N}} t^k(a, \bar{b}) = t(\bigvee_{k \in \mathbb{N}} t^k(a, \bar{b}), \bar{b}) = e(t(x, \bar{u})).$$

Hence $\Sigma \subseteq \ker(e)$ and we obtain $\Sigma \not\models_{\mathcal{V}} \varepsilon(y, z, \bar{u})$, completing the proof of $(\star)$.

Since $\mathcal{V}$ is coherent, there exists a finite set of equations $\Delta(y, z, \bar{u})$ such that for any equation $\varepsilon(y, z, \bar{u})$,

$$\Sigma \models_{\mathcal{V}} \varepsilon(y, z, \bar{u}) \iff \Delta \models_{\mathcal{V}} \varepsilon(y, z, \bar{u}).$$

So $\Sigma \models_{\mathcal{V}} \Delta$, and, by $(\star)$, also $\Pi \models_{\mathcal{V}} \Delta$. Using the compactness of $\models_{\mathcal{V}}$ (see [23]) and the fact that Δ is finite, $\Pi' \models_{\mathcal{V}} \Delta$ for some finite $\Pi' \subseteq \Pi$. But also $\{t^{k+1}(y, \bar{u}) \leq z\} \models_{\mathcal{V}} t^k(y, \bar{u}) \leq z$ for each $k \in \mathbb{N}$, and hence for some $n \in \mathbb{N}$,

$$\{t^n(y, \bar{u}) \leq z\} \models_{\mathcal{V}} \Delta.$$

Since $\Sigma \models_{\mathcal{V}} t^{n+1}(y, \bar{u}) \leq z$, also $\Delta \models_{\mathcal{V}} t^{n+1}(y, \bar{u}) \leq z$. Hence, combining these consequences, $\{t^n(y, \bar{u}) \leq z\} \models_{\mathcal{V}} t^{n+1}(y, \bar{u}) \leq z$. Finally, substituting z with $t^n(y, \bar{u})$ and y with x, we obtain $\mathcal{V} \models t^{n+1}(x, \bar{u}) \leq t^n(x, \bar{u})$. □

A less general version of this theorem was used in [19] to establish the failure of coherence for broad families of varieties of Boolean algebras with operators and residuated lattices, as well as the varieties of double-Heyting algebras (algebraic semantics for bi-intuitionistic logic) and lattices.

4 Modal Logics

In this section, we apply the general criterion from Section 3 to a wide range of normal modal logics. Since the central notion of coherence is algebraic in nature and our Theorem 3.1 is formulated algebraically, let us for convenience call a normal modal logic L *coherent* if the variety of modal algebras $\mathcal{V}_\mathsf{L}$ providing an equivalent algebraic semantics for L is coherent. Our definition of consequence in a variety $\mathcal{V}$, stated in Section 2, corresponds to *global* consequence in modal logic, so our use of modal logic terminology here should always be taken in its global meaning. Moreover, we will only consider normal modal logics, so in this section, *logic* is synonymous with *normal modal logic.*

Recall that a logic L is *strongly Kripke complete* if for every set of formulas $\Gamma \cup \{\varphi\}$, whenever $\Gamma \nvdash_\mathsf{L} \varphi$, there exists a Kripke frame $\mathfrak{F}$ and a valuation v, such that $\mathfrak{F} \models_v \Gamma$ and $\mathfrak{F} \not\models_v \varphi$. Wolter showed in [33] (see also [2]) that L is strongly Kripke complete if and only if each at most countably generated $\mathbf{A} \in \mathcal{V}_\mathsf{L}$ embeds into a complex algebra $\mathfrak{G}^+ \in \mathcal{V}_\mathsf{L}$ of some Kripke frame $\mathfrak{G}$. A variety with this property is said to be *ω-complex* (see [2]).

Let us also recall that a logic L is *n-transitive* for $n \in \mathbb{N}$ if $\vdash_\mathsf{L} \boxdot^n x \to \boxdot^{n+1} x$, or, equivalently, if the reflexive closure R of the accessibility relation in Kripke frames for L satisfies $R^{n+1} \subseteq R^n$. A logic L is called *weakly transitive* if it is n-transitive for some $n \in \mathbb{N}$. Equivalently, L is weakly transitive if $\mathcal{V}_\mathsf{L}$ has *equationally definable principal congruences (EDPC)* (see, e.g., [18]).

4.1 Non-Weakly Transitive Logics

In [19] it was shown (Theorem 5.2) that any canonical modal logic that is coherent must be weakly transitive. Hence coherence fails for all canonical modal logics that are not weakly transitive.[5] The class of canonical non-weakly transitive modal logics is already rather large, including standard logics such as K, KT, KD, KB, KTB, and many others. However, the next result shows that the assumption of canonicity used in [19] can be replaced by the weaker assumption that the modal logic is strongly Kripke complete.

Theorem 4.1 *Any coherent strongly Kripke complete modal logic is weakly transitive.*

Proof. Let L be a coherent strongly Kripke complete modal logic. Consider the term $t(x) = \Diamond x \vee x$. Clearly, t is monotone and increasing; moreover, it defines an operator, so t is completely additive in complex algebras. Let $\mathbf{A} \in \mathcal{V}_\mathsf{L}$ be finitely generated, and let $a \in A$. Then $\mathbf{A}$ is at most countable, so by the ω-complexity of $\mathcal{V}_\mathsf{L}$, it embeds into a complex algebra $\mathfrak{G}^+ \in \mathcal{V}_\mathsf{L}$ of some Kripke frame $\mathfrak{G}$ for L. By complete additivity, we have that in $\mathfrak{G}^+$,

$$\bigvee_{k \in \mathbb{N}} t^k(a) = t\big(\bigvee_{k \in \mathbb{N}} t^k(a)\big).$$

[5] In fact, a more general result was established in [19]: if a variety $\mathcal{V}$ of Boolean Algebras with Operators (BAOs) is both canonical and coherent, then several of its term reducts, including $\mathcal{V}$ itself and various reducts to a single operator, have EDPC, so coherence fails for all canonical varieties of BAOs without EDPC.

Hence $\mathcal{V}_{\mathsf{L}}$ satisfies the fixpoint embedding condition (FE) with respect to t. But then by Theorem 3.1, we have $\mathcal{V}_{\mathsf{L}} \models t^{n+1}(x) \approx t^n(x)$ for some $n \in \mathbb{N}$, so L is weakly transitive. $\square$

All canonical logics are strongly Kripke complete, but the converse does not hold; a counterexample can be obtained by applying Thomason simulation to the tense logic of the real line (see, e.g., [2] for details). Hence Theorem 4.1 is slightly stronger than the results stated in [19]. Let us remark also that although the requirements of Theorem 3.1 are satisfied whenever each countable $\mathbf{A} \in \mathcal{V}_{\mathsf{L}}$ embeds into a direct product of finite algebras from $\mathcal{V}_{\mathsf{L}}$, this property is too strong to produce interesting results. An embedding $\mathbf{A} \hookrightarrow \prod_{i \in I} \mathbf{B}_i$ can always be taken to be subdirect with subdirectly irreducible factors, which implies that each countable subdirectly irreducible algebra in $\mathcal{V}_{\mathsf{L}}$ is finite. For modal algebras (but not in general), this further implies that $\mathcal{V}_{\mathsf{L}}$ is locally finite and hence coherent.

4.2 Weakly Transitive Logics

Clearly, Theorem 4.1 cannot be used to show failures of coherence for weakly transitive logics. Also, its proof relies on the fact that t defines an operator, so weakening the assumption of canonicity narrows the scope of applications of the method. Indeed, to extend our approach to weakly transitive logics, we require a term that does not define an operator. For example, to prove that the canonical logic $\mathsf{K4}$ is not coherent, the term $\Diamond\Box x$ (which does not define an operator) can be used with Theorem 3.1. This approach also works for some other weakly transitive logics, but not for the archetypal transitive logic $\mathsf{S4}$. To establish the failure of coherence for $\mathsf{S4}$ using Theorem 3.1, a unary term t will probably not suffice, and a *positive* unary term will certainly not suffice, since the one-generated free positive interior algebra is finite (see [24]).

We therefore make use here of the ternary term

$$t(x, y, z) = \Box(y \vee \Box(z \vee x)) \vee x.$$

This term does not define an operator, and for any variety $\mathcal{V}$ of modal algebras,

$$\mathcal{V} \models x \leq t(x, y, z) \quad \text{and} \quad \mathcal{V} \models x \leq x' \ \Rightarrow\ t(x, y, z) \leq t(x', y, z).$$

We will now show that for any modal logic L whose Kripke frames include finite chains of arbitrary length, $\mathcal{V}_{\mathsf{L}} \not\models t^n(x, y, z) \approx t^{n+1}(x, y, z)$ for all $n \in \mathbb{N}$. By a *finite chain* we mean here any frame $\mathfrak{C}_n = (C, R)$ such that $|C| = n$ for some $n \in \mathbb{N}$, and the reflexive closure of R is a total order on C. Note that, according to this definition, a finite chain is not uniquely determined by the number of its elements; indeed, there are precisely 2^n finite chains with n elements, one for each choice of (a subset of) reflexive points. We say then that a logic L *admits finite chains*, if for each $n \in \mathbb{N}$ there exists at least one n-element chain that is a frame for L.

Lemma 4.2 *Let* L *be a modal logic admitting finite chains, and let* $t(x, y, z)$ *be as defined above. Then* $\mathcal{V}_{\mathsf{L}} \not\models t^n(x, y, z) \approx t^{n+1}(x, y, z)$ *for all* $n \in \mathbb{N}$.

Proof. Fix $n \in \mathbb{N}$. By the construction of t, $\mathcal{V}_{\mathsf{L}} \models t^n(x,y,z) \leq t^{n+1}(x,y,z)$, so we need to show that the converse inequality fails in $\mathcal{V}_{\mathsf{L}}$. Following common practice, in what follows we write just t^m for $t^m(x,y,z)$ $(m \in \mathbb{N})$.

Since L admits finite chains, let $\mathfrak{C}_n = (C_n, R_n)$ be a $2n+1$-element chain that is a frame for L. Without loss of generality, $C_n = \{0,1,\ldots,2n\}$, and R_n extends the natural strict order on C_n by making some points reflexive. Let v be a valuation extending the map $v\colon \{x,y,z\} \to \mathcal{P}(\{0,1,\ldots,2n\})$ given by

$$v(x) = v(z) = \{2i+1 \mid 0 \leq i \leq n-1\},$$
$$v(y) = \{2i \mid 0 \leq i \leq n\},$$

as illustrated by the following diagram:

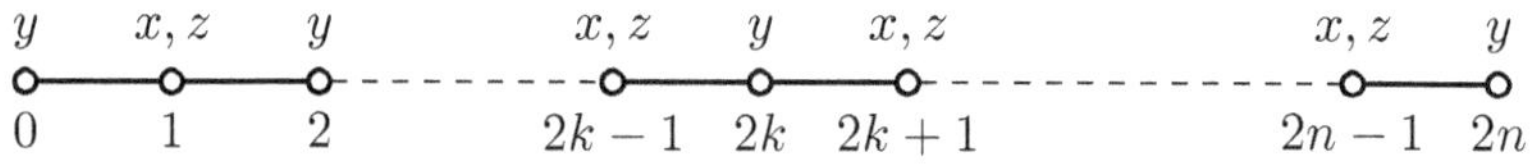

Define $s_0 = \bot$ and $s_{m+1} = \Box(y \vee \Box(z \vee s_m))$. Then $\mathcal{V}_{\mathsf{L}} \models s_m \leq t^m$ for all $m \in \mathbb{N}$. Just observe that, inductively, $\mathcal{V}_{\mathsf{L}} \models s_0 \leq x$, and since $t^{m+1} = \Box(y \vee \Box(z \vee t^m)) \vee t^m$, we obtain $\mathcal{V}_{\mathsf{L}} \models s_m \leq t^m$ by the induction hypothesis, and then $\mathcal{V}_{\mathsf{L}} \models s_{m+1} \leq t^{m+1}$ as required. By the construction of s_m, it can only fail at a point a (that is, $a \models_{\langle \mathfrak{C}_n, v\rangle} \neg s_m$) if there is a path $a = a_0 R a_1 R \ldots R a_{2m-1} R a_{2m}$ of (not necessarily distinct) points $a_0, a_1, \ldots, a_{2m}$ such that $\neg y$ holds at a_i for all odd i, and $\neg z$ holds at a_i for all even i. Inspecting $\mathfrak{C}_n$ we see that such a path does not exist for any $m > n$. So s_m holds at every point of $\mathfrak{C}_n$, for any $m > n$. Since $\mathcal{V}_{\mathsf{L}} \models s_{n+1} \leq t^{n+1}$, also t^{n+1} holds at every point of $\mathfrak{C}_n$.

Claim. For any $\ell \leq n$, we have that $2k \models_{\langle \mathfrak{C}_n, v\rangle} \neg t^\ell$ for all $k \leq n - \ell$.

Proof of Claim. Induction on ℓ. For $\ell = 0$, it is immediate that for every $k \leq n$, we have $2k \models_{\langle \mathfrak{C}_n, v\rangle} \neg x$. For $\ell + 1 \leq n$, by the induction hypothesis, t^ℓ fails at $2k$ for all $k \leq n - \ell$, so we have the following:

$$\begin{aligned}
2(n-\ell) &\models_{\langle \mathfrak{C}_n, v\rangle} \neg t^\ell, \neg z \\
2(n-\ell)-1 &\models_{\langle \mathfrak{C}_n, v\rangle} \neg y \\
2(n-\ell-1) &\models_{\langle \mathfrak{C}_n, v\rangle} \neg t^\ell, \neg z \\
&\vdots \\
2 &\models_{\langle \mathfrak{C}_n, v\rangle} \neg t^\ell, \neg z \\
1 &\models_{\langle \mathfrak{C}_n, v\rangle} \neg y \\
0 &\models_{\langle \mathfrak{C}_n, v\rangle} \neg t^\ell, \neg z.
\end{aligned}$$

Since $t^{\ell+1} = \Box(y \vee \Box(z \vee t^\ell)) \vee t^\ell$, we have that $t^{\ell+1}$ fails at $2(n-\ell-1)$ and at all even points below it. This proves the claim. □

Finally, taking $\ell = n$ in the above claim, we obtain $0 \not\models_{\langle \mathfrak{C}_n, v\rangle} t^n$, and $0 \models_{\langle \mathfrak{C}_n, v\rangle} t^{n+1}$. Therefore, $\mathfrak{C}_n^+ \models t^{n+1} \not\leq t^n$ holds for the complex algebra $\mathfrak{C}_n^+$ of the frame $\mathfrak{C}_n$, and $\mathcal{V}_{\mathsf{L}} \not\models t^n \approx t^{n+1}$, as required. □

Combining Theorem 3.1 with Lemma 4.2, we obtain immediately the following sufficient condition for the failure of coherence in a modal logic.

Theorem 4.3 *Let* L *be a modal logic admitting finite chains. If* $\mathcal{V}_{\mathsf{L}}$ *satisfies the fixpoint embedding condition* (FE) *with respect to the term* $t(x, y, z)$ *defined above, then* $\mathcal{V}_{\mathsf{L}}$ *is not coherent.*

The main obstacle to applying Theorem 4.3 is the satisfaction of (FE) with respect to the term $t(x, y, z)$. Below, we explain why this condition is satisfied for all modal logics L such that canonical extensions of countable members of $\mathcal{V}_{\mathsf{L}}$ themselves belong to $\mathcal{V}_{\mathsf{L}}$. If a variety $\mathcal{V}$ of Boolean Algebras with Operators (BAOs) possesses this property, it is said to be *countably canonical.* It is an open question whether countable canonicity implies canonicity.

We will assume basic knowledge about canonicity in algebraic form; in particular, we assume familiarity with the notion of a *canonical extension* $\mathbf{A}^\sigma$ of a BAO $\mathbf{A}$. We refer the reader to [10] for the general theory of canonical extensions, and to [17] for the necessary background on canonical extensions of BAOs. To keep the presentation smooth, we recall some terminology and facts, mostly from [17]. Let $\mathbf{A}$ be a BAO, and t a term in the signature of $\mathbf{A}$. If t is a fundamental operation of $\mathbf{A}$, the interpretation of t in $\mathbf{A}^\sigma$ is defined to be the canonical extension of $t^{\mathbf{A}}$ as a map: put succinctly, $(t^{\mathbf{A}})^\sigma = t^{\mathbf{A}^\sigma}$. Although this equality is a definition for the fundamental operations, it is in general not preserved under composition. Hence, an arbitrary term-operation $t^{\mathbf{A}}$ can be extended to an operation on $\mathbf{A}^\sigma$ in two ways: as $(t^{\mathbf{A}})^\sigma$, or as $t^{\mathbf{A}^\sigma}$. A term t is called *stable* if these two ways always coincide, that is, if for every BAO $\mathbf{A}$ of appropriate signature, $(t^{\mathbf{A}})^\sigma = t^{\mathbf{A}^\sigma}$. Terms defined by composing operators and lattice operations, or dual operators and lattice operations, are stable. In particular, the term $t(x, y, z) = \Box(y \vee \Box(z \vee x)) \vee x$ is stable.

We will also need two lemmas spelling out certain fixpoint properties of canonical extensions. The first of these is a reformulation in terms of canonical extensions of Esakia's Lemma, first proved in [8]. (For a thorough treatment of Esakia's Lemma, and its connections to canonical extensions we refer the reader to [9].)

Lemma 4.4 *Let* $\mathbf{L}$ *be a bounded lattice and let* $f \colon L \to L$ *be an order-preserving map. If* $X \subseteq L$ *is upward directed and closed under* f*, and* f *is increasing on* X*, then* $f^\sigma(\bigvee X) = \bigvee X$ *in* $\mathbf{L}^\sigma$*.*

The second is a reformulation of Lemma 4.10, from [19], where it was shown to hold, in a slightly different form, for a more general class of algebras. Here we state it for BAOs.

Lemma 4.5 *Let* $\mathcal{V}$ *be a variety of Boolean Algebras with Operators and let* $t(x, \bar{u})$ *be a stable term such that*

$$\mathcal{V} \models x \leq t(x, \bar{u}) \quad \text{and} \quad \mathcal{V} \models x \leq y \Rightarrow t(x, \bar{u}) \leq t(y, \bar{u}).$$

Let $\mathbf{A} \in \mathcal{V}$*,* $a, \bar{b} \in A$*, and* $X = \{(t^k)^{\mathbf{A}}(a, \bar{b}) \mid k \in \mathbb{N}\}$*. Then* $\bigvee X$ *exists in* $\mathbf{A}^\sigma$ *and* $t^{\mathbf{A}^\sigma}(\bigvee X, \bar{b}) = \bigvee X$*.*

Proof. Let $f\colon A \to A$ be the map defined by $f(x) = t^{\mathbf{A}}(x, \bar{b})$. Then f is order-preserving, X is upward directed, closed under f, and f is increasing on X. Hence, by Lemma 4.4, we have $(t^{\mathbf{A}})^{\sigma}(\bigvee X, \bar{b}) = \bigvee X$ in $\mathbf{A}^{\sigma}$. But t is stable, so $t^{\mathbf{A}^{\sigma}}(\bigvee X, \bar{b}) = \bigvee X$ in $\mathbf{A}^{\sigma}$, as required. □

Theorem 4.6 *Let* L *be a modal logic admitting finite chains, and such that* $\mathcal{V}_{\mathsf{L}}$ *is countably canonical. Then*

(a) $\mathcal{V}_{\mathsf{L}}$ *is not coherent;*

(b) $\mathcal{V}_{\mathsf{L}}$ *does not admit right uniform deductive interpolation;*

(c) *the first-order theory of* $\mathcal{V}_{\mathsf{L}}$ *does not have a model completion.*

Proof. By Theorem 4.3, to prove (a), it suffices to show that $\mathcal{V}_{\mathsf{L}}$ satisfies the fixpoint embedding condition (FE) with respect to $t(x, y, z)$. Let $\mathbf{A} \in \mathcal{V}_{\mathsf{L}}$ be finitely generated. Then $\mathbf{A}$ is at most countable, so $\mathbf{A}^{\sigma} \in \mathcal{V}_{\mathsf{L}}$ because $\mathcal{V}_{\mathsf{L}}$ is countably canonical. Since $t(x, y, z)$ is stable, Lemma 4.5 guarantees that $t(\bigvee_{k \in \mathbb{N}} t^k(a, b, c)) = \bigvee_{k \in \mathbb{N}} t^k(a, b, c)$ in $\mathbf{A}^{\sigma}$ for all $a, b, c \in A$. So $\mathcal{V}_{\mathsf{L}}$ satisfies (FE) as required. The statements (b) and (c) then follow using Theorem 2.1 and the fact, proved in [32], that if the first-order theory of a variety has a model completion, then the variety is coherent. □

We can also formulate a positive version of this theorem for coherent logics.

Corollary 4.7 *Let* L *be any coherent modal logic for which* $\mathcal{V}_{\mathsf{L}}$ *is countably canonical. Then for any stable term* $t(x, \bar{u})$ *where* x *occurs only positively, there exists* $n \in \mathbb{N}$ *such that* $\vdash_{\mathsf{L}} t_+^n(x, \bar{u}) \leftrightarrow t_+^{n+1}(x, \bar{u})$, *where* $t_+(x, \bar{u}) := x \vee t(x, \bar{u})$.

Theorem 4.6 implies the failure of coherence for a broad family of modal logics, including K, KT, K4, K4M (McKinsey's logic), S4, and S4.3. However, failure of coherence *does not* follow from this theorem for logics such as GL, S4Grz, and S4.3Grz that admit finite chains but are not canonical. It is known that GL (see [28]) and S4Grz (see [30]) have uniform interpolation and are therefore coherent; indeed, it was proved in [13] that the first-order theories of the varieties corresponding to these logics admit model completions. It was also proved in [13] that the first-order theory of the variety corresponding to S4.3Grz does not have a model completion, but it is not clear (at least to the present authors) if the proof given there also establishes the failure of coherence. A general negative result proved in [13] demonstrates that for any logic L extending K4 that has the finite model property and admits all finite reflexive chains and the two-element cluster, $\mathcal{V}_{\mathsf{L}}$ does not have a model completion. An analysis of this proof reveals that such logics are also not coherent. Clearly, there is a large overlap between this negative result and Theorem 4.6 (but intriguingly no inclusion either way), although our proofs are arguably simpler, since they require only finding a suitable term. Let us emphasize also that our method applies not only to modal logics but also to many other families of non-classical logics and varieties of algebras.

Let us remark finally that it would be rather easy to state the results of this section in a more general way for arbitrary varieties of BAOs. In any such variety, a great range of unary operators can be term-defined. Let $\boxtimes$ be one of them. Then we can define $t(x, y, z) = \boxtimes(y \vee \boxtimes(z \vee x)) \vee x$, reformulate the condition of admitting finite chains algebraically, and state suitable analogues of Theorem 4.3 and 4.6. Such an approach, however, would lack the simplicity and elegance of the presentation given here just for modal logics.

References

[1] Bílková, M., *Uniform interpolation and propositional quantifiers in modal logics*, Studia Logica **85** (2007), pp. 1–31.

[2] Chagrov, A., F. Wolter and M. Zakharyaschev, *Advanced modal logic*, in: *Handbook of Philosophical Logic, Vol. 3*, Kluwer, 2001 pp. 83–266.

[3] Choo, K. G., K. Y. Lam and E. Luft, *On free product of rings and the coherence property*, in: *Algebraic K-Theory, II: "Classical" Algebraic K-Theory and Connections with Arithmetic (Proc. Conf., Battelle Memorial Inst., Seattle, Wash., 1972)*, Lecture Notes in Math. **342**, Springer, 1973 pp. 135–143.

[4] Czelakowski, J., *Sentential logics and Maehara interpolation property*, Studia Logica **44** (1985), pp. 265–283.

[5] D'Agostino, G., *Uniform interpolation, bisimulation quantifiers, and fixed points*, in: *Proc. TbiLLC'05*, 2005, pp. 96–116.

[6] D'Agostino, G. and M. Hollenberg, *Logical questions concerning the μ-calculus*, J. Symbolic Logic **65** (2000), pp. 310–332.

[7] Eklof, P. and G. Sabbagh, *Model-completions and modules*, Ann. Math. Logic **2** (1971), pp. 251–295.

[8] Esakia, L., *Topological Kripke models*, Dokl. Akad. Nauk SSSR **214** (1974), pp. 298–301.

[9] Gehrke, M., *Canonical extensions, Esakia spaces, and universal models*, in: *Leo Esakia on duality in modal and intuitionistic logics*, Outst. Contrib. Log. **4**, Springer, 2014 pp. 9–41.

[10] Gehrke, M. and J. Harding, *Bounded lattice expansions*, J. Algebra **238** (2001), pp. 345–371.

[11] Ghilardi, S., C. Lutz and F. Wolter, *Did I damage my ontology? A case for conservative extensions in description logics*, in: *Proc. KR'06* (2006), pp. 187–197.

[12] Ghilardi, S. and M. Zawadowski, *Undefinability of propositional quantifiers in the modal system* S4, Studia Logica **55** (1995), pp. 259–271.

[13] Ghilardi, S. and M. Zawadowski, "Sheaves, Games, and Model Completions: A Categorical Approach to Nonclassical Propositional Logics," Springer, 2002.

[14] Gould, V., *Coherent monoids*, J. Austral. Math. Soc. Ser. A **53** (1992), pp. 166–182.

[15] Higman, G., *Subgroups of finitely presented groups*, Proc. Roy. Soc. Ser. A **262** (1961), pp. 455–475.

[16] Jónsson, B., *Extensions of relational structures*, in: *Proc. International Symposium on the Theory of Models*, 1965, pp. 146–157.

[17] Jónsson, B., *On the canonicity of Sahlqvist identities*, Studia Logica **53** (1994), pp. 473–491.

[18] Kowalski, T. and M. Kracht, *Semisimple varieties of modal algebras*, Studia Logica **83** (2006), pp. 351–363.

[19] Kowalski, T. and G. Metcalfe, *Uniform interpolation and coherence*, submitted, available to download from `https://arxiv.org/abs/1803.09116`.

[20] Lutz, C. and F. Wolter, *Foundations for uniform interpolation and forgetting in expressive description logics*, in: *Proc. IJCAI'11*, 2011, pp. 989–995.

[21] Maksimova, L., *Craig's theorem in superintuitionistic logics and amalgamable varieties of pseudo-Boolean algebras*, Algebra Logika **16** (1977), pp. 643–681.

[22] Marti, J., F. Seifan and Y. Venema, *Uniform interpolation for coalgebraic fixpoint logic*, in: *Proc. CALCO'15*, LIPIcs **35** (2015), pp. 238–252.

[23] Metcalfe, G., F. Montagna and C. Tsinakis, *Amalgamation and interpolation in ordered algebras*, J. Algebra **402** (2014), pp. 21–82.
[24] Moraschini, T., *Varieties of positive modal algebras and structural completeness*, submitted.
[25] Pigozzi, D., *Amalgamations, congruence-extension, and interpolation properties in algebras*, Algebra Universalis **1** (1972), pp. 269–349.
[26] Pitts, A., *On an interpretation of second-order quantification in first-order intuitionistic propositional logic*, J. Symbolic Logic **57** (1992), pp. 33–52.
[27] Schmidt, P., *Algebraically complete lattices*, Algebra Universalis **17** (1983), pp. 135–142.
[28] Shavrukov, V. Y., *Subalgebras of diagonalizable algebras of theories containing arithmetic*, Dissertationes Math. (Rozprawy Mat.) **323** (1993), p. 82.
[29] van Gool, S., G. Metcalfe and C. Tsinakis, *Uniform interpolation and compact congruences*, Ann. Pure Appl. Logic **168** (2017), pp. 1827–1948.
[30] Visser, A., *Uniform interpolation and layered bisimulation*, in: *Gödel '96: Logical Foundations of Mathematics, Computer Science and Physics—Kurt Gdel's legacy, Brno, Czech Republic, August 1996*, pp. 139–164.
[31] Wheeler, W., *Model-companions and definability in existentially complete structures*, Israel J. Math. **25** (1976), pp. 305–330.
[32] Wheeler, W., *A characterization of companionable, universal theories*, J. Symbolic Logic **43** (1978), pp. 402–429.
[33] Wolter, F., "Lattices of Modal Logics," Ph.D. thesis, Freie Universität Berlin (1993).

Interpolation for Intermediate Logics via Hyper- and Linear Nested Sequents

Roman Kuznets[a 1] Björn Lellmann[a 2]

[a] *TU Wien, Austria*

Abstract

The goal of this paper is extending to intermediate logics the constructive proof-theoretic method of proving Craig and Lyndon interpolation via hypersequents and nested sequents developed earlier for classical modal logics. While both Jankov and Gödel logics possess hypersequent systems, we show that our method can only be applied to the former. To tackle the latter, we switch to linear nested sequents, demonstrate syntactic cut elimination for them, and use it to prove interpolation for Gödel logic. Thereby, we answer in the positive the open question of whether Gödel logic enjoys the Lyndon interpolation property.

Keywords: Intermediate logics, hypersequents, linear nested sequents, interpolation, cut elimination, Gödel logic, Lyndon interpolation

1 Introduction

The *Craig Interpolation Property* (CIP) is one of the fundamental properties of for logics, alongside decidability, compactness, etc. It states that for any theorem $A \to B$ of the logic, there must exist an *interpolant* C that only uses propositional variables common to A and B such that both $A \to C$ and $C \to B$ are theorems. The *Lyndon Interpolation Property* (LIP) strengthens the common language requirement by demanding that each variable in C occur in both A and B with the same polarity as it does in C. One of the more robust *constructive* methods of proving interpolation, which unlike many other methods, can yield both CIP and LIP, is the so-called *proof-theoretic method*, whereby an interpolant is constructed by induction on a given analytic sequent derivation of (a representation of) $A \to B$. Unfortunately, many modal and *intermediate logics*, including Jankov and Gödel logics (intermediate in the sense of being between intuitionistic and classical propositional logics, the former denoted Int), do not possess a known cut-free sequent calculus. Various extensions of sequent calculi were developed to address this situation, including *hypersequents* and nested sequents. Already hypersequents are sufficient to capture

[1] Funded by FWF projects S 11405 (RiSE) and Y 544-N23.

[2] Funded by WWTF project MA16-28

both Jankov and Gödel logics. A method for using such advanced calculi to prove interpolation was only recently developed and applied to classical-based modal logics in [9,12,14].

The goal of this paper is to finally extend the method to intermediate logics. It is a classical result [16] by Maksimova that exactly seven intermediate logics are interpolable (have CIP). It is also known that five of these logics, including the intuitionistic and Jankov logics, have LIP [18]. This paper is devoted to proving interpolation using intuitionistic hypersequents, i.e., hypersequents with at most one formula in the consequent of each sequent component. It turned out, however, that the existence of such a hypersequent calculus does not yet guarantee that the CIP for an interpolable intermediate logic can be proved using the method. Our counterexample for Gödel logic (see the proof of Theorem 4.6) demonstrates that the special interpolation property at the core of our method is strictly stronger that the CIP. In order to overcome this difficulty, we develop an alternative formalism for Gödel logic, that of *linear nested sequents*. In particular, we use it to solve the open problem [4,10,18] of Lyndon interpolation for Gödel logic.

The paper is structured as follows. After preliminaries in Section 2, we describe the method for hypersequents in Section 3 and use it in Section 4 to prove interpolation for Int and for Jankov logic LQ. In Section 5, we introduce linear nested sequents for Gödel logic G and provide a surprisingly intricate proof of syntactic cut elimination for them. In Section 6, we explain how to modify the method from hypersequents to linear nested sequents. Finally, in Section 7, we summarize the obtained results.

2 Intermediate Logics and Hypersequents

We consider the language $A ::= p \mid \bot \mid (A \wedge A) \mid (A \vee A) \mid (A \to A)$, where p is taken from a countably infinite set Prop of *propositional variables*. The Boolean constant $\top$ and connective $\neg$ are defined in the standard way. For all intermediate logics we consider, we use *intuitionistic Kripke models*, adapted to individual logics by frame conditions.

Definition 2.1 A *Kripke frame*, or simply a *frame*, is a pair $(W, \leqslant)$ of a set $W \neq \varnothing$ of *worlds* and a partial order $\leqslant$ on W.[3] A *Kripke model*, or simply a *model*, is a triple $(W, \leqslant, V)$ where $(W, \leqslant)$ is a frame and $V \colon W \to 2^{\mathsf{Prop}}$ is a *valuation* that is monotone w.r.t. $\leqslant$, i.e., $w \leqslant u$ implies $V(w) \subseteq V(u)$.

Definition 2.2 For a model $\mathcal{M} = (W, \leqslant, V)$, the *forcing relation* $\Vdash$ between $w \in W$ and formulas is defined by $\mathcal{M}, w \Vdash p$ iff $p \in V(w)$ for each $p \in \mathsf{Prop}$; $\mathcal{M}, w \nVdash \bot$; $\mathcal{M}, w \Vdash A \wedge B$ iff $\mathcal{M}, w \Vdash A$ and $\mathcal{M}, w \Vdash B$; $\mathcal{M}, w \Vdash A \vee B$ iff $\mathcal{M}, w \Vdash A$ or $\mathcal{M}, w \Vdash B$; $\mathcal{M}, w \Vdash A \to B$ iff $\mathcal{M}, v \nVdash A$ or $\mathcal{M}, v \Vdash B$ for all $v \geqslant w$. A formula A is *valid in* $\mathcal{M}$, written $\mathcal{M} \Vdash A$, if $(\forall w \in W)\, \mathcal{M}, w \Vdash A$. A formula A is *valid in a class* $\mathcal{C}$ *of models*, written $\mathcal{C} \Vdash A$, if $(\forall \mathcal{M} \in \mathcal{C})\, \mathcal{M} \Vdash A$.

[3] A *partial order* is a reflexive, transitive, and antisymmetric binary relation.

$$\mathsf{id}\,\frac{}{A \Rightarrow A} \qquad \mathsf{id}_\bot\,\frac{}{\bot \Rightarrow} \qquad \mathsf{Ex}\,\frac{\mathcal{G} \mid \Gamma \Rightarrow \Delta \mid \Lambda \Rightarrow \Theta \mid \mathcal{H}}{\mathcal{G} \mid \Lambda \Rightarrow \Theta \mid \Gamma \Rightarrow \Delta \mid \mathcal{H}} \qquad \mathsf{EC}\,\frac{\mathcal{G} \mid \Gamma \Rightarrow \Delta \mid \Gamma \Rightarrow \Delta}{\mathcal{G} \mid \Gamma \Rightarrow \Delta}$$

$$\wedge\!\Rightarrow\,\frac{\mathcal{G} \mid \Gamma, A_i \Rightarrow \Delta}{\mathcal{G} \mid \Gamma, A_1 \wedge A_2 \Rightarrow \Delta} \qquad \Rightarrow\!\wedge\,\frac{\mathcal{G} \mid \Gamma \Rightarrow A \quad \mathcal{G} \mid \Gamma \Rightarrow B}{\mathcal{G} \mid \Gamma \Rightarrow A \wedge B} \qquad \to\!\Rightarrow\,\frac{\mathcal{G} \mid \Gamma \Rightarrow A \quad \mathcal{G} \mid \Gamma, B \Rightarrow \Delta}{\mathcal{G} \mid \Gamma, A \to B \Rightarrow \Delta}$$

$$\vee\!\Rightarrow\,\frac{\mathcal{G} \mid \Gamma, A \Rightarrow \Delta \quad \mathcal{G} \mid \Gamma, B \Rightarrow \Delta}{\mathcal{G} \mid \Gamma, A \vee B \Rightarrow \Delta} \qquad \Rightarrow\!\vee\,\frac{\mathcal{G} \mid \Gamma \Rightarrow A_i}{\mathcal{G} \mid \Gamma \Rightarrow A_1 \vee A_2} \qquad \Rightarrow\!\to\,\frac{\mathcal{G} \mid \Gamma, A \Rightarrow B}{\mathcal{G} \mid \Gamma \Rightarrow A \to B}$$

$$\mathsf{W}\!\Rightarrow\,\frac{\mathcal{G} \mid \Gamma \Rightarrow \Delta}{\mathcal{G} \mid \Gamma, A \Rightarrow \Delta} \qquad \Rightarrow\!\mathsf{W}\,\frac{\mathcal{G} \mid \Gamma \Rightarrow}{\mathcal{G} \mid \Gamma \Rightarrow A} \qquad \mathsf{EW}\,\frac{\mathcal{G}}{\mathcal{G} \mid \Gamma \Rightarrow \Delta} \qquad \mathsf{C}\!\Rightarrow\,\frac{\mathcal{G} \mid \Gamma, A, A \Rightarrow \Delta}{\mathcal{G} \mid \Gamma, A \Rightarrow \Delta}$$

Fig. 1. Hypersequent calculus HInt for Int ([6]).

Lemma 2.3 (Monotonicity [20]) *$\mathcal{M}, w \Vdash A$ implies $\mathcal{M}, v \Vdash A$ whenever $w \leqslant v$ for any model $\mathcal{M} = (W, \leqslant, V)$.*

Theorem 2.4 (Completeness [20,4,5]) *Intuitionistic logic* Int *is sound and complete w.r.t. the class of all models. Jankov logic* $\mathsf{LQ} = \mathsf{Int} + \neg A \vee \neg\neg A$ *and Gödel logic* $\mathsf{G} = \mathsf{Int} + (A \to B) \vee (B \to A)$[4] *are sound and complete w.r.t. models with a maximum element and with linear $\leqslant$ respectively.*

Definition 2.5 A *sequent* is a figure $\Gamma \Rightarrow \Delta$ where Γ and Δ are finite multisets of formulas. It is *single-conclusion* if $|\Delta| \leqslant 1$. A *hypersequent* $\mathcal{G}$ is a finite list $\Gamma_1 \Rightarrow \Delta_1 \mid \dots \mid \Gamma_n \Rightarrow \Delta_n$ with $n > 0$ of single-conclusion sequents $\Gamma_i \Rightarrow \Delta_i$, called *components* of $\mathcal{G}$. We define the length $\|\mathcal{G}\| = n$ to be the number of components of $\mathcal{G}$. A *formula interpretation* $\iota(\mathcal{G}) := \bigvee_{i=1}^{n} (\bigwedge \Gamma_i \to \bigvee \Delta_i)$.

Definition 2.6 The hypersequent system HInt for Int is presented in Figure 1. Hypersequent systems HLQ and HG for Jankov logic LQ and Gödel logic G respectively are obtained by adding to HInt respectively the rules

$$\mathsf{lq}\,\frac{\mathcal{G} \mid \Gamma, \Lambda \Rightarrow}{\mathcal{G} \mid \Gamma \Rightarrow \mid \Lambda \Rightarrow} \qquad \text{and} \qquad \mathsf{com}\,\frac{\mathcal{G} \mid \Gamma, \Gamma' \Rightarrow \Delta \quad \mathcal{G} \mid \Lambda, \Lambda' \Rightarrow \Delta'}{\mathcal{G} \mid \Gamma, \Lambda' \Rightarrow \Delta \mid \Lambda, \Gamma' \Rightarrow \Delta'}.$$

We write $\mathsf{HL} \vdash \mathcal{G}$ if the hypersequent $\mathcal{G}$ is derivable in HL for a logic L.

Theorem 2.7 (Hypersequent completeness [6]) *For* $\mathsf{L} \in \{\mathsf{Int}, \mathsf{LQ}, \mathsf{G}\}$ *we have* $\mathsf{HL} \vdash \mathcal{G}$ *iff* $\mathsf{L} \vdash \iota(\mathcal{G})$*. In particular,* $\mathsf{HL} \vdash A \Rightarrow B$ *iff* $\mathsf{L} \vdash A \to B$.

3 Interpolation via Hypersequents

The proof-theoretic method for proving interpolation constructively using hypersequents was first introduced in [12]. There it was applied to the classical-based modal logic S5 and later extended to S4.2 in [13]. In this paper, we adapt the method to intuitionistic propositional reasoning.

[4] Also known as KC and Dummett's logic LC respectively.

Definition 3.1 A *split multiset* $\widetilde{\Gamma} = \Gamma_l; \Gamma_r$ consists of the left and right multiset parts Γ_l and Γ_r (the semicolon can be omitted for the empty multiset). $|\Gamma_l; \Gamma_r| := |\Gamma_l| + |\Gamma_r|$. A split sequent $\widetilde{\Gamma} \Rightarrow \widetilde{\Delta}$ is obtained from split multisets $\widetilde{\Gamma}$ and $\widetilde{\Delta}$ the same way sequents are obtained from multisets. A *split hypersequent* $\widetilde{\mathcal{G}} = \Gamma_1; \Pi_1 \Rightarrow \Delta_1; \Sigma_1 \mid \ldots \mid \Gamma_n; \Pi_n \Rightarrow \Delta_n; \Sigma_n$ is built from split single-conclusion sequents $\Gamma_i; \Pi_i \Rightarrow \Delta_i; \Sigma_i$ the same way hypersequents are built from sequents. The left (right) side and the *conflation* of $\widetilde{\mathcal{G}}$ are obtained by dropping all right (left) formulas and by combining the two sides of $\widetilde{\mathcal{G}}$ respectively:

$$L\widetilde{\mathcal{G}} := \Gamma_1 \Rightarrow \Delta_1 \mid \ldots \mid \Gamma_n \Rightarrow \Delta_n \qquad R\widetilde{\mathcal{G}} := \Pi_1 \Rightarrow \Sigma_1 \mid \ldots \mid \Pi_n \Rightarrow \Sigma_n$$

$$LR\widetilde{\mathcal{G}} := \Gamma_1, \Pi_1 \Rightarrow \Delta_1, \Sigma_1 \mid \ldots \mid \Gamma_n, \Pi_n \Rightarrow \Delta_n, \Sigma_n$$

As before, the length $\|\widetilde{\mathcal{G}}\| := n$. It is obvious that $\|\widetilde{\mathcal{G}}\| = \|L\widetilde{\mathcal{G}}\| = \|R\widetilde{\mathcal{G}}\| = \|LR\widetilde{\mathcal{G}}\|$.

Splits do not play any semantic role. They are used solely for interpolant construction. There is a standard way of turning a given sequent-like calculus HL into its split equivalent SHL. For each rule, one considers all possible splits of the conclusion and uses the corresponding splits of the premiss(es), i.e., side formulas remain on the same side and active formula(s) in the premiss(es) are on the same side as the principal formula. A logical rule typically produces two split variants depending on whether the principal formula is on the left or on the right. To save space, we do not present splits of the rules. Most of them can be read from Figure 2 by omitting interpolants.

Theorem 3.2 *For a logic* $\mathsf{L} \in \{\mathsf{Int}, \mathsf{LQ}, \mathsf{G}\}$ *we have* $\mathsf{SHL} \vdash \widetilde{\mathcal{G}}$ *iff* $\mathsf{HL} \vdash LR\widetilde{\mathcal{G}}$.

Proof. Both directions are by induction on derivation depth. The main observation is that each split rule of SHInt becomes a rule of HInt if one takes the union of left and right formulas separately in each antecedent and consequent. Vice versa, for each split of the conclusion of a rule of HInt, there is a split of the premiss(es) that turns this rule into a rule of SHInt. □

Corollary 3.3 *For* $\mathsf{L} \in \{\mathsf{Int}, \mathsf{LQ}, \mathsf{G}\}$ *we have* $\mathsf{SHL} \vdash A; \Rightarrow; B$ *iff* $\mathsf{L} \vdash A \rightarrow B$.

Proof. The statement follows from Theorems 2.7 and 3.2. □

We define an alternative interpolation property based on direct evaluation of hypersequents into models, as opposed to first translating them into a formula interpretation and evaluating the latter.

Definition 3.4 For a sequence $\boldsymbol{w}$ of worlds from a model $\mathcal{M} = (W, \leqslant, V)$ of length $\|\boldsymbol{w}\| = n > 0$, we denote its ith member by w_i, where $1 \leqslant i \leqslant n$. For a $v \in W$, the sequence $\boldsymbol{w}$ is *v-rooted in $\mathcal{M}$* if $v \leqslant w_i$ for all $1 \leqslant i \leqslant n$. The sequence $\boldsymbol{w}$ is *$\mathcal{M}$-rooted* if it is v-rooted for some $v \in W$.

Definition 3.5 A hypersequent $\Gamma_1 \Rightarrow \Delta_1 \mid \ldots \mid \Gamma_n \Rightarrow \Delta_n$ *componentwise holds* at a sequence $\boldsymbol{w}$ of worlds from $\mathcal{M} = (W, \leqslant, V)$ of length $\|\boldsymbol{w}\| = n$, written $\mathcal{M}, \boldsymbol{w} \models \Gamma_1 \Rightarrow \Delta_1 \mid \ldots \mid \Gamma_n \Rightarrow \Delta_n$, iff for some $i = 1, \ldots, n$

$$\mathcal{M}, w_i \not\Vdash A \text{ for some } A \in \Gamma_i \quad \text{or} \quad \mathcal{M}, w_i \Vdash B \text{ for some } B \in \Delta_i.$$

A hypersequent $\mathcal{G}$ is *componentwise valid in a class $\mathcal{C}$ of models* iff $\mathcal{M}, \boldsymbol{w} \models \mathcal{G}$ for each $\mathcal{M} \in \mathcal{C}$ and each $\mathcal{M}$-rooted sequence $\boldsymbol{w}$ of length $\|\mathcal{G}\|$.

Lemma 3.6 *A hypersequent $\mathcal{G}$ is componentwise valid in a class $\mathcal{C}$ of models iff its formula interpretation is, i.e., iff $\mathcal{C} \Vdash \iota(\mathcal{G})$.*

Proof. Let $\mathcal{G} = \Gamma_1 \Rightarrow \Delta_1 \mid \ldots \mid \Gamma_n \Rightarrow \Delta_n$. It is a simple exercise in intuitionistic Kripke semantics to prove that $\mathcal{G}$ is componentwise invalid iff $\iota(\mathcal{G})$ is. It is based on two observations: (1) if $\mathcal{M}, \boldsymbol{w} \not\models \mathcal{G}$ for a v-rooted sequence $\boldsymbol{w}$, then $\mathcal{M}, v \not\Vdash \iota(\mathcal{G})$; (2) if $\mathcal{M}, v \not\Vdash \iota(\mathcal{G})$, then there exists a v-rooted sequence $\boldsymbol{w}$ such that $\mathcal{M}, \boldsymbol{w} \not\models \mathcal{G}$. □

Compared to interpolants for classical-based modal logics ([12]), the components of interpolants need to be imbued with polarity.

Definition 3.7 A *uniformula* has the form $C^{(k)}$ or $\overline{C}^{(k)}$, where C is a propositional formula and $k \geqslant 1$ is an integer. Each uniformula is a *multiformula*. If $\mho_1$ and $\mho_2$ are multiformulas, then so are $(\mho_1 \otimes \mho_2)$ and $(\mho_1 \oslash \mho_2)$. The *arity* $\|\mho\|$ of $\mho$ is the largest number k such that either $C^{(k)}$ or $\overline{C}^{(k)}$ occurs in $\mho$.

Definition 3.8 Let $\mathcal{M} = (W, \leqslant, V)$ be a model and $\boldsymbol{w} = w_1, \ldots, w_n$ be a sequence of worlds from W of length $\|\boldsymbol{w}\| = n$. For a multiformula $\mho$ of arity $\|\mho\| \leqslant n$, we define its truth inductively: (a) $\mathcal{M}, \boldsymbol{w} \models C^{(k)}$ iff $\mathcal{M}, \boldsymbol{w} \not\models \overline{C}^{(k)}$ iff $\mathcal{M}, w_k \Vdash C$; (b) $\mathcal{M}, \boldsymbol{w} \models \mho_1 \oslash \mho_2$ iff $\mathcal{M}, \boldsymbol{w} \models \mho_1$ and $\mathcal{M}, \boldsymbol{w} \models \mho_2$; (c) $\mathcal{M}, \boldsymbol{w} \models \mho_1 \otimes \mho_2$ iff $\mathcal{M}, \boldsymbol{w} \models \mho_1$ or $\mathcal{M}, \boldsymbol{w} \models \mho_2$.

Thus, $\oslash$ and $\otimes$ are componentwise analogs of $\wedge$ and $\vee$ respectively.

Definition 3.9 Multiformulas $\mho_1$ and $\mho_2$ are *componentwise equivalent*, written $\mho_1 \dashv\models \mho_2$, iff $\mathcal{M}, \boldsymbol{w} \models \mho_1 \iff \mathcal{M}, \boldsymbol{w} \models \mho_2$ for any model $\mathcal{M}$ and any sequence $\boldsymbol{w}$ of worlds from $\mathcal{M}$ of length $\|\boldsymbol{w}\| \geqslant \|\mho_1\|, \|\mho_2\|$.

Lemma 3.10 (Normal forms) *Each multiformula $\mho$ can be effectively transformed into either special DNF or CNF (SDNF or SCNF) that is componentwise equivalent to $\mho$, i.e., a DNF/CFN w.r.t. $\oslash$ and $\otimes$ such that for each $k = 1, \ldots, K \geqslant \|\mho\|$ each disjunct in the SDNF (conjunct in the SCNF) contains exactly one uniformula of type $C^{(k)}$ and one of type $\overline{D}^{(k)}$. Moreover, this can be done without changing the set of propositional variables occurring in $\mho$ or, indeed, without changing their polarities.*

Proof. The transformation to a DNF/CNF is standard. Uniformulas of the same type are merged using componentwise equivalences of the following types: $C^{(k)} \oslash D^{(k)} \dashv\models (C \wedge D)^{(k)}$, $C^{(k)} \otimes D^{(k)} \dashv\models (C \vee D)^{(k)}$, $\overline{C}^{(k)} \oslash \overline{D}^{(k)} \dashv\models \overline{C \vee D}^{(k)}$, $\overline{C}^{(k)} \otimes \overline{D}^{(k)} \dashv\models \overline{C \wedge D}^{(k)}$, $\overline{C}^{(k)} \dashv\models \overline{C}^{(k)} \oslash \top^{(k)}$, $\overline{C}^{(k)} \dashv\models \overline{C}^{(k)} \otimes \bot^{(k)}$, etc. □

Definition 3.11 A *componentwise interpolant of a split hypersequent* $\widetilde{\mathcal{G}}$ w.r.t. a class $\mathcal{C}$ of models, written $\widetilde{\mathcal{G}} \overset{\mathcal{C}}{\longleftarrow} \mho$, is a multiformula $\mho$ of arity $\|\mho\| \leqslant \|\widetilde{\mathcal{G}}\|$ containing only propositional variables common to $L\widetilde{\mathcal{G}}$ and $R\widetilde{\mathcal{G}}$ and such that for each model $\mathcal{M} \in \mathcal{C}$ and each $\mathcal{M}$-rooted sequence $\boldsymbol{w}$ of length $\|\boldsymbol{w}\| = \|\widetilde{\mathcal{G}}\|$,

$$\mathcal{M}, \boldsymbol{w} \not\models \mho \implies \mathcal{M}, \boldsymbol{w} \models L\widetilde{\mathcal{G}} \qquad \text{and} \qquad \mathcal{M}, \boldsymbol{w} \models \mho \implies \mathcal{M}, \boldsymbol{w} \models R\widetilde{\mathcal{G}}.$$

We call the existence of such $\mho$ for $\widetilde{\mathcal{G}}$ the *interpolation statement for* $\widetilde{\mathcal{G}}$.

Lemma 3.12 *Let a logic* L *be sound and complete w.r.t. a class* $\mathcal{C}$ *of models and* $\mho = \bigotimes_{i=1}^{n} \left(\overline{C_i}^{(1)} \otimes D_i^{(1)}\right)$. *If* $A; \Rightarrow ; B \overset{\mathcal{C}}{\leftarrow} \mho$, *then* $I := \bigwedge_{i=1}^{n}(C_i \to D_i)$ *is a Craig interpolant of* $A \to B$ *w.r.t.* L.

Proof. Since $L(A; \Rightarrow ; B)$ is $A \Rightarrow$ and $R(A; \Rightarrow ; B)$ is $\Rightarrow B$, the formulas C_i and D_i contain only propositional variables common to A and B.

Let $\mathcal{M} = (W, \leqslant, V) \in \mathcal{C}$ and $v \in W$ be arbitrary. We need to show that $\mathcal{M}, v \Vdash A \to I$ and $\mathcal{M}, v \Vdash I \to B$. For the former, assume $\mathcal{M}, w \Vdash A$ for an arbitrary $w \geqslant v$. For any $w' \geqslant w$, by monotonicity $\mathcal{M}, w' \Vdash A$, i.e., $\mathcal{M}, w' \not\models A \Rightarrow$. By the definition of componentwise interpolation $\mathcal{M}, w' \models \mho$. In particular, $\mathcal{M}, w' \models \overline{C_i}^{(1)} \otimes D_i^{(1)}$, in other words, $\mathcal{M}, w' \nVdash C_i$ or $\mathcal{M}, w' \Vdash D_i$ for $i = 1, \ldots, n$. Thus, $\mathcal{M}, w \Vdash C_i \to D_i$ for each $i = 1, \ldots, n$, yielding $\mathcal{M}, w \Vdash I$.

To show $\mathcal{M}, v \Vdash I \to B$, assume $\mathcal{M}, w \Vdash I$ for an arbitrary $w \geqslant v$. Then, for each $i = 1, \ldots, n$, $\mathcal{M}, w \Vdash C_i \to D_i$, implying $\mathcal{M}, w \nVdash C_i$ or $\mathcal{M}, w \Vdash D_i$ and $\mathcal{M}, w \models \overline{C_i}^{(1)} \otimes D_i^{(1)}$. Thus, $\mathcal{M}, w \models \mho$. By the definition of componentwise interpolation, $\mathcal{M}, w \models \Rightarrow B$, i.e., $\mathcal{M}, w \Vdash B$. □

4 Interpolation Algorithm for Int and LQ

Definition 4.1 We define operations on multiformulas $\mho^{n+1 \mapsto n}$ that changes every superscript $(n+1)$ into (n) and $\mho^{n \leftrightarrow n+1}$ that swaps (n) and $(n+1)$. In either case everything else is left intact.

Lemma 4.2 *Let* $\mathcal{M}$ *be a model,* $\mho$ *be a multiformula, and* $\boldsymbol{w}$ *be an* $\mathcal{M}$*-rooted sequence of worlds. For* $\|\boldsymbol{w}\| = n$ *and* $\|\mho\| \leqslant n+1$, *we have* $\mathcal{M}, \boldsymbol{w} \models \mho^{n+1 \mapsto n}$ *iff* $\mathcal{M}, \boldsymbol{w}, w_n \models \mho$. *For* $\|\boldsymbol{w}\| = n+k$ *and* $\|\mho\| \leqslant n+k$ *for some* $k \geqslant 1$, *we have* $\mathcal{M}, \boldsymbol{w} \models \mho^{n \leftrightarrow n+1}$ *iff* $\mathcal{M}, w_1, \ldots, w_{n-1}, w_{n+1}, w_n, w_{n+2}, \ldots, w_{n+k} \models \mho$.

An algorithm for constructing componentwise interpolants for split hypersequents derivable in SHInt is presented via interpolant transformations for each rule of SHInt. All non-identity transformations can be found in Figure 2. In order to apply the transformations for $(\Rightarrow \to^{\mathsf{l}})$ and $(\Rightarrow \to^{\mathsf{r}})$ the interpolant of the premiss must be first transformed to the relevant normal form by Lemma 3.10.

Lemma 4.3 *Let* $\mathcal{C}$ *be a class of models. For each rule from Figure 2, if the interpolation statement(s) in the premiss(es) of the rule hold(s), so does the interpolation statement in its conclusion. For each split hypersequent rule of* Int *not present in Figure 2, which are all unary, any componentwise interpolant of the premiss w.r.t.* $\mathcal{C}$ *is also a componentwise interpolant for its conclusion w.r.t.* $\mathcal{C}$.

Proof. We only consider several complex cases, the remaining ones being analogous and simpler. We consider an arbitrarily chosen model $\mathcal{M} = (W, \leqslant, V) \in \mathcal{C}$ and $\mathcal{M}$-rooted sequence $\boldsymbol{w}$ of appropriate length. We omit $\mathcal{M}$ from the $\Vdash$ statements about formulas and from $\models$ statements about multiformulas and hypersequents. We do not discuss the common variable conditions, which are always easy to verify, either.

$$\mathsf{id}^{\mathsf{ll}}\ \frac{}{A;\Rightarrow A;\ \overset{\mathcal{C}}{\longleftarrow}\ \bot^{(1)}}\qquad \mathsf{id}^{\mathsf{rl}}\ \frac{}{;A\Rightarrow A;\ \overset{\mathcal{C}}{\longleftarrow}\ \overline{A}^{(1)}}\qquad \mathsf{id}^{\mathsf{lr}}\ \frac{}{A;\Rightarrow ;A\ \overset{\mathcal{C}}{\longleftarrow}\ A^{(1)}}$$

$$\mathsf{id}^{\mathsf{rr}}\ \frac{}{;A\Rightarrow ;A\ \overset{\mathcal{C}}{\longleftarrow}\ \top^{(1)}}\qquad \mathsf{id}^{\mathsf{l}}_{\bot}\ \frac{}{\bot;\Rightarrow\ \overset{\mathcal{C}}{\longleftarrow}\ \bot^{(1)}}\qquad \mathsf{id}^{\mathsf{r}}_{\bot}\ \frac{}{;\bot\Rightarrow\ \overset{\mathcal{C}}{\longleftarrow}\ \top^{(1)}}$$

$$\Rightarrow\wedge^{\mathsf{l}}\ \frac{\mathcal{G}\mid\widetilde{\Gamma}\Rightarrow A;\ \overset{\mathcal{C}}{\longleftarrow}\ \mho_1\quad \mathcal{G}\mid\widetilde{\Gamma}\Rightarrow B;\ \overset{\mathcal{C}}{\longleftarrow}\ \mho_2}{\mathcal{G}\mid\widetilde{\Gamma}\Rightarrow A\wedge B;\ \overset{\mathcal{C}}{\longleftarrow}\ \mho_1\ovee\mho_2}\qquad \Rightarrow\wedge^{\mathsf{r}}\ \frac{\mathcal{G}\mid\widetilde{\Gamma}\Rightarrow ;A\ \overset{\mathcal{C}}{\longleftarrow}\ \mho_1\quad \mathcal{G}\mid\widetilde{\Gamma}\Rightarrow ;B\ \overset{\mathcal{C}}{\longleftarrow}\ \mho_2}{\mathcal{G}\mid\widetilde{\Gamma}\Rightarrow ;A\wedge B\ \overset{\mathcal{C}}{\longleftarrow}\ \mho_1\owedge\mho_2}$$

$$\vee^{\mathsf{l}}\Rightarrow\ \frac{\mathcal{G}\mid\Gamma,A;\Pi\Rightarrow\widetilde{\Delta}\ \overset{\mathcal{C}}{\longleftarrow}\ \mho_1\quad \mathcal{G}\mid\Gamma,B;\Pi\Rightarrow\widetilde{\Delta}\ \overset{\mathcal{C}}{\longleftarrow}\ \mho_2}{\mathcal{G}\mid\Gamma,A\vee B;\Pi\Rightarrow\widetilde{\Delta}\ \overset{\mathcal{C}}{\longleftarrow}\ \mho_1\ovee\mho_2}$$

$$\vee^{\mathsf{r}}\Rightarrow\ \frac{\mathcal{G}\mid\Gamma;\Pi,A\Rightarrow\widetilde{\Delta}\ \overset{\mathcal{C}}{\longleftarrow}\ \mho_1\quad \mathcal{G}\mid\Gamma;\Pi,B\Rightarrow\widetilde{\Delta}\ \overset{\mathcal{C}}{\longleftarrow}\ \mho_2}{\mathcal{G}\mid\Gamma;\Pi,A\vee B\Rightarrow\widetilde{\Delta}\ \overset{\mathcal{C}}{\longleftarrow}\ \mho_1\owedge\mho_2}$$

$$\rightarrow^{\mathsf{l}}\Rightarrow\ \frac{\widetilde{\mathcal{G}}\mid\Gamma;\Pi\Rightarrow A;\ \overset{\mathcal{C}}{\longleftarrow}\ \mho_1\quad \widetilde{\mathcal{G}}\mid\Gamma,B;\Pi\Rightarrow\widetilde{\Delta}\ \overset{\mathcal{C}}{\longleftarrow}\ \mho_2}{\widetilde{\mathcal{G}}\mid\Gamma,A\rightarrow B;\Pi\Rightarrow\widetilde{\Delta}\ \overset{\mathcal{C}}{\longleftarrow}\ \mho_1\ovee\mho_2}$$

$$\rightarrow^{\mathsf{r}}\Rightarrow\ \frac{\widetilde{\mathcal{G}}\mid\Gamma;\Pi\Rightarrow ;A\ \overset{\mathcal{C}}{\longleftarrow}\ \mho_1\quad \widetilde{\mathcal{G}}\mid\Gamma;\Pi,B\Rightarrow\widetilde{\Delta}\ \overset{\mathcal{C}}{\longleftarrow}\ \mho_2}{\widetilde{\mathcal{G}}\mid\Gamma;\Pi,A\rightarrow B\Rightarrow\widetilde{\Delta}\ \overset{\mathcal{C}}{\longleftarrow}\ \mho_1\owedge\mho_2}$$

$$\Rightarrow\rightarrow^{\mathsf{l}}\ \frac{\widetilde{\mathcal{G}}\mid\Gamma,A;\Pi\Rightarrow B;\ \overset{\mathcal{C}}{\longleftarrow}\ \bigovee_{j=1}^{m}\Big(\overline{C_j}^{(n)}\owedge D_j^{(n)}\owedge\bigowedge_{l=1}^{n-1}(\overline{E_{jl}}^{(l)}\owedge F_{jl}^{(l)})\Big)}{\widetilde{\mathcal{G}}\mid\Gamma;\Pi\Rightarrow A\rightarrow B;\ \overset{\mathcal{C}}{\longleftarrow}\ \bigovee_{j=1}^{m}\Big(\overline{D_j\rightarrow C_j}^{(n)}\owedge\bigowedge_{l=1}^{n-1}(\overline{E_{jl}}^{(l)}\owedge F_{jl}^{(l)})\Big)}$$

$$\Rightarrow\rightarrow^{\mathsf{r}}\ \frac{\widetilde{\mathcal{G}}\mid\Gamma;\Pi,A\Rightarrow ;B\ \overset{\mathcal{C}}{\longleftarrow}\ \bigowedge_{j=1}^{m}\Big(\overline{C_j}^{(n)}\ovee D_j^{(n)}\ovee\bigovee_{l=1}^{n-1}(\overline{E_{jl}}^{(l)}\ovee F_{jl}^{(l)})\Big)}{\widetilde{\mathcal{G}}\mid\Gamma;\Pi\Rightarrow ;A\rightarrow B\ \overset{\mathcal{C}}{\longleftarrow}\ \bigowedge_{j=1}^{m}\Big((C_j\rightarrow D_j)^{(n)}\ovee\bigovee_{l=1}^{n-1}(\overline{E_{jl}}^{(l)}\ovee F_{jl}^{(l)})\Big)}$$

$$\mathsf{EC}\ \frac{\widetilde{\mathcal{G}}\mid\widetilde{\Gamma}\Rightarrow\widetilde{\Delta}\mid\widetilde{\Gamma}\Rightarrow\widetilde{\Delta}\ \overset{\mathcal{C}}{\longleftarrow}\ \mho}{\widetilde{\mathcal{G}}\mid\widetilde{\Gamma}\Rightarrow\widetilde{\Delta}\ \overset{\mathcal{C}}{\longleftarrow}\ \mho^{n+1\mapsto n}}\qquad \mathsf{Ex}\ \frac{\widetilde{\mathcal{G}}\mid\widetilde{\Gamma}\Rightarrow\widetilde{\Delta}\mid\widetilde{\Lambda}\Rightarrow\widetilde{\Theta}\mid\widetilde{\mathcal{H}}\ \overset{\mathcal{C}}{\longleftarrow}\ \mho}{\widetilde{\mathcal{G}}\mid\widetilde{\Lambda}\Rightarrow\widetilde{\Theta}\mid\widetilde{\Gamma}\Rightarrow\widetilde{\Delta}\mid\widetilde{\mathcal{H}}\ \overset{\mathcal{C}}{\longleftarrow}\ \mho^{n\leftrightarrow n+1}}$$

Fig. 2. Interpolation transformations for SHInt. For the unary rules of SHInt and structural rules not depicted above, any interpolant for the premiss also interpolates the conclusion. It is required that (1) $\|\mho_i\| \leqslant \|\widetilde{\mathcal{G}}\| + 1$ for $i = 1, 2$ in ($\Rightarrow\wedge^{\mathsf{l}}$), ($\Rightarrow\wedge^{\mathsf{r}}$), ($\vee^{\mathsf{l}}\Rightarrow$), ($\vee^{\mathsf{r}}\Rightarrow$), ($\rightarrow^{\mathsf{l}}\Rightarrow$), and ($\rightarrow^{\mathsf{r}}\Rightarrow$); (2) $\|\widetilde{\mathcal{G}}\| = n-1$ in ($\Rightarrow\rightarrow^{\mathsf{l}}$), ($\Rightarrow\rightarrow^{\mathsf{r}}$), (EC), and (Ex); (3) $\|\mho\| \leqslant n+1$ in (EC); (4) $\|\mho\| \leqslant n + \|\widetilde{\mathcal{H}}\| + 1$ in (Ex).

Rule ($\Rightarrow\rightarrow^{\mathsf{l}}$). Let the componentwise interpolation statement for the premiss of ($\Rightarrow\rightarrow^{\mathsf{l}}$) in Figure 2 hold, in particular, $\|\mathcal{G}\| = n-1$. Assume, for the left side, that $\boldsymbol{w} \not\Vdash \bigovee_{j=1}^{m}\Big(\overline{D_j\rightarrow C_j}^{(n)}\owedge\bigowedge_{l=1}^{n-1}(\overline{E_{jl}}^{(l)}\owedge F_{jl}^{(l)})\Big)$. Let $\boldsymbol{v} := w_1,\dots,w_{n-1}$. For each $j = 1,\dots,m$ and each $w'_n \geqslant w_n$, either $\boldsymbol{v} \not\Vdash \bigowedge_{l=1}^{n-1}(\overline{E_{jl}}^{(l)}\owedge F_{jl}^{(l)})$, or $w'_n \not\Vdash D_j$, or $w'_n \Vdash C_j$, i.e., for each $w'_n \geqslant w_n$

$$\mathsf{lqS}\ \dfrac{\widetilde{\mathcal{G}} \mid \Gamma, \Lambda; \Pi, \Phi \Rightarrow \overset{\mathcal{J}}{\longleftarrow} \bigowedge_{j=1}^{m} \left(\overline{C_j}^{(n)} \ovee D_j{}^{(n)} \ovee \bigovee_{l=1}^{n-1} (\overline{E_{jl}}^{(l)} \ovee F_{jl}{}^{(l)})\right)}{\widetilde{\mathcal{G}} \mid \Gamma; \Pi \Rightarrow \mid \Lambda; \Phi \Rightarrow \overset{\mathcal{J}}{\longleftarrow} \bigowedge_{j=1}^{m} \left(\overline{\neg(C_j \to D_j)}^{(n)} \ovee \bigovee_{l=1}^{n-1} (\overline{E_{jl}}^{(l)} \ovee F_{jl}{}^{(l)})\right)}$$

Fig. 3. Interpolation transformation for lqS. It is required that $\|\mathcal{G}\| = n-1$.

$\boldsymbol{v}, w'_n \not\models \bigovee_{j=1}^{m} \left(\overline{C_j}^{(n)} \owedge D_j{}^{(n)} \owedge \bigowedge_{l=1}^{n-1} (\overline{E_{jl}}^{(l)} \owedge F_{jl}{}^{(l)})\right)$. Since $\boldsymbol{w} = \boldsymbol{v}, w_n$ is $\mathcal{M}$-rooted, so is $\boldsymbol{v}, w'_n$ for any $w'_n \geqslant w_n$. Thus, for each $w'_n \geqslant w_n$, it follows from the premiss interpolant that $\boldsymbol{v}, w'_n \models L\widetilde{\mathcal{G}} \mid \Gamma, A \Rightarrow B$. In other words, either $\boldsymbol{v} \models L\widetilde{\mathcal{G}}$ or, for each $w'_n \geqslant w_n$, (a) $w'_n \not\Vdash G$ for some $G \in \Gamma$ or (b) $w'_n \not\Vdash A$ or $w'_n \Vdash B$. By monotonicity, (a) implies $w_n \not\Vdash G$ for some $G \in \Gamma$. Thus, $\boldsymbol{v} \models L\widetilde{\mathcal{G}}$, or $w_n \not\Vdash G$ for some $G \in \Gamma$, or, for all $w'_n \geqslant w_n$, we have $w'_n \not\Vdash A$ or $w'_n \Vdash B$. In other words, $\boldsymbol{w} \models L\widetilde{\mathcal{G}} \mid \Gamma \Rightarrow A \to B$.

For the right side, let $\boldsymbol{w} \models \bigovee_{j=1}^{m} \left(\overline{D_j \to C_j}^{(n)} \owedge \bigowedge_{l=1}^{n-1} (\overline{E_{jl}}^{(l)} \owedge F_{jl}{}^{(l)})\right)$. Then $\boldsymbol{v} \models \bigowedge_{l=1}^{n-1} (\overline{E_{jl}}^{(l)} \owedge F_{jl}{}^{(l)})$ and $w_n \not\Vdash D_j \to C_j$ for some $1 \leqslant j \leqslant m$. The latter implies that $w'_n \Vdash D_j$ and $w'_n \not\Vdash C_j$ for some $w'_n \geqslant w_n$. It follows that $\boldsymbol{v}, w'_n \models \bigovee_{j=1}^{m} \left(\overline{C_j}^{(n)} \owedge D_j{}^{(n)} \owedge \bigowedge_{l=1}^{n-1} (\overline{E_{jl}}^{(l)} \owedge F_{jl}{}^{(l)})\right)$ for this $w'_n \geqslant w_n$. It follows from the premiss interpolant that $\boldsymbol{v}, w'_n \models R\widetilde{\mathcal{G}} \mid \Pi \Rightarrow$ for this $w'_n \geqslant w_n$, and, by monotonicity, $\boldsymbol{w} \models R\widetilde{\mathcal{G}} \mid \Pi \Rightarrow$.

Rule (EC). Here $\|\widetilde{\mathcal{G}}\| + 1 = n$. By Lemma 4.2, $\boldsymbol{w} \models \mho^{n+1 \mapsto n}$ iff $\boldsymbol{w}, w_n \models \mho$. $\boldsymbol{w}, w_n \models S(\widetilde{\mathcal{G}} \mid \widetilde{\Gamma} \Rightarrow \widetilde{\Delta} \mid \widetilde{\Gamma} \Rightarrow \widetilde{\Delta})$ implies $\boldsymbol{w} \models S(\widetilde{\mathcal{G}} \mid \widetilde{\Gamma} \Rightarrow \widetilde{\Delta})$ for $S \in \{L, R\}$. □

Lemma 4.4 *Let $\mathcal{J}$ be the class of models $\mathcal{M}$ such that each $\mathcal{M}$ has a largest element ∞, i.e., with $w \leqslant \infty$ for all worlds w from $\mathcal{M}$. The interpolation statement in the premiss of* (lqS) *implies that in the conclusion.*

Proof. $\|\mathcal{G}\| = n-1$. Let $\boldsymbol{v} := w_1, \ldots, w_{n-1}$ for a given $\mathcal{M}$-rooted sequence $\boldsymbol{w}$ of length $\|\boldsymbol{w}\| = n+1$ consisting of worlds from $\mathcal{M} = (W, \leqslant, V) \in \mathcal{J}$. For the left side, assume $\boldsymbol{w} \not\models \bigowedge_{j=1}^{m} \left(\overline{\neg(C_j \to D_j)}^{(n)} \ovee \bigovee_{l=1}^{n-1} (\overline{E_{jl}}^{(l)} \ovee F_{jl}{}^{(l)})\right)$. Then there is some $1 \leqslant j \leqslant m$ such that $\boldsymbol{v} \not\models \bigovee_{l=1}^{n-1} (\overline{E_{jl}}^{(l)} \ovee F_{jl}{}^{(l)})$ and, in addition, $w_n \Vdash \neg(C_j \to D_j)$. In particular, $\infty \not\Vdash C_j \to D_j$. Given that ∞ is the largest element, by monotonicity, $\infty \Vdash C_j$ and $\infty \not\Vdash D_j$. Therefore, for the $\mathcal{M}$-rooted sequence $\boldsymbol{v}, \infty$, we have $\boldsymbol{v}, \infty \not\models \bigowedge_{j=1}^{m} \left(\overline{C_j}^{(n)} \ovee D_j{}^{(n)} \ovee \bigovee_{l=1}^{n-1} (\overline{E_{jl}}^{(l)} \ovee F_{jl}{}^{(l)})\right)$. It follows from the premiss interpolant that $\boldsymbol{v}, \infty \models L\widetilde{\mathcal{G}} \mid \Gamma, \Lambda \Rightarrow$. Since both $w_n \leqslant \infty$ and $w_{n+1} \leqslant \infty$, by monotonicity, $\boldsymbol{w} \models L\widetilde{\mathcal{G}} \mid \Gamma \Rightarrow \mid \Lambda \Rightarrow$.

Assume now $\boldsymbol{w} \models \bigowedge_{j=1}^{m} \left(\overline{\neg(C_j \to D_j)}^{(n)} \ovee \bigovee_{l=1}^{n-1} (\overline{E_{jl}}^{(l)} \ovee F_{jl}{}^{(l)})\right)$ for the right side. Then $\boldsymbol{v} \models \bigovee_{l=1}^{n-1} (\overline{E_{jl}}^{(l)} \ovee F_{jl}{}^{(l)})$ or $w_n \not\Vdash \neg(C_j \to D_j)$ for each $j = 1, \ldots, m$. Thus, for each $j = 1, \ldots, m$, either $\boldsymbol{v} \models \bigovee_{l=1}^{n-1} (\overline{E_{jl}}^{(l)} \ovee F_{jl}{}^{(l)})$ or $z_j \Vdash C_j \to D_j$ for some $z_j \geqslant w_n$. The latter implies $\infty \Vdash C_j \to D_j$ by monotonicity. Thus, for each $j = 1, \ldots, m$, $\boldsymbol{v} \models \bigovee_{l=1}^{n-1} (\overline{E_{jl}}^{(l)} \ovee F_{jl}^{(l)})$, or $\infty \not\Vdash C_j$, or $\infty \Vdash D_j$, i.e., $\boldsymbol{v}, \infty \models \bigowedge_{j=1}^{m} \Big(\overline{C_j}^{(n)} \ovee D_j^{(n)} \ovee \bigovee_{l=1}^{n-1} (\overline{E_{jl}}^{(l)} \ovee F_{jl}^{(l)})\Big)$. It follows from the premiss interpolant that $\boldsymbol{v}, \infty \models R\widetilde{\mathcal{G}} \mid \Pi, \Phi \Rightarrow$. As before, by monotonicity $\boldsymbol{w} \models R\widetilde{\mathcal{G}} \mid \Pi \Rightarrow \mid \Phi \Rightarrow$. $\square$

Theorem 4.5 Int *and* LQ *enjoy the CIP.*

Proof. Let $\mathsf{L} \vdash A \to B$ for $\mathsf{L} \in \{\mathsf{Int}, \mathsf{LQ}\}$. By Corollary 3.3, $\mathsf{SHL} \vdash A;\ \Rightarrow\ ; B$. By Lemmata 4.3 and 4.4, we can construct a componentwise interpolant $\mho$ of $A;\ \Rightarrow\ ; B$. By Lemma 3.10, this $\mho$ can be efficiently transformed to another componentwise interpolant $\mho'$ of length 1 in SCNF. By Lemma 3.12, this $\mho'$ can be efficiently transformed to a formula interpolant C of A and B. $\square$

Theorem 4.6 HSG *does not enjoy the componentwise interpolation property w.r.t. the class $\mathcal{L}in$ of all linear models, i.e., there is a derivable hypersequent for which the componentwise interpolation statement does not hold.*

Proof. The split sequent $;q \Rightarrow p;\ \mid p;\ \Rightarrow\ ;q$ is derivable in SHG from $p;\ \Rightarrow p;$ and $;q \Rightarrow\ ;q$ by the rule (comS). Since no propositional variable occurs on both sides, componentwise interpolants could only be constructed from $\bot$. Simple induction on the interpolant construction shows that any such interpolant would be componentwise equivalent to either $\top^{(1)}$ or $\bot^{(1)}$. Depending on which it were, either $\mathcal{M}, w_1, w_2 \models q \Rightarrow\ \mid\ \Rightarrow q$ for any linear model $\mathcal{M}$ and any $\mathcal{M}$-rooted sequence w_1, w_2 or $\mathcal{M}, w_1, w_2 \models\ \Rightarrow p \mid p \Rightarrow$ for any linear model $\mathcal{M}$ and any $\mathcal{M}$-rooted sequence w_1, w_2. However, both hypersequents can be easily refuted in componentwise semantics. This contradiction completes the proof. $\square$

5 Linear Nested Sequents for G

Given that the standard hypersequent system HG for G turned out to be useless as far as interpolation proofs go, we consider a different calculus in the framework of *linear nested sequents* [15], essentially a reformulation of 2-sequents [19].

Definition 5.1 A *linear nested sequent* is a finite list of multi-conclusion sequents $\mathcal{G} = \Gamma_1 \Rightarrow \Delta_1 /\!\!/ \ldots /\!\!/ \Gamma_n \Rightarrow \Delta_n$ with $n > 0$. Its *formula interpretation* is given by $\iota(\Gamma \Rightarrow \Delta) := \bigwedge \Gamma \to \bigvee \Delta$ and $\iota(\Gamma \Rightarrow \Delta /\!\!/ \mathcal{G}) := \bigwedge \Gamma \to \bigvee \Delta \vee \iota(\mathcal{G})$.

While linear nested sequents resemble hypersequents, the formula interpretation is markedly different in that it introduces nested implications, in line with the linear structure of models from $\mathcal{L}in$. The presented calculi can be seen as building on the nested sequent calculus for intuitionistic logic from [8]. The rules capturing linearity of frames are inspired by the analogous treatment of linearity in the noncommutative hypersequent calculi for temporal logics presented in [11]. See also [7] for further discussion on other calculi for G.

Definition 5.2 The rules of the calculus LNG are given in Figure 4. In the notation $\mathcal{G} /\!\!/ \Gamma \Rightarrow \Delta /\!\!/ \mathcal{H}$, none, one, or both of $\mathcal{G}$ and $\mathcal{H}$ could be empty.

$$\frac{}{\mathcal{G}/\!\!/\Gamma, p \Rightarrow \Delta, p/\!\!/\mathcal{H}}\ \mathsf{init}_1 \qquad \frac{}{\mathcal{G}/\!\!/\Gamma, p \Rightarrow \Delta/\!\!/\mathcal{H}/\!\!/\Sigma \Rightarrow \Pi, p/\!\!/\mathcal{I}}\ \mathsf{init}_2$$

$$\frac{}{\mathcal{G}/\!\!/\Gamma, \bot \Rightarrow \Delta/\!\!/\mathcal{H}}\ \bot_L \qquad \frac{\mathcal{G}/\!\!/\Gamma, A \Rightarrow \Delta/\!\!/\Sigma, A \Rightarrow \Pi/\!\!/\mathcal{H}}{\mathcal{G}/\!\!/\Gamma, A \Rightarrow \Delta/\!\!/\Sigma \Rightarrow \Pi/\!\!/\mathcal{H}}\ \mathsf{Lift}$$

$$\frac{\mathcal{G}/\!\!/\Gamma, A, B \Rightarrow \Delta/\!\!/\mathcal{H}}{\mathcal{G}/\!\!/\Gamma, A \wedge B \Rightarrow \Delta/\!\!/\mathcal{H}}\ \wedge_L \qquad \frac{\mathcal{G}/\!\!/\Gamma \Rightarrow \Delta, A/\!\!/\mathcal{H} \quad \mathcal{G}/\!\!/\Gamma \Rightarrow \Delta, B/\!\!/\mathcal{H}}{\mathcal{G}/\!\!/\Gamma \Rightarrow \Delta, A \wedge B/\!\!/\mathcal{H}}\ \wedge_R$$

$$\frac{\mathcal{G}/\!\!/\Gamma, A \Rightarrow \Delta/\!\!/\mathcal{H} \quad \mathcal{G}/\!\!/\Gamma, B \Rightarrow \Delta/\!\!/\mathcal{H}}{\mathcal{G}/\!\!/\Gamma, A \vee B \Rightarrow \Delta/\!\!/\mathcal{H}}\ \vee_L \qquad \frac{\mathcal{G}/\!\!/\Gamma \Rightarrow \Delta, A, B/\!\!/\mathcal{H}}{\mathcal{G}/\!\!/\Gamma \Rightarrow \Delta, A \vee B/\!\!/\mathcal{H}}\ \vee_R$$

$$\frac{\mathcal{G}/\!\!/\Gamma, B \Rightarrow \Delta/\!\!/\mathcal{H} \quad \mathcal{G}/\!\!/\Gamma, A \to B \Rightarrow \Delta, A/\!\!/\mathcal{H}}{\mathcal{G}/\!\!/\Gamma, A \to B \Rightarrow \Delta/\!\!/\mathcal{H}}\ \to_L \qquad \frac{\mathcal{G}/\!\!/\Gamma \Rightarrow \Delta/\!\!/A \Rightarrow B}{\mathcal{G}/\!\!/\Gamma \Rightarrow \Delta, A \to B}\ \to_R^1$$

$$\frac{\mathcal{G}/\!\!/\Gamma \Rightarrow \Delta/\!\!/A \Rightarrow B/\!\!/\Sigma \Rightarrow \Pi/\!\!/\mathcal{H} \qquad \mathcal{G}/\!\!/\Gamma \Rightarrow \Delta/\!\!/\Sigma \Rightarrow \Pi, A \to B/\!\!/\mathcal{H}}{\mathcal{G}/\!\!/\Gamma \Rightarrow \Delta, A \to B/\!\!/\Sigma \Rightarrow \Pi/\!\!/\mathcal{H}}\ \to_R^2$$

Fig. 4. Linear nested sequent calculus LNG for G.

Note that the calculus contains two rules for the implication on the right hand side. This mirrors the two possibilities in attempting to construct a countermodel for the formula $A \to B$: either we have not created a successor to the current world yet, in which case we create such a successor satisfying A and falsifying B, or the current world already has a successor. In the latter case, the witness falsifying the implication $A \to B$ is either in between the current world and that successor, giving the first premiss, or beyond it, giving the second. We modified the $(\to_L)$ rule as usual to have contraction admissible.

Theorem 5.3 (Soundness of LNG) $\mathsf{LNG} \vdash \mathcal{G}$ *implies* $\mathsf{G} \vdash \iota(\mathcal{G})$.

Proof. By induction on the derivation depth, showing for every rule that whenever the formula interpretation of its conclusion is falsifiable, then so is that of one of its premisses. E.g., for an application of $(\to_R^2)$ with conclusion $\mathcal{G}$ being

$$\Gamma_1 \Rightarrow \Delta_1/\!\!/\ldots/\!\!/\Gamma_n \Rightarrow \Delta_n/\!\!/\Gamma_{n+1} \Rightarrow \Delta_{n+1}, A \to B/\!\!/\Gamma_{n+2} \Rightarrow \Delta_{n+2}/\!\!/\ldots/\!\!/\Gamma_m \Rightarrow \Delta_m$$

where the displayed $A \to B$ is principal, suppose that $\mathcal{M}, w_0 \not\Vdash \iota(\mathcal{G})$ for some $\mathcal{M} = (W, \leqslant, V)$ and $w_0 \in W$. (We omit the model whenever safe.) Then there are $w_1, \ldots, w_m \in W$ such that $w_i \leqslant w_{i+1}$, and $w_i \Vdash \bigwedge \Gamma_i$, and $w_i \not\Vdash \bigvee \Delta_i$ for $i \geqslant 1$. Further, $w_{n+1} \not\Vdash A \to B$, meaning there is a world $u \geqslant w_{n+1}$ such that $u \Vdash A$ and $u \not\Vdash B$. Since the model is linear, either $u < w_{n+2}$ or $w_{n+2} \leqslant u$. In the former case, the sequence $w_0, \ldots, w_{n+1}, u, w_{n+2}, \ldots, w_m$ witnesses that the formula interpretation of the first premiss is falsified at w_0. In the latter case, $w_{n+2} \not\Vdash A \to B$ and, hence, the sequence $w_0, \ldots, w_{n+1}, w_{n+2}, \ldots, w_m$ witnesses that the formula interpretation of the second premiss is falsified at w_0. The remaining cases are similar and simpler and essentially follow [8]. □

While we refrain from giving a cut rule explicitly, we will show completeness via a syntactic cut elimination proof (Theorem 5.16). In preparation for this

$$\frac{\mathcal{G} /\!\!/ \Gamma \Rightarrow \Delta /\!\!/ \mathcal{H}}{\mathcal{G} /\!\!/ \Gamma, \Sigma \Rightarrow \Delta, \Pi /\!\!/ \mathcal{H}} \mathsf{W} \qquad \frac{\mathcal{G} /\!\!/ \mathcal{H}}{\mathcal{G} /\!\!/ \Rightarrow /\!\!/ \mathcal{H}} \mathsf{EW} \qquad \frac{\mathcal{G} /\!\!/ \Gamma \Rightarrow \Delta, A /\!\!/ \Sigma \Rightarrow \Pi /\!\!/ \mathcal{H}}{\mathcal{G} /\!\!/ \Gamma \Rightarrow \Delta /\!\!/ \Sigma \Rightarrow \Pi, A /\!\!/ \mathcal{H}} \mathsf{Lower}$$

$$\frac{\mathcal{G} /\!\!/ \Gamma, A, A \Rightarrow \Delta /\!\!/ \mathcal{H}}{\mathcal{G} /\!\!/ \Gamma, A \Rightarrow \Delta /\!\!/ \mathcal{H}} \mathsf{ICL} \qquad \frac{\mathcal{G} /\!\!/ \Gamma \Rightarrow \Delta, A, A /\!\!/ \mathcal{H}}{\mathcal{G} /\!\!/ \Gamma \Rightarrow \Delta, A /\!\!/ \mathcal{H}} \mathsf{ICR} \qquad \frac{\mathcal{G} /\!\!/ \Gamma \Rightarrow \Delta /\!\!/ \Sigma \Rightarrow \Pi /\!\!/ \mathcal{H}}{\mathcal{G} /\!\!/ \Gamma, \Sigma \Rightarrow \Delta, \Pi /\!\!/ \mathcal{H}} \mathsf{mrg}$$

Fig. 5. Structural rules: internal and external *weakening*, internal *contraction*, *lower*, and *merge*.

we prove a succession of lemmata concerning admissibility of certain structural rules and invertibility for the logical rules in the following sense.

Definition 5.4 A rule is *admissible* if derivability of the premiss(es) implies derivability of the conclusion. It is *depth-preserving admissible*, abbreviated *dp-admissible*, if some derivation of the conclusion has depth no greater than that of the derivations of all premiss(es). Finally, a rule is *invertible* if derivability of the conclusion implies derivability of the premisses.

Structural rules we consider are given in Figure 5. The proofs of the following lemmata are all by induction on the depth d of the derivation.

Lemma 5.5 *Weakening* (W) *is dp-admissible in* LNG.

Lemma 5.6 *External Weakening* (EW) *is admissible in* LNG.

Proof. The case of $d = 0$ is trivial. For $d = n + 1$ we distinguish cases according to the last applied rule. If the last rule was (Lift), we apply IH, followed either by one application of (Lift) or by Lemma 5.5 and two applications of (Lift). Let the last rule be $(\to^1_R)$:

$$\frac{\mathcal{G} /\!\!/ \Gamma \Rightarrow \Delta /\!\!/ A \Rightarrow B}{\mathcal{G} /\!\!/ \Gamma \Rightarrow \Delta, A \to B} \to^1_R$$

If the new component is not introduced at the very end, we use IH followed by $(\to^1_R)$. Otherwise, by IH we obtain both $\mathcal{G} /\!\!/ \Gamma \Rightarrow \Delta /\!\!/ A \Rightarrow B /\!\!/ \Rightarrow$ and $\mathcal{G} /\!\!/ \Gamma \Rightarrow \Delta /\!\!/ \Rightarrow /\!\!/ A \Rightarrow B$, and applying $(\to^1_R)$ to the latter followed by $(\to^2_R)$ we have the desired result. Finally, let the last rule be $(\to^2_R)$:

$$\frac{\mathcal{G} /\!\!/ \Gamma \Rightarrow \Delta /\!\!/ A \Rightarrow B /\!\!/ \Sigma \Rightarrow \Pi /\!\!/ \mathcal{H} \qquad \mathcal{G} /\!\!/ \Gamma \Rightarrow \Delta /\!\!/ \Sigma \Rightarrow \Pi, A \to B /\!\!/ \mathcal{H}}{\mathcal{G} /\!\!/ \Gamma \Rightarrow \Delta, A \to B /\!\!/ \Sigma \Rightarrow \Pi /\!\!/ \mathcal{H}} \to^2_R$$

If the new component is to be inserted anywhere apart from in between $\Gamma \Rightarrow \Delta, A \to B /\!\!/ \Sigma \Rightarrow \Pi$, we apply IH followed by $(\to^2_R)$. Otherwise, using IH thrice,

$$\dfrac{\mathcal{G} /\!\!/ \Gamma \Rightarrow \Delta /\!\!/ A \Rightarrow B /\!\!/ \Rightarrow /\!\!/ \Sigma \Rightarrow \Pi /\!\!/ \mathcal{H} \qquad \dfrac{\mathcal{G} /\!\!/ \Gamma \Rightarrow \Delta /\!\!/ \Rightarrow /\!\!/ A \Rightarrow B /\!\!/ \Sigma \Rightarrow \Pi /\!\!/ \mathcal{H} \quad \mathcal{G} /\!\!/ \Gamma \Rightarrow \Delta /\!\!/ \Rightarrow /\!\!/ \Sigma \Rightarrow \Pi, A \to B /\!\!/ \mathcal{H}}{\begin{array}{c}\mathcal{G} /\!\!/ \Gamma \Rightarrow \Delta /\!\!/ \Rightarrow A \to B /\!\!/ \Sigma \Rightarrow \Pi /\!\!/ \mathcal{H} \\ \vdots\end{array}} \to^2_R}{\mathcal{G} /\!\!/ \Gamma \Rightarrow \Delta, A \to B /\!\!/ \Rightarrow /\!\!/ \Sigma \Rightarrow \Pi /\!\!/ \mathcal{H}} \to^2_R$$

For other rules, IH followed by the same rule suffices. □

Lemma 5.7 (Lower) *is dp-admissible in* LNG.

Proof. If $d = 0$, then the sequent is derived using (init_1), (init_2), or $(\bot_L)$. The last two cases are trivial; in the first case, the desired sequent is an instance of (init_1) or (init_2). Let $d = n + 1$. For the last rule $(\rightarrow^1_R)$, the lower-formula is a side formula, and applying IH and $(\rightarrow^1_R)$ suffices. For the last rule $(\rightarrow^2_R)$ with the lower-formula A a side formula, the most interesting case is

$$\frac{\mathcal{G}/\!\!/\Gamma \Rightarrow \Delta, A/\!\!/C \Rightarrow D/\!\!/\Sigma \Rightarrow \Pi/\!\!/\mathcal{H} \qquad \mathcal{G}/\!\!/\Gamma \Rightarrow \Delta, A/\!\!/\Sigma \Rightarrow \Pi, C \rightarrow D/\!\!/\mathcal{H}}{\mathcal{G}/\!\!/\Gamma \Rightarrow \Delta, A, C \rightarrow D/\!\!/\Sigma \Rightarrow \Pi/\!\!/\mathcal{H}} \rightarrow^2_R$$

Applying IH twice to the left premiss and once to the right premiss yields $\mathcal{G}/\!\!/\Gamma \Rightarrow \Delta/\!\!/C \Rightarrow D/\!\!/\Sigma \Rightarrow \Pi, A/\!\!/\mathcal{H}$ and $\mathcal{G}/\!\!/\Gamma \Rightarrow \Delta/\!\!/\Sigma \Rightarrow \Pi, A, C \rightarrow D/\!\!/\mathcal{H}$. Now $(\rightarrow^2_R)$ produces the desired result. If the last rule was $(\rightarrow^2_R)$ with the principal formula being the lower-formula, then we take the right premiss. If the last rule was $(\wedge_R)$ or $(\vee_R)$ with the principal formula being the lower-formula, we apply IH to all active formulas and then the same rule. Note that we need dp-admissibility to be able to apply IH twice for $(\rightarrow^2_R)$ and $(\vee_R)$. Finally, in all remaining cases the lower-formula is a side formula of the last applied rule, and we apply IH followed by the same rule. □

Lemma 5.8 $(\wedge_R)$ *and* $(\vee_R)$ *are invertible in* LNG.

Since we absorb contraction into the (Lift) rule, we need to use a stronger form of invertibility, *m-invertibility*, for the rules $(\wedge_L)$, $(\vee_L)$, and $(\rightarrow_L)$, in a way reminiscent of the way cut is extended to multicut, as formulated in Lemma 5.9. As usual we write A^k for the multiset containing k copies of A.

Lemma 5.9 *For* $\sum_{i=1}^{n} k_i \geqslant 1$, (1) *implies* (2), *and* (3) *implies both* (4) *and* (5), *and* (6) *implies both* (7) *and* (8):

$$\mathsf{LNG} \vdash \Gamma_1, (A \wedge B)^{k_1} \Rightarrow \Delta_1/\!\!/\ldots/\!\!/\Gamma_n, (A \wedge B)^{k_n} \Rightarrow \Delta_n \tag{1}$$
$$\mathsf{LNG} \vdash \Gamma_1, A^{k_1}, B^{k_1} \Rightarrow \Delta_1/\!\!/\ldots/\!\!/\Gamma_n, A^{k_n}, B^{k_n} \Rightarrow \Delta_n \tag{2}$$
$$\mathsf{LNG} \vdash \Gamma_1, (A \vee B)^{k_1} \Rightarrow \Delta_1/\!\!/\ldots/\!\!/\Gamma_n, (A \vee B)^{k_n} \Rightarrow \Delta_n \tag{3}$$
$$\mathsf{LNG} \vdash \Gamma_1, A^{k_1} \Rightarrow \Delta_1/\!\!/\ldots/\!\!/\Gamma_n, A^{k_n} \Rightarrow \Delta_n \tag{4}$$
$$\mathsf{LNG} \vdash \Gamma_1, B^{k_1} \Rightarrow \Delta_1/\!\!/\ldots/\!\!/\Gamma_n, B^{k_n} \Rightarrow \Delta_n \tag{5}$$
$$\mathsf{LNG} \vdash \Gamma_1, (A \rightarrow B)^{k_1} \Rightarrow \Delta_1/\!\!/\ldots/\!\!/\Gamma_n, (A \rightarrow B)^{k_n} \Rightarrow \Delta_n \tag{6}$$
$$\mathsf{LNG} \vdash \Gamma_1, B^{k_1} \Rightarrow \Delta_1/\!\!/\ldots/\!\!/\Gamma_n, B^{k_n} \Rightarrow \Delta_n \tag{7}$$
$$\mathsf{LNG} \vdash \Gamma_1, (A \rightarrow B)^{k_1} \Rightarrow \Delta_1, A^{k_1}/\!\!/\ldots/\!\!/\Gamma_n, (A \rightarrow B)^{k_n} \Rightarrow \Delta_n, A^{k_n} \tag{8}$$

Proof. We prove that (1) implies (2) by induction on the derivation depth. The crucial case is when the last applied rule is (Lift). W.l.o.g. let the first 2 components be active and $k_1 > 0$:

$$\frac{\Gamma_1, (A \wedge B)^{k_1} \Rightarrow \Delta_1/\!\!/\Gamma_2, (A \wedge B)^{k_2+1} \Rightarrow \Delta_2/\!\!/\ldots/\!\!/\Gamma_n, (A \wedge B)^{k_n} \Rightarrow \Delta_n}{\Gamma_1, (A \wedge B)^{k_1} \Rightarrow \Delta_1/\!\!/\Gamma_2, (A \wedge B)^{k_2} \Rightarrow \Delta_2/\!\!/\ldots/\!\!/\Gamma_n, (A \wedge B)^{k_n} \Rightarrow \Delta_n} \,\mathsf{Lift}$$

$\Gamma_1, A^{k_1}, B^{k_1} \Rightarrow \Delta_1 /\!\!/ \Gamma_2, A^{k_2+1}, B^{k_2+1} \Rightarrow \Delta_2 /\!\!/ \ldots /\!\!/ \Gamma_n, A^{k_n}, B^{k_n} \Rightarrow \Delta_n$ by IH for the premiss. Now apply (Lift) twice. Note that this breaks depth-preserving invertibility and shows why we need m-invertibility instead of standard invertibility. In all remaining cases, apply IH followed by the same rule. The proof of (4) and (5) from (3) is similar. (8) follows from (6) by Lemma 5.5. The proof of (7) from (6) is by induction on the depth of the derivation. If all of the implications are side formulas in the last applied rule, or principal in the (Lift) rule, apply IH and the same rule. If one of the implications is principal in $(\to_L)$ and it is not the only implication, then applying IH to the left premiss suffices. □

Lemma 5.10 *The rule* $(\to_R^2)$ *is invertible in* LNG.

Proof. Invertibility w.r.t. the right premiss follows immediately from admissibility of (Lower) (Lemma 5.7). Invertibility w.r.t. the left premiss is shown by induction on the derivation depth d. If $d = 0$, the same rule clearly applies. If $d = n + 1$, we distinguish cases according to the last applied rule. If it is one of $(\wedge_L)$, $(\wedge_R)$, $(\vee_L)$, $(\vee_R)$, $(\to_L)$, (Lift), or $(\to_R^1)$, then apply IH on the premiss(es) of that rule followed by the same rule, possibly twice together with admissibility of (W) in the case of (Lift). If the last rule is $(\to_R^2)$ and the relevant formula is principal, we are done. If the relevant formula is a side formula of a side component, we apply IH followed by $(\to_R^2)$. Otherwise,

$$\frac{\mathcal{G} /\!\!/ \Gamma \Rightarrow \Delta, A \to B /\!\!/ C \Rightarrow D /\!\!/ \Sigma \Rightarrow \Pi /\!\!/ \mathcal{H} \quad \mathcal{G} /\!\!/ \Gamma \Rightarrow \Delta, A \to B /\!\!/ \Sigma \Rightarrow \Pi, C \to D /\!\!/ \mathcal{H}}{\mathcal{G} /\!\!/ \Gamma \Rightarrow \Delta, A \to B, C \to D /\!\!/ \Sigma \Rightarrow \Pi /\!\!/ \mathcal{H}} \to_R^2 \quad (9)$$

Using IH on the premisses of this we obtain the premisses of

$$\frac{\mathcal{G} /\!\!/ \Gamma \Rightarrow \Delta /\!\!/ A \Rightarrow B /\!\!/ C \Rightarrow D /\!\!/ \Sigma \Rightarrow \Pi /\!\!/ \mathcal{H} \quad \mathcal{G} /\!\!/ \Gamma \Rightarrow \Delta /\!\!/ A \Rightarrow B /\!\!/ \Sigma \Rightarrow \Pi, C \to D /\!\!/ \mathcal{H}}{\mathcal{G} /\!\!/ \Gamma \Rightarrow \Delta /\!\!/ A \Rightarrow B, C \to D /\!\!/ \Sigma \Rightarrow \Pi /\!\!/ \mathcal{H}} \to_R^2$$

Furthermore, by dp-admissibility of Lower (Lemma 5.7) and IH on the left premiss of (9) we obtain $\mathcal{G} /\!\!/ \Gamma \Rightarrow \Delta /\!\!/ C \Rightarrow D /\!\!/ A \Rightarrow B /\!\!/ \Sigma \Rightarrow \Pi /\!\!/ \mathcal{H}$ and applying $(\to_R^2)$ yields $\mathcal{G} /\!\!/ \Gamma \Rightarrow \Delta, C \to D /\!\!/ A \Rightarrow B /\!\!/ \Sigma \Rightarrow \Pi /\!\!/ \mathcal{H}$. □

Lemma 5.11 *The rule* $(\to_R^1)$ *is invertible in* LNG.

Proof. By induction on the derivation depth. The base case is easy. For the induction step we distinguish cases according to the last rule in the derivation of $\mathcal{G} /\!\!/ \Gamma \Rightarrow \Delta, A \to B$. If that rule was one of $(\wedge_L)$, $(\wedge_R)$, $(\vee_L)$, $(\vee_R)$, $(\to_L)$, (Lift), or $(\to_R^2)$, then we apply IH on its premiss(es) followed by the same rule. If the last rule was $(\to_R^1)$ with principal $A \to B$, we take the premiss and are done. Otherwise, the last rule was $(\to_R^1)$ with $A \to B$ a side formula, so

$$\frac{\mathcal{G} /\!\!/ \Gamma \Rightarrow \Delta, A \to B /\!\!/ C \Rightarrow D}{\mathcal{G} /\!\!/ \Gamma \Rightarrow \Delta, A \to B, C \to D} \to_R^1$$

Applying dp-admissibility of Lower (Lemma 5.7) and IH on the premiss yields $\mathcal{G} /\!\!/ \Gamma \Rightarrow \Delta /\!\!/ C \Rightarrow D /\!\!/ A \Rightarrow B$. Further, Lemma 5.10 and $(\to_R^1)$ on the same premiss gives $\mathcal{G} /\!\!/ \Gamma \Rightarrow \Delta /\!\!/ A \Rightarrow B, C \to D$. Now applying $(\to_R^2)$ yields the result. □

Lemma 5.12 *Left Contraction* (ICL) *is admissible in* LNG.

Proof. By induction on the pairs $(|A|, d)$ in the lexicographic ordering, where d is the depth of the derivation and $|A|$ the complexity of the contraction formula. We distinguish cases according to the main connective of A. Let $\mathsf{LNG} \vdash \mathcal{G} /\!\!/ \Gamma, A, A \Rightarrow \Delta /\!\!/ \mathcal{H}$. Contracting initial sequents is easy. If A is a principal formula of (Lift) or a side formula, we apply IH to the premiss(es) of this rule, and then the same rule. If A is a principal conjunction/disjunction in the last rule, we apply Lemma 5.9, IH twice consecutively/in parallel and the rule $(\wedge_L)/(\vee_L)$. Finally, if $A = C \to D$ is principal in $(\to_L)$, then we have

$$\frac{\mathcal{G} /\!\!/ \Gamma, C \to D, D \Rightarrow \Delta /\!\!/ \mathcal{H} \qquad \mathcal{G} /\!\!/ \Gamma, C \to D, C \to D \Rightarrow \Delta, C /\!\!/ \mathcal{H}}{\mathcal{G} /\!\!/ \Gamma, C \to D, C \to D \Rightarrow \Delta /\!\!/ \mathcal{H}} \to_L$$

Applying Lemma 5.9 and IH for D to the left premiss gives $\mathcal{G} /\!\!/ \Gamma, D \Rightarrow \Delta /\!\!/ \mathcal{H}$. By IH for $C \to D$ in the right premiss, $\mathcal{G} /\!\!/ \Gamma, C \to D \Rightarrow \Delta, C /\!\!/ \mathcal{H}$. Now $(\to_L)$ yields the desired $\mathcal{G} /\!\!/ \Gamma, C \to D \Rightarrow \Delta /\!\!/ \mathcal{H}$. □

Lemma 5.13 *Merge* (mrg) *is admissible in* LNG.

Proof. By induction on the derivation depth. For the base case, the result of applying (mrg) to an initial sequent is still an initial sequent. For the induction step we distinguish cases according to the last rule. In most cases, including if it was one of $(\wedge_L)$, $(\wedge_R)$, $(\vee_L)$, $(\vee_R)$, $(\to_L)$, or $(\to^1_R)$, we apply IH and then the same rule. If (mrg) merges the principal components of (Lift), we apply IH and admissibility of (ICL) (Lemma 5.12). Finally, if (mrg) merges the principal components of $(\to^2_R)$, it is sufficient to apply IH to the right premiss. □

Lemma 5.14 *Right Contraction* (ICR) *is admissible in* LNG.

Proof. By induction on the pairs $(|A|, d)$ in the lexicographic ordering, where $|A|$ is the complexity of the contraction formula and d is the depth of the derivation. We distinguish cases according to the main connective of A and only show those where IH and the same rule cannot be used directly. If one of the occurrences of $A = C \wedge D / C \vee D$ is principal in $(\wedge_R)/(\vee_R)$, invert it by Lemma 5.9, and then use IH as needed. If $A = C \to D$ is principal in $(\to^1_R)$,

$$\frac{\mathcal{G} /\!\!/ \Gamma \Rightarrow \Delta, C \to D /\!\!/ C \Rightarrow D}{\mathcal{G} /\!\!/ \Gamma \Rightarrow \Delta, C \to D, C \to D} \to^1_R$$

Applying invertibility of $(\to^2_R)$ (Lemma 5.10) and admissibility of (mrg) (Lemma 5.13) to the premiss yields $\mathcal{G} /\!\!/ \Gamma \Rightarrow \Delta /\!\!/ C, C \Rightarrow D, D$. Now applying IH for D and admissibility of (ICL) for C (Lemma 5.12), followed by $(\to^1_R)$ yields $\mathcal{G} /\!\!/ \Gamma \Rightarrow \Delta, C \to D$. If $A = C \to D$ is principal in $(\to^2_R)$, we have

$$\frac{\mathcal{G} /\!\!/ \Gamma \Rightarrow \Delta, C \to D /\!\!/ C \Rightarrow D /\!\!/ \Sigma \Rightarrow \Pi /\!\!/ \mathcal{H} \quad \mathcal{G} /\!\!/ \Gamma \Rightarrow \Delta, C \to D /\!\!/ \Sigma \Rightarrow \Pi, C \to D /\!\!/ \mathcal{H}}{\mathcal{G} /\!\!/ \Gamma \Rightarrow \Delta, C \to D, C \to D /\!\!/ \Sigma \Rightarrow \Pi /\!\!/ \mathcal{H}} \to^2_R$$

From the left premiss, as before by the same Lemmata 5.10 and 5.13 we can obtain $\mathcal{G} /\!\!/ \Gamma \Rightarrow \Delta /\!\!/ C, C \Rightarrow D, D /\!\!/ \Sigma \Rightarrow \Pi /\!\!/ \mathcal{H}$, and by IH for D and admissibility of (ICL) (Lemma 5.12) for C, we get $\mathcal{G}/\!\!/\Gamma \Rightarrow \Delta/\!\!/C \Rightarrow D/\!\!/\Sigma \Rightarrow \Pi/\!\!/\mathcal{H}$. Further, from the right premiss above by dp-admissibility of (Lower) (Lemma 5.7) and IH we obtain $\mathcal{G}/\!\!/\Gamma \Rightarrow \Delta/\!\!/\Sigma \Rightarrow \Pi, C \to D/\!\!/\mathcal{H}$. Now $(\to_R^2)$ gives the desired result. □

We can now prove the admissibility of the cut rule. As we absorbed contraction into (Lift), we use a more general kind of cuts similar to "one-sided multicuts." While we do not use it explicitly, the cut rule for linear nested sequents can be read off the statement of Theorem 5.16 for the case $n = k_1 = 1$.

Definition 5.15 We define the *splice* $\mathcal{G}\oplus\mathcal{H}$ of linear nested sequents $\mathcal{G}$ and $\mathcal{H}$:

$$\begin{aligned}
(\Gamma \Rightarrow \Delta) \oplus (\Sigma \Rightarrow \Pi) &:= \Gamma, \Sigma \Rightarrow \Delta, \Pi \\
(\Gamma \Rightarrow \Delta) \oplus (\Sigma \Rightarrow \Pi /\!\!/ \Omega \Rightarrow \Theta /\!\!/ \mathcal{H}) &:= \Gamma, \Sigma \Rightarrow \Delta, \Pi /\!\!/ \Omega \Rightarrow \Theta /\!\!/ \mathcal{H} \\
(\Gamma \Rightarrow \Delta /\!\!/ \Omega \Rightarrow \Theta /\!\!/ \mathcal{H}) \oplus (\Sigma \Rightarrow \Pi) &:= \Gamma, \Sigma \Rightarrow \Delta, \Pi /\!\!/ \Omega \Rightarrow \Theta /\!\!/ \mathcal{H} \\
(\Gamma \Rightarrow \Delta /\!\!/ \Omega \Rightarrow \Theta /\!\!/ \mathcal{G}) \oplus (\Sigma \Rightarrow \Pi /\!\!/ \Xi \Rightarrow \Upsilon /\!\!/ \mathcal{H}) &:= \\
&\Gamma, \Sigma \Rightarrow \Delta, \Pi /\!\!/ \big((\Omega \Rightarrow \Theta /\!\!/ \mathcal{G}) \oplus (\Xi \Rightarrow \Upsilon /\!\!/ \mathcal{H})\big)
\end{aligned}$$

Theorem 5.16 (Cut elimination) *If* $\|\mathcal{G}\| = \|\mathcal{I}\|$, *and* $\sum_{i=1}^{n} k_i \geqslant 1$, *and* $\|\mathcal{H}\| = n-1$, *then* (10) *and* (11) *imply* (12)*:*

$$\mathsf{LNG} \vdash \mathcal{G} /\!\!/ \Gamma \Rightarrow \Delta, A /\!\!/ \mathcal{H} \tag{10}$$

$$\mathsf{LNG} \vdash \mathcal{I} /\!\!/ A^{k_1}, \Sigma_1 \Rightarrow \Pi_1 /\!\!/ \ldots /\!\!/ A^{k_n}, \Sigma_n \Rightarrow \Pi_n \tag{11}$$

$$\mathsf{LNG} \vdash (\mathcal{G} \oplus \mathcal{I}) /\!\!/ \Gamma, \Sigma_1 \Rightarrow \Delta, \Pi_1 /\!\!/ \big(\mathcal{H} \oplus (\Sigma_2 \Rightarrow \Pi_2 /\!\!/ \ldots /\!\!/ \Sigma_n \Rightarrow \Pi_n)\big) \tag{12}$$

Proof. By induction on the pairs $(|A|, d)$ in the lexicographic ordering, where $|A|$ is the complexity of the cut formula and d is the depth of the derivation. If $d = 0$, then (11) is an instance of (init_1), (init_2), or $(\bot_L)$. If none of the displayed A's is principal in that rule, (12) is an instance of the same rule. If an occurrence of $A = p$ is principal in (init_1) or (init_2), then A occurs in Π_i for some $i \leqslant n$, and we obtain (12) by (cut-free) admissibility of (Lower) (Lemma 5.7) and (W) (Lemma 5.5) from $\mathcal{G}/\!\!/\Gamma \Rightarrow \Delta, A/\!\!/\mathcal{H}$. If an occurrence of $A = \bot$ is principal in $(\bot_L)$, then we use the admissibility of

$$\frac{\mathcal{G} /\!\!/ \Gamma \Rightarrow \Delta, \bot /\!\!/ \mathcal{H}}{\mathcal{G} /\!\!/ \Gamma \Rightarrow \Delta /\!\!/ \mathcal{H}}$$

which can be proved by induction on the depth of the derivation, together with admissibility of (W) (Lemma 5.5) to obtain (12) from $\mathcal{G}/\!\!/\Gamma \Rightarrow \Delta, \bot/\!\!/\mathcal{H}$.

Let $d > 0$. If none of A's is principal in the last applied rule r, we apply IH to the premiss(es) of r, followed by r itself. If r is $(\to_R^1)$ or $(\to_R^2)$ we additionally use admissibility of (EW) (Lemma 5.6), e.g., if (11) is derived by

$$\frac{\mathcal{I} /\!\!/ A^{k_1}, \Sigma_1 \Rightarrow \Pi_1' /\!\!/ E \Rightarrow F /\!\!/ A^{k_2}, \Sigma_2 \Rightarrow \Pi_2 /\!\!/ \ldots /\!\!/ A^{k_n}, \Sigma_n \Rightarrow \Pi_n \qquad \mathcal{I} /\!\!/ A^{k_1}, \Sigma_1 \Rightarrow \Pi_1' /\!\!/ A^{k_2}, \Sigma_2 \Rightarrow \Pi_2, E \to F /\!\!/ \ldots /\!\!/ A^{k_n}, \Sigma_n \Rightarrow \Pi_n}{\mathcal{I} /\!\!/ A^{k_1}, \Sigma_1 \Rightarrow \Pi_1', E \to F /\!\!/ A^{k_2}, \Sigma_2 \Rightarrow \Pi_2 /\!\!/ \ldots /\!\!/ A^{k_n}, \Sigma_n \Rightarrow \Pi_n} \to_R^2$$

then admissibility of (EW) on $\mathcal{G}/\!\!/\Gamma \Rightarrow \Delta, A/\!\!/\mathcal{H}$ yields $\mathcal{G}/\!\!/\Gamma \Rightarrow \Delta, A/\!\!/ \Rightarrow /\!\!/\mathcal{H}$, and we get $(\mathcal{G}\oplus\mathcal{I})/\!\!/\Gamma, \Sigma_1 \Rightarrow \Delta, \Pi_1'/\!\!/E \Rightarrow F/\!\!/\big(\mathcal{H}\oplus(\Sigma_2 \Rightarrow \Pi_2/\!\!/\ldots/\!\!/\Sigma_n \Rightarrow \Pi_n\big)$ by IH on this and the top premiss above. By IH on $\mathcal{G}/\!\!/\Gamma \Rightarrow \Delta, A/\!\!/\mathcal{H}$ and the bottom premiss, $(\mathcal{G} \oplus \mathcal{I})/\!\!/\Gamma, \Sigma_1 \Rightarrow \Delta, \Pi_1'/\!\!/(\mathcal{H} \oplus (\Sigma_2 \Rightarrow \Pi_2, E \to F/\!\!/\ldots/\!\!/\Sigma_n \Rightarrow \Pi_n)$. Applying $(\to_R^2)$ to these two sequents yields (12) because $\Pi_1 = \Pi_1', E \to F$.

If one of A's is a principal formula of (Lift), then IH suffices. Otherwise, we distinguish several cases according to the main connective in A. If $A = C \wedge D$, by invertibility of $(\wedge_L)$ (Lemma 5.9) and $(\wedge_R)$ (Lemma 5.8), we can derive:

$$\mathcal{G}/\!\!/\Gamma \Rightarrow \Delta, C/\!\!/\mathcal{H} \quad \mathcal{G}/\!\!/\Gamma \Rightarrow \Delta, D/\!\!/\mathcal{H} \quad \mathcal{I}/\!\!/C^{k_1}, D^{k_1}, \Sigma_1 \Rightarrow \Pi_1/\!\!/\ldots/\!\!/C^{k_n}, D^{k_n}, \Sigma_n \Rightarrow \Pi_n$$

Since $|C|, |D| < |A|$ we can apply IH twice to obtain a derivation of

$$(\mathcal{G}/\!\!/\Gamma \Rightarrow \Delta/\!\!/\mathcal{H}) \oplus (\mathcal{G}/\!\!/\Gamma \Rightarrow \Delta/\!\!/\mathcal{H}) \oplus (\mathcal{I}/\!\!/\Sigma_1 \Rightarrow \Pi_1/\!\!/\ldots/\!\!/\Sigma_n \Rightarrow \Pi_n)$$

and admissibility of (ICL) and (ICR) (Lemmata 5.12 and 5.14) yields (12). If $A = C \vee D$, we proceed analogously.

If $A = C \to D$ and none of the occurrences of A in the right premiss (11) of the cut is principal, we proceed as for the case where A is a propositional variable. If one of its occurrences in (11) is principal, we have

$$\frac{\begin{array}{c}\mathcal{I}/\!\!/(C \to D)^{k_1}, \Sigma_1 \Rightarrow \Pi_1/\!\!/\ldots/\!\!/(C \to D)^{k_m}, \Sigma_m \Rightarrow \Pi_m, C/\!\!/\ldots/\!\!/(C \to D)^{k_n}, \Sigma_n \Rightarrow \Pi_n \\ \mathcal{I}/\!\!/(C \to D)^{k_1}, \Sigma_1 \Rightarrow \Pi_1/\!\!/\ldots/\!\!/(C \to D)^{k_m-1}, D, \Sigma_m \Rightarrow \Pi_m/\!\!/\ldots/\!\!/(C \to D)^{k_n}, \Sigma_n \Rightarrow \Pi_n\end{array}}{\mathcal{I}/\!\!/(C \to D)^{k_1}, \Sigma_1 \Rightarrow \Pi_1/\!\!/\ldots/\!\!/(C \to D)^{k_m}, \Sigma_m \Rightarrow \Pi_m/\!\!/\ldots/\!\!/(C \to D)^{k_n}, \Sigma_n \Rightarrow \Pi_n} \to_L$$

Suppose that $\mathcal{G}/\!\!/\Gamma \Rightarrow \Delta, C \to D/\!\!/\mathcal{H}$ is

$$\mathcal{G}/\!\!/\Gamma_1 \Rightarrow \Delta_1, C \to D/\!\!/\Gamma_2 \Rightarrow \Delta_2/\!\!/\ldots/\!\!/\Gamma_n \Rightarrow \Delta_n. \tag{13}$$

First, we apply IH in "cross-cuts" to (13) and each premiss of $(\to_L)$ to eliminate all $C \to D$'s in the context (if $\sum_{i=1}^{n} k_i = k_m = 1$, then admissible (W) is used for the bottom premiss instead), resulting in derivations of

$$(\mathcal{G} \oplus \mathcal{I})/\!\!/\Gamma_1, \Sigma_1 \Rightarrow \Delta_1, \Pi_1/\!\!/\ldots/\!\!/\Gamma_m, \Sigma_m, D \Rightarrow \Delta_m, \Pi_m/\!\!/\ldots/\!\!/\Gamma_n, \Sigma_n \Rightarrow \Delta_n, \Pi_n \tag{14}$$
$$(\mathcal{G} \oplus \mathcal{I})/\!\!/\Gamma_1, \Sigma_1 \Rightarrow \Delta_1, \Pi_1/\!\!/\ldots/\!\!/\Gamma_m, \Sigma_m \Rightarrow \Delta_m, \Pi_m, C/\!\!/\ldots/\!\!/\Gamma_n, \Sigma_n \Rightarrow \Delta_n, \Pi_n \tag{15}$$

From (13), we obtain $\mathcal{G}/\!\!/\Gamma_1 \Rightarrow \Delta_1/\!\!/\ldots/\!\!/\Gamma_m, C \Rightarrow \Delta_m, D/\!\!/\ldots/\!\!/\Gamma_n \Rightarrow \Delta_n$ by using admissibility of (Lower) (Lemma 5.7), invertibility of $(\to_R^2)$ (Lemma 5.10) or of $(\to_R^1)$ (Lemma 5.11) depending on whether $m < n$ or $m = n$ respectively, and admissibility of (mrg) (Lemma 5.13). Since $|C|, |D| < |C \to D|$, applying IH twice to the resulting sequent and the sequents (14) and (15) yields

$$(\mathcal{G}\oplus\mathcal{G}\oplus\mathcal{G}\oplus\mathcal{I}\oplus\mathcal{I})/\!\!/(\Gamma_1)^3, (\Sigma_1)^2 \Rightarrow (\Delta_1)^3, (\Pi_1)^2/\!\!/\ldots/\!\!/(\Gamma_n)^3, (\Sigma_n)^2 \Rightarrow (\Delta_n)^3, (\Pi_n)^2$$

Finally, using admissibility of Contraction (Lemmata 5.12 and 5.14) we obtain the desired $(\mathcal{G} \oplus \mathcal{I})/\!\!/\Gamma_1, \Sigma_1 \Rightarrow \Delta_1, \Pi_1/\!\!/\ldots/\!\!/\Gamma_n, \Sigma_n \Rightarrow \Delta_n, \Pi_n$. □

Corollary 5.17 (Completeness of LNG) G $\vdash A$ *implies* LNG $\vdash\Rightarrow A$.

Proof. It is easy to derive the axioms of G, including $\Rightarrow (A \to B) \vee (B \to A)$. Modus ponens is simulated using admissibility of cut as usual, by deriving $A, A \to B \Rightarrow B$ and applying Theorem 5.16 twice to this and the linear nested sequents $\Rightarrow A$ and $\Rightarrow A \to B$ respectively. □

6 Interpolation for G via Linear Nested Sequents

We now explain modifications to the construction of interpolants via hypersequents sufficient to adapt the method to linear nested sequents. If *validity* in Definition 3.5 is defined based on all *monotone sequences* $\boldsymbol{w}$, i.e., sequences with $w_1 \leqslant w_2 \leqslant \cdots \leqslant w_{\|\boldsymbol{w}\|}$ instead of $\mathcal{M}$-rooted ones, then Lemma 3.6 can be proved by induction on $\|\boldsymbol{w}\|$ for linear nested sequents and the formula interpretation from Definition 5.1. More generally, all definitions and statements from Sections 3 and 4 that mention $\mathcal{M}$-rooted sequences must now use monotone sequences instead. Most proofs apply as is. E.g., Lemma 3.12 still holds because for singleton sequences, both $\mathcal{M}$-rootedness and monotonicity are trivial and, hence, equivalent. Definition 4.1 and Lemma 4.2 are omitted. Since Theorem 4.5 is a direct consequence of Lem 4.3, it remains to define all split versions of the rules from Figure 4, provide interpolant transformations for them, and prove their correctness.

It is clear that splits of initial linear nested sequents (init_1) and ($\bot_L$) can be interpolated the same way as the corresponding splits of initial hypersequents (id) and ($\mathsf{id}_\bot$) from Figure 1 respectively, except that the superscript should be $\|\widetilde{\mathcal{G}}\| + 1$ instead of 1, e.g., $\widetilde{\mathcal{G}} /\!\!/ \Gamma, p; \Pi \Rightarrow \Delta; \Lambda, p /\!\!/ \widetilde{\mathcal{H}} \xleftarrow{\mathcal{L}in} p^{(\|\widetilde{\mathcal{G}}\|+1)}$. It also works for ($\mathsf{init}_2$) split the same way, e.g.,

$$\widetilde{\mathcal{G}} /\!\!/ \Gamma; \Theta, p \Rightarrow \widetilde{\Delta} /\!\!/ \widetilde{\mathcal{H}} /\!\!/ \widetilde{\Sigma} \Rightarrow \Lambda, p; \Pi /\!\!/ \widetilde{\mathcal{I}} \xleftarrow{\mathcal{L}in} \overline{p}^{(\|\widetilde{\mathcal{G}}\|+1)}.$$

The transformations for splits of ($\wedge_L$), ($\wedge_R$), ($\vee_L$), ($\vee_R$), and ($\rightarrow_L$) are the same as for corresponding splits of ($\wedge \Rightarrow$), ($\Rightarrow \wedge$), ($\vee \Rightarrow$), ($\Rightarrow \vee$), and ($\rightarrow\Rightarrow$) respectively, and the proof is the same. It is also easy to see that, as with many unary rules, any interpolant for the premiss of (Lift) also works for its conclusion, mainly because $w_{k+1} \not\Vdash A$ implies $w_k \not\Vdash A$ for any monotone sequence $\boldsymbol{w}$. Finally, the transformations for ($\rightarrow^1_R$) and ($\rightarrow^2_R$) are presented in Figure 6.

Theorem 6.1 *G enjoys the CIP.*

Proof. As for hypersequents, it is sufficient to prove that all interpolant transformations in this section preserve componentwise interpolation w.r.t. $\mathcal{L}in$, which is tedious but not difficult. While, for the lack of space, we only provide an argument for ($\rightarrow^{2l}_R$) from Figure 6, it is worth mentioning that the ${D_j}^{(n)}$ and $\overline{C_j}^{(n)}$ terms in the conclusions of ($\rightarrow^{2l}_R$) and ($\rightarrow^{2r}_R$) respectively have to be added to make sure IH can be used for a world intermediate between the $(n-1)$th and nth worlds in a given sequence.

Rule ($\rightarrow^{2l}_R$) Assume the two interpolation statements in the premisses to be true w.r.t. $\mathcal{L}in$. Consider any linear model $\mathcal{M} = (W, \leqslant, V)$ and an arbitrary monotone sequence $\boldsymbol{w}$ of worlds of length $\|\boldsymbol{w}\| = n + k$, where $n = \|\widetilde{\mathcal{G}}\| + 2$ and $k = \|\widetilde{\mathcal{H}}\|$. It is clear that for any world u such that $w_{n-1} \leqslant u \leqslant w_n$ and for the sequence $\boldsymbol{v} := w_1, \dots, w_{n-1}, u, w_n, \dots, w_{n+k}$,

$$\boldsymbol{v} \models \bigotimes_{l \neq n} (\overline{E_{jl}}^{(l)} \otimes {F_{jl}}^{(l)}) \iff \boldsymbol{w} \models \bigotimes_{l=1}^{n-1} (\overline{E_{jl}}^{(l)} \otimes {F_{jl}}^{(l)}) \otimes \bigotimes_{l=n}^{n+k} (\overline{E_{j,l+1}}^{(l)} \otimes {F_{j,l+1}}^{(l)}) \quad (16)$$

$$\to^{1l}_{R}\ \frac{\widetilde{\mathcal{G}} /\!/ \widetilde{\Gamma} \Rightarrow \Delta; \Pi /\!/ A; \Rightarrow B; \ \leftarrow\ \bigovee_{j=1}^{m} \Big(\overline{C_j}^{(n)} \owedge D_j^{(n)} \owedge \bigowedge_{l=1}^{n-1} (\overline{E_{jl}}^{(l)} \owedge F_{jl}^{(l)})\Big)}{\widetilde{\mathcal{G}} /\!/ \widetilde{\Gamma} \Rightarrow \Delta, A \to B; \Pi \ \leftarrow\ \bigovee_{j=1}^{m} \Big(\overline{D_j \to C_j}^{(n-1)} \owedge \bigowedge_{l=1}^{n-1} (\overline{E_{jl}}^{(l)} \owedge F_{jl}^{(l)})\Big)}$$

$$\to^{1r}_{R}\ \frac{\widetilde{\mathcal{G}} /\!/ \widetilde{\Gamma} \Rightarrow \Delta; \Pi /\!/; A \Rightarrow; B \ \leftarrow\ \bigowedge_{j=1}^{m} \Big(\overline{C_j}^{(n)} \ovee D_j^{(n)} \ovee \bigovee_{l=1}^{n-1} (\overline{E_{jl}}^{(l)} \ovee F_{jl}^{(l)})\Big)}{\widetilde{\mathcal{G}} /\!/ \widetilde{\Gamma} \Rightarrow \Delta; \Pi, A \to B \ \leftarrow\ \bigowedge_{j=1}^{m} \Big((C_j \to D_j)^{(n-1)} \ovee \bigovee_{l=1}^{n-1} (\overline{E_{jl}}^{(l)} \ovee F_{jl}^{(l)})\Big)}$$

$$\begin{array}{l}
\widetilde{\mathcal{G}} /\!/ \widetilde{\Gamma} \Rightarrow \Delta; \Theta /\!/ \widetilde{\Sigma} \Rightarrow \Pi, A \to B; \Lambda /\!/ \widetilde{\mathcal{H}} \leftarrow \mho \\
\qquad\qquad \widetilde{\mathcal{G}} /\!/ \widetilde{\Gamma} \Rightarrow \Delta; \Theta /\!/ A; \Rightarrow B; /\!/ \widetilde{\Sigma} \Rightarrow \Pi; \Lambda /\!/ \widetilde{\mathcal{H}} \leftarrow \bigovee_{j=1}^{m} \Big(\overline{C_j}^{(n)} \owedge D_j^{(n)} \owedge \bigowedge_{l \neq n} (\overline{E_{jl}}^{(l)} \owedge F_{jl}^{(l)})\Big) \\
\hline
\to^{2l}_{R} \qquad \longleftarrow \quad \mho \ovee \bigovee_{j=1}^{m} \Big(\bigowedge_{l=1}^{n-1} (\overline{E_{jl}}^{(l)} \owedge F_{jl}^{(l)}) \owedge \overline{(D_j \to C_j)}^{(n-1)} \owedge D_j^{(n)} \owedge \bigowedge_{l=n}^{n+k} (\overline{E_{j,l+1}}^{(l)} \owedge F_{j,l+1}^{(l)})\Big) \\
\widetilde{\mathcal{G}} /\!/ \widetilde{\Gamma} \Rightarrow \Delta, A \to B; \Theta /\!/ \widetilde{\Sigma} \Rightarrow \Pi; \Lambda /\!/ \widetilde{\mathcal{H}} \qquad \longleftarrow
\end{array}$$

$$\begin{array}{l}
\widetilde{\mathcal{G}} /\!/ \widetilde{\Gamma} \Rightarrow \Delta; \Theta /\!/ \widetilde{\Sigma} \Rightarrow \Pi; \Lambda, A \to B /\!/ \widetilde{\mathcal{H}} \leftarrow \mho \\
\qquad\qquad \widetilde{\mathcal{G}} /\!/ \widetilde{\Gamma} \Rightarrow \Delta; \Theta /\!/; A \Rightarrow; B /\!/ \widetilde{\Sigma} \Rightarrow \Pi; \Lambda /\!/ \widetilde{\mathcal{H}} \leftarrow \bigowedge_{j=1}^{m} \Big(\overline{C_j}^{(n)} \ovee D_j^{(n)} \ovee \bigovee_{l \neq n} (\overline{E_{jl}}^{(l)} \ovee F_{jl}^{(l)})\Big) \\
\hline
\to^{2r}_{R} \qquad \longleftarrow \quad \mho \owedge \bigowedge_{j=1}^{m} \Big(\bigovee_{l=1}^{n-1} (\overline{E_{jl}}^{(l)} \ovee F_{jl}^{(l)}) \ovee (C_j \to D_j)^{(n-1)} \ovee \overline{C_j}^{(n)} \ovee \bigovee_{l=n}^{n+k} (\overline{E_{j,l+1}}^{(l)} \ovee F_{j,l+1}^{(l)})\Big) \\
\widetilde{\mathcal{G}} /\!/ \widetilde{\Gamma} \Rightarrow \Delta; \Theta, A \to B /\!/ \widetilde{\Sigma} \Rightarrow \Pi; \Lambda /\!/ \widetilde{\mathcal{H}} \qquad \longleftarrow
\end{array}$$

Fig. 6. Transformations for component-creating rules of LNG. All interpolation statements are w.r.t. $\mathcal{L}in$. For all 4 rules, $\|\widetilde{\mathcal{G}}\| = n - 2$. For the last 2 rules, $\|\widetilde{\mathcal{H}}\| = k$.

We consider two possibilities depending on whether the interpolant in the conclusion holds or not. Assume first that it does not hold for $\boldsymbol{w}$. In particular $\boldsymbol{w} \not\models \mho$ and, for each $j = 1, \ldots, m$, the following does not hold at $\boldsymbol{w}$:

$$\bigowedge_{l=1}^{n-1} (\overline{E_{jl}}^{(l)} \owedge F_{jl}^{(l)}) \owedge \overline{(D_j \to C_j)}^{(n-1)} \owedge D_j^{(n)} \owedge \bigowedge_{l=n}^{n+k} (\overline{E_{j,l+1}}^{(l)} \owedge F_{j,l+1}^{(l)}). \quad (17)$$

We need to prove that, whenever $\boldsymbol{w} \not\models L\widetilde{\mathcal{G}} /\!/ L\widetilde{\Gamma} \Rightarrow \Delta /\!/ L\widetilde{\Sigma} \Rightarrow \Pi /\!/ L\widetilde{\mathcal{H}}$, we have $w_{n-1} \Vdash A \to B$. Assuming the former, we conclude from $\boldsymbol{w} \not\models \mho$ and the interpolation statement for the top premiss that $w_n \Vdash A \to B$. We now use the interpolation statement for the bottom premiss to show that $u \nVdash A$ or $u \Vdash B$ for any $u \geqslant w_{n-1}$ such that $u \ngeqslant w_n$. By linearity of $\mathcal{M}$, it follows that $u < w_n$, making $\boldsymbol{v}$ above a monotone sequence. It remains to show that the interpolant of the bottom premiss, let us denote it $\mho'$, is false at $\boldsymbol{v}$. By (16), if the jth disjunct of the conclusion interpolant is false at $\boldsymbol{w}$ because of E's and F's, then the jth disjunct of $\mho'$ is false at $\boldsymbol{v}$. If $\boldsymbol{w} \not\models D_j^{(n)}$, i.e., if $w_n \nVdash D_j$, then $u \nVdash D_j$ by monotonicity, i.e., $\boldsymbol{v} \not\models D_j^{(n)}$. And if $\boldsymbol{w} \not\models \overline{(D_j \to C_j)}^{(n-1)}$, i.e., $w_{n-1} \Vdash D_j \to C_j$, then either $u \nVdash D_j$ or $u \Vdash C_j$, i.e., $\boldsymbol{v} \not\models \overline{C_j}^{(n)} \owedge D_j^{(n)}$. Thus, $\boldsymbol{v} \not\models \mho'$, completing the proof for the left side.

For the right side, assume that the conclusion interpolant is true at $\boldsymbol{w}$. If it is true because of $\mho$, then the right side $R\widetilde{\mathcal{G}} /\!/ R\widetilde{\Gamma} \Rightarrow \Theta /\!/ R\widetilde{\Sigma} \Rightarrow \Lambda /\!/ R\widetilde{\mathcal{H}}$ of the top premiss holds, which is the same as the right side of the conclusion. Otherwise, (17) is true at $\boldsymbol{w}$ for some $1 \leqslant j \leqslant m$, in particular, $w_{n-1} \not\Vdash D_j \to C_j$ and $w_n \Vdash D_j$. Thus, there exists some $u' \geqslant w_{n-1}$ such that $u' \Vdash D_j$ and $u' \not\Vdash C_j$. Let $u := u'$ if $u' \leqslant w_n$ or $u := w_n$ otherwise, i.e., if $u' > w_n$. In the latter case, $u \not\Vdash C_j$ by monotonicity. Hence, either way, $w_{n-1} \leqslant u \leqslant w_n$ and both $u \Vdash D_j$ and $u \not\Vdash C_j$. Therefore, $\boldsymbol{v}$ above is a monotone sequence and the jth disjunct makes the bottom premiss interpolant true at $\boldsymbol{v}$. By the interpolation statement for this premiss, $\boldsymbol{v} \models R\widetilde{\mathcal{G}} /\!/ R\widetilde{\Gamma} \Rightarrow \Theta /\!/ \Rightarrow /\!/ R\widetilde{\Sigma} \Rightarrow \Lambda /\!/ \widetilde{\mathcal{H}}$, which implies $\boldsymbol{w} \models R\widetilde{\mathcal{G}} /\!/ R\widetilde{\Gamma} \Rightarrow \Theta /\!/ R\widetilde{\Sigma} \Rightarrow \Lambda /\!/ R\widetilde{\mathcal{H}}$. □

Corollary 6.2 Int, LQ, *and* G *enjoy the LIP.*

Proof. All interpolant transformations preserve variable polarity. □

7 Conclusion and Discussion

We have provided constructive proofs of Craig and Lyndon interpolation using sequent-style calculi for Jankov logic and Gödel logic. In particular, Lyndon interpolation for Gödel logic was listed as an open problem in [4,10,18]. We answer this question in the positive. Its proof uses a novel calculus for G in the framework of linear nested sequents which is of independent interest.

We are grateful to one of the reviewers who suggested a possibility that there might be alternative constructive proofs of interpolation property for these logics via translation from modal logics or based on methods and calculi from [1,2,3]. While this is a very interesting direction for further research, doing so for LIP seems at the very least not trivial, as also confirmed by M. Baaz, one of the authors of the above three papers, in a private communication. As for using known results for modal logics, the Gödel translations of the three intermediate logics we considered are S4, S4.2, and S4.3. Thus, a method with a detour via modal logics could not produce any results for Gödel logic, due to S4.3 not having the CIP [17], in effect, leaving out the most interesting case. In particular, a linear nested calculus would not help prove interpolation in the modal case. The exact source of this disparity deserves further investigation.

While our method does not produce proofs of uniform interpolation, this property presents less interest for us than LIP because for intermediate logics uniform interpolation is known to be equivalent to CIP [18].

This work is part of the project of using proof-theoretic methods to show Craig and Lyndon interpolation for intermediate logics.

Acknowledgments. We would like to thank M. Baaz for many fruitful discussions, A. Ciabattoni for her general and generous support, and D. Gabbay for encouragement. We further thank the anonymous reviewers for their careful reading of the article and their useful and thought-provoking advice, which may lead to further publications, by us and/or by them.

References

[1] Baaz, M., A. Ciabattoni and C. G. Fermüller, *Cut-elimination in a sequents-of-relations calculus for Gödel logic*, in: *ISMVL 2001*, IEEE, 2001 pp. 181–186.

[2] Baaz, M., A. Ciabattoni and C. G. Fermüller, *Hypersequent calculi for Gödel logics—a survey*, Journal of Logic and Computation **13** (2003), pp. 835–861.

[3] Baaz, M. and H. Veith, *An axiomatization of quantified propositional Gödel logic using the Takeuti–Titani rule*, in: *Logic Colloquium 1998*, LNL **13**, ASL, 2000 pp. 91–104.

[4] Chagrov, A. and M. Zakharyaschev, "Modal Logic," Clarendon Press, 1997.

[5] Ciabattoni, A. and M. Ferrari, *Hypertableau and path-hypertableau calculi for some families of intermediate logics*, in: R. Dyckhoff, editor, *TABLEAUX 2000*, LNAI **1847**, Springer, 2000 pp. 160–174.

[6] Ciabattoni, A., R. Ramanayake and H. Wansing, *Hypersequent and display calculi—a unified perspective*, Studia Logica **102** (2014), pp. 1245–1294.

[7] Dyckhoff, R. and S. Negri, *Decision methods for linearly ordered Heyting algebras*, Archive for Mathematical Logic **45** (2006), pp. 411–422.

[8] Fitting, M., *Nested sequents for intuitionistic logics*, Notre Dame Journal of Formal Logic **55** (2014), pp. 41–61.

[9] Fitting, M. and R. Kuznets, *Modal interpolation via nested sequents*, Annals of Pure and Applied Logic **166** (2015), pp. 274–305.

[10] Gabbay, D. M. and L. Maksimova, "Interpolation and Definability: Modal and Intuitionistic Logic," Clarendon Press, 2005.

[11] Indrzejczak, A., *Linear time in hypersequent framework*, Bulletin of Symbolic Logic **22** (2016), pp. 121–144.

[12] Kuznets, R., *Craig interpolation via hypersequents*, in: D. Probst and P. Schuster, editors, *Concepts of Proof in Mathematics, Philosophy, and Computer Science*, Ontos Mathematical Logic **6**, De Gruyter, 2016 pp. 193–214.

[13] Kuznets, R., *Multicomponent proof-theoretic method for proving interpolation property*, Annals of Pure and Applied Logic **To appear** (2018).

[14] Kuznets, R. and B. Lellmann, *Grafting hypersequents onto nested sequents*, Logic Journal of IGPL **24** (2016), pp. 375–423.

[15] Lellmann, B., *Linear nested sequents, 2-sequents and hypersequents*, in: H. De Nivelle, editor, *TABLEAUX 2015*, LNAI **9323**, Springer, 2015 pp. 135–150.

[16] Maksimova, L. L., *Craig's theorem in superintuitionistic logics and amalgamable varieties of pseudo-Boolean algebras*, Algebra and Logic **16** (1977), pp. 427–455.

[17] Maksimova, L. L., *Absence of the interpolation property in the consistent normal modal extensions of the Dummett logic*, Algebra and Logic **21** (1982), pp. 460–463.

[18] Maksimova, L. L., *The Lyndon property and uniform interpolation over the Grzegorczyk logic*, Siberian Mathematical Journal **55** (2014), pp. 118–124.

[19] Masini, A., *2-sequent calculus: A proof theory of modalities*, Annals of Pure and Applied Logic **58** (1992), pp. 229–246.

[20] Moschovakis, J., *Intuitionistic logic*, in: *Stanford Encyclopedia of Philosophy*, 2015.

*-Continuity vs. Induction: Divide and Conquer

Stepan Kuznetsov [1]

Steklov Mathematical Institute of the RAS
8 Gubkina St., Moscow, Russia

Abstract

The Kleene star can be axiomatised in two ways: inductively, as a fixpoint, or as the ω-iteration of multiplications. The latter is called *-continuity and is stronger than the former: not every Kleene algebra is *-continuous. In the language of only multiplication, union, and Kleene star, however, the (in)equational atomic theory (logic) of *-continuous Kleene algebras coincides with the one of all Kleene algebras (Kozen, 1994). The situation changes dramatically when one adds division operations. As shown by Buszkowski (2007), then the logic with *-continuity becomes Π_1^0-hard and therefore strictly stronger than the inductive one. This result, however, is not constructive, i.e., does not yield a formula distingushing them.
Our contribution is threefold. First, we present an example of Kleene algebra with divisions and intersection, which is not *-continuous. Second, we present a formula which makes Buszkowski's result constructive (see above). Third, we show that the calculus for *-continuity is incomplete w.r.t. more specific relational and language models, in the fragment with divisions, multiplication, intersection, and Kleene star. The choice of this fragment is natural, since union or the unit constant are known to yield incompleteness even without Kleene star.

Keywords: Kleene star, infinitary action logic, action logic, residuated Kleene lattice, *-continuity, algebra of formal languages, relational algebra

1 Introduction

1.1 Residuated Kleene Lattices

We start with the definition of residuated Kleene lattice (RKL), or action lattice, following Kozen [17] and Buszkowski [5]. The notion of RKL is a combination of action algebras, or action semilattices, by Pratt [32], and residuated lattices studied by Ono [27] and others as models for substructural logics (*i.e.,* logics lacking structural rules of contraction, weakening, and permutation). In comparison with RKL's, action algebras lack meet ($\wedge$) and residuated lattices

[1] e-mail: `sk@mi.ras.ru`
This work was supported by the Russian Science Foundation under grant 16-11-10252.

lack the most intriguing operation, the Kleene star. In RKL's, we have the full set of operations.

As a matter of notation, we follow the tradition of Lambek [21] and denote residuals as divisions ($\backslash$, $/$), rather than directed implications ($\rightarrow$, $\leftarrow$). For simplicity, we also do not use the zero constant in our algebraic definitions and logical calculi (examples of RKL's we present here, however, have a zero).

Definition 1.1 A structure $(\mathfrak{A}; \preceq, \vee, \wedge, \cdot, \mathbf{1}, \backslash, /, ^*)$ is a *residuated Kleene lattice* (RKL), if the following holds.

(i) $(\mathfrak{A}; \preceq, \vee, \wedge)$ is a lattice (*i.e.*, $\preceq$ is a preorder, $x \vee y = \sup\{x, y\}$, $x \wedge y = \inf\{x, y\}$, where sup and inf are taken w.r.t. $\preceq$ and should exist).

(ii) $(\mathfrak{A}; \cdot, \mathbf{1})$ is a monoid (*i.e.*, $\cdot$ is an associative operation and $\mathbf{1}$ is its unit).

(iii) $\backslash$ and $/$ are residuals for $\cdot$ w.r.t. $\preceq$, *i.e.*,

$$y \preceq x \backslash z \iff x \cdot y \preceq z \iff x \preceq z / y$$

In other words, $x \backslash z = \max\{y \mid x \cdot y \preceq z\}$ and $z / y = \max\{x \mid x \cdot y \preceq z\}$, where max is taken w.r.t. $\preceq$. The structure $\mathfrak{A}$ is residuated if all these maxima exist. (Partially ordered residuated semigroups are algebraic models for the Lambek calculus [21].)

(iv) a^* is the smallest fixpoint of $x \mapsto \mathbf{1} \vee a \cdot x$, *i.e.*, $\mathbf{1} \preceq a^*$, $a \cdot a^* \preceq a^*$, and if $\mathbf{1} \preceq b$ and $a \cdot b \preceq b$, then $a^* \preceq b$.

As shown by Pratt [32], in the residuated situation there is no difference between left and right RKL's (cf. Kozen [15] for an example where a^* is a fixpoint for $x \mapsto \mathbf{1} \vee a \cdot x$, but not for $x \mapsto \mathbf{1} \vee x \cdot a$).

Moreover, Pratt [32] shows that in the presence of division operations the fixpoint condition (iv) can be reformulated. Namely, the 'smallest' half of (iv) is replaced by one axiom which Pratt called *pure induction:* $(a \backslash a)^* = a \backslash a$, and the monotonicity principle: if $a \preceq b$, then $a^* \preceq b^*$.[2] Pure induction is particularly easy to check. For the other half of (iv) Pratt suggests a symmetric version: $\mathbf{1} \preceq a^*$, $a \preceq a^*$, $a^* \cdot a^* \preceq a^*$.

Further we use the notation x^n, defined inductively: $x^0 = \mathbf{1}$, $x^{n+1} = x^n \cdot x$.

Definition 1.2 An RKL $\mathfrak{A}$ is **-continuous*, if $x^* = \sup\{x^n \mid n \in \omega\}$, where supremum is taken w.r.t. $\preceq$ and ω denotes the set of all natural numbers. (In particular, this implies that all such suprema exist.)

Notice that here the usual definition of *-continuity, $y \cdot x^* \cdot z = \sup\{y \cdot x^n \cdot z \mid n \in \omega\}$, here can be simpified, since in the presence of division operations multiplication distributes over infinite joins (suprema). In presence of *-continuity, axiom (iv) becomes redundant.

---

[2] Since in lattices $a \preceq b$ iff $b = a \vee b$, monotonicity can be reformulated as an (in)equation: $a^* \preceq (a \vee b)^*$. Pratt also does provides such a reformulation of Lambek-style axioms for division operations. This yields a purely (in)equational axiomatisation of action algebras, *i.e.*, the fact that the class of action algebras is a finitely based variety [32, Theorem 7].

1.2 Action Logic and Infinitary Action Logic

In this section we define logical calculi which correspond to algebraic structures defined before. We start with **MALC**, the multiplicative-additive Lambek calculus,[3] which is obtained from the Lambek calculus with the unit [22] by extending it with additive conjunction and disjunction from Girard's linear logic [9], corresponding to meet and join.

Formulae of **MALC** are built from a countable set of variables $\{p, q, r, \ldots\}$ and the unit constant $\mathbf{1}$ using five binary connectives: $\cdot$, $\backslash$, $/$, $\wedge$, $\vee$. Sequents of **MALC** are expressions of the form $\Pi \to A$, where A is a formula and Π is a finite (possibly empty) linearly ordered sequence of formulae.

Axioms and rules of **MALC** are as follows:

$$\frac{}{A \to A}\ (\mathrm{id}) \qquad \frac{\Gamma, \Delta \to C}{\Gamma, \mathbf{1}, \Delta \to C}\ (\mathbf{1} \to) \qquad \frac{}{\to \mathbf{1}}\ (\to \mathbf{1})$$

$$\frac{\Gamma, A, B, \Delta \to C}{\Gamma, A \cdot B, \Delta \to C}\ (\cdot \to) \qquad \frac{\Gamma \to A \quad \Delta \to B}{\Gamma, \Delta \to A \cdot B}\ (\to \cdot)$$

$$\frac{\Pi \to A \quad \Gamma, B, \Delta \to C}{\Gamma, \Pi, A \backslash B, \Delta \to C}\ (\backslash \to) \qquad \frac{A, \Pi \to B}{\Pi \to A \backslash B}\ (\to \backslash)$$

$$\frac{\Pi \to A \quad \Gamma, B, \Delta \to C}{\Gamma, B / A, \Pi, \Delta \to C}\ (/ \to) \qquad \frac{\Pi, A \to B}{\Pi \to B / A}\ (\to /)$$

$$\frac{\Gamma, A_1, \Delta \to C \quad \Gamma, A_2, \Delta \to C}{\Gamma, A_1 \vee A_2, \Delta \to C}\ (\vee \to) \qquad \frac{\Pi \to A_i}{\Pi \to A_1 \vee A_2}\ (\to \vee)_i,\ i = 1, 2$$

$$\frac{\Gamma, A_i, \Delta \to C}{\Gamma, A_1 \wedge A_2, \Delta \to C}\ (\wedge \to)_i,\ i = 1, 2 \qquad \frac{\Pi \to A_1 \quad \Pi \to A_2}{\Pi \to A_1 \wedge A_2}\ (\to \wedge)$$

$$\frac{\Pi \to A \quad \Gamma, A, \Delta \to C}{\Gamma, \Pi, \Delta \to C}\ (\mathrm{cut})$$

(The cut rule is eliminable by straightforward induction, as in the original Lambek calculus [21].)

This system is extended with the Kleene star * (as a unary connective) in two ways, depending on whether we want *-continuity.

The *-continuous case corresponds to *infinitary action logic* [6], $\mathbf{ACT}_\omega$, which is obtained from **MALC** by adding the following rules:

[3] Sometimes **MALC** is called "full Lambek calculus," which sounds a bit offensive for the original, multiplicative-only system.

$$\frac{(\Gamma, A^n, \Delta \to C)_{n \geq 0}}{\Gamma, A^*, \Delta \to C}\ (^* \to)_\omega \qquad \frac{\Pi_1 \to A \quad \ldots \quad \Pi_n \to A}{\Pi_1, \ldots, \Pi_n \to A^*}\ (\to {}^*)_n,\ n \geq 0$$

Derivations in $\mathbf{ACT}_\omega$ are infinite, but well-founded trees with infinite branching at instances of $(^* \to)_\omega$. Cut elimination for $\mathbf{ACT}_\omega$ was shown by Palka [29] using transfinite induction.

The system corresponding to the inductive definition of *, *action logic* **ACT**, is obtained from **MALC** (with (cut)) by adding the following rules [5]:

$$\frac{}{\to A^*} \qquad \frac{}{A, A^* \to A^*} \qquad \frac{}{A^*, A \to A^*}$$

$$\frac{A, B \to B}{A^*, B \to B} \qquad \frac{B, A \to B}{B, A^* \to B}$$

These rules are not good Gentzen-style sequent rules, and cut in this formulation of **ACT** is not eliminable. In fact, no cut-free sequent calculus for **ACT** is known. (An attempt was taken by Jipsen [12], but Buszkowski [5] showed that in Jipsen's system cut is not eliminable.)

Besides Kleene star, we also consider positive iteration, defined as $A^+ = A \cdot A^*$. One can prove in **ACT** (and therefore in $\mathbf{ACT}_\omega$), that A^* is equivalent to $\mathbf{1} \vee A^+$.

Positive iteration becomes important if we consider systems with Lambek's restriction, where antecedents of all sequents are required to be non-empty (as in the original paper by Lambek [21]). With Lambek's restriction, standard Kleene star becomes unavailable, and is replaced by positive iteration. Algebraically, this constraint corresponds to considering semigroups instead of monoids. Lambek's restriction is motivated by linguistic applications of the Lambek calculus and yields a system which is not a conservative fragment of the system without this restriction. Indeed, even some sequents with non-empty antecedents, like $(p \mathbin{\backslash} p) \mathbin{\backslash} q \to q$, require empty antecedents during the derivation. Thus, the study of systems with and without Lambek's restriction go in parallel. For more details, we refer to [20], and in this paper continue (for simplicity) using action logic without Lambek's restriction imposed.

Thanks to cut elimination, $\mathbf{ACT}_\omega$'s fragments with restricted sets of connectives are obtained by merely taking the appropriate subset of rules. In particular, removing rules for Kleene star yields **MALC** as a conservative fragment of $\mathbf{ACT}_\omega$. For **ACT**, due to the lack of a cut-free sequent calculus, the question of conservative fragments is non-trivial. However, if a sequent $\Pi \to A$ does not include * and is derivable in **ACT**, then it is derivable in $\mathbf{ACT}_\omega$, and therefore in **MALC** by conservativity of $\mathbf{ACT}_\omega$ over **MALC**. Thus, **MALC** is a conservative fragment of both **ACT** and $\mathbf{ACT}_\omega$.

1.3 Complexity and Compactness Arguments

Kozen's completeness theorem [16] shows that for Kleene algebras, *i.e.*, in the language of $\cdot$, $\vee$, and *, though non-*-continuous Kleene algebras do exist [15], the logics for *-continuous Kleene algebras and all Kleene algebras coincide. Division operations change things dramatically. Namely, Buszkowski [5] shows that the derivability problem in $\mathbf{ACT}_\omega$ is Π^0_1-complete. On the other hand, **ACT** is clearly recursively enumerable. This has two corollaries:

(i) there exists an RKL which is not *-continuous;

(ii) there exists a sequent provable in $\mathbf{ACT}_\omega$, but not in **ACT** (in other words, true in all *-continuous RKL's, but not in all RKL's).

Buszkowski's argument, however, is not constructive, giving neither an example of a non-*-continuous RKL, nor a concrete sequent distinguishing $\mathbf{ACT}_\omega$ from **ACT**. We fill these gaps in Sections 2 and 3.

Before going further, let us mention other ways of proving statements (i) and (ii) above. The first statement can be proved by applying the well-known Gödel–Maltsev [10,23,24] compactness theorem [4]. Namely, one can write a first-order theory in the signature $\Omega = (\cdot, \backslash, /, \wedge, \vee, \mathbf{1}, {}^*; =, \preceq)$ whose models are exactly RKL's [5]. For example, Lambek's axiom for / becomes

$$\forall x \forall y \forall z \, (x \cdot y \preceq z \leftrightarrow x \preceq z \,/\, y);$$

for Kleene star one can take Pratt's equational axiomatisation (based on pure induction), *etc.* Notice that the strict version of the order, $x \prec y$, is expressible as $(x \preceq y) \,\&\, \neg(x = y)$. Now we add two new constant symbols, a and b, to the signature and the following axioms to the theory: $b \prec a^*$ and $a^n \prec b$ for each $n \geq 0$. Any finite subset of this theory includes only a finite number of such axioms and can be easily satisfied on an RKL. For example, one can take the powerset of the set of natural numbers with the set-theoretic lattice structure, $\cdot$ for elementwise addition, the Kleene star defined *-continuously, and $a = \{1\}$. Then take $b = \{0, 1, \ldots, n_0, n_0 + 1\}$, where n_0 is the biggest value of n appearing in the axioms of the finite subset taken. By compactness theorem, the whole theory also has a model. This model is an RKL, but fails to be *-continuous: b is explicitly stated to be an upper bound for all a^n, which is smaller than a^*. This way of showing existence of non-*-continuous RKL's is, however, also not constructive.

As for (ii), there exists a way of making the complexity argument presented above in a sense constructive, using the notion of *productive function* [6]. Following the notation from Soare's book [34], let W_x be the x-th r.e. set (*i.e.*, x is the natural number encoding the algorithm enumerating this set). A set P is called *productive*, if there exists a computable partial function ψ, such that if $W_x \subseteq P$, then $\psi(x)$ is defined and is an element of the set difference $P - W_x$. The function ψ itself is called *productive function.* The set $\overline{K} = \{x \mid x \notin W_x\}$ is clearly a productive one, with the trivial productive function $\psi_{\mathrm{id}}(x) = x$.

[4] Suggested by one of the anonymous reviewers.

[5] The $\wedge$ and $\vee$ *algebraic* operations here should not be confused with *logical* conjunction and disjunction used in first-order logic. For logical conjunction we use & instead of $\wedge$.

[6] Suggested by Fedor Pakhomov and Scott Weinstein.

Next, we use the following theorem [8, Theorem 2.1][26, Theorem 5][34, Theorem 4.5iii]: if P_1 is productive and it is m-reducible to P_2, then P_2 is also productive. Let P be the set of theorems of $\mathbf{ACT}_\omega$. By the Π_1^0-completeness result by Buszkowski [5], $\overline{K}$ is m-reducible to P. Therefore, P is a productive set with some computable productive function ψ. On the other hand, the set of theorems of **ACT** is r.e., *i.e.*, it is a W_y for some y; $W_y \subseteq P$, since **ACT** is weaker than $\mathbf{ACT}_\omega$. Thus, $\psi(y)$ is an element of P, but not W_y, in other words, gives an example of a sequent provable in $\mathbf{ACT}_\omega$, but not **ACT**.

Theoretically, this example can be explicitly extracted from the reasoning presented above. In order to do so, one needs to track the m-reduction of $\overline{K}$ to P used by Buszkowski (via the totality problem for context-free grammars) and transform it to a translation of the productive function from ψ_{id} to ψ. This yields a concrete algorithm for ψ, which can then be applied to y, which is the code of the algorithm for enumerating theorems of **ACT**; $\psi(y)$ is guaranteed to exist and it is the necessary example. In practice, however, performing this procedure is quite problematic. In Section 3 we give a much shorter example, which, moreover, exhibits some interesting structural features of proofs in action logic.

1.4 Models for ACT and $\mathbf{ACT}_\omega$

As mentioned by Buszkowski [5], standard Lindenbaum algebra construction shows that **ACT** and $\mathbf{ACT}_\omega$ are complete, respectively, w.r.t. the class of all RKL's and the class of *-continuous RKL's.

There are two more specific classes of RKL's which are usually considered as standard classes of models for the Lambek calculus and action logic.

The first one is the algebra $\mathcal{P}(\Sigma^*)$ of formal languages over a given alphabet Σ (here and further $\mathcal{P}(X)$ stands for the set of all subsets of X). The preorder is set-theoretical inclusion, and multiplication is defined as pairwise concatenation. It is well-known that this algebra is residuated. Language-theoretic division operations are defined as follows:

$$x \,/\, y = \max\{z \mid z \cdot y \subseteq x\} = \{u \in \Sigma^* \mid (\forall v \in y)\, uv \in x\};$$

$$y \setminus x = \max\{z \mid y \cdot z \subseteq x\} = \{u \in \Sigma^* \mid (\forall v \in y)\, vu \in x\}.$$

The preorder, $\subseteq$, enjoys arbitrary meets and joins and therefore induces a lattice structure. Finally, Kleene star is defined in the *-continuous way:

$$x^* = \sup\{x^n \mid n \geq 0\} = \{u_1 \ldots u_n \mid n \geq 0, u_i \in x\}.$$

Models of the Lambek calculus and action logic on such algebras are called L-models.

The other class is formed by relational models, of the form $\mathcal{P}(W \times W)$, where W is a non-empty set. Elements of $\mathcal{P}(W \times W)$ are binary relations of W. Multiplication is defined as composition of relations. The lattice structure is set-theoretic. Relational algebras are also residuated. Kleene star x^* is defined as the reflexive-transitive closure of relation x. Models from this class are called R-models.

Both L- and R-models are *-continuous, so they form natural classes of models for $\mathbf{ACT}_\omega$. It is well-known, however, that there is an obstacle to completeness connected with distributivity. L- and R-models are distributive lattices, *i.e.*, enjoy $x \wedge (y \vee z) = (x \wedge y) \vee (x \wedge z)$. [7] On the other hand, one of the distributivity laws, namely $(A \vee B) \wedge (A \vee C) \to A \vee (B \wedge C)$, is not derivable in **MALC**, and therefore in $\mathbf{ACT}_\omega$. (The fact that distributive law is invalid in substructural logic was noticed by Ono and Komori [28].)

Another obstacle is connected with the unit constant, $\mathbf{1}$. Dividing the unit by something, $\mathbf{1} \,/\, x$, yields a trivialisation of the interpretation and therefore extra sequents which are true, but not derivable. For both L- and R-models, such a sequent is, for example, $\mathbf{1} \,/(p \,/\, p) \to (\mathbf{1} \,/(p \,/\, p)) \cdot (\mathbf{1} \,/(p \,/\, p))$ [4,25,18].

These incompleteness issues actually have nothing to do with the Kleene star. In Section 4 we give a more fine-grained incompleteness result for language and relational semantics of $\mathbf{ACT}_\omega$, where the Kleene star plays a crucial role.

Despite incompleteness, the lower complexity bounds (Π^0_1-hardness), as shown by Buszkowski [4], are also valid for deciding general truth in R- and L-models. Essentially this comes from the fact that a fragment of $\mathbf{ACT}_\omega$ sufficient for encoding a Π^0_1-hard problem is indeed R- and L-complete (see footnote on page 508).

2 Example of an RKL which is not *-continuous

Kozen [15] presents an example of a Kleene algebra which is not *-continuous. In his construction, $\mathfrak{A} = \{\bot\} \cup (\mathbb{N} \times \mathbb{N}) \cup \{\top\}$, where $\mathbb{N} \times \mathbb{N}$ is ordered lexicographically, and $\bot$ and $\top$ are artificially added minimum and maximum. The following picture depicts the order on $\mathfrak{A}$:

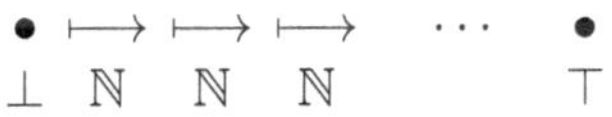

Multiplication on $\mathbb{N}$ is componentwise addition: $(a_1, b_1) \cdot (a_2, b_2) = (a_1 + b_1, a_2 + b_2)$; $\bot \cdot x = x \cdot \bot = \bot$ and $\top \cdot y = y \cdot \top = \top$ for $y \neq \bot$. The Kleene star is defined as follows: $\bot^* = (0,0)^* = (0,0)$, which is the unit element, and for $x \succ (0,0)$ we have $x^* = \top$. One can see that $(0,1)^* = \top$, while $\sup\{(0,1)^n \mid n \geq 0\} = (1,0)$, which falsifies *-continuity.

This algebra has two extra properties: it is commutative and its order is linear.

Unfortunately, it is not residuated: for example, for any z of the form $(0,i)$ we have $z \cdot (1,1) = (1, i+1) \prec (2,0)$, but $(1,0) \cdot (1,1) = (2,1) \not\preceq (2,0)$. Thus, there is no $(2,0) \,/(1,1) = \max\{z \mid z \cdot (1,1) \preceq (2,0)\}$.

This issue is quite a deep one, due to the following result which Restall [33] attributes to Pratt, calling it *Pratt's normality theorem* [33, Theorem 9.44]: in an RKL, if there exists $\sup\{a^n \mid n \geq 0\}$, then it coincides with a^*. In other words, whenever the *-continuous definition of Kleene star is available, we cannot choose another, non-standard version of Kleene star.

---

[7] As a basic fact of lattice theory, this also yields the dual: $x \vee (y \wedge z) = (x \vee y) \wedge (x \vee z)$.

This gives us a key how to fix Kozen's construction and make it residuated: we should avoid suprema (least upper bounds) in our model. Let $\mathfrak{A} = \{\bot\} \cup (\{0\} \times \mathbb{N}) \cup (\{1, 2, \dots\} \times \mathbb{Z}) \cup \{\top\}$, with the following linear order:

$$\begin{array}{cccccc} \bullet & \longmapsto & \longleftrightarrow & \longleftrightarrow & \cdots & \bullet \\ \bot & \mathbb{N} & \mathbb{Z} & \mathbb{Z} & & \top \end{array}$$

Multiplication and Kleene star are defined exactly as in Kozen's model. The same reasoning shows that Kleene star is not *-continuous; $\top$ is still the smallest fixpoint, and the supremum merely does not exist, since $\mathbb{Z}$ has no minimal element. The only thing we need to check is that division (due to commutativity, we have only one division) here is correctly defined:

Lemma 2.1 *For any $x, y \in \mathfrak{A}$, there exists* $\max\{z \mid z \cdot y \preceq x\}$ *(which we denote by $x \,/\, y = y \setminus x$).*

Proof. Easily, $x \,/\, \bot = \top$, $\top \,/\, y = \top$, $\bot \,/\, y = \bot$ (for $y \neq \bot$), $x \,/\, \top = \bot$ (for $x \neq \top$).

Now let x and y be pairs of numbers, $x = (a_1, a_2)$, $y = (b_1, b_2)$. Let us call the fragment $\{i\} \times \mathbb{Z}$ ($i > 0$) the i-th galaxy of $\mathfrak{A}$; the 0-th galaxy is truncated, $\{0\} \times \mathbb{N}$. There are four possible cases:

- $a_1 < b_1$. Then $b_1 - a_1 > 0$, and therefore the $(b_1 - a_1)$-th galaxy is not truncated, the second component there allows unrestricted subtraction, and we have $x \,/\, y = (b_1 - a_1, b_2 - a_2)$.
- $a_1 = b_1$, $a_2 \leq b_2$. Then $x \,/\, y = (0, b_2 - a_2)$.
- $a_1 = b_1$, $a_2 > b_2$. Then $x \,/\, y = \bot$, since for no pair $(c_1, c_2) \in \mathfrak{A}$ we can have $(a_1 + c_1, a_2 + c_2) \preceq (b_1, b_2)$.
- $a_1 > b_1$. Then $x \,/\, y = \bot$.

□

This example of non-*-continuous RKL inherits two extra properties of Kozen's example: commutativity and linearity of the order.

3 Example of a Sequent Distinguishing ACT and $\mathbf{ACT}_\omega$

In this section we present a concrete example of a sequent provable in $\mathbf{ACT}_\omega$, but not in **ACT**. This is obtained by presenting a new induction principle for action logic, which is admissible in the *-continuous situation, but does not follow from the fixpoint axiomatisation of Kleene star (Pratt's pure induction principle). We call it the "induction-in-the-middle" rule:

Lemma 3.1 *The following rule*

$$\frac{\to B \quad A \to B \quad A, B, A \to B}{A^* \to B} \ (^* \to)_{\mathrm{mid}}$$

is admissible in $\mathbf{ACT}_\omega$.

Proof. By induction on n we show that all $A^n \to B$ are derivable from the premises using cut. Then apply the ω-rule. $\square$

The next theorem shows, however, that $(^* \to)_{\mathrm{mid}}$ is not valid in **ACT**. This gives a concrete example of a sequent that distinguishes **ACT** and $\mathbf{ACT}_\omega$.

Theorem 3.2 *The sequent*

$$(p \wedge q \wedge (p \,/\, q) \wedge (p \setminus q))^+ \to p$$

is provable in $\mathbf{ACT} + (^* \to)_{\mathrm{mid}}$ *(and therefore in* $\mathbf{ACT}_\omega$*), but not in* **ACT**.

Proof. Proving $(p \wedge q \wedge (p \,/\, q) \wedge (p \setminus q))^+ \to p$ in $\mathbf{ACT} + (^* \to)_{\mathrm{mid}}$ is easy. First we establish a variant of the induction-in-the-middle rule for $^+$:

$$\frac{A \to B \quad A^2 \to B \quad A, B, A \to B}{A^+ \to B}\ (^+ \to)_{\mathrm{mid}}$$

This new rule is obtained from $(^* \to)_{\mathrm{mid}}$ by the following derivation (recall that $A^+ = A \cdot A^*$):

$$\dfrac{\dfrac{\dfrac{A \to B}{\to A \setminus B}\,(\to \setminus) \quad \dfrac{A^2 \to B}{A \to A \setminus B}\,(\to \setminus) \quad \dfrac{\dfrac{A \to A \quad A, B, A \to B}{A, A, A \setminus B, A \to B}\,(\setminus \to)}{A, A \setminus B, A \to A \setminus B}\,(\to \setminus)}{A^* \to A \setminus B}\,(^* \to)_{\mathrm{mid}} \quad \dfrac{A \to A \quad B \to B}{A, A \setminus B \to B}\,(\setminus \to)}{\dfrac{A, A^* \to B}{A \cdot A^* \to B}\,(\cdot \to)}\,(\mathrm{cut})$$

Next, since $p \to p$, and $p \,/\, q, q \to p$, and $p \,/\, q, p, p \setminus q \to p$ are derivable in the Lambek calculus, for $A = (p \wedge q \wedge (p \,/\, q) \wedge (p \setminus q))$ we have $A \to p$, and $A^2 \to p$, and $A, p, A \to p$, by applying $(\wedge \to)$ several times. Now the $(^+ \to)_{\mathrm{mid}}$ rule yields the necessary sequent $A^+ \to p$.

In order to show that this sequent is not derivable in **ACT**, we construct a counter-model, *i.e.*, an RKL in which this sequent is false. [8]

Fix a two-letter alphabet $\Sigma = \{a, c\}$ and consider two families of formal languages, $\mathfrak{L}_1$ and $\mathfrak{L}_2$, defined as follows:

(i) $\mathfrak{L}_1$ is the family of all finite subsets of $\{a^n \mid n \geq 0\}$, including $\varnothing$ and $\{\varepsilon\}$ (ε is the empty word), which will be the zero and the unit of our RKL;

(ii) $\mathfrak{L}_2$ is the family of all languages of the form

$$A \cup \bigcup_{h \geq 0} \{a^i c a^{i+h} \mid i \geq f_{\mathrm{R}}(h)\} \cup \bigcup_{h > 0} \{a^{i+h} c a^i \mid i \geq f_{\mathrm{L}}(h)\},$$

where A is a cofinite subset of $\{a^n \mid n \geq 0\}$ and $f_{\mathrm{L}}, f_{\mathrm{R}} \colon \mathbb{N} \to \mathbb{N}$ are functions of at least linear growth. Throughout this paper, "f is a function of at least linear growth" means that, for all h, $f(h) \geq \alpha h + \beta$ for some rational α and β, $\alpha > 0$.

[8] We cannot use the non-*-continuous RKL from Section 2 here, since it is commutative, and in the commutative case $(^* \to)_{\mathrm{mid}}$ becomes admissible in **ACT**.

Let $\mathfrak{L} = \mathfrak{L}_1 \cup \mathfrak{L}_2$ and ∞ be an extra object (fresh constant) not belonging to $\mathfrak{L}$. Now let us define an RKL structure on $\mathfrak{A} = \mathfrak{L} \cup \{\infty\}$.

(i) The preorder, $\preceq$, is defined as set-theoretic inclusion, $\subseteq$, on $\mathfrak{L}$, and ∞ is declared the maximal element. Since $\mathfrak{L}$ is closed under (finite) unions and intersections (for $x, y \in \mathfrak{L}_2$ it follows from the fact that pointwise minimum and maximum of two functions of at least linear growth are again functions of at least linear growth; other cases are obvious), this preorder forms a lattice structure on $\mathfrak{A}$.

(ii) Multiplication, $x \cdot y$, is defined as follows:
- pairwise concatenation (as in L-models), if x and y are both in $\mathfrak{L}_1$ or if one of the arguments is in $\mathfrak{L}_1$ and the other is in $\mathfrak{L}_2$ (for the second case correctness follows from Lemma 3.3 below);
- ∞, if x and y are both in $\mathfrak{L}_2$;
- $\infty \cdot \varnothing = \varnothing \cdot \infty = \varnothing$;
- $\infty \cdot x = x \cdot \infty = \infty$, if $x \neq \varnothing$.

Checking associativity is routine. The unit is $\{\varepsilon\}$.

(iii) Left and right divisions are correctly defined due to Lemma 3.4 below.

(iv) Finally, Kleene star is defined as follows: $\varnothing^* = \{\varepsilon\}^* = \{\varepsilon\}$; $x^* = \infty$ for $x \neq \varnothing, \{\varepsilon\}$. We show that this definition is correct by verifying Pratt's pure induction condition, $(x \,/\, x)^* = x \,/\, x$. For $x = \varnothing$ and $x = \infty$ we have $x \,/\, x = \infty$, $\infty^* = \infty$. For $x \neq \varnothing, \infty$ we use Lemma 3.5 below showing that $x \,/\, x = \{\varepsilon\}$; $\{\varepsilon\}^* = \{\varepsilon\}$. Other conditions accompanying pure induction in Pratt's system are obvious: our definition of Kleene star is clearly monotone, for any x we have $\mathbf{1} \preceq x^*$ and $x \preceq x^*$, and since x^* is either $\{\varepsilon\}$ or ∞, we also have $x^* \cdot x^* \preceq x^*$.

Below we state and prove several lemmata supporting correctness of this definition.

Lemma 3.3 *If $x \in \mathfrak{L}_1$, $x \neq \varnothing$, and $y \in \mathfrak{L}_2$, then $x \cdot y$ and $y \cdot x$, where multiplication is defined as pairwise concatenation, are both in $\mathfrak{L}_2$. (If $x = \varnothing$, then $x \cdot y = y \cdot x = \varnothing \in \mathfrak{L}_1$.)*

Proof. We show that $x \cdot y \in \mathfrak{L}_2$ (the statement for $y \cdot x$ is symmetric). Since finite unions of languages from $\mathfrak{L}_2$ belong to $\mathfrak{L}_2$, it is sufficient to consider $x = \{a^k\}$ and show that $\{a^k\} \cdot y \in \mathfrak{L}_2$. Recall that

$$y = A \cup \bigcup_{h \geq 0} \{a^i c a^{i+h} \mid i \geq f_{\mathrm{R}}(h)\} \cup \bigcup_{h > 0} \{a^{i+h} c a^i \mid i \geq f_{\mathrm{L}}(h)\},$$

whence

$$\{a^k\} \cdot y = \{a^k\} \cdot A \cup \bigcup_{h \geq 0} \{a^{i+k} c a^{i+h} \mid i \geq f_{\mathrm{R}}(h)\} \cup \bigcup_{h > 0} \{a^{i+h+k} c a^i \mid i \geq f_{\mathrm{L}}(h)\}.$$

Apply the following transformations:

$$\bigcup_{h > 0} \{a^{i+h+k} c a^i \mid i \geq f_{\mathrm{L}}(h)\} = \bigcup_{\ell_1 > k} \{a^{i+\ell_1} c a^i \mid i \geq f_{\mathrm{L}}(\ell_1 - k)\},$$

where $\ell_1 = h + k$;

$$\bigcup_{h \geq k} \{a^{i+k} c a^{i+h} \mid i \geq f_{\mathrm{R}}(h)\} = \bigcup_{\ell_2 \geq 0} \{a^{j_2} c a^{j_2+\ell_2} \mid j_2 \geq f_{\mathrm{R}}(\ell_2 + k) + k\},$$

where $\ell_2 = h - k$, $j_2 = i + k$;

$$\bigcup_{0 \leq h < k} \{a^{i+k} c a^{i+h} \mid i \geq f_{\mathrm{R}}(h)\} = \bigcup_{0 < \ell_3 \leq k} \{a^{j_3+\ell_3} c a^{j_3} \mid f_{\mathrm{R}}(k - \ell_3) + k - \ell_3\},$$

where $\ell_3 = k - h$, $j_3 = i + h$.

This yields

$$\{a^k\} \cdot y = \widetilde{A} \cup \bigcup_{\ell \geq 0} \{a^i c a^{i+\ell} \mid i \geq \tilde{f}_{\mathrm{R}}(\ell)\} \cup \bigcup_{\ell > 0} \{a^{i+\ell} c a^i \mid i \geq \tilde{f}_{\mathrm{L}}(\ell)\},$$

where $\widetilde{A} = \{a^k\} \cdot A$ is a cofinite subset of $\{a^n \mid n \geq 0\}$ and $\tilde{f}_{\mathrm{R}}$ and $\tilde{f}_{\mathrm{L}}$, defined as follows

$$\tilde{f}_{\mathrm{R}}(\ell) = f_{\mathrm{R}}(\ell + k) + k;$$
$$\tilde{f}_{\mathrm{L}}(\ell) = \begin{cases} f_{\mathrm{L}}(\ell - k) & \text{for } \ell > k, \\ f_{\mathrm{R}}(k - \ell) + k - \ell & \text{for } 0 < \ell \leq k, \end{cases}$$

are both functions of at least linear growth[9] ($\tilde{f}_{\mathrm{L}}$ can actually decrease on $h \leq k$, but at least linear growth is an asymptotic property). Therefore $\{a^k\} \cdot y \in \mathfrak{L}_2$.$\square$

Lemma 3.4 *For any elements $x, y \in \mathfrak{A}$ there exist $x \,/\, y = \max\{z \mid z \cdot y \preceq x\}$ and $y \setminus x = \max\{z \mid y \cdot z \preceq x\}$.*

Proof. Consider only $x \,/\, y$ ($y \setminus x$ is symmetric). First we handle some degenerate cases:

- $y = \varnothing$: since $z \cdot \varnothing = \varnothing$ for any z, we have $x \,/\, \varnothing = \infty$. In particular, $\varnothing \,/\, \varnothing = \infty$.
- $x = \infty$: since $z \cdot y \preceq \infty$ for any z, we have $\infty \,/\, y = \infty$. In particular, $\infty \,/\, \infty = \infty$.
- $y = \infty$, $x \neq \infty$: since $z \cdot \infty \preceq x$ holds only for $z = \varnothing$ (in this case we get $\varnothing \preceq x$, otherwise $\infty \not\preceq x$), we have $x \,/\, \infty = \varnothing$.

Thus, now we have only the interesting case of $x, y \in \mathfrak{L}$, $y \neq \varnothing$. First we show that in this case $x \,/\, y$ can be defined exactly as in L-models (Subsection 1.4). Namely, if the language $z_0 = \{u \in \{a, c\}^* \mid (\forall v \in y)\, uv \in x\}$ (the language-theoretic division of x by y) belongs to $\mathfrak{L}$, then $z_0 = x \,/\, y = \max\{z \cdot y \preceq x\}$ in $\mathfrak{A}$ (*i.e.*, maximum is taken over all elements of $\mathfrak{A}$, including ∞, and w.r.t. $\preceq$).

Indeed, $z_0 \cdot y \subseteq x$, and therefore $z_0 \cdot y \preceq x$. Now let $z \cdot y \preceq x$ for some other z. Since $y \neq \varnothing$, z cannot be ∞ ($\infty \cdot y = \infty \not\preceq x$). Hence, $z \in \mathfrak{L}$, $z \cdot y \subseteq x$, and

[9] Recall that k is constant.

since z_0 is the language-theoretic division, $z \subseteq z_0$ and therefore $z \preceq z_0$. Thus, z_0 is the maximum in $\mathfrak{A}$.

Now it is sufficient to show that $\mathfrak{L}$ is closed under language-theoretic division operations (with non-zero denominator). Consider four possible cases:

- $x, y \in \mathfrak{L}_1$. The class of finite language over an alphabet is closed under language-theoretic division, provided the denominator is not $\varnothing$.
- $x \in \mathfrak{L}_1$, $y \in \mathfrak{L}_2$. In this case $x \,/\, y = \varnothing$, since there exists a word $v_0 = a^{i_0} c a^{i_0} \in y$, and for any $u \in \{a, c\}^*$ the word uv_0 contains c, and therefore cannot belong to x.
- $x \in \mathfrak{L}_2$, $y \in \mathfrak{L}_1$. Since in L-models $x \,/ (y_1 \cup \ldots \cup y_n) = (x \,/\, y_1) \cap \ldots \cap (x \,/\, y_n)$, and y, being a non-empty finite set, is a finite union of singletons, it is sufficient to show that $z_0 = x \,/ \{a^k\} = \{u \in \{a, c\}^* \mid ua^k \in x\}$ belongs to $\mathfrak{L}_2$. Recall that x, being a language from $\mathfrak{L}_2$, has the form

$$x = A \cup \bigcup_{h \geq 0} \{a^i c a^{i+h} \mid i \geq f_{\mathrm{R}}(h)\} \cup \bigcup_{h > 0} \{a^{i+h} c a^i \mid i \geq f_{\mathrm{L}}(h)\}.$$

Next, since in L-models for $x = \bigcup_\gamma x_\gamma$ (x is an infinite union) we have $x \,/ \{a^k\} = \bigcup_\gamma (x_\gamma \,/ \{a^k\})$, we can divide each component of x by $\{a^k\}$ independently:

$$\bigcup_{h \geq k} \{a^i c a^{i+h} \mid i \geq f_{\mathrm{R}}(h)\} \,/ \{a^k\} = \bigcup_{\ell_1 \geq 0} \{a^i c a^{i+\ell_1} \mid i \geq f_{\mathrm{R}}(\ell_1 + k)\},$$

where $\ell_1 = h - k$;

$$\bigcup_{0 \leq h < k} \{a^i c a^{i+h} \mid i \geq f_{\mathrm{R}}(h)\} \,/ \{a^k\} =$$
$$\bigcup_{0 < \ell_2 \leq k} \{a^{j_2 + \ell_2} c a^{j_2} \mid j_2 \geq \max\{0, f_{\mathrm{R}}(k - \ell_2) - \ell_2\}\},$$

where $\ell_2 = k - h, j_2 = i + h - k = i - \ell_2$;

$$\bigcup_{h > 0} \{a^{i+h} c a^i \mid i \geq f_{\mathrm{L}}(h)\} \,/ \{a^k\} =$$
$$\bigcup_{\ell_3 > k} \{a^{j_3 + \ell_3} c a^{j_3} \mid j_3 \geq \max\{0, f_{\mathrm{L}}(\ell_3 - k) - k\}\},$$

where $\ell_3 = h + k$, $j_3 = i - k$.

Thus,

$$x \,/ \{a^k\} = \widetilde{A} \cup \bigcup_{h \geq 0} \{a^i c a^{i+h} \mid i \geq \tilde{f}_{\mathrm{R}}(h)\} \cup \bigcup_{h > 0} \{a^{i+h} c a^i \mid i \geq \tilde{f}_{\mathrm{L}}(h)\},$$

where $\widetilde{A} = A \,/ \{a^k\}$ is a cofinite subset of $\{a^n \mid n \geq 0\}$, and $\tilde{f}_{\mathrm{R}}$ and $\tilde{f}_{\mathrm{L}}$, defined as follows

$$\tilde{f}_{\mathrm{R}}(\ell) = f_{\mathrm{R}}(\ell + k);$$
$$\tilde{f}_{\mathrm{L}}(\ell) = \begin{cases} \max\{0, f_{\mathrm{R}}(k-\ell) - \ell\} & \text{for } 0 < h \le k, \\ \max\{0, f_{\mathrm{L}}(\ell - k) - k\} & \text{for } h > k, \end{cases}$$

are functions of linear growth (again, for $\tilde{f}_{\mathrm{L}}$ growth starts from $h > k$).

- $x, y \in \mathfrak{L}_2$. We show that in this case $z_0 = \{u \in \{a,c\}^* \mid (\forall v \in y)\, uv \in x\}$ belongs to $\mathfrak{L}_1$. Indeed, if u includes at least one letter c, then for $v_0 = a^{i_0}ca^{i_0} \in y$ the word uv_0 includes at least two c's, and cannot belong to y. Thus, $z_0 \subseteq \{a^n \mid n \ge 0\}$, and it remains to show that z_0 is finite. Let a^k be an element of z_0. Fix an arbitrary $a^ica^i \in y$ (such an element exists by definition of $\mathfrak{L}_2$) and multiply it by a^k. We get $a^ica^{i+k} \in x$. This yields $i \ge f_{\mathrm{R}}(k)$, where f_{R} is taken from the $\mathfrak{L}_2$ representation of x, and by growth condition $i \ge \alpha k + \beta$. Since $\alpha > 0$, we get $k \le (i - \beta)/\alpha$, which establishes a global boundary for possible values of k: recall that i, α, and β were taken independently from k. Therefore, $z_0 \subseteq \{a^k \mid 0 \le k \le (i-\beta)/\alpha\}$ is finite and belongs to $\mathfrak{L}_1$.

□

Lemma 3.5 *For any x, except $\varnothing$ and ∞, $x \,/\, x = \{\varepsilon\}$.*

Proof. Since $\{\varepsilon\}$ is the unit of $\mathfrak{A}$, $\{\varepsilon\} \cdot x = x$, therefore $x \,/\, x \succeq \{\varepsilon\}$. Suppose there exists $z \succ \{\varepsilon\}$ such that $z \cdot x \preceq x$. Then $\{a^k\} \preceq z$ for some $k > 0$, and by monotonicity $\{a^k\} \cdot x \preceq x$ (monotonicity of $\cdot$ w.r.t. $\preceq$ holds in all residuated lattices, since the corresponding logical rule is admissible in the Lambek calculus [21]). Consider two cases:

- $x \in \mathfrak{L}_1$, $x \ne \varnothing$. Then let a^m be the element of x with the greatest m. Clearly, $a^ka^m \notin x$, therefore $\{a^k\} \cdot x \not\preceq x$.
- $x \in \mathfrak{L}_2$. Take $a^{i_0}ca^{i_0} \in x$ (exists by definition of $\mathfrak{L}_2$). Since $\{a^k\} \cdot x \preceq x$, we have $a^{i_0+k}ca^{i_0} \in x$, $a^{i_0+2k}ca^{i_0} \in x$, ..., $a^{i_0+mk}ca^{i_0} \in x$, ... Thus, $f_{\mathrm{L}}(mk) \le i_0$ for arbitrary big m, which contradicts with the growth condition for f_{L}.

□

Now we finish the proof of Theorem 3.2 by falsifying $(p \wedge q \wedge (p \,/\, q) \wedge (p \setminus q))^+ \to p$ in the newly constructed RKL:

Lemma 3.6 *If p is interpreted as the language from $\mathfrak{L}_2$ with $A = \{a^n \mid n \ge 0\}$ and $f_{\mathrm{L}}(h) = f_{\mathrm{R}}(h) = 2h$, and $q = p \cdot \{a\}$, then $p \wedge q \wedge (p \,/\, q) \wedge (p \setminus q) = \{a\}$. Thus, since $\{a\}^+ = \infty \not\preceq p$, the sequent $(p \wedge q \wedge (p \,/\, q) \wedge (p \setminus q))^+ \to p$ is not true under this intepretation.*

Proof. Clearly, $a \in p$, and, since $\varepsilon \in p$, also $a \in q$. Next, $p \cdot \{a\} \preceq q$ yields $\{a\} \preceq p \setminus q$.

Next, let us show that $p \,/\, q = \{a\}$. As follows from the proof of Lemma 3.4, since both p and q are elements of $\mathfrak{L}_2$, $p \,/\, q \in \mathfrak{L}_1$, *i.e.*, it is a finite subset of

$\{a^n \mid n \geq 0\}$. This means that $p\,/\,q = \{a^n \mid (\forall v \in q)\, a^n v \in p\}$, and since any $v \in q$ is of the form ua, where $u \in p$, we have $p\,/\,q = \{a^n \mid (\forall u \in p)\, a^n ua \in p\}$.

For $n = 1$ we indeed have $aua \in p$ for any $u \in p$ (all three components in the $\mathfrak{L}_2$ representation of p are upwardly closed under multiplication by a on both sides).

For $n = 0$, take $u = c$. Since $f_{\mathrm{R}}(0) = 0$, it is an element of p. The word $ua = ca$, however, is of the form $a^i ca^{i+h}$ for $i = 0$ and $h = 1$, and since $f_{\mathrm{R}}(1) = 2 > 0$, this word does not belong to p. Therefore, $\varepsilon = a^0 \notin p\,/\,q$.

Analogous reasoning applies to $n > 1$. Again, take $u = c \in p$ and consider the word $a^n ua = a^n ca$, which is of the form $a^{i+h} ca^i$ for $i = 1$ and $h = n-1 > 0$. Having $f_{\mathrm{L}}(n-1) = 2(n-1) \geq 2 > 1$, we show that this word is not in p, therefore $a^n \notin p\,/\,q$ for $n > 1$.

Finally, having $p\,/\,q = \{a\}$ and p, q, $p \setminus q \succeq \{a\}$, we obtain $p \wedge q \wedge (p\,/\,q) \wedge (p \setminus q) = \{a\}$. □

This finishes the proof of Theorem 3.2. □

We conclude this section with some general remarks. The search for concrete formulae witnessing the fact that ω-rules are more powerful than induction has a long history, starting from Gödel's Second Incompleteness Theorem [11]. The formula there is the well-known consistency statement: "for all n, n does not encode a proof of contradiction." Later, other arithmetical statements that are true, but not provable in Peano arithmetic (*i.e.*, using the standard induction principle), of more combinatorial nature, were discovered. These include Hercules vs. Hydra by Kirby and Paris [13], Beklemishev's worm [2], *etc.* An example closer to our discussion was discovered by Kozen for PDL [14]. The key feature of our example is that it is formulated in a propositional language, while the formulae mentioned above are first-order ones. On the other hand, our example actually shows not the weakness of induction in general, but rather the fact that in action logic there exist induction principles different from (and not following from) Pratt's pure induction, but yet not transfinite.

4 Incompleteness of $\mathbf{ACT}_\omega(\cdot, \setminus, /, \wedge, {}^*)$ w.r.t. R- and L-models

Though relational and language models, both being *-continuous, are natural classes of interpretations for $\mathbf{ACT}_\omega$, there are well-known obstacles to completeness connected with distributivity and the unit constant (see Subsection 1.4). Without these problematic connectives, $\vee$ and $\mathbf{1}$, there is a hope for R- and L-completeness. This hope is supported by completeness results for the corresponding fragment without *, $\mathbf{MALC}(\cdot, \setminus, /, \wedge)$: see Andréka and Mikulás [1] for R-models of $\mathbf{MALC}(\cdot, \setminus, /, \wedge)$, Buszkowski [3] for L-models of the product-free system $\mathbf{MALC}(\setminus, /, \wedge)$, Pentus [30,31] for L-models of the Lambek calculus (in our notation, $\mathbf{MALC}(\cdot, \setminus, /)$). L-completeness of $\mathbf{MALC}(\cdot, \setminus, /, \wedge)$, with both multiplication and intersection, is still an open problem, and there are no arguments against the positive answer.

This motivates us to consider the fragment $\mathbf{ACT}_\omega(\cdot, \setminus, /, \wedge, {}^*)$ and conjecture its R- and L-completeness. Disjunction, however, is hidden inside Kleene star: $a^* = \mathbf{1} \vee a^+$, and distributivity can shoot around the corner, making $\mathbf{ACT}_\omega$ incomplete even in this restricted fragment.

Theorem 4.1 *The sequent*

$$(s \,/(r \,/\, r)) \wedge (s \,/(p^+ \wedge q^+)) \to s \,/(p^* \wedge q^*)$$

is true in all distributive RKL's, but not provable in $\mathbf{ACT}_\omega$.

Proof. First we show that this sequent is true in all distributive RKL's. Recall that $A^* \leftrightarrow \mathbf{1} \vee A^+$. By distributivity and monotonicity of $\wedge$, we get

$$p^* \wedge q^* \to (\mathbf{1} \vee p^+) \wedge (\mathbf{1} \vee q^+) \to \mathbf{1} \vee (p^+ \wedge q^+) \to (r \,/\, r) \vee (p^+ \wedge q^+).$$

Next, we put this under $s \,/ \ldots$ (the direction of the arrow changes), which allows us to replace $\vee$ by $\wedge$ using $(A \,/\, B) \wedge (A \,/\, C) \to A \,/(B \vee C)$:

$$(s \,/(r \,/\, r)) \wedge (s \,/(p^+ \wedge q^+)) \to s \,/((r \,/\, r) \vee (p^+ \wedge q^+)) \to s \,/(p^* \wedge q^*).$$

The reasoning above can be done in $\mathbf{ACT}$ + distributivity axiom. Thus, the goal sequent is true in all distributive RKL's.

For the second part, non-derivability in $\mathbf{ACT}_\omega$, we recall that cut is eliminable in this calculus [29] and perform exhaustive proof search. Since the $(\to /)$ rule is invertible, we can suppose that it was applied as the last step of the derivation:

$$\frac{(s \,/(r \,/\, r)) \wedge (s \,/(p^+ \wedge q^+)), p^* \wedge q^* \to s}{(s \,/(r \,/\, r)) \wedge (s \,/(p^+ \wedge q^+)) \to s \,/(p^* \wedge q^*)}\ (\to /)$$

Now we have four options for applying the $(\wedge \to)$ rule, with the following premises:

(i) $s \,/(r \,/\, r), p^* \wedge q^* \to s$;

(ii) $s \,/(p^+ \wedge q^+), p^* \wedge q^* \to s$;

(iii) $(s \,/(r \,/\, r)) \wedge (s \,/(p^+ \wedge q^+)), p^* \to s$;

(iv) $(s \,/(r \,/\, r)) \wedge (s \,/(p^+ \wedge q^+)), q^* \to s$.

These sequents are not generally true in L-models, and therefore are not derivable in $\mathbf{ACT}_\omega$. The counter-interpretations are as follows:

(i) $s = r = \{\varepsilon\}$, $p = q = \{a\}$;

(ii) $s = \{a\}^+$, $p = q = \{a\}$;

(iii) $s = r = \{\varepsilon\}$, $p = \{a\}$, $q = \varnothing$;

(iv) $s = r = \{\varepsilon\}$, $p = \varnothing$, $q = \{a\}$.

□

5 Conclusions and Future Work

The counter-model used in the proof of Theorem 3.2 is quite an *ad hoc* invention to falsify one particular sequent. On the other hand, it is very closely related to L-models (which are all *-continuous, and therefore useless in connection to Theorem 3.2). It looks interesting to perform a more systematic study of variants of L-models with Kleene star, and possibly find natural classes of models which are not *-continuous (*i.e.*, give semantics for **ACT**, but not $\mathbf{ACT}_\omega$). One possible starting point for such a study is the interpretation of **MALC** on syntactic concept lattices (SCL's) by Wurm [35]. SCL's are much like L-models, but can be, for example, non-distributive, thus avoiding incompleteness issues for meet and join.

Our $({}^* \to)_{\mathrm{mid}}$ rule is actually one of an infinite series of induction-in-the-middle rules arising from circular proof systems [20]:

$$\frac{\to B \quad A \to B \quad \ldots \quad A^{m+k-1} \to B \quad A^m, B, A^k \to B}{A^* \to B}$$

All these rules are admissible in $\mathbf{ACT}_\omega$, but even if we add all of them, we still get a system weaker than $\mathbf{ACT}_\omega$ (due to complexity reasons: it is still r.e.). The question is how do these rules interact with each other: for example, is there a finite subfamily of these rules which derives all of them?

Our example of a sequent derivable in $\mathbf{ACT}_\omega$, but not in **ACT**, contains additive conjunction, $\wedge$. Buszkowski [5], however, also proves Π_1^0-completeness for the fragment with $\vee$ instead of $\wedge$. Moreover, in the view of the Π_1^0-completeness result [20] for a closely related system, with Lambek's restriction, it is highly likely that even in the multiplicative-only fragment, including $\cdot$, $\backslash$, $/$, and *, the infinitary system $\mathbf{ACT}_\omega$ is stronger than **ACT**. The task of finding concrete examples of sequents to distinguish the two systems in these restricted fragments is left open for future research.

As for L- and R-completeness, the question is still open whether the purely multiplicative Lambek calculus with iteration (in our notation, $\mathbf{ACT}_\omega(\cdot, \backslash, /, {}^*)$) is L-complete (R-complete).[10]

Finally, we still have the old problem of constructing a good (cut-free) Gentzen-style system for **ACT**. For Kleene algebra, there is a recent approach by Das and Pous [7], who present a cut-free hypersequential calculus with circular proofs. Unfortunately, their approach cannot be directly generalised to the residuated case.

---

[10] L-completeness is known for two fragments of this calculus. In the first fragment, Kleene star is allowed only on the top level in formulae of the antecedent. In this case completeness follows from that of the original Lambek calculus due to invertibility of $({}^* \to)$. The second fragment is the product-free Lambek calculus with * allowed only in denominators of $\backslash$ and $/$ (*i.e.*, in subformulae of the form $A^* \backslash B$ and B / A^*). L-completeness for the variant of this fragment, with Lambek's restriction and positive iteration in place of Kleene star, is stated in [19].

Acknowledgments

The author is grateful to Fedor Pakhomov, Daniyar Shamkanov, Andre Scedrov, and Scott Weinstein for fruitful discussions. The author is also grateful to the anonymous referees for their deep comments and helpful suggestions. Being a Young Russian Mathematics award winner, the author would like to thank its sponsors and jury for this high honour. (The work on this particular paper was funded from another source, as mentioned on the first page.)

References

[1] Andréka, H. and S. Mikulás, *Lambek calculus and its relational semantics: Completeness and incompleteness*, Journal of Logic, Language, and Information **3** (1994), pp. 1–37.

[2] Beklemishev, L. D., *Provability algebras and proof-theoretic ordinals, I*, Annals of Pure and Applied Logic **128** (2004), pp. 103–123.

[3] Buszkowski, W., *Compatibility of a categorial grammar with an associated category system*, Zeitschrift für mathematische Logik und Grundlagen der Mathematik **28** (1982), pp. 229–237.

[4] Buszkowski, W., *On the complexity of the equational theory of relational action algebras*, in: R. A. Schmidt, editor, *RelMiCS 2006: Relations and Kleene Algebra in Computer Science*, Lecture Notes in Computer Science **4136** (2006), pp. 106–119.

[5] Buszkowski, W., *On action logic: Equational theories of action algebras*, Journal of Logic and Computation **17** (2007), pp. 199–217.

[6] Buszkowski, W. and E. Palka, *Infinitary action logic: Complexity, models and grammars*, Studia Logica **89** (2008), pp. 1–18.

[7] Das, A. and D. Pous, *A cut-free cyclic proof system for Kleene algebra*, in: R. A. Schmidt and C. Nalon, editors, *TABLEAUX 2017: Automated Reasoning with Analytic Tableaux and Related Methods*, Lecture Notes in Computer Science (Lecture Notes in Artificial Intelligence) **10501** (2017), pp. 261–277.

[8] Dekker, J. C. E., *Two notes on recursively enumerable sets*, Proceedings of the American Mathematical Society **4** (1953), pp. 495–501.

[9] Girard, J.-Y., *Linear logic*, Theoretical Computer Science **50** (1987), pp. 1–101.

[10] Gödel, K., *Die Vollständigkeit der Axiome des logischen Functionenkalküls*, Monatshefte für Mathematik und Physik **37** (1930), pp. 349–360.

[11] Gödel, K., *Über formal unentscheidbare Sätze der Principia Mathematica und verwandter Systeme, I*, Monatshefte für Mathematik und Physik **38** (1931), pp. 173–198.

[12] Jipsen, P., *From semirings to residuated Kleene lattices*, Studia Logica **76** (2004), pp. 291–303.

[13] Kirby, L. and J. Paris, *Accessible independence results for Peano arithmetic*, Bulletin of the London Mathematical Society **14** (1982), pp. 285–293.

[14] Kozen, D., *On induction vs. *-continuity*, in: D. Kozen, editor, *Logic of Programs 1981: Logics of Programs*, Lecture Notes in Computer Science **131** (1982), pp. 167–176.

[15] Kozen, D., *On Kleene algebras and closed semirings*, in: B. Rovan, editor, *MFCS 1990: International Symposium on Mathematical Foundations of Computer Science*, Lecture Notes in Computer Science **452** (1990), pp. 26–47.

[16] Kozen, D., *A completeness theorem for Kleene algebras and the algebra of regular events*, Information and Computation **110** (1994), pp. 366–390.

[17] Kozen, D., *On action algebras*, in: J. van Eijck and A. Visser, editors, *Logic and Information Flow*, MIT Press, 1994 pp. 78–88.

[18] Kuznetsov, S., *Trivalent logics arising from the Lambek calculus with constants*, Journal of Applied Non-Classical Logics **24** (2014), pp. 132–137.

[19] Kuznetsov, S. L. and N. S. Ryzhkova, *Fragment isčislenija Lambeka s iteracijej [A fragment of the Lambek calculus with iteration]* (in Russian), in: *Mal'tsev Meeting. International Conference Dedicated to 75th Anniversary of Yu. L. Ershov. Collection of Abstracts*, Novosibirsk, 2015, p. 213.
[20] Kuznetsov, S., *The Lambek calculus with iteration: Two variants*, in: J. Kennedy and R. de Queiroz, editors, *WoLLIC 2017: Logic, Language, Information, and Computation*, Lecture Notes in Computer Science **10388** (2017), pp. 182–198.
[21] Lambek, J., *The mathematics of sentence structure*, American Mathematical Monthly **65** (1958), pp. 154–170.
[22] Lambek, J., *Deductive systems and categories II. Standard constructions and closed categories*, in: P. J. Hilton, editor, *Category Theory, Homology Theory and their Applications I*, Lecture Notes in Mathematics **86** (1969), pp. 76–122.
[23] Malcev, A., *Untersuchungen aus dem Gebeite der mathematische Logik*, Rec. Math. [Mat. Sbornik] N.S. **1(43)** (1936), pp. 323–336.
[24] Mal'cev, A. I., *Ob odnom obščem metode polučenija lokalnyh teorem teorii grupp [On a general method for obtaining local theorems in group theory]* (in Russian), Ivanov. Gos. Ped. Inst. Uč. Zap. Fiz.-Mat. Fak. **1** (1941), pp. 3–9.
[25] Morrill, G. V., "Categorial Grammar: Logical Syntax, Semantics, and Processing," Oxford University Press, 2011.
[26] Myhill, J., *Creative sets*, Zeitschift für mathematische Logik and Grundlagen der Mathematik **1** (1955), pp. 97–108.
[27] Ono, H., *Semantics for substructural logics*, in: P. Schroeder-Heister and K. Došen, editors, *Substructural Logics*, Studies in Logic and Computation **2**, Clarendon Press, Oxford, 1993 pp. 259–291.
[28] Ono, H. and Y. Komori, *Logics without contraction rule*, Journal of Symbolic Logic **50** (1985), pp. 169–201.
[29] Palka, E., *An infinitary sequent system for the equational theory of *-continuous action lattices*, Fundamenta Informaticae **78** (2007), pp. 295–309.
[30] Pentus, M., *Models for the Lambek calculus*, Annals of Pure and Applied Logic **75** (1995), pp. 179–213.
[31] Pentus, M., *Free monoid completeness of the Lambek calculus allowing empty premises*, in: J. M. Larrazabal, D. Lascar and G. Mints, editors, *Logic Colloquium '96*, Lecture Notes in Logic **12** (1998), pp. 171–209.
[32] Pratt, V., *Action logic and pure induction*, in: J. van Eijck, editor, *JELIA 1990: Logics in AI*, Lecture Notes in Computer Science (Lecture Notes in Artificial Intelligence) **478** (1991), pp. 97–120.
[33] Restall, G., "An Introduction to Substructural Logics," Routledge, 2000.
[34] Soare, R. I., "Recursively Enumerable Sets and Degrees: A Study of Computable Functions and Computably Generated Sets," Springer-Verlag, 1987.
[35] Wurm, C., *Language-theoretic and finite relation models for the (full) Lambek calculus*, Journal of Logic, Language, and Information **26** (2017), pp. 179–214.

The Internalized Disjunction Property for Intuitionistic Justification Logic

Michel Marti [1]

Loria
Campus Scientifique, BP 239, 54506 Vandoeuvre-lès-Nancy
France

Thomas Studer [2]

Institute of Computer Science
University of Bern, Neubrückstrasse 10, 3012 Bern
Switzerland

Abstract

In intuitionistic justification logic, evidence terms represent intuitionistic proofs, that is a formula r:A means r is an intuitionistic proof of A. A natural principle in this context is the internalized disjunction property (IDP), which is: for each term r there exists a term s such that r:(A or B) implies s:A or s:B.
We introduce a light extension of iJT4, in which IDP is valid. Our proof relies on a model construction that enforces sharp evidence relations and a tight connection between syntax and semantics. This makes it possible to switch between proofs and models, which will be the key to proving IDP.

Keywords: Justification logic, intuitionistic logic, disjunction property, sharp model.

1 Introduction

Justification logics feature formulas of the form $t : A$ meaning A *is known for reason* t. The evidence term t can stand for a formal proof of A (say in Peano Arithmetic) [2,21] or t can represent an informal justification to believe that A [14]. This second reading turned out to be very useful for analyzing a variety of epistemic situations, see, e.g., [4,5,8,9,11,12,17].

It is a distinguishing feature of justification logics that they internalize their own notion of proof. If a formula A is provable, then there exists a term t such that $t : A$ is also provable and, additionally, the term t is a blueprint of the proof of A.

[1] michel.marti@loria.fr

[2] tstuder@inf.unibe.ch

Supported by the Swiss National Science Foundation grant 200021_165549.

In an intuitionistic setting, it is natural to ask whether an internalized version of the disjunction property holds, i.e., for each term t there exists a term s such that

$$t : (A \vee B) \to (s : A \vee s : B) \tag{IDP}$$

is valid. In this paper we introduce an intuitionistic justification logic $\mathsf{iJT4}^+$ and show that $\mathsf{iJT4}^+$ proves (IDP).

Our logic $\mathsf{iJT4}^+$ is a light extension of $\mathsf{iJT4}$, which is an intuitionistic version of the Logic of Proofs. Artemov introduced $\mathsf{iJT4}$ (then called $\mathcal{ILP}$) in one of the early papers on justification logic [3] with the aim of unifying the semantics of modalities and λ-calculus. He defined $\mathsf{iJT4}$ by changing the propositional base of LP to intuitionistic logic while keeping the other axioms of LP. Artemov then showed that $\mathsf{iJT4}$ can realize a λ-calculus for intuitionistic $\mathsf{S4}$.

Marti and Studer [22] recently developed a possible world semantics for $\mathsf{iJT4}$. They introduced so-called modular models for $\mathsf{iJT4}$, which feature the principle *justification yields belief* [6,19].

Steren and Bonelli [24] provide an alternative term system for $\mathsf{iJT4}$ that is based on natural deduction and hypothetical reasoning. Their aim is to reformulate the Logic of Proofs in order to explore applications in programming languages.

The axiomatization of $\mathsf{iJT4}$ does not yield a proper intuitionistic provability semantics, which means to interpret $t : A$ as *t encodes a proof of A in Heyting Arithmetic.* Artemov and Iemhoff [7] extended $\mathsf{iJT4}$ by axioms that introduce novel proof terms to internalize certain admissible rules of intuitionistic logic. The arithmetical completeness of that system was later established by Dashkov [13].

The above mentioned intuitionistic justification logics all study explicit versions of the $\Box$-modality. Kuznets, Marin, and Straßburger [18] provide a treatment of the intuitionistic $\Diamond$-modality in the style of justification logic. To do so, they introduce a new type of evidence terms that justifies consistency. Hence they obtain justification analogues of several constructive modal logics and establish a realization theorem for them.

The structure of the paper is as follows. In the next section we will introduce the logic $\mathsf{iJT4}^+$, which is a light extension of the intuitionistic justification logic $\mathsf{iJT4}$. Then we present generated models for $\mathsf{iJT4}^+$ and establish soundness of $\mathsf{iJT4}^+$. In Section 4 we define point-generated models, which are needed to prove the disjunction property. Section 5 studies atomic models, i.e. models that are completely defined by the evaluations of atomic propositions and atomic justifications. The name *atomic model* goes back to Kashev [16]; Artemov [1] uses *sharp model.* In Section 6 we define the canonical model and establish completeness of $\mathsf{iJT4}^+$ with respect to atomic models. Finally, in Section 7, we prove the internalized disjunction property for $\mathsf{iJT4}^+$. In the last section we discuss future work and give some hints about the realization of Hirai's intuitionistic modal logic [15] in $\mathsf{iJT4}^+$. The appendix contains proofs of some technical lemmas.

2 The Logic $\mathsf{iJT4}^+_{\mathsf{CS}}$

We start with a countable set of atomic propositions Prop and a countable set of term constants Const. We define terms and formulas of our language simultaneously by the following grammar:

$$t ::= c \mid ?_n \mid t+t \mid !t \mid t \cdot_A t$$
$$A ::= \bot \mid P \mid A \wedge A \mid A \vee A \mid A \rightarrow A \mid t : A$$

where $c \in \mathsf{Const}$, $P \in \mathsf{Prop}$, and $n \in \mathbb{N}$.

The set of terms is denoted by Tm, the set of formulas by Fm. We will use P, Q, R for propositions, c, d, e for constants, A, B, C, D for arbitrary formulas, and t, r, s for terms. As is usual in intuitionistic logic, $\neg A$ is defined as $A \rightarrow \bot$.

The system $\mathsf{iJT4}^+$ has the following axioms for all $t, s \in \mathsf{Tm}$ and all $A, B \in \mathsf{Fm}$

(i) all axioms for intuitionistic propositional logic

(ii) $t : A \wedge s : (A \rightarrow B) \rightarrow (s \cdot_A t) : B$ (j)

(iii) $(s \cdot_A t) : B \rightarrow t : A \wedge s : (A \rightarrow B)$ (invj)

(iv) $t : A \vee s : A \rightarrow (t+s) : A$ (+)

(v) $(t+s) : A \rightarrow t : A \vee s : A$ (inv+)

(vi) $t : A \rightarrow !t : (t : A)$ (!)

(vii) $\neg(!t : A)$ if A is not of the form $t : B$ for any formula $B \in \mathsf{Fm}$ (inv!)

(viii) $(t : A) \rightarrow A$ truth property (t)

Definition 2.1 A *constant specification* CS is any subset

$$\mathsf{CS} \subseteq \{(c, A) \mid c \in \mathsf{Const} \text{ and } A \in \mathsf{Fm} \text{ is an axiom of } \mathsf{iJT4}^+\}.$$

A constant specification CS is called *axiomatically appropriate* if for each axiom A of $\mathsf{iJT4}^+$, there is a constant c such that $(c, A) \in \mathsf{CS}$.

Definition 2.2 [$?_n$-form] For each natural number $n \in \mathbb{N}$, we define what it means for a formula to be of $?_n$-form as follows:

- for $n = 0$, A is of $?_0$-form iff $A = \neg(c : B)$ for some formula B with $B \notin \mathsf{Prop}$ and $(c, B) \notin \mathsf{CS}$
- A is of $?_{(n+1)}$-form iff $A = \neg ?_n : B$ for some formula B which is not of $?_n$-form.

Definition 2.3 For a constant specification CS the deductive system $\mathsf{iJT4}^+_{\mathsf{CS}}$ is the Hilbert system given by the axioms (axiom schemes) above and by the following three rules:

$$\frac{A \quad A \rightarrow B}{B}\ (\mathrm{MP}) \qquad \frac{(c, A) \in \mathsf{CS}}{c : A}\ (\mathrm{AN})_{\mathsf{CS}} \qquad \frac{A \text{ is of } ?_n\text{-form}}{?_n : A}\ (?_n)$$

The system $\mathsf{iJT4}^+_{\mathsf{CS}}$ defines a family of logics parameterized by the constant specification CS. Hence we should denote the derivability relation of $\mathsf{iJT4}^+_{\mathsf{CS}}$

by $\vdash_{\mathsf{CS}}$. For simplicity, however, we will only write $\vdash$ when CS is clear from the context.

The system $\mathsf{iJT4}^{+}_{\mathsf{CS}}$ is tailored towards minimal evidence relations. This explains why we included inverse axioms (invj) and (inv+) for application and sum, respectively. In order to formulate (invj) we need a labelled application operator, which goes back to Renne [23]. These inverse axioms make it possible to reduce a justification with an application to simpler justifications, see, e.g., [20].

Our aim is to prove an internalized version of the disjunction property. In order to achieve this, we have to guarantee that a constant term can only justify an axiom (as given in the constant specification) or an atomic proposition. This could be achieved with a rule of the form

$$\frac{(c, B) \notin \mathsf{CS} \text{ and } B \notin \mathsf{Prop}}{\neg(c : B)} .$$

However, since we want to have the Lifting Lemma 2.6, we need terms to justify derivations with this rule. This is the role of the $?_n$-terms and explains the rule $(?_n)$, which can be thought of as a form of negative introspection for the constant specification.

Remark 2.4 Recall that our language does *not* include justification variables. The reason is that variables stand for arbitrary terms, which does not fit our setting. Suppose that we have variables and assume $(c, A) \in \mathsf{CS}$. Then

$$\neg(!c : x : A) \tag{1}$$

is an instance of axiom (inv!) since syntactically the term c is different from the term x. However, if we substitute c for x in (1), then we obtain $\neg(!c : c : A)$, which is provable false. So by substitution, we can transform a valid formula into a provably false formula. To avoid this, we do not include variables in our language (but we will come back to this issue in the Conclusion).

As usual in justification logic, the Deduction Theorem holds.

Theorem 2.5 (Deduction Theorem) *For any set of formulas Γ and formulas A, B we have*

$$\Gamma, A \vdash B \quad \Longleftrightarrow \quad \Gamma \vdash A \to B$$

Since the rule $(?_n)$ only introduces formulas of the form $t : A$, we can prove the following form of the Lifting Lemma as usual.

Lemma 2.6 *Let CS be an axiomatically appropriate constant specification. For all terms $t_1, \ldots, t_n \in \mathsf{Tm}$ and all formulas $A_1, \ldots, A_n, C \in \mathsf{Fm}$, if*

$$t_1 : A_1, \ldots, t_n : A_n \vdash C,$$

then there is a term $t \in \mathsf{Tm}$ such that

$$t_1 : A_1, \ldots, t_n : A_n \vdash t : C.$$

3 Generated Models

We are now going to introduce a semantics for $\mathsf{iJT4}^{+}_{\mathsf{CS}}$. A *frame* $\mathcal{F} = (W, \leq)$ is a pair consisting of a set W of states and a reflexive and transitive relation $\leq$ on W. A model for $\mathsf{iJT4}^{+}_{\mathsf{CS}}$ is essentially a frame with a suitable minimal basic modular model assigned to each state. These models are called *generated* since minimality is achieved by inductively generating the models [20,25,26].

Definition 3.1 [Basis] Given a frame $\mathcal{F} = (W, \leq)$, a basis for $\mathcal{F}$ is a familiy of sets $\mathcal{B} = (\mathcal{B}_w)_{w \in W}$ such that

- $\mathcal{B}_w \subseteq \mathsf{Tm} \times \mathsf{Fm}$ for each $w \in W$
- $w \leq v \implies \mathcal{B}_w \subseteq \mathcal{B}_v$ (Monotonicity)
- $(!t, A) \notin \mathcal{B}_w$ if A is not of the form $t : B$ for any formula $B \in \mathsf{Fm}$

$\mathcal{B}_w$ must be downwards closed, i.e. for each $w \in W$:

- $(t + s, A) \in \mathcal{B}_w \implies (t, A) \in \mathcal{B}_w$ or $(s, A) \in \mathcal{B}_w$
- $((t \cdot_A s), B) \in \mathcal{B}_w \implies (s, A) \in \mathcal{B}_w$ and $(t, A \to B) \in \mathcal{B}_w$
- $(!t, t : A) \in \mathcal{B}_w \implies (t, A) \in \mathcal{B}_w$

Definition 3.2 [Evidence Closure] Let $B \subseteq \mathsf{Tm} \times \mathsf{Fm}$. For a set $X \subseteq \mathsf{Tm} \times \mathsf{Fm}$ we define $\mathsf{cl}_B(X)$ by

(i) if $(t, A) \in B$, then $(t, A) \in \mathsf{cl}_B(X)$

(ii) if $(s, A) \in X$ and $(t, A \to B) \in X$, then $((t \cdot_A s), B) \in \mathsf{cl}_B(X)$

(iii) if $(s, A) \in X$, then $((s + t), A) \in \mathsf{cl}_B(X)$

(iv) if $(t, A) \in X$, then $((s + t), A) \in \mathsf{cl}_B(X)$

(v) if $(t, A) \in X$, then $(!t, t : A) \in \mathsf{cl}_B(X)$

For any $B \subseteq \mathsf{Tm} \times \mathsf{Fm}$, the operator cl_B is monotone. Therefore, cl_B has a least fixed point, which we denote *the evidence relation induced by B*.

Definition 3.3 [Evidence Relation] Let $B \subseteq \mathsf{Tm} \times \mathrm{Fm}$. Then $\mathcal{E}(B)$ is defined as the least fixed point of cl_B.

We need the following immediate properties about evidence relations.

Lemma 3.4 (Constants are in the Basis) *For all formulas A and constants c we have:*

$$(c, A) \in \mathcal{E}(\mathcal{B}_w) \implies (c, A) \in \mathcal{B}_w.$$

Lemma 3.5 *If A is not of the form t : B, then*

$$(!t, A) \notin \mathcal{E}(\mathcal{B}_w).$$

Definition 3.6 [Generated Models] Let CS be a constant specification. A *generated model* is a structure $\mathfrak{M} = (W, \leq, V, (\mathcal{B}_w)_{w \in W})$, where

(i) $(W, \leq)$ is a frame

(ii) $(\mathcal{B}_w)_{w \in W}$ is a basis on $(W, \leq)$

(iii) $V : W \to \mathcal{P}(\mathsf{Prop})$ such that $w \le v \Longrightarrow V(w) \subseteq V(v)$ (Monotonicity)

The model $\mathfrak{M}$ is called a CS-*model* if

(i) $\mathsf{CS} \subseteq \mathcal{B}_w$ for all $w \in W$

(ii) $(?_n, A) \in \mathcal{B}_w$ for each formula A that is of $?_n$-form

(iii) $(?_n, A) \notin \mathcal{B}_w$ for each formula A that is not of $?_n$-form

Definition 3.7 [Truth in Generated Model] Let CS be a constant specification, $\mathfrak{M} = (W, \le, V, (\mathcal{B}_w)_{w \in W})$ a generated model and $A \in \mathsf{Fm}$ be a formula. We define the relation $(\mathfrak{M}, w) \vDash A$ by

(i) $(\mathfrak{M}, w) \nvDash \bot$

(ii) $(\mathfrak{M}, w) \vDash P \quad$ iff $\quad P \in V(w)$

(iii) $(\mathfrak{M}, w) \vDash B \wedge C \quad$ iff $\quad (\mathfrak{M}, w) \vDash B$ and $(\mathfrak{M}, w) \vDash C$

(iv) $(\mathfrak{M}, w) \vDash B \vee C \quad$ iff $\quad (\mathfrak{M}, w) \vDash B$ or $(\mathfrak{M}, w) \vDash C$

(v) $(\mathfrak{M}, w) \vDash B \to C \quad$ iff $\quad (\mathfrak{M}, v) \vDash C$ for all $v \ge w$ with $(\mathfrak{M}, v) \vDash B$

(vi) $(\mathfrak{M}, w) \vDash t : B \quad$ iff $\quad (t, B) \in \mathcal{E}(\mathcal{B}_w)$

Definition 3.8 We call a generated model *factive* iff for each term $t \in \mathsf{Tm}$ and each formula $A \in \mathsf{Fm}$:

$$(t, A) \in \mathcal{E}(\mathcal{B}_w) \quad \Longrightarrow \quad (\mathfrak{M}, w) \vDash A.$$

It is immediately clear from the truth definition that factive models satisfy Axiom (t) of $\mathsf{iJT4}^+_{\mathsf{CS}}$. Using the fact that in generated models, evidence relations are constructed as least fixed points, we can show soundness of $\mathsf{iJT4}^+_{\mathsf{CS}}$.

Theorem 3.9 (Soundness) *The logic* $\mathsf{iJT4}^+_{\mathsf{CS}}$ *is sound with respect to factive generated* CS-*models, i.e.*

$$\vdash A \quad \Longrightarrow \quad \vDash_{\textit{factive generated } \mathsf{CS}\textit{-models}} A$$

4 Point-Generated Models

In this section we introduce point-generated models and show that each model has an equivalent point-generated model. These models are needed for showing the disjunction property via glueing.

Definition 4.1 [Point-Generated Model] Let $\mathfrak{M}$ be a generated model and $w \in W$. The model generated by w, denoted as $\mathfrak{M}_w = (W_w, \le_w, V_w, \mathcal{B}(w))$ is defined by

$$\begin{aligned} W_w &:= \{v \in W \mid v \ge w\} \\ \le_w &:= \le \cap (W_w \times W_w) \\ V_w &:= V\restriction_{W_w} (= V \cap (W_w \times \mathcal{P}(\mathsf{Prop}))) \\ \mathcal{B}(w)_v &:= \mathcal{B}_v \text{ for all } v \in W_w. \end{aligned}$$

A proof of the following lemma can be found in the appendix.

Lemma 4.2 (Invariance for Point-Generated Models) *Let $\mathfrak{M}$ be a model, $w \in W$, and $\mathfrak{M}_w$ the model generated by w. Then we have for each formula $A \in \mathsf{Fm}$ and each $v \in W_w$:*

$$(\mathfrak{M}, v) \vDash A \iff (\mathfrak{M}_w, v) \vDash A.$$

Remark 4.3 $(W_w, \leq)$ is a tree with root w.

5 Atomic Generated Models

The idea behind atomic models is that at each state, some new formulas get justifications, but those either have to be atomic, or have to do with the constant specification. More precisely, a new formula getting a justification has to be of the form $c : P$ where c is a constant and P is a propositional variable or the justification and the formula are part of the constant specification or they are of the form $?_n : A$, where A is of $?_n$-form. You may think of $c : P$ as some bit of atomic evidence justifying some atomic fact, like a simple observation or a computation establishing some basic numeric fact.

Definition 5.1 [Atomic Model] We call a basis *atomic* iff for all $w \in W$

$$\mathcal{B}_w \subseteq (\mathsf{Const} \times \mathsf{Prop}) \cup \mathsf{CS} \cup \{(?_n, A) \mid A \text{ is of } ?_n\text{-form}\}$$

We call a generated model *atomic* iff its basis is atomic.

Lemma 5.2 *Let* CS *be a constant specification and $\mathfrak{M}$ a factive* CS*-model. If $A \notin \mathsf{Prop}$, $c \in \mathsf{Const}$, and $(c, A) \notin \mathsf{CS}$, then for all states w of $\mathfrak{M}$,*

$$(c, A) \notin \mathcal{B}_w.$$

Theorem 5.3 (Atomization) *Let* CS *be a constant specification and $\mathfrak{M}$ a factive* CS*-model with basis $\mathcal{B}$. Then there is an equivalent atomic* CS*-basis $\mathcal{B}'$, i.e.*

$$\mathcal{E}(\mathcal{B}_w) = \mathcal{E}(\mathcal{B}'_w) \quad \textit{for all } w \in W.$$

Proof. The corresponding atomic basis is simply defined as the atomic part of the original basis, i.e.

$$\mathcal{B}'_w := \{(c, P) \in \mathsf{Const} \times \mathsf{Prop} \mid (c, P) \in \mathcal{B}_w\} \cup \mathsf{CS} \cup \{(?_n, A) \mid A \text{ is of } ?_n\text{-form}\}$$

Then it follows immediately that

- $w \leq v \Longrightarrow \mathcal{B}'_w \subseteq \mathcal{B}'_v$ (Monotonicity)
- $\mathcal{B}'$ is atomic
- $\mathcal{B}'_w \subseteq \mathcal{B}_w$ and therefore $\mathcal{E}(\mathcal{B}'_w) \subseteq \mathcal{E}(\mathcal{B}_w)$

It remains to show that $\mathcal{E}(\mathcal{B}_w) \subseteq \mathcal{E}(\mathcal{B}'_w)$. We show for each pair $(t, A) \in \mathcal{E}(\mathcal{B}_w)$, that we have $(t, A) \in \mathcal{E}(\mathcal{B}'_w)$ and proceed by induction on the build-up of $\mathcal{E}(\mathcal{B}_w)$.

Base case. $(t, A) \in \mathcal{B}_w$. We continue by induction on t.

- If $t = c \in \mathsf{Const}$, then we distinguish the following three cases:

· $A = P \in \mathsf{Prop}$. Then by the definition of $\mathcal{B}'$ we have that $(t, A) = (c, P) \in \mathcal{B}'_w \subseteq \mathcal{E}(\mathcal{B}'_w)$.
· $(c, A) \in \mathsf{CS}$. Then we have $(c, A) \in \mathcal{B}'_w \subseteq \mathcal{E}(\mathcal{B}'_w)$.
· $A \notin \mathsf{Prop}$ and $(c, A) \notin \mathsf{CS}$. By the previous lemma, this case does not happen.

- $t = ?_n$. Since $\mathfrak{M}$ is a CS-model, A has to be of $?_n$-form. It follows by the definition of $\mathcal{B}'$ that $(?_n, A) \in \mathcal{B}'_w \subseteq \mathcal{E}(\mathcal{B}'_w)$.
- $t = !s$. By Lemma 3.5, A has to be of the form $s : B$ for some formula B. So we have $(!s, s : B) \in \mathcal{B}_w$, and since a basis is downwards closed, it follows that $(s, B) \in \mathcal{B}_w$. It follows by the I.H. that $(s, B) \in \mathcal{E}(\mathcal{B}'_w)$ and therefore $(!s, s : B) \in \mathcal{E}(\mathcal{B}'_w)$.
- $t = r + s$, so $(r + s, A) \in \mathcal{B}_w$. Since a basis is downwards closed, it follows that $(r, A) \in \mathcal{B}_w$ or $(s, A) \in \mathcal{B}_w$. It follows by the I.H. that $(r, A) \in \mathcal{E}(\mathcal{B}'_w)$ or $(s, A) \in \mathcal{E}(\mathcal{B}'_w)$, and therefore $(r + s, A) \in \mathcal{E}(\mathcal{B}'_w)$.
- $t = r \cdot_B s$. Again, since a basis is downwards closed, we have $(r, B \to A) \in \mathcal{B}_w$ and $(s, B) \in \mathcal{B}_w$. It follows by the I.H. that $(r, B \to A) \in \mathcal{E}(\mathcal{B}'_w)$ and $(s, B) \in \mathcal{E}(\mathcal{B}'_w)$ and therefore $(r \cdot_B s, A) \in \mathcal{E}(\mathcal{B}'_w)$.

Inductive step.

- $t = r+s$, and $(r, A) \in \mathcal{E}(\mathcal{B}_w)$ or $(s, A) \in \mathcal{E}(\mathcal{B}_w)$. It follows by the I.H. (outer induction on the build-up of $\mathcal{E}(\mathcal{B}_w)$) that $(r, A) \in \mathcal{E}(\mathcal{B}'_w)$ or $(s, A) \in \mathcal{E}(\mathcal{B}'_w)$ and therefore $(r + s, A) \in \mathcal{E}(\mathcal{B}'_w)$.
- $t = !s$, $A = s : B$ and $(s, B) \in \mathcal{E}(\mathcal{B}_w)$. It follows by the I.H. that $(s, B) \in \mathcal{E}(\mathcal{B}'_w)$ and therefore $(!s, s : B) \in \mathcal{E}(\mathcal{B}'_w)$.
- $t = r \cdot_B s$, $(r, B \to A) \in \mathcal{E}(\mathcal{B}_w)$ and $(s, B) \in \mathcal{E}(\mathcal{B}_w)$. It follows by the I.H. that $(r, B \to A) \in \mathcal{E}(\mathcal{B}'_w)$ and $(s, B) \in \mathcal{E}(\mathcal{B}'_w)$ and therefore $(r \cdot_B s, A) \in \mathcal{E}(\mathcal{B}'_w)$.

□

Corollary 5.4 *For each factive generated model, there is an equivalent atomic factive generated model.*

Definition 5.5 [Subformulas] For a formula A, its set of subformulas $\mathrm{subf}(A)$ is defined inductively as follows:

- $\mathrm{subf}(\bot) = \{\bot\}$
- $\mathrm{subf}(P) = \{P\}$ for $P \in \mathsf{Prop}$
- $\mathrm{subf}(A \odot B) := \{A \odot B\} \cup \mathrm{subf}(A) \cup \mathrm{subf}(B)$ for each $\odot \in \{\wedge, \vee, \to\}$
- $\mathrm{subf}(c : A) := \{c : A\} \cup \mathrm{subf}(A)$
- $\mathrm{subf}(!t : A) := \{!t : A\} \cup \mathrm{subf}(A)$
- $\mathrm{subf}(?_n : A) := \{?_n : A\} \cup \mathrm{subf}(A)$
- $\mathrm{subf}(t + s : A) := \{t + s : A\} \cup \mathrm{subf}(t : A) \cup \mathrm{subf}(s : A)$
- $\mathrm{subf}(t \cdot_B s : A) := \{t \cdot_B s : A\} \cup \mathrm{subf}(t : (B \to A)) \cup \mathrm{subf}(s : B)$

Observe how we use the annotation in the application operator to catch the subformulas of B. By induction on the structure of formulas, we show that the set of subformulas of any given formula is finite.

Lemma 5.6 (Set of Subformulas is Finite) *For each formula A, its set of subformulas* $\mathrm{subf}(A)$ *is finite.*

The next lemma allows us to connect the truth of a justification formula at a point of an atomic model with the provability of that formula from the atomic information of that state. The proof of this lemma is in the appendix.

Lemma 5.7 (Connection Lemma) *Let* CS *be a constant specification,* $\mathfrak{M}$ *an atomic model and* $w \in W$. *Then we have for all formulas* $A, B \in \mathsf{Fm}$ *and each term* $t \in \mathsf{Tm}$

(i) $(\mathfrak{M}, w) \vDash t : A \quad \Longrightarrow \quad \{c : P \in \mathrm{subf}(t : A) \mid w \vDash c : P\} \vdash t : A$

(ii) *If* $\mathfrak{M}$ *is a* CS*-model, then*
$\{c : P \in \mathrm{subf}(A) \mid w \vDash c : P\} \cup \{P \in \mathrm{subf}(A) \mid w \vDash P\} \vdash B \quad \Longrightarrow$
$(\mathfrak{M}, w) \vDash B$

6 Canonical Model

In this section we perform a canonical model construction to establish completeness of $\mathsf{iJT4}^{+}_{\mathsf{CS}}$. We only state the lemmas needed to obtain the final completeness result. The proofs are straightforward and left to the reader.

Definition 6.1 Given a constant specification CS, we call a set of formulas $\Delta \subseteq \mathsf{Fm}$ *prime* iff it satisfies the following conditions:

(i) Δ has the disjunction property, i.e., $B \vee C \in \Delta \Longrightarrow B \in \Delta$ or $C \in \Delta$

(ii) Δ is deductively closed with respect to CS, i.e., for any formula B, if $\Delta \vdash B$, then $B \in \Delta$

(iii) Δ is consistent, i.e., $\bot \notin \Delta$.

From now on, we will use Σ, Δ, Γ for prime sets of formulas.

We need a prime lemma relative to a fixed constant specification CS, since our relevant notion of provability involves a constant specification by using the rule of axiom necessitation and the $(?_n)$-rule.

Theorem 6.2 (Prime Lemma) *Let* CS *be a constant specification,* $B \in \mathsf{Fm}$ *be a formula ,* $N \subseteq \mathsf{Fm}$ *a set of formulas such that* $N \nvdash B$. *Then there exists a prime set* $\Delta \subseteq \mathsf{Fm}$ *with* $N \subseteq \Delta$ *and* $\Delta \nvdash B$.

Definition 6.3 [Canonical Generated Model] We define the canonical generated model $\mathfrak{M} = (W, \subseteq, V, (\mathcal{B}_\Delta)_{\Delta \in W})$ as follows:

(i) $W := \{\Delta \subseteq \mathrm{Fm} \mid \Delta \text{ is prime }\}$

(ii) $V(\Delta) := \Delta \cap \mathsf{Prop}$

(iii) for every term $t \in \mathsf{Tm}$ we set $(t, A) \in \mathcal{B}_\Delta \quad :\Longleftrightarrow \quad t : A \in \Delta$

Lemma 6.4 $\mathcal{B}_\Delta$ *is downwards closed.*

Lemma 6.5 *The canonical generated model is a generated model indeed.*

Since our prime sets are deductively closed, it follows that the canonical basis is identical to its closure.

Lemma 6.6 *We have*

$$\mathcal{B}_\Delta = \mathcal{E}(\mathcal{B}_\Delta).$$

Lemma 6.7 (Canonical Evidence) *For all terms* $t \in \mathsf{Tm}$ *and all prime sets* Δ*, we have*

$$t : A \in \Delta \quad \Longleftrightarrow \quad (t, A) \in \mathcal{E}(\mathcal{B}_\Delta).$$

Lemma 6.8 (Truth Lemma) *For each formula* $A \in \mathsf{Fm}$ *and each prime set* $\Delta \subseteq \mathsf{Fm}$*:*

$$(\mathfrak{M}, \Delta) \vDash A \quad \Longleftrightarrow \quad A \in \Delta.$$

Lemma 6.9 *The canonical model is a* CS*-model.*

Lemma 6.10 *The canonical model is a factive generated model, i.e., for all* $t \in \mathsf{Tm}$ *and all* $A \in \mathsf{Fm}$*:*

$$(t, A) \in \mathcal{B}_\Delta \quad \Longrightarrow \quad \Delta \vDash A.$$

Theorem 6.11 (Completeness w.r.t Factive Generated Models) *Let* CS *be a constant specification. The logic* $\mathsf{iJT4}^+_{\mathsf{CS}}$ *is complete with respect to factive generated* CS*-models, i.e. for every formula* $A \in \mathsf{Fm}$ *we have*

$$\vDash_{\textit{factive generated } \mathsf{CS}\textit{-models}} A \quad \Longrightarrow \quad \vdash A.$$

For each factive generated CS-model, we find an equivalent atomic model by Corollary 5.4. It follows that there is an atomic model that is equivalent to the canonical one, and, therefore, we have completeness with respect to atomic factive generated CS-models.

Corollary 6.12 *Let* CS *be a constant specification. The logic* $\mathsf{iJT4}^+_{\mathsf{CS}}$ *is complete with respect to atomic factive generated* CS*-models, i.e., for every formula* A *we have*

$$\vDash_{\textit{atomic factive generated } \mathsf{CS}\textit{-models}} A \quad \Longrightarrow \quad \vdash A.$$

7 Disjunction Properties

Our justification logic has the disjunction property. To prove this, we adapt the usual glueing technique, see, e.g., [10], to the framework of justification logic. In order to internalize the disjunction property later, we now have to consider derivations from a finite set of assumptions of the form $c : P$.

Theorem 7.1 (Disjunction Property) *Let* CS *be any constant specification and* A, B *formulas. Then we have:*

$$\begin{aligned} &c_1 : P_1, \ldots, c_n : P_n \vdash A \vee B \\ \Longrightarrow \quad &c_1 : P_1, \ldots, c_n : P_n \vdash A \quad \textit{or} \quad c_1 : P_1, \ldots, c_n : P_n \vdash B. \end{aligned}$$

Proof. We proceed by contraposition. Suppose that

$$c_1 : P_1, \dots, c_n : P_n \nvdash A \quad \text{and} \quad c_1 : P_1, \dots, c_n : P_n \nvdash B.$$

By the Deduction Theorem we obtain

$$\nvdash \bigwedge_{i=1}^{n} (c_i : P_i) \to A \quad \text{and} \quad \nvdash \bigwedge_{i=1}^{n} (c_i : P_i) \to B.$$

By completeness for factive generated CS-models, we find that there are factive generated CS-models

$$\mathfrak{M}^A = (W^A, V_A, \leq_A, (\mathcal{B}_{A,w})_{w \in W^A}), \; \mathfrak{M}^B = (W^B, V_B, \leq_B, (\mathcal{B}_{B,w})_{w \in W^B})$$

with states w_A and w_B such that

$$(\mathfrak{M}^A, w_A) \nvDash \bigwedge_{i=1}^{n} (c_i : P_i) \to A \quad \text{and} \quad (\mathfrak{M}^B, w_B) \nvDash \bigwedge_{i=1}^{n} (c_i : P_i) \to B$$

which means that there are states $v_A \geq_A w_A$ and $v_B \geq_B w_B$ such that

$$(\mathfrak{M}^A, v_A) \vDash \bigwedge_{i=1}^{n} c_i : P_i \quad \text{and} \quad (\mathfrak{M}^A, v_A) \nvDash A$$

and

$$(\mathfrak{M}^B, v_B) \vDash \bigwedge_{i=1}^{n} c_i : P_i \quad \text{and} \quad (\mathfrak{M}^B, v_B) \nvDash B.$$

Now we consider the submodels of $\mathfrak{M}^A$ and $\mathfrak{M}^B$ generated by the points v_A and v_B, respectively. Call them $\mathfrak{M}^A_{v_A}$ and $\mathfrak{M}^B_{v_B}$. These models are factive CS-models; further they are trees with roots v_A and v_B and agree with the original models on all their states.

Next we construct a new model $\mathfrak{M}$ by

$$W := \{(0,0)\} \cup \{(1,a) \mid a \in W^A_{v_A}\} \cup \{(2,b) \mid b \in W^B_{v_B}\}$$

$$(x_1, y_1) \leq_{\mathfrak{M}} (x_2, y_2) :\Longleftrightarrow \begin{cases} x_1 = 0, \\ x_1 = x_2 = 1 \text{ and } y_1 \leq_A y_2, \\ x_1 = x_2 = 2 \text{ and } y_1 \leq_B y_2. \end{cases}$$

$$V(x,y) := \begin{cases} \{P_1, \dots, P_n\}, \text{ if } (x,y) = (0,0), \\ V_A(y), \text{ if } x = 1, \\ V_B(y), \text{ if } x = 2. \end{cases}$$

and its basis is defined by

$$\mathcal{B}_{(x,y)} := \begin{cases} \{(c_1, P_1), \dots, (c_n, P_n)\} \cup \mathsf{CS} \cup \\ \qquad \{(?_n, A) \mid A \text{ is of } ?_n\text{-form}\}, \text{ if } (x,y) = (0,0), \\ \mathcal{B}_{A,y}, \text{ if } x = 1, \\ \mathcal{B}_{B,y}, \text{ if } x = 2. \end{cases}$$

Claim: $\mathfrak{M}$ is a factive generated CS-model. We have to check that our glueing does not violate the monotonicity conditions on the valuation and on the basis. We have that

$$(\mathfrak{M}^A, v_A) \models c_i : P_i \quad \text{and} \quad (\mathfrak{M}^B, v_B) \models c_i : P_i \qquad \text{for } i = 1, \ldots, n,$$

which means

$$(c_i, P_i) \in \mathcal{E}(\mathcal{B}_{A,v_A}) \quad \text{and} \quad (c_i, P_i) \in \mathcal{E}(\mathcal{B}_{B,v_B}) \qquad \text{for } i = 1, \ldots, n. \tag{2}$$

It follows by Lemma 3.4 that

$$\{(c_i, P_i) \mid i = 1, \ldots, n\} \subseteq \mathcal{B}_{A,v_A} \text{ and } \{(c_i, P_i) \mid i = 1, \ldots, n\} \subseteq \mathcal{B}_{B,v_B}.$$

Hence $\mathcal{B}_{(0,0)} \subseteq \mathcal{B}_{(x,y)}$ for each $(x, y) \geq_{\mathfrak{M}} (0, 0)$, so the monotonicity for the basis is established.

Also the monotonicity condition for the valuation is satisfied. Indeed, since $\mathfrak{M}^A$ and $\mathfrak{M}^B$ are factive, (2) implies

$$(\mathfrak{M}^A, v_A) \models P_i \quad \text{and} \quad (\mathfrak{M}^B, v_B) \models P_i \qquad \text{for } i = 1, \ldots, n.$$

Thus we find

$$V(0,0) = \{P_1, \ldots, P_n\} \subseteq V(x,y) \quad \text{if } (0,0) \leq_{\mathfrak{M}} (x,y).$$

Last but not least observe that $\mathfrak{M}$ is factive since the new state $(0,0)$ satisfies factivity by definition.

Now we show that $(\mathfrak{M}, (0,0)) \not\models (\bigwedge_{i=1}^n c_i : P_i) \to (A \vee B)$. By the definition of the basis of $\mathfrak{M}$ we immediately have

$$(\mathfrak{M}, (0,0)) \models \bigwedge_{i=1}^n c_i : P_i.$$

By the monotonicity of $\leq_{\mathfrak{M}}$ we find $(\mathfrak{M}, (0,0)) \not\models A$ and $(\mathfrak{M}, (0,0)) \not\models B$, which implies $(\mathfrak{M}, (0,0)) \not\models A \vee B$. Therefore,

$$(\mathfrak{M}, (0,0)) \not\models (\bigwedge_{i=1}^n c_i : P_i) \to (A \vee B).$$

By soundness we get

$$\nvdash (\bigwedge_{i=1}^n c_i : P_i) \to (A \vee B)$$

and finally, by the Deduction Theorem,

$$c_1 : P_1, \ldots, c_n : P_n \nvdash A \vee B$$

□

Now we can prove a first version of the internalized disjunction property, which we call *local*. It is local in the sense that it is shown semantically for a given state in the model and the term s depends on that state.

Lemma 7.2 (Local Internalized Disjunction Property) *Let* CS *be an axiomatically appropriate constant specification. Let* $\mathfrak{M}$ *be an atomic* CS*-model,* $w \in W$. *For each term* $t \in \mathsf{Tm}$ *and all formulas* A, B, *there exists a term* $s = s_{t,A,B,w}$ *such that*

$$(\mathfrak{M}, w) \vDash t : (A \vee B) \rightarrow (s : A \vee s : B).$$

Proof. Assume that $(\mathfrak{M}, w) \vDash t : (A \vee B)$. By the first part of the Connection Lemma 5.7 we get

$$\{c : P \in \mathrm{subf}(t : (A \vee B)) \mid w \vDash c : P\} \vdash t : (A \vee B).$$

Using the axiom (t), we obtain

$$\{c : P \in \mathrm{subf}(t : (A \vee B)) \mid w \vDash c : P\} \vdash A \vee B.$$

By Lemma 5.6, the set $\mathrm{subf}(t : (A \vee B))$ is finite. Hence by the disjunction property, Theorem 7.1, we get

$$\begin{array}{ll} \{c : P \in \mathrm{subf}(t : (A \vee B)) \mid w \vDash c : P\} \vdash A & \text{or} \\ \{c : P \in \mathrm{subf}(t : (A \vee B)) \mid w \vDash c : P\} \vdash B. & \end{array} \tag{3}$$

Assume that $\{c : P \in \mathrm{subf}(t : (A \vee B)) \mid w \vDash c : P\} \vdash A$, the other case being similar. We now apply the Lifting Lemma 2.6 to find an $s \in \mathsf{Tm}$ such that

$$\{c : P \in \mathrm{subf}(t : (A \vee B)) \mid w \vDash c : P\} \vdash s : A.$$

Therefore,

$$\{c : P \in \mathrm{subf}(t : (A \vee B)) \mid w \vDash c : P)\} \cup \{P \in \mathrm{subf}(t : (A \vee B)) \mid w \vDash P\} \vdash s : A.$$

Using the second part of the Connection Lemma 5.7, we get $(\mathfrak{M}, w) \vDash s : A$. Finally, combining the two cases of (3) yields $(\mathfrak{M}, w) \vDash s : A \vee s : B$. □

The above version of the internalization property is local in the sense that the term s depends on the world w. In the next lemma, we provide a global version of the internalized disjunction property, which does not have this dependency. In its proof, we make crucial use of the $+$ operator to collect all possible justification terms for the disjuncts.

Lemma 7.3 (Global Internalized Disjunction Property) *Let* CS *be an axiomatically appropriate constant specification. For each term* $t \in \mathsf{Tm}$ *and all formulas* $A, B \in \mathsf{Fm}$, *there exists a term* $s = s_{t,A,B} \in \mathsf{Tm}$ *such that for each atomic* CS*-model* $\mathfrak{M}$ *and each* $w \in W$:

$$(\mathfrak{M}, w) \vDash t : (A \vee B) \rightarrow (s : A \vee s : B).$$

Proof. In the proof of the above lemma, the term s only depends on which of the (finitely many) $c_i : P_i \in \mathrm{subf}(t : (A \vee B))$ do hold at w. Hence there are finitely many terms $s_{A,1}, \ldots, s_{A,m}$ and $s_{B,1}, \ldots, s_{B,m}$ such that for each w there exists $1 \leq j \leq m$ with

$$(\mathfrak{M}, w) \vDash s_{A,j} : A \quad \text{or} \quad (\mathfrak{M}, w) \vDash s_{B,j} : B.$$

Therefore, for all w we have

$$(\mathfrak{M}, w) \vDash s_{A,1} : A \vee \cdots \vee s_{A,m} : A \quad \text{or} \quad (\mathfrak{M}, w) \vDash s_{B,1} : B \vee \cdots \vee s_{B,m} : B.$$

Now we let

$$s := s_{A,1} + \cdots + s_{A,m} + s_{B,1} + \cdots + s_{B,m}$$

and obtain that for all w,

$$(\mathfrak{M}, w) \vDash s : A \vee s : B.$$

□

By completeness for atomic models, we immediately get the following corollary.

Corollary 7.4 (Internalized Disjunction Property) *Let* CS *be an axiomatically appropriate constant specification. For all formulas* $A, B \in \mathsf{Fm}$ *and all* $t \in \mathsf{Tm}$ *there exists* $s = s_{t,A,B} \in \mathsf{Tm}$ *such that*

$$\vdash t : (A \vee B) \to (s : A \vee s : B).$$

8 Conclusion and further work

We introduced the intuitionistic justification logic $\mathsf{iJT4}^+$, which is a light extension of $\mathsf{iJT4}$. We defined atomic models and established completeness of $\mathsf{iJT4}^+$ with respect to that class of models. This made it possible to prove the internalized disjunction property for $\mathsf{iJT4}^+$.

Hirai [15] introduced an intuitionistic modal logic based on $\mathsf{S4}$ with the additional axiom

$$\Box(A \vee B) \to (\Box A \vee \Box B),$$

which is the modal version of the internalized disjunction property. So it is a natural question whether we can realize his logic into $\mathsf{iJT4}^+$, i.e., given a modal formula A provable in Hirai's logic, is there a realization A^r provable in $\mathsf{iJT4}^+$ where A^r is A with □-operators replaced by suitable terms.

It turns out that such a realization is possible but it requires a heavy technical apparatus. First of all, we need a cut-free sequent system for Hirai's logic. Then we can perform Artemov's syntactic realization procedure to obtain the realization A^r. To do so, we need justification variables to realize □-operators occurring in negative positions. However, this is problematic as we have observed in Remark 2.4. The solution is to add variables to the language but to keep the formulation for the axioms, i.e., axioms are stated only for variable free terms. Instead we define provability of a formula with variables as provability of all its (variable-free) substitution instances, i.e.,

$$\vdash A[x_1, \ldots, x_n] \quad :\Longleftrightarrow \quad \vdash A[t_1, \ldots, t_n] \text{ for all variable-free terms } t_1, \ldots, t_n.$$

The realization result will be published in full detail somewhere else.

Appendix

A Proof of Lemma 4.2

Lemma 4.2 (Invariance for Point-Generated Models) Let $\mathfrak{M}$ be a model, $w \in W$, and $\mathfrak{M}_w$ the model generated by w. Then we have for each formula $A \in \mathsf{Fm}$ and each $v \in W_w$:

$$(\mathfrak{M}, v) \vDash A \quad \Longleftrightarrow \quad (\mathfrak{M}_w, v) \vDash A.$$

Proof. By induction on A.

(i) $A = \bot$. Follows immediately.

(ii) $A = P \in \mathsf{Prop}$. Then the claim follows immediately by the definition of $V(w)$.

(iii) $A = B \wedge C$ or $A = B \vee C$. Then the claim follows immediately by the I.H.

(iv) $A = B \to C$. For the direction from left to right, assume that $(\mathfrak{M}, v) \vDash B \to C$. We have to show that $(\mathfrak{M}_w, v) \vDash B \to C$, so let $u \geq_w v$ with $(\mathfrak{M}_w, u) \vDash B$. We have that $u \in W_w$, so it follows by the I.H. that $(\mathfrak{M}, u) \vDash B$, so $(\mathfrak{M}, u) \vDash C$. Applying the I.H. again, we get that $(\mathfrak{M}_w, u) \vDash C$. Since u was arbitrary, it follows that $(\mathfrak{M}_w, v) \vDash B \to C$.

For the direction from right to left, assume that $(\mathfrak{M}_w, v) \vDash B \to C$. We have to show that $(\mathfrak{M}, v) \vDash B \to C$, so let $u \geq v$ with $(\mathfrak{M}, u) \vDash B$. Since $u \geq v \geq w$, it follows by the transitivity of $\leq$ that $u \geq w$, so by definition we have that $u \in W_w$. It follows by the I.H. that $(\mathfrak{M}_w, u) \vDash B$. It also follows that $u \geq_w v$, and therefore $(\mathfrak{M}_w, u) \vDash C$. Applying the I.H. again, we obtain $(\mathfrak{M}, u) \vDash C$. Since u was arbitrary, it follows that $(\mathfrak{M}, v) \vDash B \to C$.

(v) $A = t : B$. We just observe that since $\mathcal{B}(w)_v = \mathcal{B}_v$ for all $v \in W_w$, we have $\mathcal{E}(\mathcal{B}(w)_v) = \mathcal{E}(\mathcal{B}_v)$ for all $v \in W_w$. Then the claim follows immediately by the definition of truth in a model.

□

B Proof of Lemma 5.2

Lemma 5.2 Let CS be a constant specification and $\mathfrak{M}$ a factive CS-model. If $A \notin \mathsf{Prop}$, $c \in \mathsf{Const}$, and $(c, A) \notin \mathsf{CS}$, then for all states w of $\mathfrak{M}$,

$$(c, A) \notin \mathcal{B}_w.$$

Proof. If $A \notin \mathsf{Prop}$, $c \in \mathsf{Const}$, and $(c, A) \notin \mathsf{CS}$, then by definition, $\neg(c : A)$ is of $?_0$-form. By the definition of CS-model, we have that

$$(?_0, \neg c : A) \in \mathcal{B}_w \subseteq \mathcal{E}(\mathcal{B}_w).$$

Since the model is factive, it follows that $(\mathfrak{M}, w) \vDash \neg c : A$. Thus $(\mathfrak{M}, w) \nvDash c : A$, which finally is

$$(c, A) \notin \mathcal{B}_w.$$

□

C Proof of Lemma 5.7

Lemma 5.7 (Connection Lemma) Let CS be a constant specification, $\mathfrak{M}$ an atomic model and $w \in W$. Then we have for all formulas $A, B \in \mathsf{Fm}$ and each term $t \in \mathsf{Tm}$

(i) $(\mathfrak{M}, w) \models t : A \quad \Longrightarrow \quad \{c : P \in \text{subf}(t : A) \mid w \models c : P\} \vdash t : A$

(ii) If $\mathfrak{M}$ is a CS-model, then
$\{c : P \in \text{subf}(A) \mid w \models c : P\} \cup \{P \in \text{subf}(A) \mid w \models P\} \vdash B \quad \Longrightarrow$ $(\mathfrak{M}, w) \models B$

Proof. First statement. By the truth definition,

$$(\mathfrak{M}, w) \models t : A \quad \Longleftrightarrow \quad (t, A) \in \mathcal{E}(\mathcal{B}_w)$$

so we proceed by induction on $\mathcal{E}(\mathcal{B}_w)$.

Base case. $(t, A) \in \mathcal{B}_w$. Since $\mathcal{B}$ is atomic, there are three subcases.

(i) (t, A) is of the form (c, P). Then the claim follows immediately.

(ii) (t, A) is of the form $(c, A) \in \mathsf{CS}$. Again, the claim follows immediately.

(iii) $(t, A) = (?_n, A)$ and A is of $?_n$-form. Again, the claim follows immediately.

Inductive step.

- $t = r + s$, and

$$(r, A) \in \mathcal{E}(\mathcal{B}_w) \quad \text{or} \quad (s, A) \in \mathcal{E}(\mathcal{B}_w)$$

It follows by the I.H. that

$$\{c : P \in \text{subf}(r : A) \mid w \models c : P\} \vdash r : A$$

or

$$\{c : P \in \text{subf}(s : A) \mid w \models c : P\} \vdash s : A$$

it follows that

$$\{c : P \in \text{subf}(r + s : A) \mid w \models c : P\} \vdash r : A$$

or

$$\{c : P \in \text{subf}(r + s : A) \mid w \models c : P\} \vdash s : A$$

it follows by propositional reasoning that

$$\{c : P \in \text{subf}(r + s : A) \mid w \models c : P\} \vdash r : A \vee s : A$$

and by the axiom $(+)$ and some more propositional reasoning we get

$$\{c : P \in \text{subf}(r + s : A) \mid w \models c : P\} \vdash (r + s) : A.$$

- $t = s \cdot_B r$, and

$$(s, B \to A) \in \mathcal{E}(\mathcal{B}_w) \quad \text{and} \quad (r, B) \in \mathcal{E}(\mathcal{B}_w)$$

It follows by the I.H. that

$$\{c : P \in \mathrm{subf}(s : (B \to A)) \mid w \models c : P\} \vdash s : (B \to A)$$

and

$$\{c : P \in \mathrm{subf}(r : B) \mid w \models c : P\} \vdash r : B$$

and therefore by the axiom (j)

$$\{c : P \in \mathrm{subf}(s : (B \to A)) \cup \mathrm{subf}(r : B) \mid w \models c : P\}$$
$$\vdash (s \cdot_B r) : A$$

and since

$$\{c : P \in \mathrm{subf}(s : (B \to A)) \ \cup \ \mathrm{subf}(r : B)\} = \{c : P \in \mathrm{subf}((s \cdot_B r) : A)\}$$

we have that

$$\{c : P \in \mathrm{subf}(s \cdot_B r : A) \mid w \models c : P\} \vdash (s \cdot_B r) : A.$$

- $t = !s$, $A = s : B$, and
$$(s, B) \in \mathcal{E}(\mathcal{B}_w).$$
Then it follows by the I.H. that
$$\{c : P \in \mathrm{subf}(s : B) \mid w \models c : P\} \vdash s : B$$
so it follows by the axiom (!) that
$$\{c : P \in \mathrm{subf}(s : B) \mid w \models c : P\} \vdash !s : (s : B)$$
i.e.
$$\{c : P \in \mathrm{subf}(s : B) \mid w \models c : P\} \vdash t : A.$$

Second Statement. We proceed by induction on the derivation of B.

- B is an axiom. Then the claim follows by soundness.
- $B \in \{c : P \in \mathrm{subf}(A) \mid w \models c : P\} \cup \{P \in \mathrm{subf}(A) \mid w \models P\}$. Then the claim follows immediately.
- B is of the form $c : D$ with $(c, D) \in \mathsf{CS}$ and was derived by $(\mathrm{AN})_{\mathsf{CS}}$. Since $\mathfrak{M}$ is a CS-model, we have by definition that $\mathsf{CS} \subseteq \mathcal{B}_w \subseteq \mathcal{E}(\mathcal{B}_w)$, and therefore $(\mathfrak{M}, w) \models c : D$.
- B was derived by the rule $(?_n)$. Then $B = ?_n : C$ for some formula C of $?_n$-form. Again, since $\mathfrak{M}$ is a CS-model, we have that $(?_n, C) \in \mathcal{B}_w \subseteq \mathcal{E}(\mathcal{B}_w)$ and therefore $(\mathfrak{M}, w) \models ?_n : C$.

- B was derived by (MP). Then there is a formula C such that

$$\{c : P \in \mathrm{subf}(A) \mid w \vDash c : P\} \cup \{P \in \mathrm{subf}(A) \mid w \vDash P\} \vdash C$$

and

$$\{c : P \in \mathrm{subf}(A) \mid w \vDash c : P\} \cup \{P \in \mathrm{subf}(A) \mid w \vDash P\} \vdash C \to B$$

Then it follows by the I.H. (since these derivations are shorter) that

$$(\mathfrak{M}, w) \vDash C \to B \quad \text{and} \quad (\mathfrak{M}, w) \vDash C, \quad \text{so } (\mathfrak{M}, w) \vDash B.$$

□

References

[1] Artemov, S., *Justification awareness models*, in: S. Artemov and A. Nerode, editors, *Logical Foundations of Computer Science* (2018), pp. 22–36.

[2] Artemov, S. N., *Explicit provability and constructive semantics*, BSL **7** (2001), pp. 1–36.

[3] Artemov, S. N., *Unified semantics for modality and λ-terms via proof polynomials*, in: K. Vermeulen and A. Copestake, editors, *Algebras, Diagrams and Decisions in Language, Logic and Computation*, CSLI Lecture Notes **144**, CSLI Publications, Stanford, 2002 pp. 89–118.

[4] Artemov, S. N., *Justified common knowledge*, TCS **357** (2006), pp. 4–22.

[5] Artemov, S. N., *The logic of justification*, RSL **1** (2008), pp. 477–513.

[6] Artemov, S. N., *The ontology of justifications in the logical setting*, Technical Report TR-2011008, CUNY Ph.D. Program in Computer Science (2011).

[7] Artemov, S. N. and R. Iemhoff, *The basic intuitionistic logic of proofs*, JSL **72** (2007), pp. 439–451.

[8] Artemov, S. N. and E. Nogina, *Introducing justification into epistemic logic*, Journal of Logic and Computation **15** (2005), pp. 1059–1073.

[9] Baltag, A., B. Renne and S. Smets, *The logic of justified belief, explicit knowledge, and conclusive evidence*, APAL **165** (2014), pp. 49–81, published online in August 2013.

[10] Blackburn, P., M. de Rijke and Y. Venema, "Modal Logic," Cambridge Tracts in Theoretical Computer Science **53**, Cambridge University Press, 2002.

[11] Bucheli, S., R. Kuznets and T. Studer, *Justifications for common knowledge*, Applied Non-Classical Logics **21** (2011), pp. 35–60.

[12] Bucheli, S., R. Kuznets and T. Studer, *Realizing public announcements by justifications*, Journal of Computer and System Sciences **80** (2014), pp. 1046–1066.

[13] Dashkov, E., *Arithmetical completeness of the intuitionistic logic of proofs*, Journal of Logic and Computation **21** (2011), pp. 665–682, published online August 2009.

[14] Fitting, M., *The logic of proofs, semantically*, APAL **132** (2005), pp. 1–25.

[15] Hirai, Y., *An intuitionistic epistemic logic for sequential consistency on shared memory*, in: *Proceedings of the 16th International Conference on Logic for Programming, Artificial Intelligence, and Reasoning*, LPAR'10 (2010), pp. 272–289.
URL `http://dl.acm.org/citation.cfm?id=1939141.1939157`

[16] Kashev, A., "Justification with Nominals," Ph.D. thesis, University of Bern (2016).
URL `http://www.iam.unibe.ch/ltgpub/2016/kas16.pdf`

[17] Kokkinis, I., Z. Ognjanović and T. Studer, *Probabilistic justification logic*, in: S. Artemov and A. Nerode, editors, *Logical Foundations of Computer Science* (2016), pp. 174–186.

[18] Kuznets, R., S. Marin and L. Straßburger, *Justification logic for constructive modal logic*, IMLA 2017 - 7th Workshop on Intuitionistic Modal Logic and Applications (2017).
URL `https://hal.inria.fr/hal-01614707`

[19] Kuznets, R. and T. Studer, *Justifications, ontology, and conservativity*, in: T. Bolander, T. Braüner, S. Ghilardi and L. Moss, editors, *Advances in Modal Logic, Volume 9*, College Publications, 2012 pp. 437–458.

[20] Kuznets, R. and T. Studer, *Update as evidence: Belief expansion*, in: S. N. Artemov and A. Nerode, editors, *LFCS 2013*, LNCS **7734**, Springer, 2013 pp. 266–279.

[21] Kuznets, R. and T. Studer, *Weak arithmetical interpretations for the logic of proofs*, Logic Journal of the IGPL (2016).

[22] Marti, M. and T. Studer, *Intuitionistic modal logic made explicit*, IfCoLog Journal of Logics and their Applications **3** (2016), pp. 877–901.

[23] Renne, B., *Evidence elimination in multi-agent justification logic*, in: A. Heifetz, editor, *Theoretical Aspects of Rationality and Knowledge, Proceedings of the Twelfth Conference (TARK 2009)* (2009), pp. 227–236.

[24] Steren, G. and E. Bonelli, *Intuitionistic hypothetical logic of proofs*, in: V. de Paiva, M. Benevides, V. Nigam and E. Pimentel, editors, *Proceedings of the 6th Workshop on Intuitionistic Modal Logic and Applications (IMLA 2013) in association with UNILOG 2013, Rio de Janeiro, Brazil*, Number 300 in Electronic Notes in Theoretical Computer Science, Elsevier, 2014 pp. 89–103.

[25] Studer, T., *Lectures on justification logic* (2012), Manuscript.

[26] Studer, T., *Decidability for some justification logics with negative introspection*, Journal of Symbolic Logic **78** (2013), pp. 388–402.

A Recursively Enumerable Kripke Complete First-Order Logic Not Complete with Respect to a First-Order Definable Class of Frames

Mikhail Rybakov

Department of Mathematics, Tver State University, Tver, Russia
and
School of Computer Science and Applied Mathematics,
University of the Witwatersrand, Johannesburg, South Africa

Dmitry Shkatov

School of Computer Science and Applied Mathematics,
University of the Witwatersrand, Johannesburg, South Africa

Abstract

It is well-known that every quantified modal logic complete with respect to a first-order definable class of Kripke frames is recursively enumerable. Numerous examples are also known of "natural" quantified modal logics complete with respect to a class of frames defined by an essentially second-order condition which are not recursively enumerable. It is not, however, known if these examples are instances of a pattern, i.e., whether every recursively enumerable, Kripke complete quantified modal logic can be characterized by a first-order definable class of frames. While the question remains open for normal logics, we show that, in the context of quasi-normal logics, this is not so, by exhibiting an example of a recursively enumerable, Kripke complete quasi-normal logic that is not complete with respect to any first-order definable class of (pointed) frames.

Keywords: first-order modal logic, recursive enumerability, Kripke completeness, first-order definability

1 Introduction

Some important (first-order) quantified modal logics are based on propositional logics characterized by classes of frames defined by essentially second-order conditions on their accessibility relations. Among them are the quantified provability logics **QGL** (Quantified Gödel-Löb) and **QGrz** (Quantified Grzegorczyk), as well as their "linear" counterparts **QGL.3** and **QGrz.3**; quantified counterparts **QPDL**, **QCLT**, and **CTL*** of propositional logics **PDL**, **CTL**, and **CTL***; quantified epistemic logics with the common knowledge operator; and the quantified logic of finite Kripke frames.

A Kripke complete propositional modal logic can be extended to a (first-order) quantified one in essentially two ways. Given a propositional logic L complete with respect to a class of frames $\mathfrak{C}$, we can either consider the set of quantified formulas true on all frames from $\mathfrak{C}$, or alternatively, add to L, considered as a logical calculus, axioms and inference rules of the classical first-order logic. If class $\mathfrak{C}$ is defined by an essentially second-order condition, then in either case, we obtain quantified logics with undesirable properties. If we consider the set of quantified formulas true on all the frames of a propositional logic with essentially second-order Kripke semantics, we obtain logics that are not recursively enumerable, and thus cannot be represented as logical calculi—this holds, for example, for logics of frames with the condition of non-existence of infinite ascending chains, such as **QGL**, **QGrz**, **QGL.3**, and **QGrz.3** [4]; for logics of frames where one binary relation is the reflexive and transitive closure of another, such as **QPDL** and **QCLT** [7]; and for quantified logics of finite Kripke frames [5]. If, on the other hand, we consider extending such a propositional logic with classical first-order axioms and rules of inference, we obtain logics that are Kripke incomplete, i.e., are not complete with respect to any class of Kripke frames,—the proofs for **QGL**, **QGrz**, **QGL.3**, and **QGrz.3** can be found in [3], [4]; similar arguments apply to all the other logics mentioned above.

It would thus appear that quantified extensions of propositional logics with essentially second-order Kripke semantics are either Kripke incomplete or not recursively enumerable. In other words, Kripke completeness with respect to semantics with essentially second-order conditions and recursive enumerability do not seem to sit well together for quantified modal logics. Whether this is indeed so has, however, not been established. More precisely, it has not been established whether every recursively enumerable, Kripke complete quantified modal logic can be characterized by a class of frames defined by a classical first-order condition. We note here that the converse is known to be true, i.e., every quantified modal logic Kripke complete with respect to a class of frames defined by a classical first-order condition is recursively enumerable,—this is a straightforward consequence of the fact that such a logic can be embedded into the classical first-order logic through the so-called standard translation (see, e.g., [2]).

For normal logics, the above question still remains open. In the present paper, we show that for logics that are not required to be normal, i.e., closed under necessitation, the answer is negative,—there do exist quasi-normal quantified modal logics that are both recursively enumerable and Kripke complete, but are not complete with respect to any first-order definable class of (pointed) frames.

The paper is structured as follows. In section 2, we briefly introduce quantified modal logic and the associated Kripke semantics. In section 3, we present an example of a recursively enumerable, Kripke complete quasi-normal logic not complete with respect to any first-order definable class of pointed frames. We conclude in section 4.

2 Quantified modal logic

A (first-order) quantified modal language contains countably many individual variables; countably many predicate letters of every arity; Boolean connectives $\wedge$ and $\neg$; a modal connective $\Box$; and a quantifier $\forall$. Formulas as well as the symbols $\vee$, $\rightarrow$, $\exists$, and $\Diamond$ are defined in the usual way.

For every formula φ, we denote by $\mathbf{md}(\varphi)$ the modal depth of φ, which is defined inductively, as follows:

$$\begin{aligned}\mathbf{md}(P(y_1,\ldots,y_n)) &= 0;\\ \mathbf{md}(\varphi_1 \wedge \varphi_2) &= \max\{\mathbf{md}(\varphi_1), \mathbf{md}(\varphi_2)\};\\ \mathbf{md}(\neg\varphi_1) &= \mathbf{md}(\varphi_1);\\ \mathbf{md}(\forall x\, \varphi_1) &= \mathbf{md}(\varphi_1);\\ \mathbf{md}(\Box\varphi_1) &= \mathbf{md}(\varphi_1) + 1.\end{aligned}$$

Modal formulas can be interpreted using Kripke semantics. A (Kripke) frame is a tuple $\mathfrak{F} = \langle W, R\rangle$, where W is a non-empty set (of worlds) and R is a binary (accessibility) relation on W. A predicate (Kripke) frame is a tuple $\mathfrak{F}_D = \langle W, R, D\rangle$, where $\langle W, R\rangle$ is a frame and D is a function from W into a set of non-empty subsets of some set (the domain of $\mathfrak{F}_D$), satisfying the condition that wRw' implies $D(w) \subseteq D(w')$. We call the set $D(w)$ the domain of w. If a predicate frame satisfies the condition that wRw' implies $D(w) = D(w')$, we refer to it as a frame with constant domains.

A (Kripke) model is a tuple $\mathfrak{M} = \langle W, R, D, I\rangle$, where $\langle W, R, D\rangle$ is a predicate Kripke frame and I is a function assigning to a world $w \in W$ and an n-ary predicate letter P an n-ary relation $I(w, P)$ on $D(w)$. We refer to I as the interpretation of predicate letters with respect to worlds in W.

An assignment in a model is a function g associating with every individual variable y an element of the domain of the underlying frame.

The truth of a formula φ at a world w of a model $\mathfrak{M}$ under an assignment g is inductively defined as follows:

- $\mathfrak{M}, w \models^g P(y_1,\ldots,y_n)$ if $\langle g(y_1),\ldots,g(y_n)\rangle \in I(w, P)$;
- $\mathfrak{M}, w \models^g \varphi_1 \wedge \varphi_2$ if $\mathfrak{M}, w \models^g \varphi_1$ and $\mathfrak{M}, w \models^g \varphi_2$;
- $\mathfrak{M}, w \models^g \neg\varphi_1$ if $\mathfrak{M}, w \not\models^g \varphi_1$;
- $\mathfrak{M}, w \models^g \Box\varphi_1$ if wRw' implies $\mathfrak{M}, w' \models^g \varphi_1$, for every $w' \in W$;
- $\mathfrak{M}, w \models^g \forall y\, \varphi_1$ if $\mathfrak{M}, w \models^{g'} \varphi_1$, for every assignment g' such that g' differs from g in at most the value of y and such that $g'(y) \in D(w)$.

Note that, given a Kripke model $\mathfrak{M} = \langle W, R, D, I\rangle$ and $w \in W$, the tuple $\mathfrak{M}_w = \langle D_w, I_w\rangle$, where $D_w = D(w)$ and $I_w(P) = I(w, P)$, is a classical predicate model.

We say that φ is true at world w of model $\mathfrak{M}$ and write $\mathfrak{M}, w \models \varphi$ if $\mathfrak{M}, w \models^g \varphi$ holds for every g assigning to free variables of φ elements of $D(w)$. We say that φ is true in $\mathfrak{M}$ and write $\mathfrak{M} \models \varphi$ if $\mathfrak{M}, w \models \varphi$ holds for every world w of $\mathfrak{M}$. We say that φ is true on predicate frame $\mathfrak{F}_D$ and write $\mathfrak{F}_D \models \varphi$ if φ is true in every model based on $\mathfrak{F}_D$. We say that φ is true on frame $\mathfrak{F}$ and write $\mathfrak{F} \models \varphi$ if φ is true on every predicate frame of the form $\mathfrak{F}_D$. Finally, we say that a formula is true on a class of frames if it is true on every frame from the class.

Let $\mathfrak{M} = \langle W, R, D, I \rangle$ be a model, $w \in W$, and $a_1, \ldots, a_n \in D(w)$. Let $\varphi(y_1, \ldots, y_n)$ be a formula whose free variables are among $y_1, \ldots, y_n$. We write $\mathfrak{M}, w \models \varphi[a_1, \ldots, a_n]$ to mean $\mathfrak{M}, w \models^g \varphi(y_1, \ldots, y_n)$, where $g(y_1) = a_1, \ldots, g(y_n) = a_n$.

Sometimes, semantics based on pointed frames, rather than frames, is useful. A pointed Kripke frame is a tuple $(\mathfrak{F}, w_0)$, where $\mathfrak{F} = \langle W, R \rangle$ is a Kripke frame and $w_0 \in W$ is a distinguished world. A formula φ is true on a pointed frame $(\mathfrak{F}, w_0)$ if it is true at w_0 in every model based on $\mathfrak{F}$.

A (first-order) quantified (quasi-normal) modal logic is a set L of formulas containing the validities of the classical first-order logic as well as the formula $\Box(p \to q) \to (\Box p \to \Box q)$, and closed under predicate substitution, modus ponens, and generalization (if $\varphi \in L$, then $\forall x\, \varphi \in L$). A normal modal logic is a quasi-normal modal logic L that is closed under necessitation (if $\varphi \in L$, then $\Box\varphi \in L$).

If $\mathfrak{C}$ is a class of frames, then the set of formulas true on every frame in $\mathfrak{C}$ is denoted by $L(\mathfrak{C})$. If $\mathfrak{C}$ is a class of pointed frames, then the set of formulas true at the distinguished world of every frame in $\mathfrak{C}$ is denoted by $rL(\mathfrak{C})$. If $\mathfrak{C}$ is a class of frames, then $L(\mathfrak{C})$ is a normal modal logic; if $\mathfrak{C}$ is a class of pointed frames, then $rL(\mathfrak{C})$ is a quasi-normal modal logic.

A quasi-normal logic is sound and complete with respect to a class of pointed frames $\mathfrak{C}$ if $L = rL(\mathfrak{C})$ for some class of pointed Kripke frames. We say that a quasi-normal logic is Kripke complete if it is sound and complete with some class of pointed frames. Analogously for normal logics and Kripke frames.

A class $\mathfrak{C}$ of (pointed) frames is first-order definable if there exists a first-order formula φ (for frames, φ contains no free variables; for pointed frames, a single free variable), containing binary predicate letters R and $=$ (and no other predicate letters), such that $\mathfrak{F} \in \mathfrak{C}$ if, and only if, φ is true in $\mathfrak{F}$ considered as a classical model (for pointed frames, the free variable of φ is interpreted as the distinguished world).

Remark 2.1 If a class $\mathfrak{C}$ of pointed Kripke frames is first-order definable, say, by formula $\varphi(x)$, then $\mathfrak{C}$ considered as a class of frames—i.e., disregarding the roots of the frames—is also first-order definable, namely by the formula $\exists x\, \varphi(x)$.

The following proposition is well-known (see, for example, [2], Proposition 3.12.8).

Proposition 2.2 *Let $\mathfrak{C}$ be a first-order definable class of frames. Then, $L(\mathfrak{C})$ is recursively enumerable.*

A similar proposition holds for pointed frames.

Proposition 2.3 *Let $\mathfrak{C}$ be a first-order definable class of pointed frames. Then, $rL(\mathfrak{C})$ is recursively enumerable.*

In the strict sense, the converses of Proposition 2.2 and Proposition 2.3 are known not to be true, as some recursively enumerable logics are not complete with respect to any class of frames (i.e., are Kripke incomplete), and thus, not complete with respect to any first-order definable class. Examples for normal logics have been mentioned in the Introduction. An example of a Kripke incomplete quasi-normal logic that is not normal is the quantified counterpart of Solovay's logic (see, e.g., [1]), which is obtained from the syntactically defined **QGL** by adding the axiom $\Box p \to p$ and removing the requirement of closure under necessitation. A more interesting question, therefore, as noted above, is whether every recursively enumerable Kripke complete logic is a logic of a first-order definable class of frames. The main contribution of this paper is to show, which we do in the next section, that this is not so for quasi-normal logics—namely, we exhibit an example of a recursively enumerable Kripke complete quasi-normal logic that is not complete with respect to any first-order definable class of pointed frames.

3 Construction of the main counterexample

In this section, we present an example of a quasi-normal quantified modal logic L_0 that is recursively enumerable, Kripke complete, but not complete with respect to any first-order definable class of frames. The logic L_0 is defined as the set of formulas true at the distinguished world of the following pointed Kripke frame $\mathfrak{F}$. Let

$$\begin{aligned} W_0 &= \{w_0^0\}; \\ W_{k+1} &= W_k \cup \{w_0^{k+1}, \ldots, w_{k+1}^{k+1}\}. \end{aligned}$$

For every $n \in \mathbb{N}$, let R_n be a binary relation on W_n such that, for every $w_m^k, w_s^t \in W_n$,

$$w_m^k R_n w_s^t \iff t = k+1 \text{ and } m = 0.$$

Let $\mathfrak{F}_n$ denote the frame $\langle W_n, R_n \rangle$. Finally, let $\mathfrak{F} = \langle W, R \rangle$, where

$$W = \bigcup_{i=0}^{\infty} W_i, \quad R = \bigcup_{i=0}^{\infty} R_i.$$

The frame $\mathfrak{F}$ is depicted in Fig. 1. We define L_0 to be the set of formulas that are true at w_0^0 in $\mathfrak{F}$; thus, L_0 is a quasi-normal modal logic. By definition, L_0 is complete with respect to a class of pointed Kripke frames (namely, the class containing a single frame, $\mathfrak{F}$). We next show that L_0 is recursively enumerable and not complete with respect to any first-order definable class of frames.

To show that L_0 is recursively enumerable, we effectively embed it into the classical first-order logic with equality **QClE**.

First, notice that, since $\mathfrak{F}_n$ is a finite frame, we can effectively construct a classical first-order formula F_n describing $\mathfrak{F}_n$,—all we need to do is say what worlds exist in $\mathfrak{F}_n$, that those worlds are pairwise distinct, that there are no other worlds in $\mathfrak{F}_n$, and describe which worlds are related by the accessibility relation.

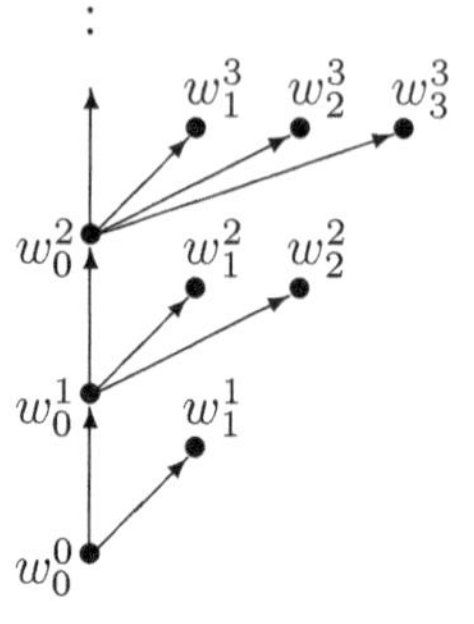

Fig. 1. The pointed frame $\mathfrak{F}$

Now, let R and D be binary, and W unary, predicate letters not occurring in φ; intuitively, $W(x)$ means "x is a world", $D(x, y)$ means "y is an element of the domain of world x", and $R(x, y)$ means "y is accessible from x". Note that w_0^0 is the only world in $\mathfrak{F}_n$ that satisfies the property $Root(x)$, defined as follows:

$$Root(x) \;=\; \forall y\, \neg R(y, x).$$

Let $ST_x(\varphi)$ be the standard translation of the formula φ into classical first-order logic, defined as follows:

$$\begin{aligned} ST_x(P(y_1, \ldots, y_m)) &= P'(y_1, \ldots, y_m, x); \\ ST_x(\varphi_1 \wedge \varphi_2) &= ST_x(\varphi_1) \wedge ST_x(\varphi_2); \\ ST_x(\neg\varphi_1) &= \neg ST_x(\varphi_1); \\ ST_x(\Box\varphi_1) &= \forall y\, (W(y) \wedge R(x, y) \to ST_y(\varphi_1)); \\ ST_x(\forall y\, \varphi_1) &= \forall y\, (\neg W(y) \wedge D(x, y) \to ST_x(\varphi_1)), \end{aligned}$$

where the arity of P' is one greater than P, letter P' is distinct from letter Q' if, and only if, P is distinct from Q, and all the newly introduced individual variables are distinct from the previously used ones. Let M be the formula

$$\begin{aligned} M \;=\; & \exists x\, W(x) \wedge \forall x\, [W(x) \to \exists y\, D(x, y)] \wedge \\ & \wedge \forall x \forall y \forall z\, [W(x) \wedge W(y) \wedge \neg W(z) \wedge D(x, z) \wedge R(x, y) \to D(y, z)]. \end{aligned}$$

Intuitively, M describes general properties of predicate Kripke frames.

Lastly, for an arbitrary classical first-order formula with equality θ, inductively define the formula θ^* as follows:

$$\begin{aligned} (x = y)^* &= (x = y); \\ (R(x, y))^* &= R(x, y); \\ (\theta_1 \wedge \theta_2)^* &= \theta_1^* \wedge \theta_2^*; \\ (\neg\theta_1)^* &= \neg\theta_1^*; \\ (\forall x\, \theta_1)^* &= \forall x\, (W(x) \to \theta_1^*). \end{aligned}$$

Lemma 3.1 *For every closed modal formula φ with $\mathbf{md}(\varphi) = n$, the following holds:*

$$(\mathfrak{F}_n, w_0^0) \models \varphi \iff M \wedge F_n^* \to \forall x\, [W(x) \wedge Root(x) \to ST_x(\varphi)] \in \mathbf{QClE}.$$

Proof. It is well-known (see, e.g., [2]) that the standard translation has the property that $\mathfrak{M}, w \models \varphi$ if, and only if, $\mathfrak{M} \models ST_x(\varphi)[w]$. Thus, if $(\mathfrak{F}_n, w)$ is a (finite) pointed frame, then $(\mathfrak{F}_n, w) \models \varphi$ if, and only if, $\mathfrak{M} \models ST_x(\varphi)[w]$ holds for all first-order models "based on" $\mathfrak{F}_n$ (i.e., their domain and the interpretation of the binary relation R coincides with the set of worlds and the accessibility relation, respectively, of $\mathfrak{F}_n$). In turn, the latter holds if, and only if, the formula $M \wedge F_n^* \to \forall x\, [W(x) \wedge Root(x) \to ST_x(\varphi)]$, which claims that $ST_x(\varphi)$ holds provided we evaluate it in a model that looks like $\mathfrak{F}_n$ with x assigned to the "root" node, is valid. □

Proposition 3.2 *L_0 is recursively enumerable.*

Proof. As, for a modal formula φ with $\mathbf{md}(\varphi) = n$,

$$(\mathfrak{F}, w_0^0) \models \varphi \iff (\mathfrak{F}_n, w_0^0) \models \varphi,$$

it immediately follows from Lemma 3.1 that, given an arbitrary $n \in \mathbb{N}$, the set of theorems of L_0 with modal depth n is recursively enumerable. Thus, using the standard technique from recursion theory, we can recursively enumerate all the theorems of L_0. □

It remains to show that L_0 is not complete with respect to any first-order definable class of pointed Kripke frames. To prove this, we need an auxiliary Lemma, whose statement is a slight modification of a result from [5].

Lemma 3.3 *Let L be a normal modal logic that is sound and complete with respect to a class $\mathfrak{C}$ of frames that satisfies the following conditions:*

(i) *if $\mathfrak{F} \in \mathfrak{C}$, then every world in $\mathfrak{F}$ can see only finitely many worlds;*

(ii) *for every $n \in \mathbb{N}$, there exist $\mathfrak{F} = \langle W, R\rangle$ in $\mathfrak{C}$ and $w \in W$ such that w can see at least n worlds.*

Then, L is not recursively enumerable.

Proof. Let φ be an arbitrary classical first-order formula and let T be a unary, and E a binary, predicate letter not occurring in φ. Let $Congr$ be a formula saying that E is a congruence relation with respect to all predicate letters in φ and let

$$A = \forall x\, \Diamond T(x) \wedge \forall x \forall y\, (\Diamond(T(x) \wedge T(y)) \to E(x, y)).$$

Let **QClFin** be the classical first-order logic of finite models and let $\varphi^* = (Congr \wedge A) \to \varphi$. We can then show that

$$\varphi \in \mathbf{QClFin} \iff \varphi^* \in L.$$

Indeed, assume that $\varphi \notin \mathbf{QClFin}$; that is, there exists a classical model $\mathfrak{M}$ with a finite domain $D = \{a_1, \ldots, a_n\}$ such that $\mathfrak{M} \not\models \varphi$. We construct a model $\mathfrak{M}^*$, based on a frame from $\mathfrak{C}$, falsifying φ^*. By assumption, $\mathfrak{C}$ contains a frame $\mathfrak{F} = \langle W, R\rangle$ such that some $w_0 \in W$ can see at least n worlds; select exactly n out of those and bijectively map them to the elements of D; let the world $\bar{a}_i$ correspond to element a_i, where $i \in \{1, \ldots, n\}$. Let $D^*(w) = D$ for every $w \in W$. Let $\mathfrak{M}, w \models T[a_i]$ if, and only if, $w = \bar{a}_i$, for $i \in \{1, \ldots, n\}$. Let $I^*(w, E)$, for every $w \in W$, be the identity relation and let all the predicate letters in φ be defined in every $w \in W$ exactly as they are defined in $\mathfrak{M}$. It is then easy to check that $\mathfrak{M}^*, w_0 \not\models \varphi^*$. As $\mathfrak{M}^*$ is based on a frame from $\mathfrak{C}$, we conclude that $\varphi^* \notin L$.

Assume, on the other hand, that $\varphi^* \notin L$; that is, there exists a frame $\mathfrak{F} = \langle W, R\rangle$ in $\mathfrak{C}$, $w_0 \in W$, and a model $\mathfrak{M}^*$ based on $\mathfrak{F}$ such that $\mathfrak{M}^*, w_0 \not\models \varphi^*$. We construct a finite classical model $\mathfrak{M}$ falsifying φ. By assumption, w_0 can see only finitely many worlds, say $w_1, \ldots, w_n$. As $\mathfrak{M}^*, w_0 \models Congr \wedge A$, for every $b \in D^*(w_0)$, we have $w' \models T[b]$ for at least one w' accessible from w_0, and for every w accessible from w_0, T holds for the elements of only one equivalence class with respect to E. As w_0 can see only finitely many worlds, $D^*(w_0)$ contains only finitely many equivalence classes $a_1, \ldots, a_n$ with respect to E. Let $D = \{a_1, \ldots, a_n\}$ and let $I(P) = I^*(w_0, P)$ for every predicate letter P occurring in φ. Let $\mathfrak{M} = \langle D, I\rangle$. As $\mathfrak{M} \not\models \varphi$ and D is finite, $\varphi \notin$ **QClFin**.

As **QClFin** is not recursively enumerable [6], the statement of the Lemma follows.

□

Corollary 3.4 *Let $\mathfrak{C}$ be a class of frames satisfying the following conditions:*

(i) *if $\mathfrak{F} \in \mathfrak{C}$, then every world in $\mathfrak{F}$ can see only finitely many worlds;*

(ii) *for every $n \in \mathbb{N}$, there exist $\mathfrak{F} = \langle W, R\rangle$ in $\mathfrak{C}$ and $w \in W$ such that w can see at least n worlds.*

Then, $\mathfrak{C}$ is not first-order definable.

Proof. Immediately follows from Lemma 3.3 and Proposition 2.2. □

Proposition 3.5 *Let $\mathfrak{C}$ be a class of pointed frames such that L_0 is sound and complete with respect to $\mathfrak{C}$. Then, $\mathfrak{C}$ is not first-order definable.*

Proof. Let $rL(\mathfrak{C})$ be the set of formulas true at the distinguished world of every frame in $\mathfrak{C}$. By the statement of the proposition, $rL(\mathfrak{C}) = L_0$. For every $i, n \in \mathbb{N}^+$ such that $i \leqslant n$, let

$$\alpha_n^i = \Diamond(p_1 \wedge \ldots \wedge p_{i-1} \wedge \neg p_i \wedge p_{i+1} \wedge \ldots \wedge p_n).$$

For every $n \in \mathbb{N}$, let

$$\beta_n = \Box^n \neg(\alpha_{n+3}^1 \wedge \ldots \wedge \alpha_{n+3}^{n+3}).$$

As from every world w_k^n of $\mathfrak{F}$, we can reach either 0 (if $k > 0$) or $n+2$ (if $k = 0$) worlds, we have $(\mathfrak{F}, w_0^0) \models \beta_n$ and, thus, $\beta_n \in L_0$. As $rL(\mathfrak{C}) = L_0$, for every pointed frame $(\mathfrak{F}', w) \in \mathfrak{C}$, we have $(\mathfrak{F}', w) \models \beta_n$. Therefore, each world in $\mathfrak{F}'$ reachable from w in n steps can see no more than $n+2$ worlds. As $\mathfrak{F}'$ is a pointed frame with distinguished world w, every world in $\mathfrak{F}'$ can thus see only finitely many worlds, and hence $\mathfrak{C}$ satisfies the first condition of Corollary 3.4.

On the other hand, $(\mathfrak{F}, w_0^0) \not\models \Box\beta_n$; hence, $\Box\beta_n \notin L_0$, and thus $\mathfrak{C}$ contains a pointed frame $(\mathfrak{F}', w)$ such that $(\mathfrak{F}', w) \not\models \Box\beta_n$. Therefore, $\mathfrak{F}'$ contains a world

w' that can see at least $n+3$ worlds, and hence $\mathfrak{C}$ satisfies the second condition of Corollary 3.4.

Thus, in view of Corollary 3.4, $\mathfrak{C}$ considered as a class of frames is not first-order definable. Then, in view of Remark 2.1, $\mathfrak{C}$ considered as a class of pointed frames is not first-order definable, either, which concludes the proof.□

4 Discussion

We have exhibited an example of a quasi-normal quantified modal logic L_0 such that (1) L_0 is recursively enumerable, (2) L_0 is Kripke complete, and (3) L_0 is not complete with respect to any class of pointed frames defined by a classical first-order condition. The logic L_0 was defined as the logic of a particular pointed Kripke frame, $\mathfrak{F}$; it is not, however, an isolated example. Recall that the frame $\mathfrak{F}$ is a tree where, for every $n \in \mathbb{N}$, the nth level contains $n+1$ worlds $w_0^n, \ldots, w_{n+1}^n$ and where the world w_0^n can see all the worlds on level $n+1$. This construction can be generalized to use an arbitrary computable function f not bound above by any $n \in \mathbb{N}$ so that the nth level of the tree contains $f(n)$ worlds, thus giving us countably many quasi-normal quantified logics satisfying properties (1) through (3). We could also work with logics whose Kripke semantics involves constant, rather than varying, domains.

The most important question for future research remains the one posed at the beginning of the present paper—whether there exists a normal quantified modal logic satisfying properties (1) through (3). It is not clear whether the technique used in the present paper is transferable to normal logics.

Acknowledgments

This work has been supported by the Russian Foundation for Basic Research, projects 17-03-00818 and 18-011-00869.

References

[1] Feys, R., "Modal Logics," E. Nauwelaerts, 1965.

[2] Gabbay, D., V. Shehtman and D. Skvortsov, "Quantification in Nonclassical Logic, Volume 1," Elsevier, 2009.

[3] Montagna, F., *The predicate modal logic of provability*, Notre Dame Journal of Formal Logic **25** (1984), pp. 179–189.

[4] Rybakov, M., *Enumerability of modal predicate logics and the condition of non-existence of infinite ascending chains*, Logicheskiye Issledovaniya **8** (2001), pp. 155–167, in Russian.

[5] Skvortsov, D., *On the predicate logics of finite Kripke frames*, Studia Logica **54** (1995), pp. 79–88.

[6] Trakhtenbrot, B., *The impossibility of an algorithm for the decidability problem on finite classes*, Proceedings of the USSR Academy of Sciences (1950), in Russian.

[7] Wolter, F. and M. Zakharyaschev, *Decidable fragments of first-order modal logics*, Journal of Symbolic Logic **66** (2001), pp. 1415–1438.

Truth-Preserving Operations on Sums of Kripke Frames

Ilya Shapirovsky [1] [2]

Steklov Mathematical Institute of Russian Academy of Sciences

Institute for Information Transmission Problems of Russian Academy of Sciences

Abstract

The operation of sum of a family $(\mathsf{F}_i \mid i \text{ in } \mathsf{I})$ of Kripke frames indexed by elements of another frame I provides a natural way to construct expressive polymodal logics with good semantic and algorithmic properties. This operation has had several important applications over the last decade: it was used by L. Beklemishev in the context of polymodal provability logic; two ways of combining modal logics, the *refinement of modal logics* introduced by S. Babenyshev and V. Rybakov, and the *lexicographic product of modal logics* proposed by Ph. Balbiani, can be defined in terms of sums of frames. This paper provides some general truth-preserving tools for operating with sums of Kripke frames, and then applies them to study properties of resulting modal logics, in particular, to investigate the finite model property.

Keywords: combinations of modal logics, sum of Kripke frames, finite model property, universal modality, polymodal provability logic, refinement of modal logics, lexicographic product of modal logics

1 Introduction

This paper contributes to the area of combining modal logics [9,12].

Given a family $(\mathsf{F}_i \mid i \text{ in } \mathsf{I})$ of frames indexed by elements of another frame I (of the same signature), the *sum of the frames* F_i*'s over* I is obtained from their disjoint union by connecting elements of i-th and j-th distinct components according to the relations in I (this operation is a particular case of *generalized sum of models* introduced by S. Shelah in [15]). Given a class $\mathcal{F}$ of frames-summands and a class $\mathcal{I}$ of frames-indices, we consider the logic of the class $\sum_{\mathcal{I}} \mathcal{F}$ of all possible sums of F_i's in $\mathcal{F}$ over I in $\mathcal{I}$. In a particular case when $\mathcal{F}$ is the class $\mathrm{Fr}\, L_1$ of all the frames of a logic L_1, and $\mathcal{I}$ is $\mathrm{Fr}\, L_2$ for another logic L_2, we obtain a natural operation on Kripke-complete logics.

Over the last decade, sums of Kripke frames have had several important applications in modal logic. In [6], L. Beklemishev used (iterated) sums over

[1] This work is supported by the Russian Science Foundation under grant 16-11-10252 and performed at Steklov Mathematical Institute of Russian Academy of Sciences

[2] shapir@iitp.ru

Noetherian orders to construct models of the polymodal provability logic (this was probably the first application of sums in the context of polymodal logics). Then in [14] it was noted that sums can be a useful tool for studying computational complexity of modal satisfiability problems. At the same time in [1], S. Babenyshev and V. Rybakov considered an operation on frames and logics called *refinement*, and showed that under a very general condition this operation preserves the finite model property and decidability; refinements of frames can be considered as special instances of sums. The *lexicographic product of modal logics*, introduced by Ph. Balbiani in [2] (and then considered in [3,5,4]), is another example of an operation that can be defined via sums of frames.

This paper presents several general tools for studying modal logics of sums. Section 3 provides some basic observations on how sums interact with operations of p-morphism, generated subframe, and disjoint union. In Section 4 we address the following question: given a class of sums $\sum_{\mathcal{I}} \mathcal{F}$, when can we replace $\mathcal{F}$ with some other class of frames $\mathcal{F}'$, preserving the logic of sums? In particular, if the logic of summands $\mathcal{F}$ has the finite model property, can we replace $\mathcal{F}$ with a class of finite frames? Theorem 4.11 provides the following answer: if $\mathcal{F}$ and $\mathcal{F}'$ have the same logic in the language enriched by the universal modality (such classes are said to be *interchangeable*), then the logics of sums $\sum_{\mathcal{I}} \mathcal{F}$ and $\sum_{\mathcal{I}} \mathcal{F}'$ are equal; moreover, these classes of sums are interchangeable again, thus we have $\mathrm{Log} \sum_{\mathcal{J}} (\sum_{\mathcal{I}} \mathcal{F}) = \mathrm{Log} \sum_{\mathcal{J}} (\sum_{\mathcal{I}} \mathcal{F}')$ for any other class of frames-indices $\mathcal{J}$, and so on. Then we apply this theorem and show that the finite model property of the logic $\mathrm{Log}\,\mathcal{F}$ of summands transfers to logics of (iterated) sums over Noetherian orders. Finally, we consider several applications to refinements and lexicographic products.

2 Preliminaries

We assume the reader is familiar with the basic notions of modal logics [7,8,9].

Let A be a set (an alphabet of indices for modalities).

The set ML_A of *modal formulas over* A (or A-*formulas*, for short) is built from a countable set of *variables* $\mathrm{PV} = \{p_0, p_1, \ldots\}$ using Boolean connectives $\bot, \to$ and unary connectives $\Diamond_a$, $a \in A$ (*modalities*). The connectives $\vee, \wedge, \neg, \top, \Box_a$ are defined as abbreviations in the standard way, in particular $\Box_a \varphi$ is $\neg \Diamond_a \neg \varphi$.

An (A-)frame is a structure $\mathsf{F} = (W, (R_a)_{a \in A})$, where $W \neq \varnothing$ and $R_a \subseteq W \times W$ for $a \in A$. A *model on* F is a pair $\mathsf{M} = (\mathsf{F}, \theta)$, where $\theta : \mathrm{PV} \to 2^W$. We write $\mathrm{dom}(\mathsf{F})$ for W, which is called the *domain* of F. For u, v in F, u is *a-accessible from b in* F if $uR_a v$. We write $u \in \mathsf{F}$ for $u \in \mathrm{dom}(\mathsf{F})$. Likewise for models. For $u \in W$, $V \subseteq W$, we put $R_a(u) = \{v \mid uR_a v\}$, $R_a[V] = \cup_{v \in V} R_a(v)$.

The *truth relation* $\mathsf{M}, w \models \varphi$ is defined in the usual way, in particular $\mathsf{M}, w \models \Diamond_a \varphi$ means that $\mathsf{M}, v \models \varphi$ for some v in $R_a(w)$. A formula φ is *satisfiable in a model* M if $\mathsf{M}, w \models \varphi$ for some w in M. For a class $\mathcal{F}$ of frames, let $\mathrm{Mod}\,\mathcal{F}$ be the class of all models (F, θ) with $\mathsf{F} \in \mathcal{F}$. A formula is *satisfiable in a frame* F (*in a class* $\mathcal{F}$ *of frames*) if it is satisfiable in some model on F (in some model in $\mathrm{Mod}\,\mathcal{F}$). φ is valid in a frame F (in a class $\mathcal{F}$ of frames) if $\neg\varphi$ is not satisfiable in F (in $\mathcal{F}$). Validity of a set of formulas means validity of every formula in this set.

A (*propositional normal modal*) *logic* is a set L of formulas that contains all classical tautologies, the axioms $\neg\Diamond_a\bot$ and $\Diamond_a(p_0 \vee p_1) \to \Diamond_a p_0 \vee \Diamond_a p_1$ for each a in A, and is closed under the rules of modus ponens, substitution and monotonicity (if $\varphi \to \psi \in L$, then $\Diamond_a\varphi \to \Diamond_a\psi \in L$, for each a in A). In particular, the set $\operatorname{Log}\mathcal{F}$ of all formulas valid in $\mathcal{F}$ is a logic; it is called the *logic of* $\mathcal{F}$; such logics are called *Kripke complete.* A logic has the *finite model property* if it is the logic of a class of finite frames (a frame is finite, if its domain is). Let $\operatorname{Fr} L$ and $\operatorname{Fr}_f L$ be the classes of all frames and all finite frames validating L respectively.

The notions of p-morphism, generated subframe or submodel are defined in the standard way, see e.g. [9, Section 1.4]. We write $\mathsf{F} \twoheadrightarrow \mathsf{G}$, if G is a p-morphic image of F. The notation $\mathsf{F} \cong \mathsf{G}$ means that F and G are isomorphic. We write $\mathsf{F}[w]$ for the subframe of F generated by the singleton $\{w\}$; such frames are called *cones in* F.

The cardinality of a set V is denoted by $|V|$. Natural numbers are considered as finite ordinals. Given a sequence $\boldsymbol{v} = (v_0, v_1, \ldots)$, we write $\boldsymbol{v}(i)$ for v_i.

3 Sums

We fix $N \leq \omega$ for the alphabet and consider the language ML_N.

Consider a non-empty family $(\mathsf{F}_i)_{i\in I}$ of N-frames $\mathsf{F}_i = (W_i, (R_{i,a})_{a\in N})$. The *disjoint union* of these frames is the N-frame $\bigsqcup_{i\in I}\mathsf{F}_i = (\bigsqcup_{i\in I} W_i, (R_a)_{a\in N})$, where $\bigsqcup_{i\in I} W_i = \bigcup_{i\in I}(\{i\} \times W_i)$ is the *disjoint union of sets* W_i, and

$$(i,w)R_a(j,v) \quad \text{iff} \quad i = j \,\&\, wR_{i,a}v.$$

Suppose that I is the domain of another N-frame $\mathsf{I} = (I, (S_a)_{a\in N})$.

Definition 3.1 The *sum of the family* $(\mathsf{F}_i)_{i\in \mathsf{I}}$ *of* N*-frames over the* N*-frame* I is the N-frame $\sum_{i\in\mathsf{I}}\mathsf{F}_i = (\bigsqcup_{i\in I} W_i, (R^{\Sigma}_a)_{a\in N})$, where

$$(i,w)R^{\Sigma}_a(j,v) \quad \text{iff} \quad i = j \,\&\, wR_{i,a}v \text{ or } i \neq j \,\&\, iS_a j.$$

The sum of models $\sum_{i\in\mathsf{I}}(\mathsf{F}_i, \theta_i)$ is the model $(\sum_{i\in\mathsf{I}}\mathsf{F}_i, \theta)$, where $(i,w) \in \theta(p)$ iff $w \in \theta_i(p)$.

For classes $\mathcal{I}$, $\mathcal{F}$ of N-frames, let $\sum_{\mathcal{I}}\mathcal{F}$ be the class of all sums $\sum_{i\in\mathsf{I}}\mathsf{F}_i$ such that $\mathsf{I} \in \mathcal{I}$ and $\mathsf{F}_i \in \mathcal{F}$ for every i in I.

Remark 3.2 We do not require that S_a's are partial orders or even transitive relations.

The relations R^{Σ}_a are independent of reflexivity of the relations S_a: if $\mathsf{I}' = (I, (S'_a)_{a\in N})$, where S'_a is the reflexive closure of S_a for each $a \in N$, then $\sum_{i\in\mathsf{I}}\mathsf{F}_i = \sum_{i\in\mathsf{I}'}\mathsf{F}_i$.

We shall be mainly interested in the polymodal case. For a simple illustration of the definition let us first consider the following unimodal examples.

Let $\mathsf{F} = (W, R)$ be a preorder. The *(irreflexive) skeleton* of F is the strict partial order $\mathrm{sk}\mathsf{F} = (\overline{W}, <_R)$, where $\overline{W}$ is the quotient set of W by the equivalence $R \cap R^{-1}$, and for $C, D \in \overline{W}$, $C <_R D$ iff $C \neq D$ and $\exists w \in C\, \exists v \in D\; wRv$. Elements of $\overline{W}$ are called *clusters in* F. Then F is isomorphic to the sum $\sum_{C \in \mathrm{sk}\mathsf{F}} (C, C \times C)$ of its clusters over its skeleton.

For another example suppose that $\mathsf{F} = (W, R)$ satisfies the property of *weak transitivity* $xRzRy \Rightarrow xRy \vee x = y$. Then F is isomorphic to a sum $\sum_{i \in \mathsf{I}} \mathsf{F}_i$, where I is a partial order and in every F_i we have $xR_iy \vee x = y$.

The propositions below show how sums interact with p-morphisms, generated subframes, and disjoint unions.

The following fact is immediate from Definition 3.1.

Proposition 3.3 *If* J *is a generated subframe of* I, *then* $\sum_{i \in \mathsf{J}} \mathsf{F}_i$ *is a generated subframe of* $\sum_{i \in \mathsf{I}} \mathsf{F}_i$.

Proposition 3.4 *Consider* N*-frames* I, J, *and two families of* N*-frames* $(\mathsf{F}_i)_{i \in \mathsf{I}}$, $(\mathsf{G}_j)_{j \in \mathsf{J}}$. *Assume that all the relations in* J *are irreflexive.*

(i) *If* $f : \mathsf{I} \twoheadrightarrow \mathsf{J}$ *and* $\mathsf{F}_i \twoheadrightarrow \mathsf{G}_{f(i)}$ *for all* i *in* I, *then* $\sum_{i \in \mathsf{I}} \mathsf{F}_i \twoheadrightarrow \sum_{j \in \mathsf{J}} \mathsf{G}_j$.

(ii) *If* $\mathsf{I} = \mathsf{J}$ *and* $\mathsf{F}_i \twoheadrightarrow \mathsf{G}_i$ *for all* i *in* I, *then* $\sum_{i \in \mathsf{I}} \mathsf{F}_i \twoheadrightarrow \sum_{i \in \mathsf{I}} \mathsf{G}_i$.

(iii) *If* $f : \mathsf{I} \twoheadrightarrow \mathsf{J}$, *then* $\sum_{i \in \mathsf{I}} \mathsf{G}_{f(i)} \twoheadrightarrow \sum_{j \in \mathsf{J}} \mathsf{G}_j$.

Proof. (i) The required p-morphism is defined as $g(i, w) = (f(i), g_i(w))$, where $g_i : \mathsf{F}_i \twoheadrightarrow \mathsf{G}_{f(i)}$ for each i in I. (ii) and (iii) are special cases of (i): in (ii), f is the identity map on I; in (iii), $\mathsf{F}_i = \mathsf{G}_{f(i)}$ for each i in I. □

Lemma 3.5 *Consider an* N*-frame* I, *a family* $(\mathsf{J}_i)_{i \in \mathsf{I}}$ *of* N*-frames, and a family* $(\mathsf{F}_{ij})_{i \in \mathsf{I}, j \in \mathsf{J}_i}$ *of* N*-frames. Then*

$$\sum_{i \in \mathsf{I}} \sum_{j \in \mathsf{J}_i} \mathsf{F}_{ij} \quad \cong \quad \sum_{(i,j) \in \sum_{k \in \mathsf{I}} \mathsf{J}_k} \mathsf{F}_{ij}.$$

The proof of this lemma is straightforward from the definition; the detailed verification is given in Appendix.

Let $(\varnothing)_N$ denote the sequence of length N in which every element is the empty set. Disjoint unions are special cases of sums: if I is a frame with empty relations $(I, (\varnothing)_N)$, then $\bigsqcup_{i \in I} \mathsf{F}_i = \sum_{i \in \mathsf{I}} \mathsf{F}_i$.

Proposition 3.6 *For a non-empty set* I, *a family* $(\mathsf{J}_i)_{i \in \mathsf{I}}$ *of* N*-frames, and a family* $(\mathsf{F}_{ij})_{i \in I, j \in \mathsf{J}_i}$ *of* N*-frames,*

$$\bigsqcup_{i \in I} \sum_{j \in \mathsf{J}_i} \mathsf{F}_{ij} \quad \cong \quad \sum_{(i,j) \in \bigsqcup_{k \in I} \mathsf{J}_k} \mathsf{F}_{ij}.$$

Proof. This is a special case of Lemma 3.5 in which $\mathsf{I} = (I, (\varnothing)_N)$. □

Proposition 3.7 *For an* N*-frame* I, *a family* $(J_i)_{i \in \mathsf{I}}$ *of non-empty sets, and a family* $(\mathsf{F}_{ij})_{i \in \mathsf{I}, j \in J_i}$ *of* N*-frames,*

$$\sum_{i \in \mathsf{I}} \bigsqcup_{j \in J_i} \mathsf{F}_{ij} \quad \cong \quad \sum_{(i,j) \in \sum_{k \in \mathsf{I}} (J_k, (\varnothing)_N)} \mathsf{F}_{ij}.$$

Proof. Follows from Lemma 3.5: let J_i be $(J_i, (\varnothing)_N)$. □

4 Replacing summands

In this section we introduce the notion of *interchangeable classes of frames* and prove the following: if $\mathcal{F}$ and $\mathcal{G}$ are interchangeable, then they have the same logic, and, for any class $\mathcal{I}$ of frames of the same signature, the classes $\sum_{\mathcal{I}} \mathcal{F}$ and $\sum_{\mathcal{I}} \mathcal{G}$ are interchangeable again. Then we show that classes are interchangeable iff they have the same logic in the language enriched by the universal modality.

4.1 Interchangeable classes

Definition 4.1 A sequence $\mathbf{\Gamma} = (\Gamma_a)_{a \in N}$, where Γ_a are sets of N-formulas, is called a *condition* (in the language ML_N).

Consider a model $\mathsf{M} = (W, (R_a)_{a \in N}, \theta)$, w in M. By induction on the length of an N-formula φ, we define the relation $\mathsf{M}, w \models_{\mathbf{\Gamma}} \varphi$ (*"under the condition* $\mathbf{\Gamma}$, φ *is true at* w *in* M"): as usual, $\mathsf{M}, w \not\models_{\mathbf{\Gamma}} \bot$, $\mathsf{M}, w \models_{\mathbf{\Gamma}} p$ iff $\mathsf{M}, w \models p$ for a variable p, $\mathsf{M}, w \models_{\mathbf{\Gamma}} \varphi \to \psi$ iff $\mathsf{M}, w \not\models_{\mathbf{\Gamma}} \varphi$ or $\mathsf{M}, w \models_{\mathbf{\Gamma}} \psi$; for $a \in N$,

$$\mathsf{M}, w \models_{\mathbf{\Gamma}} \Diamond_a \varphi \quad \text{iff} \quad \varphi \in \Gamma_a \text{ or } \exists v \in R_a(w) \;\; \mathsf{M}, v \models_{\mathbf{\Gamma}} \varphi.$$

In particular, if all Γ_a are empty, then we have the standard notion of truth in a Kripke model:

$$\mathsf{M}, w \models_{(\varnothing)_N} \varphi \quad \text{iff} \quad \mathsf{M}, w \models \varphi.$$

Let $\mathrm{sub}(\varphi)$ be the set of all subformulas of φ, and let $\mathrm{sub}(\varphi; \mathsf{M}, \mathbf{\Gamma})$ be the set $\{\psi \in \mathrm{sub}(\varphi) \mid \mathsf{M}, v \models_{\mathbf{\Gamma}} \psi \text{ for some } v\}$. In particular, $\mathrm{sub}(\varphi; \mathsf{M}, (\varnothing)_N)$ is the set of all subformulas of φ satisfiable in M. Models M and M' are said to be $(\varphi, \mathbf{\Gamma})$-*equivalent* if $\mathrm{sub}(\varphi; \mathsf{M}, \mathbf{\Gamma}) = \mathrm{sub}(\varphi; \mathsf{M}', \mathbf{\Gamma})$.

A triple $(\varphi, \Phi, \mathbf{\Gamma})$, where $\Phi \subseteq \mathrm{sub}(\varphi)$, is called a *tie*. A tie $(\varphi, \Phi, \mathbf{\Gamma})$ is *satisfiable* in a frame F (in a class $\mathcal{F}$ of frames) if there exists a model M on F (in $\mathrm{Mod}\,\mathcal{F}$) such that $\Phi = \mathrm{sub}(\varphi; \mathsf{M}, \mathbf{\Gamma})$.

We put $\mathcal{F} \preccurlyeq_\varphi \mathcal{G}$ if every tie of form $(\varphi, \Phi, \mathbf{\Gamma})$, which is satisfiable in $\mathcal{F}$, is satisfiable in $\mathcal{G}$. (Equivalently, $\mathcal{F} \preccurlyeq_\varphi \mathcal{G}$ if for every condition $\mathbf{\Gamma}$ and every model $\mathsf{M} \in \mathrm{Mod}\,\mathcal{F}$, there exists a model $\mathsf{M}' \in \mathrm{Mod}\,\mathcal{G}$ such that M and M' are $(\varphi, \mathbf{\Gamma})$-equivalent.)

If $\mathcal{F} \preccurlyeq_\varphi \mathcal{G}$ and $\mathcal{G} \preccurlyeq_\varphi \mathcal{F}$, then we put $\mathcal{F} \equiv_\varphi \mathcal{G}$. We put $\mathcal{F} \preccurlyeq \mathcal{G}$ if $\mathcal{F} \preccurlyeq_\varphi \mathcal{G}$ for all N-formulas φ. The classes $\mathcal{F}$ and $\mathcal{G}$ are *interchangeable*, denoted $\mathcal{F} \equiv \mathcal{G}$, if $\mathcal{F} \preccurlyeq \mathcal{G}$ and $\mathcal{G} \preccurlyeq \mathcal{F}$.

Proposition 4.2

(i) *If* $\mathcal{F} \preccurlyeq_\varphi \mathcal{G}$ *and* φ *is satisfiable in* $\mathcal{F}$, *then* φ *is satisfiable in* $\mathcal{G}$.

(ii) *If* $\mathcal{F} \equiv \mathcal{G}$, *then* $\mathrm{Log}\,\mathcal{F} = \mathrm{Log}\,\mathcal{G}$.

Proof. Follows from the following observation: if $\mathcal{C}$ is a class of N-frames, then φ is satisfiable in $\mathcal{C}$ iff there exists $\Phi \subseteq \mathrm{sub}(\varphi)$ such that $\varphi \in \Phi$ and the tie $(\varphi, \Phi, (\varnothing)_N)$ is satisfiable in $\mathcal{C}$. □

Theorem 4.3 *Let $\mathcal{I}$, $\mathcal{F}$, $\mathcal{G}$ be classes of N-frames.*

(i) *For every N-formula φ, if $\mathcal{F} \preccurlyeq_\varphi \mathcal{G}$, then $\sum_\mathcal{I} \mathcal{F} \preccurlyeq_\varphi \sum_\mathcal{I} \mathcal{G}$.*

(ii) *If $\mathcal{F} \equiv \mathcal{G}$, then $\sum_\mathcal{I} \mathcal{F} \equiv \sum_\mathcal{I} \mathcal{G}$.*

The proof is based on Lemmas 4.5 and 4.6 below. In what follows, $\mathbf{\Gamma}$ is a condition, φ is a formula, $\mathsf{M} = (W, (R_a)_{a\in N}, \theta)$ is a model.

Definition 4.4 Let V be a set of elements of M. Given φ and $\mathbf{\Gamma}$, let $\mathbf{\Delta}$ be the condition defined as follows: for $a \in N$,

$$\mathbf{\Delta}(a) = \mathbf{\Gamma}(a) \cup \{\chi \in \mathrm{sub}(\varphi) \mid \exists w \in R_a[V]\backslash V\ \mathsf{M}, w \models_\mathbf{\Gamma} \chi\}.$$

$\mathbf{\Delta}$ is called the *external condition of V in M with respect to φ and $\mathbf{\Gamma}$.*

We write $\mathsf{M}{\restriction}V$ for the restriction of M to V, i.e., $\mathsf{M}{\restriction}V = (V, (R_a{\restriction}V)_{a\in N}, \theta')$, where $R_a{\restriction}V = R_a \cap (V \times V)$, and $\theta'(p) = \theta(p) \cap V$ for $p \in \mathrm{PV}$.

Lemma 4.5 *Consider a sum of models $\mathsf{M} = \sum_\mathsf{I} \mathsf{M}_i$, i in I, and the set $V = \{i\} \times \mathrm{dom}(\mathsf{M}_i)$. If $\mathbf{\Delta}$ is the external condition of V in M with respect to some given φ, $\mathbf{\Gamma}$, then for all v in M_i, χ in $\mathrm{sub}(\varphi)$,*

$$\mathsf{M}, (i, v) \models_\mathbf{\Gamma} \chi \quad \textit{iff} \quad \mathsf{M}_i, v \models_\mathbf{\Delta} \chi. \tag{1}$$

Proof. By induction on the length of χ. Consider the case $\chi = \Diamond_a \psi$.

Suppose that $\psi \in \mathbf{\Gamma}(a)$. Then $\psi \in \mathbf{\Delta}(a)$, and both sides of (1) are true.

Suppose now that $\psi \notin \mathbf{\Gamma}(a)$.

Assume that $\mathsf{M}, (i, v) \models_\mathbf{\Gamma} \Diamond_a \psi$. Then we have $\mathsf{M}, (i, u) \models_\mathbf{\Gamma} \psi$ for some pair (j, u) such that $(i, v) R_a (j, u)$. If $i = j$, then by induction hypothesis, $\mathsf{M}_i, u \models_\mathbf{\Delta} \psi$; since u is a-accessible from v in M_i, we have $\mathsf{M}_i, v \models_\mathbf{\Delta} \Diamond_a \psi$. If $i \neq j$, then $\psi \in \mathbf{\Delta}(a)$, and we have $\mathsf{M}_i, v \models_\mathbf{\Delta} \Diamond_a \psi$ again.

Conversely, let $\mathsf{M}_i, v \models_\mathbf{\Delta} \Diamond_a \psi$. There are two cases. First, suppose $\mathsf{M}_i, u \models_\mathbf{\Delta} \psi$ for some u, which is a-accessible from v in M_i. Then $\mathsf{M}, (i, u) \models_\mathbf{\Gamma} \psi$ by induction hypothesis, and so $\mathsf{M}, (i, v) \models_\mathbf{\Gamma} \Diamond_a \psi$. Second, suppose $\psi \in \mathbf{\Delta}(a)$. Then since $\psi \notin \mathbf{\Gamma}(a)$, it follows that $\mathbf{\Gamma}(a) \neq \mathbf{\Delta}(a)$. By the definition of $\mathbf{\Delta}$, we have $\mathsf{M}, (j, u) \models_\mathbf{\Gamma} \psi$ for some pair (j, u) in $R_a[V]\backslash V$. It follows that j is a-accessible from i in I, so $(i, v) R_a (j, u)$. Hence $\mathsf{M}, v \models_\mathbf{\Gamma} \Diamond_a \psi$. □

Lemma 4.6 *Consider $\varphi, \mathbf{\Gamma}$, a frame I, and two sums of models $\mathsf{M} = \sum_\mathsf{I} \mathsf{M}_i$, $\mathsf{M}' = \sum_\mathsf{I} \mathsf{M}'_i$. For i in I, let $\mathbf{\Delta}_i$ be the external condition of the set $\{i\} \times \mathrm{dom}(\mathsf{M}_i)$ in M with respect to φ and $\mathbf{\Gamma}$. If the models M_i and M'_i are $(\varphi, \mathbf{\Delta}_i)$-equivalent for each i in I, then the sums M and M' are $(\varphi, \mathbf{\Gamma})$-equivalent.*

Proof. We show that for all i in I, w in M'_i, and χ in $\mathrm{sub}(\varphi)$,

$$\mathsf{M}', (i, w) \models_\mathbf{\Gamma} \chi \quad \text{iff} \quad \mathsf{M}'_i, w \models_{\mathbf{\Delta}_i} \chi. \tag{2}$$

By induction on the length of χ. The only non-trivial case is $\chi = \Diamond_a \psi$.

If $\psi \in \mathbf{\Gamma}(a)$, then $\psi \in \mathbf{\Delta}_i(a)$, and both sides of (2) are true.

Suppose that $\psi \notin \mathbf{\Gamma}(a)$.

Let $\mathsf{M}', (i, w) \models_{\mathbf{\Gamma}} \Diamond_a \psi$. Then $\mathsf{M}', (k, u) \models_{\mathbf{\Gamma}} \psi$ for some pair (k, u) which is a-accessible from (i, w) in M'. By induction hypothesis, $\mathsf{M}'_k, u \models_{\mathbf{\Delta}_k} \psi$. There are two cases: $k = i$ and $k \in S_a(i)\backslash\{i\}$, where S_a is the a-th relation in I. If $k = i$, then u is a-accessible from w in M'_i, and the right-hand side of (2) follows by Definition 4.1. Now let $k \in S_a(i)\backslash\{i\}$. We have $\psi \in \mathrm{sub}(\varphi; \mathsf{M}'_k, \mathbf{\Delta}_k)$, and since M'_k and M_k are $(\varphi, \mathbf{\Delta}_k)$-equivalent, we have $\psi \in \mathrm{sub}(\varphi; \mathsf{M}_k, \mathbf{\Delta}_k)$. It follows that $\mathsf{M}_k, u' \models_{\mathbf{\Delta}_k} \psi$ for some u' in M_k. By Lemma 4.5, $\mathsf{M}, (k, u') \models_{\mathbf{\Gamma}} \psi$. Hence $\psi \in \mathbf{\Delta}_i(a)$, and we have $\mathsf{M}'_i, w \models_{\mathbf{\Delta}_i} \Diamond_a \psi$, as required.

Conversely, let $\mathsf{M}'_i, w \models_{\mathbf{\Delta}_i} \Diamond_a \psi$. If $\mathsf{M}'_i, u \models_{\mathbf{\Delta}_i} \psi$ for some u, which is a-accessible from w in M', then the left-hand side of (2) follows from induction hypothesis. Suppose $\psi \in \mathbf{\Delta}_i(a)$. Since $\psi \notin \mathbf{\Gamma}(a)$, by the definition of $\mathbf{\Delta}_i$ we have $\mathsf{M}, (k, u) \models_{\mathbf{\Gamma}} \psi$ for some $k \in S_a(i)\backslash\{i\}$, $u \in \mathrm{dom}(\mathsf{M}_k)$. By Lemma 4.5, $\mathsf{M}_k, u \models_{\mathbf{\Delta}_k} \psi$. The models M_k and M'_k are $(\varphi, \mathbf{\Delta}_k)$-equivalent, so $\mathsf{M}'_k, u' \models_{\mathbf{\Delta}_k} \psi$ for some u' in M'_k. By induction hypothesis, $\mathsf{M}', (k, u') \models_{\mathbf{\Gamma}} \psi$. Then since $k \in S_a(i)\backslash\{i\}$, it follows that $\mathsf{M}', (i, w) \models_{\mathbf{\Gamma}} \Diamond_a \psi$.

Thus, (2) is proved. It remains only to observe that

$$\mathrm{sub}(\varphi; \mathsf{M}, \mathbf{\Gamma}) = \bigcup_{i \in \mathsf{I}} \mathrm{sub}(\varphi; \mathsf{M}_i, \mathbf{\Delta}_i) = \bigcup_{i \in \mathsf{I}} \mathrm{sub}(\varphi; \mathsf{M}'_i, \mathbf{\Delta}_i) = \mathrm{sub}(\varphi; \mathsf{M}', \mathbf{\Gamma}).$$

Indeed, the first equality holds by Lemma 4.5, the third — by (2), and the second one holds because M_i and M'_i are $(\varphi, \mathbf{\Delta}_i)$-equivalent for all i in I. □

Proof of Theorem 4.3. The first statement follows from Lemma 4.6: for $\mathsf{I} \in \mathcal{I}$, a sum $\sum_{\mathsf{I}} \mathsf{M}_i$ of models in $\mathrm{Mod}\,\mathcal{F}$, and a tie $(\varphi, \Phi, \mathbf{\Gamma})$, we choose models M'_i in $\mathrm{Mod}\,\mathcal{G}$ in such a way that $\sum_{\mathsf{I}} \mathsf{M}'_i$ is $(\varphi, \mathbf{\Gamma})$-equivalent to the initial sum. The second statement immediately follows from the first. □

It follows that $\mathcal{F} \equiv \mathcal{G}$ implies $\mathrm{Log} \sum_{\mathcal{I}} \mathcal{F} = \mathrm{Log} \sum_{\mathcal{I}} \mathcal{G}$. When $\mathcal{F} \equiv \mathcal{G}$?

4.2 Criterion of interchangeability

We shall show that classes of frames are interchangeable iff they have the same logic in the language endowed with the universal modality.

Given a condition $\mathbf{\Gamma}$, by induction on the length of φ we define $[\varphi]^{\mathbf{\Gamma}}$: $[\bot]^{\mathbf{\Gamma}} = \bot$, $[p]^{\mathbf{\Gamma}} = p$ for variables, $[\varphi_1 \to \varphi_2]^{\mathbf{\Gamma}} = [\varphi_1]^{\mathbf{\Gamma}} \to [\varphi_2]^{\mathbf{\Gamma}}$,

$$[\Diamond_a \varphi]^{\mathbf{\Gamma}} = \begin{cases} \top, & \text{if } \varphi \in \mathbf{\Gamma}(a), \\ \Diamond_a [\varphi]^{\mathbf{\Gamma}} & \text{otherwise.} \end{cases}$$

Lemma 4.7 $\mathsf{M}, w \models_{\mathbf{\Gamma}} \varphi$ *iff* $\mathsf{M}, w \models [\varphi]^{\mathbf{\Gamma}}$.

Proof. By induction on the length of φ. Consider the case $\varphi = \Diamond_a \psi$.

Suppose that $\psi \in \mathbf{\Gamma}(a)$. In this case, we have $[\Diamond_a \psi]^{\mathbf{\Gamma}} = \top$; by Definition 4.1, $\mathsf{M}, w \models [\Diamond_a \psi]^{\mathbf{\Gamma}}$ for all w in M.

Now suppose that $\psi \notin \mathbf{\Gamma}$. In this case $\mathsf{M}, w \models_{\mathbf{\Gamma}} \Diamond_a \psi$ means that $\mathsf{M}, v \models_{\mathbf{\Gamma}} \psi$ for some $v \in R_a(w)$, which is equivalent to $\mathsf{M}, w \models \Diamond_a [\psi]^{\mathbf{\Gamma}}$ by induction hypothesis. It remains to observe that in this case $\Diamond_a [\psi]^{\mathbf{\Gamma}} = [\Diamond_a \psi]^{\mathbf{\Gamma}}$. □

We fix some $u \notin N$ and consider the alphabet $N' = N \cup \{u\}$. For an N-frame $\mathsf{G} = (W, (R_a)_{a \in N})$, let G^u be the N'-frame $(W, (R_a)_{a \in N'})$, where $R_u = W \times W$;

likewise for models. For a class $\mathcal{F}$ of N-frames, $\mathcal{F}^u = \{\mathsf{F}^u \mid \mathsf{F} \in \mathcal{F}\}$. For a tie $(\varphi, \Psi, \mathbf{\Gamma})$, where φ is an N-formula, put

$$\delta(\varphi, \Psi, \mathbf{\Gamma}) = \bigwedge_{\psi \in \Psi} \Diamond_u [\psi]^{\mathbf{\Gamma}} \wedge \bigwedge_{\psi \in \mathrm{sub}(\varphi) \setminus \Psi} \neg \Diamond_u [\psi]^{\mathbf{\Gamma}} \tag{3}$$

Lemma 4.8 *$(\varphi, \Psi, \mathbf{\Gamma})$ is satisfiable in $\mathcal{F}$ iff $\delta(\varphi, \Psi, \mathbf{\Gamma})$ is satisfiable in $\mathcal{F}^u$.*

Proof. By Lemma 4.7, for any model M we have: $\Psi = \mathrm{sub}(\varphi; \mathsf{M}, \mathbf{\Gamma})$ iff the formula $\delta(\varphi, \Psi, \mathbf{\Gamma})$ is true (at any point) in the model M^u. □

Lemma 4.9 *If $\mathcal{F} \preccurlyeq \mathcal{G}$ and α is satisfiable in $\mathcal{F}^u$, then α is satisfiable in $\mathcal{G}^u$.*

Proof. Let $\mathcal{C}$ be the class of all N-frames. By [10, Theorem 3.7], there exists an N'-formula $\alpha' = \Box_u \chi \wedge \psi \wedge \bigwedge_{i<l} \Diamond_u \psi_i$ such that χ, ψ, ψ_i $(i < l)$ are N-formulas, and $\alpha \leftrightarrow \alpha'$ is valid in $\mathcal{C}^u$. Assume that α is satisfiable in M^u for some $\mathsf{M} \in \mathrm{Mod}\,\mathcal{F}$. Consider an N-formula φ containing $\neg\chi$, ψ, and all ψ_i as subformulas. Put $\Psi = \mathrm{sub}(\varphi; \mathsf{M}, (\varnothing)_N)$. Then ψ, ψ_i $(i < l)$ are in Ψ, and $\neg\chi \notin \Psi$. Since $\mathcal{F} \preccurlyeq \mathcal{G}$, for some $\mathsf{M}' \in \mathrm{Mod}\,\mathcal{G}$ we have $\Psi = \mathrm{sub}(\varphi; \mathsf{M}', (\varnothing)_N)$. It follows that α' is true at some point in M'^u. Thus α is satisfiable in $\mathcal{G}^u$. □

From Lemmas 4.8 and 4.9 we obtain the following simple characterization of interchangeable classes.

Proposition 4.10 *$\mathcal{F} \equiv \mathcal{G}$ iff $\mathrm{Log}\,\mathcal{F}^u = \mathrm{Log}\,\mathcal{G}^u$.*

Now from Theorem 4.3 and Proposition 4.10 we obtain the main result of this section:

Theorem 4.11 *Let $\mathcal{I}$, $\mathcal{F}$, $\mathcal{G}$ be classes on N-frames. If $\mathrm{Log}\,\mathcal{F}^u = \mathrm{Log}\,\mathcal{G}^u$, then $\mathrm{Log}(\sum_{\mathcal{I}} \mathcal{F})^u = \mathrm{Log}(\sum_{\mathcal{I}} \mathcal{G})^u$, and in particular $\mathrm{Log} \sum_{\mathcal{I}} \mathcal{F} = \mathrm{Log} \sum_{\mathcal{I}} \mathcal{G}$.*

The rest of this section provides some more tools for interchangeable classes.

Proposition 4.12 *If $\mathcal{F} \equiv \mathcal{G}$, then $\mathcal{F}^u \equiv \mathcal{G}^u$.*

Proof. If $\mathrm{Log}\,\mathcal{F}^u = \mathrm{Log}\,\mathcal{G}^u$, then trivially $\mathrm{Log}\left((\mathcal{F}^u)^u\right) = \mathrm{Log}\left((\mathcal{G}^u)^u\right)$ (another universal relation does nothing). Now we use Proposition 4.10. □

Proposition 4.13 *For frames F, G, if $\mathsf{F} \twoheadrightarrow \mathsf{G}$, then any tie that is satisfiable in G is satisfiable in F.*

Proof. This follows from Lemma 4.8, because $\mathsf{F} \twoheadrightarrow \mathsf{G}$ implies $\mathsf{F}^u \twoheadrightarrow \mathsf{G}^u$. □

Definition 4.14 Let $\mathsf{M} = (W, (R_a)_{a \in N}, \theta)$ and $\mathsf{M}' = (W', (R'_a)_{a \in N}, \theta)$ be models such that $W' \subseteq W$, $R'_a \subseteq R_a$ for each $a \in N$, and $\theta'(p) = \theta(p) \cap W'$ for variables. The model M' is called a *selective filtration of* M *with respect to given φ and $\mathbf{\Gamma}$* if for all ψ, $a \in N$ such that $\Diamond_a \psi \in \mathrm{sub}(\varphi)$, and all w in M'

$$\mathsf{M}, w \models_{\mathbf{\Gamma}} \Diamond_a \psi \,\&\, \psi \notin \mathbf{\Gamma}(a) \;\Rightarrow\; \exists v\, (w R'_a v \,\&\, \mathsf{M}, v \models_{\mathbf{\Gamma}} \psi).$$

Proposition 4.15 *If M' is a selective filtration of M with respect to φ and $\mathbf{\Gamma}$, then for all $\psi \in \mathrm{sub}(\varphi)$, w in M', we have $\mathsf{M}', w \models_{\mathbf{\Gamma}} \psi$ iff $\mathsf{M}, w \models_{\mathbf{\Gamma}} \psi$.*

In our formulation of selective filtration, it is important that $\Box_a$'s are abbreviations. The proof of Proposition 4.15 is straightforward (see Appendix).

Proposition 4.16 *If* M' *is a generated submodel of* M*, then for every condition* $\mathbf{\Gamma}$*, every formula* φ*, and every* w *in* M'*, we have* $\mathsf{M}', w \models_{\mathbf{\Gamma}} \varphi$ *iff* $\mathsf{M}, w \models_{\mathbf{\Gamma}} \varphi$.

Proof. A generated submodel is a selective filtration (with respect to any φ and $\mathbf{\Gamma}$). Now we use Proposition 4.15. □

5 Applications

5.1 Sums over Noetherian orders

Definition 5.1 Consider a unimodal frame $\mathsf{I} = (I, S)$ and a family $(\mathsf{F}_i)_{i \in I}$ of N-frames (or N-models). For $a \in N$, the *a-sum* ${}^{a}\sum_{\mathsf{I}} \mathsf{F}_i$ is the sum $\sum_{\mathsf{I}'} \mathsf{F}_i$, where I' is the N-frame whose domain is I, the a-th relation is S and all the other relations are empty. If $\mathcal{F}$ is a class of N-frames, $\mathcal{I}$ is a class of 1-frames, then ${}^{a}\sum_{\mathcal{I}} \mathcal{F}$ is the class of all sums ${}^{a}\sum_{\mathsf{I}} \mathsf{F}_i$, where $\mathsf{I} \in \mathcal{I}$ and all F_i are in $\mathcal{F}$.

For $s < \omega$ and a tuple $\boldsymbol{a} = (a_0, \ldots, a_{s-1}) \in N^s$, let ${}^{\boldsymbol{a}}\sum_{\mathcal{I}} \mathcal{F}$ be the class ${}^{a_0}\sum_{\mathcal{I}} \cdots {}^{a_{s-1}}\sum_{\mathcal{I}} \mathcal{F}$ (we put ${}^{\boldsymbol{a}}\sum_{\mathcal{I}} \mathcal{F} = \mathcal{F}$ if $\boldsymbol{a}$ is the empty sequence).

A strict partial order $(I, <)$ is *Noetherian* if it has no infinite ascending chain. Let NPO and PO_f be the classes of all non-empty Noetherian partial orders and all finite non-empty strict partial orders respectively (we say that a partial order is non-empty, if its domain is).

Sums over Noetherian orders play a significant role in the context of provability logics. In [6], L. Beklemishev introduced a system J, a Kripke complete approximation of the well-known polymodal provability logic GLP [11]. Semantically, J was characterised as the logic of frames called *stratified* in [6]. In our notation, this can be formulated as follows: for $N < \omega$, the N-modal fragment of J is the logic of the class ${}^{\boldsymbol{a}_N}\sum_{\mathrm{NPO}} \{\mathsf{S}_N\}$, where S_N is a singleton with N empty relations, and $\boldsymbol{a}_N = (0, \ldots, N-1)$. From [6] it follows that

$$\mathrm{Log}\ {}^{\boldsymbol{a}_N}\sum_{\mathrm{NPO}} \{\mathsf{S}_N\} = \mathrm{Log}\ {}^{\boldsymbol{a}_N}\sum_{\mathrm{PO}_f} \{\mathsf{S}_N\}. \tag{4}$$

We are going to generalize this fact in the following ways: in (4), we may replace $\{\mathsf{S}_N\}$ with an arbitrary class $\mathcal{F}$ of N-frames; if, moreover, the logic of the class $\mathcal{F}^u$ has the finite model property, then in the right-hand side of the equation we may replace $\mathcal{F}$ with the class of finite frames of its logic.

A strict partial order $(I, <)$ is called a (*transitive irreflexive*) *tree* if it has a least element (the *root*) and for all $i \in I$ the set $\{j \mid j < i\}$ is a finite chain. Let Tr_f and NTr be the classes of all finite trees and Noetherian trees respectively.

Consider a finite tree $\mathsf{I} = (I, <)$. The *branching of* i *in* I, denoted by $br(i, \mathsf{I})$, is the number of immediate successors of i (j *is an immediate successor of* i, if $i < j$ and there is no k such that $i < k < j$); the *branching of* I, denoted by $br(\mathsf{I})$, is $\max\{br(i, \mathsf{I}) \mid i \text{ in } \mathsf{I}\}$. The *height of* I, denoted by $ht(\mathsf{I})$, is $\max\{|V| \mid V \text{ is a chain in } \mathsf{I}\}$. For $n \in \omega$, let $\mathrm{Tr}(n)$ be the class of all finite trees with height and branching $\leq n$: $\mathrm{Tr}(n) = \{\mathsf{I} \in \mathrm{Tr}_f \mid ht(\mathsf{I}) \leq n \,\&\, br(\mathsf{I}) \leq n\}$.

Let $\bigsqcup\mathcal{F}$ be the class of all disjoint unions $\bigsqcup_I \mathsf{F}_i$, where I is a non-empty set and all F_i are in $\mathcal{F}$, and $\bigsqcup_{\leq k}\mathcal{F}$ the class of all such frames where $|I| \leq k$.

Let $\sharp\varphi$ be the number of subformulas of φ.

Theorem 5.2 *Let $\mathcal{F}$ be a class of N-frames, $s < \omega$, $\boldsymbol{a} = (a_0, \dots a_{s-1}) \in N^s$, $\mathrm{Tr}_f \subseteq \mathcal{I}_0, \dots, \mathcal{I}_{s-1} \subseteq \mathrm{NPO}$, $\mathcal{G} = {}^{a_0}\sum_{\mathcal{I}_0} \cdots {}^{a_{s-1}}\sum_{\mathcal{I}_{s-1}} \mathcal{F}$.*

(i) *If $s > 0$, then for every φ we have ${}^{\boldsymbol{a}}\sum_{\mathrm{NPO}} \mathcal{F} \equiv_\varphi \bigsqcup_{\leq \sharp\varphi} {}^{\boldsymbol{a}}\sum_{\mathrm{Tr}(\sharp\varphi)} \mathcal{F}$.*

(ii) $\mathrm{Log}\,\mathcal{G} = \mathrm{Log}\ {}^{\boldsymbol{a}}\sum_{\mathrm{Tr}_f} \mathcal{F}$*; moreover, a formula φ is satisfiable in $\mathcal{G}$ iff φ is satisfiable in ${}^{\boldsymbol{a}}\sum_{\mathrm{Tr}(\sharp\varphi)} \mathcal{F}$.*

(iii) *If $\mathrm{Log}\,\mathcal{F}^u$ has the finite model property, then so does $\mathrm{Log}\,\mathcal{G}$:*

$$\mathrm{Log}\,\mathcal{G} = \mathrm{Log}\ {}^{\boldsymbol{a}}\sum_{\mathrm{Tr}_f} \mathrm{Fr}_f\,\mathrm{Log}\,\mathcal{F}.$$

The proof of this theorem is based on the following statements.

Lemma 5.3 *Let $a \in N$. Every frame in ${}^{a}\sum_{\mathrm{NPO}} \bigsqcup\mathcal{F}$ is isomorphic to a frame in ${}^{a}\sum_{\mathrm{NPO}} \mathcal{F}$.*

Proof. By Proposition 3.7, a sum of form ${}^{a}\sum_{i \in \mathsf{I}} \bigsqcup_{j \in J_i} \mathsf{F}_{ij}$ is isomorphic to ${}^{a}\sum_{(i,j) \in \sum_{k \in \mathsf{I}} (J_k, \varnothing)} \mathsf{F}_{ij}$. If I is Noetherian, then the sum $\sum_{k \in \mathsf{I}} (J_k, \varnothing)$ is. □

Proposition 5.4 *Let $(I, <)$ be a Noetherian tree, $\mathcal{V}$ a finite family of subsets of I, $i_0 \in I$. Then there exists $J \subseteq I$ such that $ht(J, <) \leq |\mathcal{V}| + 1$, $br(J, <) \leq |\mathcal{V}|$, i_0 is the root of $(J, <)$, and for all $V \in \mathcal{V}$, $j \in J$ we have*

$$\exists i > j\ i \in V \ \Rightarrow\ \exists i > j\ i \in V \cap J. \tag{5}$$

The proof of this fact is by a standard 'step-by-step' construction, the details are given in Appendix. We shall use it in the following lemma, which is the crucial technical step in the proof of Theorem 5.2.

Lemma 5.5 *Let $a \in N$. Consider a model $\mathsf{M} \in \mathrm{Mod}\ {}^{a}\sum_{\mathrm{NTr}} \mathcal{F}$. For every φ, $\mathbf{\Gamma}$, and x in M, there exists a model $\mathsf{M}' \in \mathrm{Mod}\ {}^{a}\sum_{\mathrm{Tr}(\sharp\varphi)} \mathcal{F}$ which contains x and is a selective filtration of M with respect to φ and $\mathbf{\Gamma}$.*

Proof. Let $\mathsf{M} = {}^{a}\sum_{i \in \mathsf{I}} \mathsf{M}_i$, where $\mathsf{I} = (I, <)$ is a Noetherian tree. Consider the family $\mathcal{V} = \{P(\alpha) \mid \Diamond_a \alpha \in \mathrm{sub}(\varphi)\}$, where

$$P(\alpha) = \{i \in I \mid \mathsf{M}, (i, w) \models_{\mathbf{\Gamma}} \alpha \text{ for some } w\}.$$

Assume that $x = (i_0, w_0)$. By Proposition 5.4, there exists a restriction $\mathsf{J} = (J, <)$ of I such that $\mathsf{J} \in \mathrm{Tr}(|\mathcal{V}| + 1)$, $i_0 \in J$, and for all $j \in J$, $V \in \mathcal{V}$ we have (5).

Put $\mathsf{M}' = {}^{a}\sum_{i\in \mathsf{J}} \mathsf{M}_i$ and show that M' is the required selective filtration.

Let $b \in N$, $\Diamond_b\alpha \in \mathrm{sub}(\varphi)$, $\alpha \notin \mathbf{\Gamma}(b)$ and $\mathsf{M}, (j, w) \models_{\mathbf{\Gamma}} \Diamond_b\alpha$ for some j in J and some w in M_j. Let R_b be the b-th relation in M. Since $\alpha \notin \mathbf{\Gamma}(b)$, we have $\mathsf{M}, (k, u) \models_{\mathbf{\Gamma}} \alpha$ for some k in I and u in M_k such that $(j, w)R_b(k, u)$. Our aim is to choose i in J and v in M_i such that $\mathsf{M}, (i, v) \models_{\mathbf{\Gamma}} \alpha$ and $(j, w)R_b(i, v)$.

If $j = k$, we can put $i = k$ and $v = u$.

Assume that $j \neq k$. In this case $a = b$ and $k > j$. Then $k \in P(\alpha)$, and by (5) there exists $i > j$ such that $i \in J$ and $i \in P(\alpha)$. By the definition of $P(\alpha)$, we have $\mathsf{M}, (i, v) \models_{\mathbf{\Gamma}} \alpha$ for some v in M_i. Since $i > j$, we have $(j, w)R_a(i, v)$. □

Lemma 5.6 *For $a \in N$,* ${}^{a}\sum_{\mathrm{NPO}} \mathcal{F} \equiv_\varphi \bigsqcup_{\leq \sharp\varphi} {}^{a}\sum_{\mathrm{Tr}(\sharp\varphi)} \mathcal{F}$.

Proof. First, we claim that the classes ${}^{a}\sum_{\mathrm{NPO}} \mathcal{F}$ and $\bigsqcup {}^{a}\sum_{\mathrm{NTr}} \mathcal{F}$ are interchangeable. By standard unravelling arguments, if a non-empty Noetherian order J has a least element, then it is a p-morhpic image of a Noetherian tree $\mathsf{T}(\mathsf{J})$. Every frame is a p-morphic image of the disjoint union of its cones. Thus, for a non-empty Noetherian order I we have

$$\bigsqcup_{i\in\mathsf{I}} \mathsf{T}(\mathsf{I}[i]) \twoheadrightarrow \bigsqcup_{i\in\mathsf{I}} \mathsf{I}[i] \twoheadrightarrow \mathsf{I};$$

so I is a p-morhpic image of a disjoint union of Noetherian trees. Now by Propositions 3.4 and 4.13 we obtain

$${}^{a}\sum_{\mathrm{NPO}} \mathcal{F} \preccurlyeq {}^{a}\sum_{\bigsqcup \mathrm{NTr}} \mathcal{F}.$$

Since $\bigsqcup \mathrm{NTr} \subseteq \mathrm{NPO}$, we have

$${}^{a}\sum_{\bigsqcup \mathrm{NTr}} \mathcal{F} \preccurlyeq {}^{a}\sum_{\mathrm{NPO}} \mathcal{F};$$

it follows that these classes are interchangeable. By Proposition 3.6,

$${}^{a}\sum_{\bigsqcup \mathrm{NTr}} \mathcal{F} \equiv \bigsqcup {}^{a}\sum_{\mathrm{NTr}} \mathcal{F},$$

which proves the claim.

Trivially,

$$\bigsqcup_{\leq \sharp\varphi} {}^{a}\sum_{\mathrm{Tr}(\sharp\varphi)} \mathcal{F} \preccurlyeq_\varphi \bigsqcup {}^{a}\sum_{\mathrm{NTr}} \mathcal{F}.$$

To prove the converse, consider a model $\mathsf{M} = \bigsqcup_{i\in I} \mathsf{M}_i$, where I is a set and all M_i are in $\mathrm{Mod}\ {}^{a}\sum_{\mathrm{NTr}} \mathcal{F}$. Let $\Psi = \mathrm{sub}(\varphi; \mathsf{M}, \mathbf{\Gamma})$ for a given $\mathbf{\Gamma}$. For each ψ in Ψ we chose some j in I and x_j in M_j such that $\mathsf{M}_j, x_j \models_{\mathbf{\Gamma}} \psi$. Let J be the set of all these j's (if Ψ is empty, let $J = \{j\}$ for some arbitrary $j \in I$, and x_j be an arbitrary element of M_j). By Lemma 5.5, for each $j \in J$ there exists a model $\mathsf{M}'_j \in \mathrm{Mod}\ {}^{a}\sum_{\mathrm{Tr}(\sharp\varphi)} \mathcal{F}$ which contains x_j and is a selective filtration of M_j with respect to φ and $\mathbf{\Gamma}$; it follows that $\mathsf{M}'_j, x_j \models \psi$ by Proposition 4.15. On the other hand, for each $j \in J$, $\mathrm{sub}(\varphi; \mathsf{M}'_j, \mathbf{\Gamma}) \subseteq \mathrm{sub}(\varphi; \mathsf{M}_j, \mathbf{\Gamma})$ by Proposition 4.15, and $\mathrm{sub}(\varphi; \mathsf{M}_j, \mathbf{\Gamma}) \subseteq \Psi$ by Proposition 4.16. It follows that $\Psi = \bigcup_{j\in J} \mathrm{sub}(\varphi; \mathsf{M}'_j, \mathbf{\Gamma})$. By Proposition 4.16 again, we have $\mathrm{sub}(\varphi; \bigsqcup_{j\in J} \mathsf{M}'_j, \mathbf{\Gamma}) = \bigcup_{j\in J} \mathrm{sub}(\varphi; \mathsf{M}'_j, \mathbf{\Gamma})$. Thus $\bigsqcup_{j\in J} \mathsf{M}'_j$ and M are $(\varphi, \mathbf{\Gamma})$-equivalent. □

Proof of Theorem 5.2. (i) By induction on s. The case $s = 1$ is given by Lemma 5.6. For $s > 1$, we put $\boldsymbol{b} = (a_1, \dots, a_{s-1})$, $\mathcal{G} = {}^{\boldsymbol{b}}\sum_{\mathrm{NPO}} \mathcal{F}$, $\mathcal{H} = {}^{\boldsymbol{b}}\sum_{\mathrm{Tr}(\sharp\varphi)} \mathcal{F}$. We have

$$ {}^{a_0}\sum_{\mathrm{NPO}} \mathcal{G} \equiv_\varphi {}^{a_0}\sum_{\mathrm{NPO}} \bigsqcup_{\leq \sharp\varphi} \mathcal{H} \equiv {}^{a_0}\sum_{\mathrm{NPO}} \mathcal{H} \equiv_\varphi \bigsqcup_{\leq \sharp\varphi} {}^{a_0}\sum_{\mathrm{Tr}(\sharp\varphi)} \mathcal{H}; $$

the first equivalence holds by induction hypothesis and Theorem 4.3; the next step is immediate from Lemma 5.3; finally, we apply Lemma 5.6 again.

(ii) Since $\mathcal{G}$ contains ${}^{\boldsymbol{a}}\sum_{\mathrm{Tr}_f} \mathcal{F}$, we only have to check that if φ is satisfiable in $\mathcal{G}$, then φ is satisfiable in ${}^{\boldsymbol{a}}\sum_{\mathrm{Tr}(\sharp\varphi)} \mathcal{F}$. The class $\mathcal{G}$ is contained in ${}^{\boldsymbol{a}}\sum_{\mathrm{NPO}} \mathcal{F}$. Now (ii) follows from (i) and Proposition 4.2.

(iii) follows from (ii) and Theorem 4.11. □

5.2 Refinements and lexicographic products

The following construction was introduced in [1] by S. Babenyshev and V. Rybakov.

Definition 5.7 Let $\mathsf{F} = (W, R)$ be a preorder, $\mathrm{sk}F = (\overline{W}, <)$ its skeleton. Consider a family $(\mathsf{F}_C)_{C\in\overline{W}}$ of N-frames such that $\mathrm{dom}(\mathsf{F}_C) = C$ for all $C \in \overline{W}$. The *refinement* of F *by* $(\mathsf{F}_C)_{C\in\overline{W}}$ is the $(1+N)$-frame $(W, R, (R_a^{\triangleright})_{a\in N})$, where

$$ R_a^{\triangleright} \subseteq \bigcup_{C\in\overline{W}} C \times C \quad \text{for all } a \in N, \tag{6} $$

$$ (W, (R_a^{\triangleright})_{a\in N}) \restriction C = \mathsf{F}_C \quad \text{for all } C \in \overline{W}. \tag{7} $$

For a class $\mathcal{I}$ of preorders and a class $\mathcal{G}$ of N-frames let $\mathrm{Ref}(\mathcal{I}, \mathcal{F})$ be the class of all refinements of frames from $\mathcal{I}$ by frames in $\mathcal{F}$. For logics $L_1 \supseteq \mathrm{S4}, L_2$, we put $\mathrm{Ref}(L_1, L_2) = \mathrm{Log}\,\mathrm{Ref}(\mathrm{Fr}\, L_1, \mathrm{Fr}\, L_2)$.

Remark 5.8 In [1], refinements are defined in a more general way — for the cases when F is a K-frame ($K \leq \omega$) with transitive relations.

In [1] it was shown that in many cases the refinement operation preserves the finite model property. In particular, if both L_1 and L_2 *admit filtration* (in the sense of Lemmon and Scott [13]), then $\mathrm{Ref}(L_1, L_2)$ is the logic of the class $\mathrm{Ref}(\mathrm{Fr}_f\, L_1, \mathrm{Fr}_f\, L_2)$. Moreover, from the proof it follows that if L_2 admits filtration, then $\mathrm{Ref}(L_1, L_2)$ is the logic of $\mathrm{Ref}(\mathrm{Fr}\, L_1, \mathrm{Fr}_f\, L_2)$ ([1, Lemma 3.3]).

We consider refinements of frames as sums and provide another condition for the latter equality.

Let us make the convention that the universal modality comes first in the language and shifts other modalities: for an N-frame $\mathsf{G} = (W, R_0, R_1, \ldots)$, G^u is the $(1+N)$-frame $(W, W \times W, R_0, R_1, \ldots)$.

Proposition 5.9 *If* $\mathsf{F}^{\triangleright}$ *is the refinement of a preorder* F *by the frames* $(\mathsf{F}_C)_{C \in \mathrm{sk}\mathsf{F}}$, *then*

$$\mathsf{F}^{\triangleright} \cong {}^{0}\!\!\sum_{C \in \mathrm{sk}\mathsf{F}} \mathsf{F}_C^u.$$

Proof. The required isomorphism is defined as $w \mapsto (C, w)$, where $w \in C$. □

For a logic L, let L^u be the logic of the class $(\mathrm{Fr}\, L)^u$.

Theorem 5.10 *Let* L_1 *be a unimodal logic containing* S4. *For every logic* L_2 *such that* L_2^u *has the finite model property, we have*

$$\mathrm{Ref}(L_1, L_2) = \mathrm{Log}\,\mathrm{Ref}(\mathrm{Fr}\, L_1, \mathrm{Fr}_f\, L_2).$$

Proof. Suppose that a formula φ is satisfiable in $\mathrm{Ref}(\mathrm{Fr}\, L_1, \mathrm{Fr}\, L_2)$ and show that it is satisfiable in $\mathrm{Ref}(\mathrm{Fr}\, L_1, \mathrm{Fr}_f\, L_2)$. By Proposition 5.9, φ is satisfiable in a model $\mathsf{M} = {}^{0}\sum_{C \in \mathrm{sk}\mathsf{F}} \mathsf{M}_C^u$, where $\mathsf{F} \models L_1$ and for every $C \in \mathrm{sk}\mathsf{F}$, M_C is a model on a frame validating L_2. The classes $(\mathrm{Fr}\, L_2)^u$ and $(\mathrm{Fr}_f\, L_2)^u$ have the same logic L_2^u, since it has the finite model property. Hence by Proposition 4.10, $\mathrm{Fr}\, L_2 = \mathrm{Fr}_f\, L_2$. Then by Proposition 4.12, $(\mathrm{Fr}\, L_2)^u \equiv (\mathrm{Fr}_f\, L_2)^u$. We consider the condition $\mathbf{\Gamma} = (\varnothing)_{N+1}$ and use Lemma 4.6 to construct models M'_C $(C \in \mathrm{sk}\mathsf{F})$ such that

- the sums M and $\mathsf{M}' = {}^{0}\sum_{C \in \mathrm{sk}\mathsf{F}} {\mathsf{M}'_C}^u$ are $(\varphi, \mathbf{\Gamma})$-equivalent,
- every M'_C is based on a finite frame validating L_2, and
- $\mathsf{M}_C = \mathsf{M}'_C$ whenever C is finite.

Thus φ is satisfiable in M'. For $C \in \mathrm{sk}\mathsf{F}$, we put $C' = \mathrm{dom}(\mathsf{M}'_C)$. The frame of M' is the refinement of the preorder $\mathsf{G} = \sum_{C \in \mathrm{sk}\mathsf{F}} (C', C' \times C')$ by the frames of models M'_C. It follows that $\mathsf{F} \twoheadrightarrow \mathsf{G}$ (indeed, the preorder F is isomorphic to $\sum_{C \in \mathrm{sk}\mathsf{F}} (C, C \times C)$, and for each C in $\mathrm{sk}\mathsf{F}$ we have $|\,\mathrm{dom}(\mathsf{M}'_i)| \leq |C|$). It follows that G validates L_1. Thus, the frame of M' is in $\mathrm{Ref}(\mathrm{Fr}\, L_1, \mathrm{Fr}_f\, L_2)$ as required. □

Another sum-based operation is the *lexicographic product of logics*, introduced in [2] by Ph. Balbiani. Fix $N, K < \omega$.

Definition 5.11 Consider frames $\mathsf{I} = (I, (S_a)_{a \in K})$ and $\mathsf{F} = (W, (R_b)_{b \in N})$. The *l-product* $\mathsf{I} \leftthreetimes \mathsf{F}$ is the $(K+N)$-frame $(I \times W, (S_a^{\leftthreetimes})_{a \in K}, (R_b^{\leftthreetimes})_{b \in N})$, where

$$(i, w) S_a^{\leftthreetimes} (j, u) \quad \text{iff} \quad i S_a j,$$
$$(i, w) R_b^{\leftthreetimes} (j, u) \quad \text{iff} \quad i = j \;\&\; w R_b u.$$

For a class $\mathcal{I}$ of K-frames and a class $\mathcal{G}$ of N-frames, the class $\mathcal{I} \leftthreetimes \mathcal{F}$ is the class of all products $\mathsf{I} \leftthreetimes \mathsf{F}$ such that $\mathsf{I} \in \mathcal{I}$ and $\mathsf{F} \in \mathcal{F}$. For logics L_1, L_2, we put $L_1 \leftthreetimes L_2 = \mathrm{Log}\,(\mathrm{Fr}\, L_1 \leftthreetimes \mathrm{Fr}\, L_2)$.

From the definitions we have

Proposition 5.12 $\mathsf{I} \leftthreetimes \mathsf{F} = \sum_{\mathsf{I}'} \mathsf{F}_i$, *where* $\mathsf{I}' = (I, (S_a)_{a\in K}, (\varnothing)_N)$, *and for each* $i \in I$, $\mathsf{F}_i = (W, (S_{i,a})_{a\in K}, (R_b)_{b\in N})$ *with* $S_{i,a} = W \times W$ *if* $iS_a i$, *and* $S_{i,a} = \varnothing$ *otherwise.*

Let QO and QO_f be the classes of all non-empty preorders and finite non-empty preorders respectively.

Theorem 5.13 $\mathrm{S4} \leftthreetimes \mathrm{S4} = \mathrm{Ref}(\mathrm{S4}, \mathrm{S4}) = \mathrm{Log}\, {}^0\!\sum_{\mathrm{Tr}_f} {\mathrm{QO}_f}^u$.

Proof. First, we show that every product $\mathsf{I} \leftthreetimes \mathsf{F}$ of preorders is in $\mathrm{Ref}(\mathrm{QO}, \mathrm{QO})$. Let $\mathsf{I} \leftthreetimes \mathsf{F} = (W, R_0, R_1)$, $\mathsf{H} = (W, R_0)$. Notice that H is a preorder. Then $\mathsf{I} \leftthreetimes \mathsf{F}$ is the refinement of H by the family $(\mathsf{G}_C)_{C\in \mathrm{sk}\mathsf{H}}$, where G_C is the restriction of (W, R_1) to C. Each G_C is a disjoint union of copies of F, thus G_C is a preorder. Hence $\mathsf{I} \leftthreetimes \mathsf{F} \in \mathrm{Ref}(\mathrm{QO}, \mathrm{QO})$.

By the definition, $\mathrm{Ref}(S4, S4) = \mathrm{Log}\,\mathrm{Ref}(\mathrm{QO}, \mathrm{QO})$. It follows that

$$\mathrm{S4} \leftthreetimes \mathrm{S4} \supseteq \mathrm{Ref}(\mathrm{S4}, \mathrm{S4}).$$

Suppose that φ is satisfiable in $\mathrm{Ref}(\mathrm{QO}, \mathrm{QO})$. In [1], it was shown that $\mathrm{Ref}(S4, S4) = \mathrm{Log}\,\mathrm{Ref}(\mathrm{QO}_f, \mathrm{QO}_f)$. Thus φ is satisfiable in $\mathrm{Ref}(\mathrm{QO}_f, \mathrm{QO}_f)$. Hence by Proposition 5.9, φ is satisfiable in ${}^0\!\sum_{\mathrm{PO}_f} {\mathrm{QO}_f}^u$. By Theorem 5.2, φ is satisfiable in ${}^0\!\sum_{\mathrm{Tr}_f} {\mathrm{QO}_f}^u$. Thus

$$\mathrm{Ref}(\mathrm{S4}, \mathrm{S4}) \supseteq \mathrm{Log}\, {}^0\!\sum_{\mathrm{Tr}_f} {\mathrm{QO}_f}^u.$$

In [2], it was shown that $\mathrm{S4} \leftthreetimes \mathrm{S4}$ is the least logic containing the axioms of S4 for $\Diamond_0, \Diamond_1$ and the formulas

$$\Diamond_0\Diamond_1 p \to \Diamond_0 p, \quad \Diamond_1\Diamond_0 p \to \Diamond_0 p, \quad \Diamond_0 p \to \Box_1\Diamond_0 p.$$

They are valid in ${}^0\!\sum_{\mathrm{Tr}_f} {\mathrm{QO}_f}^u$, thus $\mathrm{Log}\, {}^0\!\sum_{\mathrm{Tr}_f} {\mathrm{QO}_f}^u \supseteq \mathrm{S4} \leftthreetimes \mathrm{S4}$. □

As another example, we consider the logic $\mathrm{GL} \leftthreetimes \mathrm{S4}$, where GL is the Gödel-Löb logic. The next theorem shows that $\mathrm{GL} \leftthreetimes \mathrm{S4}$ is approximable by finite products and sums. Let $\mathcal{C}$ be the class of finite frames of form $(C, \varnothing, C \times C)$.

Theorem 5.14 $\mathrm{GL} \leftthreetimes \mathrm{S4} = \mathrm{Log}\,(\mathrm{Tr}_f \leftthreetimes \mathrm{QO}_f) = \mathrm{Log}\, {}^0\!\sum_{\mathrm{Tr}_f} {}^1\!\sum_{\mathrm{Tr}_f} \mathcal{C}$.

Proof. For a frame $\mathsf{F} = (W, R)$ let $\mathsf{F}^{[\varnothing]}$ be the 2-frame $(W, \varnothing, R)$; for a class $\mathcal{F}$ of 1-frames we put $\mathcal{F}^{[\varnothing]} = \{\mathsf{F}^{[\varnothing]} \mid \mathsf{F} \in \mathcal{F}\}$.

Since NPO $=$ Fr GL, by the definition, GL $\leftthreetimes$ S4 is the logic of the class NPO $\leftthreetimes$ QO. By Proposition 5.12, NPO $\leftthreetimes$ QO $\subseteq {}^{0}\sum_{\mathrm{NPO}} \mathrm{QO}^{[\varnothing]}$. It follows that

$$\mathrm{Log}\ {}^{0}\sum_{\mathrm{NPO}} \mathrm{QO}^{[\varnothing]} \subseteq \mathrm{GL} \leftthreetimes \mathrm{S4}.$$

Consider the class $(\mathrm{QO}^{[\varnothing]})^u = \{(W, W \times W, \varnothing, R) \mid (W, R) \in \mathrm{QO}\}$. It is a standard fact the the logic of this class has the finite model property (e.g., it follows from [10, Theorem 5.9]). By Theorem 5.2 we obtain

$$\mathrm{Log}\ {}^{0}\sum_{\mathrm{NPO}} \mathrm{QO}^{[\varnothing]} = \mathrm{Log}\ {}^{0}\sum_{\mathrm{Tr}_f} {\mathrm{QO}_f}^{[\varnothing]}.$$

We shall now prove that

$$\mathrm{Log}\ {}^{0}\sum_{\mathrm{Tr}_f} {\mathrm{QO}_f}^{[\varnothing]} = \mathrm{Log}\,(\mathrm{Tr}_f \leftthreetimes \mathrm{QO}_f).$$

If φ is satisfiable in $\mathrm{Tr}_f \leftthreetimes \mathrm{QO}_f$, then φ is satisfiable in ${}^{0}\sum_{\mathrm{Tr}_f} {\mathrm{QO}_f}^{[\varnothing]}$ by Proposition 5.12. Conversely, suppose that φ is satisfiable in a sum ${}^{0}\sum_{\mathsf{I}} \mathsf{F}_i^{[\varnothing]}$, where I is a finite tree and F_i are finite preorders. Consider the Cartesian product G of the preorders $(\mathsf{F}_i)_{i \in I}$. It is easy to see that $\mathsf{G} \twoheadrightarrow \mathsf{F}_i$ and so $\mathsf{G}^{[\varnothing]} \twoheadrightarrow \mathsf{F}_i^{[\varnothing]}$ for each i in I. Now it follows from Propositions 3.4 and 5.12 that $\mathsf{I} \leftthreetimes \mathsf{G} \twoheadrightarrow {}^{0}\sum_{\mathsf{I}} \mathsf{F}_i^{[\varnothing]}$. Since G is a finite preorder, φ is satisfiable in $\mathrm{Tr}_f \leftthreetimes \mathrm{QO}_f$.

Altogether we have proved

$$\mathrm{Log}\,(\mathrm{Tr}_f \leftthreetimes \mathrm{QO}_f) \;=\; \mathrm{Log}\ {}^{0}\sum_{\mathrm{Tr}_f} {\mathrm{QO}_f}^{[\varnothing]} \;=\; \mathrm{Log}\ {}^{0}\sum_{\mathrm{NPO}} \mathrm{QO}^{[\varnothing]} \;\subseteq\; \mathrm{GL} \leftthreetimes \mathrm{S4}.$$

It follows that these four logics coincide: indeed, GL $\leftthreetimes$ S4 is contained in the logic of the class $\mathrm{Tr}_f \leftthreetimes \mathrm{QO}_f$, since this class is contained in NPO $\leftthreetimes$ QO.

Every finite preorder is (up to isomorphism) the sum of finite frames of form $(C, C \times C)$ over a finite partial order, and vice versa. Thus, the classes ${\mathrm{QO}_f}^{[\varnothing]}$ and ${}^{1}\sum_{\mathrm{PO}_f} \mathcal{C}$ coincide up to isomorphisms. It follows that GL $\leftthreetimes$ S4 is the logic of the class ${}^{0}\sum_{\mathrm{Tr}_f} {}^{1}\sum_{\mathrm{PO}_f} \mathcal{C}$. Finally, we have

$$\mathrm{Log}\ {}^{0}\sum_{\mathrm{Tr}_f} {}^{1}\sum_{\mathrm{PO}_f} \mathcal{C} = \mathrm{Log}\ {}^{0}\sum_{\mathrm{Tr}_f} {}^{1}\sum_{\mathrm{Tr}_f} \mathcal{C}$$

by Theorem 5.2. □

6 Further results

For classes $\mathcal{I}$ and $\mathcal{F}$, let $\mathcal{I}_f = \mathrm{Fr}_f \operatorname{Log}\mathcal{I}$, and $\mathcal{F}_f = \mathrm{Fr}_f \operatorname{Log}\mathcal{F}$. When do we have $\operatorname{Log}\sum_{\mathcal{I}}\mathcal{F} = \operatorname{Log}\sum_{\mathcal{I}_f}\mathcal{F}_f$? Finite summands can be obtained by Theorem 4.11. Theorem 5.2 allows us to obtain finite indices in the case of sums over Noetherian orders. The proof of Theorem 5.2 is based on selective filtration. Another way is to use filtration in the sense of Lemmon and Scott [13]: this approach was successfully used in [1] to obtain the finite model property for refinements in numerous cases. The methods developed in [1] in a combination with Theorem 4.11 suggest the following conjecture: in the case of finitely many modalities, if $\operatorname{Log}\mathcal{F}^u$ has the finite model property, and $\operatorname{Log}\mathcal{I}$ admits filtration, then the classes $\sum_{\mathcal{I}}\mathcal{F}$ and $\sum_{\mathcal{I}_f}\mathcal{F}_f$ are interchangeable.

Theorem 5.2 can be used to obtain complexity results for logics of sums over Noetherian orders, in particular – over finite orders. Let $\operatorname{Sat}\mathcal{F}$ denote the satisfiability problem for $\mathcal{F}$.

Theorem 6.1 *Let $\mathcal{F}$ be a non-empty class of N-frames, $a \in N$, $\mathcal{G} = {}^{a}\!\sum_{\mathcal{I}}\mathcal{F}$, where $\mathrm{Tr}_f \subseteq \mathcal{I} \subseteq \mathrm{NPO}$. If $\operatorname{Sat}\mathcal{F}^u$ is in PSPACE, then $\operatorname{Sat}\mathcal{G}^u$ is PSPACE-complete.*

This result generalizes [14, Theorem 35]; the proof will be given in a forthcoming paper. In particular, in view of Theorems 5.13 and 5.14, it follows that the logics $\mathrm{S4} \rightthreetimes \mathrm{S4}$ and $\mathrm{GL} \rightthreetimes \mathrm{S4}$ are PSPACE-complete.

Acknowledgements

This work is supported by the Russian Science Foundation under grant 16-11-10252 and performed at Steklov Mathematical Institute of Russian Academy of Sciences.

I would like to thank Sergey Babenyshev, Philippe Balbiani, Lev Beklemishev, and Vladimir Rybakov for their comments on an earlier version of the paper. I would also like to thank the anonymous reviewers for their suggestions.

Appendix

Proof of Lemma 3.5. Let $\mathsf{I} = (I, (S_a)_{a\in N})$. For $i \in I$, let $\mathsf{J}_i = (J_i, (S_{i,a})_{a\in N})$, and for $j \in J_i$, $\mathsf{F}_{ij} = (W_{ij}, (R_{ij,a})_{a\in N})$. Let W be the set of all triples (i, j, w) such that $i \in I$, $j \in J_i$, and $w \in W_{ij}$. By the definition, the domain of $\sum_{i\in\mathsf{I}}\sum_{j\in\mathsf{J}_i}\mathsf{F}_{ij}$ is the set of all the pairs $(i, (j, w))$ such that $(i, j, w) \in W$. Likewise, the domain of $\sum_{(i,j)\in\sum_{k\in\mathsf{I}}\mathsf{J}_k}\mathsf{F}_{ij}$ consists of all $((i, j), w)$ such that $(i, j, w) \in W$.

For $a \in N$, let R'_a, R''_a be respectively the a-th relations in $\sum_{i\in\mathsf{I}}\sum_{j\in\mathsf{J}_i}\mathsf{F}_{ij}$ and $\sum_{(i,j)\in\sum_{k\in\mathsf{I}}\mathsf{J}_k}\mathsf{F}_{ij}$. We claim that for all $(i, j, w), (i', j', w') \in W$, $a \in N$,

$$(i, (j, w))R'_a(i', (j', w')) \quad \text{iff} \quad ((i, j), w)R''_a((i', j'), w'). \tag{A.1}$$

By the definition, $(i, (j, w))R'_a(i', (j', w'))$ iff

$$i \neq i' \;\&\; iS_a i' \quad \text{or} \quad i = i' \;\&\; (j \neq j' \;\&\; jS_{i,a}j' \text{ or } j = j' \;\&\; wR_{ij,a}w'). \tag{A.2}$$

Likewise, $((i,j),w)R''_a((i',j'),w')$ iff

$$(i,j) \neq (i',j') \;\&\; (i \neq i' \;\&\; iS_a i' \text{ or } i = i' \;\&\; jS_{i,a}j') \quad \text{or}$$
$$(i,j) = (i',j') \;\&\; wR_{ij,a}w'. \quad \text{(A.3)}$$

It is straightforward that (A.2) and (A.3) are equivalent. □

Proof of Proposition 4.15. By induction on the length of ψ. Consider the case $\psi = \Diamond_a \chi \in \text{sub}(\varphi)$.

If $\chi \in \mathbf{\Gamma}(a)$, then, by Definition 4.1, $\mathsf{M}', w \models_{\mathbf{\Gamma}} \Diamond_a \chi$ and $\mathsf{M}, w \models_{\mathbf{\Gamma}} \Diamond_a \chi$.

Assume that $\chi \notin \mathbf{\Gamma}(a)$.

If $\mathsf{M}', w \models_{\mathbf{\Gamma}} \Diamond_a \chi$, then for some $v \in R'_a(w)$ we have $\mathsf{M}', v \models_{\mathbf{\Gamma}} \chi$, which is equivalent to $\mathsf{M}, v \models_{\mathbf{\Gamma}} \chi$ by induction hypothesis; since $R'_a \subseteq R_a$, we have $\mathsf{M}, w \models_{\mathbf{\Gamma}} \Diamond_a \chi$.

Conversely, assume that $\mathsf{M}, w \models_{\mathbf{\Gamma}} \Diamond_a \chi$. By Definition 4.14, $\mathsf{M}, v \models_{\mathbf{\Gamma}} \chi$ for some $v \in R'_a(w)$; by induction hypothesis, $\mathsf{M}', v \models_{\mathbf{\Gamma}} \chi$, and so $\mathsf{M}', w \models_{\mathbf{\Gamma}} \Diamond_a \chi$.□

Proof of Proposition 5.4. For $V \subseteq I$, let V' be all maximal elements of V, $\Diamond V = \{j \mid \exists i > j \; i \in V\}$. Since $(I, >)$ is well-founded, we have

$$\Diamond V = \Diamond V'. \quad \text{(A.4)}$$

Put $K = \{i_0\} \cup \bigcup\{V' \mid V \in \mathcal{V}\}$, $\mathsf{K} = (K, <)$. The height of K is not greater than $|\mathcal{V}| + 1$: indeed, if $i \in U'$, $j \in V'$, and $i < j$, then $U \neq V$.

Let h be the height of the cone $\mathsf{K}[i_0]$ (the *depth* of i_0 in K). We construct the required $J \subseteq K$ by induction on h.

If $h = 1$, then $\mathsf{K}[i_0] = (\{i_0\}, \varnothing)$, and we put $\mathsf{J} = \mathsf{K}[i_0]$; then (5) is trivial, the branching of J is 0.

Assume that $h > 1$. Consider the family

$$\mathcal{U} = \{U \in \mathcal{V} \mid i_0 \in \Diamond U\}.$$

Let $U \in \mathcal{U}$. By (A.4), we have $i_0 < j$ for some $j \in U' \subseteq K$; the height of K is finite, thus for some immediate successor i_U of i_0 in K we have

$$i_U \in U' \cup \Diamond U'. \quad \text{(A.5)}$$

In K, the depth of i_U is less than the depth of i_0. By induction hypothesis, i_U is the root of a tree $(J(i_U), <)$ whose branching is not greater than $|\mathcal{V}|$ and

$$\forall V \in \mathcal{V} \; \forall j \in J(i_U) \; (j \in \Diamond V \;\Rightarrow\; j \in \Diamond V \cap J(i_U)). \quad \text{(A.6)}$$

We put $J = \{i_0\} \cup \bigcup\{J(i_U) \mid U \in \mathcal{U}\}$. The branching of i_0 in $(J, <)$ is not greater than the cardinality of $\mathcal{U} \subseteq \mathcal{V}$, thus $br(J,<) \leq |\mathcal{V}|$. Since $J \subseteq K$, $ht(J,<) \leq ht(\mathsf{K}) \leq |\mathcal{V}| + 1$. By (A.5) and (A.6) we have (5). □

References

[1] Babenyshev, S. and V. Rybakov, *Logics of Kripke meta-models*, Logic Journal of the IGPL **18** (2010), pp. 823–836.

[2] Balbiani, P., *Axiomatization and completeness of lexicographic products of modal logics*, in: S. Ghilardi and R. Sebastiani, editors, *Frontiers of Combining Systems*, Lecture Notes in Computer Science **5749**, Springer, 2009 pp. 165–180.
[3] Balbiani, P., *Axiomatizing the temporal logic defined over the class of all lexicographic products of dense linear orders without endpoints*, in: *2010 17th International Symposium on Temporal Representation and Reasoning*, 2010, pp. 19–26.
[4] Balbiani, P. and D. Fernández-Duque, *Axiomatizing the lexicographic products of modal logics with linear temporal logic*, in: L. Beklemishev, S. Demri and A. Máté, editors, *Advances in Modal Logic*, **11** (2016), pp. 78–96.
[5] Balbiani, P. and S. Mikulás, *Decidability and complexity via mosaics of the temporal logic of the lexicographic products of unbounded dense linear orders*, in: P. Fontaine, C. Ringeissen and R. A. Schmidt, editors, *Frontiers of Combining Systems* (2013), pp. 151–164.
[6] Beklemishev, L. D., *Kripke semantics for provability logic GLP*, Annals of Pure and Applied Logic **161** (2010), pp. 756–774.
[7] Blackburn, P., M. de Rijke and Y. Venema, "Modal Logic," Cambridge Tracts in Theoretical Computer Science **53**, Cambridge University Press, 2002.
[8] Chagrov, A. and M. Zakharyaschev, "Modal Logic," Oxford Logic Guides **35**, Oxford University Press, 1997.
[9] Gabbay, D., A. Kurucz, F. Wolter and M. Zakharyaschev, "Many-Dimensional Modal Logics: Theory and Applications," Studies in Logic and the Foundations of Mathematics, North Holland Publishing Company, 2003.
[10] Goranko, V. and S. Passy, *Using universal modality: Gains and questions*, Journal of Logic and Computation **2** (1992), pp. 5–30.
[11] Japaridze, G. K., "The Modal Logical Means of Investigation of Provability," Ph.D. thesis (in Russian), Moscow (1986).
[12] Kurucz, A., *Combining modal logics*, in: *Handbook of Modal Logic*, Elsevier, NY, USA, 2006 pp. 869–924.
[13] Lemmon, E. and D. Scott, "An Introduction to Modal Logic: The Lemmon Notes," American Philosophical Quarterly Monograph Series, B. Blackwell, 1977.
[14] Shapirovsky, I., *Pspace-decidability of Japaridze's polymodal logic*, in: *Advances in Modal Logic*, **7** (2008), pp. 289–304.
[15] Shelah, S., *The monadic theory of order*, Annals of Mathematics **102** (1975), pp. 379–419.

On Kripke Completeness of Some Modal Predicate Logics with the Density Axiom

Valentin Shehtman [1] [2]

Steklov Mathematical Institute, Russian Academy of Sciences
Institute for Information Transmission Problems, Russian Academy of Sciences
National Research University Higher School of Economics, Moscow, Russia
Moscow State University

Abstract

We prove completeness for some normal modal predicate logics in the standard Kripke semantics with expanding domains. We consider quantified versions of propositional logics with the axiom of density plus some others (transitivity, confluence).
The method of proof modifies the technique developed for other cases (without density) by S. Ghilardi, G. Corsi and D. Skvorstov; but now we arrange the whole construction in a game-theoretic style.

Keywords: modal predicate logic, Kripke semantics, Kripke completeness, canonical model, model construction games, density axiom.

1 Modal logics and Kripke frames

Let us recall some basic definitions and notation; most of them are the same as in the book [3].

Atomic formulas are constructed from predicate letters P_k^n (countably many for each arity $n \geq 0$) and a countable set of individual variables Var, without constants and function letters. Also we do not use equality. *Modal (predicate) formulas* are obtained from atomic formulas by applying classical propositional connectives $(\supset, \bot)$, the quantifier $\forall$ and the modal operator $\Box$. All other connectives (and $\exists$) are derived.

In *modal propositional formulas* only the proposition letters (P_k^0) are used as atoms.

A *modal propositional logic* is a set of modal propositional formulas containing classical propositional tautologies, the axiom of **K** ($\Box(p \supset q) \supset (\Box p \supset \Box q)$, where p, q are proposition letters) and closed under the basic inference rules: Modus Ponens, $\Box$-introduction, and (propositional) Substitution.

[1] This work is supported by the Russian Science Foundation under Project No. 16-11-10252 and was carried out at Steklov Mathematical Institute, Russian Academy of Sciences.

[2] shehtman@netscape.net

As usual **K** denotes the minimal modal propositional logic, $\Lambda + A$ is the smallest logic containing a logic Λ and a formula A, and $\mathbf{K4} := \mathbf{K} + \Box \mathbf{p} \supset \Box\Box \mathbf{p}$.

Recall that Kripke semantics for propositional modal logics is given by *(propositional) Kripke frames* of the form (W, R), where $W \neq \varnothing$, $R \subseteq W \times W$. The set of all propositional formulas valid in a frame F (the *modal logic of* F) is denoted by $\mathbf{ML}(\mathbf{F})$. The class of all frames validating a propositional logic Λ (*Λ-frames*) is denoted by $\mathbf{V}(\Lambda)$.

A *p-morphism* from (W, R) onto (W', R') is a surjective map $f : W \longrightarrow W'$ such that for any $x \in W$ $f[R(x)] = R'(f(x))$. In this case $\mathbf{ML}(W, R) \subseteq \mathbf{ML}(W', R')$ (the *p-morphism lemma*).

A *cone in* $F = (W, R)$ *with root* u (denoted by $F{\uparrow}u$) is the restriction of F to the smallest subset V containing u and such that $R(V) \subseteq V$; obviously, $V = R(u) \cup \{u\}$ if R is transitive. If $F = F{\uparrow}u$, F itself is called *rooted* (or a *cone*). So a transitive frame (W, R) is rooted with root u if $W = R(u)$, or equivalently, if it has a first cluster.

A *modal predicate logic* is a set of modal predicate formulas containing classical predicate axioms, the axiom of **K** and closed under Modus Ponens, Generalization, $\Box$-introduction, and (predicate) Substitution.

$\mathbf{Q}\Lambda$ denotes the smallest predicate logic containing the propositional logic Λ (*the predicate version* of Λ).

For predicate formulas we use the standard Kripke semantics. Recall that a *predicate Kripke frame* over a propositional Kripke frame $F = (W, R)$ is a pair $\mathbf{F} = (\mathbf{F}, \mathbf{D})$, in which $D = (D_u)_{u \in W}$, $D_u \neq \varnothing$ and such that $D_u \subseteq D_v$ whenever uRv.

For a class of propositional frames $\mathcal{C}$, the class of all predicate frames (F, D) with $F \in \mathcal{C}$ is denoted by $\mathcal{KC}$.

A *valuation* ξ in $\mathbf{F}$ is a function sending every predicate letter P_k^n to a family of n-ary relations on the domains:

$$\xi(P_k^n) = (\xi_u(P_k^n))_{u \in W},$$

where $\xi_u(P_k^n) \subseteq D_u^n$ for $n = 0$ and $\xi_u(P_k^0) \in \{0, 1\}$.

The pair $M = (\mathbf{F}, \xi)$ is a *Kripke model* over $\mathbf{F}$. The definition of truth in a Kripke model is standard. So at every point $u \in W$ we evaluate *modal D_u-sentences*, i.e., modal formulas, in which all parameters (free variables) are replaced with elements of D_u; $M, u \vDash A$ means that A is true at u in M. Then

$$\begin{aligned}
&M, u \vDash P_k^n(a_1, \ldots, a_n) \text{ iff } (a_1, \ldots, a_n) \in \xi_u(P_k^n),\\
&M, u \vDash P_k^0 \text{ iff } \xi_u(P_k^0) = 1,\\
&M, u \vDash A \supset B \text{ iff } (M, u \nvDash A \text{ or } M, u \vDash B),\\
&M, u \nvDash \bot,\\
&M, u \vDash \forall x A(x) \text{ iff } \forall a \in D_u\ M, u \vDash A(a),\\
&M, u \vDash \Box A \text{ iff } \forall v \in R(u)\ M, v \vDash A.
\end{aligned}$$

A modal formula $A(x_1, \ldots, x_n)$ is called *true in* M (in symbols, $M \vDash A(x_1, \ldots, x_n)$) if $M, u \vDash A(\mathbf{a})$ for every $u \in W$ and $\mathbf{a} \in D_u^n$.

A modal formula A is *valid* in a frame $\mathbf{F}$ (in symbols, $\mathbf{F} \models \mathbf{A}$) if it is true in every Kripke model over $\mathbf{F}$. $\mathbf{ML}(\mathbf{F}) := \{\mathbf{A} \mid \mathbf{F} \models \mathbf{A}\}$ is the *modal logic of* $\mathbf{F}$.

The *modal logic of a class of frames* $\mathcal{C}$ (or the logic *determined by* $\mathcal{C}$) is $\mathbf{ML}(\mathcal{C}) := \bigcap\{\mathbf{ML}(\mathbf{F}) \mid \mathbf{F} \in \mathcal{C}\}$. Logics of this form are called *Kripke complete.*

A modal predicate logic L is *strongly Kripke complete* if every L-consistent theory (a set of sentences) is satisfied at a point of some Kripke model over a frame validating L.

Similar definitions are given for modal propositional logics. Also recall that a modal propositional logic *has the finite model property (fmp)* if it is determined by some class of finite frames.

From the definitions it follows that for a predicate frame (F, D) and a propositional formula A,

$$(F, D) \models A \text{ iff } F \models A.$$

So for a propositional logic Λ and a predicate frame $\mathbf{F}$

$$\mathbf{F} \models \mathbf{Q}\Lambda \text{ iff } \mathbf{F} \in \mathcal{K}\mathbf{V}(\Lambda).$$

2 Completeness and incompleteness in modal predicate logic

In modal predicate logic there are too many examples of incompleteness, and proofs of completeness can be rather nontrivial. For instance, for a propositional modal logic $\Lambda \supseteq \mathbf{S4}$, $\mathbf{Q}\Lambda$ is complete only if $\mathbf{S5} \subseteq \Lambda$ or $\Lambda \subseteq \mathbf{S4.3}$ (cf. [5]). Still some logics $\mathbf{Q}\Lambda$ are complete, in particular, for the well-known modal logics $\Lambda = \mathbf{K}$, $\mathbf{K4}$, $\mathbf{S4}$, $\mathbf{S5}$, $\mathbf{S4.2}$, $\mathbf{S4.3}$ (cf. [3], theorems 6.1.29, 6.6.7, 6.7.12). These results were obtained by different authors — S. Kripke, D. Gabbay, S. Ghilardi, G. Corsi and others.

In this paper we are mainly interested in the logic $\mathbf{K4}Ad := \mathbf{K4}+Ad$, where

$$Ad := \Box\Box p \supset \Box p$$

is the axiom of density; $(W, R) \models Ad$ iff R is dense, i.e., $R \subseteq R \circ R$.

An extension of $\mathbf{K4}Ad$ is $\mathbf{D4.3}Ad$ obtained by adding the axiom of non-branching (.3) and seriality ($\Diamond\top$). It is well-known that $\mathbf{D4.3}Ad = \mathbf{ML}(\mathbb{Q}, <)$, where $\mathbb{Q}$ denotes the set of rationals. Moreover, completeness transfers to the predicate version [1]:

$$\mathbf{Q}(\mathbf{D4.3Ad}) = \mathbf{ML}(\mathcal{K}(\mathbb{Q}, <)).$$

3 Trees and unravelling

A *tree* is a frame (W, R) with a root u_0 such that $R^{-1}(u_0) = \varnothing$ and $R^{-1}(x)$ is a singleton for any $x \neq u_0$. A *transitive tree* is a transitive closure of a tree, so it is a strictly ordered set $(W, <)$ with the least element such that every subset $\{y \mid y < x\}$ is linearly ordered and finite.

Lemma 3.1 *Every rooted transitive frame is a p-morphic image of a transitive tree.*

A well-known proof is by unravelling: for a rooted frame $F = (W, R)$ with root u we construct a tree $F^\sharp = (W^\sharp, <)$, where $W^\sharp$ is the set of all finite paths from u to points of W (i.e., finite sequences $x_0 x_1 \dots x_n$ such that $x_0 = u$ and $x_i R x_{i+1}$ for any $i < n$), and $\alpha < \beta$ iff β prolongs α. The required p-morphism sends every path to its last point.

Hence we have

Proposition 3.2 **K4** *is determined by the class of all (at most) countable trees.*

This follows from lemma 3.1, the p-morphism lemma and the fmp of **K4**; note that unravelling of a finite frame is finite or countable.

Definition 3.3 Let $(W, <)$ be a tree, and consider a frame $(W, <')$, in which $<'$ is obtained from $<$ by making some points reflexive. Then $(W, <')$ is called a *semireflexive tree.*

One can easily check that a semireflexive tree $(W, <')$ validates Ad iff its irreflexive points can have only reflexive immediate successors.[3] Such a semireflexive tree is called *dense.*

Proposition 3.4 **K4***Ad is determined by the class of all (at most) countable dense semireflexive trees.*

Proof. A standard filtration argument shows that **K4**Ad has the fmp, so it is determined by finite rooted **K4**Ad-frames (cf. [6]). Finite **K4**-frames consist of clusters, some of which can be degenerate (i.e., irreflexive singletons), while in finite **K4**Ad-frames successors of degenerate clusters are non-degenerate.

Now let us unravel a finite **K4**Ad-frame $F = (W, R)$ with root u more carefully than in lemma 3.1. Call a path $x_0 \dots x_n$ *long* if

$$\forall i < n \, \forall y \in F (x_i R y R x_{i+1} \Rightarrow y R x_i \vee x_{i+1} R y).$$

Consider the set W_1 of all long paths from u to points in F and take the restriction $F_1 := F^\sharp | W_1$. This frame is a tree, and the map f sending a path to its last point is still a p-morphism $F_1 \longrightarrow F$. This is because every two R-related points can be connected by a long path.

Now we extend the relation in F_1 by making reflexive every point a such that $f(a)$ is reflexive. We obtain a semireflexive tree F_2 and again f is a p-morphism $F_2 \longrightarrow F$.

F_2 is a dense semireflexive tree. In fact, if in F_2 we have an irreflexive a and its successor b, then a is a long path in F ending at an irreflexive point $f(a)$, and the cluster of $f(b)$ is a successor of $f(a)$. So $f(b)$ is reflexive, and thus b is reflexive in F_2. ■

To obtain a class of irreflexive transitive frames determining **K4**Ad we can use Segerberg's bulldozing method (cf. [6]). Viz., given a dense semireflexive tree F_2, we can replace each its reflexive point with a strict dense linear order

[3] Henceforth by a 'successor' we mean an 'immediate successor'.

without the last element (e.g., the non-negative rationals $\mathbf{Q}_+$). Then we obtain $\mathbf{K4}Ad$-frame F_3, and there is a p-morphism from F_3 sending every irreflexive point from F_2 to itself and every copy of $\mathbf{Q}_+$ to the corresponding reflexive point in F_2. We call such a frame F_3 a *sprouting tree*. So we have

Proposition 3.5 *$\mathbf{K4}Ad$ is determined by the class of sprouting trees.*

Remark 3.6 It is not clear if predicate frames over sprouting trees determine the predicate logic **QK4Ad**. The completeness proof proposed below yields more complicated frames.

4 Completeness of QK4Ad

To prove completeness for **QK4Ad** we use a method originating from G. Corsi's paper [1] and further developed by D. Skvortsov [8]; also cf. [3], sec. 6.4.

The main idea is to extract an appropriate submodel from a canonical model of a given logic L and to make a sort of unravelling which leads to a frame validating L. More exactly, this frame is obtained as a direct limit of a sequence of finite trees. This sequence can be constructed by induction, or equivalently, by playing a game.

First we recall some definitions from [3], sections 6.1, 6.3, with little changes.

We fix a denumerable set of extra constants S^*. A subset $S' \subseteq S^*$ is called *small* if the complement $(S^* - S')$ is infinite.

Definition 4.1 For a modal predicate logic L, an *L-place* is a maximal L-consistent theory (i.e, a set of sentences) Γ in the basic language with extra constants from S^* with the *Henkin property*: for any formula $\varphi(x)$ with at most one parameter x there exists a constant c such that $(\exists x\varphi(x) \supset \varphi(c)) \in \Gamma$. An L-place is *small* if the set of its constants is small.

It is well-known that every L-consistent theory with a small set of constants can be extended to a small L-place ([3], Lemma 6.1.9).

Definition 4.2 The *canonical model* VM_L is (VP_L, R_L, D_L, ξ_L), where

- VP_L is the set of all small L-places,
- $\Gamma R_L \Delta$ iff $\Box^-\Gamma \subseteq \Delta$, where $\Box^-\Gamma := \{A \mid \Box A \in \Gamma\}$,
- $(D_L)_\Gamma$ (also denoted by D_Γ) is the set of constants occurring in Γ,
- $(\xi_L)_\Gamma(P_k^m) := \{\mathbf{c} \in (\mathbf{D}_\Gamma)^\mathbf{m} \mid \mathbf{P}_\mathbf{k}^\mathbf{m}(\mathbf{c}) \in \Gamma\}$
for $m > 0$, and
$(\xi_L)_\Gamma(P_k^0) := 1$ iff $P_k^0 \in \Gamma$.

Note that $\Box^-\Gamma \subseteq \Delta$ implies $D_\Gamma \subseteq D_\Delta$; this holds, since $\Box(P_1^1(c) \supset P_1^1(c)) \in \Gamma$ for any $c \in D_\Gamma$, so $(P_1^1(c) \supset P_1^1(c)) \in \Delta$.

Then for any D_Γ-sentence A

$$VM_L, \Gamma \models A \text{ iff } A \in \Gamma$$

(the *Canonical model theorem*).

Note that for arbitrary L-places an analogue of this theorem does not hold, but we still need them for further considerations. So put $VM_L^+ := (VP_L^+, R_L, D_L, \xi_L)$, where VP_L^+ is the set of all L-places, and R_L, D_L, ξ_L are the same as above.[4] This VM_L^+ is actually a submodel of a canonical model for some larger set of extra constants.

Definition 4.3 Let L be a predicate logic, $F = (W, R)$ a propositional frame. An *L-network over F* is a monotonic map from F to (VP_L^+, R_L), i.e. a map $h : W \longrightarrow VP_L^+$ such that for any $u, v \in W$

$$uRv \Rightarrow h(u)R_L h(v).$$

The frame F is denoted by $dom(h)$ and called the *domain* of h. An L-network h is *small* if every $h(u)$ is small and *transitive* if $dom(h)$ is transitive.

With every L-network h we associate a predicate Kripke frame $\mathbf{F(h)} := (\mathbf{dom(h)}, \mathbf{D})$, where $D_u = (D_L)_{h(u)}$ for $u \in W$, and a Kripke model $M(h) := (\mathbf{F(h)}, \xi(\mathbf{h}))$, where

$$\xi(h)_u(P_k^m) := \{\mathbf{c} \in \mathbf{D_u^m} \mid \mathbf{P_k^m(c)} \in \mathbf{h(u)}\}$$

for $m > 0$ and

$$\xi(h)_u(P_k^0) := 1 \text{ iff } P_k^0 \in h(u).$$

We define the partial order on networks.

$$h \leq h' := dom(h) \text{ is a subframe of } dom(h') \text{ and } \forall u \in dom(h)\ h(u) \subseteq h'(u).$$

Definition 4.4 A *defect* in a network h over a frame (W, R) is a pair (u, A) such that $u \in W$ and $\Diamond A \in h(u)$. A defect (u, A) is *eliminated* in h if there exists $v \in R(u)$ such that $A \in h(v)$.

Henceforth in this section we assume that L contains **QK4**, so L-frames are transitive.

We will call a transitive L-network h *finite* if it is small and $dom(h)$ is a finite transitive tree.

Lemma 4.5 *(On elimination of defects) Let h be a finite L-network with a defect (u, A). Then there is a finite L-network $h' \geq h$ eliminating this defect.*

Proof. If h eliminates (u, A), take $h' = h$. Otherwise extend $dom(h)$ by adding a new successor v of u (such that v has no successors). Since $\Diamond A \in h(u)$, by the properties of the canonical model VM_L, there exists a small L-place Γ such that $A \in \Gamma$ and $h(u)R_L\Gamma$. So we can put $h'(v) := \Gamma$. ■

If Γ, Δ are L-places, $\Gamma \restriction \Delta$ denotes the restriction of Γ to the language of Δ.

Lemma 4.6 *(Skvortsov's extension lemma)*

---

[4] More exactly, R_L is extended to $VP_L^+ \times VP_L^+$, etc.

(1) *Let* Γ, Δ *be L-places,* $\Gamma_0 = \Gamma \upharpoonright \Delta$ *and suppose that* $\Box^-\Gamma_0 \subseteq \Delta$. *Then there exists an L-place* $\Delta' \supseteq \Delta$ *such that* $\Gamma R_L \Delta'$. Δ' *can be chosen small if* Γ, Δ *are small.*

(2) *Let h be a finite L-network over a transitive tree F with root v, and let* Γ *be an L-place,* $\Gamma_0 = \Gamma \upharpoonright h(v)$, *and suppose that* $\Box^-\Gamma_0 \subseteq h(v)$. *Let* F' *be the transitive tree obtained by adding a root u below F. Then there exists a finite L-network* $h' \geq h$ *over* F' *such that* $\Gamma = h'(u)$.

Proof. This is a reformulation of Lemma 6.4.28 from [3], and the proof follows the same lines.

(1) The assumptions imply that the theory $\Box^-\Gamma \cup \Delta$ is consistent (see the details in [3]); so it extends to an L-place Δ'.

(2) We can argue by induction on the cardinality of F. By (1) there exists an L-place $\Delta' \supseteq h(v)$ such that $\Gamma R_L \Delta'$. If v has no successors (i.e., F is a singleton), we are done: take h' defined on the chain $\{u, v\}$ such that $h'(u) = \Gamma,\ h'(v) = \Delta'$.

Suppose v has successors $v_1, \ldots v_n$, $F_i = F{\uparrow}v_i$. h_i is the restriction of h to F_i. Since we can rename the constants from $D_{\Delta'} - D_{h(v)}$, we may assume that they do not occur in any $h(v_i)$; thus $h(v) = \Delta' \upharpoonright h(v_i)$, and $\Box^-h(v) \subseteq h(v_i)$. Now by IH there exists $h'_i \geq h_i$ defined on the tree F_i with the added bottom element v such that $h'_i(v) = \Delta'$. Then we define the following network h' over F':

$$h'(u) = \Gamma,\ h'(v) = \Delta',\ h'|F_i = h'_i.$$

■

Now we assume that L contains **QK4Ad**.

Lemma 4.7 *(On inserts) Let h be a finite L-network, and let v be a successor of u in dom(h). Then there exists a finite L-network* $h' > h$ *such that v is not a successor of u in dom*(h').

Proof. Suppose $h(u) = \Gamma,\ h(v) = \Delta$, and let $\Delta_0 = \Delta \upharpoonright \Gamma$. It follows that the set $\Gamma' := \Box^-\Gamma \cup \{\Diamond A \mid A \in \Delta_0\}$ is L-consistent. In fact, otherwise there exist $B \in \Box^-\Gamma$ and $A \in \Delta_0$ such that $\{B, \Diamond A\}$ is inconsistent (since the sets $\Box^-\Gamma$, Δ_0 are closed under conjunction and $\Diamond(A_1 \wedge A_2)$ implies $\Diamond A_1 \wedge \Diamond A_2$). So $\vdash_L B \supset \neg\Diamond A$, or equivalently, $\vdash_L B \supset \Box\neg A$. Hence by the monotonicity of $\Box$, $\vdash_L \Box B \supset \Box\Box\neg A$; thus $\vdash_L \Box B \supset \Box\neg A$ by Ad. Since $\Box B \in \Gamma$ and A is in the language of Γ, this implies $\Box\neg A \in \Gamma$. Since $\Gamma R_L \Delta$, it follows that $\neg A \in \Delta$, which is a contradiction.

Then Γ' can be extended to an L-place Θ (with new unused constants). Let $\Theta_0 = \Theta \upharpoonright \Delta$ $(= \Theta \upharpoonright \Delta_0$, since new constants of Θ do not occur in $\Delta)$.

It follows that $\Box^-\Theta_0 \subseteq \Delta_0$. In fact, $\neg A \in \Delta_0$ implies $\Diamond\neg A \in \Gamma' \subseteq \Theta$, so $\Box A \notin \Theta_0,\ A \notin \Box^-\Theta_0$.

Consider the tree F' obtained from $F = dom(h)$ by adding a new point z between u and v. By Lemma 4.6 there exists a finite network h^1 over $F'{\uparrow}z$ such that $h^1(z) = \Theta$ and $h^1 \geq h$ on $F{\uparrow}v$. Now we can define h' on F', which coincides with h^1 on $F'{\uparrow}z$ and coincides with h at all other points. This is a network, since $\Box^-\Gamma \subseteq \Theta$, i.e., $h'(u)R_Lh'(z)$. ∎

Definition 4.8 Let Γ_0 be a small L-place. The *selective game* $SG_L(\Gamma_0)$ is played by two players, $\forall$ (the first) and $\exists$ (the second). A position after the n-th turn is a finite network h_n over a transitive tree $F_n = (W_n, R_n)$. We also assume [5] that $W_n \subseteq \omega$.

At the initial position F_0 is an irreflexive singleton 0 and $h_0(0) = \Gamma_0$.

For the $(n+1)$-th move the player $\forall$ has two options.

1. Selecting a *defect*, i.e., a pair (u, A) such that $u \in W_n$ and $\Diamond A \in h_n(u)$.
2. A query for an *insert*, i.e., a pair (u, v) such that uR_nv and there are no points between u and v.

The player $\exists$ should respond with a network $h_{n+1} \geq h_n$ such that

1. If the move of $\forall$ was a defect (u, A), then there exists v such that $uR_{n+1}v$ and $A \in h_{n+1}(v)$.
2. If the move of $\forall$ was a query for an insert (u, v), then then there exists w such that $uR_{n+1}wR_{n+1}v$.

The player $\exists$ wins if the play continues infinitely or $\forall$ cannot make his move.

Note that $\forall$ cannot make the $(n+1)$th move in the only case when $n = 0$ and h_0 has no defects. This happens if Γ_0 is an endpoint in VM_L, i.e., $R_L(\Gamma_0) = \varnothing$.

Every infinite play of the game generates a sequence of networks $h_0 \leq h_1 \leq \ldots$ Then we define the resulting network h_ω, with $dom(h_\omega) = F_\omega := (W_\omega, R_\omega)$, $W_\omega := \bigcup_n W_n$, $R_\omega := \bigcup_n R_n$, $h_\omega(u) := \bigcup_{n\geq m} h_n(u)$ for $u \in W_m$. One can easily check that this is really a network (not necessarily finite or small).

Lemma 4.9 *$\exists$ has a winning strategy in $SG_L(\Gamma_0)$.*

Proof. If $\forall$ cannot make the first move, there is nothing to prove. If the $(n+1)$-th move of $\forall$ is a defect, $\exists$ can eliminate it by her next move according to Lemma 4.5. If the move of $\forall$ is a query for an insert, $\exists$ can respond according to Lemma 4.7. ∎

Lemma 4.10 *If Γ_0 is not an endpoint in VM_L, then there exists a play of $SG_L(\Gamma_0)$ generating a sequence of networks such that $F_\omega \vDash$ **K4**Ad and for any u, for any $D_{h_\omega(u)}$-sentence A*

$$M(h_\omega), u \vDash A \text{ iff } A \in h_\omega(u).$$

Proof. A *dense tree* is a rooted strictly ordered set $(W, \prec)$, in which every subset $\{u \mid u \prec w\}$ is a dense chain. Let us construct an infinite play such that F_ω is a dense tree.

At the initial position $F_0 = (0, \varnothing)$ and $h_0(0) = \Gamma_0$.

Let us choose the further strategy for $\forall$ as follows. Fix an enumeration of the countable set $\omega \times \omega$, and an enumeration of $\omega \times \Phi$, where Φ is the set of all modal sentences with constants from S^*. An odd move $(n+1)$ of $\forall$ chooses

[5] This technical detail is needed for the further proofs.

the first new pair (u, A), which is a defect in h_n. An even move $(n+1)$ of $\forall$ chooses the first new pair $(u, v) \in \omega \times \omega$, which is a query for an insert in h_n.

By lemma 4.9 there is a winning strategy for $\exists$. For the resulting network we have

$$M(h_\omega), u \models A \text{ iff } A \in h_\omega(u).$$

This is checked by induction. The atomic case holds by the definition of $\xi(h)$; the cases of propositional connectives and quantifiers hold by the properties of L-places.

Let us consider the case $A = \Box B$. Suppose $M(h_\omega), u \not\models A$; then $M(h_\omega), v \not\models B$ for some $v \in R_\omega(u)$. Since A is in the language of $h_\omega(u)$ and h_ω is a network, we have $h_\omega(u) R_L h_\omega(v)$, so A (and B) is also in the language of $h_\omega(v)$. By IH it follows that $B \notin h_\omega(v)$; hence $A = \Box B \notin h_\omega(u)$ by the definition of R_L.

The other way round, suppose $A \notin h_\omega(u)$; then $\Diamond\neg B \in h_\omega(u)$, so $\Diamond\neg B \in h_n(u)$ (i.e., $(u, \Diamond\neg B)$ is a defect in h_n) for some finite n. Choose the minimal such n; so $(u, \Diamond\neg B)$ is a defect in h_m for all $m > n$. Since the defects subsequently appear as odd moves of $\forall$, there exists m such that $(u, \Diamond\neg B)$ is his $(m+1)$-th move. By the response of $\exists$, we have $\neg B \in h_{m+1}(v)$ for some $v \in R_{m+1}(u)$. Hence $\neg B \in h_\omega(v)$, $v \in R_\omega(u)$. By IH, we have $M(h_\omega), v \not\models B$. Thus $M(h_\omega), u \not\models A$.

To check the density for F_ω, we can use even moves. In fact, if $u R_\omega v$, there exists n such that $u R_n v$. If v is a successor of u in R_n, the pair (u, v) must show up as a later even move of $\forall$. By the response of $\exists$ we obtain w such that $u R_\omega w R_\omega v$. ■

Definition 4.11 A modal predicate logic L is strongly Kripke complete if every L-consistent set of sentences is satisfiable at some point of a Kripke model over a frame validating L.

Theorem 4.12 QK4Ad *is strongly Kripke complete.*

Proof. Every L-consistent theory Γ without constants can be extended to a small L-place Γ_0. If Γ_0 is an endpoint in VM_L, then for any A in its language

$$VM_L, \Gamma_0 \models A \text{ iff } A \in \Gamma_0$$

by the canonical model theorem. Since Γ_0 is an endpoint, the truth at this point reduces to the truth in a model over an irreflexive singleton.

In all other cases we can apply lemma 4.10. So there exists a model $M(h_\omega)$ such that $M(h_\omega), u_0 \models \Gamma_0$ and $F_\omega \models \mathbf{K4}Ad$. Hence $\mathbf{F}(\mathbf{h}_\omega) \models \mathbf{L}$. ■

Theorem 4.13 *If Π is a set of closed (i.e., constructed only from $\bot$, $\Box$ and $\supset$) propositional formulas, then* **QK4Ad** $+ \Pi$ *is strongly Kripke complete.*

Proof. By the same argument as in the previous theorem. In this case $\Pi \subset \Gamma$ for all L-places Γ (where $L := \mathbf{QK4Ad} + \Pi$), so $M(h_\omega) \models \Pi$. Hence $F_\omega \models \Pi$, and thus $F(h_\omega) \models \Pi$. ■

5 Logics with n-density

Let us first notice that for the logic **QKAd** $:=$ **QK**$+Ad$ one can use the same method as in the previous section. Now we only need finite networks over non-transitive frames. If (W,R) is a tree, R^+ is the transitive closure of R and $R \subseteq R_1 \subseteq R^+$, then (W,R_1) is called an *almost transitive tree.* Lemmas 4.5, 4.6, 4.7 are transferred to almost transitive trees and proved by the same arguments.

The analogue of lemma 4.10 also holds for **QKAd**. The same proof constructs a frame F_ω validating **KAd** (but this frame should not be called a "dense tree").

Thus we obtain

Theorem 5.1 *If* Π *is a set of closed propositional formulas, then* **QKAd** $+\Pi$ *is strongly Kripke complete.*

Now recall the n-density axiom Ad_n generalizing Ad:

$$Ad_n := \bigwedge_{i=1}^{n} \Diamond p_i \supset \Diamond(\bigwedge_{i=1}^{n} \Diamond p_i).$$

This is a Sahlqvist formula, so for the logic $\mathbf{KAd_n} := \mathbf{K}+Ad_n$ we have

Proposition 5.2 $\mathbf{KAd_n}$ *is canonical and determined by the following first-order condition on frames:*

$$\forall x, y_1, \ldots, y_n \,(\bigwedge_{i=1}^{n} xRy_i \supset \exists z\,(xRz \wedge \bigwedge_{i=1}^{n} zRy_i)).$$

Lemma 5.3 *(On inserts) For L containing* $\mathbf{QKAd_n}$ *let h be a finite L-network over a frame (W,R) and suppose $uRv_1, \ldots, uRv_n$. Then there exists a finite L-network $h' > h$ and z such that $uR'z$, $zR'v_1, \ldots, zR'v_n$, where R' is the relation in dom(h').*

Proof. The same argument as in 4.7, with slight changes.

Let $h(u) = \Gamma$, $h(v_i) = \Delta_i$, $\Delta_{i0} = \Delta_i \restriction \Gamma$. Then the set

$$\Gamma' := \Box^- \Gamma \cup \bigcup_{i=1}^{n} \{\Diamond A \mid A \in \Delta_{i0}\}$$

is L-consistent.

For, otherwise there exist $B \in \Box^-\Gamma$ and $A_i \in \Delta_{i0}$ such that $\{B, \Diamond A_1, \ldots, \Diamond A_n\}$ is L-inconsistent, i.e., $\vdash_L B \supset \neg \bigwedge_i \Diamond A_i$. Hence

$$\vdash_L \Box B \supset \Box \neg \bigwedge_i \Diamond A_i;$$

thus

$$\vdash_L \Box B \supset \neg \Diamond \bigwedge_i \Diamond A_i,$$

and

$$\vdash_L \Box B \supset \neg \bigwedge_i \Diamond A_i,$$

by Ad_n. However, $\Box B \in \Gamma$, so $\neg \bigwedge_i \Diamond A_i \in \Gamma$. On the other hand, every A_i is in the language of Γ, $A_i \in \Delta_i$, and $\Gamma R_L \Delta_i$, which implies $\Diamond A_i \in \Gamma$. Hence $\bigwedge_i \Diamond A_i \in \Gamma$, which is a contradiction.

Then Γ' can be extended to an L-place Θ such that $D_\Theta - D_{\Gamma'}$ contains only new constants. So we have $\Gamma' = \Theta \upharpoonright \Delta_i$, $\Box^- \Gamma' \subseteq \Delta_i$.

Consider the tree F' obtained from $F = dom(h)$ by adding a new unique successor z of u below all the v_i. Let $F_i := F{\uparrow}v_i$, $h_i := h|F_i$. Since $\Box^-(\Theta \upharpoonright \Delta_i) \subseteq \Delta_i$, by Lemma 4.6 there exists a finite network $h'_i \geq h_i$ defined on F_i with the added root z such that $\Theta = h'_i(z)$. Then we can define the finite network h' over F' such that $h'(z) = \Theta$, $h'|F_i = h_i$ and $h'(x) = h(x)$ for all $x \notin R(u)$. This is a network, since $\Box^-\Gamma \subseteq \Theta$, i.e., $h'(u) = h(u)R_L h'(z)$. ■

Now let $L = \mathbf{QKAd_n} + \Pi$, where Π is a set of closed propositional formulas.

Definition 5.4 The selective game $SG_L(\Gamma_0)$ is defined as in definition 4.8, but now a query for an insert at the $(m+1)$-th move is a tuple $(u, v_1, \ldots, v_n)$ such that $uR_m v_1, \ldots, uR_m v_n$ and there is no z with $uR_m zR_m v_i$ for all i.

In a response for this move there must be w such that

$$uR_{m+1}w,\ wR_{m+1}v_1, \ldots, wR_{m+1}v_n.$$

Now we have analogues of lemmas 4.9, 4.10.

Lemma 5.5 *$\exists$ has a winning strategy in $SG_L(\Gamma_0)$.*

Proof. By applying lemmas 4.5, 5.3. ■

Lemma 5.6 *If Γ_0 is not an endpoint in VM_L, then there exists a play of $SG_L(\Gamma_0)$ generating a sequence of networks such that $\mathbf{F}(\mathbf{h}_\omega) \models \mathbf{L}$ and for any u, for any $D_{h_\omega(u)}$-sentence A*

$$M(h_\omega), u \models A \text{ iff } A \in h_\omega(u).$$

Proof. The same as for lemma 4.10, with the following change.

An odd move $(m+1)$ of $\forall$ is the first new tuple from ω^{n+1} which is an insert query in h_m. These moves guarantee the n-density for F_ω. ■

Theorem 5.7 *If Π is a set of closed propositional formulas, the logic $\mathbf{QKAd_n} + \Pi$ is strongly Kripke complete.*

Proof. Similar to theorem 4.13. If an L-place Γ_0 is not an endpoint in the canonical model, we apply lemma 5.6 to obtain a model $M(h_\omega)$ satisfying Γ_0, with $\mathbf{F}(\mathbf{h}_\omega) \models \mathbf{L}$. ■

A similar result holds for the transitive case; note that $\mathbf{K4} + Ad_2 \vdash Ad_n$ for any n.

Theorem 5.8 *If Π is a set of closed propositional formulas, the logic $\mathbf{QK4}$+$Ad_2 + \Pi$ is strongly Kripke complete.*

6 Logics with confluence and density

Now let us consider logics containing the confluence ("Church–Rosser") axiom

$$A2 := \Diamond\Box p \supset \Box\Diamond p.$$

The semantical characterization of $A2$ is well-known:

Proposition 6.1 *The logic* $\mathbf{K2} := \mathbf{K}+A2$ *is canonical and determined by the following condition on frames:*

$$\forall x, y, z\,(xRy \wedge xRz \supset \exists u\,(yRu \wedge zRu)).$$

For completeness proofs in this section we also need transitivity. So we will consider extensions of $\mathbf{QK4.2} := \mathbf{QK4}+A2$.

Lemma 6.2 $\mathbf{K2} \vdash \Box\Diamond\top$.

Proof. On the one hand, it is clear that $\mathbf{K} \vdash \Diamond\top \supset \Diamond\Box\top$, so $\mathbf{K2} \vdash \Diamond\top \supset \Box\Diamond\top$.

On the other hand, $\mathbf{K} \vdash \Box\bot \supset \Box\Diamond\top$; hence the statement follows. ■

In this section we deal with finite networks over transitive trees and infinite networks over other frames (sums of trees).

Definition 6.3 A finite network h over a transitive tree (W, R) is called *rich* if its satisfies the following condition.

Let $u_1, \ldots, u_n$ be R-incomparable, and let v be their maximal common predecessor. Then the sets $D_{h(u_i)} - D_{h(v)}$ are disjoint.

Lemma 6.4 *Let* $\Delta, \Gamma_1, \Gamma_2$ *be* L*-places for* $L \supseteq \mathbf{QK4.2}$ *such that* $\Delta R_L \Gamma_1$, $\Delta R_L \Gamma_2$ *and* $D_{\Gamma_1} \cap D_{\Gamma_2} = D_\Delta$. *Then the set* $\Box^-\Gamma_1 \cup \Box^-\Gamma_2$ *is* L*-consistent.*

Proof. Suppose not. Since $\Box$ distributes over conjunction, then there exist $\Box B_1 \in \Gamma_1$, $B_2 \in \Gamma_2$ such that $\vdash_L \neg(B_1 \wedge B_2)$. Every B_i can be presented as $A_i(\mathbf{a_i}, \mathbf{b})$ for a list $\mathbf{a_i}$ of constants from $D_{\Gamma_i} - D_\Delta$, and a list $\mathbf{b}$ of constants from D_Δ. By assumption, $\mathbf{a_1}, \mathbf{a_2}$ are disjoint.

By predicate logic, it follows that

$$\vdash_L \forall\mathbf{x_1}\forall\mathbf{x_2}\neg(\mathbf{A_1}(\mathbf{x_1}, \mathbf{b}) \wedge \mathbf{A_2}(\mathbf{x_2}, \mathbf{b}))$$

for disjoint lists of variables $\mathbf{x_1}, \mathbf{x_2}$; hence

$$\vdash_L \neg(\exists\mathbf{x_1}\mathbf{A_1}(\mathbf{x_i}, \mathbf{b}) \wedge \exists\mathbf{x_2}\mathbf{A_2}(\mathbf{x_2}, \mathbf{b})).$$

CLAIM The rule $A/\Box\Diamond A$ is admissible in L.

In fact, $\vdash_L A$ implies $\vdash_L \top \supset A$, and thus $\vdash_L \Box\Diamond\top \supset \Box\Diamond A$, and finally $\vdash_L \Box\Diamond A$ by lemma 6.2.

Now by the Claim we have

$$\vdash_L \Box\Diamond\neg(\exists\mathbf{x_1}\mathbf{A_1}(\mathbf{x_i}, \mathbf{b}) \wedge \exists\mathbf{x_2}\mathbf{A_2}(\mathbf{x_2}, \mathbf{b})),$$

and so

$$\vdash_L \neg\Diamond\Box(\exists \mathbf{x_1 A_1}(\mathbf{x_i}, \mathbf{b}) \wedge \exists \mathbf{x_2 A_2}(\mathbf{x_2}, \mathbf{b})),$$

$$\neg\Diamond\Box(\exists \mathbf{x_1 A_1}(\mathbf{x_i}, \mathbf{b}) \wedge \exists \mathbf{x_2 A_2}(\mathbf{x_2}, \mathbf{b})) \in \Delta. \qquad (*)$$

On the other hand, by confluence and transitivity we have

$$\mathbf{K4.2} \vdash \Diamond\Box p_1 \wedge \Diamond\Box p_2 \supset \Diamond\Box(p_1 \wedge p_2),$$

thus

$$\vdash_L \Diamond\Box\exists \mathbf{x_1 A_1}(\mathbf{x_1}, \mathbf{b}) \wedge \Diamond\Box\exists \mathbf{x_2 A_2}(\mathbf{x_2}, \mathbf{b}) \supset \Diamond\Box(\exists \mathbf{x_1 A_1}(\mathbf{x_1}, \mathbf{b}) \wedge \exists \mathbf{x_2 A_2}(\mathbf{x_2}, \mathbf{b})). \qquad (**)$$

Since $\Delta R_L \Gamma_i$ and $\Box A_i(\mathbf{a_i}, \mathbf{b}) \in \Gamma_\mathbf{i}$, it also follows that $\Box\exists \mathbf{x_i A_i}(\mathbf{x_i}, \mathbf{b}) \in \Gamma_\mathbf{i}$, $\Diamond\Box\exists \mathbf{x_i A_i}(\mathbf{x_i}) \in \Delta$, so from (**) we obtain

$$\Diamond\Box(\exists \mathbf{x_1 A_1}(\mathbf{x_1}, \mathbf{b}) \wedge \exists \mathbf{x_2 A_2}(\mathbf{x_2}, \mathbf{b})) \in \Delta$$

contradicting (*). ■

Lemma 6.5 *(Cf. [3], Lemma 6.6.5) Let h be a rich finite small L-network over a nontrivial tree (W, R) for $L \supseteq$ **QK4.2**. Then there exists a small Θ such that $h(w)R_L\Theta$ for any $w \in W$.*

Proof. By induction on the cardinality of W.

If (W, R) is a two-element chain: $W = \{u, v\}$, uRv, then we can apply lemma 6.4 to $\Delta = h(u)$, $\Gamma_1 = \Gamma_2 = h(v)$ and construct $\Theta \supseteq \Box^- h(v)$.

The same argument goes through for any finite chain with the first element u and the last element v.

So for the induction step we may assume that (W, R) has maximal points $u_1, \ldots, u_n$, $n > 1$. By IH there exists Θ such that $h(u_1)R_L\Theta, \ldots, h(u_{n-1})R_L\Theta$, and by renaming constants we may also assume that all new constants in D_Θ do not occur in $h(u_n)$. Let $v = inf(u_1, \ldots, u_n)$. Since h is rich, it follows that $D_\Theta \cap D_{h(u_n)} = D_{h(v)}$. So lemma 6.4 is applicable, which gives us a small Θ' such that $\Theta R_L \Theta'$, $h(u_n)R_L\Theta'$. It remains to note the R_L is transitive. ■

Now let us consider the logics $L :=$ **QK4.2**+Ad or $L :=$ **QK4.2**+Ad+$\Diamond\top$.[6]

To define an appropriate game we need an increasing sequence $(S_n)_{n\geq 1}$ of subsets of the set of constants S^* such that S_1 and all the sets $(S_{n+1} - S_n)$ are infinite.

Definition 6.6 Let Γ_0 be an S_1-small L-place. The selective game $SG_L(\Gamma_0)$ is defined as in 4.8, with the some changes.

1. The length of the game is ω^2.
2. A position after the turn $\alpha = \omega \cdot m + n$ is a rich network h_α over a finite tree $F_\alpha = (W_\alpha, R_\alpha)$ such that $W_\alpha \subseteq \omega$ and all L-places $h_\alpha(u)$ are S_{m+1}-small.
3. Every tree $F_{\omega \cdot m}$ is an irreflexive singleton 0, $h_0(0) = \Gamma_0$.

[6] The method also works for the logic **QS4.2** (its completeness was first proved in [2]).

4. If $\alpha = \omega \cdot m + n$, the player $\forall$ has the same two options for the move $\alpha + 1$ as in definition 4.8: selecting a defect or a query for an insert in h_α. A response of $\exists$ is also described in 4.8; it yields a network $h_{\alpha+1} \geq h_\alpha$.

5. A limit move $\alpha = \omega \cdot (m+1)$ of the player $\forall$ is just waiting for the response of $\exists$. For the response $\exists$ should construct the limit network h^*_α over $F^*_\alpha := (W^*_\alpha, R^*_\alpha)$, where

$$W^*_\alpha := \bigcup_n W_{\omega\cdot m+n},\ R^*_\alpha := \bigcup_n R_{\omega\cdot m+n},\ h^*_\alpha(u) := \bigcup_{n\geq k} h_{\omega\cdot m+n}(u) \text{ for } u \in W_{\omega\cdot m+k};$$

then she should choose an S_{m+2}-small L-place Γ_α such that $h^*_\alpha(u) R_L \Gamma_\alpha$ for any $u \in W^*_\alpha$. The resulting position would be the network $h_\alpha : 0 \mapsto \Gamma_\alpha$.

6. The player $\exists$ wins if the play is of length ω^2 or if $\forall$ cannot make one of his moves.

In this game a position, at which $\forall$ cannot make the next move, may occur only at the stage 0 if $\Box\bot \in \Gamma_0$. In fact, otherwise at every non-limit stage we have $\Diamond\top \in h_\alpha(0)$ and also $\Diamond\top \in h_\alpha(u)$ for any $u \neq 0$ (since $\Box\Diamond\top \in \Gamma_0$ by lemma 6.2 and $h_\alpha(0) R_L h_\alpha(u)$); so $\forall$ can select a defect.

An ω^2-play generates a sequence of networks $h^*_\omega \leq h^*_{\omega\cdot 2} \leq \ldots$.

The resulting network h^+ is then defined as the sum $\sum_{m\in\omega} h^*_{\omega\cdot(m+1)}$. So $dom(h^+) = F^+ = (W^+, R^+) := \sum_{m\in\omega} F^*_{\omega\cdot(m+1)}$ (the ordered sum), i.e.,

$$W^+ := \bigcup_{m\geq 1} W^*_{\omega\cdot m} \times \{m\},\ (x,m) R^+ (y,l) \text{ iff } (m < l \vee m = l \ \&\ x R^*_{\omega\cdot m} y),$$

and

$$h^+(x,m) := h^*_{\omega\cdot m}(x) \text{ for } x \in W^*_{\omega\cdot m}.$$

One can easily see that h^+ is really a network. In fact, it coincides with $h^*_{\omega\cdot m}$ on each component. To show that $(x,m) R^+ (x,l)$ implies $h^+(x,m) R_L h^+(y,l)$ for $m < l$, it is sufficient to consider the case $l = m+1$. In this case we have

$$h^+(x,m) = h^*_{\omega\cdot m}(x) R_L h^*_{\omega\cdot(m+1)}(0) = h^+(0, m+1) R_L h^+(y, m+1).$$

Lemma 6.7 *$\exists$ has a winning strategy in $SG_L(\Gamma_0)$.*

Proof. For non-limit moves use lemmas 4.5, 4.7 with an extra observation that the networks can be always kept rich by choosing new constants.

For a limit move $\alpha = \omega \cdot (m+1)$ a response of $\exists$ also exists. In fact, F^*_α has the root 0, so $h^*_\alpha(0) R_L h^*_\alpha(u)$ for any $u \in W^*_\alpha$, $u \neq 0$. All these L-places $h^*_\alpha(u)$ are S_{m+1}-small.

We claim that the theory

$$\Sigma := \bigcup \{\Box^- h(u) \mid u \in W^*_\alpha,\ u \neq 0\}$$

is L-consistent.

In fact, otherwise the set

$$S := \Box^{-}h(u_1) \cup \ldots \cup \Box^{-}h(u_n)$$

is L-inconsistent for some finite n. Then there exist $\alpha = \omega \cdot m + k$ such that $u_1, \ldots, u_n \in domh^*_\alpha$. The network h^*_α is finite and rich, so by lemma 6.5 there exists Θ such that $h(u_i)R_L\Theta$ for every i. So Θ contains S, which is a contradiction.

Note that the set of constants of Σ is S_{m+2}-small, so this theory can be extended to an S_{m+2}-small L-place Γ_α. It follows that $h^*_\alpha(u)R_L\Gamma_\alpha$ for any $u \in W^*_\alpha$, $u \neq 0$, and $h^*_\alpha(0)R_L\Gamma_\alpha$ by transitivity. ∎

Lemma 6.8 *If Γ_0 is not an endpoint in VM_L, then there exists a play of $SG_L(\Gamma_0)$ of length ω^2 generating a network h^+ such that $\mathbf{F}(\mathbf{h}^+) \models \mathbf{L}$ and for any u, for any $D_{h^+(u)}$-sentence A*

$$M(h^+), u \models A \textit{ iff } A \in h^+(u).$$

Proof. Similar to lemma 4.10. Such a play is provided by the winning strategy of $\exists$ used against the following strategy of $\forall$.

At the initial position $F_0 = (0, \varnothing)$ and $h_0(0) = \Gamma_0$.

The further strategy for $\forall$ will be the same as in lemma 4.10 for every ω-sequence of moves $\omega \cdot m + 1, \omega \cdot m + 2, \ldots$.

So we fix an enumeration of the countable set $\omega \times \omega$, and an enumeration of $\omega \times \Phi$, where Φ is the set of all S_{m+1}-sentences.

An odd move $(\omega \cdot m + n + 1)$ of $\forall$ chooses the first new (for this sequence of moves) pair (u, A), which is a defect in $h_{\omega \cdot m+n}$. An even move $(\omega \cdot m + n + 1)$ of $\forall$ chooses the first new (again for this sequence) pair $(u, v) \in \omega \times \omega$, which is a query for an insert in $h_{\omega \cdot m+n}$.

Let $\exists$ apply her winning strategy (lemma 6.7). We claim that the resulting network h^+ satisfies the statement of lemma 6.8.

In fact, the equivalence

$$u \models A \text{ iff } A \in h^+(u)$$

is again checked by induction. In the case $A = \Box B$ 'if' follows easily, since h^+ is a network.

For 'only if' suppose $A \notin h^+(u)$, $u = (x, m)$, $x \in W^*_{\omega \cdot m}$. Then the defect $(u, \Diamond\neg B)$ appears as some move $\omega \cdot m + n$ of $\forall$, and by the strategy of $\exists$ we obtain $v \in R^+(u)$ such that $\neg B \in h^+(v)$. Then $v \not\models B$ by IH, so $u \not\models A$.

The density of F^+ in every its component $F^*_{\omega \cdot m}$ is provided by even moves. For the points $u = (x, m)$, $v = (y, m')$ in different components $(m < m')$ we have uR^+v, and there is always an intermediate point — any point accessible from u in the same m-th component.

F^+ is confluent, since the points $(x, m), (y, m')$ with $m \leq m'$ both see the root $(0, m' + 1)$ of a later component. ∎

Theorem 6.9 *The logics* **QK4.2**$+Ad$, **QK4.2**$+Ad+\Diamond\top$ *are strongly Kripke complete.*

Proof. As above, either an L-place Γ_0 is an endpoint in VM_L or by lemma 6.8 we can construct $M(h^+)$ satisfying Γ_0. ∎

Theorem 6.10 *The logics* **QK4.2**, **QK4.2** $+ \Diamond\top$ *are strongly Kripke complete.*

Proof. By applying the same method as in the previous theorem. The game $SG_L(\Gamma_0)$ is the same as in definition 6.6, but now at non-limit moves $\forall$ can only select defects. An analogue of lemma 6.7 still holds, so we can construct an appropriate network h^+. ∎

Theorem 6.11 *The logics* **QK4.2** $+ Ad_2$, **QK4.2** $+ \Diamond\top + Ad_2$ *are strongly Kripke complete.*

Proof. We can use the same method. Now definition 6.6 changes for even moves — they are queries for 2-inserts (cf. definition 5.4 for $n = 2$).

Then the resulting frame $\mathbf{F}(\mathbf{h}^+)$ is 2-dense: the 2-density of each component is guaranteed by even moves, and points in a later component $F^*_{\omega\cdot m}$ have a common predecessor, the root $(0, m)$. ∎

7 Final remarks

Axiomatizing modal predicate logics of specific frames is usually a nontrivial problem. In particular, we can be interested in predicate logics of relativistic time. The only clear case is the following.

Theorem 7.1 *Let F be the Minkowski lower halfspace with the causal future relation: aRb iff a signal can be sent from a to b. Them* $\mathbf{ML}(\mathcal{K}F) = \mathbf{QS4}$.

Proof. Every cone in F can be mapped p-morphically onto the infinite reflexive binary tree IT_2 [6]. It is also well-known that $\mathbf{ML}(\mathcal{K}IT_2) = \mathbf{QS4}$ (cf. [3], section 6.4). Hence the claim follows. ∎

However, the method does not work for the logic of chronological future. Its propositional version was axiomatized in [7], this is $\Lambda = \mathbf{K4.2} + Ad_2 + \Diamond\top$. It is hardly probable that $\mathbf{Q}\Lambda$ fits for the predicate case, and we do not know how to play a game constructing a chronological order on the Minkowski plane.

Also note that our method is inapplicable to the case of constant domains. Moreover, the corresponding logic $L' := \mathbf{QK4Ad}{+}Ba$, where

$$Ba := \forall x \Box P(x) \supset \Box \forall x P(x)$$

is the Barcan axiom, may be Kripke incomplete. In fact, incompleteness is known for the logic $\mathbf{QKAd} + \Diamond\top{+}Ba$ (cf. [4]), and it probably extends to L' (although the proof from [4] does not fit for L', because of transitivity).

I would like to thank an anonymous referee for very useful comments on the first version of this paper

References

[1] Corsi, G., *Quantified modal logics of positive rational numbers and some related systems*, Notre Dame Journal of Formal Logic (1993), pp. 263–283.

[2] Corsi, G. and S. Ghilardi, *Directed frames*, Archive for Math. Logic (1999), pp. 263–283.

[3] Gabbay, D., V. Shehtman and D. Skvortsov, "Quantification in Nonclassical Logic Vol.1," Studies in Logic and the Foundation of Mathematics **153**, Elsevier, 2009.

[4] Gasquet, O., *A new incompleteness result in Kripke semantics*, Fundamenta Informatica (1995), pp. 407–415.

[5] Ghilardi, S., *Incompleteness results in Kripke semantics*, J. Symbolic Logic (1991), pp. 517–538.

[6] Goldblatt, R., "Logics of Time and Computation," 2nd edition, CSLI Lecture Notes **7**, Stanford University, 1992.

[7] Shapirovsky, I. and V. Shehtman, *Chronological future modality in Minkowski spacetime*, Advances in Modal Logic **4** (2003), pp. 437–459.

[8] Skvortsov, D., *On the predicate logic of linear Kripke frames and some of its extensions*, Studia Logica (2005), pp. 261–282.

A Remark on the Superintuitionistic Predicate Logic of Kripke Frames of Finite Height with Constant Domains: A Simpler Kripke Complete Logic That Is Not Strongly Complete

Dmitrij Skvortsov [1]

Federal Research Center for Computer Science and Control (FRCCSC)
Molodogvardejskaja 22, korp.3, kv.29, Moscow, RUSSIA, 121351
email: skvortsovd@yandex.ru

Abstract

We show that the superintuitionistic predicate logic characterized by all Kripke frames of finite height with constant domains is not strongly Kripke complete (as well as some its extensions). This gives new examples of Kripke complete logics that are not strongly complete, cf. Problem 1 in [5]; the previous examples of such logics found by Takano were Π^1_1-hard, while ours are Π_2-arithmetical.

Keywords: Superintuitionistic predicate logics, Kripke semantics, Kripke completeness, strong Kripke completeness.

H. Ono (the talk at L.E.J. Brouwer Centenary Symposium held in 1981, cf. [5, Problem 1 (P34)]) asked if every Kripke complete intermediate predicate logic is strongly Kripke complete. M. Takano found a counterexample, mentioned by Ono in [5]: namely, *the logic of any Kripke frame with a denumerable constant domain, whose set of worlds is an infinite ordinal, is not strongly Kripke complete.* Note that all these logics are Π^1_1-hard (cf. [9,11,12]).

Here we consider the intermediate predicate logics $\mathbf{LP}_\infty$ and $\mathbf{L}^c\mathbf{P}_\infty$ characterized by all Kripke frames of finite height (with expanding and with constant domains respectively). We show that the logic $\mathbf{L}^c\mathbf{P}_\infty$ (with constant domains) is not strongly Kripke complete; moreover, a slightly weakened version of strong completeness fails as well (for this logic and for many its extensions). The similar question for the logic $\mathbf{LP}_\infty$ (with expanding domains) remains open. Note that the logics $\mathbf{LP}_\infty$ and $\mathbf{L}^c\mathbf{P}_\infty$ are not recursively axiomatizable (see [11]); on the other hand, they are obviously Π_2-arithmetical. Namely, $\mathbf{L}^c\mathbf{P}_\infty = \bigcap\limits_n \mathbf{L}^c\mathbf{P}_n$, where $\mathbf{L}^c\mathbf{P}_n$ is a finitely axiomatizable logic of Kripke frames of height n (with constant domains); and similarly for the case with expanding domains.

[1] The research presented in this paper was supported by the RFBR project 16-01-00615.

We do not know *if every recursively axiomatizable (or at least every finitely axiomatizable) Kripke complete predicate logic is strongly Kripke complete.*

The main result of the paper was announced in [13].

Remark 0. By the way, H. Ono formulated a similar question for intermediate propositional logics [5, Problem 1′ (P35)]. T. Shimura [8] found a family of propositional counterexamples; namely, he obtained the following result:

(Sh): *Intuitionistic logic is the only strongly complete intermediate propositional logic weaker than* $\mathbf{GJ}_2$ (Gabbay – de Jongh's logic of finite binary trees [1]).

This implies the following consequence for the predicate case:[2]

(Sh)′ : *Every intermediate predicate logic, the propositional fragment of which is included in* $\mathbf{GJ}_2$ *and is non-intuitionistic* (i.e., it is not equal to intuitionistic logic), *is not strongly Kripke complete.*

In particular, we can see that for any $n \geq 2$, the predicate logics, characterized by all Kripke frames (with constant and with expanding domains) over finite n-ary trees, are not strongly Kripke complete (note that these logics are Π_2-arithmetical, and they are not RE by [10]). On the other hand, the claim **(Sh)**, as well as any similar result for the propositional case, obviously does not give anything for predicate logics with the intuitionistic propositional fragment (like $\mathbf{L}^c\mathbf{P}_\infty$, $\mathbf{LP}_\infty$, and many other logics considered in the present paper).

By the way, V. Shehtman [6] showed that every Kripke complete intermediate propositional logic is strongly Kripke complete in the topological semantics. We do not know if this result transfers to predicate logics.

Section 1. Preliminary notions

We consider superintuitionistic predicate logics without equality and function symbols (called in this paper *predicate logics*, or sometimes even *logics*, for short[3]). These are defined as extensions of intuitionistic predicate logic **QH** closed under modus ponens, generalization, and substitution of arbitrary formulas for atomic ones (cf. e.g. [2, Definition 2.6.3]; the book contains the basic notions in the field). For these logics we use the standard predicate Kripke semantics. Let us recall the corresponding definitions.

1.1 A *predicate Kripke frame with expanding domains* (or a *Kripke frame*, or even a *frame*, for short) is a pair (M, U) with M a non-empty partially ordered set (poset) and U a domain map defined on M such that, for any $u, v \in M$:

$(i)\ U(u) \neq \varnothing$, and $(ii)\ u \leq v \Rightarrow U(u) \subseteq U(v)$.

We say that (M, U) is a Kripke frame *over* the poset M. If U is a constant mapping on M such that $U(u) = X$ for all $u \in M$, then we write $(M, \lambda u.X)$ for (M, U) and call it *a Kripke frame with a constant domain.*[4]

---

[2] In [8] this straightforward corollary was not mentioned and Ono's Problem 1 (P34) was not addressed.

[3] in essence, we do not consider other sorts of logics

[4] The notation $(M, \lambda u.X)$ was introduced in [10], while in [11] we used the notation $(M; X)$.

The notions of a valuation and of validity of a predicate formula on a Kripke frame are defined in a usual way (cf. e.g. [15]). Namely, a *valuation* on (M,U) is a forcing relation $u \vDash A$ between points $u \in M$ and formulas A (with parameters replaced by elements of $U(u)$), satisfying *monotonicity* : $u \leq v,\ u \vDash A \Rightarrow v \vDash A$ and the following inductive conditions:

$$u \vDash (B\&C) \Leftrightarrow (u \vDash B)\,\&\,(u \vDash C); \quad u \vDash (B \vee C) \Leftrightarrow (u \vDash B) \vee (u \vDash C);$$
$$u \vDash (B \supset C) \Leftrightarrow \forall v \geq u\,[\,(v \vDash B) \Rightarrow (v \vDash C)\,]; \qquad u \nvDash \bot;$$
$$u \vDash \forall x B(x) \Leftrightarrow \forall v \geq u\ \forall c \in U(v)\,[\,v \vDash B(c)\,];$$
$$u \vDash \exists x B(x) \Leftrightarrow \exists c \in U(u)\,[\,u \vDash B(c)\,].$$

As usual, to obtain a valuation, it is sufficient to know $u \vDash A$ only for atomic A, and by induction, the monotonicity condition for atomic A implies the monotonicity for arbitrary (non-atomic) formulas A.

A Kripke frame with a valuation is called a *Kripke model.*

A predicate formula $A(x_1, \ldots, x_n)$ is said to be *true* (under a valuation $\vDash$ on a frame (M,U)) if $u \vDash A(a_1, \ldots, a_n)$ for any $u \in M$ and $a_1, \ldots, a_n \in U(u)$. A formula A is *valid* on a Kripke frame (M,U) if it is true under any valuation on (M,U). The *predicate logic* $\mathbf{L}(M,U)$ *of a Kripke frame* (M,U) is the set of all formulas valid on (M,U). It is well known that this set is indeed a superintuituionistic logic.

1.2 The *predicate logic of a class* $\mathbf{Z}$ *of Kripke frames* is

$$\mathbf{L}[\mathbf{Z}] = \bigcap (\,\mathbf{L}(M,U) : (M,U) \in \mathbf{Z}\,).$$

A predicate logic $\mathbf{L}$ is *Kripke complete* if $\mathbf{L} = \mathbf{L}[\mathbf{Z}]$ for some class $\mathbf{Z}$ of Kripke frames. The *Kripke completion* of a logic $\mathbf{L}$ is $\mathbf{L}^+ = \bigcap(\mathbf{L}(M,U) : \mathbf{L} \subseteq \mathbf{L}(M,U)\,)$, the smallest (w.r.t. the inclusion) Kripke complete extension of $\mathbf{L}$.

The *predicate logic of a poset* M is $\quad \mathbf{L}M = \bigcap\limits_{U} \mathbf{L}(M,U)$

and the *predicate logic of a class* $\mathbf{Y}$ *of posets* is $\quad \mathbf{LY} = \bigcap(\,\mathbf{L}M : M \in \mathbf{Y}\,)$. Analogously, we define the *predicate logic with constant domains of a class* $\mathbf{Y}$:

$$\mathbf{L}^c\mathbf{Y} = \bigcap (\,\mathbf{L}(M, \lambda u.X) : M \in \mathbf{Y},\ X \neq \varnothing\,).$$

A poset M with the least element 0_M is called *rooted.* A *cone* in a poset M is $M^u = \{v \in M \mid u \leq v\}$ (for $u \in M$). The *cone* (M^u, U) of a Kripke frame (M,U) is its restriction to M^u.

It is well known that $\quad (I) \quad \mathbf{L}(M,U) = \bigcap (\,\mathbf{L}(M^u,U) : u \in M\,)$,

hence $\quad (II) \quad \mathbf{LY} = \mathbf{L}\{M^u : M \in \mathbf{Y},\ u \in M\}$ for a class $\mathbf{Y}$ of posets (and similarly with $\mathbf{L}^c\mathbf{Y}$). This means that rooted Kripke frames (and rooted posets) are sufficient for the Kripke semantics of superintuitionistic logics.

1.3 Let us consider the *constant domain principle*:

$$D = \forall x (Q(x) \vee p) \supset \forall x Q(x) \vee p,$$

where p is a propositional symbol and Q is a unary predicate symbol. It is well known that D is valid on any Kripke frame with a constant domain (i.e., on any frame of the form $(M, \lambda u.X)$ with $X \neq \varnothing$); hence $D \in \mathbf{L}^c M$ for any poset M and $D \in \mathbf{L}^c \mathbf{Y}$ for any class $\mathbf{Y}$ of posets. Moreover, for a rooted poset M the following equivalence holds:

$$D \in \mathbf{L}(M, U) \text{ iff a frame } (M, U) \text{ has a constant domain.} \qquad (\delta)$$

Note that for non-rooted M, this equivalence does not hold in general; e.g., D is valid on a disjoint union of two frames with different constant domains.

1.4 Recall that the *height* $h[M]$ of a poset M is the supremum of cardinalities of chains (i.e., linearly ordered subsets) in M. Similarly, the *width* $w[M]$ of a rooted poset M is the supremum of cardinalities of antichains (i.e., sets of pairwise incomparable elements) in M. The width of an arbitrary poset is $w[M] = \sup(w[M^u] : u \in M))$.

Let $\mathbf{P}_n$ be the class of all posets of height $h[M] \leq n$ (for $n \in \omega$, $n > 0$) and let $\mathbf{P}_\infty = \bigcup_n \mathbf{P}_n$ be the class of posets of finite height. Analogously, one can introduce the class $\mathbf{W}_n$ of all posets of width $w[M] \leq n$ (and $\mathbf{W}_\infty = \bigcup_n \mathbf{W}_n$, the class of posets of finite width).

Let S_n be an n-element chain, $n > 0$; clearly, its height is n and its width is 1. Denote $\mathbf{S}_\infty = \{S_n : n \in \omega,\ n > 0\}$. Then $\mathbf{L}\mathbf{S}_\infty = \bigcap_n \mathbf{L}S_n$ and $\mathbf{L}^c\mathbf{S}_\infty = \bigcap_n \mathbf{L}^c S_n$; similarly $\mathbf{L}\mathbf{P}_\infty = \bigcap_n \mathbf{L}\mathbf{P}_n$ and $\mathbf{L}^c\mathbf{P}_\infty = \bigcap_n \mathbf{L}^c\mathbf{P}_n$, etc. Finally, let $\mathbf{Fin}$ be the class of all finite posets. Obviously $\mathbf{S}_\infty = \mathbf{P}_\infty \cap \mathbf{W}_1$ and $\mathbf{Fin} = \mathbf{P}_\infty \cap \mathbf{W}_\infty$; so $\mathbf{S}_\infty \subset \mathbf{Fin} \subset \mathbf{P}_\infty$.

It can be easily shown (applying [3, Theorem 3.4]) that

$$\mathbf{L}M \subseteq \mathbf{L}S_n \text{ and } \mathbf{L}^c M \subseteq \mathbf{L}^c S_n \text{ for any poset } M \text{ of height } h[M] \geq n, \qquad (\sigma)$$

because the n-element chain S_n is a p-morphic image of any poset M of height $h[M] \geq n$ (note that p-morphisms were called embeddings in [3, Section 3]).

Actually moreover: $\mathbf{L}M \subseteq \mathbf{L}S_n$ *iff* $\mathbf{L}^c M \subseteq \mathbf{L}^c S_n$ *iff* $h[M] \geq n$, because $P_{n-1} \in \mathbf{L}M \setminus \mathbf{L}^c S_n$ if $h[M] < n$ (the formulas P_n of height n are defined in Section 2).

Section 2. Main result

2.1 Let $\mathbf{L}$ be a logic and Γ, Δ be two sets of sentences. A pair (Γ, Δ) is called $\mathbf{L}$-*inconsistent* if $\mathbf{L} \vdash (\&\Gamma_0 \supset \bigvee \Delta_0)$ for some finite subsets $\Gamma_0 \subseteq \Gamma$, $\Delta_0 \subseteq \Delta$. A pair (Γ, Δ) is *satisfiable* in a Kripke frame (M, U) if there exists a valuation in (M, U) and a world $u \in M$ such that $u \models A$ for all formulas $A \in \Gamma$ and $u \not\models B$ for all formulas $B \in \Delta$. We say that a predicate logic $\mathbf{L}$ is *strongly Kripke complete without parameters* if every $\mathbf{L}$-consistent pair (Γ, Δ) of sets of sentences is satisfiable in a Kripke frame validating $\mathbf{L}$. The usual notion of *strong Kripke completeness* (given in [5]) is slightly stronger: namely, formulas with parameters (not necessarily sentences) are allowed in Γ and Δ; naturally, these parameters would be evaluated by individuals taken from the corresponding domain $U(u)$.

Lemma 1 (**Main Lemma**) *There exists a predicate sentence* $A^*_{\mathbf{P}_\infty}$ (or A^* for short) *such that for every Kripke frame* (M,U): $A^*_{\mathbf{P}_\infty} \in \mathbf{L}(M,U)$ *iff*

$$\textit{for all } u \textit{ in } M \text{ [} \textit{the height } h[M^u] \textit{ is finite, or the domain } U(u) \textit{ is finite } \text{]}.$$

Hence for rooted Kripke frames with constant domains we have:[5]

$$A^*_{\mathbf{P}_\infty} \in \mathbf{L}(M,\lambda u.X) \textit{ iff } [\textit{ the height } h[M] \textit{ is finite, or the domain } X \textit{ is finite }].$$

Thus, clearly $A^*_{\mathbf{P}_\infty} \in \mathbf{LP}_\infty$, so $A^*_{\mathbf{P}_\infty}\&D \in \mathbf{L}^c\mathbf{P}_\infty$.

We present the proof of this lemma in Section 4.

2.2 Now we apply Main Lemma to show that strong Kripke completeness fails for $\mathbf{L}^c\mathbf{P}_\infty$; more precisely, we prove:

Theorem 1 *The logic* $\mathbf{L}^c\mathbf{P}_\infty$ *of frames of finite height with constant domains is not strongly Kripke complete without parameters.*

Take the following propositional formulas of finite heights:

$$P_0 = \bot \text{ and } P_{n+1} = p_n \vee (p_n \supset P_n) \text{ for } n \in \omega,$$

where $p_0,\ldots,p_n,\ldots$ are different propositional symbols. It is well known that: $P_n \in \mathbf{L}(M,U)$ iff $h[M] \leq n$. Hence $P_n \in \mathbf{LP}_n$ and $(P_n\&D) \in \mathbf{L}^c\mathbf{P}_n$ for any $n>0$. By the way, note that $\mathbf{L}^c\mathbf{P}_n = [\mathbf{QH} + P_n\&D]$ (cf. [4, Theorem 3.3]), while $\mathbf{LP}_n \neq [\mathbf{QH} + P_n]$ (i.e., the logics $[\mathbf{QH} + P_n]$ are Kripke incomplete) for $n \geq 2$ (see [4, Theorem 3.2]); by definition, $[\mathbf{QH} + P_1]$ is classical logic.[6]

Also take the sentences

$$C_m = \forall x_1,\ldots,\forall x_m\,[\mathop{\&}_i Q_i(x_i) \supset \bigvee_{i<j} Q_i(x_j)]$$

for $m>1$. Clearly, $C_m \in \mathbf{L}(M,\lambda u.X)$ iff $\mathrm{card}(X) < m$.

Take the set $\Delta^* = \{P_n : n>0\} \cup \{C_m : m>1\}$. Then the pair $(\varnothing,\Delta^*)$ is $\mathbf{L}^c\mathbf{P}_\infty$-consistent (and moreover, it is $\mathbf{L}^c\mathbf{S}_\infty$-consistent). Indeed, any finite $\Delta_0 \subseteq \Delta^*$ is included in $\{P_n : n<n_0\} \cup \{C_m : m \leq m_0\}$ for some n_0, m_0; then the corresponding disjunction $\bigvee \Delta_0$ is falsified in every Kripke frame of height n_0 (in particular, in S_{n_0}) with an m_0-element constant domain.

On the other hand, the subsequent claim shows that the pair $(\varnothing,\Delta^*)$ is not satisfiable in $\mathbf{L}^c\mathbf{P}_\infty$-frames; hence the logic $\mathbf{L}^c\mathbf{P}_\infty$ is not strongly Kripke complete without parameters.

---

[5] For non-rooted M, this equivalence does not hold in general; e.g., $A^*_{\mathbf{P}_\infty}$ is valid on a disjoint union M of all finite chains S_n (with an infinite constant domain), while $h[M]$ is infinite.

[6] Note that the logics $\mathbf{LP}_n$ are finitely axiomatizable for all $n>0$; their axioms are P_n^+, which are essentially predicate formulas similar to P_n (see [16]).

Claim *Let (M, U) be a rooted Kripke frame validating* $\mathbf{L}^c\mathbf{P}_\infty$.
Then $\mathbf{L}(M, U) \cap \Delta^* \neq \varnothing$.

Proof. Clearly, $D \in \mathbf{L}^c\mathbf{P}_\infty \subseteq \mathbf{L}(M, U)$, and so (M, U) has a constant domain (due to (δ), see in Section 1). Now, if $\mathbf{L}(M, U) \cap \Delta^* = \varnothing$, i.e., all P_n and C_m are falsified in (M, U), then its domain is infinite and the height $h(M)$ is infinite as well. Thus $A^*_{\mathbf{P}_\infty} \notin \mathbf{L}(M, U)$ and so $\mathbf{L}^c\mathbf{P}_\infty \not\subseteq \mathbf{L}(M, U)$ (since $A^*_{\mathbf{P}_\infty} \in \mathbf{L}^c\mathbf{P}_\infty$). □

Actually, our argument gives a more general result:

Theorem 2 *Let* $\mathbf{L}$ *be a predicate logic such that* $\mathbf{L}^c\mathbf{P}_\infty \subseteq \mathbf{L} \subseteq \mathbf{L}^c\mathbf{S}_\infty$. *Then* $\mathbf{L}$ *is not strongly Kripke-complete without parameters.*

Indeed, if $\mathbf{L} \subseteq \mathbf{L}^c\mathbf{S}_\infty$, then the pair $(\varnothing, \Delta^*)$ is $\mathbf{L}$-consistent. And if $\mathbf{L}^c\mathbf{P}_\infty \subseteq \mathbf{L}$, then $(\varnothing, \Delta^*)$ is not satisfiable in $\mathbf{L}$-frames. □

Corollary 1 *Let* $\mathbf{Y} \subseteq \mathbf{P}_\infty$ *be a class of posets of finite height such that* $\forall n \in \omega(\mathbf{Y} \not\subseteq \mathbf{P}_n)$ *(i.e.,* $\mathbf{Y}$ *contains posets of arbitrarily large heights). Then the Kripke complete logic* $\mathbf{L}^c\mathbf{Y}$ *is not strongly Kripke complete (even without parameters).*

Indeed, if $\mathbf{Y} \not\subseteq \mathbf{P}_n$, then $\mathbf{L}^c\mathbf{Y} \subseteq \mathbf{L}^c\mathbf{S}_n$, due to (σ) (see the end of Section 1). □

Therefore, we conclude that the following Kripke complete logics (and many other ones) are not strongly Kripke complete (without parameters): [7]

$$\mathbf{L}^c\mathbf{P}_\infty, \mathbf{L}^c\mathbf{Fin}, \mathbf{L}^c\mathbf{S}_\infty = \mathbf{L}^c(\mathbf{P}_\infty \cap \mathbf{W}_1), \mathbf{L}^c(\mathbf{P}_\infty \cap \mathbf{W}_m) = \mathbf{L}^c(\mathbf{Fin} \cap \mathbf{W}_m),$$
$$\mathbf{L}^c(\mathbf{P}_n \cup (\mathbf{P}_\infty \cap \mathbf{W}_m)) = \mathbf{L}^c\mathbf{P}_n \cap \mathbf{L}^c(\mathbf{Fin} \cap \mathbf{W}_m) \text{ for every } m, n \in (\omega \setminus \{0\}). \quad (\lambda)$$

Note that all logics $\mathbf{L}$ mentioned in Theorem 2 (in particular, all logics listed in (λ)) are not recursively axiomatizable, see [11, Theorem 1.2]. On the other hand, all logics listed in (λ) are Π_2-arithmetical, because the logics $\mathbf{L}^c\mathbf{P}_n$ for $n < \omega$ are finitely axiomatizable (see in Section 1) and all logics $\mathbf{L}^c M$ for finite posets M are recursively axiomatizable (in a uniform way), see e.g. [10]. [8]

Section 3. A short discussion and open questions

3.1 Main Lemma shows that, for the Kripke semantics with constant domains, the formula $A^*_{\mathbf{P}_\infty}$ (or, more precisely, $A^*_{\mathbf{P}_\infty} \& D$) describes the finiteness of height, up to a minor additional exception involving (arbitrary) Kripke frames with finite constant domain. Now we will explain why this addition is inevitable.

Let $\mathbf{Kr}^c_m = \{(M, \lambda u.X) \mid M \neq \varnothing, \mathrm{card}(X) = m\}$ be the class of Kripke frames with m-element constant domain (for $m \in \omega, m > 0$). Now, let

$$\mathbf{Kr}^c_\infty = \bigcup_m \mathbf{Kr}^c_m = \{ (M, \lambda u.X) \mid M \neq \varnothing, X \text{ is finite} \}$$

[7] Note that all these examples are not covered by the claim $(\mathbf{Sh})'$ (see Remark 0 at the beginning of the paper), because the propositional fragments of these logics are either intuitionistic or not included in $\mathbf{GJ}_2$.

[8] Moreover, they are finitely axiomatizable by [7, Theorem 3.7] (cf. our Proposition 2 in Section 3).

be the class of Kripke frames with finite constant domains.

Similarly, we introduce the classes of frames with finite M:
$\mathbf{Fin}^c_m = \{(M, \lambda u.X) \mid M \text{ is finite}, \mathrm{card}(X) = m\} = \{(M, \lambda u.X) \in \mathbf{Kr}^c_m \mid M \in \mathbf{Fin}\}$
and $\mathbf{Fin}^c_\infty = \bigcup_m \mathbf{Fin}^c_m = \{(M, \lambda u.X) \mid M \text{ and } X \text{ are finite } \} =$
$= \{(M, \lambda u.X) \in \mathbf{Kr}^c_\infty \mid M \in \mathbf{Fin}\}$.

Finally, we have analogous classes of frames of finite height:
$\mathbf{P}^c_{\infty,m} = \{(M, \lambda u.X) \mid M \in \mathbf{P}_\infty, \mathrm{card}(X) = m\} = \{(M, \lambda u.X) \in \mathbf{Kr}^c_m \mid M \in \mathbf{P}_\infty\}$
and $\mathbf{P}^c_{\infty,\infty} = \bigcup_m \mathbf{P}^c_{\infty,m} = \{ (M, \lambda u.X) \mid h[M] \text{ and } X \text{ are finite } \} =$
$= \{(M, \lambda u.X) \in \mathbf{Kr}^c_\infty \mid M \in \mathbf{P}_\infty\}$.

Clearly, $\mathbf{Fin}^c_m \subset \mathbf{P}^c_{\infty,m} \subset \mathbf{Kr}^c_m$ for any m, and so $\mathbf{Fin}^c_\infty \subset \mathbf{P}^c_{\infty,\infty} \subset \mathbf{Kr}^c_\infty$. Also $\mathbf{L}[\mathbf{Kr}^c_\infty] = \bigcap_m \mathbf{L}[\mathbf{Kr}^c_m]$ and similarly with $\mathbf{L}[\mathbf{Fin}^c_\infty]$, $\mathbf{L}[\mathbf{P}^c_{\infty,\infty}]$.

Lemma 2 $\mathbf{L}[\mathbf{Kr}^c_m] = \mathbf{L}[\mathbf{Fin}^c_m]$,
and so $\mathbf{L}[\mathbf{Kr}^c_m] = \mathbf{L}[\mathbf{P}^c_{\infty,m}] = \mathbf{L}[\mathbf{Fin}^c_m]$ *for every* $m \in \omega, m > 0$.

Therefore, $\mathbf{L}[\mathbf{Kr}^c_\infty] = \mathbf{L}[\mathbf{P}^c_{\infty,\infty}] = \mathbf{L}[\mathbf{Fin}^c_\infty]$.

In other words, for the Kripke semantics with finite constant domains, we have:

every finitely valid formula is generally valid, i.e.,
every formula valid on all finite posets is valid on all (non-empty) posets M.

Hence we obtain:

Corollary 2 $\mathbf{L}^c\mathbf{Fin} \subset \mathbf{L}[\mathbf{Kr}^c_\infty]$, *and thus* $\mathbf{L}^c\mathbf{P}_\infty \subset \mathbf{L}^c\mathbf{Fin} \subset \mathbf{L}[\mathbf{Kr}^c_\infty]$.

This means that any formula valid (in the semantics with constant domains) on all frames (with arbitrary domains) over finite posets M (in particular, any formula valid on all frames of finite height) is necessarily valid on all frames with finite domains (with arbitrary posets M). That is why our sentence $A^*_{\mathbf{P}\infty}$ (presented in Lemma 1), which is valid on all frames of finite height, is inevitably valid on <u>all</u> frames with finite constant domains (with arbitrary M).

Remark 1 By the way, the inclusion $\mathbf{L}^c\mathbf{Fin} \subset \mathbf{L}[\mathbf{Kr}^c_\infty]$ is definitely proper. Indeed, let A be a well-known formula (with one binary predicate symbol R) that is classically valid on all finite domains and is not valid on infinite domains (e.g. $A = \neg A_0$ for the formula A_0 defined at the beginning of Section 4). Put $A' = [\forall x, y(R(x,y) \vee \neg R(x,y)) \supset A]$. Then A' belongs to $\mathbf{L}[\mathbf{Kr}^c_\infty] \backslash \mathbf{QC}$, where $\mathbf{QC}$ is classical predicate logic; so all the more it belongs to $\mathbf{L}[\mathbf{Kr}^c_\infty] \backslash \mathbf{L}^c\mathbf{Fin}$.

Remark 2 The inclusion $\mathbf{L}^c\mathbf{P}_\infty \subset \mathbf{L}^c\mathbf{Fin}$ is proper as well. Indeed, there exists a formula $A^*_{\mathbf{Fin}}$ (introduced in [10] and denoted by F' there) such that (cf. our Lemma 1):

$$A^*_{\mathbf{Fin}} \in \mathbf{L}(M, U) \textit{ iff } \forall u \in M\, [\, M^u \textit{ or } U(u) \textit{ is finite}]. \quad (\varphi)$$

Then clearly $A^*_{\mathbf{Fin}} \in \mathbf{LFin} \backslash \mathbf{L}^c\mathbf{P}_\infty$, and so $A^*_{\mathbf{Fin}} \in \mathbf{L}^c\mathbf{Fin} \backslash \mathbf{L}^c\mathbf{P}_\infty$. Moreover, it is easily seen that $\mathbf{LFin} \not\subseteq \mathbf{L}^c\mathbf{Y}$ for any class $\mathbf{Y}$ of rooted posets that contains an infinite poset, and so $\mathbf{LY} \subset \mathbf{LFin}$ and $\mathbf{L}^c\mathbf{Y} \subset \mathbf{L}^c\mathbf{Fin}$ for any class $\mathbf{Y} \supset \mathbf{Fin}$ that contains a poset with an infinite cone. The similar statement holds for $\mathbf{LP}_\infty$ as well, due to our formula $A^*_{\mathbf{P}_\infty}$.

Now we prove Lemma 2.

Proof. Let $X_m = \{1, \dots, m\}$ be an m-element domain. To every k-ary predicate symbol P and all $j_1, \dots, j_k \in X_m$, we assign a unique propositional symbol $\bar{P}^{(j_1,\dots,j_k)}$. For a predicate formula $A(i_1, \dots, i_n)$ with parameters replaced by elements of X_m one can easily construct a propositional formula $\bar{A}_{(m)}^{(i_1,\dots,i_n)}$, which simulates A in a natural way; namely, we replace predicate atoms $P(j_1, \dots, j_k)$ with propositional atoms $\bar{P}^{(j_1,\dots,j_k)}$ and replace quantifiers $\forall$ and $\exists$ with the conjunction and disjunction over all elements of X_m. One can easily show (by induction) that the truth of $A(i_1, \dots, i_n)$ in $(M, \lambda u.X_m)$ (at a point $v \in M$) is equivalent to the propositional truth of $\bar{A}_{(m)}^{(i_1,\dots,i_n)}$ in M at the same point v (under the corresponding valuations of symbols P in $(M, \lambda u.X_m)$ and $\bar{P}^{(i_1,\dots,i_k)}$ in M). Therefore, we conclude that $A \in \mathbf{L}(M, \lambda u.X_m)$ iff $\bar{A}_{(m)} \in \mathbf{L}(M)$ for any sentence A and a poset (i.e., a propositional Kripke frame) M.

Finally, we use the following well-known fact: any propositional formula valid on all finite posets M is intuitionistically provable, and so it is valid on all (non-empty) M as well. □

3.2 Hence we obtain:

Proposition 1 *Let (M, U) be a rooted Kripke frame. Then the following conditions are equivalent:*

(1) $\mathbf{L}^c\mathbf{P}_\infty \subseteq \mathbf{L}(M, U)$;

(2) $(\mathbf{LP}_\infty + D) \subseteq \mathbf{L}(M, U)$;

(3) $(A^*_{\mathbf{P}_\infty} \& D) \in \mathbf{L}(M, U)$;

(4) *(M, U) is a Kripke frame with constant domain X (i.e., it is $(M, \lambda u.X)$) and [the height $h(M)$ or the domain X is finite].*

Proof. The implication (1) ⇒ (2) is obvious (since $D \in \mathbf{L}^c\mathbf{P}_\infty$, see in Section 1). The implications (2) ⇒ (3) ⇒ (4) readily follow from Lemma 1 and (δ) (again see in Section 1). And the implication (4) ⇒ (1) follows from Corollary 2. □

Therefore, $\mathbf{L}^c\mathbf{P}_\infty$ *is the Kripke completion of* $(\mathbf{QH} + A^*_{\mathbf{P}_\infty} \& D)$.

Note that the logic $(\mathbf{QH} + A^*_{\mathbf{P}_\infty} \& D)$ is Kripke incomplete, because its Kripke completion $\mathbf{L}^c\mathbf{P}_\infty$ is not recursively axiomatizable, as it was mentioned in Section 2 (cf. [11]). The logic $(\mathbf{QH} + A^*_{\mathbf{P}_\infty})$ is Kripke incomplete as well; its Kripke completion described by Lemma 1 is not RE (see [11, Theorem 2.2]).[9]

Now let us state two related open questions:

Question 1 Is $(\mathbf{LP}_\infty + D)$ equal to $\mathbf{L}^c\mathbf{P}_\infty$?

In other words, is the logic $(\mathbf{LP}_\infty + D)$ Kripke complete?

---

[9] By the way, analogously, $\mathbf{L}^c\mathbf{Fin}$ is the Kripke completion of the Kripke incomplete logic $(\mathbf{QH} + A^*_{\mathbf{Fin}} \& D)$, where $A^*_{\mathbf{Fin}}$ is the formula mentioned in Remark 2 (and introduced in [10]). The logic $(\mathbf{QH} + A^*_{\mathbf{Fin}})$ is again Kripke incomplete; its Kripke completion is $\mathbf{L}[\mathbf{Fin}^*]$, where $\mathbf{Fin}^*$ is the class of frames described by condition (φ) from our Remark 2, and this completion is not RE as well (cf. [10, Corollary 4]).

Question 2 Is the logic $\mathbf{LP}_\infty$ strongly Kripke complete?

Note that we are still unable to *directly* transfer our proof for $\mathbf{L}^c\mathbf{P}_\infty$ to $\mathbf{LP}_\infty$. Indeed, there exist rooted Kripke frames (M,U) with expanding domains such that: (i) $(\varnothing,\Delta^*)$ is satisfiable in (M,U) (in the root of M), i.e., $\mathbf{L}(M,U)\cap\Delta^*=\varnothing$,[10] and (ii) (M,U) validates $A^*_{\mathbf{P}_\infty}$ (i.e., it satisfies the condition described in Lemma 1). Namely, one can take a disjoint union of frames of all finite heights (e.g. the disjoint union of all finite chains S_n) with infinite domains (or with n-element domains X_n for all $n>0$) and add the root u_0 whose domain is finite (e.g. one-element). This means that unlike the proof of Claim (see in Section 2), now definitely one cannot guarantee that every frame (M,U) satisfying $(\varnothing,\Delta^*)$ falsifies $\mathbf{L}^c\mathbf{P}_\infty$ 'due to the formula $A^*_{\mathbf{P}_\infty}$'. On the other hand, we do not know if such Kripke frames (satisfying $(\varnothing,\Delta^*)$ and validating $A^*_{\mathbf{P}_\infty}$) validate the whole logic $\mathbf{LP}_\infty$ as well. In other words, we do not know, if $A^*_{\mathbf{P}_\infty}\in\mathbf{L}(M,U)$ implies $\mathbf{LP}_\infty\subseteq\mathbf{L}(M,U)$ (cf. the equivalence $(1)\Leftrightarrow(3)$ in Proposition 1).

This example is slightly related to the following open question:

Question 3 How to transfer (in a reasonable way) Lemma 2 and Corollary 2 to the case with expanding domains.[11]

3.3 To conclude this section, note that the formula $A^*_{\mathbf{Fin}}$ (see Remark 2) allows us to obtain the following variant of Theorem 2:

Theorem 3 *Let* $\mathbf{L}$ *be a predicate logic such that*

$$\mathbf{L}^c\mathbf{Fin}\subseteq\mathbf{L}\subseteq\bigcup(\,\mathbf{L}[\mathbf{Z}]:\mathbf{Z}\textit{ is an unrestricted class of Kripke frames}\,)$$

(here $\bigcup$ denotes the set-theoretical union, but not the sum of logics!).

Then $\mathbf{L}$ *is not strongly Kripke complete without parameters.*

Here a class $\mathbf{Z}$ of Kripke frames is called *unrestricted* (cf. [10, Section 1.2] or [11, Section 2.1]), if

$$\forall n\in\omega\ \exists(M,U)\in Z\ \exists u\in M\,(\mathrm{card}(M^u)\geq n,\ \mathrm{card}(U(u))\geq n).$$

The proof is similar to the proof of Theorem 2 given in Section 2; we use

[10] More precisely, in order to satisfy the pair $(\varnothing,\Delta)$ in a frame (M,U), it is required to falsify all formulas from Δ by a single valuation in (M,U). However, this is not a serious problem. Our formulas C_m (for $m>1$) look rather regularly, and so, if cardinalities of domains are sufficiently large, it is easy to construct a valuation, which falsifies all these formulas (and similarly with P_n: $n>0$).

[11] While preparing the final text of this paper, we found an answer to this question. The proof of the corresponding claims for two different versions of Kripke semantics with finite expanding domains will be presented in [14]; on the other hand, for the third natural version of the semantics these properties do not hold.

the formula $A^*_{\mathbf{Fin}}$ for $A^*_{\mathbf{P}_\infty}$ and use the propositional formulas

$$\Phi_n = \bigvee_{i<j} (p_i \equiv p_j)$$

(in symbols $p_0, p_1, \ldots, p_n$) instead of P_n. □

Again all logics mentioned in Theorem 3 are not recursively axiomatizable, see Theorem from [10] and Theorem 2.1(2) from [11]; many of these logics are Π_2-arithmetical (cf. Section 2).

Also, for Kripke semantics with constant domains, we obtain the subsequent criterion of strong Kripke completeness for logics characterized by classes of finite posets (i.e., for logics of the form $\mathbf{L}^c\mathbf{Y}$ with $\mathbf{Y} \subseteq \mathbf{Fin}$); cf. our Corollary 1 in Section 2 and [10, Corollary 3]:

Proposition 2 *Let* $\mathbf{Y}$ *be a class of finite posets.*

(I) The following conditions are equivalent:

(1) *the logic* $\mathbf{L}^c\mathbf{Y}$ *is strongly Kripke complete (with or without parameters);*

(2) *the logic* $\mathbf{L}^c\mathbf{Y}$ *is recursively axiomatizable;*

(3) *the logic* $\mathbf{L}^c\mathbf{Y}$ *is finitely axiomatizable;*

(4) *the logic* $\mathbf{L}^c\mathbf{Y}$ *is 'tabular', i.e., it equals* $\mathbf{L}^c M$ *for a finite poset* M*;*

(5) $\mathbf{Y} \subseteq \mathbf{P}_n \cap \mathbf{W}_n$ *for some* $n > 0$.

(II) If $\mathbf{Y}$ *is a class of pairwise non-isomorphic finite rooted posets, then the mentioned conditions (1) – (5) are equivalent to*

(6) $\mathbf{Y}$ *is finite.*

Proof. The implication (1) ⇒ (5) readily follows from Theorem 3: namely, if $\forall n\,[\,\mathbf{Y} \not\subseteq \mathbf{P}_n \cap \mathbf{W}_n\,]$, then the class $\mathbf{Z}$ of all frames with constant domain over posets from $\mathbf{Y}$ is unrestricted. Similarly, the implication (2) ⇒ (5) is a consequence of Theorem from [10].

Clearly, (5) implies (4), because the family of (non-isomorphic) rooted posets from $\mathbf{P}_n \cap \mathbf{W}_n$ is finite, and $\mathbf{L}^c\mathbf{Y} = \mathbf{L}^c\mathbf{Y}' = \mathbf{L}^c M$, where $\mathbf{Y}'$ is the (finite) family of cones in posets from $\mathbf{Y}$ and M is the disjoint union of posets from $\mathbf{Y}'$.

Now, the implications (4) ⇒ (1) and (4) ⇒ (3) follow from Shimura's result [7, Theorem 3.7]. Indeed, let A be a formula axiomatizing the (finitely axiomatizable) propositional superintuitionistic logic of a finite poset M, and let $\mathbf{L} = [\mathbf{QH}{+}D\&A]$. The mentioned Shimura's theorem claims the strong Kripke completeness[12] of $\mathbf{L}$. From its proof we also obtain that $\mathbf{L} = \mathbf{L}^c M$.

Finally, the implication (3) ⇒ (2) is obvious. □

[12] naturally, in the usual sense, i.e., with parameters (cf. the beginning of Section 2)

Section 4. The proof of Main Lemma

Take a binary, unary, and 0-ary (i.e., propositional) symbols R, Q, p respectively. Take the following formulas:

$$\begin{aligned}
A_0 &= \forall x \neg R(x,x) \,\&\, \forall x \exists y R(x,y) \,\&\, \forall x \forall y\, (R(x,y) \vee \neg R(x,y)) \,\&\\
&\quad \&\, \forall x \forall y \forall z\, (R(x,y) \& R(y,z) \supset R(x,z)),\\
A_1 &= \forall x \forall y\, [(\neg R(x,y) \vee p) \equiv ((Q(x) \supset Q(y)) \vee p)],\\
A_2 &= \forall x \forall y\, [(\neg R(x,y) \vee p) \equiv ((Q(y) \supset Q(x)) \vee p)],\\
A_3 &= \forall x \forall y\, [R(x,y) \& (Q(x) \supset Q(y)) \supset Q(x)],\\
A_4 &= \forall x \forall y\, [R(x,y) \& (Q(y) \supset Q(x)) \supset Q(y)],\\
A' &= A_0 \,\&\, [(A_1 \& A_3) \vee (A_2 \& A_4)];\\
A^* &= A' \supset p.
\end{aligned}$$

4.1 IF PART. Suppose that $A^* \notin \mathbf{L}(M,U)$, i.e., $u \vDash A'$ and $u \nvDash p$ for some $u \in M$ and a valuation in (M^u, U); then we show that both $U(u)$ and $h[M^u]$ are infinite. Let us define the relation $(a < b) \Leftrightarrow (u \vDash R(a,b))$ on $U(u)$. By A_0, the relation is transitive, irreflexive, and there exists a sequence of different elements $a_0 < a_1 < \ldots$ in $U(u)$; hence $U(u)$ is infinite. Also $(a \not< b) \Leftrightarrow u \vDash \neg R(a,b)$ for $a, b \in U(u)$.

Now, let us assume that $h[M^u] = n$ is finite. Put $\Theta_i = \{v \in M^u \mid v \vDash Q(a_i)\}$ for $i \in \omega$.

First, let $u \vDash A_1 \& A_3$. Then, by A_1, for $i, j \in \omega$ we have:

$$(\Theta_i \subseteq \Theta_j) \Leftrightarrow u \vDash (Q(a_i) \supset Q(a_j)) \Leftrightarrow u \vDash \neg R(a_i, a_j) \Leftrightarrow a_i \not< a_j \Leftrightarrow j \leq i$$

(recall that $u \nvDash p$). Also, by A_3, if $i < j < k$, then

$$\forall v \in \Theta_i \backslash \Theta_j \, \exists w \in \Theta_j \backslash \Theta_k \, [v \leq w].$$

Indeed, if $j < k$ and $v \in \Theta_i \backslash \Theta_j$, then $v \nvDash Q(a_j)$ and $v \vDash R(a_j, a_k)$, so $v \nvDash (Q(a_j) \supset Q(a_k))$, i.e., $v \leq w$ for some $w \in \Theta_j \backslash \Theta_k$.

Hence we obtain an $(n+1)$-element chain $v_0 < v_1 < \ldots < v_n$ in M^u such that $v_i \in \Theta_i \backslash \Theta_{i+1}$ for $i < n$. This is a contradiction.

Second, let $u \vDash A_2 \& A_4$. Then, similarly, by A_2,

$$(\Theta_i \subseteq \Theta_j) \Leftrightarrow u \vDash \neg R(a_j, a_i) \Leftrightarrow i \leq j,$$

and by A_4, for $i < j < k$ we have

$$\forall v \in \Theta_k \backslash \Theta_j \, \exists w \in \Theta_j \backslash \Theta_i \, [v \leq w].$$

Hence we obtain an $(n+1)$-element chain $v_n < \ldots < v_0$, where $v_i \in \Theta_{i+1} \backslash \Theta_i$.

4.2 ONLY IF PART. Suppose that there exists $u \in M$ such that both $U(u)$ and $h[M^u]$ are infinite; then we show that $u \nvDash A^*$ for a suitable valuation on the cone (M^u, U).

Take a denumerable subset $X_0 = \{a_i \mid i \in \omega\}$ of $U(u)$. Let $\nu(a_i) = i$ for elements of X_0 and $\nu(a) = 0$ for all other elements from all $U(v)$, $v \geq u$.

First, assume that the cone M^u contains an ω-chain $u = u_0 < u_1 < \ldots$. Then we define the following valuation on (M^u, U):

$$\begin{aligned} v &\models R(a,b) \Leftrightarrow \nu(a) < \nu(b); \\ v &\models Q(a) \Leftrightarrow v \not\leq u_{\nu(a)}; \\ v &\models p \Leftrightarrow v \neq u. \end{aligned}$$

Clearly, $u \models A_0$ and $u \not\models p$. Also $u \models A_1$, because

$$u \models \neg R(a,b) \Leftrightarrow (\nu(b) \leq \nu(a)) \Leftrightarrow \forall v \geq u\, [v \models Q(a) \Rightarrow v \models Q(b)] \Leftrightarrow u \models (Q(a) \supset Q(b)),$$

and $v \models p$ for all $v > u$.

Now, show that $u \models A_3$. Assume that $a, b \in U(v)$, $v \models R(a,b) \& (Q(a) \supset Q(b))$, $v \not\models Q(a)$. Then $\nu(a) < \nu(b)$ and $v \leq u_{\nu(a)} < u_{\nu(b)}$. Hence $v \not\models (Q(a) \supset Q(b))$, since $u_{\nu(b)} \models Q(a)$, $u_{\nu(b)} \not\models Q(b)$. This is a contradiction.

Therefore, $A^* \notin \mathbf{L}(M, U)$.

Second, assume, otherwise, that M^u does not contain ω-chains, i.e., M^u is a dually well-founded poset. Take $\Theta_i = \{v \in M^u \mid h[M^v] \leq i\}$ for $i < \omega$. Clearly, $\varnothing = \Theta_0 \subset \Theta_1 \subset \ldots \subset \Theta_i \subset \Theta_{i+1} \subset \ldots$ (recall that $h[M^u]$ is infinite). Take the following valuation on (M^u, U):

$$\begin{aligned} v &\models R(a,b) \Leftrightarrow \nu(a) < \nu(b); \\ v &\models Q(a) \Leftrightarrow v \in \Theta_{\nu(a)}; \\ v &\models p \Leftrightarrow v \neq u. \end{aligned}$$

Again $u \models A_0$ and $u \not\models p$. Also $u \models A_2$, because

$$u \models (Q(b) \supset Q(a)) \Leftrightarrow (\Theta_{\nu(b)} \subseteq \Theta_{\nu(a)}) \Leftrightarrow (\nu(b) \leq \nu(a)) \Leftrightarrow u \models \neg R(a,b).$$

Finally, $u \models A_4$. Indeed, assume that $a, b \in U(v)$, $v \models R(a,b) \& (Q(b) \supset Q(a))$, $v \not\models Q(b)$ for some $v \geq u$. Then $\nu(a) < \nu(b)$ and $v \notin \Theta_{\nu(b)}$, i.e., $h[M^v] > \nu(b)$. Then there exists $w \in M^v$ such that $h[M^w] = \nu(b)$ and so $w \in \Theta_{\nu(b)} \setminus \Theta_{\nu(a)}$, i.e., $w \models Q(b)$ and $w \not\models Q(a)$. This contradicts $v \models (Q(b) \supset Q(a))$.

Hence again $A^* \notin \mathbf{L}(M, U)$. □

4.3 In conclusion, we can easily obtain the following variation of Lemma 1:

Lemma 3 *There exists a predicate sentence $A^*_{\mathbf{WF}_d}$ such that for every Kripke frame (M, U): $A^*_{\mathbf{WF}_d} \in \mathbf{L}(M, U)$ iff*

$\forall u \in M$ [the domain $U(u)$ is finite, or the cone M^u is dually well-founded (i.e., it does not contain ω-chains)].

Namely, put $A^*_{\mathbf{WF}_d} = A'' \supset p$, where $A'' = A_0 \& A_1 \& A_3$ (i.e., we drop the disjunct $A_2 \& A_4$ in the premise A' of $A^* = A^*_{\mathbf{P}_\infty}$).

The proof of this lemma repeats the proof for Main Lemma, with obvious changes. Namely, in the IF PART, using $A_1 \& A_3$, we can obtain an ω-chain

$v_0 < \ldots < v_i < \ldots$ in M^u such that $v_i \in \Theta_i \backslash \Theta_{i+1}$ for all $i \in \omega$ (the argument with $A_2 \& A_4$ is omitted now). And in the ONLY IF PART, we suppose that, for some $u \in M$, the domain $U(u)$ is infinite and the cone M^u contains an ω-chain $u = u_0 < u_1 < \ldots$; the second case, with dually well-founded M^u, is omitted. □

In particular, for Kripke frames with constant domains we have:

$$A^*_{\mathbf{WF}_d} \in \mathbf{L}(M, \lambda u.X) \quad \textit{iff} \quad [\ M \textit{ is dually well-founded, or } X \textit{ is finite}\ ].$$

Note that here the restriction that M must be rooted (cf. footnote 5 in Section 2) is not required, because the class $\mathbf{WF}_d$ of dually well-founded posets, unlike $\mathbf{P}_\infty$ (and **Fin**), has the following convenient property:

$$\textit{Every poset } W \notin \mathbf{WF}_d \textit{ contains a cone } W^u \notin \mathbf{WF}_d. \qquad (c)$$

Clearly, $A^*_{\mathbf{WF}_d} \in \mathbf{L}\,\mathbf{WF}_d$, and so $A^*_{\mathbf{WF}_d} \& D \in \mathbf{L}^c\mathbf{WF}_d$.

Finally, Lemmas 1 and 3, together with [10, Lemma 1] (cf. our Remark 2 in Section 3), give the following chains of proper inclusions (here **PO** is the class of all posets):

Corollary 3 $\mathbf{QH} = \mathbf{L}\,\mathbf{PO} \subset \mathbf{L}\,\mathbf{WF}_d \subset \mathbf{L}\,\mathbf{P}_\infty \subset \mathbf{L}\,\mathbf{Fin}$ *and*
$[\mathbf{QH}+D] = \mathbf{L}^c\mathbf{PO} \subset \mathbf{L}^c\mathbf{WF}_d \subset \mathbf{L}^c\mathbf{P}_\infty \subset \mathbf{L}^c\mathbf{Fin}.$

In fact, $\mathbf{LWF}_d \not\subseteq [\mathbf{QH}+D]$, $\mathbf{LP}_\infty \not\subseteq \mathbf{L}^c\mathbf{WF}_d$, etc. Moreover, $\mathbf{LWF}_d \not\subseteq \mathbf{L}^c\mathbf{Y}$ for any class of posets $\mathbf{Y} \not\subseteq \mathbf{WF}_d$; thus $\mathbf{LY} \subset \mathbf{LWF}_d$ and $\mathbf{L}^c\mathbf{Y} \subset \mathbf{L}^c\mathbf{WF}_d$ for any class $\mathbf{Y} \supset \mathbf{WF}_d$ (the corresponding statements for **LFin** and for $\mathbf{LP}_\infty$ are given in Remark 2; they involve additional restrictions related to cones or to the rootedness, since the property (c) fails for **Fin** and for $\mathbf{P}_\infty$).

Remark 3 By the way, note that

$$\mathbf{L}\,\mathbf{WF} = \mathbf{L}\,\mathbf{PO} = \mathbf{QH} \quad \textit{and} \quad \mathbf{L}^c\mathbf{WF} = \mathbf{L}^c\mathbf{PO} = [\mathbf{QH}+D],$$

where **WF** is the class of well-founded posets, because the standard tree ω^* (the ω-branching tree of height ω) is well-founded, and **QH** (resp., $[\mathbf{QH}+D]$) is complete w.r.t. frames over ω^* (with expanding domains and with constant domains respectively), cf. [2, Theorem 6.4.17(1) and Proposition 7.6.15(1)].

Acknowledgements
The author would like to thank the anonymous referees, Valentin Shehtman, and Evgeny Zolin for useful and fruitful comments, which helped improve the exposition of the paper.

References

[1] Gabbay, D. and D. de Jongh, *Sequence of decidable finitely axiomatizable intermediate logics with the disjunction property*, Journal of Symbolic Logic **39**, No.1 (1974), pp. 67–78.

[2] Gabbay, D., V. Shehtman and D. Skvortsov, "Quantification in Nonclassical Logic Vol.1," Studies in Logic and the Foundation of Mathematics **153**, Elsevier, 2009.

[3] Ono, H., *A study of intermediate predicate logics*, Publications of RIMS, Kyoto Univ. **8**, No.3 (1972-1973), pp. 619–649.
[4] Ono, H., *Model extension theorem and Craig's interpolation theorem for intermediate predicate logics*, Reports on Math. Logic **15** (1983), pp. 41–58.
[5] Ono, H., *Some problems in intermediate predicate logics*, Reports on Math. Logic **21** (1987), pp. 55–67.
[6] Shehtman, V., *On strong neighbourhood completeness of modal and intermediate propositional logics, 1*, in: *Advances in Modal Logic '96*, M. Kracht, M. De Rijke, H. Wansing, M. Zakharyaschev, eds., CSLI Publications, 1997, pp. 209–222.
[7] Shimura, T., *Kripke completeness of some intermediate predicate logics with the axiom of constant domain and a variant of canonical formulas*, Studia Logica **52**, No.1 (1993), pp. 23–40.
[8] Shimura, T., *On completeness of intermediate predicate logics with respect to Kripke semantics*, Bulletin of the Section of Logic **24**, No.1 (1995), pp. 41–45.
[9] Skvortsov, D., *On axiomatizability of some intermediate predicate logics (summary)*, Reports on Math. Logic **22** (1988), pp. 115–116.
[10] Skvortsov, D., *The predicate logic of finite Kripke frames is not recursively axiomatizable*, Journal of Symbolic Logic **70** (2005), pp. 451–459.
[11] Skvortsov, D., *On non-axiomatizability of superintuitionistic predicate logics of some classes of well-founded and dually well-founded Kripke frames*, Journal of Logic and Computation **16**, No.5 (2006), pp. 685–695.
[12] Skvortsov, D., *On non-axiomatizability of some superintuitionistic predicate logics, I: Predicate logics of well-ordered and dually well-ordered Kripke frames*, in preparation.
[13] Skvortsov, D., *A new (simpler) solution to Ono's problem on the strong completeness for intermediate predicate logics*, in: *Third St.Petersburg Days of Logic and Computability*, St.Petersburg, Russia, Abstracts, 2015, p. 20.
[14] Skvortsov, D., *On superintuitionistic predicate logics of Kripke frames with finite domains, I: Basic notions and the finite model property*, in preparation.
[15] Smoryński, C., *Applications of Kripke models*, in: *Metamathematical Investigation of Intuitionistic Arithmetic and Analysis* (ed. A.S. Troelstra), Lecture Notes Math. **344**, Springer, 1973, pp. 324–391.
[16] Yokota, S., *Axiomatization of the first-order intermediate logics of bounded Kripkean heights, I*, Zeitschr. für math. Logik und Grundl. der Math. **35**, No.5 (1989), pp. 415–421.

Pointwise Intersection in Neighbourhood Modal Logic

Frederik Van De Putte [1]

Ghent University
Blandijnberg 2, 9000 Gent, Belgium
frederik.vandeputte@ugent.be

Dominik Klein [2]

University of Bayreuth & University of Bamberg
Universitätsstraße 30 95447 Bayreuth, Germany
dominik.klein@uni-bayreuth.de

Abstract

We study the logic of neighbourhood models with *pointwise intersection*, as a means to characterize multi-modal logics. Pointwise intersection takes us from a set of neighbourhood sets $\mathcal{N}_i$ (one for each member i of a set G, used to interpret the modality $\Box_i$) to a new neighbourhood set $\mathcal{N}_G$, which in turn allows us to interpret the operator $\Box_G$. Here, X is in the neighbourhood for G if and only if X equals the intersection of some $\mathcal{Y} = \{Y_i \mid i \in G\}$. We show that the notion of pointwise intersection has various applications in epistemic and doxastic logic, deontic logic, coalition logic, and evidence logic. We then establish sound and strongly complete axiomatizations for the weakest logic characterized by pointwise intersection and for a number of variants, using a new and generally applicable technique for canonical model construction.

Keywords: modal logic, neighbourhood semantics, group operators, distributed belief

---

[1] Post-doctoral fellow of the Flemish Research Foundation – FWO Vlaanderen. We are indebted to Olivier Roy and Eric Pacuit for comments on preparatory notes for this paper and to three AiML reviewers for valuable and detailed feedback.

[2] The work of DK was partially supported by the Deutsche Forschungsgemeinschaft (DFG) and Agence Nationale de la Recherche (ANR) as part of the joint project Collective Attitude Formation [RO 4548/8-1] and by DFG and Grantová Agentura České Republiky (GAČR) as part of the joint project From Shared Evidence to Group Attitudes [RO 4548/6-1].

1 Introduction

Neighbourhood semantics is a well-established tool to study generalizations and variants of Kripke-semantics for modal logic. [3] They have been successfully applied to i.a. the logic of ability [4, 24], the dynamics of evidence and beliefs [28], conflict-tolerant deontic logic [13], and the analysis of (descriptive or normative) conditionals [5, 20].

Formally, a neighbourhood function $\mathcal{N} : W \to \wp(\wp(W))$ yields a set of accessible sets $X_1, X_2, \ldots$ of worlds for every given world w in a possible worlds model. $\Box\varphi$ is then true iff there is some such X in the neighbourhood set $\mathcal{N}(w)$, that coincides with the truth set of φ (cf. Definitions 1.1 and 3.1 below).

The move from Kripke semantics to neighbourhood semantics allows us to invalidate certain schemata that are problematic for a given interpretation of the modal operator $\Box$, but also to include other schemata that would trivialize any normal modal logic. [4] Apart from that, neighbourhood models can also be used as a purely technical vehicle in order to arrive at completeness or incompleteness w.r.t. less abstract possible worlds semantics. [5]

Many applications in philosophy and AI require a multitude of modal operators $\Box_1, \Box_2, \ldots$, where the indices may represent agents (logic of agency, doxastic or epistemic logic), non-logical axioms or reasons (logic of provability or normative reasoning), or sources of a norm (deontic logic) or of evidence (doxastic logic once more). Just as for Kripke-semantics, the step from the setting with only one modal operator to a multi-indexed one is easily made, as long as no interaction among the various operators, resp. neighbourhood functions is presupposed. However, the logic of neighbourhood models where certain neighbourhood functions are obtained by operations on (one or several) other neighbourhood functions is still largely unknown. This stands in sharp contrast to the current situation in Kripke-semantics, cf. the literature on Dynamic Logic [16] and on Boolean Modal Logic [10, 11].

The current paper is a first step towards filling this gap. In particular, we study logics that are interpreted in terms of the *pointwise intersection* of neighbourhoods. This concept is defined as follows, for a fixed (finite or infinite) index set $I = \{1, 2, \ldots\}$ and a fixed set of atomic propositions $\mathfrak{P}$.

Definition 1.1 A *model* $\mathfrak{M}$ is a triple $\langle W, \langle \mathcal{N}_i \rangle_{i \in I}, V \rangle$, where $W \neq \emptyset$ is the *domain* of $\mathfrak{M}$, for every $i \in I$, $\mathcal{N}_i : W \to \wp(\wp(W))$ is a *neighbourhood function for* i, and $V : \mathfrak{P} \to \wp(W)$ is a *valuation function.*

Where $\mathfrak{M} = \langle W, \langle \mathcal{N}_i \rangle_{i \in I}, V \rangle$ is a model and $G = \{i_1, \ldots, i_n\} \subseteq I$ is non-empty, the *neighbourhood function for* G is given by

---

[3] Scott [25] and Montague [21] are often seen as the inventors of neighbourhood models; Chellas [6] and Segerberg [26] are usually cited as the main figures in their development.

[4] See Table 1 in Section 5.1 for examples.

[5] One prototypical example of a completeness proof via neighbourhood semantics is [20]. In [15], neighbourhood semantics are used to prove the incompleteness of Elgesem's modal logic of agency [7]. We refer to [23] for a critical introduction to the many forms, uses and advantages of neighbourhood semantics.

$$\mathcal{N}_G(w) = \{X_{i_1} \cap \ldots \cap X_{i_n} \mid \text{ each } X_{i_j} \in \mathcal{N}_{i_j}(w)\}$$

So, in the context of neighbourhood semantics, pointwise intersection takes as input any intersection of neighbourhoods, one for each agent $i \in G$, to form the new neighbourhood set for G. This new neighbourhood set is then used to interpret expressions of the type $\Box_G\varphi$, by means of the standard semantic clause, plugging in the neighbourhood function $\mathcal{N}_G$.[6]

Beside its mathematical interest, pointwise intersection has many potential applications. In Section 2, we briefly point out a few of these. Sections 3–5 form the technical core of the paper, providing (strong) soundness and completeness results for a number of logics interpreted in terms of models with pointwise intersection. We conclude with a summary and some open questions for future work.

2 Applications

What follows is a non-exhaustive list of (potential) applications of logics with pointwise intersection. We leave the full elaboration of these ideas for later occasions, and whenever possible, provide pointers to the literature for more background information.

Epistemic and Doxastic Logic The *distributed knowledge* of a group of agents G can be conceived as the knowledge that would be obtained if some third agent combined the individual knowledge of all group members G and closed the result under logical consequence [1]. The logic of this notion is then defined as an extension of a multi-agent version of **S5**, where each operator $\Box_G$ $(G \subseteq I)$ is interpreted in terms of the intersection of the equivalence relations R_i $(i \in G)$.[7] Analogously, one can study *distributed beliefs* of a group G as the result of aggregating (or pooling) all the beliefs of the members of G. Formally, distributed belief can be seen as all combinations of pieces of belief, one for each agent. When beliefs are conceived as neighbourhoods, the operation of pooling one's beliefs corresponds to a pointwise intersection.

In his [27] Robert Stalnaker has proposed a combined epistemic-doxastic logic that interprets belief as the mental component of knowledge. In the framework, he abandons the assumption that knowledge is negatively introspective. Also positive introspection has been heavily critisized on philosophical grounds. Correspondingly, [19] propose two logics that weaken Stalnaker's framework further by also omitting positive introspection. It turns out that this renders belief a non-normal modality: belief is closed under weakening but

[6] One obvious question, especially if we do not assume that the neighbourhood sets $\mathcal{N}_i(w)$ are closed under intersection, is whether we can also have pointwise intersection of a neighbourhood set with itself. The short answer is: yes, we can, but this takes us beyond the scope of this conference paper. We return to this point in our concluding section.

[7] The logic of distributed knowledge is investigated in the seminal work [8]. A small warning is in place here though. As Gerbrandy [12] shows, the notion of distributed knowledge has both a syntactic and a semantic reading, which are not entirely equivalent. Fagin and co-authors [8], and most others in the field focus on the semantically driven view.

not under intersection, i.e. the agent can believe φ and ψ without believing $\varphi \wedge \psi$.

This is but one example of a non-normal logic for knowledge and belief. All such logics raise the question of defining group attitudes for non-normal modal logics akin to the distributed knowledge and belief defined above. Our results show that group versions of non-normal knowledge and belief can be easily axiomatized, leading to a counterpart of the axiomatization of normal distributed belief provided in [1]. Rather than focussing on one particular axiomatization of non-normal knowledge or belief, we provide a general tool for axiomatizing the corresponding distributed attitudes. To use a slogan: we can throw away the normal modal logic bathwater, while keeping the distributed knowledge/belief baby.

Evidence Logic The framework of Evidence Logic was proposed in [28] to study the way beliefs (of a given agent) are grounded in (possibly conflicting) evidence. Technically, evidence logics are obtained by adding a monotonic operator[8] E for "the agent has evidence for ..." and a belief operator B of the type **KD45** to classical logic. E is characterized semantically in terms of a neighbourhood function $\mathcal{N}$, where $X \in \mathcal{N}(w)$ expresses that at w, the agent has evidence for X. The belief state at a world w is interpreted as the union of all intersections $\bigcap \mathcal{X}$, where $\mathcal{X}$ is a maximal set of evidence such that $\bigcap \mathcal{X} \neq \emptyset$.

Going multi-agent with this framework is fairly straightforward. Here, our results can e.g. be used to study the piecemeal aggregation of evidence from various different sources, and how diverging strategies to do so impact the resulting belief set. Formally, $X \in \mathcal{N}_{\{i,j,k\}}(w)$ indicates that X is a result of aggregating pieces of evidence of the sources i, j, and k. One interesting epistemological question – that can now be studied at a logical level – is whether it makes a difference if one first aggregates the evidence among the sources, before computing a set of beliefs, rather than using the evidence in its original form (ignoring the sources) to ground the beliefs. Our current results are an important step towards answering this question, since we provide completeness results for the "evidence fragment" of such logics.

Deontic Logic Neighbourhood semantics have been used in Deontic Logic to model (non-explosive) conflict-tolerant normative reasoning [13,14]. Here, $\Box_i\varphi$ can e.g. be used to express that there is at least one norm in the normative system S_i that makes φ obligatory; the presence of two conflicting norms in S_i can then account for the truth of a deontic conflict of the type $\Box_i\varphi \wedge \Box_i\neg\varphi$. In this context, pointwise intersection can be interpreted as the piecemeal aggregation of norms from different normative systems; a formula such as $\Box_{\{1,2\}}p$ then expresses that there are two norms, one in S_1, the other in S_2, such that obeying both norms entails that p is the case.

An altogether different application of the formal framework developed here consists in reading the indices as *reasons* for one's obligations. On this view,

[8] A modal operator $\Box$ is *monotonic* in a given system iff it satisfies the rule: from $\varphi \vdash \psi$, to infer $\Box\varphi \vdash \Box\psi$.

$\Box_r\varphi$ expresseses that r is a reason for φ to be obligatory, and one can then aggregate reasons alongside with obligations: $\Box_r\varphi \wedge \Box_{r'}\psi$ yields $\Box_{\{r,r'\}}(\varphi \wedge \psi)$. As argued in [9,22], reasons play an important, but often neglected role in our normative reasoning; a thorough logical investigation of their interaction and aggregation in deontic logic is still largely lacking.

Coalition Logic, group abilities As shown in [3], Pauly's Coalition Logic [24] corresponds to the ability-fragment of **STIT** logic [2,18]. Moreover, this fragment is known to be decidable, in contrast to full **STIT** logic for groups [17]. In Coalition Logic, $\Box_G\varphi$ expresses that "the group of agents G has the ability to ensure that φ is the case", or in more game-theoretic terminology, "G is α-effective for φ". The modality $\Box_G$ is monotonic, meaning that we can only express what one of the group's choices *necessitates* – not what defines that choice. With the results of the current paper, we can now also obtain sound and complete logics for *exact* ability, where $\Box_G\varphi$ means that "G can make a choice that is defined by φ", or in more mundane terms: "G can do exactly φ".

3 Basic intersection logic

In the remainder we use $\mathfrak{M}$, $\mathfrak{M}'$ to refer to arbitrary models as given by Definition 1.1. $X, Y, \ldots$ are used to refer to sets of worlds in a model, and $w, w', \ldots$ for single worlds. We write $G \subseteq_f I$ to denote that G is a finite subset of I.

Let $\mathfrak{L}$ be the language obtained by closing a countable set of propositional variables $\mathfrak{P} = \{p, q, \ldots\}$ and the logical constants $\bot, \top$ under the classical connectives and all unary modal operators of the type $\Box_G$, where $G \subseteq_f I$. We use $\varphi, \psi, \ldots$ as metavariables for formulas and $\Gamma, \Delta, \ldots$ as metavariables for sets of formulas. To interpret $\mathfrak{L}$, we use the models given by Definition 1.1 together with the following (standard) semantic clauses: [9]

Definition 3.1 Where $\mathfrak{M} = \langle W, \langle \mathcal{N}_i \rangle_{i \in I}, V \rangle$ is a model, $w \in W$, $\varphi, \psi \in \mathfrak{L}$, and $G \subseteq_f I$:

0. $\mathfrak{M}, w \not\models \bot$
1. $\mathfrak{M}, w \models \varphi$ iff $w \in V(\varphi)$ for all $\varphi \in \mathfrak{P}$
2. $\mathfrak{M}, w \models \neg\varphi$ iff $\mathfrak{M}, w \not\models \varphi$
3. $\mathfrak{M}, w \models \varphi \vee \psi$ iff $\mathfrak{M}, w \models \varphi$ or $\mathfrak{M}, w \models \psi$
4. $\mathfrak{M}, w \models \Box_G\varphi$ iff $\|\varphi\|^{\mathfrak{M}} \in \mathcal{N}_G(w)$

where $\|\varphi\|^{\mathfrak{M}} = \{w \in W \mid \mathfrak{M}, w \models \varphi\}$.

Validity($\Vdash \varphi$) and semantic consequence ($\Gamma \Vdash \varphi$), for a given class of models, are defined in the standard way, viz. as truth, resp. truth-preservation at all worlds in all models in that class.

In the remainder of this paper, we will consider various logics that are obtained by imposing certain frame conditions on the models defined above. We start with what we call *basic intersection logic*, i.e. the logic characterized

[9] We treat $\bot, \neg, \vee$ as primitive; the other connectives and $\top$ are defined in the standard way.

by the class of all models. To characterize this logic syntactically, we will need the following axioms in addition to classical propositional logic (henceforth, **CL**):

$$\text{where } G \cap H = \emptyset : (\Box_G \varphi \wedge \Box_H \psi) \to \Box_{G \cup H}(\varphi \wedge \psi) \tag{B1}$$

$$\Box_{G \cup H} \top \to \Box_G \top \tag{B2}$$

$$(\Box_G \varphi \wedge \Box_{G \cup H \cup J} \varphi) \to \Box_{G \cup H} \varphi \tag{B3}$$

$$(\Box_G \varphi \wedge \Box_H (\varphi \vee \psi)) \to \Box_{G \cup H} \varphi \tag{B4}$$

and, as usual, replacement of equivalents and modus ponens:

$$\text{if } \varphi \vdash \psi \text{ and } \psi \vdash \varphi, \text{ then } \Box_G \varphi \vdash \Box_G \psi \tag{RE}$$

$$\text{if } \vdash \varphi \text{ and } \vdash \varphi \to \psi, \text{ then } \vdash \psi \tag{MP}$$

Let us quickly offer some interpretations of these axioms. (B1) is an obvious syntactic consequence of taking intersections: If $\|\varphi\|^{\mathfrak{M}}$ is in G's neighbourhood and $\|\psi\|^{\mathfrak{M}}$ is in H's neighbourhood, then $\|\varphi \wedge \psi\|^{\mathfrak{M}}$ is in their intersection neighbourhood whenever G and H are disjoint. Note that the latter restriction is required; without it, the axiom is not sound for basic intersection logic.[10] Axiom (B2) states that W can only be in G's intersection neighbourhood if it is in the neighbourhood of each member of G. (B3) expresses a property of convex closure: if $X \in \mathcal{N}_G(w)$ and $X \in \mathcal{N}_{G \cup H \cup J}(w)$, then for all $i \in H$, there must be a $Y_i \in \mathcal{N}_i(w)$ such that $X \subseteq Y_i$. Consequentially, also $X \in \mathcal{N}_{G \cup H}(w)$. (B4) follows the same reasoning as (B3) but is logically independent. In the appendix we prove the following:

Lemma 3.2 *Axioms (B1)-(B4) are logically independent from each other.*

Before we move to the completeness proof, some terminological remarks are needed. In this and the next section, we use Hilbert-style axiomatizations, with (MP) and (RE) as our only rules. We work with axiom schemata; an axiom is any instance of an axiom schema in $\mathfrak{L}$. Every formula in $\mathfrak{L}$ that can be derived by the axioms and rules is a theorem of the logic. Finally, consequence relations are defined from the respective axiomatizations as follows: $\Gamma \vdash \varphi$ iff there are $\psi_1, \ldots, \psi_n \in \Gamma$ such that $(\psi_1 \wedge \ldots \wedge \psi_n) \to \varphi$ is a theorem. Note that this means that the syntactic consequence relation of the defined logics is by definition compact.

[10] To see why, note that neighbourhood functions are not generally assumed to be closed under intersection: $X, Y \in \mathcal{N}_i(w)$ does not imply $X \cap Y \in \mathcal{N}(w)$. The unrestricted version of (B1) includes the case where $G = H = \{i\}$, which is only sound if neighbourhoods are closed under intersection. We return to this point in Section 5.5.

4 Strong Completeness for basic intersection logic

In this section, we prove the following:

Theorem 4.1 (Strong Completeness for basic intersection logic) *A sound and strongly complete axiomatization of basic intersection logic is obtained by adding (B1), (B2), (B3), and (B4) to any sound and complete axiomatization of* **CL**, *and closing the result under (RE) and (MP).*

The proof of soundness is a matter of routine; it suffices to check that all the axioms are sound with respect to the class of all neighbourhood models. For the completeness proof, we need to construct a canonical model $\mathfrak{M}^c$, in which every world corresponds to a maximal consistent set (MCS) of formulas $\Lambda \subseteq \mathfrak{L}$. The main difficulty here is to construct the $\mathcal{N}_i$ in such a way that (a) if a given formula $\Box_G\varphi$ has to be true at a world w, then the pointwise intersection of the neighbourhoods $\mathcal{N}_i(w)$ for $i \in G$ will contain $\|\varphi\|^{\mathfrak{M}}$, but also (b) if $\neg\Box_G\varphi$ is to be true at world w, then no pointwise intersection of sets in $\mathcal{N}_i(w)$ for $i \in G$ will generate $\|\varphi\|^{\mathfrak{M}}$, i.e. we don't create too many intersection sets.

Let us illustrate this point with a simple example. Note that there are relatively few constraints in picking which formulas $\Box_{\{1,2\}}\varphi$ should be contained in some MCS Λ and which should not. So suppose that $\Box_{\{1,2\}}\varphi \in \Lambda$ for some $\varphi \in \mathfrak{L}$ and some MCS Λ that corresponds to a given world w in $\mathfrak{M}^c$. This means that, to arrive at (a), there should be two sets $X \in \mathcal{N}_1(w)$ and $Y \in \mathcal{N}_2(w)$, such that $X \cap Y = \|\varphi\|^{\mathfrak{M}^c}$. However, to also guarantee (b), these X and Y should be chosen in such a way that neither of them can be combined with other $Z \in \mathcal{N}_i(w)$ (for some $i \in I$) in such a way that this makes additional formulas of the type $\Box_G\psi$ true (with $1 \in G$ or $2 \in G$). Hence, X and Y should be constructed such that they are witnesses to the fact that w verifies $\Box_{\{1,2\}}\varphi$, but that (apart from certain trivial cases) they cannot be used to arrive at any further definable intersections. More specifically, we will ensure that any intersection $Z_1 \cap \ldots \cap Z_n$ that contains only one of X and Y will not yield a definable set that could not be constructed without X or Y.

To be able to have sufficiently many distinguished witnesses, we need to make copies of each MCS Λ. To simplify our construction, we index the various copies of each Λ with functions that encode which neighborhoods that MCS will be part of. The neighborhoods we require are defined by three parameters. The first two are a group of agents G and a formula φ for which the corresponding $\Box_G\varphi$ can be true or false at some MCS Λ. The third parameter is an agent $i \in G$, denoting what this particular agent should contribute to ensure that $\Box_G\varphi$ holds whenever $\Box_G\varphi \in \Lambda$. To be precise, each world w in $\mathfrak{M}^c$ will be defined as a couple, consisting of a MCS Λ and a function f that maps every pair $\langle G, \varphi\rangle$ to a unique member of that G. On the set of these w, we will define neighborhoods $X_i^{G,\varphi}$ that help ensure that $\mathfrak{M}^c, (\Lambda, f) \models \Box_G\varphi \Leftrightarrow \Box_G\varphi \in \Lambda$. These neighborhoods will be constructed like fitting pieces of a jigsaw puzzle. For a given f the set $X_i^{G,\varphi}$ either contains (Λ, f) for all Λ (if $f(G,\varphi) \neq i$) or only those (Λ, f) with $\varphi \in \Lambda$ (if $f(G,\varphi) = i$). The crucial part of the proof is to show that these neighborhoods satisfy (a) and (b). The former is relatively straightforward, the latter is the content of Lemma 4.4 below.

We now make the above ideas exact and turn to the actual proof of completeness.

Definition 4.2 Let $\mathbb{G} = \{G \mid G \subseteq_f I\}$. Let $\mathbb{F}$ denote the set of all functions $f : \mathbb{G} \times \mathfrak{L} \to I$ such that, for all $G \in \mathbb{G}$ and all $\varphi \in \mathfrak{L}$, $f(G, \varphi) \in G$.

The *canonical model for basic intersection logic* is $\mathfrak{M}^c = \langle W^c, \langle \mathcal{N}_i^c \rangle_{i \in I}, V^c \rangle$, where

1. $W^c = \{(\Lambda, f) \mid \Lambda$ is a MCS in $\mathfrak{L}$ and $f \in \mathbb{F}\}$;
2. For all $\varphi \in \mathfrak{P}$, $V^c(\varphi) = \{(\Lambda, f) \in W^c \mid \varphi \in \Lambda\}$
3. for all $i \in I$, $\mathcal{N}_i^c(\Lambda, f) = \{X_i^{G,\varphi} \mid \Box_G \varphi \in \Lambda, i \in G \subseteq_f I\}$ where,
4. for all $(G, \varphi) \in \mathbb{G} \times \mathfrak{L}$ and $i \in G$,

$$X_i^{G,\varphi} = \{(\Lambda, f) \in W^c \mid \varphi \in \Lambda \text{ or } f(G, \varphi) \neq i\}$$

It is not hard to check that $\mathfrak{M}^c$ is well-defined; it suffices to show that W^c is non-empty, which holds in view of the soundness of basic intersection logic, and by a standard Lindenbaum construction.

The real difficulty consists in proving the truth lemma (Lemma 4.6 below). We first observe an important fact about the neighbourhoods constructed in Definition 4.2.4 and prove two auxiliary lemmata that correspond, roughly, to the desiderata (a) and (b) that we discussed above.

Fact 4.3 *If* $\not\vdash \psi$, $|H| \geq 2$ *and* $(H, \psi, j) \neq (H', \psi', j')$, *then* $X_j^{H,\psi} \neq X_{j'}^{H',\psi'}$.

Lemma 4.4 *If* $\mathcal{Y} = \{X_i^{G,\varphi} \mid i \in G\}$, *then* $\bigcap \mathcal{Y} = \{(\Lambda, f) \in W^c \mid \varphi \in \Lambda\}$.

Proof: By Definition 4.2.4,

$$\bigcap_{i \in G} X_i^{G,\varphi} = \bigcap_{i \in G} \{(\Lambda, f) \in W^c \mid \varphi \in \Lambda \text{ or } f(G, \varphi) \neq i\} \tag{1}$$

In view of the definition of $\mathbb{F}$, we know that for every $i \in G$, there is some $f' \in \mathbb{F}$ such that $f'(G, \varphi) = i$. Hence,

$$\bigcap_{i \in G} \{(\Lambda, f) \in W^c \mid \varphi \in \Lambda \text{ or } f(G, \varphi) \neq i\} = \{(\Lambda, f) \in W^c \mid \varphi \in \Lambda\} \tag{2}$$

QED

Lemma 4.5 *Let* $\mathcal{Y}$ *be a set of sets* $X_i^{G,\psi}$ *with* $i \in G$ *and* $(G, \psi) \in \mathbb{G} \times \mathfrak{L}$, *such that for no* (G, φ), $\{X_i^{G,\varphi} \mid i \in G\} \subseteq \mathcal{Y}$. *Then there is an* $f' \in \mathbb{F}$ *such that*

$$\{(\Lambda, f') \in W^c\} \subseteq \bigcap \mathcal{Y} \tag{3}$$

Proof: Suppose the antecedent holds. Let $f' \in \mathbb{F}$ be such that, for every $X_i^{G,\psi} \in \mathcal{Y}$, $f'(G, \psi) = i^{G,\psi}$ for some $i^{G,\psi} \in G$ such that $X_{i^{G,\psi}}^{G,\psi} \notin \mathcal{Y}$. In view of the supposition, there is at least one such f'. Note that, for all $X_i^{G,\psi} \in \mathcal{Y}$, $f'(G,\psi) \neq i$. By Definition 4.2, for all $X_i^{G,\psi} \in \mathcal{Y}$ and all MCS Λ, $(\Lambda, f') \in X_i^{G,\psi}$. Consequently, for all MCS Λ, $(\Lambda, f') \in \bigcap \mathcal{Y}$. **QED**

Lemma 4.6 (Truth Lemma) *For all* $(\Lambda, f) \in W^c$ *and all* $\varphi \in \mathfrak{L}$*:* $\mathfrak{M}^c, (\Lambda, f) \models \varphi$ *iff* $\varphi \in \Lambda$.

Proof: By an induction on the complexity of φ. The base case and the induction step for the classical connectives are safely left to the reader. So it remains to prove that

$$\mathfrak{M}^c, (\Lambda, f) \models \Box_G\varphi \text{ iff } \Box_G\varphi \in \Lambda \tag{TL$\Box$}$$

Right to left direction of (TL$\Box$). Suppose that $\Box_G\varphi \in \Lambda$. By Lemma 4.4,

$$\bigcap_{i\in G} X_i^{G,\varphi} = \{(\Lambda', f') \in W^c \mid \varphi \in \Lambda'\} \tag{4}$$

So by the induction hypothesis (IH), we obtain:

$$\bigcap_{i\in G} X_i^{G,\varphi} = \|\varphi\|^{\mathfrak{M}^c} \tag{5}$$

Moreover, by Definition 4.2.3, for every $i \in G$, $X_i^{G,\varphi} \in \mathcal{N}_i^c(\Lambda, f)$. By Definition 1.1, $\bigcap_{i\in G} X_i^{G,\varphi} \in \mathcal{N}_G(\Lambda, f)$. By Definition 3.1, $\mathfrak{M}^c, (\Lambda, f) \models \Box_G\varphi$.

Left to right direction of (TL$\Box$). Suppose that $\mathfrak{M}^c, (\Lambda, f) \models \Box_G\varphi$. For every $i \in G$, fix a $X_i^{H_i,\psi_i} \in \mathcal{N}_i^c(\Lambda, f)$ such that

$$\bigcap\{X_i^{H_i,\psi_i} \mid i \in G\} = \|\varphi\|^{\mathfrak{M}^c} \tag{6}$$

Let $\mathcal{A}$ be the set of $X_i^{H_i,\psi_i}$ thus fixed. This $\mathcal{A}$ is finite. There are two cases:
Case 1: φ is a tautology of basic intersection logic. By the IH, $\|\varphi\|^{\mathfrak{M}^c} = W$. Hence $X = W$ for all $X \in \mathcal{A}$. In view of Definition 4.2.4, for all $i \in G$, ψ_i is also a tautology and hence, by (RE), $\Box_{H_i}\top \in \Lambda$ for all $i \in G$. By (B2), for all $i \in G$, $\Box_i\top \in \Lambda$. Since G is finite, we can derive $\Box_G\top$ using (B1) finitely many times. By (RE), $\Box_G\varphi \in \Lambda$.
Case 2: φ is not a tautology of basic intersection logic. We first prove that, for some $K \subseteq G$, $\Box_K\varphi \in \Lambda$. Let $\mathcal{B} = \{X_i^{H,\psi} \in \mathcal{A} \mid X_i^{H,\psi} \neq W$ and for all $j \in H : X_j^{H,\psi} \in \mathcal{A}\}$. By Fact 4.3, $\mathcal{B}$ is well-defined, as for each $X \in \mathcal{B}$ either $|H| = 1$ or the H, i, ψ such that $X = X_i^{H,\psi}$ are unique. Hence also $\mathcal{B}^{Grp} := \{H \mid \exists\psi, i : X_i^{H,\psi} \in \mathcal{B}\}$ is finite. Note that for $|H| = 1$, $X_i^{H,\psi} = X_i^{H,\varphi}$ implies that $\vdash \varphi \leftrightarrow \psi$. Since again the H, i, ψ such that $X = X_i^{H,\psi}$ are unique, we can pick a finite $\mathcal{B}^{Fml} \subseteq \{\psi \mid \exists H, i : X_i^{H,\psi} \in \mathcal{B}\}$ such that for each φ in the latter set there is $\psi \in \mathcal{B}^{Fml}$ with $\vdash \varphi \leftrightarrow \psi$. Note that we have

$$\bigcup_{H \in \mathcal{B}^{Grp}} H \subseteq G \tag{7}$$

$$\text{For all } H, H' \in \mathcal{B}^{Grp} : \; H = H' \text{ or } H \cap H' = \emptyset \tag{8}$$

We can now rewrite the intersection of the elements of $\mathcal{A}$ as follows:

$$\bigcap_{X \in \mathcal{A}} X = \bigcap_{X \in \mathcal{A} \setminus \mathcal{B}} X \cap \bigcap_{X \in \mathcal{B}} X \tag{9}$$

By Lemma 4.5, there is an $f' \in \mathbb{F}$ such that

$$\{(\Lambda', f') \in W^c\} \subseteq \bigcap_{X \in \mathcal{A} \setminus \mathcal{B}} X \tag{10}$$

Let $X_i^{H,\psi} \in \mathcal{B}$ be arbitrary. By Lemma 4.4,

$$\bigcap_{j \in H} X_j^{H,\psi} \cap \{(\Lambda', f') \in W^c\} = \{(\Lambda', f') \in W^c \mid \psi \in \Lambda'\} \tag{11}$$

and hence,

$$\bigcap_{X \in \mathcal{B}} X \cap \{(\Lambda', f') \in W^c\} = \bigcap_{\psi \in \mathcal{B}^{Fml}} \{(\Lambda', f') \in W^c \mid \psi \in \Lambda'\} \tag{12}$$

By (9), (10), (12), and the IH,

$$\bigcap_{\psi \in \mathcal{B}^{Fml}} \{(\Lambda', f') \in W^c \mid \psi \in \Lambda'\} = \{(\Lambda', f') \in W^c \mid \varphi \in \Lambda'\} \tag{13}$$

Hence, every MCS that contains every member of $\mathcal{B}^{Fml}$ also contains φ, and vice versa. Since $\mathcal{B}^{Fml}$ is finite, this amounts to:

$$\vdash \bigwedge_{\psi \in \mathcal{B}^{Fml}} \psi \leftrightarrow \varphi. \tag{14}$$

By Definition 4.2.3 and Fact 4.3, $\Box_H \psi \in \Lambda$ for all $\psi \in \mathcal{B}^{Fml}$. Let $K = \bigcup_{H \in \mathcal{B}^{Grp}} H$. Applying (B1) a suitable number of times, and relying on (8), we can derive that $\Box_K \bigwedge_{\psi \in \mathcal{B}^{Fml}} \psi \in \Lambda$. By (RE) and (14),

$$\Box_K \varphi \in \Lambda \tag{15}$$

Let now $i \in G$. Since $\bigcap_{X \in \mathcal{A}} X = \|\varphi\|^{\mathfrak{M}^c}$, it follows that $X_i^{H_i,\psi_i} \supseteq \|\varphi\|^{\mathfrak{M}^c}$. Let $f_i \in \mathbb{F}$ be such that $f_i(H_i, \psi_i) = i$. Hence, $X_i^{H_i,\psi_i} \cap \{(\Lambda, f_i) \in W^c\} = \{(\Lambda, f_i) \in W^c \mid \psi_i \in \Lambda\}$. This implies that $\{(\Lambda, f_i) \in W^c \mid \psi_i \in \Lambda\} \supseteq \{(\Lambda, f_i) \mid \varphi \in \Lambda\}$, and hence

$$\vdash \varphi \to \psi_i \tag{16}$$

Consequently,

$$\vdash \psi_i \leftrightarrow (\varphi \vee \psi_i) \tag{17}$$

By (RE), $\Box_H(\varphi \vee \psi_i) \in \Lambda$. Let $G = \{i_1, \ldots, i_n\}$. Now we apply (B4) n times to derive $\Box_{K \cup H_{i_1}} \varphi$, $\Box_{K \cup H_{i_1} \cup H_{i_2}} \varphi$, etc., untill we finally arrive at $\Box_{K \cup \bigcup_{i \in G} H} \varphi$. Note that $K \subseteq G \subseteq K \cup \bigcup_{i \in G} H$. From this and (15), we can derive that $\Box_G \varphi \in \Lambda$ by (B3). **QED**

5 Some Extensions

We now turn to a number of variants, obtained by imposing certain frame conditions on the neighbourhood functions $\mathcal{N}_i$. As will turn out, quite a number of additional frame conditions on the $\mathcal{N}_i$ do not impact the axiomatization for intersection neighbourhoods at all. Most results provided here will turn out to be relatively straightforward, building on our canonical model construction and completeness proof for basic intersection logic. The proofs of all theorems in this section are slight adaptions of the proof for Theorem 4.1. We offer some details on the proofs in the appendix.

5.1 Some Extensions on the Cheap

We first discuss some axioms, resp. frame conditions that require no changes in the construction of the canonical model, cf. Table 1. (NEC) and (P) are familiar from the study of Kripke-semantics. Adding (CONEC) to any normal modal logic will result in a trivial system; adding (COP) to any normal modal logic will result in a logic where the modal operator becomes useless (since $\Box\varphi$ will be a theorem for all φ). However, in the context of non-normal modal logics, both axioms can sometimes make sense. The axiom (CONEC) is not often mentioned; one of its concrete applications is in (non-normal) logics of agency [7]. The underlying idea is that an agent i cannot (deliberately) bring about a tautology like "the dishes are washed or they are not washed". The axiom (COP) has been used to characterize the notion of "deontic sufficiency" [29], often referred to as "strong permission". Here, $\Box\varphi$ means that every φ-world is a permissible world; the axiom then follows trivially from the fact that no world verifies $\bot$.

As far as these conditions are concerned, our results are modular, in the sense that the frame conditions can be axiomatized independently; and we can

$W \in \mathcal{N}_i(w)$	$\vdash \Box_i \top$	(NEC)
$W \notin \mathcal{N}_i(w)$	$\vdash \neg\Box_i \top$	(CONEC)
$\emptyset \notin \mathcal{N}_i(w)$	$\vdash \neg\Box_i \bot$	(P)
$\emptyset \in \mathcal{N}_i(w)$	$\vdash \Box_i \bot$	(COP)

Table 1
Some extensions on the cheap.

moreover restrict each of them to certain groups G. This means that we can e.g. model cases where only one of the operators $\Box_i$ satisfies necessitation, whereas the others do not.

Theorem 5.1 *The logic of any selection of frame conditions from Table 1 is axiomatized by adding the corresponding axioms from that table to basic intersection logic.*

We should highlight that in the cases of (NEC), (CONEC) and (COP), the corresponding frame condition also holds for the $\mathcal{N}_G(w)$. For instance, as soon as $W \in \mathcal{N}_i(w)$ for all $i \in I$, we can infer that $W \in \mathcal{N}_G(w)$ for all $G \subseteq_f I$. At the syntactic level, this is mirrored by the following property:

Lemma 5.2 *For any extension $\vdash$ of basic intersection logic: if for all $i \in I$, $\vdash \Box_i \top$ (resp. $\vdash \neg\Box_i \top$ or $\vdash \Box_i \bot$), then for all $G \subseteq_f I$, $\vdash \Box_G \top$ (resp. $\vdash \neg\Box_G \top$ or $\vdash \Box_G \bot$).*

In other words, the three mentioned frame conditions and the corresponding axioms readily transfer from single indices to groups. Consequently, imposing these frame conditions on groups rather than individual indices will not make any difference to the logic.

This is not true for (P). It is easy to construct a model with $I = \{1, 2\}$ where $\emptyset \notin \mathcal{N}_i(w)$ for $i \in \{1, 2\}$ and all w, but $\emptyset \in \mathcal{N}_{\{1,2\}}(w)$ for some (or even all) w.

Theorem 5.3 *The logic of frame condition $\emptyset \notin \mathcal{N}_G(w)$ in conjunction with any selection of frame conditions from Table 1 is axiomatized by adding to basic intersection logic the corresponding axioms from that table and all instances of the following axiom schema:*

$$\neg\Box_G \bot \qquad (\mathrm{P}_G)$$

Some combinations of the axioms from Table 1 obviously result in a trivial logic if we use the same G everywhere. Note also that adding (NEC) to basic intersection logic allows us to derive the following theorem, using (B1):

$$\Box_G \varphi \to \Box_{G \cup H} \varphi \qquad (\mathrm{SA})$$

(SA) stands for *superadditivity*, which is the common name used for this type of axiom in logics of (group) agency, (distributed) belief, and (distributed) knowledge. In the presence of (SA), the axioms (B2), (B3), and (B4) become derivable. So we obtain a very simple alternative characterization of the logic of all models where, for all $i \in I$, $W \in \mathcal{N}_i(w)$: all we need is (B1) and (NEC).

5.2 The T-schema

In the remainder of this section, we will point out a few completeness results that are less modular, in the sense that they concern frame conditions that are imposed on all the neighbourhoods $\mathcal{N}_i(w)$ for all $i \in I$ at once, rather than for a selection of them. We start with the T-schema: $\Box\varphi \to \varphi$. Let us call a neighbourhood function *reflexive* iff, for every $w \in W$ and for every $X \in \mathcal{N}(w)$, $w \in X$.

Theorem 5.4 *The logic of the class of models* $\mathfrak{M} = \langle W, \langle \mathcal{N}_i \rangle_{i \in I}, V \rangle$ *where each* $\mathcal{N}_i$ *is reflexive is axiomatized by adding to basic intersection logic all instances of the following axiom schema:*

$$\Box_G\varphi \to \varphi \qquad (\mathrm{T}_G)$$

Importantly, one cannot get a complete axiomatization of reflexivity by just adding the axioms (T_i), i.e. $\Box_i\varphi \to \varphi$ to basic intersection logic. To see this, note that all axioms $\Box_G\phi \to \phi$ for $G \subseteq_f I$ are sound with respect to reflexive frames. The following example of a non-reflexive frame shows that these axioms do not logically follow from $\Box_i\varphi \to \varphi$. We consider a simple case with $I = \{1, 2\}$. Take a model $\mathfrak{M}$ with two worlds, w and v, where all propositional formulas are true at both worlds. Suppose now that $\mathcal{N}_1(w) = \mathcal{N}_1(v) = \{\{w\}\}$ and $\mathcal{N}_2(w) = \mathcal{N}_2(v) = \{\{v\}\}$. Since neither $\{w\}$ nor $\{v\}$ correspond to the truth set of any formula φ in this model, $\Box_1\varphi$ and $\Box_2\varphi$ will be false for every φ, and hence (T_1) and (T_2) will be trivially valid in this model. However, this model does not validate $(\mathrm{T}_{\{1,2\}})$, since $\Box_{\{1,2\}}\bot$ is true at w and at v. So the model satisfies all formulas of the form $\Box_i\phi \to \phi$ together with (B1)-(B2), but not $\Box_G\phi \to \phi$.

5.3 Binary Consistency

In any normal modal logic, the axiom (P) is equivalent to the following axiom:

$$\Box\varphi \to \neg\Box\neg\varphi \qquad (\mathrm{D})$$

However, in neighbourhood models, the two axioms are non-equivalent. Whereas (P) expresses that $W \notin \mathcal{N}(w)$, (D) expresses that if $X \in \mathcal{N}(w)$, then $W \setminus X \notin \mathcal{N}(w)$. By adding indexed variants of the (D)-axiom, we get a complete logic for all frames that satisfy the following frame condition:

Binary consistency: for all $i \in I$: if $X \in \mathcal{N}_i(w)$, then $W \setminus X \notin \mathcal{N}_i(w)$

Theorem 5.5 *The logic of the class of models $\mathfrak{M} = \langle W, \langle \mathcal{N}_i \rangle_{i \in I}, V \rangle$ that satisfy binary consistency is axiomatized by adding to basic intersection logic all instances of the following axiom schema, for all $i \in I$:*

$$\Box_i \varphi \to \neg \Box_i \neg \varphi \tag{D$_i$}$$

5.4 Closure under Supersets

A model $\mathfrak{M} = \langle W, \langle \mathcal{N}_i \rangle_{i \in I}, V \rangle$ is called *monotone* iff, for all $i \in I$ and all $w \in W$, $\mathcal{N}_i(w)$ is closed under supersets. This means that for all $X \in \mathcal{N}_i(w)$, for all $Y \subseteq W$ with $X \subseteq Y$, also $Y \in \mathcal{N}_i(w)$.

Theorem 5.6 *The logic of the class of all monotone models is axiomatized by adding to basic intersection logic all instances of the following axiom schema:*

$$\Box_G \varphi \to \Box_G(\varphi \vee \psi) \tag{RM$_G$}$$

Here we slightly deviate from our standard canonical model construction (Definition 4.2). To ensure that the canonical model falls in the class of monotone models, we need to close all neighbourhoods under supersets (cf. Definition A.1 in the Appendix).

Note that in the presence of (RM$_G$), (RE) becomes a derived rule. Also, it can easily be observed that if we add any (consistent) combination of the axioms (T), (P), (P$_G$), (NEC) to basic intersection logic + (RM$_G$), then we can prove that the associated canonical model $\mathfrak{M}^c_\uparrow$ will be monotone and satisfy the associated frame condition.

5.5 Closure under Finite and Infinite Intersections

In regular neighbourhood modal logic with one modality $\Box$, closure of the neighbourhood function under finite intersections yields the axiom of aggregation (C): $(\Box\varphi \wedge \Box\psi) \to \Box(\varphi \wedge \psi)$. In fact, the logic obtained by adding (RE) and (C) to classical logic is complete for both, the class of frames where the neighbourhood function is closed under finite intersections, and the class of frames where the neighbourhood function is closed under arbitrary intersections. We now generalize this fact to neighbourhood models with pointwise intersection:

Theorem 5.7 *The logic of the class of all models where each $\mathcal{N}_i(w)$ is closed under arbitrary intersections is axiomatized by replacing, in basic intersection logic, the axiom (B1) with its unrestricted counterpart:*

$$(\Box_G \varphi \wedge \Box_H \psi) \to \Box_{G \cup H}(\varphi \wedge \psi) \tag{C$_G$}$$

For the proof, again, we have to deviate slightly from our canonical model construction for basic intersection logic, by closing neighbourhoods under arbitrary intersection. Note that (C$_G$) is also sound for the class of models where the neighbourhood sets are closed under finite intersection. We obtain:

Corollary 5.8 *The logic of the class of all models where each $\mathcal{N}_i(w)$ is closed under finite intersections is axiomatized by adding to basic intersection logic all instances of (C$_G$).*

6 Summary and Outlook

In this paper, we axiomatized the logic of neighbourhood models with pointwise intersection and various extensions obtained by imposing standard frame conditions on the neighbourhoods for the individual indexes. For the canonical model construction in our completeness proof we made use of a specific (new) copying technique. In forthcoming work, we generalize these results, including the operation of pointwise intersection of a neighbourhood set with *itself* and establishing the finite model property for the resulting classes of logics.

Some obvious open questions concern the other (standard) frame conditions that correspond to well-known axioms such as the (4)-axiom, the (5)-axiom, and other "usual suspects" in modal logic. Also, one may consider adding a universal modality to the logics or having multi-modal logics where only some of the individual operators satisfy certain principles (e.g. one non-normal operator for ability, and another normal operator for belief or knowledge), and check to what extent our current techniques can be applied to such extensions.

Our definition of the canonical model, we conjecture, can be easily generalized to axiomatize other operations on neighbourhood functions. One may e.g. define *pointwise union* in an analogous fashion, replacing every occurrence of $\Cap$ in Definition 1.1 with $\Cup$. Drawing inspiration from Dynamic Logic, one may also define operations of sequential composition of neighbourhoods. In sum, we believe that the perspective we have tried to sketch here allows for a plethora of fascinating new logical investigations and philosophical applications.

Appendix

For convenience, we restate theorems and lemmas before proving them.

Lemma 3.2 *Axioms (B1)-(B4) are logically independent from each other.*

Proof: We sketch the argument that (B1)-(B4) are mutually independent. In view of the soundness of these axioms w.r.t. models with pointwise intersection, we can only falsify those axioms in models of a more general type, i.e. where each of the neighbourhood functions $\mathcal{N}_G$ are primitive. We stick to the semantic clauses from Definition 3.1. All our examples work with a set of agents $I = \{1, 2, 3\}$ and a set of worlds $W = \{w_p, w_q, w_r\}$, where the atoms p, q, r are true at w_p, w_q and w_r respectively. The models we construct only differ in their neighbourhood functions. In the following, whenever a neighbourhood $\mathcal{N}_G(w)$ for $G \subseteq \{1, 2, 3\}$ remains unspecified, we assume that $\mathcal{N}_G(w) = \{\emptyset\}$. Moreover, all neighbourhood functions are assumed constant, i.e. $\mathcal{N}_G(w) = \mathcal{N}_G(w')$ for all $w, w' \in W$. We will write $\mathcal{N}$ instead of $\mathcal{N}(w)$.

To see that (B1) is independent of (B2)-(B4), we define model $\mathfrak{M}_1$ as follows: Let $\mathcal{N}_{\{1\}} = \{\{w_p, w_r\}, \emptyset\}$ and $\mathcal{N}_{\{2\}} = \{\{w_q, w_r\}, \emptyset\}$. It is easy to check that (B2)-(B4) are valid on this model. First, the antecedent of (B2) is always false. Second, for (B3) and (B4), the antecedent can only be true if $\|\varphi\|^{\mathfrak{M}_1} = \emptyset$; under this condition, the consequent is easily verified. However, we have that $\mathfrak{M}_1, w_p \models \Box_{\{1\}}(p \vee r) \wedge \Box_{\{2\}}(q \vee r)$ but $\mathfrak{M}_1, w_p \not\models \Box_{\{1,2\}}((p \vee r) \wedge (q \vee r))$, contradicting (B1).

Next, to show that (B2) is independent of (B1),(B3) and (B4), define the model $\mathfrak{M}_2$ by taking neighbourhoods to be $\mathcal{N}_{\{i\}} = \wp(W) \setminus \{1,2,3\}$ for all singletons $\{i\}$ and $\mathcal{N}_G = \wp(W)$ for all $G \subseteq I$ of cardinality at least 2. Note that for all φ and all groups G with cardinality at least 2, $\Box_G\varphi$ is true at all worlds in $\mathfrak{M}_2$. From this one can easily infer that (B1),(B3) and (B4) are valid in $\mathfrak{M}_2$. However, we have $\mathfrak{M}_2, w_p \models \Box_{\{1,2\}}\top \wedge \neg\Box_{\{1\}}\top$ contradicting (B2).

To see that (B3) is independent from (B1),(B2) and (B4) consider model $\mathfrak{M}_3$ with neighbourhood $N_{\{1\}} = N_{\{1,2,3\}} = \{\{w_p\},\emptyset\}$. Again it's easy to see that this model satisfies (B1),(B2) and (B4), but not (B3) as $\mathfrak{M}_3, w_p \models \Box_{\{1\}}p \wedge \Box_{\{1,2,3\}}p$ but $\mathfrak{M}_3, w_p \not\models \Box_{\{1,2\}}p$.

To see that (B4) is independent of (B1)-(B3) consider model $\mathfrak{M}_4$ with neighbourhoods $N_{\{1,3\}} = \{\{w_p\},\emptyset\}$ and $N_{\{1,2\}} = N_{\{1,2,3\}} = \{\{w_p, w_q\},\emptyset\}$. It is easy to see that this model satisfies (B1)-(B3), but not (B4), as $\mathfrak{M}_4, w_p \models \Box_{\{1,3\}}p \wedge \Box_{\{1,2\}}p \vee q$, but $\mathfrak{M}_4, w_p \not\models \Box_{\{1,2,3\}}p$. **QED**

Lemma 5.2 *For any extension $\vdash$ of basic intersection logic: if for all $i \in I$, $\vdash \Box_i\top$ (resp. $\vdash \neg\Box_i\top$ or $\vdash \Box_i\bot$), then for all $G \subseteq_f I$, $\vdash \Box_G\top$ (resp. $\vdash \neg\Box_G\top$ or $\vdash \Box_G\bot$).*

Proof: Assume $\vdash \Box_i\top$ for all $i \in I$ and let $G \subseteq_f I$. Then an iterated application of (B1) yields $\vdash \Box_G\top$. Likewise, if $\vdash \Box_i\bot$, an iterated application of (B1) yields $\vdash \Box_G\bot$. Finally for $\vdash \neg\Box_i\top$ note that by (B2), we have that $\vdash \Box_G\top \rightarrow \Box_j\top$ for all $G \subseteq_f I$ and $j \in G$. Hence. $\vdash \bigwedge_{j\in G}\neg\Box_j\top \rightarrow \neg\Box_G\top$. **QED**

Theorem 5.1 *The logic of any selection of frame conditions from Table 1 is axiomatized by adding the corresponding axioms from that table to basic intersection logic.*

Proof: This is a straightforward adaptation of the original proof. The only additional thing to show is that the $X_i^{G,\phi}$ do not violate any of the four frame conditions. For (NEC) and (COP) this is immediate. For (P) it follows from the fact that $\emptyset \neq X_i^{G,\phi}$ whenever $\not\vdash \bot \leftrightarrow \phi$ or $|G| \geq 2$, together with $\Box_i\bot \notin \Lambda$ for any Λ. For (CONEC) it follows from the fact that $X_i^{G,\phi} \neq W^c$ whenever $\not\vdash \phi \leftrightarrow \top$ together with Lemma 5.2.**QED**

Theorem 5.3 *The logic of frame condition $\emptyset \notin \mathcal{N}_G(w)$ in conjunction with any selection of frame conditions from Table 1 is axiomatized by adding to basic intersection logic the corresponding axioms from that table and all instances of (P_G).*

Proof: Soundness is a matter of routine: one simply checks that the axiom is valid whenever the corresponding frame condition holds.

We briefly sketch the completeness proof for (P); for each of the other three axioms the reasoning is completely analogous. First, we construct the canonical model according to Definition 4.2, with the only difference that our maximal consistent sets are constructed using the stronger logic that also contains the (P)-axiom. We then prove the auxiliary lemmata and the truth lemma, just as before (see Lemmas 4.5, 4.4, and 4.6). By the Truth Lemma, we obtain that $\emptyset \in \mathcal{N}_G(\Lambda, f)$ iff $\Box_G\bot \in \Lambda$. However, for all MCS Λ, we also know that $\neg\Box_G\bot \in \Lambda$. Hence, since every such Λ is consistent, we can infer that $\emptyset \notin \mathcal{N}_G(\Lambda, f)$. **QED**

Theorem 5.4 *The logic of the class of models $\mathfrak{M} = \langle W, \langle \mathcal{N}_i \rangle_{i\in I}, V\rangle$ where each $\mathcal{N}_i$ is reflexive is axiomatized by adding to basic intersection logic all instances of the following axiom schema:*

$$\Box_G\varphi \rightarrow \varphi \tag{T$_G$}$$

Proof: Soundness is again a matter of routine. For completeness, we can again use the canonical model construction from Definition 4.2. The auxiliary lemmata and the truth lemma are proven as before; it suffices to show that the frame condition for (T$_G$) is satisfied. So suppose that $X_i^{G,\varphi}$ is a member of $\mathcal{N}_i(\Lambda, f)$. In view of the construction, (a) $X_i^{G,\varphi}$ is a superset of the set $\{(\Lambda', f') \in W^c \mid \varphi \in \Lambda\}$ and (b) $\Box_G\varphi \in \Lambda$. By (b) and the axiom (T$_G$), also $\varphi \in \Lambda$, and hence by (a), for all $f'' \in \mathbb{F}$, $(\Lambda, f'') \in X_i^{G,\varphi}$. Consequently, $(\Lambda, f) \in X_i^{G,\varphi}$. **QED**

Theorem 5.5 *The logic of the class of models $\mathfrak{M} = \langle W, \langle \mathcal{N}_i \rangle_{i\in I}, V\rangle$ that satisfy binary consistency is axiomatized by adding to basic intersection logic all instances of the following axiom schema, for all $i \in I$:*

$$\Box_i\varphi \rightarrow \neg\Box_i\neg\varphi \tag{D$_i$}$$

Proof: We can again use the same canonical model construction. It suffices to show that in the presence of (D$_i$), this model will satisfy binary consistency. So suppose that $i \in I$, $(\Lambda, f) \in W^c$ and $X \subseteq W^c$ are such that $X, Y \in \mathcal{N}_i(\Lambda, f)$ where $Y = W \setminus X$. Case 1: X is definable, i.e. there is some φ such that $X = \|\varphi\|^{\mathfrak{M}^c}$. In that case, by the truth lemma, $\Box_i\varphi \wedge \Box_i\neg\varphi \in \Lambda$, contradicting the supposition that Λ is consistent and closed under (D$_i$).

Case 2: X and Y are not definable. Note that by the construction of $\mathfrak{M}^c$, $X = X_i^{G,\varphi}$ and $Y = Y_i^{H,\psi}$, with $i \in G \cap H$. Suppose first that $G = \{i\}$ or $H = \{i\}$. Then by Definition 4.2 and the truth lemma, $X = \|\varphi\|^{\mathfrak{M}^c}$ or $Y = \|\psi\|^{\mathfrak{M}^c}$, contradicting the assumption that neither X nor Y are definable. So there are j, k such that $j \in G \setminus \{i\}$ and $k \in H \setminus \{i\}$. Let now $f' \in \mathbb{F}$ be such that $f'(G, \varphi) = j$ and $f'(H, \varphi) = k$, and let Λ be an arbitrary MCS. Note that $(\Lambda, f') \in X_i^{G,\varphi} \cap Y_i^{H,\psi}$ by the construction of $\mathfrak{M}^c$. Hence, $X \cap Y \neq \emptyset$, contradicting the supposition that $Y = W \setminus X$. **QED**

Theorem 5.6 *The logic of the class of all monotone models is axiomatized by adding to basic intersection logic all instances of the following axiom schema:*

$$\Box_G\varphi \rightarrow \Box_G(\varphi \vee \psi) \tag{RM$_G$}$$

Proof: Soundness is a matter of routine. For completeness, we need a slightly different construction. Let $\mathfrak{M}^c = \langle W^c, \langle \mathcal{N}_i^c \rangle_{i\in I}, V^c\rangle$ be defined as before – see Definition 4.2. Now, define $\mathfrak{M}^c_\uparrow$ as follows:

Definition A.1 $\mathfrak{M}^c_\uparrow = \langle W^c, \langle \mathcal{N}_i^{c\uparrow}\rangle_{i\in I}, V^c\rangle$, where for all $(\Lambda, f) \in W^c$, $\mathcal{N}_i^{c\uparrow}(\Lambda, f)$ is the closure of $\mathcal{N}_i^c(\Lambda, f)$ under supersets: $\mathcal{N}_i^{c\uparrow}(\Lambda, f) = \{Y \subseteq W^c \mid$ for an $X \in \mathcal{N}_i^c(\Lambda, f), X \subseteq Y\}$.

Note that lemmas 4.5 and 4.4 are preserved, since these only concern the sets $X_i^{G,\varphi}$ that are used in the construction of each $\mathcal{N}_i$. The truth lemma however needs to be proved anew. Again, the crucial point is to prove the induction step for $\Box_G$:

$$\mathfrak{M}^c_\uparrow, (\Lambda, f) \models \Box_G\varphi \text{ iff } \Box_G\varphi \in \Lambda \qquad (\mathrm{TL}\Box\uparrow)$$

For the right-to-left direction of (TL□ ↑), we can simply repeat the proof of the right-to-left direction of (TL□). For left-to-right, some small changes are required, which we spell out here.

Suppose that $\mathfrak{M}^c_\uparrow, (\Lambda, f) \models \Box_G\varphi$. By the semantic clause for $\Box_G$, there is a $\mathcal{Z} = \{Z_i \mid i \in G\}$ such that each $Z_i \in \mathcal{N}_i^{c\uparrow}(\Lambda, f)$ and $\bigcap_{i\in G} Z_i = \|\varphi\|^{\mathfrak{M}^c_\uparrow}$. By the construction, for each $Z_i \in \mathcal{Z}$ there is an $X_i \in \mathcal{N}_i(\Lambda, f)$ such that $X_i \subseteq Z_i$. Hence,

$$\bigcap_{i\in G} X_i \subseteq \|\varphi\|^{\mathfrak{M}^c_\uparrow} \qquad (\mathrm{A.1})$$

We define $\mathcal{A}$ as before. Note that, in view of the preceding, each $\mathcal{N}_i^{c\uparrow}(\Lambda, f)$ with $i \in G$ is non-empty. This implies that for all $i \in G$, there is some ψ_i and some G_i that contains i, such that $\Box_{G_i}\psi_i \in \Lambda$. By (B2), $\Box_i\psi_i \in \Lambda$ and hence by (RM_G), also

$$\Box_i\top \in \Lambda \text{ for all } i \in G \qquad (\mathrm{A.2})$$

Case 1: φ is a tautology. By (A.2), using (B1) $\Box_G\top \in \Lambda$. By (RE), $\Box_G\varphi \in \Lambda$.

Case 2: φ is not a tautology. Define $\mathcal{B}$ as before. We can now reason just as before, but instead of deriving an identity, we get at the following set inclusion:

$$\bigcap_{(H,\psi)\in\mathcal{B}} \{(\Lambda', f') \in W^c \mid \psi \in \Lambda'\} \subseteq \{(\Lambda', f') \in W^c \mid \varphi \in \Lambda'\} \qquad (\mathrm{A.3})$$

Hence, every MCS that contains every member of $\{\psi \mid (H, \psi) \in \mathcal{B}\}$ also contains φ. Since $\mathcal{B}$ is finite, this gives us:

$$\vdash \bigwedge_{(H,\psi)\in\mathcal{B}} \psi \to \varphi. \qquad (\mathrm{A.4})$$

By Definition 4.2.3, $\Box_H\psi \in \Lambda$ for all $(H,\psi) \in \mathcal{B}$. Let $K = \bigcup_{(H,\psi)\in\mathcal{B}} H$. Note that, by (7), $K \subseteq G$. Applying (B1) a suitable number of times, we can derive that $\Box_K \bigwedge_{(H,\psi)\in\mathcal{B}} \psi \in \Lambda$. By (RM$_G$) and (A.4),

$$\Box_K\varphi \in \Lambda \tag{A.5}$$

From there, we can follow the exact same reasoning as that in the proof for basic intersection logic, starting after equation (15). **QED**

Theorem 5.7 *The logic of the class of all models where each $\mathcal{N}_i(w)$ is closed under arbitrary intersections is axiomatized by replacing, in basic intersection logic, the axiom (B1) with its unrestricted counterpart:*

$$(\Box_G\varphi \wedge \Box_H\psi) \rightarrow \Box_{G\cup H}(\varphi \wedge \psi) \tag{C$_G$}$$

Proof: Soundness is again a matter of routine. For completeness we close all the neighbourhood functions of the canonical model for basic intersection logic under intersection:

Definition A.2 $\mathfrak{M}^c_\cap = \langle W^c, \langle \mathcal{N}_i^{c\cap}\rangle_{i\in I}, V^c\rangle$, where for all $(\Lambda, f) \in W^c$, $\mathcal{N}_i^{c\cap}(\Lambda, f)$ is the closure of $\mathcal{N}_i^c(\Lambda, f)$ under (possibly infinite) intersections: $\mathcal{N}_i^{c\cap}(\Lambda, f) = \{\bigcap \mathcal{Y} \mid \mathcal{Y} \subseteq \mathcal{N}_i^c(\Lambda, f)\}$.

Again, right-to-left of the truth lemma for $\Box_G$ is easy, since we only added neighbourhoods to the original canonical model. For left-to-right, we need a slightly different reasoning. Suppose that $\mathfrak{M}^{c\cap}, (\Lambda, f) \models \Box_G\varphi$. So there is a $\mathcal{Z} = \{Z_i \mid i \in G\}$ such that each $Z_i \in \mathcal{N}_i^{c\cap}(\Lambda, f)$, and $\bigcap \mathcal{Z} = \|\varphi\|^{\mathfrak{M}^{c\cap}}$. By the definition of $\mathfrak{M}^{c\cap}$, for every $Z_i \in \mathcal{Z}$ there is a $\mathcal{X}_i \subseteq_f \mathcal{N}_i(\Lambda, f)$ such that $Z_i = \bigcap \mathcal{X}_i$. Let $\mathcal{X} = \bigcup_{i\in G} \mathcal{X}_i$. Note that $\bigcap \mathcal{X} = \bigcap \mathcal{Z}$. Let $\mathcal{A} = \{(H,\psi) \in \mathbb{G} \times \mathfrak{L} \mid X_i^{H,\psi} \in \mathcal{X}$ for some $i \in H\}$ and let $\mathcal{B} = \{(H,\psi) \in \mathbb{G} \times \mathfrak{L} \mid X_i^{H,\psi} \in \mathcal{X}$ for all $i \in H\}$. Note that for all $(H,\psi) \in \mathcal{B}$, $H \subseteq G$.

We now reason as before, deriving the following equation:

$$\bigcap_{(H,\psi)\in\mathcal{B}} \{(\Lambda, f) \in W^c \mid \psi \in \Lambda\} = \{(\Lambda, f) \in W^c \mid \varphi \in \Lambda\} \tag{A.6}$$

In other words, every maximal consistent set that contains all ψ for $(H,\psi) \in \mathcal{B}$ also contains φ, and vice versa. Note however that $\mathcal{B}$ needn't be finite. By the compactness of our syntactic consequence relation however, it follows that there is a finite $\mathcal{C} \subseteq \mathcal{B}$ such that:

$$\bigwedge_{(H,\psi)\in\mathcal{C}} \psi \leftrightarrow \varphi \tag{A.7}$$

Put $K = \bigcup_{(H,\psi)\in\mathcal{C}} H$. In view of the preceding, $K \subseteq G$. From there, we reason as before, deriving that $\Box_K\varphi \in \Lambda$, and finally also that $\Box_G\varphi \in \Lambda$. **QED**

References

[1] Ågotnes, T. and Y. N. Wáng, *Resolving distributed knowledge*, Artificial Intelligence **252** (2017), pp. 1 – 21.

[2] Belnap, N., P. M., X. M. and B. P., "Facing the Future: Agents and Choice in Our Indeterminist World," Oxford University Press, 2001.

[3] Broersen, J., A. Herzig and N. Troquard, *A normal simulation of coalition logic and an epistemic extension*, in: D. Samet, editor, *Proceedings of the 11th Conference on Theoretical Aspects of Rationality and Knowledge*, TARK '07 (2007), pp. 92–101.

[4] Brown, M., *On the logic of ability*, Journal of Philosophical Logic **17** (1988), pp. 1–26.

[5] Carmo, J. M. C. L. M. and A. J. I. Jones, *Deontic logic and contrary-to-duties*, in D. Gabbay and F. Guenthner, editors, Handbook of Philosophical Logic, Vol. **8**, Kluwer Academic Publishers, 2002, 2nd edition, pp. 147–264.

[6] Chellas, B., "Modal Logic: an Introduction," Cambridge: Cambridge University Press, 1980.

[7] Elgesem, D., *The modal logic of agency*, Nordic J. Philos. Logic **2** (1997), p. 146.

[8] Fagin, R., J. Y. Halpern, Y. Moses and M. Y. Vardi, "Reasoning About Knowledge," MIT Press, Cambridge, Massachusetts, 2003.

[9] Faroldi, F. L. G. and T. Protopopescu, *Hyperintensional logics of reasons* (2017), manuscript.

[10] Gargov, G. and S. Passy, *A note on boolean modal logic*, in: P. P. Petkov, editor, *Mathematical Logic*, Springer US, 1990 pp. 299–309.

[11] Gargov, G., S. Passy and T. Tinchev, *Modal environment for boolean speculations*, in: *Mathematical Logic and its Applications*, Plenum, New York, 1987, pp. 253–263 .

[12] Gerbrandy, J., *Distributed knowledge*, in: J. Hulstijn and A. Nijholt, editors, *Twendial 98: Formal Semantics and Pragmatics of Dialogue, TWLT 13*, Universiteit Twente, Enschede, 1998, pp. 111–124.

[13] Goble, L., *A logic for deontic dilemmas*, Journal of Applied Logic **3** (2005), pp. 461–483.

[14] Goble, L., *Prima facie norms, normative conflicts, and dilemmas*, in D. Gabbay, L. van der Torre, J. Horty, and X. Parent, editors, Handbook of Deontic Logic and Normative Systems, Vol. 1, College Publications, 2013 pp. 241–351.

[15] Governatori, G. and A. Rotolo, *On the axiomatisation of Elgesem's logic of agency and ability*, Journal of Philosophical Logic **34** (2005), pp. 403–431.

[16] Harel, D., D. Kozen and J. Tiuryn, "Dynamic Logic," Cambridge, MA: MIT Press, 2000.

[17] Herzig, A. and F. Schwarzentruber, *Properties of logics of individual and group agency*, in: C. Areces and R. Gobldblatt, editors, *Advances in Modal Logic* (2008).

[18] Horty, J. F., "Agency and Deontic Logic," Oxford University Press, New York, 2001.

[19] Klein, D., O. Roy and N. Gratzl, *Knowledge, belief, normality, and introspection*, Synthese (2017), pp. 1–30.

[20] Lewis, D., "Counterfactuals," Harvard University Press, Cambridge, Mass., 1973.

[21] Montague, R., *Universal grammar*, Theoria **36** (1970), pp. 373–398.

[22] Nair, S. and J. F. Horty, "The Oxford Handbook of Reasons and Normativity," USA: Oxford University Press, forthcoming.

[23] Pacuit, E., "Neighbourhood Semantics for Modal Logic," Springer, 2017.

[24] Pauly, M., *A modal logic for coalitional power in games*, Journal of Logic and Computation **1** (2002), pp. 149–166.

[25] Scott, D., *Advice on Modal Logic,* in: *Philosophical Problems in Logic. Some Recent Developments*, Reidel, Dordrecht, 1970 pp. 143–173.

[26] Segerberg, K., "An Essay in Classical Modal Logic", Filosofiska Studier, No. 13, Uppsala Universitet, Uppsala, 1971.

[27] Stalnaker, R., *On logics of knowledge and belief*, Philosophical Studies **128** (2006), pp. 169–199.

[28] van Benthem, J. and E. Pacuit, *Dynamic logics of evidence-based beliefs*, Studia Logica **99** (2011), pp. 61–92.

[29] Van De Putte, F., *That will do: Logics of deontic necessity and sufficiency*, Erkenntnis **82** (2017), pp. 473–511.

When Names Are Not Commonly Known: Epistemic Logic with Assignments

Yanjing Wang

Department of Philosophy, Peking University

Jeremy Seligman

Department of Philosophy, University of Auckland

Abstract

In standard epistemic logic, agent names are usually assumed to be common knowledge implicitly. This is unreasonable for various applications. Inspired by term modal logic and assignment operators in dynamic logic, we introduce a lightweight modal predicate logic where names can be non-rigid. The language can handle various *de dicto* / *de re* distinctions in a natural way. The main technical result is a complete axiomatisation of this logic over S5 models.

Keywords: term modal logic, axiomatisation, non-rigid constants, dynamic logic

1 Introduction

One dark and stormy night, Adam was attacked and killed. His assailant, Bob, ran away, but was seen by a passer-by, Charles, who witnessed the crime from start to finish. This led quickly to Bob's arrest. Local news picked up the story, and that is how Dave heard it the next day, over breakfast. Now, in one sense we can say that both Charles and Dave know that Bob killed Adam. But there is a difference in what they know about just this fact. Although Charles witnessed the crime, and was able to identify the murderer and victim to the police, he might have no idea about their names. If asked "Did Bob kill Adam?" he may not know. Yet this is a question that Dave could easily answer, despite not knowing who Adam and Bob are—he is very unlikely to be able to identify them in a line-up.

The distinction between these *de re* and *de dicto* readings of "knowing Bob killed Adam" is hard to make in standard epistemic logic, where it is implicitly assumed that the names of agents are rigid designators and thus that it's common knowledge to whom they refer. But in many cases, the distinction is central to our understanding. On the internet, for example, users of websites and other online applications typically have multiple identities, and may even be anonymous. Names are rarely a matter of common knowledge and distinctions as to who knows who is whom are of great interest.

Further complexities arise with higher-order knowledge and belief. In [7], Grove gives an interesting example of a robot with a mechanical problem calling out for help (perhaps in a Matrix-like future with robots ruling the world unaided by humans). To plan further actions, the broken robot, called a, needs to know if its request has been heard by the maintenance robot, called b. But how to state exactly what a *needs to know*? In English we would probably write it as:

($\star$) a knows that b knows that a needs help.

A naive formulation in standard (predicate) epistemic logic is $\mathsf{K}_a\mathsf{K}_bH(a)$. But without the assumption that the robots' names are both commonly known, there are various ambiguities. For example, if b does not know which robot is named 'a' then neither does b know whom to help nor has a any confidence of being helped. On the other hand, a may not know that 'b' is the name of the maintenance robot, thus merely knowing b knows a needs help is not enough for a to be sure it will be helped. The authors of [4] list several possible readings of ($\star$), which we will elaborate as follows: a, the broken robot, knows that

(i) the robot named 'b' knows that the robot named 'a' needs help, or

(ii) the robot named 'b' knows that it, i.e. the broken robot, needs help, or

(iii) the maintenance robot knows that the robot named 'a' needs help, or

(iv) the maintenance robot knows that it, i.e. the broken robot, needs help.

It is impossible to distinguish the above readings in standard epistemic logic. In the literature [8,7,4,5,11], various approaches are proposed. In [7], Grove correctly pinpoints the problems of *scope* and *manner of reference* in giving various *de re* / *de dicto* readings for higher-order knowledge, and proposes a new semantics for 2-sorted first-order modal logic that is based on world-agent pairs, so as to cope with indexicals like "`me`". A special predicate symbol '`In`' is introduced to capture scope explicitly: $\texttt{In}(a, b, n)$ holds at a world w iff b is someone named n by a in w. In [5,20], an intensional first-order modal logic uses predicate abstraction to capture different readings. $(\lambda x.\mathsf{K}_bHx)(a)$ says that agent b knows *de re* that a is in need of help, whether or not b knows that agent is named 'a', whereas $\mathsf{K}_bH(a)$ says that b knows *de dicto* that someone called 'a' needs help, whether or not b knows who a is. The authors of [4] propose a very general framework with complex operators based on counterpart semantics.[1] Without going into details, the formula $|t : {}^{t_1 \ldots t_n}_{x_1 \ldots x_n}|\varphi(x_1 \ldots x_n)$ means, roughly, that the agent named by term t knows *de re* that φ of the things denoted by terms $t_1 \ldots t_n$. Holliday and Perry also bring the alethic modality into the picture together with the doxastic modality, and highlight the use of *roles* to capture subtle readings in [11], where the multi-agent cases are handled by perspective switching based on a single-agent framework.

[1] The counterpart semantics helps to handle the situation in which one agent is mistakenly considered as two people, as illustrated in [4] by the story of the double agent in Julian Symon's novel *The Man who killed himself.*

In this paper, we follow the *dynamic term modal logic* approach proposed by Kooi [13], based on *term modal logic* proposed in [6]. Term modal logic uses terms to index modalities which can also be quantified, so that $\mathsf{K}_{f(a)}\neg\forall x\mathsf{K}_x\varphi$ says that a's father knows that not everyone knows φ. The accessibility relation used in the semantics of K_t, where t is a term, is then relative to the world w at which this formula is evaluated: it is the one labeled by the agent denoted by term t in w. Based on this, Kooi [13] borrows dynamic assignment modalities from (first-order) dynamic logic so as to adjust the denotation of names, now assumed to be non-rigid in general, in contrast to the usual *constants* of first-order modal logic which are assumed rigid.

Full first-order term modal language is clearly undecidable. In [16], it is shown that even its propositional fragment is undecidable, [2] and the addition of the program modalities in dynamic logic makes things worse. As Kooi remarks in [13], the combination of term modal logic and dynamic assignment logic is not even recursively enumerable. A closely related study is the doctoral thesis of Thalmann [20], which provides many results including sequent calculi and tableaux systems for both term modal logic (with quantifiers) and quantifier-free dynamic assignment logic (with regular program constructions). But the two logics are studied *separately*, leaving their combination as future work. [3] It is shown that the quantifier-free part of dynamic assignment logic is undecidable with both (Kleene) star operator and (rigid) function symbols but it is decidable if there is no star operator. [4] A rich treatment of various issues of 'semantic competence' with names that uses term modal logic is given by Rendsvig in [18].

In this paper, we take a minimalist approach, introducing only the basic assignment modalities from dynamic logic combined with a quantifier-free term modal logic, without function symbols, to obtain a small fragment of the logic in [13], which we conjecture to be decidable over S5 models (see discussions at the end of the paper). However, as we will soon see, it is already a very powerful tool for expressing various *de re/de dicto* distinctions, as well as a kind of *knowing who*, which was discussed by Hintikka [10] at the very inception of epistemic logic. [5] The language is very simple and intuitive to use as a genuine multi-agent epistemic logic that does not presuppose common knowledge of names.

Before the formal details, let us first illustrate the ideas. As in predicate epistemic logic more generally, the formula $\mathsf{K}_a Pb$ says that a knows *de dicto* that b is P, whereas $\mathsf{K}_a Px$ says that a knows *de re* of x that it is P. The formula $[x := b]Px$ says of b that it is P, which is equivalent to Pb, but

---

[2] Only the monodic fragment is decidable [16].

[3] Thalmann predicts in his conclusion that "*Using the scoping operator instead of the quantifiers in term-modal logic, should lead to many interesting decidable fragments of term-modal logic.*"

[4] The later is only stated without a proof.

[5] See [22] for a summary of related works on knowing-wh.

combining operators we get $[x := b]\mathsf{K}_a Px$, which says that a knows *de re* of b that it is P. More precisely, our semantics is based on first-order Kripke models with a constant domain of agents (not names) with formulas evaluated with respect to both a world w and a variable assignment function σ. Formula $[x := t]\varphi$ is then true iff φ is true at w when we change σ so that it assigns to x the agent named by t in w, and $\mathsf{K}_t\varphi$ is true iff φ is true at all worlds indistinguishable from w by the agent named t in w. (This is in line with the *innermost-scope* semantics of [7].)

Returning to Grove's poor broken robot a, the various readings of 'a knows that b knows that a needs help' can be expressed as follows:

(i) $\mathsf{K}_a\mathsf{K}_b H(a)$, a knows that the robot named 'b' knows that the robot named 'a' needs help,

(ii) $[x := a]\mathsf{K}_a\mathsf{K}_b H(x)$, a knows that the robot named 'b' knows it (the broken robot a) needs help,

(iii) $[y := b]\mathsf{K}_a\mathsf{K}_y H(a)$, a knows that it (the maintenance robot b) knows the robot named 'a' needs help,

(iv) $[x := a][y := b]\mathsf{K}_a\mathsf{K}_y H(x)$, a knows that it (the maintenance robot b) knows that it (the broken robot a) needs help.

Moreover, since names are non-rigid, we can express *a knowing who b is* by $[x := b]\mathsf{K}_a(x \approx b)$ which says that a identifies the right person with name b on all relevant possible worlds. This we abbreviate as $\mathsf{K}_a b$.[6] Thus we are able to express the following:

(v) $\neg\mathsf{K}_a a$: a does not know he is called a (c.f., "the most foolish person may not know that he is the most foolish person" in [13]).

(vi) $b \approx c \wedge \mathsf{K}_a b \wedge \neg\mathsf{K}_a c$: a knows who b is but does not know who c is, although they are just two names of the same person.

(vii) $[x := b][y := a](\mathsf{K}_c M(x, y) \wedge \neg\mathsf{K}_c(a \approx x \wedge y \approx b))$: Charles knows who killed whom that night but does not know the names of the murderer and the victim.

(viii) $\mathsf{K}_d M(b, a) \wedge \neg\mathsf{K}_d a \wedge \neg\mathsf{K}_d b$: Dave knows that a person named Bob murdered a person named Adam without knowing who they are.

The innocent look of our logical language belies some technical complexity. The main technical result is a complete axiomatisation of the logic over epistemic (S5) models (Section 3 and 4), requires much work to handle the constant domain without Barcan-like formulas. We conclude with discussions on the issues of decidability of our logic (Section 5).

---

[6] There are a lot of different readings of *knowing who*. E.g., knowing who went to the party may be formalized as $\forall x(\mathsf{K}W(x) \vee \mathsf{K}\neg W(x))$ under an exhaustive interpretation [21]. See [1,2] for a very powerful treatment using *conceptual covers* to give different interpretations. [17] also contains related discussions.

2 Preliminaries

In this section we introduce formally the language and semantics of our logic.

Definition 2.1 (Epistemic language with assignments) *Given a denumerable set of names* $\boldsymbol{N}$*, a denumerable set of variables* $\mathbf{X}$*, and a denumerable set* $\boldsymbol{P}$ *of predicate symbols, the language* **ELAS** *is defined as:*

$$t ::= x \mid a$$

$$\varphi ::= (t \approx t) \mid P\mathbf{t} \mid \neg\varphi \mid (\varphi \wedge \varphi) \mid \mathsf{K}_t\varphi \mid [x := t]\varphi$$

where $a \in \boldsymbol{N}$*,* $P \in \boldsymbol{P}$*, and* $\mathbf{t}$ *is a vector of terms of length equal to the arity of predicate* P*. We write* $\widehat{\mathsf{K}}_t\varphi$ *as the the abbreviation of* $\neg\mathsf{K}_t\neg\varphi$ *and write* $\langle x := t\rangle\varphi$ *as the abbreviation of* $\neg[x := t]\neg\varphi$*.*[7] *We call the* $[x := t]$*-free fragment* **EL**.

We define the semantics of **ELAS** over first-order Kripke models.

Definition 2.2 *A constant domain Kripke model* $\mathcal{M}$ *for* **ELAS** *is a tuple* $\langle W, I, R, \rho, \eta\rangle$ *where:*

- W *is a non-empty set of possible worlds.*
- I *is a non-empty set of agents.*
- $R : I \to 2^{W\times W}$ *assign a binary relation* $R(i)$ *(also written* R_i*) between worlds, to each agent* i*.*
- $\rho : \boldsymbol{P} \times W \to \bigcup_{n\in\omega} 2^{I^n}$ *assigns an* n*-ary relation* $\rho(P, w)$ *between agents to each* n*-ary predicate* P *at each world* w*.*
- $\eta : \boldsymbol{N} \times W \to I$ *assigns an agent* $\eta(n, w)$ *to each name* n *at each world* w*.*

We call $\mathcal{M}$ *an* epistemic model *if* R_i *is an equivalence relation for each* $i \in I$*.*

Note that the interpretations of predicates and names are not required to be rigid, and there may be worlds in which an agent has no name or multiple names. To interpret free variables, we need a variable assignment $\sigma : \mathbf{X} \to I$. Formulas are interpreted on pointed models $\mathcal{M}, w$ with variable assignments σ. Given an assignment σ and a world $w \in W$, let $\sigma_w(a) = \eta(a, w)$ and $\sigma_w(x) = \sigma(x)$. So although names may not be rigid, variables are.

The truth conditions are given w.r.t. pointed Kripke models with assignments $\mathcal{M}, w, \sigma$.

Definition 2.3

$$\begin{array}{rcl}
\mathcal{M}, w, \sigma \vDash t \approx t' & \Leftrightarrow & \sigma_w(t) = \sigma_w(t') \\
\mathcal{M}, w, \sigma \vDash P(t_1 \cdots t_n) & \Leftrightarrow & (\sigma_w(t_1), \cdots, \sigma_w(t_n)) \in \rho(P, w) \\
\mathcal{M}, w, \sigma \vDash \neg\varphi & \Leftrightarrow & \mathcal{M}, w, \sigma \nvDash \varphi \\
\mathcal{M}, w, \sigma \vDash (\varphi \wedge \psi) & \Leftrightarrow & \mathcal{M}, w, \sigma \vDash \varphi \textit{ and } \mathcal{M}, w, \sigma \vDash \psi \\
\mathcal{M}, w, \sigma \vDash \mathsf{K}_t\varphi & \Leftrightarrow & \mathcal{M}, v, \sigma \vDash \varphi \textit{ for all } v \textit{ s.t. } wR_{\sigma_w(t)}v \\
\mathcal{M}, w, \sigma \vDash [x := t]\varphi & \Leftrightarrow & \mathcal{M}, w, \sigma[x \mapsto \sigma_w(t)] \vDash \varphi
\end{array}$$

[7] This is for comparison with other modal logics; in fact, the assignment modality is self-dual.

An **ELAS** *formula is valid (over epistemic models) if it holds on all the (epistemic) models with assignments $\mathcal{M}, s, \sigma$.*

We can translate **ELAS** into the corresponding (2-sorted) first-order language with not only the equality symbol but also a ternary relation symbol R for the accessibility relation, a function symbol f^a for each name a, and an $n+1$-ary relation symbol Q^P for each predicate symbol P. The non-trivial clauses are for K_t and $[x := t]\varphi$ based on translation for terms:

$$\begin{array}{l}
\mathsf{Tr}_w(x) = x \qquad \mathsf{Tr}_w(a) = f^a(w) \\
\mathsf{Tr}_w(t \approx t') = \mathsf{Tr}_w(t) \approx \mathsf{Tr}_w(t') \qquad \mathsf{Tr}_w(P\boldsymbol{t}) = Q^P(w, \mathsf{Tr}_w(\boldsymbol{t})) \\
\mathsf{Tr}_w(\neg\psi) = \neg\mathsf{Tr}_w(\psi) \qquad \mathsf{Tr}_w(\varphi \wedge \psi) = \mathsf{Tr}_w(\varphi) \wedge \mathsf{Tr}_w(\psi). \\
\mathsf{Tr}_w(\mathsf{K}_t\psi) = \forall v(R(w, v, \mathsf{Tr}_w(t)) \to \mathsf{Tr}_v(\psi)) \\
\mathsf{Tr}_w([x := t]\psi) = \begin{cases} \exists x(x \approx \mathsf{Tr}_w(t) \wedge \mathsf{Tr}_w(\psi)) & \text{if } t \neq x \\ \mathsf{Tr}_w(\psi) & \text{if } t = x \end{cases}
\end{array}$$

Note that when $t \neq x$ we can also (equivalently) define $\mathsf{Tr}_w([x := t]\psi) = \forall x(x \approx \mathsf{Tr}_w(t) \to \mathsf{Tr}_w(\psi))$, since there is one and only one value of $\mathsf{Tr}_w(t)$. When $x = t$, then $[x := x]\psi$ is equivalent to ψ according to the semantics.

In the light of this translation, we can define the free and bound occurrences of a variable in an **ELAS**-formula by viewing $[x := t]$ in $[x := t]\varphi$ as a quantifier for x binding φ. Note that the t in $[x := t]$ is not bound in $[x := t]\varphi$, even when $t = x$. The set of free variables $\mathsf{Fv}(\varphi)$ in φ is defined as follows (where $\mathsf{Var}(\boldsymbol{t})$ is the set of variables in the terms $\boldsymbol{t}$):

$$\begin{array}{ll}
\mathsf{Fv}(P\boldsymbol{t}) = \mathsf{Var}(\boldsymbol{t}) & \\
\mathsf{Fv}(\neg\varphi) = \mathsf{Fv}(\varphi) & \mathsf{Fv}(\varphi \wedge \psi) = \mathsf{Fv}(\varphi) \cup \mathsf{Fv}(\psi) \\
\mathsf{Fv}(\mathsf{K}_t\varphi) = \mathsf{Var}(t) \cup \mathsf{Fv}(\varphi) & \mathsf{Fv}([x := t]\varphi) = (\mathsf{Fv}(\varphi) \setminus \{x\}) \cup \mathsf{Var}(t)
\end{array}$$

We use $\varphi[y/x]$ to denote the result of substituting y for all the free occurrences of x in φ. We say $\varphi[y/x]$ is *admissible* if all the occurrences of y by replacing free occurrences of x in φ are also free.

We first show that **ELAS** is indeed more expressive than **EL**.

Proposition 2.4 *The assignment operator $[x := t]$ cannot be eliminated over (epistemic) models with variable assignments.*

Proof. Consider the following two (epistemic) models (reflexive arrows omitted) with a fixed domain $I = \{i, j\}$, worlds $W = \{s_1, s_2\}$ and a fixed assignment $\sigma(x) = i$ for all $x \in \mathbf{X}$:

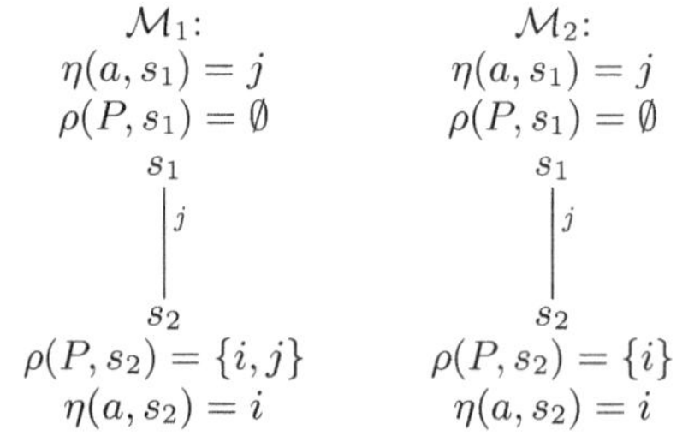

$[x := a]\widehat{\mathsf{K}}_a Px$ can distinguish $\mathcal{M}_1, s_1$ and $\mathcal{M}_2, s_1$ given σ. But the only atomic formulas other than identities are Px, Pa and $a \approx x$, which are all false at s_1 and all true at s_2 in both models. Also note that K_a and K_x have exactly the same interpretation on the corresponding worlds in the two models. Based on these observations, a simple inductive proof on the structure of formulas would show that **EL** cannot distinguish the two models given σ. □

Interested readers may also wonder whether we can eliminate $[x := t]$ in each **ELAS** formula to obtain an **EL** formulas which is *equally satisfiable.* However, the naive idea of translating $[x := t]\varphi$ into $z \approx t \wedge \varphi[z/x]$ with fresh z will not work in formulas like $\mathsf{K}_a[x := c]\widehat{\mathsf{K}}_b x \not\approx c$ since the name c is not rigid.

To better understand the semantics, the reader is invited to examine the following valid and invalid formulas over epistemic models:

1	valid	$x \approx y \to \mathsf{K}_t x \approx y$, $\quad x \not\approx y \to \mathsf{K}_t x \not\approx y$
	invalid	$x \approx a \to \mathsf{K}_t x \approx a$, $\quad x \not\approx a \to \mathsf{K}_t x \not\approx a$, $\quad a \approx b \to \mathsf{K}_t a \approx b$
2	valid	$\mathsf{K}_x\varphi \to \mathsf{K}_x\mathsf{K}_x\varphi$, $\quad \neg\mathsf{K}_x\varphi \to \mathsf{K}_x\neg\mathsf{K}_x\varphi$, $\quad \mathsf{K}_t\varphi \to \varphi$.
	invalid	$\mathsf{K}_t\varphi \to \mathsf{K}_t\mathsf{K}_t\varphi$, $\quad \neg\mathsf{K}_t\varphi \to \mathsf{K}_t\neg\mathsf{K}_t\varphi$
3	valid	$[x := y]\varphi \to \varphi[y/x]$ ($\varphi[y/x]$ is admissible)
	invalid	$[x := a]\mathsf{K}_t Px \to \mathsf{K}_t Pa$
4	valid	$x \approx a \to (\mathsf{K}_x\varphi \to \mathsf{K}_a\varphi)$, $\quad a \approx b \to (Pa \to Pb)$
	invalid	$x \approx a \to (\mathsf{K}_b Px \to \mathsf{K}_a Pa)$, $\quad a \approx b \to (\mathsf{K}_c Pa \to \mathsf{K}_c Pb)$
5	valid	$[x := y]\mathsf{K}_a\varphi \to \mathsf{K}_a[x := y]\varphi$
	invalid	$[x := b]\mathsf{K}_a Px \to \mathsf{K}_a[x := b]Px$

Remark 2.5 Here are some brief explanations:

1: It shows the distinction between (rigid) variables and (non-rigid) names. The invalid formula shows that although two names co-refer, you may not know it (recall Frege's puzzle).

2: Axioms 4 and 5 do not work for names in general, since a may not know that he is named 'a'. On the other hand, positive and negative introspection hold when the index is a variable. The T axiom works in general.

3: It also demonstrates the non-rigidity of names. $[x := a]\mathsf{K}_b Px$ does not imply $\mathsf{K}_b Pa$ since b may consider a world possible where Pa does not hold since a on that world does not refer to the actual person named by a in the real world.

4: This shows that it is fine to do the first-level substitutions for the equal names but not in the scope of other modalities.

5: The last pair also demonstrates the distinction between rigid variables and non-rigid names. In particular, the analog of *Barcan formula* $[x := t]\mathsf{K}_s\varphi \to \mathsf{K}_s[x := t]\varphi$ is not in general valid, if t is a name.

3 Axiomatisation

In this section we give a complete axiomatisation of valid **ELAS**-formulas over epistemic models. The axioms and rules can be categorised into several classes:

- For normal propositional modal logic: TAUT, DISTK, MP, NECK;
- Axiom for epistemic conditions: Tx, 4x, 5x;
- Axioms for equality and first-level substitutability: ID, SUBP, SUBK, SUBAS;
- Axioms capturing rigidity of variables: RIGIDP and RIGIDN;
- Properties of assignment operator: KAS (normality), DETAS (determinacy), DAS (executability), and EFAS (the effect of the assignment).
- Quantifications: SUB2AS and NECAS, as in the usual first-order setting (viewing assignments as quantifiers).

System SELAS

Axioms

TAUT	Propositional tautologies	SUBAS	$t \approx t' \to$ $([x := t]\varphi \leftrightarrow [x := t']\varphi)$
DISTK	$\mathsf{K}_t(\varphi \to \psi) \to (\mathsf{K}_t\varphi \to \mathsf{K}_t\psi)$		
Tx	$\mathsf{K}_x\varphi \to \varphi$	RIGIDP	$x \approx y \to \mathsf{K}_t x \approx y$
4x	$\mathsf{K}_x\varphi \to \mathsf{K}_x\mathsf{K}_x\varphi$	RIGIDN	$x \not\approx y \to \mathsf{K}_t x \not\approx y$
5x	$\neg\mathsf{K}_x\varphi \to \mathsf{K}_x\neg\mathsf{K}_x\varphi$	KAS	$[x := t](\varphi \to \psi) \to$ $([x := t]\varphi \to [x := t]\psi)$
ID	$t \approx t$		
SUBP	$\boldsymbol{t} \approx \boldsymbol{t'} \to (P\boldsymbol{t} \leftrightarrow P\boldsymbol{t'})$ (P can be $\approx$)	DETAS	$\langle x := t\rangle\varphi \to [x := t]\varphi$
		DAS	$\langle x := t\rangle\top$
SUBK	$t \approx t' \to (\mathsf{K}_t\varphi \leftrightarrow \mathsf{K}_{t'}\varphi)$	EFAS	$[x := t]x \approx t$
		SUB2AS	$\varphi[y/x] \to [x := y]\varphi$ ($\varphi[y/x]$ is admissible)

Rules:

MP $\dfrac{\varphi, \varphi \to \psi}{\psi}$ NECK $\dfrac{\vdash \varphi}{\vdash \mathsf{K}_t\varphi}$ NECAS $\dfrac{\vdash \varphi \to \psi}{\vdash \varphi \to [x := t]\psi}$ $(x \notin \mathsf{Fv}(\varphi))$

where $\boldsymbol{t} \approx \boldsymbol{t'}$ means point-wise equivalence for sequences of terms $\boldsymbol{t}$ and $\boldsymbol{t'}$ such that $|\boldsymbol{t}| = |\boldsymbol{t'}|$. It is straightforward to verify the soundness of the system.

Theorem 3.1 (Soundness) SELAS *is sound over epistemic models with assignments.*

Proposition 3.2 *The following are derivable in the above proof system (where $\varphi[y/x]$ is admissible below in* SUBASEQ*) :*

SYM	$t \approx t' \to t' \approx t$	TRANS	$t \approx t' \land t' \approx t'' \to t \approx t''$
DBASEQ	$\langle x := t\rangle\varphi \leftrightarrow [x := t]\varphi$	SUBASEQ	$\varphi[y/x] \leftrightarrow [x := y]\varphi$
EAS	$[x := t]\varphi \leftrightarrow \varphi$ $(x \notin \mathsf{Fv}(\varphi))$	T	$\mathsf{K}_t\varphi \to \varphi$
CNECAS	$\dfrac{\vdash \varphi \to \psi}{\vdash [x := t]\varphi \to \psi}$ $(x \notin \mathsf{Fv}(\psi))$	NECAS'	$\dfrac{\vdash \varphi}{\vdash [x := t]\varphi}$
EX	$[x := x]\varphi \leftrightarrow \varphi$		

Proof. (Sketch) SYM and TRANS are trivial based on ID and SUBP. DBASEQ is based on DETAS and DAS. SUBASEQ is due to the contrapositive of SUB2AS and DBASEQ. CNECAS is due to NECAS and DBASEQ for contrapositive. EAS is based on NECAS and CNECAS (taking $\psi = \varphi$). EX is a special case of SUB2AS and NECAS′

is a special case of NECAS. As a more detailed example, let us look at the proof (sketch) of T (we omit the rountine steps using the normality of $[x := t]$):

$$\begin{array}{ll}
(1) \vdash \mathsf{K}_t\varphi \to (z \approx t \to \mathsf{K}_z\varphi) & (\texttt{SUBK},\ z \text{ is fresh})\\
(2) \vdash \mathsf{K}_t\varphi \to [z := t](z \approx t \to \mathsf{K}_z\varphi) & (\texttt{NECAS(1)})\\
(3) \vdash \mathsf{K}_t\varphi \to [z := t](\mathsf{K}_z\varphi) & (\texttt{EFAS(2)})\\
(4) \vdash [z := t](\mathsf{K}_z\varphi \to \varphi) & (\texttt{NECAS}', \texttt{Tx})\\
(5) \vdash \mathsf{K}_t\varphi \to [z := t]\varphi & (\text{normality of } [z{:=}t] \text{ and } \texttt{MP})\\
(6) \vdash \mathsf{K}_t\varphi \to \varphi & (\texttt{EAS}, \texttt{MP})
\end{array}$$

□

Based on the above result, we can reletter the bound variables in any **ELAS** formula like in first-order logic.

Proposition 3.3 (Relettering) *Let z be a fresh variable not in φ and t, then*

$$[x := t]\varphi \leftrightarrow [z := t]\varphi[z/x]$$

.

Proof. Since z is fresh, $\varphi[z/x]$ is admissible. We have the following proof (sketch):

$$\begin{array}{ll}
(1) \vdash \varphi[z/x] \leftrightarrow [x := z]\varphi & (\texttt{SUBASEQ})\\
(2) \vdash [z := t]\varphi[z/x] \leftrightarrow [z := t][x := z]\varphi & (\text{normality of } [z := t])\\
(3) \vdash [z := t]\varphi[z/x] \leftrightarrow [z := t](z \approx t \wedge [x := z]\varphi) & (\texttt{EFAS})\\
(4) \vdash [z := t]\varphi[z/x] \leftrightarrow [z := t](z \approx t \wedge [x := t]\varphi) & (\texttt{SUBAS})\\
(5) \vdash [z := t]\varphi[z/x] \leftrightarrow [z := t][x := t]\varphi & (\texttt{EFAS})\\
(6) \vdash [z := t]\varphi[z/x] \leftrightarrow [x := t]\varphi & (\texttt{EAS})
\end{array}$$

□

4 Completeness

To prove the completeness, besides the treatments of $[x := t]$ and termed modality K_t, the major difficulty is the lack of the Barcan-like formulas in **ELAS**, which are often used to capture the condition of the constant domain. As in standard first-order logic, we need to provide witnesses for each name, and the Barcan formula can make sure we can always find one when building a successor of some maximal consistent set with enough witnesses. On the other hand, we can build an increasing domain pseudo model without such a formula using the techniques in [12]. Inspired by the techniques in [3], to obtain a constant domain model, when building the successors in the increasing domain pseudo model, we only create a new witness if all the old ones are not available, and we make sure by formulas in the maximal consistent sets that the new one is not equal to any old ones (throughout the whole model). In this way, there will not be any conflicts between the witnesses when we collect all of them together. We may then create a constant domain by considering the equivalence classes of all the witnesses occurring in the pseudo model with an increasing domain.

Here is the general proof strategy:

- Extend the language with countably many new variables.
- Build a pseudo canonical frame using maximal consistent sets for various sublanguages of the extended language, with witnesses for the names.
- Given a maximal consistent set, cut out its generated subframe from the pseudo frame, and build a constant-domain canonical model, by taking certain equivalence classes of variables as the domain.
- Show that the truth lemma holds for the canonical model.
- Take the reflexive symmetric transitive closure of the relations in pseudo model and show that the truth of the formulas in the original language are preserved.
- Extend each consistent set of the original model to a maximal consistent set with witnesses.

We first extend the language **ELAS** with countably infinitely many new variables, and call the new language $\mathbf{ELAS}^+$ with the variable set $\mathbf{X}^+$. We say a language L is an *infinitely proper sublanguage* of another language L' if:

- L and L' only differ in their sets of variables,
- $L \subseteq L'$,
- there are infinitely many new variables in L' that are not in L.

We use maximal consistent sets w.r.t. different infinitely proper sublanguages of $\mathbf{ELAS}^+$ that are extensions of **ELAS** to build a pseudo canonical frame.

Definition 4.1 (Pseudo canonical frame) *The pseudo canonical frame* $\mathcal{F}^c = \langle W, R \rangle$ *is defined as follows:*

- W *is the set of MCS* Δ *w.r.t. some infinitely proper sublanguages* L_Δ *of* $\mathbf{ELAS}^+$ *such that for each* $\Delta \in W$*:*
  - · $\mathbf{ELAS} \subseteq L_\Delta$,
  - · *For each* $a \in \boldsymbol{N}$ *there is a variable* x *in* L_Δ *(notation:* $x \in \mathsf{Var}(\Delta)$*) such that* $x \approx a \in \Delta$ *(call it* $\exists$*-property)*
- *For each* $x \in \boldsymbol{X}^+$, $\Delta R_x \Theta$ *iff the following three conditions hold:*

(i) x *in* $\mathsf{Var}(\Delta)$*, the set of variables in* L_Δ*.*
(ii) $\{\varphi \mid \mathsf{K}_x\varphi \in \Delta\} \subseteq \Theta$.
(iii) *if* $y \in \mathsf{Var}(\Theta) \setminus \mathsf{Var}(\Delta)$ *then* $y \not\approx z \in \Theta$ *for all* $z \in \mathsf{Var}(\Theta)$ *such that* $z \neq y$.

Observation The last condition for R_x makes sure that every new variable in the successor is distinguished from any other variables by inequalities. It is also easy to see that if $t \in L_\Delta$ then there is $x \in \mathsf{Var}(\Delta)$ such that $x \approx t \in \Delta$ by $\exists$-property and `ID`.

Proposition 4.2 *If* $\Delta R_x \Theta$ *in* $\mathcal{F}^c$*, then:*

- L_Δ *is a sublanguage of* L_Θ
- *for any* $y \neq z \in \mathsf{Var}(\mathbf{ELAS}^+)$*:* $y \approx z \in \Delta$ *iff* $y \approx z \in \Theta$.

Proof. For the first: For all $y \in \mathsf{Var}(\Delta)$, $y \approx y \in \Delta$ therefore $\mathsf{K}_x(y \approx y) \in \Delta$

by RIGIDP, thus $y \approx y \in \Theta$.

For the second:

- Suppose $y, z \in \mathsf{Var}(\Delta)$
  - If $y \approx z \in \Delta$, then $\mathsf{K}_x y \approx z \in \Delta$ by RIGIDP, thus $y \approx z \in \Theta$.
  - If $y \approx z \notin \Delta$ then $y \not\approx z \in \Delta$ since Δ is an L_Δ-MCS and $y, z \in \mathsf{Var}(\Delta)$. Then $\mathsf{K}_x y \not\approx z \in \Delta$ by RIGIDN, thus $y \approx z \notin \Theta$.
- Suppose w.o.l.g. $y \notin \mathsf{Var}(\Delta)$ thus $y \approx z \notin \Delta$.
  - If $y \notin \mathsf{Var}(\Theta)$ or $z \notin \mathsf{Var}(\Theta)$ then then $y \approx z \in \Delta$ iff $y \approx z \in \Theta$ trivially holds.
  - If $y \in \mathsf{Var}(\Theta)$ and $z \in \mathsf{Var}(\Theta)$ then $y \not\approx z \in \Theta$ due to the third condition of R_x. Therefore $y \approx z \notin \Theta$ since Θ is consistent. Thus $y \approx z \in \Delta$ iff $y \approx z \in \Theta$.

□

The second part of the above proposition makes sure that we do not have conflicting equalities in different states which are accessible from one to another.

Lemma 4.3 (Existence lemma) *If $\Delta \in W$ and $\widehat{\mathsf{K}}_t\varphi \in \Delta$ then there is a $\Theta \in W$ and an $x \in \mathsf{Var}(L_\Delta)$ such that $\varphi \in \Theta$, $x \approx t \in \Delta$, and $\Delta R_x \Theta$.*

Proof. If $\widehat{\mathsf{K}}_t\varphi \in \Delta$ then there is $x \approx t \in \Delta$ for some x, due to the fact that Δ has the $\exists$-property. Let $\Theta^{--} = \{\psi \mid \mathsf{K}_x\psi \in \Delta\} \cup \{\varphi\}$. We first show that Θ^{--} is consistent by DISTK and NECK (routine). Next we show that it can be extended to a state in W. We can select an infinitely proper sublanguage L of $\mathbf{ELAS}^+$ such that L_Δ is an infinitely proper sublanguage of L. We can list the new variables in L but not in L_Δ by $y_0, y_1, y_2, \dots$. We also list the names in $\mathbf{N}$ as $a_0, a_1, \dots$. In the following, we add the witness to the names by building Θ_i as follows:

- $\Theta_0 = \Theta^{--}$
- $\Theta_{k+1} = \begin{cases} \Theta_k & \text{if } x \approx a_k \text{ is in } \Theta_k \text{ for some } x \in \mathsf{Var}(\Delta) \ (1) \\ \Theta_k \cup \{x_i \approx a_k\} & \text{if (1) does not hold but } \{x \approx a_k\} \cup \Theta_k \\ & \quad \text{is consistent for some } x \in \mathsf{Var}(\Delta), \\ & \quad \text{and } x_i \text{ is the first such } x \text{ according to} \\ & \quad \text{a fixed enumeration of } \mathsf{Var}(\Delta) \qquad (2) \\ \Theta_k \cup \{y_j \approx a_k\} \cup & \text{if neither (1) nor (2) holds and} \\ \{y_j \not\approx z \mid z \in \mathsf{Var}(\Theta_k)\} & \quad y_j \text{ is the first in the enumeration} \\ & \quad \text{of the new variables not in } \Theta_k \qquad (3) \end{cases}$

We can show that Θ_k is always consistent. Note that Θ_0 is consistent, we just need to show if Θ_k is consistent and (1), (2) do not hold, then Θ_{k+1} is consistent too. Suppose for contradiction that $\Theta_k \cup \{y_j \approx a_k\} \cup \{y_j \not\approx z \mid z \in \mathsf{Var}(\Theta_k)\}$ is not consistent then there are fomulas $\psi_1 \dots \psi_n \in \Theta_k$, and $z_{i_1} \dots z_{i_m} \in \mathsf{Var}(\Theta_k)$ such that:

$$\vdash \psi_1 \wedge \cdots \wedge \psi_n \wedge y_j \approx a_k \to \bigvee_{i \in \{i_1, \dots, i_m\}} y_j \approx z_i \quad (\star)$$

First note that since y_j is not in Θ_k, $\Theta_k \cup \{y_j \approx a_k\}$ is consistent, for otherwise there are $\psi_1 \dots \psi_n \in \Theta_k$ such that $\vdash \bigwedge_{i \leq n} \psi_i \to y_j \not\approx a_k$, then by NECAS, we have $\vdash \bigwedge_{i \leq n} \psi_i \to [y_j := a_k] y_j \not\approx a_k$ thus by EFAS we have $\vdash \bigwedge_{i \leq n} \psi_i \to [y_j := a_k](a_k \approx y_j \wedge a_k \not\approx y_j)$, contradicting to the consistency of Θ_k (by DETAS). By ($\star$), $\Theta_k \cup \{y_j \approx a_k\}$ is consistent with one of $y_j \approx z_i$ for some z_i in $\mathsf{Var}(\Theta_k)$. Thus $\Theta_k \cup \{z_i \approx a_k\}$ is also consistent which contradicts the assumption that condition (2) and (1) do not hold.

Then we define Θ^- to be the union of all Θ_k. Clearly, Θ^- has the $\exists$-property. We build the language L' based on $\mathsf{Var}(\Theta^-)$. Note that L' is still an infinitely proper sublanguage of $\mathbf{ELAS}^+$.

Finally, we extend Θ^- into an MCS w.r.t. L' and it is not hard to show $\Delta R_x \Theta$ by verifying the third condition: when we introduce a new variable we always make sure it is differentiated with the previous one in the construction of Θ. □

Given a state Γ in $\mathcal{F}^c$, we can define an equivalence relation $\sim_\Gamma$: $x \sim_\Gamma y$ iff $x \approx y \in \Gamma$ or $x = y$ (note that $x \approx x$ is *not* in Γ if $x \notin L_\Gamma$). Due to ID, SYM, TRANS, $\sim_\Gamma$ is indeed an equivalence relation. When Γ is fixed, we write $|x|$ for the equivalence class of x w.r.t. $\sim_\Gamma$. By definition, for all $x \notin \mathsf{Var}(\Gamma)$, $|x|$ is a singleton.

Now we are ready to build the canonical model.

Definition 4.4 (Canonical model) *Given a Γ in $\mathcal{F}^c$ we define the canonical model $\mathcal{M}_\Gamma = \langle W_\Gamma, I^c, R^c, \rho^c, \eta^c \rangle$ based on the psuedo canonical frame $\langle W, R \rangle$*

- *W_Γ is the subset of W generated from Γ w.r.t. the relations R_x.*
- *$I^c = \{|x| \mid x \in \mathsf{Var}(W_\Gamma)\}$ where $\mathsf{Var}(W_\Gamma)$ is the set of all the variables appearing in W_Γ.*
- *$\Delta R^c_{|x|} \Theta$ iff $\Delta R_x \Theta$, for any $\Delta, \Theta \in W_\Gamma$.*
- *$\eta^c(a, \Delta) = |x|$ iff $a \approx x \in \Delta$.*
- *$\rho^c(P, \Delta) = \{|\boldsymbol{x}| \mid P\boldsymbol{x} \in \Delta\}$.*

Here is a handy observation.

Proposition 4.5 *If $y \in \mathsf{Var}(\Delta) \setminus \mathsf{Var}(\Gamma)$ then $y \not\approx z \in \Delta$ for all $z \in \mathsf{Var}(\Delta)$ such that $z \neq y$.*

Proof. Due to the condition 3 of the relation R_x in $\mathcal{F}^c$ and Proposition 4.2, and the fact that W_Γ is generated from Γ. □

Proposition 4.6 *The canonical model is well-defined.*

Proof.

- For $R_{|x|}$: We show that the choice of the representative in $|x|$ does not change the definition. Suppose $x \sim_\Gamma y$ then either $x = y$ or $x \approx y \in \Gamma$. In the first case, $\Delta R_x \Theta$ iff $\Delta R_y \Theta$. In the second case, suppose $\Delta R_x \Theta$. We show that the three conditions for $\Delta R_y \Theta$ hold. For condition 1, $y \in \mathsf{Var}(\Delta)$ since $y \in \mathsf{Var}(\Gamma)$ and Δ is generated from Γ by R. For condition 2, we just need to note that $\vdash y \approx x \to (\mathsf{K}_x\varphi \leftrightarrow \mathsf{K}_y\varphi)$ by SUBK. And condition 3 is given directly by condition 3 for $\Delta R_x \Theta$.

- For $\eta(a, \Delta)$: We first show that the choice of the representative in $|x|$ does not change the definition by $\vdash x \approx y \to (a \approx x \leftrightarrow a \approx y)$ (TRANS). Then we need to show that $\eta^c(a, \Delta)$ is unique. Note that due to the $\exists$-property, there is always some x such that $x \approx a$ in Δ. Suppose towards contradiction that $a \approx x \in \Delta$, $a \approx y \in \Delta$ and $x \not\sim_\Gamma y$ then clearly x, y cannot be both in $\mathsf{Var}(\Gamma)$ for otherwise $x \not\approx y \in \Delta$. Suppose w.l.o.g. x is not in $\mathsf{Var}(\Gamma)$ then we should have $x \not\approx y \in \Delta$ due to Proposition 4.5, contradicting the assumption that Δ is consistent.

□

Proposition 4.7 $R_{|x|}$ *is transitive.*

Proof. Suppose $\Delta R_{|x|} \Theta$ and $\Theta R_{|x|} \Lambda$ then in $\mathcal{F}^c$ $\Delta R_x \Theta$ and $\Theta R_x \Lambda$ (note that the representative of $|x|$ does not really matter since $R_{|x|}$ is well-defined). We have to show the three conditions for $\Delta R_x \Lambda$. For condition 1, $x \in \mathsf{Var}(\Delta)$ since $\Delta R_x \Theta$. For condition 2, by Axiom **4x**, we have for any φ such that $\mathsf{K}_x\varphi \in \Delta$ we have $\mathsf{K}_x\mathsf{K}_x\varphi \in \Delta$ thus $\mathsf{K}_x\varphi \in \Theta$ thus $\varphi \in \Lambda$, by the definition of R_x. For condition 3, suppose $y \in \mathsf{Var}(\Lambda) \setminus \mathsf{Var}(\Delta)$. Then since $\Delta \in W$, $y \notin \mathsf{Var}(\Gamma)$, so by Proposition 4.5 we are done. □

Before proving the truth lemma, we have two simple observations:

Proposition 4.8 .

(1) If $x \approx y$ is in some $\Delta \in W_\Gamma$ then $x \sim_\Gamma y$.

(2) If $x \sim_\Gamma y$ then $x = y$ or $x \approx y$ in all the $\Delta \in W_\Gamma$.

Proof. For the first, suppose $x \approx y$ is in some $\Delta \in W_\Gamma$. We just need to consider the case when $x \neq y$ for if $x = y$ then $x \sim_\Gamma y$ by definition. By Proposition 4.5, x and y must be both in $\mathsf{Var}(\Gamma)$, thus by RIGIDN and the fact that Δ is connected to Γ, $x \approx y \in \Gamma$.

The second is immediate by the definition of $\sim_\Gamma$: if $x \approx y \in \Gamma$ then $x \approx y \in \Delta$ due to RIGIDP and the fact that all the $\Delta \in W_\Gamma$ are connected to Γ.

□

Although $R_{|x|}$ is transitive, the model $\mathcal{M}_\Gamma$ is not reflexive nor symmetric in general. For the failure of reflexivity, note that some x may not be in the language of some state. For the failure of symmetry: We may have $\Delta R_{|x|} \Theta$ and $L_\Delta \subset L_\Theta$ thus it is not the case that $\Theta R_{|x|} \Delta$ by Proposition 4.2. We will turn this model into an S5 model later on. Before that we first prove a (conditional) truth lemma w.r.t. $\mathcal{M}_\Gamma$ and the canonical assignment σ^* such that $\sigma^*(x) = |x|$ for all $x \in \mathsf{Var}(W_\Gamma)$.

Lemma 4.9 (Truth lemma) *For any $\varphi \in$ **ELAS**$^+$ and any $\Delta \in W$, if $\varphi \in L_\Delta$ then:*

$$\mathcal{M}_\Gamma, \Delta, \sigma^* \vDash \varphi \Leftrightarrow \varphi \in \Delta$$

Proof. We do induction on the structure of the formulas.

For the case of $t \approx t' \in L_\Delta$, by $\exists$-property we have some $x, y \in \mathsf{Var}(L_\Delta)$ such that $t \approx x \in \Delta, t' \approx y \in \Delta$.

- Suppose $t \approx t' \in \Delta$. Since $t \approx x \in \Delta, t' \approx y \in \Delta$, by TRANS, $x \approx y \in \Delta$. Now by Proposition 4.8, $x \sim_\Gamma y$. Thus $\sigma^*(t, \Delta) = |x| = |y| = \sigma^*(t', \Delta)$, then $\mathcal{M}_\Gamma, \Delta, \sigma^* \vDash t \approx t'$.
- If $\mathcal{M}_\Gamma, \Delta, \sigma^* \vDash t \approx t'$ then $\sigma^*(t, \Delta) = \sigma^*(t', \Delta)$. If t and t' are variables x, y, then $|x| = |y|$ i.e., $x \sim_\Gamma y$. By Proposition 4.8, either $x = y$ or $x \approx y \in \Delta$. Actually, even if $x = y$, since x is in $\mathsf{Var}(\Delta)$, $x \approx x \in \Delta$ by ID. If t and t' are both in $\mathbf{N}$, then by the definition of η^c, there are $t \approx x$ and $t \approx y$ in Δ and $y \in |x|$, which means $x \sim_\Gamma y$. By Proposition 4.8 and ID again, $x \approx y \in \Delta$ therefore $t \approx t' \in \Delta$. Finally, w.l.o.g. if $t \in \mathsf{Var}(\Delta)$ and $t' \in \mathbf{N}$, then by definition of η, $t' \approx x \in \Delta$ for some x and $t \sim_\Gamma x$. Again, since $x, t \in \mathsf{Var}(\Delta)$, $x \approx t \in \Delta$ therefore $t \approx t' \in \Delta$.

For the case of $Pt \in L_\Delta$.

- If $Pt \in \Delta$, then by $\exists$-property, there are $\boldsymbol{x}$ in $\mathsf{Var}(\Delta)$ such that $\boldsymbol{x} \approx \boldsymbol{t} \in \Delta$. Then by SUBP we have $P\boldsymbol{x} \in \Delta$. Thus by the definition of ρ^c, $|\boldsymbol{x}| \in \rho^c(P, \Delta)$. By the definitions of σ^* and η^c, $\boldsymbol{\sigma^*(t, \Delta)} = |\boldsymbol{x}|$. Therefore $\mathcal{M}_\Gamma, \Delta, \sigma^* \vDash P\boldsymbol{t}$.
- If $\mathcal{M}_\Gamma, \Delta, \sigma^* \vDash P\boldsymbol{t}$, then the vector $\boldsymbol{\sigma^*(t, \Delta)} \in \rho^c(P, \Delta)$. It means that $\boldsymbol{\sigma^*(t, \Delta)} = |\boldsymbol{x}|$ (coordinate-wise) for some $P\boldsymbol{x} \in \Delta$ such that $\boldsymbol{t} \approx \boldsymbol{y} \in \Delta$ for some $\boldsymbol{y}$ such that $\boldsymbol{x} \sim_\Gamma \boldsymbol{y}$. Note that since $P\boldsymbol{x} \in \Delta$, $\boldsymbol{x} \in \mathsf{Var}(\Delta)$. It is not hard to show that $\boldsymbol{x} \approx \boldsymbol{y} \in \Delta$ by Proposition 4.8. Now based on SUBP, $P\boldsymbol{t} \in \Delta$.

The boolean cases are routine.

For the case of $\mathsf{K}_t\psi \in L_\Delta$:

- Suppose $\mathsf{K}_t\psi \notin \Delta$, then $\widehat{\mathsf{K}}_t\neg\psi \in \Delta$. By Lemma 4.3 there is some variable x and $\Theta \in W_\Gamma$ such that $\Delta R_{|x|}\Theta$, $x \approx t \in \Delta$ and $\neg\psi \in \Theta$. Therefore, by the induction hypothesis, $\mathcal{M}_\Gamma, \Theta, \sigma^* \nvDash \psi$ and so $\mathcal{M}_\Gamma, \Delta, \sigma^* \nvDash \mathsf{K}_x\psi$. If t is a variable then $x \sim_\Gamma t$ by Proposition 4.8, thus $\sigma^*(\Delta, t) = |x|$. If t is a name then by definition $\eta^c(\Delta, t) = |x|$. Therefore in either case we have $\mathcal{M}_\Gamma, \Delta, \sigma^* \nvDash \mathsf{K}_t\psi$.
- Suppose $\mathsf{K}_t\psi \in \Delta$, then by $\exists$-property, there is an $x \in \mathsf{Var}(\Delta)$ such that $x \approx t \in \Delta$, thus $\mathsf{K}_x\psi \in \Delta$ by SUBK and $\sigma^*(t, \Delta) = |x|$. By induction hypothesis, $\mathcal{M}_\Gamma, \Delta, \sigma^* \vDash x \approx t$. Now consider any $R_{|x|}$-successor Θ of Δ, it is clear that $\psi \in \Theta$ by definition of $R_{|x|}$. Now by induction hypothesis again, $\mathcal{M}_\Gamma, \Theta, \sigma^* \vDash \psi$. Therefore, $\mathcal{M}_\Gamma, \Delta, \sigma^* \vDash \mathsf{K}_t\psi$.

For the case of $[x := t]\psi \in L_\Delta$:

- Suppose $\mathcal{M}_\Gamma, \Delta, \sigma^* \vDash [x := t]\psi$.
  - If $t \in \mathbf{N}$, by $\exists$-property, we have $y \approx t \in \Delta$ for some $y \in \mathsf{Var}(\Delta)$. By induction hypothesis, $\mathcal{M}_\Gamma, \Delta, \sigma^* \vDash y \approx t$. Therefore $\sigma^*(\Delta, t) = |y|$ thus $\mathcal{M}_\Gamma, \Delta, \sigma^*[x \mapsto |y|] \vDash \psi$. Now if $\psi[y/x]$ is admissible then we have $\mathcal{M}_\Gamma, \Delta, \sigma^* \vDash \psi[y/x]$. By IH, $\psi[y/x] \in \Delta$. Thus $[x := y]\psi \in \Delta$ by SUB2AS. Since $t \approx y \in \Delta$, thus $[x := t]\psi \in \Delta$ by SUBAS. Note that if $\psi[y/x]$ is not

admissible, then we can reletter ψ to have an equivalent formula $\psi' \in L(\Delta)$ such that $\psi'[y/x]$ is admissible. Then the above proof still works to show that $[x := t]\psi' \in \Delta$. Since relettering can be done in the proof system by Proposition 3.3, we have $[x := t]\psi \in \Delta$.

· If t is a variable y, then $\mathcal{M}_\Gamma, \Delta, \sigma^*[x \mapsto |y|] \vDash \psi$. From here a similar (but easier) proof like the above suffices.

- Supposing $[x := t]\psi \in \Delta$, by the $\exists$-property of Δ, we have some $y \in \mathsf{Var}(\Delta)$ such that $t \approx y \in \Delta$. Like the proof above we can assume w.l.o.g. that $\psi[y/x]$ is admissible, for otherwise we can reletter ψ first. Thus $[x := y]\psi \in \Delta$ by SUBAS. Then by SUBASEQ, $\psi[y/x] \in \Delta$. By IH, $\mathcal{M}_\Gamma, \Delta, \sigma^* \vDash \psi[y/x] \wedge t \approx y$. By the semantics and the assumption that $\psi[y/x]$ admissible, $\mathcal{M}_\Gamma, \Delta, \sigma^* \vDash [x := t]\psi$.

□

Now we will transform the canonical model into a proper S5 model by taking the reflexive, symmetric and transitive closure of each $R_{|x|}$ in $\mathcal{M}_\Gamma$. Note that although $\mathcal{M}_\Gamma$ is a transitive model, the symmetric closure will break the transitivity. Actually, it can be done in one go by taking the reflexive transitive closure via undirected paths. More precisely, let $\mathcal{N}_\Gamma$ be the model like $\mathcal{M}_\Gamma$ but with the revised relation $R^*_{|x|}$ for each $x \in \mathsf{Var}(W_\Gamma)$, defined as:

$$\Delta R^*_{|x|}\Theta \iff \text{either } \Delta = \Theta \text{ or there are some } \Delta_1 \dots \Delta_n \text{ for some } n \geq 0 \text{ such that } \Delta_k R_{|x|} \Delta_{k+1} \text{ or } \Delta_{k+1} R_{|x|} \Delta_k \text{ for each } 0 \leq k \leq n \text{ where } \Delta_0 = \Delta \text{ and } \Delta_{n+1} = \Theta.$$

We will show that it preserves the truth value of **ELAS** formulas.

Lemma 4.10 (Preservation lemma) *For all* $\varphi \in$ **ELAS** :

$$\mathcal{N}_\Gamma, \Delta, \sigma^* \vDash \varphi \Leftrightarrow \varphi \in \Delta$$

Proof. Since we only altered the relations, We just need to check $\mathsf{K}_t\psi \in$ **ELAS**. Note that then $\mathsf{K}_t\psi$ is in all the local language L_Δ.

- If $\mathcal{N}^\Gamma, \Delta, \sigma^* \vDash \mathsf{K}_t\psi$ then since the closure only adds relations then we know $\mathcal{M}_\Gamma, \Delta, \sigma^* \vDash \mathsf{K}_t\psi$ by induction hypothesis and Lemma 4.9. Now by Lemma 4.9 again $\mathsf{K}_t\psi \in \Delta$.

- Suppose $\mathsf{K}_t\psi \in \Delta$. Since Δ has $\exists$-property, there is some $x \in \mathsf{Var}(\Delta)$ such that $x \approx t \in \Delta$ thus $\mathsf{K}_x\psi \in \Delta$. Now consider an arbitrary $R^*_{|x|}$-successor Θ in $\mathcal{N}^\Gamma$. If $\Delta = \Theta$ then by KT it is trivial to show that $\psi \in \Delta$. Now by the definition of $R^*_{|x|}$, suppose there are some $\Delta_1 \dots \Delta_n$ such that $\Delta_k R_{|x|} \Delta_{k+1}$ or $\Delta_{k+1} R_{|x|} \Delta_k$ for each $0 \leq k \leq n$ where $\Delta = \Delta_0$ and $\Theta = \Delta_{n+1}$. Now we do induction on n to show that $\mathsf{K}_x\psi \in \Delta_k$ for all those $k \leq n+1$. Note that if the claim is correct then by KT we have $\psi \in \Delta_{k+1}$ thus by IH we have $\mathcal{N}^\Gamma, \Delta, \sigma^* \vDash \mathsf{K}_t\psi$.

· $n = 0$: Then there are two cases:

$\Delta R_{|x|}\Theta$ in $\mathcal{M}_\Gamma$: by 4x, $\mathsf{K}_x\mathsf{K}_x\psi \in \Delta$ and then $\mathsf{K}_x\psi \in \Theta$ by the definition of $R_{|x|}$.

$\Theta R_{|x|}\Delta$ in $\mathcal{M}_\Gamma$: First note that there is some $y \in |x|$ such that $y \in \mathsf{Var}(\Theta)$ by the definition of $R_{|x|}$. If $y \neq x$ then by Proposition 4.8, we have $y \approx x \in \Theta$, therefore $x \in \mathsf{Var}(\Theta)$. Towards contradiction suppose $\neg\mathsf{K}_x\psi \in \Theta$. By 5x, $\mathsf{K}_x\neg\mathsf{K}_x\psi \in \Theta$. By definition of $R_{|x|}$, $\neg\mathsf{K}_x\psi \in \Delta$. Contradiction.

- $n = k+1$: Supposing that the claim holds for $n = k$, i.e., $\mathsf{K}_x\psi \in \Delta_k$. There are again two cases: $\Delta_k R_{|x|}\Delta_{k+1}$ or $\Delta_{k+1}R_{|x|}\Delta_k$ and they can be proved as above.

In sum, $\mathcal{N}^\Gamma, \Theta, \sigma^* \vDash \psi$ for any Θ such that $\Delta R_{|x|}\Theta$. Therefore, $\mathcal{N}^\Gamma, \Delta, \sigma^* \vDash \mathsf{K}_t\psi$.

□

It can be easily checked that:

Lemma 4.11 *$\mathcal{N}^\Gamma$ is an epistemic model, i.e., all the $R_{|x|}$ are equivalence relations.*

The following is straightforward by using some new variables but leaving infinitely many new variables still unused.

Lemma 4.12 *Each* SELAS*-consistent set Γ^{--} can be extended to a consistent set Γ^- w.r.t. some infinitely proper sublanguage L of* **ELAS**$^+$ *such that for each $a \in \boldsymbol{N}$ there is an $x \in \mathsf{Var}(L)$ such that $x \approx a \in \Gamma^-$. Finally we can extend it to an MCS Γ w.r.t. L.*

Theorem 4.13 SELAS *is sound and strongly complete over epistemic Kripke models with assignments.*

Proof. Soundness is from Theorem 3.1. Then given a consistent set Γ^-, using the above proposition we have a Γ. By the Truth Lemma 4.9 we have a model satisfying Γ and hence Γ^-. □

From the above proof, it is not hard to see that we can obtain the completeness of SELAS without `Tx`, `4x`, `5x` over arbitrary models by some minor modifications of the proof.

5 Discussions and future work

In this paper, we proposed a lightweight epistemic language with assignment operators from dynamic logic, which can express various *de re*/*de dicto* readings of knowledge statements when the references of the names are not commonly known. We gave a complete axiomatisation of the logic over epistemic models with constant domain of agents.

The complexity of the epistemic logic SELAS is currently unknown to us though we conjecture it is decidable due to the very limited use of quantifiers. Under the translation in Section 2, the name-free fragment can be viewed as a guarded fragment of first-order logic with transtive guards [19], which implies decidability. However, the non-rigid names translate into function symbols in first-order language, which may cause troubles since the guarded fragment with function symbols in general yields an undecidable logic [9]. We are not that far from the decidability boundary, if not on the wrong side.

To actually design a tableaux method in pursuing the decidability of our logic, we have to handle the difficulties from various sources:

- S5 frame conditions
- equalities
- constant domain
- non-rigid names
- termed modalities
- assignment operators

Some of the issues are already complicated on their own based on the knowledge of existing work. The biggest hurdle for the termination in a tableau method for S5-based logic like the ones proposed in [5,14], is to ensure loops in finite steps. This requires us in our setting to show that given a satisfiable formula, we can bound the number of necessary elements in the domain (for non-rigid names) and the number of subformulas we may encounter when building the tableau. The S5 condition and the assignment operator may ask us to always introduce new elements in the domain when creating new successors, while the new elements can essentially create new subformulas, if we add new symbols for them in the tableaux. On the other hand, without the transitivity and symmetry conditions, it is possible to bound the number of new elements in the domain to obtain decidability via some finite model property. We leave the details to a future occasion as well as the exploration of other ideas for decidability such as filtering the canonical model.

Below we list a few other further directions:

- Model theoretical issues of **ELAS**.
- Extension with function symbols.
- Extension with a (termed) common knowledge operator.
- Extension with limited quantifications over agents as in [15].
- Extension to varying domain models, where the existence of all the agents is not commonly known.

Finally, as a general direction, it would be interesting to consider what happens if we replace the standard epistemic logic with our **ELAS** in various existing logical framework extending the standard one.

Acknowledgement The authors thank Johan van Benthem, Rasmus Rendsvig, and Dominik Klein for pointers on related work. The authors are also grateful to the anonymous reviewers of AiML, whose comments helped in improving the presentation of the paper. [8] The research for this work was supported by the New Zealand Centre at Peking University and Major Program of the National Social Science Foundation of China (NO. 17ZDA026).

[8] Including the suggestion to change the previous title 'Call me by your name' of the paper.

References

[1] Aloni, M., "Quantification under Conceptual Covers," Ph.D. thesis, University of Amsterdam (2001).

[2] Aloni, M., *Knowing-who in quantified epistemic logic*, in: H. van Ditmarsch and G. Sandu, editors, *Jaakko Hintikka on Knowledge and Game-Theoretical Semantics*, Springer, 2018 pp. 109–129.

[3] Corsi, G., *A unified completeness theorem for quantified modal logics*, Journal of Symbolic Logic **67** (2002), pp. 1483–1510.

[4] Corsi, G. and E. Orlandelli, *Free quantified epistemic logics*, Studia Logica **101** (2013), pp. 1159–1183.

[5] Fitting, M. and R. L. Mendelsohn, "First-Order Modal Logic," Synthese Library, Springer, 1998.

[6] Fitting, M., L. Thalmann and A. Voronkov, *Term-modal logics*, Studia Logica **69** (2001), pp. 133–169.

[7] Grove, A. J., *Naming and identity in epistemic logic. II. A first-order logic for naming*, Artificial Intelligence **74** (1995), pp. 311–350.

[8] Grove, A. J. and J. Y. Halpern, *Naming and identity in epistemic logics. I. The propositional case*, Journal of Logic and Computation **3** (1993), pp. 345–378.

[9] Grädel, E., *On the restraining power of guards*, Journal of Symbolic Logic **64** (1998), pp. 1719–1742.

[10] Hintikka, J., "Knowledge and Belief: An Introduction to the Logic of the Two Notions," Cornell University Press, Ithaca N.Y., 1962.

[11] Holliday, W. H. and J. Perry, *Roles, rigidity, and quantification in epistemic logic*, in: A. Baltag and S. Smets, editors, *Johan van Benthem on Logic and Information Dynamics*, Springer, 2014 pp. 591–629.

[12] Hughes, G. E. and M. J. Cresswell, "A New Introduction to Modal Logic," Routledge, 1996.

[13] Kooi, B., *Dynamic term-modal logic*, in: *Proceedings of LORI-I*, 2007, pp. 173–186.

[14] Massacci, F., *Strongly analytic tableaux for normal modal logics*, in: A. Bundy, editor, *Automated Deduction — CADE-12* (1994), pp. 723–737.

[15] Naumov, P. and J. Tao, *Everyone knows that someone knows: Quantifiers over epistemic agents*, The Review of Symbolic Logic (2018).

[16] Padmanabha, A. and R. Ramanujam, *The monodic fragment of propositional term modal logic*, Studia Logica (2018).

[17] Rendsvig, R., "Towards a Theory of Semantic Competence," Master's thesis, Roskilde University (2011).

[18] Rendsvig, R., *Modeling semantic competence: A critical review of Frege's puzzle about identity*, in: L. D. and S. M., editors, *New Directions in Logic, Language and Computation, ESSLLI 2010, ESSLLI 2011*, LNCS **7415** (2012), pp. 140–157.

[19] Szwast, W. and L. Tendera, *On the decision problem for the guarded fragment with transitivity*, in: *Proceedings of the 16th Annual IEEE Symposium on Logic in Computer Science*, LICS '01 (2001), pp. 147–156.

[20] Thalmann, L., "Term-Modal Logic and Quantifier-Free Dynamic Assignment Logic," Ph.D. thesis, Uppsala University (2000).

[21] Wang, Y., *A new modal framework for epistemic logic*, in: *Proceedings Sixteenth Conference on Theoretical Aspects of Rationality and Knowledge, TARK 2017, Liverpool, UK*, 2017, pp. 515–534.

[22] Wang, Y., *Beyond knowing that: A new generation of epistemic logics*, in: H. van Ditmarsch and G. Sandu, editors, *Jaakko Hintikka on Knowledge and Game-Theoretical Semantics*, Springer, 2018 pp. 499–533.

www.ingramcontent.com/pod-product-compliance
Lightning Source LLC
Chambersburg PA
CBHW070817230426
R18182600001B/R181826PG43662CBX00001B/1
* 9 7 8 1 8 4 8 9 0 2 5 5 8 *